Principles and Practice of X-Ray Spectrometric Analysis

Second Edition

Wilhelm Conrad Röntgen
(1845-1923)
Discoverer of x-rays

Henry Gwyn Jeffreys Moseley
(1887-1915)
Founder of x-ray spectrometry

For biographies of Röntgen and Moseley, see References *(19)* and *(20)*, respectively.

Principles and Practice of X-Ray Spectrometric Analysis

Second Edition

Eugene P. Bertin

RCA Laboratories
David Sarnoff Research Center
Princeton, New Jersey

Plenum Press · New York–London

Library of Congress Cataloging in Publication Data

Bertin, Eugene P 1921-
Principles and practice of X-ray spectrometric analysis.

Bibliography: p.
Includes index.
1. X-ray spectroscopy. I. Title. [DNLM: 1. Radiation. 2. Spectrum analysis. QC481 B544p]
QD96.X2B47 1975 543'.085 74-28043
ISBN 0-306-30809-6

Photograph of W. C. Röntgen from Otto Glasser, *Dr. W. C. Röntgen* (2d ed.; Springfield, Illinois: Charles C Thomas, 1958); courtesy of Mrs. Otto Glasser and the publisher.

Photograph of H. G. J. Moseley from Mary Elvira Weeks, *Discovery of the Elements* (6th ed.; Easton, Pennsylvania: Journal of Chemical Education, 1960); courtesy of the publisher.

First Printing – April 1975
Second Printing – August 1978

A Division of Plenum Publishing Corporation
227 West 17th Street, New York, N.Y. 10011

Printed in the United States of America

Preface to the Second Edition

Since the first edition of this book was published early in 1970, three major developments have occurred in the field of x-ray spectrochemical analysis.

First, wavelength-dispersive spectrometry, in 1970 already securely established among instrumental analytical methods, has matured. Highly sophisticated, miniaturized, modular, solid-state circuitry has replaced electron-tube circuitry in the readout system. Computers are now widely used to program and control fully automated spectrometers and to store, process, and compute analytical concentrations directly and immediately from accumulated count data. Matrix effects have largely yielded to mathematical treatment. The problems associated with the ultralong-wavelength region have been largely surmounted. Indirect (association) methods have extended the applicability of x-ray spectrometry to the entire periodic table and even to certain classes of compounds. Modern commercial, computerized, automatic, simultaneous x-ray spectrometers can index up to 60 specimens in turn into the measurement position and for each collect count data for up to 30 elements and read out the analytical results in 1–4 min—all corrected for absorption-enhancement and particle-size or surface-texture effects—and wholly unattended. Sample preparation has long been the time-limiting step in x-ray spectrochemical analysis.

Second, energy-dispersive spectrometry, in 1970 only beginning to assume its place among instrumental analytical methods, has undergone phenomenal development and application and, some believe, may supplant wavelength spectrometry for most applications in the foreseeable future. There has been an amazing proliferation of highly sophisticated, computer-controlled, energy-dispersive instrumentation using lithium-drifted silicon detectors, multichannel analyzers, and low-power x-ray tube and radioisotope sources. Modern commercial, computerized, energy-dispersive x-ray spectrometers can substantially match the performance cited above for simultaneous wavelength-dispersive instruments; they cannot measure quite as many elements at once due to somewhat poorer resolution, and they must count a little longer due to lower maximum count rate.

Finally, a decreasing role of experimentation has accompanied the increasing role of mathematics, computers, electronics, and mechanics. More and more aspects of x-ray spectrometric analysis that previously required experimental ingenuity are now dealt with mathematically and automatically.

Great benefit derives from these developments. Modern computer-controlled automated instruments with their direct analytical readout enable us to get much more work done much more conveniently and in much less time. But we must forego the challenge, satisfaction—and fun!—of circumventing difficulties by experimental ingenuity. To some x-ray spectrochemists, this may seem a small price indeed. Others may share the writer's feeling that, to paraphrase the television commercial, we are analyzing more, but enjoying it less!

Analytical x-ray spectrometry continues to develop at a vigorous, wholesome pace. However, one senses an abatement in the development of instrumentation, theory, basic methods, and techniques. This trend is evident even in the younger discipline of energy-dispersion, where present activity is mostly in evaluation of performance, comparison with wavelength-dispersion, improvement of sensitivity and resolution, mathematical methods of quantitative analysis, and applications. The same applies to the ancillary field of electron-probe microanalysis. Perhaps the most significant indication of the maturity of x-ray spectrochemical analysis is the increasing attention of x-ray spectrochemists to other disciplines. One need but scan the programs of the major national and international meetings to note that the sessions previously devoted exclusively to x-ray fluorescence spectrometry or electron-probe microanalysis are now increasingly concerned with the related fields of proton and heavy-ion excitation and "electron spectrometry for chemical analysis" (ESCA), including photo- and Auger-electron spectrometry. One also finds increasing numbers of papers on the actually unrelated fields of ion-scattering spectrometry (ISS) and secondary-ion mass spectrometry (SIMS).

Unaccountably, whereas comprehensive reference treatises have appeared for all other established instrumental analytical disciplines as they matured, no such book has appeared in x-ray spectrochemistry. In undertaking this new edition, the writer was motivated by the need for such a manual, convinced that it would remain timely for many years, and encouraged by the excellent reception of the first edition.

All material in the first edition has been retained in the second, but the book has been revised, updated, and greatly expanded. The magnitude of the expansion may be conveyed in terms of the 401 pages, 23 tables, 62 figures, 221 equations, and 362 references added to the 703 pages, 47 tables, 196 figures, 344 equations and 565 references, in the first edition; three of the new figures are in color. The updating includes significant publications through 1973 and even a substantial number from 1974. The first edition cited publications through late 1968.

Substantially all sections in the book have been expanded and/or updated, but by far the most attention has been given to energy dispersion and related subjects, including low-power x-ray tubes; radioisotope sources; multichannel analyzers; lithium-drifted silicon detectors; computer-controlled multichannel energy-dispersive spectrometer systems; processing, readout, and display of energy-spectral data; and energy-dispersive diffractometry-spectrometry (Giessen–Gordon method). Other especially expanded chapters and sections include those on pulse-height selection; matrix effects (absorption-enhancement, heterogeneity, particle-size, and surface-texture); mathematical methods, including the variable-takeoff-angle method; specimen preparation, especially the borax-glass fusion method; trace analysis; radioactive specimens; and electron-probe microanalysis.

Appendixes on photon energies of principal spectral lines, absorption-edge jump ratios, and fluorescent yields have been added to the existing ones on wavelengths of principal K, L, and M spectral lines and absorption edges; K, L, and M excitation potentials; and mass-absorption coefficients. The existing table of data for 26 analyzer crystals and multilayer heavy-metal soap films has been expanded to 72. A comprehensive glossary of symbols has been added. Because problems of organization and retrieval increase with the length of the book, the detailed table of contents of the first edition has been retained and updated, the index has been made much more detailed, and the text is generously provided with cross-references.

However, the basic objectives and organization of the book remain the same as those outlined in the preface of the first edition, and the same 21 chapters in the same seven major divisions are retained. The book is still directed to those who are not content simply to operate the instrument and perform the analyses, but who want some understanding of the underlying principles. Thus, the book still emphasizes the practical experimental aspects, but also explains the principles underlying the instrumentation and analytical methods and techniques. These explanations are given in primarily descriptive, rather than mathematical, terms, but all essential mathematics is included. Rather than dealing with specific applications, the book still strives to present the widest range of analytical methods, techniques, concepts, and ideas possible in the available space, wherever possible with references for further study. Accordingly, the book does not deal *specifically* with the currently cogent applications of x-ray spectrochemical analysis to the environmental, biomedical, and forensic problems that beset our society; however, readers concerned with these fields will find all applicable techniques described and documented.

The book is still concerned primarily with x-ray fluorescence spectrometry as applied on commercial flat-crystal spectrometers and accessories. However, extensive treatment is also given to other x-ray analytical methods that can be conducted on such instruments and to ancillary methods requiring their own instrumentation. Energy-dispersion and electron-probe microanalysis are given substantial treatment.

The errors in the first edition have been corrected, hopefully without the addition of too many new ones.

Eugene P. Bertin

Harrison, New Jersey
February 1975

Preface to the First Edition

The first commercial x-ray secondary-emission (fluorescence) spectrometer became available about 20 years ago. Within a decade, x-ray spectrometry had become a firmly established method of instrumental analysis for elements having atomic number down to 12 (magnesium) in concentrations down to a few tenths of a percent. During the last decade, several thousand commercial instruments have been placed in service in laboratories and factories, and the applicability of the method has been extended to trace and micro analysis and to substantially the entire periodic table. However, although many colleges and universities offer full courses in optical and electrical methods of instrumental analysis and in x-ray diffraction, very few offer courses in x-ray spectrometry. Proficiency in this method must be acquired by self-instruction, on-the-job training and experience, the workshops held by the instrument manufacturers, and the one- or two-week courses offered by a few universities.

This book is intended primarily as a textbook for three groups of readers: (1) students; (2) technicians newly assigned to the x-ray spectrometric laboratory; and (3) technicians having practical experience, but little or no formal instruction, and therefore lacking a basic understanding of the method. The book is intended particularly for those who are not content simply to operate the instrument and perform the required analyses, but who want to understand the method and instrument. It is hoped that the book will also be useful to expert x-ray spectrochemists and to scientists and technicians in other disciplines who want to evaluate the applicability of x-ray spectrometry to their fields.

Pursuant to these objectives, the book emphasizes the practical aspects of x-ray secondary-emission spectrometry, but also gives the principles underlying the method, instrumentation, and analytical techniques. Explanations are given primarily in descriptive rather than mathematical terms, but all equations of practical or tutorial value are included.

The applications of x-ray spectrometry are so numerous and varied that it would be hopeless to attempt even to summarize them in a book of reasonable length. Moreover, the writer has always believed that an understanding of the principles and instrumentation of the method leads to proficiency in application of the method to specific problems. Consequently, rather than dealing with specific applications and materials, the book presents in general terms the widest range of analytical methods, techniques, concepts, and ideas possible in the available space. Specific examples are used only for illustration.

The book is concerned primarily with conventional x-ray secondary-emission spectrometry as applied on commercially available spectrometers and accessories. However, extensive treatment is also given to unconventional modes of operation of the instrument and to other x-ray analytical methods that can be conducted on commercial instruments. Related x-ray methods requiring their own instrumentation are also discussed.

An understanding of the instrumentation is necessary to full realization of its potentialities. Accordingly, the x-ray tube and generator, spectrogoniometer, and detection, measurement, and readout components are described schematically in considerable detail. However, specific commercial instruments are not described, except for special features.

The 21 chapters of the book fall into seven groups as follows.

Chapters 1 and 2 deal with *x-ray physics*—the origin and nature of continuous and characteristic x-ray spectra, primary and secondary excitation, fluorescent yield, absorption, scatter, and diffraction.

Chapters 3–9 constitute a detailed description of *the x-ray spectrometer, its components, and their operation.* A general description of the instrument and of the method and its scope, advantages, and limitations is given in Chapter 3. Chapter 4 (Excitation) gives the principles of secondary excitation and detailed descriptions of the x-ray tube and generator. Chapter 5 (Dispersion) is concerned primarily with conventional flat-crystal dispersion. However, most of the other flat- and curved-crystal dispersion arrangements and instruments based on them are also described. Chapter 6 (Detection) considers the construction, function, performance, properties, and principles of operation of proportional and scintillation counters. However, the relatively new solid-state detectors and other less common detectors are also described. Chapter 7 (Measurement) describes the instrumentation and methods for measurement of line and background intensities. Chapter 8 is devoted to pulse-height analysis and "nondispersive" methods. Chapter 9 describes complete conventional manual and automatic spectrometers as well as special instruments having primary or radioisotope excitation and instruments for the ultralong-wavelength region.

Chapter 10 deals with *qualitative and semiquantitative analysis.*

Chapters 11–13 consider *performance criteria* and other features of the method: error, precision, and counting statistics (Chapter 11); specific, nonspecific, special, and unusual absorption-enhancement effects (Chapter 12); and sensitivity, resolution, and spectral-line interference (Chapter 13).

Chapters 14 and 15 present the *basic methods of quantitative x-ray spectrometric analysis*: standard addition and dilution; calibration standardization; internal standardization; standardization with scattered x-rays; matrix dilution; thin-film methods; special experimental methods; and mathematical correction of absorption-enhancement effects. Chapter 15 is devoted entirely to this last method.

Chapters 16 and 17 are devoted to *specimen preparation and presentation* of solids, small parts, powders, briquets, fusion products, liquids, and supported specimens.

The remaining chapters describe *unconventional modes of operation* of the spectrometer and *related x-ray methods* of analysis. Chapter 18 deals with measurement of composition and thickness of films and platings. Chapter 19 treats the analysis of small specimens and of small selected areas on large specimens by "x-ray probe" techniques. Chapter 20 describes polychromatic and monochromatic absorptiometry, absorption-edge spectrometry, absorption-edge fine structure, microradiography, and analytical methods based on scattered x-rays. All these methods are treated with emphasis on their applicability to commercial spectrometers. This chapter also describes Auger-electron and x-ray-excited photoelectron spectrometry and x-ray-excited optical fluorescence. Chapter 21 is a brief description of the instrumentation and method of electron-probe microanalysis.

Each chapter is substantially self-contained, with many cross references to other pertinent sections and with some replication of material covered in other chapters.

Some readers may be disappointed by the absence of a chapter on micro and trace analysis. However, these subjects are dealt with in Chapters 16–19.

The Appendixes contain tables of wavelengths of spectral lines and absorption edges, excitation potentials, and mass-absorption coefficients. The Bibliography provides a list of books, reviews, bibliographies, and tables that should be available in the x-ray spectrometric laboratory, and the 513 papers and reports referred to throughout the book. In general, papers published after 1968 are not included.

The writer gratefully acknowledges that in his task he benefited substantially from access to the excellent books on x-ray spectrometry already in print, especially those by Adler (*1*), Birks (*8*), Liebhafsky *et al.* (*26*), and, particularly, Jenkins and de Vries (*22*). Finally, the writer is most grateful to Thomas J. Cullen, who read the entire manuscript and made many valuable suggestions.

Eugene P. Bertin

Harrison, New Jersey
November 1969

Contents

PART I. X-RAY PHYSICS

Chapter 1. Excitation and Nature of X-Rays; X-Ray Spectra

Chapter 2. Properties of X-Rays

PART II. THE X-RAY SPECTROMETER, ITS COMPONENTS, AND THEIR OPERATION

Chapter 3. X-Ray Secondary-Emission (Fluorescence) Spectrometry; General Introduction

Chapter 4. Excitation

Chapter 5. Dispersion

Chapter 6. Detection

Chapter 7. Measurement

Chapter 9. Laboratory, Automated, and Special X-Ray Spectrometers

PART III. QUALITATIVE AND SEMIQUANTITATIVE ANALYSIS

Chapter 10. Qualitative and Semiquantitative Analysis

PART IV. PERFORMANCE CRITERIA AND OTHER FEATURES

Chapter 11. Precision and Error; Counting Statistics

PART V. QUANTITATIVE ANALYSIS

Chapter 14. Methods of Quantitative Analysis

Chapter 15. Mathematical Correction of Absorption-Enhancement Effects

PART VI. SPECIMEN PREPARATION AND PRESENTATION

Chapter 16. Specimen Preparation and Presentation—General; Solids, Powders, Briquets, Fusion Products

Chapter 17. Specimen Preparation and Presentation—Liquids; Supported Specimens

PART VII. UNCONVENTIONAL MODES OF OPERATION; RELATED X-RAY METHODS OF ANALYSIS

Chapter 21. Electron-Probe Microanalysis

PART VIII. APPENDIXES, BIBLIOGRAPHY

Appendixes

Bibliography

Part I

X-Ray Physics

Chapter 1

Excitation and Nature of X-Rays; X-Ray Spectra

1.1. HISTORICAL

By the late 1800's, many eminent physicists had come to believe that the nature of the physical world was substantially understood and that there remained only to acquire detail and refine data. This complacency was decisively shattered by an unprecedented series of truly epochal discoveries, all within a period of about two decades: radio (Hertz, 1887); an entire, wholly unsuspected group of inert chemical elements (Ramsay, Rayleigh, and Travers, 1895–98); x-rays (Roentgen, 1895); radioactivity (Becquerel, 1896; the Curies, 1898); the electron (Thomson, 1897); quantum theory (Planck, 1900; Einstein, 1901); relativity theory (Einstein, 1905); and cosmic rays (Hess, 1910).

Roentgen's discovery was as significant as any of these, and the invaluable potentialities of x-rays in the theory and practice of science, technology, and medicine were realized immediately. X-rays were widely applied in medical (including dental) and industrial radiography and fluoroscopy within a year of their discovery. X-ray diffraction and spectrometry were applied in many university laboratories within two decades; however, the present widespread application of these disciplines had to await the development and commercial availability of reliable x-ray tubes and generators and electronic detection and readout equipment.

Moseley established the basis for qualitative and quantitative x-ray spectrochemical analysis in 1913 (*M57*). Figure 1.1, his first published

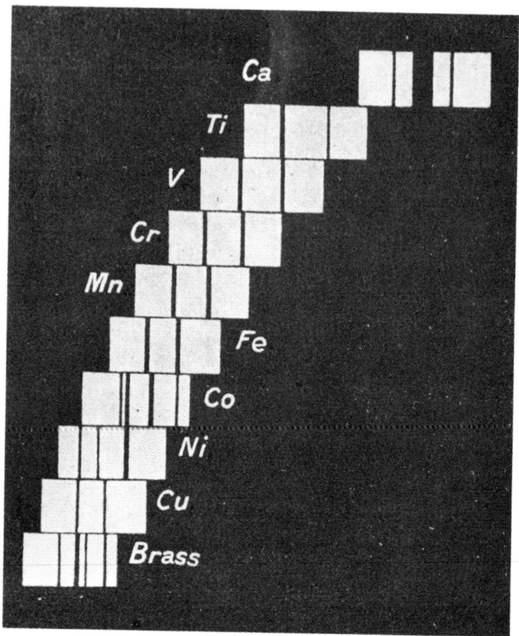

FIG. 1.1. Moseley's first published photograph of x-ray spectra, showing lines of calcium through zinc. [H. G. J. Moseley, *Philosophical Magazine* **26**, 1024 (1913); Taylor and Francis, Ltd., London; courtesy of the publisher.]

photograph of x-ray spectra, shows (1) the relationship between wavelength of x-ray spectral lines and atomic number; (2) contamination of cobalt by iron and nickel; and (3) greater intensity of copper lines than zinc lines in the x-ray spectrum of brass (70Cu–30Zn). Moseley also foresaw the use of x-ray spectra in the discovery of missing chemical elements by prediction of the wavelengths of their characteristic x-ray spectral lines.

Other outstanding contributions in the development of x-ray spectrometry are listed chronologically in Table 1.1, which shows that the Friedman–Birks instrument (*F27*)—the prototype of the first commercial instrument—did not appear until 1948. However, since then, the development of x-ray spectrometric instrumentation and technique has proceeded at a phenomenal rate and continues unabated today. It has been estimated (*49*) that whereas in 1958 there were only ~50 x-ray spectrometers in use throughout the world, by 1971 the number had increased to ~10,000.

TABLE 1.1. Chronological Development of X-Ray Spectrometry

Year	Development
1895	W. C. Roentgen[a,b] (*19*) discovered x-rays (*R19–21*)
1896	J. Perrin measured x-ray intensity, using an air ionization chamber
1909	C. G. Barkla noted evidence of absorption edges (*B8*)
1911	C. G. Barkla noted evidence of emission line series, which he designated *K*, *L*, *M*, *N*, etc. (*B7*)
1912	M. von Laue, W. Friedrich, and E. P. Knipping demonstrated diffraction of x-rays by crystals (*F29*)
1913	W. L. and W. H. Bragg built the Bragg x-ray spectrometer (*B78*)
1913	H. G. J. Moseley[a] (*20*) showed the relationship between wavelength of x-ray spectral lines and atomic number, noted that copper lines are stronger than zinc lines in the x-ray spectrum of brass; thereby established the basis of qualitative and quantitative x-ray spectrochemical analysis (*M57*)
1913	W. D. Coolidge introduced the hot-filament, high-vacuum x-ray tube (*C59*)
1913–23	M. Siegbahn did his classic work of measuring wavelengths of x-ray spectra of the chemical elements
1922	A. Hadding first applied x-ray spectra specifically to chemical analysis (of minerals) (*H1*)
1923	D. Coster and G. von Hevesy discovered hafnium, the first element to be identified by its x-ray spectrum (*C62*, *C63*)
1923	G. von Hevesy proposed quantitative analysis by secondary excitation of x-ray spectra
1923	R. Glocker and W. Frohnmeyer applied x-ray absorption-edge spectrometry (*G19*, *G20*)
1924	W. Soller constructed an x-ray spectrometer using parallel-foil collimators (*S43*)
1928	R. Glocker and H. Schreiber[c] applied x-ray secondary-emission (fluorescence) spectrometry (*G21*)
1928	H. Geiger and W. Muller developed the gas-filled detector tube to a high degree of reliability (*G6*)
1948	H. Friedman and L. S. Birks built the prototype of the first commercial x-ray secondary-emission spectrometer (*F27*)
1949	R. Castaing and A. Guinier built the first electron-probe x-ray primary-emission spectrometer

[a] See Frontispiece.
[b] In 1901, Roentgen received the first Nobel Prize in Physics for his discovery.
[c] Schreiber coined the term x-ray "fluorescence" in 1929 (*S16*).

1.2. DEFINITION OF X-RAYS

Radiation may be defined as energy, in the form of waves or particles, emanating from its source through space in divergent straight lines; however, radiation in the form of charged particles may be deflected from its linear path by electric and/or magnetic fields.

All types of radiation have a dual nature in that they exhibit some properties best explained in terms of particles, and other properties best explained in terms of waves. Nevertheless, most radiation has either *predominantly* corpuscular or wave properties, and it is convenient to classify all types of radiation in these two categories.

The corpuscular radiation includes alpha (α) rays, or helium nuclei (He^{2+}); beta (β^-) rays, or electrons (e^-); positrons (β^+), or positively charged electrons (e^+); neutrons (n); and the primary cosmic rays, which consist mostly of high-energy protons (p^+), which are hydrogen nuclei (H^+).

Wave radiation comprises the electromagnetic spectrum, which is divided into overlapping regions, as shown in Figure 1.2. The visible region (4000–7500 Å) is defined by human visual response. The gamma (γ) region comprises high-energy electromagnetic radiation originating in the nuclei of atoms undergoing radioactive decay. The secondary cosmic radiation results from interaction of the primary cosmic radiation (corpuscular) with terrestrial matter. The other spectral regions are defined somewhat arbitrarily on the basis of wavelength or, more realistically, the technology used to generate, transmit, detect, and apply the radiation in each region.

X-rays may be defined as electromagnetic radiation of wavelength $\sim 10^{-5}$ to ~ 100 Å produced by deceleration of high-energy electrons and/or by electron transitions in the inner orbits of atoms. Both these processes are discussed in detail below. The 10^{-5}-Å radiation is produced in betatrons operating at ~ 1 GV; 100 Å represents the *K*-band spectra of the lightest elements. In conventional x-ray spectrometry, the spectral region of interest is ~ 0.1 Å (U $K\alpha$) to ~ 20 Å (F $K\alpha$). In ultrasoft x-ray spectrometry, the region of interest is ~ 10 Å to ~ 100 Å (Be $K\alpha$).

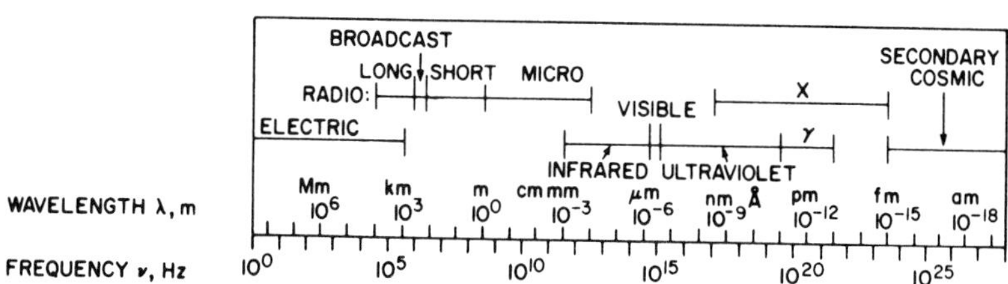

FIG. 1.2. Electromagnetic spectrum.

TABLE 1.2. Properties of X-Rays and Interaction Phenomena of X-Rays with Matter

General properties

- Occur as continuous spectra
- Occur as characteristic line spectra
- Occur as characteristic band spectra
- Produce characteristic absorption spectra
- Propagate at velocity of light
- Propagate in straight lines
- Propagate without transfer of matter
- Are undeflected and otherwise unaffected by electric and magnetic fields
- Are invisible and otherwise undetected by human senses

On encountering matter, x-rays may undergo any of the following phenomena:

- Transmission unaffected
- Reflection
- Refraction
- Polarization
- Diffraction (by pinhole, slit, or crystal)
- Coherent scatter
- Incoherent scatter
- Photoelectric absorption
- Pair production (at energy >1.02 MeV)

On absorption of x-rays, matter may undergo any of the following phenomena:

- Increase in temperature
- Alteration of dielectric properties
- Alteration of electric properties
- Ionization (especially gases and liquids)
- Radiolysis (photolysis), with consequent decomposition, chemical reaction, precipitation, gas evolution, free-radical production, etc.
- Photographic effect
- Production of color centers and other lattice defects
- Radiation damage
- Fluorescence and/or phosphorescence (infrared, visible, ultraviolet)
- Excitation of secondary characteristic x-ray line and band spectra
- Excitation of photo, Auger, and recoil electrons
- Stimulation, damage, genetic change, or death (biological tissue)

Interaction of x-rays and matter may result in emergence of any of the following types of radiation:

Corpuscular

- Ions
- Photoelectrons excited by x-rays (primary and secondary)
- Photoelectrons excited by visible and ultraviolet fluorescence
- Auger electrons
- Recoil electrons
- Electron–positron pairs (from x-rays having energy >1.02 MeV)

Electromagnetic

Incident x-radiation (continuous or line)

- Transmitted (uneffected or imprinted with the absorption spectrum of the specimen)
- Reflected
- Refracted
- Polarized
- Diffracted
- Coherently scattered
- Incoherently scattered

Emitted x rays

- Diagram spectral lines or bands
- Nondiagram (satellite) spectral lines
- Continuum (*Bremsstrahlung*) excited by photo, Auger, and recoil electrons

Other

- Infrared (thermal)
- Infrared, visible, and ultraviolet fluorescence and phosphorescence
- Annihilation radiation (0.51-MeV photons) produced by recombination of electron–positron pairs
- Cerenkov radiation[a] from very-high-speed electrons produced by very-high-energy x-rays

[a] Cerenkov radiation is faint visible light produced when a charged particle moves through a medium at a velocity greater than the velocity of light in that medium; it may be regarded as the optical analog of the sonic boom.

1.3. PROPERTIES OF X-RAYS

Table 1.2 summarizes the properties of x-rays, the phenomena occurring on interaction of x-rays with matter, and the types of radiation resulting from this interaction. X-rays exhibit the corpuscle–wave duality mentioned above: corpuscular properties include photoelectric absorption, incoherent scatter, gas ionization, and scintillation production; wave properties include velocity, reflection, refraction, diffraction, polarization, and coherent scatter. Wavelength and energy dispersion of x-rays are based on the wave and corpuscular aspects of x-rays, respectively. The wave–particle duality of x-rays is represented by Equations (1.16)–(1.18). When electromagnetic radiation exhibits the corpuscular aspect of its dual nature, the individual "particles" are termed *photons* or *quanta*.

1.4. UNITS OF X-RAY MEASUREMENT

The x-ray spectrochemist is concerned with four physical quantities in measurement of x-rays: frequency ν (nu), wavelength λ (lambda), energy E, and intensity I.

1.4.1. Frequency

Frequency is expressed in vibrations per second (s^{-1}) or, preferably, hertzes (Hz); 1 Hz $= 1\ s^{-1}$. Other units, used principally in optical spectroscopy, are wave number $\bar{\nu}$, which is the number of vibrations per centimeter (cm^{-1}), and the fresnel, which is 10^{12} Hz.

1.4.2. Wavelength

The earliest x-ray spectrometers used rock salt (NaCl) analyzer crystals cut on the (200) plane to measure x-ray wavelength. The interplanar spacing of the (200) NaCl planes $d_{(200)}$ was calculated as follows:

$$d_{(200)} = d_{(100)}/2 \tag{1.1}$$

and, since NaCl is cubic,

$$d_{(100)} = v^{1/3} = \left(\frac{\sum A/N}{\varrho}\right)^{1/3} \tag{1.2}$$

where $d_{(100)}$ is the lattice parameter (cm); v is the volume of the crystallographic unit cell (cm^3); $\sum A$ is the sum of the atomic weights of all the atoms in the unit cell (4Na + 4Cl); N is the Avogadro number (6.02×10^{23} atoms/mol); and ϱ is density (g/cm^3). The value of $d_{(200)}$ for NaCl determined in this way was 2.814×10^{-8} cm and was used in the Bragg equation [Equation (2.42)] to measure x-ray wavelength.

Wavelengths measured in this way are relative and no more accurate than the accuracy of the value of $d_{(200)}$. However, it was possible to measure wavelengths in terms of the NaCl $d_{(200)}$ spacing with much more accuracy than that with which the spacing itself was known. Therefore, it was decided to define a unit of x-ray wavelength, the x-unit (xu):

$$1 \text{ xu} \equiv d_{(200)}(\text{NaCl})/2814.00 \tag{1.3}$$

It was later found that the cleavage plane (200) of calcite ($CaCO_3$) provides a better crystal for x-ray wavelength measurements, and the x-unit was redefined:

$$1 \text{ xu} \equiv d_{(200)}(\text{calcite})/3029.45 \tag{1.4}$$

Actually, for convenience, the kilo-x-unit (kxu) was more commonly used,

$$1 \text{ kxu} \equiv 1000 \text{ xu} \tag{1.5}$$

The x-unit and kilo-x-unit are very nearly equal to 10^{-11} and 10^{-8} cm, respectively.

Later it was found possible to make absolute measurements of x-ray wavelength with ruled gratings, the spacing of which is not dependent on Avogadro's number. It was then decided to give x-ray wavelengths in terms of angstroms (Å),

$$1 \text{ Å} \equiv 10^{-10} \text{ m} \equiv 10^{-8} \text{ cm} \tag{1.6}$$

The kxu-to-Å conversion factor Λ is

$$\Lambda = \lambda_{\text{Å}}/\lambda_{\text{kxu}} = 1.00202 \qquad (\text{Pt–Ir m}) \tag{1.7}$$

At that time, the meter was defined in terms of the platinum–iridium international prototype meter. More recently, the meter has been redefined as 1,650,763.73 wavelengths of a certain spectral line in the optical emission spectrum of an isotope of krypton, ^{86}Kr. The new Λ is

$$\Lambda = \lambda_{\text{Å}}/\lambda_{\text{kxu}} = 1.0020764 \qquad (^{86}\text{Kr m}) \tag{1.8}$$

At present, Bearden (*59, 60*) proposes the use of the wavelength of the W $K\alpha_1$ x-ray spectral line (~0.209 Å) as the x-ray wavelength standard and proposes redefinition of the angstrom and a new symbol as follows:

$$1\ \overset{*}{\text{Å}} = \lambda_{\text{W}K\alpha_1}/0.2090100 \tag{1.9}$$

The $\overset{*}{\text{Å}}$-to-Å conversion factor Λ^* is

$$\Lambda^* = \lambda_{\text{Å}}/\lambda_{\overset{*}{\text{Å}}} = 1.0000197 \tag{1.10}$$

The x-ray spectrochemist rarely requires wavelengths so exact that he need be concerned with these distinctions. Throughout this book, the symbol λ indicates wavelength in angstroms (Å), except where another unit is indicated by subscript. The symbols λ_{min} and $\lambda_{I_{\text{max}}}$ indicate the short-wavelength limit and the wavelength of maximum intensity, respectively, in the continuous spectrum; $\lambda_{ZK\alpha}$, $\lambda_{ZL\alpha_1}$, etc. indicate the wavelengths of the $K\alpha$, $L\alpha_1$, etc. x-ray spectral lines of element Z; $\lambda_{ZK_{ab}}$, $\lambda_{ZL\text{III}_{ab}}$, etc. indicate the wavelengths of the K, LIII, etc. absorption edges of element Z.

The angstrom is a nonsystematic unit, and wavelengths should be expressed in nanometers (nm); 1 nm = 10^{-9} m = 10 Å. However, the angstrom is so well established that its abandonment in favor of the nanometer is most unlikely. Another unit of wavelength, commonly used in optical spectroscopy, is the millimicron (mμ). This term is to be avoided in favor of the systematic nanometer, to which it is equal.

In addition to those given above, other relationships among units of wavelength and frequency are:

$$\lambda_{\text{cm}} = c/\nu \tag{1.11}$$

$$\lambda_{\text{Å}} = (c/\nu)\times 10^8 \tag{1.12}$$

$$\nu = c/\lambda_{\text{cm}} \quad \text{Hz, s}^{-1} \tag{1.13}$$

$$\bar{\nu} = 1/\lambda_{\text{cm}} \quad \text{cm}^{-1} \tag{1.14}$$

$$\nu_{\text{fresnel}} = \nu \times 10^{-12} \tag{1.15}$$

All these quantities are defined in the foregoing text except c, which is the velocity of light (3×10^{10} cm/s).

X-ray wavelength regions are sometimes designated as *ultrahard* (<0.1 Å), *hard* (0.1–1 Å), *soft* (1–10 Å), and *ultrasoft* (>10 Å).

1.4.3. Energy

Energy of an x-ray photon in ergs is given by

$$E_{\text{erg}} = h\nu = hc/\lambda_{\text{cm}} \tag{1.16}$$

where c, ν, and λ_{cm} have been defined already; E_{erg} is photon energy in ergs, and h is Planck's constant. Of greater utility is x-ray photon energy in electron volts (eV), given by

$$E_{eV} = hc/\lambda_{cm}e \tag{1.17}$$

where h, c, and λ_{cm} have been defined already; E_{eV} is photon energy in electron volts, and e is electron charge in electrostatic units. Substituting values for h (6.6×10^{-27} erg·s), c (3×10^{10} cm/s), and e (4.8×10^{-10} esu), and inserting conversion factors 10^8 Å/cm and 1/300 esu/V, one gets

$$E_{eV} = 12{,}396/\lambda \tag{1.18}$$

where λ is in angstroms.

Throughout this book, the symbol E_x indicates photon energy in electron volts, except where another unit is indicated by subscript. Incidentally, many writers use the symbol E for electric potential ("voltage"). In this book, V (italic) indicates potential, V (roman) the unit of potential, the volt; V_{ZK}, V_{ZLIII}, etc. indicate the K, LIII, etc. excitation potentials of element Z.

1.4.4. Intensity

Intensity of an x-ray beam is defined in physics in terms of energy per unit area per unit time, usually ergs per square centimeter per second. However, in x-ray spectrometric analysis, intensities are almost always given in "counts" per unit time—that is, x-ray photons per unit area per unit time. The unit area is usually the useful area of the detector and, being constant for all measurements in a group, is not included in the unit. The unit time is usually the second, the minute, or the preset counting time. For monochromatic x-rays, intensity in ergs per second is readily calculated from intensity in counts per second by multiplying the latter by the photon energy in ergs [Equation (1.16)]:

$$I_{erg} = I_{counts/s}E_{erg} = I_{counts/s}h(c/\lambda_{cm}) \tag{1.19}$$

Many workers use the symbol R (for count *rate*) for intensity in counts per second or minute. Throughout this book, the following symbols and definitions are used: The symbol I indicates intensity in counts per second, except in a few cases where another unit is indicated by subscript. The intensity measured at the x-ray spectral line is the *peak* intensity I_P and consists of the *line* or *net* intensity I_L and *background* intensity I_B, contributed by the spectral line and background, respectively; that is, $I_L = I_P - I_B$. Some workers use the term *total* intensity I_T for peak intensity. Incidentally, many workers use the symbol I for electric current; in this book, i indicates current.

The symbol R is also used for *relative* or *normalized* intensity, that is, the ratio of the analyte-line intensities measured from the specimen and pure analyte. When it is impractical to make measurements on the pure element (gaseous elements, alkali metals, bromine, etc.), a simple stable compound can be used, for example, sodium chloride (NaCl) for sodium and chlorine, calcium carbonate ($CaCO_3$) for calcium. This ratio is substantially unaffected by the instrument. It is used almost exclusively in quantitative electron-probe microanalysis, and some workers recommend its wider use in x-ray secondary-emission spectrometric analysis. In electron-probe microanalysis, the symbol K or k, rather than R, is used for relative intensity.

Use of the terms *relative* or *normalized* intensity for this ratio is rather unfortunate because both these terms are used more commonly in other senses. *Relative intensity* is used to express the relative intensities of: (1) the analyte (or other) line from a series of measurements from the same or different specimens; (2) different x-ray spectral lines of the same element; and (3) the same x-ray spectral line (say, $K\alpha$) of different elements (Section 1.6.2.6). In this book, the symbol I_{rel} is used for relative intensity in these senses. *Normalized intensity* is used to express intensities proportionally adjusted with respect to some standard or previously established value. Normalization of intensity may also be misinterpreted as a procedure analogous to proportional normalization of analytical concentrations of the elements in a sample so that they add up to 100%. In view of the possible misinterpretation of these terms, the ratio R should either be given a new term or referred to as analyte-line intensity ratio from specimen and pure analyte.

Still another intensity concept is *intensity fraction* or *count fraction* F (*T3*). It is the intensity I, or accumulated count in unit time N, of the measured line of element i divided by the sum of the intensities, or counts, of the measured lines of all the elements in the specimen, including i, that is,

$$F_i = \frac{I_i}{\sum I_j} = \frac{N_i}{\sum N_j} \tag{1.20}$$

where j includes i. F may be expressed as a fraction or percent. An application of the intensity fraction concept is given in Section 15.2.

1.5. THE CONTINUOUS SPECTRUM

1.5.1. Nature

The continuous spectrum, also referred to as the *general* spectrum, *white* spectrum, *continuum*, and *Bremsstrahlung*, is characterized by four features: a continuous range of wavelengths (analogous to white light) having an abrupt short-wavelength limit $\lambda_{\min}$, rising to a maximum intensity $\lambda_{I_{\max}}$, and falling off gradually at longer wavelengths. The $\lambda_{I_{\max}}$ occurs at $\sim 1.5\lambda_{\min}$ and, for practical purposes, may be regarded as the *effective wavelength* of the continuum, that is, a single wavelength having substantially the same absorption in a given absorber as the continuum.

The continuum also has two less prominent features (*G13*). A second intensity maximum, very poorly defined compared with $\lambda_{I_{\max}}$, occurs at a wavelength greater than $\lambda_{I_{\max}}$ and different for each target element—for example, 1.5, 0.9, and 0.8 Å for tungsten, copper, and chromium, respectively. The phenomenon is of unknown origin, and is independent of applied potential. The other feature is an abrupt intensity discontinuity occurring at the wavelength of the absorption edge of the target element, the intensity increasing by a factor of 2–5 on the long-wavelength side of the discontinuity. The phenomenon results from the abrupt change in absorption of the target for continuous radiation excited below the surface and emerging through the overlying target layer. The magnitude of the discontinuity increases as the path length of the x-rays in emerging from the target increases, that is, as the angle between the incident electrons and target approaches 90° and as the angle between the emergent x-rays and target decreases.

The continuum may be recorded by directing the primary beam into the x-ray spectrometer. This may be done by turning the x-ray tube about its axis or, more conveniently, by placing a piece of paraffin or polyethylene in the specimen chamber and scattering some of the primary radiation into the spectrometer.

1.5.2. Generation

The electrons incident upon an x-ray tube target may interact with the target in any of several ways:

1. They may backscatter from the target toward the general direction from which they arrived. The fraction so scattered increases with the atomic number of the target—about half are scattered by the heaviest elements, very few by the lightest.

2. They may scatter within the target surface, interacting with the outermost electrons in the target atoms or with the plasma—the electron "gas" that permeates metals. Many of these valence and plasma electrons are ejected from the target as low-energy (<50-eV) secondary electrons. Each such interaction extracts 10–100 eV from the incident electron. Most of the incident electrons that do not backscatter undergo this process.

3. They may undergo Rutherford scatter in the high Coulombic field near the nuclei of the target atoms. Most such interactions are elastic, that is, do not result in loss of energy.

4. They may interact with the inner electrons in the target atoms. The probability of such interactions is small compared with that for process 2 above. This is the process that gives rise to the characteristic line spectrum of the target element and is the subject of Section 1.6.

5. They may undergo inelastic Rutherford scatter in passing near target atoms without collision, giving up some or all of their energy (speed) as x-ray photons. At the potentials involved in practical x-ray spectrometry (<100 kV), only 0.5–1% of the electrons bombarding the target undergo this process. This is the process that gives rise to the continuum and is the subject of this section.

The effect of acceleration potential and target atomic number on the extent of electron scatter within the target surface is discussed in Section 21.7 and shown in Figure 21.14.

The continuous spectrum, then, arises from the relatively few electrons that undergo stepwise inelastic nuclear scatter and deceleration in the target.

The continuous spectrum arises when high-speed electrons undergo stepwise deceleration in matter (hence the German term *Bremsstrahlung*, which means, literally, "braking radiation"). β-ray, internal-conversion, photo, Compton-recoil, and Auger electrons all produce continuous x-rays, but x-ray tubes are the principal source. X-ray tubes are discussed later; let it suffice here to say that electrons from a hot tungsten filament are accelerated by a positive potential of up to 100 kV to a metal target, where they undergo deceleration with consequent generation of continuous x-radiation. Deceleration of other high-speed particles, such as protons, deuterons, tritons, α-particles, and heavier ions, does not generate observable continuum (Section 1.8). However, a very low-intensity continuum may be generated by electrons ejected from target atoms by ion collision.

The continuum cannot be generated by secondary excitation, that is, by irradiation of matter with high-energy x-rays, because photons do not undergo stepwise loss of energy. However, continuum may appear in secondary x-ray beams due to scatter of incident continuum from the specimen and to generation by photo, recoil, and Auger electrons resulting from interaction of the incident x-rays with the specimen.

1.5.3. Short-Wavelength Limit

Consider an electron moving from filament to target in an x-ray tube operating at (peak) potential V. The shortest wavelength λ_{min} that this electron can possibly generate is emitted if the electron decelerates to zero velocity in a single step, giving up all its energy as one x-ray photon. The energy E_e gained by the electron between filament and target is

$$E_e = eV \tag{1.21}$$

where e is the electronic charge in electrostatic units, and V is potential in volts. The energy of the x-ray photon E_x is [see Equation (1.16)]

$$E_x = hc/\lambda_{\mathrm{cm}}$$

If the electron gives all its energy to the photon,

$$E_x = E_e \tag{1.22}$$

and

$$hc/\lambda_{\mathrm{cm}} = eV \tag{1.23}$$

giving

$$\lambda_{\mathrm{cm}} = hc/eV \tag{1.24}$$

Substituting the same values and conversion factors as were used in Equation (1.17), one gets

$$\lambda = 12{,}396/V \tag{1.25}$$

where λ is in angstroms, V in (practical) volts.

Equation (1.25) is the Duane–Hunt law (*D28*). It permits the calculation of the Duane–Hunt limit or short-wavelength limit λ_{min} (angstroms) for an x-ray tube operating at any specified potential V (volts), and of the minimum potential at which an x-ray tube must be operated to produce a specified wavelength. However, if this wavelength is to be generated at high intensity, the x-ray tube must be operated at a potential several times that indicated by the equation (Section 1.5.5). When Equation (1.25) is applied to an x-ray tube operating from a full-wave supply, V is peak potential. The analogy of Equations (1.25) and (1.18) is evident. Many workers use the symbol λ_0 or λ_{SWL} (for short-wavelength limit) instead of λ_{min}.

1.5.4. Origin of the Continuum

Stepwise deceleration accounts for the generation of a continuum, rather than a single wavelength λ_{min} corresponding to the x-ray tube potential. Most electrons give up their energy not in a single step, but rather in numerous unequal increments ΔV, each resulting in an x-ray photon of wavelength $12{,}396/\Delta V$ [Equation (1.25)]. The net effect of large numbers of electrons undergoing such stepwise deceleration is a continuous band of wavelengths $\geq \lambda_{min}$.

Many other phenomena contribute to the formation of the continuum. Consider a strictly monoenergetic electron beam bombarding a monatomic-layer target. Many of the electrons would pass through unaffected, and the others would undergo only one deceleration. But these decelerations need not necessarily be to zero velocity, so a continuum would appear, like that shown for layer 1 in Figure 1.3. An infinitely thick target may be regarded as a stack of such thin targets. The electrons lose energy as they penetrate deeper into the target, so that each successive layer receives electrons having average energy slightly less than those incident upon the layer above. Thus, λ_{min} and the entire spectral distribution from each successively deeper layer

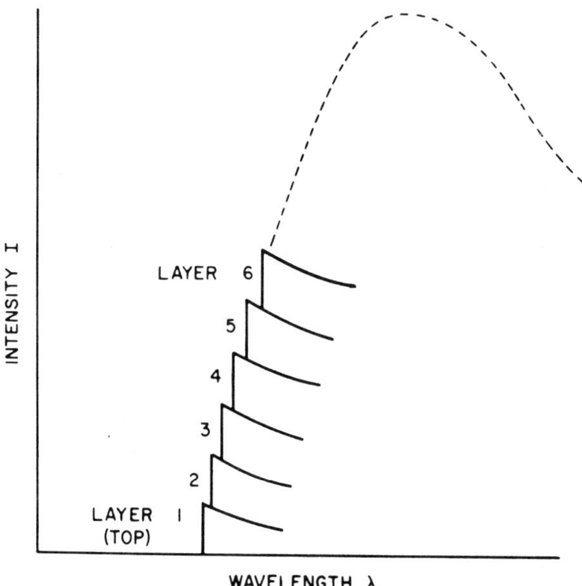

FIG. 1.3. Continuous spectrum from an infinitely thick target regarded as the summation of the individual continua from a stack of very thin targets.

are displaced to longer wavelength, as shown in Figure 1.3. Moreover, because the electrons undergo decelerations of various magnitudes, each layer receives electrons having a greater energy spread than those incident upon the layer above. Further complexity arises from lateral scatter of electrons in the target.

In practice, the electrons arriving at the target are far from monoenergetic. With full-wave rectification, V varies from zero to the peak value each half cycle. With constant potential, V fluctuates slightly with residual ripple, but even with ripple-free constant potential, the electrons have some energy spread because they leave the filament with a range of thermal energies and in random directions. Finally, x-rays generated in successively deeper layers have greater difficulty in emerging from the target surface because (1) being of slightly longer wavelength, they are more readily absorbed, and (2) they have a greater thickness of overlying target to penetrate.

Incidentally, continuum cannot be *generated* by secondary excitation, because x-ray photons, unlike electrons, cannot undergo decremental loss of energy (except for incoherent scatter, Section 2.2.2), but lose all their energy at once or not at all. However, a continuum does appear in secondary x-ray spectra due principally to scatter of the primary-beam continuum by the specimen. Photo, recoil, and Auger electrons (Sections 1.6.5.2, 2.2.2, and 2.5.1, respectively), generated by the primary x-ray beam, also contribute a small amount of continuum.

1.5.5. Effect of X-Ray Tube Current, Potential, and Target

Of the several expressions relating the spectral distribution in the continuum to excitation conditions, perhaps the most useful are the following by, respectively, Kulenkampff (*K31*), Kramers (*K29*), and Beatty (*B21*):

$$I_\lambda = CZ\left(\frac{1}{\lambda^2}\right)\left(\frac{1}{\lambda_{\min}} - \frac{1}{\lambda}\right) + BZ^2\frac{1}{\lambda^2} \tag{1.26}$$

$$I_\lambda \propto iZ\left(\frac{\lambda}{\lambda_{\min}} - 1\right)\frac{1}{\lambda^2} \tag{1.27}$$

$$I_{\text{int}} = (1.4\times 10^{-9})iZV^2 \tag{1.28}$$

where I_λ is the intensity at any wavelength λ; I_{int} is the integrated intensity of the continuum (W); Z is the atomic number of the x-ray tube target; V and i are x-ray tube potential (V) and current (A), respectively; $\lambda_{\min}$ is

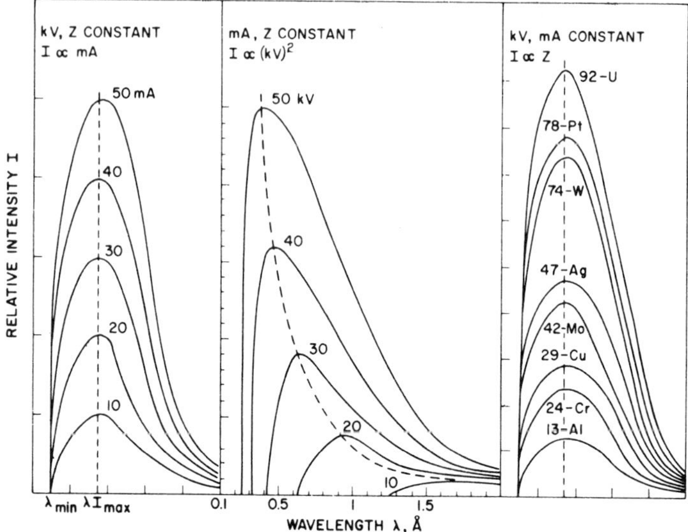

FIG. 1.4. Effect of x-ray tube current, potential, and target atomic number on the continuous spectrum.

the short-wavelength limit; and B and C are constants. In Equation (1.26), $C \gg B$, so that the second term on the right is relatively insignificant. In the region 8–14 Å, the continuum is better described by Equation (1.26) without the second term on the right and with $1/\lambda^5$ instead of $1/\lambda^2$ in the first term. Incidentally, it is differentiation of the first term on the right in Equation (1.26) with respect to λ that gives the relationship $\lambda_{I_{max}} \approx 1.5\lambda_{min}$ (Section 1.5.1). In Equation (1.28), $(1.4 \times 10^{-9})ZV$ may be regarded as the efficiency factor of an x-ray tube operated at input power iV watts.

Figure 1.4 shows the effects of x-ray tube current, potential, and target on the continuum. It is evident that none of these parameters substantially affects the general profile of the continuum. A change in x-ray tube current causes a proportional change in intensity of the continuum because the number of electrons arriving at the target from the filament is directly proportional to current. The atomic number Z of the x-ray tube target has much the same effect, because the number of orbital electrons in each target atom—and therefore the total number of electrons in the target—is proportional to Z. It is these electrons that cause the stepwise deceleration of the filament electrons. However, it is evident that the wavelengths in the continuum are in no way characteristic of the target element.

X-ray tube potential V has by far the most marked effect. As V is increased: (1) The intensity at any wavelength already present is increased because the filament electrons are accelerated to higher velocity and can undergo more decelerations; (2) $\lambda_{\min}$ and, consequently, $\lambda_{I_{\max}}$ are displaced to progressively shorter wavelengths in accordance with Equation (1.25). and (3) the intensity at and near $\lambda_{I_{\max}}$ increases rapidly with V, as much as 16-fold by doubling V. It must be remembered that with full-wave rectification the continuum shifts back and forth in wavelength each half cycle and attains its shortest $\lambda_{\min}$ and $\lambda_{I_{\max}}$ only at the peak. With constant potential, the continuum remains substantially constant with time.

1.5.6. Significance

In x-ray spectrochemical analysis, the continuum provides the principal source of specimen excitation and of background. The continuum may also provide incident radiation for polychromatic and monochromatic absorption and bracketing wavelengths for absorption-edge spectrometry. A wavelength from the continuum may be used for the background-ratio technique All these applications are discussed in detail in later sections.

1.6. THE CHARACTERISTIC LINE SPECTRUM

1.6.1. Atomic Structure

An understanding of the nature and origin of characteristic x-ray line spectra requires some knowledge of atomic structure. Every atom consists of a dense central nucleus containing all of its Z protons and all of its $M - Z$ neutrons, with all of its Z electrons revolving around this nucleus like planets around the sun. (Z and M are atomic number and mass number, respectively.) The electrons are grouped in shells designated K, L, M, N, etc. in order of increasing distance from the nucleus. The shells were thus designated by Barkla (*B7*) to allow for possible shells inside the innermost shell known at the time. The electrons in each shell are classified further with respect to angular momentum and direction of spin. Each of these parameters—shell, momentum, spin, etc.—is designated by a quantum number, which may have only certain values, and no two electrons in an atom may have an identical set of quantum numbers (Pauli exclusion principle).

Table 1.3 lists these quantum numbers, their significance, and the allowed values that each may assume for an electron in an atom. The magnetic quantum number m has little significance in this discussion and

TABLE 1.3. Quantum Numbers

Symbol	Name	Significance	Allowed values	Selection rules
n	Principal	Principal binding energy; indicates shell	$1, 2, \ldots, n$ $K, L, M, N, \ldots$	$\Delta n \neq 0$
l	Azimuthal	Orbital angular momentum; determines shape of orbital	$0, 1, \ldots, (n-1)$ $s, p, d, f, \ldots$	$\Delta l = \pm 1$
m	Magnetic	Projection of angular momentum (l) on magnetic field; indicates orientation of orbital in a magnetic field	$-l, \ldots, 0, \ldots, +l$	—
s	Spin	Direction of spin—clockwise or counterclockwise	$\pm\frac{1}{2}$	—
j	Inner precession	Vector sum of l and s	$l \pm \frac{1}{2}$, except $j \neq 0 - \frac{1}{2}$	$\Delta j = \pm 1$ or 0

is included only for completeness. Figure 1.6 lists the n, l, and j quantum numbers for all possible electrons in the first four atom shells (K, L, M, N) and in the first five subgroups of the O shell. It is evident that on the basis of these n, l, and j quantum numbers, these electrons fall into subgroups—one K, three L, five M, and seven N. These are the subgroups of principal significance in x-ray spectrometry. The number of subgroups in each shell corresponds to the allowed values of j in that shell: 1/2, 3/2, 5/2, 7/2, etc. (Table 1.3). Figure 1.6 also lists the number of electrons in each subgroup for gold atoms, and these numbers happen also to be the maximum numbers of electrons in each subgroup for any atom. Gold atoms also have one electron in the PI subgroup. Of course, if *all* the quantum numbers of each electron were considered, it would be seen that no two electrons in an atom have the same complete set of quantum numbers.

1.6.2. Nature and Origin

1.6.2.1. General

The characteristic line spectrum consists of a series of discrete wavelengths—x-ray spectral lines—characteristic of the emitting element and having various relative intensities. The line spectrum of an element originates when electrons are expelled from the inner levels of its atoms, and electrons from levels farther out fall into the vacancies. Each such transition constitutes an energy loss which appears as an x-ray photon. For example, on creation of a K-shell vacancy, a succession of spontaneous electron transitions follows; each fills a vacancy in a lower level, with resultant emission of an x-ray photon, but also creates a vacancy in a level farther out. Figure 1.5 shows two of the many possible series of transitions that may follow creation of a K-shell vacancy. The result of such processes in large numbers of atoms is the simultaneous generation of the K, L, M, etc. series of x-ray spectra of that element. Since these electron transitions correspond precisely to the difference in energy between the two atomic orbitals involved, the emitted x-ray photon has energy characteristic of this difference and thereby of the atom. The transitions are substantially instantaneous, occurring within 10^{-12}–10^{-14} s of the creation of the electron vacancy.

Figure 1.6 shows schematically all K, L, M, and N levels and the first five O levels, giving for each the n, l, and j quantum numbers and the optical spectroscopy notation. The up and down arrows show the electron transitions giving rise, respectively, to vacancies in each level, and to all K, L, and M lines of relative intensity $\sim$1 or more. (See Section 1.6.2.3.) Each

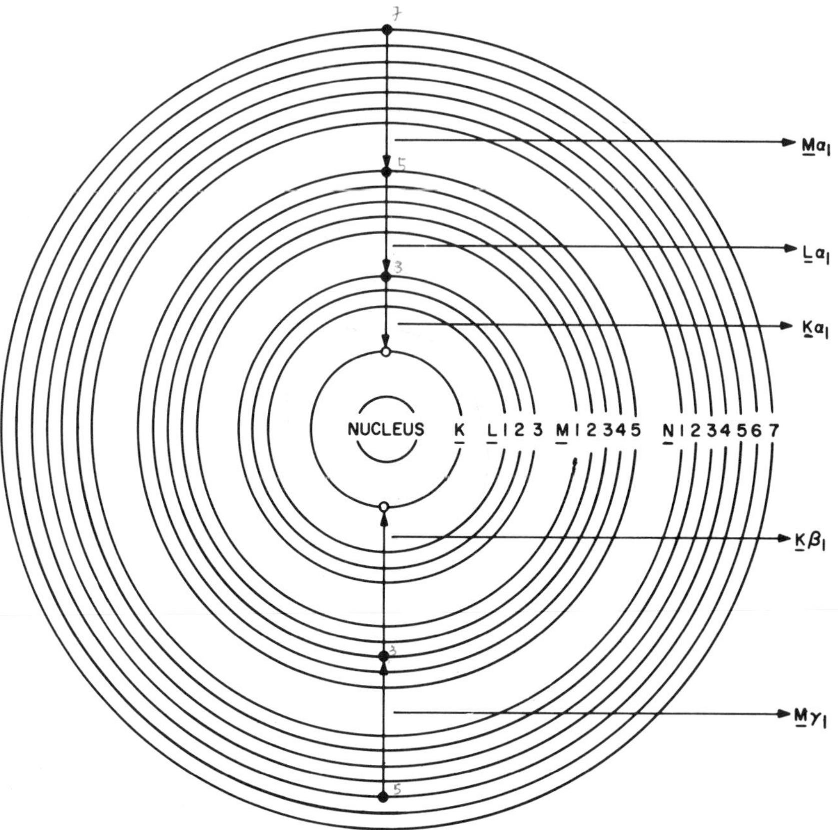

FIG. 1.5. Typical series of electron transitions that may follow creation of a *K*-shell vacancy.

transition is labeled with the symbol of the resultant line. The weakest of these lines are rarely encountered in x-ray secondary-emission spectrometry. However, in the electron-probe x-ray microanalyzer, where primary excitation and curved-crystal optics combine to give high efficiency, some of these lines may appear. At the top of the figure are the approximate relative intensities of the lines and their wavelengths and energies for gold. Physicists would prefer Figure 1.6 with the arrows pointing in the opposite direction, indicating transitions in atom state, rather than electron transitions. For example, the figure shows the $K\alpha_1$ line as originating from an *electron* transition from the *L*III level to the *K* level; physicists regard it as originating from an *atom* transition from the *K* state (or *K* quantum state) to the *L*III state.

X-ray spectral lines are grouped in series (*K*, *L*, *M*, *N*); all lines in a series result from electron transitions from various levels to the same shell.

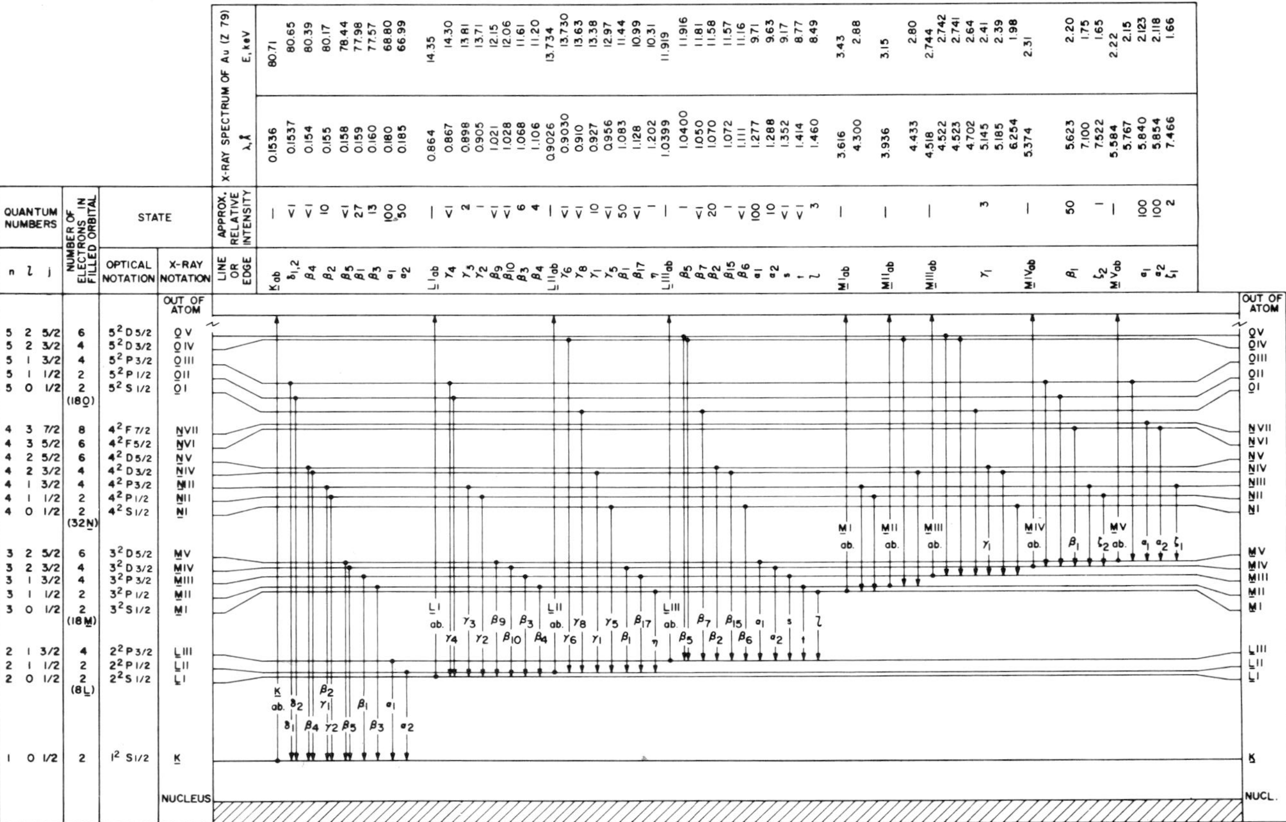

QUANTUM NUMBERS n	l	j	NUMBER OF ELECTRONS IN FILLED ORBITAL	STATE: OPTICAL NOTATION	STATE: X-RAY NOTATION
					OUT OF ATOM
5	2	5/2	6	$5^2D_{5/2}$	OV
5	2	3/2	4	$5^2D_{3/2}$	OIV
5	1	3/2	4	$5^2P_{3/2}$	OIII
5	1	1/2	2	$5^2P_{1/2}$	OII
5	0	1/2	2 (18O)	$5^2S_{1/2}$	OI
4	3	7/2	8	$4^2F_{7/2}$	NVII
4	3	5/2	6	$4^2F_{5/2}$	NVI
4	2	5/2	6	$4^2D_{5/2}$	NV
4	2	3/2	4	$4^2D_{3/2}$	NIV
4	1	3/2	4	$4^2P_{3/2}$	NIII
4	1	1/2	2	$4^2P_{1/2}$	NII
4	0	1/2	2 (32N)	$4^2S_{1/2}$	NI
3	2	5/2	6	$3^2D_{5/2}$	MV
3	2	3/2	4	$3^2D_{3/2}$	MIV
3	1	3/2	4	$3^2P_{3/2}$	MIII
3	1	1/2	2	$3^2P_{1/2}$	MII
3	0	1/2	2 (18M)	$3^2S_{1/2}$	MI
2	1	3/2	4	$2^2P_{3/2}$	LIII
2	1	1/2	2	$2^2P_{1/2}$	LII
2	0	1/2	2 (8L)	$2^2S_{1/2}$	LI
1	0	1/2	2	$1^2S_{1/2}$	K
					NUCLEUS

LINE OR EDGE	APPROX. RELATIVE INTENSITY	X-RAY SPECTRUM OF Au (Z 79) λ, Å	E, keV
K ab	—	0.1536	80.71
$\delta_{1,2}$	<1	0.1537	80.65
β_4	<1	0.154	80.39
β_2	10	0.155	80.17
β_5	<1	0.158	78.44
β_1	27	0.159	77.98
β_3	13	0.160	77.57
α_1	100	0.180	68.80
α_2	50	0.185	66.99
LI ab	—	0.864	14.35
γ_4	<1	0.867	14.30
γ_3	2	0.898	13.81
γ_2	1	0.905	13.71
β_9	<1	1.021	12.15
β_{10}	<1	1.028	12.06
β_3	6	1.068	11.61
β_4	4	1.106	11.20
LII ab	—	0.9026	13.734
γ_6	<1	0.9030	13.730
γ_8	<1	0.910	13.63
γ_1	10	0.927	13.38
γ_5	<1	0.956	12.97
β_1	50	1.083	11.44
β_{17}	<1	1.128	10.99
η	1	1.202	10.31
LIII ab	—	1.0399	11.919
β_5	1	1.0400	11.916
β_7	<1	1.050	11.81
β_2	20	1.070	11.58
β_{15}	1	1.072	11.57
β_6	<1	1.111	11.16
α_1	100	1.277	9.71
α_2	10	1.288	9.63
s	<1	1.352	9.17
t	<1	1.414	8.77
l	3	1.460	8.49
MI ab	—	3.616	3.43
		4.300	2.88
MII ab	—	3.936	3.15
		4.433	2.80
MIII ab	—	4.518	2.744
		4.522	2.742
		4.523	2.741
		4.702	2.64
γ_1	3	5.145	2.41
		5.185	2.39
		6.254	1.98
MIV ab	—	5.374	2.31
β_1	50	5.623	2.20
		7.100	1.75
ζ_2	1	7.522	1.65
MV ab	—	5.584	2.22
		5.767	2.15
α_1	100	5.840	2.123
α_2	100	5.854	2.118
ζ_1	2	7.466	1.66

FIG. 1.6. Electron transitions producing the x-ray spectrum.

(Physicists would say *atom* transitions from the same initial state to various final states.) Sometimes, series are further divided into subseries, such as *L*I, *L*II, *L*III; all lines in a subseries result from electron transitions from various levels to the same level.

Of incidental interest in x-ray spectrochemical analysis are x-ray spectra from plasmas. A *plasma* is a highly ionized gas containing equal numbers of positive and negative charges. A plasma may be formed, among other ways, by directing a high-power laser beam* on a specimen of matter or by "exploding" a fine wire by passing a high-current electric pulse through it. The instantaneous temperature in such plasmas may reach 10^6–10^7 °C, and the atoms may be stripped of many of their electrons. In this way, x-ray-emitting plasmas arise. However, unlike the x-ray spectra described above, these x-ray spectra are very complex because they contain lines from all the various multiply ionized states of the atoms in the plasma, rather than from singly ionized *K*-, *L*-, and *M*-state atoms. Plasma x-ray spectra give information about nonequilibrium atom states.

1.6.2.2. Band Spectra

Electron transitions originating from sharply defined levels emit sharp spectral lines having breadth at half height ~0.001 Å. However, the outermost levels in atoms in solids may be broadened under the influence of neighboring atoms. Electron transitions originating from such levels give rise to broad band spectra. The *K* spectra of the lightest elements are of this type. These phenomena are shown in Figure 1.7.

1.6.2.3. Selection Rules

Electron transitions cannot occur from *any* higher to *any* lower level. Only certain transitions are "permitted" by the selection rules, which are given in the last column of Table 1.3. For example, an *L*III → *L*I transition violates the first rule, $\Delta n \neq 0$. An *N*VI$(l = 3) \rightarrow$ *L*I$(l = 0)$ violates the second rule, $\Delta l = \pm 1$; *M*V$(j = 5/2) \rightarrow$ *L*II$(j = 1/2)$ violates the third rule, $\Delta j = \pm 1$ or 0. Spectral lines that "obey" the selection rules are referred to as *diagram lines*, those that violate them as *forbidden lines*.

1.6.2.4. Notation

In the Siegbahn designation system, the symbol of an x-ray spectral line (for example, Ni $K\alpha_1$, Au $L\beta_2$, W $L\eta$, U Ll) consists of: (1) the symbol

* Reference added in proof: C. M. Dozier, P. G. Burkhalter, B. M. Klein, D. J. Nagel, and R. R. Whitlock, X-Ray Emission from Laser-Produced Plasmas, *Advan. X-Ray Anal.* **17**, 423–35 (1974).

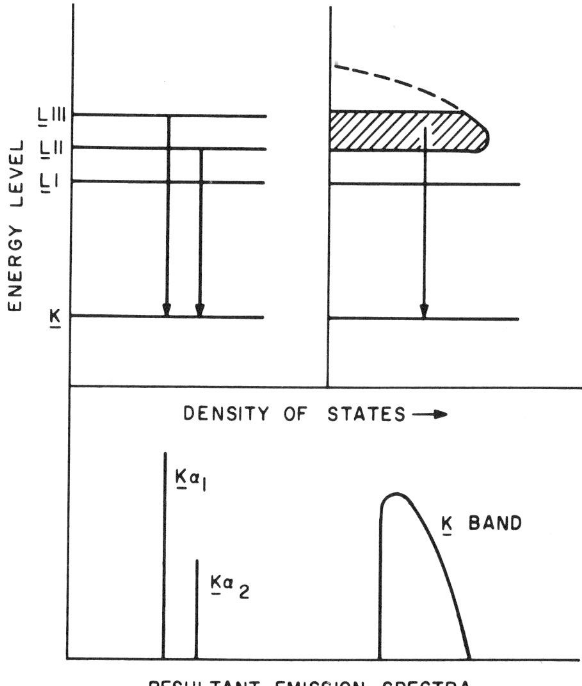

FIG. 1.7. Origin of the line and band spectra.

of the chemical element; (2) the symbol of the series (K, L, M, N, etc.); the line originates from the filling of a vacancy in, or an electron transition to, the indicated shell; and (3) a lower-case Greek letter, usually with a numerical subscript, or a lower-case italic letter denoting the particular line in the series.

The α_1 line is usually the strongest in the series, and α lines arise from $\Delta n = 1$ transitions. The β_1 line is usually the second strongest, except that $K\alpha_2$ is stronger than $K\beta_1$. The β and γ lines usually arise from $\Delta n = 1$ or 2 transitions. Aside from these conventions, the notation is not systematic. However, the line resulting from a given transition in any element is always given the same symbol.

The symbol $K\alpha$, sometimes $K\alpha_{1,2}$, indicates the $K\alpha_1\alpha_2$ pair and has wavelength equal to the average of the individual lines weighted for their relative intensities, $K\alpha_1{:}K\alpha_2 = 2{:}1$,

$$\lambda K\alpha = \frac{2\lambda K\alpha_1 + \lambda K\alpha_2}{3} \tag{1.29}$$

The symbol $K\beta$ indicates the $K\beta_1\beta_3$ pair.

Diffraction orders are indicated in any of several ways; for example, second-order Cu $K\alpha$ may be indicated Cu $K\alpha(2)$ (the notation used in this book), Cu $K\alpha(n = 2)$, Cu $2K\alpha$, or Cu $K\alpha$ II.

Another system of notation for x-ray spectral lines is the level-designation system, which gives the atomic transitions resulting in the lines. Thus, the $K\alpha_1$ line may be designated K,LIII (or K,$L3$), indicating that it arises from an *atom* transition from the K to the LIII *state*—that is, an *electron* transition from the LIII to the K *orbital*. Similarly, the $L\alpha_1$ line may be designated LIII,MV (or $L3$,$M5$), etc. Some of the less prominent L lines and most of the M and N lines have not been assigned Greek or italic symbols. These lines are indicated only by their level designations.

The prefix S before the symbol of an x-ray spectral line, for example, Al $SK\alpha_3$, indicates that it is a satellite or nondiagram line, rather than a diagram line (Section 2.5.3).

Taking note of the "archaic and unsystematic" state of the present system of notation for x-ray spectral lines, Jenkins (*J13*) has proposed what appears to be a simple, logical notation system. For the series, he would retain the K, L, M, etc. designations for all lines arising from electron transitions to orbitals having principal quantum numbers $n = 1, 2, 3$, etc., respectively. Within a series, he would designate as α, β, γ, δ, etc. all lines arising from electron transitions for which $\Delta n = 1, 2, 3, 4$, etc., respectively. For lines having the same n and Δn transitions, he would add numerical subscripts 1, 2, 3, etc. for electron transitions from the I, II, III, etc. levels, respectively, of the orbital involved. An additional subscript s or f would indicate satellite and forbidden lines, respectively.

1.6.2.5. Wavelength

The photon energy of a spectral line is the difference in energy ΔE between the initial and final levels involved in the electron transition. For example, in Figure 1.6, for gold,

K state →	LIII state +	$K\alpha_1$
80.77	11.92	68.79 keV
K state →	MIII state +	$K\beta_1$
80.77	2.74	77.97 keV
LI state →	MV state +	$L\beta_9$
14.35	2.22	12.15 keV

(The slight discrepancies in the energies are not real and are due to small errors in the table values.) Because of such relationships as these, x-ray spectrometry is extremely valuable in elucidation of atomic structure.

Wavelength is inversely proportional to energy and is calculated from Equation (1.18). The energy differences among *K*, *L*, *M*, and *N* levels of atoms are such that the wavelengths of the *K* and *L* x-ray spectra lie in the region ~0.1–100 Å. Figure 1.8 shows the strongest *K*, *L*, and *M* lines of all the chemical elements. The absorption edges and higher orders also shown are discussed later.

For a given element, wavelengths of lines in a series decrease as ΔE of the energy transition increases; thus,

$$\lambda K\beta_2(N \rightarrow K) < \lambda K\beta_1(M \rightarrow K) < \lambda K\alpha(L \rightarrow K)$$

$$\lambda L\alpha_1(M\text{V} \rightarrow L\text{III}) < \lambda L\alpha_2(M\text{IV} \rightarrow L\text{III}) < \lambda Ll(M\text{I} \rightarrow L\text{III})$$

Also, for a given element, wavelengths of lines decrease from series *M* to *L* to *K* because electrons must "fall" farther to fill vacancies in levels successively closer to the nucleus. Figure 1.6 illustrates these considerations for gold.

For the same spectral line, wavelength decreases as atomic number increases because positive nuclear charge, binding of orbital electrons, and interorbital energy differences all increase. Figure 1.8 illustrates this property.

The difference in wavelength between, say, the $K\alpha$ lines of adjacent elements increases as atomic number decreases (Figure 1.8), whereas the difference in photon energy decreases as atomic number decreases, as shown in Table 1.4. This means that wavelength (crystal) dispersion and resolution of x-ray lines of adjacent elements become easier as atomic number decreases, and energy dispersion and resolution become more difficult.

Because x-ray spectra originate in inner orbitals, which are substantially unaffected by valence, variation of line wavelength with *Z* is not periodic, as in optical spectroscopy, but uniformly progressive (Figure 1.8). Also, the wavelengths of x-ray lines are substantially—but not entirely—independent of the chemical state of the atom. The proximity of the nucleus to these inner orbitals explains the strong dependence of wavelength on *Z*, as defined by the Moseley law,

$$\sqrt{\nu} = (c/\lambda_{\text{cm}})^{1/2} = k_1(Z - k_2) \tag{1.30}$$

where ν, λ_{cm}, and *Z* are frequency, wavelength, and atomic number, respectively, and k_1 and k_2 are constants different for each line; k_2 is a screening

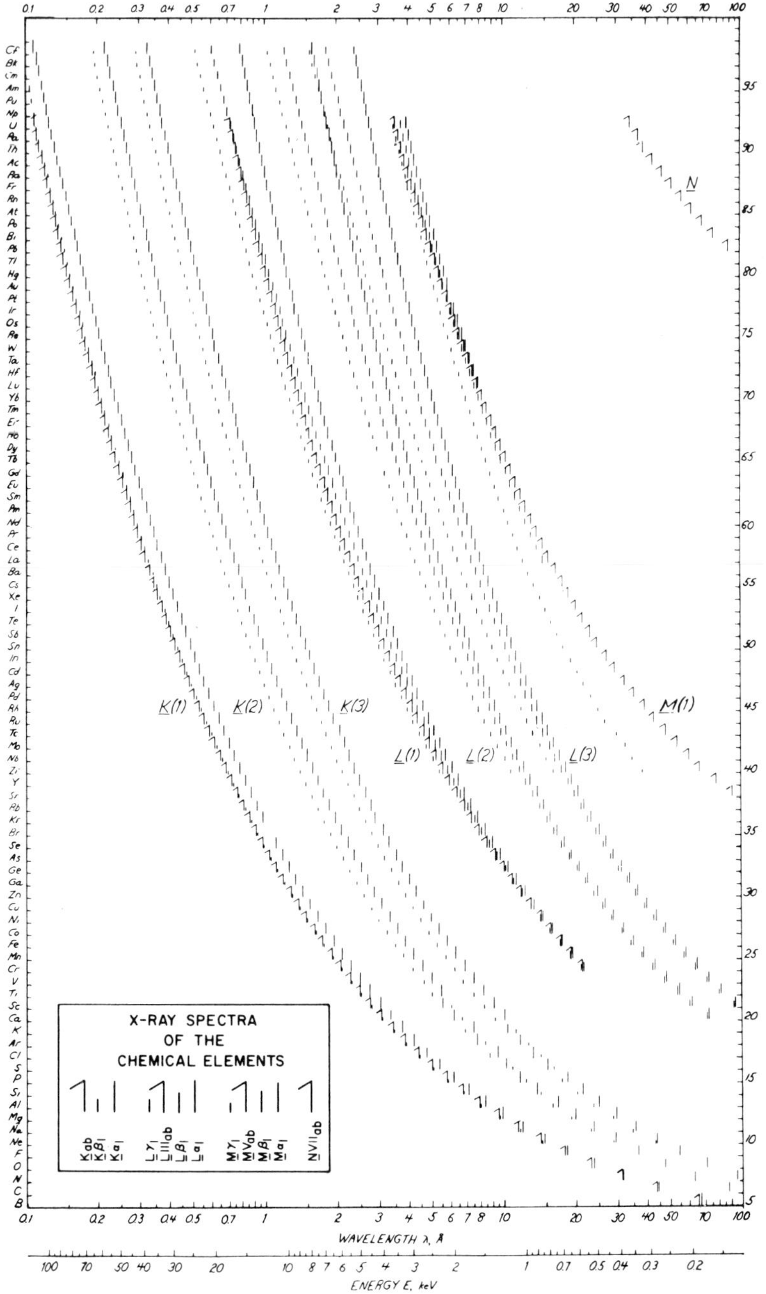

FIG. 1.8. X-ray spectra of the chemical elements.

TABLE 1.4. Wavelength and Photon-Energy Differences for $K\alpha$ Lines of Adjacent Chemical Elements

Elements	$\Delta\lambda$, Å	ΔE_x, keV
$_{46}$Pd, $_{47}$Ag	0.026	~1
$_{28}$Ni, $_{29}$Cu	0.117	~0.5
$_{13}$Al, $_{14}$Si	1.213	~0.25
$_{6}$C, $_{7}$N	13.	~0.1
$_{3}$Li, $_{4}$Be	127.	~0.05

constant to correct the effect of orbital electrons in reducing the effective nuclear charge, that is, the effective atomic number. Figure 1.9 is a Moseley-law plot for some prominent lines. The curves are linear except for a small deviation at high Z due to failure of k_2 to correct for the screening effect of the large number of electrons. In general, for practical purposes, Equation

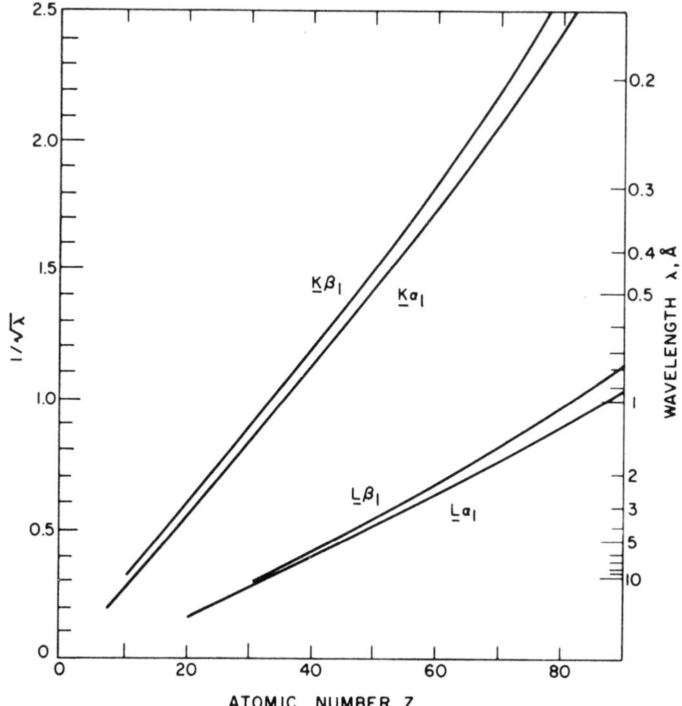

FIG. 1.9. Moseley-law diagram for selected x-ray spectral lines.

(1.30) may be reduced to the following simple relationship:

$$\lambda \propto 1/Z^2 \tag{1.31}$$

1.6.2.6. Intensity

The relative intensities of the strongest lines in the *K*, *L*, *M*, etc. x-ray spectral-line series of an element depend on the relative probabilities of expulsion of electrons from the respective shells of atoms of that element. If the primary x-radiation is of sufficiently short wavelength, it excites all the series of the irradiated element, that is, it simultaneously expels electrons from the *K*, *L*, *M*, etc. shells of the atoms of that element. However, such expulsions are most probable from the innermost shell that the primary x-rays can excite—the *K* shell in this case, so that *K*-series lines are produced with highest intensity. The probability of *K*-shell excitation and the intensities of the *K* lines increase as the primary wavelength approaches the short-wavelength side of the *K*-absorption edge, and both fall to zero at wavelength just longer than the *K* edge. Thereupon, *L*-shell excitation assumes highest probability and increases as primary wavelength approaches the *L*-absorption edges.

In general, the intensity of a line series is proportional to the jump ratio of the absorption edge associated with that series (Section 2.1.4). In a specified wavelength region, the *K*-edge jump is greater than the *L*III, which, in turn, is greater than the *M*V; thus, the $K\alpha$ line is more intense than the $L\alpha_1$, which is more intense than the $M\alpha$. As a general guide, for the same element, the *K*, *L*, and *M* series lines have approximate intensity ratios 100:10:1; and in a specified spectral region, first-order *K*, *L*, and *M* lines have approximate intensity ratios 100:10:1 for elements having the same concentration and excited at the same conditions.

The relative intensities of lines *within* a series depend on the relative probabilities of their respective electron transitions. The principal lines of analytical interest and their typical relative intensities are as follows:

$K\alpha_1$, 100	$L\alpha_1$, 100	$M\alpha_1$, 100
$K\alpha_2$, 50	$L\beta_1$, 75	$M\alpha_2$, 100
$K\alpha_{1,2}$, 150	$L\beta_2$, 30	$M\beta_1$, 50
$K\beta_1$, 15	$L\gamma_1$, 10	$M\gamma_1$, 5
$K\beta_2$, 2	$L\beta_3$, 5	
	$L\beta_4$, 3	
	Ll, 3	

(The relative intensities of the *series* are discussed above.) Some of these electron-transition probabilities—and therefore some of the relative intensities—change with atomic number. For example, although the $K\alpha_1$:$K\alpha_2$ ratio is always 2:1, the $K\alpha_1$:$K\beta$ ratio is 25:1 for $_{13}$Al, 5:1 for $_{29}$Cu, and 3:1 for $_{50}$Sn. This is because the $K\beta$ line arises from an $M \rightarrow K$ transition, and the M shell is not filled in the lighter elements, so that the probability of this transition is reduced. For similar reasons, L lines do not occur for the light elements, nor M lines for the light and intermediate elements. However, the relative intensities given above and in Figure 1.6 may serve as guidelines.

The foregoing discussion applies to the theoretical relative intensities based on atomic electron-transition probabilities. The relative intensities actually measured depend on many other factors, including x-ray tube target and potential, crystal reflectivity, absorption in the radiation path, and detector quantum efficiency, not to mention the x-ray spectral lines and absorption edges of the elements associated with the analyte in the specimen. Moreover, measured relative intensities for x-ray and electron excitation will differ for the reasons given in Sections 1.7.2. and 21.7.

Since spectral-line wavelength varies inversely as the square of the atomic number [Equation (1.31)], the separation between lines of the same series from adjacent elements decreases rapidly with increasing atomic number. This phenomenon is evident in Figure 1.8.

1.6.3. Excitation—General

We have seen that x-ray spectra arise when vacancies occur in the inner orbitals of atoms. X-ray spectral-line excitation, then, consists in creation of these vacancies in sufficiently large numbers. These vacancies are created principally in seven ways: (1) bombardment by electrons; (2) bombardment by protons, deuterons, α-particles, and other ions from particle accelerators; (3) irradiation by primary x-rays from high- or low-power x-ray tubes; (4) irradiation by α-, β-, γ-, and/or x-rays from radioisotopes; (5) irradiation by secondary x-rays from a target element (radiator) having a strong spectral line at wavelength just shorter than that of the analyte absorption edge; such secondary targets are themselves excited by primary x-rays from x-ray tubes or by radioisotopes; (6) "self-excitation" or "auto-excitation," in which no external radiation is applied to the specimen (Sections 8.2.2.3 and 14.8.4); this category may include (7) spontaneous radioactive phenomena. Most of these processes are illustrated in Figure 1.10.

In practical x-ray spectrochemical analysis, excitation of characteristic x-ray spectra may be classified as: (1) *primary* or *secondary*, depending on

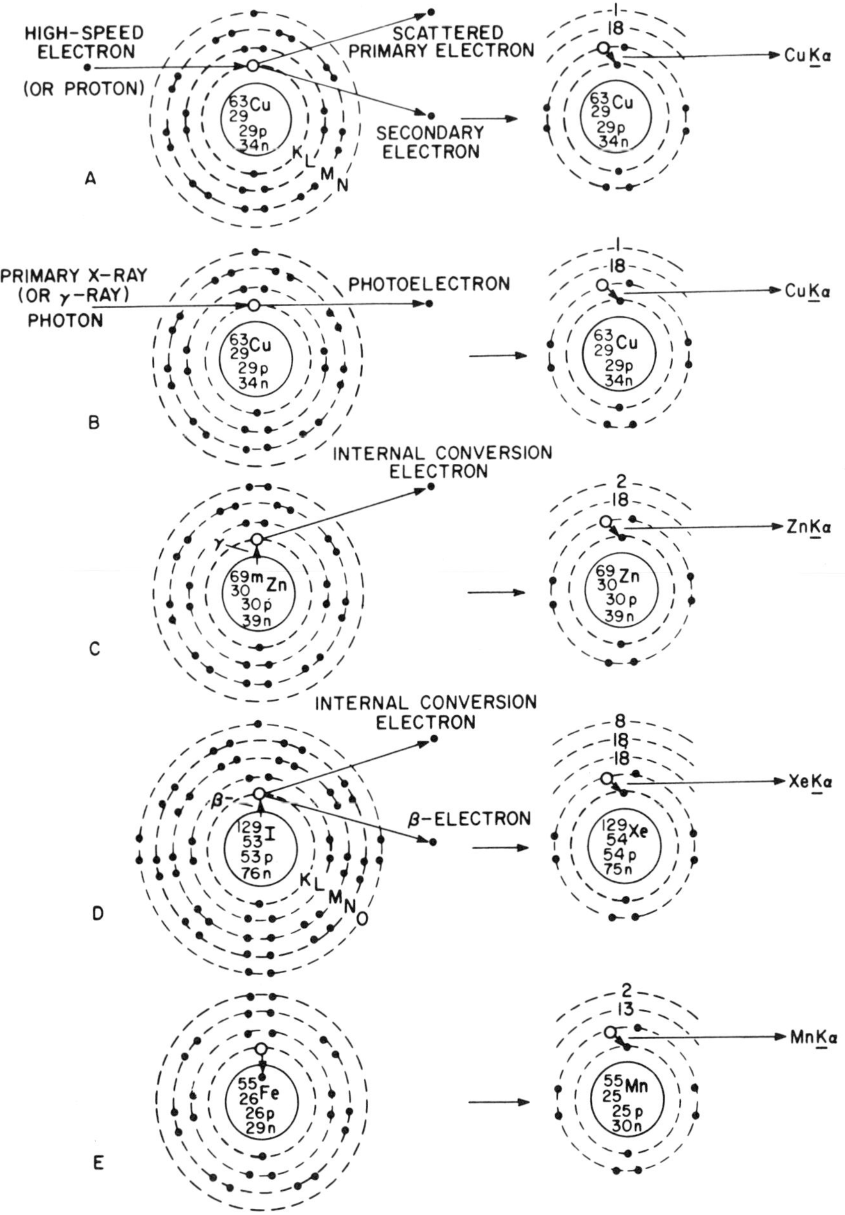

FIG. 1.10. Excitation of characteristic x-ray line spectra. A. Primary excitation. B. Secondary excitation (fluorescence). C. Internal γ conversion. D. Internal β conversion. E. Orbital-electron capture.

whether excitation is by electrons or ions or by x- or γ-ray photons, respectively; or (2) *x-ray tube*, *electron*, *ion*, or *radioisotope*, depending on the means of effecting the excitation.

Recently, an extremely interesting x-ray excitation phenomenon has been observed to occur when a target is bombarded by heavy ions (*M31*). As an incident ion approaches a target atom, the electron orbitals of the two atoms undergo various transformations and eventually may assume, instantaneously, the configuration of the atom having atomic number equal to the sum of the atomic numbers of the two colliding atoms. Thus, during bombardment of argon (Z 18) with argon ions, x-ray spectral lines of krypton (Z 36) have been observed. This has led some physicists to speculate that the x-ray spectra of superheavy elements may be observable—before the elements are produced in the usual sense! For example, bombardment of uranium (Z 92) with argon ions (Z 18) may result in production of x-ray lines of element 110!

The phenomenon just described is not to be confused with the similar and related phenomenon of *Pauli excitation* (*B80*). In this process, K- and L-shell ionization by high-speed, heavy-ion bombardment may give x-ray excitation cross sections for both projectile and target elements up to 10^3–10^5 those for Coulomb excitation. The phenomenon is attributed to the Pauli exclusion principle (Section 1.6.1) operating in the overlapping electron clouds of the interpenetrating atoms, forcing their electrons into transient quasimolecular configurations. When the atoms separate, they are in excited states and emit x-ray spectra.

Incidentally, in the Apollo-15 lunar mission, solar x-radiation was used as the excitation source for x-ray spectrometric measurement of the aluminum/silicon ratio in lunar rocks (*A13*)!

In primary, or direct, excitation (Figure 1.10A), the vacancies are created by bombarding matter with high-speed electrons. The deflected incident electron becomes a scattered electron, and the ejected electron is termed a secondary electron. High-speed protons, α-rays, or ions from radioisotopes or particle accelerators have the same effect. Primary excitation occurs in x-ray tubes, electron-probe x-ray microanalyzers, and certain instruments in which the specimen is bombarded by electrons in vacuum or outside a Lenard window, and on irradiation of matter with α or β radioisotopes.

In secondary, or fluorescence, excitation (Figure 1.10B), the vacancies are created by irradiating matter with high-energy photons, usually x-rays from an x-ray tube. The incident photon is absorbed, giving up its energy in expelling and imparting kinetic energy to the orbital electron. This process is termed photoelectric absorption, and the ejected electron is a

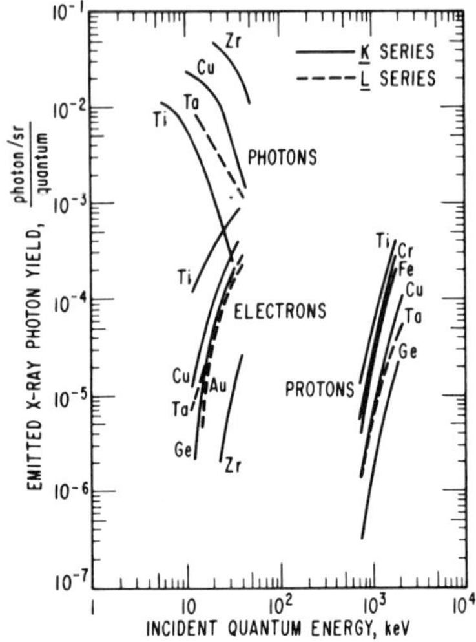

FIG. 1.11. X-ray spectral-line photon yield for bulk samples for excitation by x-ray photons, electrons, and protons. The curves give numbers of x-ray photons/steradian per incident quantum—that is, incident photon, electron, or proton—as functions of atomic number of the emitter and incident quantum energy. [L. S. Birks, R. E. Seebold, A. P. Batt, and J. S. Grosso, *Journal of Applied Physics* **35**, 2578 (1964); courtesy of the authors and the American Institute of Physics.]

photoelectron. Gamma rays have the same effect. Secondary excitation occurs in x-ray fluorescence spectrometers and on irradiation of matter with γ radioisotopes.

Figure 1.11 gives the x-ray photon yield for excitation by x-ray photons, electrons, and protons in terms of emitted x-ray photons/steradian per incident quantum—as functions of atomic number and incident quantum energy (*B65*).

X-rays generated by electron or proton bombardment, that is, by primary excitation, are termed *primary x-rays* and may be continuous or characteristic. X-rays generated by primary x-rays or perhaps by γ-rays, that is, by secondary excitation, are termed *secondary x-rays* and may be only characteristic (Section 1.5.4). X-ray generation by secondary x-rays is termed *tertiary excitation* and the radiation *tertiary x-rays*, which also may be only characteristic (Sections 1.6.5.5, 8.3.1, and 12.1). Sometimes these terms are used in a somewhat different sense: For example, in the chromium–iron–nickel system, Cr $K\alpha$ is excited by both Fe $K\alpha$ and Ni $K\alpha$, but Fe $K\alpha$ is excited only by Ni $K\alpha$. Some workers regard the excitation of Cr $K\alpha$ by primary-beam-excited Fe $K\alpha$ and Ni $K\alpha$ as secondary, and excitation of Cr $K\alpha$ by Fe $K\alpha$ excited by Ni $K\alpha$ as tertiary. Jenkins terms these two phenomena *direct enhancement* and the *third-element effect*, respectively. By the definitions given above, they would be *tertiary* and *quaternary excitation*, respectively.

Radioactive isotopes that decay by γ-emission may undergo internal conversion (Figure 1.10C). The γ photon is absorbed within the atom of its origin, giving up its energy in expelling and imparting kinetic energy to an orbital electron, which then becomes an internal-conversion electron. Internal γ-conversion may be regarded as secondary excitation in which the ionizing photon originates within the atom. The γ-ray may also escape its own atom and cause secondary excitation in another atom (Figure 1.10B).

Radioactive isotopes that decay by β-emission may also undergo internal conversion (Figure 1.10D) or cause primary excitation (Figure 1.10A) of other atoms. Internal β-conversion may be regarded as primary excitation in which the ionizing electron originates within the atom. However, the emission of a β-particle (electron, e^-) results from the conversion of a neutron to a proton ($n \rightarrow p^+ + e^-$), and thereby increases the atomic number by one. Then the internal conversion results in the emission of x-rays characteristic of this new element. In the example in Figure 1.10D,

$$^{129}_{53}\mathrm{I} \rightarrow {}^{129}_{54}\mathrm{Xe} + \beta^-$$

Thus, an ^{129}I source emits the xenon x-ray spectrum.

In radioactive isotopes that decay by orbital-electron capture (Figure 1.10E), a *K* or *L* electron actually falls into the nucleus, thereby neutralizing a proton ($p^+ + e^- \rightarrow n$) and decreasing the atomic number by one. In the example given,

$$^{55}_{26}\mathrm{Fe} \xrightarrow{K\ \text{capture}} {}^{55}_{25}\mathrm{Mn}$$

Thus, an ^{55}Fe source emits the manganese x-ray spectrum.

Radioactive isotopes are often used instead of x-ray tubes to excite the characteristic x-ray spectrum of the specimen. This mode of excitation is discussed in Section 8.2.2.3.

In primary, including β-ray, excitation, the characteristic line spectrum is superposed on strong, continuous radiation generated simultaneously. In secondary, including γ-ray, excitation, only line x-ray spectra are *emitted.* However, these line spectra are superposed on *scattered* primary continuum or the unabsorbed γ-ray spectrum. Radioisotopes that decay only by orbital-electron capture produce very nearly pure line spectra. In all the various modes of excitation of line spectra, some feeble continuum is generated by photo, internal-conversion, recoil, and Auger electrons.

Throughout this section, *K* excitation is used as the example, but the principles apply as well to *L*, *M*, and *N* excitation. Thus, there can be primary or secondary *L* excitation, internal conversion of *L* electrons, and *L*-orbital electron capture.

In any of the modes of excitation, the ionizing agent—incident electron or photon, β- or γ-ray, etc.—must be sufficiently energetic to expel the orbital electrons, and more energy is required the more tightly the electrons are bound. This means that electron expulsion from the same level—*e.g.*, the *K*—requires more energy as *Z*, and therefore the positive nuclear charge, increases; and expulsion from any one atom requires more energy the closer the level lies to the nucleus. Of course, the line wavelengths of an element are the same regardless of how the vacancies originate.

We have seen that the orbital electron vacancies necessary for generation of characteristic line spectra arise principally in two ways—by primary or secondary (fluorescence) excitation, that is, by electron bombardment or x-irradiation. We now consider each of these modes in detail.

1.6.4. Primary Excitation

The minimum x-ray tube potential that can accelerate electrons from filament to target with enough energy to expel an electron from a given level in an atom of a given target element is known as the *excitation potential* or *critical absorption energy* of that level of that element. In fact, if the numerical value of the excitation potential is expressed in electron volts (or kiloelectron volts) instead of volts (or kilovolts), one gets the corresponding electron energy. Each element has many excitation potentials: one K (V_K), three L (V_{LI}, V_{LII}, V_{LIII}), five M, seven N, etc. For each element, the excitation potential increases for levels progressively closer to the nucleus: $V_K > V_{LI} > V_{LII} > V_{LIII} > V_M > \ldots$. For the same level, excitation potential increases with Z. The K, L, and M excitation potentials of gold are represented in Figure 1.6 by the arrows directed upward from each level, and the potentials are given at the top of the figure. The K, L, and M excitation potentials of the elements are plotted in Figure 1.12 and listed in Appendix 6.

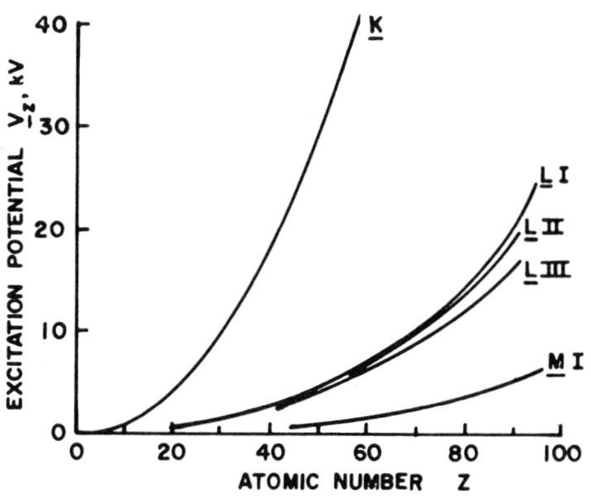

FIG. 1.12. X-ray excitation potentials of the chemical elements. (See Appendix 6.)

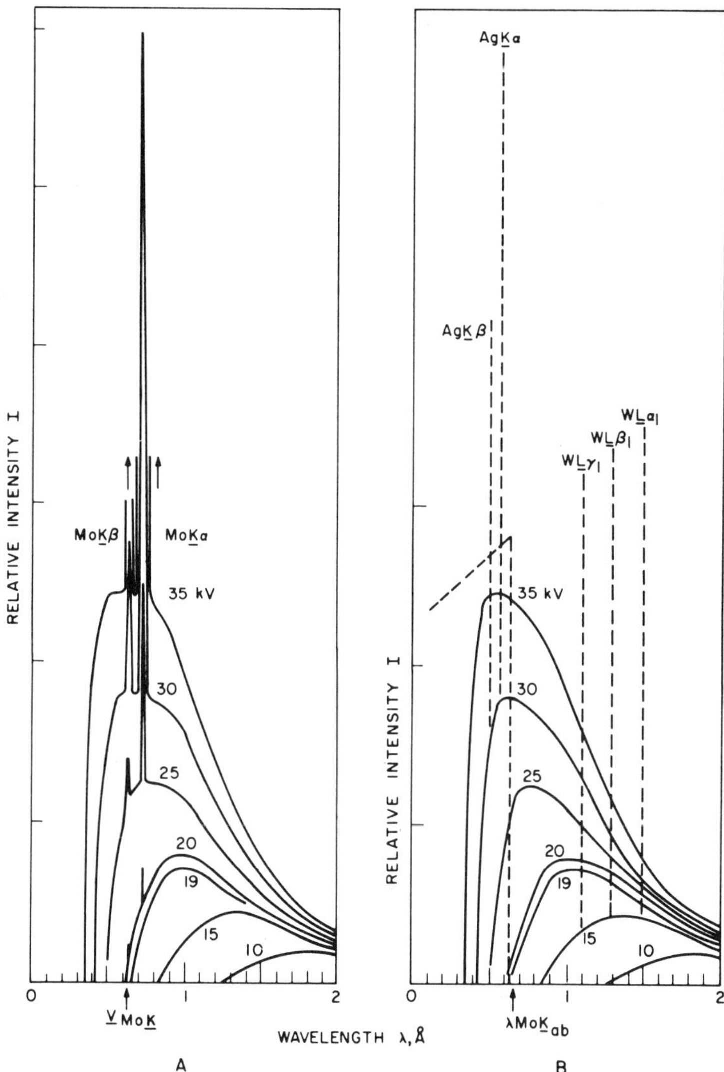

FIG. 1.13. (A) Primary and (B) secondary excitation of the molybdenum *K* spectrum.

Figure 1.13A illustrates primary excitation using the molybdenum *K* spectrum as an example, but the principles are general. (Figure 1.13B is discussed later.) The figure shows spectra from an x-ray tube having a molybdenum target and operating at various potentials. Below the molybdenum *K* excitation potential, $V_{\mathrm{Mo}K} = 20$ kV, only continuum is generated.

At 20 kV, very feeble molybdenum K lines appear superposed on the continuum. At higher potentials, the molybdenum line intensity increases strongly. In general,

$$I_K \propto i(V - V_K)^{\sim 1.7} \tag{1.32}$$

where I_K is the intensity of a specific K line; V_K is K excitation potential; and V and i are x-ray tube potential and current, respectively (*J27*). The quantity $V - V_K$ is sometimes referred to as the *overpotential* (or *overvoltage*); however, this term is more properly applied to V/V_K, the ratio of the x-ray tube and excitation potentials. For full-wave rectification at $V \geq 3V_K$, the exponent becomes $\sim$1. Analogous equations apply to the L and M levels.

Some characteristic target line radiation originates from secondary excitation of the target element by its own continuum (*C50*).

An electron that has enough energy to expel a K electron obviously can also expel any L or M electron. However, an electron having just enough energy to expel, say, an LII electron can also expel LIII, M, and N electrons, but not LI or K electrons. It follows that all spectral lines appear simultaneously that result from electron transitions to the innermost level excited and to all levels farther out. Even if only, say, LII vacancies were created by the incident electrons, vacancies farther out would result from electron transitions into the LII level (Figures 1.5 and 1.6), insofar as permitted by the selection rules.

In general, all lines of a given *level* appear at once. The exception is that lines arising from transitions to that level from levels very far from the nucleus do not appear if the atom is too small to have any electrons in that outer level.

Ordinarily, the L lines of the target would be substantially less intense than the K lines. However, the overpotential (above) can drastically alter this relationship. For example, in a thin-window rhodium-target tube operating at 45 kV(CP), the K lines (V_K 23.2 kV, $V/V_K \approx 2$) contribute $\sim$10% of the total emitted flux, the L lines ($V_{L\text{III}}$ 3.0 kV, $V/V_{L\text{III}} \approx 15$) contribute $\sim$25%.

1.6.5. Secondary Excitation

1.6.5.1. X-Ray Absorption Edges

Although x-ray absorption is discussed in detail in Chapter 2, some consideration of absorption edges is necessary here for an understanding of secondary excitation. X-ray absorption curves for several elements are shown in Figure 1.14. It is evident that in general the x-ray absorption coefficient of an element—that is, its x-ray "stopping power"—decreases with x-ray wavelength. This is to be expected because the shorter the wavelength, the greater the energy and penetrating power. However, there are abrupt discontinuities on these curves, and it is these *absorption edges* or *critical absorption wavelengths* that are significant in secondary excitation.

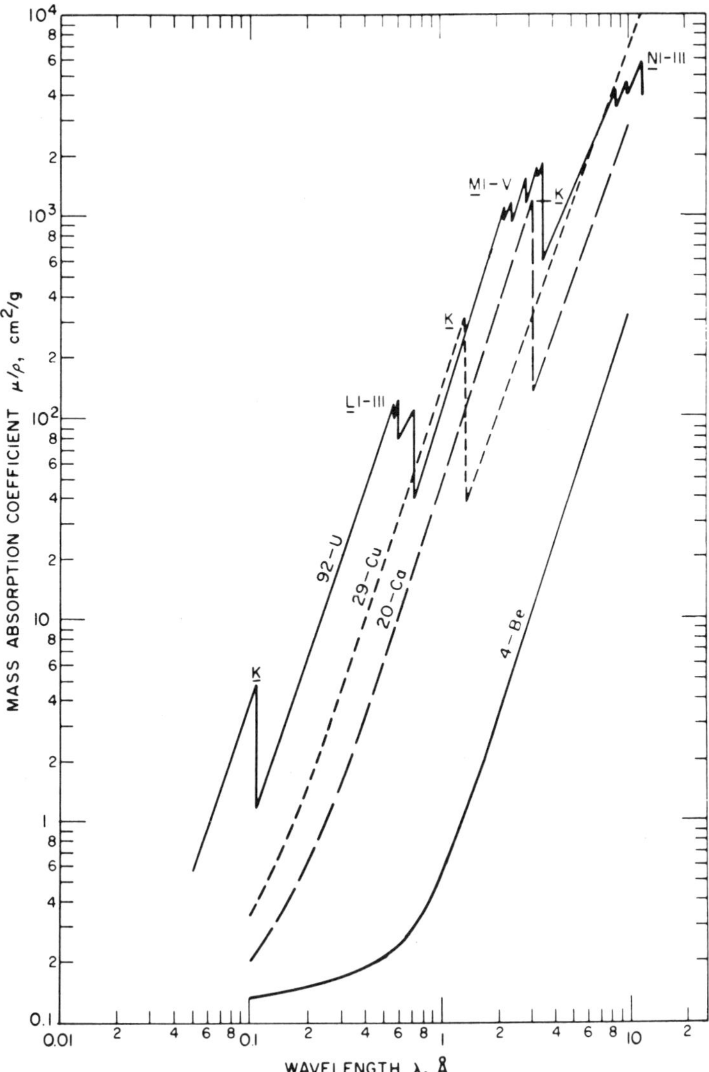

FIG. 1.14. Typical x-ray absorption curves.

The maximum wavelength (minimum photon energy) that can expel an electron from a given level in an atom of a given element is known as the *absorption edge* of that level of that element. Each element has as many absorption edges as it has excitation potentials: one K ($\lambda_{K_{ab}}$), three L ($\lambda_{LI_{ab}}$, $\lambda_{LII_{ab}}$, $\lambda_{LIII_{ab}}$), five M, seven N, etc. The most prominent edges in the multiple series are the LIII, MV, and NVII. For each element, the absorption-edge wavelength decreases (critical absorption energy increases) for levels progressively closer to the nucleus: $\lambda_{K_{ab}} < \lambda_{LI_{ab}} < \lambda_{LII_{ab}} < \lambda_{LIII_{ab}} < \lambda_{MI_{ab}} < \ldots$. For a given level, this wavelength decreases as Z increases. The K, L, and M absorption edges of gold are represented in Figure 1.6 by the arrows directed upward from each level, and the wavelengths are given at the top of the figure. The K, L, and M absorption edges of all the elements are listed in Appendix 5, and absorption curves showing K and L edges of many elements are shown in Figures 2.5 and 2.6. Figure 1.8 shows, for all elements, the K, LIII, MV, and NVII edges that occur in the wavelength region 0.1–100 Å.

1.6.5.2. Principles

Referring to the absorption curves in Figure 1.14, consider a monochromatic primary x-ray beam of wavelength λ incident on a secondary target of atomic number Z. Consider what happens as λ is made progressively shorter (more energetic).

At $\lambda > \lambda_{ZK_{ab}}$, the photons do not have enough energy to expel Z K electrons; consequently, no Z K lines appear. As λ decreases, the photons become more energetic, and the absorption coefficient decreases, that is, the secondary target becomes more transparent. At $\lambda_{ZK_{ab}}$, the photons have exactly the energy required to expel Z K electrons, and the absorption increases abruptly. The absorbed photons expel K electrons, and Z K lines are emitted. This process is *photoelectric absorption*, and the expelled electrons are *photoelectrons*. The Z K electron expulsion and line emission are most efficient at λ just less than $\lambda_{ZK_{ab}}$.

At $\lambda \ll \lambda_{ZK_{ab}}$, the photons have much more than enough energy to expel Z K electrons. However, they are so energetic that they may not be absorbed, or else, before being absorbed, they may penetrate the target to a depth from which Z K radiation cannot emerge. Thus, as λ becomes progressively shorter, line emission decreases. Note the difference in this respect between primary and secondary excitation. In primary excitation, emitted line intensity *increases* as x-ray tube potential V is made greater than the target K excitation potential V_K: $I_K \propto i(V - V_K)^{\sim 1.7}$. [Equation

(1.32)]. In secondary excitation, the emitted line intensity decreases as primary wavelength is decreased below $\lambda_{ZK_{ab}}$. This difference is evident in Figure 1.13A.

Even though at $\lambda < \lambda_{ZK_{ab}}$ photons have more than enough energy to expel Z *K* electrons, each photon that is absorbed by an electron is wholly absorbed. The surplus energy imparts kinetic energy to the photoelectron, so that the shorter the wavelength, the higher the velocity of the photoelectrons.

The preceding discussion regarding wavelengths bracketing the *K* absorption edge applies as well to wavelengths bracketing the several *L* and *M* edges.

A photon that has enough energy to expel a *K* electron obviously can also expel any *L* or *M* electron. However, a photon having wavelength between $\lambda_{LI_{ab}}$ and $\lambda_{LII_{ab}}$ (see the absorption curve for uranium in Figure 1.14) can expel *L*II, *L*III, *M*, and *N* electrons, but not *L*I or *K* electrons. The fourth and fifth paragraphs of Section 1.6.4 regarding just which line series do and do not appear simultaneously apply here as well.

1.6.5.3. Relationship of Absorption Edges and Spectral-Line Series

Figure 1.15, which complements Figure 1.6, shows all *K*, *L*, and *M* absorption edges of gold and the principal lines associated with each. The figure is largely self-explanatory, but some features warrant discussion.

It must be reemphasized that only photons having wavelength equal to or less than that of an absorption edge can excite the associated lines. It is evident in Figures 1.6 and 1.15 that there is a very narrow wavelength interval between each absorption edge and the associated line of shortest wavelength. Even a wavelength in this interval—shorter than the shortest series line, but nevertheless slightly longer than the edge—cannot excite the spectrum of this series.

Figure 1.15 shows that the lines of each series of an element occur on the long-wavelength side of the corresponding absorption edge where absorption is relatively low. It follows that an element is relatively transparent to its own spectral lines. It also follows that lines of a given series of an element cannot excite that series of that element, although they can excite a series having an absorption edge at longer wavelength. Thus, barium *K* lines excite barium *L* lines, but not barium *K* lines.

No line in a series can have wavelength shorter than that of the series edge. For example, in Figure 1.15, although some *L*I and *L*II lines are

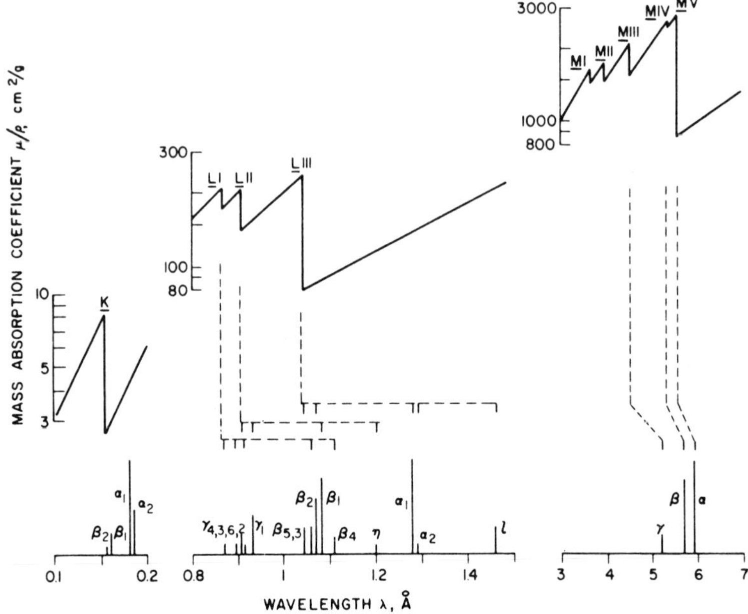

FIG. 1.15. Relationship between absorption edges and spectral-line series.

shorter than the *L*III edge, no *L*III line is shorter than the *L*III edge. Also, in Figure 1.8, the $L\gamma_1$ lines on the short side of the *L*III edges might be construed as an error in the figure. However, the $L\gamma_1$ line is associated with the *L*II edge (not shown), not the *L*III edge.

The reason lines must have less energy—longer wavelength—than their absorption edges is as follows. The wavelength of an absorption edge is equivalent to the energy *absorbed* in creating an inner-orbital vacancy, that is, in lifting an electron from that level clear out of the atom—to the conduction band or free state. The wavelength of a spectral line is equivalent to the energy *emitted* on filling that vacancy; this is effected by an electron transition, not from outside the atom, but from an outer orbital. Thus, less energy is emitted than was absorbed. The farther out the orbital from which the electron falls, the more nearly equivalent the transition to the expulsion, and the closer the line lies to the absorption edge. This can be shown in Figure 1.6 by comparing in each series the transition for the shortest line with those for the other lines. The comparison of x-ray and optical absorption in Section 2.1.5 is relevant to this discussion.

1.6.5.4. Excitation with Polychromatic X-Rays

In an x-ray fluorescence spectrometer, the primary beam is not monochromatic, but consists of the continuum and the line spectrum of the x-ray tube target element. Only that part of the primary beam having wavelength shorter than the absorption edge of an element contributes to the excitation of the spectrum of that element.

Figure 1.13B illustrates secondary excitation using the molybdenum K spectrum as an example, but the principles are general. The figure shows the spectra of the primary beam from the x-ray tube that irradiates an external molybdenum secondary target in the specimen compartment. The molybdenum K absorption edge, $\lambda_{\mathrm{Mo}K_{\mathrm{ab}}} = 0.62$ Å, is also shown.

If the molybdenum K spectrum is to be excited, the primary beam must contain x-rays having wavelength $\leq \lambda_{\mathrm{Mo}K_{\mathrm{ab}}}$. This means that the short-wavelength limit of the continuum $\lambda_{\min} \leq \lambda_{\mathrm{Mo}K_{\mathrm{ab}}}$. Consequently, by Equation (1.25), the tube must be operated at a potential

$$V = \frac{12{,}396}{\lambda_{\mathrm{Mo}K_{\mathrm{ab}}}} = V_{\mathrm{Mo}K} \tag{1.33}$$

For $\lambda_{\mathrm{Mo}K_{\mathrm{ab}}} = 0.62$ Å, $V = V_{\mathrm{Mo}K} = 20{,}000$ V, or 20 kV. Figure 1.13B shows that at potentials <20 kV, the primary spectrum contains no x-radiation having wavelength $\leq \lambda_{\mathrm{Mo}K_{\mathrm{ab}}}$, and even at 20 kV, there is very little. However, as V is increased, the amount of useful primary radiation increases sharply, and the molybdenum K spectrum is excited with great efficiency. In general, it is desirable to operate the tube at a potential such that the "hump" of the continuum ($\lambda_{I_{\max}}$) occurs at a wavelength shorter than that of the absorption edge to be excited.

So far, it has been tacitly assumed that only primary continuum excites the molybdenum K lines. However, the primary beam also contains target lines superposed on the continuum. In Figure 1.13B, if the target is tungsten, and the tube is operated at $\geq V_{\mathrm{W}L\mathrm{III}}$ (~11.5 kV), tungsten $L\alpha_1$, $L\beta_1$, and $L\gamma_1$ lines appear as shown. These lines, having wavelengths $> \lambda_{\mathrm{Mo}K_{\mathrm{ab}}}$, cannot excite the Mo K spectrum. However, if the target is silver, and the tube is operated at $\geq V_{\mathrm{Ag}K}$ (~25.5 kV), the silver $K\alpha$ and $K\beta$ lines appear as shown. These lines have wavelengths $< \lambda_{\mathrm{Mo}K_{\mathrm{ab}}}$ and contribute to the excitation of the Mo K spectrum.

1.6.5.5. Other Contributions to the Specimen Emission

Although secondary excitation by the continuum and target lines in the primary beam constitutes the principal source of characteristic line emission from the specimen, other phenomena may contribute to a smaller extent.

The spectral lines of one element may be enhanced by primary-beam-excited spectral lines of other specimen elements having wavelengths shorter than its own absorption edge. In thin-film and supported specimens, the lines of elements in the film or specimen may be enhanced by lines of elements in the substrate, and, to a smaller extent, *vice versa* (Section 18.3.4) (*P26*).

The spectral lines of an element may also be enhanced by spectral lines of other specimen elements which were themselves excited by primary-beam-excited spectral lines. Of course, each "generation" of excitation contributes substantially less than its predecessor to the emitted characteristic x-rays (Section 12.2.1) (*P24*).

Scattered primary x-rays, continuous and/or target-line, may contribute as much as 1% to the characteristic x-rays emitted from bulk specimens, and as much as 30% to those from thin films and supported specimens (*P25*).

Photo, recoil, and Auger electrons generated by the primary x-ray beam, already cited as contributing all continuum emitted by—as distinguished from scattered by—the specimen, also feebly enhance the emitted characteristic x-rays.

1.7. COMPARISON OF PRIMARY AND SECONDARY EXCITATION

1.7.1. X-Ray Tube Potential

It should be obvious that excitation potentials (which, as we have seen, are numerically equal to electron energies in electron volts or kilo-electron volts) and absorption edges are analogous quantities. Both are measures of the energy required to expel an electron from a specified level in an atom of a specified element, one applying to electrons in primary excitation, the other to photons in secondary excitation. One might say that the excitation potential is the potential or kilovolt equivalent of the absorption edge, and that the absorption edge is the wavelength or angstrom equivalent of the excitation potential. The two are related by equations having the form of Equations (1.18) and (1.25):

$$V_K = \frac{12{,}396}{\lambda_{K_{\text{ab}}}}; \qquad \lambda_{K_{\text{ab}}} = \frac{12{,}396}{V_K} \tag{1.34}$$

$$V_{LIII} = \frac{12{,}396}{\lambda_{LIII_{\text{ab}}}}; \qquad \lambda_{LIII_{\text{ab}}} = \frac{12{,}396}{V_{LIII}} \tag{1.35}$$

Thus, to excite, say, the molybdenum K spectrum, an x-ray tube must be operated at or above $V_{\mathrm{Mo}K}$. This is true regardless of whether the molybdenum is made the tube target (primary excitation) or a tube having any arbitrarily chosen target is used to irradiate the molybdenum placed external to the tube (secondary excitation). Below $V_{\mathrm{Mo}K}$, in the former case, the electrons will not be accelerated sufficiently to expel K electrons from the molybdenum target, and in the latter case, there will be in the primary beam no x-rays of wavelength shorter than $\lambda_{\mathrm{Mo}K_{\mathrm{ab}}}$.

Moreover, if the molybdenum spectrum is to be excited efficiently, the tube must be operated at $V \gg V_{\mathrm{Mo}K}$—for either mode of excitation, as shown in Figure 1.13. High x-ray tube potential in primary excitation has the effect stated in Equation (1.32); in secondary excitation, it enriches the primary beam in the wavelength region below $\lambda_{\mathrm{Mo}K_{\mathrm{ab}}}$.

1.7.2. Features

In primary excitation, as the incident electrons penetrate the specimen, they decrementally lose energy and thereby their ability to excite spectra having successively lower excitation potentials. Thus, spectra having successively higher excitation potentials originate in successively thinner surface layers. Figures 21.14B and C show the phenomenon. Conversely, in secondary excitation, as the incident photons penetrate the specimen, they retain their full photon energy right up to the instant when they are totally absorbed, regardless of how deep they penetrate. Thus, monochromatic incident x-rays excite all spectra throughout their depth of penetration regardless of excitation potential, provided only that the excitation potential is less than the photon energy. However, this is not to be confused with the phenomenon that *polychromatic* incident x-rays do decrease in effective wavelength—increase in effective photon energy—as they penetrate because longer-wavelength x-rays are absorbed preferentially in the upper layers. Thus, spectra having successively higher excitation potentials originate in successively thicker surface layers, the opposite of the case with primary excitation.

With electron excitation, the effective specimen layer contributing the measured analyte-line x-rays is determined by the depth *to* which the electrons penetrate with sufficient energy to excite the analyte line. As the electrons diffuse into the specimen surface, their initial energy decreases decrementally and successively falls below the K, L, and M excitation potentials of the *same* element, and/or below the K (or L or M) excitation potentials of elements of successively lower atomic number (Section 21.7,

Figure 21.14). With x-ray excitation, the primary beam from an x-ray tube operating at 50 kV contains wavelengths as short as 0.25 Å [Equation (1.25)]. Also, the angle between the primary beam and specimen surface is usually about twice the takeoff angle. These conditions result in primary-beam penetration and analyte-line excitation—in a relatively thick layer on the specimen surface. However, only analyte-line x-rays actually leaving the specimen can be measured. Thus, the effective layer is determined by the depth *from* which the analyte-line x-rays can emerge (Section 16.1.5). Thus, the shorter the analyte-line wavelength, the deeper is the effective layer—the reverse of the case with electron excitation.

Primary excitation gives a high-intensity line spectrum on a high-intensity continuous spectrum. The intensity ratio of target line and continuum radiation depends on x-ray tube potential, excitation potential, and whether full-wave or constant potential is used. Recent work indicates that this ratio is greater than previously believed (*G13*). Ratios of 1:3 and 3:1 have been observed at 45 kV (CP) for tungsten (*L* lines) and chromium (*K* lines) targets, respectively. Total intensity of the continuum increases in proportion to target atomic number and the square of the x-ray tube potential [Equation (1.28)]. Target-line intensity increases in proportion to the 1.7 power of the difference between operating and excitation potentials [Equation (1.32)].

Secondary excitation gives only a line emission spectrum because primary x-ray photons cannot lose energy incrementally, as electrons do. Most continuum in the secondary spectrum is scattered primary continuum; the remainder is generated by photo, recoil, and Auger electrons. Peak-to-background ratios with secondary excitation may be as high as 10,000:1, whereas with primary excitation they are of the order 1000:1 to 300:1 or less.

For a given x-ray tube power (kV, mA), the line spectrum is ~100 times as intense with primary excitation. Only a relatively small portion of absorbed primary x-radiation is reemitted as secondary line radiation, and this is emitted in all directions. Secondary radiation emitted too far below the emitter surface is wholly absorbed in the emitter.

The phenomenon of self-absorption directs attention to an interesting difference between primary and secondary x-ray sources. It is possible to obtain a small, intense *apparent* source of x-rays by viewing a primary-excited x-ray source at a small angle. The electrons penetrate the surface to a depth of only about a micrometer, so that the x-ray source is a layer so thin that it is substantially transparent, even to x-rays emitted at glancing angles. This phenomenon is applied to great practical advantage in x-ray diffraction. X-ray tubes for diffraction have linear focal spots. The takeoff angle of the emergent x-ray beam is made very small relative to the target surface to provide a very small, extremely intense line or point source, depending on whether the focal spot is viewed from the side or end. The phenomenon is also applied in the takeoff geometry of the x-ray spectrometer tube, as shown in Figure 4.4. However, it is not possible to obtain such a small apparent source by viewing a *secondary*-emitting source of x-rays at a small angle. The primary x-rays penetrate the emitting surface to a depth of 10–100 μm, and secondary x-rays emitted at small angles cannot emerge through the long path within the surface. Consequently, a loss, rather than a gain, in "brightness" occurs at small angles. Essentially, a secondary-emitting x-ray surface is a Lambert-type emitter.

The advantages of electron excitation include: (1) better sensitivity for light elements because of the greater excitation efficiency; (2) reduced absorption-enhancement effects because the electrons penetrate only a thin surface layer; (3) better sensitivity for thin films because the electrons are substantially absorbed in the film, whereas x-rays penetrate, causing little photoionization; and (4) ease of focusing the electron beam to small diameter for use with curved-crystal spectrometers and for analysis of small selected areas. The disadvantages include: (1) higher background because of the continuum; (2) low penetration of the specimen surface; (3) generally reduced sensitivity, except for low-Z elements; (4) the requirement for high vacuum; (5) possible alteration of specimen composition by the electron beam; (6) the fact that liquid specimens cannot be used, except with Lenard-window instruments, because they cannot be held in vacuum (a Lenard window is a thin metal-foil window through which electrons can emerge from a vacuum enclosure); and (7) the fact that the specimen surface must be an electrical conductor, or it must be coated with a conductive film; this feature may be especially troublesome with powders, which may disperse on charging.

Instrumentation for primary excitation is described in Section 9.4.2.

1.8. EXCITATION BY ION BOMBARDMENT (*B24*, *C1*, *S61*, *U1*)

The instrumentation for characteristic x-ray spectral-line excitation by ion bombardment is simple in principle. Some type of ion accelerator produces an ion beam of energy of the order 10 keV to 10 MeV, which is directed on the analytical target in a vacuum chamber. The emitted characteristic x-rays are detected by a lithium-drifted silicon detector used with an energy-dispersive spectrometer system. The accelerator may be of a relatively small laboratory type, or the x-ray spectrometer may be set up at the beam-exit port of a large accelerator. The ion-bombarded specimen area is typically of the order 1 mm^2, and the ion-current density and ion dose are typically of the order 1 $\mu A/cm^2$ and 0.1 μC, respectively.

Protons are perhaps the most commonly used ion projectiles; high fluence (number/cm^2 per s) and/or high-energy proton beams are obtained relatively easily, and their excitation mechanism is fairly well understood and, in fact, is similar to that of electrons. Deuteron, triton, and α-particle beams are also readily obtained. Even heavier ions, such as $_{10}Ne^+$, $_{18}Ar^+$, and $_{36}Kr^+$ are also receiving increased application. When the bombarding and target atoms have about the same atomic number, a special phenomenon known as *Pauli excitation* occurs in which inner-shell ionization and x-ray excitation are greatly enhanced by interaction between the electronic structures of the bombarding and target atoms. A similar phenomenon is described briefly in paragraph 3 of Section 1.6.3.

In general, the x-ray yield from ion bombardment is much lower than from electron or even x-ray excitation, but the yield increases with incident ion energy (Figure 1.11) (*N14*). The accelerator must be more elaborate, the higher the mass, energy, and fluence of the ion beam. Other limitations of ion excitation are the following. The emitted x-ray spectrum may be complicated by the x-ray spectrum of the bombarding ions arising from previously implanted projectile ions; the x-ray excitation cross section for such "symmetrical" collisions is very high. Heavy ions tend to remove target atoms by sputtering. This phenomenon can sometimes be used to advantage to strip the surface layer by layer for determination of the analyte depth distribution.

Ion excitation has several very valuable features:

1. The continuous spectrum is substantially nonexistent. For example, the continuum with proton excitation is weaker than for electron excitation by a factor equal to the square of the ratio of the electron and proton masses, or $\sim(1/1800)^2$. The continuum is accordingly successively less for deuterons, tritons, α-particles, and heavier ions. The reduced background favors sensitivity compared with electron and x-ray excitation, where there is, respectively, emitted and scattered continuum.

2. Ion excitation is applicable to trace analysis of the top several monolayers of the specimen surface by use of low energy or high mass to reduce penetration.

3. Heavy ions, such as Ne^+, Ar^+, and Kr^+, have higher x-ray excitation cross sections than expected, and also well-defined excitation threshold energies. These thresholds are functions of the atomic numbers of both the bombarding and target atoms and often permit "selective excitation" of a specific analyte to the exclusion of matrix elements (*C4*). For example, 100-kV Kr^+ ions bombarding an antimony-implanted silicon target excite Sb *M* x-rays, but not Si *K* x-rays.

4. The distribution of an analyte as a function of depth can be mapped with high resolution near the specimen surface.

A good review of applications of ion-bombardment x-ray excitation is given by Duggan *et al.* (*D29*).

Chapter 2

Properties of X-Rays

In this chapter, we consider several x-ray properties that are of particular significance in x-ray spectrometry.

2.1. ABSORPTION

2.1.1. X-Ray Absorption Coefficients

In Figure 2.1, a perfectly collinear monochromatic x-ray beam of intensity I_0 is directed on an absorber having uniform thickness t cm and density ϱ g/cm³. X-ray-opaque masks having windows of the same area in perfect register are placed on both sides of the absorber. The emergent collinear beam has intensity I, which is always less than I_0, that is, the x-rays undergo *absorption* or *attenuation* in passing through matter. The

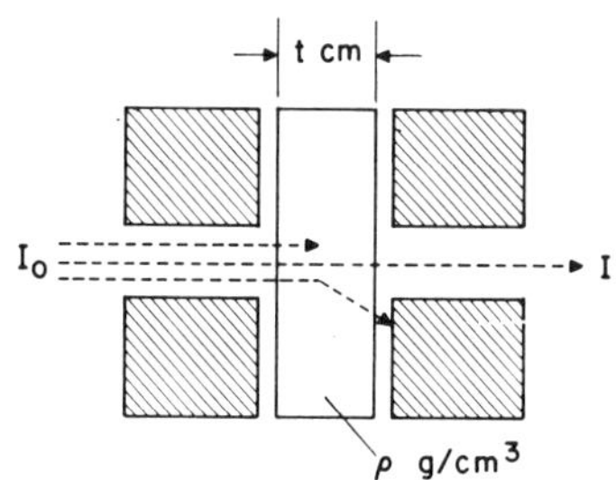

FIG. 2.1. Arrangement for x-ray absorption. The top, center, and bottom rays undergo true photoelectric absorption, transmission, and scatter, respectively (Section 2.1.2).

emergent intensity is given by

$$dI/I = -\mu \, dt \tag{2.1}$$

where μ is a proportionality constant known as the *linear absorption coefficient* and has the unit reciprocal centimeters (cm^{-1}). The negative sign indicates that the intensity always decreases on passing through matter. If the value of μ is independent of t, integration of Equation (2.1) gives the Lambert law,

$$I = I_0 \exp(-\mu t) \tag{2.2}$$

Four x-ray *absorption coefficients* (or *attenuation coefficients*) are derived from this equation:

1. *Linear absorption coefficient* μ gives absorption per unit thickness per unit area, that is, the fraction of energy absorbed from an incident x-ray beam of unit cross section in passing through unit thickness. From Equation (2.2),

$$I/I_0 = \exp(-\mu t) \tag{2.3}$$

Taking natural logarithms, one gets

$$\log_e(I/I_0) = -\mu t; \qquad \log_e(I_0/I) = \mu t \tag{2.4}$$

from which

$$\mu = \frac{\log_e(I_0/I)}{t} \quad \text{cm}^{-1} \tag{2.5}$$

2. *Mass-absorption coefficient* μ_m gives absorption per unit mass per unit area:

$$\mu_m = \mu/\varrho \quad \text{cm}^2/\text{g} \tag{2.6}$$

3. *Atomic absorption coefficient* μ_a gives absorption per atom per unit area:

$$\mu_a = \frac{\mu}{\varrho} \frac{A}{N} = \frac{\mu}{n} \quad \frac{\text{cm}^2}{\text{atom}} \tag{2.7}$$

4. *Molar absorption coefficient* μ_{mol} gives absorption per mole per unit area, that is, per gram-atom per unit area:

$$\mu_{mol} = (\mu/\varrho)A \quad \text{cm}^2/\text{mol} \tag{2.8}$$

The four coefficients are related as follows:

$$\mu = \mu_m \varrho = \mu_a \varrho (N/A) = \mu_{\text{mol}} \varrho / A \tag{2.9}$$

In Equations (2.3)–(2.9), ϱ is density (g/cm^3); A is atomic weight (g/mol); N is the Avogadro number (6.02×10^{23} atoms/mol); n is number of atoms per cubic centimeter; and N/A is number of atoms per gram.

Of these four coefficients, mass-absorption coefficient μ/ϱ ("mu over rho") is by far the most useful. It is an atomic property of each element independent—for practical purposes—of state of chemical or physical aggregation. It is a function only of wavelength and atomic number, and the function for each is relatively simple. Values of μ/ϱ for different substances are directly comparable, and μ/ϱ for a compound, solution, or mixture is readily calculated from the values for the constituents. None of these advantages applies to the linear absorption coefficient, as is strikingly illustrated by a hypothetical experiment described by Sproull [p. 72 in (*37*)]: "A beam of x-rays passing from the ceiling to the floor of a chamber filled with hydrogen and oxygen may be [say] 10% absorbed, or 90% of it will reach the floor. If a spark explodes the hydrogen and oxygen, filling the chamber with steam, 90% of the x-rays will still reach the floor. Then if the chamber is chilled so that the steam condenses to a thin layer of water or ice on the floor, 90% of the x-rays will still reach the floor. This is not true for light or ultraviolet or infrared radiation, and it explains why the mass absorption coefficient of x-rays is commonly used, whereas the linear absorption coefficient is ordinarily used in optics."

Equation (2.2) may be rewritten in terms of μ/ϱ as follows:

$$I = I_0 \exp[-(\mu/\varrho)\varrho t] \tag{2.10}$$

where ϱ is density (g/cm^3), and t is thickness (cm); ϱt is area concentration (g/cm^2).

Throughout this book, the symbol μ/ϱ indicates the mass-absorption coefficient in square centimeters per gram, and $(\mu/\varrho)_{Z,\lambda}$ indicates mass-absorption coefficient of element Z for wavelength λ. However, because the absorption coefficient used in x-ray fluorescence spectrometry is almost exclusively the mass coefficient, many workers use the symbol μ (or μ_m) for simplicity and indicate the linear coefficient by a subscript, μ_l, for example. Tables of mass-absorption coefficients are given in Appendixes 7A and 7B, and references to other tables are given in the Bibliography (*62, 65, 67, 69, 71–75, 78, 79, 81–84*). Methods for calculation of these coefficients have been published and are discussed at the end of this section.

Absorption curves of most of the chemical elements are shown in Figures 2.5 and 2.6 (*M59*).

The mass-absorption coefficient of a compound, solution, or mixture of elements A, B, C, ... in weight fractions W_A, W_B, W_C, ... is given by

$$(\mu/\varrho)_{ABC\ldots} = \sum W_i(\mu/\varrho)_i = W_A(\mu/\varrho)_A + W_B(\mu/\varrho)_B + \cdots \quad (2.11)$$

In some cases, a more useful form may be the following:

$$(\mu/\varrho)_X = W_A(\mu/\varrho)_A + (1 - W_A)(\mu/\varrho)_{X-A} \quad (2.12)$$

where X is the specimen, A the analyte, and X–A the matrix.

Equations (2.11) and (2.12) are valid for monochromatic x-rays and, to some extent, for polychromatic x-rays for which an effective wavelength can be designated (Section 4.1.3).

In Chapters 14, 15, and 18, application is made of two mass-absorption coefficients for both the primary and analyte-line x-rays: $(\mu/\varrho)'$ is the total mass-absorption coefficient of a specified element or substance for both the primary x-rays λ_{pri} and a specified analyte line λ_A, that is,

$$(\mu/\varrho)' = (\mu/\varrho)_{\lambda_{pri}} + (\mu/\varrho)_{\lambda_A} \quad (2.13)$$

$\overline{(\mu/\varrho)}$ is substantially $(\mu/\varrho)'$ corrected for geometry, that is, for path lengths of the primary beam ($\csc\phi$) and the analyte line ($\csc\psi$) in the specimen:

$$\overline{(\mu/\varrho)} = (\mu/\varrho)_{\lambda_{pri}} \csc\phi + (\mu/\varrho)_{\lambda_A} \csc\psi \quad (2.14)$$

$$= [(\mu/\varrho)_{\lambda_{pri}}/\sin\phi] + [(\mu/\varrho)_{\lambda_A}/\sin\psi] \quad (2.15)$$

Figure 4.1 shows the significance of angles ϕ and ψ. In general, $(\mu/\varrho)'$ is used when approximate results are sufficient, $\overline{(\mu/\varrho)}$ when highest accuracy is required.

Empirical methods for calculation of mass-absorption coefficients and curves have been published. Victoreen (*V12*) derives the equation

$$(\mu/\varrho)_{Z,\lambda} = C\lambda^3 - D\lambda^4 + \sigma_e Z(N/A) \quad (2.16)$$

where $(\mu/\varrho)_{Z,\lambda}$ is mass-absorption coefficient of element Z at wavelength λ, Z is atomic number, σ_e is x-ray scattering coefficient per electron, C and D are constants, and N and A are defined above. The constants C and D are different for each element and, for a specified element, for each segment of the absorption curve (μ/ϱ *versus* λ), that is, C and D are slightly different between each two consecutive absorption edges. Victoreen gives equations for calculation of C and D and tables of C and D for all elements at wavelengths below the K absorption edge and for selected elements up to the MI edge. Leroux (*L18*, *L19*) gives values of constants c and n in the equation

$$(\mu/\varrho)_{Z,\lambda} = c\lambda^n \tag{2.17}$$

for calculation of μ/ϱ for all elements except hydrogen at all wavelengths between 0.178 and 10 Å. These constants, like those of Victoreen, also differ for each element and between absorption edges. Dewey, Mapes, and Reynolds (*16*) derive a general equation for calculation of $(\mu/\varrho)_{Z,\lambda}$ from two constants K_1 and K_2, two exponents m and n, and the wavelength $\lambda_{\rm ab}$ of the nearest absorption edge of the absorber element Z longer than λ:

$$(\mu/\varrho)_{Z,\lambda} = \frac{K_1\lambda^n}{K_2\lambda_{\rm ab}^m} \approx \frac{3805\lambda^{2.80}}{20.44\lambda_{\rm ab}^{1.27}} \tag{2.18}$$

If the calculated value of $(\mu/\varrho)_{Z,\lambda}$ is <1, a correction is required, and a curve for doing this is given. These workers give a table of absorption coefficients of all elements of atomic number 4–92 (beryllium to uranium) at the $K\alpha$ lines of all these elements and at the $L\alpha_1$ lines of elements 37–92 (rubidium to uranium). These methods permit calculation of absorption coefficients at wavelengths in the gaps in the published compilations of experimental and calculated data, especially between the LI and LIII and MI and MV edges.

Heinrich (*69*) has critically reviewed the status of mass-absorption coefficients and gives equations for their calculation. He also gives tables of mass-absorption coefficients of elements 3–94 (lithium to plutonium) for $K\alpha$ and $K\beta$ lines of elements 11–42 (sodium to molybdenum), $L\alpha_1$ and $L\beta_1$ lines of elements 31–95 (gallium to americium), and $M\alpha$ and $M\beta$ lines of elements 70–92 (ytterbium to uranium).

2.1.2. X-Ray Absorption Phenomena

The linear and mass-absorption coefficients discussed in the preceding section are measures of that portion of the incident collinear beam that does not appear in the emergent collinear beam, regardless of reason (Figure 2.1). They are *total* absorption coefficients and are the result of three phenomena, each having its own linear and mass-absorption coefficients:

$$\mu = \tau + \sigma + \pi \tag{2.19}$$

$$\mu/\varrho = (\tau/\varrho) + (\sigma/\varrho) + (\pi/\varrho) \tag{2.20}$$

where τ and τ/ϱ are *true* or *photoelectric* absorption coefficients; σ and σ/ϱ are *scatter* coefficients; and π and π/ϱ are *pair-production* coefficients.

In true or photoelectric absorption, photons are really absorbed, giving their energy to expelling and imparting kinetic energy to orbital electrons with consequent emission of x-ray spectral lines. This process is discussed in Sections 1.6.3 and 1.6.5.1–3.

In the scatter process, photons are not really absorbed, but deflected from their collinear path in the absorber (Figure 2.1), in effect, disappearing from the emergent beam. This process is discussed in Section 2.2. Photoelectric absorption predominates over scatter except for the lightest elements at very short wavelengths. In the absorption curves of beryllium in Figure 1.14 and of all the light elements at short wavelengths in the curves in Figure 2.5, the marked departure from linearity is due to predominance of scatter over photoelectric absorption.

In pair production, photons passing close to atomic nuclei give their energy to creating and imparting kinetic energy to two charged particles, an electron (e^-) and a positron (e^+):

$$\text{x or } \gamma \text{ photon} \rightarrow e^- + e^+$$

This is an example of conversion of energy to matter in accordance with the Einstein equation,

$$E = mc^2 \tag{2.21}$$

where E is energy (ergs); m is mass (g); and c is velocity of light (cm/s). If we let m be the rest mass of the electron, substitute values for m (9.11×10^{-28} g) and c (3×10^{10} cm/s), and insert the conversion factor 1.6×10^{-6} erg/MeV, the equation gives the energy equivalent of the electron, 0.51 MeV. It follows that the threshold photon energy for pair production is 1.02 MeV, and that the phenomenon has no significance in x-ray spectrochemistry.

To summarize, of the principal phenomena that arise on "absorption" of x-rays in matter: (1) x-ray-excited photoelectrons, characteristic diagram and satellite lines, and Auger electrons arise from true photoelectric absorption (τ); (2) unmodified and modified scatter and Compton recoil electrons arise from scatter (σ); and (3) electron–positron pairs arise from pair-production (π).

McMaster (*78*) gives photoelectric and coherent and incoherent scatter cross sections at 1 keV to 1 MeV for elements 1–83, 86, 90, 92, and 94. The corresponding absorption coefficients can be calculated from the cross sections with equations of the form of Equation (2.34).

2.1.3. Relationship of μ/ρ, λ, and Z

Mass-absorption coefficient μ/ϱ is a property of a substance; at a specified wavelength, it is different for every substance, and, for the same substance, it is different for every wavelength. The relationship of absorption and wavelength for a specified element is best expressed in the form of a curve, plotted on a log-log grid, of mass-absorption coefficient *versus* wavelength. At a given wavelength, $(\mu/\varrho)_Z$ should increase with Z because the "heavier" the element, the greater is its stopping power. For a given element, $(\mu/\varrho)_\lambda$ should increase with wavelength because as wavelength becomes longer, x-rays become less energetic and less penetrating. Figures 1.14 and 2.2 show x-ray absorption curves for several elements, and it is evident that the above predictions are realized to a large extent and would be valid generally were it not for the absorption edges (Sections 1.6.5.1–1.6.5.3).

Many examples of departures from predicted regularity appear in Figure 2.2. The μ/ϱ of tin increases regularly with wavelength up to $\lambda_{\mathrm{Sn}K_{\mathrm{ab}}}$, then falls abruptly and increases regularly again. Below $\lambda_{\mathrm{Pb}K_{\mathrm{ab}}}$, μ/ϱ increases with Z, as it should; however, just above $\lambda_{\mathrm{Pb}L\mathrm{III}_{\mathrm{ab}}}$, copper and zinc ($Z$ 29 and 30) have higher absorption than tin and lead (Z 50 and 82), and tin and lead (ΔZ 32) differ in absorption no more than copper and zinc (ΔZ 1). The μ/ϱ of lead is greater than that of tin below $\lambda_{\mathrm{Pb}K_{\mathrm{ab}}}$, less between $\lambda_{\mathrm{Pb}K_{\mathrm{ab}}}$ and $\lambda_{\mathrm{Sn}K_{\mathrm{ab}}}$, greater between $\lambda_{\mathrm{Sn}K_{\mathrm{ab}}}$ and $\lambda_{\mathrm{Pb}L\mathrm{III}_{\mathrm{ab}}}$, then slightly less at longer wavelengths. It is evident that absorption cannot be estimated simply on the basis of atomic number and wavelength.

We have seen that three phenomena—photoelectric absorption, scatter, and pair production—constitute total absorption. In the wavelength region of x-ray spectrochemistry, pair production does not occur, and photoelectric absorption predominates over scatter. Consequently, μ/ϱ is largely determined by τ/ϱ. The coefficient τ/ϱ is itself the sum of a series of coefficients representing photon absorption due to electron expulsion from each of the atomic levels:

$$(\tau/\varrho)_\lambda = \sum (\tau/\varrho)_{\lambda,i} = (\tau/\varrho)_{\lambda,K} + (\tau/\varrho)_{\lambda,L\mathrm{I}} + (\tau/\varrho)_{\lambda,L\mathrm{II}} + (\tau/\varrho)_{\lambda,L\mathrm{III}} + (\tau/\varrho)_{\lambda,M\mathrm{I}} + \cdots \quad (2.22)$$

where $(\tau/\varrho)_\lambda$ is the photoelectric mass-absorption coefficient at λ, and $(\tau/\varrho)_{\lambda,i}$ is the photoelectric mass-absorption coefficient at λ for i-level electrons. As wavelength is increased, the absorption edge of each level is exceeded in turn and that term drops from the equation, causing an abrupt

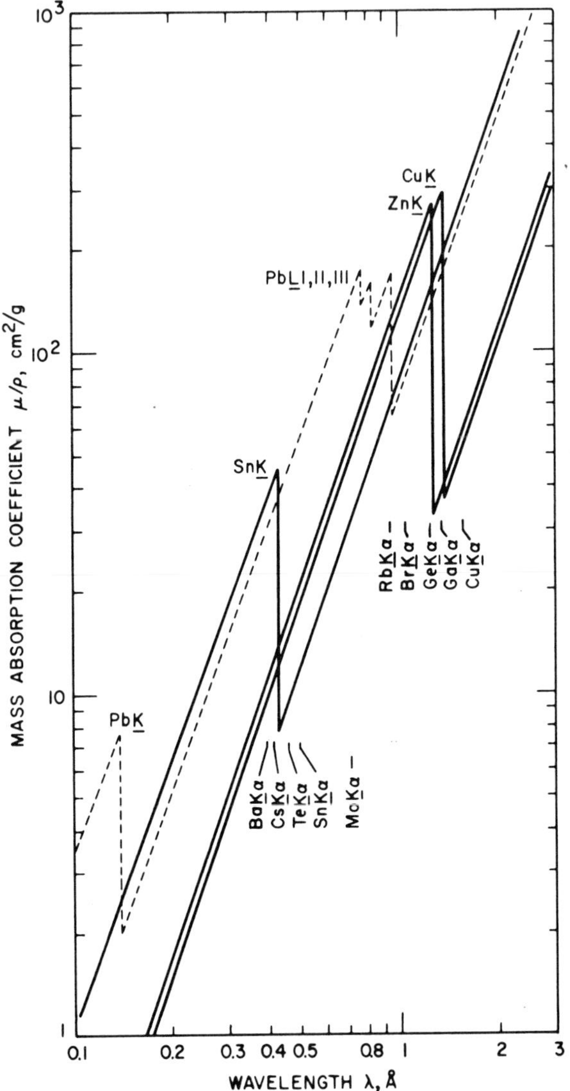

FIG. 2.2. X-ray absorption curves of copper, zinc, tin, and lead.

drop in τ/ϱ. Between absorption edges,

$$\tau/\varrho = K'(N/A)Z^4\lambda^3_{\text{cm}} \tag{2.23}$$

Here, K' is a constant that changes at each absorption edge, N is the Avogadro number (atoms/mol), and A is atomic weight (g/mol), so that N/A

is number of atoms per gram. For any given substance, N/A is constant and may be combined with K', giving the Bragg–Pierce law:

$$\mu/\varrho = KZ^4\lambda^3_{\mathrm{cm}} \tag{2.24}$$

where K still changes at each absorption edge. Thus, in a log-log plot of μ/ϱ *versus* λ, the segments of the curve between the edges are linear and substantially mutually parallel, as shown in Figures 1.14, 2.2, 2.5, and 2.6.

2.1.4. Absorption Edges

Absorption edges or *critical absorption wavelengths* are discussed in Sections 1.6.5.1–1.6.5.3. Here, we need only add a few details. A multi-element substance exhibits all the absorption edges of all its constituent elements. Figure 2.3 shows the wavelengths of the K, LIII, and MV edges as a function of atomic number.

Absorption-edge jump ratios r and jump differences δ are measures of that portion of the total absorbed x-radiation that is absorbed by a specified

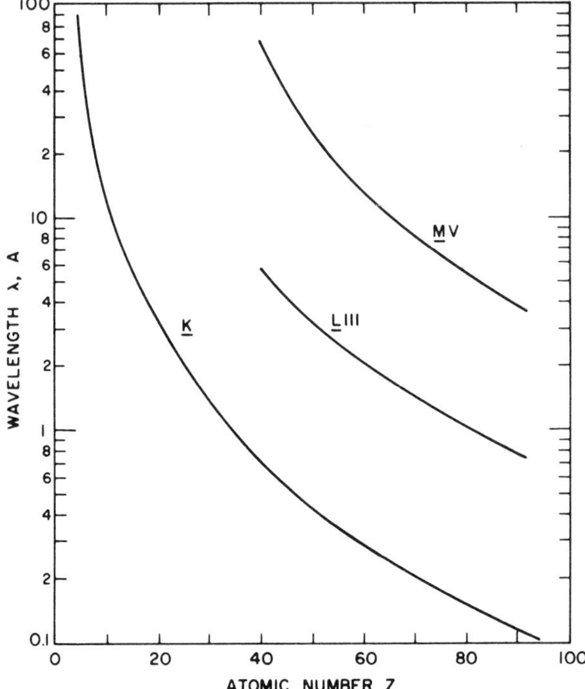

FIG. 2.3. Wavelengths of *K*, *L*III, and *M*V absorption edges.

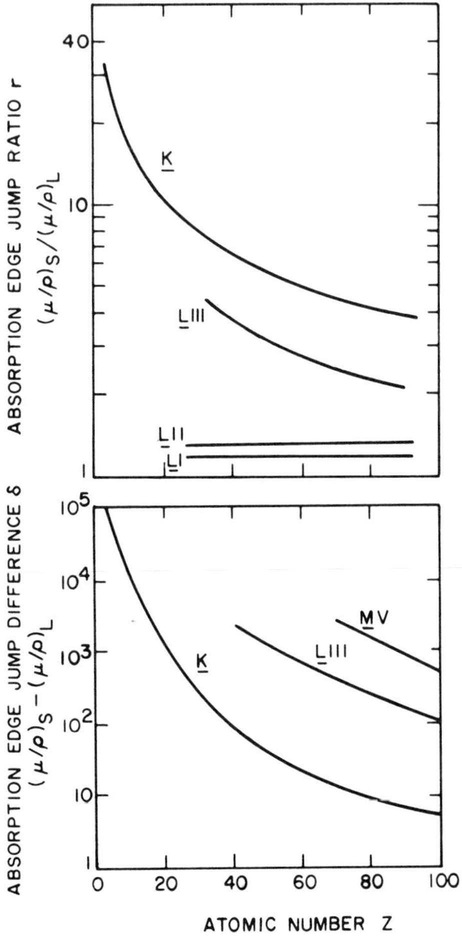

FIG. 2.4. Absorption-edge jump ratios (see Appendix 8) and differences.

atomic energy level. For example, K and LIII jump ratios are defined by

$$r_K = \frac{(\mu/\varrho)_K + (\mu/\varrho)_{LI} + (\mu/\varrho)_{LII} + (\mu/\varrho)_{LIII} + \cdots}{(\mu/\varrho)_{LI} + (\mu/\varrho)_{LII} + (\mu/\varrho)_{LIII} + \cdots} \tag{2.25}$$

$$r_{LIII} = \frac{(\mu/\varrho)_{LIII} + (\mu/\varrho)_{MI} + (\mu/\varrho)_{MII} + \cdots}{(\mu/\varrho)_{MI} + (\mu/\varrho)_{MII} + \cdots} \tag{2.26}$$

K and LIII jump differences are defined by the difference between the numerator and denominator in the respective equations. More simply,

$$r = (\mu/\varrho)_S/(\mu/\varrho)_L \tag{2.27}$$

$$\delta = (\mu/\varrho)_S - (\mu/\varrho)_L \tag{2.28}$$

where S and L refer, respectively, to the short- and long-wavelength sides of the edge, that is, the "top" and "bottom," or maximum and minimum values of μ/ϱ. The actual fraction of the total number of photoionizations that occurs in, say, the K shell is given by

$$[(\mu/\varrho)_S - (\mu/\varrho)_L]/(\mu/\varrho)_S = 1 - (1/r_K) = (r_K - 1)/r_K \qquad (2.29)$$

Figure 2.4 shows various absorption-edge jump ratios r and differences δ as functions of atomic number. Use is made of these functions in Chapters 4 and 15.

The K, L, and M absorption edges of all the elements are listed in Appendix 5, and absorption curves showing K and L edges of many elements are shown in Figures 2.5 and 2.6 (*M59*). Figure 1.8 shows, for all elements, the K, LIII, MV, and NVII edges that occur in the 0.1–100-Å region. Values of r for the K and LIII absorption edges are given in Appendix 8. McMaster *et al.* (*78*) give values of $(\mu/\varrho)_S$ and $(\mu/\varrho)_L$ for the K, LI, LII, and LIII edges of all applicable elements and for all five M edges of elements having atomic number 61 (promethium) and above. Values of r for the M edges and values of δ for all edges can be calculated from these absorption coefficients.

2.1.5. Comparison of X-Ray and Optical Absorption

X-ray and optical (visible and near-infrared and ultraviolet) absorption differ essentially only in that they occur in inner and outer orbitals of atoms, respectively. Nevertheless, their absorption spectra are strikingly dissimilar.

Optical absorption and emission spectra are complementary, the former consisting of dark lines (or bands), the latter bright lines (or bands) at the same wavelengths. X-ray absorption increases gradually with wavelength, falls abruptly at a wavelength shorter than that of the emission lines, then gradually rises again. Figure 2.7 shows this difference. X-ray photoelectric absorption is explained in Sections 1.6.5.1–1.6.5.3. In optical absorption, outer orbital electrons are elevated to higher energy levels—not clear out of the atom, as in x-ray absorption. Such *resonance* excitation requires absorption of photons having exactly the same energy as those emitted when the excited electrons return to their original levels. Thus, optical absorption and emission spectra both consist of the same discrete wavelengths.

One might wonder why x-ray absorption *lines*, analogous to optical absorption lines, do not occur. For example, why can there not be a Cu $K\alpha$

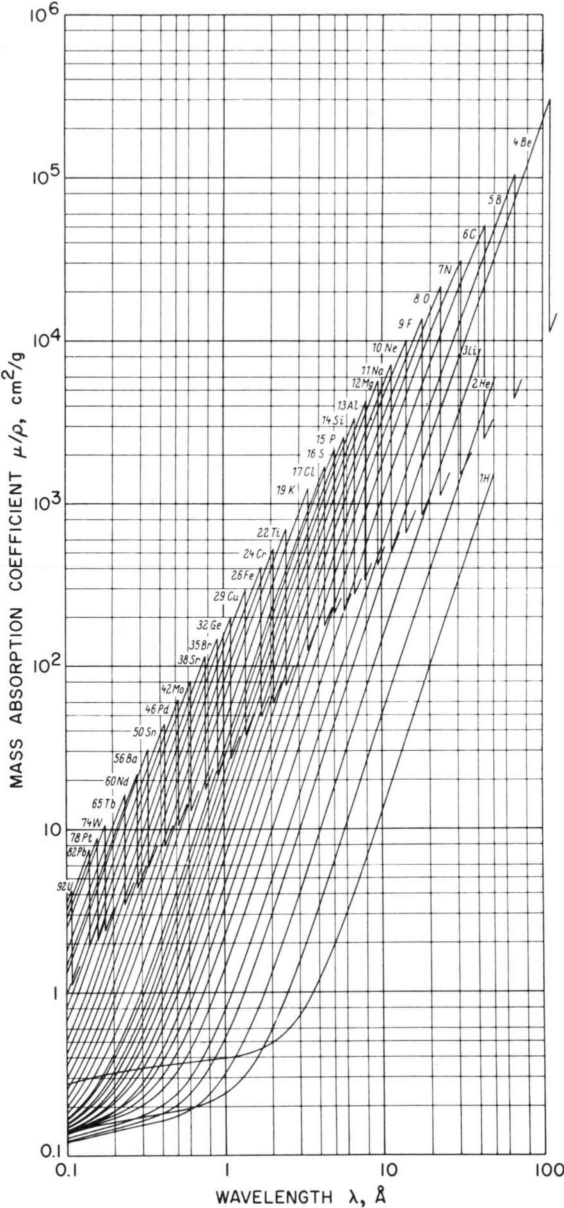

FIG. 2.5. X-ray absorption curves of the chemical elements showing *K* absorption edges. [E. A. W. Müller, *Archiv fuer Technisches Messen und Industrielle Messtechnik,* R. Oldenbourg, Munich (1956); courtesy of the author and publisher.]

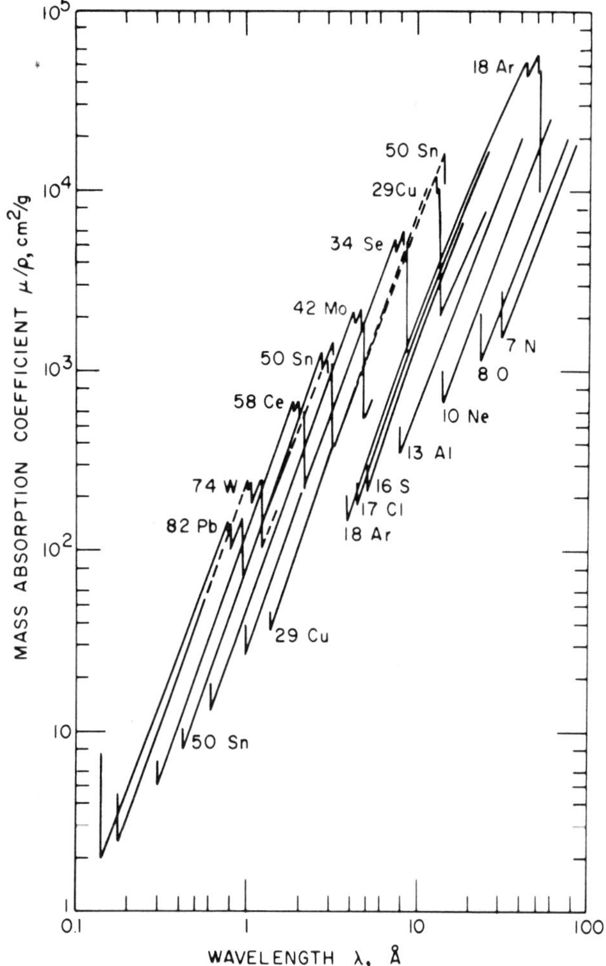

FIG. 2.6. X-ray absorption curves of selected chemical elements showing *L* absorption edges. [E. A. W. Müller, *Archiv fuer Technisches Messen und Industrielle Messtechnik*, R. Oldenbourg, Munich (1956); courtesy of the author and publisher.]

absorption line at the same wavelength as the Cu $K\alpha$ emission line, caused by elevation of a Cu K electron to the Cu LIII shell—the reverse of the transition that causes the Cu $K\alpha$ emission line? The reason is that ordinarily the Cu LIII shell is full and cannot receive the Cu K electron.

With certain exceptions, x-ray absorption, unlike optical absorption, is substantially independent of chemical state. A striking example is provided

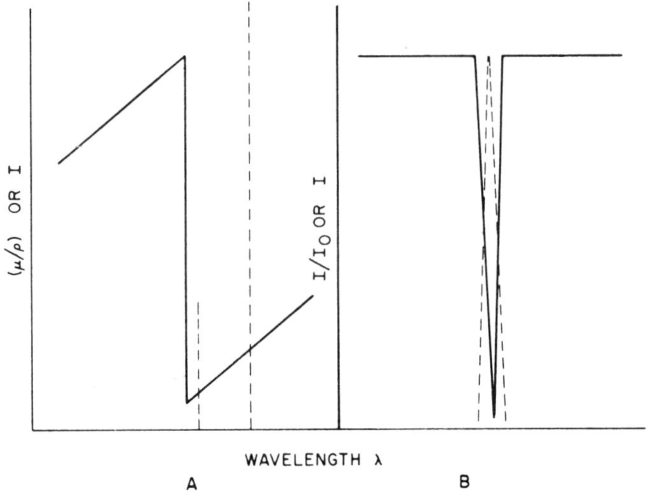

FIG. 2.7. Comparison of (A) x-ray and (B) optical absorption.

by diamond and graphite, which are, respectively, transparent and opaque to light, but have the same x-ray mass absorption coefficient. Sproull's hypothetical experiment (Section 2.1.1) further illustrates this point. Equation (2.11) shows the simple additive nature of the mass absorption coefficient of an element in a compound, solution, or mixture. No such equation is possible for optical absorption.

For other comparisons of x-ray and optical spectra, see Sections 3.2.2 and 3.3.1.1.

2.1.6. Significance

Secondary excitation of characteristic x-ray spectra is itself based on photoelectric absorption, and in almost every component of the instrument, absorption is of great significance, sometimes beneficially, sometimes detrimentally.

Absorption of primay x-rays in the x-ray tube window becomes significant at the long wavelengths required to excite elements of low atomic number. In the specimen, absorption of primary and analyte-line radiation and absorption-enhancement effects determine sensitivity and dictate the analytical strategy. Absorption in liquid and powder cell windows becomes significant in determination of low-Z elements. Photoelectric absorption of secondary x-rays in some crystals results in line emission from the crystal. Absorption of secondary x-rays in the radiation path may require use of helium or vacuum, and for the lighter elements, even these may have significant absorption. Absorption of analyte-line radiation in the detector tube window may require use of windows of thin Mylar or polypropylene or ultrathin Formvar. The gas-ionization process in gas-filled detectors and the scintillation process in scintillation counters are absorption phenomena. All filters are based on absorption: primary-beam filters to remove target lines or improve peak-to-background ratio, and secondary-beam filters to reduce spectral-line interference or permit "nondispersive" analysis. Enhancement radiators are based on photoelectric absorption by the radiator and by the analyte.

Absorption is also the basis of several analytical methods—polychromatic and monochromatic absorption, absorption-edge spectrometry, and emission-absorption methods.

2.1.7. Half-Thickness and Absorption Cross Section

Another way to express x-ray attenuation is the *half-value thickness* or *half-thickness* $t_{1/2}$, the thickness required to reduce the incident intensity by one-half. From Equations (2.2)–(2.4),

$$t = \frac{\log_e(I_0/I)}{\mu} \tag{2.30}$$

If $I = 0.5I_0$,

$$t = t_{1/2} = 0.693/\mu \tag{2.31}$$

where 0.693 is $\log_e 2$ and μ is linear absorption coefficient (cm^{-1}). Table 2.1 lists the half-thicknesses of some selected materials at several wavelengths. Half-thickness has little application in x-ray spectrometry, but is useful for expressing attenuation of polychromatic x-radiation.

Still another way to express x-ray attenuation is the *x-ray absorption cross section* σ, which is a measure of the probability that an atom will absorb an incident x-ray photon. Stated otherwise, it is the imaginary target area that an atom presents to incident x-ray photons. If Equation (2.2) for absorption of a perfectly collinear monochromatic x-ray beam is written in terms of the *numbers* of x-ray photons incident upon (n_0) and transmitted by (n) an absorber having thickness t cm and linear absorption coefficient μ cm^{-1}, one gets

$$n = n_0 \exp(-\mu t) \tag{2.32}$$

TABLE 2.1. Half-Thicknesses

Wavelength λ, Å	Half-thickness $t_{1/2}$, cm				
	Air[a]	Cellophane	$_{13}$Al	$_{29}$Cu	$_{82}$Pb
0.1	—	4.3	1.6	0.21	0.016
0.7	410	0.4	0.050	0.0016	0.00044
1.5	62	0.11	0.0056	0.0016	—
2.0	26	0.049	0.0025	0.00071	—

[a] 0°C, 760 torrs.

If the volume of absorber traversed by the x-ray photons contains n_{at} atoms/cm³, each of which presents an imaginary target area σ cm² to the photons, it can be shown that

$$\mu = n_{at}\sigma \qquad \text{or} \qquad \sigma = \mu/n_{at} \qquad (2.33)$$
$$\text{cm}^2/\text{atom} \; [=] \; \text{cm}^{-1}/(\text{atoms/cm}^3)$$

where σ is the x-ray absorption cross section and is usually given in barns/atom (1 barn $= 10^{-24}$ cm²). Like the x-ray absorption coefficient, the x-ray absorption cross section is different for each element at a specified wavelength and different at each wavelength for a specified element. Thus, symbol $\sigma_{Z,\lambda}$, analogous to $(\mu/\varrho)_{Z,\lambda}$, is appropriate. Also, like the absorption coefficient, the absorption cross section is the sum of photoelectric, scatter, and (>1.02 MeV) pair-production cross sections (Section 2.1.2). The cross section and mass-absorption coefficient are related as follows:

$$\mu/\varrho = \sigma(N/A) \qquad (2.34)$$
$$\frac{\text{cm}^2}{\text{g}} \; [=] \; \frac{\text{cm}^2}{\text{atom}} \; \frac{\text{atoms/mol}}{\text{g/mol}}$$

where N and A are the Avogadro number and atomic weight, respectively.

The cross section has little application in x-ray fluorescence spectrometry. Tables of x-ray absorption cross sections have been compiled by McMaster *et al.* (*78*).

2.1.8. Inverse-Square Law

Another form of x-ray attenuation is the reduction in intensity per unit area resulting from divergence and distance. X-rays emanating spherically from a point—or very small—source follow the inverse-square law, as shown in Figure 2.8 and the following equation:

$$I_2 = I_1(D_1^2/D_2^2) \tag{2.35}$$

where I_1 and I_2 are intensities per unit area at, respectively, distances D_1 and D_2 from the x-ray source. The inverse-square law does not apply to collinear beams. Soller collimators are an intermediate case, in that divergence occurs parallel to, but not perpendicular to, the foils.

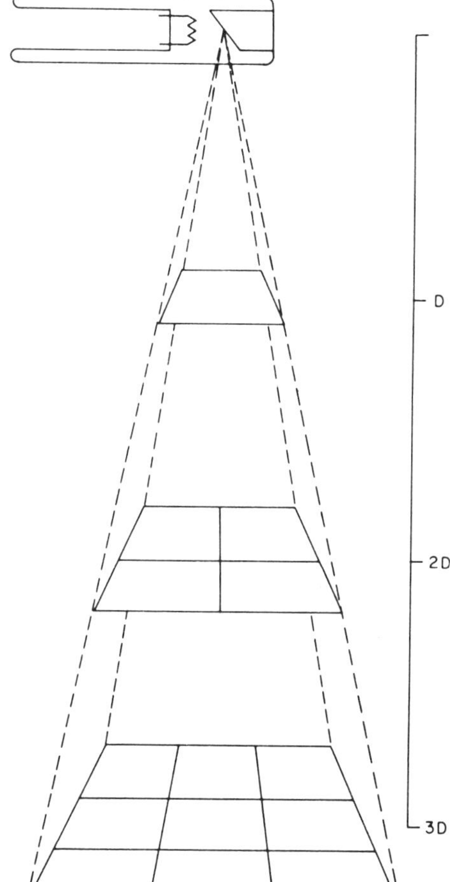

FIG. 2.8. The inverse-square law.

2.2. SCATTER

2.2.1. General

Four types of x-ray scatter must be distinguished: *Unmodified* (*elastic*, *Rayleigh*) scatter involves no change in wavelength. *Modified* (*inelastic*, *Compton*) scatter involves an increase in wavelength, that is, a decrease in photon energy. *Coherent* scatter occurs in such a way that there is a phase relationship between the incident and scattered x-rays. *Incoherent* scatter occurs without any such phase relationship. It is almost universal practice in the literature of x-ray spectrometric analysis to regard modified and incoherent scatter as synonymous. For practical purposes, this leads to no difficulty because modified scatter is necessarily incoherent, However, all incoherent scatter is not necessarily modified.

The linear scatter coefficient is given by

$$\sigma = Zf^2 + (1 - f^2) \tag{2.36}$$

where Z is atomic number, and f is the electronic structure factor, that is, the scatter power of an electron. The two terms on the right represent, respectively, the contributions of coherent and incoherent scatter.

In coherent scatter, the incident x-rays cause the electrons in the irradiated matter to oscillate at the same frequency as the x-rays. The oscillating electrons emit x-rays in all directions, again at the same frequency. Coherent scatter increases with atomic number simply because the number of electrons present increases with atomic number.

Diffraction of x-rays by crystals arises from unmodified coherent scatter and is discussed in Section 2.3. The remainder of this section is devoted mostly to unmodified (Rayleigh) and modified incoherent (Compton) scatter. X-ray scatter is discussed further in Section 14.5.

2.2.2. Modified (Compton) Scatter

The mechanism of modified scatter (*C56*) is shown in Figure 2.9. The incident x-ray photon collides with a loosely bound electron in an outer orbit of an atom. The electron recoils under the impact, leaving the atom and carrying away some of the energy of the photon, which is deflected with corresponding loss of energy or increase in wavelength. The sum of the energies of the scattered photon and recoil electron equals the incident photon energy. The encounter is described by the laws of conservation of

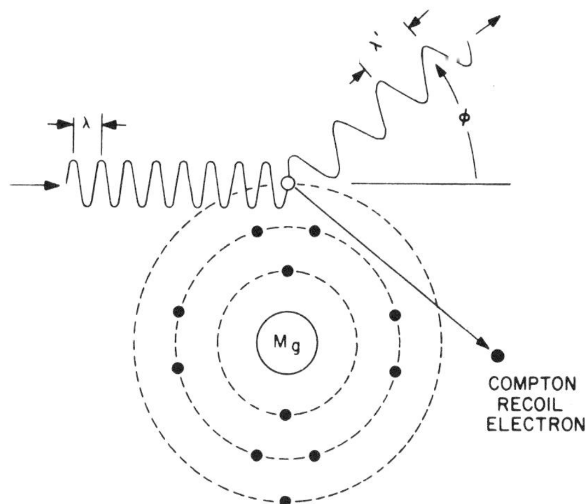

FIG. 2.9. Modified (Compton) scatter of an x-ray photon.

energy and momentum. The *recoil* or *Compton* electrons have the predicted direction and velocity, and the x-rays the predicted change in wavelength:

$$\lambda'_{\text{cm}} - \lambda_{\text{cm}} = \Delta\lambda_{\text{cm}} = (h/m_e c)(1 - \cos\phi) \tag{2.37}$$

where λ and λ' are wavelengths (cm) of the incident and modified scattered x-rays, respectively; h is Planck's constant (6.6×10^{-27} erg·s); m_e is the rest mass of the electron (9.11×10^{-28} g); c is the velocity of light (3×10^{10} cm/s); and ϕ is the angle between unscattered and scattered x-rays. Substitution of these values and insertion of the conversion factor 10^8 Å/cm gives

$$\Delta\lambda = 0.0243(1 - \cos\phi) \tag{2.38}$$

where $\Delta\lambda$ is in angstroms; 0.0243 Å is referred to as the *Compton wavelength.*

For energy-dispersive analysis, the Compton-shift equation is more conveniently applied in the following form:

$$E' = E/\{1 + [(E/m_e c^2)(1 - \cos\phi)]\} \tag{2.39}$$

$$= E/[1 + 0.001957E(1 - \cos\phi)] \tag{2.40}$$

where E and E' are the photon energies (keV) of the incident and scattered photons, respectively. To obtain Equation (2.40) from (2.39), the values of m_e and c given above are used and the term E/m_ec^2 is multiplied by the conversion factor 1.6×10^{-9} erg/keV. At primary energies less than ~50 keV, minimum scattered *energy* [E' minimum in Equations (2.39) and (2.40)] occurs at a scatter angle of 180° (straight back); however, minimum scattered *intensity*—number of scattered photons—occurs at a scatter angle of 90°, normal to the primary radiation, where it is half that at 180°.

The foregoing explanation is based on the corpuscular nature of both x-rays and electrons. Schrödinger explains the phenomenon on the basis of the wave nature of both x-rays and electrons. The initial and final states of the recoil electron mutually interfere to form a space grating of standing *matter* waves that reflects the incident x-radiation in much the same way as the space grating of atom layers in a crystal. The change in wavelength is attributed to the Doppler shift caused by the recoil velocity of the reflecting space grating.

Every wavelength in the scattered primary continuum is subject to both unmodified and modified scatter. However, in x-ray spectrometry, the principal manifestation of modified scatter is a broader peak on the long-wavelength side of each unmodified scattered x-ray tube target line and separated from it by $\Delta\lambda$.

The $\Delta\lambda$ is independent of both x-ray wavelength and atomic number of the scatterer, and varies only with ϕ. In most commercial x-ray spectrometers, the axis of the primary beam is perpendicular to the axis of the source collimator. Thus, the collimator–crystal–detector system "sees" mostly scattered primary radiation that leaves the specimen at an angle about 90° to the primary beam. Then $\phi = 90°$, $\cos \phi = 0$, and, from Equation (2.38), $\Delta\lambda \sim 0.024$ Å. As the scatter direction becomes more and more nearly backward, that is, as ϕ approaches 180°, the energy imparted to the recoil electron and, therefore, the energy retained by the scattered x-ray photon approach limiting values.

In general, modified scattered target lines are broader than unmodified scattered lines for two reasons. The primary beam has the form of a cone, and the specimen plane is usually inclined to the axis of this cone. So primary x-rays arrive at the specimen surface at various angles, and since the collimator permits divergence in the direction parallel to the foils, the spectrometer accepts secondary rays over a substantial angle. Thus, ϕ, and therefore $\Delta\lambda$, may have a range of values. Moreover, the recoil electrons may have a range of velocities and recoil directions, taking a range of energies from the scattered photons. Unless the incident x-rays have extremely short wavelength, recoil electrons have much less energy than x-ray photoelectrons.

2.2.3. Relationship of Unmodified and Modified Scatter

The proportion of modified scatter increases as: (1) the energy of the x-rays increases, that is, as wavelength decreases; (2) the binding of the orbital electrons decreases, that is, as atomic number of the scatterer decreases; and (3) ϕ increases. Modified scatter predominates when the x-ray energy greatly exceeds the binding energy of the orbital electrons. When this binding energy approaches or exceeds the x-ray energy, unmodified scatter predominates. The variation of modified scatter with wavelength and atomic number is shown in Table 2.2 and Figure 2.10 (*J20*). The figure shows primary *K* lines from copper and molybdenum x-ray tube targets scattered from four scatter targets having various effective atomic numbers from heavy ($BaSO_4$) to light (paraffin). For the longer wavelengths, Cu $K\alpha,\beta$ (~1.5 Å), Compton scatter (*C*) is absent from the two heaviest scatterers and is substantially weaker than Rayleigh scatter (*R*) even from the lightest. For the shorter wavelengths, Mo $K\alpha,\beta$ (~0.7 Å), prominent Compton scatter is present from even the heaviest scatterer and is more than twice as intense as Rayleigh scatter from the two lightest scatterers.

An interesting example of modified scatter is the frequent appearance in scattered primary spectra from x-ray tubes of a weak line corresponding to Pd $K\alpha$ (0.587 Å). Actually, this line is Compton-scattered Ag $K\alpha$ (0.561 Å) from silver solder used in construction of the tube. Note that 0.587 Å − 0.561 Å = 0.026 Å, in agreement with the $\Delta\lambda$ predicted above. The longer-wavelength lines of iron, nickel, and copper target contaminants are mostly Rayleigh-scattered, except from very light specimens.

TABLE 2.2. Ratio of Modified to Unmodified Scatter[a]

Element	I_{mod}/I_{unmod}
$_{3}Li$	All modified
$_{6}C$	5.5
$_{16}S$	1.9
$_{26}Fe$	0.5
$_{29}Cu$	0.2
$_{82}Pb$	All unmodified

[a] Page 23 in (*1*).

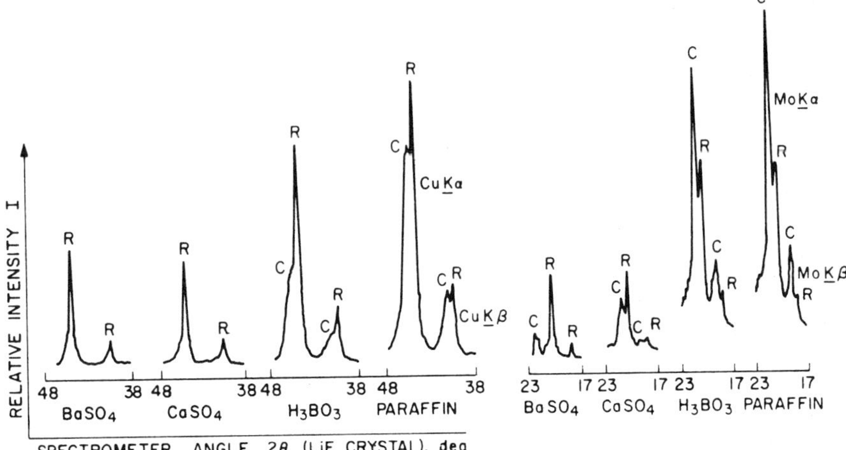

FIG. 2.10. Rayleigh (*R*) and Compton (*C*) scatter of copper and molybdenum *K* lines from heavy, medium, and light scatterers. [C. M. Johnson and P. R. Stout, *Analytical Chemistry* **30**, 1921 (1958); courtesy of the authors and the American Chemical Society.]

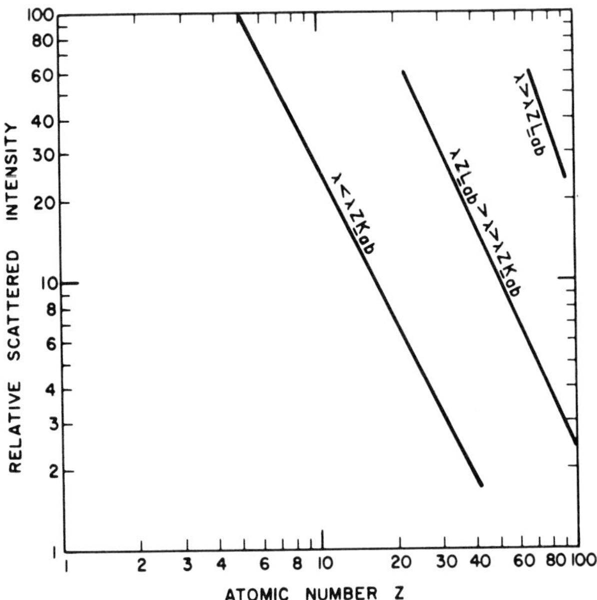

FIG. 2.11. Scattered x-ray intensity as a function of atomic number. [G. Andermann and J. W. Kemp, *Analytical Chemistry* **30**, 1306 (1958); courtesy of the authors and the American Chemical Society.]

2.2.4. Significance

Formerly, scatter was regarded only as a nuisance (*J20*). Background is mostly scattered primary radiation, and both unmodified and modified scattered target lines complicate the spectrum and increase the possibility of spectral-line interference. However, scattered x-rays may be beneficial in many ways. Sometimes, there is a question as to whether a low-intensity peak is a trace analyte line emitted from the specimen or a target line scattered from the specimen. The presence of an accompanying modified scattered peak conclusively indicates the latter. Diffraction, a form of coherent scatter, is the basis of dispersion by crystals. More recently, several techniques have been developed for using scattered x-rays to correct absorption-enhancement and surface-texture effects and other difficulties (Section 14.5). Finally, several analytical methods are based on x-ray scatter rather than emission (Section 20.2), including a method for determining carbon in hydrocarbons (Sections 14.8.3 and 20.2.2). Figure 2.11 shows scattered x-ray intensity as a function of atomic number for wavelengths shorter than the K absorption edge, between the K and L edges, and longer than the L edge (*A19*).

2.3. DIFFRACTION BY CRYSTALS

In the wavelength-dispersive x-ray spectrometer, the analyzer crystal disperses the secondary characteristic x-ray beam so that each wavelength may be measured individually. An understanding of the function of the crystal requires an elementary knowledge of the diffraction of x-rays by crystals. Diffraction is also significant in that certain spurious effects are caused by diffraction of the primary beam by crystalline specimens and of the secondary beam by Soller collimator foils.

Diffraction arises from interference of unmodified coherently scattered x-rays. Figure 2.12 shows interference of two rays of the same wavelength for three phase relationships. In Figure 2.12A, the rays are one half-wavelength (180°) out of phase ("peak to trough and trough to peak") and undergo complete destructive interference or cancellation. In Figure 2.12B, the rays are in phase ("peak to peak and trough to trough") and undergo complete constructive interference or reinforcement. The resultant ray has the same wavelength as the individual rays and the sum of their peak amplitudes. Figure 2.12C represents all intermediate phase relationships. The resultant ray has the same wavelength as the individual rays and the

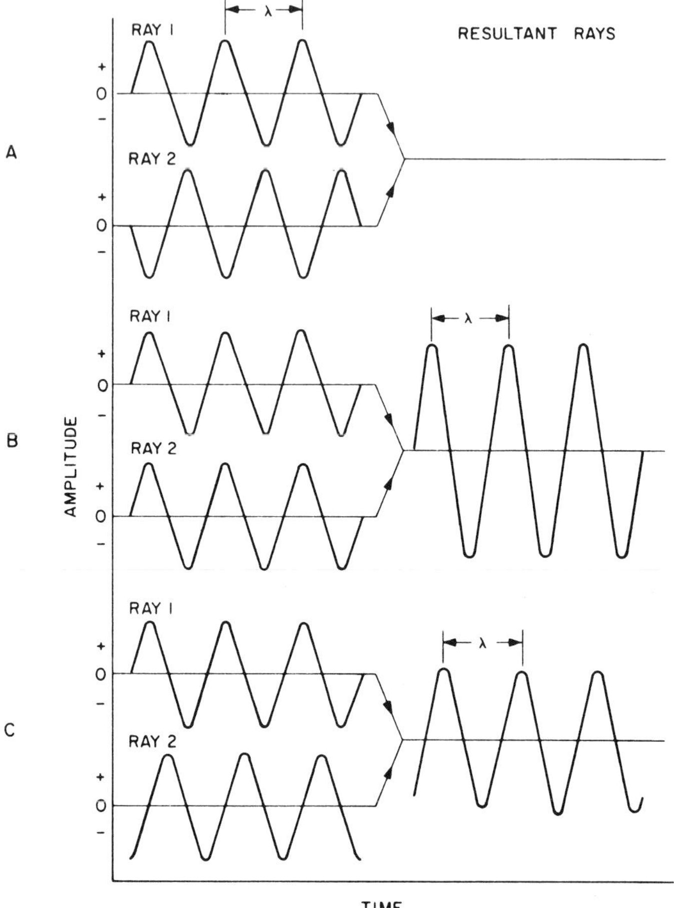

FIG. 2.12. (A) Destructive, (B) constructive, and (C) intermediate wave interference.

vector sum of their amplitudes—greater than zero, but less than the sum of the peak amplitudes. The more nearly in or out of phase the two interfering rays are, the more nearly they approach the extremes of Figures 2.12B and 2.12A, respectively.

One version of diffraction of x-rays by crystals is that of Laue (*F29*), who regards the phenomenon in terms of scatter from the individual lattice sites of the crystal. In Figure 2.13, a collimated monochromatic x-ray beam of wavelength λ and frequency ν is directed at angle θ on a set of crystal planes (hkl) of interplanar spacing d. The atoms in the crystal lattice are excited to oscillate at frequency ν and act as spherical radiators of x-rays of the same wavelength as the incident x-rays. This is the coherent scatter process. In most directions, the scattered rays are out of phase, and undergo destructive interference. However, in certain directions, they are in phase, and mutually reinforce. The group of such reinforced rays in a given direction constitutes a diffracted x-ray beam. The conditions for diffraction are as follows: (1) the angles made by the diffracting planes (hkl) with the incident and diffracted beams are equal; (2) the directions of the incident and diffracted beams and the normal to the diffracting planes are coplanar; and (3) the waves emitted by individual atoms in the direction of a diffracted beam are in phase.

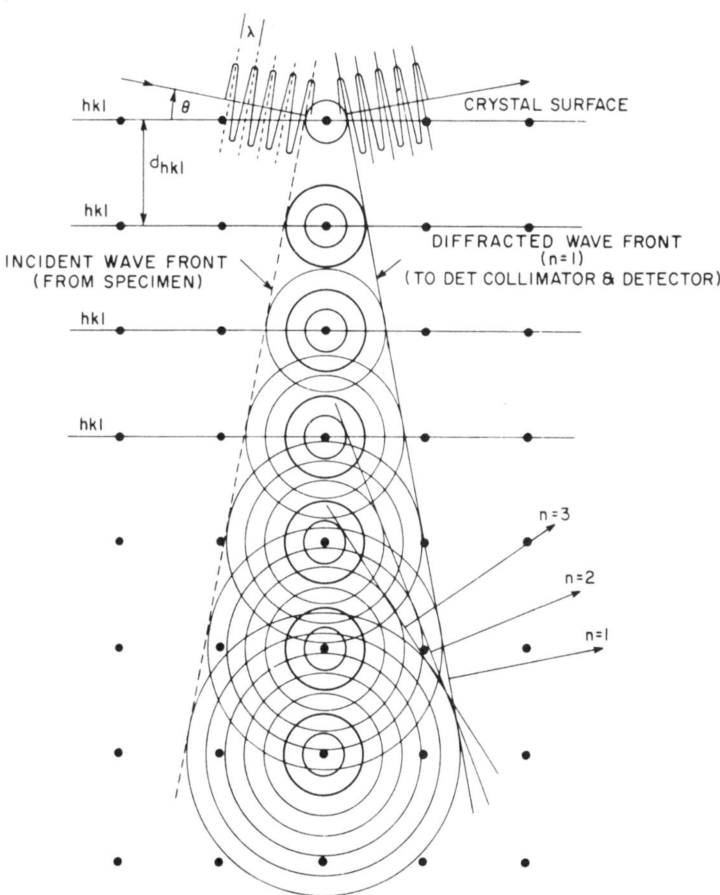

FIG. 2.13. Diffraction by crystals—Laue version.

Figure 2.13 shows the condition for first-order diffraction of λ ($n = 1$). The first-order wavefront consists of the first wave from the first (top) atom, the second wave from the second atom, etc.; that is, waves from successive atoms differ by 1. If θ is increased to appropriate discrete values, the crystal diffracts the second order ($n = 2$), where waves from successive atoms differ by 2, the third ($n = 3$), where they differ by 3, etc.

A more useful version of diffraction is that of Bragg (*B79*), who regards the phenomenon in terms of reflection from a stack of crystal planes. In Figure 2.14, a collimated monochromatic x-ray beam of wavelength λ is directed at angle θ on a set of crystal planes (hkl) of interplanar spacing d. Again the incident beam is scattered in all directions from each plane, but only in certain directions does the scatter reinforce. The conditions for diffraction are the same as for the Laue version, but the third condition might be more appropriately restated as follows: Reflected rays from successive *planes* differ in path length by an integral number of wavelengths.

In Figure 2.14, rays 1 and 2 are directed on two successive planes and scatter in all directions. Of the scattered rays, only $1'$ and $2'$ fill the conditions for diffraction. Rays 1 and 2 travel equal distances to AC, and rays $1'$ and $2'$ travel equal distances beyond AD. The path difference between rays $1A1'$ and $2B2'$ is then

$$CBD = CB + BD = 2AB \sin\theta = 2d \sin\theta \tag{2.41}$$

If these rays are to be in phase, this path difference must be an integral number of wavelengths, that is,

$$n\lambda = 2d \sin\theta \tag{2.42}$$

This is the Bragg law, where n is the order of the diffracted beam and is numerically equal to the path difference, in wavelengths, for successive planes; d is the interplanar spacing of the diffracting planes (Å); and θ, the Bragg angle, is the angle between the incident x-rays and the diffracting planes (deg). X-ray spectrometers are usually calibrated in terms of 2θ, the angle between the diffracted beam and the undeflected incident beam (Figures 2.14 and 2.15).

We have seen that complete destructive interference occurs only for two rays out of phase by exactly $\frac{1}{2}\lambda$. Two rays scattered from successive planes at an angle $\theta + \Delta\theta$ such that they differ in path length by only, say, $\frac{1}{4}\lambda$, do not cancel, but combine to a ray of smaller amplitude than if they were in phase (Figure 2.12). However, the ray scattered at $\theta + \Delta\theta$ from the third plane down in the crystal is $\frac{1}{4}\lambda$ out of phase with that from the second plane

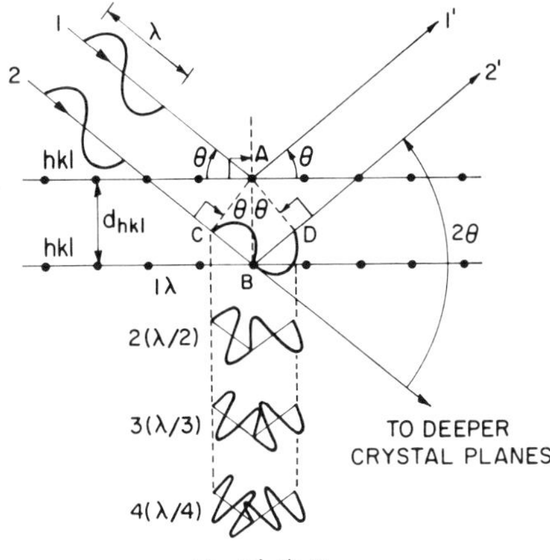

FIG. 2.14. Diffraction by crystals—Bragg version.

and $\frac{1}{2}\lambda$ out of phase with that from the first plane. Consequently, the $\theta + \Delta\theta$ rays from the first and third, second and fourth, fifth and seventh, etc. planes throughout the crystal cancel one another. Consider now rays scattered from successive planes at an angle $\theta + \Delta\theta$ such that they differ in path length by only, say, 0.005λ. The ray to cancel this ray from the first plane originates deep within the crystal, $0.5\lambda/0.005\lambda = 100$ planes down, that is, the 101st plane. Thus, the $\theta + \Delta\theta$ rays from planes 101 to 200 respectively cancel those from planes 1 to 100. Thus, diffraction occurs only at the Bragg angle. Incidentally, if the crystal is so thin that it does not have a 101st plane, the ray from the first plane is not cancelled, and the diffracted beam is broadened accordingly. This phenomenon provides the basis for measurement of single-crystal film thickness and polycrystal particle size by x-ray diffraction line broadening.

The analogy of the Bragg version of diffraction with optical reflection is evident, but the two phenomena are quite different. Optical reflection is a wholly surface effect and occurs at all incident angles greater than a certain small critical angle.

Figure 2.14 also illustrates the phenomenon of diffraction orders. Note that at the same θ angle at which wavelength λ is diffracted in first order ($n = 1$, one wavelength path difference between successive planes), wavelength $\lambda/2$ is diffracted in second order ($n = 2$, two wavelengths path difference), $\lambda/3$ in third order, etc. Thus the method of crystal diffraction cannot separate two wavelengths λ_1 and λ_2 related by the equation

$$\lambda_2 = \lambda_1/n \tag{2.43}$$

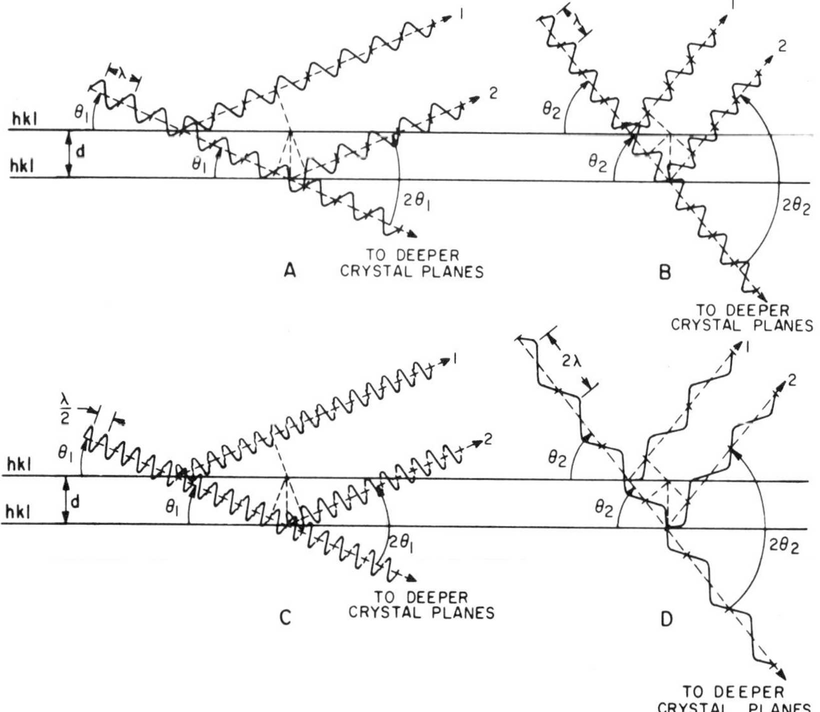

FIG. 2.15. Origin of diffraction orders.

Figure 2.15 further illustrates the phenomenon of orders: Figure 2.15A shows first-order diffraction of wavelength λ at θ_1 with path difference 1λ for rays 1 and 2. Figure 2.15B shows second-order diffraction of λ at θ_2 with path difference 2λ. Figure 2.15C shows second-order diffraction of $\lambda/2$ at θ_1, the same angle at which first-order diffraction of λ occurs. Figure 2.15D shows first-order diffraction of 2λ at θ_2, the same angle at which second-order diffraction of λ occurs. Thus, the same wavelength may be diffracted by the same planes d_{hkl} of the same crystal at more than one Bragg angle, and different wavelengths may be diffracted at the same angle.

If θ is increased progressively, the wavelength required to satisfy Equation (2.42) also increases progressively, and a polychromatic spectrum may be diffracted in sequence. At fixed θ, λ, $\lambda/2$, $\lambda/3$, etc. satisfy the equation in orders 1, 2, 3, etc.

The Bragg law disregards refraction of x-rays. The resultant error is usually negligible because the index of refraction of most substances for x-rays is of the order $1 - 10^{-6}$. The effect is greatest at the longer wavelengths, where the Bragg law may be in error by as much as $0.02° \; 2\theta$.

The application of crystals to dispersion is discussed in detail in chapter 5. Specific x-ray diffraction methods—Laue, moving-crystal, Debye–Scherrer, and divergent-beam—are described briefly in Section 8.2.3.

2.4. SPECULAR REFLECTION; DIFFRACTION BY GRATINGS

2.4.1. Specular Reflection

The index of refraction of a substance for x-rays is given by

$$n_x = 1 - \frac{N\varrho Z e^2 \lambda_{\text{cm}}^2}{2\pi A m_e c^2} = 1 - \delta \tag{2.44}$$

where N is the Avogadro number (atoms/mol); ϱ is density (g/cm^3); Z is atomic number; e is the charge of the electron (esu); λ is x-ray wavelength (cm); π is 3.1416...; A is atomic weight (g/mol); m_e is the rest mass of the electron (g); and c is the velocity of light (cm/s). In this expression,

$$N\varrho Z/A = n \tag{2.45}$$

where n is the number of electrons per cubic centimeter of the refracting medium.

In Equation (2.44), δ is of the order 10^{-6}, so that the index of refraction for x-rays is of the order $1 - 10^{-6}$, and x-rays in air or vacuum are totally reflected from any medium if incident upon it at a very small glancing angle ϕ less than ϕ_C, the critical glancing angle. By the Snell law,

$$\cos\phi_C = 1 - \delta; \qquad 1 - \cos\phi_C = \delta \tag{2.46}$$

By trigonometric identity,

$$1 - \cos 2x = 2\sin^2 x \tag{2.47}$$

from which,

$$1 - \cos\phi_C = 2\sin^2(\tfrac{1}{2}\phi_C) = \delta \tag{2.48}$$

Since $\delta \approx 10^{-6}$, the sine may be replaced by its angle in radians, giving

$$\delta = 2(\tfrac{1}{2}\phi_C)^2 = \tfrac{1}{2}\phi_C{}^2 \tag{2.49}$$

from which

$$\phi_C = \sqrt{2\delta} \tag{2.50}$$

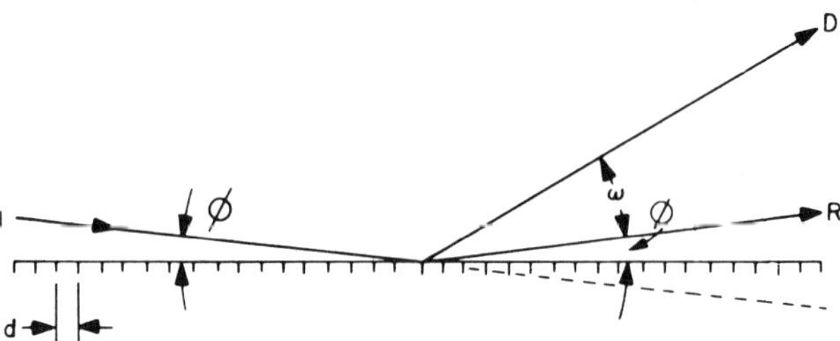

FIG. 2.16. Specular reflection from a smooth surface and diffraction by a grating.

In Figure 2.16, the highly collinear monochromatic x-ray beam I is directed at angle $\phi < \phi_C$ on a surface that for now may bc rcgarded as plane and smooth, giving reflected ray R. The incident and reflected rays make the same angle ϕ with the reflecting surface. For a typical example, for Cu $K\alpha$ (λ 1.54 Å) reflected from aluminum, δ is 8.32×10^{-6}, and ϕ_C is 4.08 milliradians or 13.0 min of arc.

Reflection has been applied to wavelength dispersion in the ultralong-wavelength region, and it occurs on the foils of Soller collimators.

2.4.2. Diffraction by Gratings

In Figure 2.16, a highly collinear monochromatic x-ray beam I having wavelength λ is directed at angle ϕ less than the critical angle for reflection on a ruled reflection grating having line spacing d. Here, R is the specularly reflected beam, and D is the first-order diffracted beam. Diffraction occurs if

$$n\lambda = 2d \sin \tfrac{1}{2}(2\phi + \omega) \sin \tfrac{1}{2}\omega \tag{2.51}$$

where n is the order; ϕ is the angle between the reflected (or incident) beam and the reflecting surface; and ω is the angle between the diffracted and reflected beams. For example, if carbon $K\alpha$ is diffracted by a grating having 12,000 lines/cm:

$$\lambda_{CK} = 44.4 \text{ Å}, \qquad d = \frac{10^8 \text{ Å/cm}}{12{,}000 \text{ lines/cm}} = 8333 \text{ Å}$$

$$\phi = 0.75°, \qquad \omega + \phi = 6.3°$$

The angle between the direction of the incident x-rays and the first-order diffracted beam D corresponds to 2θ in crystal diffraction and is $\sim 7°$

[pp. 318–19 in (*26*)]. Reflection gratings have been applied to wavelength dispersion in the ultralong-wavelength region.

Gratings may be of the simple Siegbahn type, consisting of symmetrical grooves of V-shaped cross section, or "blazed," consisting of grooves having a saw-tooth cross section. The latter may be designed, by variation of the spacing and angles of the ascending and descending slopes, to direct a high proportion of the total diffracted intensity into a specified diffraction order.

2.5. AUGER EFFECT; FLUORESCENT YIELD

2.5.1. Auger Effect

When an atom is ionized in one of its inner shells, it rearranges by filling the vacancy with an electron from a higher orbit and releasing the energy of the transition. The energy may be released as an x-ray photon; this process, of course, is x-ray emission. Alternatively, the energy may be released as an electron; this process is known variously as *radiationless transition*, *internal conversion*, or the *Auger effect* (French pronunciation oh-jay′), and the electron is an *internal-conversion* or *Auger electron* (*A25–27*). Either way, the energy of the x-ray photon or of the Auger electron is characteristic of the emitting element.

The Auger effect may be regarded in either of two ways. Consider an atom in which an initial K-shell vacancy is filled by an L electron and which then undergoes the Auger effect. It may be assumed that the atom releases the energy of the electron transition from the L to the K shell by emission of an L or M electron. Or it may be assumed that the L-to-K shell electron transition results in the production of a $K\alpha$ photon in the usual way. However, in this case, the photon does not leave the atom of its origin, but rather is absorbed within the atom with consequent expulsion of an L or M electron. This process may be regarded as internal photoelectric absorption and is analogous to the internal γ-conversion described in Section 1.6.3. Either version results in a doubly ionized atom having two vacancies, the one created by filling of the initial vacancy and the one ejected by the Auger process. These atoms are said to be in, for example, the LL or LM state. The second version of the Auger effect and the resultant production of a doubly ionized (LL) atom are illustrated in Figure 2.17. Auger-type radiationless transitions in which the original vacancy and the internal conversion occur wholly within the three L subshells, or the five M subshells, or the seven N subshells, etc. are known as *Coster–Kronig* transitions.

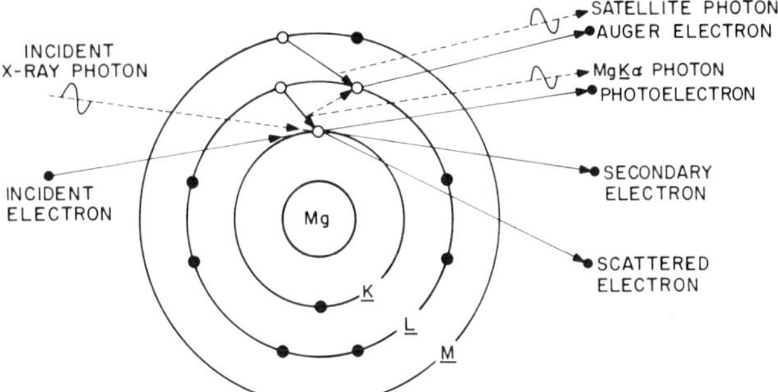

FIG. 2.17. The Auger effect—internal photoelectric absorption. If the initial K-shell vacancy is produced by an incident electron, the incident electron becomes a *scattered electron* after ejecting the K-electron, which becomes a *secondary electron*. If the initial K-shell vacancy is produced by an incident photon, the incident photon is wholly absorbed (disappears) and the ejected K-electron becomes a *photoelectron*. Filling of the K-shell vacancy by, say, an L-electron produces a Mg $K\alpha$ photon, which may be emitted (long dashed line) or absorbed internally (short dashed line). Internal absorption of the Mg $K\alpha$ photon produces, say, a second L-shell vacancy, resulting in an LL-state atom. Filling of one of these L vacancies by, say, an M electron produces a *satellite-line photon* and results in an LM-state atom.

The Auger effect is more common in elements of low atomic number because the electrons are more loosely bound and the characteristic photons more readily absorbed. The effect is more marked for the L series than for the K series for the same reasons. One can regard the phenomenon in another way: consider a $K\alpha$ photon just produced in an atom. The probability that the photon will emerge from the atom decreases the closer the wavelength of the photon lies to that of the absorption edge of the next electron shell out, LI. The wavelength of K x-ray photons approaches that of the L absorption edges as atomic number decreases.

In Figure 2.17, the initial vacancy is created by secondary excitation, but the Auger effect also occurs with primary excitation. In fact, the effect is more marked with primary excitation, but only because, as already explained, electrons are about 100 times as efficient as x-ray photons in exciting x-ray spectra. The more efficient the excitation, the greater is the opportunity for Auger absorption.

The Auger effect provides the basis for fluorescent yield, satellite lines, and the analytical method of Auger-electron spectrometry (Section 20.4.3).

2.5.2. Fluorescent Yield

One consequence of the Auger effect is that the lines in a given series are not as intense as would be predicted from the number of vacancies created in the associated orbital. The *K fluorescent yield* or *K characteristic*

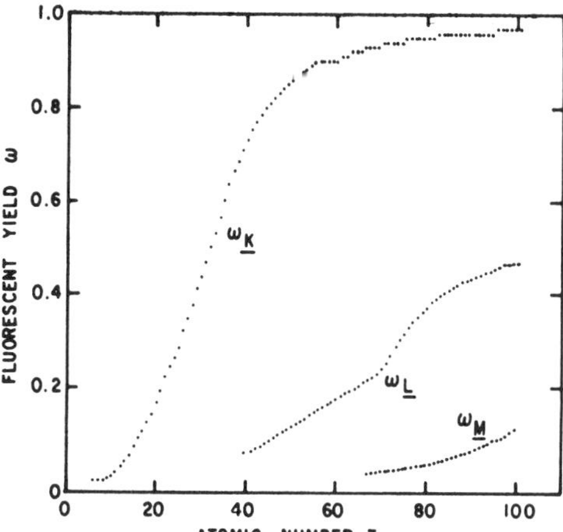

FIG. 2.18. *K*, *L*, and *M* fluorescent yields. (See Appendix 9.)

photon yield ω_K is the number of photons of all lines in the *K* series emitted in unit time divided by the number of *K*-shell vacancies formed during the same time:

$$\omega_K = \frac{\sum (n_K)_i}{N_K} = \frac{n_{K\alpha_1} + n_{K\alpha_2} + n_{K\beta_1} + \cdots}{N_K} \tag{2.52}$$

where ω_K is *K* fluorescent yield; N_K is the rate at which *K*-shell vacancies are produced; and $(n_K)_i$ is the rate at which photons of a spectral line *i* are emitted. The *L* and *M* fluorescent yields ω_L and ω_M are defined similarly. *Auger yield*, or *Auger-electron yield*, is the ratio of the numbers of Auger electrons and orbital electron vacancies produced in the same time and is equal to $1 - \omega$. Were it not for the Auger effect, ω would always be 1. Actually, ω varies with atomic number and series, as shown in Figure 2.18 and Table 2.3. Fluorescent yield is approximated by

$$\omega = Z^4/(A + Z^4) \tag{2.53}$$

TABLE 2.3. Fluorescent Yield

Element	ω_K	ω_L
$_{8}O$	0.003	—
$_{19}K$	0.118	—
$_{29}Cu$	0.425	0.006
$_{42}Mo$	0.749	0.039
$_{62}Sm$	0.915	0.180
$_{92}U$	0.960	0.478

TABLE 2.4. Constants for Calculation of Fluorescent Yields[a]

Constant [Equation (2.54)]	X-ray line series		
	K	L	M
A	−0.03795	−0.11107	−0.00036
B	0.03426	0.01368	0.00386
C	-0.1163×10^{-5}	-0.2177×10^{-6}	0.20101×10^{-6}

[a] From Colby (*C53*).

where Z is atomic number and A is a constant having values 10^6 and 10^8 for K and L x-rays, respectively. A more accurate equation is the following (*B89*):

$$\left(\frac{\omega}{1-\omega}\right)^{1/4} = A + BZ + CZ^3 \tag{2.54}$$

where the constants A, B, and C have the values given in Table 2.4; these values are not the original ones (*B89*), but updated ones (*C53*) based on recent fluorescent yield data (*F13*).

Fluorescent yield is a major limitation of sensitivity for elements of low atomic number.

Recent critical compilations of fluorescent yields are given in References (*B4*) and (*F13*), and practical values of the K, L, and M fluorescent yields of all the chemical elements are listed in Appendix 9.

Note that the *fluorescent* yield is the probability that the filling of a vacancy in a specified shell will result in emission of a characteristic x-ray photon—*regardless* of whether the vacancy arose from primary or secondary excitation. The term was originally defined on the basis of secondary x-ray intensity produced on irradiation of a target with x-rays, but now has the more general significance.

2.5.3. Satellite Lines

Another consequence of the Auger effect is the production of doubly ionized atoms, that is, atoms with two inner-orbital vacancies. For example, suppose that (1) a K-shell vacancy is created by the primary x-rays; (2) an L-shell electron falls into this vacancy, producing a $K\alpha$ photon; and (3) this photon undergoes Auger absorption with consequent expulsion of another L electron. The result is an atom in the LL state, that is, with two L-shell vacancies, one created by the L–K electron transition, the other by the expulsion of the Auger electron (Figure 2.17). If an M electron falls into one of the L vacancies and an L-series photon is emitted, the atom is left in the LM state. Doubly ionized atoms can also be produced without the Auger effect if the incident electrons (primary excitation) or x-ray photons (secondary excitation) expel more than one orbital electron in one atom. (Doubly K-ionized—KK—atoms can arise only by this process.)

Electron transitions in such doubly ionized atoms result in emission of lines having wavelengths slightly different from those emitted by the corresponding electron transitions in singly charged atoms. Such lines are known as *satellite lines* or *nondiagram lines*. For elements of high and intermediate atomic number, satellite lines usually have very low intensity and are of little significance in x-ray spectrochemistry. However, for low-Z elements, they may be relatively intense; for example, Al $SK\alpha_3$, which results from an LK–LL transition, has $\sim$10% of the intensity of Al $K\alpha_{1,2}$. The influence of chemical combination and electron distribution in the outermost orbits on the wavelength and intensity is more marked for satellite lines than for diagram lines. Satellite lines are listed by White *et al.* (*87*); typical pages of their work are reproduced in Figures 10.3 and 10.6; the satellite lines are those having symbols prefixed by the letter S. An interesting application of satellite lines for indirect detection of very light elements (boron to fluorine) is given in Section 14.8.3.

Chapters 1 and 2 show that many parameters of importance in x-ray spectrometry are functions of atomic number: wavelengths of spectral lines and absorption edges, photon energies of spectral lines, excitation potentials, mass-absorption coefficients, fluorescent yields, x-ray scattering power, ratio of Compton and Rayleigh scatter, etc. Thus, the x-ray spectrochemist should strive to think of chemical elements in terms of their atomic numbers to aid him in making quick mental comparisons of x-ray properties of specified elements of interest. Throughout this book, wherever a series of elements is listed in text or tables, they are listed in order of increasing or decreasing atomic number.

Part II

The X-Ray Spectrometer, Its Components, and Their Operation

Chapter 3

X-Ray Secondary-Emission (Fluorescence) Spectrometry; General Introduction

Having considered the excitation, nature, and properties of x-rays and x-ray spectra, we now begin our study of the application of x-ray spectra to chemical analysis. However, first we define some terms used throughout the book.

3.1. NOMENCLATURE

"What's in a name?"

Imagine going to the laboratory chemical cabinet and being confronted with shelves of bottles labeled *lye*, *caustic potash*, *lunar caustic*, *tartar emetic*, *cream of tartar*, *lime*, *calomel*, *alum*, *Epsom salt*, *Glauber's salt*, *Rochelle salt*, *blue vitriol*, *green vitriol*, *copperas*, *oil of vitriol*, *niter*, *saltpeter*, *sal ammoniac*, *corrosive sublimate*, ...* Fortunately, chemists have long since

* For the benefit of the curious: lye, sodium hydroxide (NaOH); caustic potash, potassium hydroxide (KOH); lunar caustic, silver nitrate ($AgNO_3$); tartar emetic, potassium antimonyl tartrate [$(KOOC)(CHOH)_2(COO\cdot OSb)$]; cream of tartar, potassium hydrogen tartrate [$(KOOC)(CHOH)_2(COOH)$]; lime, calcium oxide (CaO); calomel, mercury (I) chloride (HgCl); alum, potassium aluminum sulfate dodecahydrate [$KAl(SO_4)_2\cdot 12H_2O$]; Epsom salt, magnesium sulfate heptahydrate ($MgSO_4\cdot 7H_2O$); Glauber's salt, sodium sulfate decahydrate ($Na_2SO_4\cdot 10H_2O$); Rochelle salt, potassium sodium tartrate [$(KOOC)(CHOH)_2(COONa)$]; blue vitriol, copper (II) sulfate pentahydrate ($CuSO_4\cdot 5H_2O$); green

realized the vital importance of systematic nomenclature. More recently, units of physical measurement have been systematized. However, no such order exists in the nomenclature of analytical chemical instruments and methods.

The name of an instrument (or method) should: (1) classify it, showing its relationship to similar instruments; (2) distinguish it from all others in the class; (3) be appropriate and descriptive, briefly conveying its nature; and (4) be specific, that is, applicable only to that instrument. The term "atomic absorption spectrophotometer" is an example of very poor nomenclature. It obscures the relationship of the instrument to its complement —the optical flame emission spectrometer (who would ever guess from the terms that "atomic absorption" and "flame emission" spectrometry were complementary methods!), and it does not indicate the applicable spectral region. "Atomic" absorption is certainly not limited to the optical region; as we have seen, x-ray absorption is "atomic" absorption in every sense of the word. The common name for the method—simply "atomic absorption"—even obscures the fact that the method is a spectrometric one. Another example of very poor nomenclature is found on many of the printed programs of technical meetings. The sessions on optical emission spectrography and x-ray secondary-emission (fluorescence) spectrometry are distinguished by the terms *emission spectrography* and *x-ray spectrography* —as if x-ray fluorescence spectrometry were not also an emission method! The need for systematic nomenclature for analytical instruments and methods increases as new instruments and methods proliferate, which is just what is happening.

Nomenclature for a spectrometric instrument should classify the instrument as a spectro*scope* (which provides for *visual* observation of the spectrum), spectro*meter* (which provides for *measurement* of wavelength), or spectro*graph* (which *records* the spectrum). The name should then distinguish the instrument with respect to type of spectrum (emission, absorption, Raman, nuclear magnetic resonance, etc.) and/or radiation or spectral region (microwave, infrared, visible, ultraviolet, optical, x-ray, γ-ray, β-ray, mass, etc.). In the present case, another adjective is required to distinguish primary or secondary excitation.

For the instrument dealt with in this book, the writer has always preferred the term *x-ray secondary-emission spectrometer*; the term *spectrograph* is also acceptable because of the ratemeter-recorder. The term *x-ray fluo-*

---

vitriol and copperas, iron (II) sulfate heptahydrate ($FeSO_4 \cdot 7H_2O$); oil of vitrol, concentrated sulfuric acid (H_2SO_4); niter and saltpeter, potassium nitrate (KNO_3); sal ammoniac, ammonium chloride (NH_4Cl); corrosive sublimate, mercury (II) chloride ($HgCl_2$).

rescence, coined by Schreiber in 1929 (*S16*), is unfortunate, but so widely used that its abandonment seems unlikely. However, it should be used only as a synonym for secondary characteristic spectral-line emission. Thus, the instrument should be referred to as an *x-ray secondary-emission spectrometer* or *x-ray fluorescence spectrometer*, and the method as *x-ray secondary-emission spectrometry* or *x-ray fluorescence spectrometry*—but never simply *x-ray fluorescence*. In no case should the term *fluorescence* be used as a synonym for primary-excited characteristic line spectra, such as the target line spectrum from an x-ray tube or the specimen line spectrum in an electron-probe microanalyzer.

The term *x-ray emission spectrometry* may be used to distinguish x-ray emission and absorption spectrometry. However, the term is indefinite in that it can refer to either primary or secondary emission. The term *x-ray spectrochemical analysis* is a good general term for all the various analytical methods based on x-ray emission and absorption spectra, even including electron-probe microanalysis. Throughout this book, whenever there is no question that x-ray secondary-emission spectrometry is referred to, the simplified terms *x-ray spectrometry*, *x-ray spectrometer*, and *x-ray spectrometric analysis* may be used for brevity. Acronyms and other alphabet combinations used as names of methods—XRFS, ESCA, AES, etc.—are not used in this book, but are defined at appropriate places and listed in the index.

While on the subject of nomenclature, we shall define some other terms used in the book. There has long been the need for a word to replace those cumbersome terms inevitable in all papers and books on chemical analysis—"element sought," "element to be determined," "element of interest," "analytical element," etc. In this book the term *analyte* is used. The term *specimen* indicates the object placed in the instrument specimen compartment and on which measurements are made. The specimen may be: (1) a *sample*—a material of unknown composition to be analyzed or a preparation derived from it; (2) a *standard*—a similar material or preparation having known composition for calibration purposes; (3) an *intensity-reference standard*—a specimen having high chemical and physical uniformity and stability on which intensity is measured to permit adjustment of the intensities measured from samples and standards, or of the excitation conditions, to duplicate previous results; or (4) a *test specimen* for some special purpose, for example, a piece of paraffin to permit recording of the scattered primary x-ray beam (Section 4.2.4.4). The terms *major*, *minor*, and *trace* constituents refer to concentrations of the order $\gtrsim 10$, 1, and $\lesssim 0.01$ wt %, respectively. *Qualitative*, *semiquantitative*, and *quantitative* analyses consist of *detection*, *estimation*, and *determination* of the analyte, respectively. Terms such as "quantitative determination," although widely used, are redundant because determinations are necessarily quantitative.

Perhaps our nomenclature would be less chaotic if we were to refrain from devising new terms when entirely satisfactory terms already exist. The classic example is the renaming of the multiplier phototube as the "photomultiplier." The original term properly classifies the device as a phototube and gives the feature that distinguishes it from other types of phototubes—that is, the photoelectron multiplier. Conversely, the term "photomultiplier," which means literally "light multiplier," is a complete misnomer and obscures the class relationship as a phototube! Actually, the term "photomultiplier" should rather be applied to the class of electron tubes known as *image intensifiers*, which really do increase light levels. Another example is the term ESCA (electron spectrometry for chemical analysis). This term was originally "coined" to apply collectively to all types of electron spectrometry when used for *chemical* investigations and included—but was not restricted to—photo, thermionic, Auger, and beta-ray electrons. The term so used served a real need and useful purpose. However, more recently, the term ESCA seems to be restricted to photoelectron spectrometry, to the exclusion of Auger and other electron spectrometric methods. The term then simply becomes an unnecessary synonym for photoelectron spectrometry. A final example is the new term "appearance-potential" spectrometry (Section 20.7), for which the familiar term "excitation-potential" spectrometry would have been just as appropriate.

It is very encouraging that the International Union of Pure and Applied Chemistry (IUPAC) is preparing a definitive document on the nomenclature of x-ray spectrometry (*I2*). A glossary of the symbols and notation used most frequently in this book is given in Appendix 11A.

This writer does not share—and, in fact, is rather dismayed by—Birks' view [Reference (*53*) for 1974] of the establishment of standardized nomenclature in x-ray spectrometry as a "mixed blessing" and his gleeful recollections of "those heated exchanges at meetings where we *enjoyed* and employed *the happy confusion* that results from using the same symbol with different meanings" [italics mine]!

3.2. PRINCIPLE AND INSTRUMENT

This chapter is limited to wavelength-dispersive x-ray spectrometry. Wavelength and energy dispersion are distinguished in Section 5.1. The principles, advantages, and limitations of energy-dispersive x-ray spectrometry are given in Section 8.2.1, the instrumentation in Section 8.2.2.

3.2.1. Principle

X-ray secondary-emission spectrometry, or x-ray fluorescence spectrometry (XRFS), is a nondestructive instrumental method of qualitative and quantitative analysis for chemical elements based on measurement of the wavelengths and intensities of their x-ray spectral lines emitted by secondary excitation. The primary beam from an x-ray tube irradiates the specimen (sample or standard), exciting each chemical element to emit secondary spectral lines having wavelengths characteristic of that element (basis of qualitative analysis) and intensities related to its concentration (basis of quantitative analysis). The method is applicable to all chemical elements down to atomic number 4 (beryllium), but standard commercial instruments are limited to atomic number 9 (fluorine). *In favorable cases*, the method, practiced on standard commercial flat-crystal spectrometers, is applicable to: concentrations down to ~0.0001 wt % in solids and ~0.1 μg/ml (ppm) in liquids; isolated masses down to ~1 ng (10^{-9} g); and films as thin as ~0.01 monolayer. The specimen can have practically any form, including solid, powder, briquet, fusion product, film, liquid, slurry, and fabricated forms (such as rod and wire) and parts of any form and size. The useful specimen area for flat-crystal instruments is usually 5–10 cm^2, but areas as small as a few square micrometers can also be analyzed, especially with curved-crystal instruments. The precision of an individual determination may be of the order 0.1% relative, and the accuracy generally lies between $\sigma = 0.01\sqrt{C}$ and $0.05\sqrt{C}$, where σ and C are standard deviation and analyte concentration, respectively. The principles and practice of the method have been summarized in several excellent reviews (*43–53*, *F28*, *K24*, *L5*, *P11*).

3.2.2. The X-Ray Spectrogoniometer

The essential features of an x-ray fluorescence crystal spectrometer are shown in Figure 3.1. The instrument consists of three principal sections which effect excitation, dispersion, and detection and readout, respectively. The primary x-ray beam irradiates the specimen *A* held in some type of drawer or holder *B* in a specimen compartment *C*. A leaded shutter mechanism (not shown) is provided to block the primary x-ray beam at the x-ray tube window when the specimen is inserted and removed from its compartment. The secondary radiation thus excited contains many wavelengths and is emitted in all directions, but only those rays directed substantially parallel to the foils of the source Soller collimator *D* emerge from the specimen compartment and reach the analyzer crystal *E*. The crystal is mounted on a stage and may be rotated, the axis of rotation of the crystal being also the goniometer axis. As the crystal rotates, the angle θ it presents to the secondary beam varies, and each wavelength in the beam in turn satisfies the Bragg law and is diffracted. In this way, the crystal disperses the secondary spectrum, in effect causing each wavelength to go in a different direction.

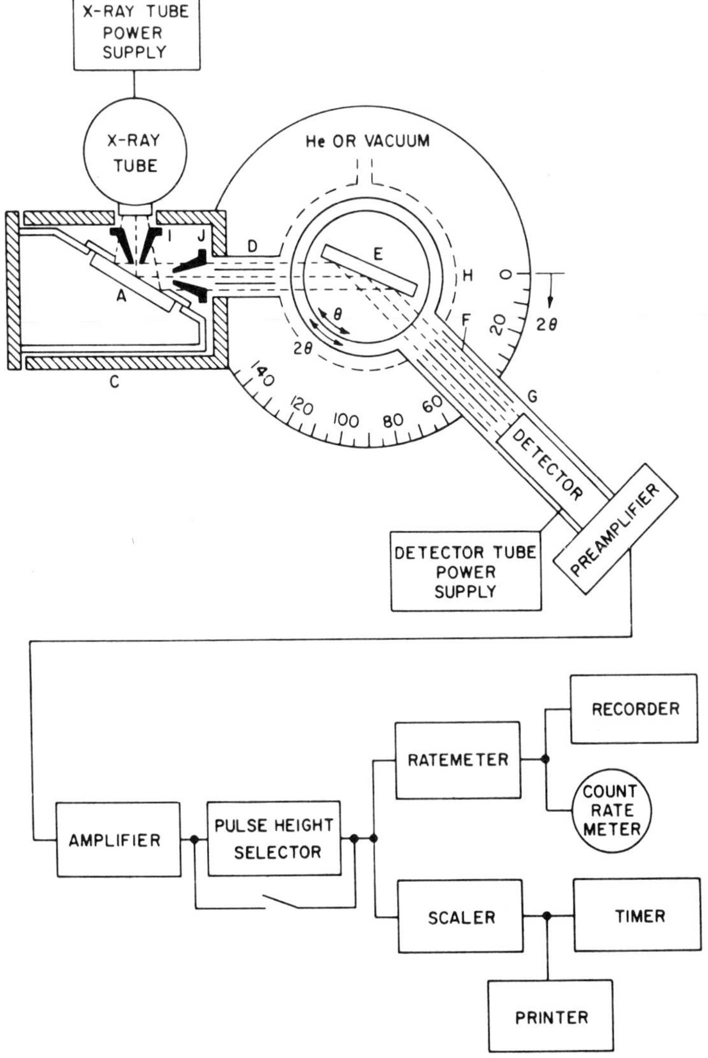

FIG. 3.1. X-ray secondary-emission (fluorescence) spectrometer.

Note the difference in this respect between wavelength-dispersive x-ray spectrometry and optical emission spectrography. In x-ray spectrometry, the spectral lines are diffracted individually in sequence. In optical spectrography, the entire spectrum is produced at once, either by a diffraction grating or refraction prism. However, in *energy*-dispersive x-ray spectrometry (Section 8.2), the entire spectrum *is* produced simultaneously.

A second detector collimator F and the detector are mounted on a common arm G that rotates on the same axis as the crystal and is geared to it with a "two-to-one" ratio, that is, as the crystal turns through an angle θ, the detector arm turns through twice that angle, 2θ. Thus, the detector and incident secondary beam always present equal angles to the crystal, and the detector is always in the correct position to receive any diffracted x-rays (Figures 2.14 and 2.15). Incidentally, x-ray spectrogoniometers are graduated in degrees 2θ, rather than degrees θ, because the detector is always at angle 2θ to the direction of the secondary beam incident on the crystal. The detector converts each x-ray photon it absorbs into a pulse of electric current; in most detectors, these pulses have amplitudes proportional to the energy of the x-ray photons. These pulses are amplified by the preamplifier, which is mounted close to the detector on the detector arm.

When wavelengths greater than ~2 Å are measured, it is necessary to reduce the absorption of the radiation path (x-ray tube window to detector window) with a helium or vacuum enclosure H. When very small specimens or small selected areas on large specimens are measured, apertures I, J may be placed in the primary or secondary x-ray beam (Chapter 19).

The spectrometer may be evaluated in terms of four criteria: analyte-line intensity (Section 13.1), resolution (Section 13.2), peak-to-background ratio (Section 11.2.4), and mechanical arrangement (Sections 5.5 and 5.6). The first three are performance criteria and depend on the analyzer crystal, collimator or slit, and detector. The fourth depends on the design and construction of the instrument, but affects the performance criteria.

The instrument just described—consisting of the specimen chamber, crystal, source and detector collimators, helium or vacuum enclosure, detector, preamplifier, and the drive mechanism for the crystal and detector arm—is known as the *goniometer* (angle-measuring device). When used for spectrometry in this way, the goniometer is referred to as a *spectrogoniometer* or *spectrometer*. The alternative mode of operation, described briefly below, is as a *diffractometer*. The crystal and detector arm may rotate in a horizontal or vertical plane—that is, the goniometer axis may be vertical or horizontal. The Picker goniometer included the x-ray tube as an integral part and could be mounted in any position.

TABLE 3.1. Comparison of the X-Ray Spectrometer and Diffractometer

Bragg law	$n\lambda =$	$2d$	$\sin\theta$
Spectrometer	Calculated	Known	Measured
Diffractometer	Known	Calculated	Measured

This type of spectrogoniometer is typical of the earlier commercial instruments, which were simply converted diffractometers. Today, all manufacturers offer instruments designed specifically for spectrometry. However, the older goniometers having specimen drawers are still available and very useful. In small laboratories and schools, they perform both spectrometric and diffractometric functions economically and with relative ease of interchangeability. They are excellent for teaching purposes because the components and operation are readily observable. They also provide versatility in choice of modes of operation, components, accessories, and specimen presentation not feasible with the newer versions.

Although both the spectrometer and diffractometer are based on the Bragg law, their modes of operation are entirely different. In a diffractometer, the specimen drawer *B* (Figure 3.1) and compartment *C* are omitted, and the x-ray tube is rotated about its axis so that the primary beam is directed through the collimator *D*. The single crystal *E* is replaced with the polycrystalline sample to be investigated. Thus, a known wavelength—the filtered $K\alpha$ line of the x-ray tube target—is directed on the sample, and as the sample is rotated through successive θ values, each set of crystal planes (hkl) having its own interplanar spacing d_{hkl} in turn satisfies the Bragg law and diffracts the incident wavelength. The functions of the spectrometer and diffractometer may be contrasted in terms of the Bragg law as shown in Table 3.1. Both instruments produce charts consisting of peaks on an intensity *versus* 2θ ("two theta") scale. On the spectrometer chart, each peak represents a different wavelength diffracted by the same set of planes in the analyzer crystal. On the diffractometer chart, each peak represents the same wavelength—the target $K\alpha$ line—but diffracted by a different set of planes in the crystalline sample.

3.2.3. Electronic Readout Components

The preamplifier is connected by cable to the amplifier, which further amplifies the detector output pulses. As we have seen, the crystal can diffract more than one wavelength at a given Bragg angle, and when this happens, each wavelength gives detector output pulses of different amplitude. The pulse-height selector sorts these pulses, passing only those having a certain preselected height. The pulse-height selector may be by-passed, as shown schematically in Figure 3.1. The amplified pulses—with or without selection—may be registered by the ratemeter, which integrates them and displays on a meter or strip-chart recorder the rate at which pulses enter the ratemeter, and thus the rate at which x-ray photons are absorbed in the detector. Alternatively, the pulses may be individually counted by the scaler–timer, the data from which may be recorded manually or by a printer. All these components and the detector-tube power supply are mounted in the electronic-circuit panel or console.

The x-ray tube and power supply, spectrogoniometer, and electronic-circuit panel collectively constitute the x-ray secondary-emission spectrometer. The particular instrument just described is a single-channel manual spectrometer—that is, it can measure only one wavelength at a time, and it is operated manually. Dual and multichannel and automatic instruments are described in Sections 5.5.1 and 9.3.3.

3.2.4. Qualitative, Semiquantitative, and Quantitative Analysis

In qualitative analysis, the spectrometer is made to scan through the entire 2θ region as detected x-ray intensity is recorded by the ratemeter–recorder. The goniometer motor drive rotates the crystal and detector arm at rates proportional to θ and 2θ, respectively, per unit time. The elements present are identified by the 2θ angles—or wavelengths—at which peaks occur on the chart, and their concentrations may be estimated from the peak heights—or intensities. Analyte peaks may be recorded from standards for comparison. Qualitative and semiquantitative analyses are described in Chapter 10.

In quantitative analysis, the intensities of analyte lines are measured from samples and standards. The measurements are almost always made with the scaler and timer, although the ratemeter may be used. Calibration curves or mathematical relationships are established from the standard data. Quantitative analysis is described in Chapter 14.

3.2.5. Phases of a Quantitative X-Ray Spectrometric Analysis

X-ray spectrochemical analyses involve five essential phases as follows:

1. Selection of the analytical method (Chapters 14 and 15) and specimen preparation and presentation (Chapters 16 and 17).

2. Excitation of the characteristic line spectrum of the analyte, involving the x-ray tube and power supply (Chapter 4).

3. Dispersion of the secondary spectrum to isolate the analyte lines and permit their individual measurement, usually effected in either of two ways or by a combination of both: wavelength dispersion (crystal or spacial dispersion), involving the crystal–collimator system (Chapter 5); and energy dispersion, involving the pulse-height analyzer (Chapter 8).

4. Detection, measurement, and readout of the analyte-line intensity, involving the detector, amplifiers, pulse-height analyzer, ratemeter and recorder, scaler and timer, and printer (Chapters 6 and 7).

5. Reduction of x-ray intensity data to analyte concentration, involving the calibration or calculation technique (Chapter 14) and/or correction for absorption-enhancement effects (Chapter 15).

The first of these phases is effected on the specimen before placing it in the instrument; the second, third, and fourth phases are effected by the spectrometer; the last may be done manually or by a computer.

3.3. APPRAISAL

This section applies principally to manual, single-channel, flat-crystal spectrometers. Advantages and limitations of semiautomatic and automatic instruments are given in Sections 9.3.2 and 9.3.3, and of energy-dispersive instruments in Sections 8.2.1.2 and 8.2.1.3.

3.3.1. Advantages

3.3.1.1. X-Ray Spectra

Unlike optical emission spectra, x-ray spectra are simple and regular. In general, each element has substantially fewer x-ray lines than optical lines. The selection rules exclude many electron transitions, and many of the permitted transitions result in lines so weak as to be insignificant. However, weak lines from major constituents may interfere with determination of minor or trace constituents, especially in the electron-probe microanalyzer, where electron excitation and curved-crystal optics result in high efficiency. In general, each element has the same K, L, and M lines, and the wavelength of each line varies regularly with atomic number, regardless of the nature of the element—metal, nonmetal, liquid, or gas. For example, Ar K lines from occluded argon frequently appear in the x-ray spectra from

FIG. 3.2. Optical emission spectra of iron, an Alnico alloy (53Fe–17Ni–13Co–9.5Al–6Cu–0.5Ti–0.5Zr), and copper. Compare with x-ray spectrum of Alnico in Figure 10.1.

metal films sputtered in argon atmospheres and from mineral oils, and Ar *K* pulse distributions occur in energy-dispersive analyses conducted in air path. The relative simplicity of x-ray spectra is illustrated by a comparison of the optical emission spectra of iron, copper, and an Alnico alloy shown in Figure 3.2 with the x-ray spectra shown in Figure 10.1.

Other advantages of x-ray spectra arise from their relative independence, compared with optical spectra, of chemical state. Although small displacements of wavelength of x-ray lines may occur, especially in light elements, the effect is much more severe in optical spectra. Several other undesirable effects present in optical spectra are completely absent in x-ray spectra: Differences in volatility of analyte elements affect relative intensities of optical lines. If the analyte volatilizes as a molecular, rather than atomic, species, the strength of the chemical bond affects the optical line intensity. Doppler line broadening arises from the rapid motion of the atoms in the arc or spark.

Internal standardization, almost always required in optical emission spectrography, is usually not necessary in x-ray spectrometry. However, when it is required, the simplicity of x-ray spectra and their relationship to atomic number make selection of an internal standard element much easier than it is in optical spectrography.

The simplicity and uniform regular relationship of x-ray spectra and atomic number are shown in Figure 1.8. No such simple diagram can be drawn for selected optical emission lines.

Incidentally, x-ray spectra from high-temperature plasmas do resemble optical spectra and may contain very large numbers of lines. (A *plasma* is a highly ionized gas containing substantially equal numbers of positive charges and electrons.) These spectra arise from atoms stripped of many of their electrons. Electrons are excited to unfilled states within the atoms —rather than out of the atoms—then return directly to their original states.

3.3.1.2. Excitation and Absorption

These properties vary with atomic number in the same uniform, regular manner as x-ray spectra, and this regularity is also uninterrupted by the physical and chemical state of the element. Elements having about the same atomic number have very similar excitation and absorption properties. An interesting example of this is provided by the determination of occluded argon in argon-sputtered silicon dioxide (SiO_2) films (*H56, L31*). Argon calibration standards would be extremely difficult to prepare. However, the analysis is readily calibrated by use of films of potassium chloride (KCl). Argon (*Z* 18) lies between chlorine (*Z* 17) and potassium (*Z* 19) in the periodic table.

3.3.1.3. Absorption-Enhancement Effects

The elements occurring with the analyte in the specimen may cause absorption-enhancement effects ("matrix" effects) in both x-ray and optical emission. In optical emission, these effects are largely unpredictable, except by experience. In x-ray emission, they are systematic, predictable, and readily evaluated. Moreover, there are many methods of dealing with them, including the following: (1) standard addition and dilution; (2) calibration standards; (3) internal standardization; (4) dilution; (5) thin-film methods; (6) methods based on x-ray scatter; (7) emission–absorption methods; (8) association (indirect) methods; and (9) mathematical correction. In addition, there are many special experimental x-ray spectrometric techniques, and the instrument can be arranged for x-ray absorption-edge spectrometry, in which absorption-enhancement effects largely disappear.

3.3.1.4. Spectral-Line Interference

Due to the simplicity of x-ray spectra, spectral-line interference is relatively infrequent. However, when it does occur, there are many ways to deal with it: (1) choice of an alternative spectral line or a higher order; (2) prevention or reduction of excitation of the interfering line by operating the x-ray tube below the interferant excitation potential, selection of the x-ray tube target, or monochromatic excitation; (3) use of a blank to measure the interfering line intensity; (4) increased collimation; (5) selection of the crystal for greater dispersion or resolution or to eliminate even-order interfering lines; (6) filtration; (7) use of air path to eliminate long-wavelength interfering lines; (8) selection of the detector for maximum efficiency for the analyte line or minimum efficiency for the interfering line;

(9) pulse-height selection; (10) dilution; and (11) mathematical correction, including unfolding (stripping) of overlapped lines.

3.3.1.5. Nondestruction of Specimen

X-ray secondary-emission spectrometry is nondestructive in the sense that the specimen placed in the instrument remains substantially unchanged during the analysis. The method is also nondestructive in the sense that, frequently, sampling is not required; the instrument can be arranged to accommodate very large objects, which then need not be damaged by cutting samples from them. These features are advantageous in many ways. Specimen composition does not change during analysis, as it frequently does during arcing in optical spectrography. Since the instrument can accommodate the specimen in nearly any form, little or no specimen preparation is required in many cases. Other analyses and tests can be made, and questionable analyses can be repeated on the same sample. Samples can be subjected to additional treatment or processing, then reanalyzed. Standards, test specimens, expensive or precious samples, criminal evidence, etc. are not consumed. Fabricated parts, art objects, antiquities, etc. are not damaged (*C29*).

However, biological, organic, and liquid specimens may undergo radiolysis; minerals, ceramics, glasses, and certain other inorganic substances may become colored, usually temporarily; standards may undergo change (Section 16.1.4.2); and plastics and electronically active materials may sustain permanent radiation damage. These effects are minimized in energy-dispersive instruments using low-power x-ray tube (Sections 4.2.5.5 and 4.2.5.6), secondary-emission (Section 8.3.1), and radioisotope (Section 8.2.2.3) sources.

3.3.1.6. Specimen Versatility

No analytical method can deal with as wide a variety of specimen forms as the x-ray spectrometer. The physical form may be solid, briquet, powder, paste, liquid, or even gas. The material may be metal, mineral, ceramic, plastic, rubber, textile, paper, or practically anything else. The size may be anything from a barely visible speck or a film having less material than a monolayer, to a very large, massive object. The shape may be plane, cylindrical, filamentary, or irregular, and the method is applicable to small, fabricated parts of all shapes. The analyte may be distributed on filter paper, Mylar film, cellulose tape, ion-exchange resin, etc. The method is applicable to special conditions such as high and low temperatures, special atmospheres, and dynamic conditions. The method is particularly useful for analysis for elements difficult to separate chemically; classic examples are mixtures of rare-earth elements, zirconium and hafnium, and niobium and tantalum.

3.3.1.7. Operational Versatility

The x-ray spectrometer can be fitted with a variety of components and accessories, and operated in a variety of modes under a variety of conditions, all selected to give optimum results in a given analysis. The selectable components include: (1) the x-ray tube target—about ten are available commercially, including a dual target (tungsten–chromium); (2) collimators of various lengths and foil spacings; (3) crystals, selected for "reflectivity," dispersion, resolution, or elimination of orders; and (4) detectors—sealed and flow proportional counters having various gas fillings, scintillation counters, or Geiger counters. The selectable accessories include rotating specimen stages, inverted optics, bulk-specimen holders, selected-area apertures, curved-crystal accessories, and helium or vacuum enclosures. It is fair to state that an x-ray spectrometer can usually be fitted with accessories optimal for a specified analysis more conveniently than any other analytical chemical instrument. The modes of operation, in addition to x-ray secondary-emission spectrometry, include polychromatic and monochromatic absorption; absorption-edge spectrometry; methods based on scattered, rather than emitted, x-rays; energy-dispersive analysis; and even polychromatic and monochromatic contact microradiography.

The work of Cowgill (*C64*) with lake core samples provides an outstanding example of versatility of operation and analytical strategy (see below) in x-ray spectrometry. The material was dried and ground to -200 mesh, then briqueted at 15,000 lb/inch^2. The briquets were used to determine 41 elements ranging in atomic number from 11 to 82 (sodium to lead) and in concentration from <1 ppm to $\sim 50\%$. The work involved three x-ray tube targets (chromium, molybdenum, platinum), eight crystals (KHP, ADP, EDDT, germanium, silicon, NaCl, LiF, topaz), and two detectors (flow and scintillation). Three basic analytical methods were used—comparison with calibration standards and the methods of Mitchell (Section 15.4) and Reynolds and Ryland (Section 14.5.5).

3.3.1.8. Versatility of Analytical Strategy

X-ray spectrometry lends itself to a variety of analytical methods and techniques of specimen preparation and presentation so large that it would be futile to attempt to summarize them here. They are described in detail in Chapters 14, 16, and 17.

3.3.1.9. Selected-Area Analysis

Small selected areas can usually be analyzed in place so that contamination and loss by chemical or physical separation are avoided. Uniformity and homogeneity of large-area samples and uniformity of thickness and composition of films and platings can be evaluated.

3.3.1.10. Semiquantitative Estimations

These can be made from data taken without standards other than the pure analyte to within a factor of two or three in x-ray spectrometry, to within an order in optical spectrography. Simple calculations based on absorption coefficients can improve the x-ray estimates.

3.3.1.11. Concentration Range

X-ray spectrometry is applicable over an extremely wide concentration range, from 100% for any element above fluorine to 0.0001% for sensitive elements in favorable matrixes. Small differences in concentration of major constituents are readily measured. For most elements, the method is applicable to micro and trace analysis. In favorable cases, traces can be determined without separation, so that contamination or loss by chemical or physical separation is avoided. Quantities $<0.1\ \mu g$ and films "thinner" than one monolayer can be detected for most elements above magnesium on modern commercial flat-crystal spectrometers.

3.3.1.12. Sensitivity

The sensitivity, although in general not as good as that of optical emission spectrography, is still very high. Instrument components, accessories, and conditions may be chosen for optimal sensitivity for a given analysis. With a given x-ray tube target, crystal, collimator system, and detector, sensitivity of the pure analyte is a relatively simple function of atomic number. Elements having about the same atomic number are likely to have about the same sensitivity in a given system. No such simplicity exists in optical emission spectrography. In Table 3.2, the sensitivities of x-ray fluorescence spectrometry and other analytical chemical methods applicable to trace and micro analysis are compared in terms of minimum detectable mass of element under *most favorable conditions.*

TABLE 3.2. Ultimate Sensitivities of Microanalytical Methods for the Most Sensitive Elements under the Most Favorable Conditions

Method	Minimum detectable amount g
Microchemistry	10^{-9}
Polarography (volt-ammetry)	10^{-9}
X-ray fluorescence spectrometry (XRFS)	10^{-9}
Optical absorption spectrometry	10^{-10}
Optical emission spectrography	10^{-10}
Auger-electron spectrometry (AES)	10^{-11}
Flame ("atomic") absorption spectrometry ("AA")	10^{-11}
Ion-scattering spectrometry (ISS)	10^{-11}
Flame emission spectrometry	10^{-12}
Ion-induced x-ray spectrometry	10^{-12}
Mass spectrometry (MS)	10^{-13}
Neutron-activation analysis (NA)	10^{-13}
Optical fluorescence microscopy	10^{-14}
Electron-probe microanalysis (EPMA)	10^{-15}
Radioactive tracer analysis	10^{-18}
Secondary-ion mass spectrometry (SIMS)	10^{-18}

3.3.1.13. Precision and Accuracy

These qualities are generally very good. Precision may usually be made as great as desired by appropriate selection of the accumulated count, and the relative standard deviation [Equation (11.8)] is typically of the order 0.1%. The accuracy compares favorably with that of other instrumental analytical methods. In conventional x-ray spectrometry, the specimen area is usually several square centimeters, so that inhomogeneity, although still a serious problem, is not as serious as in optical spectrography, where the sample area is only a very few square millimeters.

3.3.1.14. Excitation

Electronic and/or magnetic line, kV, and mA stabilization result in extremely well-regulated excitation conditions. Except for very small variations due to slight instability of x-ray tube operating conditions, x-ray emission intensity is constant with time. Optical emission intensity from an arc or spark may vary markedly with time.

3.3.1.15. Speed and Convenience

Because wavelength and intensity are measured directly on the instrument and because of the nondestructive feature, x-ray spectrometry is extremely rapid and convenient and readily applicable to automation. Typical x-ray measurement times are of the order 20–60 s, and a multichannel instrument can measure up to 20 or 30 elements in a minute or less. In fact, it is rarely the spectrometer that limits the specimen throughput rate. In general, no analytical method can deal as quickly and conveniently with very large numbers of samples when they can be analyzed in the as-received form. Even when specimen preparation is required, the method is often more rapid and convenient than other methods. X-ray spectrometry permits extremely rapid confirmation, identification, or selection of known alternatives on the basis of the presence or absence, or high or low concentration, of some element or the intensity ratios of two or more elements. For example, Kovar and stainless steel parts can be sorted on the basis of presence or absence of cobalt; 52-alloy and Invar parts (48–52 and 65–35 iron–nickel alloys, respectively) can be distinguished on the basis of relative intensities of Fe $K\alpha$ and Ni $K\alpha$ lines. With energy-dispersive spectrometers, such identifications can be made almost immediately even without any prior knowledge of the specimen whatever. Errors, anomalous results, out-of-specification results, etc. can often be noted while data is being taken, and checked immediately.

3.3.1.16. Operating Cost

Although the initial cost of the equipment is substantial, and expensive x-ray tubes must be replaced from time to time, operating costs are low because there are no consumed supplies.

3.3.1.17. Automation

Great ingenuity has been shown in designing accessories to index large numbers of samples through the specimen compartment of standard commercial x-ray spectrometers. Other instruments are available commercially for automatic sequential or simultaneous determination of up to 30 elements in a sample. Some instruments may be programmed to index a large number of samples through the specimen compartment in turn and for each to perform a series of determinations by automatic selection of kV, mA, 2θ, collimator, crystal, detector, pulse-height selector settings, and the number of counts scaled. The intensity data are printed out, or intensity data from samples and standards may be processed by a computer and analyte concentrations printed out directly. Some of these instruments are described in Chapter 9.

3.3.1.18. Process Control

Applications to on-line plating thickness gaging and chemical process control have been successful (Chapter 9).

3.3.1.19. Use with Other Methods

X-ray spectrometry has been used to complement x-ray diffraction and absorption, infrared absorption spectrometry, electrophoresis, chromatography, gas chromatography, radioactivity, and other analytical methods (Section 14.8.4).

3.3.2. Disadvantages

3.3.2.1. Light Elements

Standard commercial x-ray spectrometers are subject to certain difficulties and inconveniences for elements having atomic number 22 (titanium) or less; these difficulties become progressively more severe as atomic number decreases, and are quite serious for fluorine through silicon (atomic number 9–14). Present commercial instruments are substantially useless for elements of atomic number less than 9. However, specially built ultralong-wavelength x-ray secondary-emission spectrometers and commercial electron-probe x-ray microanalyzers are capable of measuring K spectra of elements down to beryllium (Z 4).

3.3.2.2. Penetration

Only a relatively thin surface layer (≤ 0.1 mm) contributes to the measured x-ray line intensity, so that the method is sensitive to surface texture and gives average bulk composition only for homogeneous substances. The effect becomes more severe as line wavelength increases.

3.3.2.3. Absorption-Enhancement Effects

The relationship between intensity and concentration may be seriously affected by absorption-enhancement effects caused by the matrix elements. In general, the slope of the intensity *versus* concentration calibration curve is inversely proportional to the matrix absorption for the analyte line. However, these effects are systematic and predictable, and there are many ways of dealing with them.

3.3.2.4. Sensitivity

In general, x-ray spectrometry is less sensitive than optical spectrography by two or more orders. However, there are many exceptions to this, especially the nonmetallic elements.

3.3.2.5. Qualitative Analysis

Recording one x-ray spectrum with the ratemeter–recorder may require $\frac{1}{2}$h or more, during which time a whole plate of optical spectra may be recorded and developed. However, rapid methods for recording x-ray spectra have been developed (Section 10.3), and of course, x-ray spectra are recorded in 30 s to several minutes by use of energy dispersion.

3.3.2.6. Standards

For quantitative analysis, the x-ray spectrometric method is a comparative one, like all instrumental analytical methods, and suitable standards of accurately known composition are required for calibration. The standards must have as nearly as possible the same chemical composition as the samples in both x-ray and optical spectrochemical analysis. However, in x-ray spectrometry, standards must also have the same physical form as the samples: solid metal (blanks, sheet, and foil $\geq$0.1 mm thick) can be analyzed only with solid standards, powders packed in cells only with similarly packed powder standards, briquets only with briquets, etc. The same applies to analysis of analytes distributed on filter paper, Mylar film, ion-exchange resins, etc. Moreover, solid standards must have the same surface texture as the samples, powders the same particle size distribution and packing density, etc. If the analyte line is subject to chemical effects, the analyte should be present in the standards in the same chemical state as in the samples. No such restrictions apply in optical spectrography, where, for example, metal drillings may often be analyzed with synthetic standards consisting of mixed metal powders—or even metal oxide powders.

This limitation is particularly inconvenient when the same substance must be analyzed in a number of physical forms—for example, metal blanks, drillings, shot, filings, powder, thin rod, wire, and small fabricated parts. In optical spectrography, one set of standards would suffice—and each could be in a different form, but not in x-ray spectrometry.

When standards are not available in the same physical form as the samples, and when the same substance must be analyzed in several forms, all standards and samples must be reduced to a common form—powder, fusion product, solution, etc. In such cases, much of the advantage of speed and convenience is lost.

There are exceptions to this requirement for close similarity of form of x-ray spectrometric standards and samples; some examples follow. Extremely fine particles of air pollutants collected on filter disks have been analyzed by use of standards prepared by evaporation of standard analyte solutions on similar disks. Powders dusted on Mylar have been analyzed with standards consisting of metals evaporated on Mylar (*G29*). Trace, micro, and thin-film samples in general can often be analyzed by use of standards having different matrixes and even somewhat different forms (Section 12.2.1). Organic-liquid samples have been analyzed with solid plastic-disk standards. Heavy elements in organometallic compounds supported on filter paper have been analyzed with standards of similarly supported inorganic compounds (*M19*).

3.3.2.7. Instrument Preparation

The availability of a wide selection of x-ray tube targets, crystals, collimators, and detectors has already been cited as advantageous. However, it must be admitted that in a laboratory where the variety of samples requires frequent change and adjustment of components, much time can be lost. Of course, much of this time is regained by the rapidity of x-ray spectrometric analyses, and the problem is largely eliminated in automatic instruments.

3.3.2.8. Components

Frequently, all analytes of interest in a series of specimens cannot be measured optimally with the same x-ray tube target, crystal, collimators, and/or detector. In such cases, if all measurements are to be made with the same set of components, a compromise must be made in the interest of convenience and economy at the expense of intensity. Usually, loss of intensity can be afforded except for trace constituents. The alternative is to measure one or more analytes in all specimens with one set of components, then one or more others with another set of components, etc. The difficulty may be minimized by use of a dual-target (tungsten–chromium) tube, crystal changer, and dual detector, or by providing two spectrogoniometers, each fitted with a different set of components. The difficulty is almost completely eliminated in some of the modern automatic spectrometers described in Chapter 9.

3.3.2.9. Instrument Settings

Frequently, all analytes of interest in a series of specimens cannot be measured optimally with the same settings of x-ray tube potential and current, detector-tube potential, amplifier gain, and pulse-height selector baseline and window. Of course, the individual settings may be made for each analyte on each specimen; some instruments make provision to preset several values of kV and mA and to select them with selector switches. Or, here again, the analytes may all be determined at the same settings representing a compromise between convenience and intensity, or they may be determined in groups, each group at different settings. A manual single-channel instrument provides no other alternatives, but automatic instruments change these settings automatically for each analyte in each specimen.

3.3.2.10. Error

In the operation of single-channel manual instruments, there are many opportunities for error and reduction of precision when two or more analytes are to be measured in each of many specimens. The likelihood increases as the number of settings to be changed increases—2θ (line and background), kV, mA, detector tube potential, amplifier gain, pulse-height analyzer baseline and window, and preset time or count. Finally, one must read the scaler or timer at least once for each analyte in each specimen, and usually twice (line and background). A printer eliminates this last source of error, and automatic instruments eliminate many of the others.

3.3.2.11. Tedium

In the operation of a single-channel manual instrument, data-taking may be very tedious. A printer may record the data, but specimens must be loaded and settings made manually, so that constant attention is required throughout the analysis. Here again, automatic instruments minimize the problem.

3.3.2.12. Cost

Initial cost of an x-ray spectrometer—especially an automatic one—is high, and considerable additional equipment may be required for specimen preparation (Section 16.1.6). The instrument should preferably be housed in a separate room, especially in a factory environment, and this room should preferably be air-conditioned. Certain radiation-safety requirements must be met. Operating costs are relatively low, but x-ray tubes must be replaced from time to time.

3.4. TRENDS IN X-RAY SPECTROCHEMICAL ANALYSIS

In closing this chapter on the instrumentation and method of x-ray fluorescence spectrometry, it is fitting to consider the status of, and speculate on trends in, analytical x-ray spectrometry in general and certain related analytical methods.

In general, it appears that x-ray spectrochemical analysis has evolved to the point where experimentation is declining and mathematics, computer programing, and electronic and mechanical engineering are assuming predominant roles. More and more aspects of x-ray spectrometric analyses that previously required experimental ingenuity are now dealt with mathematically and automatically. The sophisticated instrumentation for energy-dispersive spectrochemical analysis (Section 8.2.2.4) is probably the outstanding example; even particle-size, heterogeneity, and surface-texture effects have yielded to mathematical treatment (Section 12.3).

This trend is only to be expected because it has occurred in most other instrumental analytical disciplines. Moreover, great benefit derives from these developments. The modern computer-controlled automated instruments, with their sophisticated readout electronics and computerized conversion of intensities to analytical concentrations, enable us to get much more work done more conveniently and in less time. But we must forego the challenge, satisfaction—and fun!—of circumventing difficulties by experimental ingenuity. To some x-ray spectrochemists, this may seem a small price indeed. Others may share the writer's feeling that, to paraphrase the television commercial, we are analyzing more, but enjoying it less!

Among the most significant recent and continuing trends are the following.

1. *Speed and convenience.* The conversion from manual to semiautomatic and fully automatic operation is substantially complete. Modern

semiautomatic instruments have front-panel selection of crystals, collimators, detectors, and perhaps even x-ray tube targets; multiple-specimen changers are also common. High-intensity x-ray tubes and small target-specimen distances give high statistical precision in short counting times. Modern fully automatic simultaneous instruments permit determination of up to 30 elements in a series of up to 60 samples in a few minutes—all automatically and unattended.

2. *Solid-state circuitry*. In substantially all modern instruments, electron-tube circuitry has been replaced by all-solid-state circuitry, in the United States, mostly in the form of NIM (Nuclear Instruments Manufacturers) standardized modular units. The space and power requirements are greatly reduced and the degree of sophistication increased.

3. *Computers*. There is widespread and increasing use of computers to program and control automatic instruments and to store, process, and compute analytical concentrations directly and immediately from measured intensity data. The computers may be small and "dedicated" (that is, an integral part of the instrument) or large, central, time-shared units.

4. *Long- and ultralong-wavelength regions*. Most of the problems attending determination of the light elements (Z 9–22, fluorine to titanium) have been surmounted (Section 16.1.3), and laboratory and commercial instrumentation is available for determination of elements 4–8 (beryllium to oxygen) (Section 9.4.3). Spectrometers in these wavelength regions are used not only for elemental analysis, but for studies of chemical bonding, etc.

5. *Energy-dispersion*. There has been a phenomenal proliferation of highly sophisticated instrumentation for energy-dispersive analysis (Section 8.2). By one account, energy-dispersive spectrometers now outnumber wavelength-dispersive ones by four to one! This instrumentation has been made possible by the development of lithium-drifted silicon detectors (Section 6.4), multichannel pulse-height analyzers (Section 8.1.2.3), and small digital computers. This instrumentation has in turn made possible the increased use of low-power x-ray tube (Sections 4.2.5.5–4.2.5.6) and radioisotope (Section 8.2.2.3) sources. Energy-dispersive x-ray spectrometers are used not only in the form of self-contained laboratory instruments (Section 8.2.2.4), but as accessories for electron-probe microanalyzers and scanning electron microscopes, for field use, and for applications in otherwise inaccessible places, such as drill holes.

6. *Indirect (association) methods*. By stoichiometric combination with elements that are easily determinable by x-ray spectrometry, it is possible to determine elements ordinarily determinable by x-ray spectrometry only

with difficulty or not at all, and even certain classes of compounds (Section 14.8.3). These methods extend the applicability of x-ray spectrometry to the entire periodic table, to organic chemistry, and to certain specific materials—all with standard commercial instrumentation and accessories and with conventional preparation and presentation techniques.

7. *Electron-probe microanalysis* (Chapter 21). During the past decade, this method has received widespread application, and the number of electron-probe microanalyzers in use has increased from 10 or so to several hundred. Energy-dispersive spectrometers have complemented or even replaced wavelength-dispersive spectrometers on many of these instruments.

8. *Excitation by electrons, protons, and heavy ions.* The high efficiency of electron excitation, especially for low-Z elements, and the extremely low surface penetration and low continuous background of proton and ion excitation have made these sources increasingly attractive (Sections 1.8 and 4.1.1).

9. *Auger- and x-ray-excited photoelectron spectrometry.* There has been a "rediscovery" and rapid growth of these disciplines (Section 20.4), which constitute two of the several methods known collectively as "electron spectrometry for chemical analysis" (ESCA). These methods, like proton- and heavy-ion excitation, are also extremely useful for surface analysis.

10. *X-ray spectra from plasmas, colliding atoms, and solid-state energy bands.* Birks [Reference (*53*) for 1974] sees great promise in three applications of x-ray spectrometry in physical—as distinguished from analytical—chemistry. Two of these applications are investigations of the x-ray spectra arising from plasmas (Section 1.6.2.1) and from interaction of the electron shells of colliding projectile and target atoms on bombardment of matter with high-speed heavy ions (Section 1.6.3). Both these types of spectra constitute powerful means for elucidation of atomic structure, and the latter for elucidation of molecular—or, at least, quasimolecular—orbital theory. The third application is investigation of the changes occurring in x-ray spectra arising from the solid-state environment of the emitting atoms. These changes constitute means for elucidation of the band theory of solids.

Chapter 4

Excitation

4.1. PRINCIPLES

The second phase of any x-ray spectrometric analysis (Section 3.2.5) is excitation of the characteristic line spectrum of the analyte. There are many ways to accomplish this (Section 1.6.3), including special secondary-excitation techniques, such as selective, preferential, and differential excitation (Section 8.3). However, this chapter deals only with conventional secondary excitation by primary x-radiation from x-ray tubes. Excitation for energy-dispersive x-ray spectrometry is discussed in Sections 8.2.2.2 and 8.2.2.3.

In quantitative x-ray spectrometric analysis, it is the secondary spectral-line intensity emitted by the analyte that is measured and used to determine analyte concentration. The useful emitted intensity is affected by: (1) the spectral distribution (both continuous and target-line spectra) of the primary x-ray beam; (2) the absorption of the analyte and matrix for the primary x-rays; (3) the excitation probability and fluorescent yield of the analyte line; (4) the absorption of the analyte and matrix for the analyte line; and (5) the geometry of the x-ray spectrometer. In Section 4.1, equations based on these parameters are derived for the relationship between emitted analyte-line intensity and concentration.

4.1.1. General

In Figure 4.1, the number n of analyte-line photons of wavelength λ_L that enter the x-ray optical system (collimator–crystal–collimator–detector)

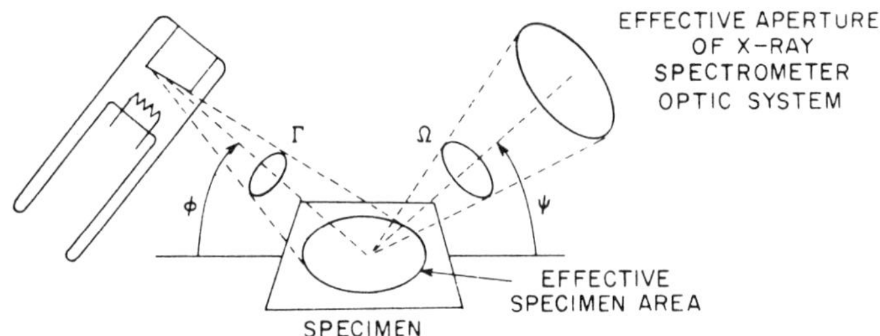

FIG. 4.1. Geometry of secondary excitation.

is (*S46*)

$$n = \frac{\Gamma\Omega}{4\pi} \int_{\lambda_{\min}}^{\lambda_{\mathrm{ab}}} T(\lambda_{\mathrm{pri}}, \lambda_L)\, J(\lambda_{\mathrm{pri}})\, d\lambda_{\mathrm{pri}} \tag{4.1}$$

In the figure and equation:

Γ	is the effective solid angle subtended at the x-ray tube target by the irradiated area of the specimen.
Ω	is the effective solid angle subtended at the specimen by the x-ray optical system.
ϕ	is the angle between the central ray of the primary cone and the specimen plane.
ψ	is the angle between the central ray of the secondary cone and the specimen plane—the takeoff angle.
λ_{pri}	is the wavelength of the primary x-rays.
λ_L	is the wavelength of the measured analyte line.
$\lambda_{\min}$	is the shortest wavelength present in the primary beam.
λ_{ab}	is the absorption edge associated with the analyte line.
$T(\lambda_{\mathrm{pri}}, \lambda_L)$	is the conversion efficiency of primary wavelength λ_{pri} to analyte line λ_L—the number of λ_L photons emitted per incident λ_{pri} photon.
$J(\lambda_{\mathrm{pri}})$	is the spectral distribution of the primary x-rays, that is, the number of photons of each primary wavelength λ_{pri} per unit solid angle Γ.

The number n depends upon: (1) the geometry of the x-ray spectrogoniometer; (2) the primary beam intensity and spectral distribution, which, in turn, depend upon x-ray tube target, potential, and current; and (3) the conversion efficiency $T(\lambda_{\mathrm{pri}}, \lambda_L)$.

Conversion efficiency is given by (*V15*)

$$T(\lambda_{\mathrm{pri}}, \lambda_L) = \tau\omega g G/\mu_{\lambda_{\mathrm{pri}}} \tag{4.2}$$

in which

$$G = \frac{1}{1 + (\mu_{\lambda_L}/\mu_{\lambda_{\text{pri}}})(\sin \phi/\sin \psi)} \tag{4.3}$$

where $\mu_{\lambda_{\text{pri}}}$ and μ_{λ_L} are linear absorption coefficients of the specimen for the primary and analyte-line radiation, respectively; τ is that portion of $\mu_{\lambda_{\text{pri}}}$ due to photoelectric absorption in the atom shell (K, L, M) associated with the analyte line; ω is the fluorescent yield of the analyte for the spectral-line series of the analyte line; and g is the fractional value or relative intensity of the analyte line in its series, or the probability of the orbital electron transition causing the analyte line, or the fraction of, say, the total K-series x-ray photons emitted by the analyte that are $K\alpha$ photons; for example,

$$g_{K\alpha} = \frac{I_{K\alpha_1} + I_{K\alpha_2}}{\sum I_K} \tag{4.4}$$

where $\sum I_K$ is the sum of the intensities of all the analyte K lines, including $K\alpha_1$ and $K\alpha_2$.

The next two sections discuss secondary excitation in somewhat more detail. The treatment follows closely that of Jenkins and de Vries [pp. 19–25 in (*22*)]. Further discussion of excitation equations is given in Sections 14.8.2 and 15.5.2.

4.1.2. Excitation by Monochromatic X-Rays

The process of secondary excitation of an analyte line is most easily understood for the case of a monochromatic primary beam (*B50*). In Figure 4.1, consider an incremental layer of thickness Δt at depth t in a specimen of density ϱ. The incident angle of the primary beam is ϕ, and the takeoff angle of the secondary beam is ψ. Now consider a single wavelength λ_{pri} having intensity $I_{0,\lambda_{\text{pri}}}$ in the incident primary beam. If the specimen has mass-absorption coefficient $(\mu/\varrho)_{\text{M},\lambda_{\text{pri}}}$ for λ_{pri}, the fraction of incident photons penetrating to depth t and arriving at layer Δt is

$$\exp -[(\mu/\varrho)_{\text{M},\lambda_{\text{pri}}}\, \varrho t \csc \phi]$$

Of the photons arriving at layer Δt, the fraction absorbed in the layer is

$$(\mu/\varrho)_{\text{M},\lambda_{\text{pri}}}\, \varrho\, \Delta t \csc \phi$$

Of the photons absorbed in layer Δt, the fraction absorbed by the analyte A having concentration C_{A} (wt fraction) and mass-absorption coefficient

$(\mu/\varrho)_{A,\lambda_{pri}}$ for λ_{pri} is

$$\frac{C_A(\mu/\varrho)_{A,\lambda_{pri}}}{(\mu/\varrho)_{M,\lambda_{pri}}}$$

Of the photons absorbed by the analyte in layer Δt, the fraction that create electron vacancies in the atom shell corresponding to the series of the analyte line is related to the absorption-edge jump ratio of the analyte for that shell r_A; the fraction is

$$1 - (1/r_A) = (r_A - 1)/r_A$$

Of the photons causing such vacancies, the fraction that lead to actual emission of photons of the spectral lines in that series is given by the fluorescent yield of the analyte for that series ω_A.

Of the primary photons causing such secondary photon emission, the fraction that lead to emission of the particular analyte line L to be measured is given by the probability of the orbital electron transition resulting in that line in analyte atoms g_L [Equation (4.4)].

Of the analyte-line photons emitted by layer Δt, only that fraction directed toward the source collimator can contribute to the measured line intensity; this fraction is given by $d\Omega/4\pi$ (Figure 4.1).

Of the analyte-line photons directed toward the collimator, the fraction that emerges from the overlying layer t having mass-absorption coefficient $(\mu/\varrho)_{M,\lambda_L}$ for the analyte line L is

$$\exp -[(\mu/\varrho)_{M,\lambda_L}\varrho t \csc \psi]$$

The foregoing processes may be grouped for convenience into five factors as follows:

1. Incident photon intensity: $I_{0,\lambda_{pri}}$.
2. Attenuation of the primary beam in reaching layer Δt:

$$\exp -[(\mu/\varrho)_{M,\lambda_{pri}}\varrho t \csc \phi]$$

3. Probability of excitation of analyte line L:

$$\frac{C_A(\mu/\varrho)_{A,\lambda_{pri}}}{(\mu/\varrho)_{M,\lambda_{pri}}} \; \frac{r_A - 1}{r_A} \; \omega_A g_L \left[\left(\frac{\mu}{\varrho}\right)_{M,\lambda_{pri}} \varrho \, \Delta t \csc \phi\right]$$

4. Fraction of analyte-line photons emitted toward source collimator: $d\Omega/4\pi$.
5. Attenuation of this analyte-line radiation in emerging from the specimen:

$$\exp -[(\mu/\varrho)_{M,\lambda_L}\varrho t \csc \psi]$$

Combination of all these expressions gives the analyte-line intensity ΔI_L emitted from layer Δt that actually enters the source collimator:

$$\Delta I_L = I_{0,\lambda_{\text{pri}}}\left\{\exp -\left[\left(\frac{\mu}{\varrho}\right)_{M,\lambda_{\text{pri}}} \varrho t \csc \phi\right]\right\}\left(\frac{\mu}{\varrho}\right)_{M,\lambda_{\text{pri}}} \varrho \, \Delta t \csc \phi$$
$$\times \frac{C_A(\mu/\varrho)_{A,\lambda_{\text{pri}}}}{(\mu/\varrho)_{M,\lambda_{\text{pri}}}} \frac{r_A - 1}{r_A} \omega_A g_L \frac{d\Omega}{4\pi} \exp -\left[\left(\frac{\mu}{\varrho}\right)_{M,\lambda_L} \varrho t \csc \psi\right] \quad (4.5)$$

Simplification of Equation (4.5) gives

$$\Delta I_L = I_{0,\lambda_{\text{pri}}}\left\{\exp -\left[\left(\frac{\mu}{\varrho}\right)_{M,\lambda_{\text{pri}}} \varrho t \csc \phi + \left(\frac{\mu}{\varrho}\right)_{M,\lambda_L} \varrho t \csc \psi\right]\right\} \varrho \, \Delta t \csc \phi$$
$$\times C_A\left(\frac{\mu}{\varrho}\right)_{A,\lambda_{\text{pri}}} \frac{r_A - 1}{r_A} \omega_A g_L \frac{d\Omega}{4\pi} \quad (4.6)$$

Integration of Equation (4.6) from $t = 0$ to $t = \infty$ gives

$$I_L = I_{0,\lambda_{\text{pri}}} \omega_A g_L \frac{r_A - 1}{r_A} \frac{d\Omega}{4\pi} \frac{C_A(\mu/\varrho)_{A,\lambda_{\text{pri}}} \varrho \csc \phi}{(\mu/\varrho)_{M,\lambda_{\text{pri}}} \varrho \csc \phi + (\mu/\varrho)_{M,\lambda_L} \varrho \csc \psi} \quad (4.7)$$

If we define

$$P_A = \omega_A g_L \frac{r_A - 1}{r_A} \frac{d\Omega}{4\pi} \quad (4.8)$$

and

$$A = \sin \phi / \sin \psi \quad (4.9)$$

Equation (4.7) becomes

$$I_L = P_A I_{0,\lambda_{\text{pri}}} C_A \frac{(\mu/\varrho)_{A,\lambda_{\text{pri}}}}{(\mu/\varrho)_{M,\lambda_{\text{pri}}} + A(\mu/\varrho)_{M,\lambda_L}} \quad (4.10)$$

Equations (4.10) and (4.7) may be regarded as the fundamental excitation equations (intensity formulas) for secondary x-ray excitation with monochromatic primary x-radiation. These equations are valid only for smooth homogeneous specimens, for collinear primary beams, and in the absence of multiple scatter and enhancement effects. (See the last paragraph of Section 4.1.3.)

In Equation (4.10), P_A may be regarded as an excitation factor, A as a geometric factor, and the fractional term as an efficiency factor. P_A is constant for a given analyte A, matrix, and spectrometer, including a specified set of components. A influences the relative importance of absorption of the primary and secondary (analyte-line) x-rays. At low takeoff angles ψ, A becomes large, and secondary absorption predominates. This relationship can be useful in excitation by continuous spectra (Section 4.1.3) because secondary absorption effects apply to the monochromatic analyte line λ_L, so calculations are much simpler than when absorption of the primary continuum predominates.

Values of ω_K and ω_L are given in Appendix 9. Values of r and $(r-1)/r$ for K and L absorption edges are given in Appendix 8. Values of g may be calculated from Equation (4.4) and the relative intensities of spectral lines given in Reference (*86*).

4.1.3. Excitation by Continuous Spectra

If, now, instead of a monochromatic beam of wavelength λ_{pri}, a continuous spectrum excites the analyte line, it becomes necessary to consider the following (*B30*, *G13*, *S45*):

1. Analyte-line radiation λ_L is excited by all primary wavelengths between the short-wavelength limit $\lambda_{\min}$ [Equation (1.25)] and the absorption edge λ_{ab} associated with λ_L. The total effective primary-beam intensity is then

$$I_\lambda = \int_{\lambda_{\min}}^{\lambda_{\text{ab}}} J(\lambda_{\text{pri}})\, d\lambda \tag{4.11}$$

$J(\lambda_{\text{pri}})$ represents the primary spectral distribution, which consists of the continuous spectrum and characteristic target lines, and depends on the x-ray tube target and potential.

There are four basic ways to deal with primary spectral distribution $J(\lambda_{\text{pri}})$: It can be calculated, replaced with a monochromatic effective wavelength, measured, or corrected for empirically by means of influence coefficients.

In calculation of $J(\lambda_{\text{pri}})$, the primary spectrum between $\lambda_{\min}$ and λ_{ab} may be, in effect, divided into a number of incremental wavelength intervals $\Delta\lambda_i$, and the intensities of the increments calculated and summed, that is,

$$\int_{\lambda_{\min}}^{\lambda_{\text{ab}}} J(\lambda_{\text{pri}})\, d\lambda = \sum (\Delta\lambda_i I_{\Delta\lambda_i} D_{\Delta\lambda_i}) \tag{4.12}$$

where $D_{\Delta\lambda_i}$ has the value 0 or 1 for wavelengths $>\lambda_{\text{ab}}$ and $<\lambda_{\text{ab}}$, respectively; and $I_{\Delta\lambda_i}$ is given by the Kramers equation [Equation (1.27)]:

$$I_{\Delta\lambda_i} \propto iZ\left(\frac{\lambda_i}{\lambda_{\min}} - 1\right)\frac{1}{\lambda_i^2}$$

where λ_i is the average wavelength in interval $\Delta\lambda_i$, and i and Z are x-ray tube current and target atomic number, respectively. The intensities of the principal target lines are given by equations having the form of Equation (1.32):

$$I_K \propto i(V - V_K)^{\sim 1.7}$$

In many applications, $J(\lambda_{\text{pri}})$ can be replaced by a single wavelength having substantially the same primary absorption and analyte excitation and enhancement efficiency as the primary beam. When excitation is by continuum, this *effective wavelength* is given by

$$\lambda_{\text{eff}} \approx \tfrac{2}{3}\lambda_{\text{ab}} \tag{4.13}$$

When excitation is predominantly by target line $\lambda_{L_{\text{tgt}}}$,

$$\lambda_{\text{eff}} \approx \lambda_{L_{\text{tgt}}} \tag{4.14}$$

Fortunately, the value of λ_{eff} is not critical, and differences of 25% or so have relatively little effect on the results.

Effective wavelength can be evaluated by measuring the incident I_0 and transmitted I intensities for a known thickness t of an element for which mass-absorption coefficients are known for a wide range of wavelengths. The effective mass-absorption coefficient is calculated from Equation (2.10), and the effective wavelength is that for which the element has the same mass-absorption coefficient.

The effective-wavelength approach is least applicable to systems where primary absorption predominates over secondary absorption (Section 12.2.1). Such systems include binary systems of elements differing in atomic number by 1–3. The effective wavelength may vary with substantial changes in specimen composition.

Measured spectral distributions and the feasibility of use of published spectral distributions are discussed in Section 15.6. Empirical correction for spectral distribution by means of influence coefficients is discussed in great detail in Section 15.5.

2. Mass-absorption coefficients must be considered for all elements in the specimen matrix for all effective primary wavelengths and for the analyte line. Expressions of the following form are obtained from Equation (2.11):

$$(\mu/\varrho)_{\text{M},\lambda_{\text{pri}}} = \sum C_i(\mu/\varrho)_{i,\lambda_{\text{pri}}} \tag{4.15}$$

$$(\mu/\varrho)_{\text{M},\lambda_L} = \sum C_i(\mu/\varrho)_{i,\lambda_L} \tag{4.16}$$

where $(\mu/\varrho)_{M,\lambda_{pri}}$ and $(\mu/\varrho)_{M,\lambda_L}$ are mass-absorption coefficients of the specimen for a specific primary wavelength λ_{pri} and the analyte line λ_L; $(\mu/\varrho)_{i,\lambda_{pri}}$ and $(\mu/\varrho)_{i,\lambda_L}$ are mass-absorption coefficients of a specific element i for λ_{pri} and λ_L; and C_i is concentration (wt fraction) of element i. In these equations, i includes A, the analyte.

Substitution of Equations (4.11), (4.15), and (4.16) in Equation (4.10) gives a general expression for I_L:

$$I_L = P_A C_A \int_{\lambda_{min}}^{\lambda_{ab}} J(\lambda_{pri}) \left(\frac{\mu}{\varrho}\right)_{A,\lambda_{pri}} \frac{1}{\sum C_i[(\mu/\varrho)_{i,\lambda_{pri}} + A(\mu/\varrho)_{i,\lambda_L}]} \, d\lambda \tag{4.17}$$

P_A and A are defined by Equations (4.8) and (4.9), respectively; their significance is given following Equation (4.10), where sources of values of g, $(r-1)/r$, and ω, contained in P_A, are also given. For a discussion of $J(\lambda_{pri})$, see the beginning of this section and Section 15.6.

Equation (4.17) may be regarded as the fundamental excitation equation (intensity formula) for secondary x-ray excitation with polychromatic primary x-radiation. Like Equations (4.10) and (4.7), this one is valid only for smooth homogeneous specimens, for collinear primary beams, and in the absence of multiple scatter and enhancement effects.

Müller (*29*) gives concise derivations of variations of the excitation equations (intensity formulas) for a pure element and an element in a two-component and a multicomponent system, for monochromatic and polychromatic x-rays, and for parallel and divergent primary beams.

The efficiency of a specified primary wavelength λ_{pri} for exciting analyte line λ_L may be defined as follows:

$$\frac{I_L}{I_{0,\lambda_{pri}}} = P_A \frac{C_A(\mu/\varrho)_{A,\lambda_{pri}}}{\sum C_i[(\mu/\varrho)_{i,\lambda_{pri}} + A(\mu/\varrho)_{i,\lambda_L}]} \tag{4.18}$$

Then, for a given analyte, matrix, and spectrometer arrangement, an efficiency factor $T(\lambda_{pri}, \lambda_L)$ for primary wavelength λ_{pri} in exciting analyte line λ_L may be defined as follows:

$$T(\lambda_{pri}, \lambda_L) = \frac{(\mu/\varrho)_{A,\lambda_{pri}}}{\sum C_i[(\mu/\varrho)_{i,\lambda_{pri}} + A(\mu/\varrho)_{i,\lambda_L}]} \tag{4.19}$$

For a pure element A, Equation (4.19) reduces to

$$T(\lambda_{pri}, \lambda_L) = \frac{(\mu/\varrho)_{A,\lambda_{pri}}}{(\mu/\varrho)_{A,\lambda_{pri}} + A(\mu/\varrho)_{A,\lambda_L}} \tag{4.20}$$

4.2. THE X-RAY TUBE

The conventional x-ray generator, shown schematically in Figure 4.4, consists of the high-power x-ray tube and its power supply. We now consider each of these in turn.

4.2.1. Function and Requirements

The x-ray tube generates the primary x-ray beam that irradiates the specimen and excites the characteristic line spectrum of its elements. The primary beam consists of the continuum and the spectral lines of the target and its contaminants. Formerly, x-ray tubes designed for other applications were used in spectrometers, but today tubes are designed specifically for spectrometers, where the following features are required: (1) high energy output, effected by high (2–5-kW) power loading, small target-to-window distance, and thin beryllium windows; (2) constancy of energy output; short-term constancy is effected by suitable design to minimize variation of internal geometry; long-term constancy is realized by prevention of target pitting and sublimation of metals onto the inside surface of the window; (3) cool operating temperature, especially important with liquid specimens; and (4) long life, with respect to filament burnout and to deterioration of energy output and spectral purity.

4.2.2. Construction

The construction of an x-ray spectrometer tube is shown in Figure 4.4 (*R22*). The external features consist of a heavy rayproof metal head *c* that confines the x-radiation, a thin beryllium window *d* that permits emergence of the useful beam, and a reentrant glass envelope *i* that retains the vacuum and provides a long electrical insulation path. This vacuum enclosure houses the x-ray generating structure.

The tungsten filament *g* is heated to incandescence by an electric current passed through the external filament terminals *h*. The hot filament emits electrons (thermionic emission) that are roughly focused by a concave focusing electrode *f* and attracted and accelerated to the target (anode) *b*, which is operated at a high positive potential. The target consists of a thin disk or plating of the target metal *b* imbedded in or plated on a heavy, hollow copper block *a*, which serves to conduct heat away from the focal spot. The focal spot is the area bombarded by the filament electrons; it is the source of the primary x-rays, and "target-to-window" and "target-to-specimen" distances are measured from the center of the focal spot. The x-rays are emitted in all directions, but emerge from the metal head *c* only through the window *d*. The internal structure is surrounded by a cylindrical metal shield *e* having an opening in register with the window. This shield intercepts improperly focused electrons from the filament and scattered electrons from the target, and minimizes metallization of the window by tungsten sublimed from the filament and by metal sputtered from the target.

The angle between the normal to the window, that is, the axis of the useful emergent x-ray cone, and the target surface is the *target angle* or *anode angle*. This angle is usually $\sim 20°$, so that the solid angle of the useful cone of x-rays is $\sim 40°$.

In practice, the target, which is an integral part of the tube head, is operated at ground potential and the filament at up to 50 (or 75 or 100) kV negative. This mode of operation permits contact with the tube head and specimen compartment, and water-cooling without precaution against grounding through the water line. A rayproof housing must be provided to prevent access to the glass end of the operating tube, both because of the poor x-ray shielding at this end and because of the electric-shock hazard. The glass envelope may be encased in an oil-filled metal cylinder that seals to the adjoining rim of the tube head. This arrangement, known as shockproof mounting, permits operation of the tube in any position and remote from the generator. The opposite arrangement, the *reverse-potential* tube, is discussed briefly in Section 4.2.5.4.

4.2.3. Design Considerations

The focal spot in an x-ray spectrometer tube, unlike that in a diffractometer tube, is not itself a part of the x-ray optical system, and, consequently, may be relatively large—10×5 mm, compared with 10×1 mm in a diffraction tube. The larger size permits greater power dissipation—2.5–3.5 kW (or kVA), compared with ~ 1 kW for a diffraction tube. However, the area and position of the focal spot must be constant throughout the analysis so that primary beam intensity and intensity distribution over the specimen area are constant.

An x-ray tube is extremely inefficient. Only $\sim 1\%$ of the power dissipated by the target—2.5 kW at 50 kV, 50 mA—is converted to x-rays, the other 99% to heat. Consequently, the target is supported on, and in intimate contact with, a block of copper, which is an excellent heat conductor. This block is water-cooled, as shown in Figure 4.4. Unless the target metal is itself a good heat conductor, it must be very thin, either a thin disk or plating. This presents no problem because electrons penetrate only an extremely thin surface layer.

The focal spot is relatively small, so the primary beam is highly divergent. The useful cone subtends at the target an angle of ~20° from the central ray. Thus, the primary beam "obeys" the inverse-square law, and if the specimen is to receive high primary intensity, the target-to-specimen —and therefore the target-to-window—distance must be as small as possible. Different manufacturers accomplish this by reducing the diameter of the tube head, flattening the window side of the head, or placing the focal spot off axis in the direction of the window (asymmetrically positioned target).

The target surface is inclined to the tube axis toward the window, as shown in Figure 4.4. The angle between the target surface and the central ray of the useful primary x-ray cone emerging from the window is the *target angle* or *anode angle* and is typically ~20°.

The window must be as transparent to x-rays as possible, especially at the longer wavelengths, which are generated only inefficiently to begin with, and which are most efficient in exciting long-wavelength spectra. Beryllium, the most transparent practical window material, is none too good: 0.25-mm (0.010-inch) beryllium transmits only 10% at 5 Å, only 0.1% at 8 Å; even 25-μm (0.001-inch) beryllium transmits only 20% at 10 Å. Beryllium thinner than 25 μm and free of pinholes is difficult to make. The window is heated by electrons scattered by the target. As the atomic number of the target increases, not only does the fraction of incident electrons that scatter increase, but the average angle between the scattered electrons and the target decreases so that more of them actually strike the window. The temperature of the window may rise to several hundred degrees centrigrade, and, since beryllium is a relatively poor heat conductor, large temperature gradients may develop across its diameter. If the window is too thin, it may rupture. The higher the atomic number of the target and the smaller the target-to-window distance, the thicker the window must be. In practice, this works out satisfactorily. The high-Z targets—molybdenum, tungsten, platinum, gold—have relatively thick (1 mm, 0.040 inch) windows. But these tubes are used to excite medium- and short-wavelength spectra, and the thick windows transmit these wavelengths efficiently. The chromium (Z 24) target is used to excite long-wavelength spectra, and a 0.010-inch beryllium window is feasible. Window bombardment by scattered electrons is precluded in the reverse-potential tube (Section 4.2.5.4) and also by deflecting the electrons away from the window by means of small permanent magnets mounted outside the tube near the window.

Each x-ray tube design has its advantages and limitations. Thin-window tubes have high output flux, especially in the long-wavelength region; for example, S $K\alpha$ intensity is ~25 times as great when excited by a tube having a 0.25-mm (0.010-inch) beryllium window compared with that when excited by a tube with a 1-mm (0.040-inch) window. However, these tubes are more fragile, shorter lived, and expensive, and have little advantage for the *K* lines of elements heavier than chromium. Tubes having off-axis focal spots also have high output flux. However, they must be operated at lower power to avoid overheating the window from bombardment by electrons scattered from the nearby focal spot. Also, intensity decreases more rapidly with life due to increased deposition of tungsten on the inside of the window. The dual-target tube costs about twice as much as a conventional tube, and a filament burnout or other failure in effect results in loss of two tubes. There is also a high initial installation cost. Other advantages and limitations of the dual-target tube are given in Section 4.2.5.1.

4.2.4. Practical Considerations

4.2.4.1. Excitation Efficiency (*B30*, *S46*)

X-ray spectral lines of a given series are excited most efficiently by primary radiation having wavelength near the short-wavelength side of the absorption edge associated with that series. If the x-ray tube target has intense lines near this wavelength, they make the dominant contribution to the excitation; otherwise, the continuum makes the dominant contribution. Target lines constitute a greater portion of the primary spectrum than was previously believed (*G13*). Recent measurements indicate that W *L* lines constitute ~25% of the spectrum of a tungsten target, Cr *K* lines ~75% of the spectrum of a chromium target. There are only about eight targets available commercially (although others may be ordered specially), so it is unlikely that an optimal target would be available for all analytes. Moreover, it is impractical to change targets too frequently, certainly for different analytes in the same specimen. Consequently, the continuum is frequently relied on for excitation, and, when this is the case, for given excitation conditions (kV, mA), the higher the atomic number, the more intense is the continuum (Figure 1.4).

Incidentally, it has been contended that at a specified x-ray tube current, the continuum emerging from, say, tungsten (*Z* 74) and chromium (*Z* 24) targets should be about the same, rather than proportional to *Z*. The reasoning is as follows. The continuum intensity is indeed proportional to the atomic number of the target for a specified electron current from *filament to target*. However, x-ray tube current is actually the total electron current flowing within the tube from filament to all components connected to ground—target, target block, internal shield, window, and tube head (Figure 4.4). As the atomic number of the target increases, an increasing fraction of the electrons arriving at the target is *scattered*, rather than collected, by the target, so that a decreasing fraction is effective in generating continuum. The scattered electrons are collected by the block, shield, window, and head. The *target current* and *scattered electron current* together constitute the *x-ray tube current* indicated on the panel mA meter. Thus, for a high-Z target, a smaller fraction of the electrons is collected by the target, but these electrons generate continuum efficiently (in proportion to Z); for a low-Z target, a larger fraction is collected, but these electrons generate continuum inefficiently. Moreover, in a tube having a low-Z target, the window can be made more transparent to whatever continuum is generated (Section 4.2.3): Because of the reduced electron scatter, the window, bombarded by fewer scattered electrons, can be made thinner. Also, the target sublimate sputtered on the inner surface of the window, having lower Z, attenuates the emergent x-rays less. Thus one might predict that the continuum intensity should remain about the same regardless of target. However, this is not substantiated by the spectral distribution data of Birks [pp. 121–8 in (*8*)] (*G13*), nor by comparison of scattered continuum intensity from low-Z specimens for tungsten and chromium targets.

All things considered, tungsten is probably the most widely used general-purpose target. Its high atomic number (74) results in intense continuum. It is refractory (melting point 3410°C) and withstands high target loading without pitting. It is available in high purity. Finally, it has intense L lines that contribute to the excitation of K and L lines of, respectively, germanium (Z 32) and iridium (Z 77) and neighboring lighter elements. Platinum (Z 78) and gold (Z 79) are substantially equivalent to tungsten except in minor respects. They give somewhat higher continuum intensity than tungsten. Platinum preferentially excites K and L lines of, respectively, selenium (Z 34) and thallium (Z 81) and neighboring lighter elements. Gold preferentially excites K and L lines of, respectively, selenium and lead (Z 82) and neighboring lighter elements.

Molybdenum (Z 42) gives only about half the continuum intensity of tungsten, platinum, and gold, but, because of preferential excitation by its K lines, at the same excitation conditions molybdenum is advantageous to tungsten for the K lines of niobium (Z 41) to arsenic (Z 33). If molybdenum tubes were fitted with thin windows, the Mo L lines would preferentially excite P $K\alpha$ and adjacent lighter elements. However, rhodium, palladium, and silver targets (Z 45, 46, and 47, respectively) are used for this application. Actually, molybdenum targets are no longer commonly used.

Rhodium, palladium, and silver give substantially lower continuum intensity than tungsten, platinum, and gold, but much higher than chromium. Their K lines preferentially excite K lines of, respectively, ruthenium (Z 44), rhodium (Z 45), and palladium (Z 46) and neighboring lighter elements. Their L lines preferentially excite K lines of chlorine (Z 17) and neighboring lighter elements and L lines of, respectively, technetium (Z 43), ruthenium (Z 44), and rhodium (Z 45) and neighboring lighter elements.

Chromium (Z 25) gives only about one-third the continuum intensity of tungsten, platinum, and gold, and about one-half the continuum intensity of rhodium, palladium, and silver. However, chromium K lines preferentially excite K lines of vanadium (Z 23) and lighter elements.

Aluminum (Z 13) targets have been used for excitation of magnesium, sodium, and lighter elements.

Chromium tubes must have thin 0.25-mm (0.010-inch) beryllium windows, and so must rhodium, palladium, and silver tubes if their L lines are to be used.

Recently, Machlett type SEG-50H x-ray spectrometer tubes have become available having 0.125-mm (0.005-inch) beryllium windows and chromium, copper, molybdenum, rhodium, silver, tungsten, platinum, and gold targets. Other targets are available on special order. These tubes are especially advantageous for excitation in the long-wavelength region.

Excitation by target lines is discussed further in Section 4.3.3.3, including Table 4.1 and Figure 4.10.

With the widespread use of rapid, automatic x-ray spectrometers, there is a great need for a single "universal" x-ray tube suitable for efficient excitation of analytes of both high and low atomic number. The tungsten–chromium dual-target tube is described in Section 4.2.5.1; the tungsten target is used for elements having atomic number greater than 22 (Ti) or 23 (V), the chromium target for the lighter elements. Some workers regard silver as the best single target element in this respect. Silver has a higher atomic number (47) than chromium (24) and therefore gives a higher continuum intensity for excitation of high-Z elements. It has a lower Z than tungsten (74) and therefore does not scatter filament electrons as efficiently, so that a thinner beryllium window can be used, which is beneficial for excitation of low-Z elements. Silver K lines preferentially excite K lines of elements 40–46 (zirconium to palladium) and L lines of thorium, uranium, and many of the lower transuranium elements. Silver *L* lines preferentially excite Cl and S *K* lines and *L* lines of elements 42–46 (molybdenum to palladium). In short, silver is better than chromium for the heavy elements—but not as good as tungsten—and better than tungsten for the light elements—but not as good as chromium. Rhodium (Z 45) is very similar to silver except that its *L* lines are not as favorable for excitation of Cl $K\alpha$.

Figures 4.10 and 9.1 and Table 4.1 show the useful spectral regions of eight commercially available targets.

4.2.4.2. Spectral-Line Interference (*B44*, *L2*)

The primary spectrum contains the following spectral lines: (1) spectra of the target element; (2) spectra of impurities in the target; (3) spectra of contaminants sublimed or sputtered on the target during operation of the tube; these contaminants include tungsten from the filament; chromium, iron, and nickel from the stainless steel internal shield and focusing electrode; copper from plated structures and the target block; and silver from silver solder; the intensities of these lines increase with life of the tube; (4) spectra arising from primary excitation of tube structures other than the target by scattered or improperly focused electrons; and (5) spectra arising from secondary excitation of these structures. Fortunately, modern tubes designed specifically for x-ray spectrometry have high spectral purity when new and maintain it to some extent with life. Any lines present in the primary beam are scattered by the specimen into the spectrometer. The lower the effective atomic number of the specimen, the more intense is the scatter.

In determination of major constituents, especially in high-Z matrixes (which scatter inefficiently), scattered primary lines may be tolerable if a suitable way of measuring background is used. However, for determining minor or trace concentrations of a given element, a target of that element should not be used, and tubes having extraneous lines of that element should be avoided. This is a serious limitation to the determination of traces of those elements whose lines are almost always present in the output of x-ray spectrometer tubes—chromium, iron, nickel, copper, silver, and tungsten (see above). Another condition to be avoided is the possible preferential excitation by target lines of a matrix element having lines that interfere with analyte lines.

Recording of target spectra and use of filters in dealing with spectral-line interference are discussed in Sections 4.2.4.4 and 4.4, respectively.

4.2.4.3. Temperature

While in operation, the x-ray tube must be cooled by a continuous flow of water at a rate of approximately two liters per minute. Usually, tap water is passed through the tube and discharged into the drain. The tap water should be filtered to prevent accumulation of suspended matter in the tube head. Alternatively, the cooling water may be circulated through a closed system cooled by: (1) a water-to-air heat exchanger consisting of a radiator cooled by forced air; (2) a water-to-water heat exchanger cooled by continuously running tap water; or (3) a small refrigerator unit. These three methods have the advantage that distilled or deionized water—rather than tap water—passes through the tube, reducing the possibility of corrosion and obstruction. The radiator and refrigerator arrangements have the advantage of economy of tap water, and the disadvantage that the target heat is discharged into the room, rather than into the drain by the tap water; this disadvantage is avoided by provision of a duct to the outside.

Additional cooling may be provided by fitting the tube head with an external jacket through which the target cooling water is passed. The jacket should be on the inlet side, that is, the water should pass through the jacket before the tube. This additional cooling is particularly desirable when liquid specimens are to be measured.

Operation of the tube at power exceeding its maximum ratings, or within its ratings with inadequate water supply, results in accelerated deterioration of the target. Operation without cooling water for even a fraction of a minute results in complete destruction of the target.

Slight changes occur in the position and dimensions of the filament and in other internal dimensions as the tube warms up to its equilibrium temperature. These changes may affect the intensity and intensity distribution of the primary beam. Consequently, it is necessary to operate the x-ray tube with both filament and target power for at least $\frac{1}{2}$ h before making any quantitative measurements.

4.2.4.4. Evaluation of the Condition of the X-Ray Tube

The x-ray tube focal spot can be photographed at 10–20-fold magnification with a pinhole camera. The pinhole is made by piercing a hole ~0.125 mm (~0.005 inch) in diameter in a piece of lead foil ~0.125 mm thick. A piece of x-ray film is wrapped in black photographic paper or aluminum foil. The pinhole is placed near and parallel to the window and the film some distance away in line with the pinhole and focal spot. The magnification is the ratio of the pinhole-to-film and pinhole-to-target distances. Such photographs taken from time to time show the condition of the focal spot.

The target spectrum can be observed as follows. A specimen containing no elements of atomic number $\gtrsim 9$ (fluorine) and having average atomic number as low as possible is placed in the specimen compartment to scatter the primary beam into the spectrometer. Suitable scatter targets include a piece of paraffin or spectroscopically pure carbon, a briquet of starch, or a cell of distilled water covered with 3- or 6-μm (0.00012- or 0.00025-inch) Mylar. A 2θ scan is made of the scattered x-rays and recorded on the ratemeter–recorder. Such spectra should be recorded whenever a new tube is first used to identify the extraneous lines, and occasionally thereafter to observe increase of spectral contamination with life.

The specimen area irradiated by the target is observed by placing a piece of unexposed x-ray film or a glass microscope slide in the specimen plane without the specimen mask and irradiating it at full power for 3–5 min. The film or glass is colored in the irradiated area. This technique aids in aligning the x-ray tube for optimal irradiation of the specimen. A method of quantitatively mapping isointensity contour lines over the specimen plane is given in Section 19.2.1.

X-ray tube output flux at constant kV and mA decreases ~3% for each 1000 h of life due to target pitting and sublimation of tungsten on the inside of the tube window. This condition is evaluated by periodic measurements on a sturdy, inert test specimen.

4.2.5. Special X-Ray Tubes

The x-ray tube just described is typical of conventional, high-power, side-window tubes. Several other types of x-ray tubes are of interest. These are described in Sections 4.2.5.1–4.2.5.6. Scanning x-ray tubes are described briefly in Section 20.3.2. In these tubes, the electron beam is focused to a very small spot and scans a rectangular raster on the target so that the x-ray source is a "flying spot," rather than a fixed spot.

4.2.5.1. Dual-Target Tube

The construction is essentially the same as that shown in Figure 4.4 except for two features: The target has two separate areas—one tungsten, platinum, or molybdenum, the other chromium—and two filaments, each having its focal spot on one of the target areas. The appropriate filament is energized by an external selector switch. This tube permits high-efficiency excitation of both long- and short-wavelength spectra without the inconvenience of changing tubes.

Dual-target tubes may be subject to certain disadvantages. They are much more expensive than single-target tubes, and the power supply must be modified to provide for switching targets. The targets contaminate each other. Usually, one target is used more than the other, so that it becomes worn sooner and contaminates the other more rapidly. When either target has become worn or contaminated to an unacceptable degree, or when either filament burns out, the tube must be discarded, even though the other target and/or one or both filaments are still good.

A low-power, dual-target tube is described in Section 4.2.5.5.

4.2.5.2. End-Window Tube

X-ray tubes are available in which the window is in the end of the tube head, normal to the axis. These tubes give somewhat more uniform specimen irradiation (if the specimen is normal to the axis) and are especially advantageous for multichannel spectrometers. (See Figure 9.4).

4.2.5.3. Demountable Tubes

Continuously pumped demountable tubes are still used on special laboratory-built spectrometers and have many advantages: burned-out filaments are easily replaced; the target can be removed for cleaning of surface contamination or resurfacing; targets are easily interchanged, and there is no limit on the element that can be used because gas-free operation and long life are not essential—thus, oxides and other reasonably stable compounds can be used when metals are unavailable; a rotatable target wheel can be made having many sectors, each of a different element; provision can be made to position the wheel without dismantling the tube; very thin, or even open, windows can be used because of the continuous pumping and ease of replacement; and high anode power is permissible because the target can be replaced easily.

Three excitation modes are feasible with demountable tubes: (1) primary excitation with the specimen on the target (*C62*, *C63*, *H1*, *M57*), giving high analyte-line intensity, but with high background; (2) secondary excitation with the specimen external to the tube; and (3) secondary excitation with the specimen inside the tube and very near the target (*E7*, *E8*). This last mode is intermediate between the other two: it is not as efficient as primary excitation, but is much more so than external secondary excitation because of the proximity to the primary source, and it still produces no continuum other than that scattered by the specimen.

In another form of demountable tube, a thin aluminum foil serves as both x-ray target and window. A film of an element having high atomic number, such as gold or platinum, can be evaporated on the inside of the foil window to increase the efficiency of x-ray excitation or to provide intense spectral lines of required wavelength. The specimen is placed just outside the window and is subjected to secondary excitation (*H40*).

Some x-ray tubes provide both x-rays and electrons. These tubes are windowless, opening into an evacuated spectrometer chamber (*W22*). An example of such a tube is described in Section 4.2.5.5.

Demountable tubes are especially advantageous for secondary excitation of trace analytes. A target can be used that has a strong line near the short-wavelength side of the analyte absorption edge. High anode power and thin windows can be used, and the sample can be placed inside the tube.

The principal disadvantages are the need for continuous pumping and, if specimens are placed inside the tube, the need for dismantling the tube for each specimen.

4.2.5.4. Tubes for Ultralong Wavelength

Figure 4.2 shows a cross section of the head of the Henke x-ray tube for excitation of ultralong-wavelength spectra (>15 Å) (*D37*, *H33*). Electrons are emitted by the thermionic cathode. An electrostatic field estab-

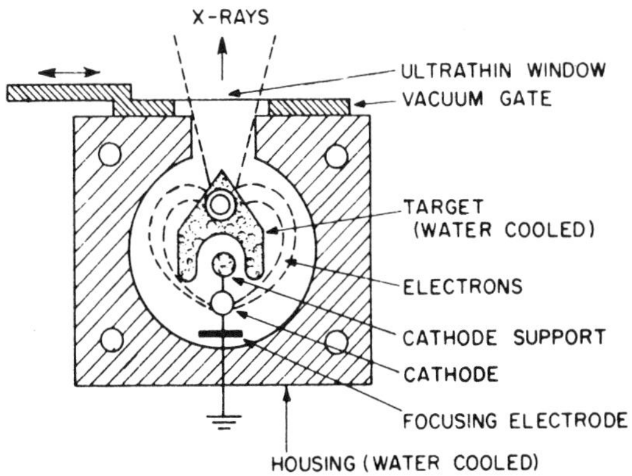

FIG. 4.2. Henke tube for excitation of ultralong-wavelength spectra. [B. L. Henke, *Advances in X-Ray Analysis* **5**, 285 (1962); courtesy of the author, University of Denver, and Plenum Press.]

lished by the focusing electrode and the positive anode constrains these electrons to move in the paths shown by the dashed lines until they strike the interchangeable anode (target) in two large (0.5×2.5 cm) focal spots normal to the page in Figure 4.2. The useful x-rays emerge through an ultrathin window. The position of the cathode behind the anode prevents sublimation of tungsten to the target and window and prevents direct thermal irradiation of the window. The anode has a large emitting area. It is easily replaced and may be selected to have characteristic lines at wavelengths just shorter than that of the analyte absorption edge. The tube is operated at several hundred volts to 15 kV and up to 300 mA, depending on the anode material. The anode is operated at positive potential with respect to the grounded cathode and tube head to prevent electron bombardment of the window. The window isolates the high, clean vacuum (10^{-6} torr) of the x-ray tube from the working vacuum (10^{-3} torr) of the spectrometer chamber, to which the x-ray tube is attached. The window consists of ultrathin Formvar (1000–2000 Å, 30–60 μg/cm^2) supported on a 200-mesh grid having 70% open area. The vacuum gate is advanced to close the exit port until both the x-ray tube and spectrometer chamber are evacuated by their separate pumps. The gate is then placed with the window over the port. This procedure prevents rupture of the window by pressure differential during evacuation. The anode and tube head are water-cooled. An ion pump and liquid-nitrogen cold finger are used to provide clean operating vacuum of 10^{-6} torr or less and to maintain a clean anode surface during many hours of operation under high-current electron bombardment. The Henke tube is available commercially (Section 9.4.3).

This tube is typical of a class known as *reverse-potential tubes*, in which all components are grounded except the anode. This arrangement is the opposite of that in conventional tubes. It has the advantage that electrons scattered from the anode are deflected away from the window, and the disadvantage that electrical leakage can occur through the cooling water, which must be distilled or deionized and, of course, supplied through plastic or rubber tubing.

Another type of x-ray tube for use at long wavelengths—a type of "focusing" x-ray tube—is shown in Figure 8.15 (Section 8.2.2.2).

4.2.5.5. Low-Power Tubes

Low-power ($<$100-W) x-ray tubes are used principally for excitation in energy-dispersive spectrometers (Section 8.2.2.2), although the windowless gas tube described below is used in a crystal spectrometer for excitation of long-wavelength (>8 Å) x-rays. Typical low-power tubes are the conventional-type single-target tube available from Applied Research Laboratories (Sunland, Calif.) and dual-target tube available from Ortec, Inc. (Oak Ridge, Tenn.); the transmission-anode, secondary-emission, and triode x-ray tubes described by Giauque, Goulding, Jaklevic, and Landis; the conic-cavity-target tube described by Herglotz; the windowless gas x-ray tube available from Compagnie Generale de Radiologie (CGR) (Issy-les-Moulineaux, France); and field-emission tubes (Section 4.2.5.6). These tubes are described briefly in the following paragraphs.

The Applied Research Laboratories *hot-filament, single-target x-ray tube* (*G1*) is essentially a small conventional tube. The tube is 12.5 cm (5 inches) long, rated at 30 kV, 1 mA (30 W) maximum, provided with side and end beryllium windows 0.25 mm (0.010 inch) thick, and available with copper, zirconium, and silver targets.

The Ortec *hot-filament, dual-target x-ray tube* (*D42*) has essentially conventional design, except for size and for the dual tungsten–molybdenum target, selected by electrostatic deflection of the filament electrons to one or the other. Other target pairs are available on special order. The tube has a 0.125-mm (0.005-inch) beryllium window. It operates at up to 50 kV, 1–200 μA ($\sim$10 W maximum), and water cooling is not required. At 7.5-cm (3-inch) target-to-specimen distance, the irradiated specimen area is an ellipse having major and minor axes 2.5 and 1.25 cm (1 and $\frac{1}{2}$ inch), respectively. Advantage can be taken of the W *L* and *M* lines and the Mo *K* and *L* lines for enhancement of the analyte line. If the lines of one target interfere with the analyte line, the other is readily selected. Incidentally, dual targets may be subject to certain disadvantages (Section 4.2.5.1), but these are likely to be less severe with low-power tubes than with high-power tubes.

The high continuum intensity produced by conventional-type tubes is advantageous in that it permits relatively efficient simultaneous excitation of substantially all elements in the specimen. However, the continuum may be disadvantageous in that it produces high scattered background intensity and thereby reduces sensitivity. The continuum may be reduced by filtration and by use of secondary targets. Usually the best filter is a thin foil or other layer of the target element itself, which passes only the strongest target lines and a narrow band of continuum having wavelength longer than that of the filter absorption edge, both at reduced intensity. The filter may be placed at the window of a conventional-design x-ray tube. The transmission-anode x-ray tube (see below) is based on the filter principle. Alternatively, the x-ray tube may irradiate a secondary target having its strongest line at wavelength just shorter than that of the analyte absorption edge; this secondary emission then excites the specimen. Only primary continuum doubly scattered from the secondary target and specimen reaches the detector. The secondary target may be placed external to the window of a conventional-design x-ray tube (Section 8.3.1) or inside a specially constructed tube near the target; the secondary-emission x-ray tube (see below) is based on this latter principle.

In the *transmission-anode x-ray tube* (*G26*, *J7*), the target is a thin film or foil, and the useful x-rays are taken from the side opposite that bombarded by the filament electrons. The target filters its own x-radiation, transmitting mostly that having wavelength just longer than its own absorption edge, including its own lines. Such tubes may be sealed off or demountable; in the latter case, the anode can be selected to provide the most efficient excitation for the analysis at hand. In either case, the anode may also serve as the x-ray tube window. A 10-W tube having a 0.1-mm (0.004-inch) molybdenum-foil anode produces ~100 times the x-ray intensity of typical radioisotope sources.

A more effective way to avoid the high continuous background of the conventional x-ray tube is provided by the *secondary-emission x-ray tube* (*J7*). The x-rays generated at the target by the filament electrons are intercepted by a secondary target mounted nearby inside the tube. These secondary x-rays then emerge from the tube to excite the external specimen. The secondary target is chosen to have its strongest spectral line on the short-wavelength side of the analyte absorption edge. For maximum x-ray tube output, the primary—electron-excited—target is chosen to have its strongest spectral line on the short-wavelength side of the secondary target's absorption edge. If this is done, the secondary-emission tube may have as much as 10% the efficiency of the transmission tube, which, however, is itself much less efficient than a conventional tube due to loss of x-rays in passing through the anode foil. However, the emergent x-rays of the secondary-emission tube are substantially free of continuum. If the primary and secondary targets are to be selectable, the tube must be demountable. Another type of secondary-emission tube is shown in Figure 8.15 (Section 8.2.2.2).

In the *triode x-ray tube* (*J6*), between the filament and target there is a third electrode in the form of a grid that can be negatively biased to cut off the electron flow between filament and target. The tube is used in energy-dispersive x-ray spectrometers to deal with dead-time problems in a novel and ingenious way. The first pulse out of the detector, in addition to being shaped, amplified, and counted in the usual way, is also used to trigger the negative grid bias on the x-ray tube. Thus the primary x-rays are immediately turned off and held off during the pulse processing time so no secondary x-rays are emitted from the specimen during this time. As soon as the first pulse has been processed and counted, the grid bias is removed and primary x-rays are generated until another pulse emerges from the detector, whereupon the bias is restored, etc. Consequently, coincidence losses and pulse pileup are completely eliminated. A very interesting condition is made possible by the triode x-ray tube system: If analyte concentration and excitation conditions are both high, it is likely that an analyte-line photon will always be emitted and detected immediately upon removal of the grid bias. The train of pulses leaving the detector is then substantially periodic, rather than random, the pulses occurring at intervals about equal to the resolving time of the detector–readout system. The effectiveness of the triode tube is limited at high analyte concentration and sensitivity when there is high probability that two or more analyte-line photons may be emitted simultaneously (Section 8.1.8.4).

The *Herglotz conic-cavity-target x-ray tube* (*H41*) produces very intense useful x-ray beams at low power by economy of electron excitation and emitted x-rays. At one end of the tube is an electron gun, of the type used in cathode-ray tubes, which produces a thin beam of electrons directed along the tube axis. At the other end is a metal block ~1 cm thick with a conic tunnel, the axis of which coincides with the tube axis, and having diameters ~3 mm and ~0.5 mm on the electron-gun side and far side, respectively. The narrow end of the tunnel is vacuum-sealed with a thin metal foil. The walls of the conic cavity and the foil constitute the target, and the foil also constitutes the x-ray window. For a given input power (kV, mA), the useful output of this conic-cavity target is very high, compared with massive or foil targets, for two reasons: The incident electrons scattered from the cavity wall and window foil are likely to fall back on the cavity wall and produce more x-rays, rather than be lost. Also, the emergent x-rays occupy a solid angle of only ~0.02 sr (steradian), compared with 2π sr for flat targets. Typical operating conditions are 30 kV, 300 μA (~9 W), and water cooling is not required.

Before W. D. Coolidge invented the high-vacuum, filament-type tube (*C59*), *gas-* (or *ion-*) *type x-ray tubes* were used for all x-ray applications. Gas tubes differ from Coolidge tubes in that the cathode is a cold, concave aluminum electrode instead of a hot thermionic tungsten filament, and a low gas pressure (~10^{-2} torr) is required. When a high dc potential is applied between the cathode (negative) and anode (positive), positive gas ions are produced and accelerated to the cathode, bombarding it and releasing electrons, which are accelerated to the target, producing x-rays. These tubes were very "temperamental," and it was difficult to achieve and maintain stable operation. Consequently, gas tubes are seldom used. However, the Compagnie Generale de Radiologie *windowless gas x-ray tube* now available operates at up to 15 kV, 10 mA (~150 W) in the residual pressure ($\sim 5 \times 10^{-2}$ torr) of a vacuum spectrometer. The specimen is excited not only by x-rays, but also by electrons scattered from the target. This combination of unfiltered, low-kV x-ray and electron excitation is very efficient for long-wavelength lines, and the absence of a filament ensures spectral purity and long life.

Sahores *et al.* (*S1*) describe a similar tube in which specimen excitation is effected entirely by electrons emitted by ion bombardment of a cold cathode, accelerated to an anode, and scattered through an open window to the specimen.

Low-power x-ray tubes, compared with high-power tubes, have the advantages that they: (1) are much more compact; (2) are much less expensive; (3) can be energized by very compact, light, inexpensive, highly regulated, solid-state power supplies; (4) require no water cooling; and (5) have much longer life.

4.2.5.6. Field-Emission Tubes

A typical commercial sealed-off miniature field-emission x-ray tube is shown in Figure 4.3 (*M20, M21*). The tube is available from Vacuum Technology Associates, Broomfield, Colo. The principal components are the cold

FIG. 4.3. Miniature field-emission x-ray tube. [Courtesy of Vacuum Technology Associates, Broomfield, Colo.]

field-emissive cathode in the form of a metal point—usually a steel sewing needle, the hemispherical target, the thin [0.025–0.125-mm (0.001–0.005-inch)] beryllium end window, and the cylindrical glass envelope. The tube is evacuated to 10^{-7} torr or less. Tubes 2.5 cm in diameter and 5–6 cm long can operate at potentials up to 30 kV, and somewhat longer tubes up to ~70 kV. Tube current is ~30 μA, giving input power of ~1 W. The applied potential can be ac, pulsating dc, or constant-potential dc. Unlike some high-power field-emission tubes for x-ray radiography, these small tubes operate in continuous, rather than pulsed, mode.

When a high potential is applied between the cathode and target (anode) with the cathode negative, an extremely high potential gradient field appears at the tip sufficient to withdraw electrons from the point, which then becomes a field emitter.

The low input power results in low x-ray intensity compared with conventional x-ray spectrometer tubes. The intensity 30 cm from the window of a miniature field-emission tube is typically 1–100 R/h (roentgens per hour), depending on the target and operating potential. By the inverse-square law, this is of the order 10^3–10^5 R/h at the window, compared with ~10^6 R/min at the window of a conventional high-power tube. Thus, use of field-emission tubes with crystal spectrometers is precluded. However, the tubes are excellent primary-beam sources for energy-dispersive and nondispersive analysis. For these applications, field-emission tubes have an impressive list of advantages over both conventional tubes and radioisotope sources.

Since the input power is only ~1 W, water-cooling is not required, and the high-potential supply can be small—1000 cm^3 in volume, 1 kg in weight. Also, the low intensity reduces the severity of high coincidence losses in the detector in energy-dispersive applications. Since the cathode is cold, no filament power supply is required, and sublimation of cathode material on the target and window does not occur. A very wide range of target elements is feasible—almost any material that can be machined, plated, or vacuum-evaporated. By suitable combination of tube targets and filters, the x-ray output can be made substantially all monochromatic (target line) or all continuum. The tubes are thus useful for providing specific wavelengths for monochromatic and differential excitation (Section 8.3.1), monochromatic absorptiometry (Section 20.1.2), and absorption-edge spectrometry (Section 20.1.4). The anode focal spot is very small [~0.125 mm (~0.005 inch)], providing a point source of x-rays for microradiography (Section 20.1.5). Excellent long-term stability of the x-ray intensity is realized, >500 h. When operated at $\lesssim$25 kV, the tube is equivalent to several curies of comparable radioisotopes. Since the tube can be turned off and constitutes no possibility of environmental contamination or ingestion, it is safer than a comparable radioisotope source. In general, the tubes are less expensive than radioisotope sources.

A serious disadvantage of field-emission tubes is that tube current cannot be varied independently of operating potential.

4.3. X-RAY POWER SUPPLY

4.3.1. Function and Requirements

The x-ray power supply provides filament power for the x-ray tube filament and high-potential power for the target. The filament power must be well regulated; the high-potential power must meet the following requirements: (1) constant-potential dc, although full-wave rectified dc is also satisfactory; (2) maximum potential at least 50 kV (peak or constant potential), although 75 or 100 kV may have some advantage; (3) maximum x-ray tube current at least 50 mA for a 50-kV supply; (4) maximum power output at least 2.5 kW (kVA) for a 50-kV supply; higher powers are preferable for 75- or 100-kV supplies; (5) regulation to ±0.05% or less provided by line, mA, and kV stabilization; (6) it must be possible to select any x-ray tube potential and current from the maximum values down to 10 kV or less and 5 mA or less; this selection should be continuous, rather than stepwise; provision should be made to preset several values of kV and mA and to select them repeatedly with selector switches when more than one analyte is to be measured in each specimen at more than one set of conditions; (7) it should be possible to operate two x-ray tubes, preferably, but not necessarily, simultaneously and at or near full power; this permits operation of two goniometers—both as spectrometers or one as a spectrometer, the other as a diffractometer.

The power supply, shown in Figure 4.4, consists of: (1) the high-potential supply; (2) the x-ray tube filament supply; (3) line, mA, and kV stabilizers; and (4) safety and protective devices. Along the bottom of the figure, the panel controls and indicators are shown. Along the top of the figure, the wave forms of the current at various key components are shown.

4.3.2. Components and Operation

In Figure 4.4, if the main power switch *A* is closed, the 220-V ac line passes through the switch and fuses *B* to the accessory outlets *C* and the line stabilizer. Auxiliary electrical equipment, such as a vacuum pump for

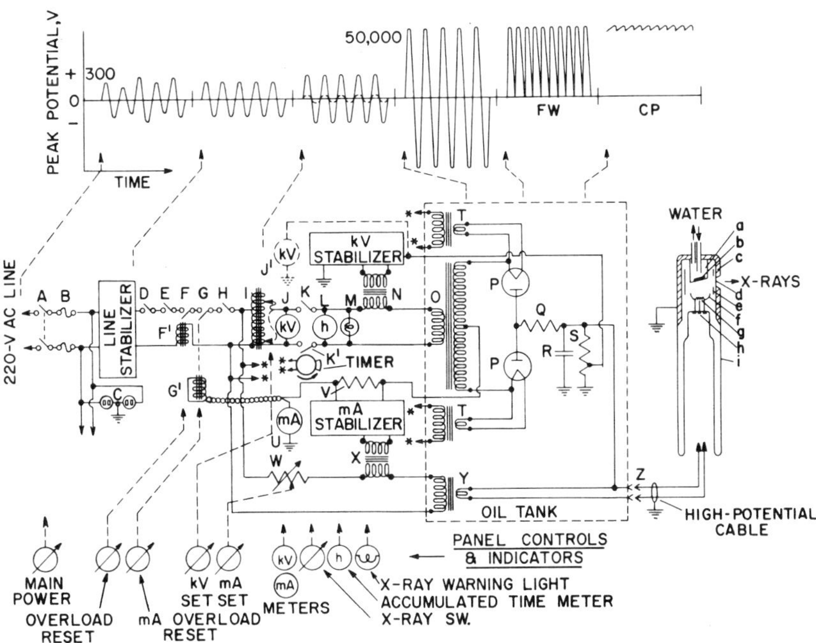

FIG. 4.4. The x-ray generator. Panel controls and indicators are shown along the bottom. Wave forms (potential *vs.* time) at selected points are shown along the top.

X-Ray Power Supply:

- *A.* Main power switch
- *B.* Fuses
- *C.* Accessory outlets
- *D.* Water switch, high-pressure
- *E.* Water switch, low-pressure
- *F.* Line-overload relay
- *F'.* Solenoid coil for *F*
- *G.* mA-overload relay
- *G'.* Solenoid coil for *G*
- *H.* Door interlocks
- *I.* Autotransformer, kV selector
- *J.* kV meter
- *J'.* kV meter, alternate position
- *K.* X-ray switch
- *K'.* Time-delay switch
- *L.* Accumulated-time meter
- *M.* X-ray warning light
- *N.* kV stabilizer reactor
- *O.* High-potential transformer
- *PP.* High-potential rectifiers
- *Q.* Resistor } Constant-potential filter
- *R.* Capacitor } Constant-potential filter
- *S.* Potential-divider resistor
- *TT.* Rectifier-filament transformers
- *U* mA meter
- *V.* Resistor for mA stabilizer
- *W.* Filament-control rheostat
- *X.* mA stabilizer reactor
- *Y.* X-ray tube filament transformer
- *Z.* High-potential socket

X-Ray Tube:

- *a.* Water-cooled target block
- *b.* Target
- *c.* Rayproof head
- *d.* Beryllium window
- *e.* Internal shield
- *f.* Focusing electrode
- *g.* Filament
- *h.* Filament terminals
- *i.* Glass envelope

a vacuum tunnel or a water pump and cooler, are also connected to the unstabilized ac line. The line stabilizer maintains a constant 220-V output for input fluctuations of 190–250 V.

If all the safety and protective devices *D–H* are closed, which they normally are, the stabilized 220-V line goes to: (1) the autotransformer *I* in the high-potential supply; (2) the x-ray tube filament transformer *Y*; (3) the rectifier filament transformers *TT*; (4) the mA and kV stabilizers; and (5) a 30–60-s time-delay switch *K′*.

4.3.2.1. High-Potential Supply

The continuously variable autotransformer *I* permits the line potential applied to the high-potential transformer to be varied from 0 to perhaps 250 V. The stabilized 220-V line is connected to the input of the autotransformer winding at the bottom and at a point a certain distance down from the top. The output is taken from the bottom and the movable contact. When the movable contact is set at the bottom of the winding (dashed arrow), the output is 0 V. As the contact is moved upward, the output potential increases until, at the point where the input is connected, output and input are equal at 220 V. If the contact is moved still higher, the output exceeds the input up to ~250 V; this is possible because of induction in the upper part of the winding.

The movable contact is operated by the kV-set knob, and the potential across the output of the autotransformer is indicated on the kV meter *J*, as shown. However, the scale of the meter is graduated 0–50 (or 75 or 100) kV, rather than 0–250 V as one would expect; how this is feasible is explained below. The output of the autotransformer is not applied to the high-potential transformer until the x-ray switch *K* and time-delay switch *K′* are closed, whereupon the accumulated-time meter *L*, x-ray warning light *M*, and primary winding of the high-potential transformer *O* are energized.

Transformers are inductive devices for increasing or decreasing an ac potential without substantial loss of power. A transformer consists of two separate windings on a laminated iron core. If an ac potential V_{pri} is applied to the primary (input) winding, an induced ac potential V_{sec} of the same frequency appears across the secondary (output) winding:

$$V_{sec} = V_{pri}(n_{sec}/n_{pri}) \tag{4.21}$$

where n_{pri} and n_{sec} are the numbers of turns of wire in the primary and secondary windings, respectively. If $n_{sec} > n_{pri}$, the device is a *step-up* transformer, and $V_{sec} > V_{pri}$; if $n_{sec} < n_{pri}$, the device is a *step-down* transformer. In the present case, if the output of the autotransformer *I*—that is, the input of transformer *O*—is 0–250 V, and the maximum output of the x-ray supply is to be 50 kV, the turns ratio must be

$$\frac{n_{sec}}{n_{pri}} = \frac{V_{sec}}{V_{pri}} \tag{4.22}$$

Then, 50,000/250 = 200/1. Thus, the secondary potential of transformer *O* is always 200 times the primary potential, that is, 200 times the output potential of the autotransformer *I*. Thus, for an autotransformer output of 50 V, the high-potential transformer output is $50 \times 200 = 10{,}000$ V (10 kV); for 125 V, it is 25 kV, etc., until, for 250 V, it is 50 kV.

It is now evident why the knob that operates the movable contact of the autotransformer is the kV-set knob. It is also evident why the panel meter *J* can be graduated in kV, since the high-potential transformer output is always 200 times the autotransformer output. Thus a small, rugged, relatively inexpensive, and easily insulated 0–250 ac voltmeter at the autotransformer output serves the same function as a bulky, expensive 0–50 ac kilovoltmeter at the output of transformer *O*. Another practical way of indicating x-ray tube potential *J′* is given below.

In Figure 4.4, transformer *O* actually has a turns ratio of 400/1 and delivers a total of 100 kV maximum. The winding is center-tapped and actually consists of two 50-kV windings. This arrangement is required for full-wave rectification, as explained below.

The transformer violates no conservation laws because, neglecting losses, the primary and secondary *powers* are the same, that is,

$$(Vi)_{sec} = (Vi)_{pri} \tag{4.23}$$

where V and i are potential and current, respectively. In the present case, if the x-ray supply is to provide 50 kV at 50 mA, the primary current, again neglecting power losses, is

$$i_{pri} = (Vi)_{sec}/V_{pri} \tag{4.24}$$

Then, $(50{,}000 \times 0.050)/250 = 10$ A.

The autotransformer *I* may be regarded as a transformer in which a single winding serves as both primary and secondary. The primary is the portion of the winding across which the 220-V line is connected. The secondary is the portion of the winding between the bottom and the movable contact.

The ac output of the high-potential transformer is rectified—that is, converted to dc—by the two rectifier tubes *PP*, each having its own filament transformer *TT*. Rectifiers are devices for converting alternating current to direct current and are available in many forms. The rectifiers in the figure are of the electron-tube type and consist of an evacuated glass tube having an electron-emissive tungsten filament surrounded by a cylindrical metal anode. Electrons can flow from the filament to the anode only when the anode is positive with respect to the filament, so the rectifier acts as a one-way electronic valve. Solid-state high-potential rectifiers are now also being used in x-ray power supplies. An incidental advantage of solid-state rectifiers is that no x-rays are generated by the rectifiers themselves. The rectifier filament transformers *TT* are step-down transformers and reduce the stabilized 220-V ac line to 5–10 V to heat the rectifier filaments. The ac input to the primary windings of these transformers is constant and in no way affected by the autotransformer *I* or the rheostat *W* for the x-ray tube filament transformer.

In full-wave supplies, the pulsating dc output of the rectifiers is connected directly to the high-potential socket *Z*. In constant-potential supplies, the full-wave rectified current is smoothed by an electric filter consisting of a resistor *Q* (or an inductor) and capacitor *R*. The potential divider *S* is a high resistance—10 MΩ or more—and allows the charge on the capacitor to leak off when the supply is turned off. Otherwise, the capacitor would retain its charge and present an electric-shock hazard, even when the supply is turned off. Other uses of the potential divider are given below.

The high-potential transformer *O*, rectifiers *PP*, constant-potential filter *QR*, potential divider *S*, rectifier filament transformers *TT*, and x-ray tube filament transformer *Y* (see below) all require high electrical insulation—up to 100 kV for a 50-kV supply. Consequently, all these components are mounted in an oil-filled tank, as shown in Figure 4.4. Incidentally, even though the three filament transformers *TTY* have primary and secondary potentials of only 220 and ~10 V, respectively, the insulation between their windings must withstand the full 100-kV potential of transformer *O*. Otherwise, the high-potential could arc through the filament transformers to the ac line and ground.

4.3.2.2. X-Ray Tube Filament Supply

The x-ray tube filament transformer *Y* is a step-down transformer similar to the rectifier filament transformers *TT* and reduces the stabilized 220-V ac line to 5–10 V to heat the x-ray tube filament. The rheostat *W* varies the primary potential in much the same way as the autotransformer *I* varies the primary potential to transformer *O*. In fact, a second autotransformer can be used in place of the rheostat. The secondary of the filament transformer is connected to the high-potential socket *Z*, thence through the high-potential cable to the x-ray tube filament terminals *h*. If the resistance of the rheostat *W* is decreased, the potential applied to the primary winding, and therefore the output of the secondary winding, increases. As a result, more current flows through the x-ray tube filament, increasing its temperature and electron emission. Thus, the x-ray tube current and x-ray output increase. It is now evident why the knob that operates the rheostat is the mA-set knob, as shown.

4.3.2.3. Operation

Closing the main power switch *A* immediately energizes the following components: (1) any equipment connected directly to the ac line (such as a water-circulating system or vacuum pump) or plugged into the outlets *C* (such as a specimen spinner); of course, any of these devices may have their own switches; (2) the line stabilizer; (3) the mA and kV stabilizers; (4) the autotransformer *I*; (5) the x-ray tube filament transformer *Y*; (6) the rectifier filament transformers *TT*; and (7) the time-delay switch *K*′. However, in order to energize the high-potential transformer *O* and thereby generate x-rays, the x-ray switch *K* must be closed. Even then, the x-ray tube is not energized unless all the safety and protective devices *D*–*H*, *K*′ are closed; these devices are discussed in Section 4.3.2.5.

We may now consider the operation of the high-potential system. Suppose that, at a particular instant, the phase of the ac at the output of the high-potential transformer *O* is such that the top of the winding is positive and the bottom is negative, that is, that the top and bottom have a deficiency and excess of electrons, respectively. Electron current always flows from the negative terminal of a source, so current flows to the filament of the lower rectifier *P*, across the vacuum to the anode, through the resistor *Q* (if present) to the socket *Z*, thence through the cable to the x-ray tube filament *g*. The x-ray tube filament electrons flow across the vacuum to the target *b*, where some generate x-rays, others flow through the copper block *a* and tube head *c* to ground. The return current flows through the ground to the milliammeter *U*, through winding *G*′ and resistance *V* (both considered later), to the center tap of the transformer *O*.

On the next half-cycle, the polarity of the secondary winding of *O* is reversed, with the top of the winding negative. Now, the same thing happens, except that the electron current flows from the top of the winding through the upper rectifier. Thus, successive half-cycles flow from alternate halves of the transformer and through alternate rectifiers, but the current applied to the x-ray tube is always in the same direction.

The electron current just described is the x-ray *tube current* and is responsible for the generation of the x-rays; it is indicated on the milli-ammeter *U*, and is usually of the order 50 mA or less. This current must be distinguished from the x-ray tube *filament current*, which flows in the circuit consisting of the x-ray tube filament and the secondary of the filament transformer *Y*, is responsible only for heating the x-ray tube filament, and is usually of the order 5 A or less.

The wave forms of the potential at key components in the system are shown at the top of Figure 4.4 on a scale of potential *versus* time. The 220-V ac line is represented by the first wave form, where line fluctuations are indicated—rather naively—by the unequal amplitudes of the alternations. The stabilized 220-V ac is represented by the second wave form, where all the alternations are of equal amplitude. The 0–250-V ac output of the auto-transformer *I* is represented by the third drawing. When the kV knob is set with the contact low on the autotransformer (dashed arrow), the output potential is low (dashed wave); when the knob is near the top of the winding (solid arrow), the output potential is high (solid wave) and may be made even higher than the 220-V input. The 50-kV ac output of the high-potential transformer *O* is shown next, followed by the 50-kV full-wave rectified pulsating output of the rectifiers. For 60-Hz ac, the full-wave pulse rate is 120 per second. In a full-wave generator, this is substantially the wave form applied to the x-ray tube. In a constant-potential generator, the full-wave pulsating current is smoothed by an electric filter *QR*, providing nearly pure direct current, as shown by the last wave form.

It is customary to state ac and pulsating dc potentials in volts *rms* (root mean square), also known as volts *effective*. This convention is followed in the x-ray generator except for the ac output of the high-potential transformer and the pulsating dc output of the rectifiers. These potentials are stated in kilovolts *peak*—kV(P)—for the following reasons. It is the peak x-ray tube potential that determines the wavelength profile of the continuous x-ray spectrum, and it is the peak potential that must exceed the excitation potential of an element if its characteristic spectrum is to be excited. For sine-wave form,

$$V_{\mathrm{peak}} = 1.414 V_{\mathrm{RMS}} \tag{4.25}$$

$$V_{\mathrm{RMS}} = 0.707 V_{\mathrm{peak}} \tag{4.26}$$

where 1.414 and 0.707 are $\sqrt{2}$ and $1/\sqrt{2}$, respectively. Constant-potential dc is stated in kilovolts constant potential—kV(CP).

In constant-potential generators, both the kilovoltmeter and milliammeter indicate substantially continuous dc, and x-ray tube power is given by

$$P_{\text{CP}} = Vi \tag{4.27}$$

where V is x-ray tube potential (V), i is x-ray tube current (A), and P_{CP} is power (W). In full-wave generators, the kilovoltmeter indicates peak kilovolts and the milliammeter rms current; x-ray tube power is given by

$$P_{\text{FW}} = 0.707 V_{\text{peak}} i_{\text{RMS}} \tag{4.28}$$

For example, an x-ray tube operating at 50 kV, 50 mA from a constant-potential generator dissipates $50{,}000 \times 0.050 = 2500$ W; the same tube operating at the same conditions from a full-wave generator dissipates $(0.707 \times 50{,}000)(0.050) = 1768$ W. About 1% of the power dissipated by the x-ray tube is converted to x-rays. It follows that, at the same nominal operating conditions (kV, mA), the x-ray output is substantially lower with full-wave than with constant-potential operation.

The constant-potential filter capacitor (R in Figure 4.4) discharges slightly between pulses as it tries to maintain the potential. Therefore, constant potential is not really constant, but has a slight ac ripple given by

$$V_{\text{ripple}} = (i/f)/C \tag{4.29}$$

where V_{ripple} is "peak-to-trough" ripple potential (V), i is the current (A) drawn from the power supply, f is the ripple frequency (Hz) (120 Hz for rectified 60-Hz ac), and C is the capacitance (F). For example, for a rectified 60-Hz constant-potential supply having a 0.2-μF capacitor and delivering 50 mA, the ripple potential is $(0.050/120)/(2 \times 10^{-7}) = 2100$ V, or $(2100/50{,}000) \times 100 = 4.2\%$.

4.3.2.4. Stabilization

Because primary x-ray intensity is proportional to iV^2 [Equation (1.28)], it is essential that the x-ray tube current and, especially, potential be highly regulated, $\pm 0.05\%$ or less. All commercial x-ray spectrometers have an ac line stabilizer and an "mA stabilizer," and some also have a "kV stabilizer." These three types of stabilizer are shown schematically in Figure 4.4, and are considered below in order.

A simple "constant-potential" or "isolation" transformer may be used between the ac line and the input terminals of the x-ray generator in laboratories where extremely wide line fluctuations are encountered. However, these transformers used alone provide only about $\pm 1\%$ output regulation for $\pm 15\%$ input fluctuation. A good line stabilizer holds the output constant at 220 V $\pm 0.1\%$ or less for an input fluctuation of 190–250 V or more, and may be electronic, electromagnetic, or a combination of both. The electronic type usually has a temperature-controlled diode. The line fluctuations cause corresponding fluctuations in the electron emission in this tube. This varying emission current generates error potentials, which are amplified and used to correct the input fluctuations.

The "mA stabilizer" functions substantially as follows. The x-ray tube current returning to the center tap of the high-potential transformer O through the milliammeter U passes through the resistor V and develops a potential across it. Fluctuations in the tube current cause corresponding fluctuations in this potential, which are amplified and, through transformer X (or a reactor), induce potentials in the primary circuit of the x-ray tube filament transformer Y in such phase as to counteract the fluctuations.

The "kV stabilizer" functions similarly. The high potential output of the power supply appears across the resistive potential divider S. A certain small fraction of this total potential is applied to the kV stabilizer, where it is compared with a fixed reference potential, thereby generating a difference potential. Fluctuations in the high potential across S cause corresponding fluctuations in the difference potential, which are amplified and, through transformer N (or a reactor), induce potentials in the primary circuit of the high-potential transformer in such phase as to counteract the fluctuations.

Incidentally, the potential taken from the divider S may also be indicated on a dc panel voltmeter J' graduated in kV, providing an alternative means of indicating x-ray tube potential. This alternative is advantageous in that the potential indicated on J' is derived directly from the potential actually applied to the x-ray tube.

4.3.2.5. Safety and Protective Devices

In Figure 4.4, K' is a 30–60-s time-delay switch that opens when the power supply is turned off and is closed by a bimetallic strip heated by a resistor or, preferably, by a small timer motor. When the main power switch A is turned on, the bimetallic strip heater or timer motor is energized and, after 30–60 s, the time-delay switch is closed. The time-delay switch prevents the application of power to the primary of the high-potential transformer O until the x-ray tube and rectifier filaments and the mA and kV stabilizers have had time to warm up. Closing switch K cannot energize transformer O unless switch K' has closed, and not even then unless all the other safety and protective devices are closed, which they normally are unless a hazardous condition exists.

Provision must be made to shut off the instrument if any difficulty develops that might present a hazard to equipment or personnel. Safety and protective devices are indicated schematically by switches D–H, shown open in Figure 4.4, but normally closed. Here, D and E represent high- and low-pressure switches in the cooling water line of the x-ray tube. F represents the line overload relay; if any of the circuits connected to the stabilized 220-V line draw excessive current, the solenoid F' is magnetized sufficiently to open F. G represents the x-ray tube current overload relay; if the x-ray tube current becomes excessive, the solenoid G' opens G. The overload relays F and G have reset buttons on the panel, as shown. H represents the several interlocks on cabinet doors and the x-ray tube housing; if any of these are disturbed while the equipment is in operation, H opens and turns off the equipment.

The fuses B will "blow" if excessive current is drawn by the line stabilizer or by any equipment connected directly to the 220-V ac line or to the outlets C, or if the protective devices fail.

4.3.3. Practical Considerations

4.3.3.1. Constant Potential

Figure 4.5 shows the wave forms of 50-kV ac, half-wave and full-wave rectified pulsating dc, and constant-potential dc. The K-excitation potentials of barium (37.4 kV), molybdenum (20.0 kV), and iron (7.1 kV) are indicated. The K lines of these elements are excited only when the x-ray tube potential exceeds their respective excitation potentials, and they are excited efficiently only when the excitation potentials are exceeded substantially. It is evident that the smaller the difference between operating and excitation potentials, the greater is the advantage of constant-potential (CP) compared with full-wave (FW) power. Adler [p. 38 in (*1*)] calculates that at 50 kV, the relative intensities for half-wave, full-wave, and constant-potential excitation are for Fe $K\alpha$ 43, 86, and 100, respectively, and for Ba $K\alpha$ 13, 26, and 100. It is also evident that constant potential is particularly advantageous for determination of minor and trace constituents having high excitation potentials. Figure 4.6 shows the intensity ratio for CP and FW supplies for excitation potentials up to 30 kV. The $\sim$30% advantage of CP

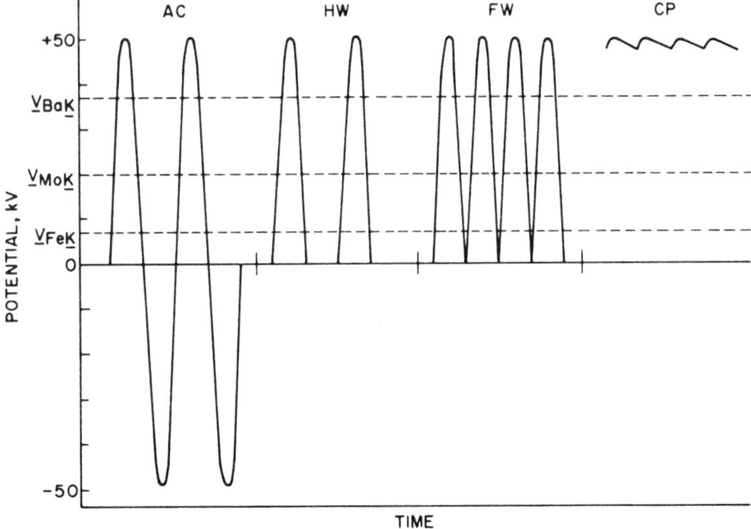

FIG. 4.5. Secondary excitation by half-wave, full-wave, and constant-potential supplies.

over FW at low excitation potentials (<10 kV) is due to the improved power factor (lead–lag phase relationship of potential and current) with CP. However, for these light elements, line and background increase at the same rate, so that line-to-background ratios with CP are no better than with FW.

If ac were applied to an x-ray tube, electrons could flow from filament to target only on the positive half-cycles when the target is positive with

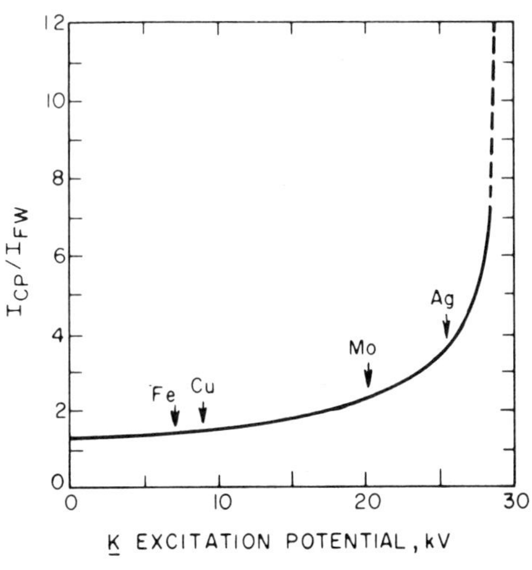

FIG. 4.6. Intensity ratio for constant-potential and full-wave excitation.

respect to the filament. Thus, in effect, the x-ray tube acts as its own rectifier, and its performance is substantially the same with ac and half-wave-rectified dc. In both cases, the tube is energized only half the time, and, for a given analyte, the secondary-emission intensity is about half that with FW. No commercial x-ray spectrometer operates on ac or half-wave-rectified dc, and they are of only academic interest.

Constant potential has another advantage when a Geiger counter is used. Suppose a Geiger counter, used on a spectrometer having a full-wave generator, gives a measured intensity at or near the upper limit of its linear region. Since the analyte is emitting photons only part of the time (Figure 4.5), it is likely that the linear limit is being exceeded during the effective part of the pulse. The error is more serious the greater the excitation potential for the measured line. No such problem arises with a CP supply. The problem is not particularly serious anyway because Geiger tubes are used relatively infrequently in x-ray spectrometers.

4.3.3.2. Maximum Target Potential

Commercial x-ray spectrometers are available with maximum high potentials of 50, 75, and 100 kV peak or CP. A potential of 50 kV excites *K* lines up to europium (*Z* 63), 75 kV up to osmium (*Z* 76), and 100 kV up to radon (*Z* 86). Of course, each of these potentials excites *K* lines efficiently only up to about five atomic numbers below those cited above. The advantage of the *K* lines of the heaviest elements is questionable. However, *K* lines are fewer in number and more intense than *L* lines, and all *K* and *L* lines excited at 50 kV are excited more efficiently at 75 or 100 kV [Equation (1.32)]. Background is also higher, but peak-to-background ratio stays about the same, so there is some net advantage. No really good crystal having both high dispersion and high "reflectivity" is available for these lines; LiF(422), LiF(420), and quartz(502) (Appendix 10) are probably the most favorable for this region. However, energy-dispersive spectrometers using lithium-drifted silicon or germanium detectors (Section 6.4) have higher resolution at short wavelengths than crystal spectrometers. Electrical insulation and radiation protection problems are more severe at 75 and 100 kV.

4.3.3.3. Operating Conditions

Choice of x-ray tube current (mA) is less a problem than choice of potential (kV) because current affects both line and background intensity linearly and therefore does not affect the line-to-background ratio. However, potential affects the two intensities very differently.

Generally, it is preferable to select the optimum potential, then set the current to the maximum permitted by the target power rating (kW or kVA). Actually, it is prudent to set the current at ~5 mA less than this value: a few milliamperes costs very little intensity, but may substantially increase the tube life.

For qualitative analysis, it is prudent to operate at maximum potential so that lines having high excitation potentials are excited. This is particularly important when the spectrum is recorded in air path, and K lines of the heavy elements must be excited because the L lines may lie in the helium or vacuum region.

For quantitative analysis, maximum potential always gives maximum intensity for all lines, regardless of whether the line is excited primarily by continuum or by target lines. The intensities of all primary wavelengths increase with x-ray tube potential. Continuum intensity increases with the square of the x-ray tube potential [Equation (1.28)]. Target-line intensities increase with approximately the 1.7 power of the difference between the x-ray tube and excitation potentials [Equation (1.32)], and for efficient excitation, the tube should be operated at or above three times the excitation potential. Maximum potential is not used when excitation of a certain element is to be avoided (selective excitation), possibly when long-wavelength spectra are to be excited, and when maximum peak-to-background ratio is to be realized. Also, it is inefficient and uneconomical to operate the tube at a potential so high that the primary x-rays are so short that they penetrate the specimen to a depth from which analyte-line radiation cannot emerge.

For long-wavelength lines, the peak-to-background ratio is smaller at higher x-ray tube potentials. As potential is increased, intensity increases in the continuum hump more rapidly than at the longer wavelengths most efficient in exciting the long-wavelength spectra. The continuum radiation scatters from the specimen and increases the background. However, pulse-height selection can eliminate this background in the long-wavelength region, so if line intensity cannot be spared, maximum x-ray tube potential can be used.

Optimal x-ray tube potential may be defined as that which gives the highest peak-to-background ratio. For excitation of short-wavelength spectra by continuum, optimal potential is usually that which places $\lambda_{I_{max}}$ (Section 1.5.1)—the continuum hump—on the short-wavelength side of the analyte absorption edge.

The considerations involved in the choice of x-ray tube target and operating potential are summarized in Figures 4.7 and 4.9.

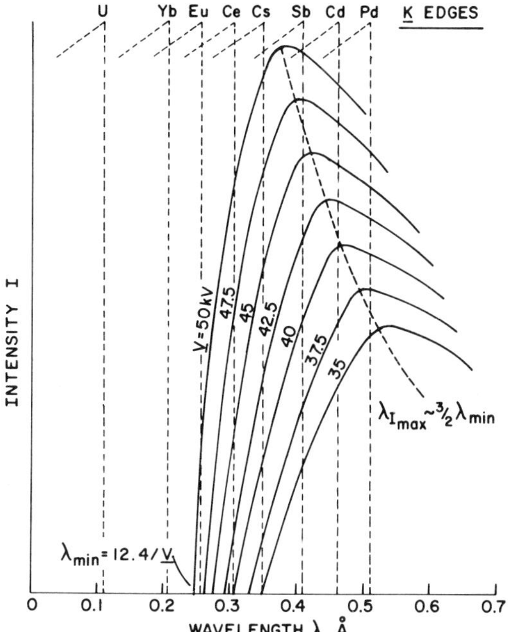

FIG. 4.7. Excitation by the primary x-ray continuum, showing the effect of x-ray tube operating potential.

Figure 4.7 illustrates excitation by the continuum. The figure shows the continuum "hump" for an x-ray tube operating at each of several potentials in the range 35–50 kV, and the *K*-absorption edges of a series of elements having atomic number 46–92 (palladium to uranium). At any specified potential: (1) only those elements having *K* edges at wavelength longer than λ_{min} are excited at all; (2) only those elements having *K* edges at wavelength substantially longer than λ_{min} are excited efficiently; and (3) those elements having *K* edges at wavelength a little longer than $\lambda_{I_{max}}$ are excited most efficiently. Of course, similar considerations apply to excitation of the *L*III edges of the elements. As x-ray tube potential is increased: (1) λ_{min} and $\lambda_{I_{max}}$ move to shorter wavelengths; (2) elements of successively higher atomic number are excited; and (3) all elements excited at lower potential are excited more efficiently.

Jenkins has suggested a convenient way to assess the efficiency of the primary continuum in exciting a specific analyte line: by plotting the primary continuum spectral distribution and analyte absorption curves superimposed on the same wavelength scale. Such plots are shown in Figure 4.8A–C for analytes having absorption edges at short, intermediate, and

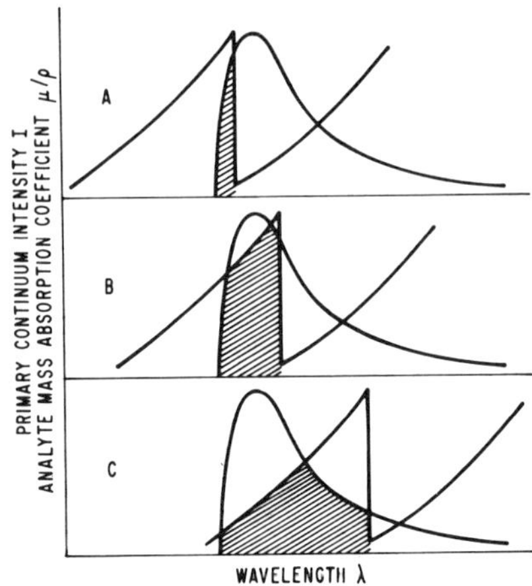

FIG. 4.8. Efficiency of excitation by the primary continuum.

long wavelengths, respectively. To a first approximation, the excitation efficiency is given by the common area under the continuum and absorption curves on the short-wavelength side of the edge—the shaded areas in the figure. Moreover, the effectiveness of an incremental band of primary wavelengths $\Delta\lambda$ is seen to depend on both the absorption coefficient and the primary intensity at that wavelength. Figure 4.8B shows that for excitation by continuum, efficiency is greatest when the "hump" lies on the short-wavelength side of the analyte absorption edge.

The primary beam invariably penetrates deeper into the specimen than the depth from which secondary analyte-line radiation can emerge. There are two principal reasons for this. The primary beam contains short-wavelength components ranging down to λ_{min}. Also, in most x-ray spectrometers, the average angle between the primary beam and specimen surface (ϕ in Figure 4.1) is usually about twice the angle between the secondary beam and specimen (ψ). Thus the path length for, say, 99.9% absorption is greater for the primary beam. It follows that although the shortest wavelength components of the primary beam excite analyte-line radiation, they may do so at a depth from which the analyte radiation cannot emerge from the specimen. Thus, the most effective primary radiation is that lying at wavelength just shorter than the absorption edge associated with the analyte line.

Figure 4.9 illustrates excitation by both continuum and target-line radiation. The figure shows the continuum and line spectrum from tungsten and chromium targets operating at 50 kV, and the K-absorption edges of a series of elements having atomic number 17–92 (chlorine to uranium). For excitation by the continuum, the tungsten target is preferable because intensity is proportional to atomic number [Equation (1.28)]. However, when a strong target line lies near the short-wavelength side of an analyte absorption edge, the target line is likely to make the predominant contribution to the excitation of that element, and the nearer the line lies to the edge, the more efficient is the excitation. Thus, Figure 4.9 shows that W $L\gamma_1$ excites the K lines of germanium and lighter elements, W $L\beta_1$ excites those of zinc and lighter elements, and W $L\alpha_1$ excites those of nickel and lighter elements. It is evident that tungsten is the optimal target for nickel because W $L\alpha_1$, $L\beta_1$, and $L\gamma_1$ all lie close to the short-wavelength side of the Ni K edge, and W $L\alpha_1$, the strongest line, lies extremely close to the edge. It is also evident why chromium is preferable to tungsten for excitation of vanadium and lighter elements: The continuum for both targets is very weak in this region, so the relative proximity of the Cr K lines to the low-Z absorption edges assumes the predominant role.

Table 4.1 and Figure 4.10 give the principal spectral lines of eight commonly used, commercially available targets and permit prediction of which analyte spectra each target will enhance. Figure 9.1 shows the useful region of each target—as well as of other spectrometer components. X-ray spectrometry slide rules that aid in the selection of x-ray tube targets, kV, and other instrument components and conditions are described in Section 9.1.

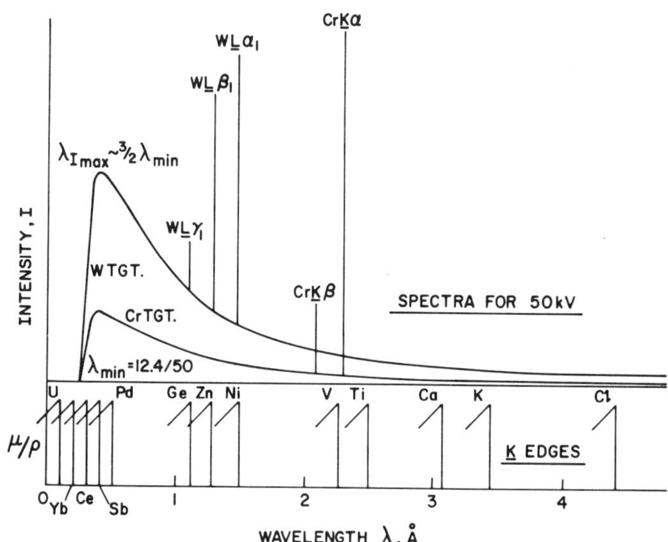

FIG. 4.9. Excitation by the continuum and spectral lines of tungsten and chromium targets.

TABLE 4.1. Excitation by X-Ray Tube Target Lines

Target	Line	Wavelength λ, Å	Relative intensity[a] (*86*)	Highest atomic number excited	
				K edge	*L*III edge
$_{79}$Au	$L\gamma_1$	0.926	10	$_{34}$Se	$_{82}$Pb
	$L\beta_2$	1.070	20	$_{32}$Ge	$_{78}$Pt
	$L\beta_1$	1.084	50	$_{32}$Ge	$_{77}$Ir
	$L\alpha_1$	1.276	100	$_{30}$Zn	$_{72}$Hf
	$M\beta$	5.624	50	$_{15}$P	$_{39}$Y
	$M\alpha$	5.847	200	$_{14}$Si	$_{39}$Y
$_{78}$Pt	$L\gamma_1$	0.958	10	$_{34}$Se	$_{81}$Tl
	$L\beta_2$	1.102	20	$_{32}$Ge	$_{77}$Ir
	$L\beta_1$	1.120	50	$_{31}$Ga	$_{76}$Os
	$L\alpha_1$	1.313	100	$_{29}$Cu	$_{71}$Lu
	$M\beta$	5.828	50	$_{14}$Si	$_{39}$Y
	$M\alpha$	6.052	200	$_{14}$Si	$_{38}$Sr
$_{74}$W	$L\gamma_1$	1.099	10	$_{32}$Ge	$_{77}$Ir
	$L\beta_2$	1.245	20	$_{30}$Zn	$_{73}$Ta
	$L\beta_1$	1.282	50	$_{30}$Zn	$_{72}$Hf
	$L\alpha_1$	1.476	100	$_{28}$Ni	$_{68}$Er
	$M\beta$	6.757	45	$_{13}$Al	$_{37}$Rb
	$M\alpha$	6.988	200	$_{13}$Al	$_{36}$Kr
$_{47}$Ag	$K\beta$	0.497	18	$_{46}$Pd	$>_{92}$U
	$K\alpha$	0.561	150	$_{43}$Tc	$>_{92}$U
	$L\gamma_1$	3.523	10	$_{18}$Ar	$_{47}$Ag
	$L\beta_2$	3.703	25	$_{18}$Ar	$_{46}$Pd
	$L\beta_1$	3.934	42	$_{17}$Cl	$_{45}$Rh
	$L\alpha_1$	4.154	100	$_{17}$Cl	$_{44}$Ru
$_{45}$Rh	$K\beta$	0.546	16	$_{44}$Ru	$>_{92}$U
	$K\alpha$	0.614	150	$_{42}$Mo	$>_{92}$U
	$L\gamma_1$	3.943	10	$_{17}$Cl	$_{45}$Rh
	$L\beta_2$	4.131	25	$_{17}$Cl	$_{44}$Ru
	$L\beta_1$	4.374	42	$_{17}$Cl	$_{43}$Tc
	$L\alpha_1$	4.597	100	$_{16}$S	$_{43}$Tc
$_{42}$Mo	$K\beta$	0.632	17	$_{41}$Nb	$>_{92}$U
	$K\alpha$	0.710	150	$_{39}$Y	$_{92}$U
	$L\gamma_1$	4.726	1	$_{16}$S	$_{42}$Mo

TABLE 4.1. *(continued)*

Target	Line	Wavelength λ, Å	Relative intensity[a] (86)	Highest atomic number excited: K edge	Highest atomic number excited: LIII edge
$_{42}$Mo	$L\beta_2$	4.923	1	$_{16}$S	$_{41}$Nb
	$L\beta_1$	5.177	45	$_{15}$P	$_{41}$Nb
	$L\alpha_1$	5.406	100	$_{15}$P	$_{40}$Zr
$_{29}$Cu	$K\beta$	1.392	20	$_{28}$Ni	$_{69}$Tm
	$K\alpha$	1.542	150	$_{27}$Co	$_{66}$Dy
	$L\beta_1$	13.079	20	$_{10}$Ne	$_{29}$Cu
	$L\alpha_1$	13.357	100	$_{10}$Ne	$_{28}$Ni
$_{24}$Cr	$K\beta$	2.085	18	$_{23}$V	$_{58}$Ce
	$K\alpha$	2.291	150	$_{22}$Ti	$_{56}$Ba
	$L\beta_1$	21.323	20	$_{8}$O	$_{24}$Cr
	$L\alpha_1$	21.714	100	$_{8}$O	$_{23}$V

[a] The relative intensities apply only within the *K*, *L*, or *M* series; for example, Au $M\alpha$ is *not* twice as intense as Au $L\alpha_1$.

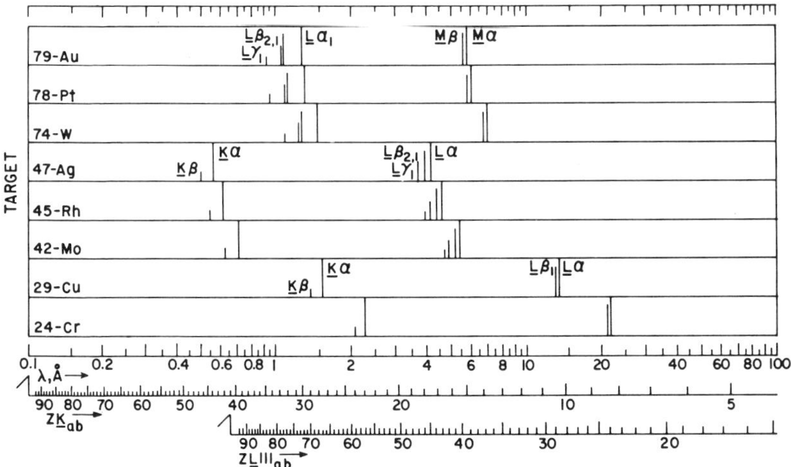

FIG. 4.10. Principal spectral lines of target elements of the common commercially available x-ray spectrometer tubes. The scales of *K* and *L*III absorption edges permit prediction of analyte-line enhancement by target lines. See also Table 4.1.

4.4. FILTERS IN SECONDARY EXCITATION

The preparation and features of x-ray transmission filters and their application in the selective-filtration method of nondispersive analysis are discussed in Section 8.3.2. Here, some applications in crystal-dispersive spectrometry are described.

4.4.1. Attenuation Filters

An attenuation filter is a thin foil or other layer placed in an x-ray beam to preferentially absorb a certain wavelength or band of wavelengths; it is based on the fact that an element absorbs wavelengths shorter than its absorption edge strongly, but is relatively transparent to longer wavelengths.

The practice of using simple (Hull) filters to remove the $K\beta$ lines from x-ray tube target spectra is well known in diffraction. The filter has an absorption edge between the $K\beta$ and $K\alpha$ lines and absorbs the $K\beta$ line very strongly. The best β filter for element Z is usually element Z − 1 or Z − 2. The filter thickness is chosen to reduce the $K\beta$:$K\alpha$ ratio from 1:5 to 1:500; this usually reduces the $K\alpha$ intensity to about half its original value.

Attenuation filters are used to remove target and other primary-beam lines when they might interfere with the analysis (*S6*). For example, Cr $K\beta$ (2.085 Å) and Cr $K\alpha$ (2.291 Å) are absorbed by a 0.1-mm titanium filter ($\lambda_{K_{ab}}$ 2.497 Å); W $L\gamma_1$ (1.099 Å), W $L\beta_1$ (1.282 Å), and W $L\alpha_1$ (1.476 Å) are absorbed by a 45-μm nickel filter ($\lambda_{K_{ab}}$ 1.488 Å), or, better, by a nickel and brass filter ($\lambda_{\mathrm{Cu}K_{ab}}$ 1.380 Å, $\lambda_{\mathrm{Zn}K_{ab}}$ 1.283 Å); if iron ($\lambda_{K_{ab}}$ 1.743 Å) is added to the filter, Cu $K\alpha$ (1.542 Å) and Ni $K\alpha$ (1.659 Å) extraneous lines are also removed.

In a multielement analysis, if the excitation conditions (kV, mA) are set to give maximum intensity for weak lines, the strongest lines are likely to be so intense as to undergo severe coincidence losses in the detector–readout system. In such cases, instead of decreasing and increasing the excitation for strong and weak lines, respectively, it may be easier to insert and remove filters in the primary or secondary beam.

4.4.2. Enhancement Filters

An enhancement filter is a thin foil or other layer placed in the primary x-ray beam to improve the analyte-line-to-background ratio. However, the analyte-line intensity is actually smaller than it would be without the filter.

Gunn (*G38*) determined traces of nickel in hydrocarbon oils using the intensity ratio of Ni $K\alpha$ to adjacent background as a function of nickel concentration. Zinc, gallium, germanium, selenium, bromine, and strontium, as powdered element or compound deposited on aluminum foil, were used as enhancement filters. Zinc ($\lambda_{K\alpha}$ 1.436 Å), predictably the optimum enhancement filter for nickel ($\lambda_{K_{ab}}$ 1.488 Å), simply reduced Ni $K\alpha$ intensity without increasing the Ni $K\alpha$:background intensity ratio. Germanium increased the ratio by a factor of 1.6, but reduced the intensity to ~17% of its unfiltered value. Strontium was most effective, increasing the ratio by a factor of 1.5 while reducing the intensity to only ~40%.

The enhancement filter technique has been developed by Caticha-Ellis *et al.* (*C32*). Figure 4.11 shows the primary continuous spectrum with and without filtration, the analyte absorption edge and lines, and the absorption edge of an enhancement filter having wavelength slightly greater than the analyte lines. Such a filter absorbs the part of the primary spectrum that excites the analyte less than it does the part that only contributes to the scattered primary background at the analyte line. Consequently, higher line-to-background ratios are to be expected. Using a similar technique, Gilmore (*G16*) determined arsenic ($\lambda_{K\alpha}$ 1.177 Å) in hydrocarbons by use of a zinc filter ($\lambda_{K_{ab}}$ 1.283 Å) and claimed 18-fold increase in peak-to-background ratio and decrease in minimum detectable concentration by a factor of 2.7.

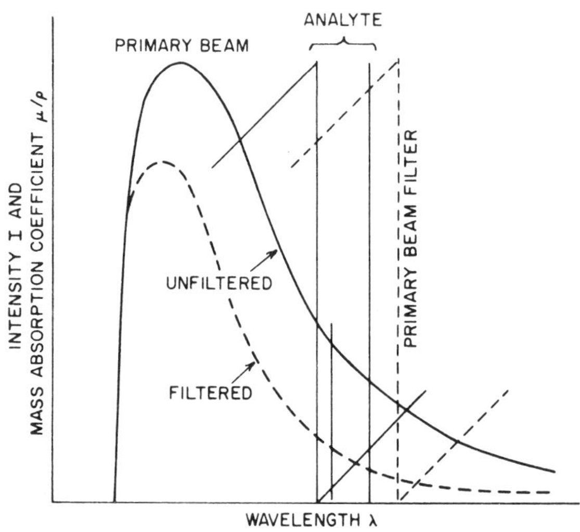

FIG. 4.11. Function of primary-beam enhancement filter.

The enhancement-filter technique is advantageous only at intermediate wavelengths. At longer wavelengths: (1) the continuum has low intensity, and filtering of the useful part is less tolerable; and (2) the scatter of secondary radiation from the crystal contributes more to the background than scatter of primary radiation from the specimen. Short-wavelength spectra lie in the region of the continuum hump, where the short wavelength and high intensity necessitate relatively thick filters to suppress the background significantly. These thick filters also substantially suppress the part of the continuum that excites the analyte. Elements having short-wavelength spectra have high excitation potentials—approaching the x-ray tube potential—so their excitation efficiency is relatively low. Consequently, high primary filtration may prohibitively reduce the analyte-line intensity.

4.4.3. Enhancement Radiators

Many microchemical and thin-film specimens are substantially radiolucent. These specimens include small specimens—specks, flakes, fibers, etc.; films on radiolucent substrates; thin layers of solid, powder, or liquid; and milligram or microgram specimens supported on filter paper, Mylar film, ion-exchange membrane, etc. A substantial portion of the primary beam passes through such specimens and is lost. Some of this wasted radiation may be retrieved by backing the specimen with a secondary radiator having a strong spectral line near the short-wavelength side of the analyte absorption edge (*H52*). The primary x-rays transmitted by the specimen excite the radiator line, which, in turn, excites additional analyte-line radiation. Enhancements of 50% have been obtained in favorable cases. This type of enhancement is treated quantitatively in References (*H52*) and (*P26*). Alternatively, the excitant may be mixed with liquid and powder specimens. Some additional excitation may be provided by simply backing x-ray-transparent specimens with beryllium or a hydrocarbon to scatter some of the transmitted primary x-rays back through the specimen. However, this technique is likely to reduce the peak-to-background ratio.

Other applications of filters are given in Section 8.3.2.

4.5. SPECIMEN PRESENTATION

Specimen presentation is discussed in detail in Section 16.1.1 and throughout Chapters 16 and 17. However, a few general comments are in order here.

The specimen plane may be horizontal below or above the x-ray tube window, inclined at 30–60° to the horizontal below or above the window, or vertical. In most x-ray spectrometers, the specimen plane is defined by a mask in the specimen chamber. The effective specimen area is the area on the specimen plane from which the x-ray optical system receives secondary emission. It is defined by the projection of the x-ray optical system on the specimen plane, or, in effect, by the projection of the source collimator (Section 5.2.2) on that plane. In general, the central rays of the primary and secondary beams meet at the center of the effective specimen area. The specimen plane is inclined at the incident angle ϕ to the central primary ray and at takeoff angle ψ to the central secondary ray. The takeoff angle is usually 30–45°, and the angle between the central primary and secondary rays is usually 90°.

For optimal performance, ϕ and, especially, ψ must be as large as possible. Large ϕ permits a smaller specimen to intercept all of the primary cone Γ (Figure 4.1), or it permits a larger portion of Γ to be intercepted by a specimen smaller than the projection of Γ on the specimen plane. Large ψ reduces self-absorption of analyte-line radiation by reducing its emergent path length within the specimen. However, large ψ also necessitates large effective areas for the collimators, crystal, and detector. Thus, in practice, ϕ may be as large as 75°, but ψ is usually limited to 30–60°.

The specimen mask material must have high mechanical rigidity and adequate absorption for the shortest wavelength emitted from the specimen area covered by the mask. Its spectral lines must lie in a region where they do not interfere with analyte lines. It must be available in high purity so that the spectrum is not complicated by spectral lines of trace constituents. High-purity aluminum 3 mm (~1/8 inch) thick is by far the most commonly used mask material. If short-wavelength lines penetrate the mask, 0.125-mm (0.005-inch) nickel or copper foil may be cemented to the under side of the mask. The foil absorbs the short-wavelength x-rays, and the aluminum prevents emission of the nickel or copper K lines. In the long-wavelength region, carbon, Lucite, nylon, or Teflon may be used; if higher-order short-wavelength lines penetrate these masks, they may be removed by pulse-height selection. Other mask materials may be used in special cases.

The specimen mask has a window of appropriate shape (usually rectangular) and dimensions, but not larger than the effective specimen area. With small windows, the area of the mask irradiated by the primary beam and, thereby, the intensity of scattered primary radiation increase. Consequently, for specimen areas of 1–50 mm^2 or so, a collimated specimen mask (*C40*, *G17*) may be used; the primary beam is conducted to the specimen and the secondary beam to the optical path in a continuous L-shaped tunnel in a block on the specimen mask. This arrangement reduces primary scatter from the mask and from air in the specimen compartment. For smaller specimen areas, selected-area techniques are used in which a pinhole or slit aperture is placed in the primary or secondary beam (Chapter 19).

Originally, all commercial x-ray spectrometers were constructed with the primary beam directed downward into a boxlike specimen compartment. Specimens were placed behind a mask in some type of drawer, which was then inserted in the compartment. When a wide variety of specimen forms must be dealt with, these drawer-type instruments are preferable because it is easy to add accessories to the drawer without altering the instrument itself. Such accessories include rotators for disk and cylindrical specimens, collimated specimen masks, selected-area accessories, and holders and manipulators for special specimen forms. For example, devices have been built to pass through the drawer continuous tapes of filter paper or other material having specimens spotted at intervals. Photographic slide changers have been arranged to index through the chamber specimens supported on 5×5-cm (2×2-inch) filter-paper squares. In more recent instruments, the specimen chamber has the form of a turret or carousel. Some can hold up to ten specimens. Others hold fewer specimens, but are fitted for continuous loading.

Chapter 5

Dispersion

5.1. INTRODUCTION

The third phase of the x-ray spectrometric analysis (Section 3.2.5) is dispersion of the secondary x-rays emitted by the specimen to permit isolation and measurement of the individual analyte lines.

In any method of chemical analysis based on measurement of characteristic x-ray spectral lines, some means must be provided to permit the lines of each analyte element in the specimen to be measured individually. There are many ways to do this. The elements may be *excited* individually (selective excitation) by potential discrimination, monochromatic excitation, or differential excitation. Alternatively, the elements may be excited all at once and some means provided to isolate the lines from each element so they can be *measured* individually; this can be done by selective diffraction (crystal or wavelength dispersion), selective filtration, selective detection (gas discrimination), or selective counting (electronic or pulse-height discrimination, energy dispersion).

Only in wavelength (crystal) dispersion are the several wavelengths emitted by the specimen actually dispersed spacially, and only in this method is it necessary to move the detector for each wavelength measured. Hence, the wavelength-dispersive method has come to be known as *dispersive* and all other methods as *nondispersive*. However, as is shown in detail in Chapter 8, the method of pulse-height analysis is also dispersive in the very real sense that the pulse distributions of the several spectral lines entering the detector are dispersed on the basis of their average pulse height and thereby their original photon energies. Thus, one should refer to the pulse-height analysis method as *energy-dispersive.*

The term *nondispersive* should then be limited to the other methods, in which the spectral lines are not really separated on the basis of either wavelength or energy: potential discrimination, monochromatic excitation, differential excitation, selective filtration, and gas discrimination.

The two dispersive modes have many synonyms. *Wavelength dispersion* (WD) or *wavelength spectrometry* (WS)—the separation of x-ray lines on the basis of their wavelengths, as indicated by the Bragg law—is also known as: (1) *crystal dispersion* when a crystal is the dispersion device; (2) *diffractive dispersion* when x-ray diffraction is the dispersion phenomenon; and (3) *geometric* or *spacial dispersion* because each wavelength leaves the disperser in a different direction (2θ). Optical gratings, specular reflectors, and other devices (Section 5.4.3) are also used for wavelength dispersion of x-rays. *Energy dispersion* (ED) or *energy spectrometry* (ES)—the separation of x-ray lines on the basis of their photon energies—is also known as: (1) *electronic dispersion* when the pulse-height analyzer is used; (2) *nondiffractive dispersion*; and (3) *nongeometric* or *nonspacial dispersion* because there is no actual separation of photons of different energies into different directions in space; this condition is the basis of the unfortunate application of the term *nondispersive* to the energy-dispersive method.

The methods described above can be applied individually or in various combinations. Energy dispersion can be applied to the output of a detector receiving a crystal-dispersed beam containing the analyte line of wavelength λ, interfering higher-order lines of wavelength $\lambda/2$, $\lambda/3$, etc., and spurious radiation arising from any of several sources. Selective filtration and detection may also supplement crystal dispersion. Pulse-height analysis is often used as an aid to the truly nondispersive methods.

Energy-dispersive and nondispersive x-ray spectrometry, as the terms are defined above, are discussed in Sections 8.2 and 8.3, respectively. This chapter is limited largely to crystal dispersion. The components of the x-ray optical system are considered first—the collimators, optical path medium, the crystal itself, and several other types of x-ray dispersion devices. Then the basic flat- and curved-crystal reflection and transmission dispersion arrangements are described. Finally, the most commonly used types of curved-crystal x-ray spectrometers are described.

5.2. COLLIMATORS

5.2.1. Function

Briefly, the function of the collimators (*W23*) is to intercept divergent x-rays so that a substantially parallel beam arrives at the crystal and detector window. The projection of the source collimator (see below) on the specimen plane defines the useful specimen area.

Each point on the specimen surface emits x-rays of many wavelengths—all the spectral lines of all the elements present at that point that can be excited by the primary beam. All these x-rays are emitted in all directions. In Figure 5.1A, consider such a point emitting a cone of x-rays containing wavelengths λ_1, λ_2, and λ_3. One ray of each wavelength arrives at the crystal at the correct Bragg angle and is diffracted, as shown. Thus, the crystal shown would diffract all wavelengths between λ_2 and λ_3. Similarly, the crystal would diffract a range of wavelengths from each point on the specimen. A pinhole or slit aperture at either of the positions shown collimates the beam and excludes all wavelengths other than λ_1. Of the x-rays emitted by the specimen, the aperture permits the crystal to receive only those that originate in the projection of the aperture on the specimen surface and that radiate in a direction θ_1 to the crystal surface. Of these x-rays, the crystal diffracts only those having wavelength λ_1 that satisfies the Bragg law for θ_1 and the d spacing of the crystal. A pinhole and slit permit these conditions to exist at a spot and a line, respectively, on the specimen surface. Figure 5.1A is essentially the arrangement of the Bragg spectrometer (*B78*, *B79*). It is extremely wasteful, in that it excludes most of the x-rays emitted by the specimen at the wavelength for which the crystal is set; this is true even for small particulate or filamentary specimens. However, this arrangement can be used with photographic registration and point or line x-ray sources—but without the pinhole or slit—to record intense short-lived spectra. [Other ways to do this are provided by the edge-crystal (Section 5.5.2.2) and twisted- and convex-crystal (Section 5.7) spectrographs.]

The Soller arrangement (*S43*), shown in Figure 5.1B, is, in effect, a series of Bragg slit systems. The multiple-foil (Soller) collimator (Figure 5.2B) allows Bragg conditions to exist for each incremental strip of a large specimen area, thereby increasing the measured intensity and averaging out specimen inhomogeneity. The foils are almost always of metal, but metal-coated plastic foils have also been used (*B48*). The detector must have a window and active volume large enough for the large diffracted beam. All flat-crystal commercial x-ray spectrometers are now of the Soller type.

The earliest Philips spectrometer used multiple-tube collimators (Figure 5.2A) and was, in effect, a series of Bragg pinhole systems. The function is the same as for the Soller arrangement, except that Bragg conditions exist for each incremental spot on the specimen area.

Some laboratory-built instruments use collimators consisting of two plates identically perforated with a number of small, regularly arrayed apertures. The plates are mounted in register some distance apart in the

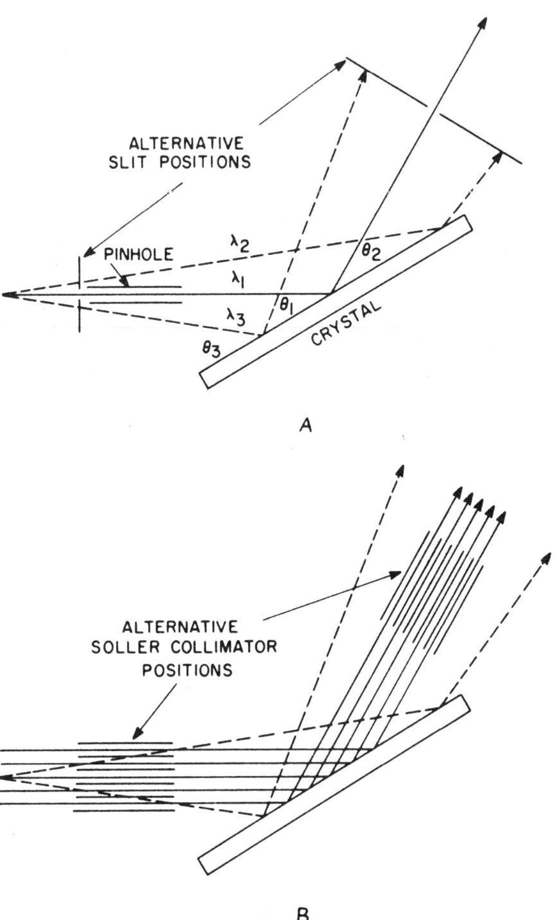

FIG. 5.1. Function of (A) pinhole and single-slit collimators and (B) multiple-slit (Soller) collimators. [H. A. Liebhafsky, H. G. Pfeiffer, E. H. Winslow, and P. D. Zemany, "X-Ray Absorption and Emission in Analytical Chemistry," John Wiley and Sons, N. Y. (1960); courtesy of the authors and publisher.]

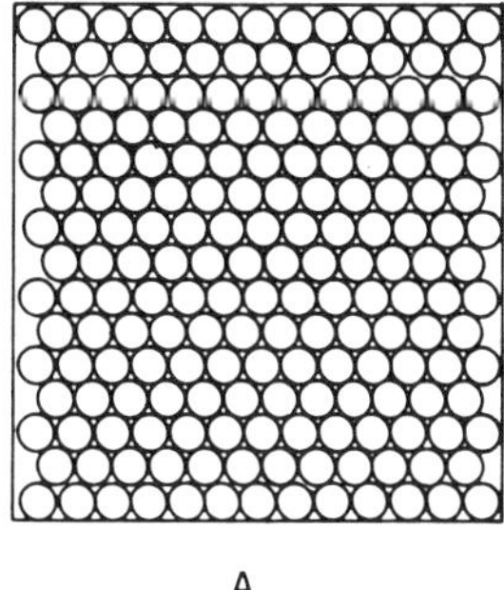

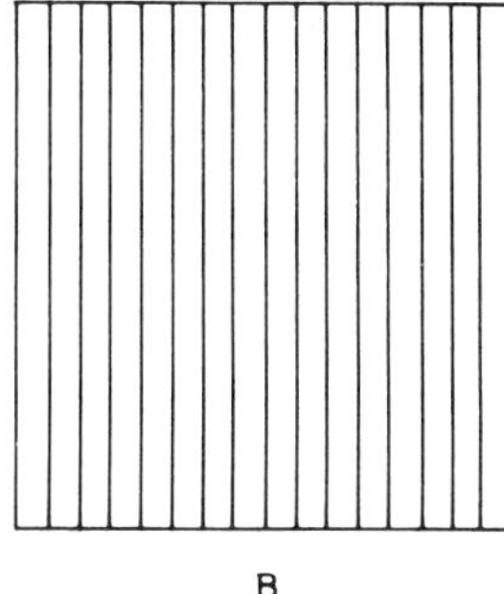

FIG. 5.2. End views of (A) multiple-tube and (B) multiple-slit (Soller) collimators.

x-ray beam. The apertures may be rows of small, square holes or uniformly spaced slits having height about the same as the plates. Such a collimator having *randomly* spaced apertures (*M25*, *U2*) is said to give better suppression of off-axis radiation and to eliminate internal reflection in the long-wavelength x-ray and extreme ultraviolet regions (2–1500 Å). The two plates are perforated with a number of small, identical, *randomly* arrayed apertures, then mounted in register a certain distance apart in the x-ray beam. As before, the apertures may be randomly arrayed, small, square holes or randomly spaced slits.

Converging focused or focusing "collimators" have foils or other channels arrayed so that converging, rather than parallel, rays are passed. Such a collimator is shown in Figure 5.24.

5.2.2. Features and Considerations

It has already been mentioned that the collimators had the form of simple pinholes or slits in the Bragg spectrometer and multiple tubes in the first commercial Philips spectrometers. Today, all flat-crystal spectrometers use multiple-foil (multiple-blade, multiple-slit, Soller) collimators.

The multiple-tube collimator transmits much lower intensity than the multiple-foil collimator for two reasons. It has a smaller cross-sectional open area, and it limits divergence in the direction parallel to the crystal axis and thereby excludes many rays that would pass a Soller collimator. The divergence perpendicular to the crystal axis is about the same for both types of collimator. The divergence parallel to the crystal axis is proportional to the tube diameter or foil height, giving the Soller collimator a substantial advantage in transmitted intensity. The advantage decreases as θ approaches 90°.

Soller collimators are placed at either or both of two positions in the optical path (Figure 3.1). The *primary*, *source*, or *divergence* collimator is placed between the specimen and crystal, and is fixed. The *secondary*, *detector*, or *receiving* collimator is placed between the crystal and detector, and moves with the detector. In principle, the two positions are equally effective for collimation of the specimen emission. However, the collimator between crystal and detector is helpful in excluding from the detector any secondary emission and spurious reflections (Section 13.3.2.1) from the crystal itself. Alternatively, the collimator between specimen and crystal makes the 2-to-1 alignment of the spectrogoniometer less critical. When collimators are used in both positions, one is usually coarse (short and/or wide-spaced foils), the other fine (long and/or close-spaced foils). When both collimators are fine, the spectrometer is difficult to align and to keep in alignment (*W23*, *W24*). Another difficulty encountered when two fine collimators are used is line splitting, or separation of a diffracted peak into two or more subpeaks. This phenomenon occurs when rays from the source collimator are blocked by foils in the detector collimator, or cross over into adjacent channels in that collimator.

In some instruments, a very short, coarse Soller collimator is placed between the front flow-proportional and rear sealed-proportional or scintillation counters of a combination (tandem) detector.

Detector collimators may be tapered, with the height of the collimator aperture and foils increasing from the crystal end to the detector end. This feature permits better utilization of the vertical divergence mentioned above, with consequent increase in transmitted intensity, and is especially advantageous for measurement of the *K* lines of the low-*Z* elements. The detector window must be high enough to receive the divergent beam. The detector collimator may be placed in actual physical contact with the thin window of a flow-proportional counter so as to support the window from explosion when the detector is used in vacuum path. This may preclude the need for a supporting grid, which would reduce the measured intensity.

In passing through the collimator, the x-rays may be specularly reflected, scattered, or diffracted by the foils, thereby producing extraneous lines. The foils may be treated to reduce these effects by etching, sandblasting, or coating with a granular material. In passing through the collimator, the x-rays may also excite the characteristic spectrum of the foil metal, thereby producing extraneous pulse-height distributions in proportional detectors.

The x-ray beam transmitted by a Soller collimator has a triangular intensity *versus* 2θ profile (Section 13.2.1.3, Figure 13.4). For a collimator of length L and foil spacing S, the width of this profile at half-maximum in degrees 2θ is

$$W_{1/2} = \tan^{-1}(S/L) \tag{5.1}$$

The spacing S is measured *between* the foils, and is not the width of the collimator aperture divided by the number of foils. Collimators are specified either in terms of length and foil spacing (for example, 100×0.250 mm) or by the angular divergence in degrees as given by Equation (5.1). Divergence is discussed in more detail in Section 13.2.

Incidentally, the divergence allowed in the direction parallel to the crystal axis by an untapered collimator having length L and height H is

$$\Delta\phi = \tan^{-1}(S/H) \tag{5.2}$$

A tapered collimator gives somewhat greater divergence.

Some of the foregoing considerations are illustrated in Figure 5.3, which shows the idealized full diffraction cones from an analyzer crystal for two wavelengths $\lambda_1 < \lambda_2$ such that the Bragg law [Equation (2.42)] is satisfied at incident angles θ_1 and θ_2, respectively. For each of these cones, the figure shows the segments passed by a Soller collimator S and by a tubular collimator T having inside tube diameter equal to the Soller foil spacing. For the Soller collimator, $2\,\Delta\theta$ is the spread allowed by the foil spacing [Equation (5.1)], and $2\,\Delta\phi$ is the spread allowed by the foil height [Equation (5.2)]. The figure shows that as θ increases, approaching 90°, a decreasing portion of the height of the Soller collimator is effective in passing the diffraction cone, and the advantage of the Soller over the tubular collimator decreases; at 90°, the two collimators are equivalent.

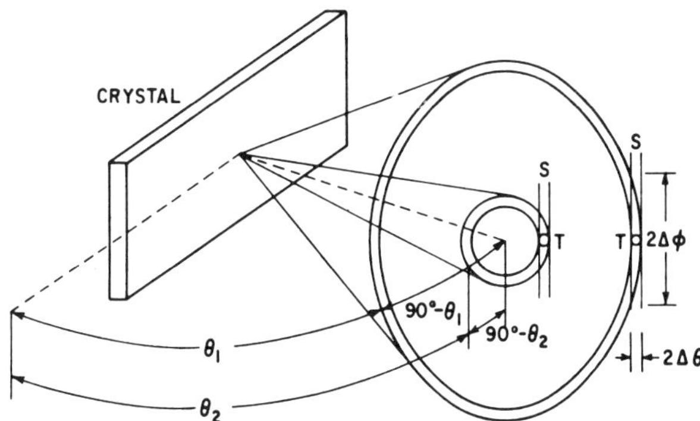

FIG. 5.3. Comparison of intensities transmitted by Soller S and tubular T collimators as θ approaches 90°.

The effects of increased collimation, that is, longer and/or closer-spaced collimators are: (1) narrower peaks; (2) narrower "tails" (or "skirts" or "wings") on the peaks; (3) increased resolution, as a consequence of (1) and (2); (4) decreased background; and (5) decreased measured analyte-line intensity. In short, the higher the collimation, the higher is the resolution and the lower is the intensity. As a rule, the shortest, most widely spaced collimator is used that gives resolution adequate for the measurements to be made. At wavelengths greater than 3 or 4 Å, the separation of the *K* lines of neighboring elements increases as atomic number decreases (Figure 1.8). Consequently, low collimation is adequate unless *M* or higher-order *K* or *L* lines of high-*Z* elements are present. This applicability of low collimation is fortunate because it partially compensates the low fluorescent yield and high absorption losses in this spectral region. Further considerations regarding choice of the crystal–collimator system are given in Section 5.4.2.3, and Figure 5.9 shows the effect of collimation for a crowded spectral region. Figure 19.1 and the associated text are also relevant to this discussion of collimators.

Collimators are subject to certain imperfections that impair their performance. Some of these defects may be present in the collimator when new; others arise with use, especially if the collimator is carelessly handled or stored and/or if the spectrometer is housed in a dusty environment. In the former group are variations in thickness or spacing of the foils from top to bottom or end to end due to nonuniformity of the foils or spacers. In the latter group are warping, sagging, or burring of the foils, damage to their ends, and accumulation of dirt in their interstices. The finer the collimator is, the more troublesome these effects will be.

A collimator may be examined by sighting through its openings toward a light, or tested for uniformity as follows (*S52*). A rigid, flat specimen having large area, homogeneity of composition, and uniformity of surface is placed in the spectrometer, which is then set at 2θ for a spectral line of an element in the specimen. The detector collimator, if any, is removed. The collimator to be tested is placed between the specimen and crystal on a special stage having a motor-drive to translate it in the direction perpendicular to the planes of the foils. Between the collimator and crystal is placed a fixed vertical slit having width 3–5 times the collimator foil spacing. Spectral-line intensity is recorded as the collimator is translated slowly across the secondary beam. The slit ensures that the same areas of the specimen, crystal, and detector are involved for all sections of the collimator. If the collimator foils and spacings are uniform and free of obstruction, the recorded intensity will be constant.

5.3. RADIATION PATH

Air path is usually satisfactory for *K* lines of elements down to atomic number 23 (vanadium) and *L* lines of elements down to atomic number 56 (barium). Air path can be used for *K* lines from relatively high concentrations of elements down to atomic number 19 (potassium). In other cases, helium or vacuum path must be used (*H64*, *M37*, *Z14*).

The measured intensities in air and helium paths are equal at wavelengths less than ~1 Å (Br $K\alpha$, Ra $L\alpha$). Helium gives some advantage at all longer wavelengths, and should always be used at wavelengths >2.5 Å (Ti $K\alpha$). Figure 5.4A shows the intensity ratio for helium and air paths for the $K\alpha$ lines of calcium to manganese.

The measured intensity for helium, hydrogen, and vacuum paths are equal at wavelengths less than ~6 Å (P $K\alpha$, Zr $L\alpha$). Vacuum gives some advantage at all longer wavelengths, and should be used at wavelengths >7 Å (Si $K\alpha$). Figure 5.4B shows transmission as a function of pressure for the $K\alpha$ lines of magnesium to chlorine. Figure 5.4C shows transmission as a function of wavelength for vacuum and helium and the intensity ratio for the two media for the $K\alpha$ lines of magnesium to calcium. Hydrogen (*B49*) is not commonly used because of the explosion hazard.

Increasing the transmission of the optical path is only one reason for using helium or vacuum path. Another is to reduce the scatter of primary

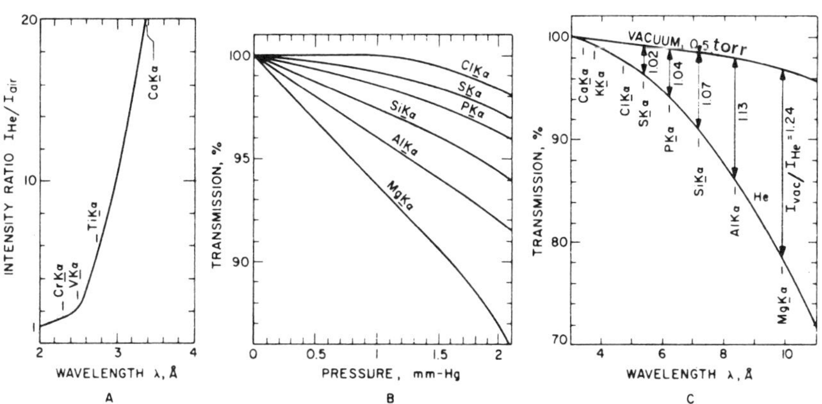

FIG. 5.4. Effect of radiation-path medium in the long-wavelength region. Data are for 25-cm path length. [J. H. Hoskins, *Norelco Reporter* **7**, 111 (1960); courtesy of the author and Philips Electronic Instruments, Inc.] A. Transmission ratio for helium and air paths. B. Transmission for $K\alpha$ lines of elements 12–17 (Mg–Cl) as a function of pressure. C. Comparison of transmission for $K\alpha$ lines of elements 12–20 (Mg–Ca) in 0.5-torr vacuum and in helium.

x-rays in the specimen compartment and thereby to reduce background. A third application of vacuum path was reported by Ashby and Proctor (*A23*). With very high stabilization of the x-ray tube kV ($\pm$0.05%) and mA ($\pm$0.02%) and temperature control of the crystal ($\pm$0.5°C), changes in density of the air path may actually limit the instrumental stability and cause observable changes in the measured intensity. Thus, if it is instrumental stability that limits the overall precision, use of vacuum path may actually improve the precision.

As might be expected, there are problems peculiar to helium and vacuum paths. Distension of the windows of liquid-specimen cells, rapid evaporation of volatile liquids, and disruptive outgassing of powders and briquets in vacuum are discussed in Chapters 16 and 17. Thin windows on gas-flow detectors used in vacuum must be supported by nylon or metal grids. Some of the earlier vacuum enclosures had limited 2θ ranges and required use of less-favorable crystals.

Probably the greatest difficulties attend the use of rubber or plastic helium jackets. Following are some of the precautions that must be taken in their use. The connections of the jacket to the collimators and crystal stage must be secure and free of gaps. The folds or pleats may not protrude into the optical path. A 0–25,000-cm^3/min (0–50-ft^3/h) gas-flow meter must be provided in the helium line. A high helium flow rate [10,000–15,000 cm^3/min (20–30 ft^3/h)] for several minutes is used initially to flush the system. If the helium jacket is attached to the crystal end of the detector collimator, a short rubber tunnel or boot must be placed between the detector end of the collimator and the detector window. During the initial helium flush, this boot should be pried open slightly to allow the helium to sweep the air out of the collimator. Alternatively, a small hole or notch may be cut in the boot to permit a slow, continuous outflow of helium. If these precautions are not taken, air may be permanently trapped in the collimator. After the initial flush, a somewhat lower flow rate [2500–10,000 cm^3/min (5–20 ft^3/h)] may be used for normal operation. The optimal flow rate is established as follows. The intensity of a spectral line having wavelength longer than that of any line to be measured is observed on the ratemeter–recorder as helium flow rate is increased incrementally until no further increase in intensity occurs. After each increase, a few minutes must be allowed for equilibration. Each time the specimen compartment is opened to air, it should remain so for as short a time as possible, and a 15–60-s flush time must be allowed before resuming the measurements. When changing specimens in drawer-type instruments, as soon as the specimen drawer is withdrawn, the next specimen, already loaded in a second, *identical* drawer, should be inserted immediately. If two identical drawers are not available, an empty drawer should be inserted in the specimen chamber while the next specimen is loaded, or the compartment should be closed in some other way. The longer the wavelength to be measured, the higher will be the required operating flow rate and the longer the flush time after loading. When 2θ must be changed to measure background or other lines, the jacket may pull and leak air or protrude into the optical path. These conditions are particularly likely to occur at very high or very low 2θ angles.

5.4. ANALYZER CRYSTALS

5.4.1. Introduction

The *analyzer crystal* or *monochromator* performs the same function in an x-ray emission spectrometer as the prism, or, more analogously, the grating, in an optical emission spectrograph. It spacially disperses the secondary x-ray beam into a wavelength spectrum. As the crystal is rotated from $0° \, 2\theta$, where it is parallel to the collimated secondary x-ray beam, through progressively larger angles, it passes successively through the Bragg angle for each wavelength present, up to a certain limit (see below) and diffracts each wavelength in turn. X-ray diffraction, the basic phenomenon involved in crystal dispersion, is discussed in Section 2.3. The principles of flat-crystal (nonfocusing) optics are reviewed in References (*25a*, *C21*, *S52*).

Many analyzer crystals are available in more than one crystallographic orientation ("cut"), each having a different crystallographic plane with different Miller indexes (hkl) and $2d$ spacing parallel to its "reflecting" surface and therefore useful in a different wavelength region. Such crystals are identified by either the Miller indexes or $2d$ spacings of their reflecting planes; for example, LiF(200), LiF(220), quartz($2d$ 8.51), and quartz($2d$ 6.69). In the Miller index notation, (hkl) and $n(hkl)$ cuts are identical; for example, LiF(100) = LiF(200) and LiF(110) = LiF(220). Also, it is common practice to omit the i in the Miller indexes of hexagonal and orthorhombic crystals $(hkil)$ since i is always equal to $-(h + k)$; for example, calcite(200) = calcite($20\bar{2}0$).

Thus, an analyzer crystal is properly specified by its name or letter symbol and the Miller indexes (hkl), or perhaps the $2d$ spacing, of its diffracting planes. Some crystals are used in only one cut, and the indexes or spacing are commonly omitted; for example, PET and KHP ("KAP").

At long wavelengths, Langmuir–Blodgett smectic, multilayer, heavy-metal, aliphatic-acid, soap films are used as analyzers (Section 5.4.3.4).

Since these are not true crystals, Miller indexes have no significance and they are adequately identified by their names or letter symbols; for example, lead stearate, LOD.

Appendix 10 gives data and shows the useful wavelength regions for all commonly used analyzer crystals and multilayer films.

A good analyzer crystal should have the following features: (1) wavelength range appropriate for the analyte lines to be measured; (2) high diffracted intensity; (3) high resolution—high dispersion and narrow diffracted peak breadth; (4) high peak-to-background ratio; (5) absence of interfering elements, that is, elements that may emit their own characteristic lines ("crystal emission"); (6) low thermal coefficient of expansion; (7) high stability in air, with use, and on prolonged exposure to x-rays; (8) good mechanical strength; (9) extremely low vapor pressure; (10) melting point $>50°C$; (11) good cleavage parallel to the crystallographic planes of interest (this feature, although preferable, is not required; if no such cleavage exists, the crystal can be cut on the required plane); and (12) availability in suitable quality and size, and at acceptable cost. Additional properties are required for crystals to be used in curved-crystal spectrometers: (1) availability in very thin plates; (2) ease of elastic or plastic curvature to specified radii; (3) ease of grinding to specified radii; this requirement applies only to Johansson crystals (Section 5.5.3.1); and (4) ability to withstand repeated flexing; this requirement applies only to crystals for spectrometers having continuously variable crystal radius (Section 5.6.2). The following sections discuss most of these features in some detail.

5.4.2. Features

5.4.2.1. Wavelength Range

In principle, the longest wavelength that a crystal can diffract is equal to twice the interplanar spacing ($2d$) of the crystal planes parallel to its surface. From the Bragg law,

$$\sin \theta = n\lambda/2d \tag{5.3}$$

If $n\lambda > 2d$, $\sin \theta > 1$, which is not possible. It follows that for dispersion of long wavelengths, long d spacing is required (*B61*). Incidentally, interplanar spacings for analyzer crystals are usually given in terms of $2d$, rather than d, because of this particular significance of $2d$.

The dispersion of a flat-crystal x-ray spectrometer increases with increasing 2θ. Consequently, if resolution is the determining factor, the optimal crystal for a specified spectral line has $2d$ spacing just longer than the line wavelength.

Crystals good for the long-wavelength region are not also good for the short-wavelength region. Such crystals usually give low diffracted intensity at short wavelengths, and, because of their large $2d$, they diffract the short wavelengths in the low-2θ region, where background is higher and the resolution lower.

Figure 5.5 shows some mechanical features that limit the useful 2θ range of the crystal (*B25*). For a collimator having breadth B in the direction perpendicular to the crystal axis, the crystal length L required to intercept the entire secondary beam at a specified diffraction angle θ is

$$L = B/\sin\theta \tag{5.4}$$

Alternatively, if a crystal of length L is used with a collimator of breadth B, the smallest diffraction angle at which the crystal can intercept the entire secondary beam is

$$\theta = \arcsin(B/L) \tag{5.5}$$

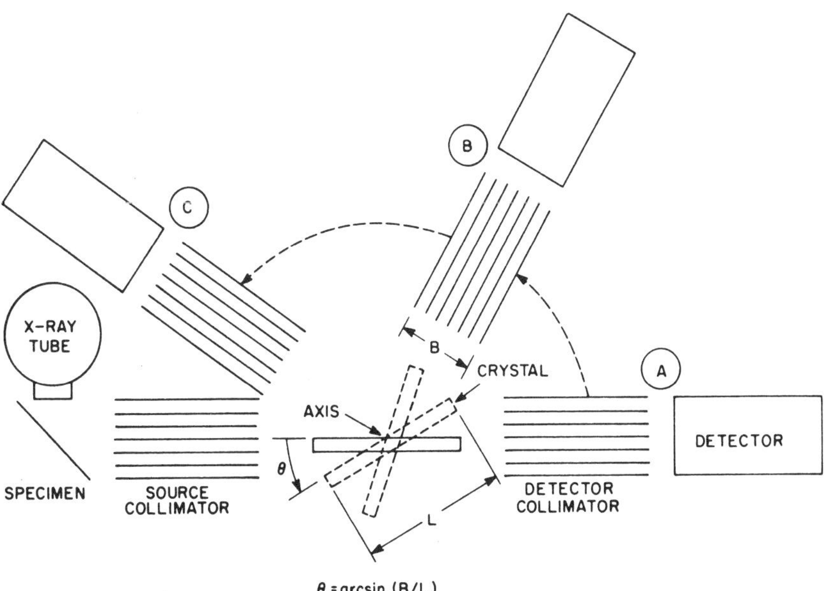

FIG. 5.5. Mechanical limitations on the 2θ range of the analyzer crystal. At the 2θ angle represented by position *B*, a crystal of length *L* just intercepts the entire useful secondary beam. At smaller 2θ angles (position *A*), some secondary rays bypass the crystal. At a certain large 2θ angle (position *C*), the detector arm strikes the x-ray tube or specimen compartment.

These are the conditions for the spectrometer in position B in Figure 5.5. As 2θ is made progressively smaller, the crystal approaches parallelism with the Soller foils, and an increasing portion of the secondary beam either strikes the edge of the crystal or passes by the crystal with resultant loss of diffracted intensity. When 2θ becomes very small, an increasing portion of this passed radiation enters the detector directly, until at 0° (position A), the entire secondary beam passes the crystal and enters the detector. Moreover, at small angles, an increasing portion of the x-rays scattered from the crystal enters the detector. Typically, B is ~1.25 cm and L is ~7.5 cm, giving $\theta \approx 10°$ ($2\theta \approx 20°$). The Bragg law [Equation (5.3)] shows that for a specified wavelength, the smaller the $2d$ spacing of the crystal, the larger is 2θ. If practicable, the crystal should be chosen to have $2d$ small enough to diffract the shortest wavelength to be measured at a minimum of 10–15° 2θ.

The high-2θ limit is set by contact of the detector arm with the tube head, specimen compartment, or some other structure, as shown in position C of Figure 5.5. However, peak breadth increases and peak intensity decreases at high 2θ angles, and, if practicable, a crystal should not be used at angles much above ~120° 2θ.

Rubber or plastic helium jackets may be difficult to use at very high and very low 2θ angles (see above), and, if practicable, a crystal should be chosen to permit use of intermediate 2θ angles.

A practical useful 2θ range for the crystal is 10–145° 2θ, and Appendix 10 gives the wavelengths and $K\alpha$ and $L\alpha_1$ lines corresponding to these limits.

5.4.2.2. Diffracted Intensity

High diffracted intensity is desirable to give high statistical precision in relatively short counting time, to give low minimum detectable analyte concentration, and to permit the use of rapid 2θ scanning rates. Figure 5.6 shows the substantially Gaussian profile of a diffracted peak and indicates peak P and integrated R intensities. These parameters may vary widely even among crystals of the same material cut on the same crystallographic plane. Diffracted intensity is affected by many factors, principally the following:

1. *Crystallographic perfection* (see below). A truly perfect crystal gives low diffracted intensity because of primary extinction. A certain amount of mosaic misorientation is desirable.

2. *Natural "reflectivity."* The coefficient of reflection for the first order depends in part on the structure factor, which, in turn, depends on atomic number.

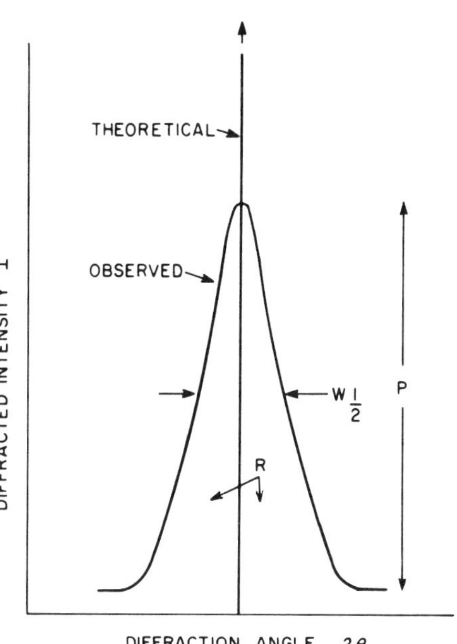

FIG. 5.6. Profile of the diffracted peak of a monochromatic x-ray beam, showing crystal parameters. [L. S. Birks, "Electron-Probe Microanalysis," Interscience Publishers, N. Y. (1963); courtesy of the author and publisher.] P: peak intensity; R: integrated intensity (area under the curve); $W_{1/2}$: rocking angle.

3. *Crystallographic plane.* The reflectivity is different for different crystallographic planes in the same crystal.

4. *Order.* The second order is usually only 5–20% as intense as the first order, and is wholly or nearly absent in some crystals, such as Si(111), Ge(111), and CaF_2(111). In general, intensity decreases as order increases, although this is certainly not always the case; for example, mica(002) has strong odd orders, weak even orders. A crystal may be selected purposely to eliminate an even-order spectral-line interference.

5. *Wavelength.* Diffracted intensity varies with wavelength, as is illustrated by the following diffracted intensity ratios for topaz and LiF: Cu $K\alpha$ (1.54 Å), 0.2; Mo $K\alpha$ (0.71 Å), 0.3; Ba $K\alpha$ (0.4 Å), 0.4.

6. *Surface.* Etched surfaces are more efficient for short wavelengths, polished surfaces for long wavelengths; the long wavelengths are absorbed by a rough surface, especially at low angles. It follows that, for example, an EDDT crystal treated to give high Al $K\alpha$ intensity will give low intensity for, say, Cu $K\alpha$, and *vice versa.*

7. *Treatment.* By means of suitable treatment, the diffracted intensity from a crystal can be increased 2–20 times without prohibitive increase in peak breadth. Birks notes that whereas a twofold increase in primary intensity would be difficult and expensive to obtain, a twofold increase in

diffracted intensity is relatively easy to obtain and is just as beneficial. The treatments (*B60, W11*) include polishing; etching; dry abrasion; wet abrasion with 15–20-μm Carborundum on glass; heating; heating to 80–300°C, then quenching in water; curving and flattening at ~300°C; and stroking with the tip of a cold ultrasonic soldering iron (*P29*). The treatment to be used depends on the nature of the crystal. Of course, there is the risk of ruining a crystal, and these techniques must be applied judiciously. Birks and Seal (*B60*) report that the diffracting power of LiF for Mo $K\alpha$ is increased fourfold, sevenfold, and tenfold by abrading, quenching, and bending, respectively; the diffracting power for Cu $K\alpha$ is increased threefold by all three techniques.

5.4.2.3. Resolution

High resolution is desirable to minimize spectral interference. The resolution of a crystal is its ability to distinguish or recognize as separate two spectral lines of nearly the same wavelength. Resolution is the combined effect of the angular dispersion—the separation of the two lines in 2θ—and the divergence—the breadth of the diffracted lines in 2θ.

Dispersion increases as $2d$ decreases, as indicated by differentiation of the Bragg equation:

$$\frac{d\theta}{d\lambda} = \frac{n}{2d\cos\theta} \tag{5.6}$$

As $2d$ decreases, $d\theta/d\lambda$ increases. For a specified wavelength, best dispersion occurs at a large θ angle with a crystal having $2d$ only slightly greater than the wavelength. Resolution is poorest at small θ angles. Dispersion is shown as a function of $2d$ in Figure 5.7 and in Table 5.1. Figure 5.8 shows the effect of $2d$ on resolution in an even more striking way. The $K\alpha$ lines of all the lanthanon elements are resolved much better by LiF(220) ($2d$ 2.848 Å) than by LiF(200) ($2d$ 4.028 Å) (*P15*).

An increase in $\Delta\theta/\Delta\lambda$ does not necessarily result in good resolution if the diffracted peaks are too broad and have broad tailings. The line breadth is the combined effect of the natural breadth of the line and the rocking curve of the crystal. The greater the line breadth, the lower is the intensity at the peak and the more difficult is the resolution of the line from adjacent lines.

Natural line width is of the order 10^{-3}–10^{-4} Å for short-wavelength lines, but increases and becomes asymmetrical at long wavelengths, due to the increased effect of valence electrons.

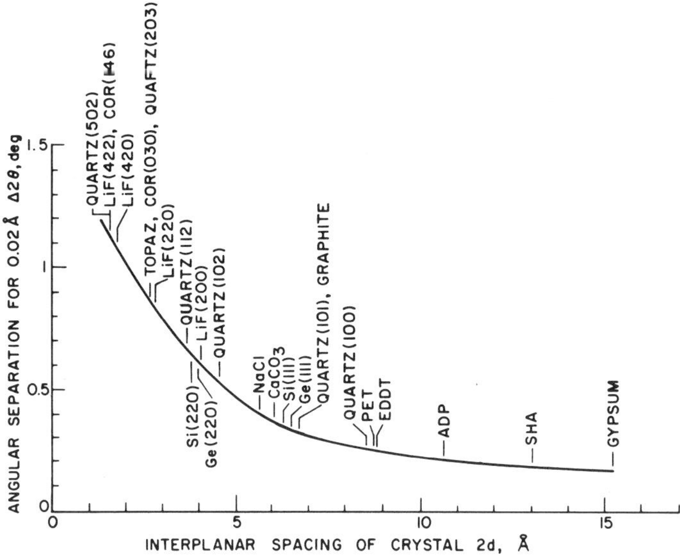

FIG. 5.7. Dispersion of common analyzer crystals as a function of interplanar spacing ($2d$).

The rocking curve is shown in Figure 5.6, together with a line free of broadening. The rocking curve is a measure of the angular range over which the diffracted monochromatic wavelength is spread due to crystal imperfection. It is expressed numerically as the width in degrees 2θ of the diffracted peak at half-maximum—full width at half-maximum (FWHM), or half-width ($W_{1/2}$). The rocking curve is approximately Gaussian and is

TABLE 5.1. Crystal Dispersion as a Function of $2d$
Dispersion of Mn $K\alpha$ (2.103 Å) and Cr $K\beta$ (2.085 Å), $\Delta\lambda$ 0.018 Å

Crystal	$2d$, Å	$\Delta 2\theta$, deg
Topaz	2.712	1.21
LiF	4.028	0.61
NaCl	5.641	0.40
EDDT	8.808	0.24
ADP	10.640	0.21
KHP	26.632	0.08
Clinochlore	28.392	0.07

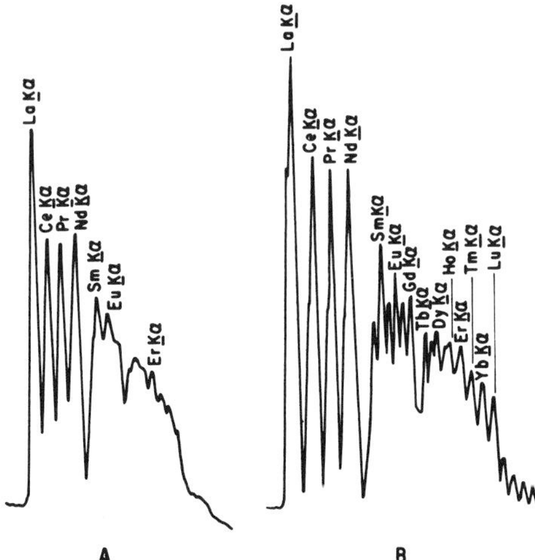

FIG. 5.8. Comparison of resolution of the Kα lines of the lanthanon elements in an aluminum oxide (Al_2O_3) matrix by LiF(200) (2*d* 4.027 Å) (A) and LiF(220) (2*d* 2.848 Å) (B) analyzer crystals. [K. W. Payne, "Proceedings of the 5th Conference on X-Ray Analytical Methods," Swansea, Wales, U. K. (1966); published by N. V. Philips Gloeilampenfabrieken, Scientific and Analytical Equipment Dept., Eindhoven, Neth.; courtesy of the author and publisher; British Crown copyright, reproduced by permission.]

indicative of the degree of crystal imperfection. It is the combined effect of mosaic misalignment, imperfections, strain, surface texture, etc.

A crystal consists of small blocks ~500 Å on a side arranged nearly perfectly, but slightly misoriented with respect to one another by up to a few minutes of arc. Thus, there are regions in the crystal that diffract at angles slightly displaced from the Bragg angle. The greater the mosaic misorientation, the greater is the breadth of the diffracted peak. A truly perfect crystal would give extremely narrow diffracted lines, but very low intensity, because of primary extinction, and would be a very poor analyzer crystal.

The rocking angles of most common analyzer crystals are in the range 0.2–0.4° 2θ, but may be much less; for example, the rocking curve of quartz is ~0.05° 2θ. A crystal may be chosen to have a small $2d$ and/or a small rocking angle to improve resolution, minimize spectral interference, and permit accurate measurement of wavelength. Figure 5.9 shows 2θ scans of a crowded spectral region made with several combinations of coarse and fine collimators and LiF ($2d$ 4.028 Å) and topaz ($2d$ 2.712 Å) crystals. The improved resolution provided by the topaz is evident. Quartz and calcite crystals are often used for very accurate measurement of wavelength because of their narrow rocking curves.

5.4.2.4. Peak-to-Background Ratio; Crystal Emission

The crystal scatters incident secondary x-rays, and the elements of which it is composed are excited by the incident x-rays to emit their own

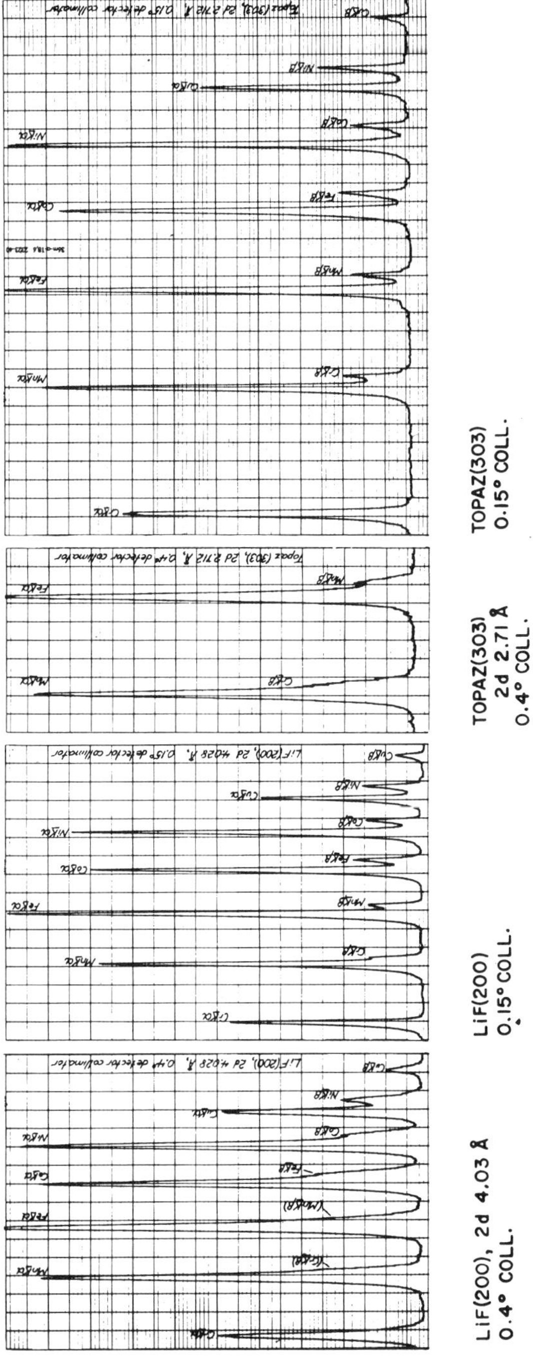

FIG. 5.9. Effect of 2*d* spacing of the analyzer crystal and of collimation on dispersion and resolution. The specimen was a mixture of oxides of Cr, Mn, Fe, Co, Ni, and Cu.

characteristic spectra. This scattered and emitted radiation emanates in all directions, contributes to the continuous background, reduces the peak-to-background ratios of the measured analyte lines, and may cause interference if the pulse-height analyzer is used. When the spectrometer is used in air path, K lines of all crystal elements up to chlorine are absorbed to insignificance, and K lines of potassium and calcium are reduced to very low intensity. However, when helium or vacuum path is used, this radiation reaches the detector, higher background results, and extraneous pulse-height distributions occur which may interfere directly with those of the analyte(s) if pulse-height analysis is used.

Organic crystals, such as EDDT and PET, are free from such emission. For example, NaCl, PET, and EDDT have diffracted intensities for $S\ K\alpha$ in ratio 4.7:2:1, respectively. In the determination of sulfur in hydrocarbons, the crystal receives, in addition to the spectra of sulfur and other impurities, only the scattered primary radiation, which would excite Cl $K\alpha$ in an NaCl crystal only inefficiently. In this case, the NaCl crystal would be chosen. However, in the determination of sulfur in iron, the iron K radiation would strongly excite Cl $K\alpha$, substantially increase the background, and create a pulse distribution close to that of S $K\alpha$. In this case, PET or EDDT would be preferable.

5.4.2.5. Thermal Expansion

An increase in temperature increases the $2d$ spacings in the crystal, causing 2θ to decrease with consequent error in measured wavelength or intensity (*L14*). The angular displacement is given by

$$-\frac{\Delta\theta}{\Delta t} = \alpha \tan\theta = \left(\frac{\Delta d}{d}\right)\left(\frac{1}{\Delta t}\right) \tan\theta \qquad (5.7)$$

where α is thermal coefficient of linear expansion ($°C^{-1}$) in the direction perpendicular to a specified set of planes (hkl); d is the interplanar spacing of those planes (Å); t is temperature (°C); and θ is diffraction angle.

Figure 5.10 shows $-\Delta 2\theta/°C$ as a function of 2θ for several common crystals. Table 5.2 gives the change in intensity due to thermal expansion of the crystal as a function of 2θ, collimator divergence $W_{1/2}$, and temperature for topaz, LiF(200), and ADP crystals. It is evident that the error increases as α, θ, and temperature increase, and as collimator divergence decreases. The increased effect of thermal expansion at high θ angles is partially compensated by the increase in width of the diffracted peak as θ

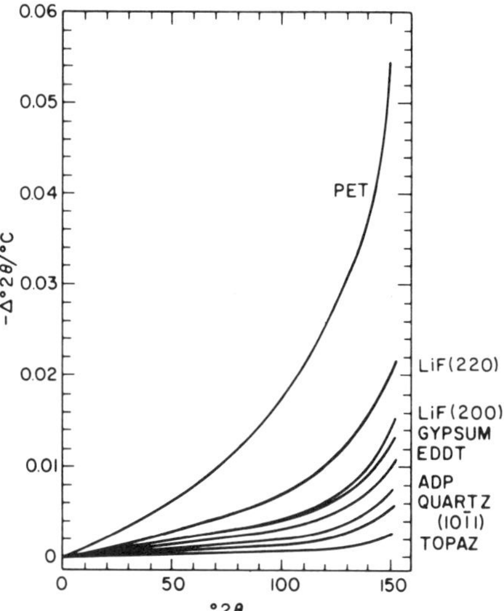

FIG. 5.10. Curves of temperature coefficient of linear expansion for some common crystals. [R. Jenkins and J. L. de Vries, "Practical X-Ray Spectrometry," ed. 2, Springer-Verlag N. Y. (1969); courtesy of the authors, publisher, and N. V. Philips Gloeilampenfabrieken.]

increases, so that angular reproducibility is less critical at high diffraction angles.

Some modern instruments provide constant-temperature chambers for the crystal maintained at $\pm 0.5°C$. These accessories are most useful for precise, high-resolution work, such as measurement of line profiles, absorption-edge fine structure, and chemical effects. However, air-conditioning to maintain laboratory temperature within ± 0.5–$1°C$ of a mean is to be recommended as standard practice.

5.4.2.6. Miscellaneous Features

Crystals should be stable in air and mechanically strong. Most of them are, but there are exceptions. For example, gypsum may effloresce in low humidity and in vacuum, and PET is very soft and fragile—about like paraffin. Some crystals, such as PET, tend to deteriorate with time and on prolonged exposure to x-rays. All crystals should be handled carefully and only by the edges. They should be stored in individual dust-free boxes when not in place on the instrument. Great care must be taken not to mar the diffracting surface and to keep it clean.

Crystals are available in several forms. Many instruments use unmounted flat crystals in the form of $7.5 \times 2.5 \times 0.3$-cm ($3 \times 1 \times 1/8$-inch) bars. Other instruments, particularly those with crystal changers, use mounted crystals of about the same length and width, but $\sim$1 mm thick cemented

TABLE 5.2. Change in Intensity as a Function of Bragg Angle, Collimation, and Temperature[a]

Crystal	α, °C^{-1}	2θ, deg 2θ	$W_{1/2}$, deg 2θ	ΔI, %				
			ΔT, °C →	1	2	3	4	5
Topaz	7.5×10^{-6}	30	0.15	0.00	0.00	0.02	0.03	0.05
		60	0.15	0.00	0.03	0.08	0.15	0.24
		90	0.15	0.01	0.08	0.23	0.44	0.73
		120	0.15	0.03	0.24	0.68	1.32	2.18
		120	0.35	0.01	0.05	0.13	0.24	0.40
		120	0.55	0.00	0.02	0.05	0.10	0.16
		120	0.75	0.00	0.01	0.03	0.05	0.09
ADP	17×10^{-6}	30	0.15	0.00	0.03	0.08	0.17	0.27
		60	0.15	0.02	0.14	0.39	0.76	1.26
		90	0.15	0.05	0.42	1.17	2.27	3.73
		120	0.15	0.14	1.26	3.46	6.66	10.80
		120	0.35	0.03	0.23	0.64	1.26	2.07
		120	0.55	0.01	0.09	0.26	0.51	0.84
		120	0.75	0.01	0.05	0.14	0.28	0.46
LiF	34×10^{-6}	30	0.15	0.01	0.12	0.34	0.66	1.09
		60	0.15	0.06	0.56	1.55	3.02	4.94
		90	0.15	0.19	1.67	4.58	8.78	14.1
		120	0.15	0.56	4.94	13.1	24.1	36.6
		120	0.35	0.10	0.93	2.55	4.94	8.03
		120	0.55	0.04	0.38	1.04	2.03	3.33
		120	0.75	0.02	0.20	0.56	1.10	1.81

[a] From F. S. Lee and W. J. Campbell (*L14*).

in accurately machined metal holders, which are different for different instruments.

Crystals usually have one finished side that is clearly intended to be the diffracting surface. The other side is usually rougher, labeled, or cemented to a mount. When a crystal is first acquired, spectral lines should be recorded in the low-, medium-, and high-2θ regions with the crystal in both its end-for-end orientations to compare performance with respect to intensity and peak breadth. If either orientation appears to be superior, the good end should be indicated on the back. However, crystals are now of excellent quality, and it is unlikely that any significant difference will be found. The crystal surface usually lies within 5 arcmin of the nominal crystallographic plane.

In manual single-channel instruments, whenever the crystal or collimator is changed, it is usually necessary to adjust the crystal so that the spectral lines occur at or near their nominal 2θ angles. An adjustment is provided for this purpose on the crystal stage. When several analyte lines are to be measured, it may not be convenient—or even possible!—to adjust the instrument so that all lines are diffracted at their correct 2θ angles. In such cases, the instrument is adjusted for a line of intermediate wavelength, and the actual position of each peak is established prior to beginning the analysis. Several techniques for doing this are described in Section 5.4.2.7. In automatic instruments and manual instruments provided with crystal changers, the crystals are of the mounted type and are aligned when installed on the changer. Thereafter, the alignment requires only occasional checking.

Several curved crystals are available commercially for Applied Research Laboratories spectrometers. Several companies will, on special order, supply any specified crystal curved, or curved and ground, to any specified dimensions and radii—any crystal, that is, that can be so treated. Some crystals can be curved, but not ground. Some can be cleaved thin enough for use in spectrometers applying transmission or variable-radius modes. Some crystals can be curved plastically and retain their curvature, for example, quartz; others curve elastically and must be held in curvature, for example, mica, aluminum, LiF, and NaCl. Fixed-radius reflecting crystals, even plastically curved ones, are always mounted on a concave metal holder of the required radius. Many workers prefer to prepare their own curved crystals, and instructions for curving and grinding crystals have been published [pp. 127–31 in ed. 1 of (*8*); pp. 185–94 in (*9*); (*B10*, *B60*, *E12*, *G12*, *P13*)]. Curved-crystal dispersion arrangements and complete curved-crystal spectrometers are described in Sections 5.5.3 and 5.6, respectively.

The analyzer crystal may be the source of extraneous spectral lines. The phenomenon is discussed in Section 13.3.2 and illustrated in Figure 13.6.

5.4.2.7. Aligning and Peaking the Goniometer

On spectrometers having crystal changers, crystals are aligned when installed on the changer and thereafter may be selected without affecting spectrometer alignment. The same applies when provision is made to select collimators mechanically. However, on manual instruments, when the crystal or collimator is changed, it is usually necessary to adjust ("peak") the crystal so that spectral lines are diffracted at their correct 2θ angles.

It may not be possible to achieve correct alignment over the full 2θ range. If alignment is correct at, say, the low-2θ end, it may progressively diverge as 2θ increases, or if it is correct at mid-scale, it may diverge toward both ends. In such cases, when multielement specimens are to be measured, the 2θ angles at which all divergent lines fall must be established experimentally. Alternatively, the deviation of observed from tabulated 2θ values may be determined for several lines distributed over the goniometer range and a curve plotted of divergence *versus* 2θ. Sometimes this can be done with several orders of a single strong line. Such curves may have lasting value with crystal changers; otherwise they must be replotted each time a crystal is changed.

The procedure for precise alignment of the spectrogoniometer mechanism is given in the instruction manual. However, for routine "peaking" after changing a crystal or collimator, or when installing crystals on a changer, the following simple procedure may be used. It is assumed that the detector is receiving the line at which alignment is to be made at a relatively high intensity—several thousand counts per second. The crystal rocking screw is set near the center of its range. The goniometer is set at the correct 2θ angle for the line. While intensity is observed on the panel ratemeter or recorder, the crystal is slowly "rocked" until maximum response is obtained. The line is then accurately located in 2θ by one of the methods given below, and the magnitude and direction of the discrepancy are noted. The crystal rocking screw is then readjusted in the direction to correct the error and the line position checked again. This procedure is repeated until the line falls satisfactorily near its correct 2θ value.

Some goniometers have a -2θ region 10–20° beyond zero. A line in this region can be peaked on both sides of zero to give very accurate alignment.

After the goniometer has been aligned, any of the following techniques may be used to determine the exact 2θ angle at which each of the analyte lines is to be measured, or to plot the divergence *versus* 2θ curves mentioned above. In all the following methods, it is assumed that the detector is receiving the line at a relatively high intensity—several thousand counts per second.

1. While intensity is observed on the panel ratemeter or recorder, the goniometer (not the crystal alone) is manually slowly "rocked" repeatedly through the peak until maximum response is observed.

2. The excitation is adjusted to give full-scale, or perhaps slightly off-scale, deflection at the peak 2θ angle established in method 1 above. While

the panel ratemeter or recorder is observed, 2θ is increased until the intensity falls to some arbitrary value—say, 8 on a 0–10 meter or recorder scale—on the high-2θ side of the peak, then decreased to 8 on the low-2θ side. The two angles are noted, and the angle midway between them may be taken as the peak. Recorded peaks are usually asymmetrical, so the intensity should not be decreased too much on the two sides of the peak; 7–9 on a 0–10 scale is usually satisfactory.

3. The goniometer is set at, say, 0.5° below the 2θ angle found in method 1, then advanced manually in 0.1° or 0.2° increments through the peak. The recorder is allowed to trace the intensity for 15–20 s at each position. The point of maximum intensity is taken as the peak.

4. Method 3 is more conveniently and accurately done by scaling, rather than recording, the intensity at each position. A step-scanner accessory does this automatically.

5. The goniometer may be made to slow-scan the top of the peak starting, say, 0.2° below the peak at, say, 0.05° 2θ/min. The position of the peak is read from the broad profile.

Regardless of which procedure is used, several precautions should be taken:

1. In general, a specimen of pure element should always be used to preclude the possibility of misalignment on a nearby line of some other element. An oxide or other compound having no interfering element may do as well. However, for the K lines of the lightest elements and for certain L lines of intermediate elements, the wavelength may be slightly different from pure element and from specimens having that element in various chemical states.

2. Linear, rather than logarithmic, ratemeter response should be used, especially for resolving an unresolved doublet (such as $K\alpha_{1,2}$) or one of the lines in a partially resolved doublet. Logarithmic ratemeters reduce the difference in response between the two individual members of the pair, and thereby in the peak profile.

3. The line should have intensity high enough to minimize ratemeter or recorder fluctuations.

4. Lag due to ratemeter time constant and recorder response must be allowed for in adjusting the crystal or goniometer. If the adjusting is done too rapidly, the panel meter or recorder response cannot follow.

5. If the goniometer may be badly out of alignment, care must be taken not to use the wrong line. Before locating the $K\alpha$ line, one should verify that it really is the $K\alpha$ line by verifying that the weaker $K\beta$ line occurs

at a lower 2θ. Before locating the $L\beta_1$ line, one should verify the presence of the (usually) stronger $L\alpha_1$ line at higher 2θ and the weaker $L\gamma_1$ line at lower 2θ.

6. The final settings of the crystal rocking screw and goniometer should always be approached from the same direction.

7. It may be wise to use a single line, such as $K\beta$, rather than an unresolved multiple line, such as $K\alpha_{1/2}$, because the latter may have an asymmetrical peak profile. This is especially important in higher orders where the α_1–α_2 separation is greater. The $L\alpha_1$ line is satisfactory because $L\alpha_2$ is relatively very weak.

8. When peaking a long-wavelength line using vacuum or, especially, helium path, it may be convenient to make preliminary adjustments in air with a higher-order, shorter-wavelength line that falls at about the same 2θ. In this way, possible difficulties with the vacuum enclosure or helium jacket (Section 5.3) are precluded. Once preliminary adjustment is effected, final peaking is done with the analyte line in vacuum or helium path. For example, if an EDDT crystal is to be peaked for Al $K\alpha$ (142.57° 2θ), preliminary adjustment may be done in air with fourth-order Mn $K\alpha$ (145.73° 2θ).

5.4.3. Other Dispersion Devices

For dispersion of very long wavelengths, analyzer crystals having very long interplanar spacing are required. Thus far, no such crystals having suitable properties have been developed, with the possible exception of octadecyl hydrogen maleate (OHM), which has $2d$ 63.5 Å. The next four sections describe other devices for dispersion of x-rays of long and ultralong wavelength.

5.4.3.1. Gratings and Specular Reflectors

Total specular reflectors (Section 2.4.1) (*D5*, *F24*) and ruled optical gratings (Section 2.4.2) (*F23*, *N16*, *S11*) have been applied to x-ray dispersion in the long- and ultralong-wavelength region. However, these devices require specially built, highly precise instruments because the critical angle for reflection and the incident and diffraction angles for gratings are very small. Moreover, the diffracted intensity for gratings is about an order lower than that from multilayer soap films (see below). Certainly, neither device can be used simply to replace the analyzer crystal on a conventional x-ray spectrometer.

5.4.3.2. Multilayer Metal Films

Dinklage and Frerichs (*D19*) prepared a multilayer analyzer consisting of 140 layers of lead 2.5 Å thick separated by magnesium layers 26.6 Å thick, by alternate vacuum evaporation of the two metals on a glass plate. The "*d* spacing" of this "crystal" was ~29 Å, so that 2*d* was ~58 Å. The analyzer was very satisfactory when first prepared, but its performance was seriously impaired by diffusion, which reduced the diffracted intensity at the peak by half in two days at room temperature. Gold-magnesium layered analyzers interdiffused similarly.

More recently (*D18*), much better performance has been realized with 130 layers of iron 14 Å thick separated by magnesium layers 16 Å thick (2*d* 60 Å) and by 100 layers of iron 13 Å thick separated by magnesium layers 39 Å thick (2*d* 104 Å). These analyzers give performance comparable with—but not as good as—lead stearate in the region 8–24 Å (Al $K\alpha$ to O $K\alpha$). The peak-to-background ratio is ~80. No detectable interdiffusion occurs at room temperature.

5.4.3.3. Metal Disulfide–Organic Intercalation Complexes

Many inorganic substances—such as graphite, molybdenum disulfide, mica, and clays—have a highly laminar crystal structure in which the chemical bonds within each layer are primarily covalent and strong, those between layers primarily van der Waals and weak. Certain chemical compounds can penetrate between the layers of such laminar crystals to form highly ordered, stoichiometric, and relatively stable new crystal structures consisting of alternating layers of the host and penetrant. Such substances are known as *intercalation complexes* and the penetrants as *intercalates*.

The transition-element dichalcogenides have this laminar structure, for example, the disulfides of titanium, zirconium, niobium, molybdenum, and tantalum. (The *chalcogens* are the group VI elements in the periodic table: oxygen, sulfur, selenium, tellurium, and polonium.) Relatively large single-crystal plates of these compounds can be grown, by the iodine-vapor-transport (van Arkel) method, for example. If such plates are subjected to prolonged immersion in certain organic liquids, such as pyridine (C_5H_5N), substituted pyridines, and aliphatic amines [$CH_3(CH_2)_nNH_2$], intercalation occurs, causing the crystals to swell physically and increase in interplanar spacing crystallographically (*G2*, *G3*). The reaction time for complete intercalation may be several hours to several weeks and proceeds more rapidly at 100–200°C than at room temperature.

Such metal disulfide–organic intercalation complexes have been prepared in areas >1 cm^2 and thicknesses from several micrometers to a substantial fraction of a millimeter. The "crystals" are mounted on suitable supports, which may be curved. The $2d$ spacing may be 6–120 Å or even more, depending on the length of the organic molecule. A series of such pseudocrystals can be prepared having a graded series of $2d$ spacings by use of a suitable series of organic compounds. The x-ray diffraction power and absorption are determined largely by the atomic number of the metal in the disulfide (or the mean atomic number of the metal dichalcogenide). Thus these intercalation complexes may have value as x-ray dispersion devices, and several laboratories are evaluating them for this application.

5.4.3.4. Multilayer Soap Films

Mesomorphic pseudocrystals consisting of multiple smectic layers of Langmuir–Blodgett (*B70*) heavy-metal soap films (*E9*, *E11*, *H35*, *N16*) have been very successful for dispersion of ultralong-wavelength x-rays. The most commonly used films are divalent-metal salts of saturated, straight-chain, aliphatic ("fatty") organic acids having the general formula $[CH_3(CH_2)_nCOO]_2M$, where M is a divalent heavy-metal cation, usually lead, barium, or strontium. The *repeat interval*, which corresponds to the interplanar spacing d of true crystals, and the diffracted intensity can be "tailored" by selection of the cation, aliphatic acid, and number of layers. By far the most commonly used cation is lead. The corresponding barium salts have substantially the same $2d$, but lower diffracted intensity. Mabis and Knapp (*M1*) claim that as the wavelength of the analyte line increases, the strontium salts become more efficient than those of lead. Multilayer films having $2d$ spacings 20–160 Å have been prepared from the 6–30-carbon aliphatic acids. True crystals are available having $2d$ spacings up to ~100 Å (Appendix 10), but these crystals are neither readily available in suitable form nor of proved reliability. Consequently, the multilayer films are widely used at the long wavelengths. Appendix 10 gives data for the most commonly used of these analyzers, many of which are available commercially. The films can be prepared in any reasonable dimension, flat or curved. A description of the structure of the pseudocrystals and one version of the mechanism of their preparation (*E9*, *H35*, *N16*) are given below in terms of lead stearate (lead octadecanoate), $[CH_3(CH_2)_{16}COO]_2Pb$, as an example.

A solution of stearic acid (octadecanoic acid, $C_{17}H_{35}COOH$) in benzene is added dropwise to the surface of a dilute water solution of a lead salt.

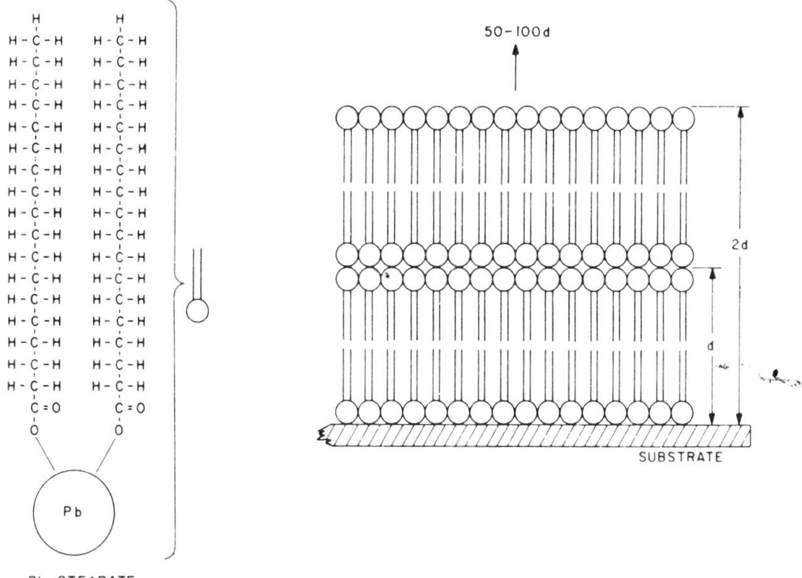

FIG. 5.11. Lead stearate molecule and the first four layers (first two repetition intervals *d*, first 2*d* interval) of a Langmuir–Blodgett, multilayer, smectic, lead stearate pseudocrystal.

The benzene spreads over the water surface and evaporates, leaving a monolayer of lead stearate molecules (Figure 5.11) with their hydrophilic (lead) ends toward the water and their hydrophobic (hydrocarbon) ends upward. By means of a floating straight-edged barrier, a two-dimensional pressure is applied to the lead stearate film parallel to the liquid surface, compressing the film and close-packing the molecules. The method of preparation of the multilayers is shown in Figure 5.12. A clean hydrophilic plate (usually glass or mica) is immersed in the solution at near normal incidence. The lead stearate does not adhere to the clean plate on immersion, but on withdrawal (pass 1), a monolayer of lead stearate adheres with the lead ends toward the plate, as shown. On reimmersion (pass 2), a second layer coheres to the first, but with opposite orientation. By successive immersion and withdrawal of the plate, a series of alternately oriented monolayers is built up, the first four of which are shown in Figure 5.11. After each pass, the floating barrier is readjusted to maintain the close-packing of the monomolecular film. If the substrate is first made hydrophobic, by application of a thin coat of stearic acid, for example, the film does adhere on the first downward pass, with the hydrocarbon ends toward the sub-

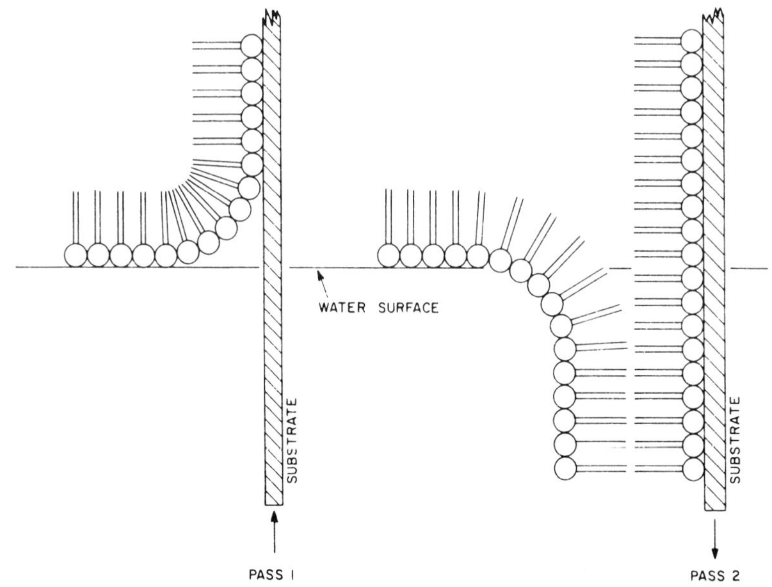

FIG. 5.12. Preparation of multilayer, smectic, heavy-element, soap-film pseudocrystals, showing the first two passes.

strate and the lead atoms outward. On withdrawal, a second layer coheres with lead to lead, etc. The resulting multilayers are identical. The stearate chain is ~25 Å long, so the *d* spacing (repeat interval, Pb to Pb) is ~50 Å, giving $2d \approx 100$ Å.

Enough monolayers should be deposited to make the film infinitely thick for the shortest-wavelength analyte line to be measured. If there are too few layers, four adverse effects may result: loss of diffracted analyte-line intensity from the film; loss of resolution—that is, increased half-width of the diffracted peak; and diffraction and/or emission from the substrate. For lead soap films, ~100 monolayers are sufficient for F $K\alpha$, ~30 layers for Be $K\alpha$. In general, films thicker than ~100 layers are difficult to prepare and give little improvement in performance.

For curved-crystal spectrometers, the films are deposited either on a rigid, curved, highly polished metal or glass blank or on a flexible blank —such as mica—that can be curved and cemented to a rigid, curved frame or other form. Only Johann geometry is feasible since the thin soap layers cannot be ground to the radius of the focusing circle (Section 5.5.3.3).

In a novel application of an analyzer consisting of a lead stearate film on a mica substrate, the film was used for the ultralong-wavelength region, and the underlying mica for the shorter wavelengths (*S22*).

Incidentally, these pseudocrystals are to be distinguished from true crystals of organic and metal-organic compounds having large *d* spacings. These crystals may be classified in four groups as follows, with examples taken from Appendix 10: (1) crystalline organic compounds (PET, EDDT, sucrose, etc.); (2) amides of long-chain aliphatic acids (tetradecanoamide, etc.); (3) diesters of dibasic organic acids (OAO, OTO, etc.), and hydrogen esters of such acids (OHM, HHM, BHM, etc.); and (4) metal salts of organic acids (silver acetate, etc.) and alkali-metal (including ammonium and thallium) hydrogen salts of dibasic organic acids (KHP, RbHP, etc.). Smectic soap multilayers and true organic crystals are distinguished in column 3 of Appendix 10.

5.4.3.5. Pyrolytic Graphite

Natural-mineral single-crystal graphite is not available in dimensions and quality suitable for use as x-ray analyzers. However, polycrystalline analyzers of highly oriented pyrolytic graphite are prepared by pyrolytic decomposition of hydrocarbon vapors on heated substrates, followed by annealing at high temperature ($>3000°C$) to effect crystallite growth and a high degree of preferred orientation (*C24*). The (0002) planes ($2d$ 6.708 Å) orient parallel to the surface. The crystallite size is ~1000 Å parallel to (0002) and ~400 Å perpendicular to (0002). However, the diffracted intensity and resolution are not as good as for good conventional analyzer crystals of comparable $2d$ spacing [quartz($2d$ 6.69 and 8.82), PET, EDDT, and ADP($2d$ 10.6)].

More recently, polycrystalline graphite analyzers have been prepared by hot-pressing graphite flakes (*G25*, *J12*). These analyzers have reflectivity >5 times that of PET for P $K\alpha$ and are especially good for the K lines of phosphorus, sulfur, and chlorine. These graphite "crystals" are very rugged and can be prepared in any form, including cylindrically or torically curved. In fact, the graphite layers are formed directly on their flat or curved support. These graphite crystals also have large rocking angles due to their large mosaic spread.

5.5. BASIC CRYSTAL-DISPERSION ARRANGEMENTS

Crystal-dispersion arrangements may be broadly classified as flat-crystal (nonfocusing) or curved-crystal (focusing). Actually, curved crystals do not focus x-rays in the true optical sense, but diffract them in such a way that they converge to a line or point.

5.5.1. Multichannel Spectrometers

Most x-ray spectrometers are single-channel instruments in the sense that they have only one crystal–collimator–detector system and thus can measure only one wavelength at a time. However, the advantage of measuring more than one wavelength at a time was soon realized, and many approaches to this goal have been devised.

Possibly the simplest method is to set the detector at a fixed 2θ, then select two or three different crystals that will diffract two or three analyte lines at the *same* angular position. A crystal changer can slide the crystals into position in sequence. Rather surprisingly, it is claimed that sets of two or three crystals can usually be found for most sets of two or three analyte lines! Another use for this arrangement is in the measurement of a single line: Two identical crystals can be mounted in the changer, one peaked for the line, the other for adjacent background.

In the Laue spectrometer, one crystal simultaneously diffracts different wavelengths at different angles from different crystallographic planes into several detectors arrayed at appropriate positions. A single-channel spectrometer having a LiF(200) crystal ($2d$ 4.03 Å) is readily converted to a two-channel single-crystal spectrometer (*B87*): A second detector is clamped on the existing detector arm at a suitable fixed angle to receive reflections from the LiF(420) planes ($2d$ 1.80 Å). Such an instrument has been applied to the simultaneous measurement of Pb $L\alpha_1$ and background intensities from specimens containing traces of lead. More elaborate Laue spectrometers are described in Section 5.5.2.3.

If the secondary-emission beam from the specimen is divergent and falls on a fixed flat crystal, different rays strike different points on the crystal at different angles. Thus the crystallographic planes parallel to the crystal surface diffract different wavelengths at different angles into different detectors placed at appropriate fixed angles. This arrangement has poor resolution, but it has been applied successfully to simultaneous measurement of the $K\alpha$ lines of titanium, chromium, and nickel.

Carson (*C28*) added a parallel flat-crystal–collimator–detector system to the Philips wide-range spectrogoniometer to permit simultaneous measurement of plate-metal and substrate lines. Jenkins realized the same objective by cutting the crystal in half laterally and mounting the two halves edge-to-edge but tipped with respect to one another so that they present different angles θ_1 and θ_2 to the collimated secondary beam. Each half-crystal could be set independently to diffract a different line. Two detectors were provided, one for each crystal. Wytzes (*W25*) also used a dual analyzer, but with sections of two different crystals, each independently aligned. Two suitably oriented detectors permit simultaneous measurement of two spectral lines.

In the Suoninen reflection–transmission spectrometer (*S68*), the same thin, flat LiF crystal is used to measure two wavelengths simultaneously, one by reflection, the other by transmission. Selection rules are given for use of this type of dual-mode crystal.

Adler and Axelrod built a two-channel curved-crystal instrument (*A10*) for simultaneous measurement of analyte and internal-standard lines, and later a four-channel instrument (*A28*) for simultaneous measurement of two analyte–internal-standard line pairs (Rb/Sr and Cs/I $K\alpha$ line ratios). Birks and Brooks (*B55*) built a three-channel curved-crystal instrument for simultaneous measurement of chromium, nickel, and molybdenum lines from steel. Modern commercial multichannel instruments are described in Chapter 9.

5.5.2. Flat-Crystal Dispersion Arrangements

5.5.2.1. Bragg and Soller

The Bragg and Soller flat-crystal dispersion systems are described in Section 5.2.1 and shown in Figure 5.1. Another feature of these arrangements may now be considered.

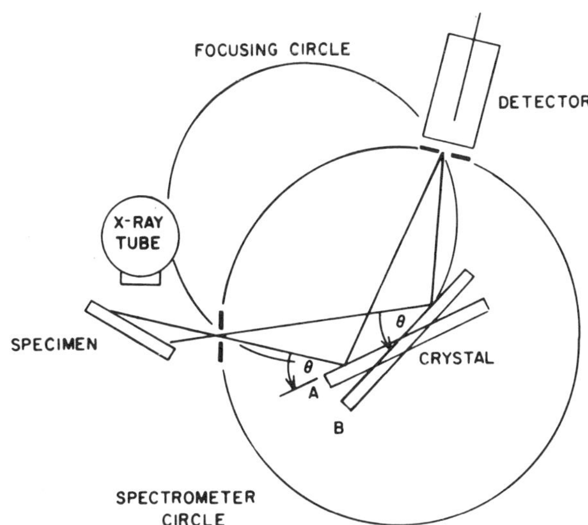

FIG. 5.13. Bragg–Brentano parafocusing.

Consider the spectrometer shown in Figure 5.13 with the crystal in position *A*. For every wavelength present, a cone of divergent rays from the x-ray point or line source intersects the crystal surface at a range of angles. Consider now the cone for a specific wavelength. One of the rays is likely to intersect the crystal at the Bragg angle and be diffracted, as shown for the lower ray in the figure. If now the crystal is rotated slightly ("rocked") to position *B*, some other ray forms the Bragg angle and is diffracted, as shown for the upper ray. Both these rays are diffracted in turn to the same point on the spectrometer circle. In fact, as the crystal is turned from *A* to *B*, a series of different rays in the cone is diffracted, all to the same point. Each wavelength present undergoes the same phenomenon at its own Bragg angle and is diffracted to a different point on the spectrometer circle, resulting in dispersion of the spectrum on the basis of wavelength. The diameter of the focusing circle is larger the shorter the wavelength. For large-area sources, a slit or pinhole must be placed at the source or detector for the purpose already explained in Section 5.2.1.

The phenomenon just described is not true focusing—even in the x-ray definition of the term—because (1) it results in *successive*, rather than *simultaneous*, diffraction of a number of rays to the same point or line, and (2) it does not increase instantaneous measured intensity. Brentano characterized this phenomenon and suggested a term for it; it is known as Bragg–Brentano *parafocusing* (*B81*).

5.5.2.2. Edge-Crystal

The edge-crystal arrangement is shown in Figure 5.14 (*B56*). For each secondary wavelength λ_1, λ_2, λ_3, . . . emitted by the specimen, a certain

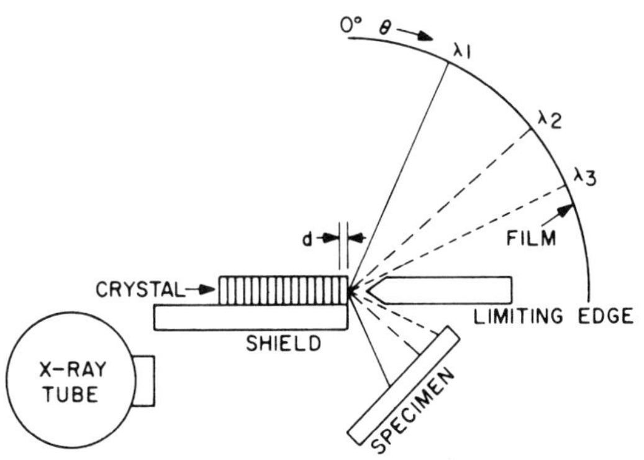

FIG. 5.14. Edge-crystal x-ray spectrograph.

series of rays lying in a line perpendicular to the figure and parallel to the crystal edge strikes the crystal edge at the Bragg angle and is diffracted. Thus the entire line spectrum is diffracted simultaneously. It can be recorded on a film in an exposure time of ~1 h, or a slit backed by a detector can be arranged to make a 2θ scan along the arc represented by the film. The crystal is ~0.1 mm thick, and the breadth of each spectral line is the projection of the crystal on the film. Various crystals may be used to record all wavelengths of interest. It must be remembered that the effective $2d$ spacing is that of the edge, rather than the major face, of the crystal. The instrument is small and mounts on an x-ray diffraction unit. The film version has no moving parts, and in the detector version, only the detector moves, so the instrument is simple in structure and convenient to operate. It may be flushed with helium or evacuated. Spectra may be recorded on the same film from series of samples and/or standards by displacing the film and changing the specimen after each exposure. Because each wavelength originates at a different place on the specimen surface (Figure 5.14), homogeneous specimens are required for semiquantitative and quantitative analysis. For accurate quantitative work, film calibration is required to convert photographic density to intensity (Section 6.6.1). Due to the small effective area of the crystal, intensities are low. An inexpensive film version is available commercially.

5.5.2.3. Laue

Figure 5.15 shows the one-crystal multichannel Laue arrangement having a flat crystal placed so that several wavelengths are diffracted simultaneously, each from a different crystallographic plane (*L3*). The planes must all be of the same crystallographic zone having zone axis parallel to the crystal axis. The figure shows a quartz crystal cut parallel to the $(10\bar{1}\bar{2})$ planes reflecting five $K\alpha$ lines, each from a different plane $(hkil)$ as indicated in the figure. Five detectors are arrayed about the crystal, each placed to receive one of the diffracted beams. Soller collimators are provided for the source beam and for each detector. The different planes have different "reflectivities"; the figure gives, for each $K\alpha$ line, the reflecting plane $(hkil)$ and the ratio of the intensity from the indicated plane to the intensity of the same line from the $(10\bar{1}\bar{2})$ plane. Since $(10\bar{1}\bar{2})$ is the surface plane, this ratio is, in effect, the ratio of the intensities from the Laue and conventional Soller spectrometers, both using a quartz $(10\bar{1}\bar{2})$ crystal.

The Laue arrangement has many advantages. It provides many of the favorable features of a multichannel instrument with one crystal, while

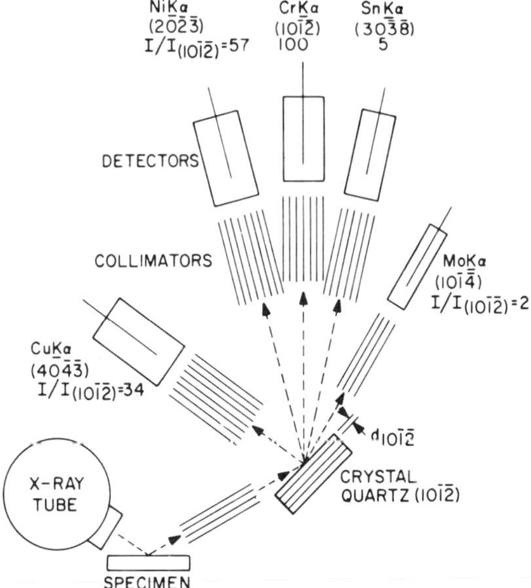

FIG. 5.15. Multichannel Laue spectrometer having five collimator–detector channels arranged to receive five different analyte lines, each from a different crystallographic plane (*hkil*) from the same quartz crystal cut parallel to the $(10\overline{12})$ planes. For each channel, the figure shows: (1) the measured spectral line; (2) the diffracting crystallographic plane; (3) the intensity ratio of that line measured from that plane and of the same line measured from the $(10\overline{12})$ plane.

retaining many of the favorable features of a single-channel instrument. A spectrometer of the type described above can be built on a standard flat-crystal spectrogoniometer. The source geometry can be optimized because the same secondary beam from the specimen is used for all channels. Minimum distance between the x-ray tube target and specimen and optimum takeoff angle may be used. The cost of the collimators can be reduced by use of a fine source collimator and relatively short, wide-spaced detector collimators. Normalization of the measured intensities, if required, is easy because it involves the relative reflectivities of crystallographic planes in the same crystal. The measured intensity of a specified analyte line can be increased by measuring it with more than one channel, that is, from more than one crystallographic plane.

The unfavorable features include the following. It is often difficult or impossible to find a crystal having lattice planes such that the crystal can be oriented to diffract a specified group of lines simultaneously. The calculations required to establish the crystallographic planes and instrument arrangement are rather complex. The intensities from the individual planes are usually lower than those measured from the same crystal in Soller mode. High absorption losses occur in the beams diffracted from planes inclined at large angles to the crystal surface. Large-area detectors and collimators may be required to intercept the full useful beams at high 2θ angles.

The Laue spectrometer may be operated in other modes, as follows.

a. Sequential. The crystal is placed in the proper position with respect to the specimen, and the crystal θ-drive is disengaged so the crystal will remain fixed while leaving the detector free to move. The spectrum may be recorded or the individual analyte lines scaled in the usual manner. This mode of operation is readily applicable to a standard flat-crystal instrument.

b. Transmission. If thin crystals are used, the spectrometer can use transmitted diffracted beams or both transmitted and reflected beams.

c. Focusing. If a curved crystal is used, a one-crystal multichannel focusing spectrometer is feasible. Again, only crystallographic planes of the same zone having zone axis parallel to the crystal axis can be used. The instrument can be arranged so that the focal points of the diffracted beams lie on a focal circle (see below). A slit aperture is provided at the specimen.

d. Scanning. A one-crystal multichannel scanning spectrometer has been described in which simultaneous 2θ scans of the short-, intermediate-, and long-wavelength regions can be made (*S51*). The three wavelength regions are recorded from separate crystallographic planes having, respectively, small, intermediate, and long $2d$ spacings, and each channel is provided with accessories optimal for its wavelength region.

5.5.2.4. Other Flat-Crystal Arrangements

Flat crystals may also be used in transmission mode with the diffracting planes perpendicular to the crystal surfaces, as shown in Figure 5.16 [p. 29 in (*8*)]. In general, this arrangement is much less satisfactory because the crystal must be thin (usually <0.025 cm) to avoid prohibitive absorption of the diffracted x-rays. This geometry is advantageous at short wavelengths where 2θ is small and a reflecting crystal of practical length does not intercept the entire secondary beam passed by the collimator (Figure 5.5). For such short wavelengths, the crystal may be a little thicker.

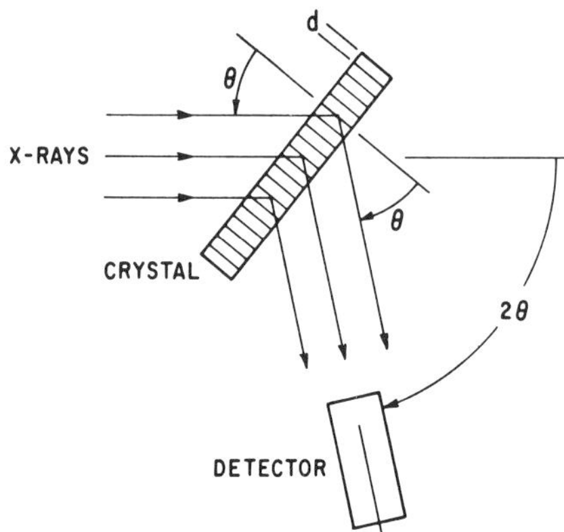

FIG. 5.16. Flat-crystal transmission optical system.

In the Suoninen *reflection–transmission spectrometer*, the same thin, flat LiF crystal is used to measure two wavelengths simultaneously, one by reflection, one by transmission (*S68*).

The use of pinhole or slit apertures to effect monochromatic diffraction from a flat crystal receiving a divergent beam from a point or line source is described in Section 5.2.1 and shown in Figure 5.1. The *two-crystal spectrometers* shown in Figure 5.17 achieve the same result without limiting apertures. In any of the arrangements in Figure 5.17, if the x-rays incident on crystal 1 are divergent, a range of wavelengths is diffracted, depending

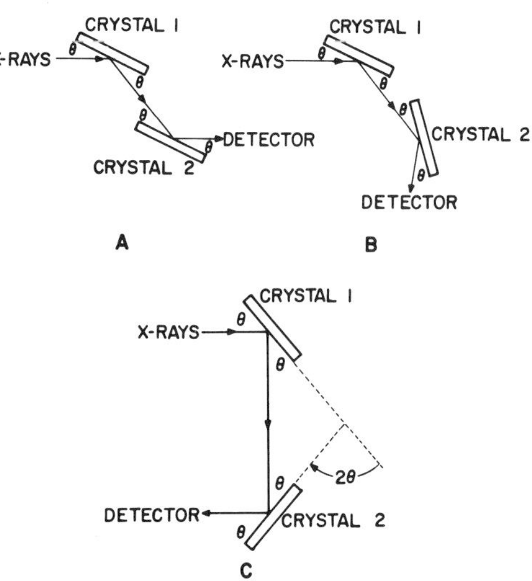

FIG. 5.17. Parallel (A), antiparallel (B), and hinge-movement (C) two-crystal x-ray spectrometers.

on the divergence between source and crystal, as shown in Figure 5.1A. However, of the wavelengths diffracted by crystal 1, only one satisfies Bragg conditions at any one setting of crystal 2. Thus, crystal 2, coupled in θ–2θ ratio with a detector, can disperse the x-rays from crystal 1 in the usual manner. In general, only a relatively small range of wavelengths is diffracted at any one setting of crystal 1, and, to diffract a new range, crystal 1 must be rotated slightly.

Several two-crystal arrangements are possible; three are shown in Figure 5.17. The parallel arrangement of Figure 5.17A has the following features: (1) the two crystals are always parallel; (2) diffraction occurs from opposite faces of the two crystals; (3) both are inclined at the Bragg angle θ to the direction of the central ray of the incident beam; (4) the crystals are coupled and driven to rotate together at the same rate, but in opposite directions; and finally, (5) the incident beam and the emergent, doubly diffracted beam are parallel. In a simpler arrangement, essentially similar to the one just described, crystal 1 is not coupled to crystal 2, but is mounted on an adjustable rotatable stage. At any specified setting of crystal 1, crystal 2 disperses the range of wavelengths diffracted by crystal 1. In the antiparallel arrangement of Figure 5.17B, the crystals are coupled to rotate together at the same rate and in the same direction, and diffraction occurs from the same faces of the two crystals. In all three arrangements (*B62*), crystal 2 not only rotates, but translates with respect to crystal 1, and the detector is coupled in θ–2θ ratio with crystal 2.

Accessories can be made to convert any standard flat-crystal spectrometer into any of the two-crystal arrangements described above. The accessories mount on the detector arm, which moves in 2θ–θ relationship with the existing crystal, which then becomes crystal 1. The accessories hold crystal 2 and provide for movement of the detector about it in 2θ–θ relationship.

A more sophisticated arrangement is shown in Figure 5.17C, in which the two crystals move with respect to one another like the two leaves of a hinge. The detector moves about the second crystal in 2θ–θ relationship as before (*G22*).

Compared with single flat-crystal spectrometers, the two-crystal instrument has higher resolution ($\sim 10\times$) and lower background ($\sim \frac{1}{15}\times$) at the expense of acceptable reduction in spectral-line intensity ($\sim \frac{1}{3}\times$). They are used for accurate measurement of wavelength, plotting of crystal rocking curves, resolution of spectral-line interference, trace analysis (because of the high peak-to-background ratio) (*K19*), and studies of wavelength shift due to chemical effects.

Two-crystal spectrometers are also feasible with flat-transmission, curved-reflection, and curved-transmission crystals. One form of an instrument having two curved-transmission crystals is described briefly in Section 5.6.5. Three-crystal spectrometers have been built in which the first two crystals are operated as in a two-crystal spectrometer, and at each setting of the first two crystals, the third crystal is swept through their narrow transmitted spectrum (*D2*).

The *multicrystal spectrometer* (*D31*) achieves curved-crystal focusing with flat crystals. Fifty small, flat calcite crystals are arranged with their reflecting surfaces arrayed similarly to reflecting planes in the curved crystal of the Cauchois arrangement (Figure 5.18), except that the radius of the array is half the diameter of the focusing circle ($R/2$), rather than R as in the Cauchois instrument.

5.5.3. Curved-Crystal Dispersion Arrangements

5.5.3.1. General (*9, 25a*)

It has already been mentioned that curved crystals do not focus x-rays in the true optical sense, but diffract them in such a way that they converge to a line or point. Consider a flat-crystal (nonfocusing) spectrometer having single- or multiple-slit collimation and set at the Bragg angle for a specified wavelength. Of all the x-rays of that wavelength emitted by each point on the useful specimen surface, only a V-shaped, thin sheet of rays arrives at the crystal at the correct Bragg angle and is diffracted. The point of the V lies at the point on the specimen, its ends lie at the top and bottom of the crystal, and its plane is vertical, that is, parallel to the crystal axis. All other x-rays emitted from that point at the wavelength for which the crystal is set are wasted. Curvature of the crystal permits rays diverging laterally also to arrive at the crystal at the Bragg angle, so that more x-rays are "collected" and diffracted. The curvature also results in convergence of the diffracted rays toward the detector.

The basic curved-crystal dispersion arrangements are shown in Figure 5.18. The first four crystals are cylindrically curved, that is, constitute sections of the surfaces of cylinders of radius R. The doubly curved crystal is torically curved, that is, constitutes a section of the surface of a tore or torus ("donut"); this crystal can have forms analogous to the Johann and Johansson forms of cylindrically curved crystals, that is, simply tangent to the tore or conforming with its surface at all points. Not shown in Figure 5.18 are the logarithmically curved crystal, which is curved to conform to

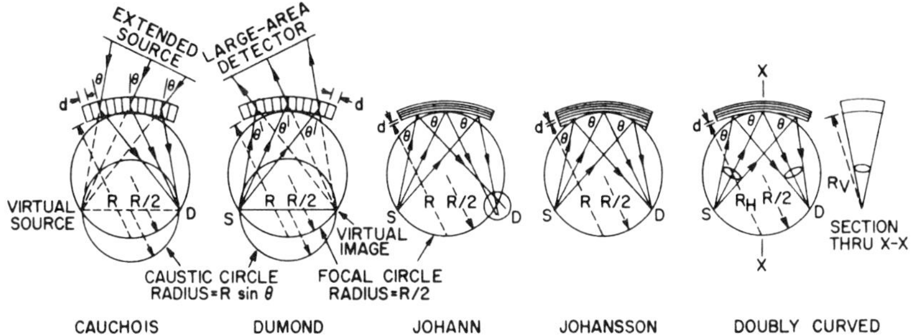

FIG. 5.18. Basic curved-crystal dispersion arrangements. Here, R is the radius of curvature of the crystal.

a logarithmic spiral, and the spherically curved crystal, which constitutes a section of the surface of a sphere of radius R. All these arrangements are described in a little more detail below.

In Figure 5.18 and the remainder of this chapter, the following terminology and symbols are used.

The *focal circle*, or *focusing circle*, is analogous to the Rowland circle in optics. For x-ray focusing, it is necessary that the secondary-emissive x-ray source (real or virtual), its focused image (real or virtual), and the crystal all lie on this common circle having radius $R/2$ (see below). A compromise must often be drawn in selection of the diameter of the focusing circle. If the circle is too small, resolution is impaired; if it is too large, the spectrometer becomes cumbersome.

The *caustic circle* has as its diameter the distance between the source (real or virtual) and its focused image (real or virtual). It has radius $R \sin \theta$.

R is the radius of cylindrical curvature (bending) of the crystal. For the doubly curved crystal, R_H and R_V are radii of curvature in the horizontal and vertical directions, respectively.

S is the optical source of the secondary characteristic x-rays, and may have one of three forms: (1) a small-area particulate or filamentary specimen on the focal circle; (2) a pinhole or slit aperture on the focal circle over an extensive specimen outside the circle; or (3) a virtual source, that is, the point on the focal circle at which the x-rays appear to originate, as shown in the Cauchois arrangement in Figure 5.18.

D is the optical focus of the source, and may have either of two forms: (1) a true line focus of the source at S (or point focus for the doubly curved crystal); or (2) a virtual focus, that is, the point on the focal circle at which the x-rays appear to focus, as shown in the DuMond arrangement in Figure 5.18. For true images, the detector slit (or circular aperture for the doubly curved crystal) is placed on the focal circle at D, behind which, outside the focal circle, lies the detector window.

θ is the Bragg angle, that is, the angle between the incident x-ray and the diffracting crystal plane in either reflection or transmission mode.

d is the interplanar spacing of the diffracting planes in either reflection or transmission mode.

L is the distance from the center of the crystal or from the crystal axis to the source S or detector slit D.

Slits at S and D and filamentary specimens at S are perpendicular to the plane of the figure.

The relationships between the source and focal image for the various curved-crystal arrangements are as follows:

Cauchois: area to line.
DuMond: line or point to area.
Johann: line or point to line.
Johansson: line or point to line.
Logarithmic: line or point to line.
Toric (doubly curved): point to point (but only at one wavelength, point to line at other wavelengths); area or line to line.
Spherical: area, line, or point to line.
Von Hamos: area to area.

5.5.3.2. Transmission

In the Cauchois arrangement (*C33*), with a polychromatic extensive source outside the focal circle and a fixed-radius crystal, each wavelength is diffracted to a different point on the focal circle. Thus, the crystal–detector distance L is different for each wavelength. If L is to be constant, provision must be made to vary the crystal radius R continuously with θ:

$$R = L/\sin\theta \tag{5.8}$$

The DuMond arrangement (*D30*) is the reverse of the Cauchois arrangement. With a polychromatic line or point source on the focal circle and a fixed-radius crystal, only one wavelength is diffracted through the crystal with the source at a given point. Thus, the source–crystal distance L is different for each wavelength. Again, if L is to be constant, provision must be made to vary the crystal radius with θ. A large-area detector or two or three conventional detectors may be required to receive the entire useful transmitted divergent beam. The DuMond arrangement gives substantially greater measured intensity than the Cauchois, and it is sometimes used in electron-probe microanalyzers because of its applicability to point sources.

Both the Cauchois and DuMond methods require crystals that can be cleaved very thin, yet remain strong enough to curve elastically; mica, aluminum, LiF, and NaCl are commonly used. In both arrangements, the diffracting planes are perpendicular to the crystal surface. A transmission crystal intercepts a greater solid angle of x-radiation than a reflecting crystal by a factor $(\cos\theta)/(\sin\theta)$. However, some of the advantage is lost by attenuation in the crystal, even at optimal thickness, which is that thickness that reduces the transmitted intensity to $1/e$ (1/2.72) of its incident value. Two or three crystals of different thickness are required for optimal performance over the full wavelength range. The crystals are usually cemented to thin spring-metal frames. Another advantage of the transmission arrangements is that measurements can be made at 2θ angles approaching 0°, where use of reflecting crystals is precluded by mechanical limitations (Section 5.4.2.1, Figure 5.5). Both transmission arrangements give excellent spectral resolution.

Cauchois has described a universal curved-crystal transmission–reflection spectrometer (*C34*) in which the secondary-emission specimen can be made to occupy the Cauchois or Johann position (see below).

5.5.3.3. Reflection

In the Johann arrangement (*J18*), the crystal is simply cylindrically curved to radius R equal to the diameter of the focal circle. The crystal–source and crystal–detector distances L are equal,

$$L = n\lambda R/2d = R\sin\theta \tag{5.9}$$

where n, λ, and d are diffraction order, wavelength, and crystal interplanar spacing, respectively.

In semifocusing arrangements, L and R are constant at all values of θ, but for optimal performance, either R is kept constant and L varies with θ, or L is kept constant and R varies with θ. Since L increases as θ increases, it may become inconveniently large at large angles for a crystal of given radius. If L is to have convenient length at all angles, two or three crystals having different radii may be required.

A Johann crystal gives optimal focusing only at its vertical axis, where it is tangent to the focusing circle, so the focusing is not sharp, as shown in exaggerated scale at D in Figure 5.18. The defect is most serious at very small 2θ angles, but decreases as 2θ increases and becomes insignificant at $2\theta > 90°$. The Johann arrangement is useful for crystals that are not readily ground for the Johansson arrangement (see below) and for spectrometers in which the crystal radius is varied continuously with θ.

"Perfect" x-ray focusing free of the defect noted above requires that two conditions be met which, on first consideration, appear to be mutually exclusive [p. 219 in *(27)*]: (1) The *Rowland condition* requires that the x-ray point or line source, crystal, and focused image all lie on a circle of radius $R/2$; to meet this requirement, the crystal must have radius R. (2) The *Bragg condition* requires that the crystal must lie tangent to the above Rowland circle at all points; to meet this requirement, the crystal must have radius $R/2$. These conditions are indeed mutually exclusive for optical gratings, which are ruled on the surface. However, with three-dimensional gratings, which, in effect, is what crystals are, the conditions are readily met, at least for physically suitable crystals.

The Johansson arrangement (*J19*) is substantially the same as the Johann, except that the focusing defect is corrected by bringing the inner surface of the crystal into contact with the focal circle. This is done by cylindrically curving the crystal to radius R, then grinding it to radius $R/2$, as shown in Figure 5.18. In effect, the crystal *planes* are curved to radius R, and the crystal *surface* is ground to radius $R/2$. Actually, it makes no difference whether the crystal is prepared this way or ground to R and then curved to $R/2$. The method is not applicable to crystals that cannot be ground, nor to spectrometers in which the crystal radius is varied continuously with θ. Equation (5.9) applies to the Johansson arrangement, as does the requirement for two or three crystals of different radii to keep L of convenient length at all angles.

In both the Johann and Johansson arrangements, the x-rays are diffracted to a line focus, so only a simple scatter slit is required at the detector.

In the doubly (torically) curved arrangement (*B10*, *H3*), the crystal is first cylindrically curved to radius R_H, the diameter of the focal circle. It is then curved to conform to the surface formed by rotation of the focal circle about the chord joining S and D, or to R_V. The surface of the resultant doubly curved crystal is a portion of a tore. Doubly curved crystals focus point sources to point images, but any one crystal gives point-to-point focusing for only one wavelength; at other wavelengths, the second (vertical) focus degenerates to a line. At its correct wavelength, a doubly curved crystal may give diffracted intensity up to 16 times that from a cylindrically curved crystal. Doubly curved crystals are applied mostly for x-ray diffraction, where monochromatic x-rays are commonly used.

The spherically curved crystal, unlike the toric crystal, does not focus a point source to a point image, and it focuses area, line, and point sources to line images inferior to those from toric crystals. However, the spherical crystal is easier to curve and grind and, unlike the toric crystal, can have continuously variable curvature (Section 5.6.6).

Recently, the logarithmically curved crystal was introduced. A horizontal section of such a crystal is a segment of a logarithmic spiral, whereas a horizontal section of a cylindrically curved crystal is an arc of a circle. The crystals are cemented to logarithmically curved metal blanks. The logarithmic form does not require grinding and so, unlike the Johansson form, is applicable to any analyzer, including mica, soft organic crystals (such as PET), and multilayer soap films. The focusing is better than that of a Johann crystal, but not as good as that of a Johansson. However, the focusing defect is easily corrected by use of a somewhat wider detector slit. Logarithmically curved crystals are used in the Siemens MRS-3 multichannel x-ray spectrometer. Three basic monochromator units having equal source slit-to-crystal and crystal-to-detector slit distances of 8, 13, and 20 cm cover the 2θ regions 13–49°, 43–77°, and 75–145°, respectively. Each channel in the multichannel spectrometer is fitted at the factory with the appropriate basic monochromator having the optimal crystal—LiF(200), PET, KHP, etc.—correctly logarithmically curved for the specific x-ray line to be measured by that channel.

5.5.3.4. Von Hamos Image Spectrograph

Von Hamos describes an *x-ray image spectrograph* or *x-ray emission micrograph* (*V14–16*) which permits photography of the area distribution of chemical elements over a flat specimen surface. The method would also be useful for mapping variations in composition and/or thickness of thin films and platings on flat substrates. The instrumental arrangement is shown in Figure 5.19. The specimen is irradiated by the primary beam from the x-ray tube, and each chemical element in the specimen surface is excited to emit its characteristic x-ray spectrum, just as in a conventional x-ray spectrometer. For each secondary wavelength λ_1, λ_2, λ_3,... emitted by the specimen, the intensity distribution over the specimen surface corresponds to the concentration distribution of the element producing it. In thin films and platings, the intensity distribution would also correspond to the point-to-point thickness. A cylindrically curved crystal is arranged to form a series

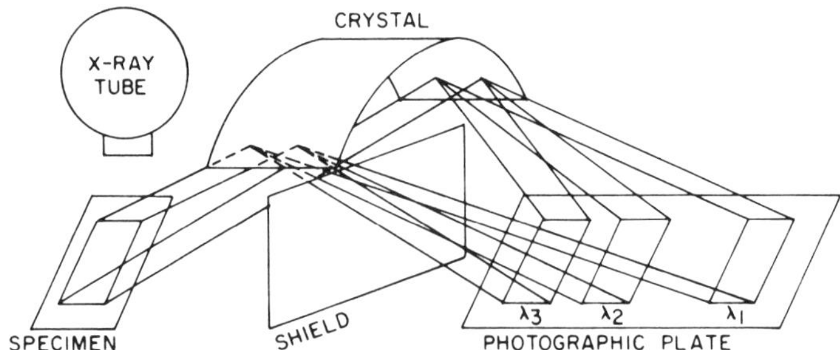

FIG. 5.19. Von Hamos image spectrograph.

of separate true monochromatic images of the specimen surface on a plane fine-grain photographic plate. Each image shows the intensity distribution of a certain spectral line over the specimen surface—and thereby the concentration or thickness distribution of the element emitting that line. Figure 5.20A shows a conventional light photograph of the surface of a copper-bearing mineral, and Figure 5.20B shows the corresponding Cu $K\alpha$ image spectrum.

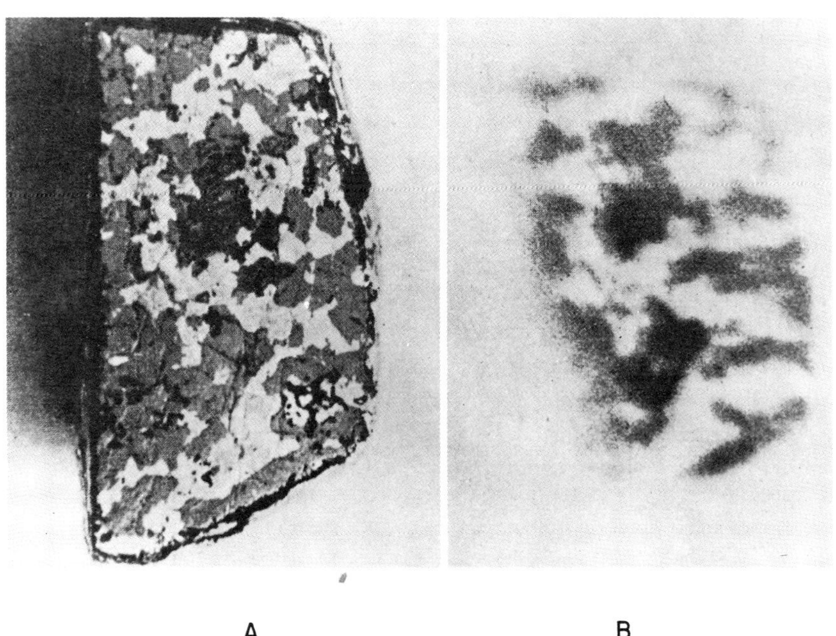

FIG. 5.20. (A) Light photograph of a polished plane surface of a copper-bearing mineral and (B) Cu $K\alpha$ image of that surface. [H. Von Hamos, *Transactions of the Royal Institute of Technology* (Stockholm); courtesy of the author and Institute.]

Estimations of surface concentrations or film thickness are made by comparing the image spectrum of the sample with spectra of standards having known composition or thickness. Determinations of surface concentrations or film thickness are made by measuring photographic densities of the image spectra of samples and standards with a microdensitometer. A photographic emulsion calibration curve is prepared to permit conversion of densities to intensities.

The useful specimen area may be as much as $\sim$1 cm^2 and the spatial resolution $\leq$0.1 mm. An idea of the sensitivity may be derived from the following datum: 0.02 μg Co in 0.01 mm^2 gave a useful photographic density (0.12) in a 1-h exposure.

5.6. CURVED-CRYSTAL SPECTROMETERS

The crystal-dispersion arrangement is combined with the x-ray tube, specimen-presentation system, detector, and drive mechanism to form the complete x-ray spectrometer, which may take any of several forms. Crystal spectrometers may be classified as being of flat-crystal or curved-crystal type. All flat-crystal instruments are nonfocusing (unless one considers parafocusing, Section 5.5.2.1). Curved-crystal instruments may be semifocusing or full-focusing. The term *focusing* is used in the sense given in paragraph 1 of Section 5.5. This section describes the principal types of curved-crystal spectrometers in common use.

5.6.1. Semifocusing Spectrometer

In semifocusing spectrometers, the flat crystal of the nonfocusing arrangement is simply replaced with a fixed-radius curved crystal, and the Soller collimators are removed. If the specimens have the form of small particles or filaments, no further modification may be required. Otherwise, simple slit or pinhole apertures are provided at the specimen and detector window. The crystal radius and axis are fixed. The crystal–source and crystal–detector distances are constant and should be equal, although this may not be true for a converted commercial flat-crystal instrument. Figure 5.21 can represent a semifocusing spectrometer at two wavelength settings if the crystal radius is assumed to remain the same as the diameter of the focal circle changes with θ.

The crystal and coupled detector arm rotate about the crystal axis at rates proportional to θ and 2θ, respectively, as in a conventional spectrom-

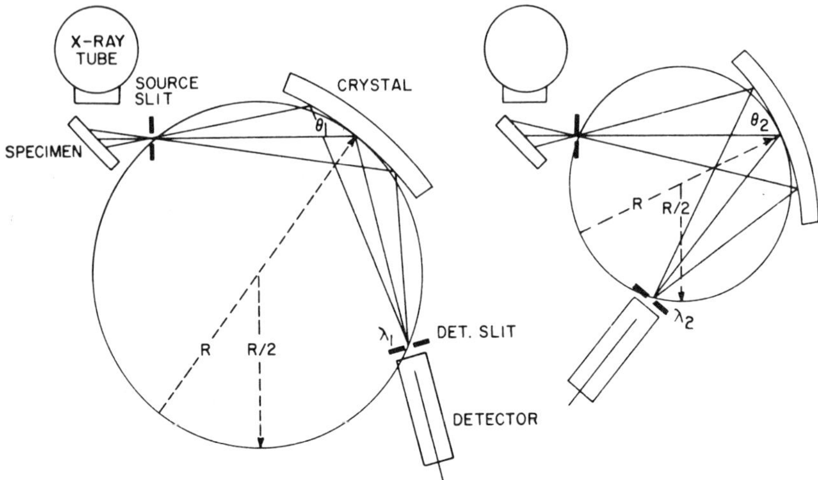

FIG. 5.21. Reflection curved-crystal x-ray spectrometer having continuously variable crystal radius, shown at two wavelength settings $\lambda_1 < \lambda_2$. The figure may also represent a semifocusing spectrometer if the crystal radius R is assumed to be the same at both settings.

eter. As the crystal rotates, the tangent to its center presents a series of Bragg angles to the incident secondary beam. The idea is that even though the crystal may not have the optimal radius, at a given Bragg angle, it collects and diffracts more of the x-rays having wavelength corresponding to that angle than a flat crystal does.

The fixed-radius crystal is likely to have the correct radius for the specimen–crystal–detector distances at a certain wavelength. As the instrument is set at wavelengths progressively farther from that wavelength, the sharpness of the focusing progressively deteriorates. Nevertheless, if the crystal radius is optimal near the center of the 2θ range, and if the spectrum is not crowded, one crystal can be used for the entire 2θ range. Otherwise, two or three crystals having different radii may be used to obtain good focusing in the low-, medium-, and high-2θ regions. Such a set of two or three curved LiF crystals of different radii and some other curved crystal for the longer wavelength region may be mounted on a three- or four-crystal changer.

In full-focusing spectrometers, provision is made to vary the crystal radius or the specimen–crystal–detector distances continuously as a function of θ to maintain optimal focusing geometry over the full 2θ range. The following sections describe several full-focusing curved-crystal spectrometers.

5.6.2. Continuously Variable Crystal Radius

An instrument that maintains optimal focusing by continuous variation of the crystal radius with θ is shown in Figure 5.21 (*E15*). The specimen aperture and crystal axis are fixed. The distances L from the crystal axis to the specimen and detector apertures are equal and constant. The crystal and detector rotate about the crystal axis at rates proportional to θ and 2θ, respectively, as in a conventional spectrometer. Provision is made to vary the crystal radius R as a function of θ:

$$R = \frac{L}{n\lambda/2d} = \frac{L}{\sin\theta} \tag{5.10}$$

The takeoff angle, the angle between the central ray of the secondary beam and the specimen plane, is constant.

The arrangement is readily adaptable to standard spectrometers because of the constant L and fixed crystal axis. It is readily applicable to selected-area work because of the constant takeoff angle. One crystal serves for the full 2θ range, but some defocusing is unavoidable. Because R varies with θ, this type of instrument is applicable only to Johann-type crystals (curved only), and only relatively few crystals can be elastically curved repeatedly in this manner, for example, LiF, NaCl, mica, aluminum.

Parker has built an x-ray spectrometer having a spherically curved crystal with continuously variable radius (Section 5.6.6).

5.6.3. Naval Research Laboratory Design

The essential features of this instrument are shown at two wavelength settings in Figure 5.22 (*B55*, *B63*). The specimen-presentation system may have any of three forms: (1) an extensive specimen lying outside the focal circle, excited by an x-ray tube in the position indicated by the solid outline, and having a pinhole or slit aperture (not shown) on the focal circle; (2) an extensive specimen lying on the focal circle, but receiving primary irradiation through an aperture from an x-ray tube in the alternate position indicated by the dashed outline; and (3) a small-area particulate or filamentary specimen lying on the focal circle and irradiated in the same way as (2) above, or even without the aperture.

The crystal radius R is constant. The diameter of the focal circle is constant, and its center is fixed and serves as the axis of the goniometer drive mechanism. Provision is made to vary the crystal–source and crystal–

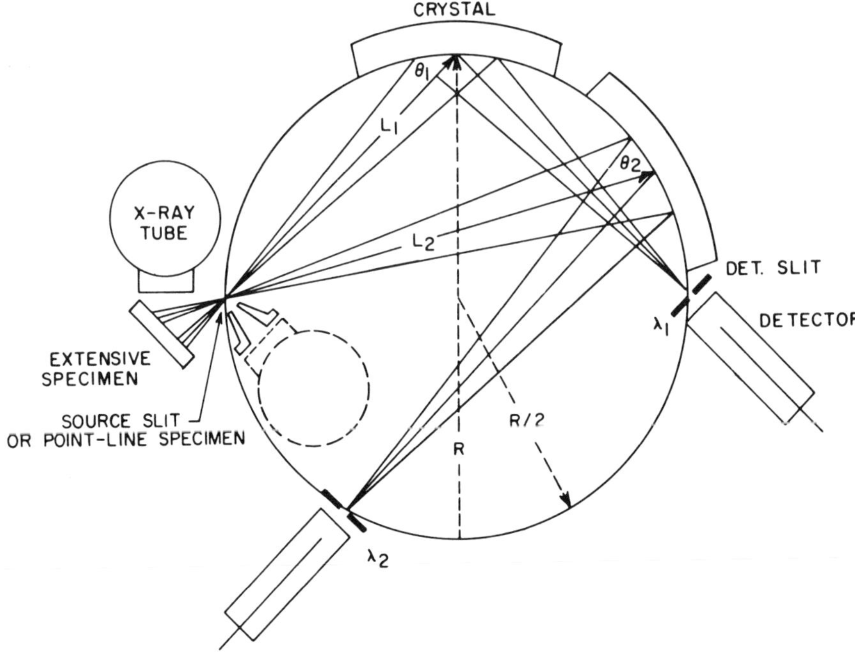

FIG. 5.22. Reflection curved-crystal x-ray spectrometer of the Naval Research Laboratory design, shown at two wavelength settings $\lambda_1 < \lambda_2$. Also shown are two alternative positions of the x-ray tube.

detector distances L in accordance with Equation (5.9). The crystal and detector are mounted on separate arms of equal length pivoted on concentric shafts at the center of the focal circle and rotated at rates proportional to θ and 2θ, respectively. Provision must also be made to keep the detector window directed toward the center of the crystal as θ varies, rather than tangent to the focal circle. The instrument drive mechanism is very simple and compact. Since $L = R \sin \theta$, if L is to remain of convenient length as θ varies from 0 to 90°, two or three crystals having different radii must be used. Since R is constant at all values of θ, this type of instrument is applicable to both Johann and Johansson crystals.

For extensive specimens lying outside the focal circle, the takeoff angle varies with θ as shown, and homogeneous specimens are required for semiquantitative and quantitative analysis. However, the instrument is also applicable to small-area specimens and to small selected areas on extensive specimens lying on the circle and irradiated through an aperture.

5.6.4. Applied Research Laboratories Design

The essential features of this instrument are shown at three wavelength settings in Figure 5.23 (*K9*). The crystal radius is constant. Provision is made to effect the following movements as a function of θ: (1) The crystal moves in a straight line away from the source aperture, thereby increasing the crystal–source distance in accordance with Equation (5.9). (2) The detector moves in a path constituting one loop (shown) of an analytic geometric curve known as an *epicycloid* (*rosette* or *four-leaf rose*) and having equation

$$r = R \sin 2\theta \tag{5.11}$$

where r is the direct distance between the source and detector aperture, as shown in the figure. (3) Finally, the detector window is continuously directed toward the center of the crystal. As a result of these movements, the center of the focal circle moves with θ in a curved path, as shown in the

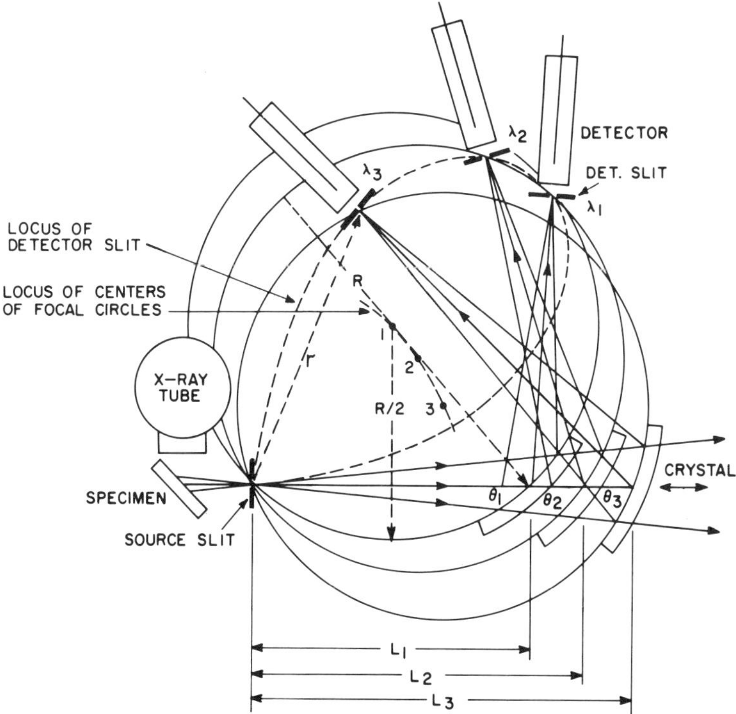

FIG. 5.23. Reflection curved-crystal x-ray spectrometer of the Applied Research Laboratories design, shown at three wavelength settings $\lambda_1 < \lambda_2 < \lambda_3$.

figure. In a more compact version of this arrangement, the detector swings about the crystal at a fixed distance, rather than about the source slit on a loop, as shown in Figure 5.23. In this arrangement, the slit aperture must be wider because the crystal-to-detector distance is shorter and the diffracted beam will not have completely converged to a line image by the time it reaches the detector. The wider detector slit increases the background intensity, which is undesirable, but can be tolerated in some applications. The takeoff angle is constant, so the instrument is applicable to selected-area work. Just as in the NRL instrument, crystals of both Johann and Johansson type are applicable, and if L is to have convenient length at all values of θ, two or three crystals of different radii are required. ARL spectrometers are used in the ARL line of automatic simultaneous instruments (Chapter 9). This type of instrument is also widely used in electron-probe microanalyzers (Chapter 21).

5.6.5. Cauchois Spectrometer

A Cauchois spectrometer (*C33*) is shown at two wavelength settings in Figure 5.24. The crystal axis is fixed, and the distance L from the crystal axis to the detector aperture is constant. A selected wavelength from an extensive source outside the focal circle is focused to a line at the detector slit on the focal circle. If L is to be constant with θ, provision must be made

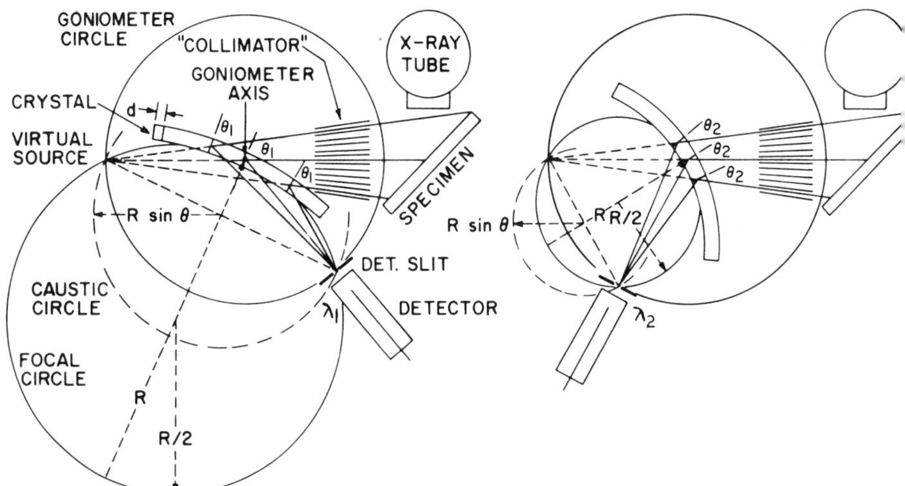

FIG. 5.24. Transmission curved-crystal x-ray spectrometer of the Cauchois type, shown at two wavelength settings $\lambda_1 < \lambda_2$.

to vary R in accordance with Equation (5.10). Although the crystal axis is fixed, the center of the focal circle moves with θ as shown. The convergent multiple-foil "collimator" shown excludes directly transmitted wavelengths other than that for which the instrument is set and prevents their reaching the detector. This type of collimator is known as a *focusing* or *focused* collimator. The foils are arrayed so that converging, rather than parallel, rays are passed. The original General Electric model XRD-3/SPG-1 spectrogoniometer was of the Cauchois type. The fixed crystal axis, constant crystal–detector distance, and specimen-presentation system permitted easy conversion to flat-crystal Soller optics.

A two-curved-crystal transmission spectrometer has been built consisting of two Cauchois-type spectrometers having radii in 2:1 ratio in series, the smaller-radius crystal having its focal point on the Rowland circle of the larger-radius crystal (*D2*).

5.6.6. Spherically Curved-Crystal Spectrometers

In Figures 5.22 and 5.23, it is evident that the cylindrically curved crystal intercepts a constant angular divergence of the x-ray beam in the plane of the figure. Consequently, in this plane, increasing the path length L affects intensity only through increasing absorption in the path. However, in the plane perpendicular to the figure, the fraction of the diffracted radiation that passes through a slit of finite height varies inversely with L. In order to collect this vertically divergent radiation, a doubly curved, point-focusing crystal is required. Ideally, the vertical radius of curvature of such a crystal is equal to the perpendicular distance from the center of the crystal to the chord of the focusing circle joining the source and detector slits (Figure 5.18) and increases with L (Figures 5.22 and 5.23).

It would be extremely difficult to continuously vary two mutually perpendicular crystal radii. A practical compromise is a spherically curved crystal that is first curved cylindrically to radius R (the diameter of the focusing circle), then ground so that its reflecting surface is a concave section of a sphere of radius $R/2$ (the radius of the focusing circle). Such a crystal partially compensates for vertical divergence and increasingly approximates true point focusing as the Bragg angle increases.

A spectrometer using such spherically curved crystals has been described in detail and evaluated (*E12*). The technique of preparation of the crystals is also given.

Parker (*P8*) has built an x-ray spectrometer having a spherically curved crystal with continuously variable radius. A thin single-crystal disk capable of elastic deformation and having diffracting planes parallel ot its surface is laid on an O-ring on the end of a cylindrical vacuum chuck. The crystal is spherically curved by application of a partial vacuum to its bottom side. As the crystal is rotated in θ, the degree of vacuum and thereby the radius of curvature are varied continuously and automatically by means of an electronic partial-vacuum regulator (*P7*). The θ–2θ drive and vacuum regulator are coupled, so the instrument may be set at a specific Bragg angle or scanned continuously. The required differential pressure (between the diffracting and bottom sides of the crystal) is a substantially linear function of θ, and as θ varies from 0 to 60°, the differential pressure varies from 0 to 600 torrs. For work at longer wavelengths, the curvature may be effected by applying the controlled partial vacuum to the spectrometer chamber and a high vacuum to the bottom side of the crystal.

Although any crystal can be used that can be cleaved or cut into thin plates capable of elastic deformation, Parker used only Si(111) wafers 25–35 mm in diameter and ~0.25 mm thick, of the type widely used in the semiconductor electronics industry.

The instrument provides three-dimensional focusing of x-rays from a point source on the focusing circle—or from an extensive source between the crystal and a virtual point source on the circle—to a line image on the circle. A slit is placed at the line image, and behind the slit, the detector. Because of the variable radius, the source–crystal and crystal–image distances are equal and constant for all values of θ, so the θ–2θ drive mechanism of a conventional flat-crystal spectrometer can be used. On Parker's instrument, this distance is 22 cm.

The potential applications of the instrument include the following: (1) for trace analysis, where the high peak-to-background ratio at low intensities is advantageous; (2) with point-source emitters, such as electron-probe microanalyzers; and (3) as a high-intensity, linear, monochromatic x-ray source, when used with high-power, sealed-off x-ray tubes.

5.7. PHOTOGRAPHIC X-RAY SPECTROGRAPHS

X-ray spectrographs used to record x-ray line spectra on photographic film must provide some means of varying the angle between the analyzing crystal and the x-rays incident upon it to be dispersed. There are four basic ways to do this:

1. The x-rays may originate in a point or line source so that they diverge and arrive at the fixed crystal over an angular range. Certain rays

of each wavelength present arrive at the crystal at the correct Bragg angle to be diffracted. This arrangement is described in some detail in Section 5.2.1.

2. The crystal (or specular reflector) may be rotated to diffract each wavelength in turn to a different point on the film. The Bragg spectrograph described in Section 5.2.1 uses this method.

3. The specimen may have a relatively large area so that for each wavelength emitted by the specimen, certain rays arrive at the fixed crystal at the correct Bragg angle to be diffracted. The edge-crystal spectrograph (Section 5.5.2.2) and the Von Hamos image spectrograph (Section 5.5.3.4) are in this class. Figures 5.25 and 5.26 show curved-crystal transmission and reflection spectrographs that operate on the same principle (*K23*, *P30*). It is evident that in all such instruments, each recorded wavelength originates at a different area on the specimen surface, so homogeneous specimens are required for quantitative analysis.

The curved-crystal spectrographs shown in Figure 5.25 use curved films to effect sharp focus at all diffracted wavelengths. By sacrificing this sharp focusing, a particularly small, rugged, convenient, and inexpensive spectrograph camera can be made. Such an instrument is shown in Figure 5.26 (*K23*), and is somewhat similar to the edge-crystal spectrograph described in Section 5.5.2.2. The instrument has no moving parts and can be

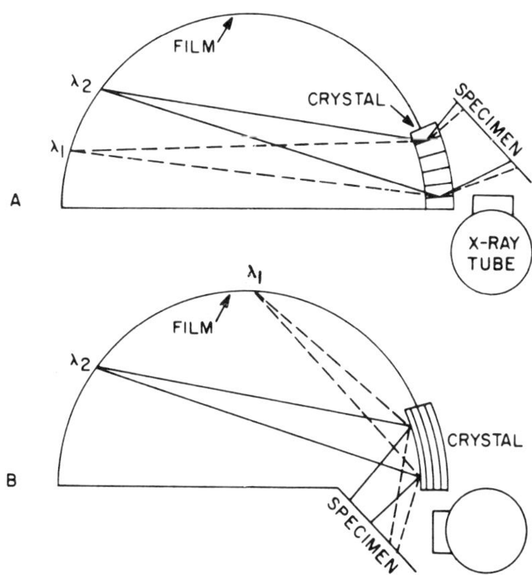

FIG. 5.25. Curved-crystal (A) transmission and (B) reflection x-ray spectrographs having photographic film registration. In both figures, $\lambda_1 < \lambda_2$.

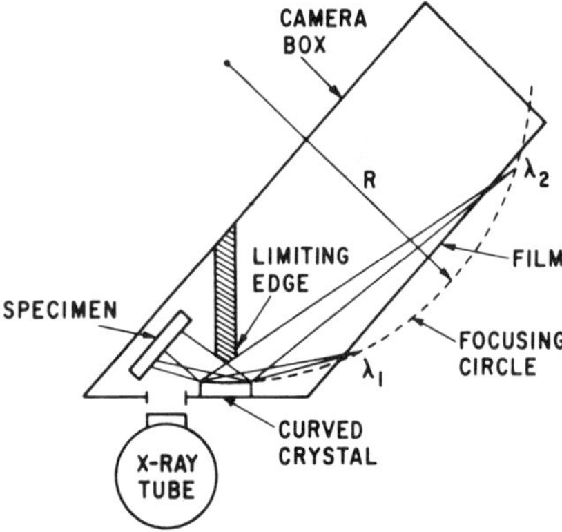

FIG. 5.26. Curved-crystal x-ray spectrograph camera (*K*23).

mounted on either an x-ray diffraction or spectrometer unit. Close coupling between the x-ray tube, specimen, and curved crystal combine with the crystal focusing to give high spectral sensitivity. Bulk or microgram specimens can be used. Several spectral regions can be recorded side-by-side simultaneously from different analyzing crystals, or spectra can be recorded side-by-side simultaneously from several small specimens. The flat-film casette is placed on a chord of the focal circle, so the recorded spectral lines are slightly defocused except at the two points where the chord intersects the focusing circle, as shown in Figure 5.26. However, this defocusing is rarely troublesome. Kodak No-Screen x-ray film or a Polaroid-Land Type 57-1 casette having a phosphor intensifier screen and using ASA-3000 Polaroid film can be used. Exposure times are 1–60 min, depending mostly on the amount of specimen.

4. The crystal itself may be distorted in such a way as to present a varying range of angles to the incident beam to be dispersed. Three examples follow.

In the *twisted-crystal spectrograph* (*F21*), the crystal is held at its top and bottom edges and twisted about its vertical axis by several degrees to form a slightly spiral diffracting surface having a continuous range of radius of curvature from top to bottom. The crystal is maintained in this twisted state. The diffracted wavelength range $\Delta\lambda$ depends on the initial angle θ between the incident x-rays and diffracting planes, and, if the crystal is twisted by angle $\Delta\theta$, is given by

$$\Delta\lambda = 2d \cos\theta \, \Delta\theta \tag{5.12}$$

Silicon and germanium crystals twisted 4–6° permit simultaneous recording of a spectral range 1.25–1.75 Å.

In the *convex-crystal spectrograph* [pp. 32–35 in (*8*)] (*B51*), a similar effect is realized by use of a convex-curved crystal. A small square of LiF 0.125 mm thick is convexly curved to a 3-mm radius. Each wavelength in a pinhole-collimated x-ray beam finds an increment of area on the convex surface that presents the correct angle to satisfy the Bragg law for that particular wavelength. Thus a complete spectrum can be recorded simultaneously on a semicylindrical film.

A third novel concept in crystal dispersion of x-rays is based on the *inverse piezoelectric effect*. This effect is the phenomenon, shown by certain asymmetrical crystals, whereby application of an electric potential in a certain crystallographic direction results in mechanical deformation of the crystal with consequent change in interplanar spacings. The change in d spacing is proportional to the applied potential. Thus, in a spectrometer having a fixed crystal and photographic plate (or detector), the wavelength dispersion can be effected by variation of an electric potential applied to the crystal. Such an arrangement (*N18*) may be useful for investigation of fine structure over a small 2θ range.

The second basic method given above (rotated crystal) and the piezoelectric technique are "sequential" methods in that the x-ray wavelengths are diffracted in turn. The first basic method (divergent beam), third basic method (edge-crystal, Von Hamos, etc.), and the twisted- and convex-crystal techniques are "simultaneous" methods in that all wavelengths are diffracted at once. The simultaneous methods are of particular value for recording x-ray spectra from short-lived or time-variable phenomena, such as spark spectra and spectra from plasmas, solar flares, etc.

Chapter 6

Detection

6.1. INTRODUCTION

The fourth phase of the x-ray spectrometric analysis (Section 3.2.5) is the detection and measurement of the characteristic secondary x-rays. Detection is the subject of this chapter; measurement is considered in Chapter 7. The detector transduces the x-rays into an electrical, luminous, or other analog that can be measured or otherwise used. Most commonly used x-ray detectors are based on the indirect effect of x-rays on the electronic structure of matter. X-rays having photon energy <1 MeV interact with matter by one of three processes: photoelectric absorption and coherent and incoherent (Rayleigh and Compton) scatter. At energies <50 keV, the initial interaction is usually photoelectric absorption, each x-ray photon producing a photoelectron having kinetic energy equal to the incident photon energy minus the binding energy of the electron. These photoelectrons expend their energy in various ways depending on their environment: In *photographic emulsion*, they render grains of silver halide developable—that is, labile to chemical reduction to silver—by an electronic process not fully understood. In *fluorescent screens* and *scintillation phosphors*, they elevate valence-band electrons to higher levels; when these electrons return to their unexcited state, the energy is reemitted in the visible and ultraviolet regions. In *gas-filled detectors*, they ionize gas atoms, producing ion–electron pairs. In *solid-state semiconductor detectors*, they elevate valence-band electrons to the conduction band, creating electron–hole pairs.

Detection may take place after crystal dispersion, or the undispersed secondary beam may be detected and electronic dispersion applied to the detector output. In the latter case, the detector itself plays an essential part in the electronic dispersion. Electronic detectors may be operated as *analog* or *continuous-current detectors*, producing a continuous output current proportional to the incident x-ray intensity, or as *digital* or *photon counters*, producing a discrete pulse of current for each detected photon. The advantages of the photon-counter mode include the following: (1) feasibility of digital data processing and display; (2) feasibility of pulse-height selection; (3) improved sensitivity to low incident intensities; (4) reduced sensitivity to instrument drift; (5) reduced difficulty from dark current; and (6) improved signal-to-noise ratio. The detector readout system may be *instantaneous*, indicating the instantaneous intensity directly, or *integrating* (or *accumulative*), accumulating the electrical analog of the x-rays for a selected time. Instantaneous and integrating detector–readout systems are functionally analogous to ammeters and coulombmeters or to speedometers and odometers, respectively. The respective readout components are the ratemeter and scaler. The common x-ray detectors—Geiger, proportional, and scintillation—may be operated in either mode by use with a ratemeter or scaler.

In analytical x-ray spectrometry, photographic film is sometimes used, but the detector is usually an electronic device—a gas-filled Geiger or proportional counter, a scintillation counter, or a semiconductor detector—in which the x-ray photons are converted to pulses of electric current. These detectors are discussed in Sections 6.2–6.4. Some less commonly used detectors are described in Section 6.6. Readout components are considered in Chapter 7. The subject of x-ray detection is reviewed in (*25d*, *K14*, *P10*, *P12*).

6.2. GAS-FILLED DETECTORS

6.2.1. Structure

6.2.1.1. Components, Classifications

The essential features of the gas-filled x-ray detector are shown in Figure 6.1. The active components are the cylindrical cathode, axial wire anode, and gas filling. In operation, the cathode and anode are operated at negative and positive potential, respectively, by an external dc power supply, as shown. Physically, the detector consists of a cylindrical metal envelope, which: (1) usually serves as the cathode, but, if not, houses,

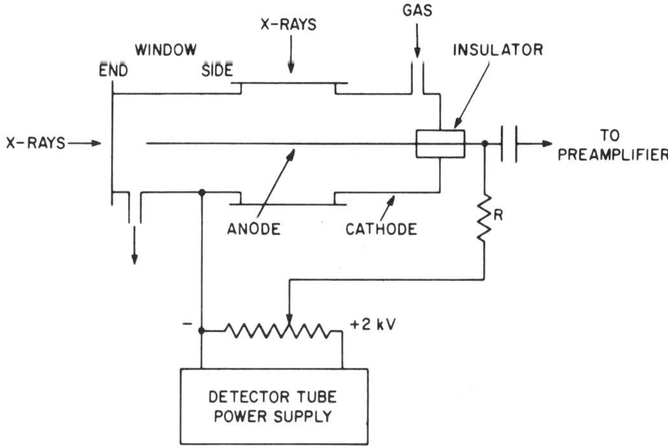

FIG. 6.1. Structure of the gas-filled x-ray detector. The tube is shown to have both end and side windows for illustration only.

supports, and provides for external connection to an enclosed cylindrical cathode; (2) supports the insulator, which, in turn, supports and provides for external connection to the thin [0.25-mm (0.010-inch) or less] wire anode; (3) retains the gas filling and excludes air; (4) provides windows for entrance and, sometimes, exit of the x-rays; and (5) may provide an inlet and outlet for continuous flow of detector gas.

Gas-filled detectors are classified in terms of at least six criteria: (1) by mode of operation—ionization chamber, proportional counter, or Geiger counter; (2) as sealed or gas-flow (*H32*) counters; (3) by window position —end or side; (4) by window thickness or material; (5) by kind of gas filling; and (6) by certain special features. The mode of operation is determined essentially by the operating potential and is discussed in detail in Section 6.2.2. The other criteria are considered in the next two sections.

6.2.1.2. Windows

Sealed detectors have sealed-on gas-tight beryllium, mica, or Lindemann-glass windows that retain the gas filling and exclude air more or less permanently. Gas-flow detectors have thin [$\leq$6-μm ($\leq$0.00025-inch)] windows of organic polymer film to permit measurement of x-rays of long wavelength. Such windows cannot be permanently sealed and are not impervious to air and the detector gas. Consequently, the gas must be passed through the detector continuously while it is in use (Section 6.2.1.3). Some ionization-chamber detectors cannot be classified as either the sealed or gas-flow type. The thin window simply isolates a volume of air at atmospheric pressure.

The detector may have a single window in the end or side, or two diametrically opposite side windows, as shown in Figure 6.1. However, the detector never has both end and side windows, as shown in the figure for illustration only. The two-window arrangement serves two purposes. Unabsorbed x-rays that might excite secondary x-ray or photoelectron emission from the back wall of the detector are allowed to emerge from the tube. When a thin-window flow counter is used to detect long-wavelength x-rays, it may be backed by a second detector more suitable for short-wavelength x-rays. The disadvantage of the side window is that the x-ray path in the active detector volume is shorter, and a gas filling having higher absorption coefficient is required for adequate absorption at short wavelengths.

When the detector is to be operated as a proportional counter (*L34*) and its output is to be subjected to pulse-height analysis, uniformity of the electric field within the active detector volume is essential to: (1) improve the proportionality of the pulse height to the photon energy; (2) reduce the width of the pulse-height distribution; and (3) reduce the slope of the plateau. (All these features are described below.) Two structural features are essential to the realization of this uniformity: use of the side-window configuration and provision of a conductive coating on the window. The side-window configuration has a more uniform electric field because the active volume does not include the dead space (end effect) and distorted electric field (clamping effect) at the ends of the anode wire. A two-window detector also permits emergence of unabsorbed x-rays that might otherwise excite secondary x-ray and photoelectron emission from the rear wall. The end and clamping effects may also be compensated by external electronic circuits (*C52*, *R15*). If the detector is to be used with pulse-height analysis, the inside surface of the window should be vacuum-coated with ~200 Å of aluminum, beryllium, or carbon to give a more uniform electric field than is possible with an insulating window. When coated thin-film windows are mounted, the coated side is identified by use of an ohmmeter. Of course, the coating, especially aluminum, reduces the window transmission at the long wavelengths. If pulse-height analysis is not to be used, end and side, and coated and uncoated windows perform about equally satisfactorily.

Some of the disadvantages of the end-window proportional counter when used for pulse-height analysis are minimized in the variable-geometry and point-anode proportional counters (Section 6.5.3).

Sealed windows are usually of 0.5–1-mm beryllium, 12–15-μm mica, or thin Lindemann glass. Materials for flow-counter windows include 3- or

6-μm (0.00012- or 0.00025-inch) Mylar, 25-μm (0.001-inch) polypropylene or polyethylene stretched to 1–6 μm (0.00005–0.00025-inch), and 500–5000-Å Formvar. Windows 2–6 μm thick are regarded as "thin," those $\lesssim$1 μm thick as "ultrathin." Sometimes two ultrathin films are used as a double layer to cover pinholes in the individual layers. Ultrathin films are often supported on some kind of grid, such as 200-mesh nickel or nylon having 75% open area.

Ultrathin collodion (nitrocellulose) films (*H35*) are prepared by placing a drop of a 2–10% (wt/vol) solution of collodion in amyl or butyl acetate on a water surface, allowing it to spread and evaporate, and scooping up the film on a wire hoop. Much stronger Formvar (polyvinyl formaldehyde) films are prepared the same way from 2–10% (wt/vol) solutions of Formvar in ethylene dichloride or dioxane (*M30*). Alternatively, the Formvar films can be prepared by dipping an extremely clean microscope slide in the solution, allowing the solution to drain and evaporate, scoring the film around the edge of the slide, then immersing the slide gently and obliquely into water; the film floats off and is scooped up as before. Detector windows made of these nitrocellulose or Formvar films may have very limited life—perhaps as little as a month.

Polypropylene is stretched as follows (*C31*). A sheet of 25-μm (0.001-inch) polypropylene film is stretched over the top of a cylindrical vacuum chamber ~30 cm in diameter and secured between two flanges fitted with *O*-rings. The chamber is evacuated by a mechanical vacuum pump (~0.1 torr), whereupon the film stretches inward *irreversibly* into a paraboloid having thickness ~1 μm at its center over a circular area ~12 cm in diameter. Air is then admitted to the chamber very slowly. A vacuum chamber of this type for stretching plastic film is available from Scientific Specialties Co., Dayton, Ohio. The 1-μm film is placed between sheets of heavy paper for storage, handling, and cutting. The stretched film is substantially free of pinholes. Counter windows supported on a 2.5 × 15-mm slit have lasted longer than three years and withstood more than 1000 pumpdowns without initial differential pumping between the spectrometer and detector (see below). Transmission of 1-μm polypropylene for Al, C, B, and Be *K* lines is about 80, 75, 70, and 35%, respectively.

Mylar (polyethylene terephthalate) contains carbon, hydrogen, and oxygen and has a higher x-ray absorption coefficient than polyethylene and polypropylene, which are hydrocarbons. Consequently, for a given thickness, Mylar has lower transmission, or, for a given transmission, polyethylene and polypropylene may be thicker and therefore less fragile. Macrofol (trade name, Hoechst Chemicals, Ltd.) (*B82*) is a polycarbonate ester film having properties intermediate between those of Mylar and polypropylene. It combines high transmission (but not as high as polypropylene) with high strength (but not as high as Mylar). For measurement of F *Kα*, the ideal window would be Teflon [$(CF_2)_n$] film $\lesssim$ 1 μm thick, because fluorine is relatively transparent to its own *Kα* line. Nitrolucid (trade name, Biodynamics Research Corp.), a plastic film having high nitrogen content, has transmission $\gtrsim$59% for the *K* bands of all elements down to atomic number 4 (beryllium) (*T15*). For measurement of Be and B *K* bands, a thin film of beryllium or boron, respectively, on the detector window absorbs background in the region of the *K* band and increases the peak-to-background ratio (*C3*). Figure 6.2 shows transmission as a function of wavelength for several window materials. When a flow counter is to be backed by another detector for short wavelengths, the front window must be thin enough to have adequate transmission for the longest wavelength to be measured. However, the back window may be much thicker.

Since the gas-flow detector usually discharges into the atmosphere—either directly or through a manostat—the gas filling is at atmospheric pressure. Thus, the windows of flow counters used with helium path—also at atmospheric pressure—experience no pressure differential if care is taken not to use excessive helium flush rates. However, windows of flow counters used with vacuum path are subjected to an outward pressure differential of up to 1 atm and may require support. There are many ways to provide such support, or otherwise prevent window explosion, including the following.

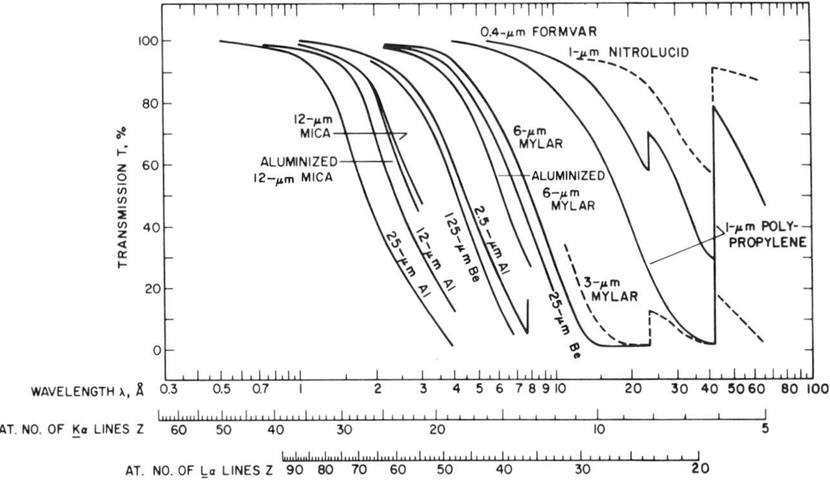

FIG. 6.2. Transmission of some typical x-ray detector window materials.

1. The window may be supported by a nylon or nickel grid. Usually, such grids are 100–200 mesh. Although meshes having up to 75% open area are available, even these result in a 25% decrease in effective window area. It is claimed that 0.1-μm Formvar and 1-μm polypropylene windows supported this way withstand differential pressures of 15 and 760 torrs (1 atm), respectively.

2. Some flow counters intended for use only in focusing curved-crystal spectrometers have window openings consisting of a narrow slit in the cathode cylinder. The film can be semipermanently cemented over this slit and a curved metal frame, also having a slit, semipermanently cemented over the film with the two slits in register.

3. The window may consist of a metal disk having a slit aperture, say, 2×20 mm for focusing spectrometers or several such parallel slits for nonfocusing spectrometers; for example, five such 2×20-mm slits have an effective aperture of 2 cm^2. A thin film of vacuum grease is applied at the periphery of one side of the disk, and the film is laid over this side and stretched to remove wrinkles. The window assembly is then placed on the detector with the film side inward and resting on an *O*-ring in a groove in the detector body, then secured in some suitable way.

4. The end of the detector collimator may lie in contact with the detector window so that the ends of the Soller foils support the window without interposition of additional material. In such cases, the film should be oriented so that the most likely splitting direction, which is usually parallel to the length of the rolled film, is transverse to the collimator foils. Also, the ends of the foils must be smooth and free of burrs.

5. Some detector bodies consist of an axial cylindrical hole in solid aluminum stock having square or hexagonal cross section. The window is machined through one of the flat outside surfaces; also machined in this surface is a track that retains a sliding metal shutter. The film, possibly supported on a mesh, is mounted over an opening in this slide. The detector and spectrometer are evacuated through separate pump lines, and while evacuation is in progress, the slide is placed to close the window, isolating the detector from the spectrometer. When both chambers are evacuated, the slide is placed with the film over the window. Alternatively, the film may be mounted directly on the window in the detector body and the slide placed to cover the window during pumping and admit x-rays to it when both chambers are evacuated.

6. A "differential pumping" system can be provided in which the detector, vacuum pump, and detector gas line are interconnected by a special valving arrangement. When air is admitted to the spectrometer,

the valve shuts off the detector gas line and simultaneously admits air to the detector. When the spectrometer is evacuated, the valve performs the following operations in sequence: (a) evacuation of the detector and spectrometer simultaneously; (b) isolation of the detector from the vacuum pump when operational vacuum is attained; and (c) admission of detector gas at reduced pressure to the detector and discharge of the effluent gas into the fore-vacuum line.

6.2.1.3. Gas Fillings

The function of the gas filling is described in detail in Sections 6.2.2 and 6.2.3. Let it suffice here to state that the gas converts the energy of the x-ray photons to electric charge and contributes to the quenching process. (Section 6.2.2.4). The two functions may be performed by the same gas, but more often a mixture of two gases is used, usually a group 0 gas—helium, neon, argon, krypton, or xenon—for detection, and an organic vapor or halogen for quenching. In the remainder of this chapter, the terms *detector gas* and *quench gas* are used to differentiate these two functions. Only those photons that are absorbed in the gas are "counted," so it is necessary for the detector gas to have absorption coefficient sufficiently high to absorb the incident x-rays in the path length of the active volume. In general, this means that the shorter the wavelength to be measured, the higher must be the atomic number, or effective atomic number, of the detector gas. In the detection of long-wavelength x-rays, it is necessary that the absorption coefficient of the gas not be so high that the photons are wholly absorbed in a thin layer of gas just inside the window.

Another consideration in the choice of detector gas is the liability of the gas to escape-peak formation (Sections 8.1.8.4a and 13.3.2.2).

Sealed detectors usually have fillings of argon, krypton, or xenon at pressure 10–60 torrs together with a halogen or organic quench gas; xenon is the most commonly used sealed-detector gas (Section 8.1.8.4a). Probably the most common flow-counter gas is "P10" gas (90% argon + 10% methane), although many other fillings have been used, especially for the long- and ultralong-wavelength region (*F15*). These gases include the following: pure methane (CH_4), ethane (CH_3CH_3), propane ($CH_3CH_2CH_3$), butane [$CH_3(CH_2)_2CH_3$], and isobutane [$(CH_3)_2CHCH_3$]; 98% argon + 2% carbon dioxide, 97% argon + 3% isobutane, 25% argon + 75% methane, and 10% argon + 90% methane ("P90" gas); 70% neon + 30% methane (*L32*); 96–97% helium + 4–3% butane or isobutane, 95% helium + 5% methane, and 88% helium + 12% carbon dioxide (*L32*). All these concentrations are vol/vol %.

One of the principal causes of deterioration of resolution in flow counters is the gradual deposition of carbon and hydrocarbons on the anode wire, probably from decomposition and polymerization of methane in the P10 gas (*D13*). Possibly use of 90% argon + 10% carbon dioxide instead of P10 would minimize this difficulty.

The helium and neon mixtures and the pure hydrocarbons cited above are especially advantageous for the ultralong-wavelength region and for the *K* lines of elements 9–17 (fluorine through chlorine). They absorb the long-wavelength analyte lines strongly, but are highly transparent to the short-wavelength background and higher-order lines of high-*Z* elements. They give narrow pulse-height distributions and do not produce escape peaks (*J10*, *L32*). Methane and propane also act, to a certain extent, as their own quench gas. Figures 6.17 and 9.1 show the useful ranges of detectors having several different gas fillings.

For the same flow counter at the same gas pressure—say, 760 torrs (1 atm)—the higher the hydrocarbon concentration of the gas, or the lower the group 0 gas concentration, the higher will be the operating potential. Potentials of 2–3 kV are not uncommon for pure hydrocarbon gases. However, a flow counter operating at reduced pressure with a high-hydrocarbon gas—such as pure propane or P90 (90% methane + 10% argon)—may require a lower operating potential than the same detector operating at atmospheric pressure with, say, P10 gas (10% methane + 90% argon). The same considerations apply to gases having high carbon dioxide concentrations.

The counter-gas cylinder is fitted with a two-gage reducing valve, the output gage of which is set at ~1 kg/cm^2 (~10 lb/inch^2). The gas line to the detector is throttled with a needle valve or with lengths of capillary tubing. If the detector discharges into the atmosphere, either directly or through a manostat, a flow meter, 0–50 cm^3/min (0–0.1 ft^3/h) or so, may be provided in the input gas line. Alternatively, the outlet line may pass through a small bubbler containing water or, preferably, light vacuum oil. The operator soon learns the bubble rate necessary for optimal detector performance. If the detector is used in a vacuum spectrometer, a trap must be provided between the detector and bubbler to prevent suction of the water or oil into the detector if its window should rupture. Special gas-line arrangements may be provided for ultrathin-window detectors (Section 6.2.1.2). Because average pulse height may change substantially with the pressure and temperature of the detector gas, a manostat may be provided in the gas line and the inlet may be water-cooled (Section 8.1.8.2b).

One is wise to adopt the habit of observing the detector gas cylinder gage at least once each day to note whether it is falling at an abnormally rapid rate. If it is, a perforation or other leak in the gas line or detector window may be indicated. A check valve may be placed at the output of the regulator on the gas cylinder; then if the detector gas line opens—perforates, breaks, pulls off, etc.—or if the detector window ruptures, atmospheric pressure will close the check valve and prevent rapid emptying of the cylinder.

The optimal gas flow rate varies with the particular detector, but is usually of the order 25–50 cm^3/min (0.05–0.1 ft^3/h). The gas must be allowed to flow for at least 30 min prior to making the first measurements, and many laboratories never interrupt the flow at all. The cathode cylinder, anode wire, and conductive window coating must be chemically inert to the gas filling; in general, this is a matter for concern only in the presence of halogen quench gases.

6.2.2. Operation

In this section, the operation of the gas detector is considered in general terms. Section 6.2.3 gives a more detailed discussion of the operation of the proportional counter.

6.2.2.1. Phenomena in the Detector Gas Volume

The basic initial process in a gas-filled x-ray detector is gas ionization or ion–electron pair production (not to be confused with electron–positron pair production, Section 2.1.2). It is this process that transduces the x-ray photon into a measurable analog (electric charge), initiates the other phenomena that occur in the gas volume, and provides the basis for proportionality of x-ray photon energy and pulse height. As used here, the term *ion–electron pair*, or simply *ion pair*, signifies a positively charged ion —a detector gas atom having an electron vacancy in one of its outermost shells—and the missing negative electron.

Photons and electrons differ in the way in which they lose energy. An electron can lose energy incrementally, but, except for incoherent scatter (Section 2.2.2), a photon loses all of its energy at once or not at all. Incidentally, this is the reason why a continuum is generated in primary, but not secondary, x-ray excitation.

The paths of photons and electrons in the detector gas also differ. The photon path is perfectly linear up to the point of its photoelectric absorption, where it disappears. The electron undergoes deflection with each ionization, that is, with each ion pair it forms, and its path is tortuous, becoming increasingly so as it loses energy.

In the detector, the incident x-ray photon is wholly absorbed by a gas atom to produce an ion pair. This process is *photoionization* or *photoelectric absorption*, and the electron so produced is a *photoelectron*. Since the ionization potential of the gas atom is only $\sim$30 eV, substantially all of the x-ray photon energy, usually $\gg$1 keV, imparts kinetic energy, that is, velocity, to the photoelectron, which then has substantially the same energy as did the x-ray photon.

This high-speed photoelectron then expends its energy in producing ion–electron pairs. This process is ionization by collision, and the electrons so produced are *secondary electrons*. These secondary electrons, unlike the photoelectron produced by the x-ray photon, have substantially no kinetic energy. With each ion pair produced, the photoelectron loses energy equivalent to the ionization potential of the detector gas ($\sim$30 eV), and as it progresses, its energy decreases and it becomes increasingly prone to interaction with gas atoms. Consequently, the density of ion pairs increases toward the end of the photoelectron path and is very high at its end.

The foregoing sequence of events occurs whenever x-rays pass through any gas—in or out of an x-ray detector and with or without an electric field. This ionization involves only the outermost electrons in the detector-gas atoms and is not to be confused with expulsion of K and L electrons with resultant emission of characteristic x-ray spectral lines. Such K and L excitation of the gas atoms can and does occur, but for simplicity, the process is disregarded here and discussed in Section 6.2.3.

The electrons formed by the original photoelectron produced by the incident x-ray photon may undergo any of several processes. Since these electrons have substantially zero velocity, on production of an ion pair, rapid *recombination* occurs unless the electrons are removed from the proximity of the positive ions. (Recombination of the original photoelectron and its ion is unlikely because this electron has such high kinetic energy—x-ray photon energy minus gas ionization potential—that it speeds away from the vicinity of the ion.) This separation is effected by application of an electric potential between the cathode and anode, as shown in Figure 6.1, causing the electrons to undergo *migration* toward the positive anode wire. If the applied potential is high enough, the electrons undergo *acceleration* and may attain such velocities that they undergo *collision* with gas atoms along their path, causing ionization and additional *ion–pair production*. The additional electrons so formed may produce still more ion pairs, the electrons from which produce still more, and so on. Thus, a single electron may initiate a shower of hundreds to billions of electrons, a phenomenon known as *Townsend avalanching*. Incidentally, the ion pairs produced by the photoelectron excited by the x-ray photon are termed *primary* ion pairs to distinguish them from the ion pairs produced by avalanching. The electrons may also excite *ultraviolet emission* from the gas atoms, and, on finally striking the anode, they may cause *secondary-electron emission* and *x-ray emission* in the form of a long-wavelength continuous spectrum.

The heavier positive ions have mobility about 1000 times slower than the electrons. They undergo slow *migration* toward the cathode, eventually undergoing *recombination* or *neutralization*—that is, return to the uncharged state—by capturing a free electron or receiving an electron from a quench-gas molecule or from the negative cathode. The process of neutralization may result in *ultraviolet emission* from the detector-gas atom. The process of collision with the cathode wall may result in *electron emission by ion bombardment*. The ultraviolet radiation permeates the entire gas volume, and, by *photoionization* of gas atoms, initiates new *avalanching*. The electrons expelled from the cathode wall also initiate *avalanching*.

Quenching, another process that occurs, at least in part, in the gas, is discussed in Section 6.2.2.4.

6.2.2.2. Proportionality in Gas Detectors

If the energy of the incident x-ray photon is imparted to a photoelectron which in turn expends its energy in producing ion pairs, and if each ion pair requires the same amount of energy, it follows that the number of primary ion pairs should be proportional to the photon energy. In fact, the number of primary ion pairs should be simply the x-ray photon energy divided by the ionization potential of the detector gas. This would probably be true if only singly charged ions were formed. However, more than one electron may be expelled from a gas atom, and the ionization potential increases for each successive ionization; for example, the first five ionization potentials for xenon are 12, 21, 32, 46, and 76 eV, respectively. Thus, each detector gas has an *effective* ionization potential somewhat larger than its first ionization potential, the relative difference between the two increasing as atomic number of the gas increases. The *average* number of primary ion pairs produced is given by the x-ray photon energy divided by the effective ionization potential V_i, that is,

$$n_0 = E_x / V_i \tag{6.1}$$

TABLE 6.1. Basis of Proportionality in Gas and Scintillation X-Ray Detectors

Detector gas	First ionization potential, eV	Effective ionization potential,[a] eV	Average number of ion pairs produced by each x-ray photon[b]	
			Cu $K\alpha$ 1,54 Å, 8040 eV	Mo $K\alpha$ 0.71 Å, 17,440 eV
$_{2}$He	24.5	27.8	289	628
$_{10}$Ne	21.5	27.4	293	637
$_{18}$Ar	15.7	26.4	304	660
$_{36}$Kr	13.9	22.8	352	765
$_{54}$Xe	12.1	20.8	386	838
Scintillator	**Light-photon energy, eV**	**Effective light-photon energy,[c] eV**	**Average number of light photons produced by each x-ray photon[b]**	
			Cu $K\alpha$	Mo $K\alpha$
NaI(Tl)	2	40	201	436

[a] Average energy actually required to produce one ion pair.
[b] X-ray photon energy divided by the effective ionization potential or effective light-photon energy.
[c] For definition, see Section 6.3.2.1.

Table 6.1 shows, for each of the five stable group 0 gases, the first ionization potential, the effective ionization potential, and the average number of ion pairs produced by Cu $K\alpha$ (1.54 Å, 8040 eV) and Mo $K\alpha$ (0.71 Å, 17,440 eV) photons.

It must be emphasized that even with a strictly monochromatic x-ray beam, the numbers of ion pairs produced by the individual photons have a wide Gaussian distribution having a mean value $\bar{n}$ and a deviation $\sqrt{\bar{n}}$. However, Table 6.1 shows that Mo $K\alpha$ photons, having about twice the energy of Cu $K\alpha$ photons, do indeed produce about twice as many ion pairs. Thus, the gas ionization process provides the basis for proportionality of output pulse height and incident x-ray photon energy.

6.2.2.3. Gas Amplification; Types of Gas Detectors

Consider an x-ray photon entering the detector shown in Figure 6.1 and passing through the active gas volume. Eventually, it undergoes photoelectric absorption by a gas atom and imparts its energy to a photoelectron, which produces ion pairs along its path until its energy is expended. The resulting electrons and ions may undergo various processes (Section 6.2.2.1). It is the potential applied to the detector that largely determines the specific behavior of the electrons and ions, and thereby determines whether the detector functions as an ionization chamber, proportional counter, or Geiger counter. The *gas amplification factor* or *gas gain* G is the number of electrons n collected on the anode wire for each electron produced by the photoelectron excited by the x-ray photon; from this definition and Equation (6.1),

$$G = n/n_0 = n(V_i/E_x) \tag{6.2}$$

Figure 6.3 shows the gas amplification factor as a function of detector potential. The curve is naturally divided by its inflections into a number of regions. Let us consider in detail the phenomena in the detector as the anode potential is increased from zero (potentiometer arm at extreme left in Figure 6.1) to progressively higher positive values.

a. Zero Applied Potential. At zero potential, the ion pairs quickly recombine, and no charge is collected on the anode. The gas amplification is zero.

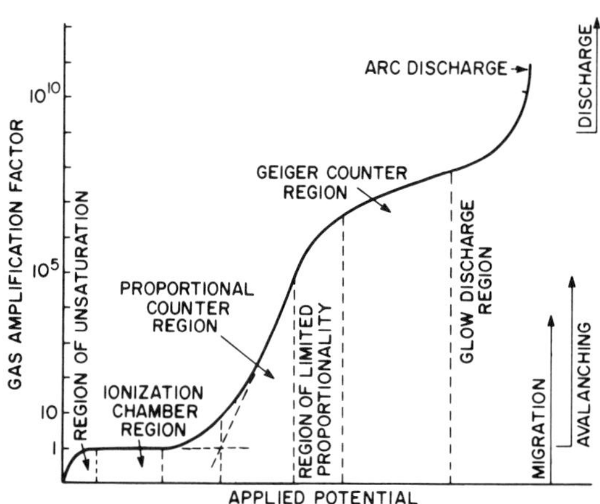

FIG. 6.3. Gas-amplification factor as a function of applied potential for the gas-filled x-ray detector.

b. Region of Unsaturation. The applied potential is so low that many electrons recombine with positive ions before they can reach the anode. The collected current is unsaturated, that is, fewer electrons are collected than were formed by the x-ray photon, and the gas amplification is <1. However, as the potential is increased, electron acceleration increases, and progressively more electrons escape recombination and reach the anode.

c. Ionization-Chamber Region. The potential is just sufficient to overcome recombination completely, all electrons are collected (saturation) and the gas amplification is 1 and remains constant over a substantial interval of applied potential. However, no secondary ionization occurs in the gas or at the electrodes. Ionization-chamber detectors operate in this region. In practice, ionization chambers are usually constructed with the two electrodes in the form of large-area parallel plates separated by the gas. The electrons and ions are all collected on the positive and negative plates, respectively. Ionization chambers have little application in x-ray spectrometry.

d. Intermediate Region. The critical detector potential for onset of gas amplification is that at which electrons from the primary ion pairs are accelerated just to the ionization potential of the detector gas by the time they reach the anode wire. At any higher potential, the electrons are able to initiate some additional gas ionization along their path. The critical potential for gas amplification is shown in Figure 6.3 as the intersection of the extensions of the approximately linear segments of the gas amplification curve in the proportional counter and ionization chamber regions. In this intermediate region, detector potential is high enough to initiate mild avalanching, and the gas amplification becomes >1. This region is of no significance for x-ray detection.

e. Proportional-Counter Region. The potential is high enough for marked avalanching to occur, and the gas amplification is 10^2–10^5. However, this avalanching is limited in extent. It is localized in a region within a few wire diameters of the anode, and, in general, each electron produced by the x-ray photon initiates only one avalanche, and the avalanches are largely free of any interaction. The number of avalanches is substantially the same as the number of initial ionizations, and, since all electrons are collected, the total collected charge is proportional to the x-ray photon energy.

The gas amplification G in the proportional region is given approximately by the following equation:

$$\log_e G \approx 1.3\left(Pi \frac{V}{V_i} \log_e \frac{r_c}{r_a}\right)^{1/2} \left[\left(\frac{V}{V_c}\right)^{1/2} - 1\right] \qquad (6.3)$$

where P is detector-gas pressure (atm); i is average ionizing power of the electron at 1 atm; V is detector potential (V); V_c is critical potential for gas amplification (V) (Section 6.2.2.3d); V_i is ionization potential of the detector gas (eV); and r_a and r_c are radii of the anode wire and cathode cylinder, respectively. Although this equation is not very accurate, it does show the strong dependence of gas amplification on P, V/V_c, and r_c/r_a.

The proportional region has another favorable feature. Since the electron mobilities are ~1000 times faster than the ion mobilities, the electrons reach the anode and are collected before the ions move very far. As the electrons are collected, the region surrounding the wire becomes depleted of negative charge, and a positive-ion space-charge sheath rapidly forms concentric with the wire. The charge density and diameter of this sheath are very similar for each successive x-ray photon counted, and the sheath tends to limit the charge accumulated—and thereby the output pulse height—for successive photons. This effect results in greater uniformity of pulse height and narrower pulse-height distribution. In general, the electrons produced by the avalanching initiated by the primary electrons produced by the incident photon are all collected within ~0.1–0.2 μs, but the positive space-charge sheath requires ~1–2 μs to decay. Thus, the output pulse has a sharp rise time and a recovery time of 1–2 μs (Figure 6.11, Sections 6.5.1.1 and 6.5.1.4).

The proportional counter is discussed in more detail in Section 6.2.3.

f. Region of Limited Proportionality. As potential is increased still further, avalanching becomes more generalized. The number of avalanches per unit volume increases, and they begin to interact. The same electron initiates multiple avalanches. The region in which avalanching occurs spreads outward from the wire. Proportionality deteriorates as potential is increased. In this region, gas amplification becomes a function of incident photon energy, and pulses of low initial amplitude are amplified more than those of high initial amplitude. The amplification is also strongly dependent on detector potential. This region, like the intermediate region, is also of little value for x-ray detection.

g. Geiger-Counter Region. The potential is now so high that the avalanching becomes generalized, and the other phenomena noted in Section 6.2.2.1 occur. The gas ions and atoms are excited in the gas volume and at the cathode wall to emit ultraviolet radiation. Secondary electrons and low-energy x-rays are emitted when the electrons strike the anode, and electrons are expelled by ion bombardment of the cathode. These emitted electrons initiate new avalanches. The ultraviolet radiation permeates the entire gas volume, causing photoionization, the photoelectrons from which also initiate avalanches. Thus, in the Geiger region, the initial ionizing event merely triggers the general discharge of the entire tube. All vestiges of proportionality disappear, and the output pulses are of substantially the same high amplitude—$\sim$1000 times the amplitude of pulses in the proportional region. The Geiger, or Geiger–Muller, counter operates in this region (*F26*, *G6*).

h. Glow-Discharge Region. The discharge is now so severe that it does not quench (Section 6.2.2.4) after each incident photon. The photon triggers not one output pulse, but a continuous, sustained discharge. If the tube window is transparent, the gas volume can be seen to glow. Operation in this condition for even a short time is likely to ruin the tube.

i. Arc Discharge. An arc discharge occurs between the cathode and anode, and the wire is likely to "burn out."

6.2.2.4. Quenching

The duration of the discharge initiated by the x-ray photon is *quenched*, that is, terminated, by the action of the quench gas, the positive ions, and the external resistor (*R* in Figure 6.1).

The quench gas is mixed in relatively low concentration with the detector gas and may be an organic vapor or a halogen. The organic gases used include methane (CH_4), ethane (CH_3CH_3), propane ($CH_3CH_2CH_3$), butane [$CH_3(CH_2)_2CH_3$], and isobutane [$(CH_3)_2CHCH_3$]; methyl chloride (CH_3Cl), methyl bromide (CH_3Br), and methyl iodide (CH_3I); and ethanol (CH_3CH_2OH). Molecules of these compounds undergo irreversible decomposition as a result of their quenching action and limit the useful life of the detector to $\sim 10^9$ counts. The halogens used include bromine (Br_2) and bromine chloride ($BrCl$). Molecules of these substances dissociate to halogen atoms, but recombine to restore the molecule. Detectors having halogen quench gas have substantially unlimited life (*L28*).

The quench gas functions in several ways to terminate the discharge. Quench-gas molecules neutralize some of the positive detector-gas ions by donating electrons to them. This is possible because the ionization potential of the quench gas is much lower than that of the detector gas. Thus, a slow, positive detector-gas ion is replaced by an even slower, positive quench-gas ion, which is less likely to expel electrons from the cathode wall. Moreover, the difference between the work functions of the detector and quench gases is much smaller than the difference between the work functions of the detector gas and cathode wall. Thus, no ultraviolet emission accompanies neutralization by the quench gas. The quench gas also absorbs secondary electrons and ultraviolet radiation produced by neutralization of detector-gas ions at the cathode.

The positive ions and the external resistor both tend to reduce the *effective* potential between cathode and anode to the extent that avalanching is no longer sustained. Since the positive-ion mobilities are $\sim$1000 times slower than the electron mobilities, the electrons are all collected before the ions have migrated very far. Thus, the ions form a positive space-charge sheath around the anode wire. This sheath migrates outward toward the cathode and reduces the effective electric field potential near the anode. The external resistor (R in Figure 6.1) prevents the electrons arriving at the cathode from leaking off as fast as they are collected. This causes the anode to become more negative (more accurately, less positive) and reduces the effective potential across the tube.

Sometimes, an external electronic quenching circuit is provided. As already explained, during a pulse, the effective potential across the detector drops. Quenching circuits reduce the effective potential even further, hold it down until the detector is cleared of ions, then restore it to equivalence with the applied potential. Special quenching circuits for Geiger counters reduce or even reverse the applied potential soon after initiation of the discharge to reduce the number of ions reaching the cathode, and they do this so quickly that the discharge cannot spread throughout the tube.

6.2.3. Proportional Counters (*A22*, *L6*, *M61*)

6.2.3.1. Phenomena in the Detector Gas Volume

It has been explained that: (1) the output of a gas detector consists of a pulse of electric current for each x-ray photon absorbed; (2) the numbers of primary ion pairs produced by the individual photons of a strictly monochromatic x-ray beam are not the same, but have a Gaussian distribution about a mean value; and (3) in proportional and Geiger counters, these ion pairs initiate a complex gas-amplification process, resulting in an output pulse having 10^2–10^{10} times as many electrons as were produced by the x-ray photon. Consequently, it is not surprising that the amplitudes of the detector output pulses also have a Gaussian distribution. In a Geiger counter, the average pulse height is the same regardless of x-ray photon energy. In a proportional counter, the average pulse height is proportional to the photon energy.

We are now in a position to consider the operation of the proportional counter in more detail. The diagrams in Figure 6.4 show five processes that an x-ray photon may initiate in the active volume of a proportional counter. Column 1 shows each process in a simplified schematic way. Column 2 shows the individual detector output pulse resulting from each process. Column 3 gives the output pulse height for each process in terms of incident x-ray photon energy E_x. Column 4 shows, for each process, the pulse-height distribution curve for a detector that is receiving a continuous monochromatic x-ray beam and in which only that process occurs. The "peaks" on these curves are *not* profiles of individual pulses; such profiles would be plotted on a grid of pulse height in volts *versus* time. Nor are they to be confused with the familiar 2θ scan of intensity *versus* 2θ angle. Rather, they show the numbers of pulses of each pulse height present in the detector output in unit time. It is also necessary to distinguish *pulse height*—the amplitude (in volts) of the individual pulse triggered by an individual x-ray photon—and *intensity*—the number of pulses per second. Pulse height for a detector is the maximum potential that appears across resistor R in Figure 6.1 during the life of that pulse. A careful study of Figure 8.4 may clarify these relationships.

Let us consider the five processes in detail. In Figure 6.4 and the following discussion, argon is assumed to be the detector gas, but the principles apply to any detector gas.

a. Transmission (Figure 6.4A). The x-ray photon may pass through the gas wholly unabsorbed, to be absorbed in the detector wall or to emerge through a rear window. Of course, no output pulse is produced. This process is more likely the shorter the x-ray wavelength and the lower the absorption coefficient of the detector gas. In this way, short-wavelength x-rays pass through a two-window detector optimal for long-wavelength x-rays to a backing scintillation or sealed proportional counter.

b. Ion-Pair Production (Figure 6.4B). The x-ray photon may cause outer-shell photoionization of a detector gas atom, imparting its energy to the photoelectron, which in turn ionizes the detector gas atoms along its path to produce Ar^+, e^- pairs. Actually, the ion pairs are not uniformly distributed along the electron path, but become denser as the electron progressively loses energy. The output pulse height is proportional to the photon energy. If only this process occurred in a detector receiving a monochromatic beam, the output would consist of a single pulse-height distribution (peak) of average pulse height proportional to the photon energy.

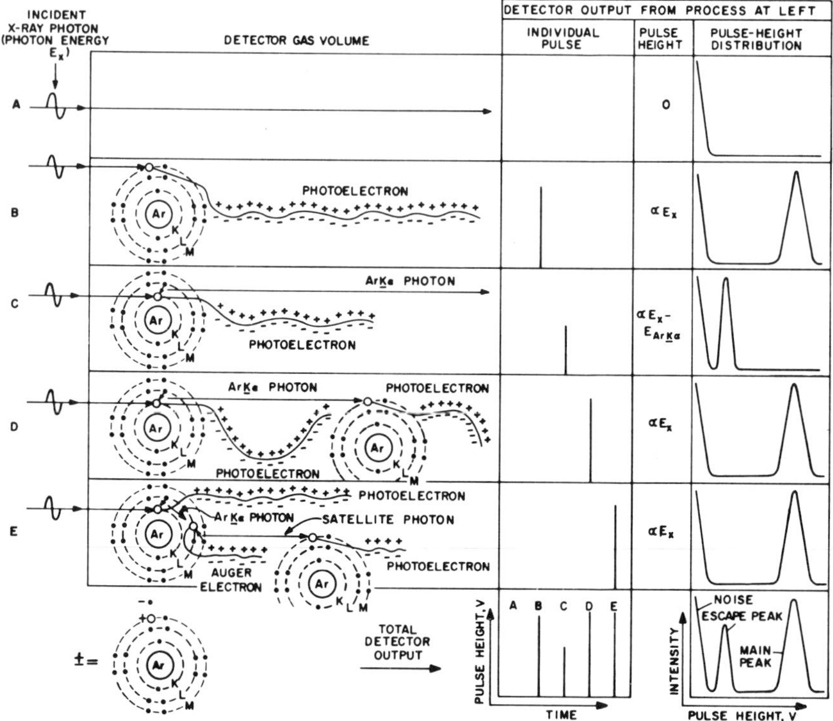

FIG. 6.4. Phenomena in the active volume of an argon-filled proportional counter, showing, for each phenomenon, the photon–atom interaction, an individual pulse, the pulse height, and the pulse-height distribution; E_x is the energy of the incident x-ray photon.

A. The x-ray photon is not absorbed.

B. The x-ray photon undergoes photoelectric absorption in the outermost shell of a gas atom; the resulting photoelectron expends its energy producing Ar^+,e^- pairs.

C. The x-ray photon undergoes photoelectric absorption in the K shell of a gas atom; the resulting photoelectron expends its energy producing Ar^+,e^- pairs; the Ar $K\alpha$ photon "escapes" the gas volume unabsorbed.

D. The x-ray photon undergoes photoelectric absorption in the K shell; the resulting photoelectron behaves as in process C; the Ar $K\alpha$ photon behaves as the incident x-ray photon in process B.

E. The x-ray photon undergoes photoelectric absorption in the K shell; the resulting photoelectron behaves as in process C; the Ar $K\alpha$ photon undergoes internal absorption (Auger effect), producing an Auger electron and a satellite-line x-ray photon; the Auger electron expends its energy producing Ar^+,e^- pairs; the satellite photon behaves as the incident x-ray photon in process B.

It has been assumed so far that photoionization of the outer shells of the detector gas atoms is the only process by which x-ray photons are converted to electric charge. In fact, it would be preferable if this were true. However, if the x-ray photon has wavelength shorter than that of the argon K-absorption edge, it may undergo photoelectric absorption in the K shell of an argon atom (Figure 6.4C,D,E), expelling an argon K electron and expending the remainder of its energy in imparting kinetic energy (velocity) to the photoelectron. The K-excitation potential of argon (3.2 keV) is much higher than the ionization potential of the outermost electron in argon atoms (15.7 eV). Consequently, the incident x-ray photon loses much more energy in K-shell ionization than in outer-shell ionization. It follows that photoelectrons produced by K-shell ionization have much less energy (E_x − 3.2 keV) than those produced by the process in Figure 6.4B (E_x − 15.7 eV). The photoelectron expends its energy in producing Ar^+, e^- pairs as shown. However, the Ar $K\alpha$ (or $K\beta$) photon resulting from the filling of the K-shell vacancy can undergo one of three processes as follows.

c. Argon K Excitation Followed by Escape of the Ar Kα Photon (Figure 6.4C). The Ar $K\alpha$ (or $K\beta$) photon may escape the active volume of the detector. This is not an unlikely process because an element is relatively transparent to its own characteristic radiation. The output pulse height is proportional to the difference between the energies of the incident and Ar $K\alpha$ photons. If only this process occurred in a detector receiving a monochromatic beam, the output would consist of a single pulse-height distribution of correspondingly lower average pulse-height than that in Figure 6.4B. This distribution is known as an *escape peak* because it is produced when some of the energy carried into the detector by the x-ray photon "escapes" the active volume.

d. Argon K Excitation Followed by Absorption of the Ar Kα Photon (Figure 6.4D). The Ar $K\alpha$ photon may undergo photoelectric absorption in the outer shell of another argon atom, producing a photoelectron which expends its energy producing Ar^+, e^- pairs. To the extent that this occurs, the escape peak is avoided. The sum of the numbers of ion pairs formed by (1) the photoelectron produced by the incident x-ray photon and (2) the photoelectron produced by the Ar $K\alpha$ photon equals the number formed in Figure 6.4B. The pulse height and pulse-height distribution are the same as in Figure 6.4B.

e. Argon K Excitation Followed by Auger Effect (Figure 6.4E). The Ar $K\alpha$ photon may undergo absorption within the atom of its origin (Auger effect, Section 2.5) with resultant emission of an Auger electron and a satellite-line x-ray photon. The Auger electron expends its energy in producing Ar^+, e^- pairs, and the satellite photon undergoes photoelectric absorption, which, in turn, produces Ar^+, e^- pairs. The sum of the numbers of ion pairs formed by (1) the photoelectron produced by the incident x-ray photon, (2) the Auger electron, and (3) the photoelectron produced by the satellite photon equals the number formed in Figure 6.4B. The pulse height and pulse-height distribution are the same as in Figure 6.4B.

6.2.3.2. Detector Output; Escape Peaks

To summarize, the output of a proportional counter consists of a series of electric-current pulses of various heights randomly distributed in time, each pulse originating from absorption of a single incident x-ray photon. In addition, there is a series of very low-amplitude detector noise pulses. For monochromatic incident x-rays having wavelength longer than that of the absorption edge of the detector gas, the output contains a single pulse-height distribution (peak) having mean pulse height proportional to the x-ray photon energy. For monochromatic incident x-rays having wavelength shorter than that of the absorption edge of the detector gas, the output may contain two pulse-height distributions (peaks). The *main* (*principal*, *natural*, or *photo*) peak has mean pulse height proportional to the x-ray photon energy. The *escape* peak has mean pulse height proportional to the difference between the photon energies of the incident x-rays and the spectral line of the detector gas. For incident x-rays containing more than one wavelength, the output contains a pulse-height distribution for each and an escape peak for each that can excite the detector gas. Since these peaks are relatively broad, they may overlap one another, and the main peak of the analyte line may be overlapped by a main or an escape peak of another element.

The intensity of the escape peak relative to the main peak is greater (1) the closer the wavelength of the x-rays lies to the short side of the absorption edge of the detector gas, and (2) the greater the fluorescent yield of the gas. This second condition is illustrated in Figure 6.5, which shows the output of a neon detector for P $K\alpha$, an argon detector for Sc $K\alpha$, a krypton detector for Mo $K\alpha$, and a xenon detector for Ce $K\beta$. Each spectral line lies close to the K-absorption edge of its detector gas and can excite its K lines very efficiently. It is evident that the intensity of the escape peak relative to the main peak increases substantially as the fluorescent yield (shown) of the detector gas increases. The escape peaks for the krypton and xenon detectors are more intense than their main peaks.

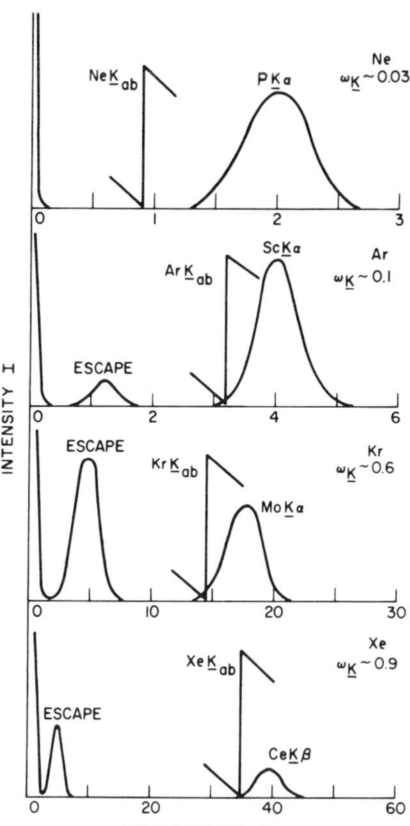

FIG. 6.5. Effect of K-fluorescent yield ω_K of the detector gas on the relative intensities of the main and escape peaks (pulse-height distributions) for detectors filled with neon, argon, krypton, and xenon, respectively. In each case, the incident x-rays have sufficient photon energy to expel K electrons from the detector gas atoms. As ω_K increases, the intensity of the escape peak relative to the main peak increases. [R. Jenkins and J. L. de Vries, "Practical X-Ray Spectrometry," Springer-Verlag N. Y. (1967); courtesy of the authors and publisher.]

6.3. SCINTILLATION COUNTERS

6.3.1. Structure

The essential features of a scintillation counter are shown in Figure 6.6 (*H57*, *N15*, *W8*). The detector consists of a scintillation crystal (phosphor) affixed to the envelope of a multiplier phototube ("photomultiplier") just outside the photocathode. Each incident x-ray photon is converted into a burst of visible-light photons, which fall on the photocathode and emit photoelectrons, which, in turn, initiate the secondary-electron multiplication process.

6.3.1.1. Scintillation Crystal

The scintillation crystal (phosphor, scintillator) must have the following characteristics: (1) The absorption coefficient should be high enough to

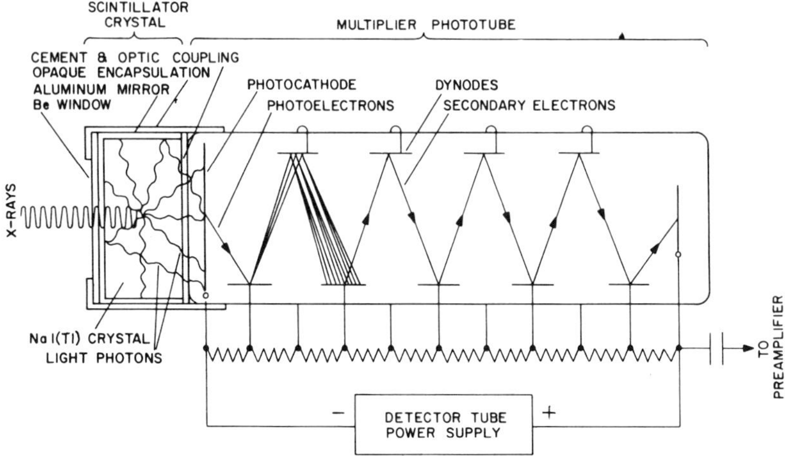

FIG. 6.6. Structure of the scintillation counter.

permit high absorption of the incident x-rays in a convenient thickness; (2) the conversion of x-ray photons to visible-light photons must be high; (3) the visible fluorescence spectrum should match the spectral sensitivity of the photocathode of the multiplier phototube; (4) the crystal should be optically clear and transparent to its own visible fluorescence; (5) the crystal should have index of refraction similar to that of the glass of the phototube envelope (or face plate) to ensure efficient transfer of light from the phosphor to the photocathode; and finally, (6) the rise and decay times of the scintillation should be short enough to permit high counting rates without serious coincidence loss.

Thallium-activated sodium iodide, NaI(Tl) or NaI:Tl, meets most of these requirements and is almost exclusively used in scintillation counters for analytical x-ray spectrometry. The heavy iodine atoms give high x-ray absorption over the entire useful spectral range of analytical x-ray spectrometry and minimize production of escape peaks (Sections 6.3.2.2 and 8.1.8.4a). An effective light photon is produced for each 50 eV of incident photon energy, and the resulting light is in the actinic blue spectral region (4100 Å). Sodium iodide can be grown synthetically in large, clear single crystals and cut to any required shape. The decay time is ~0.25 μs. The principal disadvantages are the somewhat high index of refraction, ~1.8, and, especially, the high hygroscopicity, which necessitates hermetic sealing.

The scintillator is a single-crystal disk of NaI(Tl), ~2.5 cm in diameter and 2–5 mm thick. The disk is aluminized around the edge and on the side that is to be toward the x rays to reflect scintillation light radiating away from the detector and direct it toward the phototube. The aluminized disk is sandwiched between a beryllium window ~0.2 mm thick on the x-ray side and a thin Lucite disk on the phototube side to act as a light coupler. The sandwich is then hermetically sealed and perhaps coated with opaque paint on the edge and x-ray side to exclude moisture and light, respectively. The packaged scintillator is affixed to the face of an end-window multiplier phototube just outside the photocathode. A layer of silicone is applied between the Lucite and phototube to improve optical coupling.

The composite "phoswich" scintillator is described in Section 7.1.4.3.

Two important characteristics of the scintillator crystal, *effective light-photon energy* and *crystal efficiency*, are defined in Sections 6.3.2.1 and 6.3.1.2, respectively.

6.3.1.2. Multiplier Phototube

The multiplier phototube ("photomultiplier") consists of a photocathode, a series of secondary-electron-emissive *dynodes*, and a collector sealed in an evacuated glass envelope and provided with a base having prongs to permit external connection of the electrodes. The term *dynode* arises from the dual function of collection and secondary emission of electrons. The tube is coated with optically opaque black paint to exclude external light and may be enclosed in a mu-metal shield to exclude stray magnetic fields. A regulated dc potential of 600–1000 V is applied to the tube with the photocathode negative. Each successive dynode is maintained at a higher positive potential by means of a potential divider usually consisting of a series of resistors mounted in the tube socket.

When illuminated by the light from the scintillation, the photocathode emits photoelectrons, which are attracted to the first dynode, where each expels two to four secondary electrons. These secondary electrons are attracted to the second dynode, where more secondary electrons are emitted, etc. Thus, a progressively increasing shower of secondary electrons cascades through the multiplier. The collector receives the secondary-electron current, which constitutes the output pulse from the scintillation counter.

The important characteristics of the scintillation counter, in addition to effective light-photon energy defined in Section 6.3.2.1, are the following. The *crystal efficiency* or *optical coupling efficiency* is that fraction of the light photons produced in the scintillator that actually reach the photocathode in the multiplier phototube, and is typically ~0.9. The *photocathode efficiency* is that fraction of the light photons reaching the photocathode that actually result in emission of a photoelectron, and is typically 0.1–0.2; however, with more modern tubes, the efficiency may be as high as 0.4. The *first-dynode collection efficiency* is that fraction of the photoelectrons that actually reach the first dynode, and is typically ~0.9. The *secondary-emission ratio* of the dynodes is the number of secondary electrons emitted for each electron incident on the dynode, and is typically 3–4. The *gain G* of the electron multiplier is the number of electrons arriving at the collector for each photoelectron arriving at the first dynode:

$$G = k\sigma^n \tag{6.4}$$

where k is the first dynode collection efficiency (~0.9), σ is the secondary-emission ratio (3–4), and n is the number of dynodes, usually 9–14. Thus, for a 10-stage multiplier having $\sigma \approx 4$ and $k \approx 0.9$, the gain is 0.9×4^{10} or $\sim 10^6$. The *statistically limiting process* in the scintillation counter is the process involving the least numerous events and is the collection of photoelectrons on the first dynode.

Even when the photocathode is in darkness, the multiplier phototube produces a low output current. This *dark current* arises from: (1) thermionic emission from the photocathode; (2) cold field emission from points, corners, and edges, where high potential gradients exist; (3) radioactivity (for example, from traces of ^{40}K in the glass bulb and mica and ceramic insulators); (4) cosmic radiation; (5) ionization of residual gas; and (6) electrical (ohmic) leakage over insulating surfaces. The last two of these are dc in nature and present little difficulty for a photon counter. However, the first four are pulsed; the pulses have amplitudes comparable with those from x-rays in the wavelength region >3 Å and determine the long-wavelength useful limit of the detector. The dark current may be reduced in several ways: (1) selection of the multiplier phototube for low noise; (2) operation at as low a potential as feasible; (3) cooling; the socket may be water-cooled to reduce heating by the potential divider, which usually consists of a bank of resistors mounted in the socket; the tube itself may be cooled with liquid nitrogen; and (4) use of a coincidence circuit; the outputs of two matched multiplier phototubes sharing a common phosphor may be connected to a coincidence circuit, which rejects all pulses that do not come from both tubes simultaneously. Of these techniques, only selection, low-potential operation, and water-cooling are applied on analytical spectrometers.

6.3.2. Operation

6.3.2.1. Proportionality in Scintillation Counters

The production of light photons in the scintillation crystal is analogous to the production of ion–electron pairs in the gas detector. If the energy of the incident x-ray photon is imparted to a photoelectron which in turn expends its energy in producing light photons, and if each light photon requires the same amount of energy, it follows that the number of light photons should be proportional to the x-ray photon energy. In fact, the average number of light photons produced should be—and is—simply the x-ray photon energy divided by the light photon energy, ~3 eV for 4100-Å photons. However, the overall scintillation process is much less efficient than the overall gas-ionization process. Thus, the *effective light-photon energy* may be defined as the average energy required to produce each photoelectron actually collected on the first dynode of the electron multiplier, and may be calculated as follows. Of the 3-eV light photons produced by the incident x-ray photon, ~0.9 of them reach the photocathode; of these ~0.1 produce photoelectrons; and of these ~0.9 reach the first dynode (Section 6.3.1.2). Thus $0.9 \times 0.1 \times 0.9 = 0.08$ of the light photons—that is, 8 out of each 100—are effective in initiating a secondary-electron cascade and contributing to the output pulse. The effective light-photon energy is then $(100/8) \times 3 \approx 40$ eV. Table 6.1 shows the average number of light photons produced by Cu $K\alpha$ (1.54 Å, 8040 eV) and Mo $K\alpha$ (0.71 Å, 17,440 eV) photons.

Again it must be emphasized that even with a strictly monochromatic x-ray beam, the numbers of light photons produced by individual x-ray photons have a wide Gaussian distribution having a mean value $\bar{n}$ and a deviation $\sqrt{\bar{n}}$. However, Table 6.1 shows that Mo $K\alpha$ photons, having about twice the energy of Cu $K\alpha$ photons, do indeed produce about twice as many light photons. Thus, the scintillation process provides the basis for proportionality of output pulse height and incident photon energy.

6.3.2.2. Phenomena in the Scintillation Counter

To continue the analogy of light-photon (scintillation) and ion–electron pair production, the x-ray photon in the scintillation crystal may initiate one of five processes directly analogous to those that occur in the gas volume of a proportional counter (Section 6.2.3.1, Figure 6.4):

1. The x-ray photon may pass through the crystal unabsorbed, to be absorbed in the glass envelope or internal structure of the phototube. This process is unlikely except at very short wavelengths.

2. It may produce a burst of light photons (a scintillation). The x-ray photon undergoes photoelectric absorption in an outer electron shell of an atom in the phosphor, producing a photoelectron having kinetic energy equal to the incident photon energy minus the binding energy of the orbital electron, which is only a few electron volts. This photoelectron then expends its energy producing light photons by interaction with phosphor atoms.

If the x-ray photon has wavelength shorter than that of the iodine K-absorption edge, it may undergo photoelectric absorption in the K-shell of an iodine atom, expelling an iodine K electron and expending the remainder of its energy in imparting kinetic energy to the photoelectron. The photoelectron expends its energy in producing light photons. However, the I $K\alpha$ (or $K\beta$) photon resulting from the filling of the K-shell vacancy can undergo one of the three processes listed below. Of course, analogous processes can occur when incident x-rays expel I L-shell electrons, but for simplicity the following discussion is in terms of I K-shell excitation.

3. The I $K\alpha$ (or $K\beta$) photon may escape the scintillation crystal. This is not an unlikely process, because an element is relatively transparent to its own characteristic radiation.

4. It may produce a burst of light photons by process 2 above.

5. It may undergo absorption within the atom of its origin (Auger effect) with resultant emission of an Auger electron and a satellite-line x-ray photon. The Auger electron expends its energy producing light photons. The satellite photon undergoes process 2 above.

The first of the foregoing processes produces no light photons. The second process produces a number of photons proportional to the x-ray photon energy. In the fourth process, the sum of the photons produced by the photoelectron and the I $K\alpha$ photon equals the number produced by the second process. In the fifth process, the sum of the photons produced by the photoelectron, the Auger electron, and the satellite-line photon equals the number produced by the second process. Thus, the second, fourth, and fifth processes produce a number of light photons proportional to the x-ray photon energy. However, in the third process, the number of light photons is proportional to the difference between the energies of the incident and I $K\alpha$ photons. This is, of course, the escape-peak phenomenon.

A certain fraction ($\sim$0.9) of the scintillation photons find their way to the photocathode. A certain fraction (0.1–0.2) of these result in emission of photoelectrons, a certain fraction ($\sim$0.9) of which fall upon the first dynode and initiate the secondary-electron multiplication process. If there are n dynodes having a secondary-emission ratio of σ, the output pulse contains $0.9 \times 0.1 \times 0.9 \times \sigma^n$ electrons for each photon in the scintillation.

Thus, the output of the scintillation counter, like that of the proportional counter, contains a pulse-height distribution (peak) for each incident x-ray wavelength and may contain an escape peak for each incident wavelength shorter than the iodine *K*-absorption edge. The closer the incident wavelength lies to the *K*-absorption edge, the greater is the intensity of the escape peak relative to the main peak. Escape peaks are not nearly as troublesome with scintillation counters as with argon- and krypton-filled proportional counters. The iodine *L*III absorption edge lies at 2.7 Å and the principal iodine *L* lines at ~3 Å or more. These wavelengths are at the long-wavelength limit of the usual working range of the scintillation counter. The iodine *K*-absorption edge lies at 0.37 Å. Thus, escape peaks occur only for $K\alpha$ lines of elements having atomic number 57 (lanthanum) or higher, and these elements are usually determined by measurement of their *L* lines with gas-flow proportional counters. In general, all pulse-height distributions are much broader for a scintillation counter than for a proportional counter, and the detector noise pulses have much higher intensity.

6.4. LITHIUM-DRIFTED SILICON AND GERMANIUM DETECTORS

6.4.1. Structure

Probably the most significant recent development in x-ray intensity measurement is the lithium-drifted, or lithium-compensated, semiconductor detector, principally lithium-drifted silicon, Si(Li) or Si:Li, and germanium, Ge(Li) or Ge:Li ("silly" and "jelly," respectively, in jargon) (*B75*, *B92*, *G5*, *G27*, *H26*, *S31*). The lithium-drifted detector consists of a silicon (or germanium) single-crystal semiconductor having a compensated intrinsic (*i*-type) region sandwiched between positive and negative (*p*- and *n*-type) regions. It is thus a *p–i–n*-type diode. The compensated region is formed by diffusion ("drifting") of lithium into *p*-type silicon (or germanium) under carefully controlled conditions. The lithium compensates impurities and dopants already present. The thin *p*-type layer on the detector surface is inactive—that is, does not contribute to the detection process—and is known as the *dead layer*. The two most prominent physical features of the detector are its area and thickness. Geometric efficiency—intercepted solid angle—increases, but resolution decreases with increasing area; absorption

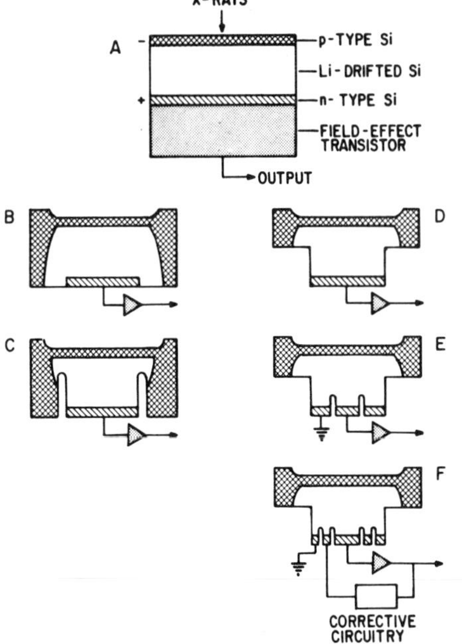

FIG. 6.7. Types of lithium-drifted silicon [Si(Li)] detectors. The detectors are circular and shown here in vertical cross section. A. Detector and field-effect transistor preamplifier (schematic). B. Planar. C. Grooved (Woo-type). D. Top hat. E. Simple guard-ring. F. Double guard-ring.

efficiency increases with thickness. Figure 6.7 shows the lithium-drifted silicon detector schematically and in several practical forms (*G27*) designed to improve charge collection and resolution.

Elcctrodcs consisting of films of gold ~200 Å thick are evaporated on the faces of the lithium-drifted silicon wafer, which, together with its field-effect transistor preamplifier (see below), is then provided with a beryllium window $\lesssim$10 μm thick and encapsulated on the back side and rim. The window and encapsulation protect the detector from: (1) surface contamination, for which, being at liquid-nitrogen temperature (see below), it constitutes a likely condensation target; (2) extraneous light and light from optically fluorescent specimens; and (3) in electron-probe microanalyzers and scanning electron microscopes, scattered electrons. Alternatively, the window and encapsulation may be omitted from the detector itself and provided by the vacuum cryostat. For measurement of very long wavelengths, such as the *K* lines of carbon, nitrogen, and oxygen, provision may be made to tip open the window when the instrument is under vacuum and close it when air is admitted.

When in use, the detector must be maintained at liquid-nitrogen temperature (−196°C, 77°K) to reduce noise and ensure optimal resolution. Moreover, lithium has an extremely high diffusion rate, even at room temperature, so that lithium-drifted silicon detectors must be maintained at liquid-nitrogen temperature at *all* times, even when not in use and when being shipped. However, ordinarily a 20-l dewar reservoir requires filling only once each week or two. It is likely that semiconductor detectors will be developed that do not require cooling when idle; in fact, some Si(Li) detectors can be "cycled"—that is, warmed to room temperature, then returned to liquid-nitrogen temperature—up to 25 times. Such detectors can also stand at room temperature for several days. However, it is unlikely that such detectors can be developed that do not require cooling when in use. Finally, operation in vacuum is usually required to permit operation at long wavelengths and to prevent condensation of moisture on the detector, and light must be excluded to prevent light-generated detector noise.

The *guard-ring detector* arrangements shown in Figure 6.7E,F were devised by Goulding *et al.* (*G27*) to reduce the principal source of continuous background in Si(Li) detectors—degradation of pulses arising from scattered primary x-rays. If this primary radiation is monochromatic, all these pulses should occupy a narrow distribution at the high-energy end of the spectrum. However, various phenomena in the detector result in incomplete charge collection and consequent clutter of the entire spectrum with abnormally small pulses. The guard-ring arrangements, used with appropriate external solid-state circuitry, effectively preclude this effect.

6.4.2. Operation

In operation, a reverse-bias potential of 300–900 V is applied across the detector. The resulting electric field depletes the compensated region of electron–hole free-charge carriers, the electrons and holes being drawn to the *p*- and *n*-type regions, respectively. The depleted region constitutes the radiation-sensitive detector volume and typically has an area of 30–100 mm^2 and a depth of 2–5 mm. Because of the carrier-depleted region, the lithium-drifted detector is a high-impedance device.

Consider an x-ray photon entering the detector shown in Figure 6.7 and passing through the active semiconductor volume—the lithium-drifted silicon region. Eventually, it undergoes photoelectric absorption by a silicon atom and imparts its energy to a photoelectron, which produces electron–hole pairs along its path until its energy is expended—just as ion–electron pairs are produced in a gas detector and light photons in a scintillation counter. This is the basis for proportionality in the Si(Li) detector. Each such ionization extracts ~3.8 eV from the photoelectron. The applied bias potential results in the collection of these electron–hole pairs, and the total charge collected Q is linearly proportional to the x-ray photon energy E_x and is given by Equation (6.1),

$$Q = (E_x/3.8)(1.6 \times 10^{-9}) \quad \text{coulomb} \tag{6.5}$$

The semiconductor detector does not have the inherent multiplication analogous to that provided by gas amplification in the proportional counter or by secondary electron multiplication in the scintillation counter. Thus the Si(Li) detector is analogous to the gas-filled ionization chamber (Section 6.2.2.3c), but with the charge carriers formed in a semiconductor solid rather than in a gas. However, work is in progress on development of the solid-state *avalanche* detector (Section 6.4.5), which does have internal electron multiplication and is the solid-state analog of the gas-filled proportional counter. Due to the absence of internal amplification, a low-noise, high-gain, charge-sensitive preamplifier is required to integrate the total collected charge into a current pulse. For this function, a field-effect transistor (FET) is used, and this must also be maintained at liquid-nitrogen temperature to reduce noise level, and is actually integral with the detector (Figure 6.7). The current pulse is subsequently amplified and converted to a potential pulse by additional preamplifier stages.

The lithium-drifted silicon (or germanium) detector, then, consists of the Si(Li) or Ge(Li) detector itself and the field-effect transistor in a liquid-nitrogen-cooled, light-tight, vacuum cryostat, several additional preamplifier stages outside the cryostat, and the liquid-nitrogen reservoir, which typically has capacity of ~20 l, a 10–14-day supply.

Detection efficiency is a function of the atomic number of the semiconductor and the depth of the depletion layer. A 3-mm layer of silicon absorbs 50% of incident 30-kV (0.4-Å) photons, but photoelectric absorption decreases rapidly with increasing photon energy. Thus, Si(Li) detectors are optimal at wavelengths >0.4 Å (Cs $K\alpha$), Ge(Li) detectors are preferable at shorter wavelengths. It is evident that the Si(Li) detector is suitable for substantially the entire spectral range of a 50-kV x-ray spectrometer. The reduced efficiency of silicon at short wavelengths is actually beneficial in that the K spectra of the light and intermediate elements and the L and M spectra of the heavy elements are less cluttered with short-wavelength peaks. Another advantage is that the silicon escape peak is small because of low fluorescent yield and usually lies outside the useful region of the silicon detector.

There are certain "escape" phenomena peculiar to the solid-state detector [pp. 2.9–13 in (*39*)]:

The first of these is *Compton escape*. The source–specimen–detector arrangement in energy-dispersive spectrometry is favorable for scatter of primary x-rays from a tube, secondary-target, or radioisotope source into the detector: The specimen is commonly supported on a low-Z substrate, and the specimen–detector distance is small. If the primary x-rays consist principally of spectral lines, the modified and unmodified scattered primary x-rays will give pulse distributions in the Si(Li) detector output. However, they also undergo scatter—mostly Compton—in the detector itself. If the Compton-scattered photon escapes from the detector, of the incident scattered primary photon energy, only the energy of the Compton recoil electron remains to produce electron–hole pairs. Moreover, because of the close proximity of the specimen and detector and the absence of collimation, the scattered primary x-rays arrive at the detector over a substantial angular range and therefore undergo a range of energy losses [Equations (2.38) and (2.39)] and produce recoil electrons having a range of energies. As a result, there appears at the low-energy end of the energy spectrum a broad, diffuse background—the *Compton-escape continuum*—having a more or less well-defined *Compton-escape edge* on its high-energy side.

Another interesting "escape" phenomenon occurs if the x-rays incident on the detector are capable of exciting Si K x-rays and are absorbed in the dead layer (Section 6.4.1). Any Si K x-rays arising from such absorption that "escape" into the active detector volume are detected in the usual way and give a Si K pulse distribution even if silicon is absent from the specimen —or an erroneously intense one if it is present.

For incident x-rays able to excite Si K x-rays ($\lambda < \lambda_{\text{Si}K_{ab}}$, 6.74 Å; $E_x > V_{\text{Si}K}$, 1.84 keV), the escape peak intensity increases as wavelength increases. This is not only because the wavelength lies progressively nearer the Si K-absorption edge, but because the photons are absorbed progressively nearer the surface of the active detector volume, and any excited Si K photons are more likely to escape.

6.4.3. Advantages

The lithium-drifted silicon detector has six principal advantages:

1. The energy conversion is very efficient. The production of one electron–hole pair requires 3.8 and 2.9 eV in silicon and germanium, respectively, compared with 20–26 eV to produce an ion–electron pair in

TABLE 6.2. Width and Energy Resolution of Pulse-Height Distributions of Proportional, Scintillation, and Semiconductor X-Ray Counters

Cu $K\alpha$ photon energy		8040 eV
Ar-filled proportional counter:		
Effective ionization potential of Ar		26.4 eV per Ar^+,e^- pair
Number of Ar^+,e^- pairs per Cu $K\alpha$ photon[a] $\bar{n}$	8040/26.4	305 pairs
Deviation for Gaussian distribution σ	$\sqrt{305}$	17.5 pairs
Half-width of distribution $W_{1/2}$	$2.35\sqrt{305}$	41.3 pairs
Increase in half-width due to fluctuations in gas gain in the avalanching process	$\sim 1.3\times$	53.7 pairs
Resolution R	$(53.7/305)\times 100$	17.6%
NaI(Tl) scintillation counter:		
NaI(Tl) phosphor efficiency		50 eV/(light) photon
Number of light photons per Cu $K\alpha$ photon	8040/50	161 photons
Number of light photons arriving at photocathode (efficiency of optical coupling between phosphor and photocathode $\sim$0.9)	0.9×161	145 photons
Number of photoelectrons emitted (efficiency of photocathode $\sim$0.1)	0.1×145	14.5 photoelectrons
Number of photoelectrons collected by the first dynode[a] (efficiency of collection $\sim$0.9), $\bar{n}$	0.9×14.5	13 photoelectrons
Deviation for Gaussian distribution σ	$\sqrt{13}$	3.6 photoelectrons
Half-width of distribution $W_{1/2}$	$2.35\sqrt{13}$	8.5 photoelectrons
Increase in half-width due to fluctuations in the secondary-electron multiplication process	$\sim 1.2\times$	10.2 photoelectrons
Resolution R	$(10.2/13)\times 100$	78%
Si(Li) semiconductor detector:		
Energy to produce one electron–hole pair		$\sim$3.8 eV (e^-,hole) pair
Number of (e^-,hole) pairs per Cu $K\alpha$ photon[a] $\bar{n}$	8040/3.8	2116 pairs
Deviation for Gaussian distribution σ	$\sqrt{2116}$	46 pairs
Half-width of distribution $W_{1/2}$	$2.35\sqrt{2116}$	108 pairs
Resolution R	$(108/2116)\times 100$	5%

[a] Statistically limiting process.

a gas detector and ~50 eV to produce an effective light photon in a scintillation detector.

2. It follows that the resolution is excellent; the FWHM of the pulse-height distribution (Section 6.5.1.12) may be as small as 130 eV for 6–8-keV photon energy, and spectral lines of the same series for adjacent atomic numbers can be resolved. Table 6.2 gives the calculation of resolution of a Si(Li) detector for Cu $K\alpha$ photons. Figure 6.8 compares the Ag $K\alpha,\beta$ pulse-height distributions for the NaI(Tl) scintillation, xenon proportional, and Si(Li) semiconductor detectors. Figure 6.9 compares the resolution of the NaI(Tl) scintillation, argon proportional, and Si(Li) detectors and the LiF(200) analyzer crystal as a function of photon energy. Figure 6.10 shows the performance of the Si(Li) detector used with a 400-channel analyzer for the K lines of some elements of high atomic number.

3. The detector, if sufficiently thick (~3 mm), like the scintillation counter, has substantially 100% quantum efficiency over much of its useful spectral region.

4. The Si(Li) detector is free from pulse-height drift and broadening at intensities up to ~10,000 counts/s.

5. Semiconductor detectors, particularly Ge(Li). permit measurement

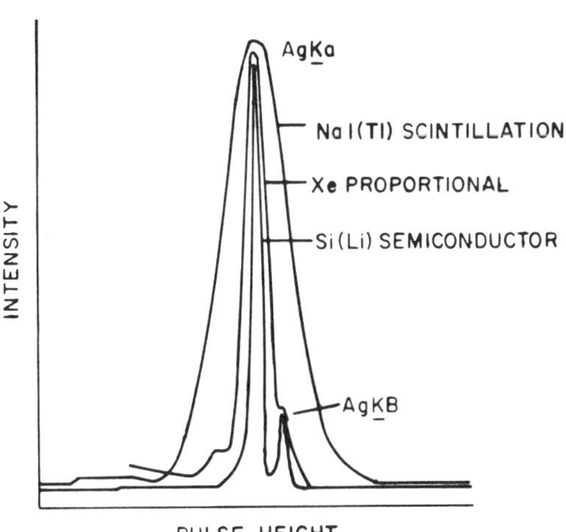

FIG. 6.8. Comparison of the Ag $K\alpha,\beta$ pulse-height distributions for the NaI(Tl) scintillation counter, Xe proportional counter, and Si(Li) semiconductor detector. [P. G. Burkhalter and W. J. Campbell, *Oak Ridge National Laboratory Report* **ORNL-IIC-10**, vol. 1 (1967); courtesy of the authors.]

of the K lines of the heaviest elements with higher efficiency and resolution than is realized with a conventional crystal spectrometer.

6. The small size of the detectors permits their use in very close proximity to the specimen and in relatively inaccessible places, such as in electron-probe microanalyzers. Geometric efficiency—that is, the fraction of emitted x-rays collected—increases with detector size, whereas resolution decreases with size; for example, at 5.9 keV (Mn $K\alpha$), 300-mm^2 and 10-mm^2 Si(Li) detectors may have resolutions of $<$350 eV and $<$170 eV, respectively. Some of the advantages of energy-dispersive methods in general (Section 8.2.1.2) are attributable to the small physical size of the semiconductor detector, for example, ability to place the detector physically close to the specimen to intercept a large solid angle of radiation, and relative independence of the detector from specimen position and orientation.

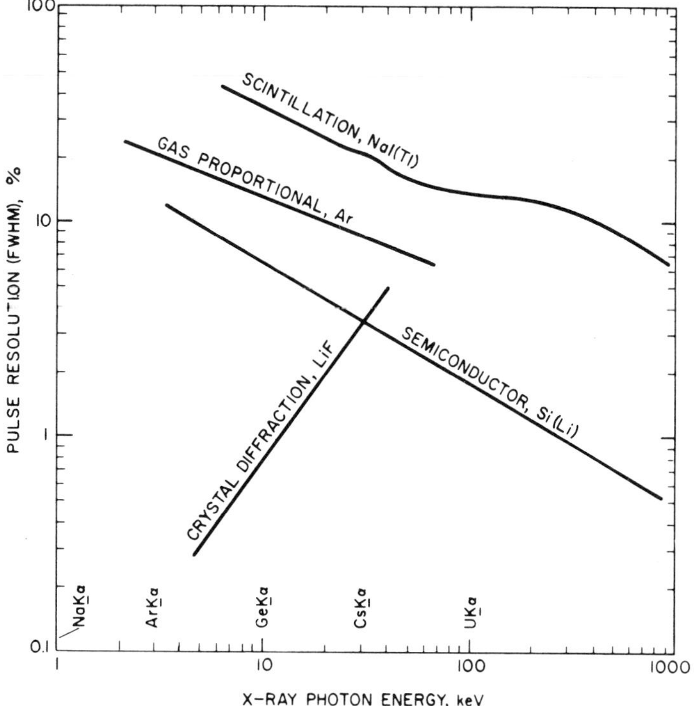

FIG. 6.9. Comparison of resolutions of the NaI(Tl) scintillation counter, Ar proportional counter, and Si(Li) semiconductor detectors, and LiF analyzer crystal. [P. G. Burkhalter and W. J. Campbell, *Oak Ridge National Laboratory Report* **ORNL-IIC-10**, vol. 1 (1967); courtesy of the authors.]

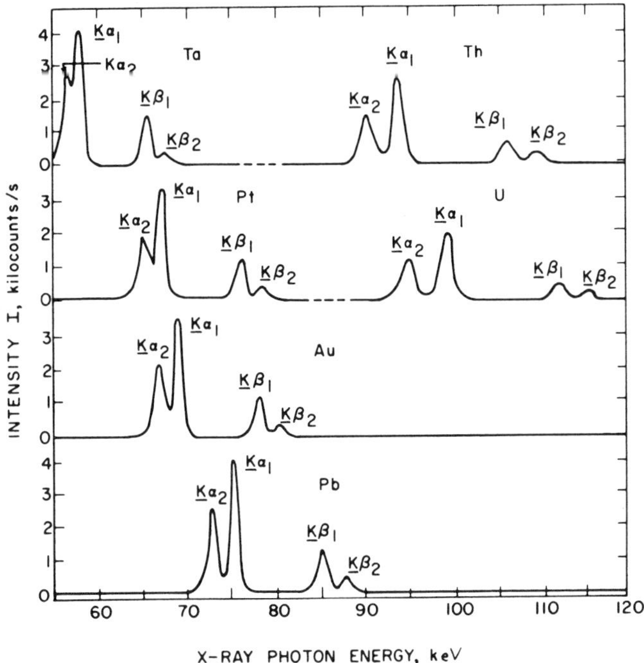

FIG. 6.10. Pulse-height distribution curves from a Si(Li) semiconductor detector used with a 400-channel analyzer. [H. R. Bowman, E. K. Hyde, S. G. Thompson, and R. C. Jared, *Science* **151**, 562 (1966); courtesy of the authors and the American Association for the Advancement of Science.]

6.4.4. Limitations

The principal limitations of the detector are the following:

1. The high mobility of the lithium causes intolerable noise levels and substantially permanent deterioration of detector characteristics at room temperature. Consequently, the detector must be maintained perpetually at liquid-nitrogen temperature.

2. Even at liquid-nitrogen temperature, the noise level is relatively high, especially for low (<6 keV) photon energies. This limits the applicability of the detector to atomic number not lower than ~11 (sodium) for ~1% analyte concentration.

3. The detector is limited to relatively low intensities—not more than ~20,000 counts/s—and resolution deteriorates rapidly as intensity increases.

This limitation is particularly serious because the detector is almost invariably used in energy-dispersive mode where the entire specimen spectrum enters the detector at once—not just the analyte line. With certain refinements, the detector can be used for total intensities as high as 50,000 counts/s. However, even at this intensity, measurement of the line intensity from an analyte present in low concentration may require long counting time for acceptable statistical precision. For example (*G14*), 50 ng/cm^2 of analyte in a sample–substrate system of 5 mg/cm^2 constitutes 10 ppm and would require accumulation of 10^5 photons for each analyte-line photon. At 20,000 counts/s, it would take a minimum live time of 500 s (~8 min) to accumulate 100 analyte-line counts for 10% σ. Real counting time may exceed live time by a factor of 2–3 because of dead time while the electronic circuits are processing a pulse. Conversely, a proportional or scintillation counter may receive intensities as high as 10^5/s, and, at high analyte concentrations, this may be substantially all analyte-line x-radiation.

4. The resolution at long wavelengths, although excellent compared with a proportional counter, is still inferior to that of a conventional crystal–proportional counter system. The best commercial Si(Li) detectors (~150 eV resolution) do not resolve the $K\alpha$ lines of element Z from the $K\beta$ lines of element Z − 1 for elements 17–28 (chlorine to nickel), and typical detectors have poor resolution for the $K\alpha$ lines of elements below chlorine. Thus elements below nickel may require mathematical unfolding (Section 8.1.9), which may be acceptable for similar intensities, but not for low intensities in the presence of high ones. Little further improvement in resolution is to be expected for commercial detectors, so it appears that the LiF(200) crystal spectrometer will retain superior resolution for wavelengths >0.8 Å, which includes the useful lines of nearly all elements [Reference (*53*) for 1972]. However, Si(Li) and Ge(Li) detectors have resolution superior to that of a crystal spectrometer for the *K* lines of elements having atomic number >40 (zirconium), and the *L* lines of the heaviest elements (Figure 6.10).

5. The effective area of a high-resolution Si(Li) detector is at present 12.5–30 mm^2, compared with several square centimeters for a typical proportional counter. The smaller area results in smaller intercepted solid angle of specimen emission, in part offsetting the advantage of close proximity of detector and specimen.

6. Although windowless Si(Li) detectors can be used for the *K* lines of elements down to atomic number 6 (carbon), because of the sensitivity of such detectors to surface contamination, they can be used only under carefully controlled conditions.

At present, the Si(Li) detector finds its most valuable applications where overall intensity—including analyte-line intensity—is low. Such applications include trace and micro analysis, selected-area analysis, electron-probe microanalysis, scanning electron microscopy, and analyses where power limitations require low-level excitation. It follows that the detector is applicable with a wide latitude of excitation sources—low-power x-ray tubes, electron beams, radioisotopes, proton and other ion beams, and high-power x-ray tubes with secondary x-ray targets.

6.4.5. Avalanche Detectors

It is mentioned above that the lithium-drifted silicon or germanium detector is the solid-state analog of the gas-filled ionization chamber—there is no internal electron multiplication. The solid-state analog of the gas-filled proportional counter is known as the *avalanche detector*. The primary electrons produced by the initial photoelectric absorption of the incident x-ray photon undergo electron multiplication—avalanching—just as the primary electrons in a gas-proportional counter. At present, the electron multiplication is not uniform across the detector. However, when these counters are perfected, they will combine small size, ruggedness, and high counting rates without coincidence loss and may well replace the gas proportional counter and perhaps the presently used solid-state semiconductor detectors.

6.5. EVALUATION OF X-RAY DETECTORS

6.5.1. Detector Characteristics

This section describes some parameters that may apply to the detector or to the detector–readout system. However, modern x-ray spectrometers are usually equipped with readout components having count-rate capability slower than that of the detector, and it is usually the readout components that limit the performance with respect to these characteristics.

6.5.1.1. Rise Time (Figure 6.11)

This is the time interval, after initiation of a pulse, for the pulse to attain 90% of its full amplitude.

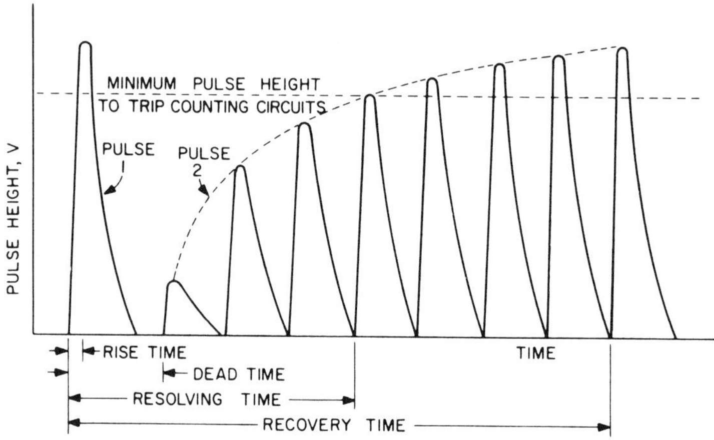

FIG. 6.11. Amplitude of a detector output pulse as a function of the time interval after initiation of the preceding pulse.

6.5.1.2. Dead Time (Figure 6.11)

This is the time interval, after initiation of a pulse of substantially full amplitude, during which the detector is completely insensitive and gives no response whatever to a second incident x-ray photon. In a gas detector, dead time is determined by the time required for the positive-ion sheath to decay to the extent that some avalanching can occur again. In proportional counters, the avalanching is localized very near the anode wire, whereas, in Geiger counters, it is generalized throughout the gas volume. Thus, the positive-ion sheath decays much more rapidly in proportional counters than in Geiger counters. In a scintillation counter, dead time is determined by the time required for the scintillation to decay.

Normally, dead time is ~200 μs for Geiger counters, ~0.2 μs for proportional counters, and ~0.25 μs for scintillation counters. However, dead time is affected by the type of x-ray generator, and by the intensity and energy of the measured x-rays. Dead time is longer if the measured x-rays are excited by a full-wave generator than if by a constant-potential generator. If the second x-ray photon follows closely after expiration of the dead time for pulse 1, pulse 2 has an abnormally small amplitude and is followed by an abnormally short dead time. Finally, dead time is a function of, and in fact varies about linearly with, pulse height.

The detector responds to an x-ray photon after the expiration of the dead time, but the pulse height is abnormally low. The longer after the expiration the photon enters the detector, the more nearly normal is the pulse height.

Actually, the term *dead time*, as commonly used, refers to the overall detector–readout system and is the time required to detect a photon and "process" its pulse. During this interval, the system cannot respond to another x-ray photon and is thus, in effect, "dead." The time after registration of a pulse but before initiation of the subsequent one is termed *live time*. In brief, dead time is the time spent in actual measurement, live time the time spent in "waiting" for x-ray photons.

Incidentally, if two or more x-ray photons enter the detector *simultaneously*, or substantially so, the effect is the same as if the detector had received a single photon having energy equal to the sum of the individual photon energies (Section 8.1.8.4b).

6.5.1.3. Resolving Time (Figure 6.11)

This is the time interval, after initiation of a pulse, after which the detector can produce a second pulse high enough to actuate the counting circuits, or, more simply, it is the minimum time interval between pulses that can be counted. In calculation of coincidence loss (see below), resolving time is often incorrectly referred to as dead time because, in effect, the detector is "dead" during the resolving time, in the sense that it cannot count a photon.

6.5.1.4. Recovery Time (Figure 6.11)

This is the time interval, after initiation of a counted pulse, after which the detector can produce a second pulse of substantially full pulse height. To summarize, after the dead time, a second pulse can occur; after the resolving time, a second pulse can be counted; after the recovery time, a second pulse has normal amplitude.

6.5.1.5. Linear Counting Range (Figure 6.12)

This is the incident x-ray intensity range over which the number of countable pulses per second from the detector is directly proportional to the number of x-ray photons arriving at the detector per second. The linear ranges of Geiger, proportional, and scintillation counters are given in Table 6.3. A detector capable of very high count rate without substantial departure from linearity is known as a *fast* counter.

6.5.1.6. Coincidence Loss (Figure 6.12)

This term refers to the x-ray photons that are "lost," that is, not counted, because they arrive at the detector too soon after the preceding

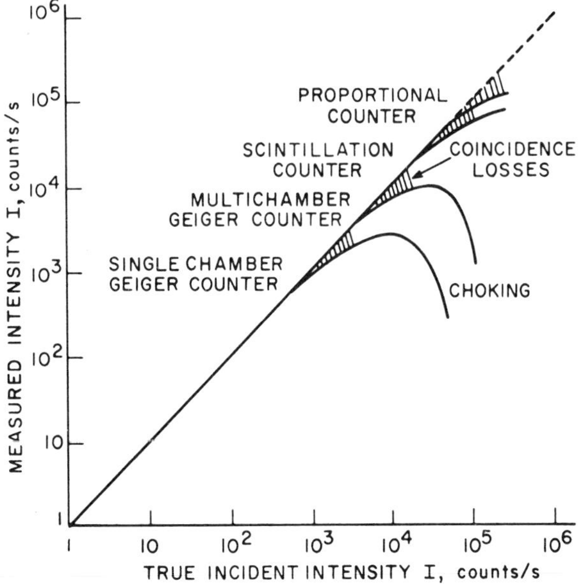

FIG. 6.12. Linear counting range, coincidence loss, and choking in commonly used x-ray detectors.

photon. The nonlinear portions of Figure 6.12 are the result of coincidence loss and occur at intensities so high that the rate at which x-ray photons enter the detector exceeds its resolving time. The following equation may be used to correct for coincidence loss or, with the multiple-foil method, to calculate resolving time:

$$I_{\text{true}} = I_{\text{obs}}/(1 - I_{\text{obs}}T_R) \tag{6.6}$$

where I_{true} and I_{obs} are true and observed intensity, respectively (counts/s), and T_R is resolving time (s). This equation is applicable to intensities measured at peaks, but not to integrated intensities, because the observed intensity, and therefore the coincidence loss, varies continuously during the integration. Also, the equation is valid only if the x-ray photons arriving at the detector have a wholly random time distribution. Consequently, the equation is much less satisfactory with full-wave excitation than with constant potential because with the former, x-rays are generated in the x-ray tube in bursts (Section 4.3.3.1).

Coincidence losses in the preset-count method (Section 7.2.2.2) are calculated with the equation

$$I_{\text{true}} = N/(T - NT_R) \tag{6.7}$$

where N is the accumulated count, T the counting time (s), and T_R the resolving time (s).

As a practical example, consider a proportional counter having a resolving time of 1 μs giving a measured intensity of 10,000 counts/s. The detector does not count photons for 1 μs after the initiation of each counted pulse. Thus, at 10,000 counts/s, the detector is inactive $(10^4)(1 \times 10^{-6}) = 0.01$ s/s, or 1% of the time, and 1% of the incident photons are lost. Thus, at intensities at or below this level, coincidence losses may usually be disregarded. Moreover, many counting systems have resolving times <1 μs.

In order to use Equation (6.6), it is necessary to know the dead time of the detector–readout system, which may be assumed to be limited by the electronic components, rather than the detector. The dead time of the electronics may be measured directly by use of a double-pulse generator by observing the closest pulse separation that is just resolved. Alternatively, one of the methods given below (*S33*) may be used to establish an empirical curve, and the value of dead time that best fits the curve is calculated. In the following methods, the measurements should be made on a monochromatic spectral line after diffraction from the analyzer crystal.

In the *multiple-foil method*, shown in Figure 6.13, a stack of metal foils, usually aluminum or nickel, is placed in the x-ray path. The foils are removed one at a time, and x-ray intensity is scaled after each removal. In the figure, the plotted data show noticeable departure from linearity at ~7000 counts/s and a true incident intensity of ~150,000 counts/s. The foils must be flat, free of pinholes, warp, and wrinkles, and of identical thickness. They must be placed precisely perpendicular to the x-ray beam, and secondary emission from the foils must be excluded from the detector.

In the *x-ray tube current method*, intensity is measured from a suitable specimen at a wide range of x-ray tube currents—say, 0–50 mA at 2-mA increments—but with constant x-ray tube potential. A curve of measured intensity *versus* current is plotted on linear or log-log paper. The method is not as good as the multiple-foil method because the milliammeter may not give an accurate indication of true effective electron current in the tube (Section 4.2.4.1).

In the *single-foil absorption-ratio method*, intensity is measured directly, I_0, and through a single foil, I, having transmission $I/I_0 \approx 0.3$–0.5. The ratio is measured at progressively increasing incident intensities I_0 of ~100–10^5 counts/s set by variation of the x-ray tube current. A curve of $\log_{10} I_0$ *versus* I/I_0 is plotted. The coincidence loss affects the higher intensity I_0 first, and thereafter the loss in I_0 will always be greater than that in I. Consequently, I/I_0 increases as coincidence loss increases, as shown in

Figure 6.14A. From the true value of I/I_0 at low intensity, a correction curve is derived showing the correction to be added for a specified observed I, as shown in Figure 6.14B. The foil must be flat and free of pinholes, warp, and wrinkles. It must be placed precisely perpendicular to the x-ray beam, and secondary emission from the foil must be excluded from the detector.

Alternatively, the ratio method may be applied to other measured intensities. The $K\beta/K\alpha$ intensity ratio may be measured from some element at a series of intensity levels and $\log_{10}I_{K\alpha}$ *versus* $I_{K\beta}/I_{K\alpha}$ plotted. The same technique may be applied to the second and first orders of the same line. The effectiveness of these methods depends on the reproducibility of the resetting of the goniometer from line to line. The ratio of the intensities of an x-ray line may be measured at a series of intensity levels from two spec-

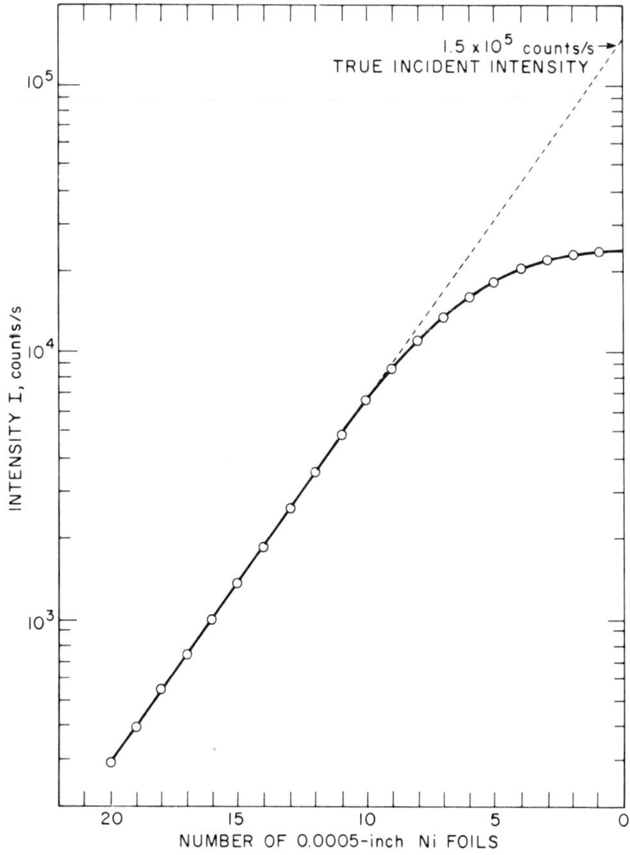

FIG. 6.13. The multiple-foil method for measurement of true intensity.

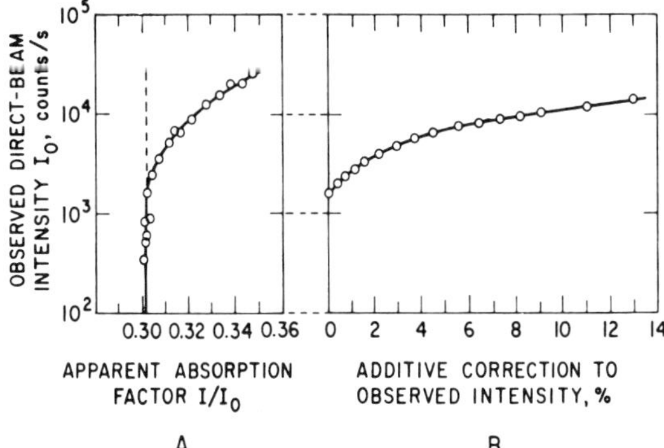

FIG. 6.14. The single-foil absorption-ratio method for correction of dead time. A. Variation of apparent I/I_0 for a 0.018-mm (0.0007-inch) nickel foil with increasing incident intensity. B. Correction curve derived from curve A.

imens containing that element in different concentrations, say 3% and 10%; the plotted function is $\log_{10} I_{10\%}$ *versus* $I_{3\%}/I_{10\%}$. Finally, the ratio of the intensities measured alternately through relatively large l and small s apertures may be used; the plotted function is $\log_{10} I_l$ *versus* I_s/I_l. It is necessary that the apertures can be reproducibly placed repeatedly and that the intensity distribution across the x-ray beam does not change substantially with change in x-ray tube current.

A simpler $K\alpha/K\beta$ ratio method is given by Birks [p. 63 in (7)]. Excitation conditions are set to give $K\alpha$ intensity of ~1000 counts/s; $K\beta$ intensity is then ~200 counts/s. Coincidence losses are negligible at these intensities. The two intensities are measured, $I_{K\alpha}$ and $I_{K\beta}$. The excitation is now increased to increase the $K\beta$ intensity to ~1000 counts/s, and the two intensities are measured again, $I'_{K\alpha}$ and $I'_{K\beta}$. The $K\beta$ intensity is still free of coincidence loss, so that $I'_{K\beta,\text{true}} = I'_{K\beta}$, but the $K\alpha$ intensity is now subject to coincidence loss, and the true $K\alpha$ intensity is given by

$$I'_{K\alpha,\text{true}} = I_{K\alpha}(I'_{K\beta}/I_{K\beta}) \tag{6.8}$$

The dead time is then given by Equation (6.6) using the true and observed $I'_{K\alpha}$ intensities.

In all the above ratio procedures, background has been neglected on the assumption that the contribution of the background remains a substantially constant fraction of the line intensity.

Other ways of dealing with coincidence loss are provided by the Harms dead-time and Lowes live-time corrections (Section 8.2.2.4). Probably the ultimate method for elimination of coincidence loss is provided by the triode x-ray tube and associated electronic circuitry described briefly in Section 4.2.5.5.

6.5.1.7. Choking (Figure 6.12)

Choking or blocking is the condition whereby an increase in incident intensity actually results in a decrease in measured intensity. In Figure 6.11, if pulse 2 occurs shortly after the dead time of pulse 1, it produces a pulse too small to be counted, and, in effect, extends the dead time following pulse 1. The higher the incident intensity, the more such small pulses occur, until, finally, all pulses are too small to be counted, that is, the dead time is extended indefinitely, and the counter completely chokes. In making a 2θ scan through an extremely intense peak, it is possible for the detector–readout system to choke to the extent that the peak profile splits symetrically. The higher the intensity, the deeper the split is until, ultimately, measured intensity—that is, detector–readout system output—may actually fall to zero at the center of the peak. This phenomenon is known as peak *reversal.*

6.5.1.8. Plateau (Figures 6.15 and 6.16)

This is the substantially straight, horizontal portion of the intensity *versus* potential characteristic of the detector tube, in which the counting rate is substantially independent of the applied potential. Suppose that monochromatic x-ray photons are arriving at the detector at some constant intensity, say 1000/s, and that amplifier gain and pulse-height analyzer baseline settings (Chapter 7) are constant. If, now, the detector tube potential is increased from zero, the observed detector output intensity follows the curve shown in Figure 6.15. The measured intensity just perceptibly rises above zero at the *starting potential*, becomes substantially constant at the *threshold potential*, remains substantially constant over a range termed the *plateau*, and finally rises again at the *discharge potential.* The *operating potential* is usually set near the center of the plateau. The interval between the threshold and operating potentials is termed the *overpotential.* The plateau curve is displaced to progressively lower potential the higher the amplifier gain, the lower the discriminator setting (that is, the lower the minimum countable pulse height), and the shorter the wavelength of the incident x-rays, as shown by the dashed lines in Figure 6.16.

Figure 6.16 shows the origin of the plateau phenomenon. It has already been explained that the output of a detector receiving a monochromatic

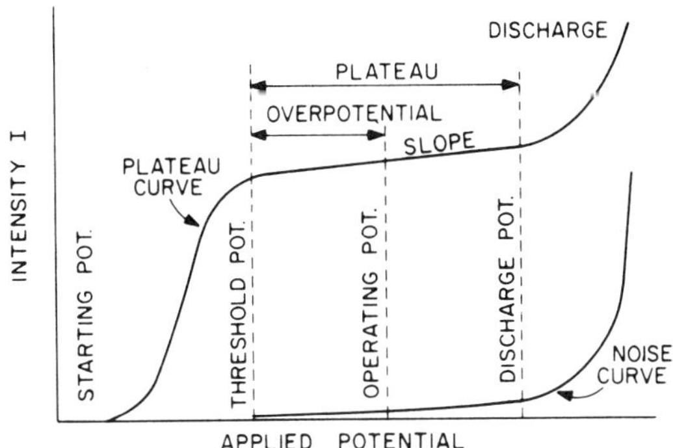

FIG. 6.15. Plateau and noise curves of an x-ray detector.

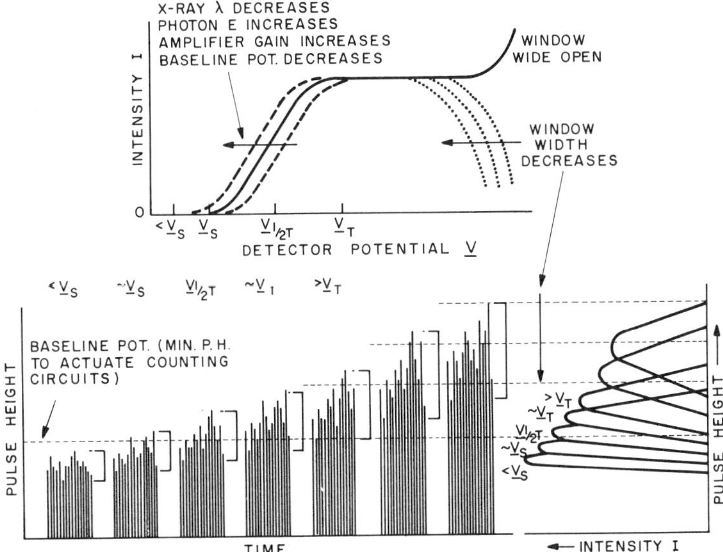

FIG. 6.16. X-ray detector plateau phenomena. As detector potential increases, average pulse height increases and progressively more—and finally all—of the pulses exceed the minimum pulse height required to actuate the counting circuits, that is, the discriminator baseline. The plateau curve is shifted to lower potentials as x-ray wavelength decreases, amplifier gain increases, and discriminator baseline potential decreases. If the window is operative, increased detector potential eventually pushes pulses out the top of the window, and the observed intensity decreases; this occurs at progressively lower detector potential the narrower the window.

x-ray beam is a pulse-height distribution. In Chapter 8, it is explained that the average amplitude of this distribution can be varied, among other ways, by varying the detector potential. Figure 6.16 shows that at detector potentials below the starting potential V_S, none of the pulses in the distribution exceeds the minimum pulse height required to actuate the counting circuits. As detector potential is increased, progressively more of the pulses exceed this minimum, until, at the threshold potential V_T, all pulses are counted and further increase in detector potential has no effect on the measured intensity. Figure 6.16 illustrates the plateau phenomenon in two ways, with plots of intensity *versus* pulse height and pulse-height *versus* time. The former figure is shown in unconventional orientation to permit correlation with the latter.

If there is an escape peak (Sections 6.2.3.1c and 6.2.3.2), the *double-plateau* phenomenon may occur, in which a step appears on the rising slope of the plateau curve between the starting and threshold potentials. An escape peak is a second pulse-height distribution having lower average pulse height than the main pulse distribution. As detector potential is increased, the main pulse distribution reaches plateau first, then the escape peak with its lower amplitude. The stepped rising slope—double plateau—is actually an integral pulse-height distribution curve (Section 8.1.2).

It is also evident in Figure 6.16 why the plateau curve shifts toward lower potentials with increased amplifier gain, decreased discriminator setting, and shorter wavelength. Increased amplifier gain increases all pulses proportionally so they exceed the minimum countable height at lower detector potential. Decreased discriminator setting lowers this minimum countable height. Shorter wavelength means higher photon energy and proportionally higher pulse amplitude.

The foregoing discussion applies to plateau curves made with the pulse-height analyzer in integral mode, that is, with only the baseline discriminator operative. If the analyzer is placed in differential mode with a window of finite width, increase in detector potential eventually results in increase of pulse height to the extent that pulses are "pushed" out the top of the window. In this case, the plateau curve falls off at higher detector potential. The narrower the window is, the lower is the potential at which this occurs and the shorter is the plateau itself, as shown by the dotted lines in Figure 6.16.

To summarize, a plateau curve is a plot of observed intensity *versus* detector potential made with constant incident intensity, amplifier gain, and discriminator setting—with or without the window. Similar plateau curves of observed intensity *versus* amplifier gain may be made with constant incident intensity, detector potential, and discriminator setting—with or without the window. Finally, it is enlightening to compare pulse-height distribution curves with plateau curves. An integral pulse-height distribution curve is a plot of observed intensity *versus* discriminator setting made with constant incident intensity, detector potential, and amplifier gain, and with infinite window. A differential pulse-height distribution curve is a plot of observed intensity *versus* discriminator setting made the same way, but with a narrow window.

6.5.1.9. Slope (Figure 6.15)

Slope is the inclination of the straight portion of the plateau curve expressed as the ratio of the change in measured intensity to the change in applied potential near the midpoint of the plateau. It may be expressed in counts per second per volt, or it may be expressed as percent change in measured intensity per volt. The slope increases and the plateau shortens as the detector deteriorates with life, and these are good criteria of the condition of the detector.

6.5.1.10. Inherent Noise and Background (Figure 6.15)

These two, taken together, constitute the intensity measured from the detector without incident x-rays, and they are evaluated by running a plateau curve without x-rays.

6.5.1.11. Quantum Efficiency (Figure 6.17)

Quantum efficiency (*T5*) is the fraction, usually expressed as percent, of incident x-ray photons that produce pulses of countable amplitude, or the number of countable output pulses from the detector divided by the number of incident x-ray photons, both in the same time interval. Quantum efficiency is measured at counting rates at which coincidence loss is not significant. Figure 6.17 shows typical quantum efficiency curves for some common detectors. However, these curves are only typical, and the same kind of detector—say, a gas-flow proportional counter having 6-μm (0.00025-inch) Mylar window and argon–methane gas—from different manufacturers may give somewhat different curves.

6.5.1.12. Resolution

The theoretical resolution R of a detector can be expressed in terms of the *half-width* $W_{1/2}$ or $\Delta V_{1/2}$, or *full width at half maximum* height (FWHM),

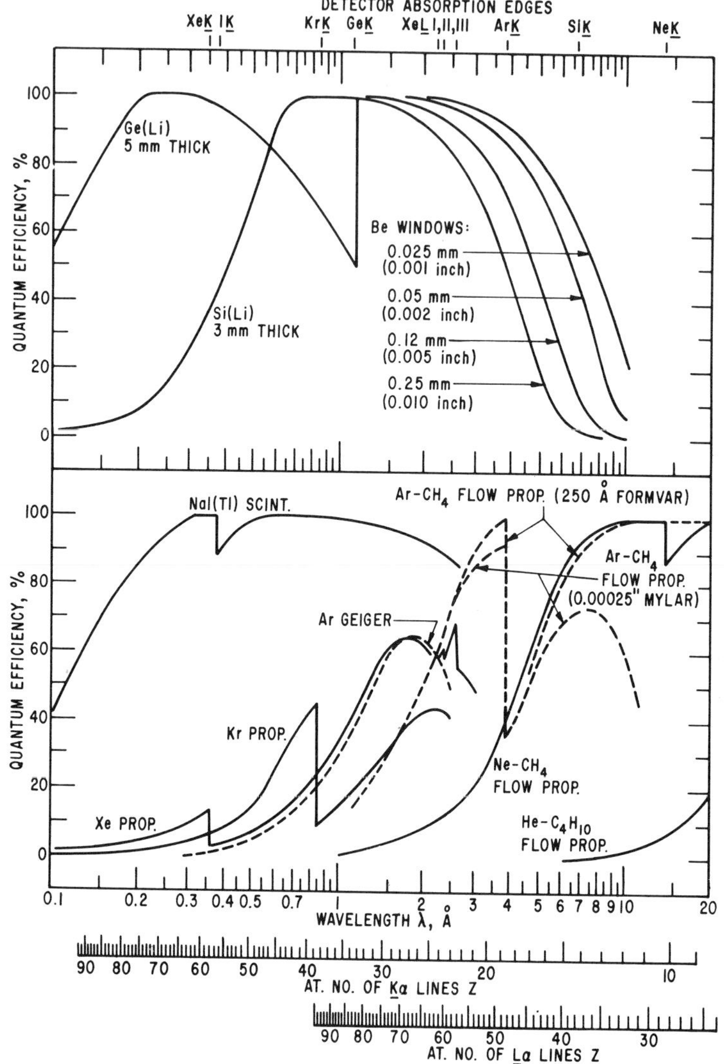

FIG. 6.17. Typical quantum efficiencies of several x-ray detectors.

of the pulse-height distribution above background. These terms are synonymous and expressed in volts. The half-width of a Gaussian distribution is 2.35 σ intervals (Section 11.1.2, Figure 11.1), so that

$$R = \Delta V_{1/2} = W_{1/2} = \text{FWHM} = 2.35\sigma \tag{6.9}$$

A more commonly used term is the *percent resolution*, or *relative resolution*, which is the half-width divided by the mean pulse height $\bar{V}$, expressed in percent, that is,

$$R = 100\Delta V_{1/2}/\bar{V} \; [=] \, \% \tag{6.10}$$

These equations can be put in terms of the x-ray photon energy E_x and the average number of electrons n in the electron population limiting the statistical precision of the detector. From Equation (11.18), $\sigma = \sqrt{n}$, or, from Equation (11.8), $\%\sigma$ (or ε) $= 100(\sqrt{n}/n)$. Equations (6.9) and (6.10) then become, respectively,

$$R = 2.35\sqrt{n} \tag{6.11}$$

$$R = 235(\sqrt{n}/n) = 235/\sqrt{n} \; [=] \, \% \tag{6.12}$$

For a gas proportional counter, n is the average number of primary ion–electron pairs generated by the photoelectron produced by the incident x-ray photon. For a semiconductor detector, n is the average number of electron–hole pairs generated by the photoelectron produced by the photon. For a scintillation counter, n is the number of photoelectrons collected by the first dynode of the multiplier phototube for each incident x-ray photon. If in all three cases one defines an *effective ionization potential* V_i as the average energy (eV) required to produce one electron in the statistically limiting process, then

$$n = E_x/V_i \tag{6.13}$$

where E_x is x-ray photon energy (eV). Equation (6.12) then becomes

$$R = \frac{235}{(E_x/V_i)^{1/2}} \; [=] \, \% \tag{6.14}$$

This equation is usually put in the form

$$R = \frac{235\sqrt{E_x V_i}}{E_x} \; [=] \, \% \tag{6.15}$$

Approximate values of effective ionization potential for various detector gases and for the NaI(Tl) scintillation counter are given in Table 6.1. For the Si(Li) detector, V_i is $\sim$3.8 eV at 77°K, the temperature of liquid nitrogen.

X-rays having photon energy $<$1 MeV interact with solids by one of three processes: photoelectric absorption and coherent and incoherent (Rayleigh and Compton) scatter. At energies $<$50 keV, the initial interaction is usually photoelectric absorption, producing a photoelectron having kinetic energy equal to the incident photon energy minus the binding energy of the electron. This photoelectron expends its energy in various ways depending on its environment. For example, in a solid-state detector, it produces electron–hole pairs by impact ionization and optical phonons (a *phonon* is a "quantum" of crystal lattice vibration); in a scintillator, it produces light photons and phonons. Due to competition among these and other possible processes, the statistical variance [Equation (11.7)] in the number of electron–hole pairs in a solid-state detector, the number of light photons in a scintillation detector, and the number of primary ion–electron pairs in a gas-filled detector is smaller than if it were governed by a Poisson or Gaussian distribution. Consequently, the observed resolution for x-ray detectors based on ionization is substantially better than is indicated by statistics alone.

Fano (*F10*) introduced the factor F (*Fano factor*) to correct the variance, that is, to bring theoretical statistical variance [Equation (11.7)] into agreement with observed variance, which becomes Fv or $F\sigma^2$. The Fano factor is an intrinsic material constant which reflects the ultimate detector resolution and correlates the observed resolution and average ionization energy. The observed resolution is then

$$R = \text{FWHM} = 2.35(FE_xV_i)^{1/2}\ [=]\ \text{V} \tag{6.16}$$

where the notation is that of Equations (6.13) and (6.14). The term $(E_xV_i)^{1/2}$ is the half-width FWHM in electron volts to be expected for a Gaussian distribution of E_x/V_i electron–hole pairs. This value is reduced by the Fano factor F, which has a value <1.

The Fano factor is evaluated by correlation of pulse distribution half-width and x-ray photon energy, and probably has a maximum value of ~ 0.125.

From Equation (6.14), it is evident that resolution is inversely related to x-ray photon energy,

$$R \propto 1/\sqrt{E_x} \tag{6.17}$$

For two x-ray photon energies E_1 and E_2 or wavelengths λ_1 and λ_2, detector resolutions are related by

$$R_1/R_2 = (E_2/E_1)^{1/2} = (\lambda_1/\lambda_2)^{1/2} \tag{6.18}$$

Incidentally, the equations in this section are for resolution of the detector only. The resolution of the overall detector–readout system is given by introducing the noise contributions of the several components (detector, preamplifier, amplifier, etc.) into Equation (11.11):

$$R = \{\sum \sigma^2 + [2.35(FV_iE_x)^{1/2}]^2\}^{1/2} \tag{6.19}$$

Mean pulse amplitude, half-width, and energy resolution are all related to photon energy E_x; they are proportional to E_x, $E_x^{1/2}$, and $E_x^{-1/2}$, respectively.

Table 6.2 gives calculations of the energy resolution for an argon proportional counter, NaI(Tl) scintillation counter, and lithium-drifted silicon semiconductor detector. The table shows that the width of the pulse-height distribution, and thereby the energy resolution, of the scintillation counter is limited by the collection of photoelectrons on the first dynode, but is also affected by statistical fluctuations in the scintillation process, light-photon collection at the photocathode, photoelectron emission, and secondary-electron multiplication. Conversely, the width of the pulse-height distribution of the proportional counter is limited by the deviation of the initial ionizing event. The only other contributing factor is statistical fluctuation in the gas amplification.

6.5.2. Comparison of Conventional Detectors *(D25)*

Geiger, proportional, and scintillation counters are compared in detail in Table 6.3, and the useful spectral regions of the scintillation counter and of neon, argon, krypton, and xenon proportional counters are shown in Figure 9.1. Probably the three most commonly used detectors on standard commercial spectrometers are the argon–methane (P10 gas) gas-flow proportional counter for the longer wavelengths, and the NaI(Tl) scintillation counter or sealed xenon proportional counter for the shorter wavelengths.

6.5.3. Modified Gas-Filled and Scintillation Detectors

Because no one detector is suitable for the entire spectral region, *combination* (*double*, *tandem*, *universal*, "piggy-back") detectors are used to facilitate selection of the optimal detector for each region. Combination detectors consist of a two-window flow-proportional counter for long wavelengths backed by a sealed krypton or xenon proportional counter or a scintillation counter for short wavelengths. The output may be taken from either detector, or the combined output from both may be used. Figure 6.18 shows the performance of a combination flow–scintillation counter. It is

TABLE 6.3. Comparison of Common X-Ray Detectors

Geiger counter	Proportional counter	Scintillation counter
Detection process		
1) Gas ionization	1) Gas ionization	1) Scintillation production in NaI(Tl) crystal
2) Townsend avalanching (generalized throughout tube)	2) Townsend avalanching (localized near anode wire)	2) Photoelectron emission from photocathode
3) Ultraviolet excitation		3) Secondary electron multiplication in dynodes
Autoamplification, output electrons/incident x-ray photon		
$\sim 10^9$	$\sim 10^6$	$\sim 10^6$
Output pulse amplitude, V		
~ 1	$\sim 10^{-3}$	$\sim 10^{-3}$
Useful wavelength region, Å		
0.3–4	Sealed: 0.3–4 Flow [6-μm (0.00025-inch) Mylar]: 0.7–10 Flow (ultrathin Formvar): 0.7–50	0.1–3 or 4
Dead time, μs		
~ 200	≥ 0.2	≥ 0.1
Maximum useful count rate without substantial coincidence loss, counts/s		
Single channel: ~ 1000 Multichannel: $\sim 10^4$	10^5	10^6

Background, counts/s		
~1	Sealed: ~0.5 Flow: ~0.2	~10
Energy resolution (width of pulse-height distribution at half height ÷ mean pulse amplitude), % (for Cu $K\alpha$)		
—	~15	~45
Life		
Practically unlimited (if halogen quench used)	Sealed: ~10^9 counts Flow: Unlimited, but windows and anode wire require occasional replacement	May be relatively short (see limitations below)
Advantages		
1) Output pulse amplitude ~1000× higher than proportional or scintillation counter; requires little amplification or associated circuitry 2) Construction simple, since no need to realize narrow pulse-height distribution 3) Plateau wide, precludes need for high regulation of detector power supply 4) Stability high 5) Quantum efficiency high 6) Background low 7) Life substantially unlimited for tubes having halogen quench gas 8) Since pulse amplitude is independent of wavelength, best detector for counting method that uses capacitor to integrate detector output	1) Pulse height proportional to x-ray photon energy; pulse-height selection can be used 2) Energy resolution better (2–3×) than scintillation counter: a) Pulse-height distribution for monochromatic x-rays is narrower b) Difference in pulse height for a given difference in x-ray photon energy is greater 3) Resolving time fast; useful at intensities up to 10^5–10^6 counts/s 4) Useful wavelength region very wide if suitable construction, window, and gas filling are used	1) Pulse-height proportional to x-ray photon energy; pulse-height selection can be used 2) Output pulse amplitude somewhat greater than proportional counter 3) Resolving time somewhat faster than proportional counter; useful at intensities up to 10^6–10^7 counts/s 4) Quantum efficiency substantially independent of wavelength and very high (~100%) over wide useful wavelength range 5) Escape peak phenomena not serious with NaI(Tl) scintillator 6) Pulse-height-distribution drift and distortion relatively uncommon

TABLE 6.3. (*continued*)

Geiger counter	Proportional counter	Scintillation counter
Limitations		
1) Pulse height independent of x-ray photon energy; pulse-height selection cannot be used	1) Cannot resolve close wavelengths: a) With crystal resolves ΔZ 1 (low Z) to 3 (high Z) b) Without crystal (nondispersive) resolves ΔZ 3 (low Z) to 6 (high Z)	1) Cannot resolve close wavelengths: a) With crystal resolves ΔZ 1–3 b) Without crystal (nondispersive) resolves ΔZ 6–10
2) Resolving time very long; maximum useful count rate much lower than proportional or scintillation counter	2) Output pulse amplitude only a few mV, requires high amplification	2) Energy resolution poorer (2–3×) than proportional counter: a) Pulse-height distribution for monochromatic x-rays is wider b) Difference in pulse height for a given difference in x-ray photon energy is smaller
3) Dead time dependent on wavelength (for full-wave generator): efficiency of excitation of a spectral line depends on excitation potential and varies continuously with time	3) Detector-tube potential must be highly stabilized (±0.1%), since gas amplification factor varies rapidly with detector potential	3) Wavelength region limited to maximum of ~4 Å
4) Quantum efficiency highly wavelength dependent: a) For a given gas filling, efficiency varies widely with wavelength b) Different gas fillings required for maximum efficiency at different wavelengths	4) Quantum efficiency highly wavelength dependent: a) For a given gas filling, efficiency varies widely with wavelength b) Different gas fillings required for maximum efficiency at different wavelengths	4) Background high due to multiplier phototube noise; increases with age and detector-tube potential
		5) Life may be limited:

5) Since maximum useful count rate is very low, poorly suited for semi-quantitative scanning of spectra where strong and weak lines occur 6) For same reason, poorly suited for integrated counts (intensity "under a peak") where intensity is very high at the peak and falls off on either side	5) Thin [6-μm (0.00025-inch)] Mylar or ultrathin (100–200-Å) Formvar windows and continuous gas flow required to realize full useful wavelength region 6) Life of sealed tubes limited to $\sim 10^9$ counts: a) Width of pulse-height distribution increases progressively with life b) Require replacement or refilling after $\sim 10^9$ counts 7) Flow counter requires occasional cleaning and replacement of windows and anode wire 8) Escape peaks more troublesome than for scintillation counter (with Ar and Kr gas fillings) 9) Pulse-height-distribution drift and distortion more prevalent than for scintillation counter; especially so for gas-flow counter 10) Flow counter requires gas cylinder and lines, may require pressure and temperature control	a) Not as rugged as gas-filled detector to shock: encapsulation may fail, crystal may crack or deteriorate, multiplier phototube may break or its electrodes become misaligned b) Phototube noise increases with age c) Phototube may be permanently damaged if applied potential is too high

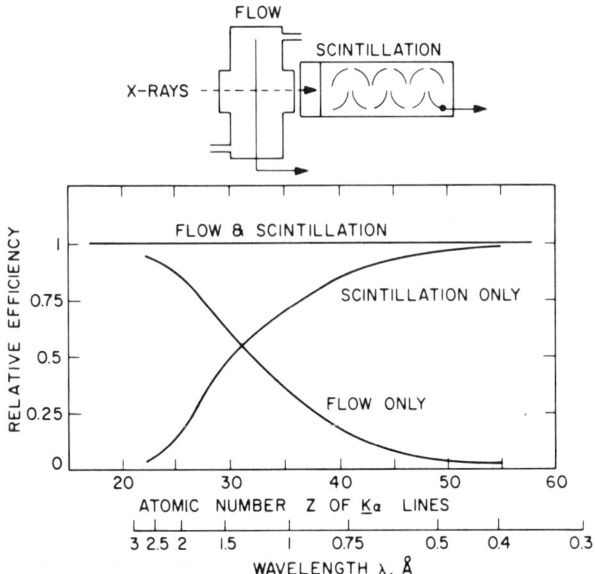

FIG. 6.18. Performance of the gas-flow and scintillation counter. [R. Jenkins and J. L. de Vries, "Practical X-Ray Spectrometry," ed. 2, Springer-Verlag N. Y. (1969); courtesy of the authors and publisher.]

evident that over half the spectral region, the combined outputs exceed the output of either detector alone. When the combined detector is used for elements of low atomic number ($\leq$26, iron), substantially all the analyte-line radiation is absorbed in the front (flow) detector, and the x-rays passing into the back detector are largely higher-order, shorter-wavelength x-rays comprising the background. In such applications, the back detector is not used.

In combination detectors, the front window of the flow counter may be thin (1–6 μm) Mylar, polyethylene, or polypropylene, or ultrathin (100–500 Å) Formvar to pass the long wavelengths. However, the rear window, which passes the short-wavelength x-rays to the rear detector, may be much heavier, so it need not be supported in vacuum nor be replaced with prolonged use. In some instruments, a short collimator is placed between the two detectors. In some vacuum instruments, the flow counter is inside the vacuum chamber, the scintillation counter outside, with the vacuum window of the chamber between. The short-wavelength x-rays measured by the back (scintillation or xenon- or krypton-filled proportional) counter are attenuated much more by the argon path through the flow counter than by the equivalent length of vacuum, helium, or air path. Consequently, in some instruments, provision is made to swing the flow counter out of the x-ray path when the back counter is used. When this is not done, an alternative is to provide means to flush the flow counter with pure air, nitrogen, or helium from a cylinder after the longer wavelengths have been measured and prior to measurement of the short wavelengths with the back detector.

Efforts have been made to extend the limited linear intensity range of the Geiger counter to take advantage of its high sensitivity and high pulse amplitude. In the *multichamber Geiger tube*, the active volume is partitioned into two or more separate compartments having a common anode wire and gas. Each chamber counts independently of the other(s), giving increased resolving time and substantially linear intensity range up to 10^4 counts/s. Use of a *mesh cathode cylinder*, rather than sheet, reduces the cathode surface and thereby the electron emission by ion bombardment. Use of a *beaded anode* (*C75*), having small glass beads spaced along the wire, limits the discharge to a small fraction of the total volume, giving a linear intensity range up to several thousand counts per second. However, the reduced sensitivity necessitates higher applied potential. Thin-window *gas-flow Geiger counters* have been used for long wavelengths; these tubes have shorter length to reduce absorption of short wavelengths. Special quenching circuits for Geiger counters have already been mentioned (Section 6.2.2.4).

The *point-anode proportional counter* (*R2*) is an end-window flow counter having improved performance, compared with conventional end-window counters, when used for pulse-height analysis. The construction is essentially similar to that of an end-window counter except that the anode wire is replaced by a thin rod, at the window end of which is a small pin chuck. This chuck holds a steel phonograph needle having ~40-μm tip radius and selected for smooth finish. Although a point-to-window distance can be found that is satisfactory for the entire useful wavelength range, there is a critical distance for optimal performance at each wavelength. Consequently, it is preferable that provision be made to vary this distance from 0 to 2 cm or so. Cathode cylinders of large diameter (~4 cm) give narrower pulse-height distributions, but the active volume is only the immediate vicinity of the point. Consequently, the window need not have a large area, and apertures 1–4 mm in diameter are placed outside the window to confine the incident x-rays to the useful region. Because of the small active volume, the point-anode proportional counter is applicable only at long wavelengths—for example, *K* lines of aluminum to beryllium—and ultrathin windows are required. The pulse-height distribution is narrower than for conventional end-window counters, but wider than for side-window counters.

In a conventional side-window flow counter, the distance between window and anode is fixed, so that if x-ray photon path length is to be favorable for different wavelengths, either the gas pressure must be varied or different gases used. Also, such detectors may be geometrically difficult to place inside small spectrometer chambers. The *variable-geometry proportional counter* (*C2*) is an end-window flow counter in which the anode–window distance can be varied, while the detector is in operation, by means of a graduated dial on the end of the detector. A preamplifier is placed inside the detector body. The paper (*C2*) includes a figure showing the effect of window–anode distance on the Cu *L* (0.93 keV) and Si *K* (1.7 keV) pulse distributions. By varying the distance, the two distributions can be detected with equal efficiency, or either can be detected to the substantial exclusion of the other.

In a novel arrangement (*W4*), silicon was determined in sodium silicates by placing each sample and a ^{55}Fe radioisotope source inside the detector itself. The detector was a demountable flow-proportional counter connected to an energy-dispersive spectrometer system.

Recently, the two-chamber gas-flow proportional counter was introduced. This detector is basically two identical flow counters in tandem (see paragraph 1 of this section). The detector body consists of a metal block about 3 cm square and 5 cm long having: (1) two longitudinal cylindrical holes 1 cm in diameter and parallel to and about equally spaced from each other and a front and back side; and (2) a rectangular side-window channel 2.5 cm high and 1 cm wide from the front side, through the front hole, and into the back hole, but not out through the back side. A gas inlet and outlet are provided. A thin Mylar or other film covers the window channel. The tops and bottoms of the holes are fitted with insulators having axial holes. Stiff anode wires 5 cm long and 0.08 mm in diameter having heads—like long pins—are inserted in the bottom insulators and extend axially through the cylindrical chambers into the top insulators. The preamplifier is mounted on the bottom of the detector. Common electrical connection is made to the two anode pins by a metal bar, which also serves to hold the pins in place. This two-channel detector has high quantum efficiency, high linear intensity range, and low detector background. The small cathode diameter (1 cm) and large anode diameter (0.08 mm) permit use of lower detector potentials, $\leq$2 keV, even with pure hydrocarbon gases, helium–carbon dioxide mixtures, etc. Also, the relatively large-diameter anode wire accumulates solid hydrocarbons and other foreign material much more slowly than a fine wire, so pulse-distribution shift and distortion develop more slowly (Section 8.1.8). These detectors are used in the Siemens MRS-3 multichannel spectrometer. For fluorine, neon, sodium, and magnesium *K* lines (and *L* and *M* lines in the same wavelength region), the anode pin is removed from the rear detector, so that only the front detector is operable. For all other wavelengths up to ~3 Å, both detectors are used, and pulses are taken from both simultaneously.

Efforts have also been made to extend the useful range of the scintillation counter farther into the long-wavelength region, beyond 3 Å. This limit is imposed partly by the absorption of the beryllium window and aluminum reflecting layer, but principally by the inherent noise pulses of the multiplier phototube, which have the same amplitude as pulses from x-rays having wavelength ~3 Å or more (Section 6.3.1.2). Fischer and Baun (*F15*) claim that the useful region of the scintillation counter can be extended to ~15 Å by removal of the scintillator window, use of a thallium-activated cesium iodide [CsI(Tl)] scintillator, and selection of the multiplier phototube for low noise.

6.6. OTHER X-RAY DETECTORS

6.6.1. Photographic Film

Instruments using photographic-film registration have been described in Sections 5.1 (the earliest Bragg spectrograph), 5.5.2.2 (edge crystal), 5.5.3.4 (Von Hamos image spectrograph), and 5.7 (curved-crystal film spectrographs). In the image spectrograph, the focal surface is plane, and photographic plate can be used. In most of the other instruments, the focal surface is curved, and x-ray diffraction film is used, although one spectrograph camera uses a Polaroid film casette.

For quantitative analysis, a film calibration curve must be established to permit correlation of photographic density and x-ray intensity. A uniform monochromatic x-ray beam is photographed through a stack of strips of thin metal foil having progressively shorter length from bottom to top, exposing a stepped wedge of graded thickness. The intensity transmitted by each step is calculated from the wavelength and intensity of the incident x-rays, the absorption coefficient of the metal, and the thickness of the step. The photographic density of the image of the step is measured with a microdensitometer. A plot of photographic density *versus* logarithm of exposure

TABLE 6.4. X-Ray Sensitivity of Kodak No-Screen X-Ray Film[a,b]

Photographic density	Exposure factor	X-ray photon energy E_x, keV	Exposure multiplier, photons/cm^2
0.1	0.09	5	2.62×10^7
0.2	0.19	6	3.32
0.3	0.27	7	3.97
0.4	0.35	8	4.64
0.5	0.45	9	5.39
0.6	0.55	10	6.29
0.7	0.66	11	7.39
0.8	0.77	12	8.75
0.9	0.88	13[c]	1.00×10^8
1.0	1.00	14[c]	7.07×10^7
1.1	1.12	15	7.80
1.2	1.23	16	8.55
1.3	1.35	17	9.32
1.4	1.46	18	1.01×10^8
1.5	1.57	19	1.09
1.6	1.69	20	1.19
1.7	1.80	21	1.25
1.8	1.92	22	1.34
1.9	2.03	23	1.42
2.0	2.15	24	1.51
2.1	2.28	25[d]	1.60
2.2	2.41	26[d]	8.38×10^7
2.3	2.55	27	8.93
2.4	2.70	28	9.50
2.5	2.85	29	1.00×10^8
2.6	3.02	30	1.06
2.7	3.21	32	1.16
2.8	3.41	34	1.25
2.9	3.62	36	1.33
3.0	3.86	38	1.40
3.1	4.13	40	1.47

TABLE 6.4. *(continued)*

Photographic density	Exposure factor	X-ray photon energy E_x, keV	Exposure multiplier, photons/cm²
3.2	4.42	42	1.54×10^8
3.3	4.74	44	1.60
3.4	5.10	46	1.66
3.5	5.50	48	1.72
3.6	5.95	50	1.77
3.7	6.45	55	1.90
3.8	7.01	60	2.01
3.9	7.64	—	—
4.0	8.36	—	—

[a] Courtesy of L. S. Birks, U.S. Naval Research Laboratory.
[b] Procedure for use of the table: Select photographic density in column 1 and photon energy in column 3; then multiply the value in column 2 opposite the selected density by the value in column 4 opposite the selected energy to get exposure in photons/cm².
[c] The bromine *K*-excitation potential is 13.48 keV.
[d] The silver *K*-excitation potential is 25.52 keV.

(*IT*, intensity × time) is plotted. The photographic densities of the spectral lines are measured and applied to the curve to derive *IT*, from which intensity is calculated from the exposure time.

Birks *et al.* [p. 33 in (*8*)] (*D26*) give tables and graphs of sensitivity of Kodak No-Screen x-ray film for x-rays having photon energies in the x-ray spectrometric region. Table 6.4 permits calculation of accumulated x-ray exposure in photons/cm² required to give photographic densities of 0.1–4.0 for x-ray photon energies 5–60 keV.

6.6.2. Photoelectric Detectors

6.6.2.1. The Phosphor-Phototube Detector

This detector is physically similar to a scintillation counter, in that it consists of a phototube or multiplier phototube on the envelope of which is mounted a piece of fluorescent screen with the fluorescent side facing the photocathode. Alternatively, a layer of phosphor may be applied directly to the bulb just outside the photocathode. This detector is used principally for continuous-current instantaneous measurements, especially in industrial x-ray thickness and filling gages. It has no application in spectrometry.

6.6.2.2. Photoelectric Detectors for the Ultralong-Wavelength Region

This section describes four detectors that are used principally in the long- and ultralong-wavelength region (*G32*). These detectors all require vacuum, but may be operated without windows if mounted in or attached to a vacuum spectrometer. The detectors are usually prepared so that their active surfaces can withstand repeated exposure to air.

The *windowless multiplier phototube* (*B22*, *F15*, *G32*, *J4*) is simply an open electron multiplier, the first electrode of which is made of beryllium–copper alloy and serves as an x-ray-sensitive photocathode. The x-rays to be measured irradiate this photocathode, thereby expelling photoelectrons, which initiate the secondary-electron multiplication process.

In the *photoelectron scintillation counter* (*G32*, *S18*) (Figure 6.19C), a series of up to six parallel, spaced, thin metal foils are arranged just above and perpendicular to a plastic or crystalline phosphor cemented to a multiplier phototube. Foils successively farther from the x-ray source are maintained at successively higher potential by a potential divider. The incident x-rays expel photoelectrons from the front and back surfaces of the foils. Some of these photoelectrons strike the phosphor and excite scintillations, which are counted by the phototube.

In the *resistance-strip magnetic photoelectron multiplier* (*G32*) (Figure 6.19A), two plane semiconductor strips, a field strip and a dynode strip, are arranged mutually parallel and a short distance apart. The dynode strip acts as a secondary-electron emitter. The electric potentials applied to the strips cause uniform potential drops along the strips in such a way as to initiate an electric field having inclined equipotential lines as shown. A uniform magnetic field is applied parallel to the planes of the strips by means of small, external, permanent magnets. The incident x-rays irradiate the photocathode, emitting photoelectrons, which initiate the secondary-electron multiplication process, as shown. In the crossed electric and magnetic fields, the motion of a photo or secondary electron is a cycloid directed perpendicular to the magnetic field and along the equipotential line on which the electron originated. Thus, each electron travels a short distance, ~1 mm, then falls back on the dynode surface, emitting additional secondary electrons.

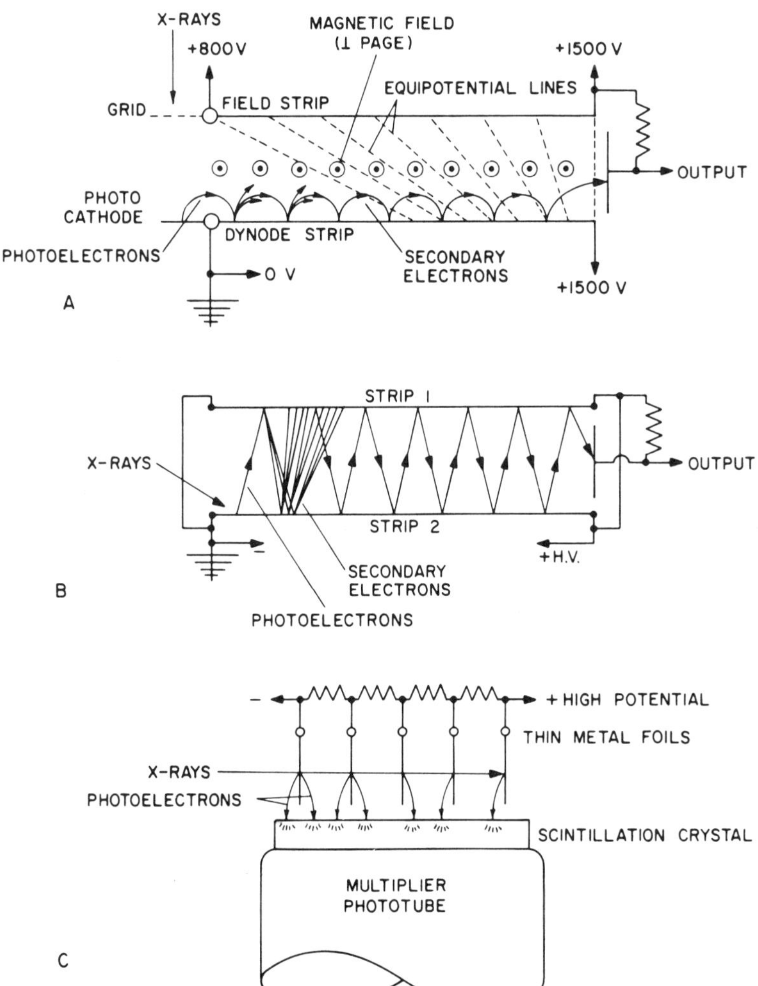

FIG. 6.19. Some photoelectric x-ray detectors for use in the ultralong-wavelength region. A. Resistance-strip magnetic photoelectron multiplier. B. Continuous-strip photoelectron multiplier. C. Photoelectron scintillation counter.

In the *continuous-strip* or *channel photoelectron multiplier* (*G32*) (Figure 6.19B), two secondary-emissive strips are deposited longitudinally and diametrically opposite one another on the inside surface of a glass tube. Alternatively, two such parallel strips may be separated and supported by insulating side walls. A potential of 500–2000 V is applied to the ends of the two strips, causing an electric field having equipotential lines perpendicular to the strips. The incident x-rays expel photoelectrons, which initiate the secondary-electron multiplication process, as shown. Many modifications of this detector have been developed (*W19*). In one, the input end is flared and the tube diameter decreases continuously to the output end. This detector may be linear or circular, like a tiny funnel or French horn, respectively. In another form, the tubular detector is coiled into a cylindrical spiral of two or more turns. Large-area (diameter $\lesssim$2.5 cm), multiple-channel detectors are made by heating cylindrical bundles of lead glass capillaries (inside diameter $\lesssim$0.25 mm) in a reducing atmosphere. The capillaries sinter together, and their inside surfaces become partially conductive and secondary-emissive due to formation of a thin film of free lead. The sintered bundle is then sectioned into disks or cylinders of appropriate length. The detector surfaces are polished and made conductive to serve as the electrodes for the applied detector potential.

6.6.3. Crystal Counters

If metal electrodes are affixed to the ends of a small single crystal of a suitable material and a potential is applied to the electrodes, the device functions as a photon counter (*F25*). Cadmium sulfide, zinc sulfide, germanium, and diamond act as crystal counters at room temperature, silver chloride and silver bromide at very low temperature. Crystal counters are very small and require relatively simple electronic equipment. They are applicable to pulse-height selection, perform well at high intensity, and have high sensitivity and signal-to-noise ratio. The disadvantages include the following: They are sensitive to the purity, crystallographic perfection, and heat treatment of the crystal; the pulse height changes with time, due to accumulation of charge in the crystal; dark current may be high, and signal current persistent. Crystal counters have found little application in analytical x-ray spectrometry.

Chapter 7

Measurement

7.1. INSTRUMENTATION

7.1.1. Introduction

In Chapter 6, it is shown that the outputs of x-ray detectors of the types used on analytical x-ray spectrometers consist of series of pulses of electric current, one pulse for each photon detected. The pulses are randomly distributed in time. In Geiger counters, all these pulses have about the same amplitude (height). In proportional, scintillation, and Si(Li) detectors, the pulses have average amplitude proportional to the photon energy.

Even for strictly monochromatic x-rays, the pulses are not all of the same height, but have a Gaussian distribution about a *mean* height proportional to the photon energy. For incident x-ray beams containing more than one wavelength, that is, more than one spectral line, the detector output contains (1) a pulse-height distribution peak for each wavelength, and (2) another escape peak for each wavelength shorter than the absorption edge of the detector gas, the iodine in the NaI(Tl) scintillator, or the silicon in the Si(Li) detector.

The electronic readout components count these pulses or measure their rate of production for analytical purposes. Ideally, these electronic components should have a response time at least as fast as the detector resolving time and preferably as fast as the recovery time. However, in practice, the linear counting range is limited by the counting circuits, rather than the detector. Of course, clearly, the counting circuits cannot improve the linear range of the detector.

In addition to the pulses that arise from the discrete incident spectral wavelengths, the detector output also contains pulses arising from background and, especially in scintillation counters, pulses arising in the detector itself.

Section 7.1 (Instrumentation) is limited to the readout instrumentation associated with wavelength-dispersive (crystal) spectrometers. Readout and display instrumentation for energy-dispersive spectrometry is discussed in Section 8.2.2.4. Sections 7.2 (Measurement of Intensity) and 7.3 (Background) are, in general, applicable in principle to both wavelength- and energy-dispersive spectrometry.

The detector and electronic readout components of an x-ray spectrometer are shown in simplified block-diagram form in Figure 7.1 and schematically in Figure 7.2. The principal panel controls are shown along the bottom of Figure 7.1 and along the left edge of Figure 7.2. Highly simplified diagrams of pulse height *versus* time are shown along the top of Figure 7.1 and at the upper right of Figure 7.2 for pulses of three amplitudes. The low pulses represent low-energy background and detector and amplifier noise. The medium and high pulses originate from two incident x-ray spectral lines having relatively low and high photon energies, respectively.

In Figure 7.2, the components are divided into three groups: the detector itself, the amplifier and pulse-height analysis components, and the readout and display components. The detector and preamplifier are mounted

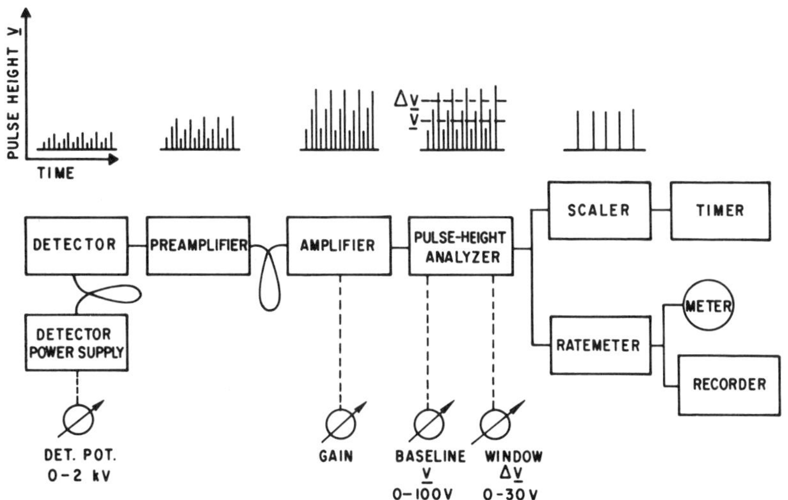

FIG. 7.1. Electronic detection and readout components of an x-ray spectrometer—simplified block diagram.

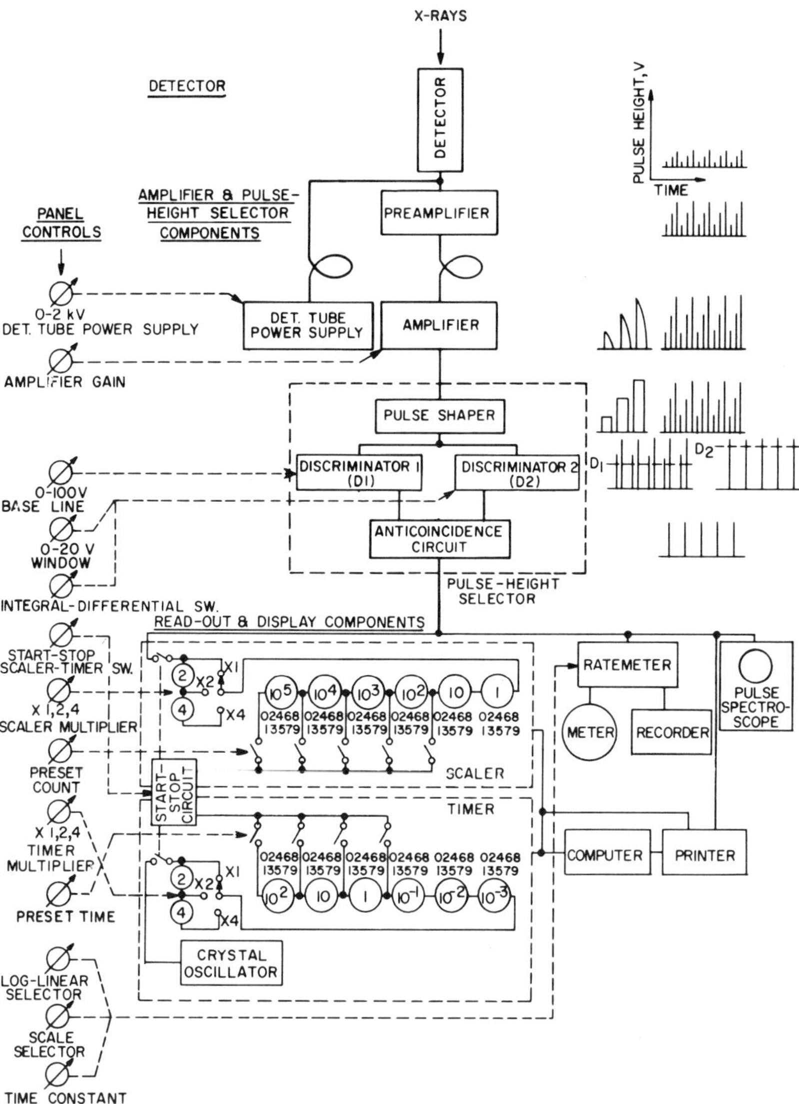

FIG. 7.2. Electronic detection and readout components of an x-ray spectrometer. Panel controls are shown along the left. Simplified wave forms (pulse height *vs.* time) are shown in the upper right.

on the spectrogoniometer. All the other components are mounted in some sort of relay rack or console, with the possible exception of the computer (if any), which may be separate, even in a location remote from the spectrometer. The detector power supply is included with the electronic components. It is a well-regulated ($\pm 0.05\%$ or so) 0–2 kV dc supply. Formerly, all the electronic components used electron tubes, and many still do. However, today, the components have been largely converted to solid-state semiconductor devices—transistors, diodes, rectifiers, and integrated circuits—with consequent reduction in size and power consumption. Also, there is increasing use of plug-in, solid-state, "NIM" (Nuclear Instruments Manufacturers) modular units having standardized dimensions, connections, potentials, etc.

7.1.2. Preamplifier

The preamplifier is mounted on the detector arm of the spectrogoniometer as an almost integral part of the detector itself. In fact, some instruments provide water-cooling of the detector mounting plate to reduce heat conduction from the preamplifier to the detector. The preamplifier output is connected by a relatively long coaxial cable to the input of the main amplifier in the electronics console. The preamplifier usually consists of one or two linear amplifier stages and a cathode-follower stage to provide optimal coupling to and minimal attenuation along the cable. The preamplifier also prevents amplifier feedback from affecting the detector. "Sine-function" preamplifiers are one means of providing automatic pulse-height selection (Section 8.1.7).

In order to conserve space, flow counters mounted inside vacuum enclosures may have only a cathode-follower mounted directly on the detector. A short cable connects this with the preamplifier mounted nearby outside the vacuum chamber.

A procedure for evaluation of preamplifier noise level is given in Section 8.1.5.1.

7.1.3. Amplifier

It is the basic function of the main amplifier to linearly amplify the detector–preamplifier output pulses to potentials high enough to actuate the pulse-height selector, ratemeter, and scaler. The amplifier is a linear, low-noise, fast-response, video-type amplifier having gain that can be varied up to ~10,000. Alternatively, many of these amplifiers have a fixed gain of, say, 10,000, and provide for *attenuation* to lower gain. The amplifier is designed to amplify preamplifier output pulses to a maximum of 5–100 V. (0–10 V in NIM modular units). Geiger, proportional, and scintillation counters require low, intermediate, and high amplification, respectively.

The amplifier should have the following characteristics: (1) linearity of amplification, that is, equal gain for all pulse amplitudes; (2) stability of gain; (3) low noise; (4) rapid overload recovery, that is, rapid return to normal baseline after receiving pulses of abnormally high amplitude or frequency; (5) good thermal stability; (6) good baseline restoration (dc restoration), that is, rapid return to normal baseline after termination of a pulse; and (7) good pole-zero cancellation; sometimes the trailing or descending side of a pulse may undershoot the baseline, that is, momentarily fall to a potential less than that of the baseline; the pole-zero cancellation precludes this, or at least rapidly returns the undershoot to the baseline. Separate baseline restoration and pole-zero cancellation circuits or modules may be provided to effect these last two functions. Such circuits, in effect, clamp the pulse baseline to some stable reference potential. Improper adjustment of these circuits reduces the valley between adjacent pulses and thereby impairs resolution. The pulse width near the baseline, say, the full width at tenth maximum (FWTM), is a useful parameter for indication and evaluation of these phenomena and of "tailing" due to pulse pileup, etc. For an ideal Gaussian peak profile, FWTM $= 1.85 \times$ FWHM (Section 6.5.1.12). The circuits are adjusted to minimize departures from this value.

A procedure for evaluation of amplifier noise threshold is given in Section 8.1.5.1. Some considerations on the setting of amplifier gain are given in Section 8.1.4.

7.1.4. Pulse-Height Selectors

7.1.4.1. Pulse-Height Selector; Discriminator

A *discriminator* passes all pulses having height greater than some preset minimum, and rejects all lower pulses. A *pulse-height selector*, or *pulse-height analyzer*, passes on to the ratemeter and scaler all pulses having height greater than one preset level but lower than a second preset level, and rejects all pulses outside this *window* or *channel*. All x-ray spectrometers, have a discriminator to reject pulses having height, after amplification, of ~5 V or less. This discriminator eliminates the detector and amplifier noise and some low-energy background. Although most modern x-ray spectrometers have pulse-height selectors, this component may be omitted, and, even if present, it may be by-passed.

The pulse-height selector is shown in Figure 7.2 enclosed in a dash-line box. The particular type shown consists of four components, the functions of which are illustrated by the pulse diagrams at the right of each component. It will be recalled that the low pulses represent detector and amplifier noise, and the medium and high pulses represent two x-ray spectral lines having medium and high photon energy, respectively. The four components are as follows.

The *pulse shaper* forms rectangular or square-wave pulses from the amplified pulse forms, as shown by the two groups of enlarged pulses, but does not alter their relative amplitudes. The shaped pulses are passed on to the inputs of both of the discriminators.

Discriminator 1 (the *lower-level discriminator*) passes all pulses having amplitude greater than level D_1. Thus, it passes the pulses of both spectral lines, but rejects the low-amplitude noise. A discriminator functioning in this way is always required, even when the pulse-height selector is absent or by-passed.

Discriminator 2 (the *upper-level discriminator*) passes all pulses having amplitude greater than level D_2. Thus, it passes only the pulses of the high-energy spectral line.

The *anticoincidence circuit* cancels all pulses that arrive simultaneously from both discriminators. Thus, it rejects the high-amplitude pulses, and passes on to the ratemeter and scaler only the pulses of intermediate amplitude.

To summarize, pulses of amplitude smaller than a selected minimum D_1 do not pass either discriminator. Pulses of amplitude greater than a selected maximum D_2 pass both discriminators and are eliminated by the anticoincidence circuit. Thus, only pulses having amplitude between the two levels pass on to be counted. The operation of the pulse-height selector is perhaps more clearly illustrated by the simplified diagram in Figure 7.1.

The potential V corresponding to the lower discriminator level is usually referred to as the *baseline*, sometimes as the *window level* or *channel level.* It represents the minimum countable pulse height referred to so frequently in Chapter 6 and shown in Figures 6.11 and 6.16. The potential interval ΔV between the two levels is usually referred to as the *window* or *channel*, sometimes as the *window width* or *channel width.* The baseline may be varied from 0 to 100 V in electron-tube instruments, and 0 to 10 V in solid-state NIM modular units. The window may be varied from 0 to 10,

20, or 30 V in electron-tube instruments. However, in most NIM modules, the window may be varied over the full baseline, 0–10 V.

The systems described above may be operated as discriminators, passing only pulses having amplitude greater than the baseline V, or as pulse-height selectors, passing only pulses having amplitude greater than the baseline V but less than the potential of the baseline plus window, $V + \Delta V$.

In addition to selection of the *input* pulses on the basis of their amplitudes, the pulse-height selector has another function. This is the conversion of the *output* pulses into standard pulses of uniform amplitude, time duration, and shape compatible with the input requirements of the ratemeter, scaler, and other readout and display components. That is, *once selected*, all pulses—regardless of height—are converted to *identical* output pulses. This function is not shown in Figure 7.2.

In wavelength-dispersive mode, the detector is mounted on a spectrogoniometer (Section 3.2.2, Figure 3.1) in which each spectral line emitted by the specimen is diffracted by the crystal to the detector *individually*, in its turn, at a specific 2θ angle. Thus, the emitted wavelengths are separated prior to detection, and, in principle, the detector output contains only one pulse distribution at a time, and pulse selection is not required. However, in practice, pulse selection is often required with wavelength dispersion, as discussed in Section 8.1.6.

In energy-dispersive mode, the spectrogoniometer is omitted or set at 0° 2θ, and all the spectral lines emitted by the specimen pass into the detector *simultaneously*. Thus, the detector output contains many pulse distributions, and pulse selection or analysis is required to isolate the analyte pulse distribution.

In either mode of operation, a discriminator is required to exclude detector and amplifier noise pulses.

The instrument described above is a *single-channel* pulse-height selector or analyzer (SCA). The *multichannel* analyzer (MCA) is described in Section 8.1.2.3. Pulse-height selection and analysis are distinguished in the introduction of Chapter 8 and described in Sections 8.1 and 8.2, respectively.

7.1.4.2. Pulse Reverter

The pulse-selection function performed by the pulse shaper, two discriminators, and anticoincidence circuit can also be performed by a single circuit known as a *pulse reverter* or, simply, *reverter*—but not as well. The reverter was used in early General Electric x-ray spectrometers. The function of the reverter is shown in Figure 7.3.

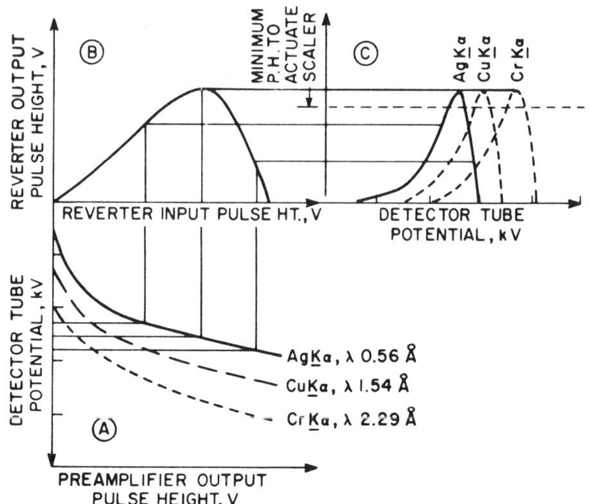

FIG. 7.3. Function of the reverter. [Courtesy of General Electric Co., X-Ray Dept.] A. Preamplifier output pulse height *vs.* detector-tube potential for a short, medium, and long wavelength. B. Reverter output pulse height *vs.* input pulse height. C. Reverter output pulse height *vs.* detector potential for a short, medium, and long wavelength.

The curves in Figure 7.3A show preamplifier output pulse height as a function of detector potential for Cr $K\alpha$, Cu $K\alpha$, and Ag $K\alpha$, relatively long-, intermediate-, and short-wavelength x-rays, respectively. The curve is shown in an unusual orientation to permit graphic correlation with the other parts of the figure. The arrows on the axes indicate directions of increasing pulse height and detector potential.

Figure 7.3B shows the reverter response characteristic and reverter output pulse height as a function of input pulse height. Because the reverter follows the preamplifier, the preamplifier output is the reverter input.

The curves in Figure 7.3C show reverter output pulse height as a function of detector potential for Cr $K\alpha$, Cu $K\alpha$, and Ag $K\alpha$. The discriminator baseline, or the minimum pulse height that will actuate the scaler, is also indicated.

It is evident that as detector potential is increased, the height of the Ag $K\alpha$ pulses increases progressively at the preamplifier output (Figure 7.3A), but increases to a maximum, then decreases at the reverter output (Figure 7.3C). Thus, only for a relatively narrow range of detector potential do Ag $K\alpha$ pulses exceed the discriminator baseline. The Cu $K\alpha$ and Cr $K\alpha$

pulses undergo the same phenomenon at successively higher detector potentials. Thus, by suitable choice of detector potential, the chromium, copper, or silver pulses may be passed individually.

7.1.4.3. Pulse-Shape Selector

Electronic components are also available that discriminate pulses on the basis of their shapes, rather than amplitudes (*H19*). A description follows of a system using such a component.

The "*phoswich*" is a composite scintillator (Section 6.3.1.1) consisting of a thin front disk of one scintillator material, such as NaI(Tl), optically coupled to a thicker back disk of a different scintillator material, such as CsI(Tl). This composite scintillator is provided with a suitable radiolucent window, encapsulated to exclude stray light and moisture, and optically coupled to a multiplier phototube. The output of this *phoswich detector* consists of pulses originating within either or both of the scintillators. With such a detector, x-ray intensity can be measured in the presence of high-energy γ-radiation. The thin front disk has substantially 100% efficiency for the relatively low-energy x-rays, but transmits the unwanted high-energy γ-rays to the back disk, where they are absorbed. The output of the phoswich detector goes to a *pulse-shape selector*, which discriminates the signals arising from the two scintillators on the basis of their phosphor rise and/or decay times, rather than their amplitudes. However, a large portion of the detector output consists of coincident pulses produced in both scintillators by high-energy incident radiation. In such cases, the back scintillator acts as an anticoincidence shield, and the electronic circuitry must be capable of rejecting pulses arising simultaneously from both disks.

Phoswich detectors and pulse-shape selectors are finding increasing application in nuclear science, but, except for the application cited above, are as yet little used in analytical x-ray spectrometry.

7.1.5. Ratemeter and Recorder

The ratemeter provides readout and display in *analog* mode, like an automobile speedometer. It indicates the mean value of the current consisting of the individual pulses from the discriminator or pulse-height analyzer accumulated over a certain selected short time interval determined by the time constant. This mean current is proportional to the instantaneous x-ray intensity incident on the detector and is displayed as intensity or count rate on a panel meter or a strip-chart recorder. Thus, the ratemeter provides instantaneous intensity readout. The ratemeter may have linear or logarithmic response. Linear response reduces tailing of the peaks and is preferable for estimation of intensity. Logarithmic response may be advantageous for recording spectra having both intense and weak peaks. The ratemeter–recorder combination is used principally to record the x-ray spectrum for qualitative or semiquantitative analysis, or preparatory to quantitative analysis. Its use and modification for these purposes are described in Section 10.3.

7.1.6. Scaler and Timer

The scaler and timer provide readout and display in *digital* mode, like an automobile odometer. (It is mentioned above that the ratemeter is analogous to a speedometer.) The *scaler* or *count register* actually counts the individual pulses passed by the discriminator or pulse-height analyzer during a certain time interval. Their rate of arrival is determined by the x-ray intensity at the detector, and their time distribution is random. The scaler thus provides integrated or accumulative intensity readout.

The *timer* or *time register* is a clock, but it keeps time by counting the positive pulses from the sine-wave output of a high-frequency crystal oscillator. Their rate of arrival is determined by the frequency of the oscillator and is, say, 1000 Hz, that is, 1000 positive pulses per second. Their time distribution is perfectly uniform, and, at 1000 Hz, they are spaced at 1-ms intervals.

Both the scaler and timer, then, are electronic binary or decimal counters, and if there is any essential difference between them, it is that the scaler must be capable of counting at extremely high rates, whereas the timer need not count faster than $\sim$1000/s. Thus, the scaler and timer are "fast" and "slow" counters, respectively.

Figure 7.2 shows, very schematically, a scaler and timer, each having a six-decade decimal counter and provision for twofold or fourfold binary multiplication.

A *binary circuit* may be regarded as one that passes every second pulse it receives; that is, it blocks pulses 1, 3, 5, ... and passes pulses 2, 4, 6, ... Usually, provision is made to indicate which condition the circuit is in. When it has received and blocked a pulse, a small neon lamp glows until the next pulse is received and passed.

A *decimal circuit* may be regarded as one that passes every 10th pulse it receives, blocking—or rather, storing—pulses 1–9. Provision is made to indicate how many pulses are accumulated in the circuit. In the E1T decimal indicator tube, a fluorescent spot advances sequentially from 0 to 9 on a numerical scale. In the Nixie indicator, wire numerals 0 to 9, mounted in a neon-filled bulb, glow in sequence. Other indicators are also in use.

In Figure 7.2, suppose that both the scaler and timer multiplier switches are at $\times 1$, that all of the preset-count and preset-time switches are open, and that both start–stop switches are simultaneously closed (start). The analyte-line pulses from the pulse-height analyzer pass directly to the first scaler decade, and the pulses from the crystal oscillator pass directly to the first timer decade.

The first nine analyte-line pulses are accumulated in and indicated by the first scaler decade. Pulse 10 passes to the second decade, causing it to indicate 1 and returning the first decade to zero. Pulses 11–19 accumulate in the first decade, and pulse 20 passes to the second. This sequence continues until both decades are filled by pulse 99. Pulse 100 then passes to the third decade, returning the first two to zero. Thus, the individual decades indicate, respectively, individual pulses, 10's, 100's, 1000's, etc. as shown.

Meanwhile, the timer is counting positive pulses from the crystal oscillator in the same way, but since these pulses arrive at exact 1-ms intervals, the timer is, in effect, counting milliseconds of elapsed time. However, the decades are labeled 0.001, 0.01, 0.1, 1, . . . *seconds*, rather than 1, 10, 100, 1000, . . . *milliseconds*.

Since the scaler and timer were started together, if they are now stopped together at any arbitrary time, the accumulated count and counting time may be read on the scaler and timer, respectively. The two instruments are stopped by opening the start–stop switches.

If it is intended to make a preset-count measurement of, say, 10^4 counts, the switch following the 10^3 scaler decade, that is, between the 10^3 and 10^4 decades, is closed. The scaler and timer are started simultaneously. The action is the same as before up to the 9999th pulse. However, the 10,000th pulse not only passes to the 10^4 decade, but also passes through the switch and actuates the stop circuit, opening both start–stop switches and stopping the scaler and timer. The counting time is then read on the timer.

Similarly, if it is intended to make a preset-time measurement of, say, 100 s, the switch following the 10 timer decade is closed. The first 99,999 millisecond pulses fill the first five timer decades. The 100,000th pulse (100.000 s) stops the scaler and timer. The accumulated count is then read on the scaler.

Now, suppose that the scaler multiplier switch is set at $\times 2$. The analyte pulses no longer pass directly to the decimal counter, but first pass through a binary stage (marked "2" in Figure 7.2), which passes only every second pulse. The action is the same as before, but the accumulated count in the scaler must be multiplied by 2. Similarly, if the multiplier switch is set at $\times 4$, the incoming pulses pass through two binary stages in series, so that only every fourth pulse reaches the counter, and the accumulated count must be multiplied by 4. Binary multipliers are also shown on the timer. Thus, the scaler and timer in Figure 7.2 have capacities of 1, 2, and 4×10^6 counts and 1, 2, and 4×10^3 s, respectively.

Provision is also made to clear the scaler and timer, that is, reduce all decades to zero, prior to starting a new measurement.

Another way to measure intensity is to allow the pulses to accumulate in a capacitor for a certain interval, then measure the accumulated charge. This method is inexpensive but has the disadvantage that the total accumulated count is not known, so that calculation of counting error is less convenient than for a scaler. This method is used in Applied Research Laboratories automatic x-ray spectrometers (Section 9.3.3).

A printer may be provided and programmed to print out the analytical data—sample number; symbol of analyte; and line, background, and net intensities—on a paper strip.

7.1.7. Computers

Computers are widely used in x-ray spectrochemical analysis for three purposes: (1) calculation of analytical concentrations from measured intensity data input to the computer some time after the measurements are made; (2) storage, processing, and calculation of analytical concentrations from accumulated counts as measured (Section 8.2.2.4); and (3) complete programing and control of fully automated spectrometer systems, including: indexing of specimens in up to 20-specimen multiple changers; for each specimen, selection of 2θ for up to 20 analyte lines and their backgrounds; for each analyte line, selection of kV, mA, crystal, detector, pulse-height analyzer settings, and all the functions listed in item 2 above (Section 9.3). The computer may be relatively small and "dedicated," that is, an integral part of the spectrometer, or a large, central, time-shared unit. The more sophisticated the function of the computer, the larger its capacity must be.

A discussion of computers is beyond the scope of this book. Several papers describe computers suitable for use in x-ray spectrometry and integrated x-ray spectrometer–computer systems (*B84*, *G24*, *J11*, *M11*, *R3*, *S38*, *S60*). Various computer programs are available for calculation of analytical concentrations from measured intensity data in x-ray fluorescence spectrometry (*24*) and electron-probe microanalysis (*B15*, *C53*).

7.2. MEASUREMENT OF INTENSITY

In principle, the *integrated intensity* under the entire analyte-line peak should be measured. In energy-dispersive spectrometry, where all x-ray photons are admitted to the detector simultaneously, integrated intensity is measured readily (Section 8.2.2.4). But in wavelength-dispersive spectrometry, where the crystal diffracts only a discrete wavelength to the detector, integrated intensity measurement is inconvenient and time-consuming (Section 7.2.2.3). However, the *peak intensity* at the 2θ angle of maximum intensity has been found to be representative of the integrated intensity, except perhaps for the K lines of elements of low atomic number. Since the peak intensity is also measured more conveniently, it is almost universal practice in wavelength-dispersive x-ray spectrometry to measure peak intensity. Another advantage of analyte-line measurement at the peak is the relative insensitivity of the measured intensity to slight error in the 2θ setting.

Before any intensity measurements are undertaken, all excitation, detection, and readout components should be allowed to warm up for 30–60 min. During the warmup period, the x-ray tube should be operated at or near the excitation conditions (kV, mA) to be used. The operating potential should be applied to the detector, and, for flow counters, the counter gas should be flowing at its proper rate. All other electronic components should be on. The gas-flow rate for flow counters is so small (usually <1 ft^3/day) that, except for special expensive gases (neon, for example), most laboratories leave the gas on at all times. Moreover, many workers advocate leaving the electronic detection and readout components on at all times. However, this is to be recommended only for instruments for which highest precision is required daily.

With the older electron-tube electronics, warmup time was determined by the time for the electronics to equilibrate. With the newer solid-state electronics, warmup time is determined by the time for the x-ray tube to equilibrate, and 30 min is more than adequate. Moreover, since very little heat is produced, operating conditions are more stable.

For accurate quantitative analysis, measured peak intensities are usually corrected for background (Section 7.3.3) and sometimes for "dead time" of the detector–readout system (Section 6.5.1.6.)

7.2.1. Ratemeter Methods

Intensity can be measured with the ratemeter panel meter when only very rough estimates are required—for example, in distinguishing 35–65 and 50–50 nickel–iron alloys.

Intensity is frequently measured with the ratemeter–recorder for semiquantitative analysis (Section 10.5) and for quantitative analysis when highest accuracy is not required. The ratemeter should have linear, rather than logarithmic, response for this application and should be carefully calibrated. In general, the accuracy of the intensity measurement is greater the more closely the following conditions are met: The 2θ scanning rate should be slow; the time constant should be fast enough to permit the recorder to indicate the true peak intensity, but slow enough to dampen excessive recorder fluctuations; insofar as practicable, peaks from all samples and standards should be recorded at the same full-scale deflection and other instrument settings; the sample and standard peaks should have as nearly as possible the same deflection.

Four ratemeter–recorder techniques are used for measurement of x-ray intensity, of which three involve recording of the entire analyte peak for each sample and standard. The least accurate technique is the measurement of peak height above background. The intensity measured this way is usually less than the scaled intensity, and the discrepancy is greater for peaks having different intensity and for different time constants and other instrument settings. A more accurate method is the measurement of peak height above background at half-width, in the manner commonly used in x-ray diffraction. An even more accurate method is the measurement of the area under the peak, but above the background, by counting squares, with a planimeter, or by actually cutting out the peak and weighing the paper (*D4*). The fourth, and possibly most accurate method, does not require recording of the full peak. The goniometer is set at the 2θ angle for the analyte line and left undisturbed. The recorder traces analyte-line intensity for several centimeters from each sample and standard in turn. A mean line is drawn through the recorder noise of each tracing. Background may be recorded the same way. In all four methods, measurements are made above background, and calibration curves or mathematical relationships are established from the standard data.

7.2.2. Scaler-Timer Methods

7.2.2.1. Preset-Time Method

In the preset-time method, the timer is set for a specified time T long enough to permit accumulation of enough counts for the required statistical precision. The timer and scaler are started simultaneously, operate until the preset time has elapsed, then stop simultaneously. The number of counts accumulated N is read on the scaler. Intensity $I = N/T$. If the same preset time is used for all samples and standards, the accumulated counts N may be used in calculations and calibrations instead of intensities. If this is done, and background corrections are made, the same preset time must be used for them also.

7.2.2.2. Preset-Count Method

In the preset-count method, the scaler is set for a specified number of counts N large enough to give the required statistical precision. The timer and scaler are started simultaneously, operate until the preset count has accumulated, then stop simultaneously. The elapsed time T is read on the timer. As before, intensity $I = N/T$. If the same preset count is used for all samples and standards, the elapsed times may be used in calculations and calibrations instead of intensities. If this is done, and background corrections are made, the same preset count must be used for them also. The preset-time and preset-count methods are compared statistically in Sections 11.2.2.2 and 11.2.3.1.

Incidentally, the terms *fixed* count and time should not be used as synonyms for *preset* count and time, as explained in Section 11.2.2.2 following paragraph f.

7.2.2.3. Integrated-Count Method

An integrated count may be made in either continuous or stepwise mode. In the continuous mode, the spectrometer is set at a 2θ angle on one side of the peak to be measured (*a* in Figure 14.4). The 2θ drive, timer, and scaler are started simultaneously, operate until the entire peak has been scanned (*a–b* in Figure 14.4), then are stopped simultaneously. The accumulated count and lapsed time are read. The integrated count or intensity may be used. In the stepwise mode, the spectrometer is moved manually or with a step-scanner in small, equal-2θ increments, and a preset-time or preset-count measurement is made at each position. The integrated-count method is used when an accurate peak profile is required—to deal with spectral overlap, for example (*B53*).

7.2.2.4. Monitor and Ratio Methods

It is evident that the preset-time, preset-count, and integrated-count methods are simply ways of defining the duration of the counting time. In the three techniques, the count is continued for the duration of, respectively, a preset time, the time to accumulate a preset count, and the time to scan through the analyte peak. Two other ways to define the counting time are the monitor and ratio methods.

When a line having very low intensity is to be measured, a very long counting time is required to accumulate enough counts to give reasonable statistical precision. During such long times, the count is also affected by fluctuations in excitation and other instrument conditions. Thus, correlation of the measurements from a series of specimens may be meaningless. A monitor provides one way to deal with this condition. A small portion of the primary x-ray beam is diverted, usually by a small scatter target, directly into a second detector–scaler system. The spectrometer and monitor counts are started simultaneously and terminated when the monitor has accumulated a preset count. All samples and standards are counted for the time for the monitor to accumulate this same preset count.

In the ratio method, the time to accumulate a specified count from the standard defines the time during which the analyte-line count is accumulated from the sample. In effect, a preset count N_S is made from the standard, during which an "up-and-down" timer "winds up" from time 0 to T_{N_S}, then "runs down" as counts accumulate from the sample. The standard count may be: (1) the analyte line measured from a calibration standard or from pure analyte; (2) an internal standard line measured from the sample itself; (3) a scattered target line or primary-continuum wavelength measured from the sample; or (4) a scattered target line or continuum wavelength measured from a separate scatter specimen. These methods are widely used on automatic spectrometers.

In the application of ratio methods in x-ray fluorescence spectrometry, it is usually preferable not to use the pure analyte, but rather a standard similar to the samples and having analyte concentration near the middle of the expected range. There are several reasons for this: The pure element may be of unsuitable physical form, chemically reactive, or unavailable in the required form or amount. Moreover, if the analyte concentration is low in the samples, excitation conditions set to give adequate analyte-line intensity from the samples are likely to give extremely high intensity from the pure element. The great disparity in matrix is likely to complicate matrix-correction procedures. Finally, the primary-absorption effect is much more severe than the secondary-absorption effect in pure elements (Section 12.2.1).

The monitor and various ratio techniques are not equally effective in compensating fluctuations in x-ray tube potential V. The integrated intensity of the primary continuum is proportional to V^2 [Equation (1.28)]. The intensity of a specific wavelength in the primary continuum is proportional to $(V_\lambda - V)$, where V_λ is the potential equivalent to the wavelength. The intensity of a secondary analyte line is proportional to $(V - V_A)^2$, where V_A is the analyte-line excitation potential [Equation (1.32)]. Thus, fluctuations in x-ray tube potential are compensated most effectively by those ratio techniques in which the count is terminated by a preset count of the analyte line from a standard or from pure analyte.

7.2.3. X-Ray Dose and Dose Rate

For purposes of radiation health and safety, x- and γ-radiation are measured in terms of *dose* or *dose rate*, which have units roentgens (R) and roentgens per unit time, respectively. The roentgen is the quantity of x- or γ-radiation that produces in 1 cm³ (0.001293 g) of dry air at 0°C and 760 torrs, ions carrying 1 esu of charge, regardless of sign. Equations for conversion of x-ray measurements in counts or counts per second into roentgens or roentgens per hour have been derived (*C6*, *K26*). Kohler and Parrish (*K26*) show that in the wavelength region 0.2–3 Å, the dose produced by one photon/cm² in air is approximated by

$$d \approx (4.8 \times 10^{-10})\lambda^2 \quad \text{R per photon/cm}^2 \tag{7.1}$$

and the number of photons per square centimeter in air required to produce a dose of 1 R is approximated by

$$n_R \approx 1/d \approx (2.1 \times 10^9)\lambda^{-2} \quad \text{photons/cm}^2 \text{ per R} \tag{7.2}$$

In both approximations, λ is in angstroms. The lower curve in Figure 7.4 is a plot of n_R as a function of wavelength.

For a total measured count N of x-rays of wavelength λ in a photon detector having effective area A cm² and quantum efficiency Q, expressed as a fraction, the dose D in roentgens is given by

$$D = k_1 N = \frac{1}{n_R A Q} N \tag{7.3}$$

$$= \frac{(4.8 \times 10^{-10})\lambda^2}{AQ} N \tag{7.4}$$

For a measured intensity I counts/s, the dose rate D/T in roentgens per

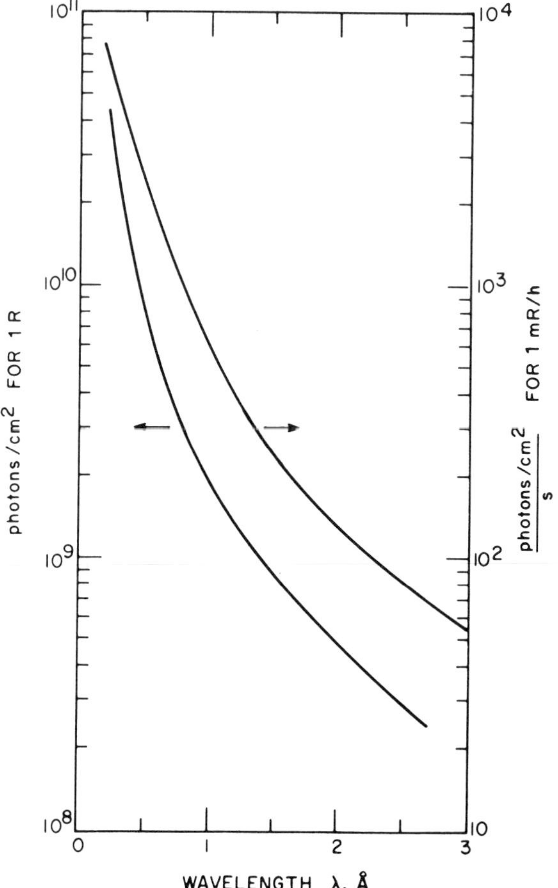

FIG. 7.4. Conversion of accumulated counts into dose in roentgens, and conversion of intensity in counts per second into dose rate in milliroentgens per hour.

hour is given by

$$\frac{D}{T} = k_2 I = \frac{3600}{n_{\mathrm{R}} A Q} I \tag{7.5}$$

$$= \frac{(1.7 \times 10^{-6})\lambda^2}{AQ} I \tag{7.6}$$

where 3600 is the number of seconds per hour. The upper curve in Figure 7.4 is a plot of intensity in photons per square centimeter per second required to give a dose rate of 1 mR/h as a function of wavelength. The constants k_1 and k_2 apply to a particular detector at a specified wavelength, and have units R/count and R/h per count/s, respectively.

7.3. BACKGROUND

7.3.1. Definition and Significance

In its broadest sense, background may be defined as the intensity that would be measured at the 2θ angle of the analyte line if the analyte were absent; strictly speaking, it cannot be measured directly. In this sense, the background consists of a more or less continuous spectrum and the contributions of spectral lines at or near the 2θ angle of the analyte line.

Background is significant in x-ray spectrometric analysis for three reasons: (1) frequently, it must be subtracted from the measured analyte and internal-standard peak intensities; (2) in the background-ratio method, it may compensate absorption errors and certain specimen and instrument errors; and (3) it limits the minimum detectable amount, which is that amount that gives a net intensity of $3\sqrt{I_B}$.

Background assumes increased significance and its correction becomes more difficult in energy-dispersive spectrometry because the energy peaks (pulse-height distributions) are much broader than the wavelength (2θ) peaks from crystal spectrometers. Thus the energy peaks cover more background, which is then likely to have a nonlinear profile and contain discontinuities due to absorption edges. The absorption edges of all elements present in the specimen may be superimposed on the continuous background. This phenomenon arises with electron excitation because continuum generated below the specimen surface must emerge through a thin surface layer, and with x-ray excitation because scattered primary continuum must penetrate and emerge from a surface layer.

7.3.2. Origin and Nature

Background radiation originates from: (1) cosmic radiation; (2) radioactive emission from the environment; (3) spurious pulses and noise arising in the detector and in the electronic amplification components; (4) radioactive emission from the specimen, if it is radioactive; (5) primary x-ray emission from the x-ray tube; and (6) secondary x-ray emission from the specimen, crystal, and other components in the x-ray path. The first two sources constitute the irreducible minimum background and contribute an intensity of about two counts/s. The detector and amplifier noise pulses have very low energy and are substantially eliminated by the discriminator. The background, then, largely results from the collective transmission, scatter, diffraction, or reflection of x-rays from the last two sources, and each component in the spectrometer contributes to the background.

Air in the specimen compartment scatters the primary continuous spectrum and the target-line and other spectral-line radiation originating in the x-ray tube. The *specimen mask* and other structures in the specimen compartment also scatter the primary spectrum and may contribute additional emission lines. The *specimen* scatters the primary spectrum and emits spectral lines that may wholly or partially overlap the analyte line (spectral-line interference); crystalline specimens may diffract specific wavelengths from the primary continuum, or even one or more spectral lines in the primary spectrum. The *crystal*, aside from its intended function of diffracting the analyte-line radiation corresponding to its θ setting: (1) diffracts any incident x-radiation having product $n\lambda$ that satisfies the Bragg law for its θ setting; (2) scatters all incident x-rays; and (3) emits the characteristic line spectrum of its own elements. The *collimators* scatter, diffract, or specularly reflect x-rays passing through them and emit their own characteristic spectrum. The *detector*, especially the multiplier phototube of the scintillation counter, contributes spurious pulses (noise) (Section 6.5.2). The *amplifier* and other electronic components also contribute noise.

In the foregoing discussion, scatter may be coherent or incoherent, diffraction may be in first and higher orders. The primary spectral lines scattered coherently and incoherently by the air, specimen compartment, and specimen are diffracted by the crystal as spectral lines, as, of course, are the interfering lines emitted by the specimen. Diffraction from the specimen, and diffraction and specular reflection from the collimator foils may have the form of very sharp "pseudolines" or "spikes." All other emission and scatter appear as continuous background unless the pulse-height analyzer is used, in which case each wavelength entering the detector produces its own pulse-height distribution. For example, without the pulse-height analyzer, emission from the analyzer crystal causes only continuous background, because the emission radiates in all directions. With the pulse-height analyzer, the pulse distributions of the spectral lines of the elements in the crystal may interfere with the distribution of the analyte line.

The continuous component of the background may have minor discontinuities at 2θ angles corresponding to the absorption edges of the x-ray tube target element, the major constituents of the specimen, and the detector gas or scintillation crystal. The pulse-height analyzer may also introduce anomalous discontinuities in the background. An example of this phenomenon is given in Section 8.1.8.4 in the next to last paragraph preceding the discussion of double or sum peaks.

Neglecting the contributions of spectral-line interference, for practical purposes, the background intensity I_B measured at the 2θ angle of the analyte line consists principally of the sum of three contributions:

$$I_B = I_{\mathrm{pri,sc}} + I_{C,\mathrm{sc}} + I_{C,\mathrm{em}} \tag{7.7}$$

where $I_{\mathrm{pri,sc}}$ is intensity of coherent and incoherent scattered primary x-rays; $I_{C,\mathrm{sc}}$ is intensity of secondary x-rays scattered from the crystal; and $I_{C,\mathrm{em}}$ is intensity of x-rays emitted from the crystal.

7.3.3. Measurement

When the specimens have a relatively wide range of composition, it may be necessary to measure background from every specimen. However, when this is not the case, it may be feasible to measure it only at the beginning and end of the series from a single typical specimen. If a large number of specimens must be measured, requiring a long time, background should also be measured at intervals during the run—perhaps each time the standards are measured.

Figure 14.3 shows one way to evaluate background intensity. Analyte-line peak intensities are measured and plotted against analyte concentration. The curve, extrapolated to zero concentration, intercepts the intensity axis at the background intensity. This extrapolation method is valid only if variation in analyte concentration does not significantly affect x-ray scatter from the specimens. Otherwise, scatter and therefore background intensity vary with analyte concentration, and background intensity is different for each specimen. Figure 14.4 shows a composite typical x-ray spectrum (2θ scan) having emitted and scattered x-rays that may be used for, or interfere with, an analysis. The significance of each feature is given in the caption. The analytical features of the figure are discussed in Section 14.3.1. Here, the figure serves to illustrate several ways to measure background. In the following discussion, the notation has the significance given in the caption of the figure.

If the analyte line A is superposed on uniform, continuous background, background is measured adjacent to the peak at B_{adj}. If the analyte line A' is superposed on nonuniform, continuous background, background is measured at equal 2θ intervals on the low-2θ B_L' and high-2θ B_H' sides of the peak. Then,

$$B_P' = (B_L' + B_H')/2 \tag{7.8}$$

If the background profile is strongly nonlinear in the region of the line, it may be necessary to measure the background at two or more places on each side of the line, plot the background profile, and read off the background at the peak. If the analyte line A'' is superposed on a scattered target line or some other scattered spectral line T, three alternatives are available: (1) B_P'' may be measured from a blank at the 2θ angle for the analyte line. (2) From measurements on blanks, the intensity ratio can be established between the interfering line T and another scattered target line that can be measured without interference, say T_R; then measurement of T_R on each sample and standard permits calculation of T. (3) A substantially constant relationship usually exists between adjacent background B_{adj}'' and the background at the peak B_P'', which, in this case, includes T. Then B_P'' may be calculated from measured B_{adj}'':

$$B_P'' = kB_{\text{adj}}'' \tag{7.9}$$

The constant k is evaluated from measurements at the 2θ angles of B_P'' and B_{adj}'' on suitable blanks containing no analyte. A variation of this technique has been used for measurement of uniform background in determination of trace elements in hydrocarbons (*D39*). From measurements on pure hydrocarbons having various hydrogen/carbon ratios, a relationship is established between background at the analyte peak and the intensity ratio of a coherently and incoherently scattered target line. Thereafter, the ratio is measured from each sample and standard and used to derive the background at the peak. The technique has the advantage that the scattered target lines from a low-Z matrix are measured more rapidly and more precisely than the less-intense, uniform, scattered background.

If the analyte line A''' is superposed on a line of a matrix constituent M, background, including M, may be measured from a blank. Alternatively, the intensity ratio can be established between the interfering line M and another line of the same matrix element that can be measured without interference M'. Then measurement of M' on each sample and standard permits calculation of M.

If the peak intensity is measured by a continuous or stepwise integrated count (a–b), background may be measured by an integrated count on an equal-2θ interval of adjacent uniform background (a'–b'), or by a count at B_{adj} for a time interval equal to that required for the integrated peak count.

Reference is made above to the use of blanks. A blank is usually defined as a specimen identical with those to be measured in composition and physical form, except that the analyte is absent. In practice, if the analyte is a trace constituent, or if its atomic number is near the average atomic number of the remaining matrix, this definition may apply rigorously. However, if the analyte is a minor or major constituent having atomic number significantly different from that of the matrix, its omission may result in a change in scattering power of the remaining matrix. The change may be to higher or lower scattered intensity, depending on the relative values of the atomic number of the analyte and effective atomic number of the matrix. In such cases, a correction factor may be applied, or the analyte may be replaced in the blank with an equal concentration of an element having about the same scattering power, but no interfering spectral lines. An element having atomic number one or two higher or lower than the analyte may be suitable. Sometimes one can use for a blank a pure chemical element having atomic number about equal to the effective atomic number of the specimens—but having no spectral lines at the analyte-line wavelength. For supported specimens (Section 17.2), the blank is usually either a clean support (substrate) or one that has been subjected to the entire specimen-preparation procedure, but without the analytical material; the latter type is known as a *reagent blank* and is preferable for trace analysis.

In general, at the peak of element Z, the background intensity due to scattered primary x-rays is substantially the same as the intensity measured at 2θ for the Z peak from the same concentration of element $Z \pm 1$ or $Z \pm 2$. Usually element Z *plus* one or two is preferable for this purpose. There is a discontinuity in the intensity of the scattered spectrum at the absorption edge of the scattering element, the intensity being slightly less on the short-wavelength side. The phenomenon is analogous to the discontinuity in the primary *emission* spectrum, as described in Section 1.5.1. If element $Z - 1$ or $Z - 2$ is used, this low-intensity region is certain to extend to the wavelength of the Z line, giving an abnormally low scattered intensity. However, the absorption edges of elements of higher Z occur at wavelengths shorter than the Z line, and the scattered intensity at the Z line will be substantially normal.

In a multielement system, the background intensity $(I_B)_i$ at the peak of analyte element i may be closely approximated by

$$(I_B)_i = W_i I_i + \sum W_j I_j \tag{7.10}$$

where W is weight fraction and I_j is the intensity measured at the analyte i peak from pure specimens of each of the other elements j in the specimen. I_i may be measured from pure analyte i adjacent to the i peak, or from element $i \pm Z$ or $i \pm 2Z$ at the i peak (see above). Of course, if for any element j, I_j has a component—such as a spectral line—other than scattered primary x-rays, an adjacent element must be used for that element. Finally, for highest accuracy, the measured I_j should be multiplied by a factor to compensate the slight increase in scattered intensity with decreasing Z or *vice versa*, that is,

$$I_Z = I_{Z'}(Z'/Z) \tag{7.11}$$

where I is the scattered primary intensity at the analyte peak, and Z and Z' are the atomic numbers of the analyte and matrix element, respectively. This last method for deriving background is sometimes used in electron-probe microanalysis, where background is emitted, rather than scattered, by the specimen. However, both scatter and primary emission intensities are functions of atomic number.

In the measurement of the low intensities from trace elements, long counting times may be required with consequent loss of precision due mostly to fluctuations in the x-ray generator. In such cases, the error is decreased substantially by simultaneous measurement of line and background intensities. This is no problem with multichannel spectrometers, but is impossible with unmodified single-channel instruments. Section 5.5.1 presents several ways to modify single-channel instruments for multichannel operation. Very long peak and background counting times may be divided into intervals, which are then alternated until the total time has been counted (Section 11.1.3.2).

Some background correction methods for energy-dispersive spectrometry are described in Section 8.2.2.4.

7.3.4. Reduction

In general, background intensity is reduced by the following conditions:

1. X-ray tube target atomic number is as low as possible.
2. Excitation conditions (kV, mA) are as low as possible.
3. Filters are used to remove target lines or to reduce continuum intensity in the region of a specific analyte line (Section 4.4).
4. The specimen compartment is evacuated or flushed with helium to eliminate air scatter.

5. The specimen compartment is fitted with baffles, primary and/or secondary apertures, or a collimated specimen mask to minimize background from air in the specimen compartment, the surface outside the specimen area of interest, or other structures in the specimen compartment.

6. Matrixes of low atomic number are avoided, such as solutions, fusion products, and glass disks.

7. Small specimens and small selected areas on large specimens are measured in a collimated specimen holder or by selected-area techniques.

8. Very small specimens are supported at the ends of fine plastic or quartz fibers in free space.

9. Milligram to submicrogram quantities of analytical material are supported on thin substrates; background, and therefore detection limit, decreases successively for filter paper, Millipore, 6- and 3-μm (0.00025- and 0.00012-inch) Mylar, and ultrathin Formvar or collodion substrates (Section 17.2.1).

10. Such thin-film substrates are supported on open frames or stretched over the tops of empty liquid cells or other cylindrical supports; primary-beam apertures or special support-cell geometries are used to prevent primary-beam irradiation of the inside cell walls to avoid scatter from the walls; and primary x-ray beam traps are used behind the specimen (opposite the x-ray tube) (Section 17.2.1).

11. Low-Z binders or diluents and plastic cell windows are avoided for powder specimens.

12. Liquid specimens are frozen or held in open-top cells, again to avoid cell-window covering.

13. High-order lines are used; these lines lie at larger 2θ angles where background is lower.

14. The crystal contains no element(s) of relatively high atomic number to produce secondary emission.

15. Collimator foils are treated to minimize reflection and scatter.

16. Fine collimation is used.

17. A slit is placed at the detector window (in curved-crystal spectrometers).

18. The crystal axis is not parallel to the specimen plane (see below).

19. A gas-filled detector is used, rather than a scintillation counter, to avoid multiplier phototube noise.

20. If a scintillation counter is used, the multiplier phototube is selected for low noise.

21. The detector is operated at relatively low potential.
22. The amplifier is operated at relatively low gain.
23. Pulse-height selection is used.

Probably the most intelligent procedure is to determine whether reduction of background is beneficial for the particular analysis at hand. If so, then the various ways to reduce it are tried in order of decreasing effectiveness, using peak-to-background ratio as a criterion of effectiveness.

It is evident that most of the techniques that reduce background also reduce analyte-line intensity, and therefore may give little or no increase in peak-to-background ratio, and may even reduce it. Peak-to-background ratio is mostly a matter of excitation conditions and is discussed in Section 4.3.3.3.

The pulse-height selector discriminates against only that portion of the background that has wavelength substantially different from that of the analyte line. At a specified spectral line (2θ), that component of the background originating from primary radiation scattered by the specimen (or by the mask or other structure in the specimen compartment) *and* diffracted by the crystal has the same *product* of order and wavelength $n\lambda$ as the analyte line. However, the pulse-height selector is still applicable if the *wavelengths* are not the same.

The effectiveness of the pulse-height selector in reducing background varies with the spectral region. In the low-2θ (short-wavelength) region, the background has high intensity and is principally first-order scattered continuum, which is why it has the characteristic "hump." It is also essentially "monochromatic" in the sense that there is only one wavelength at each increment of 2θ. In the high-2θ (long-wavelength) region, the background has low intensity and is principally higher-order scattered continuum. Thus, the effect of the pulse-height selector in rejecting background is least at low 2θ, as shown in Figure 7.5.

Birks [pp. 102–3 in ed. 1 of (7)] points out a precaution to be taken in measurement of background with the pulse-height selector (Section 8.1.1). Suppose that with the goniometer at 2θ for the analyte line, the amplifier gain is set to put the analyte-line pulse distribution in the window. If the goniometer is now moved to a nearby 2θ, the measured adjacent background may be erroneously low. Strictly speaking, the amplifier gain should be reset to move the pulse distribution of the wavelength corresponding to the new 2θ angle into the window. Of course, automatic pulse-height selection (Section 8.1.7) effects this function automatically. In practice, the

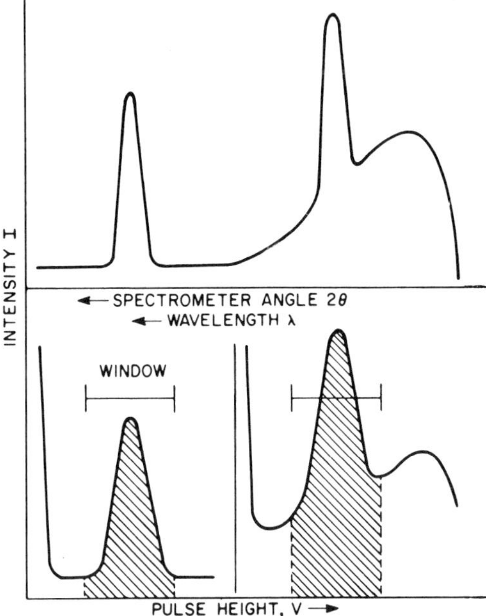

FIG. 7.5. Effectiveness of the pulse-height analyzer in reducing background. The upper curve is an idealized 2θ scan showing peaks in the high-2θ region, where background is low, and in the low-2θ region, where it is high. The lower curve shows the corresponding pulse-height distributions, and the shaded areas are the pulses passed by the windows shown.

error arising from a fixed window–amplifier combination may be negligible over a small 2θ interval, especially if background can be measured close to the peak and if the peak and background intensities are *relatively* high and low, respectively.

Anater (private communication) points out another, perhaps more likely, source of error in measurement of background with the pulse-height selector—pulse-distribution shift when the goniometer is moved from a high-intensity peak to a low-intensity background (Section 8.1.8.2d). The effect is demonstrated as follows. A high-intensity spectral-line pulse distribution is centered in the window by use of the amplifier gain. Background is then measured on both sides of the peak, using the gain to center each background wavelength in the window. If intensity shift predominates over simple energy shift as a function of wavelength (2θ), the required amplifier gain is lower for *both* backgrounds than for the peak. This is because the analyte-line pulses are abnormally low due to the intensity effect and require more than the expected gain to place them in the window. For simple photon-energy shift with wavelength, the amplifier gain would be progressively higher for low-2θ background, the peak, and high-2θ background. Automatic pulse-height selection cannot correct this effect. In fact, if background is measured on the long-wavelength (high-2θ) side of the peak, it is actually detrimental since less, not more, gain is required.

Champion and Whittem (*C36*) have pointed out that certain instrument geometries are more prone to high background than others. Background consists mostly of scattered primary x-rays. In most instruments, the scatter angle intercepted by the collimator–crystal–detector system is $\sim 90°$, so the primary x-rays scattered by the specimen are highly plane-polarized. Conversely, the characteristic x-rays emitted by the specimen are completely random in direction and unpolarized. Consequently, a substantial increase in peak-to-background ratio would be effected by a polarization analyzer to discriminate against the polarized component of the x-radiation leaving the specimen. The analyzer crystal may constitute just such a device. In spectrometers having a horizontal axis (vertical plane) (Section 9.2.2, Figure 9.3), the axis of rotation of the crystal is parallel to the specimen plane, and at 2θ angles near 90°, the crystal and specimen planes themselves are substantially parallel. This arrangement transmits the polarized component efficiently, especially near 90° 2θ, as shown in Figure 7.6A. In spectrometers having a vertical axis (horizontal plane) (Section 9.2.2, Figure 9.2), the crystal and specimen axes and planes are more or less mutually perpendicular, and the instrument discriminates against the polarized x-rays, as shown in Figure 7.6B. Figure 7.6C shows the ratio of the background intensities scattered from a cell of pure water for arrangements B (crystal and specimen planes perpendicular) and A (planes parallel).

An application of polarization to suppression of scattered background in energy-dispersive spectrometry is described briefly on pp. 2.58–9 in (*39*). In the arrangement in Figure 7.6, suppose that the specimen is replaced by a low-Z scatter target and the crystal by the analytical specimen, but that the x-ray tube and detector remain as shown. The unpolarized primary x-ray beam becomes highly plane-polarized on undergoing 90° scatter from the low-Z target ("SPECIMEN"). This polarized scattered primary beam is now used to excite the specimen ("CRYSTAL"). As in the arrangement described above, the scattered polarized primary beam cannot rescatter at 90°, so that the detector receives substantially no scattered primary x-rays. However, x-ray emission excited by the polarized primary x-rays is not polarized and reaches the detector.

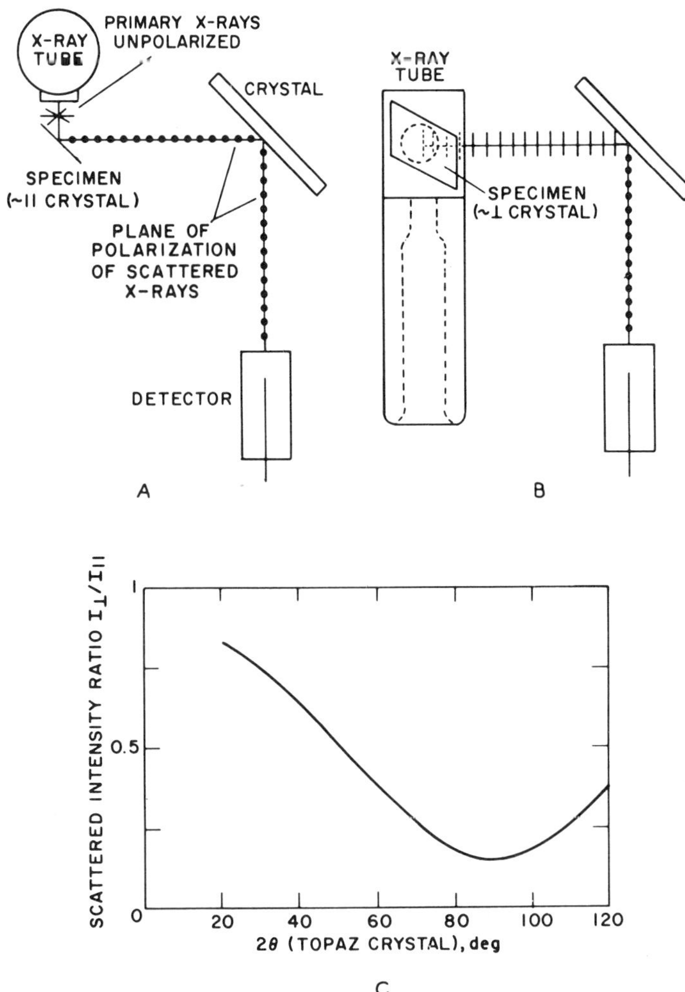

FIG. 7.6. Reduction of background by discrimination against polarized scattered primary x-rays. [K. P. Champion and R. N. Whittem, *Nature* **199,** 1082 (1963); courtesy of the authors and the Macmillan Co.] A. Crystal and specimen planes essentially parallel, especially at 90° 2θ. B. Crystal and specimen planes essentially perpendicular. C. Ratio of scattered primary intensity from water for arrangements B and A.

7.3.5. Considerations

It is by no means always necessary to correct analyte-line intensity data for background, and there are actually disadvantages to doing so. Unless background intensity can be derived from a blank or by extrapolation, at least one additional measurement is required from each specimen, with consequent increase in analysis time and cost. Moreover, background correction increases counting error because the statistical error in $I_P - I_B$ is greater than that in I_P alone.

The advantages of background correction include the following. The background measurement may serve to correct absorption-enhancement effects (Section 14.5.1) and as an indication of improper or abnormal operation of the instrument or of error in procedure. When working without a calibration curve, for example, with a single-standard or standard-addition or standard-dilution method, the net intensity is preferable.

Liebhafsky and his colleagues [pp. 212–15 in (*26*)] classify background problems in six categories as follows. In the discussion, I_P and I_B are analyte-line peak and background intensities, respectively, and C_A is analyte concentration.

1. The background is negligible, that is, I_B is so small that $(I_P - I_B) \approx I_P$, and background need not be measured.

2. The background is substantial, but need not be corrected. If I_B is constant over the entire analytical range of interest, and if the analyte-line *versus* concentration curve is linear, the function $I_P - I_B$ *versus* C_A is linear and intercepts the origin, whereas I_P *versus* C_A is linear and intercepts the intensity axis at I_B. In such cases, there is no advantage in correcting background.

3. Background correction is required, but simple; for example, if I_B can be measured from a single measurement on a blank.

4. Background correction is required, and there are no complications, but it must be measured from each specimen on one or both sides of the analyte peak.

5. Background correction is required and complex; for example, when there are interfering lines superposed on the continuous background near the analyte peak and a blank is not feasible, or the background is affected by absorption edges.

6. Background is so large with respect to the analyte line that it may seriously interfere with the analysis. In trace analysis, $I_P \approx I_B$, and statistical fluctuations in both may lead to reporting of an analyte as present when it is absent, and *vice versa*.

In general, background correction should be made only when it leads to greater analytical reliability. However, it is always prudent to exercise all practical means of reducing background.

Chapter 8

Pulse-Height Selection; Energy-Dispersive Analysis; Nondispersive Analysis

Before continuing, it is necessary to distinguish pulse-height *selection* and *analysis*. Pulse-height selection (PHS) is the separation of one or a very few pulse distributions from the analyte pulse distribution and is usually used in conjunction with a crystal spectrometer. Pulse-height analysis (PHA) is the separation of all the pulse distributions in the entire x-ray energy spectrum emitted by the specimen so that each can be measured individually. It is the most widely used method of energy-dispersive analysis. The distinction drawn in this book between energy-dispersive and nondispersive methods of x-ray spectrometric analysis is discussed in Section 5.1. In terms of the definitions given above and in Section 5.1, pulse-height selection, pulse-height analysis or energy-dispersive analysis, and nondispersive analysis are discussed in Sections 8.1, 8.2, and 8.3, respectively. The general principles of pulse-height selection are given in Section 8.1, but apply also to Section 8.2.

8.1. PULSE-HEIGHT SELECTION

The instrumentation and general function of the pulse-height selector (or pulse-height analyzer) are described in Section 7.1.4.1. This section considers the method of pulse-height selection in greater detail.

8.1.1. Principle of Pulse-Height Selection
(*F22, H27, M36, V3, V4*)

It has been explained (Sections 6.2.3.2, 6.3.2.1, and 6.4.2) that the output of a proportional, scintillation, or semiconductor counter consists of a series of electric-current pulses having mean height proportional to the incident photon energy, each pulse originating from absorption of a single incident x-ray photon. It has also been explained that, even with strictly monochromatic incident x-rays, there is a statistical spread in the numbers of ion pairs produced per x-ray photon in a proportional counter, in the numbers of useful light photons produced per x-ray photon in a scintillation counter, and in the numbers of electron–hole pairs produced per x-ray photon in a semiconductor detector. It follows that there is also a statistical spread in the amplitudes of the pulses in the detector and amplifier outputs. Thus, one properly speaks of pulse-height *distributions.*

Consider a xenon-filled proportional counter receiving three x-ray spectral lines simultaneously—Sc $K\alpha$, Cu $K\alpha$, and Nb $K\alpha$. Figure 8.1A

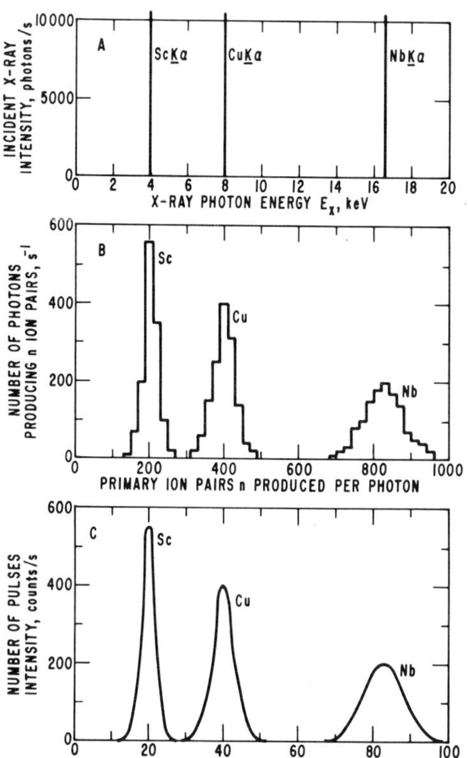

FIG. 8.1. Origin of pulse-height distributions for three x-ray spectral lines. A. X-ray photon energies. B. Distribution in numbers of primary ion–electron pairs produced in a xenon-filled proportional counter. C. Distribution in pulse amplitude of the amplifier output.

shows the photon energies E_x of the three lines—4.09, 8.04, and 16.6 keV, respectively. Figure 8.1B shows the distribution in the numbers of primary ion–electron pairs produced in the detector gas by the x-ray photons. Assuming the effective ionization potential V_i of xenon to be ~20 eV, the mean number $\bar{n}$ of primary ion pairs produced per photon is E_x/V_i [Equation (6.1)]—200, 400, and 830, respectively. The statistical spread σ in the number of pairs is $\sqrt{\bar{n}}$ [Equation (11.18)]—14, 20, and 29, respectively. Figure 8.1C shows the pulse-amplitude distribution in the amplifier output. The half-width (FWHM) of each distribution is approximately $235/\sqrt{\bar{n}}$ [Equation (6.12)]—17, 12, and 8%, respectively. Incidentally, the figure disregards differences in quantum efficiency of the detector at the three wavelengths.

The height of an individual pulse and the mean height and width of a pulse-height distribution are influenced by many parameters, principally the following: (1) photon energy; (2) the number of useful electrons produced by the initial ionizing event; (3) the internal electron multiplication (gas or secondary) in the detector; (4) the relationship of incident x-ray intensity and detector resolving and recovery times; (5) detector potential; and (6) amplifier gain. One might be inclined to wonder how proportionality can have any significance with so many variables. Of course, the answer is that if all the variables except photon energy are kept constant, the photon energy becomes the determining parameter.

Mean pulse height is given by

$$V = \frac{E_x}{E_D} A_D \frac{A_E}{L} \tag{8.1}$$

where V is mean pulse height for a specified spectral line (V); E_x is x-ray photon energy for the line (eV); E_D is the average energy to produce one useful electron in the detector (eV); this would be one ion pair in a proportional counter, one photoelectron collected at the first dynode in a scintillation counter, and one electron–hole pair in a semiconductor detector; A_D is internal detector amplification—gas amplification or secondary-electron multiplication; A_E is external amplifier gain; and L is amplifier attenuation.

Given the mean pulse height V_1 for x-rays having photon energy E_1, the mean pulse height V_2 for x-rays having energy E_2 is given by

$$V_1/V_2 = E_1/E_2 \tag{8.2}$$

In general, if all other parameters are constant, the pulse-height distribution has lower mean amplitude V, narrower half-width $W_{1/2}$ or ΔV,

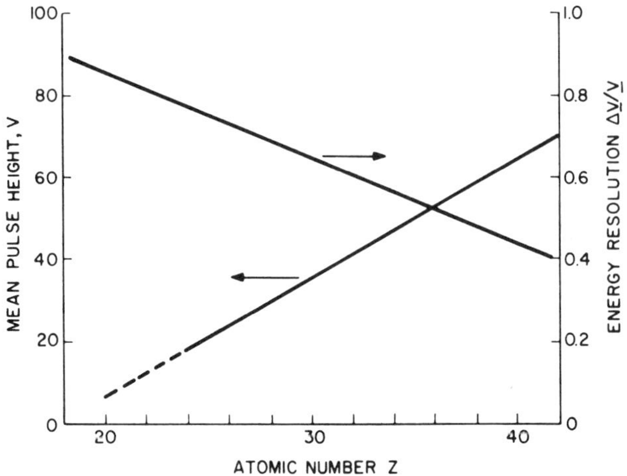

FIG. 8.2. Mean pulse height, and ratio of width of pulse-height distribution and mean pulse height, as functions of atomic number for $K\alpha$ lines.

and higher energy resolution $\Delta V/V$: (1) for proportional counters than for scintillation counters; (2) the longer the x-ray wavelength (the lower the photon energy), or, for a specified spectral line, such as $K\alpha$ or $L\alpha_1$, the lower the atomic number; (3) the lower the applied detector potential; and (4) the lower the amplification (gain). The relationship of mean pulse height, energy resolution, and atomic number for $K\alpha$ lines is shown in Figure 8.2. The last three conditions above can all be illustrated by Figure 8.3, which is explained in its caption and should be studied carefully. In all three cases, although the peak intensities decrease progressively from *A* to *C*, the integrated intensities, the areas under the peaks, remain substantially constant.

Figure 8.3 can also be used to illustrate the principle of pulse-height selection. Assume that the figure shows three pulse-height distributions of x-rays having low, medium, and high photon energy (long, medium, and short wavelength), respectively (case I in the caption). It is evident that with the window set as shown, the *B* peak is passed to the exclusion of *A* and *C*. If the baseline potential is varied, the pulse-height distributions remain fixed, but the window is moved and can be set to admit any one of them. If the detector potential and/or amplifier gain are varied, the window remains fixed, but the pulse-height distributions are moved, and any one can be moved into the window. Finally, the window width can be varied to pass as many as possible of the pulses in the analyte-line distribution while excluding as many as possible of the pulses from neighboring peaks, which may, unlike the peaks shown here, overlap. For example, it is evident that the window width shown in Figure 8.3 just admits the entire *B* peak. If the window were moved to the *A* peak, it could be made narrower without loss of *A* pulses, but if it were moved to the *C* peak, many *C* pulses would be excluded unless the window were made wider. In practice, it is customary to move the pulse-height distribution to the window, rather than the window to the distribution.

Incidentally, although emphasis is placed on proportionality of mean pulse height and photon energy—and therefore atomic number—the half-width is also proportional to photon energy and atomic number. Consequently, half-width can be used for identification of elements if the $W_{1/2}$ *versus* E relationship is previously established (*H31*).

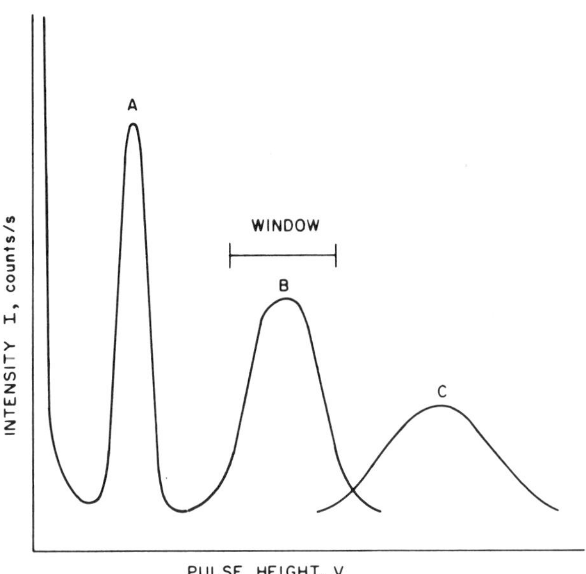

FIG. 8.3. Effect of incident x-ray photon energy, detector potential, and amplifier gain on mean pulse height, width, and peak intensity of the pulse-height distribution. The areas under the peaks are equal. The figure represents any of three cases:

I. Detector potential and amplifier gain constant: peaks *A*, *B*, and *C* represent pulse-height distributions for low, medium, and high photon energies, respectively.

II. Photon energy and amplifier gain constant: peaks *A*, *B*, and *C* represent successively higher detector potentials.

III. Photon energy and detector potential constant: peaks *A*, *B*, and *C* represent successively higher amplifier gain.

The *analyzer transmission* is the fraction of the pulses originating from a monochromatic x-ray beam that pass the pulse-height analyzer and are counted. The *detector efficiency* is the fraction of the photons of a monochromatic x-ray beam arriving at the detector window that are counted. It is equal to the product of the detector quantum efficiency (fraction of photons detected) and the analyzer transmission. Both these terms are usually expressed as percent.

8.1.2. Pulse-Height Distribution Curves

8.1.2.1. Introduction

Pulse-height distribution curves are best described in specific, rather than general, terms. A good illustrative example is the determination of silicon at concentration ~1% in iron with an EDDT crystal (*K12*). Fourth-order Fe $K\beta$ slightly overlaps Si $K\alpha$. Table 8.1 gives the pertinent data.

Actually, the two lines are separated by 2° 2θ, but the Fe $K\beta(4)$ line is so intense relative to Si $K\alpha$ that the tail of the iron line contributes intensity at the 2θ angle of the relatively weak silicon line. It must be emphasized that although the Fe $K\beta(4)$ intensity at the Fe $K\beta(4)$ peak (106° 2θ) is ~1600 counts/s, the Fe $K\beta(4)$ intensity *overlap* at the Si $K\alpha$ peak (108° 2θ) is only ~30 counts/s. Thus, if the spectrogoniometer is set to measure the Si $K\alpha$ line (108° 2θ), the detector or amplifier output contains five groups of pulses of various intensities and average amplitudes as follows: (1) detector–amplifier noise of very low amplitude and very high intensity;

TABLE 8.1. Data for Si $K\alpha$ and Fourth-Order Fe $K\beta$[a]

Spectral line	Si $K\alpha$	Fe $K\beta(4)$
Wavelength λ, Å	7.13	1.76
$n\lambda$ ($n = 4$)	—	7.04
Energy E, eV	1739	7045
Diffraction angle (EDDT) 2θ, deg	108	106
Net intensity I, counts/s		
At the peak	~200	~1600
At 108° 2θ	~200	~30
Background intensity, counts/s	~160	~160

[a] Data apply to Figures 8.4–8.8.

(2) continuous x-ray background having a wide range of amplitudes and an intensity, say, 160 counts/s; (3) Si $K\alpha$ pulses of low amplitude and intensity ~200 counts/s; (4) Fe $K\beta$ pulses of relatively high amplitude and intensity ~30 counts/s; and (5) Fe $K\beta$ escape-peak pulses (assuming an Ar-filled proportional counter); these are of no significance in the discussion to follow and are not considered further here; however, the Fe $K\beta$ escape peak is discussed in Section 13.3.2.2b and shown in A of Figure 13.7.

There are several ways to display pulse-height distribution data, three of which are shown in Figure 8.4. In Figure 8.4B, *individual* pulses are displayed on a grid of pulse height *versus* time. Three of the groups of pulses listed above are distinguishable and are marked at the left of the drawing: detector–amplifier noise, Si $K\alpha$, and Fe $K\beta$. The background pulses are not shown because they have a wide range of pulse heights and would obscure the other pulses. As expected, the high-energy Fe $K\beta$ distribution has higher average pulse height and is wider than the low-energy Si $K\alpha$ distribution. However, also as expected, at 108° 2θ, the Si $K\alpha$ pulses are more *numerous* (~200/s) than the Fe $K\beta$ pulses (~30/s). The x-ray photons incident on the detector, and therefore the output pulses, have a random distribution in time. Such a display is obtained by connecting the vertical input of a cathode-ray oscilloscope to the output of the linear amplifier. The displays at the output of the detector or preamplifier would differ only in linear amplification. Incidentally, if NIM modular units are used, the baseline pulse-height scale would be 0–10 V, not 0–100 V.

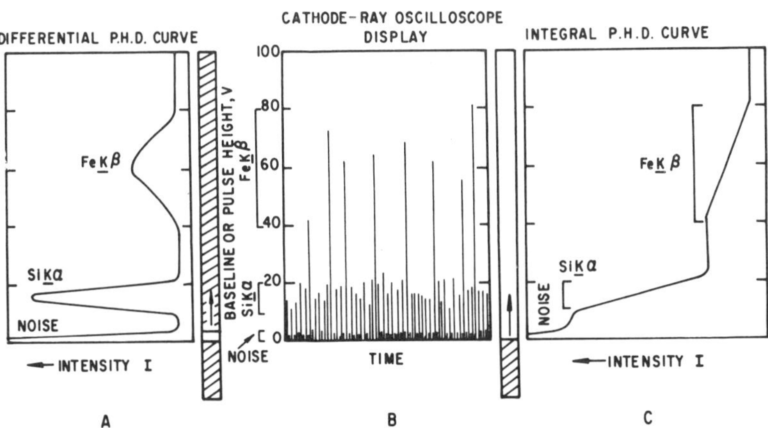

FIG. 8.4. Function of the pulse-height selector. The center figure (B) represents the amplifier output, pulse height *vs.* time, and shows three groups of pulses from noise, Si $K\alpha$, and Fe $K\beta$. The left figure (A) shows the differential pulse-height distribution curve. The right figure (C) shows the integral pulse-height distribution curve.

8.1.2.2. Single-Channel Pulse-Height Selector

The single-channel pulse-height selector or analyzer (SCA) and ratemeter–recorder may be used to display the pulse-height distribution in two ways.

A *differential pulse-height distribution curve* (Figure 8.4A) is recorded by: (1) setting the spectrogoniometer at 108° 2θ (Si $K\alpha$); (2) setting the window at an arbitrary small width ΔV, say 1 V; and (3) "scanning the baseline," that is, moving the baseline discriminator level V—and with it the window—from 0 to 100 V as intensity is recorded. At any instantaneous baseline setting, the window passes to the ratemeter only those pulses having height between V and $V + \Delta V$. The resulting curve of intensity *versus* pulse height, Figure 8.4A, shows two pulse-height distribution peaks and the detector–amplifier noise. The orientation is unconventional, to permit correlation with Figure 8.4B. At very low baseline potential, the high-intensity, low-energy noise passes the window, giving high response, which then falls to zero until the window passes through the Si $K\alpha$ distribution, giving a relatively high peak. The response then again falls to zero until the window passes the less intense Fe $K\beta$ distribution. The continuous x-ray background pulses are not shown in Figure 8.4B, and so background is also omitted in Figure 8.4A; it would appear simply as a finite—rather than zero—intensity between the noise, Si $K\alpha$, and Fe $K\beta$ peaks.

An *integral pulse-height distribution curve* (Figure 8.4C) is recorded by: (1) setting the spectrogoniometer at 108° 2θ as before; (2) making the window inoperative or, in effect, "wide open"; and (3) scanning the baseline V from 0 to 100 V as intensity is recorded. At any instantaneous baseline setting, *all* pulses having height greater than V pass to the ratemeter. The resulting curve of intensity *versus* pulse height, Figure 8.4C, shows three *steps*, instead of the *peaks* of Figure 8.4A—the noise, Si $K\alpha$, and Fe $K\beta$. Again, the orientation is unconventional, to permit correlation. At very low baseline potential, *all* pulses—noise, Si $K\alpha$, Fe $K\beta$, and background—pass to the recorder, and the response is very high. At 3–5 V, the noise is discriminated out, and the intensity decreases. At a somewhat higher potential, the Si $K\alpha$ distribution is excluded, and at still higher potential, the Fe $K\beta$. Here again, if x-ray background were shown in the figure, it would appear simply as a finite—rather than zero—intensity after the Fe $K\beta$ step.

Both types of pulse-height distribution curve convey the same information, and the choice is largely a matter of personal preference. Such curves are made prior to undertaking analysis of a new sample type. They reveal escape peaks and overlapping of neighboring pulse-height distributions with that of the analyte, and permit the choice of baseline and window settings for the analysis. The window settings for Si $K\alpha$ and Fe $K\beta$ (the latter of academic interest only) are shown in Figure 8.4B and 8.4C.

The baseline can be scanned in any of four ways: (1) The baseline control may be motor-driven; (2) the baseline may be advanced manually in equal, small increments at equal, short time intervals—say, 0.5 V every other chart division—as the recorder continuously registers intensity; (3) the baseline may be advanced incrementally, as above, and the intensity scaled at each setting (step scanners are available to perform this function automatically); and (4) the baseline may be scanned by electronic means. Incidentally, it is customary to scan the baseline from 100 to 0 V, rather than 0 to 100 V, as in the descriptions above. The reason is that it is more convenient to scan downward until the high-intensity noise is encountered, then stop.

8.1.2.3. Multichannel Pulse-Height Analyzer

In the *multichannel pulse-height analyzer* (MCA) (*V5*), the baseline in Figure 8.4 is, in effect, divided into 200, 400, 800, 1200, or even more individual equal, narrow, fixed windows, each connected to its own accumulator or memory. The analyzer receives the amplified detector output and sorts the incoming pulses by height, directing each to its proper "box" along the baseline. The process continues long enough to permit accumulation of a suitably large count—say 10^5—in the channel corresponding to the highest peak. Then the number of counts in each channel is rapidly read out electronically and indicated on an X–Y recorder, oscilloscope, or digital printer. A 400-channel analyzer, in effect, step-scans the baseline in 400 incremental steps and scales the intensity at each step—but it does it "all at once," rather than sequentially.

Figure 8.5 shows the function of a multichannel analyzer in a highly simplified schematic way, using only 30 channels for ease of illustration. Figure 8.4B, showing noise, Si $K\alpha$, and Fe $K\beta$ pulse distributions, is reproduced at the left of Figure 8.5. However, instead of "scanning the baseline" sequentially in one of the two ways described in Section 8.1.2.2, the baseline is in effect divided into 30 equal pulse-height intervals or windows (*channels*), each having its own accumulator. The instrument accumulates pulses in all channels simultaneously until the channel corresponding to the pulse height of the highest intensity (channel 4, average pulse height of the Si $K\alpha$ pulse distribution) has accumulated some arbitrarily large preset count. Then the number of counts accumulated in each

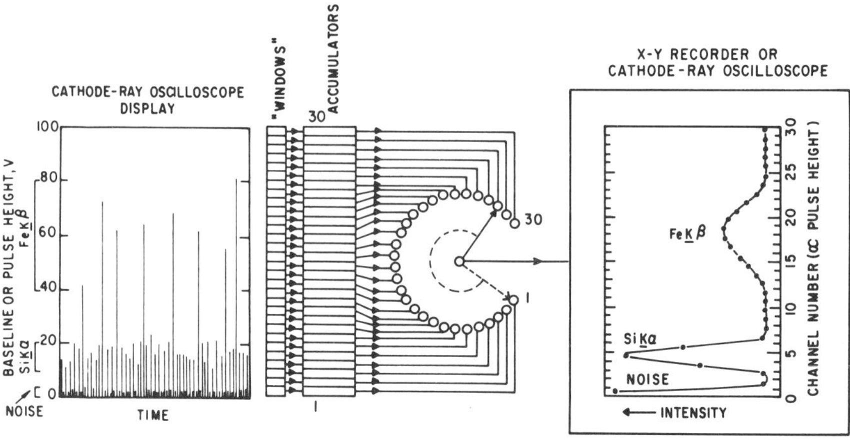

FIG. 8.5. Schematic representation of function of the multichannel analyzer.

channel is rapidly read out electronically—not mechanically, as in Figure 8.5—and displayed on an *X–Y* recorder or cathode-ray tube, as shown, or printed out digitally. If an instrument having 200 or more channels were used, the display would appear substantially as in Figure 8.4A. Alternatively, the accumulation may proceed for a preset time, or until a *group* of channels ("window") accumulates a preset count. The window may consist of a group of channels bracketing a specified peak, for example, channels 2–7 (Si $K\alpha$) in Figure 8.5.

The multichannel analyzer circuit in most common use is that due to Wilkinson (*W12*) and consists essentially of an input capacitor (rundown capacitor), address register, oscillator (address clock), and memory unit, each address in which corresponds to a specific channel in the multichannel analyzer. The peak potentials of the input pulses are used to gate the oscillator. Each pulse entering the multichannel analyzer initiates the following process: (1) The incoming pulse charges the input capacitor to the peak potential of the pulse. (2) The capacitor then discharges at a constant rate. (3) During the entire discharge time, the address register receives and counts pulses from the oscillator. (4) At the conclusion of the discharge period, the register, on the basis of the number of oscillator pulses it has received, designates which memory address (channel) is to be incremented by one count. To summarize, the greater the input pulse amplitude, the greater the potential at which the capacitor is charged, the longer the capacitor discharge time, the greater the number of oscillator pulses received by the address register, and the higher the channel number incremented by one count. At the end of the process, the next pulse is admitted, and the sequence is repeated, pulse by pulse, for the duration of the counting time. In effect, in a 1000-channel analyzer, input pulse heights of 0–10 V (for NIM modules) are converted proportionally to numbers between 0 and 1000; for example, a 2.5-V pulse would result in addition of one count to channel 250. Usually, the system just described is preceded by upper- and lower-level discriminators, say 0.5–10 V, to reject noise and abnormally high pulses. Actually, the discriminators and other components just described, except for the multichannel count register itself, constitute an analog-to-digital converter (ADC), the conversion being from pulse height to time.

It must be emphasized that the energy spectrum shown in Figure 8.5 results from multichannel analysis of the x-rays diffracted at 108° 2θ (Si $K\alpha$) by an EDDT crystal. A more likely application of the multichannel analyzer would be to eliminate the crystal and disperse directly the x-rays emitted by the specimen. This energy spectrum would show very intense Fe *K* and relatively weak Si *K* pulse-height distributions. Moreover, the broad pulse distributions shown in Figure 8.5 result from conventional scintillation or proportional counters. If a lithium-drifted silicon [Si(Li)] detector (Section 6.4) is used, very narrow pulse distributions are obtained with consequent resolution of pulse distributions for photon energies as close as 200 eV or less.

The multichannel analyzer is calibrated to correlate channel number and photon energy with known spectral lines, often from radioisotopes, ^{55}Fe, for example.

Alternatively, the instrument can be used as a *multichannel scaler* to accumulate in successive channels pulses having the *same* average pulse height from a series of related individual measurements. The series may be simple replicates, an evaluation of short- or long-term instrument stability (Section 11.3.2.2), or a point-by-point line or raster scan across the specimen by an x-ray "milliprobe" (Chapter 19) or electron-probe microanalyzer (Chapter 21). Two striking examples of this last application are shown in Figures 21.5 and 21.6.

A multichannel analyzer to be used with a Si(Li) detector in an energy-dispersive x-ray spectrometer system should have the following characteristics (*G5*): (1) 1000 channels or more; a 1000-channel analyzer used to cover the photon-energy range 0–20 keV (Mo $K\alpha$, any $L\alpha_1$) would have 20 eV/channel; (2) count capacity per channel 10^6 or more; (3) good integral linearity—that is, proportionality of channel number and energy; (4) good differential linearity—that is, uniformity of channel width; (5) low dead time; (6) dc-coupled input able to accept Gaussian pulse shapes having rise times of the order 20 μs; (7) an accurate live-time clock that also compensates dead time due to amplifier pulse widths; and (8) good stability of conversion gain and zero intercept with respect to temperature and count rate.

The principal advantages of the multichannel analyzer all result from the fact that the entire energy spectrum is accumulated simultaneously: It provides the most rapid way to display an energy spectrum and therefore to make a qualitative x-ray spectrometric analysis. Analyses can be terminated as soon as the spectrum is seen to be adequate for the purpose at hand. Also, instrumental instability has minimal effect on the simultaneously accumulated spectral-line counts.

The applications of the multichannel analyzer in x-ray spectrometry include the following: (1) rapid qualitative and quantitative energy-dispersive analysis (Section 8.2.2.4); (2) preliminary "setup" work for crystal spectrometers; (3) improvement of spectral resolution by expansion of regions of particular interest (Section 8.2.2.4); (4) rapid comparison analyses (Section 8.2.2.4); (5) unfolding of overlapped pulse-height distributions (Section 8.1.9); (6) determination of optimal excitation conditions quickly and without the long times required for a 2θ scan or for a baseline scan on a single-channel pulse-height analyzer; and (7) energy-dispersive x-ray diffraction-spectrometry (Section 8.2.3). The applications of the instrument as a multichannel scaler, in addition to those cited above, include two-dimensional printout intensity topographs in electron-probe microanalysis (Section 21.4.4, Figure 21.5).

The multichannel analyzer, combined with a high- or low-power x-ray tube, secondary-target, radioisotope, or electron excitation source, a specimen-presentation system, a Si(Li) detector and its liquid-nitrogen cryostat, and computer and memory units, constitutes an extremely compact, versatile, rapid, and convenient energy-dispersive x-ray spectrometer system. Such systems are described in Section 8.2.2.4.

8.1.3. Pulse-Height Selector Displays

It is informative at this point to summarize and compare the various ways of displaying x-ray intensity data, particularly with respect to the pulse-height selector. The silicon–iron example is used again for illustration. Figure 8.6 shows four types of ratemeter–recorder display.

Figure 8.6B is a 2θ scan of the spectral region near the Si $K\alpha$ line recorded without pulse-height selection, that is, with only a discriminator set to reject detector noise. The scan shows intensity as a function of 2θ

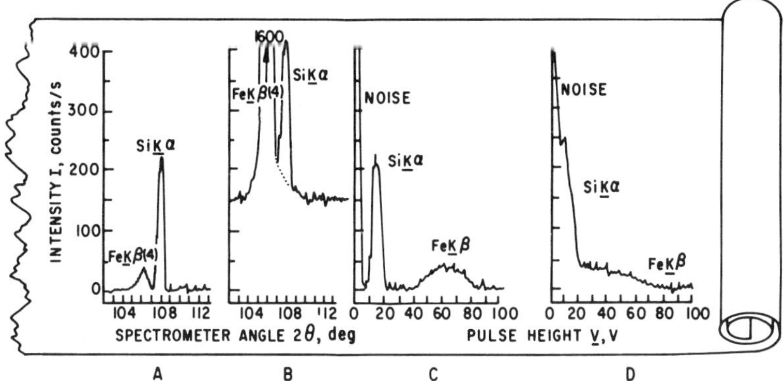

FIG. 8.6. Ratemeter–recorder readout of intensity data. A. 2θ scan (EDDT crystal) with pulse-height selection. B. 2θ scan without pulse-height selection. C. Differential pulse-height distribution curve. D. Integral pulse-height distribution curve.

(wavelength). The continuous background is very high, and the tail of the Fe $K\beta(4)$ peak contributes intensity at the 2θ angle for Si $K\alpha$. The Si $K\alpha$ line-to-background ratio, that is, the intensity ratio of Si $K\alpha$ to background plus Fe $K\beta$, is only ~1:1.

Figure 8.6A is the same 2θ scan, but with pulse-height selection. The baseline and window are set to pass the Si $K\alpha$ distribution, as shown in Figure 8.4. The continuous background has been substantially eliminated. Only a vestigial Fe $K\beta(4)$ peak remains, and it no longer contributes intensity at the Si $K\alpha$ peak. The Si $K\alpha$ net intensity is substantially the same as before, but the line-to-background ratio is now ~37:1.

Figure 8.6C is a differential pulse-height distribution curve showing intensity as a function of pulse height. It is equivalent to Figure 8.4A. The spectrogoniometer is fixed at 108° 2θ (Si $K\alpha$). The window is set at 1 V. The baseline is scanned continuously from 0 to 100 V. The curve shows the number of pulses per second passing through the narrow window at each instantaneous baseline potential. The differential pulse-height distribution curve is an x-ray *energy* spectrum directly analogous to the x-ray *wavelength* spectrum represented by a 2θ scan.

Figure 8.6D is an integral pulse-height distribution curve, also showing intensity as a function of pulse height. It is equivalent to Figure 8.4C. The spectrogoniometer is fixed at 108° 2θ as before. The window is inoperative ("wide open"). The baseline is scanned continuously from 0 to 100 V. The curve shows the number of pulses per second passing over the baseline at each instantaneous baseline potential.

Figure 8.7 shows digital printer readouts of two series of intensity

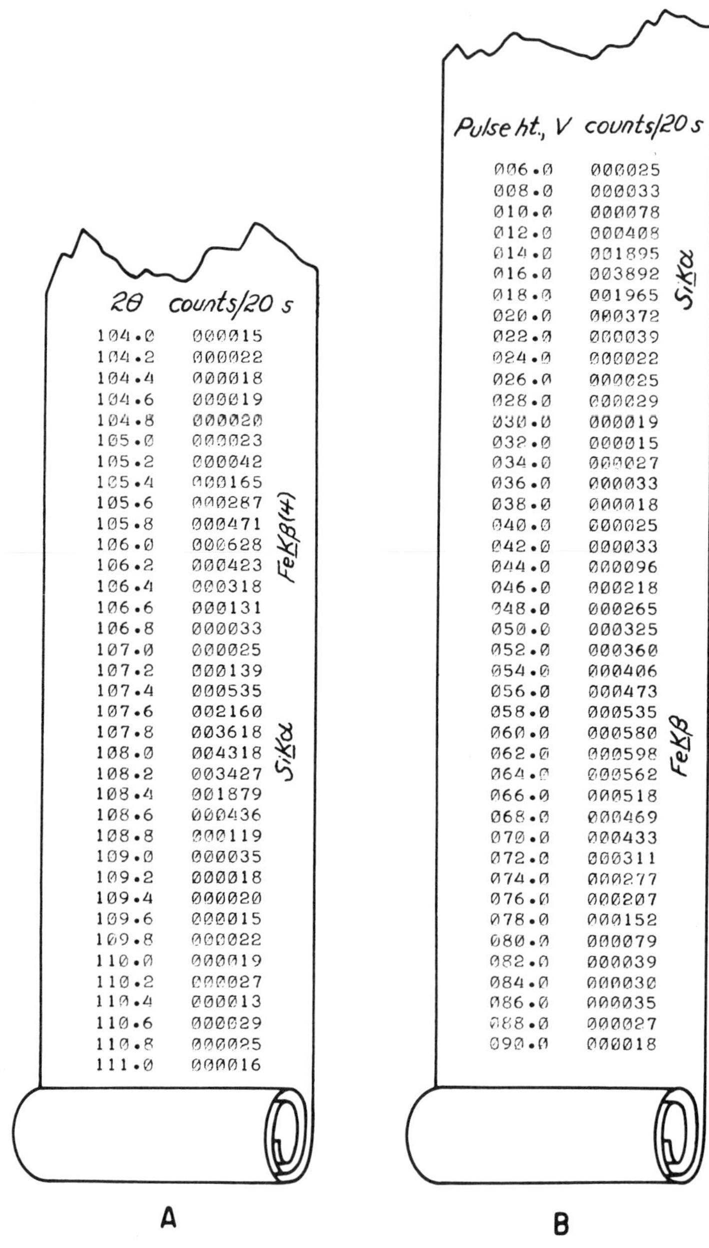

FIG. 8.7. Printer readout of intensity data. A. 2θ scan with pulse-height selection (Figure 8.6A). B. Differential pulse-height distribution curve (Figure 8.6C).

measurements. Figure 8.7A shows data for a 2θ scan of the spectral region near the Si $K\alpha$ line with pulse-height selection. The data plotted as intensity *versus* 2θ would resemble Figure 8.6A. The spectrogoniometer is manually indexed or automatically step-scanned in increments of 0.2° 2θ, and, at each position, the intensity is scaled for a preset time of 20 s. Figure 8.7B shows data for a differential pulse-height distribution curve. The data plotted as intensity *versus* pulse height would resemble Figure 8.6C. The spectrogoniometer is fixed at 108° 2θ. The window is set at 1 V. The baseline is manually or automatically scanned in increments of 2 V, and, at each potential, the intensity is scaled for a preset time of 20 s.

Figures 8.4B and 8.8 show cathode-ray oscilloscope displays, the former as a drawing. Figure 8.4B, as already mentioned, shows the individual pulses in the output of the detector, preamplifier, or amplifier on a scale of pulse height *versus* time. The spectrogoniometer is fixed at 108° 2θ, and the display shows the pulses due to detector noise, Si $K\alpha$, and Fe $K\beta$. Figure 8.8 shows a differential pulse-height distribution curve photographed from a Siemens pulse spectroscope (*B83*). The conditions are similar to those for Figure 8.6C, except that the scanning is effected electronically. The black bar on the Si $K\alpha$ peak is an electronic marker showing the window position. The marker lengthens and moves as the window is widened and shifted along the baseline.

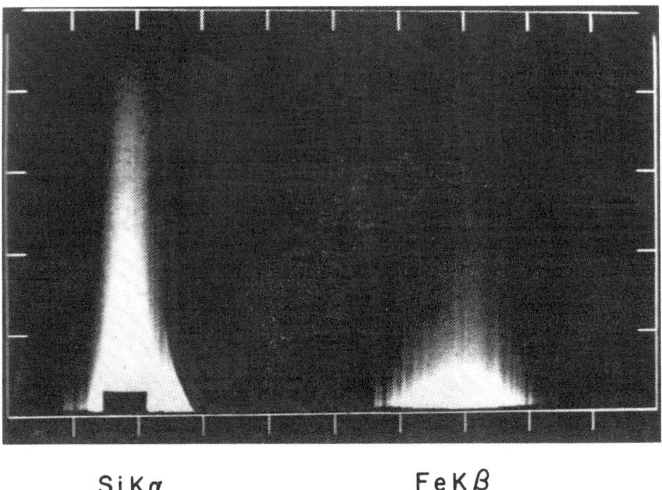

FIG. 8.8. Pulse-spectroscope display of the Si $K\alpha$ and Fe $K\beta$ pulse-height distributions. The black bar on the Si $K\alpha$ peak is an electronic window marker.

8.1.4. Pulse-Height Selector Operating Controls

The operating controls involved in the use of the pulse-height selector are the detector potential (up to ~2 kV), amplifier gain (up to ~10,000), baseline (0–100 V), window (0–10, 20, or 30 V), and differential–integral mode switch. For NIM modular units, the baseline and window potentials are smaller by a factor of 10.

The detector potential and amplifier gain both affect the mean pulse height, peak intensity, and half-width of the pulse-height distribution. Decreasing of either control decreases the mean pulse height, increases the peak intensity, and decreases the width; increasing of either control has the reverse effect (Figure 8.3). The detector potential and amplifier gain controls are used together to: (1) give optimum signal-to-noise, or peak-to-background, ratio; and (2) move the pulse-height distributions along the baseline, usually to a relatively low potential.

The maximum permissible detector potential is established from the detector noise curve, and the useful potential range from the plateau curve (Section 8.1.5.1).

With the inherently noisy scintillation counter, low detector potential and high amplifier gain must be used. With the relatively noise-free proportional counter, relatively high detector potential and intermediate amplifier gain may be used. The higher detector potential permits advantageous use of the relatively noise-free gas amplification of the detector. With the Geiger counter (which, of course, is not applicable to pulse selection), the high output pulse amplitude permits use of low amplifier gain.

Usually, the amplifier gain is set so that pulses of the lowest amplitude to be measured (that is, those originating from the longest wavelength) fall at a selected baseline setting. Then the gain is attenuated to bring successively higher-amplitude (shorter-wavelength) pulses to the selected working range.

The discriminator baseline is usually set at ~10–20 V for several reasons: (1) Detector noise is completely rejected; (2) the pulse-height distribution has high peak intensity and narrow width (Figure 8.3); (3) the energy resolution is high, reducing the interference from other distributions; (4) a narrow window can be used without substantial loss of analyte pulses; and finally, (5) the line-to-background ratio is increased because the narrow window admits fewer continuous background pulses.

In general, the window is set wide enough to pass the entire analyte pulse-height distribution and its escape peak, if any. If the detection process is preceded by crystal dispersion, the window may usually be much wider than the distribution because the crystal eliminates neighboring wavelengths, and distributions from higher orders are well separated in energy. Thus, the analyte distribution is likely to be free from neighboring distributions. A wide window is advantageous to allow for shift and distortion effects. If the optics are energy-dispersive, neighboring distributions are likely to be present, and it may be necessary to narrow the window, sacrificing analyte pulses to exclude pulses from adjacent peaks.

The baseline and window controls are usually 10-turn helical potentiometers. Provision may be made to motor-drive the baseline control continuously or stepwise. Provision may be made to couple the baseline alone or the baseline and window to the spectrogoniometer θ–2θ drive (Section 8.1.7). The window always "tracks" the baseline, that is, the window may be regarded as resting on the baseline.

In some instruments, provision is made to operate the window in either of two modes: In *fixed-width* mode, the window has the same preset width in volts at all baseline settings; in *relative-width* mode, the window width varies in proportion to the baseline setting, always maintaining a preset relative value. In either case, the width may be selected: 0–10, 20, or 30 V in fixed width, 0–20% of baseline in relative width.

The differential–integral control determines the mode of operation of the pulse-height selector. In differential mode, both discriminator levels are activated, and the selector functions as a pulse selector, passing only pulses having heights between the two levels. In integral mode, the second discriminator, and therefore the upper level, is inactivated, making the window inoperative (wide open). The selector now functions simply as a discriminator, passing all pulses having height greater than the baseline.

8.1.5. Use of the Pulse-Height Selector

8.1.5.1. Evaluation of Detector and Amplifier Characteristics

The amplifier gain is set at low, medium, or high for Geiger, proportional, or scintillation counters, respectively. The Geiger counter is not applicable to pulse selection, but does require the evaluation described in this section. All of the following evaluations are made with the pulse-height selector in integral mode.

The *amplifier noise threshold* is the maximum permissible amplifier gain and is established as follows. The preamplifier is disconnected from the amplifier input, the baseline (integral mode) is set at ~2 V (~0.5 V for NIM modules), and the ratemeter is set at its lowest range. Then, as the panel ratemeter or recorder is observed, the amplifier gain is gradually increased until a sustained response is just attained. This is the amplifier noise threshold, and further increase in gain results in a sharp increase in response. The amplifier must always be operated at gain well below this threshold.

The *preamplifier noise level* is established as follows. The detector power supply is turned off, the amplifier gain is set at just less than its noise threshold (see above), and the ratemeter is set at its lowest range. Then, as the panel ratemeter or recorder is observed, the baseline (integral mode) is gradually decreased until a sustained response is just attained. This should occur at a baseline setting of substantially less than 5 V (0.5 V on NIM modules). Otherwise, the preamplifier must be repaired.

The *detector noise level* is established as follows. With no x-rays, the baseline discriminator setting is reduced to very *low* values until a sharp increase in response is observed on the ratemeter or recorder. This is the detector noise level. The minimum baseline setting should be 2–5 V above this level to ensure adequate discrimination against the noise. Usually, a minimum baseline of 5 V is satisfactory. For pulse-height analysis, a baseline of 10–20 V is typical.

The *detector noise curve* is established as follows. Again, with no x-rays, the detector potential is increased incrementally from zero, and intensity is read on the recorder at each point. At some relatively high potential, the response increases, gradually at first, then sharply. The plot of intensity *versus* detector potential without x-rays is the detector noise curve and indicates the maximum permissible operating potential (Section 6.5.1.10, Figure 6.15).

The *detector plateau curve* is established as follows. A specimen of pure element (or its oxide) having a major spectral line in the region of interest is excited to give an intensity of ~1000 counts/s for Geiger counters and 2000–5000 counts/s for proportional and scintillation counters. The baseline is set at 5–10 V. Detector potential is increased incrementally from zero, and intensity is read on the recorder at each point. The response rises, remains constant over a substantial potential range, then rises again. The plot of intensity *versus* detector potential, with x-rays, is the detector plateau curve. The linear plateau itself is the useful range of detector potential. Geiger counters are operated near the center of the plateau. So are proportional and scintillation counters, but their potential may be varied somewhat as an operating variable to displace pulse-height distributions along the baseline. The detector potential should never be moved off the plateau. It must be recalled that the plateau curve is shifted to lower detector potential at shorter wavelengths, higher amplifier gain, and lower discriminator baseline (Section 6.5.1.8, Figure 6.15 and 6.16).

The foregoing evaluation need be made carefully for each specific detector only at intervals of several months. However, rapid checks at more frequent intervals monitor the condition of the detector and give warning of onset of deterioration.

8.1.5.2. Establishment of Pulse-Height Selector Settings

The pulse-height selector settings—baseline and window width—for determination of a specified analyte in a specified sample type (matrix) can be established in one of three ways: (1) from a pulse-height distribution curve; (2) by observing intensity as a narrow window is moved over the analyte pulse distribution; and (3) by observing intensity as the baseline is moved and window width varied on the analyte pulse distribution. These three methods are described below.

A pulse-height distribution curve should be recorded (method 1 above) if: (1) the analyte has an escape peak; and/or (2) there is any possibility that with the goniometer set at 2θ for the analyte line λ_A, the detector output may contain pulse distributions from other elements. Such extraneous distributions may arise from: (1) partial wavelength overlap, $\lambda \approx \lambda_A$ (complete wavelength overlap, $\lambda = \lambda_A$, cannot be dealt with by pulse-height selection, but other means can be used, as discussed in Section 13.3); (2) higher-order wavelength overlap, $n\lambda = \lambda_A$; (3) escape peaks, from matrix element lines, overlapping the analyte pulse distribution; and (4) crystal emission (Section 13.3.2.2a).

The pulse-height distribution curves are recorded with a sample of the type to be analyzed and usually with the excitation conditions to be used for the analysis. Very high intensities are avoided to preclude pulse-distribution shift (Section 8.1.8.2). Procedures for recording pulse-distribution curves are given briefly below, in detail in Section 8.1.2.

The differential pulse-height distribution curve is recorded as follows. With the pulse-height selector in differential mode, the window is set at 0.5 or 1 V, and, as intensity is observed on the panel ratemeter or recorder, the baseline is decreased, manually and fairly rapidly, from 100 V to the detector noise. In this way, the maximum intensity to be encountered, other than the noise, is established and the recorder set accordingly. The baseline is now set to bring the noise just on-scale and scanned by motor-drive or electronically to 100 V as intensity is recorded. The integral pulse-height distribution curve is recorded the same way except with the pulse-height analyzer in integral mode. In the absence of motor-driven or electronic baseline scanning, the baseline can be scanned manually in equal, small increments at equal, short time intervals, say 0.5 V every other chart division.

Examination of either pulse-height distribution curve reveals escape peaks and possible overlapping peaks. The curves permit selection of the optimal baseline and window settings to admit as many pulses as possible from the analyte peak and as few as possible from peaks of other elements.

If it is certain that the analyte pulse distribution is free from interference, the baseline and window settings can be established much more conveniently by either of the other two methods cited in paragraph 1 of this section. These two methods have several features in common: The specimen is pure analyte, if practicable; an oxide or other compound or alloy may be used, but no other element should be present having an observable pulse distribution—main or escape—that may be confused with that of the analyte. With the pulse-height selector in integral mode and the discriminator (baseline) set at ~5 V to exclude the noise, the excitation is set to give analyte-line intensity of several thousand counts per second. Very high intensity is avoided to preclude pulse-distribution shift. The goniometer is set at 2θ for the analyte line.

Method 2 is then conducted as follows. With the pulse-height selector in differential mode, the baseline is set at ~20 V and the window at 0.5–1 V. As intensity is observed on the meter or recorder, the detector potential and/or amplifier gain are adjusted to give maximum intensity, indicating that the peak of the analyte pulse distribution has been moved to the selected baseline potential and is more or less centered "in the window." Now, while intensity is still observed, the baseline—and therefore the position of the window—is slowly decreased manually until intensity no longer decreases, indicating that the low end of the pulse distribution has been reached. The window width is then increased until intensity no longer increases, indicating that the high end of the distribution is reached. If on decreasing the baseline, intensity continues to decrease right down to the point where it increases again due to detector–amplifier noise, it means that the low end of the pulse distribution overlaps the noise. In this case, detector potential and/or amplifier gain must be increased to move the distribution farther up the baseline.

Method 3 is conducted as follows. The excitation conditions and goniometer are set as for method 2. The baseline and window settings can now be established in either of two ways.

The baseline and window may be set to bracket the pulse-height distribution as shown on peak *A* in Figure 8.9 (assume, for the present, that peak *B* is not there). With the selector in integral mode with a low baseline, the entire analyte distribution is passed and its intensity indicated on the panel ratemeter or recorder. While intensity is observed on the meter or recorder, the baseline is moved to progressively higher potentials until the intensity decreases sharply, indicating that the baseline has begun to discriminate the low-amplitude side of the pulse-height distribution (*B*1 on peak *A*). The baseline is then "backed off" (*B*2) to allow some latitude for shift and distortion of the distribution. The selector is now switched to differential mode with the window at maximum width. The width is decreased progressively until the intensity decreases sharply, indicating that the window has begun to exclude the high-amplitude side of the distribution (*W*1 on peak *A*) and is then increased for latitude (*W*2).

If, when the baseline is first increased, the observed intensity sharply decreases immediately, the pulse distribution lies partly below the original baseline. In this case, amplifier gain or detector potential must be increased to move the pulse distribution to higher average amplitude. Similarly, if, when the window width is first decreased, the observed intensity sharply decreases immediately, the pulse distribution lies partially above the original (maximum) window. In this case, amplifier gain or detector potential must be decreased to move the pulse distribution to lower average amplitude.

Alternatively, the pulse-height distribution may be moved to the window as shown in Figure 8.9 on peak *B*, which, for the moment, assume

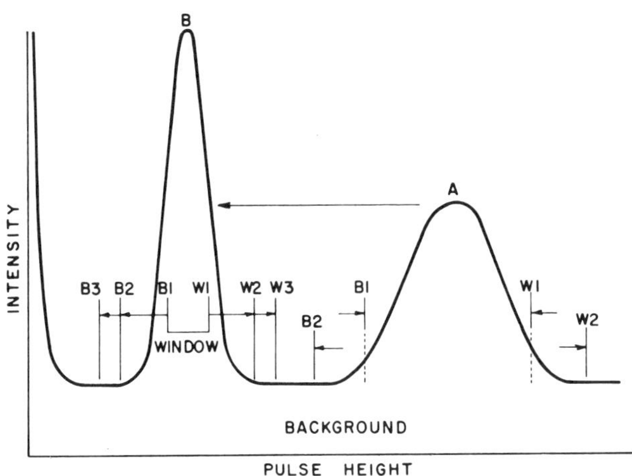

FIG. 8.9. Simplified techniques for determination of pulse-height selector baseline and window settings.

not to be there. The selector is set in differential mode with the baseline at some arbitrarily low potential (*B*1 on peak *B*), say 10–20 V, and a narrow window (*W*1), say 5–10 V. At these conditions, the pulse-height distribution is likely to fall wholly or partially outside the window (peak *A*), and the panel ratemeter or recorder indicates low intensity accordingly. While intensity is observed on the meter or recorder, the amplifier gain or detector potential is varied until the observed intensity is maximum, indicating that the peak of the pulse distribution has moved into the window—peak *A* has moved to peak *B*, as indicated by the long arrow in the figure. The baseline is now reduced decrementally while the upper level of the window is held fixed at its original absolute potential. This means that for each decrement in baseline potential, the window width is *increased* an equal increment in potential. This is repeated until no further increase in intensity is observed, indicating that the low side of the pulse distribution now lies entirely in the window (*B*2 on peak *B*); the baseline is then decreased a little farther (*B*3). The window is now widened until no further increase in intensity is observed, indicating that the high side of the distribution now lies entirely in the window (*W*2 on peak *B*), and is then widened a little more (*W*3).

Incidentally, in the foregoing discussion, it must be borne in mind that a decrease in baseline potential and/or an increase in window width *always* result in an increase in intensity; this is true even in a region where there is no pulse-height distribution, and is due to admission of increased numbers of background pulses. Similarly, an increase in baseline and/or decrease in window width *always* result in a decrease in intensity due to exclusion of background pulses. However, these intensity variations are usually very small and gradual, and the operator soon learns to distinguish them from the large, sharp intensity changes when a pulse distribution moves in or out of the window, or *vice versa*.

If the pulse-height selector is used only to reduce background, rather than to exclude another pulse distribution, the operator may want to determine whether the selector is really beneficial. A simple test follows. After the baseline and window are set by one of the methods described above, the goniometer is moved just off the analyte-line peak and the adjacent background intensity is observed on the panel ratemeter or recorder, or scaled. Then the pulse-height selector is set at integral mode and the baseline discriminator at 5–20 V, and the background intensity is measured again. If this intensity is substantially the same as before, the pulse-height selector is not doing any good and may just as well be operated as a simple discriminator.

Regardless of which of the three foregoing procedures is used, insofar as possible, the baseline and window should be set to pass the entire analyte pulse-height distribution and its escape peak, if any, with some latitude. Some other considerations are given in Sections 8.1.4 (general), 8.1.8.2 (pulse-distribution shift), and 8.1.8.4a (escape peaks).

Finally, all settings—kV, mA, detector potential, amplifier gain, baseline, and window—are recorded and always used thereafter to determine the specified analyte in the specified type of sample. The foregoing procedure must be followed for each other analyte to be determined with the aid of the pulse-height selector. If the analyte having spectral line of longest wavelength is evaluated first, it may be that the other analyte lines simply require progressively lower amplifier gains to move their pulse distributions to the window. This preliminary work is done only once for a given sample type. However, it is prudent to check conditions occasionally, especially if the detector plateau curve shows a change. The great convenience of the Siemens pulse spectroscope (Figure 8.8) for this preliminary work is obvious.

8.1.6. Applications and Limitations

Certainly the pulse-height selector is not required, perhaps not even beneficial, in many analyses. When sharp, intense spectral lines occur in a sparse spectrum having low, continuous background, there is little, if any, advantage in use of the analyzer. However, in other situations, it is extremely beneficial, even indispensable.

The principal applications of the pulse-height selector are the following: (1) elimination of higher-order spectral-line interference (harmonic overlap); (2) reduction of background by elimination of that portion having wavelength different from that of the measured line; (3) increase of detection limit in trace and micro analysis because of the reduced background; (4) higher resolution of lines that are relatively close in wavelength; (5) higher sensitivity at long wavelengths because crystal dispersion need no longer resolve higher-order interfering lines and coarser collimators may be used; (6) elimination of radioactive background from radioactive specimens; and (7) feasibility of energy-dispersive operation (Section 8.2) with resultant ~100-fold increase in measured intensity.

Probably the principal applications of pulse-height selection are the first two listed above and the last. It is very significant that two spectral lines that diffract at the same—or nearly the same—2θ angle in crystal dispersion may have widely separated pulse-height distributions. For example, Cu $K\alpha$ (1.542 Å) and third-order In $K\alpha$ (λ 0.513 Å, 3λ 1.539 Å) are inseparable by crystal dispersion, but have wide energy dispersion (photon energies 8040 and 24,140 eV, respectively). However, the pulse-height selector cannot separate pulse distributions from similar wavelengths.

The resolution of a pulse-height selector used with a proportional counter is far superior to that with a scintillation counter—but is still none too good. Pulse distributions for the same lines of elements differing in atomic number by ± 1 are not resolved; those differing by ± 2 or 3 are resolved only by sacrifice of a substantial fraction of the total intensity, with consequent loss in counting efficiency; those differing by ± 4 are resolved with little loss of counting efficiency. The ability of a counter (proportional or scintillation) to resolve pulse distributions of neighboring elements is substantially constant throughout the periodic table (*H31*). With the advent of the lithium-drifted silicon and germanium detectors (Section 6.4), resolution of lines of adjacent elements has been accomplished.

The application of the pulse-height selector in reduction of background is discussed in Section 7.3.4 and illustrated in Figures 7.5 and 8.6A,B. Again, the analyzer removes only background having wavelength substantially different from that of the analyte line.

The effectiveness of the pulse-height selector is limited by: (1) the difference in mean pulse height of the pulse-height distributions to be resolved; (2) their half-widths; (3) their profiles, that is, whether Gaussian or distorted; (4) their relative intensities; (5) the presence or absence of escape peaks and their character with respect to the first four properties; and (6) the continuous background pulses. The pulse-height analyzer is more effective: (1) the more widely separated, narrower, and more nearly Gaussian the distributions; (2) the more intense the analyte distribution and the less intense the interfering one; (3) the fewer, weaker, narrower, and more separated the escape peaks; and (4) the lower the background.

8.1.7. Automatic Pulse-Height Selection

Because mean pulse height is proportional to x-ray photon energy, for each wavelength to be measured, different settings are required for the baseline and window, or for the detector potential and/or amplifier gain. Such resetting is very inconvenient when several analyte lines are to be measured on a number of specimens, and is intolerable in automatic instruments and in recording of 2θ scans. Accordingly, methods have been devised to couple the pulse-selection function to the θ–2θ drive of the spectrogoniometer so as to vary the pulse-selection conditions automatically and synchronously with 2θ, that is, with wavelength. With such automatic pulse-height selectors, a 2θ scan gives a pure first-order spectrum.

The methods for effecting this function are of two basic types—those that vary the pulse height and those that vary the baseline, or, in effect, those that move the pulses to the window and those that move the window to the pulses. The two approaches are best discussed in terms of Equation (8.1), $V = (E_x/E_D)A_D(A_E/L)$, where V is mean pulse height; E_x is x-ray photon energy; E_D is energy to produce one useful electron in the detector (Section 8.1.1); A_D is internal electron multiplication in the detector; A_E is external amplifier gain; and L is amplifier attenuation.

In the first method, the baseline and window are constant, and, as 2θ varies, the amplifier gain and/or attenuation A_E/L is varied to amplify the pulses to different degrees and move them into the preset window. The same effect would result from variation of detector potential, and, thereby, of A_D. However, A_D is highly dependent on the specific detector, and this method is avoided. For a specified detector, E_D is constant if the detector potential and therefore A_D are constant. Then, from Equation (8.1) and the Bragg law,

$$V \propto E_x(A_E/L), \qquad E_x \propto 1/\lambda, \qquad \lambda \propto (2d \sin \theta)/n$$

Then,

$$A_E/L \propto V(2d/n) \sin \theta$$

For a specified analyzer crystal and order n of λ, $2d/n$ is constant. Then,

$$A_E/L \propto V \sin \theta$$

Consequently, if A_E/L is varied sinusoidally with θ, all pulse-height distributions occur at V. Amplifiers designed to effect this function are known as *sine-function amplifiers* (*W24*).

However, if the crystal ($2d$) or spectral-line order n is changed, all pulse-height distributions are displaced to some new constant value V'. If the original pulse-height selector settings are to be retained, the average pulse height must be moved from V' to V by changing the amplifier gain as a function of $2d/n$. Consequently, in semiautomatic and automatic spectrometers, coupling of the amplifier gain and 2θ necessitates provision for automatic change of the amplifier gain also with selection of the analyzer crystal and order. In the following examples, Ni $K\alpha$ [1.659 Å, 7.47 keV, 48.67° 2θ (LiF)] is used for illustration, and it is assumed that the amplifier gain–2θ coupling is calibrated for LiF; that is, when the goniometer is at

48.67° 2θ, the amplifier gain is set automatically to amplify pulses from 1.659-Å (7.47-keV) photons to the correct amplitude to pass the preset pulse-height selector.

Suppose now that second-order Ni $K\alpha$ [2λ 3.318 Å, 2θ (LiF) 111.00°] is to be measured. But when the goniometer is set at 111.00° 2θ, the amplifier gain is set automatically to amplify pulses originating from 3.318-Å (3.74-keV) photons, not Ni $K\alpha$ photons. Since these photons are only half as energetic as Ni $K\alpha$ photons, the amplifier will assume automatically twice the gain it would have at 48.67° 2θ (Ni $K\alpha$). Thus, provision is required to halve the amplifier gain automatically when the order selector is set at $n = 2$, or, in general, to change the gain by $1/n$ when the order selector is set at n.

Similarly, if Ni $K\alpha$ is to be measured with, say, LiF(420), PET, or gypsum crystals, the peak is diffracted at a different 2θ angle for each, as shown in line 3 of Table 8.2. Incidentally, in order to show the relationships more simply, these particular crystals were chosen because they have $2d$ spacings respectively ~1/2, ~2, and ~4 times that of LiF(200). However, at each of these 2θ angles, the amplifier gain will be set automatically to amplify pulses originating from photons of the wavelength that would be diffracted at that angle *by LiF(200)* (line 4 of the table). The photon energies for these wavelengths are given in line 5 and are different from that for Ni $K\alpha$. Thus the gain would be different from that required for Ni $K\alpha$. The photon energies are in the same ratio as the $2d$ values (line 6), and the required amplifier gain is proportional to $1/2d$ (line 7). Thus, provision

TABLE 8.2. Relationship of Required Amplifier Gain and Crystal $2d$ Spacing for Automatic Pulse-Height Selection

Crystal	LiF(420)	LiF(200)	PET	Gypsum
$2d$, Å	1.80	4.028	8.742	15.185
2θ for Ni $K\alpha$, deg	134.37	48.67	21.88	12.54
λ, Wavelength diffracted at above 2θ by LiF(200), Å	3.712	1.659	0.764	0.440
E_x, Photon energy at above wavelength, keV	3.34	7.47	16.22	28.17
R, Ratio of $2d$ (or E_x) to $2d$ (or E_x) for LiF(200)	0.447	1.00	2.17	3.77
Required relative amplifier gain for λ, $1/R$	2.24	1.00	0.461	0.265

must be made to vary the amplifier gain automatically as a function of $2d$ when the crystal changer is actuated to select a different crystal.

A disadvantage of the sine amplifier arises when it is applied over very wide x-ray wavelength regions. The sine amplifier maintains constant pulse height at the pulse selector input, regardless of 2θ, by varying the amplifier gain while leaving detector potential constant. Now for measurement of F $K\alpha$ to Cl $K\alpha$, relatively high detector potential is preferable to take advantage of the relatively noise-free internal amplification of the detector. Conversely, at shorter wavelengths, relatively lower potential is preferable to avoid running the detector into the nonproportional region. Consequently, at least one manufacturer supplies a spectrometer in which detector potential, rather than amplifier gain, is linked to the 2θ drive, so that potential varies inversely as $\sin\theta$.

In the other method of automatic pulse-height selection, as 2θ varies, the pulse-height distributions of the spectral lines are allowed to undergo their natural variation in mean height and in width. The baseline (*R15*, *S7*), and perhaps also the window width (*W5*), are varied to receive them. Again, for a specified detector and potential, E_D and A_D are constant. Then, if A_E/L is also constant,

$$V \propto E_x \propto 1/\lambda \propto 1/\sin\theta$$

Consequently, if, as θ varies, the baseline is varied in proportion to $1/\sin\theta$, the baseline follows the pulse-height distributions.

As x-ray photon energy increases, the window width required to pass the pulse distribution increases (Figure 8.3), and if the width remains constant, some analyte-line pulses may be excluded. Similarly, as photon energy decreases, the required window width decreases, and if the width remains constant, some unnecessary background pulses or pulses from an adjacent distribution may be admitted. Thus, an added refinement to any of the types of automatic pulse selector described above would be an arrangement to vary the window width to correspond with the variation in the width of the pulse distribution with 2θ. For instruments in which amplifier gain and detector potential are linked to the 2θ drive, the window width should vary as $1/(\sin\theta)^{1/2}$ and $(\sin\theta)^{1/2}$, respectively. In practice, variation as $1/\sin\theta$ and $\sin\theta$, respectively, is satisfactory. One commercial instrument provides two window widths—one narrow, one wide—which are automatically selected on selection of the flow-proportional and scintillation counter, respectively, and which can be selected manually for low and high photon energies, respectively.

8.1.8. Problems with Pulse-Height Selection

8.1.8.1. General

If the detector is receiving a monochromatic x-ray beam and the pulse-height analyzer is not used, each pulse resulting from an incident x-ray photon is counted, provided only that it is higher than the discriminator baseline. It does not matter whether the pulse is a natural or escape pulse, or whether the pulse distribution has shifted or become distorted. However, when the pulse-height analyzer is used, many conditions become troublesome that would otherwise be insignificant. Jenkins and de Vries have given a particularly good review of these problems, and much of the treatment below is derived from theirs. They classify pulse-height analysis problems in four categories: (1) pulse-amplitude shift; (2) pulse-amplitude distortion; (3) additional pulse-height distributions arising from the measured wavelength; and (4) additional pulse-height distributions arising from other wavelengths.

The fourth category comprises distributions arising from: (1) true spectral interference, where a second line is present having wavelength very close to the measured wavelength; (2) higher-order lines, where a second line is present having $n\lambda$ very close to the measured wavelength; (3) secondary characteristic emission from the analyzer crystal; (4) spurious lines diffracted from crystal planes inclined slightly to the surface planes (Figure 13.6); and (5) spurious lines that enter the detector because of its relatively large angle of acceptance. These phenomena are discussed in Section 13.3.2.

The other three classes of problems listed above are discussed in this section, where the treatment is based on that of Jenkins and de Vries [pp. 75–88 in (*22*)].

8.1.8.2. Shift of Pulse-Height Distribution

Consider a pulse-height analyzer set to pass the distribution outlined by the solid outline in Figure 8.10A. Now suppose that the distribution shifts as shown by the dashed outline. Those pulses in the shaded area now lie outside the window and are not counted, so the measured intensity decreases proportionally. Such shifts may occur during the course of an analysis, or on a long-term basis, affecting analyses in which the same long-established analyzer settings are used without occasional verification.

Pulse-amplitude shifts are always caused by a change in amplification, either internal (detector) or external (electronic). Proportional counters are more prone to shift than scintillation counters, and, in proportional count-

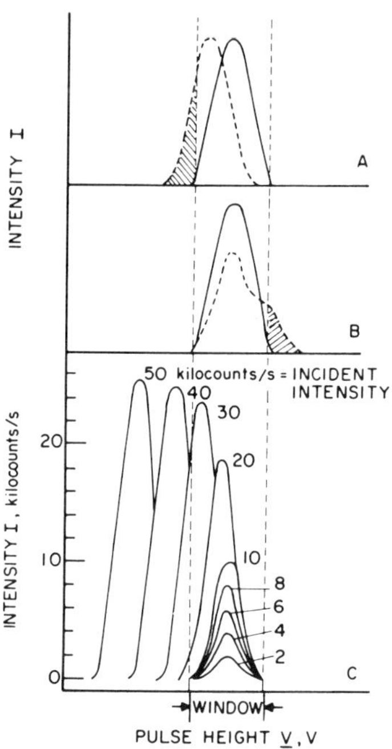

FIG. 8.10. Effect of shift and distortion of the pulse-height distribution. A. Shift. B. Distortion. C. Intensity effect—shift caused by dead-time effect. The detector is assumed to be free from dead-time losses up to 10,000 counts/s. At lower intensities, the detector output intensity decreases proportionally and without change in pulse height. At higher incident intensities, the detector output intensity increases at a progressively slower rate than the incident intensity, and the entire pulse-height distribution shifts to smaller average pulse height.

ers, the effect is greater for: (1) flow counters than for sealed counters at comparable potentials; (2) high applied potential, that is, high gas amplification; (3) high-atomic-number detector gas; (4) small anode wire diameter, that is, small detector capacitance; (5) nonuniform anode-wire diameter; and (6) stainless steel wire than for tungsten wire of the same diameter.

Pulse-amplitude shifts usually result from one or more of six causes:

a. Drift in Applied Detector Potential. Variation in detector potential affects the magnitude of the avalanching or secondary-electron multiplication process in proportional or scintillation counters, respectively, and thereby affects the magnitude of the average pulse height. A change of 1.5% in applied potential may cause a 25% change in average pulse height in a flow counter. The obvious remedy is high regulation ($\pm$0.05% or less) of the detector power supply.

b. Gas Density Effect. Increase in pressure or decrease in temperature increases the density of the detector gas, thereby decreasing the mean free path of electrons, the gas amplification, and the average pulse height. Decrease in pressure or increase in temperature has the reverse effect. Typical laboratory variations of 3% in absolute temperature and 7% in pressure may result in shift of mean pulse height by 20% and 40%, respectively. Fortunately, since the laboratory is not a sealed enclosure, temperature and pressure changes tend to compensate one another, but the net effect may still be troublesome. Typically, a 1% change in either may cause ~5% change in pulse height. The effect is dealt with in several ways. The detector may be mounted inside a temperature-controlled spectrometer chamber. The detector mount may be water-cooled to prevent thermal conduction from the preamplifier. A simple way to substantially reduce the effect of fluctuations in ambient temperature on the detector gas is to pass the gas through a metal coil immersed in a reservoir containing a few liters of water. Because the water is relatively slow to warm or cool, it has a sort of "flywheel" effect, retaining a more or less constant temperature as room temperature fluctuates. Alternatively, the reservoir can be replaced with a water jacket through which passes a slow stream of tap water. The idea is that the ground temperature of the tap water is likely to be more nearly constant than the ambient room temperature. The gas outlet of a flow counter should not discharge directly into the atmosphere, but into a gas-density compensator or manostat to maintain constant pressure in the detector with fluctuating barometric pressure and ambient temperature. A typical manostat is shown in Figure 8.11. The ambient pressure and temperature act on the fixed mass of gas inside the bellows, causing expansion or contraction against the adjustable spring load at the top. The movement actuates the needle valve and regulates the effluent gas flow to maintain

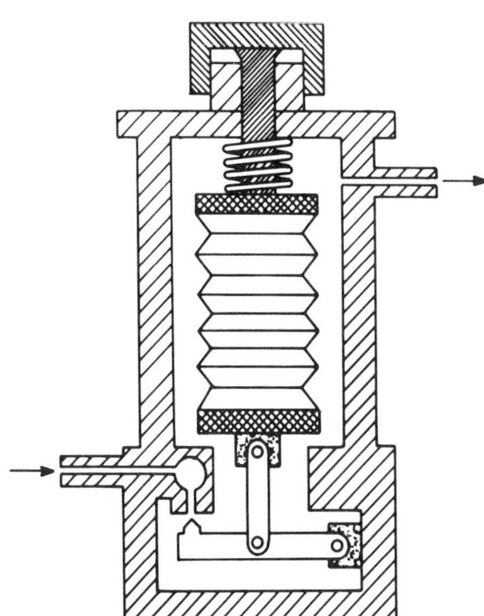

FIG. 8.11. Manostat for gas-flow proportional counter. [R. Jenkins and J. L. de Vries, "Practical X-Ray Spectrometry," Springer-Verlag N. Y. (1967); courtesy of the authors and publisher.]

constant pressure in the detector. Alternatively, the manostat can be placed between the gas cylinder and the detector, in which case the effluent gas from the detector should discharge into the atmosphere through a needle valve or capillary (*W24*). Another solution is to make the pulse-height analyzer window wide enough to accommodate a reasonable amount of shift.

c. Gas Composition Effect. The ratio of detector and quench gas may change when the cylinder is nearly empty (*J9*). A decrease of 0.55% in methane (CH_4) concentration in P10 gas (90% argon, 10% methane) may occur between a full and nearly empty cylinder. The phenomenon results from slightly more rapid flow of the lighter methane molecules (molecular weight 16) relative to argon atoms (atomic weight 40) through the needle valve and capillaries in the gas line. The difference in rate increases as the pressure remaining in the cylinder decreases. A change of 0.5% in the methane concentration may cause a 10–20% shift in pulse height. The obvious remedy is to replace the cylinder when it is 90% empty or so.

d. Intensity Effect. When the incident intensity is such that photons enter the detector between the resolving and recovery times, the pulses are higher than the baseline, but not of normal height. Thus the average pulse height is displaced. The magnitude of the effect depends not only on the counting rate, but also on the pulse decay process, and anything that slows this mechanism increases the resolving time and the pulse-height shift. The effect is serious with a proportional counter at 50,000 counts/s, but may not be serious with a scintillation counter even at 100,000 counts/s. The remedies include the following. The most obvious remedy is to reduce the analyte-line intensity by reducing the x-ray tube operating potential or current or by using a higher-order line. If a combination detector is being used, the signal may be taken from the *less* favorable one for the wavelength being measured. A more risky remedy is to use standards having analyte concentrations very close to those of the samples in the hope that the intensity effect will be similar for all specimens. To shorten the recovery time, all dirt and corrosion must be cleaned from the tube, especially from the anode wire.

The intensity effect may be visualized by reference to Figure 8.10C. Let the outline marked 10 represent the pulse-height distribution for the maximum intensity at which no significant coincidence loss occurs—say, 10,000 counts/s. If now the incident intensity is increased incrementally to 20, 30, 40, 50, etc. kilocounts/s, the pulse-height distribution undergoes two changes: (1) The intensity of the entire distribution increases, but each

10-kilocount increase gives a progressively smaller increase in measured intensity; and (2) the entire distribution shifts to progressively *smaller* average pulse height—that is, to the *left* in the figure. Conversely, if the incident intensity is decreased incrementally to 8, 6, 4, 2, etc. kilocounts/s, the distribution remains fixed in position along the pulse-height axis, but decreases in intensity in proportion to the incident intensity.

e. Pulse Pileup. This is a phenomenon that causes pulses to appear higher than they really are and occurs in the amplifier at high count rates. It consists in superposition of pulses on the trailing edges of immediately preceding pulses, so that the apparent pulse amplitude is the true amplitude plus the contribution of the preceding pulse. Of course, the pulse-height analyzer treats such pulses according to their apparent—abnormally high—amplitudes. Pulse pileup appears as abnormally high and prolonged tailing on the low-energy sides of intense peaks in the energy spectrum. It is dealt with by pulse pileup correction circuits.

f. Crystal Emission. Secondary characteristic emission from the analyzer crystal may sometimes cause an *apparent* pulse-height shift. Jenkins and de Vries give this example: If P $K\alpha$ is measured with a gypsum ($CaSO_4 \cdot 2H_2O$) crystal, a S $K\alpha$ (2.31 keV) distribution may occur with the P $K\alpha$ (2.02 keV) distribution. The two are not resolved, so the observed distribution is a composite of the two. At low P $K\alpha$ intensities, the S $K\alpha$ peak predominates, and the peak of the composite distribution lies near the peak of the sulfur distribution. As P $K\alpha$ intensity increases, its peak assumes dominance, with a resultant shift of the peak of the composite distribution.

8.1.8.3. Distortion of Pulse-Height Distribution

The distortion effect is shown in Figure 8.10B, where the pulses in the shaded area lie outside the window if the pulse-height distribution (solid outline) becomes distorted (dashed outline), that is, departs from its Gaussian profile (*D13*). The distorted pulse distribution appears as though the incident photons, although monochromatic, were creating two overlapping pulse distributions of somewhat different average pulse height. This does in fact happen when, for example, the potential field in the detector is locally distorted by, say, a dirty or corroded spot on the anode wire. Pulses of two different average heights would arise from the dirty and clean portions of the wire. Distortion, like shift, is most serious with flow-proportional counters. It is not uncommon for the resolution of a new counter, or a counter that has been cleaned and fitted with a new anode wire, to deteriorate by a factor of two or three within a month. Distortion results from an inhomogeneous electric field caused by window effects; anode-wire end and clampling effects; and dirt, corrosion, pitting, and variation in diameter of the anode wire. Side-window counters are much less prone to distortion than end-window counters, and the inside surfaces of windows must be metallized or carbonized. The principal source of dirt is the gas supply, and a filter in the gas line may substantially retard the deterioration of detector performance. The presence of traces of oxygen and/or water vapor in the gas may cause irreversible corrosion or pitting of the anode wire; preventive measures include specification of high-purity detector-grade gas and provision of chemical absorbers for oxygen and water vapor. The hydrocarbon component of the gas itself (methane, propane, etc.) may undergo polymerization and/or reduction at the anode wire, leaving a deposit of solid hydrocarbon or carbon; use of argon–carbon dioxide mixtures instead of argon–hydrocarbon mixtures has been suggested as a remedy for this condition (*D13*).

When resolution has deteriorated to an unacceptable extent, the wire may be cleaned by gentle brushing with a camel-hair brush or pipe-cleaner dipped in benzene, or replaced with a new, similarly cleaned wire.

8.1.8.4. Additional Pulse-Height Distributions Arising from the Measured Wavelength

In addition to its own natural pulse-height distribution peak, the measured wavelength may initiate *escape peaks* and *double* or *sum peaks*.

a. Escape Peaks. The origin of escape peaks (and of the term) is explained for proportional counters in Sections 6.2.3.1 and 6.2.3.2 and for scintillation counters in Section 6.3.2.2. The escape peak as a source of interference is discussed in Section 13.3.2.2b.

The ratio of the mean pulse heights of the escape and main (photo, principal, or natural) peaks is

$$\frac{V_{\mathrm{esc}}}{V_{\mathrm{main}}} = \frac{E_x - E_{\mathrm{det}}}{E_x} \tag{8.3}$$

where V_{esc} and V_{main} are mean pulse heights of the escape and main peaks, respectively; E_x is the energy of the incident x-ray photon; and E_{det} is the energy equivalent of the absorption edge of the detector gas or of the iodine in the NaI(Tl) scintillator. The intensity of the escape peak relative to the main peak increases with (1) proximity of the incident wavelength to the absorption edge of the detector gas or iodine, (2) the fluorescent yield of the detector gas or iodine; the effect of fluorescent yield is shown in Figure 6.5; (3) opacity (high μ/ϱ) of the detector to the analyte line; (4) transparency (low μ/ϱ) of the detector to its own line; and (5) shortness of the detector path. As analyte-line wavelength decreases, separation of photo and escape peaks decreases. In general, at long analyte-line wavelengths, the escape peaks lie in the detector noise, and at short wavelengths, the two peaks are close enough so that both are counted.

In general, escape peaks are not a serious problem in analytical x-ray spectrometry. In argon proportional counters, elements of atomic number 18 (argon) and below do not excite the argon *K*-absorption edge, and for elements of atomic number 28 (nickel) and above, the escape and main peaks are not resolved. Thus, escape peaks may be troublesome only for elements 19–27 (potassium through cobalt). Even for these, the low argon *K* fluorescent yield (~0.1) results in relatively weak escape peaks and the potassium and calcium escape peaks may well be lost in the detector noise. However, if these escape peaks do interfere with an analyte photo peak, use of propane detector gas may provide the remedy. For example, when an argon detector is used for determination of fluorine in the presence of calcium, the Ca *K* escape peak interferes with the F *Kα* photo peak. The interference does not occur in propane detectors.

In xenon proportional counters, the *L*III absorption edge (2.6 Å) is outside the usual spectral region of the detector. In NaI(Tl) scintillation counters, the I *L*III absorption edge (2.7 Å) is at the long-wavelength limit of the useful range. Thus, x-ray lines most efficient in exciting these edges give very low-energy escape peaks that are lost in the detector noise. Although such escape peaks are lost for counting purposes, at least they cannot interfere with other, useful pulse distributions. In both the xenon and NaI(Tl) scintillation detectors, the *K* edges (both ~0.36 Å) are excited only by the *K* lines of elements of atomic number ~57 (lanthanum) or above. The *K* lines of these elements are excited relatively inefficiently by a 50-kV generator, and the elements are usually determined by use of their *L* lines. However, the *K* fluorescent yields of xenon and iodine are high (~0.9), and, when escape peaks do occur, they have two to three times the intensity of the main peak.

In the krypton proportional counter, the *K* edge (0.86 Å) is excited by elements of atomic number ~38 (strontium) and above, and the *K* fluorescent yield is relatively high (~0.6). Thus, the krypton detector is the most prone to escape peaks of all the common detectors, but even it has a wide useful range.

Escape peaks, like main peaks, arise from the x-ray spectral line being measured, so, in general, the pulse-height analyzer window should be set to admit both. However, this is not feasible if: (1) the escape peak has such low pulse height that it is lost in the detector noise; (2) the two peaks are too widely separated; (3) too many background pulses are admitted; or (4) a main or escape peak of another element is wholly or partially admitted. Sometimes, even the last three of these difficulties can be surmounted if two pulse-height analyzers are available and set for the main and escape peaks, respectively. In this way, background and interfering peaks between the two peaks of interest are rejected.

Escape peaks may be troublesome in two ways. If, for any of the reasons given above, it is not feasible to include both the main and escape peaks in the window of a single-channel analyzer, only one can be measured. Then the measured intensity is reduced by the intensity represented by the rejected peak. Also, escape peaks complicate the energy spectrum and increase the probability of overlapping (Section 13.3.2.2b).

An interesting escape-peak phenomenon occurs when a NaI(Tl) scintillation counter is used with automatic pulse-height selection to record a 2θ scan from a low-Z specimen having high scattered background (*P9*). A discontinuity occurs in the background at $\sim$15.8° 2θ (topaz crystal), the intensity being lower at lower 2θ (shorter wavelength). The explanation follows. For topaz, 15.8° 2θ corresponds to the iodine K absorption edge. Below 15.8°, the scattered continuum is sufficiently energetic to excite the I $K\alpha$ line in the scintillator. An escape peak forms, but its amplitude lies outside the window automatically set to receive natural pulses corresponding to each 2θ angle, and it is not counted. Thus, measured background is abnormally low below 15.8° 2θ, but is normal at higher angles.

Another interesting escape-peak phenomenon occurs in the measurement of Sn L lines from tin-plated steel by use of an argon-filled detector (*C57*). The photon energy of Sn $L\alpha_1$ is 3.44 keV. The Fe $K\alpha$ (E_x 6.40 keV) escape peak in an argon detector (Ar $K\alpha$ E_x 2.96 keV) corresponds to $6.40 - 2.96 = 3.44$ keV. Thus, the iron escape peak is counted along with the tin photo peak. As tin thickness increases, the intensities of the tin and escape peaks increase and decrease, respectively, and as tin thickness decreases, the reverse effect occurs. Thus the two pulse distributions tend to compensate one another, and the measured intensity remains substantially constant as tin thickness varies. The problem is readily solved by using a different detector gas.

b. Double or Sum Peaks. If two identical x-ray photons enter the detector simultaneously, or very nearly so, a pulse of twice normal amplitude occurs. If this happens often enough, a second pulse distribution appears in the detector output at mean amplitude about twice that of the main peak. This new peak is known as a *double* or *sum* peak. Such a peak sometimes appears in pulse-height distribution curves recorded with fast counters at very high intensities. Triple and even higher sum pulses may also occur. If a pulse-height analyzer is used and the window set to admit the main peak, the number of photons that are detected but not counted is *twice* the number of double pulses. Even without the analyzer, there is a loss *equal* to the number of double pulses. Moreover, two or more simultaneous pulses of amplitude too low to be counted—that is, below the baseline—may sum to give a countable pulse and an erroneous net gain of one pulse.

Sum pulses may be difficult to understand in view of the dead-time phenomenon. However, the problem is resolved if one considers that, in effect, the two concurrent photons are equivalent to a single photon of twice the energy. A photon entering the detector after the quench process has begun is, of course, subjected to the dead-time effect.

8.1.9. Unfolding of Overlapping Pulse-Height Distributions

8.1.9.1. Principle

Pulse-height distributions from neighboring elements are unresolved or, at best, very poorly resolved by proportional and scintillation counters. Even in semiconductor detectors, resolution decreases with decreasing atomic number. Thus, total measured intensity at the mean pulse-height potential for element i is the sum of the intensity from i and the fractional contributions from the pulse-height distributions of other neighboring elements j at the position of the i peak.

Dolby (*D21*) has shown that a set of linear simultaneous equations can be written for the intensities at the mean pulse-height positions—that is, the pulse-height distribution peaks—corresponding to each element. These equations are solved algebraically by use of the intensities measured at the position of the i peak from the sample, pure i, and pure specimens of each other element j contributing to the total intensity at the i peak. These equations have the form

$$I_i = I_{ii}R_i + \sum_{1}^{j} (R_j I_{jj} I_{ji}) \qquad (8.4)$$

where I_i and I_{ii} are net intensities measured at the position of the i peak from the sample and pure i, respectively; I_{jj} is net intensity measured at the position of the j peak from pure j; I_{ji} is the fractional net intensity measured at the position of the i peak from element j; and R_i and R_j are true relative net intensities of elements i and j, that is, I_i/I_{ii} and I_j/I_{jj}, respectively. These R's are the values required for quantitative analysis and are found by solution of the simultaneous equations.

This technique of deriving the individual contributions in overlapping pulse-height distribution curves is known variously as *unfolding*, *convolution sorting*, or *stripping*. The following examples are taken from work by Birks and Batt (*B52*) [pp. 98–102 in ed. 1 of (*7*)].

The basic approach is illustrated in Figure 8.12. The three solid curves represent pulse-height distributions from three consecutive adjacent elements A, B, and C in a hypothetical specimen. The curves were drawn with Gaussian distributions and arbitrarily assigned relative peak intensities of 4.5, 2.0, and 2.5, respectively. The dashed curve is the summation of the three individual curves and represents the single, unresolved pulse-height distribution that would be observed from the hypothetical specimen. The composite curve has relative *measured* intensities at the positions of A, B, and C peaks of 6.06, 6.25, and 4.30, respectively.

The simplest approach to the unfolding is to assume that the individual distributions are Gaussian (this would not be known if only the composite

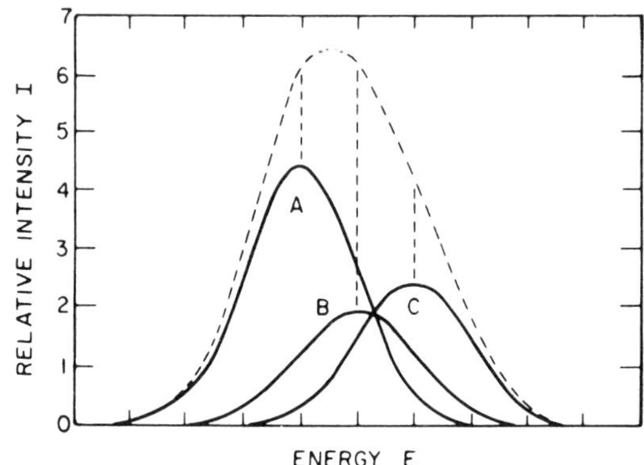

FIG. 8.12. Idealized pulse-height distribution curves of arbitrary intensities for spectral lines of adjacent elements A, B, and C in a hypothetical specimen. The dashed curve is the summation of the three individual curves. [L. S. Birks, "Electron-Probe Microanalysis," ed. 1, Interscience Publishers, N. Y. (1963); courtesy of the author and publisher.]

curve were available), and have standard deviation σ equal to the separation between peaks for adjacent atomic numbers. For a Gaussian distribution, the relative intensities at the peak, 1σ, and 2σ are 1.00, 0.607, and 0.135, respectively. Then the relative intensities of element A at its own peak and at the peaks of its first and second consecutive adjacent neighbors are 1.00, 0.607, and 0.135, respectively. These assumptions and the data for Figure 8.12 may be used to establish three simultaneous equations:

$$I_A = R_A + 0.607R_B + 0.135R_C = 6.06 \tag{8.5a}$$

$$I_B = 0.607R_A + R_B + 0.607R_C = 6.25 \tag{8.5b}$$

$$I_C = 0.135R_A + 0.607R_B + R_C = 4.30 \tag{8.5c}$$

where I indicates intensities measured from the specimen (dashed curve in Figure 8.12), each I being the sum of contributions from all three elements; and R indicates true relative intensity in the specimen.

If only the composite curve is known and it is required to calculate R_A, R_B, and R_C, the three equations are solved by use of determinants. The solution gives 4.56, 2.06, and 2.50, respectively, in excellent agreement with the starting values given above.

8.1.9.2. Application

In practice, the pulse-height distributions are not exactly Gaussian, and the peak intensity varies with atomic number due to dependence of detector sensitivity on wavelength. Consequently, the multipliers in Equations (8.5) must be determined by actual measurements on the pulse-height distributions of the pure elements. Figure 8.13 shows the pulse-height distributions for pure iron, copper, and zinc and for two samples containing these elements as minor (4–14%) constituents in a low-Z matrix. Table 8.3 gives the data derived from these curves at the positions of the iron, copper, and zinc peaks. The three simultaneous equations established from the data for sample 1 are as follows:

$$\text{Fe:}\quad 1015 = 10765R_{Fe} + (9730\times 0.077R_{Cu}) + (7845\times 0.069R_{Zn}) \tag{8.6a}$$

$$\text{Cu:}\quad 659 = (10765\times 0.017R_{Fe}) + 9730R_{Cu} + (7845\times 0.506R_{Zn}) \tag{8.6b}$$

$$\text{Zn:}\quad 680 = (10765\times 0.003R_{Fe}) + (9730\times 0.559R_{Cu}) + 7845R_{Zn} \tag{8.6c}$$

The simultaneous equations for sample 2 are identical except for the quantities to the left of the equality signs; these values for sample 2 are given in Table 8.3. The set of equations for each sample is solved for the three R values.

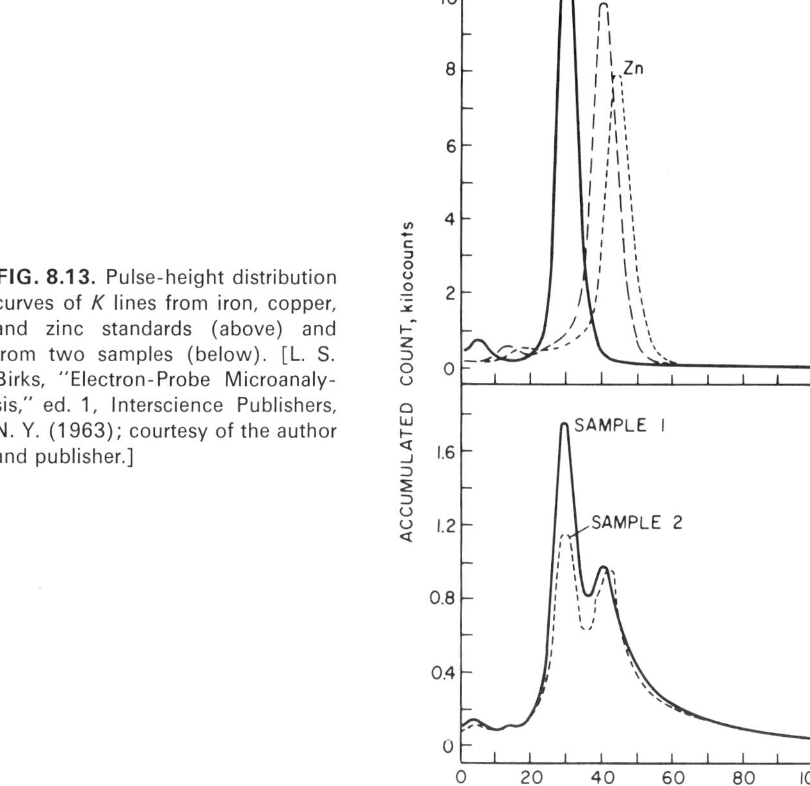

FIG. 8.13. Pulse-height distribution curves of *K* lines from iron, copper, and zinc standards (above) and from two samples (below). [L. S. Birks, "Electron-Probe Microanalysis," ed. 1, Interscience Publishers, N. Y. (1963); courtesy of the author and publisher.]

8.1.9.3. Simplified Variations

When some accuracy can be sacrificed, the unfolding procedure may be simplified. The following example is given by Birks [pp. 67–8 in (*8*)]. Figure 8.14A shows the individual Cr, Fe, and Ni $K\alpha$ pulse-height distributions and the unresolved composite distribution from a stainless steel having composition ~70Fe–18Cr–9Ni. The unfolding may be done in any of three ways:

1. Figure 8.14A: For highest accuracy, the full unfolding procedure is applied, as already described.

2. Figure 8.14B: The intensity at the peak of the unresolved distribution may be taken as the peak intensity of the major constituent, iron.

TABLE 8.3. Data for Unfolding of Pulse-Height Distribution Curves Shown in Figure 8.13[a]

Specimen	$K\alpha$ net intensity, counts/s	Pulse-height distribution peak position		
		Fe	Cu	Zn
		Fractional $K\alpha$ net intensity at peak position indicated above		
Fe	10,765	1.0	0.017	0.003
Cu	9.730	0.077	1.0	0.559
Zn	7,845	0.069	0.506	1.0
		Measured $K\alpha$ net intensity at peak position indicated above, counts/s		
Sample 1	—	1015	659	680
Sample 2	—	1610	740	630

[a] From Birks and Batt (*B52*).

Then the contribution of the iron only is subtracted from the intensities at the positions of the peaks of the chromium and nickel distributions.

3. Figure 8.14C: The intensity at the position of the peak of each element is taken as being contributed by only that element, even though this is obviously not the case.

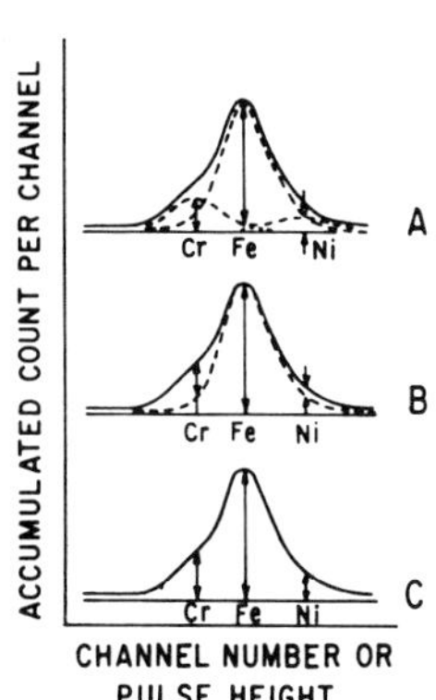

FIG. 8.14. Full and simplified methods for unfolding of overlapped pulse-height distributions of adjacent elements. [L. S. Birks, "X-Ray Spectrochemical Analysis," ed. 2, Interscience Publishers, N. Y., p. 68 (1969); courtesy of the author and publisher.] A. Complete unfolding by use of simultaneous equations. B. Subtraction of analyte-line intensity of the major element. C. Uncorrected intensities.

Methods 2 and 3 are progressively more simple—and less accurate than the full unfolding procedure. The calculations for these two methods are easily done by hand. Results of the three methods for several stainless steels are compared in Table 8.4. The second method gives results nearly as good as complete unfolding, and even the third method is suitable for semi-quantitative analysis.

TABLE 8.4. Data for Comparison of Three Methods for Unfolding of Overlapped Pulse-Height Distribution Curves of Adjacent Elements[a]

Steel	Analyte	Concentration, wt%	Concentration difference[b], wt% (x-ray – chemical)		
			Unfolding	Subtraction of Fe $K\alpha$ intensity	Uncorrected intensity ratio
301	Cr	17.9	0.4	0.5	0.7
	Fe	72.7	−1.9	−2.5	−2.5
	Ni	7.2	0.6	0.6	3.9
303	Cr	17.2	0.1	0.1	06
	Fe	71.2	−0.9	−1.6	−1.6
	Ni	8.7	0.7	0.6	3.0
304	Cr	18.6	0.1	0.2	0.1
	Fe	69.5	−1.3	−1.5	−1.5
	Ni	9.4	0.2	0.2	2.3
321	Cr	17.8	0.2	0.2	0.4
	Fe	68.2	−0.2	−0.3	−0.3
	Ni	10.8	0.2	0.2	1.5
347	Cr	17.7	0.2	0.2	0.3
	Fe	67.9	−0.8	−1.0	−1.0
	Ni	10.7	0.5	0.6	1.7
Average error in % of amount present:					
	Cr		1.3	1.4	2.4
	Fe		1.4	2.0	2.0
	Ni		5.0	4.9	27.6

[a] From Birks [p. 67 in (*8*)].

[b] Each value is the average of 10 determinations.

These unfolding techniques are made feasible by the speed and convenience of the multichannel analyzer (Section 8.1.2.3). Collection of the data in Table 8.4 requires ~1 h with a single-channel pulse-height analyzer, but only ~5 min with the multichannel analyzer. In applying the unfolding techniques, care must be taken to avoid pulse-amplitude shift as a function of intensity. Solid-state Si(Li) detectors can resolve pulse-height distributions of adjacent elements in the shorter-wavelength spectral region. Their use in this region may preclude the need for unfolding and is beneficial in any case. Incidentally, there appears to be no reason why unfolding techniques cannot be applied also to overlapped wavelength spectra (2θ scans).

The unfolding method described above may be regarded as *de*synthesis of a complex pulse distribution into its individual distributions. An alternative approach is to *synthesize* or *reconstruct* the complex distribution from the individual ones (*C58*). The composite distribution and the individual distributions from the pure contributing elements are stored in a tape-recorder memory. The composite curve is displayed on the screen of a cathode-ray tube. This composite profile is then duplicated simultaneously on the same screen by trial-and-error superimposition of the individual distributions in appropriate proportions. See also Section 8.2.2.4b.

8.2. ENERGY-DISPERSIVE ANALYSIS

8.2.1. Introduction

The distinction drawn in this book between energy-dispersive and nondispersive methods of x-ray spectrometric analysis is discussed in Section 5.1. The two methods are described in this section and Section 8.3, respectively.

8.2.1.1. Principles

In wavelength-dispersive (crystal) spectrometers, the several wavelengths emitted by the specimen are dispersed (separated) spacially on the basis of their wavelengths *prior* to detection; thus, in principle at least, the detector receives only one wavelength at a time. In energy-dispersive spectrometers (*B46*, *F16*), the detector receives the undispersed secondary beam comprising all excited lines of all specimen elements. The amplified detector output is then subjected to pulse-height selection, in which the pulse distributions arising from the several detected wavelengths are separated on the basis of their average pulse heights—thereby on the basis of the photon energies of the corresponding incident x-ray lines. By appropriate setting of the baseline and window, the pulse distribution of each analyte line can be counted individually, or by appropriate setting of the detector potential and/or amplifier gain, the analyte distributions can be moved individually into a fixed window. Alternatively, all pulse distributions present can be displayed simultaneously with a multichannel analyzer.

A qualitative x-ray spectrometric analysis by either wavelength- or energy-dispersive mode results in a series of peaks on a chart or on some other type of display. In wavelength dispersion, each peak represents an x-ray spectral line of an element in the specimen, and the chart is a plot of intensity *versus* 2θ and therefore intensity *versus* wavelength (Figure 10.1). In energy dispersion, each peak represents the pulse-height distribution of an x-ray spectral line of an element, and the chart is a display of intensity *versus* pulse height—or, with a multichannel analyzer (Section 8.1.2.3), channel number—and therefore intensity *versus* x-ray photon energy (Figures 8.6C and 8.17–8.23).

8.2.1.2. Advantages

The principal advantages of energy-dispersive analysis arise from the simplicity of the instrumentation, economy of emitted analyte-line intensity, and feasibility of simultaneous accumulation and display of the entire spectrum. Elimination of the crystal precludes the requirement for, and alignment of, a precision drive mechanism for the crystal and detector, and the loss of intensity from inefficiency of the diffraction process. The energy-dispersive spectrometer has no moving parts. Further, the recorded spectrum is less cluttered due to absence of higher orders. Also, the detector may now be placed very close to the emitting specimen, providing even greater economy of intensity by reduction of distance, increased solid angle of interception, and elimination of collimators. The resulting 100-fold or more increased sensitivity is advantageous when selective excitation and filtration are used and for trace and microspecimens. The conservation of intensity is so great that it is feasible to use radioactive isotope excitation sources, thereby eliminating the large, expensive, and power-consuming x-ray power supply and tube. Alternatively, excitation is feasible with low-power x-ray tubes (Section 4.2.5.5), field-emission tubes (Section 4.2.5.6), and secondary monochromatic radiators or "fluorescers" (Section 8.3.1) excited by standard x-ray tubes. When radioisotopes are used, many other advantages are realized (Section 8.2.2.3). In the instrument arrangement for energy-dispersive analysis, photons of all x-ray spectral lines emitted by the specimen enter the detector simultaneously. This condition provides the basis for simultaneous accumulation and display of the entire energy spectrum, including background, by means of a multichannel analyzer and cathode-ray tube display, and background and spectral interferences are clearly shown. Thus, semiconductor-detector x-ray spectrometers permit qualitative analyses to be performed much more rapidly and conveniently than crystal spectrometers. Also, energy-dispersive spectrometers are very well suited to investigation of dynamic systems. Close proximity of source, specimen, and detector, absence of the need for "optical" focusing, and simultaneous accumulation of the entire spectrum result in less sensitivity to specimen position than in a wavelength-dispersive spectrometer. An incidental advantage of energy dispersion is that the entire integrated analyte line is measured, rather than just the peak. Consequently, the effect of analyte-line wavelength shift due to chemical state is minimized. Simultaneous accumulation also reduces the effect of instrument drift, increases the statistical precision in net counts, and permits rapid and convenient processing of the spectrum in various ways (Section 8.2.2.4). Finally, simultaneous accumulation permits observation and measurement of elements other than those specified for analysis and reduces the possibility of overlooking unsuspected elements. When low-power x-ray tube, secondary emission, or radioisotope sources are used, energy dispersion is particularly advantageous for specimens subject to radiation damage—such as liquids and organics, which may undergo radiolysis, and glasses, art objects, antiquities, etc., which may undergo discoloration.

8.2.1.3. Limitations

The two principal limitations of the energy-dispersive spectrometer are imposed by the Si(Li) detector (Section 6.4) and have to do with resolution and tendency to coincidence loss. At wavelengths longer than ~0.8 Å (15 keV), the resolution of a Si(Li) energy-dispersive spectrometer is poorer than that of a crystal spectrometer. (The resolution of an energy-dispersive instrument using a proportional or scintillation counter is poorer at any wavelength.) Thus the crystal spectrometer may be preferable for the intermediate-wavelength region and must be used for the ultralong-wavelength region—for the *K* lines of elements of atomic number $\lesssim$11 (sodium). As mentioned above, in energy dispersion, all x-rays emitted and scattered by the specimen enter the detector simultaneously—not just the discrete wavelength diffracted by a crystal. Thus coincidence losses, pulse-height shift, and detector choking occur much more readily. In fact, energy-dispersive instruments should be provided with a total-intensity ratemeter to enable the operator to avoid these conditions. Higher individual intensities can be measured with crystal spectrometers because only a narrow spectral region is admitted to the detector and because gas-filled proportional and scintillation counters are capable of higher count rates than Si(Li) detectors. When the multichannel analyzer is used, it is difficult to realize high sensitivity for weak lines in the presence of strong lines; the instrument counts all lines simultaneously, and the strongest line determines the counting time and stops the count. Other disadvantages apply when radioisotope sources are used (Section 8.2.2.3).

Perhaps the best comparison of wavelength- and energy-dispersive modes to date is that by Gilfrich, Burkhalter, and Birks (*G14*). They evaluated sensitivities and detection limits for 19 elements ranging in atomic number from 13 (aluminum) to 82 (lead) in micro quantities in filter paper disks in the following spectrometer arrangements under conditions as nearly identical as possible:

1. Wavelength-dispersive spectrometer with chromium-, rhodium-, and tungsten-target x-ray tubes; LiF(200), graphite, and KHP crystals; and gas-flow proportional counter with P10 gas (90% argon, 10% methane).

2. Energy-dispersive spectrometer with Si(Li) detector and the following excitation sources: (1) x-ray tubes: molybdenum and tungsten targets at 50 kV(CP), 3 mA; (2) secondary radiators: manganese, copper, silver, and chromium–zirconium alloy, excited by a tungsten-target tube at 45 kV (CP), 20 mA; (3) radioisotopes: 7-mCi ^{55}Fe and 70-mCi ^{109}Cd; and (4) ions: 5-MeV protons and alpha particles.

These workers conclude that: (1) major concentrations can be determined in either mode with any of the excitation sources; (2) intermediate concentrations can be determined in either mode, but direct x-ray tube excitation must be used for energy dispersion; and (3) concentrations approaching the detection limit can be determined only by wavelength dispersion.

Hanson (*H15*) has compiled a table which gives, for each of up to seven principal K and L lines of 71 elements ranging from $_{11}$Na to $_{92}$U, the 1000-s count in the peak channel of a 512-channel analyzer for 40-mCi ^{55}Fe, 6-mCi ^{109}Cd, and 6-mCi ^{241}Am sources. The data are also presented as a set of graphs of intensity *versus* photon energy.

An excellent evaluation of the semiconductor-detector energy-dispersive spectrometer for trace analysis is given by Giauque *et al.* (*G9*).

In summary, the following generalizations may be made. The detection limit, all things considered, is substantially the same for both energy and wavelength dispersion. For resolution, energy dispersion is better at shorter wavelengths (higher photon energies), and wavelength dispersion is better at longer wavelengths (lower energies). For qualitative analysis, energy dispersion is generally advantageous. It is certainly more rapid and convenient, but weak peaks may be unresolved if situated near strong peaks, and resolution is generally poorer at longer wavelengths. For quantitative analysis, energy dispersion may be advantageous, compared with single-channel (manual and sequential) crystal spectrometers, for large numbers of analytes because all data are accumulated simultaneously. However, here again, weak peaks may be unresolved from closely adjacent strong ones, and resolution becomes generally poorer at longer wavelengths. Wavelength dispersion is advantageous for measurement of individual analytes because only one analyte line enters the detector at a time, to the exclusion of the remainder of the spectrum. Wavelength dispersion is also preferable to resolve weak peaks near strong ones, and is still generally required for the longer wavelengths, especially >10–12 Å (Na *K* and Zn *L* lines).

Birks [Reference (*53*) for 1974] classifies simultaneous multianalyte analyses in two categories: (1) If the analytes are known, multichannel crystal spectrometers may be advantageous because of their better resolution and detection limit for neighboring elements of widely varying concentration. (2) If the analytes are not known in advance, energy dispersion may be advantageous because of its versatility and preclusion of the need for setting for specific elements. Quantitative analysis of neighboring elements can usually be effected by unfolding of overlapped spectral lines. However, counting time may be larger because of the count-rate limit of the Si(Li) detector and the fact that all photons must be processed regardless of whether they are of analytical interest.

Incidentally, whereas x-ray spectrochemists using wavelength-dispersive spectrometers become accustomed to thinking of x-ray spectral lines and absorption edges in terms of their *wavelengths* in angstroms, those using energy-dispersive instruments think in terms of their *energies* in kiloelectron volts. For example, in wavelength terms, one would say that if Mo $K\alpha$ (λ 0.71 Å) is to be excited efficiently, the primary beam must be rich in wavelengths shorter than that of the Mo K absorption edge ($\lambda_{\mathrm{Mo}K_{\mathrm{ab}}}$

0.62 Å). In energy terms, one would say that if Mo $K\alpha$ (E_x 17.4 keV) is to be excited, the excitation radiation must be substantially more energetic than the Mo K critical absorption energy or excitation potential ($V_{\text{Mo}K}$ 20.0 keV). The photon energies of the principal K and L lines are given in Appendix 4. The K, L, and M excitation potentials are given in Appendix 6.

8.2.2. Instrumentation

8.2.2.1. General

The standard commercial x-ray fluorescence crystal spectrometer comprises the following basic components: high-power x-ray tube and power supply; specimen presentation system; spectrogoniometer (crystal, collimators, and drive mechanism); proportional or scintillation detector; and electronic readout components, including a scaler–timer and ratemeter–recorder. Excepting the specimen presentation system, any of these "standard" components can be replaced with what might be termed "nonstandard" components. In this way, a wide variety of x-ray spectrometric instrumentation is derived. First, in order to realize narrower pulse-height distributions at the shorter wavelengths, one might simply replace the proportional or scintillation detector with a lithium-drifted silicon detector (Section 6.4), retaining all other components. The next step would be to replace the electronic readout components with a multichannel analyzer (Section 8.1.2.3). An instrument thus modified can be used in both wavelength-dispersive and, at 0° 2θ, energy-dispersive modes. If now the spectrogoniometer, with its photon-wasteful crystal and collimators, is omitted entirely, one can replace the high-power x-ray tube with a low-power tube, an electron-excitation source, or a radioisotope source. These arrangements, of course, are energy-dispersive.

8.2.2.2. Excitation by X-Rays

The various nondiffractive methods are conveniently applied on a standard commercial x-ray spectrometer (*J10*). The instrument arrangement is the same as for secondary-emission spectrometry, except that the goniometer is fixed at 0° 2θ, and the crystal—preferably also the crystal stage—is removed. Coarse collimators may be used in both positions. The distance from specimen to detector may be reduced by mounting the detector directly after the source collimator or even at the exit port of the specimen compartment. In instruments having specimen drawers, a special drawer may be provided in which the specimen plane is inclined in the opposite direction, away from the crystal, toward the front of the drawer. A window may be cut in the front of the drawer and the detector mounted at this window, preferably after a short, wide-spaced collimator.

Nondiffractive methods are also applied on a standard diffractometer. The specimen is placed in the diffractometer specimen position on the axis of the goniometer, which is then set at $\sim 90° \, 2\theta$. Care is taken to avoid any lines diffracted by the specimen. Wide slits and coarse collimators may be used. The specimen is irradiated by the direct primary beam, preferably from a diffraction tube having a tungsten target.

However, most energy-dispersive work by far is done on spectrometers designed specifically for this application (Section 8.2.2.4).

Conventional high-power (2–5-kW) x-ray tubes can be used for excitation in such instruments in any of three modes: (1) direct excitation by the unfiltered primary beam in the usual way so that both continuum and target-line x-rays irradiate the specimen; (2) direct excitation by the primary beam filtered so that substantially monochromatic target-line x-rays irradiate the specimen; and (3) excitation of a secondary target having its strongest line at wavelength just shorter than that of the analyte absorption edge; the secondary x-rays in turn irradiate the specimen.

For energy-dispersive work, unfiltered direct excitation by high-power tubes has the serious disadvantages that, even at the lowest power attainable with conventional high-power generators, the tubes produce high-intensity continuum and relatively high-intensity specimen emission. The continuum causes high scattered background, reducing the sensitivity attainable, and the high total background and emission intensity may cause coincidence loss, pulse-height shift, and/or choking in the detection system (Section 8.2.1). Conversely, continuous radiation permits relatively efficient simultaneous excitation of substantially all the elements in the specimen, whereas the two monochromatic modes are much more selective. Thus, excitation by continuum is advantageous for qualitative and semiquantitative analyses, comparisons, and "signature" or "fingerprint" analyses (Section 8.2.2.4).

The disadvantages of excitation by continuum are minimized by filtered primary excitation, but the excitation is then more specific. The element having an absorption edge at wavelength just longer than that of the strongest target line is excited most efficiently, each successively lower atomic number less so. Elements six or seven atomic numbers lower are excited very inefficiently and may not be detected if present in low concentration.

Excitation by secondary emission permits selection of optimal excitation wavelength, gives more suitable specimen emission intensity, and gives lower background and consequently higher peak-to-background ratios. Only ~0.1 of the x-rays generated in the tube are intercepted by the secondary target, and the inefficient secondary-emission process results in a further reduction by a factor of 10^{-2}–10^{-3}. The only continuum background arises from double scatter of the primary continuum from both the secondary target and specimen. High-power x-ray tubes, when used for energy-dispersive excitation, are used almost exclusively in the indirect mode, and usually only high-power tubes give intensities really adequate for this mode.

Low-power (<100-W), sealed-off conventional (Section 4.2.5.5) and field-emission (Section 4.2.5.6) tubes are available for excitation in energy-dispersive spectrometers. Demountable tubes (Section 4.2.5.3) are also useful for this application.

Some energy-dispersive x-ray spectrometers combine the x-ray tube, secondary target, specimen, and detector in a compact integral unit. An instrument typical of this type is shown in Figure 8.15 (*H12*, *H13*). The circular filament *A* emits electrons *B*, which bombard the inside surface of the bottomless, saucer-shaped primary target *C* and excite intense primary x-radiation *D*. Because of the shape of *C*, the primary x-rays are directed to the interchangeable conic secondary target *E*, where they excite its characteristic x-rays *F*. Again because of the shape of the emitter, these x-rays converge on the specimen *G*. The secondary target is selected to have a strong spectral line on the short-wavelength side of the analyte absorption edge. In this way, efficient selective excitation of the analyte spectrum *H* is

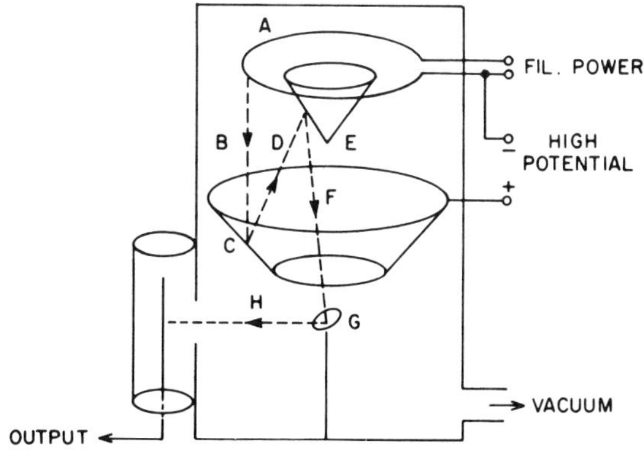

FIG. 8.15. Energy-dispersive high-intensity x-ray spectrometer.

realized. The x-rays from the specimen are detected by a proportional counter *I*, the output of which is subjected to pulse-height analysis. The instrument was designed primarily for biological applications, where the interest is principally in light elements, up to atomic number ~30. Consequently, the primary target is designed for currents up to 500 mA at relatively low potentials, up to ~15 kV.

X-ray tubes, particularly low-power tubes, have certain advantages as excitation sources for energy-dispersive spectrometry, compared with radioisotope sources. These advantages are discussed in the remainder of this subsection.

The effective wavelength and intensity of the excitation radiation can be varied much more widely with tubes than with isotopes by appropriate selection of: (1) continuous or monochromatic x-rays; (2) x-ray tube target; (3) internal or external secondary target; and (4) x-ray tube operating conditions (kV, mA). The effective wavelength and intensity of an x-ray tube output are readily varied over a wide range, those of a radioisotope source are largely limited by the specific isotope and its activity (Ci). Almost any element in some form or other can serve as a secondary target for use with an x-ray tube, so that substantially any wavelength is available to excite the specimen. The choice of materials for actual x-ray tube primary targets is somewhat more limited by the requirements for electrical and thermal conductivity and for stability in vacuum and under electron bombardment. The choice of targets for radioisotopes is limited to elements efficiently excited by the radiation emitted by available practical isotopes (Table 8.5).

The tube target or secondary radiator can be selected not only to excite the analyte line efficiently, but to avoid spectral-line interference or Compton-scatter overlap with it (last paragraph of Section 8.2.2.3).

As already mentioned, excitation by intense primary continuum, although producing high background and perhaps undesirably high total intensity, permits efficient simultaneous excitation of all elements in the specimen.

The intensity from low-power and field-emission tubes is higher than that from practical radioisotope sources, and accordingly, analysis time is shorter. For example, a 100-mCi source—a relatively high activity source—emits 3.7×10^9 photons/s, whereas an x-ray tube operating at only 100-μA tube current emits $\sim 10^{12}$ photons/s.

With power supplies having well-regulated (<0.1%) kV and mA stabilization, the long-term stability of low-power and field-emission tubes is excellent, far exceeding that of radioisotopes. For example, over a four-day period, ^{55}Fe and ^{109}Cd sources (half-lives 2.7 and 1.3 y, respectively) decay

TABLE 8.5. Commonly Used Radioisotope Sources for X-Ray Spectrometric Excitation[a]

Radioisotope source	Principal radioactive decay process[b]	Half-life	Useful radiation[c]		Typical practical source activity, Ci	X-rays excited efficiently
			Type	Energy, keV		
$^{3}_{1}H$–Ti	β^-	12.3 y	Continuum	3–10	5[d]	$_{11}$Na–$_{29}$Cu *K*
			Ti *K* x-rays	4–5		
$^{3}_{1}H$–Zr	β^-	12.3 y	Continuum	2–12	5	$\lesssim _{30}$Zn *K*
			Zr *L* x-rays	2		
$^{55}_{26}Fe$	OEC	2.7 y	Mn *K* x-rays	5.9	0.020[d]	$_{13}$Al–$_{24}$Cr *K*
$^{57}_{27}Co$	OEC	270 d	Fe *K* x-rays	6.4	0.5	$\lesssim _{98}$Cf *K*
			γ	14		
			γ	122		
			γ	136		
$^{109}_{48}Cd$	OEC	1.3 y	Ag *K* x-rays	22	0.003[d]	$_{20}$Ca–$_{43}$Tc *K*
			γ	88		$_{74}$W–$_{92}$U *L*
$^{125}_{53}I$	OEC	60 d	Te *K* x-rays	27		$\lesssim _{54}$Xe *K*
			γ	35		
$^{147}_{61}Pm$–Al	β^-	2.6 y	Continuum	12–45	0.5	$\lesssim _{60}$Nd *K*
$^{153}_{64}Gd$	OEC	236 d	Eu *K* x-rays	42	0.010[d]	$_{42}$Mo–$_{58}$Ce *K*
			γ	97		$_{69}$Tm–$_{92}$U *L*
			γ	103		
$^{210}_{82}Pb$	β^-	22 y	Bi *L* x-rays	11	0.010	$\lesssim _{62}$Sm *K*
			γ	47		
$^{238}_{94}Pu$	α	89.6 y	U *L* x-rays	15–17	0.030[d]	$_{20}$Ca–$_{35}$Br *K*
						$_{74}$W–$_{82}$Pb *L*
$^{241}_{95}Am$	α	470 y	Np *L* x-rays	11–22	0.010[d]	$_{50}$Sn–$_{69}$Tm *K*
			γ	26		
			γ	59.6		

[a] P. 64 in (*8*), (*R9–11*).
[b] OEC = orbital-electron capture.
[c] For the $^{3}_{1}H$–Ti, $^{3}_{1}H$–Zr, and $^{147}_{61}Pm$–Al sources, the useful radiation is that from the overall source, rather than from only the radioisotope.
[d] The x-ray output of each of these sources is $\sim 10^8$ photons/s.

by 0.29% and 0.59%, respectively, whereas low-power tubes may remain stable within 0.25% for one to two weeks (*D42*).

Finally, x-ray tubes are safer than radioisotopes. They can be turned off when not in use and, if desirable, even when changing specimens. They cannot contaminate the laboratory environment, and there is nothing to ingest. Also, there is no high-energy (>100-keV) radiation to be shielded from the environment—and from the detector.

Conversely, although low-power x-ray tubes do not require water-cooling and can be energized by very small, compact, light, solid-state power supplies, an x-ray tube source is much larger than a radioisotope source and does require a source of power (ac line, battery, motor-generator, etc.). These considerations place x-ray tubes at a disadvantage and may even preclude their use in certain field and space applications, and in applications involving small inaccessible places, such as wells and boreholes.

Certain advantages and disadvantages specific to low-power, field-emission, and demountable tubes are given in the sections dealing with these tubes.

To summarize, x-ray tube sources are much more versatile and flexible than radioisotope sources, but much more cumbersome, complex, and expensive. The x-ray tube may be high- or low-power and have many forms. The specimen may be excited directly by the primary x-rays or indirectly by secondary radiators having spectral lines appropriate for the specific analysis. Direct excitation may be by continuum or target lines; in the former case, excitation conditions (kV, mA) may be chosen for optimal excitation, in the latter case, the target, which may also be used with or without a filter to exclude continuum.

8.2.2.3. Excitation by Radioisotopes

The elimination of collimation and crystal dispersion and the reduction of radiation path length result in extremely low analyte photon loss between specimen and detector. This economy of analyte radiation makes excitation by radioisotopes feasible (*C13*, *K6*, *M28*, *R9–11*, *S19*).

For practical purposes, a chemical element may be defined as a substance all atoms of which contain the same number of protons in their nuclei, and therefore have the same atomic number Z. However, atoms of the same element may contain different numbers of neutrons in their nuclei, and therefore have different mass numbers M. The mass number is the sum of the numbers of protons and neutrons in the atom nucleus. An *isotope* is a form of a chemical element all atoms of which contain both the same number of protons and the same number of neutrons in their nuclei, and therefore have both the same atomic number and mass number. A *radioisotope* or *radioactive isotope* is an isotope that undergoes radioactive decay, that is, that spontaneously changes, at a certain rate, into an isotope of a different element or into a different isotope of the same element. For example, carbon (atomic number 6) has eight known isotopes: ${}^{9}_{6}C$, ${}^{10}_{6}C$, ${}^{11}_{6}C$, ${}^{12}_{6}C$, ${}^{13}_{6}C$, ${}^{14}_{6}C$, ${}^{15}_{6}C$, and ${}^{16}_{6}C$. All these isotopes have 6 protons in their atomic nuclei, but 3, 4, 5, 6, 7, 8, 9, and 10 neutrons, respectively. Of these isotopes, ${}^{12}_{6}C$ and ${}^{13}_{6}C$ are stable, the others radioactive.

The correct symbol for an isotope gives the symbol of the element, atomic number Z, and mass number M in the form ${}^{M}_{Z}El$, for example, ${}^{80}_{35}Br$. Unfortunately, the older notation ${}_{Z}El^{M}$ (${}_{35}Br^{80}$) is also still commonly used. The former symbol is preferable because it leaves the right superscript and subscript positions open for, respectively, ionic charge and number of atoms in a molecule or radical; for example, a singly charged molecule of the above bromine isotope would be ${}^{80}_{35}Br_2{}^{+}$. When the discussion warrants, either the atomic or mass number may be omitted—${}^{80}Br$ or ${}_{35}Br$; the old form of the former symbol is Br^{80}. Also, for simplicity, some authors use symbols of the form 35-Br and Br-80. In this book, the preferred notation above is used, except that the form Z-El is used in some of the tables.

A radioisotope *source* is a specified amount of a specified radioisotope fabricated in a form suitable for a specified application. The radioactive material is encapsulated to prevent its dispersion into and thereby contamination of the laboratory environment, and shielded to retain radiation, except in certain directions. Tritiated aluminum and zirconium sources are unsealed and consist simply of the metals with tritium chemisorbed in the surface layer. Some low-activity sources do not require shielding.

Radioisotope sources for x-ray spectrometric application are characterized by four principal properties as follows: (1) radioactive decay process and type of emitted radiation—α-, β-, or γ-emission, or K- or L-orbital-electron capture, which results in x-ray spectral-line emission (Section 1.6.3); (2) energy of the emitted radiation—1–150 keV; (3) activity of the source —1 mCi to 5 Ci (the *curie*, Ci, is that mass of radioisotope in which 3.7×10^{10} disintegrations occur each second); and (4) half-life, the time required for half the atoms of the radioisotope to disintegrate, that is, for the activity of the source to fall to half its initial value. If a radioisotope source has activity n_{T_0} at time T_0, the activity at time T is given by

$$n_T = n_{T_0} \exp -[\lambda(T - T_0)] \tag{8.7}$$

where n is activity, the *number* of radioactive atoms that disintegrate in unit time; and λ is the decay constant (*not* wavelength), the *fraction* of radioactive atoms that disintegrate in unit time. The decay constant is characteristic of the radioisotope and is readily calculated from the half-life $T_{1/2}$: At $T = T_{1/2}$, $n_T = n_{T_0}/2$ and Equation (8.7) becomes

$$n_{T_0}/2 = n_{T_0} \exp(-\lambda T_{1/2}) \tag{8.8}$$

The n_{T_0}'s cancel, and if the remaining terms are put in natural logarithmic form, one gets

$$\lambda = 0.693/T_{1/2} \tag{8.9}$$

where 0.693 is $\log_e 2$.

Table 8.5 lists these properties for each of the radioisotope sources commonly used for x-ray spectrometric excitation. In the tritium–titanium source ($^3_1$H–Ti), tritium, a radioactive hydrogen isotope emitting 18-keV β-rays, is absorbed in nonradioactive titanium. The β-rays excite the titanium to emit x-ray continuum and the Ti K spectrum, which, in turn, excite the specimen. The $^3_1$H–Zr and $^{147}_{61}$Pm–Al sources are of the same nature.

In addition to the parameters cited above, some other considerations determine the suitability of a radioisotope for x-ray spectrometric excitation. The radioactive emission should be as simple as possible, preferably a single intense γ- or x-ray line. It follows that radioisotopes with stable (that is, nonradioactive) decay ("daughter") products are preferable. Also, one must be aware of possible spectral interference from the emission lines from radioisotopes, especially in trace and micro analysis. For example, a ^{241}Am source emits two strong γ lines and the Np L lines. All these lines may scatter from the specimen in unmodified and modified form and generate photo and possibly escape peaks in a detector. (See also the last paragraph in Section 8.2.2.3.)

The three principal source–specimen–detector geometry arrangements are shown in Figure 8.16: the central or button source, the annular or ring source, and the source-target arrangement. In the central or annular geometries, the source irradiates and excites the specimen directly. In the source–target geometry, the source irradiates and excites the characteristic x-ray spectrum of the target, which is selected to have its principal line at wavelength just shorter than that of the analyte absorption edge. This target x-radiation then irradiates and excites the analyte. The targets may consist of pure metals in the form of rings or small thimbles (Figure 8.16, bottom) or oxide powders bonded with epoxy resin to ring or thimble supports. If

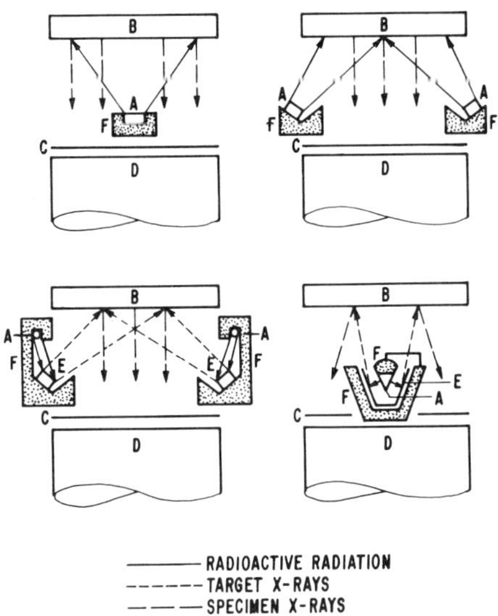

FIG. 8.16. Radioisotope source–specimen–detector geometry arrangements: center or button source (top left), annular or ring source (top right), and two types of source-target arrangements (bottom), the left of which has annular form. *A*. Radioisotope. *B*. Specimen. *C*. X-ray filter (if used); the filter has an open center for the source-target geometry. *D*. Detector—scintillation, proportional, or Si(Li) semiconductor. *E*. Target or radiator. *F*. Shielding for radioactive radiation. *A* and *F* together constitute the radioisotope source.

more than one characteristic line is required, alloys or mixed oxide powders are used. The targets are made so as to be easily interchangeable. This arrangement gives extremely pure analyte spectra having high peak-to-background ratios, but requires high-activity sources because of the inefficiency of the secondary excitation process, and even then gives relatively low analyte-line intensity. A disadvantage of the source–target arrangement is its relative specificity, exciting only elements having absorption edges at wavelengths longer than the secondary x-rays, and those with decreasing efficiency the farther away the edge is. Also, the specimen must be shielded from direct irradiation by the radioisotope. Figure 8.16 also shows the position at which filters, if used, are inserted.

The center-source geometry is most efficient with large detector windows and is most widely used with scintillation and proportional counters. The annular-source geometry is used with Si(Li) detectors, which, having a diameter of, say, 15–20 mm, would lie in the "shadow" of a center source.

Alpha rays produce no continuum and are useful to realize high peak-to-background ratios, especially for analyte lines of long wavelength. However, excitation efficiency for photon energies >14 keV (<0.88 Å) is much lower than for β-rays. Unfortunately, α-rays have extremely low penetrating power, and it is extremely difficult to prepare sealed α sources having windows thin enough to permit emergence of α-rays, yet strong enough to withstand radiation damage and normal handling.

Beta-ray excitation of x-rays is much more efficient, and the efficiency increases with β-ray energy. However, unlike α-rays, β-rays do produce continuum, and the continuum intensity also increases with β-ray energy. Thus β-excitation results in low peak-to-background ratios. Also, β-rays are second only to α-rays in low penetrating power, and it is also difficult to prepare sealed sources having windows transparent to β-rays. The principal application of β-isotopes is admixed with a "radiator" to produce x-ray continuum and the characteristic spectrum of the radiator. The radiator may be selected to have strong lines at wavelength just shorter than that of the analyte absorption edge. It is easy to prepare sealed sources having windows transparent to this x-radiation, which then excites the specimen. The tritium–titanium, tritium–zirconium, and promethium–aluminum sources in Table 8.5 are of this type.

The most effective radioisotope sources are those that emit x-rays or low-energy γ-rays at discrete wavelengths or in relatively narrow wavelength bands. Such radiation gives much lower background than sources that emit x-ray continuum. If the scattered incident radioisotope radiation and emitted analyte-line radiation can be resolved by the pulse-height analyzer or other means, these sources give very high peak-to-background ratios for medium- and low-Z analytes, especially those having absorption edges just longer than the exciting wavelengths. The ^{55}Fe, ^{57}Co, ^{109}Cd, ^{238}Pu, and ^{241}Am sources are of this type. High-energy γ-ray sources are unsuitable because it is not feasible to shield the detector from the highly penetrating radiation.

The source-target geometry (Figure 8.16) is used principally with x- or γ-ray sources, including those consisting of β-emitting isotopes admixed with radiator.

When elements having widely separated excitation potentials are to be determined, it may be feasible to irradiate the specimen with two radioisotope sources simultaneously, one efficient for excitation of high-potential elements, the other for low-potential elements.

In a novel technique, Kowalski and Isenhour (*K28*) actually generated an extremely short-lived γ-emitting radioisotope in the specimen itself. They admixed a small amount of ^{10}B with the specimen, which was then placed in a thin-walled quartz vial and inserted in a beam of thermal neutrons. Some of the stable ^{10}B atoms transmute to metastable ^{7m}Li by neutron capture and subsequent emission of an α-particle ($^4_2$He^{2+}). The ^{7m}Li decays with a half-life of 5×10^{-14} s by emission of a 477-keV γ-ray to stable ^{7}Li. The transmutation and decay are represented as follows:

$$^{10}_{5}\mathrm{B}\ (^1_0\mathrm{n},\ ^4_2\alpha)\ ^{7m}_{3}\mathrm{Li} \rightarrow {}^7_3\mathrm{Li} + \gamma(477\ \mathrm{keV})$$

The γ-rays excite the x-ray spectra of the heavy elements in the specimen. These x-rays are detected by a Ge(Li) detector connected to an energy-dispersive x-ray spectrometer. This is an example of what might be termed "self-excitation" or "autoexcitation" of x-ray spectra in that no external excitation radiation is incident upon the specimen during measurement. Another example is discussed in Section 14.8.4.

In some applications, selective excitation, filtration, or detection may permit measurement of analyte-line intensity simply by counting the detector output directly. This may be the case when the specimen contains only two or a very few elements having relatively widely spaced absorption edges and spectral lines, or when differential excitation or balanced filters are used (Section 8.3.2.1). Otherwise, the detector output must be subjected to pulse-height analysis.

When pulse selection is required, the resolution is poorer with scintillation counters than with proportional counters, and the best resolution is realized with lithium-drifted silicon semiconductor detectors, which permit resolution of lines of adjacent elements. A single-channel pulse-height analyzer may be used, but a multichannel analyzer is much more rapid and convenient (*L7*); 200 or 400 channels may be sufficient for analytical x-ray spectrometry.

The use of radioisotope sources results in some advantages in addition to those of energy-dispersive methods in general (Section 8.2.1.2). Elimination of the x-ray tube and generator results in a very small, light-weight, inexpensive, easily portable instrument. Compact source–detector assemblies (Figure 8.16) can be placed very close to specimens that would be difficult to present to a conventional crystal spectrometer. Spectrometers for lunar and planetary surface analysis may also be of this type. There is a wide choice of sources with respect to the four properties noted above to suit the particular analysis. With radioisotopes of relatively long half-life, the excitation source is extremely stable during the analysis—for practical purposes, absolutely so. Excitation conditions equivalent to a 100–150-kV generator are readily attainable so that *K* lines of even the heaviest elements can be excited. Elimination of the x-ray generator makes radioisotope excitation especially attractive for on-line process control, geological field use, etc. The low intensity of radioisotope sources, a disadvantage in some ways, can be advantageous in that it tends to prevent detector choking, which occurs readily in energy-dispersive mode (Section 8.2.1.3), and substantially eliminates radiolysis of liquids, organics, etc., and discoloration of glasses, art objects, antiquities, etc. The extreme source stability permits use of long counting times to accumulate statistically acceptable counts.

Use of α sources results in very high peak-to-background ratios; the continuum generated by α-rays is weaker than that generated by electrons by a factor equal to the square of the ratio of the electron and α-particle masses, or $\sim(1/7200)^2$. Use of the source-target geometry and of direct specimen excitation by pure x-ray emitting sources also give "clean" analyte spectra with very low background. In the source-target arrangement, scatter of radioisotope radiation from the specimen is minimal, giving greatly reduced background. Thus, even if the excitation intensity is very low compared with that from an x-ray tube, peak-to-background ratios are high and, as already mentioned, long counting times are feasible. Consequently, sensitivites of 10^{-11} g (0.01 ng) or less are attainable. Use of α- and β-emitting sources minimizes absorption-enhancement effects, due to the low penetration of these types of radiation.

The principal disadvantages of radioisotope excitation are the low intensity compared with x-ray tubes, and the personnel health hazards attendant upon the use of radioactive materials. Radioisotope sources have intensities $\sim 10^{-6}$–10^{-7} those of standard x-ray tubes, and the detection limits realized with such sources are $\sim$10–100$\times$ those realized with conventional tube excitation. It is explained above that the low intensity is usually not a serious problem because of the low background and high source stability. The radioisotope sources are usually sealed and of relatively low activity, so that ingestion is unlikely and shielding is easily provided. Another disadvantage is the gradual decrease of source intensity with decay of the radioisotope, necessitating frequent recalibration—more frequent the shorter the half-life of the radioisotope. Finally, the desired radioisotope may be accompanied by contaminating and/or daughter decay isotopes that may emit high-energy γ-rays. These γ-rays may find their way to the detector by scatter or even by penetration of shielding with resultant increase in background.

When secondary radiators are used (source-target geometry), some of the x-radiation from the target may undergo incoherent scatter from the specimen (Section 2.2.2). Care must be taken to select the radiator so that the incoherently scattered line does not overlap the analyte line. The intensity of the incoherently scattered x-rays increases as wavelength decreases, atomic number of the specimen (scatterer) decreases, and the angle ϕ between the directions of the unscattered and scattered x-rays increases. The wavelength shift is given by Equation (2.38), where ϕ in this case is the angle between the directions of the x-rays emitted by the radiator and specimen, and approaches 180°. For example, if it is required to detect lanthanum ($\lambda_{K_{ab}}$ 0.318 Å), the ideal secondary target should be samarium

($\lambda_{K\alpha}$ 0.310 Å). However, at ~180° scatter angle, the incoherently scattered Sm $K\alpha$ line lies at 0.370 Å and interferes with La $K\alpha$ at 0.372 Å. Thus, for maximum excitation efficiency, the primary radiation should have wavelength just shorter than that of the analyte absorption edge, but sufficiently shorter so that incoherently scattered primary radiation does not overlap the analyte line significantly.

8.2.2.4. Energy-Dispersive Multichannel X-Ray Spectrometer Systems (*P27*)

A modern, energy-dispersive, multichannel x-ray spectrometer system consists of the following components: (1) high-power (Sections 4.2.1–4.2.4) or low-power (Sections 4.2.5.5–4.2.5.6) x-ray tube, secondary-emission (Section 8.3.1), radioisotope (Section 8.2.2.3), electron (Section 1.6.4), or proton or other ion (Section 1.8) excitation source; (2) specimen-presentation system; (3) Si(Li) detector with its liquid-nitrogen vacuum cryostat (Section 6.4.1); (4) multichannel analyzer (Section 8.1.2.3); (5) computer and memory unit; and (6) cathode-ray tube, X–Y recorder, digital, and/or print-out display units. These systems are applicable to all elements down to atomic number ~11 (sodium), the limitation being imposed by the Si(Li) detector. Sensitivity *in favorable cases* may be <1 ppm. Specimens may have almost any form, including priceless antiquities and dynamic, continuous-flow fluids. Up to 20 specimens can be placed in automatic indexing turntable changers, or up to 100 in towerlike changers. Counting times may vary from <1 min to 15 min or more depending on whether major, minor, or trace concentrations are present, and/or whether x-ray tube, secondary-emission, electron, ion, or high- or low-activity radioisotope excitation sources are used.

These systems are extremely convenient and versatile, and usually rapid as well. This section is intended to acquaint the reader with the latitude of their functions.

The energy spectrum is accumulated in the multichannel analyzer, as already explained (Section 8.1.2.3), until one channel or window accumulates a preset count (100–10^6), or for a preset time. The accumulated spectrum may be displayed, plotted, or printed out directly on a cathode-ray tube, X–Y recorder, or printer, respectively. The accumulated spectrum may also be transferred to the computer memory store and processed in a wide variety of ways and with great convenience. If an X–Y recorder or printer is used, the number of counts accumulated in each channel is plotted or printed out on demand. If a cathode-ray display is used, the display

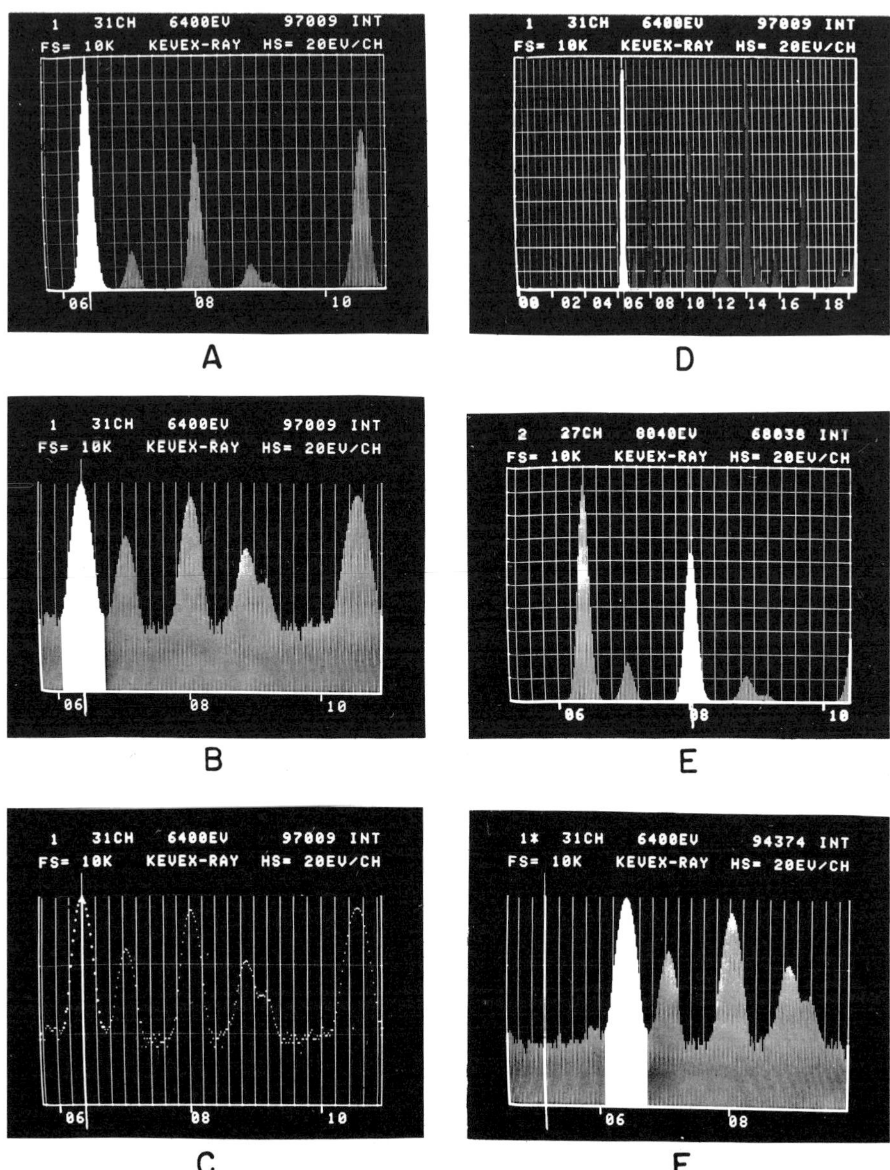

FIG. 8.17. X-ray energy spectra of a test specimen containing Fe, Cu, Sr, Mo, and Pb. [Courtes of Kevex Corp.] The alphanumeric data have the following significance: *Upper top* (from left number of brightened "window"; width of selected window, number of channels (CH); energ (EV) of peak centroid in selected window; integrated (INT) total accumulated count in the selecte window. (This position can also display net count in the selected window or total accumulate count or count rate for the entire spectrum). *Lower top* (from left): number of counts for fu

is continuous, and one can actually watch the spectrum accumulate—that is, watch the channels filling—and observe the progress and results of the processing. Usually 5-inch cathode-ray tubes are used, but for lecture and demonstration purposes, 12–21-inch television monitors can be used. Of course, the cathode-ray display can be photographed with a Polaroid camera. In the remainder of this section, a cathode-ray display is assumed.

Let us now consider the remarkable functioning of the multichannel analyzer–computer-memory–cathode-ray-display system in five categories: (a) display and labeling of the energy spectrum; (b) spectrum processing; (c) qualitative and comparison analysis; (d) quantitative analysis; and (e) automation. Although all the features described below may not be available on any one commercial instrument, most are available on all instruments, and all are available on one or another. The remainder of Section 8.2.2.4 is illustrated by Figures 8.17–8.23, which have detailed captions and should be studied carefully. It is recommended that these figures be studied together in sequence, more or less independently of the text. However, specific figures are referred to in the following text.

a. Display and Labeling of the Energy Spectrum (Figures 8.17–8.23). The data are displayed on the cathode-ray tube in three forms: bar or dot graphs, alphanumerical data, and fixed or movable bright-line markers. Also, there can be a rectangular *X–Y* grid over the entire display. The sample number and name or other identification can be displayed alpha-

---

vertical scale (FS); horizontal scale (HS) in eV/channel (EV/CH). *Bottom*: energy scale in keV.

A. X-ray energy spectrum in the region 5–11 keV, showing peaks at energies (keV) as follows (left to right): Fe $K\alpha$ (6.40), Fe $K\beta$ (7.06), Cu $K\alpha$ (8.04), Cu $K\beta$ (8.90), Pb*Ll* (9.18), and Pb $L\alpha_1$ (10.55). In this spectrum and those in B and C below, a window (No. 1) 31 channels wide is set on the Fe $K\alpha$ peak (6400 eV) and brightened, and a bright-line peak-centroid marker is visible above and below the peak.

B. Spectrum A displayed on a three-decade logarithmic vertical scale.

C. Spectrum B with dot instead of bar display.

D. X-ray energy spectrum in the region 0–20 keV, showing the peaks in spectrum A plus the following (left to right): Pb $L\beta_1$ (12.61), Sr $K\alpha$ (14.14), Pb $L\gamma_1$ (14.76), *Pb* γ_3 (15.21), Sr $K\beta$ (15.83), Mo $K\alpha$ (17.44), and Mo $K\beta$ (19.60).

E. Spectrum A slightly expanded horizontally and with a brightened window (No. 2) 27 channels wide set on the Cu $K\alpha$ peak (8040 eV).

F. Spectrum B slightly expanded horizontally with Fe $K\alpha$ background subtracted numerically. The asterisk (*) after the window number (1) indicates that the top right number (94374) is a net count rather than a total count as in spectrum B (97009). The background is obtained by multiplying the number of counts in the channel marked by the bright line by 31, the number of channels in the Fe $K\alpha$ window, and subtracting the product from the total accumulated count in the window (97009).

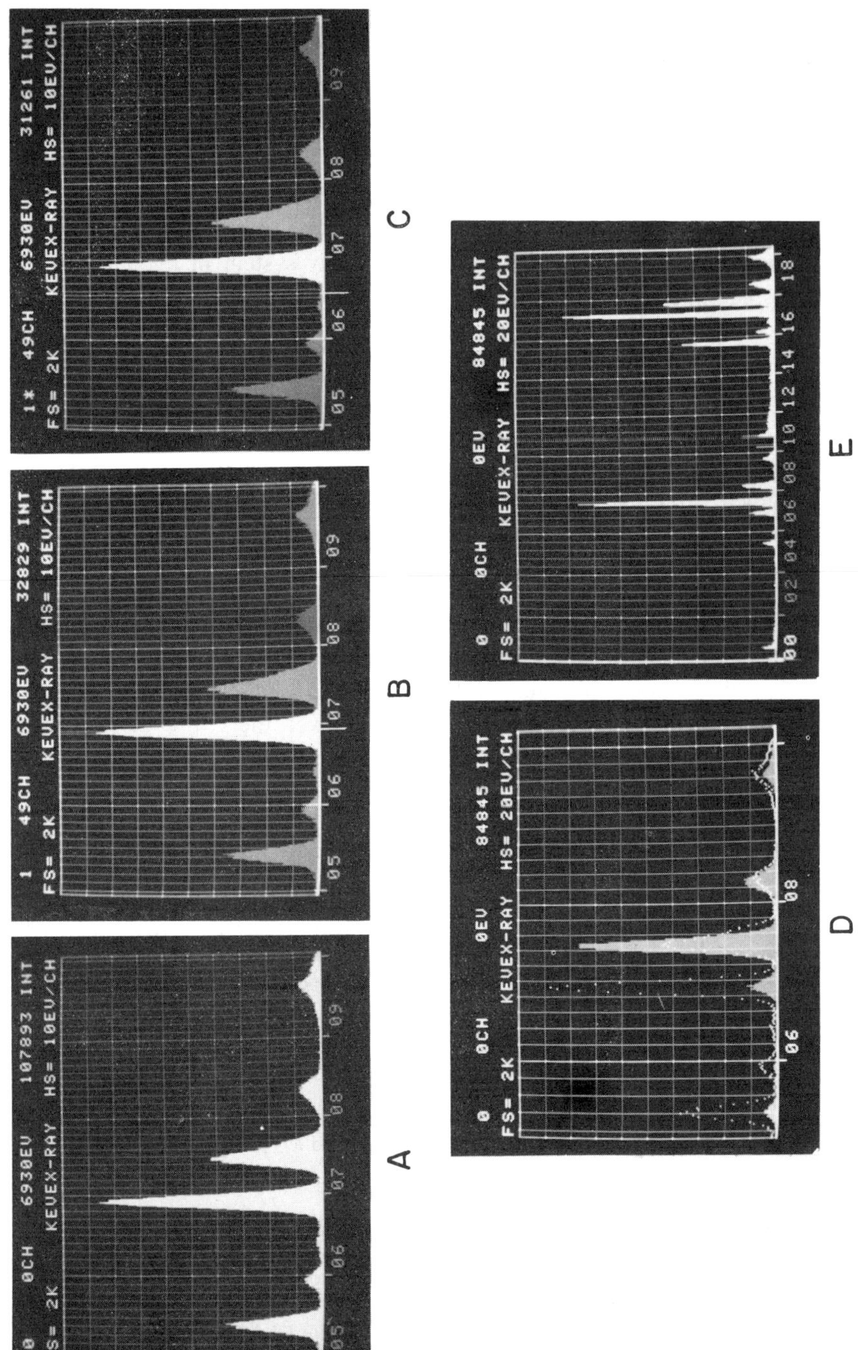
0 0CH 6930EV 107893 INT
FS= 2K KEVEX-RAY HS= 10EV/CH
05 06 07 08 09
A
1 49CH 6930EV 32829 INT
FS= 2K KEVEX-RAY HS= 10EV/CH
05 06 07 08 09
B
1* 49CH 6930EV 31261 INT
FS= 2K KEVEX-RAY HS= 10EV/CH
05 06 07 08 09
C
0 0CH 0EV 84845 INT
FS= 2K KEVEX-RAY HS= 20EV/CH
06 08
D
0 0CH 0EV 84845 INT
FS= 2K KEVEX-RAY HS= 20EV/CH
00 02 04 06 08 10 12 14 16 18
E

FIG. 8.18. X-ray energy spectra of a test specimen containing Cr, Fe, Co, Ni, and W. The alphanumeric data have the same significance as in Figure 8.17. Window No. 0 indicates that the entire spectrum is brightened. [Courtesy of Kevex Corp.]

A. X-ray energy spectrum in the region 4–10 keV, showing peaks at energies (keV) as follows (left to right): Cr $K\alpha$ (5.41), Cr $K\beta$ (5.95), Fe $K\alpha$ (6.40), Co $K\alpha$ (6.93), Ni $K\alpha$ (7.47), W $L\alpha_1$ (8.40), and W $L\beta_1$ (9.67). All channels are brightened (window No. 0).

B. Spectrum A with a brightened window (No. 1) 49 channels wide set on the Co $K\alpha$ peak (6930 eV). A bright-line peak-centroid marker is visible above and below the peak.

C. Spectrum B with Co $K\alpha$ background subtracted. The net count in the Co $K\alpha$ window (31261) is obtained in the same way as in Figure 8.17F from the total accumulated count in the window as displayed in spectrum B (32829), the accumulated count in the channel marked by the bright line, and the number of channels in the Co $K\alpha$ window (49).

D. Spectrum A (dots) compared with a superimposed reference spectrum (bars). The displayed integrated count is the total count for both spectra.

E. Spectrum A compressed into the right half of the display compared with the reference spectrum in D above compressed into the left half of the display. The displayed integrated count is the total count for both spectra. The energy scale is correct for the left spectrum, 10 keV high for the right spectrum.

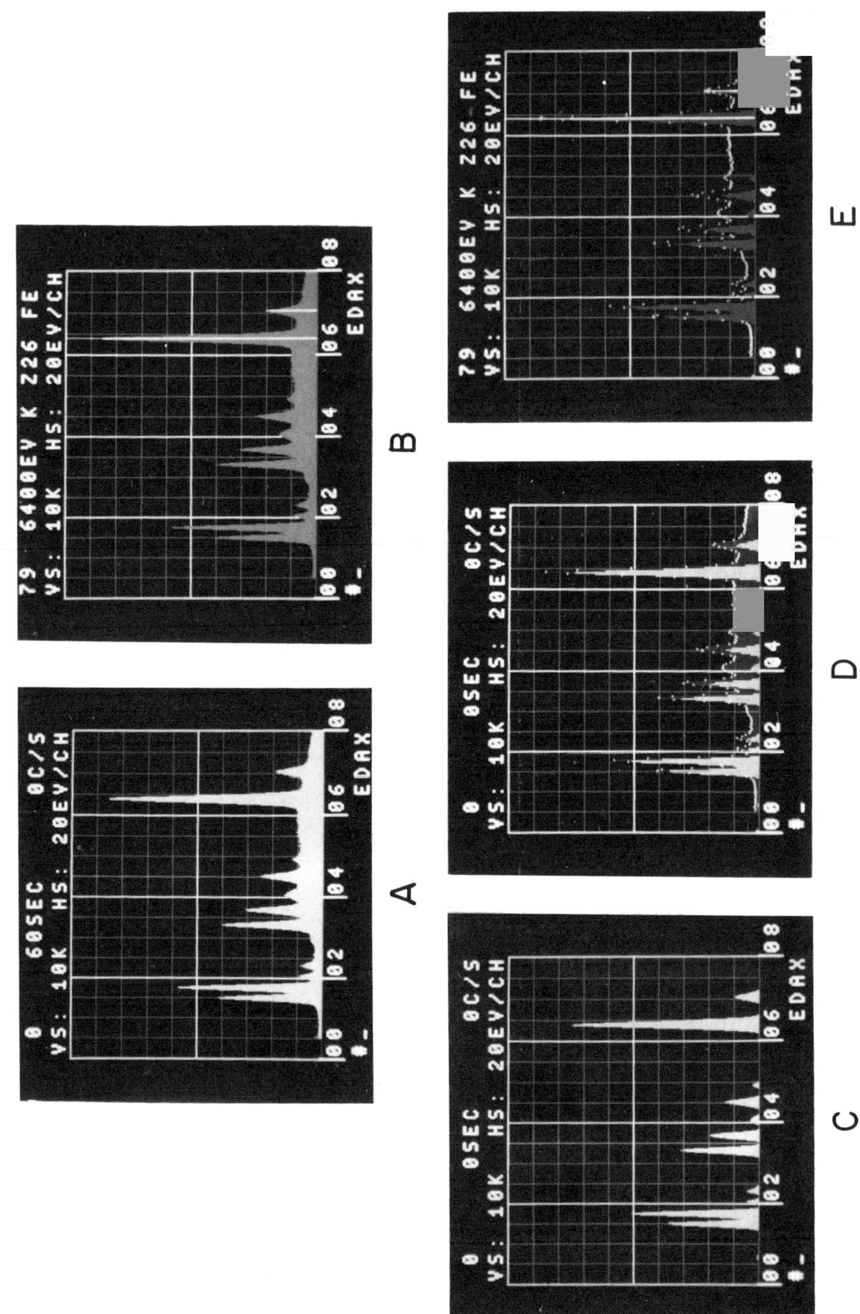
0 60SEC 0C/S
VS: 10K HS: 20EV/CH
EDAX
A
79 6400EV K Z26 FE
VS: 10K HS: 20EV/CH
EDAX
B
0 0SEC 0C/S
VS: 10K HS: 20EV/CH
EDAX
C
0 0SEC 0C/S
VS: 10K HS: 20EV/CH
D
79 6400EV K Z26 FE
VS: 10K HS: 20EV/CH
E

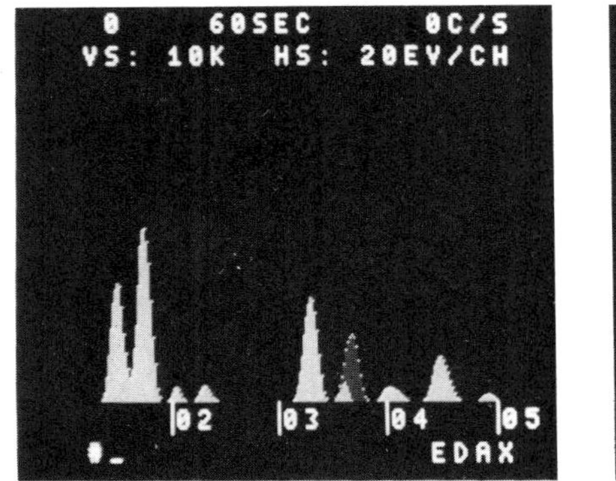

F

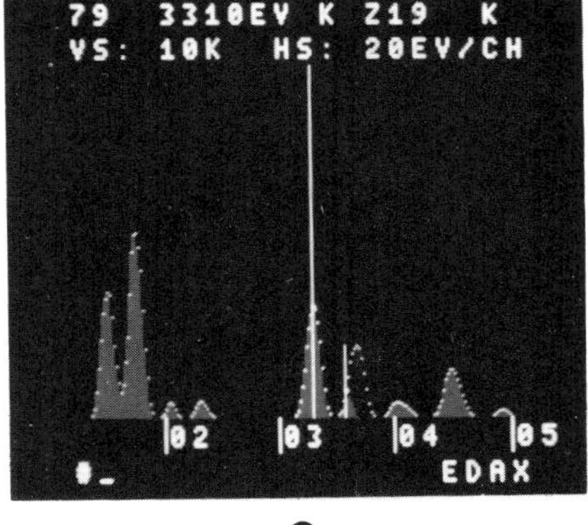

G

FIG. 8.19. X-ray energy spectra of a test specimen containing Al, Si, P, S, K, Ca, Ti, and Fe. The alphanumeric data have the same significance as in Figures 8.17 and 8.18 with the following exceptions and additions: VS (vertical scale) is used instead of FS (full scale); in A and F, 60 SEC indicates the counting time; in B and E, K Z 26 FE indicates that the bright-line markers label the *K* lines of Fe (Z 26); in G, K Z 19 K indicates that the bright-line markers label the *K* lines of K (Z 19). [Courtesy of EDAX International, Inc.]

A. Entire x-ray energy spectrum in the region 0–8 keV, showing peaks at energies (keV) as follows (left to right): Al $K\alpha$ (1.49), Si $K\alpha$ (1.74), P $K\alpha$ (2.01), S $K\alpha$ (2.31), K $K\alpha$ (3.31), Ca $K\alpha$ (3.69), Ca $K\beta$ (4.01), Ti $K\alpha$ (4.51), Ti $K\beta$ (4.93), Fe $K\alpha$ (6.40), and Fe $K\beta$ (7.06).

B. Spectrum A with bright-line markers on Fe $K\alpha$ and Fe $K\beta$.

C. Spectrum A with background subtracted.

D. Spectrum A (dots) superimposed on spectrum C (bars).

E. Similar to spectrum D with bright-line markers on Fe $K\alpha$ and Fe $K\beta$.

F. The 1–5-keV region of spectrum A expanded with the Ca $K\alpha$ peak (dots) stripped to uncover the underlying K $K\beta$ peak (bars).

G. Spectrum F with bright-line markers on K $K\alpha$ and K $K\beta$.

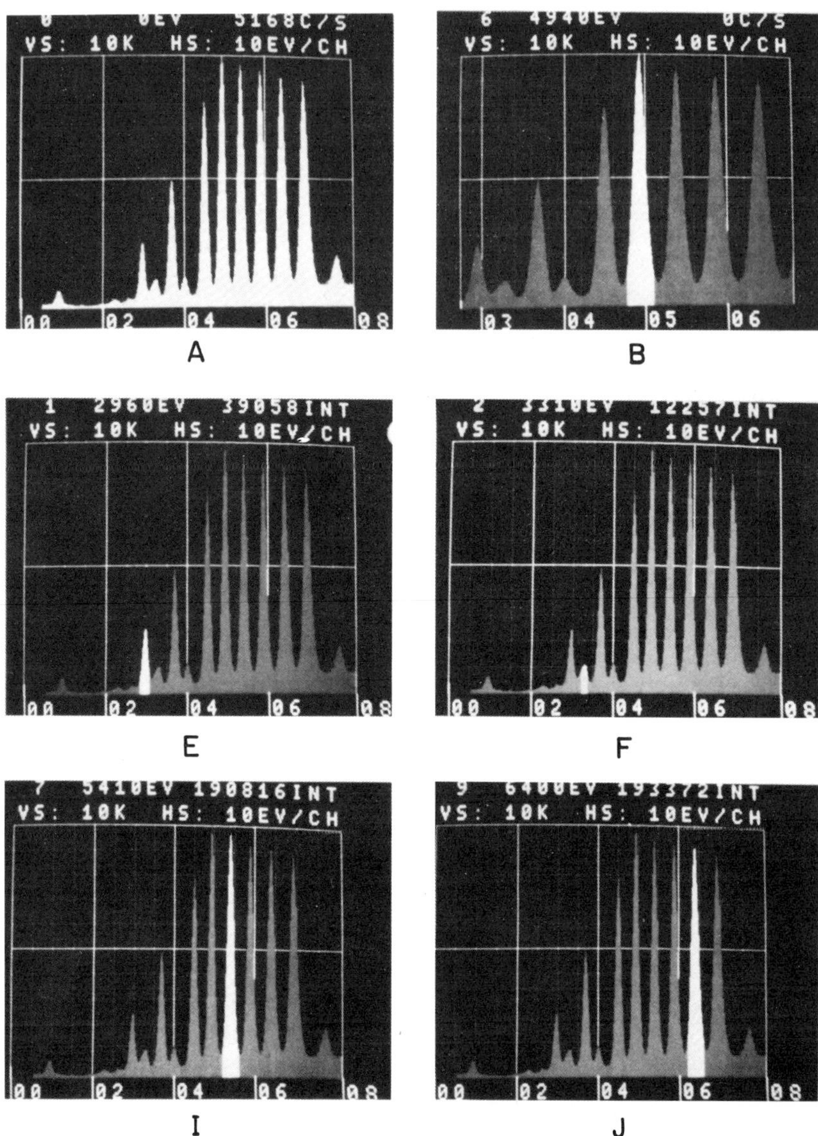

FIG. 8.20. X-ray energy spectra of a test specimen containing Na, S, Cl, K, Ca, Ti, V, Cr, Mn, Fe, Co, and occluded Ar. The alphanumeric data have the same significance as in Figures 8.17 and 8.18 except that the width of the brightened window is not displayed and VS is the same as FS, as in Figure 8.19. [Courtesy of EDAX International, Inc.]

A. Entire x-ray energy spectrum in the region 0–8 keV, showing peaks at energies (keV) as follows (left to right): Na $K\alpha$ (1.04), S $K\alpha$ (2.31), Cl $K\alpha$ (2.62), Ar $K\alpha$ (2.96), K $K\alpha$ (3.31), Ca $K\alpha$ (3.69), Ca $K\beta$ (4.01), Ti $K\alpha$ (4.51), V $K\alpha$ (4.95), Cr $K\alpha$ (5.41), Mn $K\alpha$ (5.90), Fe $K\alpha$ (6.40), Co $K\alpha$ (6.93), and Co $K\beta$ (7.65).

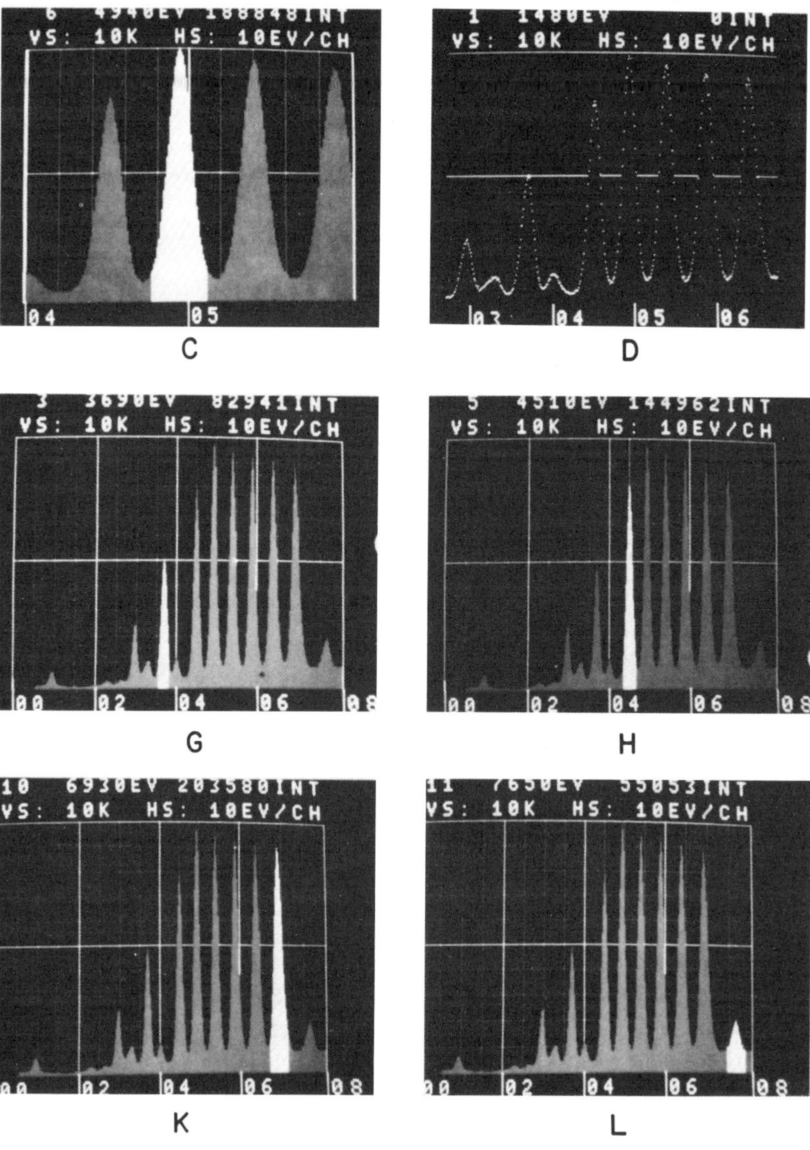

B. Spectrum A with two-fold horizontal expansion and brightening of V *Kα* (window 6).

C. Spectrum A with fourfold horizontal expansion and brightening of V *Kα*.

D. Spectrum B with dot instead of bar display and without brightening of V *Kα*.

E–L. Spectrum A with the following peaks brightened, respectively: Ar *Kα*, K *Kα*, Ca *Kα*, Ti *Kα*, Cr *Kα*, Fe *Kα*, Co *Kα*, and Co *Kβ*. For each spectrum, the window number, peak-centroid energy (eV), and peak integrated total count in the window are given alphanumerically at the top.

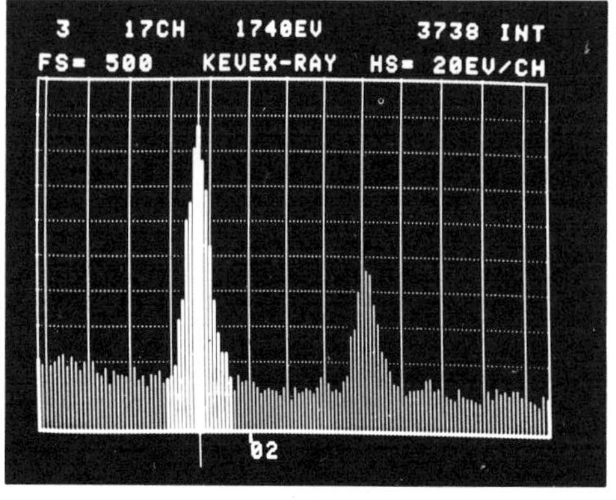

A

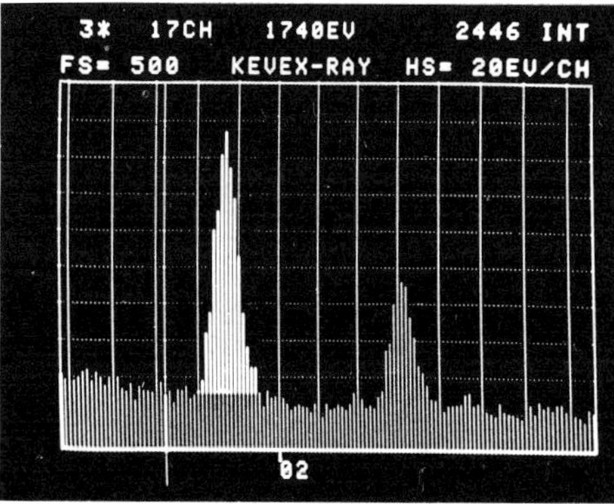

B

FIG. 8.21. Background subtraction. X-ray energy spectrum showing Si $K\alpha$ and Cl $K\alpha$ peaks at 1.74 and 2.62 keV, respectively. The alphanumeric data have the same significance as in Figure 8.17. [Courtesy of Kevex Corp.]

A. A brightened window (No. 3) 17 channels wide and a bright-line peak centroid marker are set on the Si $K\alpha$ peak (1740 eV).

B. Spectrum A with Si $K\alpha$ background subtracted. The asterisk (*) after the window number (3) indicates that the top right numer (2446) is a net count rather than a total count as in spectrum A (3738). The background is obtained by multiplying the number of counts in the channel marked by the bright line by 17, the number of channels in the Si $K\alpha$ window, and subtracting the product from the total accumulated count in the window (3738). Note that only the net count in each window channel is brightened. Compare with Figure 8.17F.

numerically. The spectrum itself can be displayed as a bar (histogram) or dot graph on a linear or logarithmic intensity scale. The display can be marked with an energy (keV) or wavelength (Å) scale in the form of a series of bright-line tics along the bottom edge, each marked numerically in keV or Å, and the calibration constant (eV or Å per channel) can be displayed alphanumerically; the scale may vary automatically as the spectrum is contracted or expanded. The positions of the $K\alpha$, $K\beta$, $L\alpha_1$, $L\beta_1$, $L\gamma_1$, $M\alpha$, and $M\beta$ lines of a specified element can be marked with bright-line tics, and the symbol of the marked element displayed (Figure 8.19B,G). All of these seven lines that lie in the displayed spectral region are thus marked simply by dialing-in the atomic number of the element. The α, β, and γ tics are long, medium, and short, respectively, suggestive of their relative intensities.

Any displayed peak can be selected with a movable bright-line cursor and its energy (keV), wavelength (Å), net peak counts, and/or net integrated counts displayed. The peak centroid of any selected peak can be accented or otherwise marked; the *centroid* is that channel, among the several constituting the peak, containing the highest accumulated count. The bars or dots in a selected number of consecutive channels can be brightened so as to stand out visually—for example, all channels in a specified peak, or several channels in the background adjacent to the peak. The integrated total counts in such brightened groups of channels can be read out for analytical purposes.

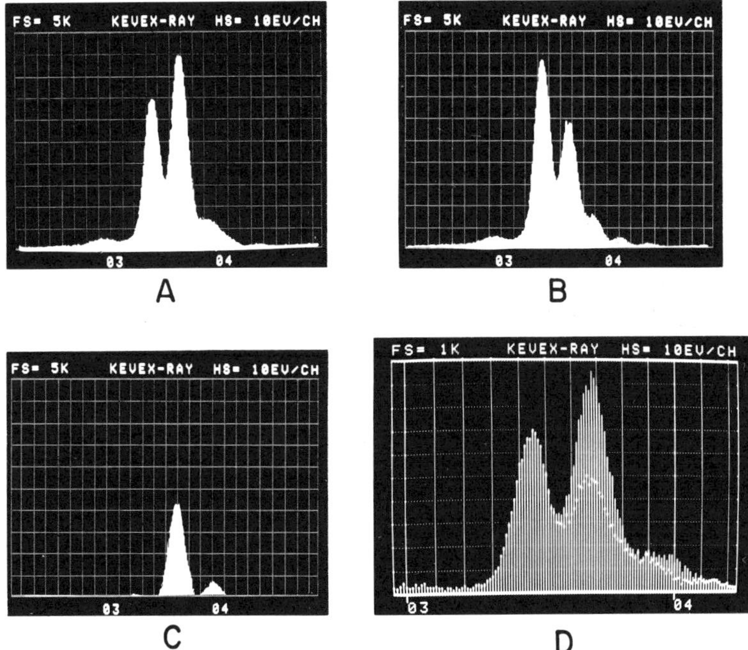

FIG. 8.22. Stripping of overlapped energy-spectral peaks. [Courtesy of Kevex Corp.]

A. X-ray energy spectrum of a specimen containing Ca and Sn, showing overlapped Ca K and Sn L peaks at energies as follows (left to right): Sn $L\alpha_1$ (3.44); Sn $L\beta_1$ (3.66) and Ca $K\alpha$ (3.69); Sn $L\beta_2$ (3.90) and Ca $K\beta$ (4.01); and Sn $L\gamma_1$ (4.13).

B. Sn L spectrum from pure tin.

C. Spectrum B, adjusted to appropriate amplitude, subtracted from spectrum A, leaving Ca $K\alpha$ and Ca $K\beta$ peaks.

D. Spectra A (as bars) and B (as dots) expanded horizontally, superimposed, and displayed on a logarithmic vertical scale.

These selected groups of channels are sometimes termed energy "windows," not to be confused with the term *window* in the sense used in Sections 7.1.4.1 and 8.1.1. In one 1024-channel commercial instrument, as many as 63 such windows can be superimposed on the spectrum. These windows are independently variable in width from 1 to 301 channels and may overlap. The window number (1–63) and width (in channels) are displayed alphanumerically. The count data—bars or graphs—in a selected window may be accented.

The total number of counts accumulated in all channels can be displayed. As is explained below, comparison spectra can be displayed simultaneously side by side with or superimposed on the analytical spectrum.

The displayed data can be color-coded (Kevex Color Video Monitor, Figure 8.23) to simplify visual conception and interpretation as follows: the spectrum—red; the peak centroid (see above)—white; the energy "window" (see above) and alphanumeric data pertaining to it—yellow; other alphanumeric data—white; etc.

After the spectrum has been subjected to smoothing, background subtraction, stripping, etc. (see below), the total number of net counts in each peak can be integrated. The instrument can then display a list of all statistically significant peaks (say, 99% confidence level) in order of energy, listing the peak energy and integrated net counts for each. Each peak can then be tested electronically for Gaussian profile, and those that are asymmetrical or possibly composite are marked for further consideration by the operator.

It is then possible to establish a "window" for each peak of interest. The windows may be defined either by the energies at the low and high edge of each peak or simply by the peak centroid. In the latter case, the window width is set automatically at 1.2 times the peak half-width (FWHM). A number, say 10, of such windows can be preset for use in analytical programs (see below).

b. Spectrum Processing. The analytical spectrum can be processed in several ways as the progress of the processing is observed on the display. The system can be programed (see below) to perform a specified series of spectrum processings automatically. The accumulation of the spectrum from the next sample can proceed while the previous spectrum is being processed in the computer. The analytical spectrum can be corrected for coincidence loss and pulse pileup, smoothed, stripped, and/or reconstructed.

Coincidence-loss correction is facilitated by the simultaneous accumulation of the spectrum, which permits application of novel and ingenious ways to compensate these losses. The *Harms dead-time correction* circuit

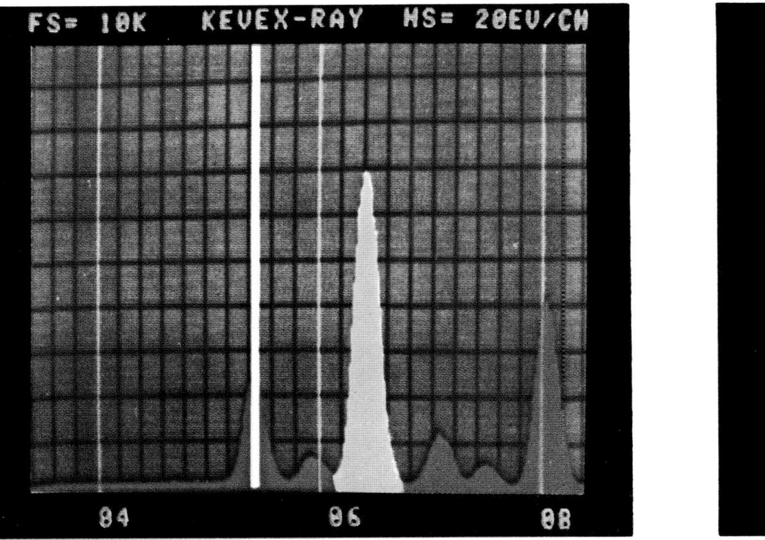

A

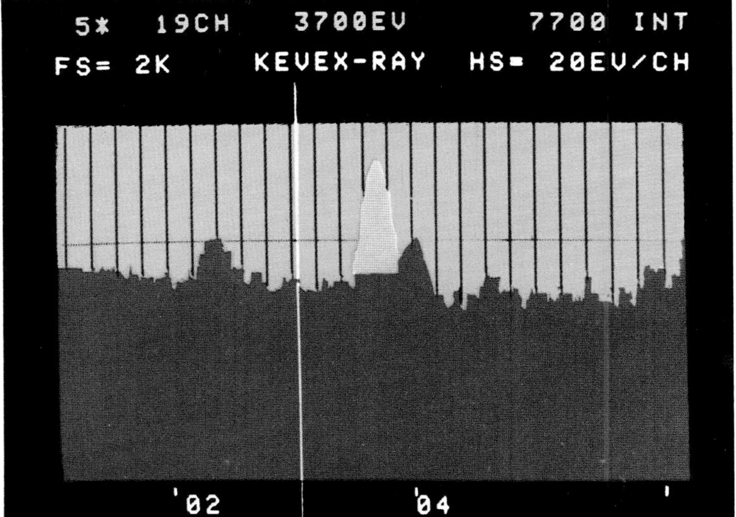

B

FIG. 8.23. Color-coded x-ray energy spectra. [Courtesy of Kevex Corp.] (A) Color-coded x-ray energy spectrum in the region 3.5–8.5 keV, showing peaks at energies (keV) as follows (left to right): Cr $K\alpha$ (5.41), Cr $K\beta$ (5.95), Fe $K\alpha$ (6.40), Fe $K\beta$ (7.06), Ni $K\alpha$ (7.47), and Cu $K\alpha$ (8.04). A bright-line marker is set on the Cr $K\alpha$ peak and a window on the Fe $K\alpha$ peak. The color code is red for the general spectrum, yellow for the window, white for the marker. (B) Color-coded logarithmic x-ray energy spectrum, showing a Ca $K\alpha$ peak at 3.69 keV. The alphanumeric data have the same significance as in Figure 8.17. A window (No. 5) 19 channels wide is set on the Ca $K\alpha$ peak (3700 eV). The asterisk (*) after the window number (5) indicates that background has been subtracted from the window and the top right number (7700) is integrated net counts in the window. The background was obtained from the number of counts in the channel marked by the bright line in the same way as in Figures 8.17F and 8.21B. Note that only the net count in each window channel is coded yellow. Compare with Figures 8.17F and 8.21B. The color code is red for the general spectrum, yellow for the window and alphanumeric data pertaining to it, and white for markers and other alphanumeric data.

(*H16*) continuously compares the number of pulses actually arriving from the detector—regardless of amplitude—with the number of pulses actually stored in the multichannel analyzer after baseline restoration and pulse pileup rejection. The circuit stores a number of pulses equal to the difference. This number corresponds to the pulses lost due to coincidence effects. At the end of the counting time, the circuit apportions these counts among the channels in the multichannel analyzer, incrementing each in proportion to its share of the total accumulated count. The *Lowes live-time correction*, in effect, estimates the dead time for each pulse from its half-width, keeps account of these dead times, then extends the counting time to allow accumulation of enough additional pulses to compensate for those lost by coincidence effects.

Probably the ultimate method for elimination of coincidence loss is provided by the triode x-ray tube and associated electronic circuitry described briefly in Section 4.2.5.5.

Smoothing reduces the channel-to-channel fluctuations due to counting statistics and is especially useful when very weak peaks are present that might otherwise be "lost in the noise."

Stripping consists of subtraction of certain data from the analytical spectrum after they have been accumulated, or even while they are actually being accumulated. The data may actually be subtracted (Figures 8.19A,C, D,F; 8.22), leaving only the net spectrum in the display, or the net data—in each channel, that portion of the bar histogram that represents net counts—can be brightened (Figures 8.21B, 8.23B). The subtracted data may be: (1) the spectrum contributed by the specimen support or matrix; (2) background; and/or (3) overlapped interfering peaks. The subtracted spectral data may be accumulated in unused portions of the multichannel analyzer from analyte-free specimen support or matrix, blanks, pure interfering elements, etc. Alternatively, the subtracted data may be entered from punched or magnetic tapes or stored in the computer memory, if it has sufficient capacity; interfering peaks to be subtracted can actually be generated electronically.

Background for specimens supported on Mylar film, Millipore filters, filter paper, ion-exchange membrane or paper, etc. may usually be corrected by subtraction of spectral data from clean support. Background for trace and, usually, minor constituents may usually be corrected by subtraction of spectral data from analyte-free blanks (Section 7.3.3). Background for minor and major constituents in more massive specimens is usually more difficult to deal with. One instrument provides two windows (Section 8.2.2.4a) in the adjacent background region bracketing the selected analyte peak; the accumulated count from these windows is summed, halved, then subtracted from the peak count.

Alternatively, the integrated count can be measured in one or more channels adjacent to the peak window on each side; background is then stripped from each channel in the window assuming linearity of background profile between the measured adjacent backgrounds. In still another instrument, one of the windows is set to just bracket the peak of interest; let us assume that the required window width is, say, 40 channels. A single channel adjudged by the operator to be representative of the background under the peak is selected in the spectrum adjacent to this window and indicated on the display by a bright-line background marker. The instrument multiplies the number of counts in this background channel by the number of channels in the analyte peak window (40) and subtracts the product from the integrated analyte-peak count in the window. The net count is displayed numerically, and the net peak can be brightened on the display (Figures 8.17F, 8.18C, 8.21B, 8.23B). A different representative background can be selected for each peak.

A still more sophisticated approach is the frequency-filtering method of Russ (*R36*). The energy spectrum has four principal frequency components: (1) very-low-frequency, continuously but nonlinearly varying, continuum background; (2) intermediate-frequency, relatively short-span energy peaks; (3) high-frequency channel-to-channel statistical fluctuations; and (4) very-high-frequency absorption-edge discontinuities in the continuous background. Components 1 and 3 are filtered out, leaving components 2 and 4 as the net energy spectrum.

In subtracting interfering peaks (Figures 8.19F, 8.22), if the subtracted data are accumulated directly from a blank, they may usually be subtracted directly. Otherwise, a scale factor is applied to adjust the subtracted peak heights. Alternatively, the progress of the subtraction can be observed on the cathode-ray tube as the amplitude of the subtracted data is varied, and, sometimes, the correct amplitude can be established visually. When only a few overlapping peaks are involved, the more accurate method of ratio subtraction can be used. The method requires that for each interfering (overlapping) peak, there be another peak of the interfering element in the analytical spectrum free of interference. The ratio of the overlapping and free peaks is established from measurements on pure element. Thereafter, the height of the peak to be subtracted is established from the height of the free peak measured on the sample, and the ratio.

Reconstruction (*C58*) is the opposite of stripping. The analytical spectrum is displayed in, say, channels 1–600. A series of peaks is generated electronically of elements known or likely to be present in the specimen, and displayed in channels 601–1200. The generated peak heights and widths are then adjusted to obtain a close visual match to the analytical spectrum.

c. Qualitative and Comparison Analysis. The multichannel analyzer may have a total of, say, 1200 channels, which may be used as a single unit, as two 600-channel units, or as four 300-channel units (Figure 8.18E). The analytical spectrum may be accumulated originally in the full 1200 channels, then later compressed into, say, channels 1–600 and a selected portion of it expanded into channels 601–1200 for closer inspection at higher resolution; or the full spectrum may be compressed into channels 1–300 and the portion of interest expanded into channels 301–1200.

In some units, any portion of the spectrum can be expanded horizontally (pulse-height, energy, or channel axis) (Figures 8.17A,D; 8.19F; 8.20A,B,C) or vertically (counts axis).

In one system, channels, say, 301–900 (the central 600 channels) can be used to expand, say, a 300-channel portion of the spectrum as the spectrum is passed slowly, by manual or automatic control, past this central portion. Thus, in effect, the entire spectrum can be swept horizontally past a sort of "magnifying glass," the unexpanded end portions displayed in channels 1–300 and 901–1200, the expanded portion in channels 301–900. The translation can be stopped at will for closer inspection, photography, etc.

Qualitative analyses may be performed in several ways: The energy (or wavelength) corresponding to each peak can be read off by use of the calibrated scale in the usual way. Or the movable bright-line cursor mentioned above can be moved to each peak in turn so that its energy (or wavelength) is displayed digitally. The peaks are then identified by use of tables in the usual way. Energy-to-line tables (Figure 10.8) are available (*77*). Finally, the atomic number of each element likely to be present can be dialed-in one at a time so as to mark the positions of its principal lines (Figures 8.19B,E,G; 8.20). This last technique is especially advantageous for searching for very weak peaks.

Comparison analyses are performed by displaying reference or standard spectra for instantaneous and direct comparison with the spectrum from the sample. The spectra may be compared side by side (Figure 8.18E) or superimposed (Figures 8.18D, 8.22). For example, in a 1200-channel analyzer, channels 1–600 may display the sample spectrum while channels 601–1200 simultaneously display, on the same scale, one at a time, various reference spectra for direct comparison. It is also possible to display the sample spectrum in channels 1–300 and, say, three comparison spectra in channels 301–600, 601–900, and 901–1200. Alternatively, the sample and comparison spectra can be superimposed, both as bar graphs, or one as a bar graph, the other as dots. Either of two such superimposed spectra can be displaced vertically relative to the other to any desired degree of overlap by means of a manual offset control. In either side-by-side or superimposed mode, the comparison spectra can be displayed while the sample spectrum is actually being accumulated, or after accumulation is complete. The standard or reference spectra may be accumulated in the unused sections of the multichannel analyzer from standards or reference specimens, or they may be entered from magnetic or punched tapes, or they may be previously stored in the computer memory, if it has sufficient capacity.

These comparison displays lead to the feasibility of a method of x-ray spectrometric qualitative analysis known as comparison, "fingerprint," or "signature" analysis, based on visual or automatic electronic comparison of the x-ray energy spectra of the unknown sample and prospective known materials. For example, suppose that a laboratory is required to identify and classify incoming metal samples as one of, say, 30 specified alloys. The x-ray energy spectra ("fingerprints" or "signatures") of the 30 alloys are recorded on punched or magnetic tapes or stored in the computer memory, if it has sufficient capacity. The spectrum of each sample is substantially instantaneously compared with the taped or stored spectra and thereby identified. This can be done visually on the cathode-ray display, or electronically by the computer. It is claimed that the latter method is so effective that subjective evaluation is virtually eliminated from the comparison. Note that this comparison analysis, unlike quantitative analysis, does not require correction for absorption-enhancement effects because, when the correct reference spectrum is found, the sample and standard are the same material—the same analytes in substantially the same concentrations in the same matrix.

Use of the multichannel analyzer as a multichannel scaler to accumulate data for mapping the distributions of elements along a line or over an area on the specimen is discussed in Sections 8.1.2.3, 21.4.3, and 21.4.4.

d. Quantitative Analysis. The analytical spectrum can be processed in many ways, the following being typical: (1) Each peak is corrected for dead time. (2) Each peak is corrected for pulse pileup and background. (3) Each peak is corrected for, or stripped of, spectral-line interference as described above. (4) The channels to be integrated for each peak to be measured and its adjacent background are selected; the selection is facilitated by brightening the dots or bars in the selected channels, as mentioned above. (5) The corrected integrated intensities are then printed out. (6) Alternatively, the intensities can be converted to analytical concentrations by the computer programmed to follow the method selected by the operator. This may be a simple intensity *versus* concentration function, in which case intensity data from standards must be accumulated, entered from tape, or stored. Or the method may make Lucas-Tooth and Pyne, Lachance–Traill, or other absorption-enhancement corrections, in which case constants for the elements present are entered, or, if the computer has sufficient capacity, constants for all elements are permanently stored. No knowledge of the mathematics of the method nor of the electronics of the instrument is required, and the operation is largely automatic.

e. Automation. It is explained above that a quantitative analysis involves selection of the peaks to be actually measured, then for each, correction for dead time, pulse pileup, background, and spectral-line interference, and selection of the channels to be integrated for peak and background. The operator establishes these conditions more or less by trial and error while observing the progress on the display screen. Once the sequence of operations is established, the operator runs through the sequence once with the instrument in "learn" mode. Thereafter, the instrument repeats the sequence on all subsequent samples of the series. Up to 50 steps can be learned and programmed in this way. If a multiple specimen holder is provided, the system can be programmed to index each specimen into position in turn, accumulate its spectrum, process the data, and print out the results.

Any accumulated spectrum or analytical procedure program can be recorded on magnetic or punched tape or in the computer memory, if its capacity is adequate.

8.2.3. Energy-Dispersive Diffractometry-Spectrometry

The condition for diffraction of x-rays by a crystal lattice can be expressed in the form of the Bragg law (Section 2.3),

$$n\lambda = 2d_{(hkl)} \sin \theta \qquad (8.10)$$

where n is the diffraction order; λ is the wavelength of the diffracted x-rays; $d_{(hkl)}$ is the interplanar spacing of the diffracting planes (hkl); and θ is the angle between the incident x-rays and the diffracting planes. Since in x-ray diffraction it is always a measurement of d that is required, the many dif-

TABLE 8.6. Comparison of X-Ray Diffraction, Spectromety, and Energy-Dispersive Diffractometry-Spectrometry

	$n\lambda$	=	$2d$	×	$\sin\theta$
Diffraction, Laue	Variable		Calculated		Fixed
Diffraction, moving-crystal	Fixed		Calculated		Variable
Diffraction, Debye—Scherrer	Fixed		Calculated		Variable
Diffraction, divergent-beam	Fixed		Calculated		Variable
Spectrometry	Calculated		Fixed		Variable
Diffractometry-spectrometry	Variable		Calculated		Fixed

fraction methods fall into two fundamental groups on the basis of the Bragg law—those that vary λ and those that vary θ. However, more practically, x-ray diffraction methods are classified in the four groups given in Table 8.6—Laue, moving-crystal, Debye–Scherrer, and divergent-beam.

In the *Laue methods*, a collimated polychromatic x-ray beam irradiates a fixed single crystal. Each set of planes (hkl) in the crystal presents a certain fixed angle $\theta_{(hkl)}$ to the incident beam and "selects" from the beam and diffracts the particular wavelength that satisfies the Bragg law for its d and θ.

In the *moving-crystal methods*, a collimated, monochromatic x-ray beam irradiates a single crystal. If the crystal were fixed, as in the Laue case, only by chance would any of the planes happen to present the required θ to the incident beam to satisfy the Bragg law. However, if the crystal is rotated or made to undergo some other appropriate motion, various sets of planes are brought in turn to the required θ's to satisfy the Bragg law for their respective values of $d_{(hkl)}$ and the incident wavelength.

In the *Debye–Scherrer* (or *powder*) *methods*, a collimated, monochromatic x-ray beam irradiates a polycrystalline specimen. Any specified set of planes (hkl) can diffract the incident wavelength only from those crystallites so oriented as to present that plane to the incident beam at the required θ. However, if the number of crystallites in the beam is large and they are randomly oriented, each plane is likely to have some crystallites properly oriented to permit it to diffract.

In the *divergent-beam methods*, a spherically divergent point source of monochromatic x-rays is generated at or just above the surface of a fixed, thin, radiolucent single crystal. A solid cone of x-rays is transmitted through the crystal. If there were no diffraction, a photographic plate a short distance below the crystal would be *uniformly* exposed by the transmitted x-rays. Actually, certain rays of the diffracted cone strike each fixed crystallographic plane at the correct angle for diffraction. The intensities of the rays in the divergent cone are increased in the directions of the diffracted rays, and decreased in the directions of the corresponding incident rays prior to diffraction. In this way, a pattern of dark and light circles, parabolas, and hyperbolas (*Kossel lines*) is formed on the photographic plate.

The method of *energy-dispersive x-ray diffractometry-spectrometry* (*G10, G11*) combines features of the Laue and Debye–Scherrer x-ray diffraction methods with energy-dispersive x-ray spectrometry. In fact, the method may be regarded as an energy-dispersive Laue method (because polychromatic x-rays are used) or as an energy-dispersive Debye–Scherrer method (because polycrystalline specimens are used). The originators of the method propose the term *energy-dispersive spectrometric powder diffractometry* (SPD) (*G11*). The technique is also known as the "Giessen–Gordon method." The instrument arrangement is shown in Figure 8.24 and is essentially that of a conventional diffractometer. An x-ray tube, Si(Li) detector, and a plane polycrystalline specimen are arranged with the incident and diffracted beams making equal angles θ_1 with the specimen surface. However, the arrangement differs from that in diffractometry in that the geometry is fixed and the x-ray tube irradiates the specimen with polychromatic, rather than monochromatic (filtered target-line) x-rays.

Diffracted x-rays from a specified set of planes (*hkl*) can be detected only from those crystallites that happen to be oriented so that those planes lie parallel to the specimen plane—that is, present angle θ_1 to the incident and diffracted rays. Those crystallites for which this is the case "select" from the incident beam and diffract the wavelength that satisfies the Bragg law for their $d_{(hkl)}$ and fixed angle θ_1. Other crystallites permit other planes to diffract other wavelengths in the same way.

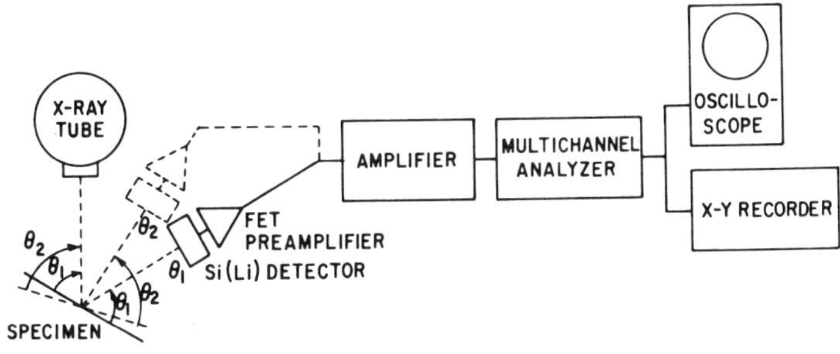

FIG. 8.24. Instrument arrangement for energy-dispersive x-ray diffractometry-spectrometry.

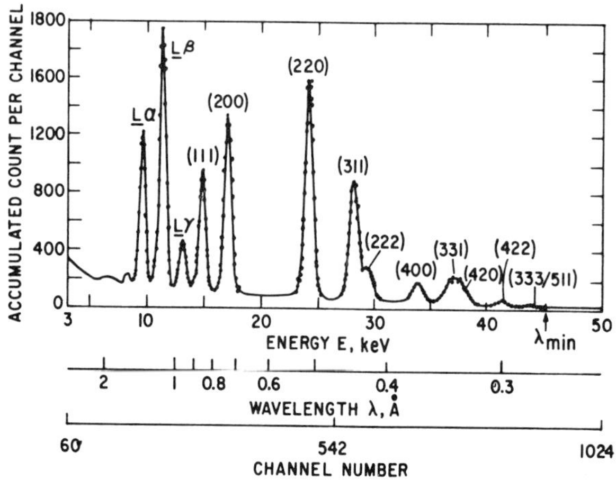

FIG. 8.25. Energy-dispersive x-ray diffractometry-spectrometry spectrum of polycrystalline platinum. [B. G. Giessen and G. E. Gordon, *Science* **159**, 973 (1968); courtesy of the authors and the American Association for the Advancement of Science.]

Actually, the specimen may also be cylindrical with its axis at the point of intersection of incident and diffracted rays and perpendicular to their plane. In this case, diffracted rays from a specified set of planes are detected only from those crystallites that happen to be oriented so that their plane normals are coplanar with and bisect the angle $(180° - 2\theta_1)$ between the incident and diffracted rays.

The x-rays entering the detector consist of: (1) diffracted rays of various wavelengths, each diffracted by a different set of planes (hkl); (2) emitted x-rays of various wavelengths arising from secondary excitation of the elements in the specimen; and (3) scattered primary x-rays—continuum and target lines. The detector output is amplified and fed to a multichannel analyzer, the accumulated output of which is displayed on a cathode-ray tube or X–Y recorder. Figure 8.25 shows a diffraction-spectrometry energy spectrum of polycrystalline platinum foil. Each peak is a narrow pulse-height distribution, and the distributions of the Pt L lines and of the several wavelengths diffracted by the crystal planes are shown.

In Figure 8.25, the Pt L peaks constitute a Pt L energy spectrum identical with that obtainable from an energy-dispersive x-ray spectrometer. However, the diffraction peaks differ from those obtainable on a conventional diffractometer: In Figure 8.25, each peak represents a *different* wavelength diffracted from a different Pt (hkl) plane at the *same* angle θ_1.

In a diffractometer record, each peak represents the *same* wavelength (the filtered target line) diffracted from a different plane at a *different* angle $\theta_{(hkl)}$.

The *emission* peaks can be identified by simple inspection if: (1) only one or a very few elements in the specimen emit lines giving pulse-height distributions in the energy region displayed; (2) these elements are known; and (3) there is no overlap of emission and diffraction pulse distributions. These conditions prevail in Figure 8.25. If there is difficulty in identifying the emission lines, a second energy spectrum is recorded at a different angle θ_2 (Figure 8.24) 5–15° away from θ_1. The pulse-height distributions of the *emitted* and *scattered* x-rays are the same in both spectra, but the distributions of the *diffracted* x-rays are different at the two angles. The emitted and scattered peaks may be eliminated by visual comparison of the two spectra, or the two multichannel analyzer outputs may be fed directly to a computer and the peaks common to both—that is, the emission and scatter peaks—electronically subtracted.

The interplanar spacing $d_{(hkl)}$ of the crystal plane giving rise to each diffraction peak is readily calculated from the Bragg law and Equation (1.18):

$$\lambda = 12.396/E_x \tag{8.11}$$

where E_x is photon energy (keV). Substitution in the Bragg equation gives

$$d_{(hkl)} = \frac{12.396/E_x}{2 \sin \theta} = \frac{6.198}{\sin \theta} \frac{1}{E_x} \tag{8.12}$$

For any specified angle θ, $\sin \theta$ may be included in the constant, giving simply $d = k/E_x$. The E_x is obtained from the calibrated scale of the multichannel analyzer.

If accurate relative intensities of the diffraction peaks are required, the detector response at each wavelength in the primary beam must be established and used to correct the observed relative intensities.

Applications and advantages of the method arise from the following three features: (1) the extreme rapidity with which the pattern is recorded, as quickly as 1 s; (2) simultaneous analysis for both elements and chemical species; and (3) elimination of the need for scanning over a wide, continuous angular range. The method is applied to studies of continuous dynamic phenomena, short-lived phenomena, very large objects, and specimens in restricted enclosures, such as high-pressure and high-temperature vessels.

The relationships, in terms of the Bragg law, among the various methods considered in this section are summarized in Table 8.6. It is evident that in the Laue, Debye–Scherrer, and divergent-beam methods, all planes diffract simultaneously, whereas in the moving-crystal method, in general, only one plane diffracts at a time. In all these methods, the diffracted rays from different planes travel in different directions. In energy-dispersive diffractometry-spectrometry, all planes diffract simultaneously and in the same direction.

An instrument designed exclusively for energy-dispersive diffractometry-spectrometry, the "Speedifrax" (TM), is available from Nuclear Equipment Corp., San Carlos, Calif. (*M14*).

8.3. NONDISPERSIVE ANALYSIS

The distinction drawn in this book between nondispersive and energy-dispersive methods of x-ray spectrometric analysis is discussed in Section 5.1. The individual nondispersive methods are discussed below.

8.3.1. Selective Excitation

The analyte line may be excited selectively—that is, to the exclusion, or substantially so, of the lines of other elements in the specimen—by potential discrimination, monochromatic excitation, or differential excitation.

In *potential discrimination*, the x-ray tube is operated at a potential below that at which the interfering line is excited. The details of this technique are given in Section 13.3.3.2. The method is applicable to simple systems having relatively few elements, and in which the analyte excitation potential is relatively low; it may be used as a supplement to crystal dispersion. Loss of analyte-line intensity results from the reduced operating potential. An example of an ideal application is selective excitation of Fe $K\alpha$ in tin-plated steel.

In *monochromatic* or *tertiary* (Section 1.6.3) *excitation*, the primary beam may be used to irradiate a secondary target (radiator or "fluorescer") having strong spectral lines on the short-wavelength side of the analyte absorption edge. The resulting spectral emission excites the analyte line. The method may be useful when intensity can be sacrificed for a "clean" spectrum, that is, one free of continuum; the only continuum background arises from double scatter of the primary continuum from both the secondary target and the specimen. The analyte-line intensity is very low, but its measurement by the nondispersive method is feasible. Monochromatic x-radiation can also be obtained by filtering the continuum from the primary beam, leaving substantially pure target-line radiation (Section 8.2.2.2), and from radioisotope sources (Section 8.2.2.3). The filtered primary radiation is more intense than either secondary-target or radioisotope sources. However, secondary targets provide substantially any required wavelength, whereas the other two sources are much more limited.

Birks [pp. 62–63 in (*8*)] describes the variations of the technique shown in Figure 8.26 and gives an example of their application to a multi-element system:

The arrangement in Figure 8.26A is used for determination of the analyte having line of longest wavelength. The primary beam excites the radiator chosen to have a strong line capable of exciting only that analyte having absorption edge—and therefore spectral line—of longest wavelength.

The arrangement in Figure 8.26B is used for determination of the analyte having line of shortest wavelength. The primary beam excites the specimen, the secondary emission from which excites the radiator chosen to have an absorption edge that is excited only by the analyte line of shortest wavelength. The intensity of the *radiator* line is a measure of the intensity of the analyte line of shortest wavelength.

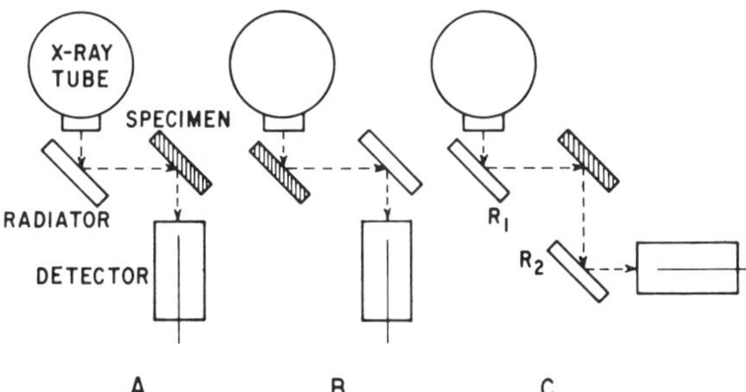

FIG. 8.26. Monochromatic excitation by use of secondary "fluorescers." [L. S. Birks, "X-Ray Spectrochemical Analysis," ed. 2, Interscience Publishers, N. Y., p. 63 (1969); courtesy of the author and publisher.] A. The radiator excites only specimen elements having absorption edges at wavelengths longer than that of its own line. B. The radiator is excited only by specimen emission having wavelength shorter than that of its own absorption edge. C. The first radiator excites only specimen elements having absorption edges at wavelengths longer than that of its own line. The second radiator is excited only by specimen emission having wavelength shorter than that of its own absorption edge.

The arrangement in Figure 8.26C is a combination of the other two and is used for determination of an analyte having a line of intermediate wavelength. The primary beam excites the first radiator chosen to have a spectral line capable of exciting only the analyte and elements having absorption edges at longer wavelengths. The secondary emission from the specimen irradiates a second radiator that is excited only by the line of the shortest wavelength present in the specimen emission, that is, the analyte line. The intensity of the *second radiator line* is a measure of the intensity of the analyte line of intermediate wavelength.

As an example, suppose that chromium, iron, and nickel are to be determined in stainless steel by monochromatic excitation. The wavelengths of the $K\alpha$ lines and K-absorption edges are as follows:

	$\lambda_{K_{ab}}$	$\lambda_{K\alpha}$
Cr	2.070	2.291
Fe	1.743	1.937
Ni	1.488	1.659

Chromium is determined with the arrangement of Figure 8.26A with iron as radiator; Fe $K\alpha$ excites chromium, but not nickel. Nickel is determined with the arrangement of Figure 8.26B with iron as radiator; Fe $K\alpha$ is excited by Ni $K\alpha$ from the specimen, but not by Cr $K\alpha$. Iron is determined with the arrangement of Figure 8.26C with nickel and chromium as first and second radiators, respectively; Ni $K\alpha$ from the first radiator excites both Cr $K\alpha$ and Fe $K\alpha$ in the specimen, but of these, only Fe $K\alpha$ excites Cr $K\alpha$ in the second radiator.

In instruments having drawer-type specimen presentation, a special drawer can be provided to accommodate the specimen and radiator(s) in proper geometry. The drawer can be arranged so that the intensity to be measured emerges from the specimen compartment in the usual direction. In this case, the goniometer is set at 0° 2θ, the crystal stage removed, and the spectrometer detector used. Alternatively, the drawer can be arranged so that the intensity emerges in the opposite direction—out the front of the drawer. In this case, a hole is cut in the drawer and the detector placed there.

When secondary radiators are used, some of the x-radiation from the radiator may undergo incoherent scatter from the specimen (Section 2.2.2). Care must be taken to select the radiator so that the incoherently scattered line does not overlap the analyte line. The situation is similar to that described in Section 8.2.2.3 for source-target radioisotope excitation. The wavelength shift is given by Equation (2.38), where ϕ in this case is the angle between the directions of the x-rays emitted by the radiator and specimen.

In *differential excitation*, the specimen is irradiated successively by two wavelengths λ_S and λ_L bracketing the analyte absorption edge on the short- and long-wavelength sides, respectively. A detector is placed to receive x-rays emitted and scattered by the specimen. The detector receives analyte-line emission at λ_S, but not at λ_L. The difference in measured intensity at the two wavelengths gives analyte-line intensity—without pulse-height analysis. Miniature field-emission tubes having different targets to provide the bracketing wavelengths are suitable for this application (Section 4.2.5.6). More details of the method, especially as applied to selected-area analysis, are given in Section 19.3.

8.3.2. Selective Filtration

8.3.2.1. Methods

One of the ways to isolate a specified spectral line from a beam containing many such lines is by use of filters (*S20*) (Figure 8.27). The choice of filter material and thickness depends on three criteria: (1) degree of attenuation of the interfering x-rays; (2) remaining intensity of the wanted x-rays after adequate attenuation of the interfering x-rays; and (3) possible increase in background intensity by secondary excitation of the filter element(s).

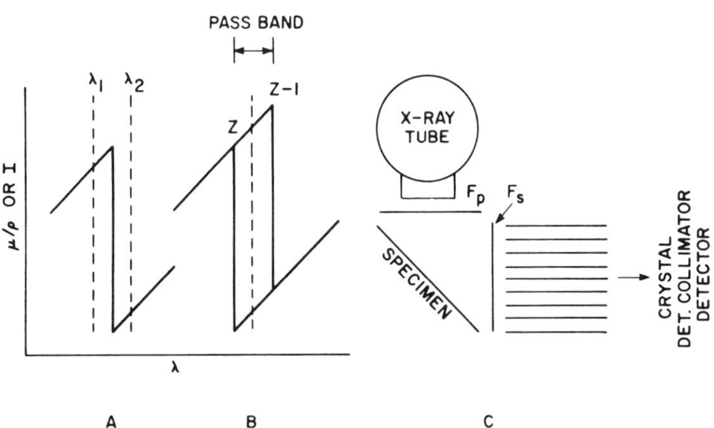

FIG. 8.27. Filter arrangements. A. Simple (Hull). B. Balanced (Ross). C. Kustner.

The principle of the *simple* (*Hull*) *filter* (*S6*) is shown in Figure 8.27A. The filter is a thin foil or other layer of an element having a K or LIII absorption edge at a wavelength just shorter than that of the analyte line λ_2. Thus, the filter has low absorption for the analyte line. All lines of shorter wavelength, such as λ_1, are strongly absorbed. The method is very simple, but gives only rough discrimination, and none for interfering lines on the long-wavelength side of the filter edge. The intensity of the passed wavelength is lower the more nearly complete the discrimination—that is, the thicker the filter. The method is most useful for removing the shorter of two interfering lines bracketing the filter absorption edge. Two examples of simple filters follow. In the analysis of brass (copper–zinc alloy), a filter of nickel ($\lambda_{K_{ab}}$ 1.49 Å) 24 μm thick almost completely suppresses the Zn $K\alpha$ line (λ 1.44 Å) while reducing the Cu $K\alpha$ (λ 1.54 Å) intensity only to ~0.35 of its original value. In the analysis of tantalum–tungsten alloys, a filter of nickel 20 μm thick reduces the W $L\alpha_1$ line (λ 1.48 Å) to ~0.02 of its original intensity while reducing the Ta $L\alpha_1$ line (λ 1.52 Å) only to ~0.44 of its original intensity.

The principle of the *balanced* (*Ross*) *filter* (*K16*, *R27*) is shown in Figure 8.27B. A balanced filter consists of two separately mounted thin foils or other layers of two elements of adjacent atomic number selected so that the wavelength to be measured lies in the *pass band* between their absorption edges. For example, nickel ($\lambda_{K_{ab}}$ 1.49 Å) and cobalt ($\lambda_{K_{ab}}$ 1.61 Å) filters can be used to measure Cu $K\alpha$ (λ 1.54 Å). The thicknesses of the filters are adjusted so that they transmit equal intensities at all wavelengths except those between their absorption edges. This is done by actual adjustment of their thickness or area density (mg/cm^2), or by stretching or tilting the filters in their frames. Another way is to add to a filter having slightly low absorption a thin filter of a third element having very low absorption, no absorption edge, and no spectral line in the wavelength region of interest—aluminum, for example.

Intensity is measured first with one filter in the x-ray beam, then the other. The difference between the two intensities is due entirely to x-rays having wavelengths between the two absorption edges, principally the analyte line. Alternatively, the x-rays emitted and scattered by the specimen can be received by two identical detectors simultaneously, each covered by one of the balanced filters. Again, the difference in the two detector outputs is due to the wavelengths in the filter pass band.

The balanced-filter method can be made substantially specific for any wavelength, but the filters are difficult to balance, and a different pair of filters is required for each analyte line (*A1*, *M26*, *T2*). The resolution of a balanced filter varies from ~300 eV at 1.5 keV (Al $K\alpha$) to ~700 eV at 10 keV (Ge $K\alpha$) to ~2 keV at 70 keV (Hg $K\alpha$) (*45*). Scintillation counters, despite their poor resolution, can be used with balanced filters because resolution is determined by the pass band of the filter, rather than by the detector.

The principle of the *Kustner* filter method (*K34, K35*) is shown in Figure 8.27C. The filter is a foil or other layer of an element having a K or LIII absorption edge at a wavelength just longer than that of the analyte line. Thus, the filter has high absorption for the analyte line. Intensity is measured first with the filter in the primary beam F_P, then in the secondary beam F_S. The intensity of the scattered x-rays reaching the detector is the same with the filter in either position, whereas the analyte-line intensity is much lower with the filter in the secondary beam. Thus, the difference between the two intensities is due to analyte-line emission from the specimen.

The Dothie "Filterscan" method (*D22, D23*) is based on the relative wavelengths of K-series (or L-series) α_1 lines and absorption edges: The $K\alpha_1$ line of the element of atomic number Z is transmitted by a filter of element $Z - n$ more than by one of element $Z - (n + 1)$. The value of n is given in Table 8.7. Dothie (*D23*) gives a table of the $K\alpha_1$, $L\alpha_1$, and $L\beta_1$ lines lying between the K-absorption edges of pairs of adjacent elements of atomic number 1–100 (hydrogen to fermium).

A series of filters of successive atomic numbers are placed in turn in the secondary beam, and the intensity transmitted through each is measured and plotted. The resulting plot is, in effect, a "filter scan" of the secondary spectrum. The method can resolve lines of elements of adjacent atomic number. The filters must have about the same area density (mg/cm^2) of element.

TABLE 8.7. Variation of n with Atomic Number of Radiating Element[a]

Atomic number Z	Elements	n
1–22	H–Ti	0
23–35	V–Br	1
36–47	Kr–Ag	2
48–59	Cd–Pr	3
60–73	Nd–Ta	4
74–92	W–U	5
93–100	Np–Fm	6

[a] From Dothie and Gale (*D23*).

In certain cases, two lines may fall between two successive absorption edges. Such cases may be dealt with by use of appropriate standards or perhaps by combinations of *K*, *L*II, and *L*III absorption edges. For example, both Ti *K*α and V *K*α fall between the scandium and titanium *K* edges. However, the tellurium and iodine *L*II edges lie between Ti *K*α and V *K*α, so that Ti *K*α lies between the scandium–tellurium or scandium–iodine filter pair, but V *K*α is excluded.

The method is applicable to any specimen form. Because the method is based on the difference in transmitted intensities, it is applicable to radioactive specimens. Dothie applied the method to secondary spectra excited by radioisotopes.

In all filter methods, *K*-absorption edges are used whenever practical. The *L*III edges of elements of higher atomic number may be used, but in a specified spectral region, *K* edges have higher jump ratios than *L*III edges (Figure 2.4, Appendix 8), and elements with *L*III edges have higher absorption than elements with *K* edges.

8.3.2.2. X-Ray Transmission Filters

X-ray transmission filters should have the following features: (1) uniform thickness; (2) optimal thickness for the application; (3) absence of warp, wrinkles, creases, cracks, and pinholes; (4) mechanical rigidity and stability; (5) chemical stability; (6) absence of interfering emission lines; and (7) absence of impurity elements, especially those having interfering absorption edges. Filters that are not self-supporting require a substrate. This substrate must be highly radiolucent in thicknesses adequate to support the film and must have no interfering x-ray spectral lines or absorption edges. Beryllium foil, Mylar film, and pyrolytic graphite (see below) make excellent substrates.

Filters are prepared in several ways. Ductile metals may be rolled to thin foil. Nonductile metals may be ground and polished with abrasive, electropolished, or chemically etched. Metals may be electroplated or vacuum-evaporated on thin, x-ray-transparent substrates. Metal or oxide powders may be slurried with an organic binder and spread on similar substrates, or spread on glass, dried, and stripped off. Powders can also be briqueted with an inert radiolucent diluent, such as starch. Filter paper may be impregnated with solutions, then dried.

Vassos *et al.* (*V8*) recommend use of thin plates of pyrolytic graphite as substrates for x-ray transmission filters prepared by thin-layer electrodeposition. The graphite is: (1) readily cleaved in thin (0.1–0.3 mm) smooth blanks; (2) rigid without support, even in such thicknesses; (3) chemically and mechanically stable; (4) available in high purity; (5) highly radiolucent; (6) undamaged by prolonged exposure to x-rays at high intensity; (7) electrically conductive; and (8) free of spectral lines and absorption edges in the useful wavelength region of commercial x-ray fluorescence spectrometers. The method of thin-layer electrodeposition produces films of uniform and accurately controllable thickness. Multiple filters can be plated on the same side of the substrate, provided the metals are plated in order of decreasing nobility. Alternatively, a different metal can be plated on each side of the substrate. The filters may be given a thin coating of lacquer or clear Krylon to prevent corrosion or abrasion. Electroplated films of mercury evaporate under irradiation by intense x-ray beams; this is also prevented by coating with lacquer or Krylon. Fortunately, in many—if not most—filter applications, the filter is subjected only to the less intense secondary beam.

Other uses of filters in x-ray spectrometry include elimination of x-ray tube target lines (Section 4.4.1); removal of primary-beam continuum to provide a substantially monochromatic target-line excitation source (Sections 8.2.2.2 and 8.3.1); improvement of peak-to-background ratio (Section 4.4.2); combination emission–absorption methods (Section 14.8.1); exclusion of radiation from radioactive specimens (Section 17.4); reduction of analyte-line intensity to preclude coincidence losses (Section 4.4.1); and reduction of spectral-line interference (Section 13.3.3.3a).

8.3.3. Selective Detection

Selective detection, or "gas discrimination," takes advantage of the dependence of quantum efficiency of the detector (Figure 6.17) on wavelength to realize high detection efficiency for the analyte line—or low efficiency for interfering lines. The method is somewhat inconvenient to apply to specific spectral lines, but may be used in special cases. Gas discrimination is most effective in the long-wavelength region. Helium, neon, methane, and propane detector gases are relatively efficient at these wavelengths, but have very low absorption for higher-order, shorter-wavelength lines of elements of higher atomic number.

8.3.4. Modulated Excitation

The principle and instrumentation of this method (*V17*) are illustrated in Figure 8.28. The x-ray tube is energized by a pure direct current of potential

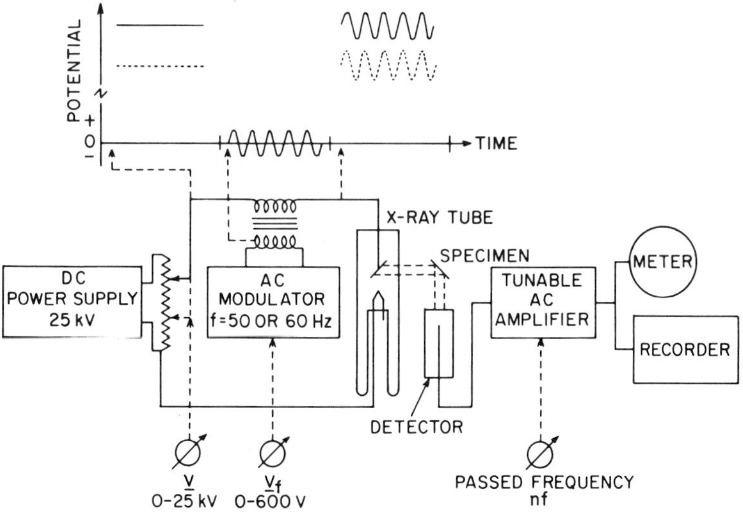

FIG. 8.28. Method of modulated excitation.

V variable from, say, 0 to 25 kV and modulated with a sinusoidal alternating potential V_f of ~600 V and having frequency f the same as the ac line (50 or 60 Hz). For any selected dc potential V, the modulated potential applied to the x-ray tube varies between $V \pm V_f$ at frequency f, that is, between $V \pm 600$ V at 50 or 60 Hz. The wave forms at the top of the figure show the potentials at three places in the apparatus.

If V is set at the analyte excitation potential, the analyte is excited during the positive half-cycles of the modulation, but not during the negative half-cycles. Thus analyte-line radiation is emitted by the specimen in pulses at frequency f and, due to the nonlinearity of the excitation, also at harmonics nf. If now V is varied continuously from 0 to 25 kV at a rate of 0.5–3.5 kV/min, the detector receives the following components as a function of time.

1. Scattered primary continuum: a direct (unvarying) component from $V - V_f$ (intensity, $\lambda_{\min}$, and $\lambda_{I_{\max}}$ all constant), and an alternating component from $V \pm V_f$ at frequencies f and $2f$ (intensity, $\lambda_{\min}$, and $\lambda_{I_{\max}}$ all alternate between fixed limits).

2. Characteristic secondary emission of all elements present in the specimen having excitation potentials $\lesseqgtr(V + V_f)$: direct components from all elements having excitation potentials $\lesseqgtr(V - V_f)$, and alternating components from elements having excitation potentials between $V \pm V_f$ at frequencies f, $2f$, and, at very low intensity, $3f$.

3. Characteristic secondary radiation of that element (if any) having excitation potential equal to V: a direct component and alternating components having frequencies f, $2f$, and also $3f$ and higher harmonics at substantial intensities.

Thus, if as V varies from 0 to 25 kV, the ac amplifier is set to pass a selected harmonic nf with $n \gtrequal 3$, the recorder traces peaks of lines having excitation potential corresponding to the instantaneous value of V as a function of V. With 600-V modulation potential and excitation potentials up to ~15 kV ($_{37}$Rb K, $_{87}$Fr L), resolution is feasible for lines of the same series from elements differing in atomic number by 2. Peak-to-background ratios vary from 2 to 20 and are favored by low atomic number, high modulation ratio (V_f/V), and high order of the measured harmonic. The peak-to-background ratio is higher at low atomic numbers because the modulation ratio is higher.

The method is also applicable to x-ray absorption analysis (Section 20.1). The primary x-rays pass *through* a thin layer of specimen directly to the detector.

Chapter 9

Laboratory, Automated, and Special X-Ray Spectrometers

9.1. INTRODUCTION

In earlier chapters, we have considered the principles of x-ray spectrometry and the individual components that comprise the instrument. In this chapter, we consider the integration of these components into complete x-ray spectrometer instrument systems. This chapter is limited to wavelength-dispersive (crystal) spectrometers. Energy-dispersive spectrometers are discussed in Section 8.2.2.

By way of a summary of the discussions of components in Chapters 4–6, Figure 9.1 shows the useful spectral regions of the principal x-ray tube targets, collimators, radiation-path media, and detectors. The figure also serves as a convenient guide to their selection. The useful spectral regions of analyzer crystals are shown in Appendix 10 (Figure A10.1).

An x-ray spectrometer slide rule is available commercially ("Spectro-Rule," Somar Laboratories, Inc., New York) (*S42*) that aids in the selection of instrument components and operating conditions. The slide rule indicates or permits calculation of the following: (1) wavelengths of the first-order (λ) and second-order (2λ) $K\alpha$, $K\beta$, $L\alpha_1$, and $L\beta_1$ lines for all elements down to atomic number 6 (carbon), and the K and LIII absorption edges for all elements down to atomic number 3 (lithium); (2) selection of optimal target (chromium, molybdenum, silver, tungsten, platinum, or gold) for a specified analyte line, or selection of optimal line ($K\alpha$ or $L\alpha_1$) for a specified target; the selection can be made on the basis of continuum or principal target

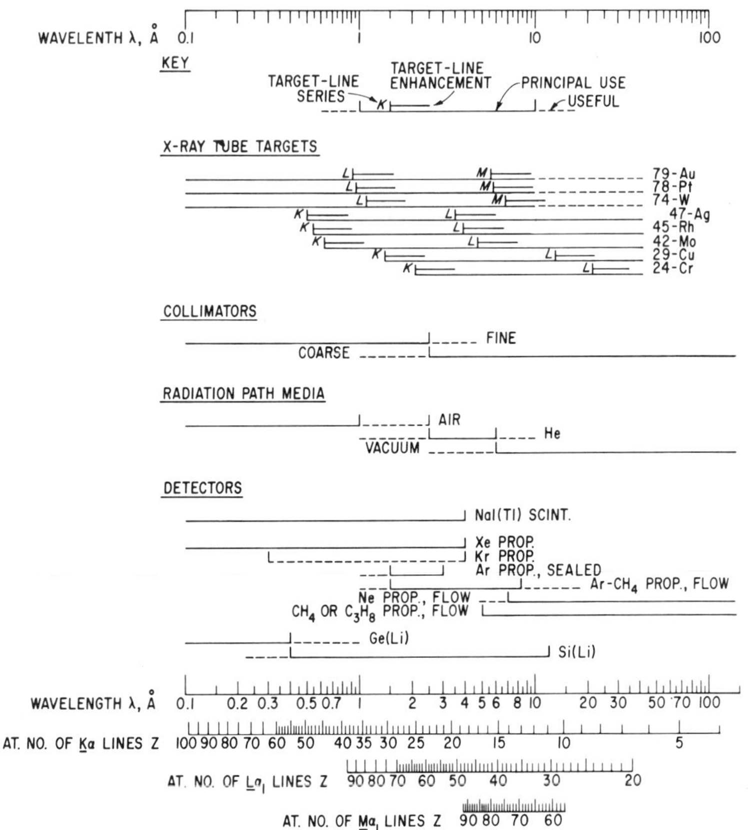

FIG. 9.1. Useful spectral regions of the most commonly used x-ray tube targets, collimators, radiation-path media, and detectors. A similar chart for analyzer crystals and multilayer Langmuir–Blodgett films is given in Appendix 10 (Figure A10.1). More detailed information on targets is given in Table 4.1 and Figure 4.10.

lines; (3) minimal and optimal operating potentials (kV) for excitation of a specified analyte line; the optimal calculation can be made on the basis of 3.5 times the excitation potential or of placing the continuum "hump" on the short-wavelength side of the analyte absorption edge; (4) selection of optimal crystal [topaz, LiF(200), NaCl, quartz(101), PET, EDDT, ADP, gypsum, KHP, or lead stearate] for a specified analyte line; (5) angle 2θ for first- and second-order reflection of a specified line ($K\alpha$, $K\beta$, $L\alpha_1$, or $L\beta_1$) of a specified analyte for a specified crystal; (6) optimal detector (scintillation or flow) for a specified analyte line; (7) identification and 2θ angles of prospective interfering lines (first- or higher-order $K\alpha$, $K\beta$, $L\alpha_1$, or $L\beta_1$) for a specified analyte line; (8) relative intensities of the continuum for each of the six targets listed in item 2 above; and (9) x-ray tube operating potential for potential discrimination of higher-order lines.

The Somar calculator has the form of a two-sided $17\frac{3}{4} \times 3\frac{5}{8}$-inch rectangular slide rule having a 1-inch-wide slide and a hairline sliding cursor, and provided with a rigid, hinged-top case with detailed instructions in the lid.

Another calculator is available from Severn Science, Ltd., Dursley, Gloucester, U. K. (*A2*). This calculator has the form of a circular slide rule consisting of two transparent circular dials 14 inches in diameter on a white cursor 14 inches square. It permits rapid identification of: (1) all principal *K*, *L*, and *M* lines, (2) in orders 1–10, (3) in the wavelength region 0.1–50 Å, (4) for elements 6–95 (carbon to americium), (5) for 13 crystals [topaz, LiF(220), LiF(200), NaCl, calcite, quartz(101), quartz(100), PET, EDDT, ADP, gypsum, mica(002), and KHP]. Although the calculator is actually marked only for these 13 crystals, it is applicable for any crystal having $2d \leq 50$ Å.

An x-ray spectrometer may be a standard or custom-built commercial instrument, or it may be wholly designed and built in the laboratory, or assembled from commercial mechanical and electronic components. X-ray spectrometers may be classified in many ways, principally the following.

The *intended application* may be for the physics laboratory, analytical chemistry laboratory, production control, process control, or portable/mobile field use.

The *mode of operation* may be manual, semiautomatic, or automatic; and automatic instruments may be further classified as sequential or simultaneous.

The *analytical mode* may be: (1) scanning, in which fixed samples are analyzed for each element sequentially; (2) multichannel, in which fixed samples are analyzed for all elements simultaneously; or (3) on-stream, in which moving or flowing samples are analyzed repetitively, either sequentially or simultaneously.

The *excitation source* may be an x-ray tube, electron gun, proton or other ion source, or radioactive isotope; and the x-ray tube power may be full-wave or constant-potential and have maximum potential of 50, 60, 75, or 100 kV.

The *optics* may be wavelength-dispersive or energy-dispersive ("nondispersive"), and the former may be further classified as nonfocusing (flat-crystal), semi- or full-focusing (curved-crystal), or double-crystal. The instrument may have single-channel, two-channel, or multichannel optical systems. The goniometer may have a horizontal or vertical plane, and may or may not have "inverted" optics; or it may be possible to orient the goniometer in any plane.

The *readout* may have many forms, including panel meter, strip chart, electronic digital display, printed paper tape or typewriter copy, punched paper tape, and photographic film. The data itself may be analyte-line intensity, or the intensity data for samples and standards may be entered into a computer and analyte concentration read out directly.

Many of these criteria and instrument types are considered in other chapters. The others are defined and discussed in this chapter, which is devoted to commercially available standard, automated, and special x-ray emission spectrometers.

Jenkins and de Vries (*49*) give a list of 10 "western world" manufacturers of x-ray spectrometers and the types of instrument they provide, and a list of the approximate cost of several types of instruments.

9.2. LABORATORY X-RAY SPECTROMETERS

9.2.1. General

Laboratory x-ray spectrometers are usually designed for application in physics or analytical chemistry. Physics laboratory instruments are designed for the greatest attainable precision and accuracy in measurement of x-ray wavelengths for applications such as cataloging of x-ray spectra and evaluation of atomic energy levels. Speed and convenience of operation are of secondary importance.

Analytical laboratory instruments are designed for use in chemical analysis and materials characterization in their broader senses. They have the following features: (1) speed and convenience of operation; (2) rugged construction to withstand continuous operation at the hands of relatively unskilled personnel; (3) favorable balance of intensity and resolution; (4) high precision and accuracy in setting wavelength (2θ) and measuring intensity; and (5) versatility of specimen form and presentation to permit application to a wide variety of analytical problems arising in science, engineering, technology, geology, medicine, etc. It is these analytical laboratory instruments that are the principal concern of this book.

All the original commercial analytical laboratory x-ray spectrometers were basically modified diffractometers. The goniometer was fitted with collimators, analyzer crystal, and a boxlike specimen compartment, which received a specimen holder having the general form of a drawer. Most manufacturers still offer instruments of this type. However, modern instrumentation is designed specifically for x-ray spectrometry; the goniometer is concealed and barely recognizable, and the specimen chamber has the form of a multiunit turret or carousel. The Philips Universal Vacuum Spectrometer represented an intermediate phase in which the x-ray tube, four-position specimen turret, collimators, crystal, dual flow-proportional and scintillation detector, and vacuum enclosure all formed a self-contained unit. The standard Philips upright goniometer provided only the θ–2θ drive. The General Electric SPG-3 Cauchois and SPG-5 vacuum spectrometers are examples of early laboratory instruments representing a complete departure from the former basic design.

9.2.2. Instrument Arrangements

The newer instrument designs are advantageous for laboratories where the principal work load is analysis of solids, briquets, powders, and liquids. However, for laboratories required to deal with a wide variety of specimens of all forms, quantities, sizes, and shapes, the older design permits a greater versatility. Moreover, these instruments are more conveniently arranged for modes of operation other than conventional secondary-emission spectrometry. These instruments have two principal forms—those having a horizontal goniometer plane (vertical axis), and those having a vertical goniometer plane (horizontal axis).

Figures 9.2 and 9.3 show various arrangements of instruments having horizontal and vertical goniometer planes, respectively. In either figure, arrangement *A* represents conventional geometry for x-ray secondary-emission spectrometry. The x-ray tube is rotated to place the window in position *W*1, where the primary beam is directed downward into a specimen compartment having the specimen plane in position 2. It is evident that horizontal-plane goniometers (Figure 9.2) have inclined specimen planes, whereas vertical-plane goniometers (Figure 9.3) have horizontal specimen planes. This arrangement can also be used for other x-ray methods, as described in the following paragraphs.

Selected-area analysis (Chapter 19) is done by placing apertures *PA* and *SA* in the primary and secondary beams, respectively.

All three of the principal absorption methods (Section 20.1) are feasible; absorption specimens are placed at position 5 or 6. Polychromatic absorptiometry is done by placing a very low-Z scatter target in position 2 and setting the goniometer at 0° 2θ. Monochromatic absorptiometry is done by placing at position 2 a secondary-emission target having a strong spectral

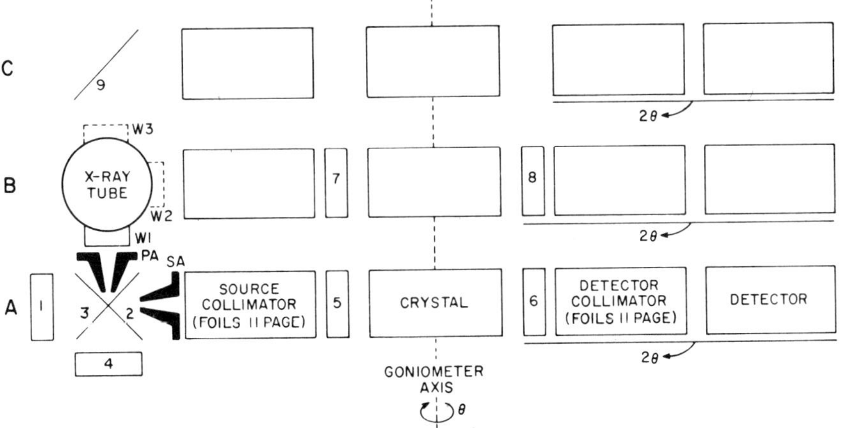

FIG. 9.2. Various arrangements of an x-ray spectrogoniometer having a horizontal plane (vertical axis).

line at the required wavelength and setting 2θ to diffract and detect that wavelength. Absorption-edge spectrometry is done the same way, except that the secondary target is selected to provide two wavelengths bracketing the analyte absorption edge—or else two targets are used in turn—and 2θ is set at each wavelength alternately.

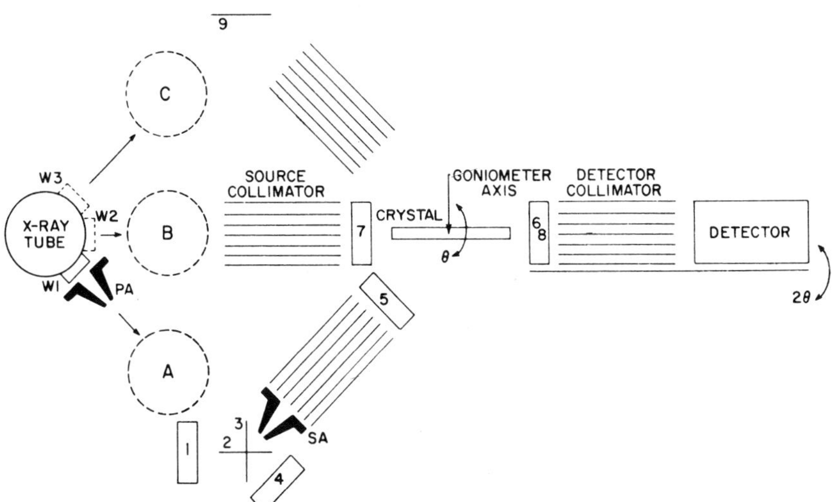

FIG. 9.3. Various arrangements of an x-ray spectrogoniometer having a vertical plane (horizontal axis).

Polychromatic contact microradiography is done by exposing the camera to the primary beam at position 2 or 4. Monochromatic contact microradiography is done by placing the camera at 5 and exposing it to the spectrum of a secondary target at 2, or the specimen drawer can be modified and the camera and secondary target placed at 1 and 3, respectively. Contact microradiography is discussed in more detail in Section 20.1.5.

Arrangement *C* (Figures 9.2 and 9.3) represents "inverted" geometry for x-ray secondary-emission spectrometry. The x-ray tube is rotated to place the window in position *W*3, where the primary beam is directed upward into a specimen compartment having the specimen plane in position 9. Here again, as in arrangement *A*, horizontal- and vertical-plane goniometers have inclined and horizontal specimen planes, respectively. The inverted horizontal plane is advantageous for liquid and loose-powder specimens. The geometry in *C* permits analysis of very large specimens (see below). The selected-area, absorption, and microradiography techniques feasible with arrangement *A* are also feasible with *C*.

Arrangement *B* represents the geometry of a diffractometer, although the components may also be those of a spectrometer. The x-ray tube window, at *W*2, is directed into the optical system. This arrangement can be used for absorption methods that use the primary beam as the source of incident x-rays. The absorption specimens are placed at position 7 or 8. For polychromatic absorptiometry, the goniometer is set at $0° \ 2\theta$, for monochromatic absorptiometry at 2θ to diffract and detect the appropriate wavelength. For absorption-edge spectrometry, the goniometer is set alternately at the 2θ angles for the wavelengths bracketing the analyte absorption edge.

It is evident that if a horizontal-plane goniometer having arrangement *A* is to be converted to arrangement *B* or *C*, the source collimator, crystal stage, and collimator–detector arm must be elevated to be in line with the x-ray beam to be measured.

Some other nonconventional modes of operation are discussed in Chapters 8 and 20.

9.2.3. Accessories

Instruments having the geometry shown in *A* of Figures 9.2 and 9.3 are converted to "inverted" geometry (*C* in the figures) by rotation of the x-ray tube about its axis to move the window from position *W*1 to *W*3 and elevation of the collimators, crystal stage, and detector on spacers to align them with the new specimen plane. On some instruments, the original specimen compartment can also be mounted on a bracket at an elevated inverted position above the x-ray tube window. However, more often, a multiple-specimen turret or large-specimen holder is provided.

The specimen drawer can be replaced by a *multiple-specimen accessory* that holds two to as many as 60 specimens in a turret arrangement that permits indexing each specimen in turn into measurement position. Such accessories may be designed for *continuous loading* so that new specimens may be added as measured ones are indexed out of position. Provision may be made to rotate the specimen in the measured position. Multiple-specimen accessories can be made for vacuum instruments, in which case it is desirable to have an air-lock prepump chamber in which the next specimen to be measured may be prepumped prior to indexing into the measurement position.

With conventional geometry, the specimen size is limited by the space between the x-ray tube window and goniometer base or instrument table top. With inverted geometry, there is no such limitation. A *large-specimen holder* consists of a stage having a specimen-mask window and hinged, lead-lined lid; it accommodates specimens up to perhaps 10 inches square and 4 inches thick. If the lid is removed and other provision made for x-ray shielding, even larger specimens and parts can be accommodated. Provision can be made to mount the standard specimen drawer—with any of its accessories—in the large-specimen holder.

The crystal stage may be replaced with a *crystal changer* having two or more crystals mounted on a turret arrangement so that each may be indexed into position as required. Each crystal is adjusted to the goniometer 2θ scale when installed and thereafter indexed into position without readjustment.

A flat-crystal spectrometer can be converted into a semifocusing curved-crystal instrument without any modification other than replacement of the flat crystal with a fixed-radius curved crystal and removal of the Soller collimators. It is usually preferable to place at the detector window a simple slit parallel to the goniometer axis. It is preferable that the width of this slit be variable, either by means of adjustable jaws or by provision of a set of slits of different widths; these slits may be separate or cut in a common movable slide. Such an arrangement is optimal at only one 2θ angle, but if this optimal angle is $\sim 50°$ 2θ, and if some line broadening can be tolerated at high and low 2θ angles, one crystal may suffice for the entire 2θ range. However, for near-optimal performance over the full 2θ range, two or three crystals having different radii are required. Such a set of crystals can be mounted on a crystal changer. Full-focusing curved-crystal accessories provide for continuous variation of crystal radius or of specimen–crystal–detector distances to maintain optimal focusing as 2θ is varied. Curved-crystal accessories of either type require primary or secondary apertures (*PA* and *SA* in Figures 9.2 and 9.3) when used with extensive specimens. They may be used with or without apertures for very small-area or filamentary specimens. Curved-crystal accessories are discussed more fully in Chapter 5 and Section 19.2.3 (Figure 19.9).

Helium and vacuum accessories are described in Chapter 5. Specimen rotators for drawer-type specimen holders are described in Chapter 16 (*B38*, *Q1*, *T1*). The newer turret or carousel holders have built-in rotators, which may be energized by a panel switch. Various specimen cells and specimen-presentation systems are described in Chapters 16 and 17. Selected-area accessories are described in Chapter 19.

A *step-scanner accessory* automatically advances the goniometer in small, equal, preset increments of 2θ and scales the intensity at each position for a preset time or count. The device is useful for measurement of integrated intensity and for accurate location of the 2θ angle of a peak.

Some goniometers have provision for moving rapidly from line to line when two or more lines in widely separated 2θ regions are to be measured on each of several specimens. For example, if Mo $K\alpha$ ($\sim 20°\ 2\theta$ for a LiF crystal) and Cr $K\alpha$ ($\sim 69°\ 2\theta$) are to be measured, it is inconvenient to have to "crank" the goniometer between these lines for each specimen. In some instruments, the 2θ drive gear can be disengaged to permit rapid manual movement of the goniometer to within a degree of the specified line. The gear is then engaged and 2θ set accurately. In other instruments, a high-speed "slewing" gear permits rapid motor-drive of the goniometer to the vicinity of the line (Section 9.3.4). The conventional gear is then engaged as before. In one commercial instrument, the slewing speed can be as high as $600°\ 2\theta/\text{min}$.

Finally, the spectrogoniometer may be fitted with accessories to reduce instrumental error by provision of *environmental control or correction* of certain critical components as follows: crystal temperature (Section 5.4.2.5); gas pressure in the flow-proportional counter (Section 8.1.8.2b); detector temperature (Section 8.1.8.2b); and variations in atmospheric density (Section 5.3).

9.3. AUTOMATED X-RAY SPECTROMETERS

9.3.1. General

Suppose that a manual, single-channel laboratory spectrometer of the type considered so far throughout this book is used for production control. It would be required to analyze large numbers of similar multicomponent samples submitted continuously throughout the day, and, in such cases, it is common for results to be required very soon after samples are submitted. Even instruments used for incoming inspection and lot approval are required to make large numbers of such analyses from time to time, perhaps with somewhat less urgency. In such analyses, if the analytes have a substantial range of atomic number, no one set of components or conditions is likely to give optimal—or even acceptable—results for all of them. Sometimes, the same components may be used for the entire multicomponent analysis with a certain amount of sacrifice of sensitivity and statistical precision; otherwise, one or more components must be changed during the course of the analysis, with substantial loss of time. In multicomponent analyses, many individual settings may be required for each analyte and perhaps for one or more internal standards—kV, mA, 2θ, preset count or time, pulse-height selector, etc.—and the accumulated count or lapsed time must be read and recorded. For large numbers of multicomponent samples, the number of settings and readings becomes very large, and, aside from the tedium, the opportunity for error increases. If the specimen compartment can accommodate only one specimen and there is no provision for continuous loading, the instrument is completely idle while one sample is removed and the next one loaded. For large numbers of samples, the cumulative loss of time can be substantial.

Consequently, it is not surprising that ever since the first commercial instruments were introduced, modifications have been made to facilitate all these operations. The first innovations were the multiple-specimen changer, which permits indexing a series of preloaded specimens into the measuring position, and the continuous specimen loader, which permits loading the next specimen while one is being measured. The dual detector, crystal changer, and dual-target tube permit convenient interchange of these components without substantial interruption of a series of multielement analyses, and the printer automatically reads out scaler–timer data. An entirely different approach was the introduction of multichannel instruments having two or more complete optical systems (Section 5.5.1). These were particularly useful for simultaneous measurement of analyte and internal-standard lines. These modifications are the direct antecedents of the sophisticated semiautomatic and fully automatic instruments available today. The components of automated instruments are essentially the same as those in the laboratory instruments, except, of course, for the electronic and mechanical automatic program and control components.

Excepting simultaneous spectrometers (Section 9.3.3), automatic x-ray spectrometers are not necessarily faster than manual ones: They must perform the same operations as a human operator. Moreover, automatic spectrometers are not necessarily more accurate than manual ones: Accuracy still depends largely on the standards and calibration function. Rather, the benefits of automatic instruments lie in their reduction of operational instrumental error and operator error, and in freeing the operator for other work while the analyses are in progress.

Kemp (*251*) classifies automated x-ray spectrometers as production-control or process-control instruments. Production-control instruments are designed for rapid, accurate, routine, repetitive analysis of the same types of samples for the same elements day after day, and for reduced dependence on the attendance and training of the operator. The precision and accuracy are comparable with those of a laboratory instrument. The equipment is usually installed in a central laboratory.

Process-control instruments (*K13*, *M23*, *M62*) are designed for continuous, on-line, largely unattended analysis of a product during some phase of its processing, or of a by-product, residual or waste material, or other material that can indicate the condition of the process. The output is read out under continuous observation, and any changes may be used to guide adjustment of the process. The ultimate objective of the process-control instrument is to use the output to automatically control the process. This has been realized, at first only for some simple processes, such as plating of metal sheet or recovery of a single element from a slurry of powdered ore. However, today instruments are in use that can analyze for up to nine elements in up to 21 "streams" in a complex process. If no material in the process has suitable form, sample preparation is required, such as finishing, grinding, drying, briqueting, dilution, fluxing, or addition of an internal standard; this preparation must also be automated. It is fair to state that the principal problem in design of process-control equipment is the design of reliable dynamic sample-presentation systems. The most common specimen forms are continuous strip, liquids and solutions, slurries, and briquets.

Process-control instruments may be classified in four groups: (1) converted laboratory or production-control instruments fitted with special dynamic sample-presentation systems for continuous strip, slurries, liquids, etc.; (2) commercial instruments designed specifically for a single, widely used industrial process; the Philips plating thickness gage is of this type; (3) commercial basic instruments designed for process control, but having a built-in flexibility for relatively convenient adaptability to a variety of specific processes; the Applied Research Laboratories Quantrol, General Electric XEG, and Philips Continuous On-Stream Process Analyzer are of this type; and (4) instruments custom-designed for the customer's application. Most manufacturers of x-ray spectrometers will undertake the design, construction, and installation of such instruments, and probably most of the process-control instruments in use today are of this category. Although many process-control x-ray spectrometers are in use, the field must still be regarded as being in an early phase of its development. No further consideration is given to process-control instruments in this book.

A situation intermediate between production and process control is provided by processes that do not require continuous analysis. In such cases, an automated process for sampling at suitable intervals may be established and a production-control instrument used.

9.3.2. Sequential Automatic Spectrometers

A sequential automatic x-ray spectrometer is basically an automated single-channel instrument programmed to index automatically to a preselected series of analyte lines, stopping at each to accumulate a preset count from the sample and from a standard. The instrument is programmed to automatically select components and conditions optimal for each analyte line. This type of instrument greatly expedites routine analyses. The operator is relieved of the tedium of making the many instrument settings and readings required for multielement analyses of large numbers of routine samples, and the opportunity for error is minimized.

The sequential x-ray spectrometers presently available commercially include the Applied Research Laboratories (ARL) MXQ, Diano XRD-704 and XRD-710, Hilger Fluoroprint, Japan Electron-Optics Laboratories (JEOL) JSX-60S$_4$ [*sic*], Philips PW-1450 (*W24*), Rigaku Geigerflex, Siemens SRS-1, and Solartron XZ-1030 (*L44*). The Philips instrument is the successor to the first commercial automatic sequential spectrometer, the Philips Autrometer (*C67*).

All these instruments provide for automatic selection of a number of the following components and conditions for each programmed analyte, but none provides for all. Of course, any component or parameter for which selection is not provided is the same for all analytes.

1. *Specimen*: Various instruments accommodate 2–30 specimens, which are automatically indexed into position in turn by means of a carousel arrangement.
2. *Specimen mask area*: Choice of two or three window sizes.
3. *Specimen motion*: Fixed or rotated.
4. *Element*: The goniometer automatically indexes to up to 15–24 2θ angles, depending on the instrument. In one instrument, the optimal 2θ

angle for, say, Al $K\alpha$ can be programmed for both the element and aluminum oxide to allow for wavelength shift due to chemical effects.

5. *X-ray tube target*: Tungsten or chromium, platinum or chromium, or molybdenum or chromium (only with a dual-target tube).

6. *Primary-beam filter*: In or out; if in, choice of up to five to remove target lines and/or to reduce analyte-line intensity and thereby preclude coincidence loss without changing x-ray tube operating conditions.

7. *X-ray tube potential*: A selection of preset values up to 50, 60, 75, or 100 kV, depending on the instrument.

8. *X-ray tube current*: A selection of preset values up to 50, 75, or 100 mA, depending on the instrument.

9. *Collimator*: Coarse, medium, or fine.

10. *Crystal*: Choice of up to six.

11. *Diffraction order*: The instrument can be instructed to measure the first- or second-order line.

12. *Path*: Air, helium, or vacuum.

13. *Detector*: Most instruments have a dual scintillation and gas-flow proportional counter, and the output can be taken from either or both.

14. *Delay time*: The instrument may be set to delay a preset time before making a measurement each time a new specimen is indexed into position to ensure adequate helium flushing or vacuum and/or to allow primary-beam equilibration after changing x-ray tube kV and mA.

15. *Amplifier gain*: A selection of several preset values.

16. *Automatic pulse-height selection*: In or out, and window width.

17. *Preset count or preset time*: Which mode, and a selection of values. For each analyte, that count (or time) is programmed that gives the required statistical precision.

18. *Repetition*: The instrument can be set to repeat a measurement a specified number of times.

Some instruments also provide for selection of analytical measurement modes as follows:

1. *Specimen or element sequence*: The instrument can be programmed to measure all analytes on the same specimen, then advance to the next specimen, or to measure the same analyte on all specimens, then advance to the next analyte.

2. *Peak or profile mode*: The instrument can be programed to "slew" to the peak 2θ, scale and print out or store the peak intensity, then slew to the next peak, etc.; alternatively, it can be made to slew to a specified low-2θ edge of the peak, step-scan or continuously scan through the peak

to a specified high-2θ edge of the peak, print out or store the profile or integrated count, then slew to the low-2θ side of the next peak, etc.

These spectrometers can be operated in any of three modes as follows.

1. *Manual.* In this mode, the instrument becomes substantially a standard single-channel scanning spectrometer useful for qualitative analysis and for quantitative analysis of individual samples.

2. *Semiautomatic.* For each sample, the analytes are selected in turn, and for each analyte, the components and conditions are selected—all by manual operation of pushbuttons and multiposition selector switches on the control panel. As the operator makes each setting, the instrument indicates that it has executed its "instructions" correctly. This mode is used for analysis of large numbers of nonroutine samples for which no program has been established.

3. *Fully Automatic.* The samples are indexed into position and, on each sample, the specified analytes are measured, each with its specified components and conditions—all automatically. This mode is used for analysis of large numbers of routine samples that warrant establishment of a program. The program may be set manually, or by means of a patchboard, switchboard, prewired plug-in casette, punched card, or punched tape specific for a certain series of analytes in a certain type of sample. Alternatively, the program can be stored in the computer memory (see below) and selected by addressing it on the computer input–output typewriter.

The spectrometer output may be "interfaced" with a computer, which may be "dedicated," that is, a permanent complement to the spectrometer, or a central "time-shared" computer. The computer converts intensity data measured from the samples directly to concentration by means of calibration functions automatically established from intensity data measured from the standards. The calibration may be based on a single standard, a linear function derived from two or more standards, or even a nonlinear function divided into regions, each considered to be linear.

The equipment is usually designed so that the basic manual instrument can be purchased first, then later expanded to a semiautomatic, then to a fully automatic, instrument by addition of controller and programming units, respectively. Finally, the computer may be added.

The output may be read out on a printer, a digital display, or the computer typewriter. A strip-chart recorder may be provided for ratemeter readout with manual operation. The output data for each analyte in each sample may be read out in any of three forms: (1) absolute accumulated count (or counting time); (2) ratio of accumulated count (or time) for the sample and standard; and (3) concentration.

The specimens may be solids, briquets, powders, slurries, or liquids. Provision is usually made for continuous loading—through an air lock if helium or vacuum path is used. Most sequential spectrometers are useful for all elements down to atomic number 9 (fluorine).

Several specific instruments have noteworthy features.

The *ARL model MXQ* has two full-focusing curved-crystal scanning channels, each with a two-crystal changer and a nondispersive ratio channel, which monitors the primary x-ray beam. The instrument can be operated manually, semiautomatically, or automatically, and, in automatic operation, the two channels can be operated serially or in parallel. In serial operation, one channel can be fitted and programmed for several lines in the short-wavelength region, the other for several lines in the long-wavelength region; then the first channel sequentially measures several analytes down to potassium, whereupon the other channel continues the analysis down to fluorine. In parallel operation, the two channels simultaneously measure two wavelengths from the specimen; the second wavelength may be a second analyte, an internal-standard line, or a scattered continuum wavelength or target line. The MXQ has features of both simultaneous and sequential instruments. The instrument can be purchased with a single channel and the other added later.

The *General Electric XRD-6SP* (*D7*), no longer available, operated unattended while up to 11 analytes were automatically measured on up to 90 samples 1/8–3/16 inch thick, or 45 samples 1/4–3/8 inch thick. The samples were placed in a vertical tower, which indexed downward, one sample at a time, discharging each sample onto an inclined belt, which, in turn, conveyed it to the specimen compartment. All measurements were made in air at a constant preset potential up to 75 kV, but any of the following parameters could be programmed automatically for each analyte line: (1) tungsten or chromium target; (2) any of four preset values of x-ray tube current up to 100 mA; (3) any of four crystals; (4) output from scintillation or flow-proportional counter or both; (5) any of six preset counting times; (6) any of six preset values of amplifier gain; (7) pulse-height selector in or out; (8) repetition of the measurement—single, duplicate, or quadruplicate. A digital printer printed sample number, channel (2θ) number, counting time; and accumulated count for each of the one, two, or four measurements. If the sample mechanism jammed, a loud alarm sounded, and if the instrument received no attention within a preset time, it shut off.

The instrument permitted unattended overnight operation. The XRD-6SP could also function with helium path, but only four samples could be programmed.

The *Hilger Fluoroprint* has a nondispersive primary-beam monitor channel, against which all analyte-line measurements are ratioed.

The *Solartron XZ-1030* (*L44*), like the ARL MXQ, has two identical dispersive channels for sample and standard, respectively. It also provides for setting of pulse-height selector window width for each analyte line.

The *Siemens SRS-1* has several notable features:

1. The instrument can be operated in two sequential modes. In the sample-sequential (conventional) mode, the instrument measures all analytes on one sample before automatically indexing to the next sample. In the element-sequential mode, the instrument measures one analyte on all samples before automatically indexing to the next analyte.

2. The system can be programmed to automatically periodically calibrate against a standard—or two or more. In this way, the instrument periodically checks and revises its calibration data, and uses the revised data to derive analyte concentration.

3. While a programmed analysis is in progress, if an urgent sample of a different type should require analysis, a "priority interrupt" feature permits the instrument to be switched to manual or semiautomatic mode, the special sample to be analyzed, then the automatic program resumed.

4. Several features are included to ensure maximum precision. Precise specimen positioning is provided. The spectrometer compartment temperature is controlled to $\pm 0.5^\circ C$ to keep the crystal temperature constant, to reduce scintillation counter noise, and to reduce change in gas amplification due to temperature. The gas-flow counter discharges into a constant-pressure manostat to reduce change in gas amplification due to pressure.

Sequential x-ray spectrometers, compared with simultaneous spectrometers (Section 9.3.3), have the advantages that: (1) they are much more compact because of the absence of a multiplicity of spectrometers arrayed about the specimen; (2) they are economical of collimators, crystals, detectors, and electronic readout components; (3) x-ray tube target, primary-beam filter, kV, and mA can be programmed parameters; (4) the measured intensity for all analytes leaves the specimen in the same direction, so that homogeneity and surface texture are not so critical; and finally (5) because there are no fixed preset spectrometers, they are readily indexed to any specified 2θ angle for measurement of any analyte line, scattered target line, or background. The principal disadvantages are that the indexing mechanism is very complex, use of curved crystals is not convenient, analysis time is much longer, and the advantages of simultaneous measurement of analyte and internal-standard lines, analyte line and background, or all analytes cannot be realized.

9.3.3. Simultaneous Automatic Spectrometers

A simultaneous x-ray spectrometer is basically a number of single-channel instruments, each having its own collimators, crystal, detector, and count scaler or integrator, but having a common specimen and readout. Some early forerunners of modern simultaneous instruments are described in Section 5.5.1. The simultaneous x-ray spectrometers presently available commercially include the Applied Research Laboratories (ARL) VXQ and VPXQ X-Ray Quantometers, Philips PW-1270, Rigaku Simultix, and Siemens MRS-1 and MRS-3.

In the ARL X-Ray Quantometers, most of the channels (monochromators) are fixed, that is, fitted, usually at the factory, with optimal components and set at 2θ for a specified analyte line, and thereafter left undisturbed. However, one or more of the channels may be scanning channels. These are conveniently set to measure elements determined less frequently than those for which the fixed channels are set, thereby precluding the necessity to disturb the fixed channels. The scanning channels can also be used to make 2θ scans for qualitative analysis. Finally, a nondispersive channel is provided to monitor the primary beam. In the ARL instruments, this channel receives x-rays from a separate semipermanent monitor specimen and has a phosphor-phototube detector. The VXQ can have up to nine analytical channels, not counting the nondispersive channel, of which three may be scanners; the VPXQ can have up to 22 channels, of which five may be scanning.

The original ARL X-Ray Quantometers (XIQ and XRQ) used flat crystals and Soller collimators (*H22*, *K9*), but more recent instruments use full-focusing curved-crystal optics and adjustable bilateral source and detector slits. The ARL curved-crystal spectrometer is shown in Figure 5.23 and described in Section 5.6.4. The crystals in these instruments are of the Johansson curved-and-ground type, and eight are available—LiF, NaCl, Ge, SiO_2, PET, EDDT, ADP, and KHP; a curved lead stearate decanoate crystal is also available. The slits are set individually for optimal balance of resolution and sensitivity for the analyte to be determined by that channel. The ARL spectrometers use end-window x-ray tubes, which are particularly well suited to instruments of this type because they occupy no space radial to the specimen plane and therefore permit arrangement of more channels

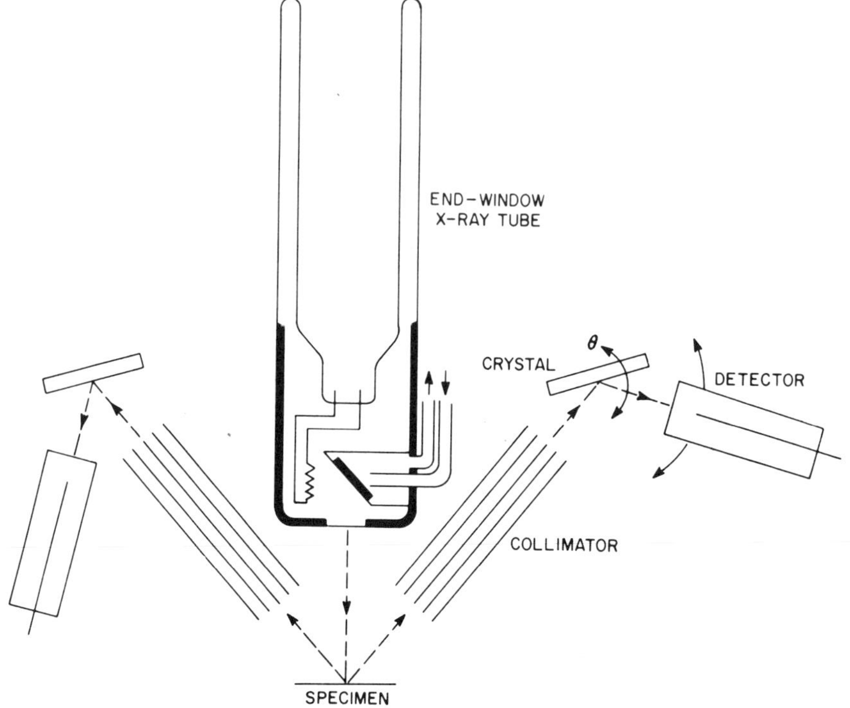

FIG. 9.4. End-window x-ray tube with two x-ray optical channels.

about the sample. Chromium, rhodium, tungsten, and platinum targets are available with ~1/8-mm beryllium windows. Figure 9.4 shows an end-window tube and two optical channels of the flat-crystal type used previously. Present-day instruments also have inverted geometry. The specimens may be solids, briquets, powders, slurries, or liquids, and may have dimensions up to 4-inch diameter and 2-inch thickness. An air lock permits loading of the next sample while one is being measured. Air, helium, and vacuum path can be used. The detectors are sealed proportional counters having aluminum or beryllium windows and xenon, krypton, argon, and neon gas fillings; a gas-flow counter having an ultrathin window is also available.

In operation, the output of the detector in every channel, including the nondispersive one, charges a separate integrating capacitor until the charge in the capacitor in the standard channel (see below) reaches a predetermined value, whereupon all detectors are disconnected from their capacitors. Then the accumulated charge in each analytical capacitor is subjected to high-speed sequential readout at a rate ~2.5 s for each. The standard channel may be the nondispersive one, which, as mentioned earlier, in effect monitors the primary beam. Alternatively, any one of the analytical channels can serve as the standard channel, and the crystal may be set to diffract a wavelength in the continuum, a scattered target line, or the line of an internal-standard element.

A computer can be used with a simultaneous spectrometer, and the output can be read out by the same components and in the same forms as for a sequential spectrometer. A significant feature of the ARL X-Ray Quantometers is that their output can be read out on the consoles of ARL Optical Emission Quantometers.

The ARL VPXQ has a unique modular construction. The basic unit can accommodate up to six channel positions arranged radially about the specimen compartment. Each position receives a metal plate on which can be placed one or two curved-crystal channels. Thus, the basic unit can accommodate up to 12 analytical channels. The single unit can be expanded to a dual unit by addition of a semicircular annex, not unlike a bay window, that can accommodate five more positions—10 more channels. Thus, the VPXQ can measure up to 22 elements simultaneously.

The Philips PW-1270 is similar in principle, but with some distinctive features. The maximum excitation power is 60 kV, 80 mA. Side-window x-ray tubes are used, and gold, tungsten, silver, molybdenum, and chromium targets are available. The analytical channels are of the flat-crystal type. The specimen carousel holds up to 10 specimens, and a 160-unit overnight holder is available. In operation, a scaler in each channel accumulates counts, instead of the capacitor integration in the ARL instruments. The instrument can be operated in any of three modes as follows: (1) absolute—all channels accumulate counts for a preset time; (2) monitor—the analytical channels accumulate counts until the monitor channel has accumulated a preset count; and (3) ratio—analyte-line counts are accumulated from a standard and sample in turn; the standard is measured for the time required to accumulate a preset count; the sample is then measured for the same time, and the sample and standard counts are ratioed (Section 7.2.2.4).

The simultaneous technique has some of the advantages of the sequential and some additional ones. Simultaneous instruments are extremely rugged because of the fixed preset spectrometers and the absence of moving parts. The preset spectrometers can be of the full-focusing curved-crystal type. The components (crystal, detector, etc.) of each channel can be optimized for its specific element. The simultaneous measurement of all analytes permits greater economy of analysis time and cost. Analysis time is limited by the counting or integrating time required to give acceptable statistical precision for the analyte having lowest intensity, plus short loading and readout times. If no element requires more than 1-min counting time, an analysis for 22 elements can require as little as 2 min for both counting and readout. A sequential instrument requires the sum of the individual counting times plus the operation time. Another advantage is that many instrument and specimen variables are compensated by simultaneous measurement of all intensities. Also, it is very convenient to be able to measure analyte and internal-standard lines, and even line and background, simultaneously. Finally, in trace analysis, higher sensitivity and statistical precision can be realized in relatively short time by setting two channels at the same trace analyte line and combining their outputs.

The principal disadvantages of simultaneous spectrometers are that: (1) they are very bulky because of the multiplicity of spectrometers; they become especially cumbersome when provision must be made for large numbers of elements; (2) they require large numbers of expensive components—collimators, crystals, and detectors; (3) the electronic readout system is elaborate and complex; (4) x-ray tube target, primary-beam filter, kV, and mA cannot be programmed parameters; (5) because each spectrometer views the specimen from a different direction, homogeneity and surface topography are critical, and specimen rotation is usually required; and (6) it is inconvenient to change a preset channel to a different spectral line. Because of this last limitation, simultaneous instruments are relatively inflexible and best suited to analysis of very large numbers of similar samples, especially in production control. For example, these instruments are ideally suited for analysis of iron-base alloys in iron foundries and steel mills, copper-base alloys in brass and bronze foundries, briqueted cements in cement factories, and hydrocarbon liquids in petroleum refineries. The flexibility of simultaneous instruments is substantially increased by inclusion of one or more "scanning"—as distinguished from preset fixed—spectrometers that can be used to record 2θ scans and can readily be set to any specified analyte line.

9.3.4. "Slewing" Goniometers

The recording of the x-ray spectrum (wavelength or 2θ scan) for qualitative or semiquantitative analysis, or preparatory to quantitative analysis, may be very time-consuming. At 2° 2θ/min, a 120° scan requires 1 h. One way to speed up the scan is to use short ratemeter time constants (0.4–0.2 s), fast full-scale recorder response times (0.5–0.25 s), and fast chart speeds (4–8 cm/min). By such means, scanning speeds of 8–16° 2θ/min are feasible without loss of resolution or of low-intensity peaks (Section 10.3). Standard recorders must be modified to incorporate these features. Of course, the most rapid way to make a qualitative analysis is to use a Si(Li) detector and multichannel analyzer.

Another means for making fast scans is provided in some automatic instruments. The goniometer can be programmed to scan the *peaks* at normal scanning speed, but pass over the intervals *between* peaks at high speeds—up to 600° 2θ/min. This rapid transit between peaks is termed *slewing* (Section 9.2.3). In some instruments, the operator can preselect the minimum peak intensity above background that will be recorded; for example, the instrument may be set to record only peaks having net intensity 50 counts/s or more. The 2θ scale is established by a pulse counter. The goniometer is started at a preselected 2θ, and the pulse counter marks 0.5° or 1° 2θ intervals on the chart during the scan. Alternatively, recorder chart speed can be varied synchronously with the scanning speed. Slewing may reduce scanning time by a factor of 10 or more. Another use of slewing is given in Section 9.2.3.

9.4. SPECIAL X-RAY SPECTROMETERS

Energy-dispersive x-ray spectrometers having low-power x-ray tube or radioisotope excitation sources are described in Section 8.2.2. Other special x-ray spectrometers are described briefly in the following sections.

9.4.1. Portable Spectrometer

The Pitchford Portaspec (*M34*, *P18*) is a portable x-ray fluorescence spectrometer consisting of two units connected by a 9-ft cable. The power unit weighs 100 lb and houses a 30-kV, 1–4-mA constant-potential supply that operates from the 110-V ac line and uses 220 W at full power. The x-ray unit weighs 23 lb and houses the x-ray tube, specimen compartment, goniometer, detector, ratemeter, and detector-tube power supply. The x-ray tube has a tungsten target and beryllium end-window. The goniometer consists of source and detector collimators, LiF crystal, and an argon-filled detector; the 2θ angle is varied by means of an external knob in the range 13–89°, thereby permitting measurement of all elements down to atomic number 22 (titanium). The ratemeter output is indicated on a panel meter. A goniometer drive motor and strip-chart recorder are available to permit making 2θ scans. A scaler–timer unit is also available. A specimen holder is available for small solids, powders, and liquids, another for bar, rod, and sheet stock. Sensitivity down to 0.01 wt % is claimed for favorable cases. The instrument can be used in the plant and warehouse for identification and sorting of metal stock, chemicals, leaded fuels, etc. With a mobile generator, it can be used in the field for geological and mining applications. The instrument has also been used to indicate the lead content of painted surfaces for investigation of sources of lead poisoning.

9.4.2. Primary Excitation

Standard x-ray tubes are relatively inefficient for excitation of long-wavelength x-ray spectra, so electron excitation (Section 1.6.4) becomes attractive in this region. However, some of the advantage is offset by the increased continuum intensity and resultant higher background. Electron penetration is much shallower than x-ray penetration, so that the effective specimen layer is only a few micrometers at most. This reduces absorption-enhancement effects, but greatly increases surface-texture effects. Also, it is necessary that the surface be representative of the bulk specimen. At the point of impact, the electron beam heats the specimen, possibly expelling volatile constituents, and also causes deposition of carbon from residual vacuum pump oil vapors. These conditions are avoided by use of a large-area electron beam, or by scanning a focused beam over a large area, $\sim$1 cm^2.

An example of a commercial instrument having electron excitation is the "Betaprobe" manufactured by TELSEC Instruments, Oxford, England, and available in the United States from Jarrell-Ash. A stabilized electron gun and an electron-beam scanning system provide an electron beam having current 0.1, 0.2, 0.5, or 1 mA and scanning an area $\sim 1.5 \times 1.5$ cm on the specimen surface. The emitted primary x-rays are intercepted by either of two interchangeable spectrometer units—a 12-channel spectrometer system or a single-channel scanning spectrometer. A metal plate is fitted with up to 12 fixed x-ray optical channels, each consisting of a source collimator, flat crystal, ultrathin-window gas-flow proportional counter, and preamplifier. The multichannel instrument is available in either sequential or simultaneous readout mode. After the counting time, the output of each channel may be applied sequentially to a single readout unit, or 12 such units may be provided for simultaneous operation. Alternatively, the 12-channel assembly can be replaced with a single-channel, flat-crystal, 2θ-scanning spectrometer. For either arrangement, the readout can be by ratemeter and recorder or by scaler and printer or digital display.

The crystals available are LiF, NaCl, KBr, PET, EDDT, and KHP. The instrument is useful for 0.005–100% concentrations of elements down to atomic number 6 (carbon). The advantages derive from the high excitation efficiency and low penetration of electrons, compared with primary x-rays. The efficiency results in high analyte-line intensity, which is especially beneficial for the low-Z elements, and short counting times, typically ~10 s, so that eight elements can be read out in <2 min. The low penetration also reduces absorption-enhancement effects.

Other instruments having direct excitation by an electron gun are described by Fox (*F20*) and by Kimoto and his co-workers (*K15*). Lenard windows provide another approach to primary excitation. The electron beam from an electron gun passes through a thin aluminum or glass window into the air (*F19*, *S37*). The specimen is placed just outside the window and is subjected to primary excitation.

9.4.3. Ultralong-Wavelength Spectrometry

(*25i*, *F15*, *H33–37*, *H60*, *H61*, *P20*)

In x-ray spectrometry, the ultralong-wavelength (ultrasoft) x-ray region extends from 15 to 150 Å, or over about 1–0.1 keV. Its short-wavelength limit lies at the long-wavelength limit of standard commercial x-ray spectrometers. The region contains the $K\alpha$ lines of elements 9 (fluorine, 18.3 Å) to 4 (beryllium, 114 Å); the $L\alpha_1$ lines of elements 26 (iron, 17.6 Å) to 14 (silicon, 136 Å); and the M spectra of elements 57 (lanthanum, $M\alpha$ 14.9 Å) to 37 (rubidium, $M\zeta$ 129 Å).

Formerly, work in the ultralong region, at least beyond ~25 Å, was done on laboratory-designed and built optical-grating spectrometers. These instruments require highly collimated beams from pinhole or slit apertures and very small incidence and takeoff angles. They require critical adjustment and are generally inconvenient to operate. However, ultralong-wavelength accessories are now available commercially for the Philips Universal-Vacuum x-ray spectrometer (*D36*, *D37*) to permit measurement of K spectra down to beryllium and L spectra down to silicon. The accessories include a special x-ray tube power supply, x-ray tube, analyzer crystal, and detector.

The power supply delivers relatively low potential, but high current—up to 20 kV, 500 mA. The x-ray tube is the demountable, continuously pumped Henke tube described in Section 4.2.5.4. The tube mounts in the same position as the conventional tube, and the eight-position specimen turret may be used.

KHP crystals permit measurement of K spectra of elements down to oxygen, but Langmuir–Blodgett multilayer smectic pseudocrystals of heavy-element soaps are required for the longer wavelengths. Compared with gratings, these crystals have several advantages. They can be used with large-area specimens and the Soller-collimator and flat-crystal optics of standard commercial instruments, and they are readily interchanged with conventional flat crystals. They can also be used with line or point sources and curved-crystal optics. The solid angle of acceptance of a grating is small, and the diffracted intensity is comparable with that from a crystal only for a line or point source. For a large-area specimen, the crystal is much more efficient because Soller-collimated x-rays are accepted by a crystal at a Bragg angle that is large compared with the glancing angle of a grating. The principal disadvantage of the crystal is that its resolving power is only 0.2–0.5 that of a grating. However, it must be remembered that the grating requires a special spectrometer, critical alignment, and less convenient operating procedures. Data for KHP and the commonly used multilayer crystals are given in Appendix 10.

The detector is a side-window gas-flow proportional counter. The window consists of ultrathin Formvar supported on a 200-mesh nickel grid having 70% open area and sandwiched between apertured metal plates. The gas may be P10 (90% argon, 10% methane) or neon, but is frequently pure methane or propane. The pressure is 1 atm and the flow rate 0.5 ft^3/h. The gas discharges into a manostat to ensure constant pressure in the active volume. Another type of detector suitable for ultralong-wavelength work is the open multiplier phototube having beryllium–copper photocathode and dynodes (Section 6.6.2.2).

Windows are required on the x-ray tube to isolate the clean high vacuum (10^{-6} torr or less) from the working vacuum in the spectrometer, and on the detector to isolate the detector gas from the spectrometer. Window materials include aluminum leaf dipped in Formvar solution in ethylene dichloride, stretched polypropylene film, Nitrolucid (TM), and ultrathin (1000–2000 Å, 30–60 $\mu g/cm^2$) collodion or Formvar (Section 6.2.1.2). The two ultrathin films are usually used as double layers, each having half the indicated thickness, to eliminate pinholes. Ultrathin films are supported on some kind of grid, such as 200-mesh nickel having 70% open area.

Table 9.1 gives some typical performance data for the modified Philips Universal-Vacuum spectrometer.

Alternatively, the eight-position specimen turret may be replaced with a continuously pumped, self-contained, primary-excitation accessory consist-

TABLE 9.1. Performance of an Ultralong-Wavelength X-Ray Spectrometer[a]

Element	Specimen	Analyte concentration, wt %	Spectral line and wavelength, Å	X-ray tube anode	Crystal[b]	Intensity, counts/s		Minimum detectable concentration,[c] wt %
						Peak	Background	
$_8O$	SiO_2	53.3% O	O K, 23.6	Cu	PbSt	13,297	759	0.03
$_7N$	BN	56.4% N	N K, 31.6	Cu	PbSt	3,840	462	0.11
$_6C$	Graphite	100% C	C K, 44.6	Cu	PbSt	21,681	696	0.04
$_5B$	BN	43.6% B	B K, 67.8	C	PbSt	12,444	108	0.01
$_4Be$	Be foil	100% Be	Be K, 114	C	PbL	1,855	132	0.2

[a] From B. L. Henke (*H37*). Philips Universal-Vacuum X-Ray Spectrometer; Henke demountable x-ray tube: 6 kV, 330 mA; Soller collimators: source, 3.7°, detector, 2.7°; detector: gas-flow proportional counter, methane gas at 1 atm, 250 cm^3/min (0.5 ft^3/h); x-ray tube and detector windows: Formvar, 60 $\mu g/cm^2$, supported on 200-mesh nickel grid, 70% transmission.

[b] PbSt: lead stearate (octadecanoate) multilayer, $2d$ 100.6 Å. PbL: lead lignocerate (tetracosanoate) multilayer, $2d$ 130 Å.

[c] $(0.3C\sqrt{I_B})/(I_P - I_B)$, where I_P and I_B are peak and background intensities, respectively (columns 7 and 8), and C is analyte concentration in the specimen (column 3).

ing of an electron gun and target. The x-rays from the target are directed into the optical system of the Universal-Vacuum spectrometer. This accessory permits analysis of specimens deposited on the target and provides a source of high-intensity, ultralong-wavelength x-rays for absorption measurements or for irradiation of materials. This accessory is not available commercially.

Ultralong-wavelength x-ray spectrometry extends the applicability of x-ray spectrometric analysis from fluorine to beryllium. Because of the long wavelengths, the method is particularly applicable to relatively thin surface layers. The method is extremely valuable for investigation of chemical state and valence-electron band structure because the spectra in the ultralong region are relatively highly influenced by chemical state (Section 11.1.3.4). The chemical effects cause the spectral lines or bands to shift in wavelength, change in profile, or split into multiple bands. Such investigations are of great significance in chemistry, solid-state physics, and semiconductor electronics.

Because ultralong-wavelength x-rays penetrate only a few micrometers of matter, surface contamination of the specimen, crystal, detector window, etc. becomes a serious problem. Even thin films of back-streamed vacuum-pump oil vapor can be detrimental if the equipment is not adequately cold-trapped (*K3*).

More recently, ultralong-wavelength x-ray spectrometry has found application in high-temperature plasma physics and in the study of solar x-rays. In matter at temperatures of the order of millions of degrees, electron and atom energies are of the order of 0.1–1 keV, and the resulting electron accelerations and atom ionizations produce x-rays in the ultralong-wavelength region. Such phenomena occur in stars and in high-temperature plasmas, which are being investigated as a possible source of energy from controlled thermonuclear fusion. Ultralong-wavelength spectrometers are used to evaluate or characterize phenomena and performance in plasmas, and they have been sent up in rockets to measure solar x-rays.

9.5. X-RAY SAFETY AND PROTECTION

It has come to be almost obligatory to include in any book on x-ray practice a section on x-ray safety and protection. Actually, a careful, alert, competent worker who is aware of the potential hazards of x-rays need have little apprehension for his personal safety in working with commercial x-ray spectrometers. These instruments are adequately shielded to confine the x-rays and prevent irradiation of the environment.

The principal sources of possible extraneous x-rays from an x-ray-tube-excited spectrometer are: (1) the interface between the x-ray tube head and the specimen chamber; (2) the specimen chamber itself when specimens are inserted and removed, especially if the x-ray shutter assembly is faulty; (3) the high-potential electron-tube rectifiers, which may radiate through the relatively thin steel of the oil tank on the peak-inverse half-cycles; solid-state rectifiers do not produce x-rays; and (4) the crystal, which diffracts, scatters, and emits x-rays in directions other than that intercepted by the detector collimator. Of course, the extraneous intensity from any source increases with increasing x-ray tube operating potential and current, and is substantially greater in 100-kV instruments than in 50-kV instruments.

The greatest possible radiation hazard in any x-ray generator is the primary beam, particularly close to the x-ray tube window. In even a 50-kV x-ray spectrometer, the primary intensity at the x-ray tube window may be of the order of megaroentgens per minute (see below), and even momentary exposure to it is almost certain to result in a lesion at the irradiated site. The sources of extraneous radiation listed above are likely to have very low intensity. However, an operator taking data for several hours each day may well receive a substantial fraction of his maximum permissible dose (see below) from this source.

Although the radioisotope sources used in energy-dispersive x-ray spectrometry emit radiation at intensities lower by several orders than that of high-power x-ray tubes, their potential hazard is actually substantially greater. An x-ray tube can be turned off when not in use and cannot contaminate the laboratory environment; also, there is nothing to ingest. None of these favorable safety features applies to radioisotope sources. Consequently, such sources must be kept shielded at all times and handled—in the prescribed way—with great care to prevent rupture of their encapsulation and dispersion of the radioactive material. Radioisotopes emit several types of radiation, principally α-, β-, x-, and γ-rays. The penetrating power increases in the order given, but ingestion of β- and, especially, α-emitting isotopes may be extremely hazardous.

The hazard from ionizing radiation in general (α, β, γ, x) is compounded by the fact that such radiation is completely insidious. It is quite possible that one could be receiving a lethal dose of ionizing radiation and be wholly unaware of any sensation. Further, the physiological effects are delayed, and symptoms may not appear for days or even weeks after the exposure. Finally, the physiological effects are cumulative if one receives radiation at a rate faster than that at which the body can repair the physiological damage, that is, faster than the maximum permissible dose rate (see below).

For purposes of radiation health and safety, radiation is measured in terms of *dose* or *dose rate*. The unit of biological radiation exposure or dose is the *rad* (*r*adiation *a*bsorption *d*ose), which is that quantity of ionizing radiation of any kind (α, β, γ, x) that results in absorption of 100 ergs of energy per gram of biological tissue, or equivalent material. The unit of x- or γ-ray dose is the *roentgen* R, which is that quantity of x- or γ-radiation that produces in 1 cm^3 of dry air at 0°C and 760 torrs (1 atm), ions carrying 1 esu of charge, regardless of sign; this is equivalent to 2.08×10^9 ion pairs/cm^3, or 84 ergs. The unit of biological radiation damage is the *rem* (*r*oentgen *e*quivalent, *m*an), which is equivalent to 1 rad of x- or γ-radiation, giving average specific ionization of 100 ion pairs per micrometer in water. Dose rate is given in terms of dose per unit time, for example, milliroentgens per hour (mR/h), millirems per hour (mrem/h). For practical purposes, for x-radiation produced in x-ray spectrometers excited by tubes operating at $\lesssim$100 kV, the doses in roentgens, rems, and rads are substantially equivalent.

The maximum permissible whole-body dose rate is 5 rems/year or 3 rems/quarter (13 weeks); this is substantially equivalent to 5 R/year or 100 mR/week; for a 40-h work week, this is 2.5 mR/h. The maximum permissible accumulated life-time whole-body dose is $5(y - 18)$ R, where y is age in years and must be >18. Certain radiologically insensitive portions of the body may receive substantially greater doses; for example, 30 R/year for the skin, 75 R/year for hands, forearms, feet, ankles, neck, and head (excluding the eyes). Background radiation at sea level from natural radioactivity in the environment and from cosmic radiation is typically 0.01–0.1 mR/h, say, 0.05 mR/h; this is $0.05 \times 24 = 1.2$ mR/day, $1.2 \times 365 \approx 500$ mR/year.

All evidence indicates that if a healthy normal adult receives x-radiation at a dose rate no greater than this, his body can easily repair the minor physiological damage. However, every effort should be made to reduce the dose rate to as little as possible, and, pursuant to this objective, certain simple precautions are prudent.

The x-ray spectrometer should be thoroughly monitored with a radiation-survey meter when it is installed and at routine 3–6-month intervals thereafter. In x-ray tube-excited spectrometers, particular attention should be given to the regions of the sources of extraneous radiation given above and to the place at which the operator sits when taking data. The region of the x-ray tube head and specimen compartment should also be monitored each time the x-ray tube is changed or the compartment disturbed. All monitoring should be done with the x-ray tube operating at maximum potential (50 or 100 kV) and maximum permissible current, and with an emission specimen having a strong spectral line of short wavelength—molybdenum, for example. The survey meter should be a model having good response for x-rays having energy up to 50 keV (or 100 keV) and should be properly calibrated to read milliroentgens per hour.

In general, the survey meter should indicate not more than 0.5 mR/h at 5 cm from any accessible surface or part of the instrument. Actually, the 5 cm should be measured to the center of the active volume of the detector in the survey meter. Thus, in practice, the survey meter should be held as close to the spectrometer as possible without danger of damaging the meter. If the reading exceeds 0.5 mR/h, the source of the "leak" should be examined to ascertain that the components involved are properly assembled and that any shielding provided by the manufacturer is securely in place. If all is proper, and a radiation leak still exists, additional shielding must be provided.

Particularly hazardous conditions may exist when the x-ray tube is operated without its housing in place, that is, with its glass end exposed, or without its shutter, or without the specimen compartment in place. Laboratory-designed and built instruments and accessories and, especially, temporary arrangements should be carefully monitored.

All x-ray laboratory personnel should be provided with film badges and/or pocket dosimeters. The person responsible for the x-ray laboratory must be familiar with all federal, state, and municipal regulations pertaining to x-ray safety and must take all necessary measures to comply with them (*A16*, *J15*, *M53*, *M54*).

Another hazard is that of electric shock. Contact with the high-potential supply while in operation may well result in electrocution. Fortunately, when the high-potential cable is secured at both the oil-tank and x-ray-tube ends, contact with the high potential is unlikely. However, even when the power is off, there may be danger of shock from charge stored in capacitors in the high-potential circuit if there is no "bleeder" resistor or if the resistor is open. Also, great care must be taken when it is necessary to work on the equipment with the cabinet door interlocks thwarted. Contact with the 220-V line power is very hazardous.

Part III

Qualitative and Semiquantitative Analysis

Chapter 10

Qualitative and Semiquantitative Analysis

10.1. GENERAL

Many analytical problems require only qualitative or semiquantitative, rather than quantitative, analysis of the sample(s)—that is, *detection* or *identification* of the elements present, or *estimation*, rather than *determination*, of their concentrations. A quantitative x-ray spectrometric analysis usually gives analyte concentration within a relatively narrow known range of accuracy as derived from precisely measured and carefully corrected intensity data, and from calibration data from standards very similar to the samples in analyte concentration and matrix composition. A semiquantitative x-ray spectrometric analysis usually gives analyte concentration as derived from intensities measured with speed and convenience, rather than precision, as primary objectives, and from standards applicable to a variety of related sample types. If samples and standards are very similar, a semiquantitative analysis may give high accuracy, but the more they differ, the poorer is the accuracy.

X-ray secondary-emission spectrometry is extremely well suited to qualitative and semiquantitative analysis. It is convenient, rapid, nondestructive, and applicable to major, minor, and trace constituents in any sample that can be placed in the instrument. When practiced on commercial instruments, it is applicable to all elements down to magnesium, sodium, or fluorine, although sensitivity falls off for elements below phosphorus. Identification of the elements is greatly facilitated by: (1) synchronism of the goniometer and chart drives so that chart divisions are directly convertible to 2θ; (2) the simplicity of x-ray spectra; and (3) the availability of excellent x-ray spectrometer tables. Intensities are easily measured with reasonable accuracy either on the recorder charts or on the scaler. Semiquantitative analyte concentrations sufficiently accurate for many purposes are readily derived from these intensities, and often these concentrations are remarkably accurate. Even more accurate concentrations W_A (weight fraction) are usually obtained from the uncorrected ratios of analyte-line intensities from the sample $I_{A,X}$ and pure analyte $I_{A,A}$

$$W_A \approx I_{A,X}/I_{A,A} \tag{10.1}$$

Another application of qualitative analysis is as the preliminary work preparatory to a quantitative analysis. A 2θ scan should be recorded for each new sample type to be analyzed. For this purpose, it is helpful to plot the positions of the absorption edges of all major and minor constituent elements present—at least their K and LIII edges. Examination of the 2θ scan, especially with absorption edges indicated, permits prediction of spectral-line interference, higher-order overlap, and absorption-enhancement effects to be expected in the analysis.

10.2. RECORDING THE SPECTRUM

A qualitative x-ray spectrometric analysis consists in recording of the x-ray spectrum of the sample, identification of the peaks, and, perhaps, classification of each element identified as a major, minor, or trace constituent on the basis of the intensity of its peaks.

The spectrum is recorded as follows. The sample is placed in the x-ray spectrometer (Figure 3.1), where it is irradiated by the primary beam and emits secondary x-rays consisting of the wavelengths of the spectral lines of its constituent elements. This secondary beam is collimated and subjected to a 2θ scan (or θ scan) as follows. The analyzer crystal and detector arm are driven about a common axis at relative speed ratio 1:2. Thus, the crystal and the normal to the detector window always present angles θ and 2θ, respectively, to the direction of the secondary beam, and the detector is always in position to receive any x-rays diffracted by the crystal. As the crystal rotates, it presents a continuous succession of θ angles to the secondary beam and, one by one, in accordance with the Bragg law, diffracts each wavelength in the beam to the detector. The detector output is amplified and applied to the ratemeter, the output of which displaces the recorder

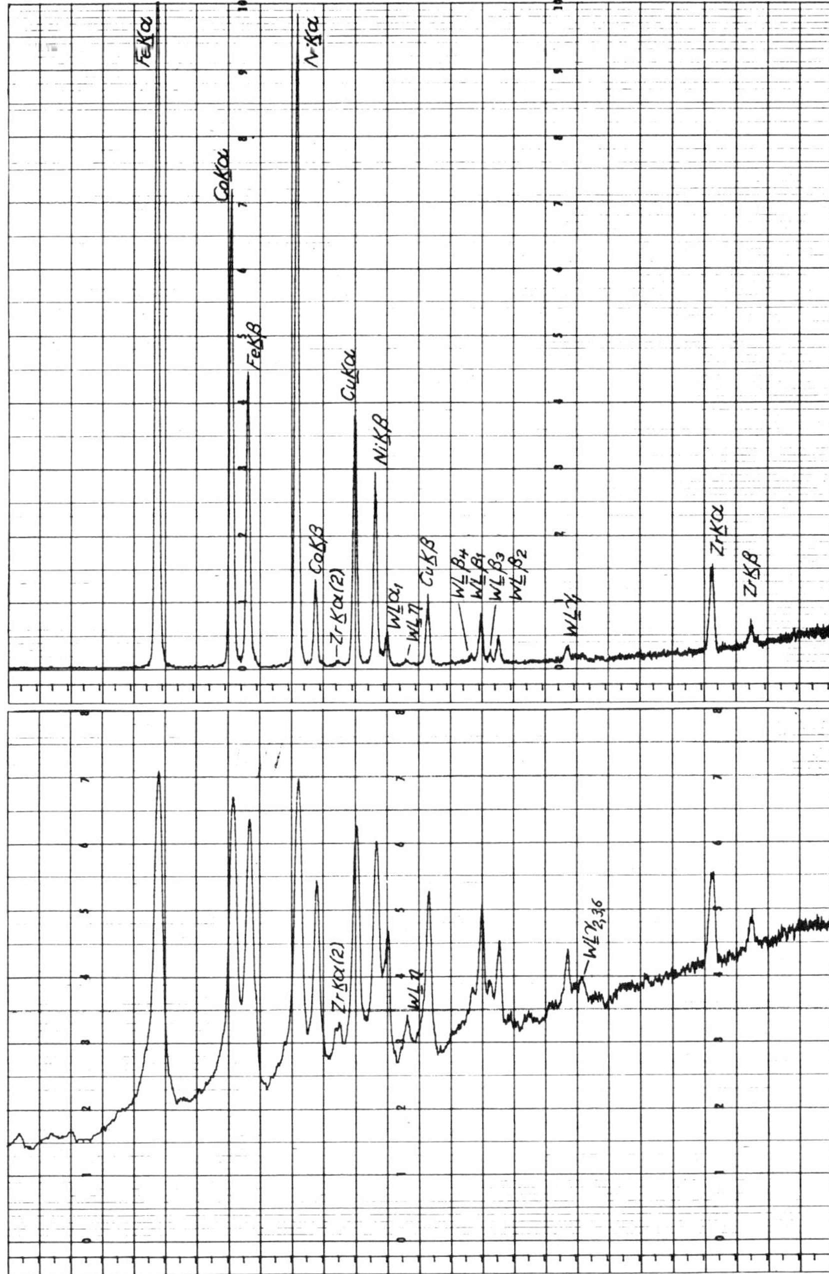

FIG. 10.1. Ratemeter–recorder charts of x-ray spectra (2θ scans) of an Alnico alloy (53Fe–17Ni–13Co–9.5Al–6Cu–0.5Ti–0.5Zr). The upper chart was recorded with linear, the lower with logarithmic ratemeter mode. Compare with optical-emission spectrum of Alnico in Figure 3.2.

pen. The recorder chart drive is synchronized with the goniometer drive. The result is a chart having the form of those shown in Figure 10.1. The chart is a plot of intensity *versus* 2θ and is, in fact, an x-ray wavelength spectrum.

10.3. INSTRUMENT CONDITIONS

Of the many instrument conditions to be considered in recording the spectrum, the following are the most important. *Spectrogoniometer conditions*: x-ray tube target, potential (kV), and current (mA); collimators; crystal; optical-path medium; detector; and goniometer scanning speed. *Readout conditions*: pulse-height analysis; ratemeter time constant (response time); ratemeter sensitivity (intensity required for full-scale recorder deflection); ratemeter mode (linear or logarithmic); recorder response time (time to reach full scale); and recorder chart speed. If there is interest in only one or a few elements, the instrument is arranged for optimal recording of these elements. If practically nothing is known of the specimen and maximum information is required, the following general considerations may guide the choice of conditions for standard commercial instruments.

The x-ray tube should be operated at maximum potential to ensure excitation of spectra of elements having high excitation potentials, and at maximum current to ensure maximum sensitivity for minor and trace elements. In general, a tungsten or platinum target is preferable because the high atomic number ensures maximum x-ray output for a given input power. If interest is limited to elements below vanadium, a chromium target is preferable. A dual tungsten–chromium target permits convenient changeover during the recording of the spectrum. A primary beam filter to remove target lines may be beneficial if the sample has low effective atomic number and therefore scatters primary radiation efficiently.

The collimators are selected to provide a balance between high intensity and high resolution. Certainly no crystal can exceed the combination of high diffracted intensity, high resolution, and wide useful range of elements (down to potassium) provided by lithium fluoride. A crystal changer permits convenient extension of the element range down to fluorine. Helium or vacuum is of some benefit for trace concentrations of elements having line wavelengths >1 Å, and is required if spectra of elements below vanadium are to be recorded. A combination detector consisting of a scintillation counter and a flow-proportional counter having an ultrathin window gives satisfactory response down to fluorine.

The goniometer 2θ scanning speed should be slow enough to permit response to weak peaks, yet fast enough to permit the spectrum to be recorded in reasonable time. A commonly used speed is $2° 2\theta$/min. The weaker the peaks of interest, the slower must be the scanning speed. The ultimate is a statistically designed point-by-point step scan. Some form of automatic pulse-height selection may be beneficial to exclude certain types of spectral-line interference and background.

If there are several major constituents, the full-scale deflection is usually set so that the strongest peak is just on scale. If there is only one major constituent, or even two, it may be preferable to set the scale so that the strongest minor constituent peak is just on scale and let the major peak(s) run off scale. In either case, the peak to be used to set the scale may be known or found by making a rapid manual 2θ scan. It must be remembered that response is smaller when the goniometer scans through a peak than when it is stationary at the peak, and response decreases the faster the scanning speed, the longer the ratemeter time constant, and the slower the recorder response time. If the scale is made too small, background "noise" fluctuations become severe, and weak peaks may be obscured.

The ratemeter time constant should be short enough to permit response to weak peaks, yet long enough to damp background noise. Typical values are 1–4 s, but, in general, the time constant is selected for compatibility with the 2θ scanning speed and the full-scale deflection. If minor and trace as well as major constituents are of interest, that is, if very weak as well as very strong peaks are to be recorded, operation of the ratemeter in logarithmic, rather than linear, mode may be preferable. Recorder chart speed is selected principally to give a chart of convenient length. The charts in Figure 10.1 were recorded at a chart speed of 1 cm/min and a goniometer speed of $2° 2\theta$/min, giving a $52° 2\theta$ spectrum on a 26-cm chart. It is evident that in the logarithmic spectrum, the peaks of lower intensity are accentuated, but so are the background and the "tails" ("skirts," "wings") of all peaks.

An inspection of the recorded spectrum may reveal cases where the presence of a peak is questionable because of poor resolution from an adjacent peak or from background noise. These parts of the spectrum may be rerun at conditions to accentuate the questionable peak. Techniques for improving resolution are given in Section 13.2.2. The most convenient ones are use of higher collimation, a crystal having a smaller d spacing, and higher-order spectra.

Figure 10.2 (*25p*) shows the effect of several instrument conditions on the recorded 2θ scans. Each pair of spectra was recorded at linear mode and

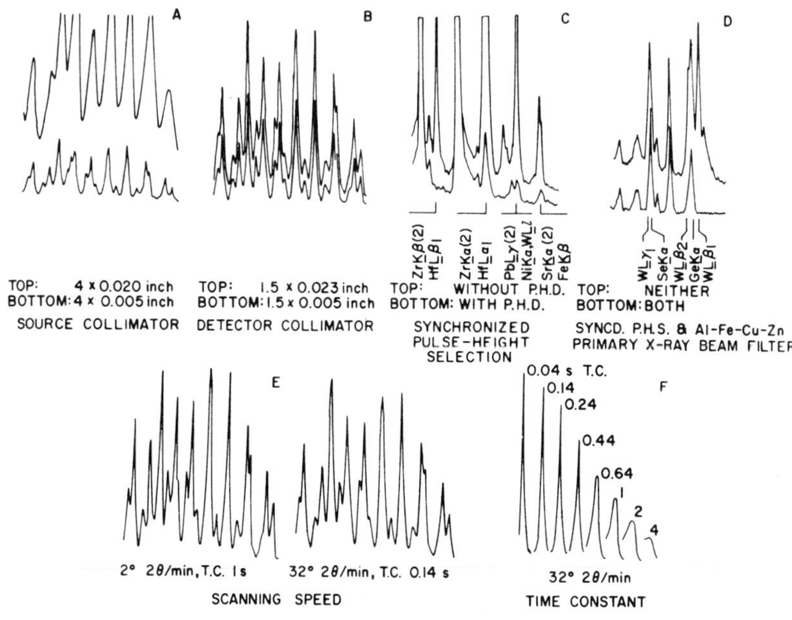

FIG. 10.2. Effect of certain instrument conditions on intensity and resolution in 2θ scans. [M. L. Salmon, in "Handbook of X-Rays," E. F. Kaelble, ed.; McGraw-Hill Book Co. (1967); courtesy of the author and publisher.]

identical conditions except for the one being evaluated. Figure 10.2A compares 10×0.05-cm (4×0.020-inch) and 10×0.0125-cm (4×0.005-inch) source collimators. The fine collimator gives only ~25% as much net peak intensity, but resolves many new peaks. Figure 10.2B compares 3.75×0.0575-cm ($1\frac{1}{2}\times0.023$-inch) and 3.75×0.0125-cm ($1\frac{1}{2}\times0.005$-inch) detector collimators. Here, the coarse collimator doubles the intensity with little loss of resolution and peak-to-background ratio. Figure 10.2C shows the effect of continuous variation of pulse-height selector baseline in synchronism with 2θ. Second-order zirconium, lead, and strontium peaks are eliminated, leaving the first-order hafnium, nickel, and iron peaks clear. Figure 10.2D shows the combined effect of continuous pulse-height selection and an Al–Fe–Cu–Zn primary-beam filter to clear the germanium and selenium K lines from overlapping scattered tungsten L lines. Figure 10.2E shows that high intensity and resolution can be obtained at very high 2θ-scanning rates. The first spectrum was recorded at $2°\ 2\theta$/min and 1-s time constant on a conventional recorder having 1-s full-scale response. The other spectrum was recorded at $32°\ 2\theta$/min and 0.14-s time constant on a recorder having 0.25-s full-scale response, and has substantially the same intensity and resolution as the slow scan. Figure 10.2F shows the effect of time constant on the same peak scanned at 32° 2θ/min. Commercial spectrometers must be modified to have such rapid 2θ scan rates, ratemeter time constants, and recorder response times.

10.4. IDENTIFICATION OF THE PEAKS

Figure 10.1 shows the x-ray spectrum of an Alnico alloy (53Fe–17Ni–13Co–9.5Al–6Cu–0.5Ti–0.5Zr) recorded with linear and logarithmic ratemeter modes. Certainly the linear spectrum is preferable for general purposes, and is always used when intensities are to be estimated from the chart. However, the very weak Zr $K\alpha(2)$, W $L\eta$, and W $L\gamma$ peaks are accentuated on the logarithmic spectrum.

10.4.1. Spectral-Line Tables

The peaks may be identified by use of transparent overlay scales (*25p*, *G28*, *S4*) showing the positions of all but the weakest spectral lines. A different overlay is required for each combination of crystal, 2θ-scanning speed, and chart speed. Peaks may also be identified by use of x-ray spectral calculators (Section 9.1). However, peaks are usually identified by use of tables (*58*, *61*, *68*, *80*, *86*, *87*).

The 2θ angle at which each peak on the chart occurs is obtained from the known 2θ scan and chart speeds or from a scale previously prepared from these parameters. Also, most recorders in x-ray spectrometers have a degree-marker pen that is actuated by a micro switch in the goniometer and makes a tick along the edge of the chart at 1° or 0.5° intervals. Degree-marker traces are shown along the bottom edges of the charts in Figure 10.1. The peaks are then identified by reference to x-ray spectrometer tables, several of which are listed in the Bibliography. These tables are of two forms: The *line-to-2θ* table enables one to find at what 2θ angle a specified order of a specified line is diffracted by a specified crystal; the *2θ-to-line* table enables one to identify a peak from the 2θ angle at which it is diffracted, knowing the crystal that was used. The 2θ-to-line table is also useful for searching for possible spectral-line interference. The analyte line is located in the table, and a search is made for 1° or 2° 2θ on either side of the line for lines of other elements known to be present in the specimen.

A portion of a page of a line-to-2θ table is shown in Figure 10.3 (*87*). Elements are listed in order of increasing atomic number, and, for each

PAGE 27.0

EL	LINE	N	I	Z	R	KEV	LAMBDA	LSC	KAP	MICA	GYPM	ADP	EDDT	PET	SIO2	CACO3	NACL	LIF
RU	KA2	3	50	44	1	19.15	1.9422	2.22	8.36	11.24	14.70	21.04	25.48	25.67	33.77	37.31	40.28	57.67
RU	KA1,2	1	150	44	1	19.23	0.6445	.74	2.77	3.72	4.87	6.95	8.39	8.46	11.06	12.19	13.12	18.42
RU	KA1,2	2	150	44	1	19.23	1.2890	1.47	5.55	7.45	9.74	13.92	16.83	16.96	22.23	24.52	26.42	37.33
RU	KA1,2	3	150	44	1	19.23	1.9335	2.21	8.33	11.19	14.63	20.94	25.36	25.56	33.61	37.14	40.09	57.39
RU	KA1	1	100	44	1	19.28	0.6430	.73	2.77	3.71	4.85	6.93	8.37	8.44	11.04	12.16	13.09	18.37
RU	KA1	2	100	44	1	19.28	1.2860	1.47	5.54	7.43	9.72	13.88	16.79	16.92	22.17	24.46	26.36	37.24
RU	KA1	3	100	44	1	19.28	1.9290	2.20	8.31	11.16	14.60	20.89	25.30	25.50	33.53	37.05	40.00	57.24
RU	SKBN	1	1	44	1	21.53	0.5757	.66	2.48	3.33	4.35	6.20	7.49	7.55	9.88	10.88	11.72	16.44
RU	SKBN	2	1	44	1	21.53	1.1513	1.32	4.96	6.65	8.70	12.42	15.02	15.14	19.83	21.86	23.56	33.22
RU	SKBN	3	1	44	1	21.53	1.7270	1.97	7.44	9.99	13.06	18.68	22.61	22.79	29.93	33.05	35.66	50.79
RU	KB3	1	15	44	1	21.63	0.5731	.65	2.47	3.31	4.33	6.17	7.46	7.52	9.83	10.83	11.66	16.36
RU	KB3	2	15	44	1	21.63	1.1461	1.31	4.93	6.62	8.66	12.37	14.95	15.07	19.74	21.76	23.45	33.07
RU	KB3	3	15	44	1	21.63	1.7192	1.96	7.40	9.94	13.00	18.60	22.51	22.68	29.79	32.90	35.49	50.54
RU	KB1	1	15	44	1	21.65	0.5725	.65	2.46	3.31	4.32	6.17	7.45	7.51	9.82	10.82	11.65	16.34
RU	KB1	2	15	44	1	21.65	1.1449	1.31	4.93	6.62	8.65	12.35	14.94	15.05	19.72	21.74	23.42	33.03
RU	KB1	3	15	44	1	21.65	1.7174	1.96	7.39	9.93	12.99	18.58	22.49	22.66	29.76	32.86	35.45	50.48
RU	SKB7	1	1	44	1	21.72	0.5707	.65	2.46	3.30	4.31	6.15	7.43	7.49	9.79	10.79	11.61	16.29
RU	SKB7	2	1	44	1	21.72	1.1413	1.30	4.91	6.60	8.62	12.32	14.89	15.00	19.65	21.67	23.35	32.93
RU	SKB7	3	1	44	1	21.72	1.7120	1.96	7.37	9.90	12.95	18.52	22.42	22.59	29.67	32.76	35.34	50.31
RU	SKB6	1	1	44	1	21.78	0.5691	.65	2.45	3.29	4.30	6.13	7.41	7.47	9.76	10.76	11.58	16.25
RU	SKB6	2	1	44	1	21.78	1.1383	1.30	4.90	6.58	8.60	12.28	14.85	14.96	19.60	21.61	23.28	32.84
RU	SKB6	3	1	44	1	21.78	1.7074	1.95	7.35	9.87	12.91	18.47	22.36	22.53	29.59	32.67	35.24	50.17
RU	KB5	1	1	44	1	21.83	0.5678	.65	2.44	3.28	4.29	6.12	7.39	7.45	9.74	10.73	11.56	16.21
RU	KB5	2	1	44	1	21.83	1.1357	1.30	4.89	6.56	8.58	12.25	14.82	14.93	19.56	21.56	23.23	32.76
RU	KB5	3	1	44	1	21.83	1.7035	1.95	7.33	9.85	12.88	18.43	22.30	22.47	29.52	32.59	35.16	50.05
RU	SKB8	1	1	44	1	21.98	0.5639	.64	2.43	3.26	4.26	6.08	7.34	7.40	9.68	10.66	11.48	16.10
RU	SKB8	2	1	44	1	21.98	1.1279	1.29	4.85	6.52	8.52	12.17	14.71	14.83	19.42	21.41	23.07	32.53
RU	SKB8	3	1	44	1	21.98	1.6918	1.93	7.28	9.78	12.79	18.30	22.15	22.32	29.31	32.36	34.91	49.68
RU	KB2	1	5	44	1	22.07	0.5616	.64	2.42	3.24	4.24	6.05	7.31	7.37	9.64	10.62	11.43	16.03
RU	KB2	2	5	44	1	22.07	1.1233	1.28	4.83	6.49	8.48	12.12	14.65	14.76	19.34	21.32	22.97	32.39
RU	KB2	3	5	44	1	22.07	1.6849	1.93	7.25	9.74	12.74	18.22	22.06	22.22	29.19	32.22	34.76	49.46
RU	KB4,	1	1	44	1	22.10	0.5608	.64	2.41	3.24	4.23	6.04	7.30	7.36	9.62	10.60	11.41	16.01
RU	KB4,	2	1	44	1	22.10	1.1217	1.28	4.83	6.48	8.47	12.10	14.63	14.74	19.31	21.29	22.94	32.34
RU	KB4,	3	1	44	1	22.10	1.6825	1.92	7.24	9.73	12.72	18.20	22.02	22.19	29.14	32.18	34.71	49.39
RU	SKB+2	1	1	44	1	22.18	0.5589	.64	2.41	3.23	4.22	6.02	7.28	7.33	9.59	10.56	11.37	15.96
RU	SKB+2	2	1	44	1	22.18	1.1179	1.28	4.81	6.46	8.44	12.06	14.58	14.69	19.25	21.22	22.86	32.23
RU	SKB+2	3	1	44	1	22.18	1.6768	1.92	7.22	9.70	12.68	18.13	21.95	22.12	29.04	32.07	34.59	49.21
RH	M3-M5	1	1	45	0	0.19	65.4500	81.47										
RH	M2-M4	1	1	45	0	0.21	59.5400	72.83										
RH	MZ1,2	1	1	45	0	0.26	47.5900	56.65										

FIG. 10.3. Portion of a page of a line-to-2θ x-ray spectrometer table giving data for all crystals together. [E. W. White, G. V. Gibbs, G. G. Johnson, Jr., and G. R. Zechman, Jr., "X-Ray Emission Line Wavelength and Two-Theta Tables," *ASTM Data Series* **DS-37**, American Society for Testing and Materials (1965); courtesy of the authors and A.S.T.M.]

element, spectral lines are listed in order of wavelength. Many tables include absorption edges, even though these do not appear on the ratemeter–recorder charts. The table shown in Figure 10.3 includes satellite lines (those having line symbols prefixed by S) and some lines to which no symbols have been assigned (those indicated by initial and final atom state, for example, M3–M5). The column headings signify, in order, the element (EL), spectral line (LINE), order (N), relative intensity (I), atomic number (Z), reference to source of data (R), photon energy (KEV), and product of order and wavelength in angstroms $n\lambda$ (LAMBDA). The wavelength of a line (1λ) is the value given for the first order of that line. The last 11 columns give the 2θ angle at which the line is diffracted by each of 11 crystals. This table lists only three orders for each line, and is a little difficult to use because it contains so many lines of no significance in conventional x-ray spectrometric analysis.

The second edition of this table (*86*) gives data for 25 crystals. Actually, only 23 crystals are listed, but the data for graphite ($2d$ 6.708 Å) apply also to quartz(101) ($2d$ 6.687 Å), and the data for lead stearate apply also to lead stearate decanoate ($2d$ for both ~100 Å). The table lists practically all x-ray spectral lines and absorption edges of wavelengths up to 160 Å—some 3400 first-order *K*, *L*, *M*, and *N* lines and edges. Also, whereas most published 2θ tables give a constant set of relative intensities for *L*-series lines of all elements, this one gives the variation of relative intensity with atomic number. Finally, this table includes a column indicating the magnitude to be expected in the wavelength shift due to chemical effects on *K* lines. The format is the same as for the first edition (*87*), except that only first-order lines are listed in the line-to-2θ portion of the table.

The form of table shown in Figure 10.3 lists 2θ values for each line for all crystals together, that is, on the same page. Other line-to-2θ tables have the forms shown in Figures 10.4 and 10.5, where the data for each crystal are separate. Figure 10.4 (*80*) shows part of a page of such a table for a LiF crystal in which each spectral line has its own column and each of the five orders its own row. Figure 10.5 (*68*) shows part of a page of a table in which each spectral line has its own row and each of the five orders its own column.

A portion of a page of a 2θ-to-line table is shown in Figure 10.6 (*87*), where the column headings have the significance given to Figure 10.3 above. The lines are listed in order of increasing $n\lambda$ or 2θ. The second edition of this table (*86*) gives data for 25 crystals in up to 10 orders for the strongest lines. Data for all 25 crystals are not given for all lines. Rather, the various crystals are phased in and out of the table over their useful 2θ regions. The format is the same as for the first edition (*87*).

ASCENDING ATOMIC NUMBER DATA FOR LIF 2D 4.028

ELEM	ORD	KA1	KA2	KB1	KB3	LA1	LB1	LB2	LY1	LB3	LB5	LL	LB4	LY6
81 TL	1	4.84	4.98	4.27	4.30	34.87	29.19	29.04	24.89	28.78	28.19	40.22	29.90	24.2
	2	9.68	9.97	8.54	8.60	73.64	60.53	60.20	51.06	59.61	58.30	86.89	62.11	49.6
	3	14.55	14.98	12.83	12.91	128.04	98.22	97.57	80.55	96.41	93.88		101.40	78.0
	4	19.44	20.02	17.13	17.25				119.08	167.48	153.90			114.1
	5	24.36	25.09	21.46	21.61									
82 PB	1	4.70	4.84	4.15		33.92	28.22	28.25	24.07	27.84	27.37	39.16	28.96	23.4
	2	9.40	9.68	8.31		71.38	58.36	58.43	49.30	57.52	56.48	84.18	60.00	47.8
	3	14.12	14.55	12.49		122.12	94.00	94.13	77.46	92.39	90.43		97.18	74.9
	4	18.86	19.44	16.67			154.41	154.93	113.06	148.42	142.30		180.00	108.4
	5	23.64	24.36	20.88										
83 BI	1	4.58	4.70	4.04	4.07	33.00	27.34	27.43	23.32	26.96	26.58	38.17	28.07	22.6
	2	9.17	9.40	8.09	8.14	69.23	56.42	56.61	47.68	55.58	54.75	81.68	58.04	46.2
	3	13.77	14.12	12.14	12.23	116.87	90.31	90.68	74.64	88.75	87.21	157.56	93.38	72.1
	4	18.40	18.86	16.21	16.33		141.95	143.01	107.87	137.65	133.72		151.96	103.5
	5	23.06	23.64	20.30	20.45									158.1

FIG. 10.4. Portion of a page of a line-to-2θ x-ray spectrometer table giving data for each crystal separately and having spectral lines in separate columns. [M. C. Powers, "X-Ray Fluorescent Spectrometer Conversion Tables," Philips Electronic Instruments, Inc. (1957); courtesy of the author and Philips Electronic Instruments, Inc.]

For use in energy-dispersive analysis (Section 8.2), *line-to-keV* and *keV-to-line* tables, analogous to line-to-2θ and 2θ-to-line tables, respectively, are required. Portions of pages of such tables are shown in Figures 10.7 and 10.8 (*77*). These tables list all lines in table (*86*) up to 50 Å (0.25 keV). The column headings are the same as for Figures 10.3 and 10.6, insofar as applicable. Inasmuch as most 2θ tables also list photon and excitation

LIF 2D 4.0267

ELEMENT	LAMBDA	ORDER	1	2	3	4	5
82 PB	.14077	K	ABS 4.01				
82 PB	.14155	KB4	4.03	8.06	12.11	16.17	20.25
82 PB	.14191	KB2	4.04	8.08	12.14	16.21	20.30
82 PB	.14494	KB5	4.13	8.26	12.40	16.56	20.74
82 PB	.14596	KB1	4.15	8.31	12.49	16.67	20.88
82 PB	.14680	KB3	4.18	8.36	12.56	16.77	21.01
82 PB	.16536	KA1	4.71	9.42	14.15	18.91	23.70
82 PB	.16701	KA	4.75	9.52	14.29	19.10	23.94
82 PB	.17029	KA2	4.85	9.70	14.58	19.48	24.41
82 PB	.78152	L I	ABS 22.38				
82 PB	.78588	LY4	22.51	45.95	71.68	102.64	154.76
82 PB	.78710	LYP4	22.54	46.03	71.81	102.87	155.56
82 PB	.80231	LY11	22.99	46.97	73.42	105.69	170.05
82 PB	.81484	LY3	23.35	47.75	74.76	108.08	
82 PB	.81508	L II	ABS 23.36				
82 PB	.81686	LY6	23.41	47.87	74.97	108.47	
82 PB	.82127	LY2	23.54	48.15	75.45	109.34	
82 PB	.82327	LYP8	23.59	48.27	75.67	109.73	
82 PB	.82368	LY8	23.61	48.30	75.71	109.81	
82 PB	.83971	LY1	24.07	49.30	77.45	113.05	
82 PB	.86645	LY5	24.85	50.98	80.41	118.79	
82 PB	.92677	LB9	26.61	54.81	87.33	134.03	
82 PB	.93383	LB10	26.82	55.27	88.17	136.14	
82 PB	.95029	LIII	ABS 27.30				

FIG. 10.5. Portion of a page of a line-to-2θ x-ray spectrometer table giving data for each crystal separately and having diffraction orders in separate columns. ["X-Ray Wavelengths for Spectrometer," General Electric Co., X-Ray Dept.]

PAGE 141.0

EL	LINE	N	I	Z	R	KEV	LAMBDA	LSC	KAP	MICA	GYPM	ADP	EDDT	PET	SIO2	CACO3	NACL	L F
PT	KA2	8	50	78	1	65.11	1.5231	1.74	6.56	8.81	11.51	16.46	19.92	20.07	26.33	29.06	31.33	44.44
HF	LN	1	1	72	1	8.14	1.5232	1.74	6.56	8.81	11.51	16.46	19.92	20.07	26.33	29.06	31.33	44.45
BI	LG4+	2	1	83	1	16.27	1.5239	1.74	6.56	8.81	11.52	16.47	19.93	20.08	26.34	29.07	31.35	44.47
IR	KB3	9	15	77	1	73.20	1.5241	1.74	6.56	8.81	11.52	16.47	19.93	20.08	26.35	29.08	31.35	44.47
ER	KA1,2	6	150	68	1	48.80	1.5241	1.74	6.56	8.81	11.52	16.47	19.93	20.08	26.35	29.08	31.35	44.47
PR	KB3	5	15	59	1	40.65	1.5246	1.74	6.56	8.81	11.52	16.48	19.94	20.09	26.36	29.09	31.36	44.49
TA	SLAS	1	1	73	1	8.13	1.5252	1.74	6.57	8.82	11.53	16.48	19.94	20.10	26.37	29.10	31.38	44.51
TH	L3-P2,	2	1	90	1	16.25	1.5253	1.74	6.57	8.82	11.53	16.48	19.94	20.10	26.37	29.10	31.38	44.51
AT	LG1	2	10	85	2	16.25	1.5258	1.74	6.57	8.82	11.53	16.49	19.95	20.10	26.38	29.11	31.39	44.53
HF	KB2	8	5	72	1	64.97	1.5263	1.74	6.57	8.82	11.54	16.49	19.96	20.11	26.39	29.12	31.40	44.54
GD	KB1	6	15	64	1	48.72	1.5265	1.74	6.57	8.83	11.54	16.50	19.96	20.11	26.39	29.12	31.40	44.55
TH	L3-P1	2	1	90	1	16.24	1.5267	1.74	6.57	8.83	11.54	16.50	19.96	20.12	26.39	29.13	31.41	44.55
TA	KB3	8	15	73	1	64.94	1.5271	1.74	6.57	8.83	11.54	16.50	19.97	20.12	26.40	29.14	31.42	44.57
IR	KA1	8	100	77	1	64.91	1.5279	1.75	6.58	8.83	11.55	16.51	19.98	20.13	26.42	29.15	31.43	44.59
PD	KB4	3	1	46	2	24.34	1.5279	1.75	6.58	8.83	11.55	16.51	19.98	20.13	26.42	29.15	31.43	44.59
CU	SKA+3	1	1	29	1	8.11	1.5280	1.75	6.58	8.83	11.55	16.51	19.98	20.13	26.42	29.15	31.43	44.59
BI	L1-N6,	2	1	83	1	16.22	1.5281	1.75	6.58	8.83	11.55	16.51	19.98	20.13	26.42	29.16	31.44	44.60
PO	LG3,6	2	3	84	1	16.22	1.5289	1.75	6.58	8.84	11.56	16.52	19.99	20.14	26.43	29.17	31.45	44.62
GD	LG3	1	2	64	1	8.11	1.5290	1.75	6.58	8.84	11.56	16.52	19.99	20.15	26.43	29.17	31.46	44.63
RN	KA2	10	50	86	2	81.07	1.5290	1.75	6.58	8.84	11.56	16.52	19.99	20.15	26.43	29.17	31.46	44.63
TH	LB5	2	1	90	1	16.21	1.5295	1.75	6.58	8.84	11.56	16.53	20.00	20.15	26.44	29.18	31.47	44.64
TB	LG1	1	10	65	1	8.10	1.5297	1.75	6.59	8.84	11.56	16.53	20.00	20.16	26.45	29.19	31.47	44.65
TM	LB1	1	50	69	1	8.10	1.5299	1.75	6.59	8.85	11.56	16.53	20.01	20.16	26.45	29.19	31.47	44.65
TH	LB1	2	50	90	1	16.20	1.5303	1.75	6.59	8.85	11.57	16.54	20.01	20.16	26.46	29.20	31.48	44.67
PD	KB2	3	5	46	1	24.29	1.5307	1.75	6.59	8.85	11.57	16.54	20.02	20.17	26.46	29.21	31.49	44.68
TL	KA1	9	100	81	1	72.86	1.5313	1.75	6.59	8.85	11.58	16.55	20.02	20.18	26.47	29.22	31.50	44.70
GD	KB3	6	15	64	1	48.57	1.5313	1.75	6.59	8.85	11.58	16.55	20.02	20.18	26.47	29.22	31.50	44.70
BI	L1-O1	2	1	83	1	16.18	1.5321	1.75	6.60	8.86	11.58	16.56	20.03	20.19	26.49	29.23	31.52	44.72
PB	KA2	9	50	82	1	72.81	1.5322	1.75	6.60	8.86	11.58	16.56	20.04	20.19	26.49	29.24	31.52	44.72
TA	LA2	1	10	73	1	8.09	1.5328	1.75	6.60	8.86	11.59	16.57	20.04	20.20	26.50	29.25	31.54	44.74
CU	SKA4	1	1	29	1	8.09	1.5330	1.75	6.60	8.86	11.59	16.57	20.05	20.20	26.50	29.25	31.54	44.75
LU	L2-M2	1	1	71	1	8.08	1.5333	1.75	6.60	8.86	11.59	16.57	20.05	20.20	26.51	29.26	31.55	44.76
CU	SKA3+	1	1	29	1	8.08	1.5339	1.75	6.60	8.87	11.60	16.58	20.06	20.21	26.52	29.27	31.56	44.78
GD	LG2	1	1	64	1	8.08	1.5341	1.75	6.60	8.87	11.60	16.58	20.06	20.21	26.52	29.27	31.56	44.78
TB	SLG9	1	1	65	1	8.08	1.5345	1.75	6.61	8.87	11.60	16.58	20.07	20.22	26.53	29.28	31.57	44.80
CU	SKA3	1	1	29	1	8.08	1.5347	1.75	6.61	8.87	11.60	16.59	20.07	20.22	26.53	29.28	31.58	44.80
OS	L3-M2	1	1	76	2	8.08	1.5347	1.75	6.61	8.87	11.60	16.59	20.07	20.22	26.54	29.28	31.58	44.80
TH	SLB2•7	2	1	90	1	16.15	1.5355	1.75	6.61	8.88	11.61	16.60	20.08	20.23	26.55	29.30	31.59	44.83
I	KB1	4	15	53	1	32.29	1.5355	1.75	6.61	8.88	11.61	16.60	20.08	20.23	26.55	29.30	31.59	44.83
CU	SKA+	1	1	29	1	8.07	1.5361	1.76	6.61	8.88	11.61	16.60	20.09	20.24	26.56	29.31	31.61	44.84

FIG. 10.6. Portion of a page of a 2θ-to-line x-ray spectrometer table. [E. W. White, G. V. Gibbs, G. G. Johnson, Jr., and G. R. Zechman, Jr., "X-Ray Emission Line Wavelength and Two-Theta Tables," *ASTM Data Series* **DS-37**, American Society for Testing and Materials (1965); courtesy of the authors and A.S.T.M.]

energies (keV), the principal difference in the energy tables is that no higher-order lines are listed. Of course, in nondiffractive systems, orders have no significance and so only confuse the search.

Another type of table is of incidental interest (*70*, *88*). Some x-ray spectrometers have 2θ-drive drums (odometers) calibrated to indicate wavelength in angstroms directly when used with a certain crystal, usually LiF or ADP. Such spectrometers are frequently used on electron-probe microanalyzers (Chapter 21). If a crystal other than the one for which it is calibrated is used on such an instrument, the odometer no longer indicates wavelength, and the indicated reading must be corrected as follows:

$$R_X = (d_C/d_X)\lambda_L \tag{10.2}$$

where R_X is odometer reading for the interchanged crystal X; d_C and d_X are interplanar spacings (Å) for the diffracting planes of, respectively, the crystal for which the odometer was calibrated C and the new crystal X; and λ_L is the wavelength of the line—or the odometer reading for that line with the original crystal. Tables have been compiled that give odometer settings for various crystals used on a spectrometer calibrated for a specified

Page 2.

El	Line	I	Z	R	KeV	Lambda
Cr	$K\alpha_1$	100	24	6	5.414	2.290
Cr	$K\beta_{1,3}$	18	24	6	5.946	2.085
Cr	$K\beta_5$	.03	24	6	5.986	2.071
Mn	Ll	2	25	6	0.556	22.290
Mn	Ln	1	25	6	0.567	21.850
Mn	$L\alpha_{1,2}$	100	25	6	0.637	19.450
Mn	$L\beta_1$	30	25	6	0.649	19.110
Mn	$L\beta_{3,4}$	.1	25	6	0.721	17.190
Mn	$K\alpha_2$	50	25	6	5.887	2.106
Mn	$K\alpha_{1,2}$	150	25	6	5.894	2.103
Mn	$K\alpha_1$	100	25	6	5.898	2.102
Mn	$K\beta_{1,3}$	20	25	6	6.489	1.910
Mn	$K\beta_5$	.03	25	6	6.534	1.897
Fe	Ll	8	26	6	0.615	20.150
Fe	Ln	2	26	6	0.628	19.750
Fe	$L\alpha_{1,2}$	100	26	6	0.705	17.590
Fe	$L\beta_1$	20	26	6	0.718	17.260
Fe	$L\beta_{3,4}$	.5	26	6	0.792	15.650
Fe	$K\alpha_2$	50	26	6	6.390	1.940
Fe	$K\alpha_{1,2}$	150	26	6	6.398	1.937
Fe	$K\alpha_1$	100	26	6	6.403	1.936
Fe	$K\beta_{1,3}$	20	26	6	7.057	1.757
Fe	$K\beta_5$	.03	26	6	7.107	1.744
Co	Ll	9	27	6	0.678	18.292
Co	Ln	2	27	6	0.694	17.870

El	Line	I	Z	R	KeV	Lambda
Ga	Ll	1	31	6	0.957	12.953
Ga	Ln	1	31	6	0.984	12.597
Ga	$L\alpha_{1,2}$	100	31	6	1.098	11.292
Ga	$L\beta_1$	35	31	6	1.125	11.023
Ga	$L\beta_{3,4}$	2	31	6	1.197	10.359
Ga	$K\alpha_2$	50	31	6	9.223	1.344
Ga	$K\alpha_{1,2}$	150	31	6	9.241	1.341
Ga	$K\alpha_1$	100	31	6	9.250	1.340
Ga	$K\beta_3$	7	31	6	10.259	1.208
Ga	$K\beta_1$	14	31	6	10.263	1.208
Ga	$K\beta_5$	.04	31	6	10.346	1.198
Ga	$K\beta_2$	.3	31	6	10.365	1.196
Ge	Ll	1	32	6	1.036	11.965
Ge	Ln	1	32	6	1.068	11.609
Ge	$L\alpha_{1,2}$	100	32	6	1.188	10.436
Ge	$L\beta_1$	35	32	6	1.218	10.175
Ge	$L\beta_4$	1	32	6	1.286	9.640
Ge	$L\beta_3$	1	32	6	1.294	9.581
Ge	$K\alpha_2$	50	32	6	9.854	1.258
Ge	$K\alpha_{1,2}$	150	32	6	9.874	1.255
Ge	$K\alpha_1$	100	32	6	9.885	1.254
Ge	$K\beta_3$	7	32	6	10.976	1.129
Ge	$K\beta_1$	14	32	6	10.980	1.129
Ge	$K\beta_5$	.05	32	6	11.073	1.119
Ge	$K\beta_2$	.5	32	6	11.099	1.117

FIG. 10.7. Portion of a page of a line-to-energy x-ray spectrometer table. [G. G. Johnson, Jr. and E. W. White, "X-Ray Emission Wavelengths and keV Tables for Nondiffractive Analysis," *ASTM Data Series* **DS-46**, American Society for Testing and Materials (1970); courtesy of the authors and A.S.T.M.]

Page 24.

El	Line	I	Z	R	KeV	Lambda	El	Line	I	Z	R	KeV	Lambda
Th	$L\gamma_2$	1	90	6	19.302	0.642	U	$L\beta_3$	6	92	6	17.452	0.710
Ru	$K\alpha_1$	100	44	6	19.276	0.643	Mo	$K\alpha_{1,2}$	150	42	6	17.441	0.711
Ru	$K\alpha_{1,2}$	150	44	6	19.233	0.645	Mo	$K\alpha_2$	50	42	6	17.371	0.714
Ra	$L\gamma_{13}$	.01	88	6	19.215	0.645	Fr	$L\gamma_1$	10	87	6	17.300	0.717
Ra	L_1-$O_{4,5}$	.01	88	6	19.165	0.647	Ra	$L\gamma_5$	.1	88	6	17.271	0.718
Ru	$K\alpha_2$	50	44	6	19.147	0.547	Pu	$L\beta_2$	20	94	6	17.252	0.719
Th	L_1-N_1	.01	90	6	19.143	0.643	U	$L\beta_1$	50	92	6	17.217	0.720
Pu	$L\beta_{10}$	.01	94	6	19.124	0.648	Pu	$L\beta_{15}$	1	94	6	17.205	0.720
Am	$L\beta_3$	6	95	6	19.103	0.649	U	L_3-$P_{4,5}$	.01	92	6	17.159	0.722
Ra	$L\gamma_4$	.1	88	6	19.081	0.650	Th	$L\beta_9$	.01	90	6	17.136	0.723
Ra	$L\gamma_4p$	.1	88	6	19.032	0.651	U	L_3-$P_{2,3}$	.01	92	6	17.115	0.724
Th	L_2-N_5	.01	90	6	19.009	0.652	U	L_3-P_1	.01	92	6	17.093	0.725
Cf	$L\beta_2$	20	98	3	18.983	0.653	U	$L\beta_5$	1	92	6	17.067	0.726
Th	$L\gamma_1$	10	90	6	18.979	0.653	Np	$L\beta_4$	4	93	6	17.058	0.727
Nb	$K\beta_4$	.01	41	6	18.978	0.653	Y	$K\beta_4$	.01	39	6	17.033	0.728
Nb	$K\beta_2$	4	41	6	18.949	0.654	Y	$K\beta_2$	4	39	6	17.013	0.729
Pa	$L\gamma_5$	.1	91	6	18.925	0.655	Th	$L\beta_{10}$	.01	90	6	16.978	0.730
Am	$L\beta_1$	50	95	6	18.849	0.653	U	L_3-O_3	.01	92	6	16.960	0.731
Th	L_2-N_3	.01	90	6	18.725	0.662	Pa	$L\beta_3$	6	91	6	16.927	0.732
Ra	$L\gamma_{11}$	.01	88	6	18.629	0.665	U	L_3-O_2	.01	92	6	16.904	0.733
Nb	$K\beta_1$	.6	41	6	18.619	0.666	Am	$L\beta_6$	.1	95	6	16.884	0.734
Nb	$K\beta_3$	7	41	6	18.603	0.665	Y	$K\beta_5$	.07	39	6	16.877	0.734
Ra	L_1-N_4	.01	88	6	18.596	0.667	U	$L\beta_7$	.1	92	6	16.842	0.736
Pu	$L\beta_3$	6	94	6	18.537	0.669	Np	$L\beta_2$	20	93	6	16.837	0.736
Bk	$L\beta_2$	20	97	3	18.529	0.669	U	Lu	.01	92	6	16.783	0.739

FIG. 10.8. Portion of a page of an energy-to-line x-ray spectrometer table. [G. G. Johnson, Jr. and E. W. White, "X-Ray Emission Wavelengths and keV Tables for Nondiffractive Analysis," *ASTM Data Series* **DS-46**, American Society for Testing and Materials (1970); courtesy of the authors and A.S.T.M.]

crystal. These tables have line-to-setting and setting-to-line forms analogous to the forms of conventional x-ray spectrometer tables. Figure 10.9 (*88*) shows a portion of a page of the setting-to-line form of such a table. The first seven column headings have the same significance as in Figure 10.3. The six columns ADP/LSD, ADP/KAP, etc. give odometer readings for LSD, KAP, etc. used on a spectrometer calibrated for ADP. The six columns LIF/LSD, LIF/KAP, etc. have the same significance for a spectrometer calibrated for LiF. A later edition of this table (*85*) gives data only for spectrometers calibrated for LiF(200) crystals. The line-to-setting table gives data for 10 interchangeable crystals. The setting-to-line table gives data for 20 crystals, which are phased in and out of the table over their useful wavelength regions.

10.4.2. Identification of Peaks

There is a systematic procedure for identification of the peaks. First, by use of a line-to-2θ table, all peaks known to be present are identified. These include peaks of elements known to be present in the specimen and coherently and incoherently scattered lines of the target and its contami-

PAGE 84.0

EL	LINE	N	I	Z	KEV	LAMBDA	ADP LSD	ADP KAP	ADP EDDT	ADP SIO2	ADP NACL	ADP LIF	LIF LSD	LIF KAP	LIF ADP	LIF EDDT	LIF SIO2	LIF NACL
TL	L3-M3	3	1	81	12.05	3.0860		1.2329	3.7278	4.9101	5.8212	8.1533		0.4666	1.1680	1.4109	1.8584	2.2033
W	LG4+	3	1	74	12.05	3.0860		1.2329	3.7278	4.9101	5.8212	8.1533		0.4666	1.1680	1.4109	1.8584	2.2033
CA	SKB•5	1	1	20	4.02	3.0864		1.2330	3.7283	4.9108	5.8219	8.1544		0.4667	1.1681	1.4111	1.8587	2.2036
EU	LG4,	2	1	63	8.03	3.0876		1.2335	3.7298	4.9127	5.8242	8.1576		0.4669	1.1686	1.4117	1.8594	2.2044
RH	KA2	5	50	45	20.07	3.0882		1.2338	3.7305	4.9136	5.8253	8.1591		0.4669	1.1688	1.4120	1.8598	2.2048
CU	KA2	2	50	29	8.03	3.0886		1.2339	3.7310	4.9143	5.8261	8.1602		0.4670	1.1690	1.4121	1.8600	2.2051
TM	LB4	2	4	69	8.03	3.0886		1.2339	3.7310	4.9143	5.8261	8.1602		0.4670	1.1690	1.4121	1.8600	2.2051
RE	SLB2+	3	1	75	12.04	3.0890		1.2341	3.7315	4.9149	5.8269	8.1613		0.4671	1.1691	1.4123	1.8602	2.2054
SM	KA1	10	100	62	40.13	3.0892		1.2342	3.7317	4.9152	5.8272	8.1618		0.4671	1.1692	1.4124	1.8604	2.2056
CA	KB1	1	15	20	4.01	3.0896		1.2343	3.7322	4.9159	5.8280	8.1629		0.4672	1.1694	1.4126	1.8606	2.2058
CF	LG1	6	10	98	24.07	3.0900		1.2345	3.7326	4.9164	5.8287	8.1638		0.4672	1.1695	1.4128	1.8608	2.2061
FR	LA1	3	100	87	12.03	3.0902		1.2346	3.7329	4.9168	5.8291	8.1644		0.4672	1.1696	1.4129	1.8610	2.2063
BA	KA1,2	8	150	56	32.07	3.0926		1.2355	3.7358	4.9206	5.8337	8.1708		0.4676	1.1705	1.4140	1.8624	2.2080
PA	LB2	4	20	91	16.02	3.0946		1.2363	3.7382	4.9238	5.8374	8.1761		0.4679	1.1713	1.4149	1.8636	2.2094
W	L1-O1	3	1	74	12.02	3.0951		1.2365	3.7388	4.9246	5.8383	8.1773		0.4680	1.1714	1.4151	1.8639	2.2097
AU	SLB2•,	3	1	79	12.01	3.0959		1.2368	3.7398	4.9259	5.8399	8.1795		0.4681	1.1718	1.4155	1.8644	2.2103
RE	LG2	3	1	75	12.01	3.0968		1.2372	3.7409	4.9273	5.8416	8.1819		0.4682	1.1721	1.4159	1.8649	2.2110
PR	KA1	9	100	59	36.03	3.0968		1.2372	3.7409	4.9273	5.8416	8.1819		0.4682	1.1721	1.4159	1.8649	2.2110
PB	L1-M1	3	1	82	12.01	3.0969		1.2372	3.7410	4.9274	5.8417	8.1821		0.4683	1.1721	1.4159	1.8650	2.2110
MO	LB10	2	1	67	8.01	3.0970		1.2373	3.7411	4.9277	5.8420	8.1824		0.4683	1.1722	1.4160	1.8651	2.2111
IN	KA2	6	50	49	24.00	3.0992		1.2382	3.7438	4.9312	5.8461	8.1883		0.4686	1.1730	1.4170	1.8664	2.2127
MG	LB3	3	6	80	11.99	3.1004		1.2386	3.7453	4.9331	5.8484	8.1914		0.4688	1.1735	1.4175	1.8671	2.2136
CA	SKB+	1	1	20	4.00	3.1011		1.2389	3.7461	4.9342	5.8497	8.1933		0.4689	1.1737	1.4179	1.8675	2.2141
AU	SLB5•,	3	1	79	11.99	3.1022		1.2394	3.7474	4.9359	5.8518	8.1962		0.4691	1.1741	1.4184	1.8682	2.2149
U	L3-M2	3	1	92	11.98	3.1037		1.2400	3.7493	4.9383	5.8546	8.2002		0.4693	1.1747	1.4191	1.8691	2.2159
BI	L2-M2	3	1	83	11.98	3.1038		1.2400	3.7493	4.9384	5.8547	8.2003		0.4693	1.1747	1.4191	1.8691	2.2160
MO	KB2	5	5	42	19.96	3.1047		1.2404	3.7505	4.9399	5.8565	8.2028		0.4694	1.1751	1.4195	1.8697	2.2166
BK	LB1	5	50	97	19.96	3.1050		1.2405	3.7508	4.9403	5.8570	8.2035		0.4695	1.1752	1.4196	1.8699	2.2168
SM	KA1,2	10	150	62	39.92	3.1052		1.2406	3.7511	4.9407	5.8575	8.2042		0.4695	1.1753	1.4197	1.8700	2.2170
I	SLA9	1	1	53	3.99	3.1055		1.2407	3.7514	4.9412	5.8580	8.2049		0.4696	1.1754	1.4199	1.8702	2.2172
BR	SKA4	3	1	35	11.97	3.1058		1.2408	3.7518	4.9417	5.8586	8.2057		0.4696	1.1755	1.4200	1.8704	2.2174
BR	SKA3,	3	1	35	11.96	3.1091		1.2421	3.7558	4.9469	5.8648	8.2145		0.4701	1.1768	1.4215	1.8724	2.2198
RE	LG6	3	1	75	11.95	3.1109		1.2428	3.7580	4.9498	5.8683	8.2192		0.4704	1.1774	1.4224	1.8735	2.2211
PR	KA1,2	9	150	59	35.86	3.1112		1.2430	3.7583	4.9503	5.8688	8.2200		0.4704	1.1776	1.4225	1.8736	2.2213
AU	SLB2•5	3	1	79	11.95	3.1124		1.2434	3.7598	4.9522	5.8711	8.2232		0.4706	1.1780	1.4230	1.8744	2.2222
AC	LB3	4	6	89	15.93	3.1128		1.2436	3.7602	4.9527	5.8717	8.2241		0.4707	1.1781	1.4232	1.8746	2.2224
CS	KB2	9	5	55	35.82	3.1148		1.2444	3.7627	4.9560	5.8756	8.2296		0.4710	1.1789	1.4241	1.8758	2.2239
IN	KB2	7	5	49	27.86	3.1149		1.2444	3.7628	4.9562	5.8758	8.2298		0.4710	1.1790	1.4242	1.8759	2.2239
SN	LB9	1	1	50	3.98	3.1150		1.2445	3.7629	4.9564	5.8760	8.2301		0.4710	1.1790	1.4242	1.8759	2.2240
I	SLA8	1	1	53	3.98	3.1150		1.2445	3.7629	4.9564	5.8760	8.2301		0.4710	1.1790	1.4242	1.8759	2.2240

FIG. 10.9. Portion of a page of an x-ray spectrometer table for use when other crystals are used on spectrometers calibrated directly in angstroms for LiF or ADP crystals. [E. W. White, G. V. Gibbs, G. G. Johnson, Jr., and G. R. Zechman, Jr., "X-Ray Wavelengths and Crystal Interchange Settings for Wavelength-Geared Curved-Crystal Spectrometers," *Mineral Industries Experiment Station Special Publication* **3-64**, Pennsylvania State University (1965); courtesy of the authors and Pennsylvania State University.]

nants. The incoherently scattered peaks appear as relatively broad peaks on the high-2θ (long-wavelength) sides of the coherently scattered peaks (Figure 2.10). All scattered peaks are more intense the lower the effective atomic number of the specimen. Then, by use of a 2θ-to-line table, all remaining peaks are identified, starting with the most intense.

Peaks may be as much as 0.5° 2θ away from their tabulated values because of slight misalignment of the goniometer or lag in ratemeter–recorder response, and this displacement may vary progressively from start to end of the spectrum. Moreover, it may be difficult to read the chart more precisely than $\pm 0.2^\circ$ 2θ.

One soon learns to recognize groups of associated lines. The K x-ray spectrum of an element consists of the strong $K\alpha$ line and the weaker (1/5 or less) $K\beta$ line at shorter wavelength (lower 2θ). In the second and higher orders, the $K\alpha$ peak splits into the $K\alpha_1\alpha_2$ doublet in 2:1 intensity ratio, the more intense $K\alpha_1$ component occurring at lower 2θ. The L x-ray spectrum of an element usually consists of at least the following three lines more or less equally spaced at progressively shorter wavelengths (lower 2θ): the strongest (usually) $L\alpha_1$, the somewhat weaker (usually) $L\beta_1$, and the still weaker ($\sim$0.4 or less) $L\gamma_1$. For elements of atomic number $\sim$82 (lead), the relatively intense $L\beta_2$ line lies very close to the $L\beta_1$ line and may be unresolved. At lower Z, $L\beta_2$ lies at shorter wavelength (lower 2θ) than $L\beta_1$, and at higher Z, at longer wavelength. If the element is present in high concentration, the many other L lines appear, most of them between $L\alpha_1$ and $L\gamma_1$, although Ll lies at longer wavelength (higher 2θ) than $L\alpha_1$, and $L\gamma_2$ lies at shorter wavelength (lower 2θ) than $L\gamma_1$.

Insofar as possible, more than one of its lines should be identified for an element to be regarded as present. If a peak is believed to be Fe $K\alpha$, the weaker ($\sim$1/5) Fe $K\beta$ should be present. If a peak is believed to be Pb $L\beta_1$, the slightly stronger Pb $L\alpha_1$ and weaker Pb $L\gamma_1$ peaks should be present. If relatively strong Mo $K\alpha$–$K\beta$ peaks are present, their second- and perhaps third-order peaks should be present. If the associated peaks cannot be found, their absence must be accounted for. Some examples follow: If a $K\alpha$ peak is very weak, the $K\beta$ peak may not be distinguishable from the background; $K\beta$ lines of elements 9–17 (fluorine to chlorine) are very weak relative to the $K\alpha$ lines; a peak may be lost in an interfering peak; and second-order peaks may undergo extinction by the crystal.

Relative intensities of peaks must be considered and anomalies accounted for. For example, suppose a strong peak is believed to be Cu $K\alpha$ but the Cu $K\beta$ peak is very weak, when it should be $\sim$1/5 as intense as Cu $K\alpha$. Perhaps another line is superposed on Cu $K\alpha$ giving it an anomalously high apparent intensity, or perhaps a matrix element is preferentially absorbing the Cu $K\beta$—nickel, for example. Then again, perhaps the strong line is not Cu $K\alpha$ after all. In considering relative intensities, one can be guided by certain general relationships: the relative intensities of the lines within a series (K, L, etc.) are given in the tables, especially (*86*), and in Figure 1.6; in the same 2θ region and for the same concentrations, first-order $K\alpha$ lines are 7–10 times as intense as first-order $L\alpha_1$ lines; and for LiF crystals, second-order lines are ~1/10 as intense as first-order lines.

A peak may appear anomalously intense if: (1) another peak—an emission line from a sample element or a coherently or incoherently scattered target line—is superimposed on it; (2) it lies on the short-wavelength side of an absorption edge of the detector gas or of the iodine in a scintillation counter phosphor; and/or (3) it is preferentially excited by a matrix-element line or a strong x-ray tube target line. A peak may appear anomalously weak if: (1) it lies on the short-wavelength side of an absorption edge of a matrix element; and/or (2) it lies on the long-wavelength side of a detector absorption edge.

10.5. GENERAL PROCEDURES FOR QUALITATIVE AND SEMIQUANTITATIVE ANALYSIS

If two specimens A and B are identical except that A has twice the, say, iron concentration of B, the Fe $K\alpha$ net intensity from A will be about twice that from B. However, this simple relationship is not generally true for *different* elements in the *same* specimen for many reasons, including the following: (1) *Excitation potential.* Each element has its own excitation potential, and the amount by which this potential is exceeded by the x-ray tube operating potential affects the excitation efficiency. (2) *Atomic weight.* If a specimen is, say, 50 wt % iron (at. wt. 56) and 50 wt % silicon (at. wt. 28), there are about twice as many silicon atoms as iron atoms. (3) *Particle size* and/or *surface texture* of the specimen affect different wavelengths to different degrees. (4) *Absorption-enhancement effects.* (5) *Crystal "reflectivity"* and *detector quantum efficiency* vary with wavelength. Thus, simple comparison of peak heights on a ratemeter–recorder chart, or of scaled intensities, does not necessarily permit semiquantitative estimation of relative concentrations, although it does permit designation of the elements present as major, minor, or trace constituents. The following two sections describe ratemeter–recorder procedures for qualitative and semiquantitative x-ray spectrometric analysis.

10.5.1. Normalization Factor Method

Consider a specimen having 10 wt % element A and 10 wt % element B in a spectrometer operating at specified standard conditions. If, say, the A $K\alpha$ peak were twice as "high" as the B $K\alpha$ peak, a normalization factor —2—for element B with respect to element A could be applied to the B $K\alpha$ peak to bring the two peak heights into agreement with the relative concentrations. Similarly, the major peak of each element of analytical interest could be normalized with respect to the A peak. If this were done, each analyte would then automatically be normalized to every other analyte, and the presence of element A would not even be required for a valid analysis. If the peaks of *all* elements present were thus normalized, the normalized heights could be totaled, and the concentration of each element would be the fraction of this total represented by its normalized peak height. The normalization factors are derived from ratemeter–recorder 2θ scans on standards of the same general composition as the samples to be analyzed.

Anater (*A17*) developed a semiquantitative method based on this principle. He normalized all elements with respect to iron because it was present in most of his samples and gives relatively high Fe $K\alpha$ intensity even in low concentration. He derived his normalization factors from NBS alloy standards by the following procedure.

A qualitative 2θ scan is made from a standard over the region 10–90° 2θ. A similar scan is made from a suitable blank to permit measurement of scattered target and other instrument residual lines. Then for each detected element of analytical interest A, the following steps are taken:

1. The major peak and adjacent background are rerecorded in duplicate, using a ratemeter–recorder scale factor giving maximum on-scale peak height. This is done for both the standard and blank.

2. The average (of the duplicate tracings) net (peak minus background) peak height is measured in units of chart divisions for both standard and blank.

3. The average net peak height (step 2) is multiplied by the scale factor to give "adjusted chart divisions," ACD, for both standard and blank.

4. The ACD (step 3) for the blank is subtracted from that for the standard to give "net adjusted chart divisions," NACD.

5. From the known composition of the standard, the ratio wt % A/wt % Fe is calculated, and from step 4, the ratio $(\text{NACD})_{\text{A}}/(\text{NACD})_{\text{Fe}}$ is calculated.

6. The weight concentration ratio is divided by the NACD ratio to give the normalization factor NF for the analyte.

Steps 1–6 are repeated for several appropriate standards, and the several normalization factors thus obtained for each analyte are averaged. These average NF's are subsequently used to derive semiquantitative analyte concentrations from the measured ratemeter peak heights from samples. The normalization factors are said to be relatively insensitive to the amount and form of the sample. Also, factors derived from alloys are applicable to solutions of the alloys and to the mixed oxides obtained by evaporation of the solutions and firing of the residues.

Anater applied his method to 27 elements having atomic number $\gtrsim 13$ (aluminum) with precision usually within ± 20 wt % of the amount present. The normalization factors varied from 80 for aluminum, the least sensitive element in the group, to 0.74 for calcium, the most sensitive.

10.5.2. Method of Salmon

The techniques developed by Salmon (*25p*, *S4*) for qualitative and semiquantitative x-ray spectrometric analysis of minerals and ores permit large numbers of accurate analyses to be performed in a remarkably short time. These techniques provide an excellent example of the speed and accuracy attainable by x-ray spectrometric techniques and well warrant detailed consideration.

All samples are reduced to at least -200 mesh by a standardized grinding procedure, loosely packed in CaPlugs, sealed with 6-μm (0.00025-inch) Mylar, and placed in the inverted position in a Philips 100-kV spectrometer. The CaPlugs are available in a range of sizes. The Mylar-sealed samples minimize contamination of the samples, instrument, and laboratory; reduce loading time; and can be stored for future reexamination. The procedure is also applicable to residues from evaporation or ashing, homogeneous bulk solids, and liquids and slurries.

Six superposed 2θ scans are made. For each, the chart is rolled back to the appropriate starting place, and the recorder pen is replaced with another having ink of a different color. All scans are made at the following conditions, except as noted below:

W target
50 kV, 33 mA
LiF crystal
10×0.0125-cm (4×0.005-inch) source collimator
3.75×0.0575-cm ($1\frac{1}{2} \times 0.023$-inch) detector collimator
Air path

Scintillation counter
Scanned region and rate: 90° 3° 2θ, 8° 2θ/min
Ratemeter time constant: 0.44 s
Full-scale recorder deflection (switched automatically):

90°–60° 2θ, 800 counts/s
60°–31° 2θ, 1600 counts/s
31°–17° 2θ, 3200 counts/s
17°–3° 2θ, 6400 counts/s

Pulse-height selection synchronized with 2θ drive

Scan 1 is made at the indicated conditions and provides complete qualitative data and intensity data for semiquantitative analysis for minor and trace constituents, except where scattered target lines interfere.

Scan 2 is made at the same conditions, but without pulse-height selection, and the goniometer, full-scale deflection, and chart position are adjusted to rerun 2θ regions where peaks are off-scale. This scan shows gross intensities at all wavelengths without discrimination against higher-order peaks. The data are used for estimation of major constituents and to supplement the data in scan 1 for minor and trace constituents.

Scan 3 is limited to the region 40°–28° 2θ with full-scale deflection 400 counts/s, and a filter is placed over the x-ray tube window to supress tungsten *L* lines. The filter consists of three foils: 50-μm (0.002-inch) aluminum, 25-μm (0.001-inch) brass (Cu–Zn), and 12.5-μm (0.0005-inch) iron. This scan permits *K* lines of gallium, germanium, arsenic, and selenium, and *L* lines of elements 73–81 (tantalum to thallium) to be observed without interference from scattered tungsten *L* lines.

Scan 4 is made only if in scan 1 there are instances of overlapping first-order peaks. Such peaks are rerun at the same conditions except that a filter having an absorption edge between the peaks is placed in the secondary beam. In this way, the peak having the shorter wavelength is absorbed.

Scan 5 is limited to the region 26°–3° 2θ with excitation power 100 kV, 16.5 mA. This scan gives improved data for estimation of minor and trace concentrations of elements 38–74 (strontium to tungsten).

Scan 6 is made with helium path. This scan gives improved intensity data for wavelengths $>$1 Å and permits more accurate estimation of minor and trace concentrations of titanium, vanadium, chromium, and manganese.

The peaks are identified by placing the chart over a transparent template mounted on an illuminated cylindrical drum. The template is prepared from the tables for the 2θ-scan and chart speeds used and shows the angles at which all orders of all lines of any interest appear. The marks on the template show through the illuminated chart.

The semiquantitative analyses are based on intensities measured from the charts. Net intensity of a peak is measured directly if interfering peaks are absent. If an analyte peak includes contributions from more than one element, the contributions of the interfering elements are estimated from their unobscured peaks, and the analyte peak is corrected.

Standards are obtained by chemical analysis of selected samples or by synthesis. Oxide powders may be mixed, or they may be added to a base mineral of known composition to adjust the concentration of the analytes. If the powders to be mixed vary substantially in density, a solution of the denser constituent is added to the lighter powder; the paste is then thoroughly mixed and dried. The standards are subjected to the same procedure as the samples to obtain intensity data from which to prepare calibration curves.

Several techniques are used to convert the intensity data to semiquantitative analytical data. The calibration curve may simply have the function intensity *versus* concentration. Other semiquantitative techniques are based on the fact that scattered-background and analyte-line intensities decrease in constant ratio as the absorption coefficient of the sample increases. Sets of calibration curves of *net* intensity *versus* concentration are prepared, each curve for a different background level. Each sample is then analyzed by use of the curve corresponding to its background level. The background-ratio method may also be used, in which the calibration curve has the form analyte line/background intensity ratio *versus* concentration. The ratioed background intensity may be measured adjacent to the peak or in the scattered continuum hump. In all these methods, intensities are in terms of chart divisions.

Intensity ratios of lines of the same series for elements of nearly the same—preferably adjacent—atomic number are usually substantially proportional to their concentration ratios, unless an absorption edge intervenes.

In the emission-absorption technique, the analyte-line intensity is measured from each sample directly and through a thin filter of the same sample placed in the secondary beam. The filter consists of a 25–500-μm layer of the finely powdered sample in a thin cell having two Mylar windows or spread with binder on a radiolucent support. For layer thickness $\lesssim$250 μm excellent uniformity is realized by mixing equal volumes of powder specimen and Vaseline and pressing the paste between spaced 25-μm (0.001-inch) Mylar films (*S4*). A rough matrix-correction factor is calculated from the transmittance measurement and used to correct the intensities derived from the direct measurement. The corrected intensities are then applied to the calibration curves. The absorption measurement can be made by recording another 2θ scan of the peaks to be corrected, this time with the sample filter in the secondary beam. Alternatively, the scaler can be used for both the direct and the filtered measurements.

While this book was in press, Jenkins [pp. 100–102 in (*21*)] presented two excellent examples of semiquantitative analysis by application of simple corrections to ratemeter–recorder chart peak heights:

Table 10.10 shows data for analysis of a mixture of oxides of selenium, zirconium, and antimony in concentrations 25, 50, and 25 wt%, respectively, containing a trace iodate impurity. This is a very favorable case: The three major heavy elements do not differ widely in atomic number ($Z = 34$, 40, and 51 for Se, Zr, and Sb, respectively); the same line ($K\beta$) is measured for each; and the only other major element is oxygen. The table shows that the relative intensities (chart peak heights above background) give a very passable semiquantitative analysis. Correction for oxide form improves the results only slightly, although presumably this refinement would be more beneficial with less favorable systems.

Table 10.11 shows data for analysis of a 50–50-wt% Sn–Pb solder containing trace antimony and bismuth impurities. This is a more complex case:

TABLE 10.1. Semiquantitative Analysis of an Oxide Mixture[a]

Element and line	Oxide	Oxide factor[b]		Relative intensity[c] I_{rel}			Composition, wt% Estimated	True
Se $K\beta$	SeO_2	1.40	×	28	=	39.2	39.2/133 = 29	25
Zr $K\beta$	ZrO_2	1.35	×	48	=	64.8	64.8/133 = 49	50
Sb $K\beta$	Sb_2O_5	1.33	×	20	=	26.6	26.6/133 = 20	25
I $K\beta$	10_3^-	1.38	×	2	=	2.8	2.8/133 = 2	Trace
						133.4	100	

[a] From Jenkins [pp. 100–102 in (*21*)].

[b] M/nA, where M is the molecular weight of the oxide, A is the atomic weight of the heavy element, and n is the number of atoms of heavy element in a molecule of oxide: $SeO_2/Se = 111/79$; $ZrO_2/Zr = 123/91$; $Sb_2O_5/2Sb = 324/244$; $IO_3^-/I = 175/127$.

[c] Ratemeter–recorder chart divisions.

TABLE 10.2. Semiquantitative Analysis of a Tin–Lead Solder[a]

Element and line	Relative intensity[b] I_{rel}	Weighted fluorescent yield[c] ω'	Excitation potential, kV V_{exc}	Excitation factor[d] F	$\omega' F$	Corrected intensity[e] $I_{rel,cor}$	Composition, wt%: Estimated[f] C_{est}	Composition, wt%: True
Sn $K\alpha$	250	0.7	29	130	91	2.75	49.2	50
Sb $K\alpha$	2.5	0.7	30	121	85	0.03	0.5	Trace
Pb $L\alpha_1$	180	0.2	13	323	64.6	2.79	49.9	50
Bi $L\alpha_1$	1.5	0.2	13	323	64.6	0.02	0.4	Trace
						5.59	100.0	

[a] From Jenkins [pp. 100–102 in *(21)*].

[b] Ratemeter–recorder chart divisions.

[c] The weighted fluorescent yield $\omega' = \omega g$, where ω and g are the fluorescent yield and probability factor [Equation (4.4)], respectively. For example, the weighted fluorescent yield for Sn $K\alpha$ is the Sn K fluorescent yield ($\omega_{\mathrm{Sn}K} = 0.845$) multiplied by $g_{\mathrm{Sn}K\alpha}$, the fraction of all Sn K-line photons that are Sn $K\alpha$ (Sn $K\alpha_1$ or Sn $K\alpha_2$) photons. This fraction is the relative intensity of Sn $K\alpha$ divided by the sum of the relative intensities of all the principal Sn K lines ($I_{rel} \gtreqless 1$):

$$\mathrm{Sn}\ K\alpha_1 : \mathrm{Sn}\ K\alpha_2 : \mathrm{Sn}\ K\beta_1 : \mathrm{Sn}\ K\beta_2 : \mathrm{Sn}\ K\beta_3 = 100:50:19:5:9$$
$$(100 + 50)/(100 + 50 + 19 + 5 + 9) = 150/183 = 0.82$$
$$0.845 \times 0.82 = 0.7$$

The best source of relative intensities is Reference *(86)*, although approximate values can be obtained from the column headings in Appendixes 1–3.

[d] $F = (V - V_{exc})^{1.7}$ [Equation (1.32)], where V is x-ray tube operating potential, 50 kV in this case.

[e] $I_{rel,cor} = I_{rel}/(\omega' F)$.

[f] $C_{est} = (I_{rel,cor}/\Sigma I_{rel,cor}) \times 100$.

The two major elements differ widely in atomic number ($Z = 50$ and 82 for Sn and Pb, respectively); lines of different series are measured for each (Sn $K\alpha$, Pb $L\alpha_1$); and the excitation potentials differ substantially ($V_{\mathrm{Sn}K} = 29$ kV, $V_{\mathrm{Pb}L\mathrm{III}} = 13$ kV). The table shows that the uncorrected relative intensities are unsuitable, and how correction is made for line series by use of "weighted fluorescent yields" and for excitation potential by use of Equation (1.32). The corrected semiquantitative analysis is in excellent agreement with the known composition.

Part IV

Performance Criteria and Other Features

Chapter 11

Precision and Error; Counting Statistics

11.1. ERROR IN X-RAY SPECTROMETRIC ANALYSIS

The principal objective of the analytical method or "strategy" (Chapters 14 and 15) is to eliminate, minimize, circumvent, or correct for absorption-enhancement effects; another objective is to minimize the effect of instrumental drift, by use of a ratio method, for example. The principal objective of the specimen preparation and presentation technique (Chapters 16 and 17) is to eliminate or minimize specimen errors. It is appropriate that the nature and sources of these errors be discussed first. This chapter defines and discusses all the various sources of error that beset a quantitative x-ray spectrometric analysis and considers counting statistics and analytical precision in detail. Absorption-enhancement effects and spectral-line interference are discussed in Chapters 12 and 13, respectively.

11.1.1. Nature of Error

The *error* in a measurement (of intensity, for example) or in an overall analytical result is the difference between the measured value and the "true" value. However, since the true value must itself be determined by measurement, it might seem impossible to evaluate the error. Strictly speaking, this may be true, but in practice, error can be evaluated satisfactorily in terms of precision and accuracy, which, although often used synonymously, are quite different.

The *precision* of a measurement or analysis is the degree of agreement among replicate determinations made under conditions as nearly identical as possible. Quantitatively, precision p_i is the difference between the individual measurement m_i or analysis and the mean $\bar{m}$ of a large number or *set* of independent replicate measurements or analyses, usually expressed relative to the mean and as percent, that is,

$$p_i = \frac{m_i - \bar{m}}{\bar{m}} \times 100 \tag{11.1}$$

Thus the greater ("better") the precision, the smaller is its numerical value. The mean $\bar{m}$ is regarded as the "best known" value, that is, the value most likely to be true. Precision can be evaluated experimentally or calculated by summation of the individual contributing errors [Equation (11.11)].

Incidentally, the difference between the highest and lowest individual values m_i in a related set of measurements is known as the *range* or *spread.*

The *accuracy* of a measurement or analysis is the degree of agreement with the "true," accepted, or most reliably known value. Quantitatively, accuracy a_i is the difference between the individual measurement m_i or analysis and the true value t, usually expressed relative to the true value and as percent, that is,

$$a_i = \frac{m_i - t}{t} \times 100 \tag{11.2}$$

Thus again, the greater ("better") the accuracy, the smaller is its numerical value.

Incidentally, the difference between the individual and mean values, $m_i - \bar{m}$ in Equation (11.1), and between the individual and true values, $m_i - t$ in Equation (11.2), is known as the *deviation.*

In short, precision is a measure of the reproducibility of the measurement or analysis, and accuracy is a measure of its correctness. Accuracy is unattainable without precision, but precision does not necessarily guarantee accuracy. A measurement or analysis may be very precise, that is, reproducible, but very inaccurate. The *reliability* of a measurement or analysis is the degree to which it possesses precision and accuracy.

In quantitative x-ray spectrometric analysis, analyte concentration is derived from analyte-line intensities measured from the sample and one or more standards. The precision of the x-ray intensity measurements is discussed in great detail in Section 11.2. The overall analytical precision depends on the magnitude of the errors affecting the x-ray intensity measurements and on the effectiveness of the techniques used to minimize them. The overall accuracy depends on: (1) the accuracy with which the composition of the standards is known; (2) the degree to which the sample represents the bulk material being analyzed, (3) the uniformity of samples and standards with respect to composition, density, particle size and distribution, surface texture, etc.; (4) the similarity of samples and standards with respect to these properties; and (5) the precision and accuracy of the intensity measurements and the validity of the calibration. Often the accuracy is as good as the precision because it can usually be assumed that: (1) the measured x-ray intensity can be adjusted to agree with the analytical value by the calibration; and (2) the analytical concentration in the standards is accurately known or at least accepted as true. Analytical precision is discussed further in Section 11.3.

Error may be expressed in absolute or relative terms. *Absolute error* is the degree by which the measured and true values may be expected to differ, expressed in the physical unit of the measured quantity. *Relative error* is the degree by which the measured and true values may be expected to differ, expressed as a fraction or percent of the measured value. Thus, in x-ray spectrometric analysis, absolute error is expressed in terms of analyte concentration in weight percent, micrograms per square centimeter, or other unit, relative error in fraction or percent "of the amount present." In either case, the degree of certainty of the indicated error must be stated, as discussed in Section 11.1.2. For example, if an analyst states that the analyte concentration is 20% ±1% absolute (with 95% certainty), he means that he is 95% sure that the true concentration lies between 19 and 21%, which is 20% ±5% *relative*. If he states that the analyte concentration is 20% ±1% relative (with 95% certainty), he means that he is 95% sure that the true concentration lies between 19.8 and 20.2%, which is 20% ±0.2% *absolute*.

The error is said to be *high* or *low*, or *positive* or *negative*, depending on whether the measured value is, respectively, greater or less than the true value.

The errors affecting an x-ray spectrometric analysis may be classified as random or systematic. *Random errors* consist of small differences in successive values of a measurement made repetitively with great care by the same competent person under conditions as nearly constant as possible. The magnitude of random errors can be evaluated, and the errors can be minimized, but not corrected for. *Systematic errors* are those that can be avoided, or at least evaluated and corrected for; they may be constant, in which case they account for the deviation or *bias* of the experimental result from the true value; or they may fluctuate about a mean value, in which case they contribute to the precision. To a first approximation, random and systematic errors limit the precision and accuracy, respectively.

11.1.2. Elementary Statistics

Each measurement or overall analytical determination results in a numerical value x. Replicate measurements or determinations result in a series of values $x_1, x_2, \ldots, x_n$, which constitute a population of n members. If the measurements are subject only to random errors, a plot of frequency of occurrence of individual values x_i *versus* the value of x will have a *Gaussian* or *normal* distribution having the form shown in Figure 11.1 and given by the equation

$$P(x) = \frac{1}{(2\pi\bar{x})^{1/2}} \exp - \frac{(x - \bar{x})^2}{2\bar{x}} \qquad (11.3)$$

where $P(x)$ is the probability of occurrence of a specified value x, $\bar{x}$ is the mean value of x in the population (see below), and $(x - \bar{x})$ is the deviation of an individual value from the mean.

A series of replicate measurements is characterized by its population (the number of measurements or determinations), range or spread (the difference between the highest and lowest values), mean, and scatter or divergence. The Gaussian distribution of the series is characterized simply by the mean and variance. The *mean* $\bar{x}$ is the best approximation of the true

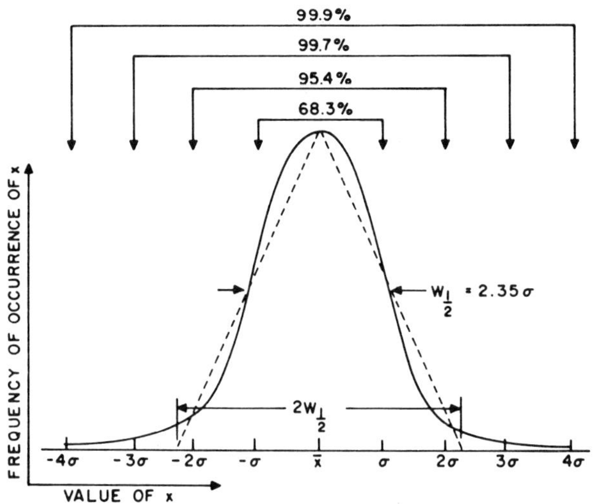

FIG. 11.1. The Gaussian (normal) distribution. The full width at half maximum (FWHM) is 2.35 σ intervals, as shown. The full width at one-tenth maximum (FWTM, not shown) is 1.823 σ intervals. A Gaussian distribution may be represented approximately by an isosceles triangle having base twice the width at half height (dashed lines); such a triangle contains ~92% (~2σ) of the total population of the distribution.

value and is given by

$$\bar{x} = (\sum x_i)/n \tag{11.4}$$

The scatter in the individual values is a measure of the precision of the measurement or analysis and is evaluated by the variance and standard deviation. The *variance* v is the sum of the squares of the deviations of the individual values x_i from their mean $\bar{x}$, divided by the number of degrees of freedom, which is one less than the number in the population:

$$v = \frac{\sum (x_i - \bar{x})^2}{n - 1} \tag{11.5}$$

The *standard deviation* σ is the square root of the sum of the squares of the individual deviations divided by the degrees of freedom:

$$\sigma = \left[\frac{\sum (x_i - \bar{x})^2}{n - 1}\right]^{1/2} = \left\{\frac{1}{n - 1}\left[\sum x_i^2 - \frac{(\sum x_i)^2}{n}\right]\right\}^{1/2} \tag{11.6}$$

Variance and standard deviation are related as follows:

$$v = \sigma^2; \qquad \sigma = \sqrt{v} \tag{11.7}$$

Actually, in statistics, the *standard deviation* σ is defined as the square root of the mean of the squares (rms) of the deviations of an infinite set of measurements, or at least of all possible measurements ν, from their arithmetic mean μ. In practice, the *estimate of the standard deviation* s is the rms deviation of a limited set of measurements n from their mean $\bar{x}$, which is itself the *estimate of the mean.* However, the practice of using the symbol σ instead of s in the x-ray literature is very common, as is the practice of referring to precision in terms of one, two, and three "sigma." Therefore, the symbol σ is used in this book.

The practical significance of the standard deviation is shown in Figure 11.1 and may be stated in various ways, including the following:

1. The probability is 68.3% ($\sim$7/10) that any individual value of x will deviate from the average $\bar{x}$ of a very large number of values by $\leq\sigma$.

2. The probability is 68.3% that any x will have a value between $\bar{x} \pm \sigma$.

3. Of a very large number of measurements of x, 68.3% will have values of x lying between $\bar{x} \pm \sigma$.

4. It follows that 68.3% of the area under the Gaussian distribution curve lies between $\bar{x} \pm \sigma$.

Similarly, the probability is 95.4% (19/20) that an individual x will lie between $\bar{x} \pm 2\sigma$, 99.7% (997/1000) that it will lie between $\bar{x} \pm 3\sigma$, and 99.9% (999/1000) that it will lie between $\bar{x} \pm 4\sigma$. Finally, the probability is 50% (1/2) that an individual x will lie between $\bar{x} \pm 0.67\sigma$; 0.67σ is known as the *probable error* or *probable fractional error.*

Figure 14.3 shows a typical calibration curve on which the associated error is indicated on each calibration point in the conventional manner. The ranges may indicate $\pm 1\sigma$ or some other specified precision. For example, if $\pm 1\sigma$ is indicated, the probability is 68.3% that the true intensity for each point is neither greater nor less than the limits shown. If the fixed-count method is used, the range is the same for each point, as shown, but if different numbers of counts are accumulated for each point, the range is also different. In addition to the random error shown for all the points, the point for the standard having second highest concentration also shows a substantial systematic error.

The larger σ is, the flatter is the Gaussian distribution curve; the width of the distribution at half its maximum height is $2(2 \log_e 2)^{1/2}$ σ intervals, or 2.35σ (Figure 11.1). The flatter the curve, the greater is the spread in replicate intensity measurements or analytical results, and the lower is the precision in the measurements or results. The larger the number n of replicate measurements, the more nearly their distribution corresponds to the Gaussian curve and the more reliable is the calculated value of σ.

The standard deviation expressed relative to the mean is the *relative standard deviation* or *coefficient of variation* ε and is often expressed as a percent, although in this chapter, for simplicity, ε is expressed as a fraction:

$$\varepsilon = \sigma/\bar{x} \qquad \text{or} \qquad \varepsilon = 100\sigma/\bar{x} \tag{11.8}$$

If n replicate determinations are made and averaged of a quantity having standard deviation σ and relative standard deviation ε, the terms for the mean are given by

$$\sigma_n = (1/\sqrt{n})\sigma \tag{11.9}$$

$$\varepsilon_n = (1/\sqrt{n})\varepsilon \tag{11.10}$$

These equations are used to calculate the standard counting error (Section 11.2.2) for replicate measurements of an intensity or accumulated count.

If it is required to calculate the standard deviation or relative standard deviation of a combination of measurements or determinations, each having its own standard deviation, the following equations apply.

For the sum or difference of two values, $x_1 + x_2$ or $x_1 - x_2$, each having its own σ and ε,

$$\sigma = (\textstyle\sum \sigma_i^2)^{1/2} \tag{11.11}$$

$$\varepsilon = \frac{[(x_1\varepsilon_1)^2 + (x_2\varepsilon_2)^2]^{1/2}}{x_1 \pm x_2} \tag{11.12}$$

TABLE 11.1. Statistical Precision

Symbol	Term	Confidence level	Probability
0.67σ	Probable error	50.0%	1/2
σ	Standard deviation, sigma	68.3%	7/10
2σ	Two sigma	95.4%	19/20
3σ	Three sigma	99.7%	997/1000
4σ	Four sigma	99.9%	999/1000

Note that Equation (11.11) applies to any number of individual values added and/or subtracted. In Equation (11.12), the denominator is $x_1 + x_2$ or $x_1 - x_2$ according to whether the calculation is for a sum or difference, respectively. These equations are used to calculate the standard counting error (Section 11.2.2) of a net intensity ($I_P - I_B$) or net count ($N_P - N_B$), where P and B refer to peak and background, respectively.

For the product of two measurements $x_1 x_2$,

$$\sigma = x_1 x_2[(\sigma_1/x_1)^2 + (\sigma_2/x_2)^2]^{1/2} \tag{11.13}$$

$$\varepsilon = (\varepsilon_1^2 + \varepsilon_2^2)^{1/2} \tag{11.14}$$

For the quotient of two measurements x_1/x_2,

$$\sigma = (x_1/x_2)[(\sigma_1/x_1)^2 + (\sigma_2/x_2)^2]^{1/2} \tag{11.15}$$

Equation (11.14) applies also for the quotient.

One final statistical term: The highest and lowest values of x between which the actual measurement is likely to fall with a specified probability are known as the *confidence limits* or *confidence levels*. For example, one may predict that a measured x will lie between $\bar{x} \pm \sigma$ at the 68.3% confidence level, and $\bar{x} + \sigma$ and $\bar{x} - \sigma$ are the 68.3% confidence limits. The most frequently cited confidence limits are given in Table 11.1.

11.1.3. Sources of Error

11.1.3.1. General

The errors besetting x-ray spectrometric analyses may be classified by source as follows:

1. *Statistical counting error* (Section 11.2) constitutes the best possible attainable precision and depends only on the total accumulated count.

2. *Instrumental errors* (Section 11.1.3.2) consist of short-term and long-term variation, instability, and drift in instrumental components, conditions, and parameters, principally the following: (a) X-ray tube potential (kV) and current (mA). (b) Intensity and distribution of the primary x-ray beam (caused by changes in dimension and position of internal components of the tube). (c) Crystal interplanar spacing (caused by changes in temperature; such changes cause slight displacement in 2θ for the spectral line). (d) Density of the air in the radiation path. (e) Gas-amplification of the proportional counter (caused by drift in detector-tube potential, changes in ambient temperature, and, in flow counters, changes in ambient pressure). (f) Secondary-emission ratio of the scintillation counter (caused by drift in detector-tube potential and changes in ambient temperature). (g) Coincidence (dead-time) losses in the detector and electronic circuitry. (h) Shift and distortion of pulse-height distributions (only when pulse-height analysis in used). (i) Electronic circuitry.

3. *Operational errors* (manipulative or resetting errors) (Section 11.1.3.2) consist of slight nonreproducibility in settings of instrument conditions, principally the following: (a) x-ray tube potential (kV); (b) x-ray tube current (mA); (c) goniometer angle (2θ); (d) amplifier gain; (e) pulse-height analyzer baseline and window; and (f) detector-tube potential.

4. *Specimen errors* (Section 11.1.3.3) arise in the specimen itself. However, specimen errors do not include *sampling errors*, that is, errors arising from failure of the submitted sample to be representative of the bulk of the material to be analyzed. Specimen errors may be classified as follows. (a) Absorption-enhancement ("matrix") effects (Chapter 12). (b) Chemical effects (Section 11.1.3.4). (c) Position effects, including variations in specimen plane, takeoff angle, position, orientation, and flatness (warp, ripple, wrinkle, etc.); position effects include the specimen-insertion effect. (d) Physical attributes, including thickness (if less than infinite); heterogeneity of composition; heterogeneity of density (porosity, voids, cracks, etc.); surface texture (roughness, grind marks, etc.); particle size (average and distribution); packing density of loose and briqueted powders; and expansion, volatility, bubble formation, radiolysis, and window distension of liquid specimens.

5. *Error in estimation of concentration from the calibration curve.*

6. *Spectral-line interference.* Inasmuch as a nearby spectral line can contribute to an intensity measurement of an analyte line, spectral-line interference may constitute a source of error (Section 13.3).

Counting error, most instrumental errors, operational errors, and errors in the estimation of analyte concentration from the calibration curve are random. Specimen errors (including absorption-enhancement effects) and spectral-line interference are systematic. Each of the foregoing groups of errors, or each individual source of error, may be expressed in terms of the standard deviation (see below) that it contributes to the precision of the measured intensity or of the analytical result. Of all the errors listed above, only the statistical counting error is easily and accurately calculated. The others must be evaluated experimentally.

In general, errors due to specimen heterogeneity are the most difficult to remedy, and instrumental errors are the easiest.

11.1.3.2. Instrumental and Operational Error

Instrumental instability is usually limited by x-ray generator instability, but may increase with the complexity of the electronic readout components. Instrumental instability consists of two components: *short-term* fluctuations and drift occurring from moment to moment and/or over the duration of a series of analytical measurements, and *long-term* drift over a period of several days to a year. Short-term instability is caused by fluctuations and drift in the ambient temperature, pressure, and humidity of the laboratory, in the ac power line, in the geometry of the x-ray tube, and in the operation of the x-ray tube power supply and electronic detection and readout components. Short-term drift may be either upward or downward, that is, toward higher or lower measured intensities, and may fluctuate upward and downward from time to time. Long-term drift is caused by aging of instrument components and is almost invariably downward. The output potential of the x-ray tube power supply gradually decreases due to aging of resistors and other components. X-ray tube output gradually decreases due to pitting of the target and sublimation of metal on the inside of the window. Crystal reflectivity gradually decreases from exposure to x-rays and the atmosphere. The quantum efficiency of the detector gradually decreases with age.

Long-term instability usually has several times the magnitude (as high as 10% in a year) of short-term instability. There are two approaches to reduction of error due to instrumental drift: one is to reduce the instability itself; the other is to choose a counting strategy that reduces the effect of any remaining instability on the intensity measurement.

Instrumental drift can be reduced in several ways, principally the following. Efficient kV and mA stabilization reduce drift in these two parameters. Efficient line stabilization ($\pm 0.05\%$ or less) and adequate warmup time (at least 30 min) reduce all forms of instrument instability except those caused by the effects of temperature and pressure on the crystal and detector.

Another form of instrumental instability is not generally recognized. When x-ray tube loading (kV, mA) is changed substantially, the primary-beam intensity may undergo a short period of instability. Consequently, if different excitation conditions are required for different analytes in a multi-element analysis, after each change of excitation, measurement should be delayed briefly to permit equilibration.

The goniometer may provide two sources of error: error in establishing the peak 2θ angle of the analyte line and error in resetting the goniometer repeatedly to this angle—the *reset error.* These errors are absent when the same analyte line is measured on all specimens without changing the goniometer setting. The reset error is always minimal at the analyte-line peak. An interesting example of the reset error is cited by Liebhafsky and his coauthors [p. 346 in *(27)*]: They made repeated intensity measurements from a specimen, resetting 2θ after each. The precision was increased 10-fold by pressing inward on the 2θ setting drum when resetting 2θ.

Drift in 2θ due to change in crystal d spacing with change in ambient temperature is minimized by avoidance of crystals with high thermal expansion coefficients (such as PET) and by temperature control of the spectrometer chamber. Temperature control also minimizes the effect of ambient temperature on the gas amplification of sealed and flow-proportional counters. The effect of ambient pressure on gas amplification in flow counters is minimized by discharging the effluent gas into a manostat. Shift and distortion of pulse-height distribution are discussed in Section 8.1.8.

The radiation path may also be a source of error. Variations in ambient temperature, barometric pressure, and humidity affect the x-ray absorption of an air path. Variations in helium flow and vacuum affect the x-ray absorption of these media.

Having reduced instrumental instability as much as possible, the operator can select a counting strategy to minimize the effects of the residual instability on the intensity measurement. These effects are minimized in ratio techniques (Chapter 14), and substantially eliminated when the two ratioed intensities are measured simultaneously. The instrumental effects are also substantially eliminated by monitor techniques (Section 7.2.2.4).

If measurements are made of a large number of specimens, total measurement time may be very long, and if maximum precision is required, it may be advisable to measure the samples and standards in random sequence to minimize the effect of any long-term drift. Adler proposes that the samples and standards be assigned serial numbers, then placed in random sequence by use of a table of random numbers. If replicate measurements or multielement analyses are to be made, the entire series of measurements should be thus randomized [p. 146 in *(1)*].

If the analyte-line intensity is very low, very long counting time is required to realize acceptable precision. In such cases, the effect of instrument drift may be minimized by dividing the line and background counting times into intervals and alternating them until the total times have been counted. Probably the ultimate example of this technique is the work of Chappell *et al.* (*C41*), who determined down to ~0.04 ppm of rubidium and strontium in rocks by continuous alternation of peak and background counts for a total counting time of ~10 h.

Operational error is substantially dependent on "slop" and backlash in the controls and on the attentiveness and care of the operator. It is a problem only when settings must be changed during an analysis, and is far less serious with stepwise controls than with continuous controls. The error is minimized by always approaching a setting from the same direction. Operational errors are substantially absent in modern semiautomatic and automatic instruments (Chapter 9).

It has been proposed that in a multielement analysis, maximum precision is realized by first measuring a certain analyte-line intensity on all specimens at $2\theta_L$, then background intensity on all specimens at $2\theta_B$. Then kV, mA, and the readout parameters are changed as required and another analyte line and background measured the same way. However, this procedure is very prone to specimen errors, and, because of the repeated reloading, briquet, powder, and liquid specimens must be handled with great care.

11.1.3.3. Specimen Error

Instrumental and operational errors may be very small in modern commercial instruments, and the counting error usually can be made very small by accumulation of large counts. Therefore, specimen errors may limit the precision of a quantitative x-ray spectrometric analysis. Techniques for reducing specimen error are discussed in detail in Chapters 16 and 17. Only general considerations are given here.

Specimen manipulative or placement error arises from the act of placing the specimen in its compartment, either repeated placement of a single specimen or placement of a series of geometrically and chemically similar specimens.

The measured surfaces of all specimens in an analytical series must lie precisely in the plane defined by the specimen mask. If the target-to-specimen distance is 25 mm along the central ray of the primary beam, a 0.5-mm shift along this line changes the analyte-line intensity by ~4%. If the takeoff angle is ~45°, a change of 1° changes analyte-line intensity ~3.5%. Solids must have plane surfaces and be seated firmly against the specimen mask. Rippling or wrinkling of sheet and foil specimens cannot be tolerated. Precautions must be taken to prevent warping of sheets, foils, and filter disks, and distension of liquid-cell windows due to heating by the primary x-rays. These and other precautions are discussed in Chapters 16 and 17. The specimen compartment and drawer must be replaced when they become worn, and even when they are in good condition, a leaf spring may be required to hold the drawer firmly in a reproducible position.

The concept of infinite thickness is discussed in Section 14.7.2. Let it suffice here to state that if the specimens are not infinitely thick, or of identical thickness, variations in thickness constitute a source of error. Moreover, in a given specimen, infinite thickness occurs at different depths for different analyte lines. The shorter the wavelength of the line, or the smaller the absorption coefficient of the specimen for the line, the deeper is the infinite thickness. It follows that a specimen may be infinitely thick for one wavelength, but not for another.

In most x-ray spectrometers, the central rays of the primary and secondary beams are mutually perpendicular, so there is the possibility of absorption and shielding of the two beams by surface texture. The surface need not have a high polish, but it must be plane and smooth, and the longer the wavelength of the analyte line or the higher the absorption coefficient of the specimen for the line, the finer must the specimen finish be. In general, surface roughness should be less than the infinite thickness for the analyte line, and must be the same for all samples and standards. If the specimens have substantially parallel grind marks, and are not rotated during measurement, the grind marks should be oriented in the same direction for all specimens. To reduce shielding, this direction should be parallel to the plane defined by the central rays of the primary and secondary beams.

Powder specimen errors are reduced by grinding the powders to 20-μm particle size or smaller. Specimen rotation reduces most specimen errors arising from inhomogeneities of one kind or another (composition, porosity, particle size, etc.) and from surface texture. Liquid specimens are discussed in Chapter 17.

A source of long-term specimen error often overlooked is the deterioration of some types of standards, especially with repeated use (Section

16.1.4.2). The principal causes of the deterioration are: (1) mechanical abrasion from repeated handling and insertion; (2) oxidation and corrosion, especially from perspiration and from ozone produced in the specimen chamber by the primary x-rays; (3) radiation damage by the high primary-beam intensity; and (4) aging. The types of standards subject to such deterioration include plated and coated metals and ceramics, and heterogeneous materials in which the phases are unequally prone to chemical or mechanical damage.

11.1.3.4. Chemical Effects (*25e*, *F5*, *H59*, *W10*, *Z5*)

In general, x-ray spectral lines of analytical interest in x-ray secondary-emission spectrometry arise from electron transitions to the *K* or *L* shell from the *L* or *M* shell. In most elements to which the method is applicable, the *L* and *M* shells are not the outermost shells and are therefore not involved in chemical phenomena. Thus, influence of chemical state of the atom on the wavelength of the analyte line is relatively uncommon and very small at most. However, for atomic numbers below 18 (argon), the 3*p* (*M*II, *M*III) orbital is unfilled. The electron transition giving the $K\beta$ line originates in this orbital, and, consequently, the $K\beta$ line of elements 9–17 (fluorine to chlorine) may show slight wavelength dependence on chemical state. The same consideration applies to the $K\beta_5$ line of elements 21–28 (scandium to nickel), which have only partially filled 3*d* orbitals, and to the $L\beta_2$ lines of elements 40–45 (zirconium to rhodium), which have unfilled 4*d* (*N*IV, *N*V) orbitals. In general, if a spectral line arises from an electron transition *from* the outermost orbital or *from* an unfilled orbital, there is probability of chemical effect.

Oxidation number is the most important chemical feature that contributes to wavelength shift of an x-ray spectral line; others include coordination number, type and electronegativity of bonds, induction effects, and the nature of the atom or radical to which the emitting atom is bonded. The magnitude of the wavelength displacement varies from 0.005 to 0.25 Å.

The chemical effect provides an excellent means for investigating the various chemical phenomena listed above. Such studies require an x-ray spectrometer capable of extremely high resolution and having a temperature-controlled crystal chamber. Although some of these studies have been done on standard commercial instruments, usually the small wavelength displacements are masked by other effects, such as 2θ shift due to thermal expansion of the crystal. However, chemical effects in the lighter elements and for certain lines of the heavier elements may constitute a source of error. If the goniometer 2θ angle for the analyte line is peaked for, say, the pure analyte, the line may occur at a slightly different 2θ for standards prepared from pure oxides, and at still different 2θ for complex mineral samples. Moreover, in complex substances, 2θ may vary from sample to sample.

11.2. COUNTING STATISTICS

It is the purpose of this section to present practical equations for calculating counting error for various x-ray spectrometric measurement techniques. The errors considered in this section are those arising *only* from the random arrival of x-ray photons and do not include possible errors arising in the x-ray source (x-ray tube and power supply) or detector (counter tube, power supply, and electronic readout components). The evaluation of other errors and of the overall analytical precision are discussed in Section 11.3.

The counting error always represents the maximum precision attainable in an x-ray spectrometric analysis. In modern x-ray spectrometers, instrumental errors are very small, and in semiautomatic and automatic instruments, operational errors are similarly reduced. Consequently, in analyses where specimen errors are absent or successfully dealt with, counting error may constitute substantially the analytical precision. A quotation from Liebhafsky and his coauthors [p. 333 in (*27*)] is appropriate here. They point out that "the counting error is *predictable* from the total count and *controllable* through it. We have here a situation highly unusual in analytical chemistry, for (if everything works out favorably) we have in x-ray emission spectrography the possibility of predicting and controlling the best precision attainable as long as the analytical-line intensity is established [that is, limited] by counting."

A particularly good discussion of counting statistics in x-ray spectrometric analysis is given by Jenkins and de Vries [Ch. 5 in (*22*)], and a substantial portion of the following treatment is based on theirs. Briefer reviews of counting statistics are given in (*M4*, *P22*).

11.2.1. Nature of the Counting Error

Consider a perfectly smooth, plane, homogeneous specimen in an x-ray spectrometer, all components of which, from the x-ray generator to the readout circuitry, are perfectly stable. Suppose that a series of equal preset-time intensity measurements 1, 2, 3, . . . , n are made of an x-ray spectral line of an element in the specimen without disturbing the specimen or any instrument setting during the entire series. Even under these ideal—and unattainable—conditions, the numbers of counts accumulated N_1, N_2, N_3, . . ., N_n during successive identical time intervals would not be equal. The reason is that emission of x-ray photons is a random process, so that the photons in an x-ray beam have a random time distribution. The fluctuation in the instantaneous value of the x-ray intensity about its mean value is known as the "noise," and the ratio of the mean value to the amplitude of the fluctuations as the *signal-to-noise ratio* (S/N). It will be shown (Section 11.2.2.1) that a measure of the fluctuation is given by $\sqrt{N}$, so that the signal-to-noise ratio is approximated by $\bar{N}/\sqrt{N}$, where $\bar{N}$ is the average number of counts accumulated in a specified time.

If a large number n of measurements are taken, the individual accumulated counts N always have a Poisson distribution:

$$P(N) = \frac{\bar{N}^N \exp -\bar{N}}{N!} \tag{11.16}$$

where $P(N)$ is the probability of occurrence of a specified accumulated count N; $\bar{N}$ is the mean of a very large number of measurements; and $N!$ is N-factorial, that is, $N \times (N-1) \times (N-2) \times \cdots \times 1$.

As $\bar{N}$ becomes progressively larger, the distribution approaches a Gaussian (normal) distribution, and for $N > 100$, the distribution may be regarded as Gaussian and represented by Equation (11.3) by setting $x = N$ and $\bar{x} = \bar{N}$,

$$P(N) = \frac{1}{(2\pi\bar{N})^{1/2}} \exp - \frac{(N-\bar{N})^2}{2\bar{N}} \tag{11.17}$$

The Poisson distribution is represented by a histogram (bar graph), the Gaussian distribution by a smooth, continuous curve. A plot of a Gaussian distribution is shown in Figure 11.1, where frequency of occurrence of a given value of N is plotted as a function of the value of N.

Gaussian distributions for sets of x-ray counts differ from Gaussian distributions for sets of other kinds of measurements or for replicate analyses in that the former are characterized simply by average total accumulated count $\bar{N}$, whereas the latter are characterized by the mean $\bar{x}$ and variance [Equation (11.5)]. Moreover, standard deviations for sets of x-ray counts (standard counting errors, Section 11.2.2.1) differ from standard deviations for other kinds of measurements or for replicate analyses in that the former involve only the average accumulated count [Equation (11.18)], whereas the latter involve the mean $\bar{x}$, divergence $x_i - \bar{x}$, and population n of the set [Equation (11.6)]. More simply, standard deviation is determined by the dispersion of the individual values about their mean; standard counting error is simply the square root of the mean. The standard deviation of replicate x-ray spectrometric analyses and the standard counting error of the corresponding replicate analyte-line counts will be the same only if the analysis is subject to no error except that arising from the random nature of x-rays, that is, from the counting error.

The distinction must be emphasized between the equations that follow in this section and Equations (11.11) and (11.75). The latter combine the different *errors* (counting, instrumental, operational, specimen, etc.) that apply to a single measurement or analysis. The equations in this section, excepting Equations (11.18) and (11.19), combine the counting errors of different measurements (peak, background, sample, standard, etc.) that apply to a single complex quantity (net count, ratio, etc.).

11.2.2. Calculation of Counting Error

11.2.2.1. Counting Error for Accumulated Counts

Following are equations for calculation of the standard deviation σ_N and relative standard deviation ε_N in terms of accumulated count N measured by various counting techniques. In this application, some workers prefer the terms *standard counting error* to standard deviation and *relative fractional counting error* to relative standard deviation. The use of these terms should be encouraged to avoid confusion with standard deviation and relative standard deviation of other errors and of the overall precision (Section 11.3).

a. Single Measurement. For a single measurement of N counts,

$$\sigma_N = \sqrt{N} \tag{11.18}$$

$$\varepsilon_N = \frac{\sigma_N}{N} = \frac{\sqrt{N}}{N} = \frac{1}{\sqrt{N}} = \frac{1}{\sqrt{IT}} \tag{11.19}$$

b. Net Count (Z9). For a measurement of N_P counts at the analyte peak corrected for a background measurement of N_B counts, both measurements taken for the same time interval, from Equation (11.11),

$$\sigma_N = (\sigma_P{}^2 + \sigma_B{}^2)^{1/2} \tag{11.20}$$

$$= (N_P + N_B)^{1/2} \tag{11.21}$$

$$\varepsilon_N = \frac{(N_P + N_B)^{1/2}}{N_P - N_B} = \frac{1}{\sqrt{N_P}} \frac{\{(N_P/N_B)[(N_P/N_B) + 1]\}^{1/2}}{(N_P/N_B) - 1} \tag{11.22}$$

Equation (11.22) is important in trace analysis in that it shows the importance of background.

c. Ratio Method (Uncorrected). In this technique, the time T_S to accumulate N_S counts from a standard S is also used to accumulate N_X counts from the sample X (Section 7.2.2.4). The N_X is measured at the analyte line; N_S may be measured at the analyte line from a standard, or it may be measured from the sample at an internal-standard line, or a scattered target line, or a scattered wavelength in the continuum. The method is very widely used in automatic spectrometers. If no background corrections are made,

$$\varepsilon_N = \left[\left(\frac{\sqrt{N}}{N}\right)_X^2 + \left(\frac{\sqrt{N}}{N}\right)_S^2\right]^{1/2} \tag{11.23}$$

$$= \left(\frac{1}{N_X} + \frac{1}{N_S}\right)^{1/2} \tag{11.24}$$

d. Ratio Method (Corrected). If the corresponding backgrounds $(N_B)_X$ and $(N_B)_S$ are measured for the same times T_S,

$$\varepsilon_N = \left[\frac{(N_P + N_B)_X}{(N_P - N_B)_X^2} + \frac{(N_P + N_B)_S}{(N_P - N_B)_S^2}\right]^{1/2} \tag{11.25}$$

e. Replicate Measurements. The foregoing equations apply to single measurements of N, $N_P - N_B$, or N_X/N_S, as the case may be. If n replicate determinations are made of each of these quantities, the standard counting error and relative fractional counting error are given by Equations (11.9) and (11.10):

$$(\sigma_N)_n = \frac{1}{\sqrt{n}}\,\sigma_N \tag{11.26}$$

$$(\varepsilon_N)_n = \frac{1}{\sqrt{n}}\,\varepsilon_N \tag{11.27}$$

Equations of this form are also used to calculate the standard deviation of the overall average of a set of n averages, each derived from a series of measurements and each having its own σ and ε.

It is evident that standard counting error decreases as N increases, and, in principle, may be made as small as required if a sufficiently long counting time is permissible. However, accumulation of a large N at low intensity requires a very long counting time; this may be disadvantageous for several reasons. Analysis time is increased. Extremely high stability is

TABLE 11.2. Standard and Relative Counting Error

Number of counts N	σ,[a] counts/s	$\varepsilon_{68.3}$,[b] %	2σ,[a] counts/s	$\varepsilon_{95.4}$,[b] %	3σ,[a] counts/s	$\varepsilon_{99.7}$,[b] %
100	10	10.0	20	20.0	30	30.0
200	14	7.1	28	14.1	42	21.2
500	22	4.5	44	9.0	66	13.4
1,000	32	3.2	64	6.3	96	9.5
2,000	45	2.2	90	4.5	135	6.7
5,000	71	1.4	142	2.8	213	4.2
10,000	100	1.0	200	2.0	300	3.0
20,000	141	0.70	282	1.4	423	2.1
50,000	224	0.45	448	0.90	672	1.3
100,000	316	0.32	632	0.63	948	0.95
200,000	447	0.22	894	0.45	1341	0.67
500,000	707	0.14	1414	0.28	2121	0.42
1,000,000	1000	0.10	2000	0.20	3000	0.30
2,000,000	1414	0.07	2828	0.14	4242	0.21
5,000,000	2236	0.05	4472	0.09	6708	0.13
10,000,000	3162	0.03	6324	0.06	9486	0.10

[a] $\sigma = \sqrt{N}$.
[b] $\varepsilon = 100(n\sigma/N)$; the subscript of ε is the confidence level; n is 1, 2, or 3.

required in the x-ray tube potential and current and in the electronic detector and readout components. Finally, difficulties arise on prolonged exposure of liquid specimens to the x-ray beam.

Table 11.2 gives values of σ, 2σ, and 3σ and ε, 2ε, and 3ε for a range of values of N. Figure 11.2 shows the correlation of ε for $N = 10^2$–10^7 counts at 0.67σ, σ, 2σ, and 3σ.

11.2.2.2. Counting Error for Intensities

This section presents substantially the same equations as Section 11.2.2.1, but in terms of intensity I rather than accumulated count N. However, from Equation (11.8),

$$\varepsilon_N = \sigma_N/N \qquad \text{and} \qquad \varepsilon_I = \sigma_I/I \tag{11.28}$$

Therefore, for values of intensity and accumulated count related to the

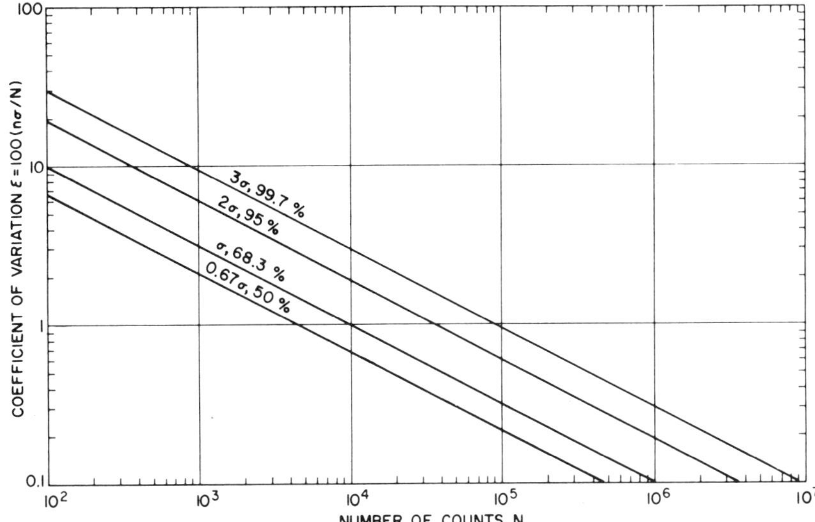

FIG. 11.2. Correlation of coefficient of variation and the accumulated count for various standard deviations and confidence levels.

same counting time T,

$$N = IT \tag{11.29}$$

Then standard and relative counting errors in terms of N and I are equal, that is,

$$\varepsilon_I = \varepsilon_N; \qquad \sigma_I/I = \sigma_N/N \tag{11.30}$$

Equation (11.30) applies only if I and T are related by Equation (11.29) and if the error associated with the measurement of T is negligible and has no distribution of its own. When high precision is required and T is small, the error in T may require consideration. The equations for standard counting error σ_I and relative fractional counting error ε_I in terms of intensity follow.

a. Uncorrected Intensity. For a single measured intensity,

$$\sigma_I = I\varepsilon_N = \frac{I}{\sqrt{N}} = \frac{I}{(IT)^{1/2}} = \left(\frac{I}{T}\right)^{1/2} \tag{11.31}$$

$$\varepsilon_I = \frac{(I/T)^{1/2}}{I} = \frac{1}{(IT)^{1/2}} = \frac{1}{\sqrt{N}} = \varepsilon_N \tag{11.32}$$

Equation (11.32) confirms Equation (11.30).

b. Net Intensity. For a measurement of I_P counts/s at the analyte peak corrected for I_B background counts,

$$(\sigma_I)_P = \left(\frac{I_P}{T_P}\right)^{1/2}; \qquad (\sigma_I)_B = \left(\frac{I_B}{T_B}\right)^{1/2} \tag{11.33}$$

From Equation (11.11),

$$\sigma_I = (\sigma_P{}^2 + \sigma_B{}^2)^{1/2} \tag{11.34}$$

$$= \left(\frac{I_P}{T_P} + \frac{I_B}{T_B}\right)^{1/2} \tag{11.35}$$

$$\varepsilon_I = \frac{[(I_P/T_P) + (I_B/T_B)]^{1/2}}{I_P - I_B} = \frac{1}{\sqrt{I_P}} \frac{\{(I_P/I_B)[(I_P/I_B) + 1]\}^{1/2}}{(I_P/I_B) - 1} \tag{11.36}$$

or, from Equation (11.28),

$$\varepsilon_I = \frac{[(I_P\varepsilon_P)^2 + (I_B\varepsilon_B)^2]^{1/2}}{I_P - I_B} = \frac{1}{T^{1/2}} \frac{1}{I_P^{1/2} - I_B^{1/2}} \tag{11.37}$$

where, of course, $I_P - I_B$ is net intensity.

c. Ratio Method (Uncorrected). For I_X/I_S, both measured at T_S (see above),

$$\varepsilon_I = \frac{1}{\sqrt{T_S}} \left(\frac{1}{I_X} + \frac{1}{I_S}\right)^{1/2} \tag{11.38}$$

$$= \frac{1}{\sqrt{N_S}} \left(1 + \frac{I_S}{I_X}\right)^{1/2} \tag{11.39}$$

If $I_S \approx I_X$,

$$\varepsilon_I \approx (2/N_S)^{1/2} \tag{11.40}$$

Equation (11.38) can be used to calculate the standard and relative fractional counting errors for the two commonly used counting methods, fixed count and fixed time, and for the optimal-fixed-time method (see below).

There are three basic methods for measurement of the peak and background intensities: *fixed-count*, in which the same total count is accumulated for both; *fixed-time*, in which the same counting time is used for both; and *optimal fixed-time*, in which the total counting time is apportioned between the two measurements so as to give minimal error in the net intensity. Equations for calculation of counting error and relative counting error for these three methods follow.

d. Fixed Count. This method is characterized by the following conditions: $I_P T_P = I_B T_B = N$, $T_P/T_B = I_B/I_P$, and $T_P + T_B = T$ (total counting time); then,

$$\sigma_{\text{FC}} = \left(\frac{1}{\sqrt{T}}\right)(I_P + I_B)^{1/2}\left(\frac{I_P}{I_B} + \frac{I_B}{I_P}\right)^{1/2} \tag{11.41}$$

e. Fixed Time. This method is characterized by the following conditions: $T_P = T_B = T/2$, and $T_P + T_B = T$; then,

$$\sigma_{\text{FT}} = \left(\frac{2}{T}\right)^{1/2}(I_P + I_B)^{1/2} \tag{11.42}$$

$$\varepsilon_{\text{FT}} = \left(\frac{2}{T}\right)^{1/2}\left(\frac{I_P + I_B}{I_P - I_B}\right)^{1/2} \tag{11.43}$$

f. Optimal Fixed Time. When the difference of two intensities is to be measured, as in a net intensity measurement, the two intensities should be known with about the same absolute error. The optimal-fixed-time method is essentially the same as the fixed-time method except that the apportionment of the total counting time T between the peak and the background is calculated to give minimum error in the measured net intensity. This condition exists when $T_P + T_B = T$ and

$$\frac{T_P}{T_B} = \left(\frac{I_P}{I_B}\right)^{1/2} \tag{11.44}$$

Then,

$$\sigma_{\text{FTO}} = \frac{1}{\sqrt{T}}\left(\sqrt{I_P} + \sqrt{I_B}\right) \tag{11.45}$$

$$\varepsilon_{\text{FTO}} = \frac{1}{\sqrt{T}}\left(\frac{\sqrt{I_P} + \sqrt{I_B}}{I_P - I_B}\right) \tag{11.46}$$

$$= \frac{1}{\sqrt{T}}\left(\frac{1}{\sqrt{I_P} - \sqrt{I_B}}\right) \tag{11.47}$$

Incidentally, an equation similar to Equation (11.44) applies when an optimal time apportionment is required for the measurement of the *ratio* of two intensities I_1/I_2:

$$\frac{T_1}{T_2} = \left(\frac{I_1}{I_2}\right)^{1/2} \tag{11.48}$$

Equations (11.26) and (11.27) for replicate measurements apply to the equations for σ_I and ε_I as well.

One must distinguish the terms *preset* time and count (Sections 7.2.2.1–7.2.2.2) and *fixed* time and count as defined above. The former refer simply to the mechanics of making the measurement, the latter to the strategy of the measurement (Section 11.2.3).

g. *Ratemeter–Recorder.* The precision of intensity measurements made with the ratemeter–recorder is also subject to statistical fluctuation. Measurement of an intensity with a ratemeter having a time constant of RC seconds is equivalent to accumulation of counts during a time interval of $2RC$ seconds. Then the relative fractional counting error is

$$\varepsilon_I = 1/[2I(RC)]^{1/2} \tag{11.49}$$

Actually, R signifies resistance in ohms (Ω), and C signifies capacitance in farads (F). Dimensionally, $1\ \Omega = 1$ s/esu-cm, and $1\ \text{F} = 1$ esu cm. Then RC [=] ΩF [=] (s/esu-cm)(esu cm) [=] s.

11.2.3. Counting Strategy

Counting strategy may be defined, perhaps with some oversimplification, as the selection of counting technique, counting time, and/or accumulated count to make a specified intensity measurement or series of measurements with a specified precision in the shortest time.

11.2.3.1. Measurement of Net Intensity

The choice of counting strategy for the measurement of a net intensity involves the choice of the fixed-count, fixed-time, or optimal fixed-time method, apportionment of the total counting time between measurement of peak and background intensities, and certain practical considerations.

The *fixed-count method* has the advantages that only 2θ need be changed between peak and background measurements and that, since $N_B = N_P$, the net counting error is only slightly greater than that for the peak alone [Equation (11.20)]. However, the accumulation of the same number of counts at the background as at the peak may require a long counting time, and the lower I_B is, the longer the counting time is. A minor advantage of the fixed-count method is that dead-time corrections, if required, are very easily made (Section 6.5.1.6).

The *fixed-time method* has three variations, one of which is not really a fixed-time method at all, and each of which has a different counting error. A discussion of these variations follows.

In case I, the counting time at the peak T_P is selected to accumulate about the same number of counts N_P as would be selected if the fixed-count method were to be used; then T_B is the same as T_P.

In case II, the counting times at the peak and background are also equal, but each is about half the *total* counting time used to accumulate the peak and background counts by the fixed-count method. For example, if the fixed-count method requires peak and background counting times of, say, 30 and 70 s, respectively, the total counting time is 100 s, and in the fixed-time method, T_P and T_B would each be 50 s.

In case III, the fixed times for peak and background are different and are apportioned as specified by Equation (11.44). This is the *optimal fixed-time* method and is discussed below. This is a preselected-time method (Section 7.2.2.1), but not truly a fixed-time method.

The true fixed-time method, then, is characterized by equality of counting times for peak and background (cases I and II above). Like the fixed-count method, it also requires only a change of 2θ between peak and background measurements, but does not require background counting times as long as the fixed-count method. The net counting error for fixed-time is greater or less than that for fixed-count, depending on whether one considers case I or case II above. The following paragraphs consider counting error for the two cases in detail.

For case I, the net counting error is the same or greater for fixed-time than for fixed-count, but never more than 10% greater, and this increase is justified by the economy of background counting time. Birks [pp. 82–83 in (*8*)] (*B57*) compares counting errors for the fixed-count and case I fixed-time methods as follows. In the fixed-time measurement, let

$$N_B/N_P = r \qquad (T_B = T_P) \tag{11.50}$$

Then the relative counting error ε_L [Equation (11.22)] in the net line count $N_P - N_B$ is

$$\varepsilon_L = \frac{(\sigma_P{}^2 + \sigma_B{}^2)^{1/2}}{N_P - N_B} \tag{11.51}$$

From Equations (11.18) and (11.50),

$$\sigma_B{}^2 = N_B = rN_P = r\sigma_P{}^2 \tag{11.52}$$

Therefore,

$$\varepsilon_{L(\mathrm{FT})} = \frac{[N_P(1 + r)]^{1/2}}{N_P(1 - r)} \tag{11.53}$$

For fixed-count measurement, the background counting time T_B is increased so that

$$N_{B(\mathrm{FC})} = N_P = N_{B(\mathrm{FT})}(T_B/T_P) \tag{11.54}$$

Then the error in the background measurement is reduced by the ratio

$$[N_{B(\mathrm{FT})}/N_{B(\mathrm{FC})}]^{1/2} = r^{1/2} \tag{11.55}$$

That is,

$$\sigma_{B(\mathrm{FC})} = r^{1/2}\sigma_{B(\mathrm{FT})} \tag{11.56}$$

and

$$\varepsilon_L = \frac{[\sigma_P{}^2 + \sigma^2_{B(\mathrm{FC})}]^{1/2}}{N_P - [N_{B(\mathrm{FC})}(T_P/T_B)]} \tag{11.57}$$

If now one lets

$$\sigma^2_{B(\mathrm{FC})} = r\sigma^2_{B(\mathrm{FT})} = rN_{B(\mathrm{FT})} = r^2 N_P \tag{11.58}$$

and

$$N_{B(\mathrm{FC})}(T_P/T_B) = N_{B(\mathrm{FT})} = rN_P \tag{11.59}$$

then

$$\varepsilon_{L(\mathrm{FC})} = \frac{[N_P(1 + r^2)]^{1/2}}{N_P(1 - r)} \tag{11.60}$$

The counting error for fixed-time compared with that for fixed-count is given by the ratio of Equations (11.53) and (11.60); the denominators cancel, giving

$$\left(\frac{\varepsilon_{\mathrm{FT}}}{\varepsilon_{\mathrm{FC}}}\right)_L = \frac{[N_P(1 + r)]^{1/2}}{[N_P(1 + r^2)]^{1/2}} = \left(\frac{1 + r}{1 + r^2}\right)^{1/2} \tag{11.61}$$

The full range for r is zero (for $N_B = 0$) to one (for $N_L = N_P - N_B = N_B$). A plot of Equation (11.61) for $r = 0$–1 is given in Figure 11.3 and shows that at worst, the counting error for fixed-time measurement is only $\sim$1.1

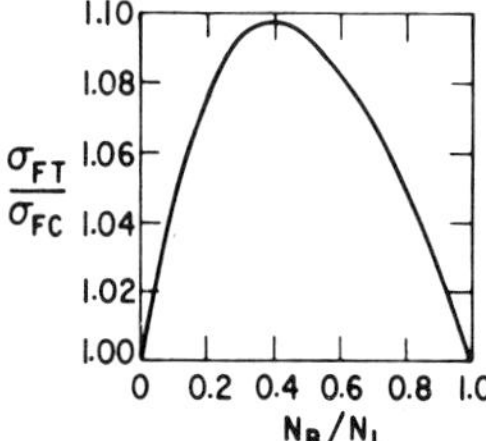

FIG. 11.3. Ratio of relative counting errors for fixed-time σ_{FT} and fixed-count σ_{FC} methods *vs.* ratio of accumulated counts for background N_B and analyte line N_L. [L. S. Birks, "X-Ray Spectrochemical Analysis," ed. 2, Interscience Publishers, N. Y., p. 82 (1969); courtesy of the author and publisher.]

times that for fixed-count measurement. This negligible difference in precision is more than offset by the convenience and shorter background counting time for the fixed-time method.

For case II—*total* counting time the same for fixed-time and fixed-count methods—the net counting error is the same or less for fixed-time than for fixed-count. Moreover, for a fair comparison, the precision of the two methods should be compared for a specified total counting time.

Gaylor (*G4*) has shown that: (1) for a specified total counting time for peak and background, the variance of the fixed-time method is always less than or equal to that of the fixed-count method; and (2) for a specified precision, the total counting time required for the fixed-time method is always less than or equal to that for the fixed-count method. It is also demonstrable that when the ratio of two intensities is required, the fixed-count and fixed-time methods give the same counting error, but when the difference of two intensities or a net intensity is required, the fixed-time method always gives a lower counting error in a specified total measuring time with peak and background counting times the same.

Jenkins and de Vries [p. 101 in (*22*)] compare counting errors for the fixed-count and case II fixed-time methods as follows. Since it is always true that I is positive and $I_P > I_B$, it follows that

(1) $(I_P - I_B) > 0$

(2) $(I_P - I_B)^2 > 0$

(3) $$\frac{I_P^2 + I_B^2}{I_P I_B} > 2$$

(4) $$\left(\frac{I_P}{I_B} + \frac{I_B}{I_P}\right) > 2$$

(5) $$\left(\frac{I_P}{I_B} + \frac{I_B}{I_P}\right)^{1/2} > \sqrt{2}$$

Hence,

$$\left(\frac{I_P}{I_B} + \frac{I_B}{I_P}\right)^{1/2}\left(\frac{1}{T}\right)^{1/2}(I_P + I_B)^{1/2} > \sqrt{2}\left(\frac{1}{T}\right)^{1/2}(I_P + I_B)^{1/2}$$

and, for case II, from Equations (11.41) and (11.42), $\sigma_{\mathrm{FC}} > \sigma_{\mathrm{FT}}$.

Aside from the improved counting statistics of the fixed-time method compared with the fixed-count method, for a specified total counting time, there are some practical advantages. Some examples follow.

Suppose that a method has been developed for the determination of a specified analyte in a specified type of sample in concentration range, say,

TABLE 11.3. Comparison of Fixed-Count and Fixed-Time Methods[a]

	Fixed-count method $N = 10{,}000$ counts	Fixed-time method $T = 100$ s
Analyte concentration 1%, $I = 100$ counts/s		
N, counts	10,000	10,000
$\%\varepsilon = 100(\sqrt{N}/N)$	1%	1%
T, s	100	100
Analytical result	$1 \pm 0.01\%$ (σ)	$1 \pm 0.01\%$ (σ)
Analyte concentration 0.1%, $I = 10$ counts/s		
N, counts	10,000	1000
$\%\varepsilon = 100(\sqrt{N}/N)$	1%	3.2%
T, s	1000	100
Analytical result	$0.1 \pm 0.001\%$ (σ)	$0.1 \pm 0.003\%$ (σ)

[a] From Jenkins and de Vries, p. 97 in (*22*).

25–30%. For samples in which the analyte concentration falls in or near this range, the two counting techniques are equally *convenient*. However, for occasional samples in which the analyte is present in very low concentration, analysis time is lost in accumulating the same fixed count to determine a concentration outside the calibration range. This consideration assumes prime importance in the programming of automatic spectrometers.

Now suppose that a calibration curve has been established for a specified analyte in concentration range 0–1%, that the slope of the curve is 100 counts/s per percent analyte, and that the required precision is $\pm 1\%$ relative for 1σ at the 1% concentration level. Table 11.3 summarizes data for fixed-count and fixed-time measurements on two samples having analyte concentrations 1% and 0.1%, respectively. In this case, the analyst would usually prefer the fixed-time method at the sacrifice of some relative precision at low concentrations, rather than count ten times as long to achieve a precision ($\pm 0.001\%$) that, relatively, is ten times that at the 1% concentration level ($\pm 0.01\%$).

The final example is the most significant in that it concerns the background measurement. In principle, background should be measured with the same precision as the analyte peak, that is, the same number of counts should be accumulated for each. However, when $I_P \gg I_B$, the absolute magnitude of the fluctuations at the peak is much greater than at the background, and high precision is not required for the background measurement. For example, a background of, say, 25 counts/s measured with a precision of $\pm 5\%$ would contribute an error of only ± 1 count/s. Thus, a fixed-count measurement may require a very long time to count the background, and much time is wasted in reducing the smaller component of the variance in the net intensity [Equation (11.34)], time better spent in reducing the larger component.

The *optimal fixed-time method* has the advantages that net counting error is minimal and that long background counting times are not required. However, in addition to 2θ, the counting time must also be changed between peak and background measurements. A more often cited disadvantage is that it is necessary to make preliminary measurements of I_P and I_B and to calculate the optimal apportionment of the counting times. These calculations are somewhat elaborate, and because of this, some workers reject the optimal fixed-time method. Other workers use the fixed-time method for small numbers of specimens, but take advantage of the economy of counting time of the optimal method for analyses involving large numbers of specimens and for trace analyses where the counting times are very long.

The counting errors of the fixed-time and optimal fixed-time methods may be compared in a way similar to that used above to compare the fixed-count and fixed-time methods [p. 102 in (*22*)]. For all values of I_P and I_B, except for $I_P = I_B$:

(1) $(I_P - I_B)^2 > 0$

(2) $I_P^2 - 2I_P I_B + I_B^2 > 0$

(3) $I_P^2 + 2I_P I_B + I_B^2 > 4I_P I_B$

(4) $(I_P + I_B)^2 > [2(I_P I_B)^{1/2}]^2$

(5) $(I_P + I_B) > 2(I_P I_B)^{1/2}$

(6) $2(I_P + I_B) > I_P + 2(I_P I_B)^{1/2} + I_B$

Hence,

$$\sqrt{2}\,(I_P + I_B)^{1/2} > \sqrt{I_P} + \sqrt{I_B}$$

and, from Equations (11.42) and (11.45), $\sigma_{\mathrm{FT}} > \sigma_{\mathrm{FTO}}$. Then, for measurement of net intensities, $\sigma_{\mathrm{FC}} > \sigma_{\mathrm{FT}} > \sigma_{\mathrm{FTO}}$ for all values of I_P and I_B, provided $I_P \neq I_B$.

The counting errors of all three methods may be compared by solution of Equations (11.41) for σ_{FC}, (11.42) for σ_{FT}, and (11.45) for σ_{FTO} for the *same* total counting time with the following typical conditions: $I_P = 10{,}000$

counts/s, $I_B = 100$ counts/s, and T (*total* counting time) $= 100$ s. The result is $(\sigma_{FC} = 100) > (\sigma_{FT} = 14) > (\sigma_{FTO} = 11)$, where the numbers are counts.

In practice, Equations (11.41), (11.42), and (11.45) can be used to calculate the total counting time T required to realize a specified standard counting error for the peak and background intensities actually encountered from the specimens. Then the optimal apportionment of T between peak I_P and background I_B intensities can be calculated from the following equations:

$$T_P + T_B = T \tag{11.62}$$

$$T_P/T_B = (I_P/I_B)^{1/2} \tag{11.63}$$

Of course, the instrument is not likely to provide the exact preset times given by these calculations, and in such cases, the closest preset times provided are used.

Another approach is to calculate the number of counts that must be accumulated at the peak N_P and at the background N_B to achieve a specified relative standard deviation and in minimal counting time. The equations (*M4*) are

$$N_P = \frac{10^4}{\varepsilon^2} \frac{(I_P/I_B)^{1/2} + 1}{[(I_P/I_B) - 1]^2} \left(\frac{I_P}{I_B}\right)^{3/2} \tag{11.64}$$

$$N_B = \frac{10^4}{\varepsilon^2} \frac{(I_P/I_B)^{1/2} + 1}{[(I_P/I_B) - 1]^2} \tag{11.65}$$

Mack and Spielberg (*M4*) have established a convenient graph for determination of the optimal numbers of counts to accumulate at the peak and background for specified peak-to-background ratios for 1, 5, or 10% statistical error at 50, 90, or 99% confidence level. The graph is derived in their paper and shown in Figure 11.4 as improved by Adler [p. 238 in (*1*)]. The lower horizontal scale is the peak-to-background ratio $R = I_P/I_B$. The upper horizontal scale gives the number of counts to accumulate at the peak N_P or background N_B for a specified error and confidence level. The vertical scale is "Z," a complex function of R derived in (*M4*). The use of the graph is illustrated by the following example.

Suppose that a preliminary measurement on a typical specimen of the type to be analyzed gives analyte peak I_P and background I_B intensities 1000 and 200 counts/s, respectively. Then $R = 1000/200 = 5$. Suppose further that 1% error is required at 90% confidence level. The total count to be accumulated at the peak N_P is derived as follows.

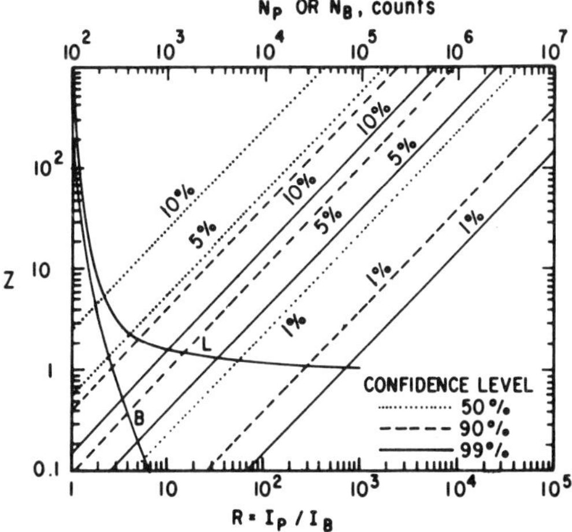

FIG. 11.4. Counting-strategy graph to permit selection of total accumulated count for peak N_P and background N_B for specified peak-to-background intensity ratio I_P/I_B and confidence level. [M. Mack and N. Spielberg, *Spectrochimica Acta* **12**, 169 (1958); courtesy of the authors and Pergamon Press; this version by I. Adler, "X-Ray Emission Spectrography in Geology," Elsevier Publishing Co., N. Y., p. 238 (1966); courtesy of the author and publisher.]

1. Locate 5 on the I_P/I_B scale.
2. Move vertically (parallel to the Z axis) to intersection with curve L (line); this occurs at $Z = 2.2$.
3. Move horizontally (parallel to the R axis) to intersection with the 1% error, 90% confidence level line.
4. Move vertically to the N scale and read $N_P = 5.4 \times 10^4$ counts.

The count to be accumulated at the background N_B is derived as follows.

1. Locate 5 on the I_P/I_B scale, as before.
2. Move vertically to intersection with curve B at $Z = 0.2$.
3. Move horizontally to intersection with the 1% error, 90% confidence level line.
4. Move vertically to the N scale and read $N_B = 5.2 \times 10^3$ counts.

The counting times are then

$$T = \frac{N}{I}; \qquad T_P = \frac{54{,}000}{1000} = 54 \text{ s}; \qquad T_B = \frac{5200}{200} = 26 \text{ s}$$

Then the total counting time required at $I_P/I_B = 5$ for 1% error at 90%

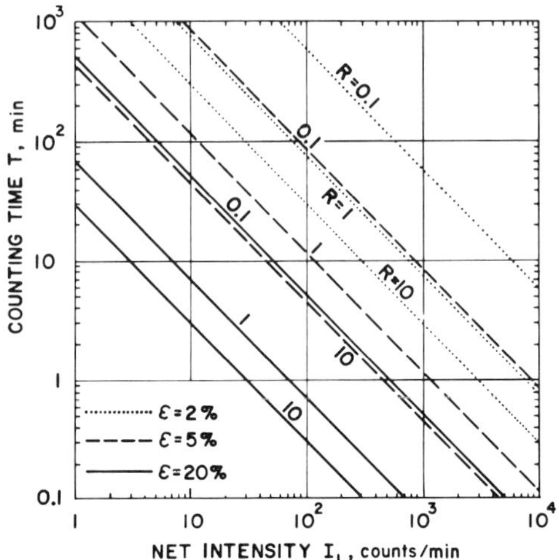

FIG. 11.5. Counting time as a function of net intensity I_L for *line*-to-background ratios $I_L/I_B = R$ of 0.1, 1, and 10 for relative standard deviations ε [Equation (11.8)] of 2, 5, and 20% (*N13*).

confidence level is 80 s. Preset-time or preset-count measurements are made using the available instrument settings closest to those indicated by the procedure above.

Neeb (*N13*) prepared the nomograph shown in Figure 11.5, which permits determination of counting times $T = T_P = T_B$ as a function of *net* analyte-line intensity I_L for *line*-to-background ratios $I_L/I_B = R$ of 0.1, 1, and 10 and for relative standard counting errors ε of 2, 5, and 20%. Neeb's graph is useful principally for trace analyses, where I_L is low, and in cases when it is necessary to measure a relatively weak line in a region of high background.

11.2.3.2. The Ratio Method

In the ratio method, one of the three counting techniques—fixed-count, fixed-time, and optimal fixed-time—is used to measure two intensities, one from the sample, the other from an internal or external standard, a scattered x-ray tube target line, or a wavelength from the scattered continuum.

It can be shown that when applied to the measurement of intensity ratios, the fixed-time and fixed-count methods yield exactly the same precision in the same measuring time:

$$\varepsilon_{\mathrm{FC}} = \varepsilon_{\mathrm{FT}} = \left[\frac{2}{T}\left(\frac{1}{I_X} + \frac{1}{I_S}\right)\right]^{1/2} \tag{11.66}$$

However, when intensity ratios are measured, the numerator and denominator should have substantially the same relative error. Decreasing the relative counting error ε in only one of the values does not improve the precision of the ratio significantly. For such applications, maximum precision is realized by use of a counting strategy such that a constant relative error is realized at all intensities; this is the case for the fixed-count method.

The optimal fixed-time method can also be applied to the ratio method (*M2*), and when it is, $\sigma_{\mathrm{FTO}} < \sigma_{\mathrm{FC}} = \sigma_{\mathrm{FT}}$:

$$\varepsilon_{\mathrm{FTO}} = \frac{1}{\sqrt{T}} \, \frac{1}{\sqrt{I_X} + \sqrt{I_S}} \tag{11.67}$$

where the notation is the same as in Equation (11.66). There remains the apportionment of the total counting time T between the two measurements. Let the net-intensity ratio R of the analyte A and standard S lines be defined as follows:

$$R = \frac{(I_P - I_B)_{\mathrm{A}}}{(I_P - I_B)_S} = \frac{(I_L)_{\mathrm{A}}}{(I_L)_S} \tag{11.68}$$

where S may be an internal or external standard or a scattered x-ray tube target line. Also, $(I_P - I_B)_{\mathrm{S}}$ and $(I_L)_{\mathrm{S}}$ may be replaced by the symbol I_{S}, indicating a scattered continuum wavelength. The optimal time apportionment is then given by Equation (11.48):

$$\frac{T_{\mathrm{A}}}{T_{\mathrm{S}}} = \frac{(T_P + T_B)_{\mathrm{A}}}{(T_P + T_B)_{\mathrm{S}}} = \frac{(I_P^{1/2} - I_B^{1/2})_{\mathrm{S}}}{(I_P^{1/2} - I_B^{1/2})_{\mathrm{A}}} \tag{11.69}$$

Then, for a specified relative standard deviation ε_R,

$$\varepsilon_{\mathrm{A}} = \frac{\varepsilon_R}{\{1 + [1/(T_{\mathrm{A}}/T_{\mathrm{S}})]\}^{1/2}} \tag{11.70}$$

$$\varepsilon_{\mathrm{S}} = \frac{\varepsilon_R}{[1 + (T_{\mathrm{A}}/T_{\mathrm{S}})]^{1/2}} \tag{11.71}$$

Having thus determined ε_{A} and ε_{S}, one can use graphs similar to Figures 11.4 and 11.5 or equations similar to (11.64) and (11.65) to find the numbers of counts to be accumulated for the four intensities, $(N_P)_{\mathrm{A}}$, $(N_B)_{\mathrm{A}}$, $(N_P)_{\mathrm{S}}$, and $(N_B)_{\mathrm{S}}$.

In automatic instruments, the time T_{S} to accumulate N_{S} counts from a standard S is often used as the measuring time to accumulate N_{X} counts from the sample X. The relative fractional counting error is given by Equation (11.39),

$$\varepsilon = \frac{1}{\sqrt{N_S}} \left(1 + \frac{I_S}{I_X}\right)^{1/2}$$

If I_S is of the same order as I_X,

$$\varepsilon \approx (2/N_S)^{1/2} \tag{11.72}$$

It is evident that this value is $\sqrt{2}$ greater than the relative fractional counting error associated with a simple measurement of N counts, $(1/N)^{1/2}$ [Equation (11.19)]. However, although the counting error is greater in the ratio method, the instrumental errors are substantially smaller. The choice between absolute or ratio method must be based on the value of ε for the two methods in the following equation:

$$\varepsilon = (\varepsilon_N^2 + \varepsilon_{\text{instr}}^2)^{1/2} \tag{11.73}$$

If $\varepsilon_{\text{instr}}$ is, say, 0.3%, the corresponding N is 10^5 counts (Figure 11.2); then it is a waste of time to accumulate more than 10^5 counts in a single measurement because above this number, instrumental errors limit the precision. If the required precision is greater than that corresponding to the instrumental error, replicate measurements must be made. For example, for a single measurement of $N = 10^6$ counts, $\sigma_N = \sqrt{10^6} = 10^3$, or $\varepsilon_N = 0.1\%$; for 10 replicate measurements of $N = 10^6$, from Equation (11.26), $\sigma_N = (1/\sqrt{10})\sqrt{10^6} \approx 300$, or $\varepsilon_N = 0.03\%$.

Some other counting strategies are discussed briefly in Section 11.1.3.2: counting of samples and standards in random sequence and alternation of portions of very long peak and background counts.

11.2.4. Figure of Merit

Peak-to-background ratio (P/B, I_P/I_B) or the product of peak-to-background ratio and analyte-line intensity [$(I_P/I_B) \times I_L$] provide commonly recognized figures of merit to be attained in setting instrument operating conditions. Peak-to-background ratios with secondary excitation may be as high as 10,000:1, whereas with primary excitation, they are at best a few thousand to one. Counting precision decreases rapidly as peak-to-background ratio decreases below ~3:1.

Jenkins and de Vries [p. 103 in (*22*)] propose other figures of merit based on Equation (11.47). For major and minor constituents (relatively high analyte-line intensity) and a properly apportioned total counting time T, if ε is to be minimum, $\sqrt{I_P} - \sqrt{I_B}$ should be maximum. For constituents having very low concentrations (low analyte-line intensity), I_P is not much greater than I_B and, in fact, approaches it at or near the detection limit. Then $(I_P I_B)^{1/2} \approx I_P$. In such cases, if ε is to be minimum, $(I_P - I_B)/\sqrt{I_B}$ should be maximum.

If $I_P - I_B$ (counts/s) is replaced by the slope m (counts/s per % analyte) of the intensity *versus* concentration calibration curve, the figure of merit is $m/\sqrt{I_B}$. This is the figure of merit proposed by Spielberg and Brandenstein (*S49*). It is most useful at low concentrations where I_P approaches I_B, and, like $(I_P - I_B)/\sqrt{I_B}$, should be as large as possible.

Another concept that may serve as a figure of merit is *percent equivalent background* P_B (*L45*):

$$P_B = I_B / I_{1\%} \tag{11.74}$$

where I_B is background intensity at the analyte line, and $I_{1\%}$ is net intensity per 1% analyte. On the calibration curve of intensity, uncorrected for background, *versus* analyte concentration, P_B is the negative intercept on the concentration axis (Figure 14.3). Both $I_{1\%}$ and I_B vary linearly with x-ray tube current, but P_B remains constant and is therefore a fairer criterion for peak-to-background ratios than I_B. The detection limit decreases as $I_{1\%}/P_B$ increases.

Figures of merit can be used not only to establish operating conditions, but to compare the precision of intensity measurements on different instruments at substantially the same conditions and of the same instrument at different conditions.

11.3. ANALYTICAL PRECISION

11.3.1. Nature of Analytical Precision

Table 11.4 gives data for 12 (n) determinations of germanium made on the same sample by timing (T) 12 preset counts of 51,200 (N) of the Ge $K\beta$ line, calculating the intensities ($I = N/T$), and applying them to a calibration curve. The curve was prepared previously from measurements on standards regarded as accurate. The specimen was reloaded without regard to orientation for each measurement, but no instrument setting was disturbed. The standard deviation σ of the set of n results x from their mean $\bar{x}$ is calculated from Equation (11.6),

$$\sigma = \left(\frac{\sum d^2}{n-1} \right)^{1/2}$$

TABLE 11.4. Precision of an X-Ray Spectrometric Analysis

i	Counting time T, s	Ge $K\beta$ net intensity I, counts/s	Ge concentration C, at % x_i	Deviation $d = x_i - \bar{x}$	Deviation squared d^2
1	26.1	1962	34.7	+0.3	0.09
2	26.0	1969	35.1	+0.7	0.49
3	26.3	1947	33.9	−0.5	0.25
4	26.2	1954	34.3	−0.1	0.01
5	26.1	1962	34.7	+0.3	0.09
6	26.2	1954	34.3	−0.1	0.01
7	26.2	1954	34.3	−0.1	0.01
8	26.2	1954	34.3	−0.1	0.01
9	26.4	1939	33.4	−1.0	1.00
10	26.0	1969	35.1	+0.7	0.49
11	26.2	1954	34.3	−0.1	0.01
$12 = n$	26.2	1954	34.3	−0.1	0.01
		$\Sigma I = 23472$	$\Sigma x = 412.7$		$\Sigma d^2 = 2.47$
		$\Sigma I/n = 1956$	$\Sigma x/n = 34.4 = \bar{x}$		$\Sigma d^2/(n-1) = 0.22$

$$\sigma = [\Sigma d^2/(n-1)]^{1/2} = 0.47 \text{ at \% Ge}$$

$$\varepsilon = 100(\sigma/\bar{x}) = 1.4\ \%$$

Standard counting error for $N = 51200$:

$$\sigma_N = \sqrt{N} = 226$$

$$\varepsilon_N = 100(\sigma_N/N) = 0.44\%$$

where d is the deviation of an individual determination x_i from the average $\bar{x}$ of the n replicate determinations. The coefficient of variation ε is then calculated from Equation (11.8):

$$\varepsilon = 100(\sigma/\bar{x})$$

The standard deviation is a measure of absolute precision; the coefficient of variation is a measure of relative precision, that is, precision relative to the "amount" (concentration) of analyte present.

In Table 11.4, the analytical result of, say, the first determination ($i = 1$) may be reported by any combination of the following statements: 34.7% Ge ±0.5% absolute, or 34.7% Ge ±1.4% relative, or 34.7% Ge ±1.4% of the amount present; followed by: for one standard deviation, or for one sigma (1σ), or at the 68.3% confidence level. This means that the probability is 68.3% (7/10) that the "true" germanium concentration, that is, the mean of a very large number of determinations, lies within ±0.5% absolute of 34.7%, or that it lies within (0.014×34.7) of 34.7%. Similarly, the result may be reported as 34.7% Ge ±0.9% absolute or ±2.8% relative for 2σ or at the 95.4% confidence level. Table 11.1 gives other degrees of precision that are used in reporting analytical results. If 12 replicate determinations are made, as shown in Table 11.4, and their mean (34.4% Ge) is reported as the analytical result, the standard deviation of this mean result is given by Equation (11.26): $\sigma = (1/\sqrt{12})(0.47) = 0.13$.

Table 11.4 also gives the standard counting error σ_N and relative counting error ε_N for the preset count N of 51,200. It is evident that the coefficient of variation of a single determination of germanium (1.4%) is nearly three times the relative counting error for a count of 51,200 (0.44%). This discrepancy shows that serious errors other than the counting error are impairing the precision of the determination. Standard counting error is—or is derived from—the square root of the mean. Standard deviation is a measure of the dispersion of a group of individual results about their mean. The two are identical in an x-ray spectrometric analysis only if the analytical process is truly random and subject to no error other than that arising from the random time distribution of the x-ray photons.

In the preceding paragraph, it is implied that in the absence of any source of error other than the counting error, the relative standard deviation of the analytical concentration should be the same as the relative counting error of the intensity or combination of intensities (difference, ratio, etc.) from which it is derived. This is substantially true only for linear calibra-

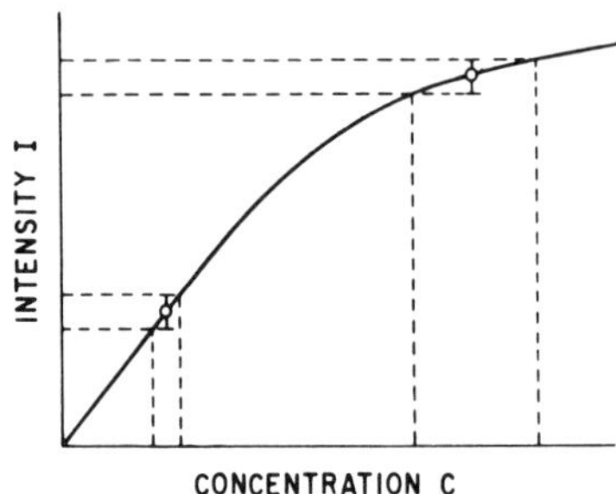

FIG. 11.6. Relationship of counting error and that portion of the concentration error contributed by the counting error. The indicated spread may represent, say, $\pm\sigma$.

tion curves that pass through the origin, but is not generally true, as shown in Figure 11.6, where σ_I is the standard counting error.

11.3.2. Evaluation of Precision

11.3.2.1. General Considerations

The standard counting error σ_N has twofold significance in x-ray spectrometric analysis: (1) It represents the maximum precision attainable, and cannot be eliminated from a process based on counting photons; and (2) it can be accurately calculated and used as a standard against which to evaluate the precision actually attained.

The precision of an x-ray spectrometric analysis is obtained by combining the variances for the individual sources of error [Equation (11.11)], so that the precision may be represented by

$$\sigma_T^2 = \sigma_N^2 + \sigma_I^2 + \sigma_O^2 + \sigma_S^2 + \sigma_M^2 \qquad (11.75)$$

where the subscripts represent, in order, total, counting, instrumental, operational, specimen, and miscellaneous errors. A similar equation can be written for each of the individual errors, expressing the variance for each as the sum of the variances of several specific sources of error, as outlined in Section 11.1.1. The following paragraphs describe the general procedure for revealing, identifying, and evaluating errors.

If the standard deviation of the result of a series of replicate determinations of an analyte does not differ significantly from the standard counting error, the analysis is free of significant error, and the precision is limited by the counting statistics. If the standard deviation and standard counting error do differ significantly, some error(s) other than the counting error is limiting the precision. If the magnitude of the error is not acceptable, the source of the error must be found. In seeking the source of error, one first investigates the most probable source and continues with other sources in order of decreasing probability until the discrepancy is accounted for.

The following paragraphs outline a general procedure for evaluation of precision of an overall analytical method. Incidentally, a computerized automatic instrument may be programmed to conduct such a diagnostic test automatically and indicate the sources of error. For example, the computer may reveal such things as contamination of a certain specimen position, nonreproducible seating of one of the collimators or of the kV or mA settings, etc.

11.3.2.2. Instrumental Instability

A plane, smooth, thick ($\geq$1 mm), rigid piece of homogeneous metal—preferably a pure element—is placed in the specimen chamber. The x-ray generator, detector, and all electronic components are allowed to warm up for 30–60 min. Then, without disturbing the specimen or any instrument settings, 30-s to 1-min preset-time measurements are made of the line intensity of a spectral line of an element in the specimen for several hours. The standard deviation σ_1 is calculated for the entire set of measurements, and the counting error σ_N is calculated for the average number of counts accumulated during the preset time. Any significant difference between the two values may be attributed to some kind of long-term instrument instability. The variance for this instability is

$$\sigma^2 = \sigma_1{}^2 - \sigma_N{}^2 \tag{11.76}$$

Short-term instability can also be evaluated by calculation of the standard deviation σ_2 for sets of 10–20 consecutive measurements from the above set and comparison of these values with σ_N in the same way. If the sets of 10–20 consecutive measurements are distributed throughout the full series—say, a set every 30 min—a comparison of their σ_2 values reveals whether the short-term instability is constant or varies during prolonged operation of the instrument.

The stability of the detector–readout system can be evaluated by similar series of measurements made with an ^{55}Fe or other radioisotope source rather than the x-ray tube-specimen system.

11.3.2.3. Operational Error

A series of measurements is made without disturbing the specimen, but this time, after each measurement, all instrument controls are reset—kV, mA, 2θ, and perhaps detector and pulse-height analyzer settings. Moreover, if a crystal changer is used, this, too, should be reset. If the standard deviation σ_3 for this series is significantly larger than that for short-term instability σ_2 (see above), some kind of operational error is indicated. If the magnitude of the error is unacceptable, each setting must be investigated individually. Settings made by stepwise selector switches are less likely to be sources of error than settings made by continuous controls.

If helium or vacuum path is to be used, the effect on precision must be evaluated, preferably with a spectral line that, although usually measured in helium or vacuum path, can also be measured in air—Ti $K\alpha$ or Ca $K\alpha$, for example. Several series of measurements may be required: Two series made without disturbing the specimen or settings, one in air, the other in helium or vacuum, permit evaluation of the constancy of the helium or vacuum path. Another series without disturbing the settings but with specimen reinsertion after each measurement permits evaluation of the effectiveness of the helium flush period or of the air lock. A final series should be made without disturbing the specimen and without disturbing any setting except the goniometer angle. After each measurement, the goniometer is displaced to a very high and a very low 2θ. This series reveals whether movement of the goniometer permits air to leak through the sliding vacuum seal (if any), or whether it pulls the helium jacket, allowing air to enter or pleats to protrude into the x-ray path.

11.3.2.4. Specimen Error

Specimen error is best evaluated with a specimen of the type to be analyzed. A series of measurements is made without disturbing specimen or settings. If the standard deviation σ_4 for this series is significantly larger than that for short-term instability σ_2 (see above) or than the counting error σ_N, deterioration of the specimen in the primary beam or wrinkling of foil specimens is indicated. Next, a series of measurements is made without disturbing any setting, but this time the specimen is reloaded, and therefore also reinserted, after each measurement. If the standard deviation σ_5 for this series is significantly larger than σ_4, some type of specimen error is indicated. The series is repeated, but this time, the specimen is only reinserted—not reloaded—care being taken to insert the specimen in the same orientation each time. A comparison of the standard deviation σ_6 of this series with σ_5 indicates how much, if any, of σ_5 is reinsertion error. If a substantial portion of σ_5 remains, other experiments are made to identify the source of the error.

For liquid specimens, a series of measurements on identical cells filled with the same liquid in the same way permits evaluation of the effect of slight differences in the degree of filling or in stretching of the Mylar cover. A series of measurements on the same undisturbed cell permits evaluation of the effect of temperature rise, bubble formation, and radiolysis on prolonged irradiation. If the cell is removed, jarred to release bubbles, then reinserted after each measurement, the effect of bubble formation can be evaluated.

Powders loosely packed in Mylar-covered cups may be shaken or jarred between measurements, and powders packed into cells may be dumped out and repacked.

Solids, glass disks, briquets, etc. may be marked at the edge and measurements made with the marks up, down, right, and left in the specimen holder. Measurements can be made on different areas of the sample, or the sample may be resurfaced between measurements. These specimen types, as well as filter disks, are best evaluated by use of a specimen rotator. Three series of measurements are made: without rotation, with rotation only between measurements, and with continuous rotation.

Finally, if specimen treatment or preparation is involved, a carefully selected homogeneous specimen should be divided into a number of portions, and each portion subjected to the preparation process. A measurement is made on each specimen, and a standard deviation σ_7 is calculated and compared with σ_5 for the reloaded single specimen (see above). A significant difference indicates that specimen preparation is contributing to the error.

The foregoing discussion is not intended to be a complete outline of techniques for evaluation of precision, nor is it to be inferred that such a comprehensive procedure must be followed for every analysis. Instrumental and operational error require only occasional evaluation—a few times a year at most. However, a suitable evaluation of specimen error is required for each combination of analyte and sample type. Once the individual errors have been identified and evaluated, measures can be taken to reduce those that contribute error of unacceptable magnitude. These errors are then evaluated again, and the overall analytical precision is calculated with an equation similar to Equation (11.75).

The precision of x-ray spectrometric analyses may be very high, with standard deviation σ [Equation (11.6)] of the order 0.001. The accuracy of any analytical method may often be expressed by the relation, $\sigma = k\sqrt{C_A}$, where C_A is analyte concentration. In x-ray spectrometric analyses, k usually has the value 0.02–0.05, compared with 0.008–0.05 for conventional chemical methods (*49*).

11.3.2.5. Evaluation of Internal Consistency of Data

When the calibration standards are of questionable accuracy, the analyst must be content to demonstrate that his data are "internally consistent," that is, consistent with his standards—however accurate they may be. A particularly convenient method for doing this for linear calibration functions is given in this section. This method provides the equation of the linear calibration curve, evaluates the internal consistency of the measured data and of the standards themselves, and gives a measure of the disparity between theoretical and observed analytical data. The method is illustrated by use of the data in Table 11.5 for standards for determination of cobalt plating thickness on nickel. The plot of intensity *versus* thickness is linear and defined by the equation

$$I = mt + b \tag{11.77}$$

where I is Co $K\alpha$ net intensity, m is slope, and b is intercept on the intensity axis. Note that b is *not* background, as in Equation (14.20), because net intensities are used here; in fact, ideally, b should be zero, but may have a finite positive or negative value. The procedure follows.

An expression having the form of Equation (11.77) is written for each data point in Table 11.5. These equations are divided into two groups such that the sums of the equations in the two groups are as nearly equal as feasible. In the example, this would lead to groups consisting of the equations for points 1–4 and 5, respectively. However, if data point 5 happened to be divergent, the final result would be impaired, so a preferable grouping consists of the equations for points 1–3 and 4 and 5, respectively. Following

TABLE 11.5. Data for Evaluation of Internal Consistency

Data point	Co thickness t, μinch	Co $K\alpha$ net intensity I, counts/min
1	0.260	965
2	0.705	3,994
3	1.63	8,031
4	2.98	14,672
5	5.80	29,981

TABLE 11.6. Evaluation of Internal Consistency

Co thickness t, μinch	$mt =$ $5105t$	$I_{calc} =$ $mt - 85$, counts/min	I_{exp} (Table 11.5), counts/min	$\Delta = I_{exp} - I_{calc}$		Δ / I_{calc}	$\Delta t = (\Delta / I) \times t$	
				(+)	(−)		μinch	Å
0.260	1,327	1,242	965	—	277	0.223	0.058	14.5
0.705	3,599	3,514	3,994	480	—	0.136	0.096	24.0
1.63	8,321	8,236	8,031	—	205	0.025	0.041	10.2
2.98	15,213	15,128	14,672	—	456	0.030	0.089	22.2
5.80	29,609	29,524	29,981	457	—	0.016	0.093	23.2
				Σ +937	−938		Av. 0.075	18.8

are the five individual equations, the sum of each group (A and B), and the sum of all five equations (C):

$$
\begin{array}{llllll}
(1) & 965 = 0.260m + b & & & \\
(2) & 3{,}994 = 0.705m + b & & & \\
(3) & 8{,}031 = 1.63m + b & \text{(A)} & 12{,}990 = & 2.595m + 3b \\
(4) & 14{,}672 = 2.98m + b & & & \\
(5) & 29{,}981 = 5.80m + b & \text{(B)} & 44{,}653 = & 8.780m + 2b \\
\hline
& & \text{(C)} & 57{,}643 = & 11.375m + 5b
\end{array}
$$

Equations (A) and (B) are multiplied by 2 and 3, respectively, then solved simultaneously for the slope m, giving the value 5105 counts/min per μinch:

$$
\begin{array}{lrl}
3\times(\text{B}): & 133{,}959 = & 26.34m + 6b \\
2\times(\text{A}): & 25{,}980 = & 5.19m + 6b \\
\hline
(-) & 107{,}979 = & 21.15m \\
 & m = & 5105
\end{array}
$$

This value of m is substituted in equation (C), which is then solved for b—the amount by which the experimental curve "misses" the origin:

$$
\begin{aligned}
57{,}643 &= (11.375\times 5105) + 5b \\
b &= (57{,}643 - 58{,}069)/5 \\
&= -85.2
\end{aligned}
$$

Substitution of these values of m and b in Equation (11.77) gives the equation for the calibration curve,

$$I = 5105t - 85.2 \tag{11.78}$$

and analytical thickness is then given by

$$t = (I + 85.2)/5105 \tag{11.79}$$

Equation (11.78) can now be used to calculate the theoretical intensity for each standard thickness, and the calculated and observed intensities are then compared. The calculations are summarized in Table 11.6. The agreement between the sums of the positive and negative deviations Δ shows the self-consistency of both the experimental data and the nominal thicknesses of the standards. The disparity between calculated and observed thickness ranges from ~0.04 to ~0.1 μinch (~10 to ~25 Å) and has average value 0.075 μinch (18.8 Å).

Chapter 12

Matrix Effects

12.1. INTRODUCTION

Consider a thick specimen free from all sources of positional and chemical specimen error in an x-ray spectrometer free from all sources of instrumental and operational error (Section 11.1.1). One would expect the intensity $I_{A,M}$ of a spectral line of element A in matrix M (see below) to be given by

$$I_{A,M} = W_{A,M} I_{A,A} \qquad \text{or} \qquad I_{A,M}/I_{A,A} = W_{A,M} \tag{12.1}$$

where $W_{A,M}$ is weight fraction of analyte A in matrix M, and $I_{A,A}$ is analyte-line intensity from pure A, and both the sample and pure analyte are infinitely thick. Unfortunately, even under these ideal—and unattainable—conditions, this simple relationship rarely applies. Instead, $I_{A,M}$ is a function of the matrix M as well as the two parameters given in Equation (12.1), that is,

$$I_{A,M} = f(W_{A,M}, I_{A,A}, \text{M}) \tag{12.2}$$

This chapter considers the effect of the matrix on analyte-line intensity.

The *matrix* consists of the entire specimen except the particular analyte under consideration. Thus, in a multielement system, the matrix of the same specimen is different for each analyte in the specimen, and each analyte constitutes a part of the matrix of every other analyte. The term *matrix* applies to the specimen as measured in the spectrometer. For example, if an alloy is to be dissolved and measured as a solution, one may consider the alloy itself as consisting of certain analytes in a certain basic matrix, but it is the solution matrix that is of analytical significance. In such cases, the alloy and solution matrixes may be referred to as the *original* and *specimen* matrixes, respectively. The same applies if an internal standard is added; the treated matrix is the significant one. Some confusion may arise in the calculation of absorption coefficients for evaluation and correction of "matrix" effects; the absorption coefficient of the entire specimen —including the analyte—must be considered.

The effects of the matrix on the measured analyte-line intensity may be classified in two categories—those arising from the chemical composition of the matrix (absorption-enhancement effects) and those arising from the particle-size and/or surface texture of the specimen. These two classes of matrix effects are discussed in Sections 12.2 and 12.3, respectively.

In general, absorption-enhancement effects become more severe and particle-size and surface-texture effects less severe the deeper the active layer thickness is, that is, the deeper the excitation radiation penetrates the specimen and, more important, the greater the depth from which analyte-line radiation can emerge—or the shorter the primary and analyte-line radiation.

In general, matrix effects are substantially the same for wavelength- and energy-dispersive and nondispersive x-ray spectrometry, provided that the excitation is by x-radiation. However, if the excitation is by electrons, ions, or α or β radioisotopes, the exciting radiation does not penetrate as deeply into the specimen surface. In these cases, absorption-enhancement effects become less severe, and surface-texture and particle-size effects more severe.

This chapter is concerned principally with the nature of matrix effects. Methods of dealing with compositional matrix effects by experimental means are discussed in Chapter 14, and by mathematical means in Chapter 15. Methods of dealing with particle-size and surface-texture effects by experimental means are discussed in Sections 16.2.3 and 16.2.4 for solids, and in Section 16.3.5.1 for powders and briquets; mathematical means are mentioned in Section 12.3.

12.2. ABSORPTION-ENHANCEMENT EFFECTS

12.2.1. General; Definitions

The effects of the matrix composition on the measured analyte-line intensity are known variously as *matrix*, *interelement*, *self-absorption*, and *absorption-enhancement* effects. Liebhafsky and his co-authors [p. 172 in

(*26*)] raise valid objections to the first three of these terms: *Matrix effect* excludes the analyte itself, which contributes to the absorption coefficient of the specimen just as any matrix element does. *Interelement effect* excludes the self-absorption effect when only one element is present, or when only one element having a short-wavelength line is present in a very light matrix. *Self-absorption* excludes enhancement effects. These authors prefer the term *absorption-enhancement* effect, and this term is used here.

Absorption-enhancement effects arise from the following phenomena:

1. The matrix absorbs primary x-rays (*primary-absorption effect*); it may have a larger or smaller absorption coefficient than the analyte for primary x-rays, and it may preferentially absorb or transmit those wavelengths that excite the analyte line most efficiently, that is, those near the short-wavelength side of the analyte absorption edge.

2. The matrix absorbs the secondary analyte-line radiation (*secondary-absorption effect*); it may have a larger or smaller absorption coefficient than the analyte for the analyte-line radiation, and it may preferentially absorb or transmit this wavelength.

3. The matrix elements emit their own characteristic lines, which may lie on the short-wavelength side of the analyte absorption edge, and thereby excite the analyte to emit line radiation in addition to that excited by the primary x-rays (*enhancement*).

Because analyte-line radiation is a discrete wavelength, whereas the primary radiation is usually a continuum, the secondary-absorption effect is usually more severe than the primary-absorption effect, but is also more easily predicted, evaluated, and corrected. The primary-absorption effect is usually more severe in simple systems, such as binary alloys and, especially, pure elements. This accounts, at least in part, for the unsuitability of pure analytes as standards in x-ray secondary-emission (fluorescence) spectrometric analysis. Enhancement effects are usually much less severe—~10% at most—than absorption effects. Only the strongest matrix lines contribute significantly, and these only when they lie close to the short-wavelength side of the analyte absorption edge. However, enhancement effects are invariably much more difficult to correct than absorption effects.

The absorption-enhancement effects may be classified in two ways: On the basis of their effect on the analyte-line intensity, they may be *positive* or *negative* absorption effects, or *true* or *apparent* enhancement effects. On the basis of their origin or general nature, they may be *nonspecific* (*general*), *specific*, *secondary* (*second order*), or *unusual* (*special*).

In the *positive absorption effect*, the matrix has a smaller absorption coefficient than the analyte for the primary and analyte-line radiation, and the analyte-line intensity for a specified analyte concentration is greater than would be predicted from Equation (12.1). In the *negative absorption effect*, the matrix has a larger absorption coefficient than the analyte, and the analyte-line intensity is smaller than would be predicted. In the *true enhancement effect*, one or more spectral lines of matrix elements have wavelengths shorter than the analyte absorption edge. Thus, the matrix actually excites analyte-line radiation in addition to that excited by the primary beam, and the intensity is greater than that predicted by Equation (12.1). The *apparent enhancement effect* is simply the positive absorption effect. The analyte-line intensity is greater, but only because matrix absorption is lower; the matrix excites no additional analyte-line radiation.

True enhancement may take either or both of two forms: *direct enhancement* and the *third-element effect* (Section 1.6.3). Consider a ternary system A–B–C in which: (1) A is the analyte; (2) the respective strongest lines λ_A, λ_B, λ_C have progressively shorter wavelength; and (3) λ_C can excite λ_B and λ_A, and λ_B can excite λ_A. In direct enhancement, λ_B and λ_C excite λ_A directly. In the third-element effect, λ_C excites λ_B, which in turn excites λ_A. For example, in the chromium–iron–nickel system, the wavelengths of the respective $K\alpha$ lines are 2.29, 1.94, and 1.66 Å, and the respective K absorption edges 2.07, 1.74, and 1.49 Å. Thus, Ni $K\alpha$ excites iron and chromium, and Fe $K\alpha$ excites chromium. The contributions to the emitted Cr $K\alpha$ intensity from primary-beam excitation, direct enhancement by Fc $K\alpha$, direct enhancement by Ni $K\alpha$, and third-element enhancement by Fe $K\alpha$ excited by Ni $K\alpha$ are 72.5, 23.5, 2.5, and 1.5%, respectively. The theory of the third-element effect is given by Pollai (*P24*).

Nonspecific or *general* absorption-enhancement effects result simply from differences in the absorption coefficients of analyte and matrix elements for the primary and, especially, analyte-line radiation. Specific absorption edges are not involved. Nonspecific effects involve only absorption—unless one considers apparent enhancement. *Specific* absorption-enhancement effects result from interaction of analyte and matrix spectral lines and absorption edges in close proximity. *Secondary* or *second-order* absorption-enhancement effects arise from the influence of the overall matrix on a nonspecific or specific effect on a certain analyte–matrix element pair. They take the form of departures from the effects predicted from the absorption coefficients and the wavelengths of spectral lines and absorption edges of the specified analyte–matrix pair. *Unusual* or *special* absorption-enhancement effects include cases in which analyte-line intensity remains substantially constant, or even decreases, as analyte concentration increases.

In the derivation of equations for mathematical correction of absorption-enhancement effects (Chapter 15) and in the interpretation of intensity *versus* concentration calibration functions influenced by enhancement effects, it is sometimes expedient to regard enhancement as the opposite of absorption, that is, as "negative absorption, $-\mu/\varrho$." However, the concept of negative absorption is not to be confused with the negative absorption *effect*, in which analyte-line intensity is reduced. If analyte-line intensity measured in a specific matrix is greater than predicted by Equation (12.1), the cause may be either of the following: (1) The matrix may have a lower absorption coefficient than the analyte for analyte-line and/or primary radiation; this is the positive absorption effect; or (2) the spectral lines of the matrix element(s) may be enhancing the analyte line; this is the true enhancement effect, but it has the same effect on analyte-line intensity as the positive absorption effect. In other words, true enhancement, regarded as negative absorption, acts as an *apparent* positive absorption effect, just as true positive absorption acts as an apparent enhancement effect, as already mentioned.

Nonspecific effects can be predicted by use of the guidelines given in Sections 12.2.2–12.2.4 and evaluated by simple calculations based on the mass-absorption coefficients of analyte and matrix elements. Specific effects can be predicted by use of the guidelines given in Sections 12.2.3 and 12.2.5 and evaluated by measurement of analyte-line intensity from mixtures of the analyte and a single matrix element in various proportions. The more complex effects can be predicted by use of the guidelines given in Section 12.2.6 and evaluated by measurement of analyte-line intensity from a specific analyte–matrix element pair as the remainder of the matrix is varied. Chapters 14 and 15 outline in detail the several experimental and mathematical methods for evaluation and correction of absorption-enhancement effects.

Whatever absorption-enhancement effects a specified analyte–matrix system may be subject to, they are most severe at and above infinite thickness, decrease in severity as thickness decreases below infinite thickness, and substantially disappear in thin films. Also, whatever absorption-enhancement effects a specified matrix element may cause, they decrease in severity as the concentration of that element decreases.

12.2.2. Effects on Calibration Curves

Figure 12.1 shows a series of calibration curves that illustrate absorption-enhancement effects. If only absorption effects are present, the curves

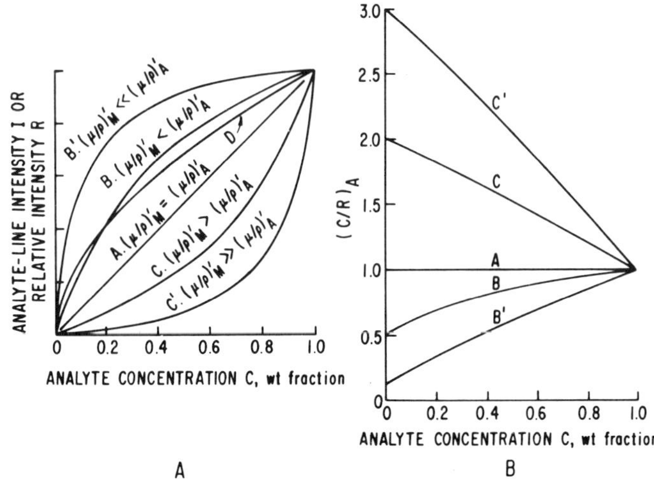

FIG. 12.1. Relationship of absorption-enhancement effects and the form of calibration curves. A. Intensity I_A or relative intensity R_A *vs.* analyte concentration. B. $(C/R)_A$ *vs.* analyte concentration.

represent various combinations of analyte and matrix absorption coefficient for the analyte line and for the primary radiation that excites it most efficiently. In the figure, $(\mu/\varrho)'$ represents the total mass-absorption coefficient for both primary and analyte-line radiation. However, similar curves are obtained when enhancement effects are present, the enhancement appearing as a positive absorption effect (curves B and B'), as explained above.

Rasberry and Heinrich (*R4*, *R5*, *Z12*) have considered the curves in Figure 12.1A in greater detail. If the curves for binary systems are plotted in terms of R, the ratio of analyte-line intensities from the specimen and pure analyte (Section 1.4.4), and if only true, simple absorption effects are present, curves B, B', C, and C' are hyperbolic and described by the equation

$$R = \frac{C}{C + a(1 - C)} \tag{12.3}$$

where R is analyte-line relative intensity, C is analyte concentration (weight fraction), and a is a positive, real-number constant. The value of a is 1, <1, or >1 for the three cases, respectively, $(\mu/\varrho)_M' = (\mu/\varrho)_A'$ (curve A), $(\mu/\varrho)_M' < (\mu/\varrho)_A'$ (curves B and B'), and $(\mu/\varrho)_M' > (\mu/\varrho)_A'$ (curves C and C'). The greater the displacement of the curve from the linear case (curve A), the greater is the divergence of a from 1. However, when true enhancement occurs, although the enhanced calibration curves may be represented in general appearance by curves of the form B and B', these curves do not exactly follow the hyperbolic function [Equation (12.3)] and actually constitute a fourth case, curve D. This is to be expected since analyte-line intensity increases with increasing concentration of both the analyte (excited element) and excitant element. Thus, when analyte concentration is near unity, analyte-line intensity is high from excitation by primary x-rays, but there is very little excitant present to cause enhancement. However, at low analyte concentrations, enhancement is increased due to high excitant concentration. The stronger the enhancement, the greater is the displacement of curve D from curve A. Case D is significant in that some derivations of influence-coefficient equations assume that enhancement may be considered simply as "negative absorption."

Calibration curves of the form shown in Figure 12.1A may also be plotted in terms of $(C/R)_A$, that is, ratio of analyte concentration (weight fraction) and analyte-line relative intensity. When this is done, the hyperbolas of Figure 12.1A become the substantially straight lines of Figure 12.1B. The curve representing no absorption-enhancement effect in Figure 12.1A, curve A, becomes the zero-slope "no-effect" line in Figure 12.1B.

The effects of absorption and enhancement on the form of calibration curves is discussed further in Sections 12.2.4 and 14.3.2.5.

12.2.3. Prediction of Absorption-Enhancement Effects

Mitchell (*M39*) gives an excellent discussion of prediction of absorption-enhancement effects, and this entire section is taken from her article. Mitchell states,

> "Quantitative analysis of a sample by x-ray spectrography is facilitated by prior qualitative examination for its major constituents. Knowledge of the sample matrix allows general prediction of the absorption effects which will occur. Intelligent selection of the appropriate method for interelement effect correction, and choice of analytical lines subject to a minimum of absorption are then possible. Interelement effects can be predicted by a study of mass-absorption coefficients and the relative positions of x-ray lines and element absorption edges."

12.2.3.1. *K* Lines

The relationships among K spectral lines and absorption edges discussed in this section are shown in Figure 14.10.

For K lines of pairs of elements adjacent to one another in the periodic table from atomic number 22 (titanium) to 72 (hafnium), only moderate absorption-enhancement effects occur, and $K\alpha$ intensities are nearly linear functions of concentration; examples are Cu $K\alpha$ and Zn $K\alpha$ from Cu–Zn alloys and Zr $K\alpha$ and Nb $K\alpha$ from Zr–Nb alloys. However, for some adjacent elements in this group, it may not be possible to resolve the $K\alpha$ line of the higher from the $K\beta$ line of the lower element; for example, $_{23}$V $K\alpha$ (2.50 Å) and $_{22}$Ti $K\beta$ (2.51 Å). One must then use the $K\beta$ line of the higher element (V $K\beta$ in the example) as its analytical line. But in this case, in addition to the lower intensity, severe absorption occurs due to the proximity of V $K\beta$ to the short-wavelength side of the Ti K-absorption edge. In general, the $K\beta$ line of the higher of any two or more adjacent elements suffers high absorption by the lighter elements, the effect being most severe by the immediately adjacent (lighter) element and decreasing in severity as Z decreases. For example, in Figure 14.10, Zn $K\beta$ from Cu–Zn alloys is strongly absorbed by copper, and from Zn–Cu–Ni alloys by both copper and nickel; Mo $K\beta$ from Mo–Nb alloys is strongly absorbed by niobium, and from Mo–Nb–Zr alloys by both niobium and zirconium. In these systems, the K lines of the lighter elements are enhanced, but not strongly, because the $K\beta$ line is relatively weak.

When the $K\alpha$ line is subject to spectral-line interference and the $K\beta$ line is strongly absorbed, the second-order $K\alpha$ line may be useful.

For K lines of triads of adjacent elements from atomic number 22 (titanium) to 35 (bromine) (Figure 14.10), no general pattern appears, and absorption-enhancement effects vary:

1. There is a gradual displacement of the $K\alpha$ lines of the higher elements so that by atomic number 35, all three $K\alpha$ lines of an adjacent triad lie on the long-wavelength side of all three of their absorption edges, thus freeing them from specific absorption.

2. In adjacent triads from atomic number 22–26 (titanium to iron), both K lines of the middle element are subject to spectral-line interference. For example, in the Ti–V–Cr triad, Ti $K\beta$ interferes with V $K\alpha$, and Cr $K\alpha$ interferes with V $K\beta$. Most spectrogoniometers and all pulse-height selectors cannot resolve these lines. In the Cr–Mn–Fe triad (not shown in Figure 14.10), the interference with the Mn K lines is less severe, and most modern spectrogoniometers can resolve both these lines. The Mn $K\alpha$ line is preferable because of its higher intensity and its location on the long-wavelength side of all three K-absorption edges.

3. Adjacent triads from atomic number 35 up are represented by the Zr–Nb–Mo and In–Sn–Sb triads. The $K\alpha$ lines of all three elements are on the long-wavelength side of all three of their K edges. However, the $K\alpha$ line

of the highest element is subject to interference by the $K\beta$ line of the lowest; for example, Mo $K\alpha$ and Zr $K\beta$ are only 0.008 Å apart.

4. Even if a 75- or 100-kV generator is available to excite the K lines of the heavier elements, such as the rare earths, very high resolution is required to separate their $K\alpha$ lines; the $K\alpha$ lines of lanthanum, cerium, and praseodymium are each 0.01 Å from the adjacent line.

Relationships among K lines and edges of adjacent elements of atomic number less than 22 are shown in Figure 14.10 for the K–Ca–Sc and Mg–Al–Si adjacent triads. This region has the simplest x-ray spectra, and combinations of two or more elements are free from mutual spectral-line interference. In adjacent pairs of triads, both K lines of the heavy element are absorbed by the lighter ones. The effect is most severe in the K–Ca–Sc triad, least in the Mg–Al–Si triad.

12.2.3.2. *L* Lines

Prediction of absorption-enhancement effects for L-series lines is more difficult because of the multiplicity of the L lines and edges. Figure 12.2 shows relationships among L spectral lines and edges for elements 57–62 (lanthanum to samarium). In a La–Ce–Pr–Nd–Sm system, the choice of analytical lines is based on considerations of spectral-line interference and absorption-enhancement effects as follows.

1. La $L\alpha_1$ is on the long-wavelength side of all the L edges. The La $L\alpha_1$ intensity is a nearly linear function of concentration; it varies generally with matrix composition because it is absorbed in decreasing degree by cerium, praseodymium, neodymium, and samarium; it is subject to non-

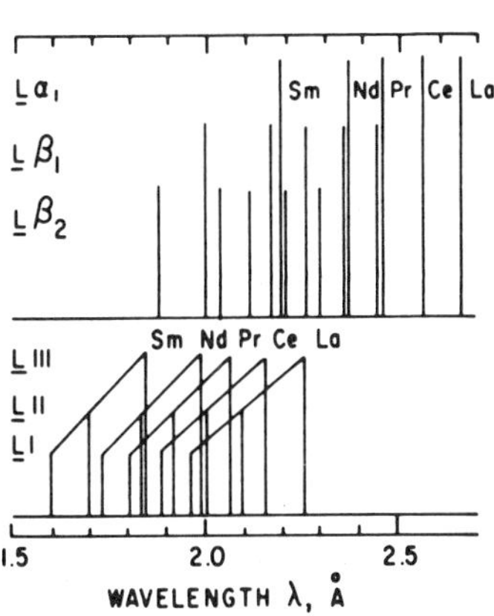

FIG. 12.2. *L*-series spectral lines and absorption edges of lanthanum, cerium, praseodymium, neodymium, and samarium (atomic numbers 57, 58, 59, 60, and 62).

specific absorption by all the other elements and to enhancement by those having lines lying on the short-wavelength side of the La *L*III edge—especially Sm $L\alpha_1$ and Pr $L\beta_1$.

2. The Ce $L\alpha_1$ intensity is also a nearly linear function of concentration; it varies generally with composition because it is absorbed less by lanthanum than by praseodymium, neodymium, and samarium.

3. For praseodymium, neodymium, and samarium, the $L\beta_1$ lines are usually used because of close spectral-line interference with their $L\alpha_1$ lines. The Nd $L\beta_1$ line lies on the short-wavelength side of the La *L*III edge and is absorbed specifically by neodymium, nonspecifically by cerium, praseodymium, and samarium. The same applies to Pr $L\beta_1$, which lies extremely close to the La *L*III edge. The Sm $L\beta_1$ line (2.00 Å) is subject to moderate absorption by La *L*II (2.10 Å) and Pr *L*III (2.08 Å) and to slight absorption by Ce *L*III (2.16 Å).

Mitchell points out that in the prediction of interelement effects, many simplifications are possible because of the regular variation of wavelength of lines and edges with atomic number. Elements adjacent to one another in the periodic table are affected by and affect certain other elements in a similar manner. For example, Nb *K* lines (*Z* 41) are specifically absorbed by $_{73}$Ta to about the same degree as by $_{74}$W or $_{72}$Hf, and are nonspecifically absorbed by $_{22}$Ti to about the same degree as by $_{23}$V. The Zr *K* lines (*Z* 40) are specifically absorbed by tantalum, tungsten, and hafnium and nonspecifically absorbed by titanium and vanadium to almost the same degree as Nb *K* lines (*Z* 41). In addition to simplifying the prediction of absorption-enhancement effects, this similarity in behavior of adjacent and neighboring elements also provides the basis of internal standardization (Section 14.4) and reduces the number of standards and/or matrix corrections required in calibration standardization (Section 15.4).

12.2.4. Nonspecific Absorption Effects

The relative intensities I_1 and I_2 measured under identical conditions from two samples having concentrations C_1 and C_2 of the same analyte in the same matrix are given by

$$\frac{I_1}{I_2} \approx \left(\frac{C_1}{C_2}\right) \frac{(\mu/\varrho)_2'}{(\mu/\varrho)_1'} \tag{12.4}$$

where $(\mu/\varrho)'$ is mass-absorption coefficient of the sample for both primary and secondary x-rays. The ratio of the slopes m_1 and m_2 of the calibration curve (Figure 14.3) at C_1 and C_2 is

$$\frac{m_1}{m_2} = \frac{(\mu/\varrho)_2'}{(\mu/\varrho)_1'} \tag{12.5}$$

Equations (12.4) and (12.5) show that as the ratio C_1/C_2 changes, I_1/I_2 also changes, but not necessarily at the same rate, as determined by $(\mu/\varrho)_2'/(\mu/\varrho)_1'$. This difference in rate may lead to positive and negative deviations from linearity of the calibration function. In special cases, when $(\mu/\varrho)'$ increases at a greater rate than C, even a negative slope may occur, as shown in Section 12.2.7.

It follows from Equation (12.4) that for the same analyte concentration in two samples 1 and 2 having different matrixes with mass-absorption coefficients $(\mu/\varrho)_1'$ and $(\mu/\varrho)_2'$ for primary and secondary x-rays, the relative analyte-line intensities I_1 and I_2 are

$$\frac{I_1}{I_2} = \frac{(\mu/\varrho)_2'}{(\mu/\varrho)_1'} \tag{12.6}$$

In Equations (12.4) and (12.6), for practical purposes, $(\mu/\varrho)'$ may be replaced with the corresponding mass-absorption coefficient for the analyte line. Then, to a first approximation, analyte-line intensity may be corrected for matrix absorption simply by use of mass-absorption coefficients of the samples. If these are neither known nor readily calculated, they may be measured; this is the basis of emission–absorption methods (Sections 10.5.2 and 14.8.1).

Figure 12.1A shows a series of calibration curves that illustrate nonspecific absorption effects and apparent enhancement effects.

In a *neutral matrix* (curve A), the absorption coefficient of the matrix $(\mu/\varrho)_M'$ and analyte $(\mu/\varrho)_A'$ are substantially the same for both the analyte-line and the primary radiation that excites it most efficiently. The primary x-rays reach each incremental specimen volume, and the excited analyte-line radiation emerges from the specimen with about the same attenuation in the matrix as in pure analyte. Thus, analyte-line intensity I_A increases at about the same rate as C_A, the I_A–C_A curve is substantially linear, and the absorption effect is very small.

Elements within one or two atomic numbers of the analyte usually constitute neutral matrixes for that analyte, at least down to atomic number 22 (titanium). For these medium-Z and high-Z elements, absorption-enhancement properties of adjacent elements are so similar that they can serve as internal standards for one another (Section 14.4.2). However, at

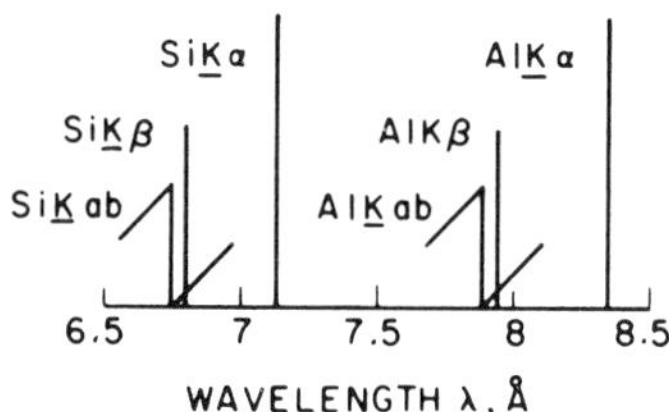

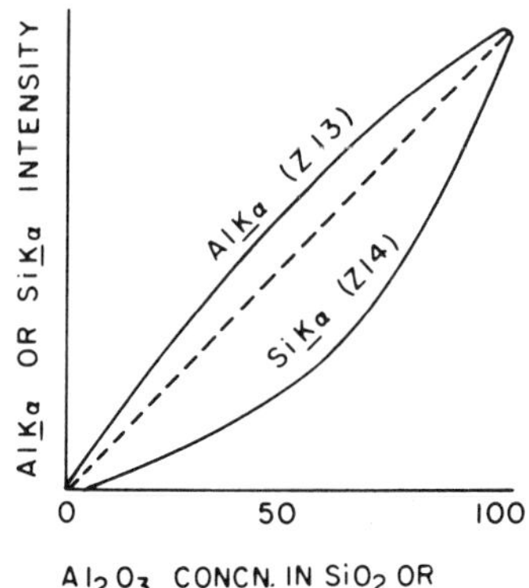

FIG. 12.3. Mutual absorption-enhancement effects of aluminum and silicon.

lower atomic numbers, the differences in wavelength (Figure 1.8) and absorption coefficient (Figure 2.5) for adjacent elements have increased to a degree such that the Z $K\alpha$ line is very strongly absorbed by element Z − 1. For example, one might expect curve *A* in Figure 12.1A to represent the curves of $K\alpha$ intensity *versus* concentration for aluminum (*Z* 13) and silicon (*Z* 14) in their mixed oxides. The actual curves and the relationships among their *K* lines and absorption edges are shown in Figure 12.3. The Al $K\alpha$ curve shows marked enhancement even though Al $K\alpha$ is absorbed more strongly by silicon ($\mu/\varrho \sim 500$) than by aluminum ($\mu/\varrho \sim 385$). The reason is that Si $K\alpha$ is absorbed very strongly by aluminum ($\mu/\varrho \sim 3500$), thereby causing true enhancement of Al $K\alpha$ sufficient to more than balance the absorption. The Si $K\alpha$ curve shows very strong absorption because of the high absorption coefficient of aluminum just mentioned. Thus, aluminum, although an adjacent element, actually constitutes a very heavy matrix for silicon. It is interesting to compare the aluminum–silicon system with the copper–zinc system, where the absorption coefficients of both copper and zinc for both Cu $K\alpha$ and Zn $K\alpha$ are all substantially the same, ranging from ~44 to 60. Only a slight enhancement of Cu $K\alpha$ by Zn $K\beta$ occurs.

In a *light matrix* (curves B and B' in Figure 12.1A), $(\mu/\varrho)_M'$ is less than $(\mu/\varrho)_A'$ for the analyte line and primary x rays. In such a matrix, the primary x-rays undergo less attenuation in reaching any incremental specimen volume and analyte-line radiation undergoes less attenuation in emerging from the specimen in the matrix than in pure analyte. Thus, I_A increases at a faster rate than C_A. At higher concentrations, the specimen composition approaches pure analyte, and the difference in rates of increase for I_A and C_A decreases. The I_A–C_A curves are nonlinear and show the positive absorption effect; the greater the difference between $(\mu/\varrho)_M'$ and $(\mu/\varrho)_A'$, the more distorted is the curve, as is evident from a comparison of curves B and B'. This positive absorption effect constitutes an apparent or pseudo enhancement effect. Examples of light matrixes are usually provided by analytes having relatively short-wavelength lines in relatively low-Z matrixes. Lead as tetraethyllead in gasoline is an example of an extreme case (curve B').

In a *heavy matrix* (curves C and C'), $(\mu/\varrho)_M'$ is greater than $(\mu/\varrho)_A'$, primary and analyte-line radiation undergo greater attenuation in the matrix than in pure analyte, I_A increases at a slower rate than C_A, and the curves show the negative absorption effect. Again, at higher analyte concentration, the difference in the two rates decreases, and the greater the difference in $(\mu/\varrho)_M'$ and $(\mu/\varrho)_A'$, the more distorted is the curve. Examples of heavy matrixes are usually provided by analytes having relatively long-wavelength lines in relatively high-Z matrixes. Determinations of magnesium, aluminum, and silicon in molybdenum alloys provide examples of extreme cases (curve C').

It must be emphasized that whether a matrix is "heavy," "light," or "neutral" for a specified analyte line is determined by the mass-absorption coefficient of that matrix for that line, not by the effective atomic number of the matrix. Admittedly, it is usually true that a high-Z element constitutes a heavy matrix for a long-wavelength line, and a low-Z element constitutes a light matrix for a short-wavelength line. It is also usually true that a high-Z element constitutes a heavy matrix for the spectral lines of low-Z elements, a low-Z element constitutes a light matrix for the lines of high-Z elements, and any element constitutes a more or less neutral matrix for lines of adjacent and neighboring elements. However, these simple guidelines are far from generally applicable for reasons given in paragraphs 1 and 2 of Section 2.1.3 and must be used with great caution. For example, lead (Z 82) is a heavy matrix (μ/ϱ 5968), as expected, for Na $K\alpha$ (Z 11, λ 11.9 Å), but is a relatively light matrix (μ/ϱ 85) for Br $K\alpha$ (Z 35, λ 1.04 Å). Similarly, aluminum (Z 13) is a light matrix (μ/ϱ 34 to 6), as expected, for the $K\alpha$ lines of elements 31–42 (gallium to molybdenum), but is a heavy matrix for Si $K\alpha$

(Z 14, λ 7.13 Å, μ/ϱ 3493) and for Zr $L\alpha_1$ (Z 40, λ 6.07 Å, μ/ϱ 2236), for which the "rules" above would predict it to be neutral and light, respectively.

12.2.5. Specific Absorption-Enhancement Effects

If the analyte line occurs at a wavelength just less than that of an absorption edge of a particular matrix element B, the A line is highly absorbed by B, and A-line intensity is reduced in proportion to B concentration. This is the case for Fe $K\alpha$ (1.94 Å) in the presence of chromium ($\lambda_{K_{ab}}$ 2.07 Å), and chromium has a similar, but progressively weaker, negative absorption effect on the $K\alpha$ lines of elements of successively higher atomic number—cobalt, nickel, copper, etc.

Conversely, if the analyte absorption edge occurs at a wavelength just greater than that of a line of a particular matrix element B, the B line is absorbed by A, and A-line intensity is enhanced in proportion to B concentration. This is the case for Fe K_{ab} (1.74 Å) and Ni $K\alpha$ (1.66 Å), and nickel has a similar, but progressively weaker, enhancement effect on the $K\alpha$ lines of elements of successively lower atomic number—manganese, chromium, vanadium, etc.

The effects just described are specific, in that they depend on the proximity of spectral lines and absorption edges of analyte and matrix elements. These specific absorption-enhancement effects are now considered in detail, with Fe $K\alpha$ serving as an example of a typical analyte line [pp. 165–6 in (*26*)]. Ideally, in the determination of iron in various matrixes, Fe $K\alpha$ intensity should be given by Equation (12.1):

$$I_{\mathrm{Fe}K\alpha,\mathrm{M}} = W_{\mathrm{Fe,M}} I_{\mathrm{Fe}K\alpha,\mathrm{Fe}} \tag{12.7}$$

That is, Fe $K\alpha$ intensity in matrix M should be given by the product of the weight fraction of iron in matrix M and the Fe $K\alpha$ intensity from pure iron. Let us consider specific absorption-enhancement effects on the Fe $K\alpha$ intensity measured from binary alloys or mixtures consisting of iron (atomic number 26) in matrixes of aluminum (13), chromium (24), manganese (25), cobalt (27), nickel (28), cerium (58), and lead (82). Figure 12.4 shows the spectral lines, absorption edge, and mass-absorption curve for each of these elements in the spectral region of the iron K absorption edge and lines.

The effect of each matrix element on Fe $K\alpha$ intensity must be considered in terms of three phenomena: (1) absorption of that portion of the primary beam that excites Fe $K\alpha$ most efficiently, that is, having wavelength near

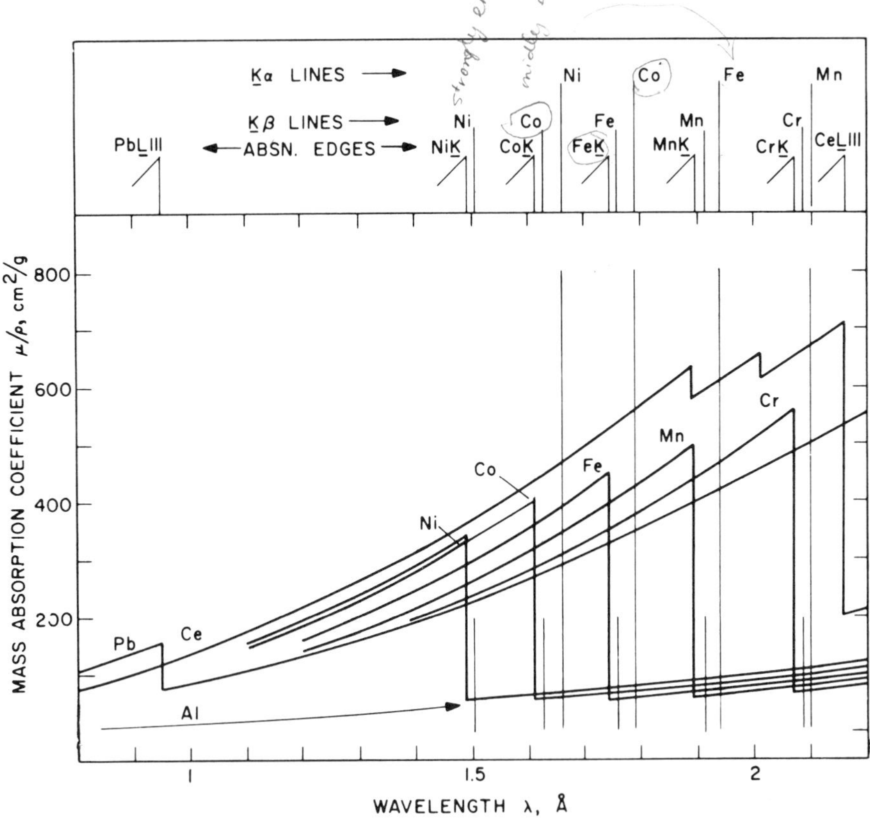

FIG. 12.4. Specific absorption-enhancement effects of several matrix elements on Fe $K\alpha$ intensity.

the short-wavelength side of the iron K absorption edge; (2) absorption of the Fe $K\alpha$ line; and (3) enhancement of the Fe $K\alpha$ line by matrix lines having wavelengths on the short side of the iron K edge.

With respect to absorption of the primary x-rays, at all wavelengths on the short side of the iron K edge, manganese absorption is about the same as that of iron, cerium absorption is substantially higher, chromium and lead absorption are substantially lower, and aluminum absorption is very low. Cobalt and nickel absorption are very low in the extremely critical region within, respectively, ~0.15 Å and ~0.25 Å of the iron K edge, but they rise to about the same value as iron at shorter wavelengths.

With respect to the much more significant absorption of Fe $K\alpha$ radiation, aluminum, manganese, iron, cobalt, and nickel all have about the same very low absorption. Lead, chromium, and, especially, cerium have much higher absorption.

With respect to enhancement, aluminum, chromium, manganese, cerium, and lead have no spectral lines at wavelengths anywhere near the short side of the iron K edge, and therefore do not enhance Fe $K\alpha$. The Co $K\beta$ and Co $K\alpha$ lines bracket the iron K edge, with Co $K\beta$ on the short side; thus, cobalt mildly enhances Fe $K\alpha$. Both Ni $K\alpha$ and Ni $K\beta$ lie on the short side of the iron K edge, and nickel strongly enhances Fe $K\alpha$.

The net effect of each of these matrix elements on Fe $K\alpha$ intensity may be presented in terms of its predicted effect based on the foregoing discussion as compared with the intensity calculated from Equation (12.7); that is,

$$\frac{(I_{\mathrm{Fe}K\alpha,\mathrm{M}})_{\mathrm{predicted}}}{W_{\mathrm{Fe},\mathrm{M}} I_{\mathrm{Fe}K\alpha,\mathrm{Fe}}} = R \tag{12.8}$$

Aluminum. $R \gg 1$ because of the very low absorption of primary and Fe $K\alpha$ radiation and despite the absence of any enhancing spectral lines (strong positive absorption or apparent enhancement effect).

Chromium. $R \ll 1$ because the high absorption of Fe $K\alpha$ outweighs the somewhat lower absorption of the primary x-rays (strong negative absorption effect).

Manganese. $R \approx 1$ because manganese has substantially the same absorption coefficient as iron for primary and Fe $K\alpha$ radiation and causes no enhancement (absorption-enhancement effects largely absent; neutral matrix).

Iron. $R = 1$.

Cobalt. R is somewhat greater than 1. The absorption of primary radiation within ~0.15 Å of the short side of the iron K edge is very low, and Co $K\beta$ enhances Fe $K\alpha$ (combination of positive absorption and mild enhancement).

Nickel. $R \gg 1$ because the absorption of primary radiation within ~0.25 Å of the short side of the iron K edge is very low, and both Ni $K\alpha$ and Ni $K\beta$ enhance Fe $K\alpha$ (combination of positive absorption and strong enhancement).

Cerium. $R \ll 1$ because of the very high absorption of both primary and Fe $K\alpha$ radiation and the absence of enhancement (strong negative absorption).

Lead. $R < 1$. The effect of lead is similar to that of cerium, except that the absorption coefficient of lead is not as high in this spectral region (relatively strong negative absorption).

An analysis of the possible absorption-enhancement effects should be made whenever an analytical method is to be developed for a new combination of analyte(s) and matrix. Often, it is sufficient to plot on a wavelength scale the principal spectral lines and absorption edges of all elements present, in the manner shown at the top of Figure 12.4. Backerud (*B1*) gives a comprehensive, largely mathematical procedure for establishing the scope and limitations of a proposed determination of major, minor, and trace elements in a complex alloy system.

12.2.6. Secondary Absorption-Enhancement Effects

12.2.6.1. General

Mitchell and Kellam (*M43*) have carried the characterization of absorption-enhancement effects a significant step farther by consideration of the overall matrix composition on the absorption-enhancement effect of a particular matrix element on a specified analyte line. They describe five secondary effects that may markedly influence the effect predicted for an analyte–matrix element pair on the basis of the nonspecific and specific effects just described. The discussion that follows is largely summarized from their excellent paper.

Figure 12.5 illustrates the secondary effects, using Fe $K\alpha$ as the analyte line. For all curves, Fe $K\alpha$ is measured from synthetic specimens having the same iron concentration (10%) in a matrix M consisting of various proportions of a specific matrix element B and one other matrix element (M − B). Figure 12.5A shows secondary absorption effects on the absorption effect of chromium (element B) on Fe $K\alpha$. Figure 12.5B shows secondary enhancement effects on the enhancement effect of copper (element B) on Fe $K\alpha$. Table 12.1 gives the data necessary for consideration of the figure. Following is a glossary of the symbols and terms used in the discussion:

A is the analyte element.

B is a specific matrix element, the effect of which on the analyte is being considered as a function of the overall matrix composition.

M is the whole matrix—all specimen constituents except the analyte; in the examples given below, the matrix consists of only two elements, B and "(M − B)."

(M − B) is the matrix excluding element B; it is the element, the effect of which on the absorption-enhancement effect of B on A is being evaluated.

μ/ϱ is the mass-absorption coefficient for the analyte line, unless specified otherwise.

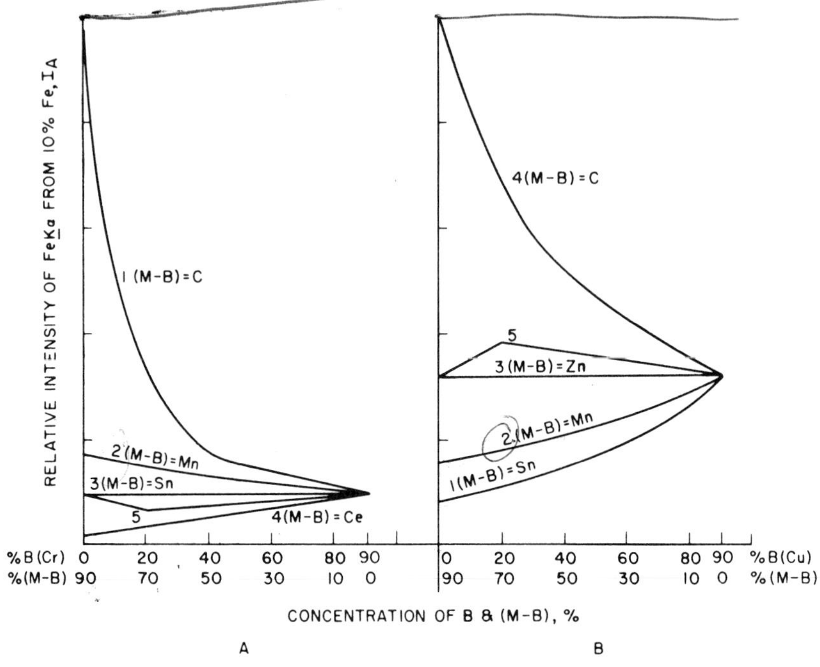

FIG. 12.5. Secondary absorption-enhancement effects. A. Secondary absorption effects of chromium on Fe $K\alpha$ intensity in the presence of several other matrix elements. B. Secondary enhancement effects of copper on Fe $K\alpha$ intensity in the presence of several other matrix elements. [B. J. Mitchell and J. E. Kellam, *Appl. Spectrosc.* **22,** 742 (1968).]

A *light matrix* is one in which $(\mu/\varrho)_{\mathrm{M-B}} < (\mu/\varrho)_{\mathrm{A}}$.

A *neutral matrix* is one in which $(\mu/\varrho)_{\mathrm{M-B}} \approx (\mu/\varrho)_{\mathrm{A}}$.

A *heavy matrix* is one in which $(\mu/\varrho)_{\mathrm{M-B}} > (\mu/\varrho)_{\mathrm{A}}$.

An *equivalent matrix* is one in which $(\mu/\varrho)_{\mathrm{M-B}} \approx (\mu/\varrho)_{\mathrm{B}}$, that is, in which element B and the remainder of the matrix have about the same absorption coefficient for the analyte line.

12.2.6.2. Secondary Absorption Effects

In Section 12.2.5, it is shown that chromium has a strong negative absorption effect on Fe $K\alpha$ intensity, that is, strongly depresses it. However, when other elements are present in the matrix, the absorption effect becomes much more complex. Figure 12.5A shows the effect of various elements (M − B) on the negative absorption effect of chromium (element B) on Fe $K\alpha$ intensity. Table 12.1 gives the supplementary data. Mitchell and Kellam define five secondary-absorption effects as follows:

1. *Exaggerated absorption* (curve 1) occurs in a very light matrix, where $(\mu/\varrho)_{M-B} \ll (\mu/\varrho)_B > (\mu/\varrho)_A$. Absorption of Fe $K\alpha$ by carbon (μ/ϱ 11) is so much less than that by chromium (μ/ϱ 490) that, as chromium progressively replaces carbon in the matrix, the absorption of Fe $K\alpha$ by chromium appears to be strongly exaggerated; that is, Fe $K\alpha$ intensity is depressed at a much faster rate than would be expected.

2. *Normal absorption* (curve 2) occurs in a neutral matrix, where $(\mu/\varrho)_{M-B} \approx (\mu/\varrho)_A$. Chromium has about the same absorption effect on Fe $K\alpha$ in the presence of manganese (μ/ϱ 64) as in the presence of iron alone (μ/ϱ 73).

3. *Cancellation of absorption* (curve 3) occurs in an equivalent matrix, where $(\mu/\varrho)_{M-B} \approx (\mu/\varrho)_B$. Replacement of tin ($\mu/\varrho$ 457) with chromium (μ/ϱ 490) has relatively little effect on Fe $K\alpha$ intensity. The expected decrease in Fe $K\alpha$ intensity on increase in chromium concentration does not occur.

4. *Reversal of absorption* (curve 4) occurs in a heavy matrix, where $(\mu/\varrho)_{M-B} \gg (\mu/\varrho)_B > (\mu/\varrho)_A$. An expected reduction in analyte-line intensity due to the presence of a specific matrix element B having a high absorption coefficient for the analyte line may actually become an apparent en-

TABLE 12.1. Secondary Absorption-Enhancement Effects[a]

Curve[b]	Matrix type[b]	Absorption effects[c] (Figure 12.5A)		Enhancement effects[d] (Figure 12.5B)	
		(M − B)[e]	$(\mu/\varrho)_{M-B,FeK\alpha}$	(M — B)[e]	$(\mu/\varrho)_{M-B,FeK\alpha}$
1	Very light	$_{6}C$	11	$_{6}C$	11
2	Neutral	$_{25}Mn$	64	$_{25}Mn$	64
3	Equivalent	$_{50}Sn$	457	$_{30}Zn$	109
4	Heavy	$_{58}Ce$	636	$_{50}Sn$	457
5	Medium heavy	$_{53}I$	527	$_{27}Co$	81

[a] From Mitchell and Kellam (*M43*). Refer to Figure 12.5.
[b] Curve numbers and matrix types apply to both Figures 12.5A and 12.5B.
[c] For absorption effects:
Analyte A = Fe, $\lambda_{FeK\alpha} = 1.94$ Å, $(\mu/\varrho)_{Fe,FeK\alpha} = 73$ cm²/g.
Matrix element B = Cr, $\lambda_{CrK_{ab}} = 2.07$ Å, $(\mu/\varrho)_{Cr,FeK\alpha} = 490$ cm²/g.
[d] For enhancement effects:
Analyte A = Fe, $\lambda_{FeK_{ab}} = 1.74$ Å, $(\mu/\varrho)_{Fe,FeK\alpha} = 73$ cm²/g.
Matrix element B = Cu, $\lambda_{CuK\alpha} = 1.54$ Å, $(\mu/\varrho)_{Cu,FeK\alpha} = 99$ cm²/g.
[e] The entire matrix M consists of elements B and (M − B).

hancement effect upon change in the overall matrix composition. This is the case when, as B concentration increases, it replaces a matrix having much higher absorption than itself. As chromium (μ/ϱ 490) replaces cerium (μ/ϱ 636), the matrix absorption of Fe $K\alpha$ actually decreases.

5. *Absorption-enhancement inflection* (curve 5) occurs in a medium-heavy matrix, where $(\mu/\varrho)_{\text{M-B}} > (\mu/\varrho)_{\text{B}} > (\mu/\varrho)_{\text{A}}$. At low B concentrations, B absorbs analyte-line radiation. As B concentration increases, it replaces a matrix having somewhat higher absorption than itself, causing a slight apparent enhancement. This phenomenon might be expected for Fe $K\alpha$ in a matrix of chromium (element B, μ/ϱ 490) and iodine (element M − B, μ/ϱ 527).

It is evident that chromium may be said to decrease measured Fe $K\alpha$ intensity only under certain conditions, and even then to a varying extent.

12.2.6.3. Secondary Enhancement Effects

In Section 12.2.5, it was shown that nickel has a strong enhancement effect on Fe $K\alpha$ intensity. The same applies to copper; both Cu $K\alpha$ and Cu $K\beta$ lie on and near the short side of the iron K edge. However, here again, when other elements are present in the matrix, the enhancement effect becomes much more complex. Figure 12.5B shows the effect of various elements (M − B) on the enhancement effect of copper (element B) on Fe $K\alpha$ intensity. Table 12.1 gives the supplementary data. Mitchell and Kellam define five secondary enhancement effects analogous to the five secondary absorption effects described in Section 12.2.6.2:

1. *Exaggerated enhancement* (curve 1) occurs in a heavy matrix, where $(\mu/\varrho)_{\text{M-B}} \gg (\mu/\varrho)_{\text{B}} \approx (\mu/\varrho)_{\text{A}}$. Replacement of tin ($\mu/\varrho$ 457) with copper (μ/ϱ 99) results not only in true enhancement by excitation of Fe $K\alpha$ by Cu $K\alpha$ and Cu $K\beta$, but additional apparent enhancement by reduction of the overall matrix absorption for Fe $K\alpha$.

2. *Normal enhancement* (curve 2) occurs in a neutral matrix, where $(\mu/\varrho)_{\text{M-B}} \approx (\mu/\varrho)_{\text{A}}$. The absorption of the analyte line (Fe $K\alpha$) is about the same for both copper (μ/ϱ 99) and manganese (μ/ϱ 64). Also, the absorption of the enhancing lines (Cu $K\alpha$ and Cu $K\beta$) is about the same for both iron (μ/ϱ for Cu $K\alpha$ is 324) and manganese (μ/ϱ for Cu $K\alpha$ is 284). Thus, copper has about the same enhancement effect on Fe $K\alpha$ in the presence of manganese as in the presence of iron alone.

3. *Cancellation of enhancement* (curve 3) occurs in an equivalent matrix, where $(\mu/\varrho)_{\text{M-B}} \approx (\mu/\varrho)_{\text{B}}$. Replacement of zinc ($\mu/\varrho$ 109) with copper

(μ/ϱ 99) has relatively little effect on absorption of Fe $K\alpha$ intensity, and Cu $K\alpha$ and Zn $K\alpha$ excite Fe $K\alpha$ with about the same efficiency, as do Cu $K\beta$ and Zn $K\beta$. The expected increase in Fe $K\alpha$ intensity on increase of copper concentration does not occur.

4. *Reversal of enhancement* (curve 4) occurs in a very light matrix, where $(\mu/\varrho)_{M-B} \ll (\mu/\varrho)_B \approx (\mu/\varrho)_A$. An expected enhancement of analyte-line intensity due to the presence of a specific matrix element B having a strong spectral line near the short-wavelength side of the analyte absorption edge may actually become an absorption effect upon change in the overall matrix composition. This is the case when, as B concentration increases, it replaces a matrix having much lower absorption than itself. As copper (μ/ϱ 99) replaces carbon (μ/ϱ 11), the matrix absorption of Fe $K\alpha$ increases so much that the enhancement by Cu $K\alpha$ is outweighed.

5. *Enhancement-absorption inflection* (curve 5) occurs in a medium-light matrix, where $(\mu/\varrho)_{M-B} < (\mu/\varrho)_B \approx (\mu/\varrho)_A$. At low B concentrations, B enhances analyte-line radiation. As B concentration increases, it replaces a matrix having somewhat lower absorption than itself, causing a slight negative absorption effect. This phenomenon might be expected for Fe $K\alpha$ in a matrix of copper (element B, μ/ϱ 99) and cobalt (element M − B, μ/ϱ 81).

It is evident that copper may be said to enhance measured Fe $K\alpha$ intensity only under certain conditions, and even then to a varying extent.

12.2.7. Unusual Absorption-Enhancement Effects

Some of the absorption-enhancement phenomena described in the preceding section may be considered unusual. However, the term is used here for cases in which analyte-line intensity remains substantially constant or even actually decreases with increasing analyte concentration. Two examples have been reported.

The x-ray secondary-emission spectrometer is generally unable to distinguish various concentrations of an analyte having a relatively short-wavelength line in an extremely light matrix. For example, it is, in general, not possible to distinguish the various oxides of an element of medium or high atomic number simply by comparing the intensities of its spectral line measured from the untreated oxides. Table 12.2 gives data for iron and its oxides [pp. 182–4 in (*26*)]. Assuming an effective primary wavelength of 1.39 Å, the oxygen is highly transparent to Fe $K\alpha$ (μ/ϱ 22) and to the primary x-rays (μ/ϱ 8), so that the iron (μ/ϱ 73 and 250, respectively) largely

TABLE 12.2. Unusual Absorption Effect for Oxides of Iron[a]

	Fe	FeO	Fe_3O_4	Fe_2O_3
Fe concentration, wt %	100	77.8	72.5	70.0
Fe $K\alpha$ intensity, relative	100	95.7	95.5	95.0
Critical thickness for Fe $K\alpha$, cm $\times 10^{-3}$	1.36	2.31	2.72	2.77

[a] From Liebhafsky *et al.* [pp. 182–4 in (*26*)].

determines the absorption coefficient of the oxide and thereby the critical thickness or depth from which Fe $K\alpha$ can emerge from the specimen. Thus, substantially the same number of iron atoms contribute to the measured Fe $K\alpha$ intensity for each oxide, and the relative intensity increases only from 95.0% to 95.7% while iron concentration increases from 70.0% to 77.8%. The difficulty is easily remedied in several ways: (1) elimination of the absorption effect by dispersion as a thin film; (2) leveling the matrix to a constant absorption by low-absorption dilution or (3) high-absorption dilution; (4) for some elements, by measurement of an L or M line instead of a K line; or (5) by use of an internal standard (*L11*).

Another example of the phenomenon is provided by the determination of lead in glass. The calibration curve is so strongly curved that no substantial increase in Pb $L\alpha_1$ intensity occurs for lead concentrations >25 wt %. Addink *et al.* (*A4*) obtained substantially linear calibration curves by admixture of copper powder with the sample glass powder in weight ratio 5:1.

The basis of this phenomenon is explained by reference to Equations (4.10) and (4.17), the fundamental excitation equations. For an analyte having a relatively high atomic number and short-wavelength analyte line, in a light matrix,

$$(\mu/\varrho)_{A,\lambda_{pri}} \gg (\mu/\varrho)_{i,\lambda_{pri}}$$
$$(\mu/\varrho)_{A,\lambda_L} \gg (\mu/\varrho)_{i,\lambda_L}$$

where i is a matrix element excluding the analyte A, and λ_L and λ_{pri} are the wavelengths of the analyte-line and primary x-rays, respectively. Then $\sum [C_i(\mu/\varrho)_{i,\lambda_{pri}}]$, again excluding the analyte, can be disregarded as compared with $C_A(\mu/\varrho)_{A,\lambda_{pri}}$, assuming only that C_A is not too small. The same applies to $\sum [C_i(\mu/\varrho)_{i,\lambda_L}]$. Then I_L, or I_L/I_0, becomes substantially independent of C_A, a calibration curve of I_L *versus* C_A becomes substantially horizontal, and the analysis is not feasible.

It is also possible for analyte-line intensity to actually decrease as analyte concentration increases. This phenomenon occurs when the following two conditions occur simultaneously: (1) a specific matrix element B has a high absorption coefficient for the analyte line, and (2) the concentration of element B increases at a faster rate than analyte concentration. The phenomenon was first noted in mixtures of aluminum-containing chromites [$FeO \cdot (Cr, Al)_2O_3$] and olivines [$2(Mg, Fe)O \cdot SiO_2$] in the composition range 5–25% iron (*B59*, *M43*), as shown in Figure 12.6. Chromium has a high mass-absorption coefficient for Fe *Kα* (490 cm^2/g) compared with that of iron (73 cm^2/g). Pure chromite ($FeO \cdot Cr_2O_3$) and olivine ($FeO \cdot MgO \cdot SiO_2$) contain 25 and 10% iron, respectively, and have mass-absorption coefficients for Fe *Kα* of 231 and 68 cm^2/g, respectively; chromite contains 46% chromium. Thus, as chromite minerals are added to olivines, chromium concentration increases at a faster rate than iron concentration. The high absorption of chromium and disproportionate rates of increase of chromium and iron concentrations result in the negative absorption effect of chromium outweighing the increased Fe *Kα* emission from the iron, and the iron calibration curve has a negative slope, as shown.

A simple calculation proves the point [Equation (12.6)]. Assume that the Fe *Kα* intensity from pure olivine is 100 counts/s per 1% iron. Then, for pure olivine (10% Fe), Fe *Kα* intensity is $10 \times 100 = 1000$ counts/s; for pure

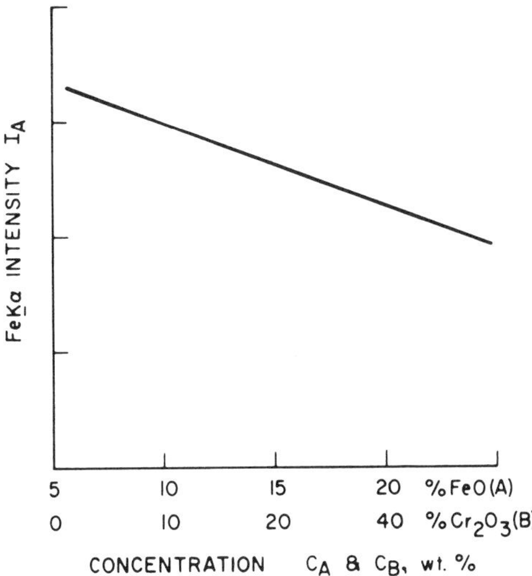

FIG. 12.6. Unusual matrix effect of chromium on Fe *Kα* intensity. [B. J. Mitchell and J. E. Kellam, *Appl. Spectrosc.* **22**, 742 (1968).]

chromite (25% Fe), Fe $K\alpha$ intensity is $(25 \times 100)(68/231) = 735$ counts/s, where 68 and 231 are the mass-absorption coefficients of olivine and chromite, respectively.

12.3. PARTICLE-SIZE, HETEROGENEITY, AND SURFACE-TEXTURE EFFECTS

Derivations of the basic excitation equations (4.10) and (4.17) (Sections 4.1.2 and 4.1.3) assume a homogeneous, smooth specimen. This means that the relatively thin surface layer that emits the measured x-ray spectral-line intensities contains all elements present in the specimen homogeneously distributed in their true concentration ratios.

A powder or polycrystalline solid is said to be homogeneous or heterogeneous according to whether all particles (grains, crystallites) have the same chemical composition or the material is a mixture of particles having two or more compositions. A powder or solid is said to have uniform or nonuniform particle size according to whether all particles have the same size or the material is a mixture of particles of different sizes. The particle-size distribution is a curve showing the relative numbers of particles of each size present—rather analogous to a pulse-height distribution.

Emitted analyte-line intensity may be affected by the particle size and particle-size distribution in a powder or briquet—or in a polycrystalline solid—even if the composition is homogeneous. Emitted analyte-line intensity may be affected by heterogeneity of composition even if the particle size is uniform. These effects arise when the specimen is nonuniform with respect to composition or particle size over distances about the same as the path length of the analyte-line x-rays. In powders having nonuniform particle size, segregation may occur by size, and in heterogeneous powders, segregation may occur by density. In either case, sifting and settling are likely to accentuate the separation.

Emitted analyte-line intensity may also be affected by the surface texture of a massive solid, even if both the composition and particle size are uniform. The emitted intensity is affected not only by the roughness of the surface finish, but by the orientation of the grind or polish marks with respect to the directions of the primary and secondary x-ray beams (Section 16.2.4, Figures 16.5–16.7).

Clearly, if compositional homogeneity, particle size, particle-size distribution, and/or surface texture vary among samples and standards, the measured analyte-line intensities are likely to be difficult to correlate with

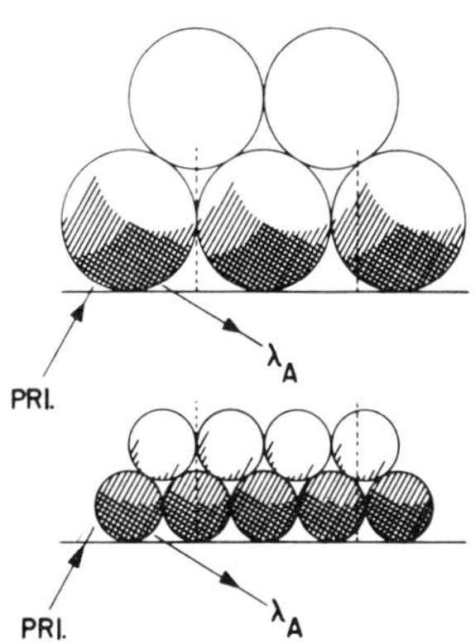

FIG. 12.7. Portions of the two outermost layers of particles in powder specimens having large and small particles, respectively, showing the volume penetrated by the primary x-rays (total shaded area) and the effective volume (cross-hatched area) from which analyte-line x-rays λ_A can emerge.

each other and with analyte concentration. The severity of the effects increases as heterogeneity and nonuniformity of the specimen and wavelength of the analyte line increase.

Figure 12.7 shows portions of the two outermost layers of particles in powder (or briquet or polycrystalline solid) specimens having large and small particles, respectively. The entire shaded area in each particle represents the volume penetrated by the primary beam—that is, the volume in which analyte-line x-rays are generated. The cross-hatched area represents the volume from which analyte-line x-rays can emerge—that is, the effective volume. For the larger particles in the figure, only ~1/3 of the volume is effective, for the smaller particles, ~2/3. For very small particles, the entire volume in the outermost layer and perhaps several deeper layers would be effective. The higher the x-ray tube operating potential, the larger will be the excited volume. The shorter the analyte-line wavelength, the more nearly will the effective volume equal the excited volume, and the longer this wavelength, the smaller will be the effective volume. In general, the particle size should be no larger than—and should be substantially smaller than—the effective volume for the longest wavelength to be measured. A similar guideline applies to surface roughness of massive solid specimens. Other experimental considerations are given in Sections 16.2.3 and 16.2.4 for solids and in Section 16.3.5.1 for powders and briquets.

De Jongh (*D11*) describes three distinct particle-size effects for a two-phase heterogeneous powder: In the *particle-size effect*, the analyte is present in only one phase, and both phases have substantially the same mass-absorption coefficient for the analyte line. In such systems, the analyte-line intensity is influenced only by the relative particle sizes of the two phases, that is, by the actual fraction of the analyte phase in the effective specimen layer. The effect is most severe when the analyte-line path length in the specimen is less than the average particle size. In the *intermineral effect*, the analyte is present in only one phase, but the two phases have substantially different absorption coefficients for the analyte line. Analyte-line intensity depends not only on particle size, but also on the relative absorption coefficients. In the *mineralogical effect*, the analyte is present in both phases, which have different absorption coefficients for the analyte line. The effect is similar to that in the intermineral effect, but more complex. The nomenclature was suggested by De Jongh. However, inasmuch as all three effects are particle-size effects and occur in all kinds of powders, more appropriate terms are to be preferred.

Particle-size and heterogeneity effects have been treated mathematically, and mathematical methods have been devised for dealing with them (*B27–29*, *C49*, *H73*, *R13*). Ebel *et al.* (*E5*) have considered the use of scattered x-rays to predict and correct for particle heterogeneity, and Claisse (*C48*) has discussed the possible use of internal standards. Myers *et al.* (*M63*) have incorporated particle-size corrections into the Lucas-Tooth method for correction of absorption-enhancement effects (Section 15.5.4.5). Somewhat more general methods have been developed by Criss (*C66*) and Berry (*B31*). Berry's method is simpler mathematically and does not require use of a computer. However, although it is satisfactory for heterogeneous powders of uniform particle size, it is less reliable for nonuniform particle sizes and distributions. Criss' method is more complex and does require a computer, but it is applicable to nonuniform particle sizes and distributions. Birks [p. 559R in (*53*) for 1972] gives guidelines for the application of Criss' and Berry's methods based on the extent of absorption of primary or analyte-line x-rays in a single particle: For $<10\%$ absorption, the specimen may be regarded as homogeneous—that is, free from particle-size effects. For 10–50% absorption, either the Criss or Berry equations give adequate correction. For 50–80% absorption, the Criss equations give better correction. For $>80\%$ absorption, the particle-size and heterogeneity effects are so severe that mathematical correction is likely to fail, and the specimen must be ground, fused, or dissolved.

Like absorption-enhancement effects, particle-size effects become less severe as specimen thickness decreases below infinite thickness. Rhodes (*R12*, *R13*) has found that particle-size effects are simple and readily dealt with in specimens meeting any one or more of three criteria: (1) the *thin-specimen criterion*, $m(\mu/\varrho)' \leq 0.1$, where m is mass per unit area (g/cm^2), and $(\mu/\varrho)'$ is the sum of the mass-absorption coefficients (cm^2/g) of the specimen for both primary and analyte-line x-rays [Equation (2.13)]; (2) the *monolayer criterion*, that is, the particles are distributed never more than one deep; and (3) the *low volume-packing-density criterion*, that is, particles distributed in a low-Z medium of given area are equivalent to no more than a monolayer of particles on that same area. The last two criteria require that each primary photon penetrating the specimen passes through not more than one particle, and each analyte-line photon must emerge through only the particle of its origin. For such specimens, analyte-line intensity I_A is given by

$$I_A = S_A P_A m_A \tag{12.9}$$

where S_A is analyte sensitivity (counts/s per $\mu g/cm^2$), m_A is mass of analyte per unit area ($\mu g/cm^2$), and P_A is a particle-size factor given by

$$P_A = \frac{1 - \exp -(\mu_A' d)}{\mu_A' d} \tag{12.10}$$

where μ_A' is the sum of the linear absorption coefficients (cm^{-1}) of the emitting particles for both primary and analyte-line x-rays, and d is the particle size (cm) of these particles. Note that P_A depends only on the size of the emitting particles—those containing analyte A—and so may be combined with S_A. Thus, interparticle effects are absent.

Equation (12.9) was verified experimentally for monolayers of mixtures of particles of dolomite ($CaCO_3 \cdot MgCO_3$) and pyrite (FeS_2) in which the individual particles varied in size from almost opaque to Mg $K\alpha$ to almost transparent to Fe $K\alpha$. Equation (12.9) is particularly useful for supported specimens consisting of particulate matter filtered from liquids or gases on filter disks, and of liquids evaporated on various supports.

Surface texture may have one or more of three principal effects: (1) The path lengths of the primary and analyte-line x-rays within the specimen may vary from point to point. (2) Because the measured analyte-line x-rays leave the surface at $\sim 90°$ to the incident primary x-rays, there is the possibility of shielding and shadowing effects. (3) Finally, extremely coarse surface topography may actually influence the effective distance between x-ray tube target and specimen. These effects become progressively more severe as: (1) the wavelength of the analyte line increases; (2) the mass-absorption coefficient of the specimen for primary and, more important, analyte-line x-rays increases; and (3) the wavelength of the x-ray tube target line increases. The third item applies to Cr $K\alpha$ and Rh, Pd, and Ag L lines, but only when the target lines make the predominant contribution to the analyte-line excitation.

Chapter 13

Sensitivity and Resolution; Spectral-Line Interference

13.1. SENSITIVITY

13.1.1. Definitions

Sensitivity in x-ray spectrometric analysis may be defined in either of two ways—in terms of minimum detectable amount of analyte, or rate of change of analyte-line intensity with change in amount of analyte.

The *minimum detectable amount, minimum detection limit* MDL, *lower limit of detectability* LLD, or *concentration at the detection limit* C_{DL} may be defined in several ways, but the definition given by Birks [p. 54 in ed. 1 of (*8*)] is probably the most widely accepted: that amount of analyte that gives a net line intensity equal to three times the square root of the background intensity for a specified counting time, or, in statistical terms, that amount that gives a net intensity equal to three times the standard counting error of the background intensity. "Amount" signifies concentration (%, mg/ml, etc.) for infinitely thick specimens, area density (mg/cm^2) for specimens less than infinitely thick, or mass (μg) for extremely small particulate or filamentary specimens. Figure 13.1 shows a ratemeter–recorder 2θ scan of a weak peak on a uniform background. Suppose that in a specified preset time, the accumulated count for the background is 100 counts. The standard counting error for the background count σ_N [Equation (11.18)] is then $\sqrt{100}$, or 10 counts. The background count remains within $\pm 3\sigma$ (± 30 counts) of its mean (100 counts) 99.7% of the time. Then, if there is to be

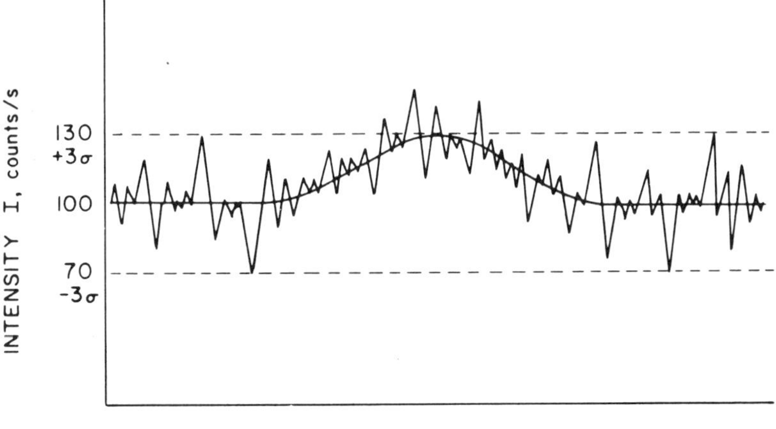

FIG. 13.1. Detection limit.

95% certainty (2σ) of detection of the peak, its accumulated count in the same preset time must exceed the background by $3\sigma_N$, or 30 counts, that is, the accumulated peak count must be 130. Conversely, if the accumulated count is 130, then there is 95% probability that there is really a peak present. Whatever concentration, area density, or isolated mass is required to give analyte-line intensity corresponding to this count is the minimum detectable amount, or limit of detectability.

The value $3\sqrt{I_B}$ is obtained as follows. From Equation (11.34), the standard counting error for a net count is calculated from the counting errors of the individual peak σ_P and background σ_B measurements:

$$\sigma = (\sigma_P{}^2 + \sigma_B{}^2)^{1/2} \tag{13.1}$$

For trace concentrations, I_P approaches I_B, and $\sigma_P \approx \sigma_B$; then,

$$\sigma = (2\sigma_B{}^2)^{1/2} = \sqrt{2}\sigma_B \tag{13.2}$$

For 95% confidence,

$$2\sigma = 2\sqrt{2}\sigma_B \approx 3\sigma_B = 3\sqrt{I_B} \tag{13.3}$$

The counting time should also be specified and is usually 100 s, that is, $T_P = T_B = 50$ s. From Equation (11.33), $\sigma_B = (I_B/T_B)^{1/2}$, so that the minimum detectable amount is $3(I_B/T_B)^{1/2}$. If now one includes the slope m (counts/s per 1% analyte) of the intensity *versus* concentration calibration curve, the minimum detectable amount becomes

$$C_{\mathrm{DL}} = (3/m)(I_B/T_B)^{1/2} \tag{13.4}$$

It is evident that the minimum detectable amount is smaller the lower the background intensity, the greater the slope, and the longer the counting time.

To return to our example (Figure 13.1), suppose that application of a net analyte-line count of 30 to a previously prepared calibration curve represents 0.01% analyte; this is the *concentration at the detection limit* or *minimum detection limit*. The precision of the analysis is given by Equation (11.22): $(130 + 100)^{1/2}/(130 - 100) = (230)^{1/2}/30 = 15.2/30 \approx 0.5$, that is, $\pm 50\%$ of the amount present or ± 0.005 wt % for 1σ.

Pantony and Hurley (*P3*) have considered the detection limit in detail and derived equations for its calculation, most notably the following:

$$C_{\mathrm{DL}} = 0.8\sqrt{I_B}/m \tag{13.5}$$

These equations were tested on 11 elements ranging in atomic number from 11 to 82 (sodium to lead), in high- and low-absorption matrixes, in three spectrometers of different makes, and under a variety of instrument conditions. Among their conclusions is that relatively large changes in background have relatively small effect on the detection limit, and efforts are more profitably directed toward increasing the slope of the calibration curve than to decreasing the background.

The lower limit of detectability or detection limit applies to *detection* of the analyte, that is, to qualitative analysis. The lower limit of *determination* or determination limit for a quantitative analysis is substantially greater and is, in fact, defined as three times the detection limit.

Alternatively, sensitivity may be stated in terms of intensity per unit analyte concentration (counts/s per %, counts/s per mg/ml, etc.) for infinitely thick specimens; intensity per unit area density (counts/s per mg/cm^2) for thin specimens; or intensity per unit mass (counts/s per μg) for very small specimens. Sometimes, sensitivity is given per unit x-ray tube current—for example, counts/s per % per mA. With radioisotope and ion-bombardment excitation, sensitivity may be stated in terms of amount of isotope in millicuries (mCi) and amount of incident ions in microcoulombs (μC)—counts/s per % per mCi and counts/s per % per μC, respectively. Actually, all these definitions constitute the slope of the calibration function—analyte-line intensity *versus* amount of analyte, and when the calibration function is nonlinear, the sensitivity varies with the concentration or amount of analyte. Although this may not be the case with thin-film and very small specimens, it is often the case with the more common solid, briquet, powder, and liquid specimens. In such cases, it is necessary to express sensitivity at a specified concentration level, for example, counts/s per % "at the 10% concentration level."

Sometimes, the true x-ray spectrometric sensitivity stated in terms of the measured sample is more conveniently stated in terms of the original sample from which the sample specimen was prepared. This is most often the case when the analytes have been preconcentrated. For example, suppose that a certain analyte, preconcentrated and distributed on a filter-paper disk, has a detection limit of 1 μg. If the material on the filter disk was separated from 10 g of solid, 10 l of water, or 10 m^3 of air, the detection limit of the analyte may be stated as 0.1 μg/g, 0.1 μg/l, or 0.1 μg/m^3, respectively.

13.1.2. Factors Affecting Sensitivity *(25o)*

Regardless of the terms used to express sensitivity, the value must be rigorously qualified. So many factors affect the sensitivity that a given figure applies only for a specified list of specimen and instrument conditions. This section lists the specimen and instrument conditions that result in increased analyte-line intensity. Sensitivity is usually, although not necessarily, improved by an increase in analyte-line intensity. Analyte-line intensity is increased by the conditions listed in Sections 13.1.2.1 13.1.2.4. Although these sections apply specifically to crystal spectrometers, most of the conditions apply as well to energy-dispersive spectrometers, including those having radioisotope excitation. For the other conditions, analogies for energy dispersion are readily inferred.

13.1.2.1. Excitation Conditions

1. Constant-potential, rather than full-wave, generator.

2. X-ray tube potential and current as high as possible; however, background also increases with x-ray tube potential and current, and it may be preferable to reduce excitation to increase peak-to-background ratio.

3. X-ray tube target lines close to the short-wavelength side of the analyte absorption edge; when inconvenience, limited choice of targets, or spectral interference by target lines precludes this condition, excitation is effected principally by the continuum, and the following two conditions assume added significance.

4. X-ray tube target atomic number as high as possible to give high continuum intensity.

5. Primary continuum "hump" close to the short-wavelength side of the analyte absorption edge.

6. X-ray tube window of beryllium and as thin as possible, especially when low-Z analytes are to be determined.

7. X-ray tube target-to-window distance as small as possible so that the target-to-specimen distance can be made small.

8. X-ray tube target-to-specimen distance as small as possible.

9. X-ray tube target surface free of pitting from long use at high power dissipation, and x-ray tube window free of high-Z sublimate.

10. Primary beam spectrally pure to minimize interference and background.

11. Primary-beam filters avoided; such filters always reduce primary-beam intensity, and therefore analyte-line intensity; however, such filters may improve peak-to-background ratio, and thereby decrease the detection limit.

12. Irradiated area on the specimen plane as large as possible. However, there is no point in irradiating an area larger than the projection of the collimator–crystal–detector system on the specimen plane.

13. If the specimen is small, it must be placed on the specimen plane at the point of highest primary intensity (Section 19.2.1).

13.1.2.2. Specimen Conditions

1. Useful specimen area equal to or slightly greater than the area of the projection of the spectrometer optical system on the specimen plane (for flat-crystal instruments).

2. Enhancement radiator (Section 4.4.3) used behind x-ray-transparent specimens.

3. Analyte absorption-edge wavelength just longer than the wavelengths of the x-ray tube target lines or just longer than the wavelength of the continuous radiation hump.

4. Matrix absorption coefficient low.

5. Matrix spectral lines enhance the analyte line.

6. Matrix spectral lines sufficiently remote from the analyte line so as not to contribute to the background.

7. Low-absorption dilution used for a matrix having high absorption coefficient.

8. Preconcentration or separation of analytes of low concentration.

9. Particle size of powder or briquet or surface texture of solid specimens <50–$100\ \mu m$.

10. Grind marks parallel to the plane defined by the central rays of the primary and secondary beams.

11. Optimal analyte line used (*C12*): For $Z < 55$ (Cs), the $K\alpha$ line is most intense. For Z 55–65 (Tb), the $L\beta_1$ line is most intense. For $Z > 65$, the $L\alpha_1$ line is most intense. However, sometimes an L line, although less intense than a K line, may give a more favorable peak-to-background ratio if it occurs in a less intense region of the scattered continuum. With a 50-kV generator, in practice, K lines are always used for elements up to atomic number 40 (zirconium), and L lines are always used above atomic number 60 (neodymium); K or L lines may be used for the intermediate elements. First-order lines have about 10 times the intensity of second-order lines with a LiF crystal.

12. It is possible for scattered primary background x-rays to originate from depths greater than that from which analyte-line x-rays can emerge from the specimen—that is, the effective layer thickness for scattered background can exceed that for the analyte line. In such cases, peak-to-background ratios can be increased by reduction of specimen thickness to about the effective layer thickness for the analyte line. Further reduction in thickness results in decreased analyte-line intensity without increase in peak-to-background ratio.

13. Small specimens supported at ends of plastic or quartz fibers or on thin or ultrathin plastic film.

14. Small specimens or small selected areas on large specimens measured in collimated specimen holder or by selected-area techniques.

13.1.2.3. Optical System

1. Source geometry favorable: specimen plane inclined at 60° or more to the primary beam, takeoff angle 30–60°.

2. Useful aperture as large as possible—that is, useful area of collimators, crystal, and detector window (*S35*).

3. Collimators short, with wide-spaced foils.

4. Axial divergence permitted, that is, divergence in the plane of, or parallel to, the goniometer axis.

5. High-reflectivity crystal used.

6. Crystal length and width sufficient to intercept all useful secondary radiation.

7. Crystal permits measurement of the analyte line at a 2θ angle at which the entire useful secondary beam is intercepted by the crystal.

8. Curved crystal used for small particulate or filamentary specimens.

9. Optical path as short as possible.

10. Optimal optical path used: helium or vacuum is beneficial for all wavelengths greater than 1 Å; vacuum is preferable to helium for the K lines of elements 9–20 (fluorine to calcium).

11. In multichannel instruments, two channels may be set to the same analyte line and their outputs combined.

13.1.2.4. Detector and Readout Conditions

1. Detector quantum efficiency for the analyte line as high as possible; this involves selection of detector type, gas filling, etc.

2. Detector window as thin as possible, especially for the lightest elements.

3. Detector resolution high to permit use of a narrow pulse-height selector window to pass all analyte pulses, yet exclude as many background pulses as possible.

4. Pulse-height selector avoided if possible, but if used, energy resolution as high as possible.

Klockenkämper (*K17*, *K18*) cites the following four spectrometer parameters as having the greatest influence on sensitivity: (1) spectral resolution and reflection factor of the crystal; (2) spacial resolution and transmission factor of the collimator(s); (3) energy resolution and quantum efficiency of the detector; and (4) energy resolution of the pulse-height selector. He gives mathematical relationships between these parameters and sensitivity and gives practical methods for their evaluation.

13.1.3. Photon Losses in the X-Ray Spectrometer

At this point, it is very informative to calculate the number of x-ray photons generated at the x-ray tube target and follow their progress to the specimen, then to calculate the number of spectral-line photons generated in the specimen and follow their progress to the detector. The calculations reveal which components or processes in the spectrometer are responsible for the greatest losses in the measured intensity.

Consider a specimen consisting of a monatomic layer of copper having area 1 cm^2. For copper, the atomic diameter is 2.55 Å (2.55×10^{-8} cm), and the density is 8.92 g/cm^3. Then the mass of the monolayer is (1 cm^2)

$(2.55 \times 10^{-8}$ cm) (8.92 g/cm^3) $= 2.3 \times 10^{-7}$ g, or 0.23 μg. Now consider the monolayer to be placed in an x-ray spectrometer and irradiated by the primary beam from an x-ray tube having a tungsten target and operating at 50 kV, 50 mA constant potential. The minimum wavelength in the primary spectrum from this tube is given by Equation (1.25):

$$\lambda_{\text{min}} = 12.4/50 \text{ kV} = 0.25 \text{ Å}$$

The effective wavelength (Section 1.5.1) is then

$$\lambda_{\text{eff}} \approx \lambda_{I_{\text{max}}} \approx 1.5\lambda_{\text{min}} \approx 0.375 \text{ Å}$$

However, it is the longer primary wavelengths nearer the copper K-absorption edge (1.38 Å) that excite Cu $K\alpha$ most efficiently. A wavelength of 1 Å represents a compromise between the most intense and the most efficient primary wavelengths, and, for simplicity, in the calculations it is assumed that the primary beam is monochromatic and has wavelength 1 Å. Moreover, this is about the effective wavelength given by Equation (4.13). Table 13.1 gives the data required for the calculations, including specimen and instrument parameters and, for beryllium, copper, and air, densities and mass-absorption coefficients for primary (1 Å) and spectral-line (Cu $K\alpha$) x-rays. The calculations follow.

1. Number of primary photons (1 Å) produced at the target focal spot: The x-ray tube power expended in producing x-rays is given by Equation (1.28):

$$\begin{aligned} I_{\text{W}} &= (1.4 \times 10^{-9}) ZiV^2 \\ &= (1.4 \times 10^{-9})(74)(0.05 \text{ A})(5 \times 10^4 \text{ V})^2 \\ &= 13 \text{ W} \end{aligned}$$

Then,

$$I_{\text{erg/s}} = (13 \text{ W})(10^7 \text{ ergs/W-s}) = 1.3 \times 10^8 \text{ ergs/s}$$

Photon energy for 1-Å x-rays is given by Equation (1.16):

$$\begin{aligned} E &= h\nu = h(c/\lambda) \\ &= (6.6 \times 10^{-27} \text{ erg-s}) \frac{3 \times 10^{10} \text{ cm/s}}{1 \times 10^{-8} \text{ cm}} \\ &= 2 \times 10^{-8} \text{ erg} \end{aligned}$$

The number of 1-Å photons per second is then

$$I = \frac{1.3 \times 10^8 \text{ ergs/s}}{2 \times 10^{-8} \text{ erg/photon}} \approx 10^{16} \text{ photons/s}$$

TABLE 13.1. Data for Calculation of Photon Losses in the Flat-Crystal X-Ray Spectrometer

Specimen data:	
Element	Cu
Area a_X	1 cm^2
Thickness t	2.55 Å, 2.55×10^{-8} cm
Spectral line	Cu $K\alpha$, 1.54 Å, 1.54×10^{-8} cm
K-absorption edge jump ratio r_K	~8
K fluorescent yield ω_K	~0.4
Instrument data:	
X-ray tube target, and target Z	W, Z74
X-ray tube potential V	50 kV, 5×10^4 V
X-ray tube current i	50 mA, 0.05 A
Target to specimen distance d_1	4 cm
X-ray tube window to specimen distance d_2	2 cm
Specimen to detector distance d_3	32 cm
Soller collimator, $L \times S$	8.75×0.025 cm (3.5×0.010 inch)
Crystal (flat)	LiF(200)
Detector	NaI(Tl) scintillation counter

Mass-absorption coefficients and densities:

Material	Density ϱ, g/cm³	Mass-absorption coefficient μ/ϱ, cm²/g	
		1 Å (primary)	1.54 Å (Cu $K\alpha$)
Be	1.85	0.57	1.35
Cu	8.92	130	53
Air	0.0012	2.45	9.56

2. Fraction of primary photons directed through the solid angle subtended by the specimen: If uniform spherical propagation of the primary x-rays is assumed, the fraction of the spherical propagation surface a_S intercepted by the specimen area a_X (1 cm^2) at the target-to-specimen distance d_1 (4 cm) is

$$\frac{a_X}{a_S} = \frac{a_X}{4\pi d_1^2} = \frac{1 \text{ cm}^2}{(4)(3.14)(4 \text{ cm})^2} = 0.005$$

where $4\pi d_1^2$ is the area of a sphere having radius d_1.

3. Fraction of primary photons transmitted by the x-ray tube window: For a beryllium window having thickness t_{Be} 0.1 cm (0.040 inch), Equation (2.10) gives

$$\begin{aligned} I/I_0 &= \exp[-(\mu/\varrho)_{\text{Be},1\text{Å}}\varrho_{\text{Be}}t_{\text{Be}}] \\ &= \exp[-(0.57\ \text{cm}^2/\text{g})(1.85\ \text{g/cm}^3)(0.1\ \text{cm})] = \exp -0.105 \\ &= 0.90 \end{aligned}$$

4. Fraction of primary photons transmitted by the air path from x-ray tube window to specimen:

$$\begin{aligned} I/I_0 &= \exp[-(\mu/\varrho)_{\text{air},1\text{Å}}\varrho_{\text{air}}d_2] \\ &= \exp[-(2.45\ \text{cm}^2/\text{g})(0.0012\ \text{g/cm}^3)(2\ \text{cm})] = \exp -0.0059 \\ &= 0.994 \end{aligned}$$

5. Fraction of primary photons absorbed in the specimen: The fraction transmitted is

$$\begin{aligned} I/I_0 &= \exp[-(\mu/\varrho)_{\text{Cu},1\text{Å}}\varrho_{\text{Cu}}t] \\ &= \exp[-(130\ \text{cm}^2/\text{g})(8.92\ \text{g/cm}^3)(2.55\times 10^{-8}\ \text{cm})] = \exp -(3\times 10^{-5}) \\ &= 0.99997 \end{aligned}$$

The fraction absorbed is $1 - 0.99997 = 3\times 10^{-5}$.

6. Fraction of absorbed primary photons that create copper K-shell vacancies: This is given by the copper K-absorption edge jump ratio r_K (Section 4.1.2, Appendix 8): $1 - (1/r_K) = 1 - (1/8) = 0.875$.

7. Fraction of copper K-shell vacancies that result in copper K photons: This is given by the copper K fluorescent yield ω_K (Section 2.5.2, Appendix 9): $\omega_K \approx 0.4$.

8. Fraction of copper K photons consisting of Cu $K\alpha_1$ and Cu $K\alpha_2$ photons: The relative intensities of the principal Cu K lines are as follows: $K\alpha_1{:}K\alpha_2{:}K\beta = 100{:}50{:}20$; then,

$$\frac{K\alpha_1 + K\alpha_2}{K\alpha_1 + K\alpha_2 + K\beta} = \frac{150}{170} = 0.88$$

9. Fraction of Cu $K\alpha$ photons transmitted by the collimator: For an 8.75×0.025-cm (3.5×0.010-inch) Soller collimator having 2.5° opening in the plane of the crystal axis and ~80% open cross-sectional area, divergence is given by Equation (13.7):

$$\begin{aligned} B &= \tan^{-1}(S/L) = \tan^{-1}(0.025/8.75) = \tan^{-1} 0.003 \\ &= 0.17^\circ \end{aligned}$$

Then,

$$\left(\frac{0.17^\circ}{360^\circ}\right)\left(\frac{2.5^\circ}{360^\circ}\right)(\pi)(0.8) = 9\times 10^{-6}$$

10. Fraction of Cu $K\alpha$ photons transmitted by the air path from specimen to detector:

$$\begin{aligned} I/I_0 &= \exp[-(\mu/\varrho)_{\mathrm{air,Cu}K\alpha}\varrho_{\mathrm{air}}d_3] \\ &= \exp[-(9.56\ \mathrm{cm^2/g})(0.0012\ \mathrm{g/cm^3})(32\ \mathrm{cm})] = \exp -0.367 \\ &= 0.692 \end{aligned}$$

11. Fraction of Cu $K\alpha$ photons diffracted by the crystal: For Cu $K\alpha$ and a LiF crystal, ~0.1.

12. Fraction of Cu $K\alpha$ photons counted by the detector: The quantum efficiency of a scintillation counter for Cu $K\alpha$ (Figure 6.17) is ~0.95.

TABLE 13.2. Summary of Photon Losses in the Flat-Crystal X-Ray Spectrometer

Calculation	Process	Fraction	Cumulative fraction	Number of photons remaining, s^{-1}
1	Primary photons emitted by target	1.0	1.0	10^{16}
2	Primary photons directed toward specimen	0.005	5×10^{-3}	5×10^{13}
3	Primary photons transmitted by x-ray tube window	0.90	4.5×10^{-3}	4.5×10^{13}
4	Primary photons transmitted by window-specimen air path	0.994	4.47×10^{-3}	4.47×10^{13}
5	Primary photons absorbed in specimen	3×10^{-5}	1.34×10^{-7}	1.34×10^{9}
6	Cu K-shell vacancies created	0.875	1.17×10^{-7}	1.17×10^{9}
7	Cu K photons emitted	0.4	4.68×10^{-8}	4.68×10^{8}
8	Cu $K\alpha$ photons emitted	0.88	4.12×10^{-8}	4.12×10^{8}
9	Cu $K\alpha$ photons transmitted by collimator	9×10^{-6}	3.71×10^{-13}	3.71×10^{3}
10	Cu $K\alpha$ photons transmitted by specimen–detector air path	0.692	2.57×10^{-13}	2.57×10^{3}
11	Cu $K\alpha$ photons diffracted by crystal	0.1	2.57×10^{-14}	2.57×10^{2}
12	Cu $K\alpha$ photons counted by detector	0.95	2.44×10^{-14}	244

Table 13.2 summarizes the calculations and permits convenient appraisal of the photon losses. In a monolayer specimen, absorption of primary x-rays is very low, and the loss due to transmitted primary photons is correspondingly great. Self-absorption of emergent Cu $K\alpha$ photons is absent in a monolayer. Coincidence losses are absent, because of the low measured intensity. Loss from rejection of pulses by the pulse-height analyzer is disregarded. It is evident that by far the greatest loss of the photons emitted by the specimen occurs in the collimator, the next greatest loss in the crystal. Both these components are eliminated in energy dispersion (Section 8.2).

13.1.4. Sensitivity Performance

Table 3.2 compares the sensitivities of x-ray fluorescence spectrometry and other analytical chemical methods applicable to micro and trace analysis in terms of the minimum detectable mass of element under most favorable conditions.

Because of the multiplicity of factors that affect sensitivity, it is difficult to cite "typical" performance. For extremely small particulate, filamentary, or residue specimens, a flat-crystal spectrometer can measure as little as 10^{-6}–10^{-7} g, a curved-crystal spectrometer as little as 10^{-8}–10^{-9} g. For comparison, the electron-probe microanalyzer can detect as little as 10^{-12}–10^{-15} g. For thin-film specimens having area 1 cm^2 or more, a flat crystal instrument can measure a monolayer of any element having atomic number 12 (magnesium) or more, and perhaps as little as 0.01 monolayer of the more sensitive elements.

A good, modern, commercial x-ray spectrometer, using optimal accessories (target, crystal, collimator, detector) and maximum excitation (50 kV, 50 mA, CP), should give typical $K\alpha$ net intensities (counts/s) for "pure element" specimens as follows: $_{9}$F (in NaF), 300; $_{11}$Na (in NaF), 3000; $_{13}$Al, 250,000; $_{42}$Mo, 10^6; and $_{56}$Ba [in $Ba(NO_3)_2$], 320,000.

Table 13.3 gives what may be near optimal sensitivity for x-ray spectrometric analyses. The measurements in columns 3, 4, and 5 were made on standard commercial flat-crystal spectrometers, those in column 6 on a full-focusing curved-crystal spectrometer (Figure 19.9). Column 3 gives sensitivity for direct determination in solid iron at the conditions given in the table. Column 4 gives sensitivities for the borax-fusion method for 1-g portions of the iron samples in column 3; the analyte is preconcentrated before being subjected to the fusion (chemical-separation borax-disk method, Section 16.4.3). Column 5 is taken from Table 13.4 (see below). For

TABLE 13.3. Optimal Sensitivity of X-Ray Spectrometric Analysis

Atomic number	Element and line	Direct determination in solid Fe (*L39*), %	Borax-fusion method 1-g Fe sample (*L39*), %	Water solution (*C10*), μg/ml	Preconcentrated in 3-mm (1/8 inch) ion-exchange membrane disks (*L39*), μg
13	Al $K\alpha$	0.012[a]	—	—	—
14	Si $K\alpha$	0.007[a]	—	—	—
15	P $K\alpha$	0.003[a]	—	—	—
16	S $K\alpha$	—	—	140	—
19	K $K\alpha$	—	—	3	0.06
20	Ca $K\alpha$	—	—	2	—
22	Ti $K\alpha$	0.0001[b]	0.0001	—	0.008
25	Mn $K\alpha$	0.0007[c]	0.0001	—	—
26	Fe $K\alpha$	—	—	2	0.005
28	Ni $K\alpha$	0.0009[d]	0.0001	—	0.002
29	Cu $K\alpha$	0.0015[d]	—	2	—
31	Ga $K\alpha$	—	—	—	0.01
32	Ge $K\alpha$	—	0.0002	—	—
33	As $K\alpha$	0.001[e]	—	2	—
37	Rb $K\alpha$	—	—	—	0.03
38	Sr $K\alpha$	—	0.0004	1.5	—
42	Mo $K\alpha$	0.001[e]	0.0009	2	—
50	Sn $K\alpha$	0.002[e]	—	10	—
56	Ba $L\alpha_1$	—	0.0003	5	—
74	W $L\alpha_1$	0.006[e]	—	—	—
82	Pb $L\alpha_1$	—	0.0006	5.5	—

Footnote	Target	kV	mA	Crystal	Collimator		Path	Detector
					cm	inches		
([a])	Ti	50	60	PET	8.75×0.050	3.5×0.020	He	Flow (Ar)
([b])	Cr	50	60	LiF	8.75×0.025	3.5×0.010	He	Flow (Ar)
([c])	W	50	75	LiF	8.75×0.025	3.5×0.010	He	Flow (Ar)
([d])	W	50	75	LiF	8.75×0.025	3.5×0.010	Air	Prop. (Xe)
([e])	W	50	75	LiF	8.75×0.025	3.5×0.010	Air	Scint.

Counting times: columns 3, 4, 6: 100 s; column 5: 2 min.

column 6, the analyte is separated from the matrix by microchemical techniques and taken up in ion-exchange membrane disks 3 mm (1/8 inch) in diameter, which are then measured in a full-focusing curved-crystal spectrometer.

Hale and King (*H9*) determined nickel in concentration 0.07–0.6 μg/ml with an absolute precision (1σ) of $\pm$0.035 μg/ml. The Ni $K\alpha$ intensity and background intensity 0.01 Å from the Ni $K\alpha$ peak were measured simultaneously on an Applied Research Laboratories multichannel spectrometer. A very narrow detector slit was used, and the detector output was integrated, rather than scaled, for up to 30 min.

Figure 13.2 (*25o*) gives sensitivity data for loosely packed powders. Table 13.4 (*C10*, *C22*) gives typical intensity and sensitivity data for water solutions, and Figure 13.3 (*L5*) gives the concentration in water solution required to give a net intensity of 10 counts/s at various conditions given in the caption. Table 9.1 gives typical performance data for the ultralong-wavelength region, showing *K*-line intensities for elements 4–8 (beryllium

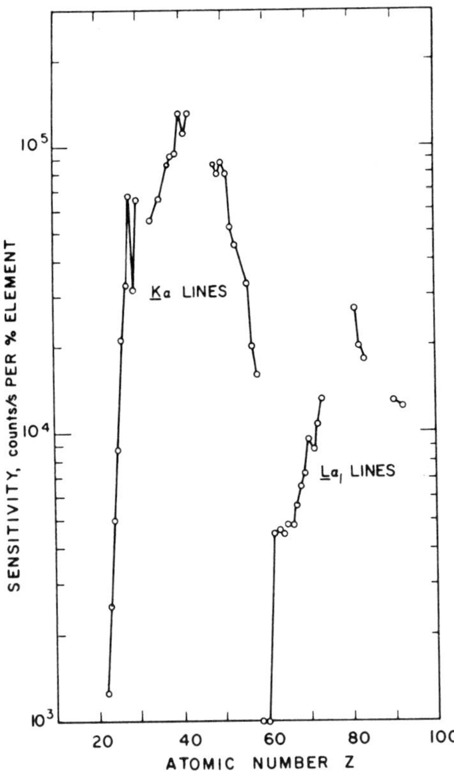

FIG. 13.2. Typical sensitivity performance of x-ray spectrometric analysis of powders. [Data from M. L. Salmon, in "*Handbook of X-Rays*," McGraw-Hill Book Co., p. **32**-7 (1967).] Powders containing up to 1% analyte in silica matrix loosely packed in cells having 4-μm (0.00014-inch) Mylar windows. Philips 100-kV spectrometer; W target, 50 kV(CP), 50 mA; 10×0.0125 cm (4×0.005-inch) collimator; LiF crystal; air path; scintillation detector.

TABLE 13.4. Typical Sensitivity of X-Ray Spectrometric Analysis in Water Solution[a]

Atomic number	Element and line	X-ray tube current,[b] mA	Crystal	Path	Detector[c]	Intensity for 1 mg/ml, counts/s: Line (net)	Intensity for 1 mg/ml, counts/s: Bkgnd.	Minimum detectable concentration,[d] μg/ml
16	S $K\alpha$	25	EDDT	He	Flow	3	2	140
17	Cl $K\alpha$	,,	EDDT	,,	,,	7	2	62
19	K $K\alpha$	,,	LiF	,,	,,	320	14	3.2
20	Ca $K\alpha$	,,	,,	,,	,,	311	7	2.3
23	V $K\alpha$	,,	,,	,,	,,	772	19	1.5
24	Cr $K\alpha$	,,	,,	,,	,,	1030	370	5.1
26	Fe $K\alpha$	,,	,,	Air	Scint.	2000	170	1.8
29	Cu $K\alpha$	10	,,	,,	,,	2100	350	2.4
33	As $K\alpha$	,,	,,	,,	,,	3140	460	1.8
38	Sr $K\alpha$	,,	,,	,,	,,	6400	1300	1.5
42	Mo $K\alpha$	,,	,,	,,	,,	8000	3300	2.0
48	Cd $K\alpha$	,,	,,	,,	,,	3200	4000	5.4
53	I $K\alpha$	,,	,,	,,	,,	1200	2800	12
56	Ba $K\alpha$	,,	,,	,,	,,	1070	3200	14
57	La $K\alpha$	,,	,,	,,	,,	635	2540	20
42	Mo $L\beta_1$	25	EDDT	He	Flow	18	11	49
48	Cd $L\beta_1$	,,	LiF	,,	,,	75	8	10
53	I $L\alpha_1$	,,	,,	,,	,,	232	13	4.2
56	Ba $L\alpha_1$	,,	,,	,,	,,	275	23	4.8
57	La $L\alpha_1$	,,	,,	,,	,,	275	29	5.3
62	Sm $L\alpha_1$	,,	,,	,,	,,	440	45	4.1
70	Yb $L\alpha_1$	,,	,,	Air	Scint.	675	285	6.8
79	Au $L\alpha_1$	10	,,	,,	,,	690	2400	19
82	Pb $L\alpha_1$	,,	,,	,,	,,	1060	460	5.5
90	Th $L\alpha_1$	,,	,,	,,	,,	1220	850	6.5

[a] From Campbell and Leon (*C10*, *C22*). The instrument was a Philips Inverted Three-Position x-ray spectrometer (*T12*); a 10 × 0.05-cm (4 × 0.020-inch) collimator and pulse-height discrimination were used for all measurements.

[b] Tungsten target, 50 kV peak.

[c] *Flow* signifies a gas-flow proportional counter having a 6-μm (0.00025-inch) Mylar window and using P10 gas (90% argon, 10% methane); *Scint.* signifies a NaI(Tl) scintillation counter.

[d] Defined as that concentration that gives a net line intensity equal to three times the square root of the background intensity for a 2-min counting time.

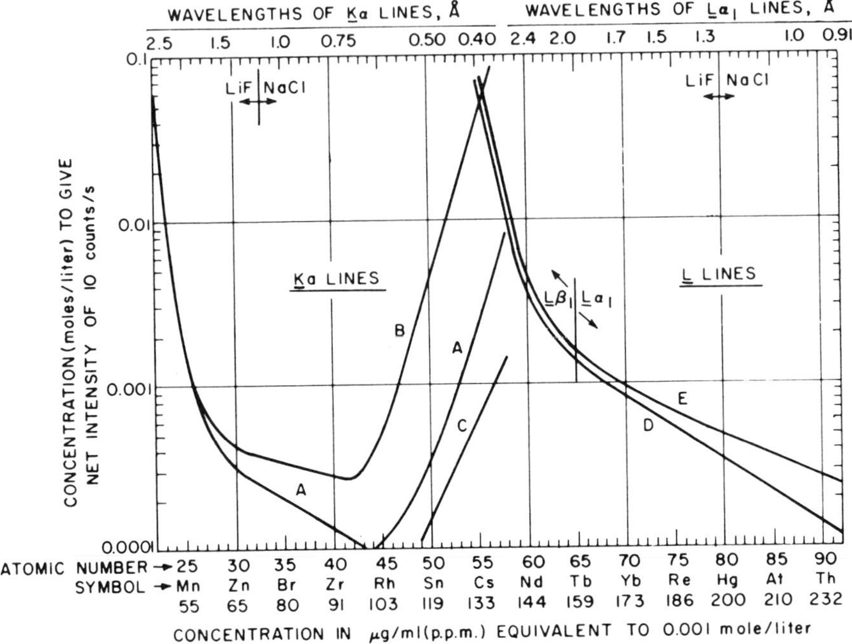

FIG. 13.3. Typical sensitivity performance of x-ray spectrometric analysis of water solutions. [Data from M. C. Lambert, *Norelco Reporter* **6**, 37 (1959).] Philips Inverted Three-Position spectrometer; W target, 50 kV (peak), 40 mA, except curve *C*, 56 kV (peak); collimation not specified; LiF or NaCl crystal, as indicated; scintillation counter for curves *A*, *B*, *D*; Ar-filled Geiger counter for curves *B*, *E*.

to oxygen) measured on a Philips Universal Vacuum Spectrometer fitted with accessories for this region.

In general, for a standard 50-kV instrument, a sensitivity curve of counts/s per 1% analyte *versus* atomic number has the general appearance of a lopsided M (Figure 13.2), and a curve of minimum detectable concentration C_{DL} *versus* atomic number has the general appearance of a lopsided W. The left half of the M or W represents elements of atomic number 9–55 (fluorine to cesium), for which $K\alpha$ lines are used; the right half represents the heavier elements, for which $L\alpha_1$ lines are used. The reasons for the shapes of the curves follow.

Sensitivity is highest (C_{DL} lowest) for elements of atomic number 20 (calcium) to 40 (zirconium) because of favorable combination of efficient excitation, high fluorescent yield, low analyte-line absorption, high-reflectivity crystals, and high-efficiency detectors. Incidentally, possibly the most sensitive element of all is titanium (Z 22).

As atomic number decreases from 20 to 9 (fluorine), the $K\alpha$ wavelength increases from 3.4 to 18 Å and sensitivity decreases (C_{DL} increases) because of: (1) poor excitation due to low transmission of the long-wavelength primary x-rays through the x-ray tube window; (2) decreasing fluorescent yield; and (3) increasing analyte-line absorption in the matrix, path, and cell and detector windows.

As atomic number increases from 40 to ~55 (cesium), the $K\alpha$ wavelength decreases from 0.8 to 0.4 Å and sensitivity decreases (C_{DL} increases) because of: (1) poor excitation due to increasing excitation potential; (2) failure of the analyzer crystal to intercept the entire secondary beam as 2θ decreases; (3) high background consisting mostly of first-order scattered radiation; and (4) decreased spectrometer resolution as 2θ decreases [Equation (13.10)].

For elements 55 to ~80 (mercury), $L\alpha_1$ lines are used and, for these elements, have wavelengths 2.9–1.2 Å, about the same as the $K\alpha$ lines of elements 20–40 above. Sensitivity increases (C_{DL} decreases) for substantially the same reasons as given for elements 20–40. However, sensitivity does not attain its magnitude for the most sensitive $K\alpha$-line elements because of the lower intensity of $L\alpha_1$ lines relative to $K\alpha$.

As atomic number increases beyond 80, sensitivity decreases again for substantially the same reasons as given for $K\alpha$-line elements 40–55 above.

Incidentally, if the detection limit of an analyte is known in a specified matrix, it may be calculated approximately for another matrix by multiplying the known limit by the ratio of the mass-absorption coefficients of the two matrixes for the analyte line. For example, suppose that the detection limit of titanium in iron is 0.0001%, and the detection limit in silver is required. The mass-absorption coefficients of iron and silver for Ti $K\alpha$ are 186 and 1042 cm^2/g, respectively. The detection limit of titanium in silver is then $(1042/186)(0.0001) \approx 0.00056\%$. Unfortunately, this simple estimation is less successful for analyte lines of short wavelength because they occur in a spectral region having high and varying background.

With energy-dispersive analysis using a 10-W x-ray tube or the equivalent amount of radioisotope for excitation, in favorable cases, sensitivity as low as 10–20 ng/cm^2 may be realized for specimens on Millipore substrates (*G26*).

With proton or other ion excitation, continuous background is extremely low (Section 1.5.2). In favorable cases, sensitivity as low as 1–2 ng/cm^2 may be realized for specimens on film substrates of evaporated carbon or ultrathin plastic having 10–20 $\mu g/cm^2$ mass thickness. Filter paper, Millipore filters, and even 3-μm (0.00012-inch) Mylar substrates are destroyed rapidly by ion beams and cannot be used.

13.2. RESOLUTION

13.2.1. Definitions

13.2.1.1. Resolution

Resolution, or resolving power, is a measure of the ability of the spectrometer to distinguish, or recognize as separate, two closely spaced spectral lines, and is a function of their dispersion and divergence—that is, their separation and breadth. Depending on whether wavelength (2θ) or energy dispersion is used, x-ray spectrometer resolution is expressed in terms of *wavelength resolution*—separation of the 2θ peaks of the two lines—or *energy resolution*—separation of their pulse-height distributions. Wavelength resolution or resolving power may be expressed in terms of the difference in wavelength $\Delta\lambda$ of two spectral lines of average wavelength λ that are just resolved, that is, $\lambda/\Delta\lambda$; $\Delta\lambda$ is placed in the denominator so that the higher the resolution, the higher is the number expressing it. Energy resolution may be expressed analogously in terms of the difference in photon energy ΔE of two spectral lines of average photon energy E that are just resolved, that is, $E/\Delta E$.

Perhaps more practically, x-ray spectrometric wavelength resolution is usually expressed as the quotient of the dispersion D of two peaks that are just resolved, by their divergence or breadth B, that is, D/B, where both terms are in degrees 2θ. Dispersion and divergence are discussed below.

Energy resolution may be expressed analogously as the quotient of the separation $V_1 - V_2$ of the peaks of two pulse-height distributions that are just resolved, by their half-widths $W_{1/2}$ or FWHM (Section 13.2.1.3), that is, $(V_1 - V_2)/W_{1/2}$, where all terms are in volts. However, energy resolution of an x-ray detector or spectrometer is usually expressed as the quotient of the half-width of an *individual* pulse-height distribution by the *average* pulse height in that distribution V_{av}, that is, $W_{1/2}/V_{av}$, usually as percent. Again, both terms are in volts. It is evident that by this definition, the higher the resolution, the lower is the number expressing it.

Detector resolution is discussed in more detail in Section 6.5.1.12.

13.2.1.2. Dispersion

Dispersion is the separation of the two lines in 2θ or of their pulse-height distributions in average pulse height, that is, $\Delta 2\theta/\Delta\lambda$ (degrees 2θ per angstrom) or $\Delta V/\Delta E$ (volts per electron volt). The dispersion of an analyzer crystal is given by differentiation of the Bragg law,

$$\frac{d\theta}{d\lambda} = \frac{n}{2d\cos\theta} \tag{13.6}$$

Crystal dispersion is greater the smaller the d spacing of the crystal. As the d spacing of the crystal decreases, $d\theta/d\lambda$ must increase, because $\cos\theta$ is always <1.

Energy dispersion is more complex, and is discussed in Section 8.1. Let it suffice here to say that energy dispersion is increased by use of a proportional, rather than a scintillation, counter, and by low detector potential and amplifier gain.

13.2.1.3. Divergence

Divergence is a measure of the intensity distribution of the diffracted peak in 2θ (wavelength divergence), or of its pulse-height distribution in baseline potential (energy divergence). It is expressed as the half-width $W_{1/2}$ or breadth B of the diffracted peak in degrees 2θ, or of the pulse-height distribution in volts, measured midway between the peak maximum and the background. The choice of symbol, $W_{1/2}$ or B, is largely a matter of personal choice. Still another term having the same significance is *full width at half-maximum*, or FWHM.

The intensity distribution of the x-ray beam diffracted by the analyzer crystal is known as its rocking curve. The distribution is Gaussian and has divergence B_D, as shown in Figure 13.4A. The divergence B_D is smaller the more nearly perfect is the mosaic structure of the crystal. The intensity distribution of the x-ray beam transmitted by a single collimator C_1 is

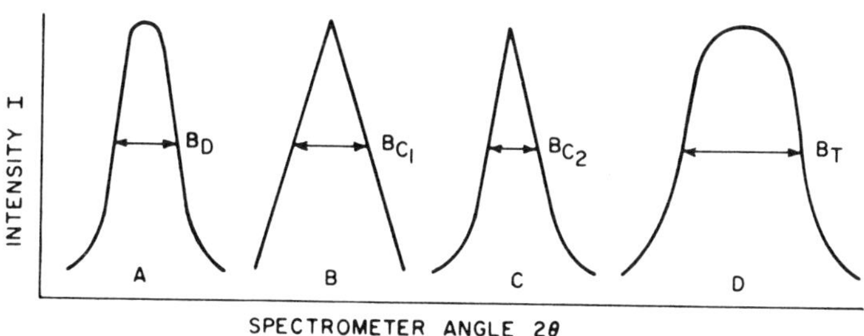

FIG. 13.4. Intensity distribution and divergence of diffracted and collimated beams. [L. S. Birks, "X-Ray Spectrochemical Analysis," Interscience Publishers, Inc., p. 22 (1959); courtesy of the author and publisher.] A. Crystal. B. One collimator. C. Two collimators having the same divergence. D. Resultant distribution of A and B.

triangular, as shown in Figure 13.4B. If the collimator has length L and foil spacing S, the divergence is given by

$$B_{C_1} = \tan^{-1}(S/L) \tag{13.7}$$

When a crystal is used with a single collimator, the resultant intensity distribution, shown in Figure 13.4D, is obtained from the convolution of the crystal rocking curve (Figure 13.4A) and the intensity distribution of the collimator (Figure 13.4B). The total divergence B_T is calculated by assuming the collimator curve to approximate a Gaussian distribution and applying the rule for combining variance [Equation (11.11)],*

$$B_T = (B_D{}^2 + B_{C_1}^2)^{1/2} \tag{13.8}$$

If a second collimator having the same divergence as C_1 is added, the intensity distribution for the two collimators, shown in Figure 13.4C, is obtained by squaring each ordinate of the curve for a single collimator (Figure 13.4B). The divergence of the squared curve $B_{C_2} \approx 0.6 B_{C_1}$. In general, for two collimators having foil spacings S_1 and S_2, respectively, and distance L' from the specimen end of the source collimator to the detector end of the detector collimator, the divergence is (*S52*)

$$B_{C_2} = (S_1 + S_2)/L' \tag{13.9}$$

In practice, the crystal rocking curve has little effect on the total divergence, even with relatively imperfect crystals, and the divergence of the crystal–collimator system is substantially that of the collimators [Equations (13.7) and (13.9)].

13.2.2. Factors Affecting Resolution

Figure 13.5 shows two closely spaced peaks having divergence B and separation $\Delta 2\theta$. Resolution may be regarded as adequate when $\Delta 2\theta = 2B$, although two peaks may often be distinguished at much smaller separation, as shown in Figure 13.5. Combining Equation (13.6) and the Bragg law, one gets

$$\frac{\lambda}{\Delta\lambda} = \frac{2\tan\theta}{\Delta 2\theta} \tag{13.10}$$

* Note added in proof: Equation (13.8) guides the matching of collimator and crystal. For example, a coarse collimator (10 cm × 0.9 mm, $B_{C_1} = 0.516°$) used with a topaz (303) crystal ($B_D \approx 0.08°$) would pass unnecessarily high background intensity without increasing analyte-line intensity. Similarly, a fine collimator (10 cm × 0.15 mm, $B_{C_1} = 0.086°$) used with a graphite crystal ($B_D \approx 0.6°$) would sacrifice analyte-line intensity without improving resolution.

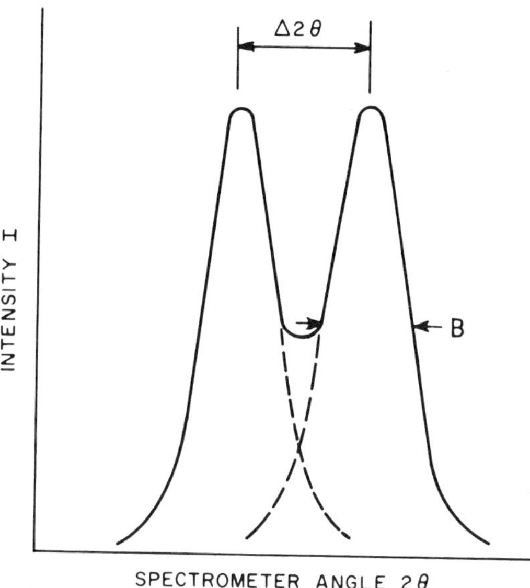

FIG. 13.5. Resolution of adjacent peaks.

Substitution of the minimum acceptable value of $\Delta 2\theta$—*i.e.*, $2B$—gives

$$\frac{\lambda}{\Delta\lambda} = \frac{\tan\theta}{B} \tag{13.11}$$

The left side of Equation (13.11) may be regarded as the required resolution, the right side as the attainable resolution, which involves θ and B. The equation shows that resolution increases rapidly with θ (2θ) because of the tangent function. From the Bragg law $n\lambda = 2d\sin\theta$ [Equation (2.42)], it is evident that the smaller the d spacing of the crystal, the larger is θ, and, thereby, the higher is the resolution. Also, for a specified crystal, successively higher orders occur at successively higher 2θ angles, and thus have successively higher resolution. The second-order spectrum of an element has approximately twice the resolution of the first-order spectrum.

For a flat crystal, in principle, the divergence depends on collimation and the degree of crystal perfection. However, as already mentioned, in practice, the divergence is determined principally by the collimation.

For a transmission curved crystal, divergence depends on the precision of the focusing and can be made very small. For a reflecting curved crystal, divergence is subject to many influences, and only the Johansson crystal has even theoretically perfect focusing; these influencing parameters include the following: (1) width of the source slit; (2) accuracy of curving and grinding; (3) mechanical tolerances in the instrument movements (these are especially serious when source-to-crystal and crystal-to-detector distances are not constant); (4) aging of the cement holding the crystal to its curved support block; (5) accentuation of the crystal mosaic structure due to the curving and grinding; and (6) penetration of the crystal by x-rays of short wavelength, leading to diffraction by planes lying off the focusing circle.

The divergence of a transmission curved crystal is smaller than that of a reflecting curved crystal, which, in turn, is substantially larger than that of a flat crystal.

When the pulse-height analyzer is used, it is the breadth and separation of the pulse-height distributions of the two lines that must be considered. Figure 13.5 could also illustrate two pulse-height distributions on intensity *versus* pulse height axes. Birks [p. 47 in ed. 1 of (*8*)] has pointed out an interesting way to improve energy resolution in cases where the escape peaks are intense. For neighboring elements, the main pulse-height distributions overlap. The escape peaks are separated by the same number of volts (energy dispersion), but their breadths are smaller because they result from smaller numbers of initial ionizations (Section 6.2.3.2, Figure 6.4). Thus, their resolution is better. However, it is still true that usually escape peaks only interfere with the measurements, and they should be avoided whenever possible.

In general, the resolution of an analyte line from an adjacent, potentially interfering line will be greater under the following conditions:

1. A second- or higher-order analyte line is measured.
2. The intensity of the interfering line is small.
3. Long, close-spaced collimators are used. The longer and the closer the foil spacing of the collimator, the more nearly parallel are the emergent x-rays. However, the measured intensity is proportional to the open cross-sectional area and decreases as foil spacing decreases.
4. A high-resolution crystal is used, that is, one having a small d spacing and a narrow rocking angle (high mosaic perfection).
5. The crystal diffracts the analyte line at a high 2θ angle. The separation of lines of neighboring elements increases with increasing 2θ angle for a specified crystal.

If pulse-height selection is used, the following additional conditions are beneficial.

6. A proportional counter is used, rather than a scintillation counter. The pulse-height distribution is narrower, and there is a greater difference in average pulse height for a given difference in photon energy.
7. Low detector-tube potential and amplifier gain are used to reduce the width of the pulse-height distributions.
8. The escape peak can be measured instead of the main peak (see above).

TABLE 13.5. Effect of Collimation[a]

Soller foil spacing S,[b] cm (inch)		Equatorial divergence α,[c] deg	Peak half-width $W_{1/2}$, deg 2θ	Relative peak intensity I_P	Relative peak/bkgnd. intensity ratio I_P/I_B
Source collimator	Detector collimator				
0.0125 (0.005)	—	0.14	0.24	53	8
0.025 (0.010)	—	0.29	0.42	91	5
0.050 (0.020)	—	0.58	0.74	100	3
0.0125 (0.005)	0.0125 (0.005)	0.04	0.15	27	100
0.0125 (0.005)	0.025 (0.010)	0.06	0.20	37	45
0.0125 (0.005)	0.050 (0.020)	0.11	0.24	40	35
0.025 (0.010)	0.025 (0.010)	0.09	0.28	57	35

[a] From Spielberg *et al.* (*S52*). All measurements made with a quartz $(10\bar{1}1)$ crystal and Cu $K\alpha$ x-rays.
[b] Length L of all collimators is 10 cm (4 inches), foil thickness 0.005 cm (0.002 inch).
[c] For one collimator, $\alpha = 2S/L$. For two collimators, $\alpha = (S_1 + S_2)/L'$; L' is 34 cm (13.4 inches), the distance from the specimen end of the source collimator to the detector end of the detector collimator.

It is evident that the conditions favorable to high resolution are unfavorable to high intensity. In practice, the intensity determines the speed and precision of the measurement, so only enough resolution should be used to give sufficient separation of adjacent lines to permit measurement of their line and background intensities. It is rarely necessary for the spectrometer to have resolution greater than that required to partially resolve the $K\alpha$ doublet. The degree of resolution required can often be established from a preliminary ratemeter–recorder 2θ scan.

Table 13.5 gives typical performance data for several collimator dimensions and arrangements.

13.3. SPECTRAL-LINE INTERFERENCE (*B44*)

13.3.1. Definition

It has already been pointed out that due to the relative simplicity of x-ray spectra, spectral-line interference is relatively infrequent, and that when it does occur, there are many ways to deal with it. Two types of x-ray spectral interference occur: (1) *wavelength* (2θ) *interference*, in which the interfering line overlaps the 2θ interval occupied by the analyte line; the interfering and analyte lines may have the same or nearly the same wavelength (λ) or product of order and wavelength ($n\lambda$); and (2) *energy interference*, in which the interfering pulse-height distribution overlaps the analyte pulse-height distribution; energy interference, as such, is observed only when pulse-height analysis is used. In either type of interference, the overlap may have any degree from slight to complete superposition.

Spectral interference, then, may be defined as that condition whereby (1) photons of a line other than the measured line enter the detector at the 2θ angle of the measured line, and/or (2) pulses caused by a line other than the measured line enter the window of the pulse-height analyzer at the settings for the measured pulse-height distribution. The measured line is usually the analyte line, but it may also be an internal-standard or internal intensity reference standard line, or a scattered target line to be used for standardization.

In this section, the above definition of spectral-line interference is broadened to include so-called "spurious" or "anomalous" lines, regardless of whether they overlap a 2θ peak or pulse distribution of a line to be measured. These lines often clutter and confuse the spectrum, defy identification, and may be falsely identified as associated with elements actually not present in the specimen.

13.3.2. Origin of Interfering Spectral Lines

13.3.2.1. Wavelength Interference

The x-ray spectrum (2θ scan) recorded without the use of the pulse-height analyzer contains "lines" of various origins as listed below. Basically, the x-ray tube emits the primary continuous and line spectra, and any object in the x-ray path, including the path medium itself, can then act as a scatterer or secondary emitter.

1. Spectral lines originating in the x-ray tube and scattered by the specimen (*L2*): (a) Primary spectrum of the target element. (b) Primary spectra of impurities in the target. (c) Primary spectra of contaminants sublimed, sputtered, or otherwise deposited on the target during operation of the tube, principally the following: tungsten from the filament; copper from the anode block and from plated components; chromium, iron, and nickel from the stainless steel focusing electrode and filament-to-anode tunnel; and silver from silver solder. (d) Primary and secondary spectra of internal structures other than the target excited by scattered or improperly focused electrons or by the primary x-rays. (e) Primary and secondary spectra of contaminants on the tube window excited by the electrons that bombard the window and/or by the primary x-rays.

These x-ray tube lines may be scattered coherently or incoherently. The intensity of the scattered spectrum and the ratio of incoherent to coherent scatter become greater as the effective atomic number of the specimen decreases. Special cases of interference by incoherent scatter are cited in Sections 8.2.2.3 and 8.3.1. The intensity of the spectra of the contaminants increases with the age of the tube and with the power at which it is usually operated. The tubes of each manufacturer have a more or less characteristic extraneous-line spectrum. Modern x-ray spectrometer tubes are designed specifically for this application and are much "cleaner" than their early predecessors. The scattered spectrum is identified by recording a 2θ scan of the x-rays scattered from a piece of paraffin or spectrographically pure graphite, or from a cell of pure water in the specimen compartment. The incoherent peaks are recognized as somewhat broader peaks on the high-2θ (long-wavelength) side of each coherently scattered peak. If reduction of total primary intensity can be afforded, any or all lines originating in the x-ray tube can be eliminated by filters placed at the tube window.

2. Secondary spectra of the specimen mask and associated structures.

3. Secondary spectra of trace impurities in the specimen support (Mylar, filter paper, etc.) or cell.

The spectra of the specimen mask and support are identified by comparison of the recorded 2θ scans of the unloaded mask and the mask loaded with the support or cell containing no specimen material.

4. Secondary spectra of the specimen itself, including feeble contributions of primary spectra excited by photo, recoil, and Auger electrons: (a) Diagram lines; of course, among these lines are the analyte lines and the lines of the internal standard or internal intensity reference standard, if any. A particularly difficult situation is the presence of two elements having mutually interfering lines, one at high concentration, the other at low concentration. (b) Nondiagram (satellite) lines; these lines result from electron transitions in doubly charged (*KL*, *LL*, *LM*, etc.) atoms. (c) Forbidden lines; these lines result from electron transitions forbidden by the selection rules; for example, in Figure 10.3, the second and third lines from the bottom (*M3–M5* and *M2–M4*), if correctly identified, are forbidden because $\Delta n = 0$ (Section 1.6.2.3).

K, *L*, *M*, or *N* spectra may be excited and recorded, depending on the x-ray tube potential, atomic numbers of the specimen elements, and spectral region recorded. The lines may appear at their "table" values, or they may be displaced as much as 0.25° 2θ by chemical effects (Section 11.1.3.4), or as much as 0.5° 2θ by positional effects (Section 19.2.1). Misalignment of thc goniometer also results in displacement of line position.

5. Diffraction lines from highly crystalline specimens (*B93*, *E6*): Single-crystalline specimens may give Laue diffraction, and polycrystalline specimens may give Debye–Scherrer diffraction (Section 8.2.3) of discrete primary wavelengths λ that meet the Bragg requirements for their d spacings and inclinations θ to the primary beam: $n\lambda = 2d \sin \theta$ [Equation (2.42)]. The diffracted radiation may consist of discrete wavelengths from the continuum or, less probably, of target lines.

Diffraction arises from lattice planes substantially parallel to the specimen surface. In polycrystalline specimens, substantial numbers of crystallites are more likely to be so oriented when there is a high degree of preferred orientation. This occurs especially in rolled metal sheet and foil and in packed and briqueted powders. In most x-ray spectrometers, the primary and useful secondary beams are perpendicular so that the specimen surface—and therefore the potentially diffracting planes—present Bragg angles of ~45° to the primary x-rays. Thus, the diffracted wavelengths would be $\sim 2d \sin 45° \approx (2d)(0.7) \approx 1.4d$ Å. Peaks at 2θ angles corresponding to these wavelengths would then appear in the x-ray spectrum.

The diffracted intensity is higher: (1) for single-crystalline specimens and polycrystalline specimens having a high degree of preferred orientation; (2) for highly "reflective" crystallographic planes; (3) for diffracted target lines, rather than continuum; and (4) if the diffracted wavelengths arise from the continuum, for high-Z targets (tungsten, platinum, gold) rather than low-Z targets (chromium) (Figure 1.4).

The diffracted beams may appear in the x-ray spectrum as more or less normal-looking peaks or as series of close-spaced, extremely sharp "spikes." They are identified as diffraction peaks by comparison of spectra from the same specimen: (1) in various orientations about its own place normal; (2) fixed and rotated about that normal; (3) tilted in the specimen holder to present different angles to the primary beam; and/or (4) in solid and powdered form. The emission lines in all these spectra remain *about* the same. Conversely, diffraction lines, especially the spikes, may be highly dependent on orientation; they may undergo marked change in intensity or disappear altogether, and new lines (wavelengths) may appear. However, the normal-looking diffraction peaks may be extremely difficult to identify if they arise from planes truly parallel to the surface of a single-crystalline or highly oriented polycrystalline specimen. If the diffracted peaks are target lines, the intensity ratios of the diffracted target line to other Rayleigh- or Compton-scattered target lines will be abnormally high. As a last resort, diffraction peaks may be identified by comparison of the spectra of the solid and fused or dissolved specimen, in which all diffraction peaks disappear.

Diffraction peaks are troublesome in that they: (1) clutter the spectrum (especially the spikes); (2) give rise to unidentifiable peaks; (3) may be identified as peaks not present in the specimen, leading to incorrect qualitative analyses; and (4) may constitute a serious variable error when the analyte and target element are the same and the analyte line and diffracted target line are the same.

Diffraction peaks may be minimized or eliminated by one or another of the following means, depending on the particular case: (1) careful orientation of the specimen, possibly including tilting in the holder; (2) specimen rotation; (3) pulverization of solids; (4) loose, random packing of powders to avoid preferred orientation; or (5) use of a different x-ray tube target. If these measures fail, fusion or dissolution may be required.

6. Additional lines introduced by the dispersion system. (a) Higher-order diffracted lines. (b) Spurious diffracted lines (see below). (c) Split lines, that is, individual peaks split into two or more peaks because of interaction of long, close-spaced source and detector collimators. (d) Lines

diffracted or specularly reflected by collimator foils. (e) Line displacement by thermal expansion of the crystal (Section 5.4.2.5).

7. Additional lines introduced by the relatively large angle of acceptance of the detector. For example, second-order Ca $K\beta$ (2λ 6.18 Å, 2θ 89.08° for EDDT) may interfere with P $K\alpha$ (λ 6.15 Å, 2θ 88.66°) when a short, wide-spaced detector collimator is used.

Normally, the detector receives only radiation diffracted from the planes parallel to the reflecting surface of the analyzer crystal. However, there may be a second set of planes only slightly inclined to the surface planes and arrayed longitudinally (as in topaz) or crosswise (as in EDDT). This condition is shown in Figure 13.6. Under certain circumstances, the detector set to receive the analyte peak from the surface planes may also receive x-rays of a different wavelength from these other planes. This phenomenon is unlikely with crystals of high symmetry, such as cubic crystals, but may occur with crystals of low symmetry, such as topaz (orthorhombic) and EDDT (monoclinic), especially when used with a short, wide-spaced detector collimator and a large-area detector window (*S47*, *S50*).

The spurious peaks are recognized by their greater width. They are further identified by use of a tandem detector, especially one having a collimator between the front and back detectors. The spurious peaks are reduced in intensity or eliminated when measured from the back detector alone because of the increased distance and collimation. The spurious peaks may also be minimized or better resolved from the analyte line by increased detector collimation, by turning the crystal end for end or top for bottom, or by pulse-height selection if the spurious and analyte-line wavelengths are substantially different.

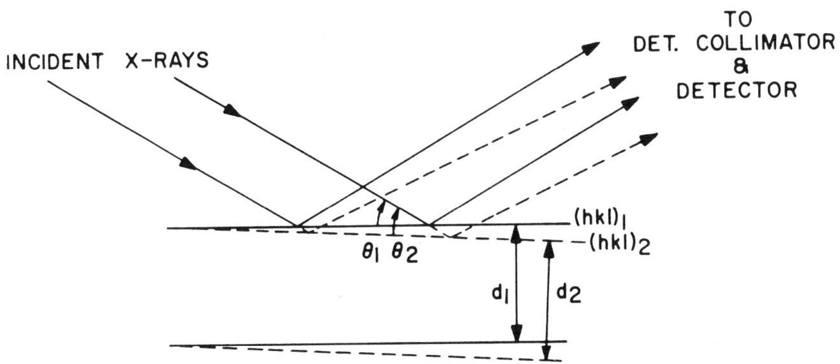

FIG. 13.6. Origin of spurious reflections from the analyzer crystal; $(hkl)_2$ is a second set of planes slightly inclined to $(hkl)_1$, on which the crystal is cut.

13.3.2.2. Energy Interference

If the pulse-height selector is not used, photons of any of the seven groups of lines listed above that enter the detector at the 2θ setting for the measured line constitute wavelength spectral interference. If the pulse-height selector is used, any such photons give rise to pulse-height distributions having average pulse height proportional to the photon energy. These distributions do not necessarily overlap the window set for the measured distribution. When they do not, the pulse-height analyzer eliminates a source of wavelength spectral interference, but when they do, spectral energy interference occurs. In addition, the pulse-height analyzer may introduce forms of energy spectral interference that have no analog in wavelength interference. The two principal forms are crystal emission and escape peaks.

a. Crystal Emission. The incident secondary x-rays excite the elements in the analyzer crystal to emit their characteristic spectra. If the crystal is purely organic (EDDT, PET, etc.), these spectra are of little significance in conventional x-ray spectrometry. If the crystal contains a heavier element, such as aluminum (topaz), silicon (quartz, silicon), phosphorus (ADP), sulfur (gypsum), chlorine (NaCl), or potassium (KHP), the spectra still do not interfere if high-Z elements are being determined in air path, or if low-Z elements are being determined in helium or vacuum *without* pulse-height analysis. In the former case, the air absorbs the crystal emission before it reaches the detector. In the latter case, since the crystal emission radiates in all directions, it merely contributes to the continuous background. However, if the pulse-height analyzer is used, that portion of the crystal emission that does enter the detector gives rise to a pulse-height distribution that may interfere with the measured distribution.

b. Escape Peaks. When the pulse-height analyzer is used, one must consider the overlap of the pulse-height distributions not only of the measured line and interfering line, but also of the measured line and the escape peak of the interfering line. Jenkins has pointed out that in using the pulse-height selector, a situation can arise where, although the main pulse-height distribution of an interfering element can be removed, the associated escape peak may still interfere with the analyte pulse-height distribution. In general, a pulse-height selector set to pass pulses from x-rays of energy E also passes pulses from x-rays of energy $E' \approx E + E_{\text{det}}$, where E_{det} is the energy of the principal spectral line of the detector gas or of the iodine in the scintillator. The reason is that the escape peak of x-rays of energy E' has the same average pulse height as the main peak of the measured line of energy E, that is, $E' - E_{\text{det}} \approx E$. For example, if a pulse-height analyzer is used with a krypton-filled proportional counter (Kr $K\alpha$: λ 0.98 Å, E 12.6 keV) to pass Cu $K\alpha$ (λ 1.54 Å, E 8.0 keV), it will also pass Rh $K\alpha$ (λ 0.62 Å, E 20.2 keV): $20.2 - 12.6 = 7.6$ keV. Partial overlap will occur for energies substantially higher and lower than 20 keV.

An interesting example of this phenomenon occurs when a flow-proportional counter with P10 gas (90% argon, 10% methane) is used to measure Ca $K\alpha$ from specimens containing tungsten (*C19*). Second-order W Ll (2λ 3.356 Å, E 7.39 keV) gives an almost exact wavelength interference with Ca $K\alpha$ (λ 3.360 Å, E 3.69 keV). The pulse-height analyzer easily rejects the main W Ll pulse distribution, but with the argon detector gas (Ar $K\alpha$: λ 4.193 Å, E 2.96 keV), the W Ll escape peak ($7.39 - 2.96 = 4.43$ keV) partially overlaps the Ca $K\alpha$ peak.

Other examples, cited by Jenkins, are shown in Figure 13.7, where A, B, and C represent resolvable, partial, and severe escape-peak interference, respectively. In all three cases, the analyte pulse distribution is the one at

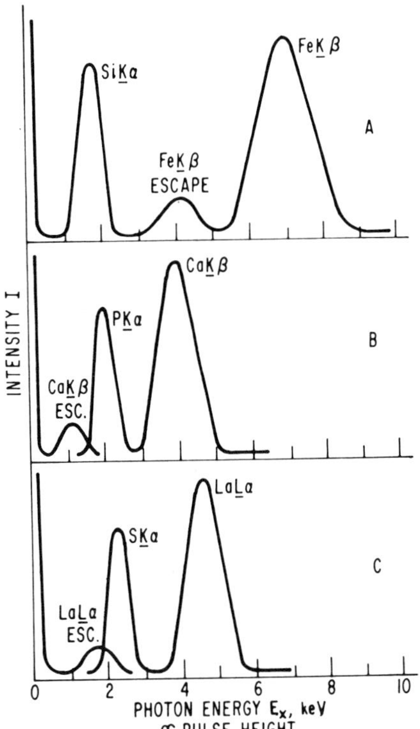

FIG. 13.7. Interference of escape peaks with analyte pulse-height distributions. The Si, P, and S $K\alpha$ peaks are the analyte pulse distributions. A, B, and C represent resolved, partial, and severe escape-peak interference, respectively.

lower photon energy (lower average pulse height) and is fully resolved from the *main* pulse distribution of the interfering line. All three examples arise in argon-filled proportional counters; the photon energy of Ar $K\alpha$ is 2.96 keV. In A, the Fe $K\beta$ escape peak (7.05 − 2.96 = 4.09 keV), like the main peak, is also resolved from the Si $K\alpha$ pulse distribution (1.74 keV) and is easily excluded. The Si $K\alpha$ and Fe $K\beta(4)$ pulse distributions are discussed in detail in Section 8.1.2. In B, the Ca $K\beta$ escape peak (4.01 − 2.96 = 1.05 keV) partially overlaps the P $K\alpha$ pulse distribution (2.01 keV). If the P $K\alpha$ intensity is high enough that some P $K\alpha$ pulses can be sacrificed, the window can be set to exclude the Ca $K\beta$ escape peak. In C, the La $L\alpha_1$ escape peak (4.66 − 2.96 = 1.70 keV) causes unresolvable interference with the S $K\alpha$ pulse distribution (2.31 keV).

13.3.2.3. Common Sources of Spectral Interference

Of course, among the lines listed above are the lines to be measured, that is, the analyte line and perhaps a line for use as an internal standard or an internal intensity reference standard. The reference line may come from the specimen (internal standard or internal intensity reference standard), the specimen mask (external intensity reference standard), or the x-ray tube (scattered target line). All other lines present in the x-rays reaching the detector are potential sources of interference.

Obviously, the higher the intensity of the measured line, and the lower the intensity and greater the separation of the interfering line, the less serious is the interference. Orders higher than the fifth, satellite lines, and M and N lines are usually troublesome in x-ray secondary-emission spectrometry only for weak measured lines, although they may be more troublesome in the electron-probe microanalyzer (Chapter 21). In general, the great majority of instances of spectral-line interference may be classified in one of the following four categories, regardless of the source of the interfering line:

1. Superposition of first-order lines of the same series (K, L, M) from adjacent or neighboring elements in the periodic table. Some examples follow. The $K\alpha$ lines of adjacent elements above technetium (Z 43) lie within 0.03 Å of one another, as do the $L\alpha_1$ lines of adjacent elements above mercury (Z 80). For titanium through cobalt (Z 22–27), the Z $K\alpha$ line lies within 0.03 Å of the $(Z-1)$ $K\beta$ line, and for rhodium through indium (Z 45–49), the Z $L\alpha_1$ line lies within 0.03 Å of the $(Z-1)$ $L\beta_1$ line.
2. Superposition of first-order lines of different series; for example, As $K\alpha$ (λ 1.177 Å) and Pb $L\alpha_1$ (λ 1.175 Å).

3. Superposition of first-order K lines and higher-order lines from elements of higher atomic number; for example, Ni $K\alpha$ (λ 1.659 Å) and Y $K\alpha$ (2λ 1.660 Å), P $K\alpha$ (λ 6.155 Å) and Cu $K\alpha$ (4λ 6.168 Å), Al $K\alpha$ (λ 8.337 Å) and Cr $K\beta$ (4λ 8.340 Å) or Ba $L\alpha_1$ (3λ 8.325 Å). This type of interference is particularly troublesome in the determination of low-Z elements in multicomponent samples having many medium-Z and high-Z elements. Higher orders of scattered target lines also cause this type of interference with low-Z elements.

4. Superposition of first-order L lines and higher-order K lines from elements of lower atomic number. The classic examples are Hf $L\alpha_1$ (λ 1.570 Å) and Zr $K\alpha$ (2λ 1.574 Å), and Ta $L\alpha_1$ (λ 1.522 Å) and Nb $K\alpha$ (2λ 1.494 Å). Similar interferences also exist between $L\beta_1$ and second-order $K\beta$ lines of these elements.

In general, intensity decreases as order increases, although this is certainly not always the case. The likelihood of an analyte line suffering interference from a higher order of a shorter-wavelength line increases as analyte-line wavelength increases. The severity of such interference is greater the weaker the analyte line is. Secondary emission lines from the specimen in orders greater than the fourth or fifth rarely cause trouble, and only when the interfering lines are very intense in the first order and the affected analyte lines relatively weak. Target lines scattered from low-Z matrixes may cause difficulty in higher orders.

Classes 1 and 2 above represent true spectral-line interference and cannot be dealt with by pulse-height selection. Classes 3 and 4 may be termed "apparent" or "pseudo" spectral-line interference, higher-order overlap, or harmonic overlap and are usually readily dealt with by pulse-height selection.

13.3.3. Reduction of Spectral Interference

13.3.3.1. General Methods

Methods for elimination of certain specific types of spectral interference have already been cited in Section 13.3.2. The methods for dealing with spectral interference may be classified in nine categories as follows:

1. Selection of an alternative analyte line that is free from interference.

2. Removal, or reduction of the concentration, of the interfering element (Chapters 16 and 17).

3. Prevention, or reduction of the efficiency, of excitation of the interfering element, or increase of the efficiency of excitation of the analyte.

4. Increase of the resolution of the spectrometer (Section 13.2).
5. Prevention, or reduction, of the transmission of the interfering line through the spectrometer, or increase of the transmission of the analyte line.
6. Prevention, or reduction of the efficiency, of detection of the interfering line, or increase of the efficiency of detection of the analyte line.
7. Prevention of the counting of detector pulses originating from the interfering line.
8. Correction for the interfering line by experimental technique.
9. Correction for the interfering line by mathematical means.

The practicability of any approach to the reduction of spectral interference is determined by the specific features of the particular case. The relative intensities of the analyte and interfering lines, the degree of their overlap, and the difference in their true wavelengths (λ, not $n\lambda$) determine the seriousness of the interference. The intensity of the analyte line determines how much intensity may be sacrificed to reduce spectral interference. If several elements are to be determined in a sample and only one or two are subject to interference, the "economy" of the overall analysis must be considered. The optimal instrument arrangement to reduce the interference may well be unfavorable for the other analyses.

In the first method listed above, the alternative spectral line may be in the same series ($K\beta$ instead of $K\alpha$) or in a different series ($L\alpha_1$ instead of $K\alpha$), or it may be a higher order of the same line. Of course, if it is the optimal line that is interfered with, the alternative line will necessarily have lower intensity or peak-to-background ratio. The seventh method, the prevention of counting of output pulses originating from the interfering line, is the function of the pulse-height selector, which is discussed in detail in Chapter 8. It must be emphasized that the pulse-height selector is useless for discrimination against lines of the same or nearly the same wavelength. Nevertheless, energy resolution is improved by use of a proportional counter with low applied potential and low amplifier gain.

13.3.3.2. Excitation of Analyte and Interferant Lines

There are many ways to apply the third method given in the preceding section.

a. Choice of X-Ray Tube Target. The target may be selected to: (1) eliminate a scattered interfering target line; (2) excite the analyte line at maximum efficiency; or (3) excite the interfering line at minimum efficiency.

The analyte line is excited most efficiently by a target having lines of wavelength just shorter than its absorption edge. The interfering line is excited least efficiently by a target having lines of wavelength just longer than its absorption edge. However, if the interferant is, say, chromium, a chromium target is a poor choice even though it excites chromium very inefficiently. The target-line radiation scattered by the specimen may more than compensate the reduced emission from the specimen, especially for low chromium concentration in a low-Z matrix.

b. Choice of X-Ray Tube Potential. The potential may be selected to: (1) prevent excitation of an interfering line (selective excitation); (2) excite the analyte line at maximum efficiency; or (3) excite the interfering line at minimum efficiency. Selective excitation is feasible if the excitation potential of the interfering line is much higher than that of the analyte line, *and* is high enough so that if the x-ray tube is operated just below this potential, the primary-beam intensity is still high enough for efficient analyte-line excitation. For example, if third-order In $K\alpha$ (3λ 1.542 Å) interferes with Cu $K\alpha$ (λ 1.542 Å), the x-ray tube can be operated below the indium K-excitation potential (28.0 kV); the x-ray tube potential is still well above the copper K-excitation potential (9.0 kV) and high enough to give high primary continuum intensity. However, the interference of fourth-order Cu $K\alpha$ with P $K\alpha$ from low phosphorus concentrations could not be dealt with this way, because operation of the x-ray tube below 9 kV would prohibitively reduce P $K\alpha$ intensity. Of course, both these examples are largely academic, because both are best dealt with by pulse-height selection.

If the analyte and interferant lines have high and low excitation potentials, respectively, analyte-line intensity is increased more than interferant-line intensity by increase of operating potential and by use of constant potential. If an L line of a high-Z element is subject to interference, use of a 75- or 100-kV generator may permit use of a K line.

Incidentally, selective excitation is much less limited in energy-dispersive spectrometry. For example, it may well be feasible to excite Ti $K\alpha$ to the exclusion of Fe $K\alpha$, which have excitation potentials 4.96 and 7.11 kV, respectively.

c. Choice of X-Ray Tube Current. If the overlap of analyte and interfering lines is only partial and if some analyte-line intensity can be sacrificed, the overlap can be reduced, perhaps to insignificance, by reduction of the x-ray tube current.

d. Monochromatic Excitation. The primary beam may be used to irradiate a secondary target having strong spectral lines on the short-wavelength side of the analyte absorption edge and the resulting spectral emission used to excite the analyte line. This method may be useful when intensity can be sacrificed for a "clean" spectrum. The analyte-line intensity would be very low, but nondispersive measurement would probably be feasible. This method is discussed in detail in Section 8.3.1.

e. Specimen Mask, Support, and Matrix. If the interfering line arises from one of these sources, judicious choice of materials is indicated (Chapters 16 and 17).

f. Shielding. The intensity of interfering lines may be reduced by appropriate use of apertures, shields, and baffles in the specimen compartment to confine the primary beam to the specimen area and to reduce scatter. With specimens of small area, selected-area techniques may be used (Chapter 19), or a collimated specimen mask having a continuous L-shaped tunnel to convey the primary beam to the specimen and the secondary beam to the optical path.

13.3.3.3. Transmission and Detection of Analyte and Interferant Lines

The fifth and sixth methods given in Section 13.3.3.1 may be applied as follows.

a. Filters. Filters are useful for reduction of spectral interference in three cases: (1) If the interfering line is a scattered target line, a filter at the x-ray tube window may reduce it to insignificance—at the expense of primary intensity; for example, a filter consisting of 25-μm (0.001-inch) foils of brass and nickel at the window of a tungsten-target tube effectively removes the tungsten L lines (*S6*). (2) If the superposed lines have very different wavelengths—as is the case for superposition of λ and $n(\lambda/n)$—a filter having an absorption edge of wavelength just longer than either line may be used, but the technique is likely to be more effective when the interfering line is the line of longer wavelength. (3) Finally, if the superposed lines have nearly the same wavelength, a filter having an absorption edge between them absorbs the line of shorter wavelength. Preparation and features of x-ray filters are discussed in Section 8.3.2.2.

b. Shielding. If the analyte-line intensity is very low, air scatter of target lines may become significant and may be reduced by appropriate use of apertures, shields, baffles, and tunnels in the specimen compartment.

c. Radiation Path. The path may be used to deal with spectral interference in two ways. Use of helium or vacuum in the specimen chamber reduces target-line scatter. The other way is best explained with an example: If phosphorus is to be determined in the presence of copper, fourth-order Cu $K\alpha$ (4λ 6.16 Å) interferes with P $K\alpha$ (λ 6.16 Å). In the absence of a pulse-height selector, the intensity at the 2θ angle for P $K\alpha$ may be measured first in helium or vacuum path, then in air. The contribution of Cu $K\alpha$ is substantially the same in both paths, but P $K\alpha$ is completely absorbed in air. Thus, the difference between the two measured intensities is P $K\alpha$ intensity.

d. Crystal. The crystal may be selected to have high extinction for an interfering higher-order line (*L42*).

e. Detector. By use of Figure 6.17, a detector can be chosen having high quantum efficiency for the analyte line or low quantum efficiency for the interfering line. The usc of gas-filled detectors in this way is known as *gas discrimination*, and is particularly effective in discriminating against high-order lines of high-Z elements when the long-wavelength lines of low-Z elements are being measured. For example, neon is preferable to argon for discrimination against wavelengths shorter than the argon K-absorption edge (3.9 Å).

13.3.3.4. Experimental and Mathematical Correction of Spectral Interference

Frequently, the net intensity of the interfering line can be measured from a blank at the 2θ angle of the analyte line, then subtracted from the net intensities of the analyte line measured from the specimens. Blanks are discussed in Section 7.3.3.

Alternatively, the contribution of the interferant line to the analyte line can be evaluated from standards. In the determination of strontium in the presence of rubidium, the Sr $K\alpha$ line (λ 0.877 Å) lies between, and is overlapped on both sides by, Rb $K\beta$ (λ 0.829 Å) and Rb $K\alpha$ (λ 0.927 Å). The contribution of rubidium to the measured Sr $K\alpha$ intensity was determined as a function of rubidium concentration from standards having the same strontium concentration with different rubidium concentrations (*H43*).

Another technique is to establish a ratio between the intensity of the interfering line at the 2θ angle for the analyte line and the intensity of another line of the interfering element that can be measured on the specimens without interference. The ratio is established from measurements on the pure interfering element or its oxide. Then, for each specimen, the interfering-line intensity can be calculated from the measured intensity of the second interferant line. The method has been used to correct U $L\alpha_1$ net intensity (λ 0.911 Å) for spectral interference by second-order In $K\beta$ (2λ 0.910 Å) (*B66*):

$$I_{\text{net, U}L\alpha_1} = I_{\text{peak, U}L\alpha_1} - I_B - \left(\frac{I_{\text{net, In}K\beta(2)}}{I_{\text{net, In}K\alpha}}\right) I_{\text{net, In}K\alpha} \tag{13.12}$$

where I_B is adjacent continuous scattered background. The net intensity ratio of In $K\beta(2)$/In $K\alpha$, measured from pure indium, is 0.0451.

Still another approach is illustrated by the following two examples, in which the interfering lines arise from the specimen and x-ray tube target, respectively. In the determination of copper, zinc, silver, and tin in dental alloys, the $K\alpha$ lines of copper, zinc, and silver give linear calibration curves, but Ag $K\beta$ interferes with Sn $K\alpha$. The interference is dealt with by plotting the intensity ratio (Sn $K\alpha$ + Ag $K\beta$)/Ag $K\alpha$ *versus* the Sn/Ag concentration ratio (*R7*). Similarly, in the determination of traces of selenium in organic chemicals, if a tungsten-target tube is used, Se $K\alpha$ is overlapped by W $L\gamma_1$. The interference is dealt with by use of the calibration function (Se $K\alpha$ + W $L\gamma_1$)/W $L\beta_1$ intensity ratio *versus* Se concentration (*O1*). Of course, use of a molybdenum target would preclude the problem entirely.

Birks and Brooks (*B53*) devised a correction technique for the determination of niobium and tantalum in their binary mixtures. Niobium was determined from the Nb $K\alpha$ line (λ 0.747 Å), which is free from interference. Tantalum may be determined from the Ta $L\alpha_1$ line (λ 1.522 Å), which is subject to interference from second-order Nb $K\alpha$ (2λ 1.494 Å), or from the Ta $L\beta_1$ line (λ 1.327 Å), which is subject to interference from second-order Nb $K\beta$ (2λ 1.332 Å). The integrated intensities of the Nb $K\alpha$ and unresolved Ta $L\beta_1$–Nb $K\beta(2)$ peaks were measured from each of several standards consisting of the mixed oxides in various proportions. A curve was plotted of the integrated intensity ratio of the Nb $K\alpha$ and Ta $L\beta_1$–Nb $K\beta(2)$ peaks *versus* tantalum concentration. Another curve was plotted of Nb $K\alpha$ integrated intensity *versus* niobium concentration. For each sample, the integrated intensities of the two peaks were measured and applied to the curves. The same technique was applied to mixtures of zirconium and hafnium oxides.

Zemany (*Z7*) determined vanadium, chromium, and manganese in steel by a correction method. The V $K\alpha$ is free of interference in steels that contain no titanium; V $K\beta$ interferes with Cr $K\alpha$, and Cr $K\beta$ interferes with Mn $K\alpha$. The procedure follows:

1. From each of several standards, the integrated intensity was measured for each of the three peaks of interest: V $K\alpha$ (I_V), unresolved Cr $K\alpha$–V $K\beta$ (I_{Cr}), and unresolved Mn $K\alpha$–Cr $K\beta$ (I_{Mn}). The measurements were made by making 100-s scans of the peaks at 2° 2θ/min with the scaler operating.

2. The $K\beta/K\alpha$ intensity ratio was established for vanadium and for chromium by measurement of the integrated intensities of V $K\alpha$ and V $K\beta$ from pure vanadium oxide (V_2O_5), and of Cr $K\alpha$ and Cr $K\beta$ from pure potassium dichromate ($K_2Cr_2O_7$). The values were V $K\beta$/V $K\alpha$ = 0.278 and Cr $K\beta$/Cr $K\alpha$ = 0.247.

3. Three curves were plotted from the standard data: I_V *versus* % V, I_{Cr} *versus* % Cr, and I_{Mn} *versus* % Mn.

4. The intercept of each curve with the intensity axis at 0% element was read and designated as the background: $I_{B,V}$, $I_{B,Cr}$, and $I_{B,Mn}$.

5. The integrated intensity corresponding to a concentration of 1% element was read from each curve: $I_{1\%V}$, $I_{1\%Cr}$, and $I_{1\%Mn}$.

6. For each sample, the three integrated intensities I_V, I_{Cr}, and I_{Mn} were measured in the same way as for the standards and used in the following equations to derive analytical concentrations, C_V, C_{Cr}, and C_{Mn}:

$$C_V = \frac{I_V - I_{B,V}}{I_{1\%V}} \tag{13.13}$$

$$C_{Cr} = \frac{I_{Cr} - I_{B,Cr} - (\text{V } K\beta/\text{V } K\alpha)(I_V - I_{B,V})}{I_{1\%Cr}} \tag{13.14}$$

$$C_{Mn} = \frac{I_{Mn} - I_{B,Mn} - (\text{Cr } K\beta/\text{Cr } K\alpha)(I_{Cr} - I_{B,Cr})}{I_{1\%Mn}} \tag{13.15}$$

See Section 14.4.4 for some other examples of mathematical correction of spectral-line interference.

The various techniques used by Birks and Brooks (*B53*) and Campbell and Carl (*C16*) in the determination of hafnium in the presence of zirconium, and tantalum in the presence of niobium, provide an example of how a number of techniques, none entirely satisfactory by itself, can be combined to reduce spectral interference. In these analyses, Zr $K\alpha(2)$ (2λ 1.58 Å) interferes with Hf $L\alpha_1$ (λ 1.57 Å), and Nb $K\alpha(2)$ (2λ 1.49 Å) interferes with Ta $L\alpha_1$ (λ 1.52 Å). Similarly, Zr $K\beta(2)$ and Nb $K\beta(2)$ interfere with Hf $L\beta_1$ and Ta $L\beta_1$, respectively. The following techniques were applied:

1. The hafnium and tantalum $L\beta_1$ lines were used instead of the $L\alpha_1$ lines. The $L\beta_1$ lines are almost as intense as the $L\alpha_1$ lines, but the $K\beta(2)$

lines that interfere with $L\beta_1$ are only ~1/5 as intense as the $K\alpha(2)$ lines that interfere with $L\alpha_1$.

2. A molybdenum target (Mo $K\alpha$ 0.71 Å) was used to excite the hafnium and tantalum L lines more efficiently (LIII absorption edges 1.30 and 1.25 Å), eliminate tungsten target-line interference, and reduce the continuous background.

3. The x-ray tube potential was decreased (in some experiments) for selective excitation of the hafnium and tantalum L lines (LIII excitation potentials 9.5 and 9.9 kV) without excitation of the zirconium or niobium K lines (excitation potentials 18 and 19 kV). Incidentally, in their discovery of hafnium, Coster and Von Hevesy used this technique to excite Hf L lines to the exclusion of Zr K lines (*C62, C63*).

4. Alternatively, the x-ray tube potential was increased to permit excitation of the hafnium and tantalum K lines (excitation potentials 65 and 67 kV), which are free of interference from zirconium and niobium.

5. High collimation, 10×0.0125 cm (4×0.005 inch), was used before and after the crystal.

6. A quartz($11\bar{2}2$) crystal ($2d$ 3.636 Å) was used to provide good dispersion and narrow peak width.

7. An argon-filled detector was used to provide high quantum efficiency for the hafnium and tantalum L lines.

8. Integrated intensities of the unresolved Hf $L\beta_1$–Zr $K\beta(2)$ and Ta $L\beta_1$–Nb $K\beta(2)$ peaks were corrected by comparison with the integrated intensities of the fully resolved Zr $K\beta$ or Nb $K\beta$ peaks (see above).

Part V

Quantitative Analysis

Chapter 14

Methods of Quantitative Analysis

14.1. INTRODUCTION

This chapter and the next are devoted to more or less general methods and techniques of x-ray spectrometric analysis. Determinations of specific elements in specific materials are cited only to illustrate such methods and techniques. Liebhafsky and his coauthors [pp. 535–49 in (*27*)] list 66 elements, plus the lanthanons as a group, and for each cite many literature references for its x-ray spectrometric determination. For some elements, the references are classified by specimen type—metals and alloys, cement materials, minerals and ores, slags, organics, etc. The bibliography has 362 references. Similar listings and bibliographies are given in the biennial reviews in *Analytical Chemistry* (*53*) and in each quarterly issue of *X-Ray Fluorescence Spectrometry Abstracts* beginning with volume 4 (1973).

The following quotation from Lucas-Tooth and Pyne (*L47*) is a most fitting preface for Chapters 14 and 15 on quantitative analysis: "The x-ray fluorescence spectrometers presently available have a very high inherent reproducibility. It has not always been possible to translate this reproducibility into analytical accuracy of the same order of magnitude. To accomplish this, it is first necessary to have a set of calibration standards whose chemical composition is known to a high degree of accuracy and secondly to be able to sort out the confusing interelement effects."

It has already been pointed out that were it not for the effect of the matrix, the analyte-line intensity $I_{A,M}$ from analyte A in a thick specimen having matrix M would be simply a function of weight fraction $W_{A,M}$ of A in M and the analyte-line intensity $I_{A,A}$ from pure analyte, that is,

$$I_{A,M} = W_{A,M} I_{A,A} \qquad (14.1)$$

It has also been pointed out that the principal objective of the analytical "method" or "strategy" is to eliminate, minimize, circumvent, or correct for the effect of the matrix, that is, for the absorption-enhancement effects. However, this is not necessarily the only objective of the analytical method.

Attempts have been made at absolute x-ray spectrometric analysis by conversion of intensity data to analytical data by wholly mathematical means. The calculations require knowledge of: (1) the intensity–wavelength distribution of the primary x-ray beam; (2) conversion efficiency of each primary wavelength to the analyte line; (3) absorption coefficient of the specimen for each primary wavelength and for the analyte line; (4) conversion efficiency of each primary wavelength to each matrix element line that can excite the analyte line; (5) absorption coefficient of the specimen for each of these lines; (6) conversion efficiencies of each of these lines to the analyte line; and (7) efficiency of the spectrometer (collimators, crystal, path, detector) for transmission, diffraction, and detection of the analyte line. Obviously, the calculations are extremely complex, and absolute methods have had only limited success. Even this has been accomplished only with simple systems having relatively few components, and then only with simplifying assumptions.

Other approaches to the absorption-enhancement problem have been more successful. Most of these methods involve the use of calibration standards. X-ray spectrometric analysis is still largely a calibration method, and intensity data are converted to analytical concentration by use of calibration curves or mathematical relationships derived from measurements on standards.

The many x-ray spectrometric analytical methods may be classified in eight categories with respect to their basic approach to the reduction of absorption-enhancement effects.

1. *Standard addition and dilution methods.* The analyte concentration is altered quantitatively in the sample itself. In a certain sense, the sample, subjected to one or more quantitative incremental concentrations or dilutions of the analyte, actually provides its own standard(s) in its own matrix.

2. *Thin-film methods.* The specimens are made so thin that absorption-enhancement effects substantially disappear.

3. *Matrix-dilution methods.* The matrix of all specimens is leveled or diluted to a composition such that the effect of the matrix is determined by the diluent.

4. *Comparison-standard methods.* The analyte-line intensity from the samples is compared with that from standards having the same form as the sample(s) and as nearly as possible the same analyte concentration and matrix.

5. *Internal standardization.* The comparison-standard method is improved by quantitative addition to all specimens of an internal-standard element having excitation, absorption, and enhancement characteristics similar to those of the analyte in the particular matrix. The calibration function involves the intensity ratio of the analyte and internal-standard lines.

6. *Standardization with scattered x-rays.* The intensity of the primary x-rays scattered by the specimen is used to correct the absorption-enhancement effects.

7. *Experimental correction.* Various special experimental techniques have been devised to minimize or compensate absorption-enhancement effects.

8. *Mathematical correction.* Absorption-enhancement effects are corrected mathematically by use of experimentally derived parameters.

Most of these methods also reduce error from other sources. Table 14.1 summarizes the effectiveness of each method in compensating the principal sources of error. Mathematical methods are discussed in Chapter 15. In this chapter, each of the other methods is discussed with respect to principles, variations, advantages and disadvantages, limitations, and general applicability. Chapters 16 and 17 discuss specific techniques of specimen preparation and presentation.

Any quantitative analysis of a new sample type should be preceded by a qualitative analysis (2θ scan), as explained in the last paragraph of Section 10.1.

Jenkins and de Vries (*49*) classify quantitative x-ray spectrometric methods in four categories and give an estimate of the fraction of all such analyses constituted by each category: (1) "in-type" methods, in which the sample is analyzed in its as-received form by use of calibration standards in the same form (70%); (2) liquid-solution and solid-solution (fusion) methods (15%); (3) absorption correction methods, including internal standardization, standardization with scattered x-rays, and ratio methods (5%); and (4) mathematical correction methods (10%). These workers also

TABLE 14.1. Effectiveness of Analytical Methods in Compensating Errors

Analytical method	Sources of error[a]										
	Instrumental		Operational	Specimen							
	Short-term	Long-term		Absorption	Enhancement	Inhomogeneity	Position	Surface	Particle size	Density	Bubbles
Standard addition and dilution	—	C	—	C	C	—	C	—	—	—	—
Thin-film method	—	—	—	C	C	—	—	C	C	—	—
Matrix-dilution											
Dry (low absorber)	—	—	—	C	C	—	—	P	P	C	—
Fusion (low absorber)	—	—	—	C	C	C	—	C	C	C	—
Solution (low absorber)	—	—	—	C	C	C	—	C	C	C	—
High absorber (dry)	—	—	—	C	C	—	—	P	P	C	—
Calibration standardization	—	—	—	X	X	—	—	X	X	X	—
Internal standardization	S	C	P	C	C	—	P	P	P	P	P
Other standardization methods											
Internal control standardization	S	C	P	C	P	—	—	—	—	—	—
Internal intensity-reference standardization	S	C	P	C	P	—	C	—	—	C	C
External standardization	S	C	P	C	P	—	—	—	C	C	—
Scattered x-ray methods											
Background-ratio method	S	C	P	C	—	—	C	C	C	C	P
Scattered target-line ratio method	S	C	P	C	—	—	C	C	C	C	P
Reynolds–Ryland method	S	C	P	C	—	—	C	C	P	C	—
Briqueting	—	—	—	—	—	—	—	—	P	C	—
Mathematical methods (geometrical, empirical, influence coefficient methods)	—	—	—	C	C	—	—	—	—	—	—

[a] C: complete or substantial correction; P: partial correction; S: correction only if both the analyte-line and standard intensities are measured

give a table of references to applications of these four general methods to analyses of 10 classes of materials: titanium-, iron-, nickel-, copper-, zirconium-, and uranium-base alloys, light metals, solders, noble metals, and slags.

14.2. STANDARD ADDITION AND DILUTION METHODS

14.2.1. Principles and Considerations

If the calibration function—analyte-line intensity I_A *versus* analyte concentration C_A—is linear in the region of analytical interest, I_A is proportional to C_A, and an incremental or decremental change in C_A gives a proportional change in I_A. If such a linear relationship is questionable, a second change and new measurements may be made to ascertain linearity. If the calibration curve is nonlinear, the slope of the curve is established by repetitive additions. Most standard addition and dilution methods also assume that the calibration curve passes through the origin. The terms *doping* or *spiking* are sometimes used for addition, and *dopant* for additive. The terms *active* and *inert* addition refer to addition of analyte and other material, respectively.

Standard addition and dilution methods usually require no standards or calibration curves, so the methods are useful for samples analyzed too infrequently to warrant preparation of standards and calibration curves, and for samples for which there is no prior knowledge or inadequate knowledge about the matrix. The methods are applicable only to trace and minor analytes (less than 5–10%) where the I_A–C_A calibration function is likely to remain linear for suitable incremental additions or dilutions. In standard-addition methods, an addition must be made for each analyte appropriate for its estimated concentration. In standard-dilution methods, analytes having the same order of concentration may be determined from the same diluted sample. However, usually a portion of the sample must be treated at a different level for each analyte or group of analytes having a different order of concentration. The methods are applicable only to samples to which additions can be made—principally powders, fusion products, and solutions.

The addition of analyte increases the analyte concentration, but, in effect, also reduces the original concentration in proportion to the dilution factor. Thus, to simplify the calculations, the actual weight of substance added should be small compared with the total sample weight. In other words, the analyte concentration in the additive should be high, and the pure analyte itself is preferable whenever its use is feasible. Intimate homogeneous admixture of the additive is required. This condition is easily attained for solutions and melts. For powders, the particle size and density of the additive must be similar to those of the samples.

The methods are most accurate when used with monochromatic primary x-rays, or with polychromatic x-rays and analytes having absorption edges at wavelengths just longer than those of the target lines.

As already mentioned, standard addition and dilution methods may be regarded as self-calibration or self-standardization methods in that the incrementally concentrated or diluted sample actually constitutes its own standard. Jenkins and de Vries regard standard-addition methods as internal standardization with the same (analyte) element.

Another form of "standard addition" may be used in conjunction with calibration standardization methods (Section 14.3) when a matrix element, present in minor and variable concentration, interferes with the determination of another trace or minor element. The interfering element may be added quantitatively to all samples and standards in an amount that overrides the variable normal concentration present; that is, the concentration of interferant is made substantially constant. For example, in the determination of lead in gasoline, interference by bromine may be eliminated by addition of ~1 vol % ethylene bromide. This represents 30–100 times the amount of bromine present in the gasoline (*P28*).

14.2.2. Methods

4.2.2.1. Standard Addition

Essentially, a weighed portion of the sample X is treated with analyte to increase the analyte concentration by ΔX, say 1% absolute, from C_X to $C_{X+\Delta X}$. The analyte-line net intensity is then measured from the untreated I_X and treated $I_{X+\Delta X}$ samples. Then,

$$\frac{I_X}{I_{X+\Delta X}} = \frac{C_X}{C_{X+\Delta X}} \tag{14.2}$$

In practice, a weighed portion of the sample X is treated with a weighed portion of a standard material S having known analyte concentration C_S to form a mixture XS containing concentrations $C_{(X)}$ and $C_{(S)}$ of *sample* and *standard*, respectively. Incidentally, $C_{(X)}$ is actually the dilution factor: weight of X/weight of X + S, or weight of X/weight of XS. In the *standard-addition method* (*C14*, *C17*, *L9*), analyte-line intensity is measured from the untreated sample I_X and mixture I_{XS}. Analyte concentration in the original sample is given by

$$C_X = \frac{(I_X/I_{XS})C_{(S)}}{1 - [(I_X/I_{XS})C_{(X)}]} \tag{14.3}$$

$$= \frac{(I_X/I_{XS})C_S}{1 + \{(w_X/w_S)[1 - (I_X/I_{XS})]\}} \tag{14.4}$$

where C is weight fraction or percent, and w_X and w_S are the weights of sample and added standard, respectively. For analysis of liquids,

$$C_X = \frac{(I_X/I_{XS})C_S}{(\varrho_S/\varrho_X) + \{(v_X/v_S)[1 - (I_X/I_{XS})]\}} \tag{14.5}$$

where C is weight/volume fraction or percent (g/ml); ϱ_X and ϱ_S are densities of sample and standard, respectively; and v_X and v_S are volumes of sample and added standard, respectively. In practice, the density ratio is usually $\sim$1.

In the *active-dilution method* (*L5*, *S26*), analyte-line intensity is measured from the untreated sample I_X, mixture I_{XS}, and standard material I_S. Analyte concentration in the original sample is given by

$$C_X = C_S\left(\frac{1 - C_{(X)}}{C_{(X)}}\right)\left[\frac{(1/I_{XS}) - (1/I_S)}{(1/I_X) - (1/I_{XS})}\right] \tag{14.6}$$

where, in place of reciprocal intensities, the times to accumulate the same preset count can be used. If the standard material and sample are mixed in ratio 1:1, $C_{(X)}$ is 0.5, the first parenthetic term in Equation (14.6) becomes unity, and the equation reduces to

$$C_X = C_S \frac{(1/I_{XS}) - (1/I_S)}{(1/I_X) - (1/I_{XS})} \tag{14.7}$$

The standard material may have a matrix different from that of the sample and may be used for the same analyte in different kinds of samples. It may be a stoichiometric oxide or other simple compound of the analyte, or it may be chosen to have relatively high absorption for the analyte line. (See the discussion of high-absorption dilution in Section 14.6.1.)

14.2.2.2. Standard Dilution (*M27, W2*)

Weighed portions of the sample X and of a standard material S having known analyte concentration C_S of the same order as that of the sample are separately mixed in the same known proportion with an inert diluent containing no analyte(s). Analyte-line intensity is measured from the untreated sample I_X and standard I_S and from the diluted sample $I_{X'}$ and standard $I_{S'}$. Analyte concentration in the untreated sample is given by

$$C_X = C_S \frac{I_S - I_{S'}}{I_X - I_{X'}} \tag{14.8}$$

14.2.2.3. Multiple Standard Addition (*R23*)

The weighed sample is dissolved, treated as required, and adjusted in volume to an analyte concentration of, say, ~50 μg/ml. To five 1-ml portions of this solution, aliquots of a pure standard analyte solution are added containing increasing amounts of analyte—20, 40, 60, 80, 100 μg—bracketing the amount expected to be present in each portion of sample solution. Each of the five treated solutions, a sixth untreated 1-ml portion of sample solution, and a reagent blank are individually absorbed in 500-mg portions of powdered chromatographic paper. The specimens are then mixed, dried, ground, and briqueted against a backing of pure chromatographic paper to form a disk having 1-inch diameter. Analyte-line intensity is measured from all disks, corrected for background measured from the blank, and plotted as a function of analyte mass, as shown in Figure 14.1. The curve is extrapolated to the analyte-mass axis to derive analyte mass in the 1-ml portions of sample solution. The two curves in the figure are for 50- and 100-mg sample weights and give analytical results of 33.5 and 66.5 μg (per 1-ml portion of sample solution), respectively.

14.2.2.4. Slope-Ratio Addition (*R23*)

This method is a refinement of the multiple standard-addition method. It is used when the variation in density and composition of the samples is such that on addition of 1 ml of sample solution to 500 mg of chromatographic paper, the sample matrix would still dominate and override the dilution effect of the paper. The sample solution volume is adjusted to an analyte concentration of, say, 500 μg/ml. To five 1-ml portions of this sample solution, aliquots of a pure standard analyte solution are added containing increasing amounts of analyte—200, 400, 600, 800, 1000 μg—

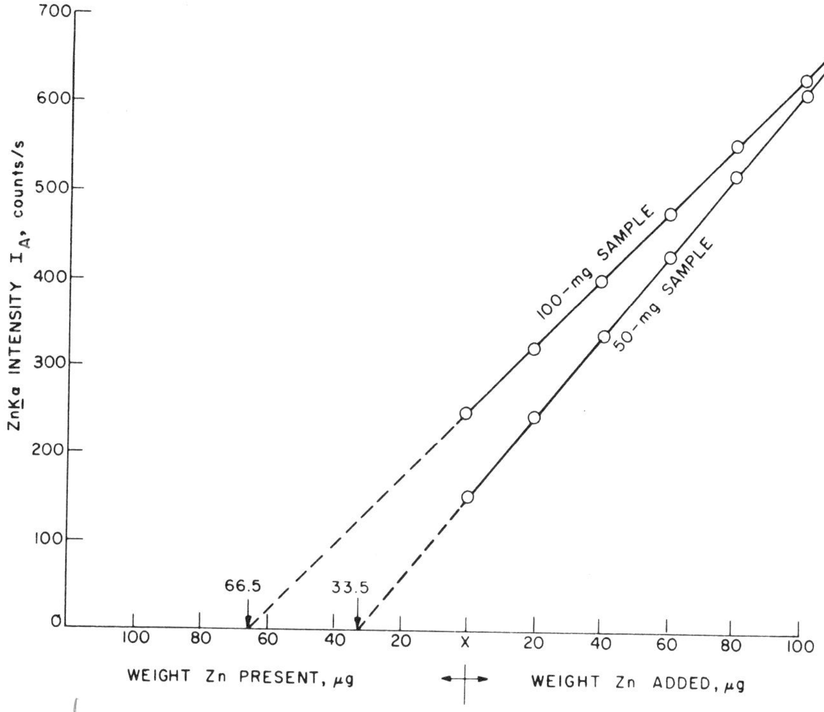

FIG. 14.1. Multiple standard-addition method with graphic solution. [H. J. Rose, Jr., and F. Cuttitta, *Advances in X-Ray Analysis* **11**, 23 (1968); courtesy of the authors, University of Denver, and Plenum Press.]

bracketing the amount expected to be present in each portion of sample solution. A sixth 1-ml portion of sample solution is left untreated. So far, the procedure is the same as that for the multiple standard-addition method. Now five more identical aliquots of pure analyte solution are added to 1-ml portions of water. Each of these 11 solutions and a reagent blank are taken up in chromatographic paper and briqueted as before. Analyte-line intensity is measured from all disks and plotted as a function of analyte mass. Figure 14.2 shows such plots for determination of bromine in brines. Curve *A* represents the pure bromine (as sodium bromide, NaBr) disks. Curves *B*, *C*, and *D* represent sets of disks for three different brine samples.

For each sample, a corrected background is calculated from the slope ratios of the sample and pure-analyte curves:

$$I_{B,X} = I_{B,S}(m_X/m_S) \tag{14.9}$$

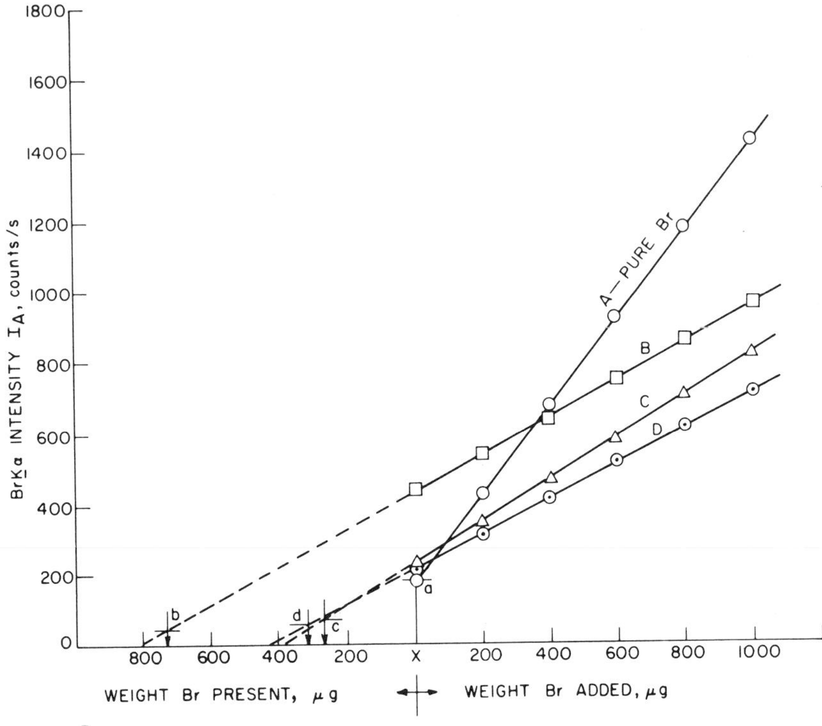

FIG. 14.2. Slope-ratio addition method applied to determination of bromine in brines. Curve *A* represents pure bromine (NaBr), curves *B*, *C*, and *D*, three brines. [H. J. Rose, Jr. and F. Cuttitta, *Advances in X-Ray Analysis* **11**, 23 (1968); courtesy of the authors, University of Denver, and Plenum Press.]

where $I_{B,X}$ is corrected background intensity for a sample; $I_{B,S}$ is background intensity measured from the blank (*a* in Figure 14.2); and m_X and m_S are the slopes of the sample and pure-analyte curves, respectively. The corrected backgrounds for curves *B*, *C*, and *D* are indicated as *b*, *c*, and *d*, respectively, in the figure. Normals dropped from these points to the analyte-mass axis give the analyte masses in the 1-ml portions of the sample solutions.

The multiple-addition and slope-ratio addition methods are discussed above in terms of the solution and paper-disk techniques used by their originators (*R23*). However, the methods are by no means limited to these techniques, and are applicable to any sample form suitable for any standard-addition method.

14.2.2.5. Double Dilution

Analyte-line intensity I_{A,C_X} for analyte A in *original-sample* concentration C_X in a fusion product, solution, or diluent is given by

$$I_{A,C_X} = \frac{C_X}{1 + \phi C_X} \tag{14.10}$$

where ϕ is a matrix factor characteristic of analyte A in that sample type in that flux, solvent, or diluent. It follows that two intensity measurements at two different concentrations C_X and C_X' permit calculation of ϕ and thereby correction of measured intensities for absorption-enhancement effects. This is the basis of Tertian's double-dilution method (*T7, T8*). In principle, the method applies to any analyte at any concentration in any matrix that can be diluted, without any assumptions or prior knowledge about the nature or composition of the sample. It requires no systematic calibration, but, unlike most standard addition and dilution methods, does require standards—at most, one for each analyte. These standards may be pure analytes, pure analyte compounds, or specimens containing analytes in accurately known concentrations. The method is especially valuable for complex and variable matrixes, such as minerals. The following presentation of the method is taken largely from Müller [pp. 194–196 in (*29*)].

The relative spectral-line intensity of element A in the original untreated sample is

$$\frac{I_A}{I_{100A}} = \frac{C_A}{C_A + [r(1 - C_A)]} = \frac{1}{r}\,\frac{C_A}{1 + \{[(1 - r)/r]C_A\}} \tag{14.11}$$

where I_A and I_{100A} are analyte-line intensities from the untreated sample and pure A, respectively; C_A is *analyte* concentration; and r is a matrix factor defined in Section 15.5.2.2 and Equation (14.13).

If now two portions of the original sample are fused, dissolved, or diluted so that the *sample* concentrations in the two fusion products, solutions, or mixtures are C_X and $2C_X$, respectively, the measured analyte-line intensities from the two treated samples, relative to the untreated sample, are

$$\frac{I_{A,C_X}}{I_A} = \frac{1}{r}\,\frac{C_X}{1 + \phi C_X} \quad \text{and} \quad \frac{I_{A,2C_X}}{I_A} = \frac{1}{r}\,\frac{2C_X}{1 + 2\phi C_X} \tag{14.12}$$

where ϕ has the significance given in Equation (14.10). Then the corresponding corrected intensities are

$$\frac{(I_{A,C_X})_{cor}}{I_A} = \frac{1}{r} C_X \qquad \text{and} \qquad \frac{(I_{A,2C_X})_{cor}}{I_A} = \frac{1}{r} 2C_X \tag{14.13}$$

The corrected and measured intensities are related as follows:

$$\begin{aligned} (I_{A,C_X})_{cor} &= I_{A,C_X}(1 + \phi C_X) \\ (I_{A,2C_X})_{cor} &= I_{A,2C_X}(1 + 2\phi C_X) \end{aligned} \tag{14.14}$$

The matrix factor ϕ is evaluated from the ratio of the measured intensities from the two treated samples:

$$\frac{I_{A,2C_X}}{I_{A,C_X}} = \frac{2C_X}{C_X} \frac{1 + \phi C_X}{1 + 2\phi C_X} \tag{14.15}$$

from which

$$\phi = \frac{1}{2C_X} \frac{2 - (I_{A,2C_X}/I_{A,C_X})}{(I_{A,2C_X}/I_{A,C_X}) - 1} \tag{14.16}$$

The corrected intensity for the $2C_X$ sample is then calculated from

$$(I_{A,2C_X})_{cor} = \frac{I_{A,2C_X}}{(I_{A,2C_X}/I_{A,C_X}) - 1} \tag{14.17}$$

Since corrected intensity is the intensity that is proportional to concentration,

$$(I_{A,C_X})_{cor} = kC_XC_A \qquad \text{and} \qquad (I_{A,2C_X})_{cor} = 2kC_XC_A \tag{14.18}$$

where k is a proportionality constant, and C_A is analyte concentration in the untreated sample:

$$C_A = \frac{(I_{A,C_X})_{cor}}{kC_X} = \frac{(I_{A,2C_X})_{cor}}{2kC_X} \tag{14.19}$$

The value of k is obtained experimentally by use of a standard, which may be pure analyte, an analyte compound, or a specimen having known analyte concentration. Two portions of the standard are processed in the same flux, solvent, or diluent, and by the same procedure as the samples. The two treated standards have the same 2:1 concentration ratio as the treated samples. Intensities are measured and corrected intensities calculated [Equations (14.14) and (14.17)]. Then, from the corrected intensities and the known value of C_A (here, analyte concentration in the *untreated stan-*

dard), k is calculated from Equation (14.19); k need be evaluated only once for each analyte in each flux, solvent, or diluent. Incidentally, ratios other than 2:1 may be used, provided, of course, the same ratio is used for both samples and standards.

14.3. CALIBRATION STANDARDIZATION

14.3.1. Principles

With the exception of the standard addition and dilution methods, all x-ray secondary-emission spectrometric analytical methods now in common use are based on comparison of analyte-line intensities measured from the sample(s) and one or more standards. This applies even to the standard addition and dilution methods if one regards the concentrated or diluted samples as standards. The standards must be similar to the samples with regard to: (1) physical form—solid, briquet, powder, fusion product, solution; or specimen supported on Mylar, filter paper, ion-exchange membrane, etc.; (2) analyte concentration; (3) matrix composition; and (4) physical features, such as surface finish, particle size, and packing density.

In fitting a smooth curve to the data points measured from the standards, it is generally agreed that the best fit is the *Cohen least-squares* curve. This is the curve for which the sum of the squares of the deviations of the individual points from the curve is minimal.

Accurate analysis is feasible with a simple calibration method—analyte-line intensity *versus* concentration—if the effect of the matrix is not serious, or, if it is serious, if the analyte concentration range is small and the standards are very similar to the samples. Otherwise, the calibration method must be supplemented by one of the other basic approaches—internal standardization, thin-film, dilution, scatter, mathematical, or special experimental methods.

Figure 14.3 shows a typical x-ray spectrometric calibration curve. If the measured intensities are not corrected for background, the curve does not pass through the origin (zero intensity and concentration), but intercepts the intensity axis at a value corresponding to the background intensity, as shown. The figure illustrates one method of measuring background —by extrapolation. Alternatively, background can be measured for each standard and net intensity plotted, in which case the curve should pass through the origin. The figure also indicates the precision of the measured intensities and shows a data point subject to strong systematic error. The

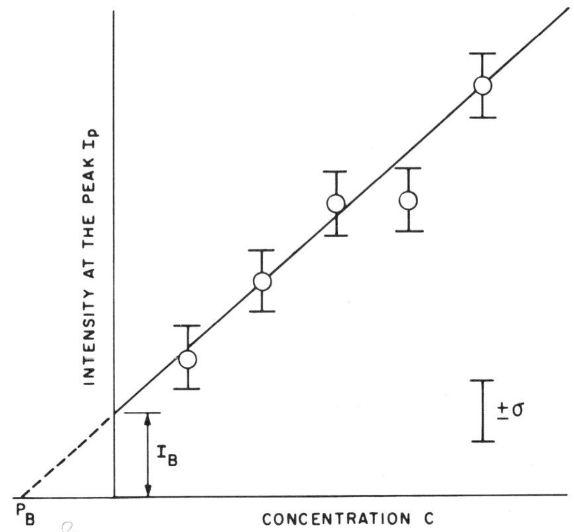

FIG. 14.3. Typical x-ray spectrometric calibration curve. For significance of P_B, equivalent background, see Section 11.2.4.

analysis may be done by measurement of the analyte-line intensity from the sample and application of this intensity to the calibration curve to derive analytical concentration graphically. Alternatively, a mathematical calibration factor may be derived from the curve or from the data itself. For example, in Figure 14.3,

$$I_P = mC + I_B \tag{14.20}$$

where m is the slope of the curve and may serve as a *calibration factor*,

$$m = \frac{I_P - I_B}{C} \tag{14.21}$$

The unit for m is counts per second per unit concentration (%, mg/ml, μg/cm², etc.). Then analytical concentration is given from the net intensity measured from the sample:

$$C = \frac{I_P - I_B}{m} \tag{14.22}$$

If the calibration curve of net intensity *versus* concentration is known to be linear and pass through the origin, in principle, one standard is enough to define it. However, for a wide concentration range, few analysts would care to proceed without at least three standards.

Linear calibration curves of analyte-line intensity *versus* concentration usually occur for analytes in low concentrations or small total amount in matrixes having low absorption coefficients. Specimen types likely to have linear calibration curves include the following: (1) thin films; (2) supported specimens having small total amounts of analytical material on filter paper, Mylar, ion-exchange resin, etc.; (3) minor and trace analytes in plastics, hydrocarbons, and other low-Z matrixes; (4) minor and perhaps even major constituents in diluted, fused, and dissolved specimens; and (5) trace analytes in their original matrixes.

The calibration curve need not be linear to be of analytical value, although certainly a linear curve is preferable. More standards are required to plot a nonlinear curve than a linear curve, and the greater the curvature, the more standards required. If the analyte concentration range of interest is small and occupies only a short segment of a nonlinear curve, and if standards closely bracketing this segment are available, it is often feasible simply to assume the useful interval to be linear.

A moderate to high ("steep") slope is desirable so that a small interval in analyte concentration gives a relatively large interval in analyte-line intensity. To a first approximation, the slope is inversely proportional to the mass-absorption coefficient of the matrix for the analyte line, that is, the higher the matrix absorption, the "shallower" the slope. Jenkins points out that slope and background level are also roughly dependent on the wavelength of the analyte line. The short-wavelength region (<1 Å) is characterized by moderate slope and high background, the long-wavelength region (3–20 Å) by low slope and background. The intermediate-wavelength region is most favorable, having high slope and low background. In general, for a linear calibration curve, the sensitivity of the analytical method is greater the steeper the slope and the lower the background intensity. Methods for reducing background are given in Section 7.3.4. Slope is increased by increasing the measured analyte-line intensity. Methods for doing this are given in Section 13.1.2.

Tertian (*T6*) has devised a method in which the tangent, at the origin, of the intensity *versus* concentration curve is used as a measure of the true intensity–concentration relationship.

Figure 14.4 shows a composite typical x-ray spectrum (2θ scan) showing emitted and scattered radiation that may be used for, or interfere with, an analysis. The significance of each feature of the spectrum is given in the caption. Some features of the figure are discussed in the section on background measurement (Section 7.3.3). Here, the figure illustrates many of the various calibration functions that may serve as the basis of an x-ray

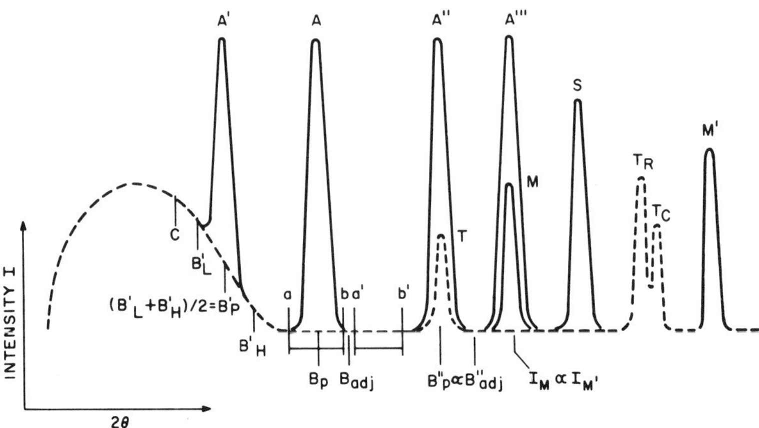

FIG. 14.4. Composite typical x-ray spectrum (2θ scan) showing emitted and scattered radiation that may be used for, or interfere with, an analysis.

Dashed curve: scattered target spectrum

Solid curve: emitted specimen spectrum

Analytical lines:

A	on uniform continuous background
A'	on nonuniform continuous background
A''	on scattered target line T
A'''	on emitted matrix line M

Background:

B_{adj}	adjacent to peak
B_P	at peak
B_L', B_H'	adjacent to peak at equal 2θ intervals on the low L and high H sides
C	intensity at a certain 2θ angle in the "hump" of the scattered continuous spectrum

Other lines:

M	line of matrix element
M'	line of same matrix element free of spectral-line interference
S	internal standard, internal-control standard, or internal intensity-reference standard line
T	scattered target line
T_C	Compton-scattered target line
T_R	Rayleigh-scattered target line
a——b	limits of integrated-count scan of peak
a'——b'	limits of integrated-count scan of background

spectrometric analysis. The intensities or intensity ratios that may be plotted as a function of analyte concentration include the following:

Intensities:

1. Analyte-line peak intensity A.
2. Analyte-line net intensity $A - B$.
3. Intensity of a line of an element stoichiometrically combined with the analyte (association analysis, Section 14.8.3).

Intensity Ratios:

4. Analyte line/internal-standard line A/S.
5. Analyte line/internal control-standard line A/S (*M44*).
6. Analyte line/internal intensity-reference standard line A/S (*J25*).
7. Analyte line/adjacent background A/B_{adj} (*A19*).
8. Analyte line/background at the peak A/B_P or A/B_P'.
9. Analyte line/background measured in the scattered continuum "hump" A/C (*A19*).
10. Analyte line/coherently scattered target line A/T_R (*C70*).
11. Analyte line/incoherently scattered target line A/T_C (*B76*).
12. Minor or trace analyte line/line of major matrix element of substantially constant concentration A/M' (*H45*).
13. Line of analyte A/line of analyte B (binary-ratio method) A/A' (*B36*).
14. Analyte line from specimen/analyte line from standard.
15. Analyte line from specimen/analyte line from pure analyte.
16. Analyte line from specimen/radiation from external standard.

Other calibration functions applicable to films and platings are given in Sections 18.1 and 18.2.2.

The intensities may be peak or net, depending on the particular system. For the last three ratios, the two ratioed intensities are not necessarily present in the spectrum simultaneously. The first two functions listed represent simple calibration methods. The other functions involve other methods of dealing with matrix effects and are discussed in subsequent sections of this chapter. If the same preset time is used for all line and background measurements, the accumulated counts may be used instead of intensities. If the same preset count is used for all measurements, the counting times may be used. In all cases, the calibration functions may be used graphically, or mathematical calibration factors may be derived.

In electron-probe microanalysis (Chapter 21), a ratio sometimes used is that of analyte-line intensity to *total* continuum intensity (*M12*). These intensities are substantially linearly proportional to mass of analyte and total mass, respectively, in the excited volume.

In general, intensity ratios give more reliable analytical results than absolute intensities, and this increased reliability usually more than compensates the increased counting error (Section 11.2.2). Liebhafsky and his coauthors go so far as to state that "a comparison of intensity ratios for standard and unknown can compensate (remedy) all difficulties in x-ray emission spectrography if standard and unknown are physically and chemically identical, a situation easiest to realize when the samples [specimens] are homogeneous."

Although there is no universally accepted convention, most workers plot x-ray spectrometric calibration curves with the intensity or intensity ratio on the ordinate (vertical axis) and the concentration or amount on the abscissa, as shown in Figure 14.3. This convention is followed in this book.

Shapes of calibration curves are discussed in Sections 12.2.2, 12.2.4, and 14.3.2.5.

14.3.2. Special Calibration Methods

14.3.2.1. Single-Standard Method

When the samples X have a very narrow range of analyte concentration C bracketing the analyte concentration of a standard S, the following simple proportionality holds reasonably well for the measured net intensities:

$$I_X/I_S = C_X/C_S \tag{14.23}$$

A refinement of this method is the *corrected single-standard method* (*S64*). Suppose that all components are to be determined in a certain type of alloy. Several chemically analyzed samples X′ are obtained, one of which is designated as the standard S. A set of calibration curves—one for each analyte—is established as follows. Analyte-line net intensity is measured for each analyte from each analyzed sample $I_{X'}$ and from the standard I_S. Then for each analyte in each analyzed sample: (1) the intensity ratio $I_{X'}/I_S$ is calculated; (2) analyte concentration $C_{X'}$ is calculated by the single-standard method [Equation (14.23)]; and (3) the ratio of the x-ray and chemical concentrations $C_{X'}/C_{\text{chem}}$ is calculated. For each of the analyzed samples, the "net density" is calculated from the $C_{X'}$ values of all the analytes A and their densities ϱ: $\sum (C_{X'}\varrho)_A$. Finally, for each analyte A, a correction curve is plotted of $(C_{X'}/C_{chem})_A$ as a function of net density. The curve for each analyte has as many points as there are analyzed samples X′.

Thereafter, the procedure for analysis of each sample X is as follows. For each analyte, I_X and I_S are measured from the sample and standard, respectively; I_X/I_S is calculated; and C_X is derived from Equation (14.23). Net density is calculated for the sample, $\sum (C_X \varrho)_A$, and applied to the correction curve for each analyte A in turn to derive $(C_{X'}/C_{chem})_A$ for each analyte. The corrected analyte concentration is then $[C_X(C_{X'}/C_{chem})]_A$.

Single-standard methods are particularly suitable for analyses made with automatic spectrometers.

14.3.2.2. Two-Standard Method (*H50*)

When the analyte concentration in the sample X lies between that in two standards S1 and S2, and the net intensity *I versus* concentration *C* function is substantially linear in the interval bracketed by the standards,

$$C_X = C_{S1} + \frac{I_X - I_{S1}}{I_{S2} - I_{S1}}(C_{S2} - C_{S1}) \tag{14.24}$$

The solid lines in Figure 14.5 show the intensity–concentration relationships for the two-standard method for the case of a linear calibration function. The dashed lines show how nonlinearity introduces error. Another form of the equation for the two-standard method is the following (*H48*, *H50*):

$$\frac{[(I_{S1}/I_{S2})(C_{S2}/C_{S1})] - 1}{C_{S2} - C_{S1}} = \frac{[(I_{S1}/I_X)(C_X/C_{S1})] - 1}{C_X - C_{S1}} \tag{14.25}$$

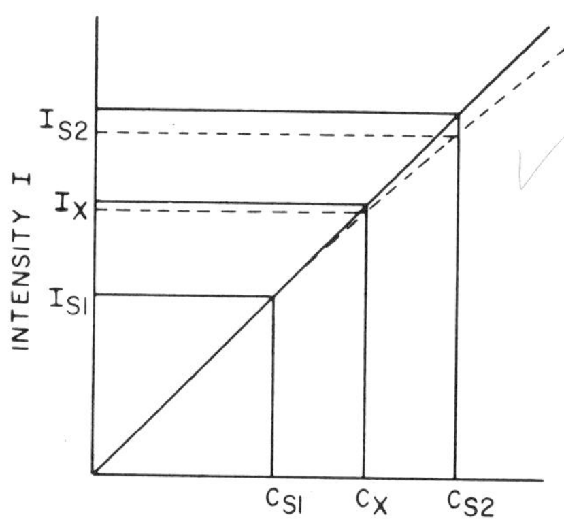

FIG. 14.5. Two-standard method.

14.3.2.3. Binary-Ratio Method (*B36, F7*)

The binary-ratio method is applicable to three classes of samples: (1) those in which only two elements A and B are present and both are determinable by x-ray spectrometry—for example, lead–tin alloys; (2) those in which only two such elements vary in an otherwise constant matrix—for example, indium gallium arsenide, (In,Ga)As; and (3) those in which only two elements having atomic number about 19 (potassium) or more are present in a very light matrix—for example, mixtures of chromium and silver oxides. Classes 2 and 3 may be termed *pseudobinary* systems, which may be defined as multielement systems in which the analyte line intensity of each of two elements depends wholly on its own concentration and that of the other element, or in which two elements vary in concentration in an otherwise substantially constant matrix. For such specimens, the net intensities of the analyte lines of elements A and B are measured from each sample and standard. From the standard data, a calibration curve is established of the function $\log_{10}(I_A/I_B)$ *versus* $\log_{10}(C_A/C_B)$, where I and C are net intensity and concentration, respectively.

The method has many advantages. It has greater sensitivity to change in concentration than the conventional I *versus* C method because I_A increases as I_B decreases, and *vice versa*. Calibration curves are always linear, regardless of how distorted the individual curves (I_A–C_A and I_B–C_B) may be, that is, regardless of how severe the mutual absorption-enhancement effects are. The method is insensitive to reasonable variations in surface texture (porosity, chipping, cracking, grind marks), and the data points usually lie on the linear binary-ratio calibration curve regardless of how widely scattered they are on the individual curves. The method is insensitive to reasonable variations in specimen area and position and to surface contour of small irregular-shaped specimens and fabricated parts. Specimens need not be masked to constant area. In many cases, the standards need not necessarily have the same physical form as the samples. No additional measurements other than the two analyte lines and their backgrounds are required. The principal disadvantage of the method is its inapplicability near the concentration extremes where C_A/C_B and I_A/I_B approach zero or infinity.

The method has been evaluated for tungsten–rhenium, bismuth–tin, and niobium–tin alloys; Cr_2O_3–Ag_2O and Fe_3O_4–CuO powder mixtures; (Ga,In)As; and solutions containing molybdenum and tungsten. Figure 14.6 shows the effectiveness of the method as applied to a relatively unfavorable system—mixtures of Cr_2O_3 and Ag_2O powders. It is evident that

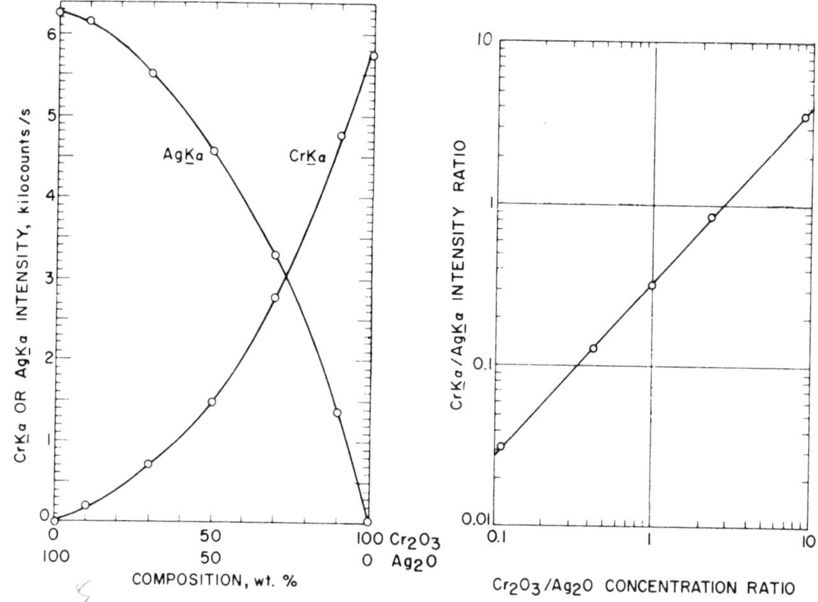

FIG. 14.6. Binary-ratio method. Intensity *vs.* concentration curves for the two individual elements (left) are compared with the binary-ratio curve (right). [E. P. Bertin, *Analytical Chemistry* **36**, 826 (1964); courtesy of the American Chemical Society.]

the individual I *versus* C curves are very distorted, but the binary-ratio curve is linear.

For adjacent elements in the periodic table, or even for elements differing by ± 2 in atomic number, the spectral-line intensity ratio is substantially equal to the concentration ratio. Such systems correspond to curve A in Figure 12.1A. Of course, the same line ($K\alpha$, $L\beta_1$, etc.) must be measured for both elements, and no specimen, crystal, or detector absorption edge may lie between the measured lines. Rubidium–strontium concentration ratios in minerals have been determined in this way for estimation of geological age (*H43*), and the method has been applied to copper–zinc (brass), tungsten–rhenium, and tungsten–osmium alloys.

In the simultaneous determination of copper and mercury in complex organic compounds, Daugherty *et al.* (*D4*) used as calibration functions the logarithm of the ratio of the Hg $L\beta_{1,2}$ and Cu $K\alpha$ peak *areas versus* mercury concentration and the logarithm of the inverse ratio *versus* copper concentration. Both curves were linear for 0–100 mol % of the metal complexes.

Eddy and Laby (*E8*) made extremely precise determinations of the copper/zinc ratio in brass by use of the ratio technique with electron excitation.

14.3.2.4. Mutual Standards Method (*C46*)

This method is applicable to substances in which: (1) the matrix is dominated by a single major constituent; (2) the other elements are known, have relatively low concentrations, and are determinable by x-ray spectrometry; and (3) it can be assumed that the sum of the concentrations of all the determined elements is 100%. The method has been applied to the determination of iron (88–100%) and seven other elements in meteoritic sulfide nodules. The nodules are "dissolved" in nitric acid, evaporated to dryness, and fired to expel sulfur trioxide and convert all eight elements to oxides. In the resulting specimens,

$$C_{\mathrm{Fe}} + \sum C_E = 100\% \tag{14.26}$$

where C is concentration as oxide (wt %), and E represents each of the seven minor and trace elements. Then,

$$C_{\mathrm{Fe}} = 100 - \sum C_E = \frac{100}{1 + \sum (C_E/C_{\mathrm{Fe}})} \tag{14.27}$$

$$= \frac{100}{1 + \dfrac{C_{\mathrm{V}}}{C_{\mathrm{Fe}}} + \dfrac{C_{\mathrm{Cr}}}{C_{\mathrm{Fe}}} + \dfrac{C_{\mathrm{Co}}}{C_{\mathrm{Fe}}} + \dfrac{C_{\mathrm{Ni}}}{C_{\mathrm{Fe}}} + \dfrac{C_{\mathrm{Cu}}}{C_{\mathrm{Fe}}} + \dfrac{C_{\mathrm{Zn}}}{C_{\mathrm{Fe}}} + \dfrac{C_{\mathrm{As}}}{C_{\mathrm{Fe}}}} \tag{14.28}$$

Seven working curves of line-intensity ratio I_E/I_{Fe} *versus* C_E were established from measurements on the seven sets of binary mixtures. An analysis consisted in measurement of the line intensities I_E of all eight elements from the sample, calculation of the seven I_E/I_{Fe} ratios, and application of these ratios to their respective working curves to derive C_E/C_{Fe} concentration ratios. These ratios were substituted in Equation (14.28), which was then solved for C_{Fe}. Finally, the seven C_E values were calculated from their respective C_E/C_{Fe} ratios.

14.3.2.5. Sets of Calibration Curves

When analyte-line intensity depends on the concentration not only of the analyte A, but also of some matrix element B, a set of calibration curves of A-line intensity *versus* A concentration may be established, each curve for a different B concentration (*D10*). An analysis for element A requires the previous determination of element B. The A-line intensity is then measured and applied to the particular I_{A}–C_{A} curve for the known B concentration. Those B concentrations not represented by a curve are dealt

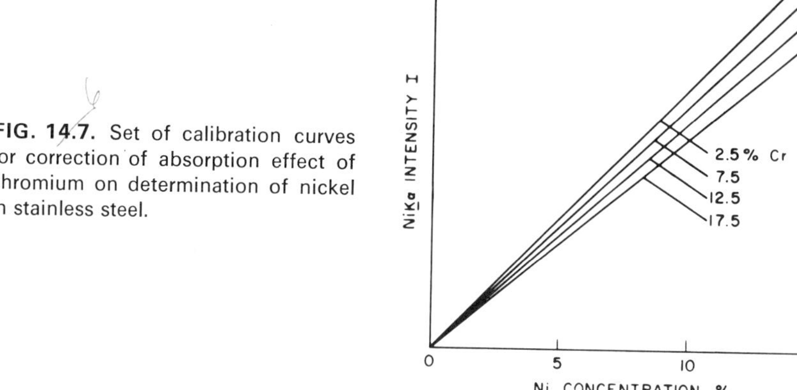

FIG. 14.7. Set of calibration curves for correction of absorption effect of chromium on determination of nickel in stainless steel.

with by interpolation. Figure 14.7 shows a typical set of curves for determination of nickel in stainless steel, each curve for a different chromium concentration. In the analysis of petroleum products, sets of curves may be used, each curve for a different base-stock composition (*K27*).

The principle of the method warrants consideration in that it elucidates the nature of calibration curves in general. The relationship of analyte-line intensity and concentration in a binary system is represented by a line. Figure 14.8A shows calibration curves for A-line intensity in the A–B, A–C, and A–D binary systems (see also Figure 15.4A). The shape and departure from linearity of the curves depend upon the relative absorption of A, B, C, and D for the primary and, especially, A-line x-rays (Section 12.2.4, Figure 12.1). As these absorption coefficients approach equality, the curves approach linearity and congruence. The relationship of analyte-line intensity and concentration in a ternary system is represented by a lenticular field. In Figure 14.8A, A-line intensities for all possible compositions in the A–B–C ternary system lie between the curves for A–B and A–C. Similarly, A-line intensities for all possible compositions in the A–B–D and A–C–D ternary systems lie between the curves for, respectively, A–B and A–D, and A–C and A–D.

Now, a set of calibration curves can be drawn for A-line intensity in, say, the A–B–C ternary system, and these curves may take either of two forms: constant B/C concentration ratio or constant C concentration. Figure 14.8B shows a set of curves of A-line intensity for 0–100% A in 100–0% of a mixture of B + C in weight ratios 8/2, 6/4, 4/6, and 2/8. Figure 14.8C shows a set of curves of A-line intensity for 0–80% A in (20% C + 80–0% B), 0–60% A in (40% C + 60–0% B), etc. Müller [p. 162

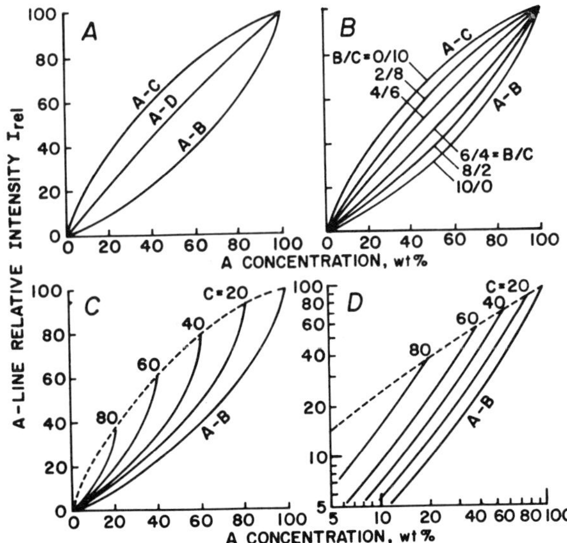

FIG. 14.8. Calibration curves for a ternary system. A. Calibration curves for the individual binary systems. B. Calibration curves for the A–B–C ternary system for constant B/C weight ratios. C. Calibration curves for the A–B–C ternary system for constant C concentration. D. The curves of C plotted on a log-log grid.

in (*29*)] points out that if such curves are plotted on a log-log grid, as shown in Figure 14.8D, rather than on a linear grid (Figure 14.8C), substantially parallel curves are obtained separated by substantially equal intervals.

Of course, the entire concentration range may not be of analytical interest. In this case, portions of the sets of curves may be established. Linear interpolation between the curves is usually permissible and is especially simple with the log-log plots (Figure 14.8D).

14.4. INTERNAL STANDARDIZATION

14.4.1. Principles (*A11*)

If an element has a spectral line having excitation, absorption, and enhancement properties identical with those of the analyte line in a particular matrix, that element would be an ideal internal standard IS for that analyte A in that matrix, and

$$C_A/C_{IS} = I_A/I_{IS} \tag{14.29}$$

where C and I are concentration and intensity, respectively. In practice, absorption-enhancement effects may be compensated by the addition to all samples and standards of a constant concentration of an internal-standard element IS that fills these requirements as closely as possible. Internal standardization also minimizes the effects of variations in packing density in loose and briqueted powders, surface ripple in glass disks, bubble formation and thermal expansion in liquids, and the inevitable variations in supported specimens. Fortuitously, a suitable internal-standard element may already be present in the specimens in a suitable constant concentration. Line intensities are measured from the analyte and internal standard from all samples and standards, and a calibration curve of I_A/I_{IS} intensity ratio *versus* analyte concentration is established from the standard data. Table 14.2 gives some typical data demonstrating the effectiveness of the method.

Alternatively, a calibration constant K may be derived from a standard having known concentrations of analyte C_A and internal standard C_{IS}:

$$I_A/I_{IS} = K(C_A/C_{IS}) \tag{14.30}$$

Thereafter, K can be used to calculate analyte concentration from the intensity ratio measured from the samples.

Section 14.5.4 describes a method in which the internal-standard line is excited in a radiolucent disk at the x-ray tube window and scattered from the specimens.

TABLE 14.2. Comparison of Ratio Methods[a]

Iron concentration,[b] mg/ml	Cu $K\alpha$ net intensity, counts/s	Ratio $I_{\mathrm{Cu}K\alpha}/I_{\mathrm{std}}$		
		Internal standard (2 mg Zn/ml)	Background ratio (0.6 Å)	Coherently scattered target line (W $L\beta_1$)
0	2117	1.045	3.261	3.541
5	1821	1.042	3.258	3.530
10	1628	1.040	3.247	3.522
15	1459	1.035	3.234	3.510
20	1308	1.032	3.221	3.503

[a] From Cullen (*C73*).
[b] Cu concentration in all solutions 2.0 mg/ml.

The effectiveness of an internal standard is evaluated by the degree of constancy of the ratio I_A/I_{IS} from specimens containing the same analyte concentration in matrixes having compositions varying over the expected range.

14.4.2. Selection of Internal-Standard Element

The basic requirements of the internal standard have already been given. In practice, the criterion is that the internal standard and analyte lines and absorption edges should have wavelengths as close as possible. Figure 14.9 shows four conditions with respect to the relative wavelengths of the spectral lines and absorption edges of the analyte A, one or two matrix elements M, and the prospective internal standard IS. In every case, the positions of the analyte and internal standard could be interchanged without altering the discussion below. Figure 14.10 shows the wavelengths of the K-absorption edges and $K\alpha$ and $K\beta$ lines of triads of adjacent elements Z + 1, Z, and Z − 1 in several regions of the periodic table. In this figure, element Z (solid lines) may be regarded as the analyte, and elements Z + 1 and Z − 1 (dashed lines) as its prospective internal standards.

In Figure 14.9, case A represents the ideal condition. The analyte and internal-standard lines are close and on the same sides of both matrix absorption edges shown, and the analyte and internal-standard absorption edges are close and on the same sides of both matrix lines shown. The analyte and internal-standard lines are absorbed and enhanced to substantially the same extent by each of the two matrix elements. Incidentally, if the two matrix lines and edges are sufficiently remote from the analyte line and edge, internal standardization may not be required. The condition in Figure 14.9A prevails for the $K\alpha$ lines of analyte elements of atomic

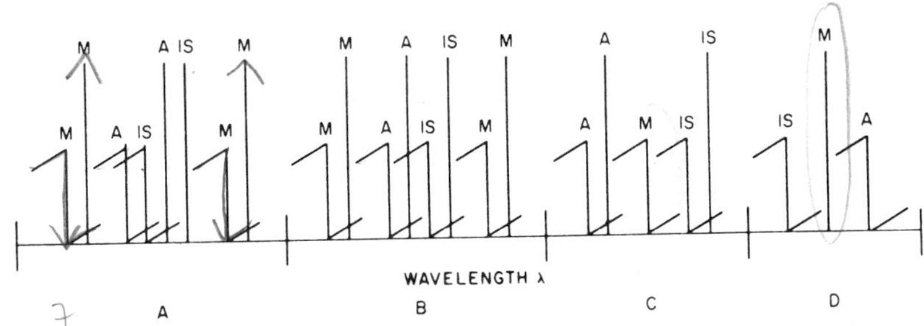

FIG. 14.9. Specific absorption-enhancement effects and the selection of an internal-standard element. The figure shows spectral lines and absorption edges of analyte A, matrix M, and prospective internal standard IS elements. Figure A is the ideal case, B is acceptable, C and D are unacceptable. In all four cases, interchange of the A and IS lines and/or edges would have no effect.

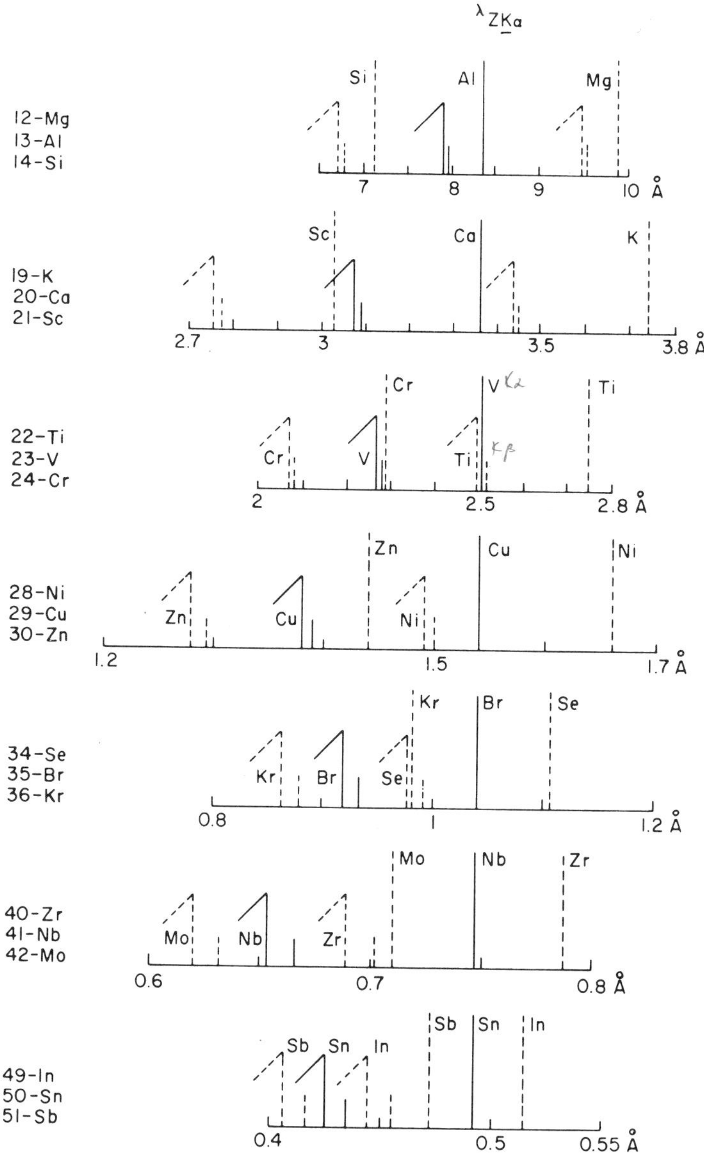

FIG. 14.10. Relationships among *K* spectral lines and absorption edges of adjacent elements in various regions of the periodic table. The long and short straight lines represent $K\alpha$ and $K\beta$ lines, respectively, the hooked lines *K*-absorption edges. The solid lines and edges represent the analyte.

number down to 23 (vanadium) and internal standards Z ± 1, as shown in Figure 14.10.

Case B is not as favorable, but is feasible, and many successful analyte–internal standard pairs are of this type. The same comments apply as for case A regarding the positions of the analyte and internal-standard lines and edges relative to the lines and edges of the matrix elements, and regarding the effect of each matrix element on the analyte and internal standard. However, here the analyte line falls between the analyte and internal-standard absorption edges, and the internal-standard edge falls between the analyte and internal-standard lines. Alternatively, the analyte and internal standard may be interchanged. One might expect case B to be unfavorable. However, difficulty is minimal because the internal-standard concentration is constant in all samples and standards and because the calibration function is based on the intensity ratio. Nevertheless, this condition should be avoided if case A is feasible. An example of case B is the use of Nb $K\beta$ as an internal standard for Mo $K\beta$ (*H6*). Figure 14.10 shows that the niobium *K*-absorption edge lies between these lines.

Case C is unacceptable: The matrix absorbs the analyte line strongly, the internal-standard line weakly. Interchanging of the analyte and internal standard would only reverse the effects and do no good. Similarly, case D is unacceptable. The matrix line is absorbed weakly by the internal standard, but strongly by the analyte, thereby enhancing the analyte line. Again, interchanging of the analyte and internal standard would do no good.

Note that the intensities of the analyte and internal-standard lines may be severely affected by the matrix. What is important is that they both be affected the same way (absorption or enhancement) and to substantially the same extent.

The selection requirements are usually best filled by an element having atomic number one or two above or below that of the analyte. Then the analyte and internal-standard lines can be of the same series and have about the same intensity for the same concentration; also, the analyte and internal standard have about the same mass-absorption coefficients for the primary and secondary x-rays. As already mentioned, and as is evident in Figure 14.10, this "rule" applies for analytes down to atomic number 23 (vanadium). Among the lighter elements, element Z − 1 strongly absorbs the $K\alpha$ line of elements having atomic number 22 (titanium) or lower, so element Z + 1 is preferable. The Z + 1 elements enhance the Z *K* lines, but the effect is relatively small in the low-*Z* region. Also, among the lighter elements, case B (Figure 14.9) is unavoidable. In Figure 14.10, for analytes calcium and aluminum, an absorption edge lies between the analyte and internal-standard lines regardless of whether element Z + 1 or Z − 1 is chosen as the internal standard.

In selecting the internal standard, one must avoid the introduction of new absorption-enhancement effects or of spectral-line interference. For example, in a three-component system A–B–C, a fourth element D may be the ideal internal standard for A, but it may strongly absorb or enhance B, and its lines may interfere with the optimal line of C.

Rather surprisingly, there are many successful applications of an L internal-standard line for a K analyte line and *vice versa*; for example, Y $K\alpha$ (0.83 Å) serves as an internal-standard line for U $L\alpha_1$ (0.91 Å). However, such selections are to be avoided if possible. An internal standard having L lines at about the same wavelength as an analyte K line would necessarily have much higher atomic number than the analyte and therefore very different absorption coefficients for the primary and secondary x-rays. An analogous appraisal can be made for an internal standard having K lines at about the same wavelength as an analyte L line. Also, for the same analyte and internal-standard concentration, K lines are 5–10 times as intense as the L lines. An internal-standard line should be chosen from a series different from that of the analyte line only when a satisfactory internal-standard element cannot be found having a line in the same series.

Of course, the less severe the absorption-enhancement effects are, the less critical are the requirements for the internal standard. It follows that the selection of the internal standard is also less critical for dilute and thin-layer specimens.

Müller [p. 191 in (*29*)] lists lines of 24 elements ranging from $_{15}$P to $_{94}$Pu and for each gives one or more internal standard lines that have been used with it and references to the original work.

14.4.3. Advantages and Limitations

The principal advantages of the internal-standard method are the following. It compensates absorption-enhancement and long-term instrument drift for all types of specimen. It partially compensates variations in density in powder and briquet specimens. It compensates density, evaporation, and bubble formation in liquids. Because the analyte and internal-standard lines are very close, their backgrounds are usually substantially the same and often need not be measured, in which case the peak-intensity ratio is used.

The principal disadvantages are the following. The method is not applicable to many types of sample, such as bulk solids, foils, and small fabricated parts. It is also not applicable to high analyte concentrations—above, say, 25%. Analysis time is greatly increased and the analysis is made much less convenient by the requirement for quantitative and thorough admixture of the internal standard and for measurement of a second line (and perhaps background) from each sample and standard. However, all measurements can be made simultaneously with a multichannel instrument. There is some loss of precision because of the additional measurement. The internal standard may be expensive; for example, the ideal internal standard for tantalum is hafnium; if large numbers of samples are to be analyzed, the cost of the highly tantalum-free hafnium would be substantial. Finally, although the method is fairly easily applied to one analyte in a multicomponent system, it is frequently impractical for more than one analyte in the same sample. However, multiple internal standardization is sometimes feasible: A single internal standard may serve for more than one analyte, or a mixture of two or more internal standards may be added to the specimens (*D40*). Some of these limitations are mitigated by the scattered internal-standard line method described in Section 14.5.4.

14.4.4. Considerations

The internal standard must be intimately and homogeneously mixed with the specimens. This is most easily accomplished with solutions and fusion products. The internal standard may be admixed with powders in several ways. The specimens may be dissolved, treated with the internal standard, then precipitated. Specimen and internal-standard powders may be mixed if they have substantially the same particle size and density; otherwise they must be mixed and ground, with or without an inert abrasive (such as silicon carbide or aluminum oxide), to a very small particle size. The powders may be ground dry or repeatedly to dryness under ether or ethanol (*A8*). The internal standard in solution may be added to the specimen powder and the resulting slurry mixed, dried, and ground (*C60*). Even greater problems arise in introduction of the internal standard to specimens supported on ion-exchange resins or membranes.

The internal standard should be added at a concentration level equivalent to about the middle of the expected analyte concentration range. It is this requirement that limits the internal-standard method to analyte concentrations up to ~25% at most. The effectiveness of the internal standard in compensating absorption-enhancement effects increases as analyte concentration decreases, or as dilution increases.

For greatest convenience and precision, it is preferable to measure analyte and internal-standard lines—and their backgrounds, if necessary—simultaneously. This is no problem with multichannel spectrometers, but is impossible with unmodified single-channel instruments. Section 5.5.1 presents several ways to modify single-channel instruments for multichannel operation.

On single-channel instruments, the method is more precise the shorter the lapsed time between measurement of the analyte and internal-standard lines.

Interaction of the analyte and internal standard has already been mentioned (Section 14.4.2, Figure 14.9B). Actually, this condition probably always exists to a certain extent. For example, even if the analyte and internal-standard $K\alpha$ lines both lie on the same side of both their absorption edges, one or the other of their $K\beta$ lines is likely to lie between the edges and cause a slight interaction. In Figure 14.10, this condition is seen to be common. The probability of interaction is even greater with the more complex L spectra.

Hakkila *et al.* (*H6*) provide an example of how to deal with variations of the slope and intercept of the calibration curve caused by spectral interference of a matrix element line with the analyte or internal-standard line. They encountered such spectral interference from uranium L lines in the determination of zirconium and molybdenum in the presence of uranium, using niobium as the internal standard. For zirconium, the curve of Zr $K\alpha$/Nb $K\alpha$ intensity ratio R *versus* zirconium concentration is linear. As uranium concentration changes, the slope changes, but the intercept on the R axis remains constant. Zirconium concentration is given by

$$C_{\mathrm{Zr}} = \frac{R - b}{a - c(d - 100)} \tag{14.31}$$

where C_{Zr} is zirconium concentration (mg/10 ml); a is the slope of the curve for solutions containing 100 mg U/10 ml; b is the constant intercept on the R axis; c is the change in slope per 1 mg U/10 ml; and d is uranium concentration (mg/10 ml). For molybdenum, the curve of Mo $K\beta$/Nb $K\beta$ intensity ratio R *versus* molybdenum concentration is also linear. As uranium concentration changes, the slope remains constant, but the intercept on the R axis changes. Molybdenum concentration is given by

$$C_{\mathrm{Mo}} = \frac{R - b + c(100 - d)}{a} \tag{14.32}$$

where C_{Mo} is molybdenum concentration (mg/10 ml); a is the constant slope of the curve; b is the intercept on the R axis for solutions containing

100 mg U/10 ml; c is the change in intercept per 1 mg U/10 ml; and d is uranium concentration (mg/10 ml).

14.4.5. Special Internal-Standardization Methods

14.4.5.1. Single-Standard Internal-Standard Method (*44*)

A weighed portion of sample X is treated with a weighed portion of a substance containing internal-standard element IS, giving a treated sample X′ having internal-standard concentration $C_{IS,X'}$. A standard S is prepared having equal concentrations of analyte and internal-standard elements in a matrix similar to that of the sample. Intensities of the analyte line I_A and internal-standard line I_{IS} are measured from X′ and S. Then analyte concentration in the untreated sample is given by

$$C_{A,X} = C_{IS,X'} D \left(\frac{I_{IS}}{I_A}\right)_S \left(\frac{I_A}{I_{IS}}\right)_{X'} \tag{14.33}$$

where D is the dilution factor, weight of X′/weight of X.

14.4.5.2. Variable Internal-Standard Method

Strasheim and Wybenga (*S66*) compensated absorption-enhancement effects in the determination of iridium, platinum, and gold by a novel *variable internal-standard* method. Intensities of the Pt $L\alpha_1$ line and its background were measured from solutions containing platinum in concentration 400 μg/ml in pure 10% hydrochloric acid and in various concentrations of various other matrix constituents. Ratios were calculated of Pt $L\alpha_1$ intensity from 10% HCl to Pt $L\alpha_1$ intensity from each of the other matrixes M: $I_{Pt,HCl}/I_{Pt,M}$. Ratios were calculated of background intensity from 10% HCl to background intensity from each of the other matrixes: $I_{B,HCl}/I_{B,M}$. A curve was plotted of the function $I_{Pt,HCl}/I_{Pt,M}$ *versus* $I_{B,HCl}/I_{B,M}$; this curve was linear. An analysis then consisted of measurements of Pt $L\alpha_1$ and background intensities from the samples and from a platinum standard in pure 10% hydrochloric acid. The ratio $I_{B,HCl}/I_{B,M}$ was calculated for each sample and applied to the curve to derive $I_{Pt,HCl}/I_{Pt,M}$, which was then used to correct the Pt $L\alpha_1$ intensity measured from the sample. The analytical platinum concentration was then obtained from the corrected Pt $L\alpha_1$ intensity for the sample and the Pt $L\alpha_1$ intensity measured from the standard.

The platinum already present in the samples was then used as a variable internal standard for determination of iridium and gold. Intensity ratios

Ir $L\alpha_1$/Pt $L\alpha_1$ and Au $L\alpha_1$/Pt $L\alpha_1$ were measured, and iridium and gold concentrations determined relative to platinum. Ruthenium, rhodium, and palladium were determined the same way (*W21*). Palladium was determined the same way as platinum, then used as a variable internal standard for determination of ruthenium and rhodium.

14.4.6. Other Standardization Methods

14.4.6.1. Internal Control-Standard Method

An internal control standard is similar to an internal standard in some ways. It is added in constant concentration to all samples and standards, unless a suitable element is already present in the samples in appropriate concentration. The calibration function is the intensity ratio of the analyte and internal control-standard lines *versus* analyte concentration. However, the control standard does not necessarily have the same excitation, absorption, and enhancement properties as the analyte in the specimen matrix, and it may serve for several analytes in the same specimen. The method has been applied to solutions where it compensates variations in volume, temperature, density, specimen plane, acid matrix, and, to a smaller degree, composition of the original matrix.

The method has been evaluated by Mitchell and O'Hear (*M44*) for the rapid, accurate, simultaneous determination of chromium, manganese, iron, cobalt, nickel, and copper in chromium and manganese alloys, Alnico, refractory alloys, ores, and slags. The sample concentration was 1 g/100 ml; analyte concentration was 0.1–99.9% of the original sample. The samples were put into solution by various acid and/or fusion treatments. All analyte lines and the internal control-standard line were measured simultaneously on an ARL X-Ray Quantometer. For each analyte in each original sample matrix, one set of standards suffices for all acid or fusion matrixes.

The method has been used to compensate the effect of pits, cracks, and gas holes in the surfaces of specimen disks cut from ingots of crude copper (*Y1*). The Cu $K\alpha$ line is used as an internal control standard for determination of antimony and lead. The intensity ratio of the analyte and Cu $K\alpha$ lines is plotted *versus* analyte concentration.

14.4.6.2. Internal Intensity-Reference Standard Method

In the determination of manganese in gasoline, Jones (*J25*) compensated variable carbon/hydrogen ratio by mounting an iron rod in the liquid-specimen cell (Figure 17.10) to serve as an internal intensity-reference standard. The calibration function was Mn $K\alpha$/Fe $K\alpha$ intensity ratio *versus* manganese concentration. Gunn (*G39*) applied the same method in the determination of lead in gasoline by using a platinum disk in the cell and the Pb $L\alpha_1$/Pt $L\alpha_1$ intensity ratio. Ideal compensation is realized with the disk at such a depth that the intensity ratio is the same for the same lead concentration in liquids having different carbon/hydrogen ratios. Gunn gives a procedure for determination of optimal depth. By selection of a true internal-standard element to be the internal intensity-reference standard, a degree of true internal standardization is realized without the inconvenience of making a quantitative addition to all specimens.

14.4.6.3. External-Standard Method

An external standard, or external intensity-reference standard, is a specimen from which some intensity value is measured to be ratioed with the analyte-line intensity, principally to compensate long-term instrument drift. The standard may be one of the samples or standards retained for this purpose, or it may be any stable specimen not necessarily bearing any relation to the samples to be analyzed (*L30*)—such as a piece of copper alloy or nickel alloy. The measured intensity may be emitted or scattered from the standard. The standard may be measured once after each sample and calibration standard, or, more likely, from time to time during a long series of measurements. A particularly interesting example of the external-standard method is reported by Hirokawa and his colleagues, who used the emission lines from specimen-mask plates of zinc or lead as external-standard lines in the determination of chromium, nickel, and tungsten in steel (*H46*).

Another way to use an external standard is to set the excitation parameters to reproduce the intensity emitted or scattered from it. For example, in the determination of aluminum, silicon, and iron in clays and bauxites, day-to-day instrument drift was compensated by adjustment of the x-ray tube current and/or potential to reproduce intensity values established at the time of the original calibration from selected aluminum and iron plates (*S67*).

14.5. STANDARDIZATION WITH SCATTERED X-RAYS

Scattered x-rays, formerly regarded only as a nuisance (*J20*), provide the basis for several extremely convenient and effective means for dealing with absorption-enhancement effects as well as particle-size and surface-texture effects, packing density, and instrument errors. The basis of the applicability to absorption effects is that the emitted analyte line and diffusely scattered intensities are both affected by the mass-absorption coefficient of the specimen. The scattered wavelength may be the background at or adjacent to the analyte line, a wavelength in the scattered continuum "hump," or a coherently or incoherently scattered target line. These scatter methods are especially advantageous for specimens having low effective atomic number, for which scattered intensity is relatively high and can be measured with good statistical precision in relatively short counting time. Although scatter methods may reduce enhancement effects in some cases, in general, they are effective principally for absorption effects. A very comprehensive mathematical treatment of scatter, especially the coherent-to-incoherent scatter ratio, is given by Ebel and Ebel (*E3*). Other x-ray analytical methods based on scatter are described in Sections 20.1.4.3 and 20.2.

14.5.1. Background-Ratio Method (*A19*, *K2*)

For a given minor or trace analyte A, the intensity ratio of the analyte line and scattered background I_A/I_B remains substantially constant as the mass-absorption coefficient of the specimen varies—that is, as the matrix varies. Andermann and Kemp (*A19*) were the first to present the theory of this phenomenon and demonstrate its great value in compensating absorption, instrumental, density, and particle-size effects. Of course, the background wavelength cannot be enhanced by the matrix, so the method is ineffective against enhancement.

The total scattered intensity at a specified wavelength λ may be represented by

$$I_{B,\lambda,\mathrm{sc}} \propto \frac{I_{\mathrm{coh}}}{2(\mu/\varrho)_\lambda} + \frac{I_{\mathrm{incoh}}}{(\mu/\varrho)_\lambda + (\mu/\varrho)_{\lambda-\Delta\lambda}} \tag{14.34}$$

where I_{coh} is the coherent (Rayleigh) scattering intensity per atom; I_{incoh} is the incoherent (Compton) scattering intensity per atom; $(\mu/\varrho)_\lambda$ is the mass-absorption coefficient of the specimen at wavelength λ; and $(\mu/\varrho)_{\lambda-\Delta\lambda}$ is the mass-absorption coefficient of the specimen at the incident wavelength $\lambda - \Delta\lambda$ that undergoes increase in wavelength $\Delta\lambda$ upon scattering incoherently [Equation (2.38)].

The emitted analyte-line intensity may be represented by

$$I_A \propto \frac{1}{(\mu/\varrho)_{\lambda_{\mathrm{pri}}} + (\mu/\varrho)_{\lambda_A}} \tag{14.35}$$

where $(\mu/\varrho)_{\lambda_{\mathrm{pri}}}$ and $(\mu/\varrho)_{\lambda_{\mathrm{A}}}$ are the mass-absorption coefficients of the specimen for the primary and emitted analyte-line radiation, respectively.

All quantities in Equations (14.34) and (14.35) vary with atomic number Z. Thus, a change in matrix composition affects both the numerator and denominator of Equation (14.34), but only the denominator of Equation (14.35). The influence of the effective atomic number of the specimen on a specified analyte-line and scattered wavelength can be estimated from the following proportionalities:

$$I_{B,\lambda,\mathrm{sc}} \propto (I_{\mathrm{coh}} + I_{\mathrm{incoh}})/(\mu/\varrho) \tag{14.36a}$$

$$I_{\mathrm{coh}} + I_{\mathrm{incoh}} \propto Z^{(1-2)} \tag{14.36b}$$

$$\mu/\varrho \propto Z^{4} \tag{14.36c}$$

$$I_{\mathrm{A}} \propto Z^{-4} \tag{14.36d}$$

Therefore,

$$I_{B,\lambda,\mathrm{sc}} \propto Z^{-(3-2)} \tag{13.36e}$$

and

$$I_{\mathrm{A}}/I_{B,\lambda,\mathrm{sc}} \propto Z^{-(1-2)} \tag{14.36f}$$

A comparison of Equations (14.36f) and (14.36d) shows that the ratio is much less dependent on atomic number—and therefore on matrix composition—than analyte-line intensity. A more rigorous mathematical treatment is given in (*T4*).

Equation (14.36f) applies when the analyte-line and scattered wavelengths are about the same. Actually, scatter and absorption are also wavelength dependent, and it may be possible to select a scattered wavelength for which the ratio $I_{\mathrm{A}}/I_{B,\lambda,\mathrm{sc}}$ is substantially independent of matrix.

Analyte-line and scattered intensities are similarly affected by specimen position and x-ray tube current. If the analyte line is excited principally by the primary continuum, rather than by the target lines, analyte-line and scattered intensities are also similarly affected by x-ray tube potential. Thus, the ratio compensates variations in these parameters.

The background-ratio method is similar to the internal-standard method, except that the calibration function is $I_{\mathrm{A}}/I_{\mathrm{B}}$ *versus* analyte concentration. Here, I_{A} may be peak or net analyte-line intensity, and I_{B} may be the intensity of: (1) the scattered background at the analyte-line peak measured from a blank or in any of the ways described in Section 7.3.3; (2) the scattered background measured adjacent to the peak; or (3) a specified wavelength in the scattered continuum "hump." The denominator of the ratio may also be the intensity of a coherently or incoherently scattered target line (Section 14.5.3). A wavelength in the hump or a scattered target line is used when: (1) the background at or adjacent to the peak is too low to be measured with acceptable statistical precision in a reasonable counting time; or (2) such a wavelength gives better standardization than the background. The background-ratio method is sometimes described as "internal standardization with scattered x-rays," and, in a sense, this is a valid description, because the scattered intensity is dependent on the matrix [Equation (14.34)].

Figure 14.11 shows the relative intensities of emitted Ni $K\alpha$ and scattered 0.6-Å x-rays measured from minerals having the same nickel concentration but 10–60% iron. Both individual intensities are strongly dependent on iron concentration, but the curves are both hyperbolic and more or less parallel, indicating that the Ni $K\alpha$/0.6 Å intensity ratio should be about constant with change in iron concentration. This prediction is strikingly confirmed in Figure 14.12. The data points on the Ni $K\alpha$ curve are widely scattered. Plotting of the same data as the intensity ratio of Ni $K\alpha$ and 0.6-Å scatter reduces the scatter in the points to insignificance.

Figure 14.13 strikingly shows the effectiveness of the method in compensating variations in x-ray tube potential and current, specimen plane, and particle size. Figures 14.11 and 14.12 are examples of an intensity ratio of an analyte line (Ni $K\alpha$, 1.66 Å) and a wavelength (0.6 Å) from the scattered continuum hump. Figure 14.13 is an example of an intensity ratio of an analyte line (Cu $K\alpha$, 1.54 Å) and its adjacent background (~1.4 Å). Table 14.2 gives some typical data for the scattered continuum method.

The background-ratio method, or, sometimes, simply the scattered intensity measured at a certain wavelength, is used to compensate varia-

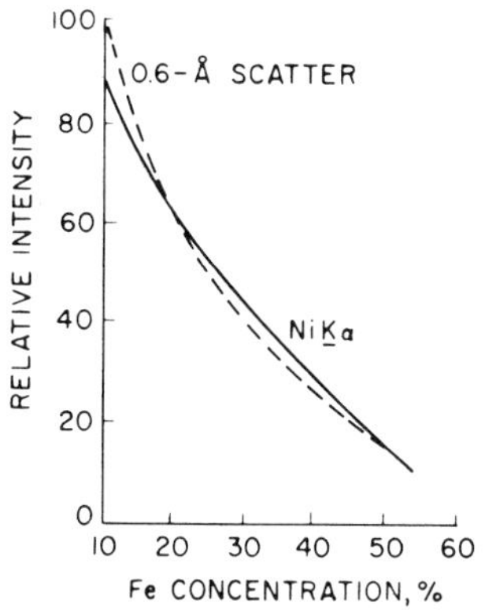

FIG. 14.11. Approximate constancy of relative intensities of Ni $K\alpha$ and 0.6-Å scattered continuum from specimens having the same nickel concentration but different iron concentrations. [G. Andermann and J. W. Kemp, *Analytical Chemistry* **30**, 1306 (1958); courtesy of the authors and the American Chemical Society.]

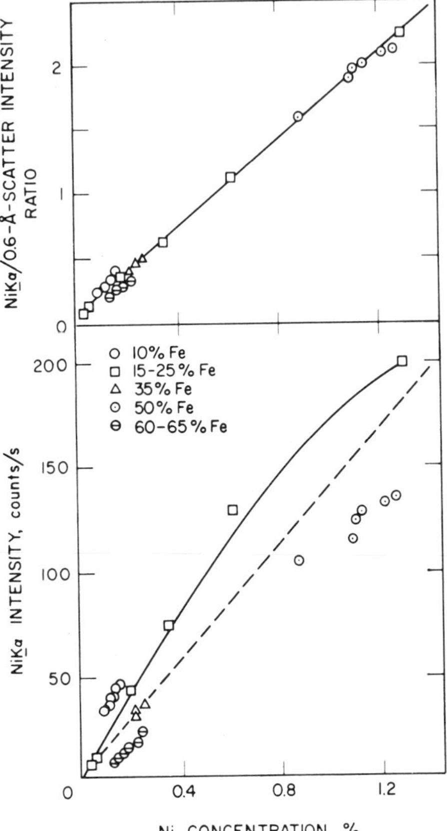

FIG. 14.12. Effectiveness of the background-ratio method in correcting for absorption-enhancement effects of iron on Ni $K\alpha$ intensity from iron–nickel minerals. [G. Andermann and J. W. Kemp, *Analytical Chemistry* **30**, 1306 (1958); courtesy of the authors and the American Chemical Society.]

tions in the carbon/hydrogen ratio in the determination of minor and trace analytes in hydrocarbons. However, the scattered target-line ratio method (Section 14.5.3) is preferable for this application. The background-ratio and scattered-intensity methods are also used to compensate slurry density in fluid specimens (Section 14.5.6) and packing density in powder and briquet specimens, especially in on-line process-control instruments.

Compared with the internal-standard method, the background-ratio method has the advantages that no addition need be made to the specimens, that it is applicable to specimens having practically any physical form, and that it is applicable to many analytes in the same specimen. The method is especially applicable to minor and trace analytes in low-Z matrixes, where the scattered intensity is high and can be measured precisely in short counting times. The method is largely ineffective against enhancement. In specific cases, it may not be as effective in compensating absorption effects as the internal-standard method, where standardization is effected by an emission line originating *in* the specimen. Also, all scattered wavelengths are not equally effective in the ratio, and principles for selection of the optimal wavelength are difficult to establish, so the choice is usually left to trial and error. No absorption edge should lie between the analyte line and the scattered standard x-rays, including that of the analyte itself. Thus, in this method, the analyte itself may actually constitute an interfering element for scattered wavelengths shorter than its own line, unless analyte concentration is low. However, there are successful analyses in which this condition does exist.

Deutsch (*D16*) has developed a method based on background standardization for determination of bromine in water, which, once calibrated, requires no standards. A series of standards S was prepared containing up to 50 μg Br/ml in pure water. From each, intensity was measured at the Br $K\alpha$ peak $I_{P,\mathrm{S}}$ and background $I_{B,\mathrm{S}}$; the same two measurements were made on pure water, $I_{P,0}$ and $I_{B,0}$. For each standard, the following calculations were made: (1) normalization factor $F_{\mathrm{S}} = I_{B,\mathrm{S}}/I_{B,0}$; (2) normalized peak intensity $I_{P,\mathrm{S}(n)} = I_{P,\mathrm{S}}/F$; and (3) normalized net intensity $I_{L,\mathrm{S}(n)} = I_{P,\mathrm{S}(n)} - I_{P,0}$. A calibration curve was established from the standard data: $I_{L,\mathrm{S}(n)}$ *versus* Br concentration; this curve was linear and passed through the origin. Thereafter, an analysis consisted in making the same two measurements on the sample(s) $I_{P,\mathrm{X}}$ and $I_{B,\mathrm{X}}$ and on pure water,

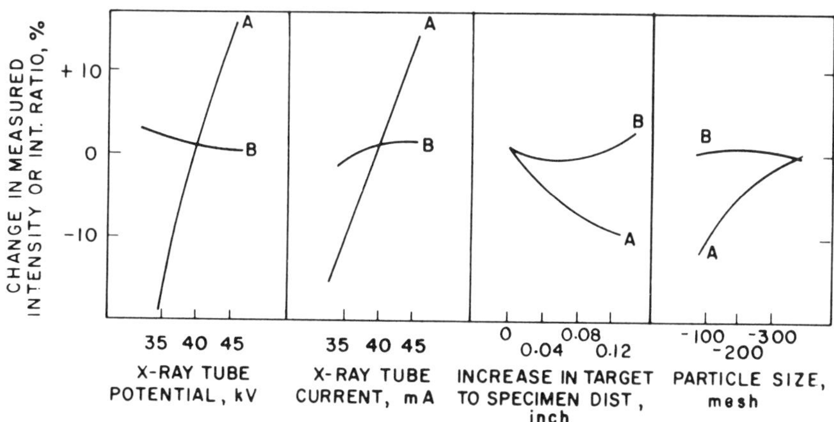

FIG. 14.13. Effectiveness of the background-ratio method in correcting for instrumental variables and the particle-size effect. Curves *A* show variation of Cu $K\alpha$ intensity. Curves *B* show variation of the intensity ratio of Cu $K\alpha$ and its adjacent background. [G. Andermann and J. W. Kemp, *Analytical Chemistry* **30**, 1306 (1958); courtesy of the authors and the American Chemical Society.]

calculating F_X, $I_{P,X(n)}$, and $I_{L,X(n)}$ for each, and reading the analytical concentration from the previously established curve. The method was applied to sea waters and brines.

14.5.2. Graphic Method

Kalman and Heller (*K2*) developed a graphic background correction method for determination of minor and trace analytes in low-Z matrixes. For each analyte, a set of calibration curves is established of analyte-line intensity I_A *versus* background intensity I_B, each curve for a different analyte concentration. All curves in the set are linear and have a more or less common origin. The curves were established from measurements on synthetic standards having matrixes consisting of various proportions of aluminum oxide (Al_2O_3), calcium carbonate ($CaCO_3$), sodium chloride (NaCl), and iron(III) oxide (Fe_2O_3) to give various scattered intensities. An analysis consists in measurement of I_A and I_B from the sample and location of their common point on the set of curves for the appropriate analyte. Analyte concentrations not represented by a curve are dealt with by interpolation.

14.5.3. Scattered Target-Line Ratio Method

This method is a special case of the background-ratio method. The calibration function is the intensity ratio of the analyte line and a scattered x-ray tube target line (*B76*, *C70*, *L41*, *S62*). Either the coherently (*C70*) or incoherently (*B76*) scattered line may be used. If only, say, the coherently scattered line is measured, Equation (14.34) reduces to

$$I_{B,\lambda,\mathrm{sc}} \propto I_{\mathrm{coh}}/[2(\mu/\varrho)_\lambda] \tag{14.37}$$

Coherent scatter I_{coh} is proportional to Z^2, and μ/ϱ is proportional to Z^4, so that $I_{B,\lambda,\mathrm{sc}} \propto Z^{-2}$. Clearly, the scattered target-line method is most effective if the target line is a good internal-standard line for the analyte line. The closer the target line lies to the analyte absorption edge, the more efficiently it enhances excitation of the analyte line. If the scattered target line has an incoherent component, the wavelength of this component more closely approximates the incident target line wavelength the longer the incident wavelength is. Finally, the closer the coherently or incoherently scattered target line lies to the analyte line, the more nearly the two lines undergo the same absorption, particle-size, and other effects in the specimen. The limited number of x-ray tube targets available commercially and convenience considerations often preclude the optimal choice. The next best alternative is the target having a strong line as close to the analyte line as possible. However, when secondary radiators are used for specimen excitation (Section 8.3.1), optimal selection becomes feasible. The target-line ratio method, like the background-ratio method, is largely ineffective against matrix enhancement. Table 14.2 gives some typical data for this method. The method is very effective in compensating variations in carbon/hydrogen ratio in determination of minor and trace analytes in hydrocarbons (*S29*).

Shenberg, Haim, and Amiel (*S25*) point out that the target-line ratio method may fail when the specimen contains one or more elements having absorption edges at wavelengths longer than that of the analyte line. Some analyte-line intensity is lost in excitation of these elements. Some target-line intensity is also lost in this way, but rarely do the two losses compensate, so the ratio is affected. An interesting example of this phenomenon occurs in the determination of uranium in solution, using a ^{241}Am–I radioisotope source-target for excitation. The U $L\alpha_1$/I $K\alpha$ and U $L\beta_2$/I $K\alpha$ intensity ratios are linear over a very wide range of uranium concentration (0.1–1000 mg U/ml), but the ratios with U $L\beta_1$ and $L\gamma_1$ lines are not. The $L\alpha_1$ and $L\beta_2$ lines lie on the long-wavelength side of the U LIII absorption edge; the $L\beta_1$ and $L\gamma_1$ lines are associated with the U LII edge and lie on the short side of the LIII edge. Thus, in this analysis, uranium can actually interfere with itself.

Burkhalter (*B91*) avoided this type of interference in the determination of silver by choosing a secondary target, tellurium, whose Compton-scattered Te $K\alpha$ line has wavelength just longer than that of the Ag K absorption edge. Thus, the target and analyte lines are close and subject to the same matrix effects. This method is applicable when a suitable target is feasible and its scattered line can be resolved from the analyte line.

Shenberg *et al.* (*S25*) developed a more general method of dealing with this type of interference and applied it to determination of 0.5–25% copper in the presence of 0.5–25% iron in solutions and powdered ores. A ^{241}Am–As radioisotope source-target was used. The procedure for each sample follows.

1. Iron concentration is estimated from a curve of Fe $K\alpha$/As $K\alpha$ intensity ratio *versus* Fe concentration established from pure iron standards. This curve is linear and substantially independent of copper concentration.

2. This iron concentration is applied to a curve of $(\text{Cu } K\alpha/\text{As } K\alpha)_{C_{\text{Fe}}}/(\text{Cu } K\alpha/\text{As } K\alpha)_{C_{\text{Fe}}=0}$ intensity ratio *versus* Fe concentration established from standards containing the same copper concentration with various iron concentrations C_{Fe} and in the absence of iron, $C_{\text{Fe}} = 0$. This curve is linear.

The appropriate correction quotient is read from the curve. These quotients vary with iron concentration, but, at a specified iron concentration, are independent of copper concentration.

3. This quotient is used to correct the Cu $K\alpha$/As $K\alpha$ intensity ratio measured from the sample.

4. Finally, the corrected ratio is applied to a curve of Cu $K\alpha$/As $K\alpha$ intensity ratio *versus* Cu concentration established from pure copper standards to obtain the analytical concentration.

Steele (*S55*) has developed a method which permits detection and determination of minor and trace analytes whose spectral lines are present in the primary x-ray beam—the target element and its contaminants, principally iron, nickel, copper, silver, and tungsten. For each of several specimen materials, a coefficient of coherent scattering is calculated from the atomic weights, weight fractions, atomic scattering factors, and mass-absorption coefficients of the elements present. The target- or contaminant-line intensity coherently scattered by the specimen is proportional to this scattering coefficient. The analytical procedure consists in measurement of the analyte-line intensity from the sample and from an analyte-free reference standard. The contribution of the scattered primary line to the total measured intensity from the sample is the product of the (scattered) analyte-line intensity from the reference standard and the ratio of the scattering coefficients of the sample and standard. This is then subtracted from the intensity measured from the sample.

14.5.4. Scattered Internal-Standard Line Method

This method, described by Vos (*V18*), combines the principles of the true internal-standard and scattered target-line ratio methods. A true internal-standard line for the specified analyte is used (Section 14.4.2). However, instead of adding the internal-standard element quantitatively to all samples and standards, a radiolucent filter disk is prepared by briqueting an oxide or other suitable compound of this element with plastic powder. This disk is placed at the x-ray tube window. The internal-standard line is excited by the primary beam passing through the disk, and scattered from the samples and standards. The intensities of the emitted analyte line and scattered internal-standard line are measured from each sample and standard, and their ratio used in the usual way.

It is relatively convenient to use a different disk for each analyte, or two or more internal-standard lines can be emitted by the same disk, provided that interference can be avoided. The disks are more or less permanent and can be used repeatedly.

The method is applicable largely to light matrixes, which include fusion products, solutions, and supported specimens, where the scattered intensity is high enough to give good statistical precision in acceptable counting times.

The technique minimizes many of the limitations of the added internal-standard method. The time and inconvenience of adding the internal standard to all samples and standards are eliminated. It is feasible to use the optimal internal-standard line for each analyte in multielement specimens. The technique permits application of the internal-standard method to specimen types to which it is impractical to add an internal standard—bulk solids, sheet and foil, small fabricated parts, etc. Internal-standard material is not consumed. The method improves on the scattered target-line ratio method in that it is convenient to use the optimal internal-standard line for each analyte in a multielement specimen.

14.5.5. Ratio of Coherent to Incoherent Scattered Intensities

For a given wavelength λ, both $(\mu/\varrho)_\lambda$ and the coherent-to-incoherent (Rayleigh-to-Compton) scatter ratio are related to atomic number, and therefore to each other. For light matrixes (*C20*),

$$I_{\mathrm{A}} = \frac{KC_{\mathrm{A}}}{(\mu/\varrho)_{\lambda_{\mathrm{pri}}} + (\mu/\varrho)_{\lambda_{\mathrm{A}}}} \tag{14.38}$$

where I_{A} and C_{A} are analyte-line intensity and concentration (wt fraction); $(\mu/\varrho)_{\lambda_{\mathrm{pri}}}$ and $(\mu/\varrho)_{\lambda_{\mathrm{A}}}$ are mass-absorption coefficients for the primary and analyte-line x-rays, respectively; and K is a constant. To a good approximation, this equation reduces to

$$I_{\mathrm{A}} \approx \frac{K'C_{\mathrm{A}}}{(\mu/\varrho)_{\lambda_{\mathrm{A}}}} \tag{14.39}$$

Ryland (*R38*) has shown that for light matrixes, μ/ϱ is a linear function of the ratio of the intensities of the coherently (Rayleigh) I_{R} and incoherently (Compton) I_{C} scattered target lines; that is,

$$(\mu/\varrho)_{\lambda_{\mathrm{A}}} \propto (I_{\mathrm{R}}/I_{\mathrm{C}}) = R \tag{14.40}$$

Then

$$I_{\mathrm{A}} \approx K''C_{\mathrm{A}}/R \tag{14.41}$$

Ryland has based a very successful absorption correction method on this equation. Analyte-line intensity is measured from the sample I_X and from a single standard I_S having known analyte concentration C_S in a matrix similar to that of the sample. Rayleigh- and Compton-scattered target-line intensities are measured from both the sample and standard and used to calculate R_X and R_S [Equation (14.40)]. Analyte concentration in the sample is then given by

$$C_X = C_S\left(\frac{I_X}{I_S}\right)\left(\frac{R_X}{R_S}\right) \tag{14.42}$$

Ryland applied the method to the determination of elements having atomic number 13–82 (aluminum to lead) in samples having average atomic number 5–10 in the form of briquets, glass disks, and water and organic solutions. Carman (*C26*) applied a similar technique to heavy-element matrixes diluted with polyethylene; he used calibration curves of $I_X R_X$ *versus* analyte concentration.

Reynolds (*R8*) used a somewhat different approach. I_R and I_C are measured from a series of standards having known or readily measured or calculated absorption coefficients $(\mu/\varrho)_\lambda$ at some wavelength near the center of the wavelength range of the spectral lines of analytical interest. Reynolds chose 0.9 Å. The data were plotted in a graph of $(\mu/\varrho)_\lambda$ *versus* R, and gave a linear curve. Analyses consisted in the measurement of analyte-line intensity from the sample I_X and from a single standard having known analyte concentration C_S, known $(\mu/\varrho)_\lambda$, and a matrix similar to that of the sample. I_R and I_C are measured from the sample, and R_X is calculated and applied to the curve to derive $(\mu/\varrho)_\lambda$. Analyte concentration in the sample is then given by

$$C_X = C_S\left(\frac{I_X}{I_S}\right)\frac{(\mu/\varrho)_{\lambda,X}}{(\mu/\varrho)_{\lambda,S}} \tag{14.43}$$

Later, Reynolds (*R8*) abandoned the ratio method and measured only the incoherently scattered target line I_C, but, except for use of I_C for R, the method is the same. The ratio method was abandoned because of difficulty in measurement of I_R due to diffraction effects from his highly crystalline powdered mineral samples. Compton-scattered radiation, being incoherent, has no fixed phase relationship with the scattering centers, or between incident and emitted radiation, and is free from diffraction effects. Ryland's samples—solutions, glasses, organic solids—are free from diffraction effects, thereby accounting for her success with the ratio method.

Ryland used a chromium-target x-ray tube; Reynolds used molybdenum. Carman used a tungsten–chromium dual-target tube; the tungsten and chromium targets were used for high-Z and low-Z elements, respectively.

These techniques, in addition to providing effective absorption correction, may be used with a single standard and without calibration curves.

14.5.6. Slurry-Density Method

X-ray spectrometric analyses of powders and briquets are limited to thin surface layers and may be seriously influenced by inhomogeneity and particle size. In such cases, if it is difficult or inconvenient to fuse or dissolve the powder, it may be advantageous to analyze it as a slurry. In on-line process control, slurries are one of the most common specimen forms. A specimen cell for laboratory analysis of slurries is described in Section 17.1.4.7. In process control, the cells are fitted for continuous flow of continuously agitated slurries.

In the analysis of slurries, variations in the slurry density, or "pulp density," must be corrected for. In laboratory analyses, this is often easily done by appropriate specimen preparation or by use of internal standardization. However, in on-line process control, other methods are required. One such method is that of Smallbone (*S40*), which involves scattered x-rays.

A series of slurries is prepared having the highest expected slurry density, say 2 g/cm^3, and covering the expected concentration ranges of all analytes. A portion of each of these slurries is diluted to give a series of slurries having the lowest expected slurry density, say 1.5 g/cm^3. Other portions of the original slurries are diluted to give series having intermediate densities, say 1.6, 1.7, 1.8, and 1.9 g/cm^3. From each of these many slurries, analyte-line intensities are measured for all analytes, and scattered continuum intensity is measured at some appropriate wavelength. In a simultaneous process-control instrument (Section 9.3.3), all these intensities are measured from a specimen at once. These measured intensities are used to establish density-correction and calibration curves as follows.

The *slurry-density correction curve* (Figure 14.14A) correlates scattered intensity and slurry density or solids content, and is used for slurry-density corrections for all analytes in this type of slurry. If the dry density of the suspended solid is known, the density axis can be calibrated in terms of solids content. Small variations in dry density do not seriously affect the

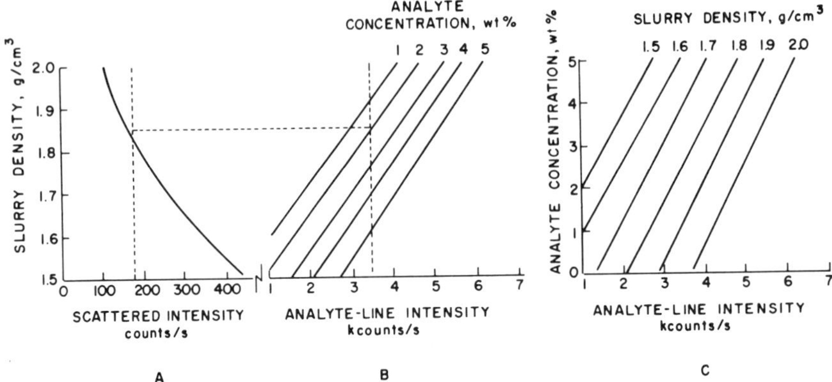

FIG. 14.14. Correction of analyte-line intensity for slurry density. A. Slurry-density correction curve. B. Element-dilution curve. C. Analytical calibration curve. The dashed lines in A and B show a graphic correction method (see text).

curve. However, large variations—say from high lead oxide (PbO) to high titanium oxide (TiO_2), densities 9.5 and 4.3 g/cm³, respectively—would necessitate different correction curves to reduce the effect of specimen matrix. The lower the dry density, the farther to the right the curve would lie, that is, the higher the scattered intensity would be.

The *element-dilution curves* (Figure 14.14B) correlate analyte-line intensity and slurry density for specified analyte concentrations. Of course, a set of curves is required for each analyte. The *analytical calibration curves* (Figure 14.14C) correlate analyte-line intensity and concentration for specified slurry densities. These curves are derived from the element-dilution curves, and, here again, a set of curves is required for each analyte.

The slurry-density and element-dilution curves may be set up to permit rapid graphic analysis, as shown by the dashed lines in A and B of Figure 14.14. Assume that the analyte-line and scattered intensities are 3500 and 175 counts/s, respectively. The intersection of 175 counts/s and the density curve indicates a slurry density of 1.85 g/cm³. The intersection of this density and 3500 counts/s on the dilution curves indicates a concentration of 2%. The same result is obtained from the intersection of 3500 counts/s and a slurry density of 1.85 g/cm³ on the calibration curve (Figure 14.14C). Of course, in all these sets of curves, values not actually falling on a curve are dealt with by interpolation.

Scattered-intensity methods for determination of dry mass of biological and medical specimens are described in Section 20.2.3.

14.6. MATRIX-DILUTION METHODS

14.6.1. Principles

The term *matrix-dilution methods* is applied to a group of methods that: (1) correct for absorption-enhancement effects by leveling the absorption coefficients of the samples and standards to a common value determined by an added diluent; and/or (2) correct for inhomogeneity and particle-size effects by dissolution or fusion of the specimens. In general, analyte-line intensities from diluted specimens are substantially proportional to analyte concentrations, and the calibration curves are substantially linear.

The efficiency of the primary beam of effective wavelength λ_{pri} for exciting analyte line λ_L is given by Equation (4.18):

$$\frac{I_L}{I_{0,\lambda_{\text{pri}}}} = P_{\text{A}} \frac{C_{\text{A}}(\mu/\varrho)_{\text{A},\lambda_{\text{pri}}}}{\sum C_i[(\mu/\varrho)_{i,\lambda_{\text{pri}}} + A(\mu/\varrho)_{i,\lambda_L}]}$$

where:

I_L	is analyte-line intensity.
$I_{0,\lambda_{\text{pri}}}$	is incident primary intensity.
C_{A}	is analyte concentration (wt fraction).
C_i	is concentration of an individual matrix element i.
$(\mu/\varrho)_{\text{A},\lambda_{\text{pri}}}$	is mass-absorption coefficient of the analyte for the primary x-rays.
$(\mu/\varrho)_{i,\lambda_{\text{pri}}}$, $(\mu/\varrho)_{i,\lambda_L}$	are mass-absorption coefficients of an individual matrix element for the primary and analyte-line x-rays, respectively.
P_{A}, A	are defined by Equations (4.8) and (4.9), respectively, and are constant for any given analysis and need not concern us here.

For the present discussion, Equation (4.18) may be expanded as follows:

$$\frac{I_L}{I_{0,\lambda_{\text{pri}}}} = P_{\text{A}} \frac{C_{\text{A}}(\mu/\varrho)_{\text{A},\lambda_{\text{pri}}}}{C_{\text{M}}[(\mu/\varrho)_{\text{M},\lambda_{\text{pri}}} + A(\mu/\varrho)_{\text{M},\lambda_L}] + C_{\text{D}}[(\mu/\varrho)_{\text{D},\lambda_{\text{pri}}} + A(\mu/\varrho)_{\text{D},\lambda_L}] + \sum C_i[(\mu/\varrho)_{i,\lambda_{\text{pri}}} + A(\mu/\varrho)_{i,\lambda_L}]} \quad (14.44)$$

where M refers to a specific matrix element, D to the added diluent, and i to the individual elements in the remainder of the matrix; M and D are discussed below.

If absorption-enhancement effects occur, variations in matrix composition among samples and standards may change the denominator in Equation (14.44) and thereby the value of $I_L/I_{0,\lambda_{\mathrm{pri}}}$ for the same analyte concentration C_{A}. Suppose that a certain matrix element M undergoes a change in concentration ΔC_{M} from specimen to specimen, causing a change in its contribution to the denominator of Equation (14.44), the first term in the denominator. If the change in this contribution is very small, the effect on the denominator—and therefore on the analyte-line intensity—may be negligible. This is the case if: (1) C_{M} is very small to begin with, so that even a *relatively* large ΔC_{M} and consequent *relatively* large change in its contribution may have little effect on the denominator; (2) ΔC_{M} is very small, so that the change in the term is small; or (3) $(\mu/\varrho)_{\mathrm{M},\lambda_{\mathrm{pri}}}$ and $(\mu/\varrho)_{\mathrm{M},\lambda_L}$ are not substantially different from the average absorption coefficient of the specimen.

However, the effect of ΔC_{M} on the analyte-line intensity is substantial when a relatively large ΔC_{M} occurs in a relatively large C_{M} of a matrix element having absorption coefficients very different—either higher or lower—from the average absorption coefficient of the specimen.

The denominator of Equation (14.44) may be made relatively insensitive to changes in concentration of the matrix element by judicious selection and addition of a diluent D in either or both of two ways: (1) A high concentration C_{D} may be added of a diluent having very low absorption coefficients $(\mu/\varrho)_{\mathrm{D},\lambda_{\mathrm{pri}}}$ and $(\mu/\varrho)_{\mathrm{D},\lambda_L}$ relative to the average absorption coefficient of the specimens (*G36*); and/or (2) a low concentration C_{D} may be added of a "diluent" having high absorption coefficients relative to the average absorption coefficients (*A4*, *C45*, *R24*). In either case, the diluent largely determines the absorption coefficient of the specimen for both primary and analyte-line radiation, and the variations in absorption coefficient caused by variations in concentrations of matrix elements are reduced, often to insignificance.

Dilution also minimizes enhancement, either by actual reduction of the concentration of the enhancing element, or by increasing the absorption of its enhancing spectral lines.

Frequently, when a low-absorption diluent (solvent or flux) is used to reduce inhomogeneity or particle-size effects, it is necessary to add a high-absorption "diluent" to level residual absorption-enhancement contributions of the original matrix (*R24*). The absorber is added in concentration such that reasonable variations in concentrations of elements in the specimen do not substantially affect the analyte-line intensity from the prepared specimen.

14.6.2. Discussion

It is evident from the preceding section that matrix-dilution methods may be classified in two ways. On the basis of whether the diluent is simply admixed with the specimen or actually destroys the original specimen form, there are, respectively, *dry-dilution methods* and *solution*, *fusion*, and *glass-disk methods*. On the basis of the relative values of the absorption coefficients of the diluent and original specimen matrix, there are *low-absorption* and *high-absorption methods*.

In a sense, methods in which the specimen is supported on filter paper, Millipore filters, or ion-exchange resin or membrane are also low-absorption dilution methods.

The technique of dilution to the extent that matrix absorption is about the same in all specimens is applicable only if, on such dilution, analyte-line intensity is still adequate. Cullen (*C73*) recommends the following procedure for evaluation of the technique as applied to solutions, although the principle is applicable to any specimen form. Two solutions are prepared having the same analyte concentration, one in pure solvent, the other in solution with the maximum expected concentrations of matrix elements. The solutions are diluted stepwise until analyte-line intensity is the same for both. It is then determined whether the analyte-line intensity is adequate for the required statistical precision in acceptable counting time. Table 14.3

TABLE 14.3. Effect of Dilution on Cu $K\alpha$ Intensity in the Presence of Iron[a]

Dilution[b]	Excitation kV, mA	Cu $K\alpha$ intensity, counts/s		Cu $K\alpha$ intensity ratio
		Cu[c]	(Cu + Fe)[d]	I_{Cu}/I_{Cu+Fe}
None	30, 20	8707	7194	1.210
1:1	30, 20	6562	6072	1.081
10:1	50, 20	4709	4568	1.033
20:1	50, 35	4866	4858	1.002
20:1	50, 50	7059	7029	1.004
100:1	50, 50	3549	3541	1.002

[a] From Cullen (*C73*).
[b] Volume of water: volume of solution (Cu or Cu + Fe).
[c] Cu concentration in undiluted solution 10 mg/ml.
[d] Cu and Fe concentrations in undiluted solution 10 mg/ml each.

gives data for such an evaluation for Cu $K\alpha$ in the presence of iron. Incidentally, the table also illustrates another feature of low-absorption dilution techniques: On dilution, analyte-line intensity decreases at a much lower rate than concentration, as is evident from comparison of the Cu $K\alpha$ intensities for 0 and 1:1 dilution, and for 20:1 and 100:1 dilution at 50 kV, 50 mA.

High-absorption dilution is applied in the high-absorption-fusion, absorption-buffer, and matrix-masking solution techniques. Among the buffers commonly used are potassium pyrosulfate ($K_2S_2O_7$), barium oxide (BaO), barium sulfate ($BaSO_4$), lanthanum oxide (La_2O_3), and tungstic acid (H_2WO_4). For addition to powders, lanthanum oxide is an excellent high-absorption diluent. It has high absorption coefficients; it is stable and available in high purity; its K lines occur at relatively short wavelengths (~0.37 Å) and are relatively inefficiently excited by a 50-kV generator—especially a full-wave generator ($V_{\mathrm{La}K}$ 39 kV); and its L lines occur at relatively long wavelengths (2–3 Å). A disadvantage of lanthanum oxide is its hygroscopicity; it can absorb up to 14% of its weight in moisture. Powder and briquet standards containing lanthanum oxide may change gradually with time, and in unfavorable cases, it may be necessary to prepare new standards with each group of samples (*F1*).

An advantage of the heavy-absorber method is that the high-Z diluent tends to accentuate secondary (analyte-line) absorption with respect to primary absorption, so that relatively simple secondary absorption corrections can be made. The heavy-absorber method has been applied especially successfully to analysis of geological samples.

A *matrix-masking agent* in the form of a salt of a noninterfering, relatively light element may be added in constant concentration to solution specimens. Cullen (*C73*) demonstrated this technique by addition of calcium nitrate [$Ca(NO_3)_2$] in concentration 10 mg/ml to solutions of copper alone and with other elements. His data are given in Table 14.4. The method is often used to reduce absorption-enhancement effects in determination of plate metals in plating solutions, which have extremely variable matrixes.

Specific dilution techniques are discussed in Chapters 16 and 17.

In addition to correction of absorption-enhancement, inhomogeneity, and particle-size effects, dilution methods are useful when the same material must be analyzed in a variety of forms—such as bar, rod, wire, foil, drillings, shot, powder, and small fabricated parts. If all forms are fused or put into solution, a single set of standards serves for all. The principal disadvantages of the method are that specimen preparation is required and that low-concentration analytes may be diluted to the extent that analyte-line in-

TABLE 14.4. Effect of Matrix-Masking Agent on Cu $K\alpha$ Intensity in the Presence of Various Matrix Elements[a]

Element added and atomic no.	Cu $K\alpha$ net intensity, counts/s	Cu $K\alpha$ net intensity ratio I_{Cu}/I_{Cu+M}
None	6143 (I_{Cu})	1.000
$_{19}$K	6094	1.008
$_{26}$Fe	5999	1.024
$_{28}$Ni	6113	1.005
$_{30}$Zn	5964	1.030
$_{47}$Ag	6005	1.023
$_{82}$Pb	5895	1.042

[a] From Cullen (*C73*). Matrix masking agent $Ca(NO_3)_2$ in concentration 10 mg/ml. Cu concentration in all solutions 2 mg/ml; concentrations of matrix elements 2 mg/ml.

tensity is prohibitively low, especially with low-Z analytes. Other advantages and disadvantages of specific techniques are given in Chapters 16 and 17.

Considerations for the selection of dry diluents are given in Reference (*L17*).

14.7. THIN-FILM METHODS

14.7.1. Principles

Absorption-enhancement effects substantially disappear in thin-film-type specimens because neither primary nor analyte-line x-rays are significantly absorbed in the extremely short path lengths in the thin layer. Each atom absorbs and emits substantially independently of the other atoms without significantly altering the incident beam. Thus, for very thin films of constant thickness, analyte-line intensity is directly proportional to analyte concentration, and for films of constant composition, intensity is proportional to thickness. These relationships provide the basis for a method free from absorption-enhancement effects for analysis of films containing more than one element, and a method for measurement of thickness of films of known composition. Of course, many samples are thin films to begin with, but for those that are not, reduction to a thin film is one way to deal with absorption-enhancement effects. In general, most supported specimens (Section 17.2) constitute thin specimens.

Gunn (*25h*) demonstrated the simple relationship between spectral-line intensity of an element and the number of atoms of that element in a thin film. The contribution dI of an incremental volume of element of constant area and thickness dt at depth t is

$$dI = k(\csc \phi)I_0 \exp -\{[(\mu/\varrho)_{\lambda_{\mathrm{pri}}} \csc \phi + (\mu/\varrho)_{\lambda_L} \csc \psi]\varrho t\}\, dt \qquad (14.45)$$

where:

I_0	is incident primary intensity.
ϕ	is the angle between the central ray of the primary cone and the specimen.
ψ	is the angle between the central ray of the secondary cone and the specimen—the takeoff angle.
λ_{pri}, λ_L	are primary and analyte-line wavelengths, respectively.
μ/ϱ	is the mass-absorption coefficient of the specimen (cm^2/g).
ϱ	is the density of the specimen (g/cm^3).
t	is film thickness (cm).
$t \csc \phi$, $t \csc \psi$	are path lengths of the primary and secondary beams, respectively, in the specimen.
k	is a constant.

If t is very small, the term inside the braces becomes negligible, and

$$dI = k(\csc \phi)I_0\, dt \qquad (14.46)$$

or, incrementally,

$$\Delta I = k(\csc \phi)I_0\, \Delta t \qquad (14.47)$$

For a constant area, $\Delta n \propto \Delta t$, where n is the number of analyte atoms in the layer. Then,

$$\Delta I = k'(\csc \phi)I_0\, \Delta n \qquad (14.48)$$

Thus, for thin films, $\Delta I \propto \Delta n$.

Birks [p. 63 in ed. 1 of (*8*)] demonstrated the absence of absorption-enhancement effects of diverse elements on the analyte-line intensity of an analyte in a thin film. The proof begins with Equation (15.32) derived in Section 15.5.2.1 for spectral-line intensity from element i in a thin film of thickness t:

$$I_{S,\lambda_L} = \frac{T_i \varrho_i' I_0}{\overline{(\mu/\varrho)}\varrho} \{1 - \exp - [\overline{(\mu/\varrho)}\varrho t]\}$$

where:

I_{S,λ_L} is secondary analyte-line intensity.

T_i is an excitation constant relating incident primary intensity and emitted analyte-line intensity of element *i*.

ϱ_i' is density of element *i in the layer* (g/cm³).

I_0, ϱ, t have the same significance as in Equation (14.45).

$\overline{(\mu/\varrho)}$ represents the term in square brackets in Equation (14.45), that is,

$$\overline{(\mu/\varrho)} = (\mu/\varrho)_{\lambda_{\text{pri}}} \csc \phi + (\mu/\varrho)_{\lambda_L} \csc \psi \qquad (14.49)$$

In Section 15.5.2.1, Equation (15.32) provides the basis for the derivation of a method for mathematical correction of absorption-enhancement effects; here, it provides the basis for proof of the absence of these effects in thin films. In Section 15.5.2.1 it is shown that

$$T_i I_0 = I_{100i} \qquad (14.50)$$

where I_{100i} is analyte-line intensity from pure *i* (100% *i*). It is also shown that the quantity $\overline{(\mu/\varrho)}$ may be expressed in terms of influence coefficients a_{ij}, each representing the overall absorption-enhancement effect of a particular matrix element *j* in the film on element *i*. Then,

$$\overline{(\mu/\varrho)}\varrho = \sum a_{ij}\varrho_j' \qquad (14.51)$$

where ϱ_j' is the density of element *j in the film*. Substitution of Equations (14.50) and (14.51) in Equation (15.32) gives

$$I_i = \frac{I_{100i}\varrho_i'}{\sum a_{ij}\varrho_j'} (1 - \exp -t \sum a_{ij}\varrho_j') \qquad (14.52)$$

where I_i is the I_{S,λ_L} of Equation (15.32). If t is very small, the exponential term may be approximated by use of the mathematical relationship

$$\exp -y \approx 1 - y \qquad (y \text{ very small}) \qquad (14.53)$$

giving

$$I_i = \frac{I_{100i}\varrho_i'}{\sum a_{ij}\varrho_j'} (1 - 1 + t \sum a_{ij}\varrho_j') \qquad (14.54)$$

This equation reduces to

$$I_i = I_{100i}\varrho_i' t \qquad (14.55)$$

Equation (14.55) shows that the analyte-line intensity from a thin film is linearly related to thickness and is, in fact, simply the product of the intensity from massive pure element I_{100i} and area density $\varrho_i't$ (g/cm^2). The intensity is unaffected by other elements j in the film, since the a_{ij} values do not appear in the equation.

14.7.2. Infinite (Critical) Thickness

Figure 14.15 shows spectral-line intensity measured from a thin film as a function of thickness. The curve has three regions. For extremely thin films, attenuation of incident primary and emergent secondary spectral-line radiation is extremely small. The curve is linear, and intensity is simply proportional to thickness. For films of intermediate thickness, attenuation of both primary and secondary x-rays increases with depth. The longer-wavelength components of the primary beam are absorbed preferentially, so that as the beam penetrates progressively deeper, it becomes both weaker and of shorter effective wavelength. The emergent secondary spectral x-rays continue to increase in intensity with thickness, but at a decreasing rate. At infinite (critical) thickness, t_∞ or t_C, secondary x-rays are excited at depths from which they cannot emerge to the surface. Further increase in thickness results in no further increase in intensity. For example, infinite thickness for chromium, iron, and nickel is 30–40 μm (*K24*).

Equation (14.45) may be used to calculate infinite thickness. From this equation, the intensity I from a film of thickness t is

$$I_t = kI_0 \csc\phi \int_0^t \exp\{-[\overline{(\mu/\varrho)}\varrho t]\}\, dt \tag{14.56}$$

$$= \frac{kI_0(\csc\phi)\{1 - \exp -[(\mu/\varrho)\varrho t]\}}{\overline{(\mu/\varrho)}\varrho} \tag{14.57}$$

where the notation has the same significance as in Equation (14.45), except $\overline{(\mu/\varrho)}$, which is defined in Equation (14.49). At infinite thickness, $t = \infty$, and

$$I_\infty = \frac{kI_0 \csc\phi}{\overline{(\mu/\varrho)}\varrho} \tag{14.58}$$

and

$$I_t/I_\infty = 1 - \exp -[\overline{(\mu/\varrho)}\varrho t] \tag{14.59}$$

or

$$\log_e [1 - (I_t/I_\infty)] = -\overline{(\mu/\varrho)}\varrho t \tag{14.60}$$

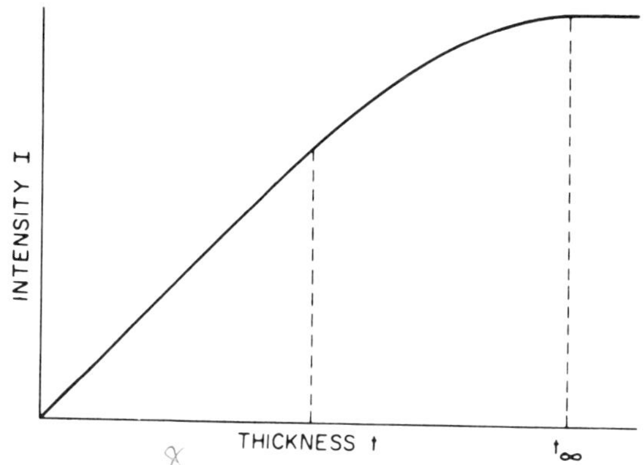

FIG. 14.15. Infinite (critical) thickness.

At the infinite (critical) thickness in the x-ray sense (Figure 14.15), $I_t/I_\infty = 1$. To calculate the thickness at which I_t just becomes I_∞, a value of, say, 0.999 is arbitrarily chosen for I_t/I_∞, and Equation (14.59) or (14.60) solved for t.

In films consisting of more than one element, the absorption coefficient of the film is different for the spectral line of each element. Thus, infinite thickness is different for each spectral line.

14.7.3. Discussion

Rhodes (*R12*) defines a "thin" specimen as one in which $m(\mu/\varrho)' \lesssim 0.1$, where m is mass per unit area (g/cm^2), and $(\mu/\varrho)'$ is the sum of the mass-absorption coefficients (cm^2/g) of the specimen for both primary and analyte-line x-rays [Equation (2.13)]. The principal advantages of thin specimens are: (1) linearity of the analyte-line intensity–concentration function over several orders of magnitude; (2) elimination of absorption-enhancement effects; (3) reduction of particle-size effects, if any; (4) increased ratio of analyte-line to scattered primary intensities, that is, increased peak-to-background ratio; and (5) low and substantially "flat" background, that is, low and constant background over a wide wavelength region.

It has already been pointed out that some specimens already have the form of thin films. These include true evaporated films, platings, corrosion layers, and sublimation deposits, as well as microgram amounts of specimen evaporated on Mylar or absorbed in filter paper or ion-exchange membrane. Extremely small particles or flakes and extremely thin fibers also act as thin specimens. Birks refers to all specimens having less than infinite thickness as "specimens of limited quantity." Alternatively, in order to reduce absorption-enhancement effects, bulk specimens may be reduced to thin films by vacuum evaporation, rolling and beating, etching, etc., or by grinding to extremely fine powder, suspension in an organic vehicle-binder, and casting to a thin film with a casting knife. Metals can be electroplated on flat copper or other electrodes, which then constitute the samples.

The intensities from extremely thin films are very low, so thin-film methods are applicable principally to major constituents. Moreover, both composition and thickness are variables. Thick films may begin to show absorption-enhancement effects. An internal standard may be added to such films by volumetric addition of a standard solution of an internal-standard element.

For specimens less than infinitely thick, as thickness increases, absorption effects increase more rapidly for emitted analyte-line x-rays than for scattered primary x-rays. Thus, sensitivity decreases as infinite thickness is approached.

Thin-film standards can be prepared by vacuum evaporation, electrodeposition, or casting, or by evaporation of small volumes of dilute solutions in filter paper or on Mylar film. Standards on specific substrates can be prepared by deposition directly on the substrate, or the film may be deposited on thin Mylar film or Scotch tape, then backed by the required analyte-free substrate (*S12*). A set of graduated thin-film standards may be prepared by deposition of the thinnest film on a very thin x-ray-transparent substrate, such as Mylar or filter paper, then stacking various numbers of layers of these films.

A novel thin-film method requiring no standards and no quantitative addition or dilution has been developed for analysis of binary systems (*F12*). Filter-paper disks are impregnated with equal volumes of the solution of the binary sample. Spectral-line intensities are measured for both elements from increasing numbers of stacked papers. A curve is plotted of the intensity ratio as a function of thickness, that is, number of stacked papers, and extrapolated to zero thickness. If the sample is a binary material, the limiting ratio is directly equivalent to the concentration ratio. For more complex samples, two methods were used: (1) an addition method in which a known amount of one of the constituents is added to a fixed amount of sample material; and (2) the internal-standard method, in which the intensity ratios of analyte and internal-standard lines are extrapolated. This multiple filter-disk method provides a means for correcting absorption-enhancement effects.

Yoneda and Horiuchi (*Y4*) point out that even thin Mylar supports may have many times the mass—and therefore give many times the scattered intensity—of micro samples. In an ingenious method for reducing this scattered background, they deposit micro samples as thin films on optically flat glass plates. A highly slit-collimated primary x-ray beam irradiates the plate at an angle of ~0.01°—less than the critical angle for total reflection from glass—and is totally reflected. Thus the primary beam excites the sample film twice—as an incident and reflected beam—and no x-rays are scattered by the support. A Ge(Li) detector, connected to an energy-dispersive spectrometer, is mounted above the specimen to receive its emitted x-rays.

Ebel has developed methods based on variation of takeoff angle for quantitative analysis of multielement films without chemical standards, and thickness (area density) of single-element and multielement films without thickness standards. These methods are described in detail in Section 14.8.2.

X-ray spectrometry has been applied to the determination of silver, chlorine, bromine, and iodine in photographic emulsions on plates, films, and papers (*B12*, *M52*). X-ray spectrometric determination of silver in developed films used for radiation dosimetry has extended their useful dosimetry range from ~100 to ~2000 R.

If the thin-film specimen is unsupported, or if it is supported on an x-ray-transparent substrate, such as Mylar or filter paper, most of the primary beam passes through the specimen and is wasted. This lost radiation may be used to excite a secondary radiator having spectral lines on the short-wavelength side of the absorption edge of the analyte, and mounted under the specimen. The secondary radiation excites additional analyte-line radiation (*H52*). This technique is discussed in greater detail in Sections 4.4.3 and 18.3.4.

X-ray spectrometric methods for determination of film and plating thickness are discussed in Chapter 18.

14.8. SPECIAL EXPERIMENTAL METHODS

This section describes a few examples of the many x-ray spectrometric methods that cannot be classified in any of the categories considered so far.

14.8.1. Emission-Absorption Methods

In emission-absorption methods (*D40*, *S3*), a portion of the sample itself is used as an absorption filter to obtain data for correction of absorption-enhancement effects on the analyte-line emission measured from another portion of the sample in the specimen chamber. A thin, uniform layer of sample is prepared to serve as an absorption filter. This filter may be mounted on the exit end of the source collimator or the entrance end of the detector collimator, as shown in Figure 14.16 (*S69*), or it may be superimposed on the emission sample itself. If a known mass of sample is formed into a briqueted disk or other uniform layer of known thickness and area, the area-density ϱt (g/cm^2) in Equation (2.10) is known directly. The mass-absorption coefficient of the sample for the primary beam is evaluated in one of three ways: The goniometer may be set at 0° 2θ and a piece of paraffin placed in the specimen chamber to scatter primary x-rays into the detector, or the x-ray tube turned so that its window faces the detector. In the latter technique, very low x-ray tube current and a small limiting aperture are used to reduce the primary intensity to a measurable value, and considerable rearrangement of the goniometer may be required (Section 9.2.2). Alternatively, one may place in the specimen compartment a secondary target having a strong spectral line at wavelength about the same as the effective wavelength of the primary beam (Section 4.1.3) and the goniometer set at 2θ for this wavelength. In any case, the intensity is measured directly and through the filter, giving incident I_0 and transmitted I intensities, respectively, in Equation (2.10). The mass-absorption coefficient for the analyte line is evaluated by measuring analyte-line intensity, emitted by the sample in the specimen chamber and diffracted by the crystal, directly and through the filter. An application of this technique is given in Section 10.5.2. In any of the foregoing methods, by placing the filter between two identical two-window flow counters, incident and transmitted intensities can be measured simultaneously from the front and back detector, respectively (Section 20.1.4.3). The combined mass-absorption coefficient for both primary and secondary x-rays is evaluated by measuring analyte-line intensity from the emission sample directly and with the filter placed over the sample so that both primary and secondary x-rays must pass through it, in opposite directions, of course.

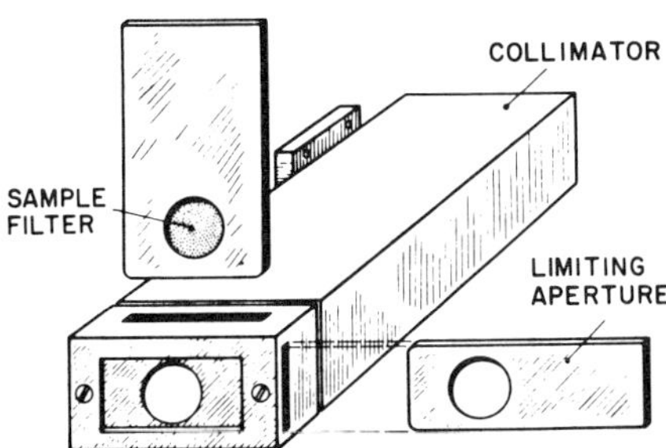

FIG. 14.16. Filter holder for the emission–absorption method. [I. R. Sweatman, K. Norrish, and R. A. Durie, Division of Coal Research, Commonwealth Scientific and Industrial Research Organization (Australia), *Miscellaneous Report* **177** (1963); courtesy of the authors and CSIRO.]

Probably the most sophisticated example of the emission–absorption method is the method of Leroux and Mahmud (*L20*), which provides correction for absorption of both primary and secondary x-rays. The method is described in detail in the following paragraphs.

Each analysis requires three briqueted specimen disks: X, the sample disk to be analyzed; S, a standard disk having known analyte concentration near the average of the expected analytical range; and R, a "radiator" disk containing analyte. The S and R disks may serve for all analyses of a given sample type. The X and S disks must be x-ray-transparent, that is, less than infinitely thick for the analyte line, so that analyte-line radiation emitted from the radiator disk can be measured after passing through an X or S disk. The weights of the X and S disks must be known. The R disk must have an analyte concentration high enough so that its analyte-line emission is measurable through the X and S disks.

Each determination requires 10 x-ray intensity measurements—five analyte-line peak intensities and their backgrounds—to obtain the following five net intensities: I_{R}, net analyte-line intensity measured from the radiator disk exposed directly to the primary x-ray beam; I_{X} and I_{S}, net analyte-line intensities measured from the sample and standard disks, respectively, each exposed alone to the primary beam; and $I_{\mathrm{X,R}}$ and $I_{\mathrm{S,R}}$, net analyte-line intensities measured from the sample and standard disks, respectively, each backed by the radiator disk. In practice, I_{R} is measured behind an open-ring spacer having thickness about the same as that of pellets X and S.

The transmittance T of the sample and standard disks to analyte-line radiation is then calculated:

$$T_{\mathrm{X}} = \frac{I_{\mathrm{X,R}} - I_{\mathrm{X}}}{I_{\mathrm{R}}} \quad \text{and} \quad T_{\mathrm{S}} = \frac{I_{\mathrm{S,R}} - I_{\mathrm{S}}}{I_{\mathrm{R}}} \tag{14.61}$$

From the values of T_{X} and T_{S}, values of the function $\phi(T)$ are calculated for the sample and standard:

$$\phi(T_{\mathrm{X}}) = \frac{-\log_e T_{\mathrm{X}}}{1 - T_{\mathrm{X}}} \quad \text{and} \quad \phi(T_{\mathrm{S}}) = \frac{-\log_e T_{\mathrm{S}}}{1 - T_{\mathrm{S}}} \tag{14.62}$$

For frequently analyzed sample types, a curve of T as a function of $\phi(T)$ may be prepared. Analyte concentration is given by solution of the following equation for C_{X}:

$$\frac{C_{\mathrm{X}}}{C_{\mathrm{S}}} = \left(\frac{I_{\mathrm{X}}}{I_{\mathrm{S}}}\right)\left(\frac{w_{\mathrm{S}}}{w_{\mathrm{X}}}\right) \frac{\phi(T_{\mathrm{X}})}{\phi(T_{\mathrm{S}})} \tag{14.63}$$

where C_X and C_S are analyte concentration in the sample and standard disks, respectively, and w_X and w_S are their weights.

Carr-Brion (*C27*) used a similar method to analyze mineral powders dispersed in wax disks 0.05 cm thick (Section 16.3.4.3). He measured analyte-line intensity from the sample disk X alone I_X, a standard disk S in the same plane as the sample disk I_S, and the sample disk backed by the standard disk $I_{X,S}$. Pure analyte was used for the standard. The intensities measured from the covered and uncovered standard are related by

$$I_{X,S} = I_S \exp -\{[(\mu/\varrho)_{X,\lambda_{pri}} \csc \phi + (\mu/\varrho)_{X,\lambda_A} \csc \psi]\varrho t\} \tag{14.64}$$

where $(\mu/\varrho)_{X,\lambda_{pri}}$ and $(\mu/\varrho)_{X,\lambda_A}$ are mass-absorption coefficients of the sample for the primary and analyte-line x-rays, respectively; ϕ and ψ are the incident and takeoff angles, respectively; and ϱ and t are the density (g/cm³) and thickness (cm) of the wax sample disk, respectively. Substituting $\overline{(\mu/\varrho)}$ [Equation (2.14)] for the term in square brackets, and substituting m/at for ϱ, where m, a, and t are, respectively, the mass (g), area (cm²) and thickness (cm) of the sample disk, one gets

$$I_{X,S} = I_S \exp -[\overline{(\mu/\varrho)}(m/a)] \tag{14.65}$$

Now, defining a sort of matrix coefficient M,

$$M = \overline{(\mu/\varrho)}/a \tag{14.66}$$

Equation (14.65) becomes

$$I_{X,S} = I_S \exp -(Mm)_X \tag{14.67}$$

Analyte-line intensity from an infinitely thick wax disk $I_{X,\infty}$ is related to that from a thin wax sample disk of thickness t cm I_X by

$$I_{X,\infty} = I_X/[1 - (I_{X,S}/I_S)] \tag{14.68}$$

It is necessary to prepare for each analyte a log-log graph of $I_{X,\infty}$ *versus* M for, say, 1 wt % analyte in various matrixes. A series of 0.05-cm-thick wax disks is prepared containing 1% analyte in various matrixes. For example, Carr-Brion prepared disks having 1% zinc, as zinc oxide, in carbon, sodium carbonate, ferric oxide, cupric sulfate, gallium oxide, molybdenum oxide, barium carbonate, mercuric oxide, and lead oxide. Each disk is weighed. Then I_X, I_S, and $I_{X,S}$ are measured from each disk. For each matrix, M is calculated from I_S, $I_{X,S}$, and the weight of the disk m

in Equation (14.67), and $I_{X,\infty}$ is calculated from I_X, I_S, and $I_{X,S}$ in Equation (14.68). A log-log plot of $I_{X,\infty}$ for 1% analyte *versus* M is substantially linear.

Thereafter, the procedure for analysis of a sample for a specified analyte is as follows. A 0.05-cm-thick disk is prepared and weighed. I_X, I_S, and $I_{X,S}$ are measured. M is calculated from Equation (14.67) and applied to the appropriate curve to derive $I_{X,\infty}$ for 1% analyte in that matrix. I_X is converted to $I_{X,\infty}$ from Equation (14.68). This value of $I_{X,\infty}$ is divided by $I_{X,\infty}$ for 1% analyte to obtain the analytical result.

Salmon (*S2*) has adapted Carr-Brion's method to analysis of microgram quantities of analyte dispersed in chromatographic paper disks.

14.8.2. Method of Variable Takeoff Angle

14.8.2.1. Basic Excitation Equations

The availability of suitable standards for quantitative x-ray spectrometric analysis is often a very serious problem, and the analyst always welcomes a method that requires no standards: The methods of standard addition and dilution (Section 14.2) and absorption-edge spectrometry (Section 20.1.4) are described elsewhere in the book. This section considers another method.

The fundamental excitation equations, such as Equation (4.7), show that analyte-line intensity depends not only on analyte concentration, primary intensity distribution, and mass-absorption coefficients of the specimen for the primary and analyte-line x-rays, but also on the incident and takeoff angles. Ebel (*E1*, *E2*, *E4*) has developed a method of quantitative x-ray spectrometric analysis based on variation of the takeoff angle. Bulk and film specimens can be analyzed without chemical standards, and film thickness (or area density) can be measured without thickness standards. The principal specimen requirements are homogeneity and a smooth, plane surface.

The geometry of the variable-takeoff-angle method is shown in Figure 14.17, in which the analyte-line intensity emitted from layer dt that actually enters the source collimator is given by

$$\Delta I_L = I_{0,\lambda_{\mathrm{pri}}}\left(\frac{d\Omega}{4\pi}\right)\omega_A g_L\left(\frac{r_A-1}{r_A}\right)\left(\frac{n_{A,X}}{n_{A,A}}\right) \times \left\{\exp-\left[\left(\frac{\mu_{X,\lambda_{\mathrm{pri}}}}{\cos\alpha}+\frac{\mu_{X,\lambda_L}}{\cos\beta}\right)t\right]\right\} d\lambda_{\mathrm{pri}}\, dt \tag{14.69}$$

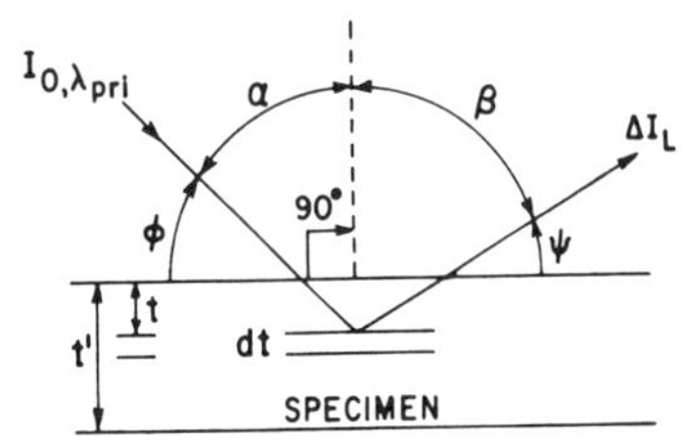

FIG. 14.17. Geometry of the method of variable takeoff angle.

where:

$d\Omega/4\pi$	is the useful solid angle of secondary analyte-line x-rays accepted by the x-ray optical system (Figure 4.1).
A	is the analyte element.
g_L	is the relative intensity of the analyte line in its series (Equation 4.4).
I_L	is the intensity of the analyte line.
$I_{0,\lambda_{pri}}$	is the intensity of the incident primary beam.
$n_{A,A}$	is the number of analyte atoms per cm^3 in pure analyte.
$n_{A,X}$	is the number of analyte atoms per cm^3 in the specimen.
r_A	is the absorption-edge jump ratio of the analyte.
t	is the depth of layer *dt* below the specimen surface.
t'	is the specimen thickness.
α	is the angle between the normal to the specimen surface and the central ray of the incident primary beam ($\alpha = 90° - \phi$).
β	is the angle between the normal to the specimen surface and the central ray of the secondary beam ($\beta = 90° - \psi$); this is the *takeoff angle* of Ebel's method.
λ_L	is the wavelength of the analyte line.
λ_{pri}	is the wavelength of the primary x-rays.
μ_{X,λ_L}	is the linear absorption coefficient of the specimen for the analyte line.
$\mu_{X,\lambda_{pri}}$	is the linear absorption coefficient of the specimen for the primary x-rays.
ω_A	is the fluorescent yield of the analyte line.

Equation (14.69) is substantially the same as Equation (4.6) except that Ebel uses linear absorption coefficients and measures incident and takeoff angles from the normal to the specimen surface (α and β, respectively), rather than from the specimen surface itself (ϕ and ψ, respectively).

The following two substitutions are now made in Equation (14.69):

$$C_{\mathrm{A}}(\varrho_{\mathrm{A}}'/\varrho_{\mathrm{A}}) = n_{\mathrm{A,X}}/n_{\mathrm{A,A}} \tag{14.70}$$

$$P_{\mathrm{A}} = \left(\frac{d\Omega}{4\pi}\right)\omega_{\mathrm{A}} g_L \left(\frac{r_{\mathrm{A}} - 1}{r_{\mathrm{A}}}\right) \tag{4.8}$$

where C_{A} is analyte concentration (wt fraction), and ϱ_{A}' and ϱ_{A} are the densities of analyte in the specimen and in pure analyte, respectively. By integration and application of the first mean-value theorem of integral calculus to the substituted equation, one gets

$$I_L = C_{\mathrm{A}}\left(\frac{\varrho_{\mathrm{A}}'}{\varrho_{\mathrm{A}}}\right) \frac{1 - \exp - \left[\left(\dfrac{\mu_{\mathrm{X},\bar{\lambda}_{\mathrm{pri}}}}{\cos\alpha} + \dfrac{\mu_{\mathrm{X},\lambda_L}}{\cos\beta}\right)t'\right]}{\dfrac{\mu_{\mathrm{X},\bar{\lambda}_{\mathrm{pri}}}}{\cos\alpha} + \dfrac{\mu_{\mathrm{X},\lambda_L}}{\cos\beta}} \int_{\lambda_{\min}}^{\lambda_{\mathrm{A_{ab}}}} I_{0,\lambda_{\mathrm{pri}}} P_{\mathrm{A}}\, d\lambda_{\mathrm{pri}} \tag{14.71}$$

where $\bar{\lambda}_{\mathrm{pri}}$ is the mean primary wavelength (see below); $\mu_{\mathrm{X},\bar{\lambda}_{\mathrm{pri}}}$ is the linear absorption coefficient of the specimen for $\bar{\lambda}_{\mathrm{pri}}$; $\lambda_{\min}$ is the short-wavelength limit of the primary beam; and $\lambda_{\mathrm{A_{ab}}}$ is the analyte absorption edge. Ebel derives four working equations from Equation (14.71):

1. Analyte-line intensity from infinitely thick ($t' = t_\infty$) multielement specimens is given by

$$I_{L,\infty} = C_{\mathrm{A}}\left(\frac{\varrho_{\mathrm{A}}'}{\varrho_{\mathrm{A}}}\right) \frac{1}{(\mu_{\mathrm{X},\bar{\lambda}_{\mathrm{pri}}}/\cos\alpha) + (\mu_{\mathrm{X},\lambda_L}/\cos\beta)} \int_{\lambda_{\min}}^{\lambda_{\mathrm{A_{ab}}}} I_{0,\lambda_{\mathrm{pri}}} P_{\mathrm{A}}\, d\lambda_{\mathrm{pri}} \tag{14.72}$$

2. For pure analyte, $C_{\mathrm{A}} = 1$ and $\varrho_{\mathrm{A}}' = \varrho_{\mathrm{A}}$, so that $C_{\mathrm{A}}(\varrho_{\mathrm{A}}'/\varrho_{\mathrm{A}}) = 1$; then analyte-line intensity from infinitely thick pure analyte is given by

$$I_{L,\infty} = \frac{1}{(\mu_{\mathrm{A},\bar{\lambda}_{\mathrm{pri}}}/\cos\alpha) + (\mu_{\mathrm{A},\lambda_L}/\cos\beta)} \int_{\lambda_{\min}}^{\lambda_{\mathrm{A_{ab}}}} I_{0,\lambda_{\mathrm{pri}}} P_{\mathrm{A}}\, d\lambda_{\mathrm{pri}} \tag{14.73}$$

where $\mu_{\mathrm{A},\bar{\lambda}_{\mathrm{pri}}}$ and $\mu_{\mathrm{A},\lambda_L}$ are the linear absorption coefficients of the analyte for $\bar{\lambda}_{\mathrm{pri}}$ and λ_L, respectively.

3. For thin-film specimens,

$$t' = m/a\varrho_{\mathrm{A}}' \tag{14.74}$$

where a and m are the area (cm^2) and mass (g) of the effective area of film from which analyte-line emission is intercepted by the x-ray optical system, and ϱ_{A}' is the density ($\mathrm{g/cm}^3$) of analyte in the film. Analyte-line intensity from multielement film specimens is given by

$$I_{L,f} = C_A\left(\frac{\varrho_A'}{\varrho_A}\right)\frac{1-\exp-\left[\left(\dfrac{\mu_{X,\bar\lambda_{pri}}}{\cos\alpha}+\dfrac{\mu_{X,\lambda_L}}{\cos\beta}\right)\dfrac{m}{a\varrho_A'}\right]}{\dfrac{\mu_{X,\bar\lambda_{pri}}}{\cos\alpha}+\dfrac{\mu_{X,\lambda_L}}{\cos\beta}}\int_{\lambda_{min}}^{\lambda_{A_{ab}}} I_{0,\lambda_{pri}}P_A\,d\lambda_{pri} \tag{14.75}$$

4. By following the same line of reasoning as for pure bulk analyte specimens (item 2 above), one gets for analyte-line intensity from pure analyte film specimens

$$I_{L,f} = \frac{1-\exp-\left[\left(\dfrac{\mu_{A,\bar\lambda_{pri}}}{\cos\alpha}+\dfrac{\mu_{A,\lambda_L}}{\cos\beta}\right)\dfrac{m}{a\varrho_A}\right]}{\dfrac{\mu_{A,\bar\lambda_{pri}}}{\cos\alpha}+\dfrac{\mu_{A,\lambda_L}}{\cos\beta}}\int_{\lambda_{min}}^{\lambda_{A_{ab}}} I_{0,\lambda_{pri}}P_A\,d\lambda_{pri} \tag{14.76}$$

In all the foregoing equations, mass-absorption coefficients μ/ϱ multiplied by the appropriate densities ϱ may be substituted for the linear coefficients μ.

14.8.2.2. Evaluation of Mean Wavelength and Excitation Integral

In Equations (14.72), (14.73), (14.75), and (14.76), mean primary wavelength $\bar\lambda_{pri}$ and the excitation integral can be either determined by variation of the takeoff angle β or eliminated by use of intensity ratios. Mean primary wavelength is different for each of the four equations and must be evaluated for each. Examples of the evaluation of $\bar\lambda_{pri}$ for pure-analyte bulk and film specimens follow.

For infinitely thick analyte, a curve of $I_{L,\infty}$ *versus* β is established experimentally for $\beta = 10$–$90°$. In Equation (14.73), only $I_{L,\infty}$, μ_{A,λ_L}, and $dI_L/d\beta$ depend on β, and the equation may be differentiated and solved for $(\mu/\varrho)_{A,\bar\lambda_{pri}}$, the mass-absorption coefficient of the analyte for the mean primary wavelength (see the last paragraph of Section 14.8.2.1):

$$\left(\frac{\mu}{\varrho}\right)_{A,\bar\lambda_{pri}} = -\left(\frac{\mu}{\varrho}\right)_{A,\lambda_L}(\cos\alpha)\left[\frac{(dI_L/d\beta)_{\beta=90°}}{I_L}+\frac{1}{\cos\beta}\right] \tag{14.77}$$

This equation may then be used to calculate $\bar\lambda_{pri}$ as a function of β.

For thin films ($t' \lesssim \sim 2000$ Å), Equation (14.76) is approximated by

$$I_{L,f} = (m/a\varrho_A)\int_{\lambda_{min}}^{\lambda_{A_{ab}}} I_{0,\lambda_{pri}}P_A\,d\lambda_{pri} \tag{14.78}$$

The intensity ratio of analyte-line intensity from film and bulk specimens of pure analyte,

$$R_1 = \frac{I_{L,f}}{I_{L,\infty}} = \frac{\text{Equation (14.78)}}{\text{Equation (14.73)}} \tag{14.79}$$

can then be solved for $(\mu/\varrho)_{A,\bar{\lambda}_{\text{pri}}}$:

$$(\mu/\varrho)_{A,\bar{\lambda}_{\text{pri}}} = \{(R_1 a/m) - [(\mu/\varrho)_{A,\lambda_L}(1/\cos\beta)]\}\cos\alpha \tag{14.80}$$

The excitation integrals, unlike the mean primary wavelengths, are all alike in Equations (14.72), (14.73), (14.75), and (14.76) since they depend only on excitation conditions and not on takeoff angle, composition, or mass per unit area. Knowledge of $(\mu/\varrho)_{A,\bar{\lambda}_{\text{pri}}}$ permits calculation of the excitation integral from Equation (14.72).

Ebel outlines the application of these equations to various determinations as described in the following sections.

14.8.2.3. Mass per Unit Area of an Element Film from Its Own Line

Consider the ratio of analyte-line intensity from film and bulk specimens of pure analyte A:

$$R_2 = \frac{I_{L,f}}{I_{L,\infty}} = \frac{\text{Equation (14.76)}}{\text{Equation (14.73)}} \tag{14.81}$$

All quantities are known or can be evaluated except m/a, the mass per unit area of the film. The problem can be solved by introduction of an *apparent* m/a:

$$\left(\frac{m}{a}\right)_{\text{ap}} = \frac{\cos\beta}{(\mu/\varrho)_{A,\lambda_L}} \log_e(1 - R_2) \tag{14.82}$$

Equation (14.82) contains only known or measurable quantities, and the function $(m/a) = f(\beta)$ is linear and can be extrapolated linearly to $\beta = 90°$, that is,

$$\lim_{\beta \to 90°} (m/a)_{\text{ap}} = m/a \tag{14.83}$$

Thus, m/a for the element film is determined without reference films by: (1) measuring analyte-line intensity I_L from the film and from pure bulk analyte at various takeoff angles β; (2) calculating $R_2 = I_{L,f}/I_{L,\infty}$ at each β; (3) plotting R_2 *versus* β; and (4) extrapolating to $\beta = 90°$.

14.8.2.4. Mass per Unit Area of an Element Film from a Substrate Line

The intensity of a line of element A in a bulk multielement specimen coated with a film of element B is obtained from Equations (14.72) and (2.10):

$$I_{L,\infty} = \text{Equation (14.72)} \times \exp -\left[\left(\frac{\mu_{B,\lambda_{pri}}}{\cos\alpha} + \frac{\mu_{B,\lambda_L}}{\cos\beta}\right)\left(\frac{m}{a\varrho}\right)_B\right] \quad (14.84)$$

where I_L and λ_L refer to element A in the bulk substrate, and subscript B refers to the film element. A ratio may now be taken of substrate-line intensity from the coated and uncoated bulk substrate:

$$R_3 = \frac{I_{L,\infty}(\text{coated})}{I_{L,\infty}(\text{uncoated})} = \frac{\text{Equation (14.84)}}{\text{Equation (14.72)}} \quad (14.85)$$

The solution of the problem now parallels that for a pure element film with its own line (above), using apparent m/a:

$$\left(\frac{m}{a}\right)_{ap} = -\frac{\cos\beta}{(\mu/\varrho)_{B,\lambda_L}} \log_e R_3 \quad (14.86)$$

Then m/a for the element film is determined without reference films by: (1) measuring substrate-line intensity I_L from the coated and uncoated bulk substrate at various takeoff angles β; (2) calculating $R_3 = I_{L,\infty(\text{coated})}/I_{L,\infty(\text{uncoated})}$ at each β; (3) plotting R_3 *versus* β; and (4) extrapolating to $\beta = 90°$.

14.8.2.5. Mass per Unit Area and Composition of Multielement Films

For this application, thin element films are required having known mass per unit area $(m/a)_i$, one such film for each element i in the multielement film. The films can be evaluated by either the film-line or substrate-line method described above. Since absorption-enhancement effects substantially disappear in thin films, the spectral-line intensity from a specified m/a of an element i is substantially the same in a pure-element or multielement film, and is given by

$$(I_{L,f})_i = (C/\varrho)_i (m/a)_i \int_{\lambda_{min}}^{\lambda_{i_{ab}}} I_{0,\lambda_{pri}} P_i \, d\lambda_{pri} \quad (14.87)$$

The ratio of analyte-line intensity from the pure-analyte film $(I_{L,f})_i$ and multielement film $(I_{L,f})_X$ is then

$$R_i = \frac{(I_{L,f})_i}{(I_{L,f})_X} = \frac{\text{Equation (14.87)}}{\text{Equation (14.78)}} = C_i \frac{(m/a)_{i,X}}{(m/a)_{i,R}} \tag{14.88}$$

where C_i is concentration of element i in the multielement film, and $(m/a)_{i,X}$ and $(m/a)_{i,R}$ are mass per unit area values for element i in the multielement sample film X and the pure-analyte reference film R, respectively. R_i is measured, and $(m/a)_{i,R}$ is known, leaving $C_i(m/a)_{i,X}$ the only unknown quantity in the equation. If the multielement film consists of n elements, n equations of the form of Equation (14.88) are required, together with

$$\sum_{i=1}^{i=n} C_i = 1 \tag{14.89}$$

Thus $n + 1$ simultaneous equations can be written to solve the n values of $C_i(m/a)_{i,X}$.

14.8.2.6. Composition of Bulk Multielement Specimens

The ratio of analyte-line intensity of element i from a bulk multielement sample X and bulk pure analyte i is

$$R_i = \frac{(I_{L,\infty})_X}{(I_{L,\infty})_i} = \frac{\text{Equation (14.72)}}{\text{Equation (14.73)}} \tag{14.90}$$

$$= C_i\left[\frac{(\mu/\varrho)_{i,\lambda_i}}{(\mu/\varrho)_{X,\lambda_i}}\right] \frac{1 + [(\mu_{i,\bar{\lambda}_{\text{pri}}}/\mu_{i,\lambda_i})(\cos\beta/\cos\alpha)]}{1 + [(\mu_{X,\bar{\lambda}_{\text{pri}}}/\mu_{X,\lambda_i})(\cos\beta/\cos\alpha)]} \tag{14.91}$$

R_i is independent of $\bar{\lambda}_{\text{pri}}$, and

$$\lim_{\beta\to 90^\circ} R_i = C_i[(\mu/\varrho)_{i,\lambda_i}/(\mu/\varrho)_{X,\lambda_i}] \tag{14.92}$$

Thus, the limiting value of Equation (14.92) permits calculation of C_i since $(\mu/\varrho)_{i,\lambda_i}$ is known and

$$(\mu/\varrho)_{X,\lambda_i} = \sum_{j=1}^{j=n} [C_j(\mu/\varrho)_{j,\lambda_i}] \qquad (j \text{ includes } i) \tag{14.93}$$

where j signifies each element in the multielement sample. If the multielement sample consists of n elements, n equations of the form of Equation (14.91) are required, together with Equation (14.89). Thus, $n + 1$ simultaneous equations can be written to solve the n values of C_i.

Thus, the concentration of each element in a multielement sample is determined without standards by the following procedure. For each of the n elements: (1) Analyte-line intensity I_L is measured from the multielement sample and from pure analyte at various takeoff angles; (2) the intensity ratio $R_i = (I_{L,\infty})_X/(I_{L,\infty})_i$ is calculated for each β; (3) R_i is plotted as a function of β; (4) the curve is extrapolated to $\beta = 90°$ to give the limiting value of R_i; and (5) an equation of the form of Equation (14.92) is established by use of the limiting value of R_i, the known value of $(\mu/\varrho)_{i,\lambda_i}$, and Equation (14.93). The n resulting equations and Equation (14.89) are solved simultaneously or by determinants for the n values of C_i.

14.8.3. Indirect (Association) Methods

An extremely significant development in x-ray spectrometric analysis is the indirect determination—or determination by association—of the following classes of analytes: (1) elements relatively inconvenient to determine by x-ray spectrometry—atomic numbers 9–17 (fluorine through chlorine); (2) elements not determinable at all on standard commercial spectrometers—atomic numbers less than 9; (3) organic compounds and other specific chemical compounds and substances; and (4) analytes that, although readily determinable, give low intensities due to a combination of small total amount and low analytical sensitivity. Such analyses are sometimes conveniently done by stoichiometric association with a readily determined element.

Examples of the first group (above) are the determination of trace chlorine in titanium by precipitation as silver chloride and measurement of the silver (*R34*); the determination of silicon by formation of silicomolybdate-8-quinolinol complex (Si:Mo = 1:12) and measurement of the molybdenum (*S41*); and the determination of sodium by precipitation of sodium zinc uranyl acetate [$NaZn(UO_2)_3(OOCCH_3)_9$] and measurement of U $L\alpha_1$ (*L52*).

A striking example of the second group is the microdetermination of nitrogen in amounts 0.2–5 μg in biological materials by converting it to ammonia, absorbing the ammonia in filter paper impregnated with Nessler's reagent, washing out excess reagent, then determining the mercury in the mercury–iodine–amine complex thus formed (*M17*). Boron can be precipitated as barium borotartrate and the Ba $L\alpha_1$ line measured (*L52*); lithium and beryllium can be determined as their phosphates or arsenates and measurement of the P $K\alpha$ or As $K\alpha$ lines (*L51*). A related example is the determination of carbon—or carbon-to-hydrogen ratio—in hydrocarbons by means of the intensity ratio of coherent to incoherent scattered target-line radiation (Section 20.2.2); of course, this is not an association method (*D38*, *D39*).

Three techniques have been evaluated for stoichiometric association of a determinable element with organic compounds (*P4*): bromination with bromine water, chelation with 5-chloro-7-iodo-8-quinolinol, and salt-formation with barium or lead. These techniques reverse the usual procedure of using an organic reagent, such as 8-quinolinol or dithizone, to specifically separate and precipitate an element prior to its determination by x-ray spectrometry.

Montmorillonite is determined in montmorillonite–kaolinite clay mixtures by prolonged agitation of the samples in strontium chloride solution, filtration, and measurement of the strontium in the recovered clay (*H44*). Only the montmorillonite undergoes ion exchange with the strontium to a significant extent. The method is calibrated by use of standards having known montmorillonite–kaolinite ratios. An alternative method would be to use strontium chloride solutions of known concentration, then determine the strontium depletion in the recovered filtrate.

The significance of these indirect methods cannot be overestimated. They extend the applicability of x-ray spectrometry to the entire periodic table, to organic chemistry, and to other specific materials—all with standard commercial instrumentation and accessories and with conventional preparation and presentation techniques. Many of these methods are applicable to micro and trace analysis.

Another type of "indirect" analysis is the method of Esmail *et al.* (*E18*) for detection of elements 5–9 (boron to fluorine). The method is based on the fact that when a heavy element combines with a light element, satellite lines appear near those heavy-element diagram lines that originate from electron transitions from its valence shell. The energy difference between the satellite and diagram lines is characteristic of the ligand and is approximately 5, 9, 14, and 20 eV for carbon, nitrogen, oxygen, and fluorine, respectively. Thus, the light element combined with a heavy element can be identified by measurement of the wavelength of the heavy-element satellite line.

14.8.4. Combinations of X-Ray Spectrometry and Other Methods

X-ray spectrometry has been used to complement many other analytical methods. Combined x-ray emission and absorption methods are discussed in Section 14.8.1. An example of the combination of x-ray emission and diffraction is provided by the measurement of the thickness of chromium and nickel double platings on brass (*G7*). The chromium (top) plating was determined by x-ray diffractometry. The underlying nickel plating was determined by measurement of Ni $K\alpha$ intensity and correction for absorption in the overlaying chromium layer.

Thorium in concentration $\sim$0.5% was determined in uranium–thorium ores by a combination of x-ray spectrometry and radiochemistry (*C15*). Total radioactivity was measured and expressed as equivalent uranium concentration. Then the U $L\alpha$/Th $L\alpha$ intensity ratio was measured and used to calculate the uranium-to-thorium weight ratio. Finally, the known relative radioactivity of uranium and thorium was used to calculate absolute uranium and thorium weight concentrations.

Another example is provided by the determination of gold by the ion-exchange filter-paper technique (*G30*) (Section 17.2.4). The retention of gold is incomplete and variable. The same amount of radioactive ^{195}Au is added as a collection monitor to all sample and standard solutions prior to taking up the gold in the ion-exchange filters. The disks are placed in the specimen compartment to measure total Au $L\alpha_1$ intensity in the usual way, then directly before the scintillation counter to measure the γ-radiation from the ^{195}Au and thereby the amount of ^{195}Au recovered. A calibration curve is plotted of the intensity ratio of Au $L\alpha_1$ x-rays and ^{195}Au γ-rays *versus* Au concentration in the standard solutions. It is not necessary that the gold recovery be complete, proportionally constant, or known.

Alkylbenzene sulfonate and residual sodium sulfate were determined in water slurries by a combination of x-ray spectrometry and infrared spectrophotometry (*K32*). Total sulfur was determined by the x-ray method, the alkylbenzene sulfonate by the infrared method. Then the sulfur equivalent to the sulfonate was calculated and subtracted from total sulfur to give sodium sulfate concentration.

Sulfur in concentration down to 0.002% was determined in gasoline by a combined x-ray spectrometric standard-addition and density method (*J26*). Two 50-ml portions A and B were taken of samples having sulfur concentration $\leq$0.15%. Portion A was treated with 5 ml of sulfur-free iso-octane, B with 5 ml of a standard solution containing 0.13 g of sulfur. Net S $K\alpha$ intensities were measured from the treated samples, and the density ϱ of the untreated sample was measured with a hydrometer. Sulfur concentration in weight percent is given by

$$C = \frac{2[0.13 I_{\mathrm{A}}/(I_{\mathrm{B}} - I_{\mathrm{A}})]}{\varrho} \tag{14.94}$$

Jackwerth and Kloppenburg (*J1*, *J2*) describe an accessory for passing chromatographic paper strips continuously through the x-ray spectrometer specimen compartment. Small chromatographic spots are irradiated as a whole; large spots are passed under a slit as emitted analyte-line intensity is recorded. The area under the recorded curve is proportional to the amount of analyte. Organic compounds can be determined if combined with an equivalent of heavy element. Houpt (*H66*) describes a scanner for thin-layer chromatograms having the form of a small cylindrical drum that fits inside a spectrometer specimen cup and is driven by a small clockworks inside the drum.

X-ray spectrometry has been used to complement gas chromatography for determination of halogens and sulfur in gas samples (*S15*). Chromatographic effluent—carrier gas and separated analytical constituents—is passed through a thin-walled capillary in the specimen compartment. With a 5-cm length of capillary, the detection limit was 0.5 μg of carbon disulfide per milliliter of carrier gas. X-ray spectrometry has also been used to complement paper electrophoresis (*W7*).

X-ray spectrometry has been combined with high-energy electron diffraction (HEED) to determine elemental and species composition and crystal structure of films evaporated on various substrates in high vacuum. An electron beam is diffracted by and excites x-ray spectra in the specimen films (*S22*).

Mantel and Amiel (*M9*, *M10*) have applied x-ray spectrometry to complement neutron activation. The specimen is subjected to neutron bombardment to convert a certain portion of each analyte into a radioisotope. However, the specimen is then analyzed by measurement of the characteristic x-ray spectra emitted by the induced radioisotopes, rather than the β- or γ-ray spectra. These x-ray spectra arise from *K*- and *L*-electron capture and internal β- and γ-ray conversion (Section 1.6.3 and Figure 1.10). The x-ray instrument consists of a Si(Li) detector and energy-dispersive spectrometer. This is an example of what might be termed "self-excitation" or "autoexcitation" of x-ray spectra in that no excitation radiation is incident upon the specimen during measurement. Another example is discussed in Section 8.2.2.3.

Chapter 15

Mathematical Correction of Absorption-Enhancement Effects

15.1. INTRODUCTION

Having considered primarily experimental methods of dealing with absorption-enhancement effects, we now consider primarily mathematical methods. Most of these methods are so complex that a comprehensive treatment cannot be given here. Moreover, no good, comprehensive, critical review of the scope, advantages, and limitations of all of the various mathematical methods has appeared, and this chapter does not provide one. However, we can set forth the principles of some of the outstanding methods.

It is explained in Section 12.1 that, disregarding particle-size, heterogeneity, and surface-texture effects, were it not for absorption-enhancement effects by matrix elements, measured analyte-line intensity would be directly proportional to analyte concentration. In fact, the ratio of analyte-line intensities from sample and pure analyte would equal the analyte concentration. In practice, this ideal condition rarely obtains, and one must deal with absorption-enhancement effects by judicious choice of the analytical method (Chapter 14) or by mathematical methods. In the latter, the analyst measures one of three quantities: (1) analyte-line intensity from the sample; (2) ratio of analyte-line intensities from the sample and pure analyte; or, preferably, (3) ratio of analyte-line intensities from the sample and a standard similar to the samples and having analyte concentration in the middle of the expected range. Analyte concentration is derived from this measured quantity mathematically by one of six basic approaches: (1) graphic calculation (Section 15.2); (2) correction for simple absorption (Section 15.3); (3) empirical calculation (Section 15.4); (4) calculation with polynomial (multiple-regression) equations (Section 15.7); (5) calculation from first principles (fundamental parameters) (Section 15.6); and (6) calculation by use of influence coefficients to correct for the effect of each matrix element on the measured analyte-line intensity ratio (Section 15.5).

15.2. GEOMETRIC METHODS

Certainly the outstanding examples of the geometric approach to the correction of absorption-enhancement effects are the methods for ternary systems. The chemical composition of a ternary system can be represented by a triangular composition diagram. Each point in the triangle uniquely represents a set of three values for the concentrations of the three components C_A, C_B, C_C. Each such point may also represent the corresponding analyte-line intensities I_A, I_B, I_C that would be measured from this composition. Figure 15.1 [Müller, pp. 164–166 in *(29)*] shows a triangular composition diagram for the TiO_2–Nb_2O_5–Ta_2O_5 ternary system; the light grid represents concentration in intervals of 5 wt %. Calibration curves of Nb $K\alpha$ intensity from the Nb_2O_5–TiO_2 and Nb_2O_5–Ta_2O_5 binary systems are constructed on the corresponding sides of the triangle. These curves permit establishment of Nb $K\alpha$ intensity scales along the two edges of the triangle, as shown. Similarly, curves of Ti $K\beta$ and Ta $L\alpha_1$ intensities from the TiO_2–Nb_2O_5, TiO_2–Ta_2O_5, and Ta_2O_5–Nb_2O_5 binary systems permit establishment of Ti $K\beta$ and Ta $L\alpha_1$ intensity scales along the appropriate sides.

Isointensity lines can then be drawn for the three spectral lines, as shown by the dark lines in Figure 15.1. If the calibration curves for the A line in A–B and A–C are very similar, these isointensity lines are substantially straight. This is the case for the Ti $K\beta$ lines in the figure; the Ti $K\beta$ calibration curves (not shown) for the TiO_2–Nb_2O_5 and TiO_2–Ta_2O_5 systems are very similar (μ/ϱ of niobium and tantalum for Ti $K\beta$ is 564 and 576 cm^2/g, respectively). However, if the curves are markedly dissimilar, the isointensity lines are curved. This is the case for Nb $K\alpha$ (μ/ϱ of titanium and tantalum for Nb $K\alpha$ is 28 and 103 cm^2/g, respectively). In such cases, a few ternary standards are required to establish the lines. In any region of analytical interest, linear interpolation is usually permissible.

Taylor *et al.* (*T3*) have developed the graphic method for ternary systems. A description of their work follows.

In a ternary system of elements A, B, and C having atomic weights A_A, A_B, and A_C in weight fractions W_A, W_B, and W_C and atom fractions

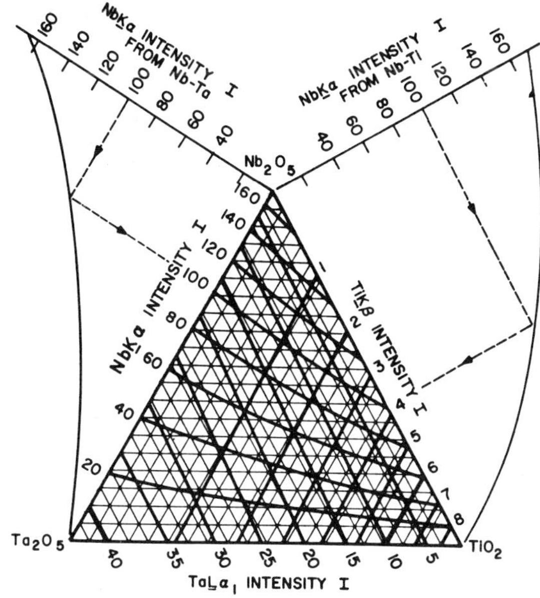

FIG. 15.1. Triangular composition diagram (light lines) for the TiO_2–Nb_2O_5–Ta_2O_5 ternary system, with superimposed isointensity lines (dark lines) for Ti $K\beta$, Nb $K\alpha$, and Ta $L\alpha_1$. The light grid represents concentration in intervals of 5 wt%. Calibration curves for Nb $K\alpha$ intensity from the Nb_2O_5–TiO_2 and Nb_2O_5–Ta_2O_5 binary systems are erected on the corresponding sides of the triangle. [R. Müller, "Spectrochemical Analysis by X-Ray Fluorescence", Plenum Press, N. Y., p. 166 (1972); courtesy of the author and publisher.]

A_A', A_B', and A_C', composition of any specimen is given by equations of the form

$$(W_A)_{ABC} = \frac{A_A' A_A}{A_A' A_A + A_B' A_B + A_C' A_C} \qquad (15.1)$$

$$(A_A')_{ABC} = \frac{W_A/A_A}{(W_A/A_A) + (W_B/A_B) + (W_C/A_C)} \qquad (15.2)$$

The corresponding relationships in the three binary systems A–B, A–C, and B–C have the forms

$$(W_A)_{AB} = \frac{A_A' A_A}{A_A' A_A + A_B' A_B} = \frac{A_A' A_A}{A_A' A_A + (1 - A_A') A_B} \qquad (15.3)$$

$$(A_A')_{AB} = \frac{W_A/A_A}{(W_A/A_A) + (W_B/A_B)} = \frac{W_A/A_A}{(W_A/A_A) + (1 - W_A)/A_B} \qquad (15.4)$$

All these equations are readily expressed in terms of percent as well as fraction.

Suppose that the triangular composition diagram for the ternary system A–B–C is plotted in terms of atom concentration. Then Equation (15.3) and the analogous equations for the A–C and B–C binary systems can be used to mark off weight concentrations along the A–B, A–C, and B–C sides of the triangle. A skew grid may be constructed on the triangle by drawing lines joining equal or complementary concentrations on the three sides—for example, 90A–10C on A–C with 90A–10B on A–B and 90B–10C on B–C, and 50A–50C on A–C with 50A–50B on A–B and 50B–50C on B–C. The grid permits reading of weight concentrations directly on the atom composition diagram. Similarly, Equation (15.4) and its analogs can be used to construct a skew grid to permit reading of atom concentrations on a weight composition diagram.

Now, if the spectral-line intensities emitted by the three elements in the ternary system A–B–C are I_A, I_B, and I_C, the "count fraction" F for each intensity is given by an equation of the form

$$(F_A)_{ABC} = I_A/(I_A + I_B + I_C) \tag{15.5}$$

In the corresponding binary systems, the count fraction equations have the form

$$(F_A)_{AB} = I_A/(I_A + I_B) \tag{15.6}$$

Like weight and atom fractions, count fractions can also be expressed in percent.

Taylor *et al.* (*T3*) have based a geometric method for correction of absorption-enhancement effects in ternary systems on the obvious similarity of the equations for weight, atom, and count fractions. Their procedure is as follows:

1. Three sets of standards are prepared, one for each of the three binary systems.

2. Three calibration curves (count percent F *versus* weight percent W) are established from the intensity data from the standards: $(F_A)_{AB}$ *versus* W_A in A–B, $(F_B)_{BC}$ *versus* W_B in B–C, and $(F_C)_{CA}$ *versus* W_C in C–A.

3. These curves are used to calibrate a triangular coordinate graph having corners $100F_A$, $100F_B$, $100F_C$ and sides graduated in linear 0–100% F scales. From the calibration data, scales of 0–100% W_A, W_B, and W_C are plotted along the A–B, B–C, and C–A sides, respectively.

4. A weight-concentration skew grid is then drawn on the triangle over the linear count-percent grid.

The result is a ternary calibration graph. An analysis then consists in: (1) measuring I_A, I_B, and I_C from the sample; (2) calculating F_A, F_B, and F_C from equations having the form of Equation (15.5); (3) finding the point having these three F values on the linear count-percent grid; and (4) reading the concentrations off the skewed composition grid.

Figure 15.2 shows the application of the method to the germanium–tellurium–bismuth system. The sides of the triangle are graduated linearly in counts percent, as shown by the equally spaced "ticks." In practice, the triangle is a standard triangular coordinate diagram, but in the figure, the triangular grid is omitted for simplicity. X-ray intensity data measured from germanium–tellurium, germanium–bismuth, and tellurium–bismuth binary

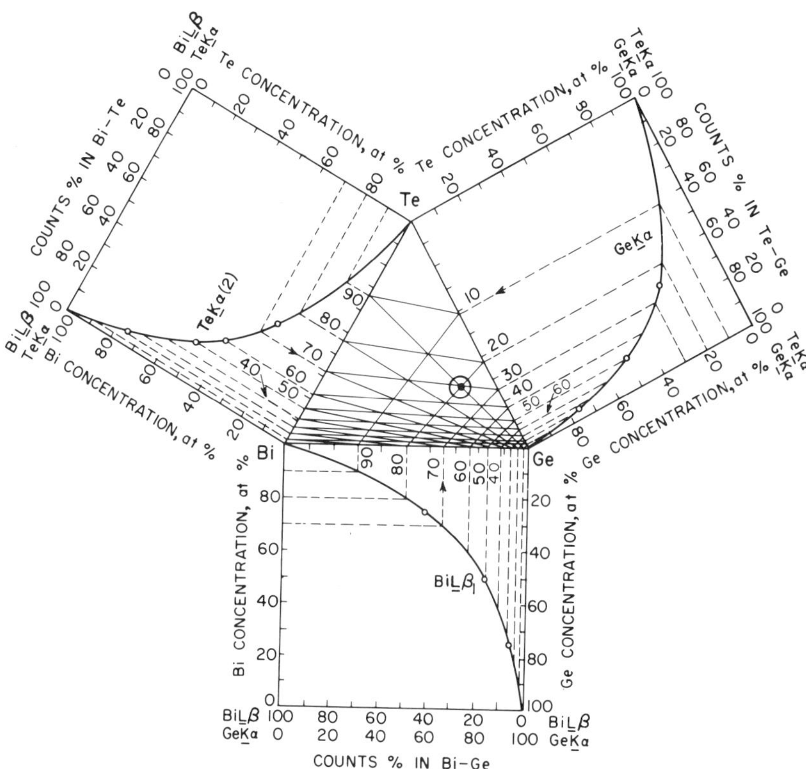

FIG. 15.2. Graphic method for analysis of ternary systems. The ticks on the sides of the triangle are linear scales of count percent. The skew grid is derived from the calibration curves constructed on the sides of the triangle and gives concentration. The dark spot is an analytical point (see text). [Data from A. Taylor, B. J. Kagle, and E. W. Beiter, *Analytical Chemistry* **33**, 1699 (1961).]

standards are used to construct calibration curves of counts percent *versus* atom percent on the sides of the triangle. These curves are used to graduate the triangle in concentration units, as shown by the dashed lines and numbers along each side. Finally, a concentration skew grid is drawn on the triangle.

An example of the application of Taylor's method follows. Consider a sample that gives count fractions for Ge $K\alpha$, Te $K\alpha(2)$, and Bi $L\beta$ of 58, 27, and 15%, respectively. This point is located on the triangular diagram of Figure 15.2 by reference to the uniformly spaced ticks along the sides of the triangle, which are linear scales of count percent. For simplicity, the point was selected to fall on an intersection on the atom concentration skew grid, and is circled on the figure. The atom concentrations of the three elements are now read off the skew grid as follows: Ge—20 at % on the Ge–Te scale; Te—70 at % on the Te–Bi scale; and Bi—10 at % on the Bi–Ge scale.

Another example of the geometric method is the application of nomographs (*M49*, *M50*).

15.3. ABSORPTION CORRECTION

Absorption-enhancement effects often consist simply of absorption effects, that is, the effects of the absorption coefficients of the matrix elements. In Table 15.1 and Figure 15.3B, the Sn $K\alpha$ intensity data from tin–lead alloys shows negative deviation from linearity (defined in Section 12.1)

TABLE 15.1. X-Ray Intensity Data for Tin–Lead Alloys[a]

Sample	Sn concn., wt %	Sn $K\alpha$ net intensity, counts/s	Pb concn., wt %	Pb $L\alpha_1$ net intensity, counts/s
1	100	497	0	4
2	74.98	314	24.97	177
3	59.97	227	39.92	260
4	40.00	144	59.89	373
5	20.18	66	79.84	473
6	0	2	100	568

[a] From Jenkins and de Vries [p. 138 in (*22*)].

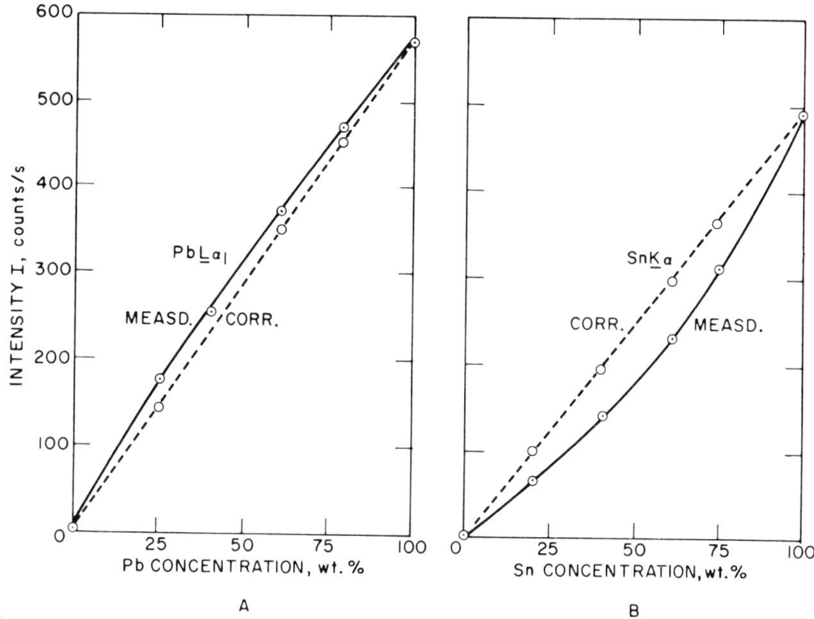

FIG. 15.3. Absorption-enhancement effects in the tin–lead system. [Data from R. Jenkins and J. L. de Vries, "Practical X-Ray Spectrometry," ed. 2, Springer-Verlag New York Inc., pp. 137–143 (1967).]

for this reason [pp. 141–143 in (*22*)]. At the primary wavelengths most efficient in exciting Sn $K\alpha$—those near the tin K-absorption edge (0.425 Å)—the mass-absorption coefficients of tin and lead are 40 and 25 cm²/g, respectively; at Sn $K\alpha$, they are 12 and 52 cm²/g. Then the total mass-absorption coefficients for both primary and Sn $K\alpha$ radiation are, for tin and lead, 52 and 77 cm²/g, respectively.

If it is assumed that only absorption effects are present,

$$I_X = C_{Sn} I_{100Sn} \frac{(\mu/\varrho)'_{100Sn}}{(\mu/\varrho)_X{}'} \quad (15.7)$$

where C_{Sn} is tin concentration; I_X and I_{100Sn} are Sn $K\alpha$ net intensities from the sample X and pure tin 100Sn, respectively; $(\mu/\varrho)_X{}'$ and $(\mu/\varrho)'_{100Sn}$ are total mass-absorption coefficients of the sample and pure tin, respectively, for both primary and Sn $K\alpha$ radiation. Equation (15.7) is derived from Equation (4.17) by substitution of a monochromatic effective wavelength for the term $J(\lambda_{pri})$, which represents the total primary spectrum. Equation (15.7) may now be used to correct the Sn $K\alpha$ data in Table 15.1. Values of

$(\mu/\varrho)_X'$ for samples 1–6 are calculated by use of the $(\mu/\varrho)'$ values for tin and lead given above. The Sn $K\alpha$ intensities from samples 2–5 are multiplied by the ratio $(\mu/\varrho)_X'/(\mu/\varrho)'_{100Sn}$, that is, $(\mu/\varrho)_X'/(\mu/\varrho)_1'$, where 1 refers to sample 1, which is pure tin. Then,

$$(I_X)_{cor} = (I_X)_{measd} \frac{(\mu/\varrho)_X'}{(\mu/\varrho)'_{100Sn}} \tag{15.8}$$

where $(I_X)_{cor}$ and $(I_X)_{measd}$ are, respectively, corrected and measured Sn $K\alpha$ net intensities from the sample. The $(I_X)_{cor}$ values are plotted in Figure 15.3B, giving the dashed linear curve. The Pb $L\alpha_1$ data in Table 15.1 and Figure 15.3A are discussed in Section 15.5.1.

Of course, $(\mu/\varrho)'$ for a sample cannot be calculated before the sample composition is known. This problem may be circumvented in several ways, including the following (see also Section 14.8.1):

1. If some accuracy may be sacrificed, an approximate analytical concentration may be derived from the uncorrected intensity data, and this value used to calculate $(\mu/\varrho)_X'$ to a first approximation.

2. Analyte-line intensity may be measured from pure analyte directly and with a sample disk of known mass and thickness placed over it. Since both primary and analyte-line radiation must pass through the sample disk, total $(\mu/\varrho)'$ for the matrix may be calculated.

3. If primary absorption can be neglected, the sample disk may be used to attenuate the analyte-line radiation from a sample of pure analyte.

4. This third technique can also be used to evaluate primary absorption if an effective wavelength can be assigned to the primary spectrum. A secondary target having a strong spectral line at this wavelength may be placed in the specimen compartment and its intensity measured through the sample disk.

5. If total (primary and secondary) absorption is not measured, the best approach is iteration. Approximate analyte concentrations are derived by assumption of a linear relationship between concentration and uncorrected analyte-line intensity. If all elements present are determined, the concentrations so derived may be normalized to 100%. These normalized concentrations are then used to calculate mass-absorption coefficients of the sample for all the analyte lines. So far, the procedure is simply that of method 1 above. These absorption coefficients are then used to correct the intensities, and a second set of concentrations is obtained and normalized. This iteration process is repeated until satisfactory convergence is realized.

Hower (*H67*) developed an absorption correction method for determination of trace heavy-element analytes in minerals. The method is based on the assumptions that: (1) of the three principal absorption-enhancement effects—primary-beam absorption, mutual enhancement, and secondary-beam absorption—only the first and last are significant for high-*Z* trace elements in minerals having light matrixes, and (2) these effects are determined by the mass-absorption coefficient of the mineral. Curves of mass-absorption coefficient as a function of wavelength were plotted for elements that are commonly major constituents in minerals. These curves showed that *relative* mass-absorption coefficient is approximately constant at all wavelengths. Curves of relative mass-absorption coefficient were plotted for various minerals relative to aluminum oxide. Matrix correction was then made for trace concentrations of nickel and heavier analytes by determination of the absorption of the sample mineral relative to aluminum oxide at a specified wavelength.

15.4. EMPIRICAL CORRECTION FACTORS

Mitchell (*M38, M40*), using empirical correction factors, successfully analyzed systems having up to seven components. The basic principle of the method is as follows. In the A–B–C–D–E–F–G system, line intensity of element A is measured, corrected for absorption-enhancement effects by C, D, E, F, and G, then applied to a calibration curve of A-line intensity *versus* A concentration in a B matrix. In other words, after correction, A is determined as if it were in an A–B binary system. Mitchell applied the method to determination of all seven metallic elements in the system TiO_2–V_2O_5–Fe_2O_3–ZrO_2–Nb_2O_5–Ta_2O_5–WO_3. The procedure follows, but for simplicity, the discussion is in terms of the elements, rather than the oxides.

The calibration procedure is as follows. For each element, one of the other six is selected as its binary complement. The complement may be the element having highest concentration or most severe absorption-enhancement effect, or it may simply be an element for which binary standards are available. The complement is not necessarily the same for all seven analytes. Table 15.2 gives Mitchell's choice of binary complements. Standards are prepared for each of the seven binary systems in the composition range of interest, and the analyte-line intensities are measured from all standards. A calibration curve of analyte-line intensity *versus* concentration is established for each of the seven systems. Then, for each analyte binary system, correction factors are derived for the other five elements, except that sim-

TABLE 15.2. Plan of Analysis of Seven-Component System[a]

Analyte	Binary complement for basic calibration curve	Interfering elements for which calculation of correction factors is required
Ti	Ta	V or Fe, Zr or Nb, W
V	Ti	Fe, Zr or Nb, Ta or W
Fe	Ta	Ti or V, Zr or Nb, W
Zr	Ta or W	Ti or V, Fe, Nb
Nb	Ta or W	Ti or V, Fe, Zr
Ta	Zr or Nb	Ti or V, Fe, W
W	Ta	Ti or V, Fe, Zr or Nb

[a] From Mitchell (*M40*).

ilarity of certain elements may preclude the need for separate factors, as shown in column 3 of Table 15.2. The derivation of the correction factors is described below.

The analytical procedure is as follows. Line intensity is measured for each of the seven elements from the sample(s). Mitchell measured the intensities for all seven elements simultaneously on an Applied Research Laboratories model XIQ multichannel x-ray quantometer. The measured line intensity of each of the seven analytes is applied to its binary calibration curve to derive its approximate concentration. This approximate concentration is then corrected by the following procedure, outlined for niobium as an example.

For each interfering element for the niobium system (Table 15.2), the correction factor is derived by applying the approximate analyte and interferant concentrations to the appropriate table or curve in the manner discussed below. The total correction for niobium in tantalum F_{NbTa} is then calculated from

$$F_{\mathrm{NbTa}} = \sum F_{\mathrm{NbTa},i} = F_{\mathrm{NbTa,Ti}} + F_{\mathrm{NbTa,V}} + F_{\mathrm{NbTa,Fe}} + F_{\mathrm{NbTa,Zr}} \quad (15.9)$$

where F_{NbTa} is the correction factor to be applied to the Nb $K\alpha$ intensity measured from the sample, and $F_{\mathrm{NbTa},i}$ is the correction factor for the effect of element i on Nb $K\alpha$ intensity in a tantalum matrix. Because of similarity of the effect of titanium and vanadium on Nb $K\alpha$ intensity in a tantalum matrix (column 3 of Table 15.2),

$$F_{\mathrm{NbTa,V}} = F_{\mathrm{NbTa,Ti}} \tag{15.10}$$

Because of similarity of the effect of tantalum and tungsten, no correction is required for tungsten, and these two elements form a common matrix. The niobium-line intensity measured from the sample is then corrected:

$$I_{\mathrm{Nb,cor}} = I_{\mathrm{Nb,measd}}/F_{\mathrm{NbTa}} \tag{15.11}$$

This corrected intensity is applied to the niobium calibration curve to derive corrected niobium concentration. This same procedure is followed for each of the other six elements.

The correction factors are derived as follows. Figure 15.4A shows calibration curves of Nb $K\alpha$ intensity *versus* Nb_2O_5 concentration in TiO_2 (or V_2O_5), Fe_2O_3, ZrO_2, and Ta_2O_5 (or WO_3) matrixes. These curves could be used for determination of Nb_2O_5 in the respective binary systems. For a Nb–Ti–Ta oxide ternary system, Nb $K\alpha$ intensities for all possible compositions lie on or between the Nb–Ti and Nb–Ta curves. Likewise, in the Nb–Fe–Zr system, all possible Nb $K\alpha$ intensities lie on or between the Nb–Fe and Nb–Zr curves. Of course, the entire "intensity area" may not be of analytical interest. Mitchell derived correction factors for multicomponent systems on the basis of such ternary relationships in two ways.

In general, for the multicomponent system A–B–C–D–· · ·, let A be the analyte and B its designated binary complement. Then the correction

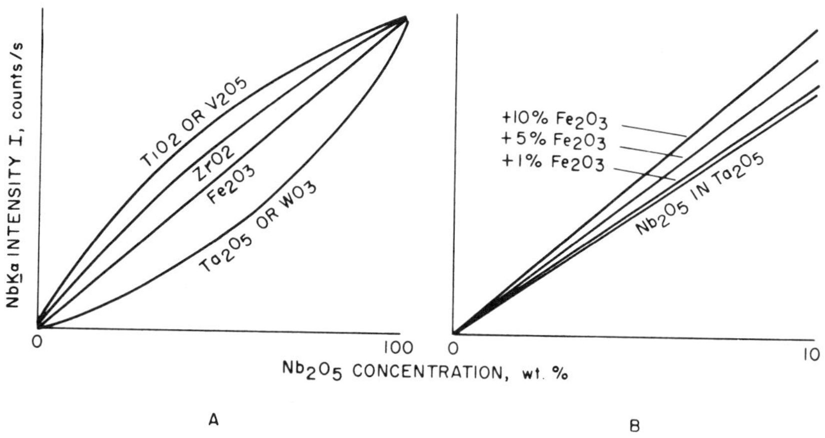

FIG. 15.4. Absorption-enhancement effects of several elements on Nb $K\alpha$ intensity.

factor for the effect of, say, C on A in a B matrix is

$$F_{\mathrm{AB,C}} = \frac{I_{\mathrm{A}} \text{ in a } B + C \text{ matrix}}{I_{\mathrm{A}} \text{ in a } B \text{ matrix}} \tag{15.12}$$

Note that the factor F is really analogous to relative intensity (Section 1.4.4) with the ternary and pure binary equivalent to the analytical sample and pure analyte, respectively. For absorption, $F < 1$; for enhancement, $F > 1$. Standards having appropriate A–B–C compositions are prepared, and the $F_{\mathrm{AB,C}}$ value is derived for each. A similar procedure is followed for A–B–D, A–B–E, etc.

The other method for derivation of correction factors is based on the fact that within a limited low-concentration region, curves of analyte-line intensity *versus* concentration are substantially linear. Figure 15.4B shows a set of curves of Nb $K\alpha$ net intensity as a function of Nb_2O_5 concentration in a Ta_2O_5 matrix having 0, 1, 5, and 10% Fe_2O_3 as the interferant (element C). These curves have the form

$$I = mC_{\mathrm{A}} + b \tag{15.13}$$

where m and b are slope and I-axis intercept, respectively. If I is net intensity, $b = 0$, as shown. Then, for a given C_{A}, correction factors are calculated for interferant concentrations 1, 2, 3, 4 (in Figure 15.4B, 0, 1, 5, 10% Fe_2O_3, respectively) as follows

$$\begin{aligned} F_1 &= m_1/m_1 = I_1/I_1 = 1 \\ F_2 &= m_2/m_1 = I_2/I_1 \\ F_3 &= m_3/m_1 = I_3/I_1 \\ F_4 &= m_4/m_1 = I_4/I_1 \end{aligned} \tag{15.14}$$

Regardless of which method is used, for each analyte binary system, a set of curves is prepared for each interfering element in the multicomponent system. These curves relate F with interferant concentration, each curve for a different analyte concentration. The curves have the form shown in Figure 15.5. Alternatively, the data may be prepared in tabular form similar to Table 15.3.

In a later paper, Mitchell (*M41*) showed that data based on the relationships of absorption coefficient, atomic number, and wavelength permit prediction of spectral-line intensity for a particular element over the full concentration range in a matrix of any atomic number on a particular

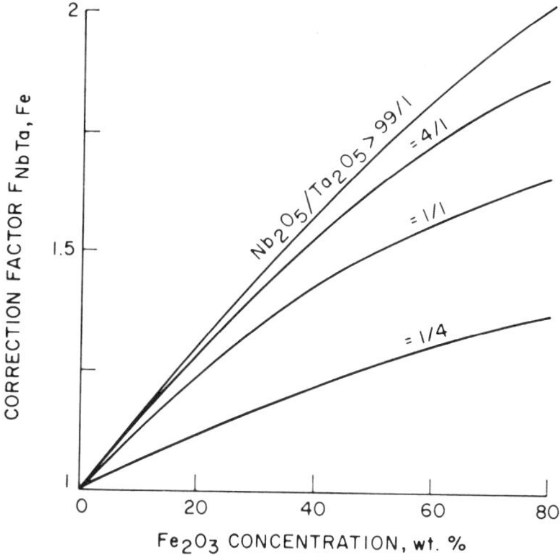

FIG. 15.5. Correction factors for iron on Nb $K\alpha$ intensity from a Ta_2O_5 matrix. [Data from B. J. Mitchell, *Analytical Chemistry* **30**, 1894 (1958).]

TABLE 15.3. Correction Factors for Effect of Fe_2O_3 on Nb $K\alpha$ Intensity from Nb_2O_5 in a Ta_2O_5 Matrix[a]

Fe_2O_3 concn., wt %	Correction factor F, for various Nb_2O_5/Ta_2O_5 weight ratios			
	>40/1	4/1	1/1	1/4
1	1.015	1.010	1.005	1.002
5	1.09	1.07	1.05	1.04
10	1.17	1.14	1.10	1.08
15	1.27	1.21	1.15	1.12
20	1.35	1.29	1.21	1.17
25	1.43	1.355	1.27	1.215
30	1.52	1.43	1.33	1.265
35	1.60	1.50	1.40	1.315
40	1.69	1.575	1.47	1.37
45	1.77	1.65	1.55	1.425
50	1.85	1.73	1.63	1.48

[a] From Mitchell (*M40*).

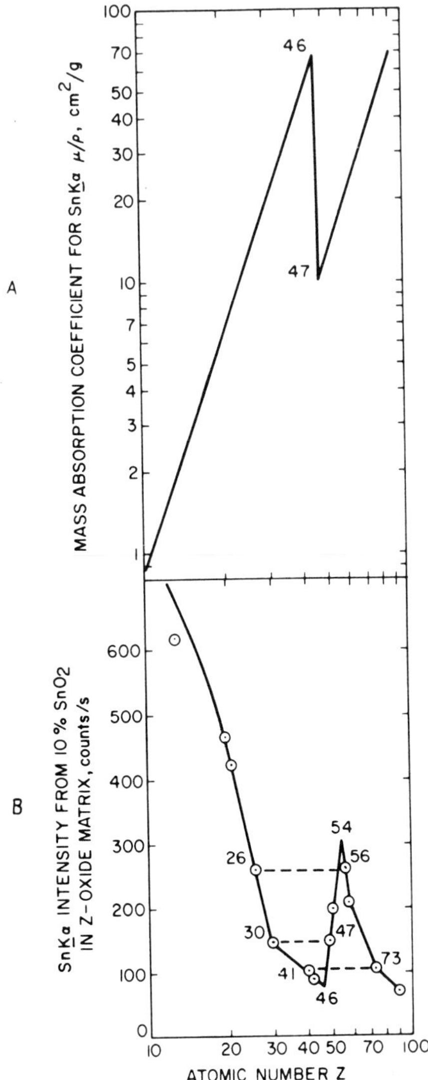

FIG. 15.6. Correlation of atomic number and (A) mass-absorption coefficient for Sn $K\alpha$ and (B) Sn $K\alpha$ intensity from 10% SnO_2. [B. J. Mitchell, *Analytical Chemistry* **33,** 917 (1961); courtesy of the author and the American Chemical Society.]

spectrometer. The spectral-line intensity of an element at each of several concentrations in each of several selected matrixes is measured and plotted. The form of the curves is shown in Figure 15.6B for 10% SnO_2 in matrixes of atomic number 10–92. The curve is essentially the reverse of the μ/ϱ *versus* Z curve for Sn $K\alpha$ (Figure 15.6A). However, the peak of the I–Z curve occurs at Z 54, rather than Z 47, as in the (μ/ϱ)–Z curve, because of enhancement of Sn $K\alpha$ by the $K\beta$ lines of elements having $Z > 51$ and by the $K\alpha$ lines of elements having $Z > 54$. This enhancement effect is more complex in a matrix emitting strong L lines. The line of interest undergoes variations in absorption due to its position with respect to the three matrix L-absorption edges, and variations in enhancement by the matrix L lines. It is evident that Sn $K\alpha$ intensity is the same for certain pairs of elements having very different atomic numbers—for example, $_{26}$Fe and $_{56}$Ba, $_{30}$Zn and $_{47}$Ag, and $_{41}$Nb and $_{73}$Ta.

From such sets of curves, intensity may be calculated for other concentrations and for other matrixes. In this way, Mo $K\alpha$ data measured at several concentrations in eight matrixes permitted the establishment of the curves for 0–100% molybdenum in all matrixes from $_{20}$Ca to $_{92}$U. Representative curves from this set are shown in Figure 15.7. Table 15.4 shows the additive nature of the correction factors for niobium in tantalum. The presence of titanium, vanadium, iron, and zirconium increases the Nb $K\alpha$ intensity relative to its intensity in a pure tantalum matrix: 1% each of TiO_2, V_2O_5, Fe_2O_3, and ZrO_2 gives an 8.7% enhancement; 10% of each gives an 86.8% enhancement. Mitchell proposes that these correction factors may be determined under standardized conditions readily attainable in any x-ray laboratory; then, in other laboratories, only a single basic calibration is required for each element in conjunction with sets of correction-factor tables or curves. In this respect, Mitchell's method is similar to that of Harvey (*H20*) in the field of optical emission spectrography.

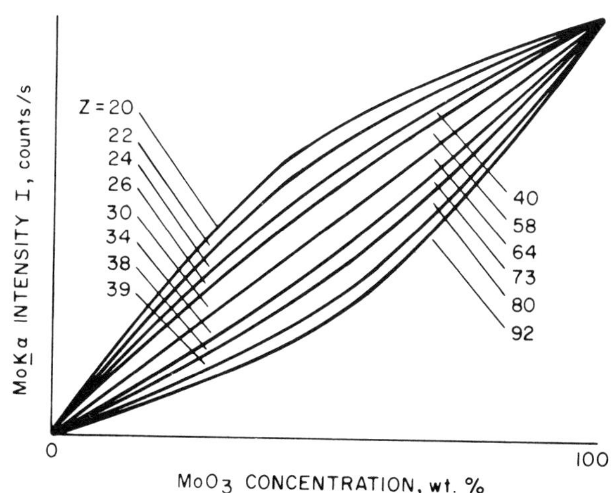

FIG. 15.7. Estimated Mo $K\alpha$ intensity patterns for matrixes of calcium to uranium (atomic number 20–92). [B. J. Mitchell, *Analytical Chemistry* **33**, 917 (1961); courtesy of the author and the American Chemical Society.]

TABLE 15.4. Additive Nature of Absorption-Enhancement Effects on Nb $K\alpha$ Intensity from Nb_2O_5 in a Ta_2O_5 Matrix[a]

Interferant and concentration, wt %	Individual effect of interferant: increase in Nb $K\alpha$ intensity,[b] %	Measured combined effect,[b] %
1% TiO_2	2.5	
1% V_2O_5	2.5	
1% Fe_2O_3	1.5	
1% ZrO_2	1.5	
TOTAL	8.0	8.7
10% TiO_2	26	
10% V_2O_5	26	
10% Fe_2O_3	17	
10% ZrO_2	18	
TOTAL	87	86.8

[a] From Mitchell (*M40*).
[b] With respect to Nb $K\alpha$ intensity from Nb_2O_5 in a Ta_2O_5 matrix in weight ratio >40/1.

15.5. INFLUENCE COEFFICIENTS

15.5.1. Introduction

Sherman (*S26–28*) attempted to derive general mathematical relationships between measured intensity and concentration in multicomponent systems. He first derived the intensity–concentration relationship for the element having highest atomic number. The *I–C* relationships for elements of successively lower atomic number were then corrected for enhancement by the spectra of elements of higher atomic number. However, the equations were successful only for limited concentration ranges in two- or three-component systems. Sherman's approach has been used also by Noakes (*N19*), Pluchery (*P23*), and Weyl (*W9*). Burnham *et al.* (*B94*) used graphic solutions to eliminate the need for solution of simultaneous equations.

The Pb $L\alpha_1$ data in Table 15.1 and Figure 15.3A provide a convenient and simple introduction to the influence-coefficient method for correction of absorption-enhancement effects [pp. 137–140 in (*22*)]. The Pb $L\alpha_1$ curve shows positive deviation from linearity (defined in Section 12.1), but for a reason different from that which explained the deviation in the Sn $K\alpha$ data (Section 15.3). The Pb $L\alpha_1$ line is not absorbed to very different degrees by tin and lead (μ/ϱ 134 and 128 cm²/g, respectively). The Pb $L\alpha_1$ line, which is associated with the Pb LIII absorption edge (0.95 Å), is not strongly enhanced by Sn $K\alpha$ (0.49 Å). The deviation is due to the difference in absorption coefficients of tin and lead for the primary wavelengths most efficient in exciting Pb $L\alpha_1$—those near Pb $L\text{III}_{\text{ab}}$; for example, at 0.9 Å, the absorption coefficients of tin and lead are 64 and 145 cm²/g, respectively; at 0.75 Å, they are 40 and 160. Thus, the excitation efficiency of the primary beam for Pb $L\alpha_1$ and the slope of the calibration curve increase with tin concentration. If it is assumed that the slope increases in proportion to tin concentration, the following equation obtains:

$$C_{\text{Pb}} = \frac{I_{\text{Pb}}}{m_{\text{Pb}}}(1 + a_{\text{PbSn}}C_{\text{Sn}}) \tag{15.15}$$

where C is concentration (wt fraction); I is net intensity; m is the slope of the curve, or Pb $L\alpha_1$ intensity per 1% lead for pure lead; and a_{PbSn} is the absorption-enhancement influence factor of tin for the Pb $L\alpha_1$ line. From this,

$$(I_{\text{Pb}})_{\text{cor}} = (I_{\text{Pb}})_{\text{measd}}(1 + a_{\text{PbSn}}C_{\text{Sn}}) \tag{15.16}$$

The value of a_{PbSn} is determined by substitution of calibration data from standards into Equation (15.15). In the present example, a_{PbSn} was -0.0021. The $(I_{\text{Pb}})_{\text{cor}}$ values are plotted in Figure 15.3A, giving the dashed linear curve. Alternatively, Equation (15.15) may be put in the form

$$C_{\text{Pb}}m_{\text{Pb}}/I_{\text{Pb}} = a_{\text{PbSn}}C_{\text{Sn}} + 1 \tag{15.17}$$

which has the form $y = mx + b$, so that a plot of $(Cm/I)_{\text{Pb}}$ *versus* C_{Sn} is substantially linear and has slope a_{PbSn}. If tin concentration in the standards is not known, Equation (15.15) may be expressed in terms of Sn $K\alpha$ net intensity I_{Sn}:

$$C_{\text{Pb}} = \frac{I_{\text{Pb}}}{m_{\text{Pb}}}(1 + a_{\text{PbSn}}I_{\text{Sn}}) \tag{15.18}$$

An equation analogous to Equation (15.16) is readily obtained from Equation (15.18), and a_{PbSn} is obtained from this equation in the same way as from Equation (15.16).

In general terms,

$$C_i = \frac{I_i}{m_i}[1 + \Sigma\,(a_{ij}C_j)] \qquad (j \neq i) \tag{15.19}$$

where C, I, and m have the significance given above; i is the analyte element; j is a matrix element; and a_{ij} is the absorption-enhancement effect of element j on the analyte i. The factor a is negative for calibration curves having negative deviation, $(\mu/\varrho)_j > (\mu/\varrho)_i$, and is positive for positive deviation, $(\mu/\varrho)_j < (\mu/\varrho)_i$. The influence factor corrects for both absorption and enhancement because the latter may be regarded as negative absorption.

For a ratio method,

$$C_{iX} = \left(\frac{I_X}{I_S}\right)_i \{1 + \sum [a_{ij}(C_X - C_S)_j]\} C_{iS} \qquad (j \neq i) \qquad (15.20)$$

where C and I are concentration and net intensity, respectively, and S and X refer to sample and standard, respectively.

Removal of the summation sign from Equation (15.20) gives the effect of a single matrix element j on analyte i. Rearrangement then gives

$$\left(\frac{C_X}{I_X}\,\frac{I_S}{C_S}\right)_i = a_{ij}(C_X - C_S)_j + 1 \qquad (15.21)$$

which has the form $y = mx + b$. Thus, a plot of $[(C_X/I_X)(I_S/C_S)]_i$ *versus* $(C_X - C_S)_j$ should be linear, intercept the ordinate at 1, and have slope a_{ij}. This serves as a test of the validity of a_{ij} values (see Section 15.5.3.5).

Beattie and Brissey (*B20*) analyzed three- and four-component systems by establishing linear simultaneous equations involving empirical absorption parameters or influence coefficients a relating intensities and mass concentrations for pairs of elements. For a binary system A–B, these coefficients have the form

$$a_{AB} = \frac{W_A}{W_B}(R_A' - 1) \qquad (15.22)$$

$$a_{BA} = \frac{W_B}{W_A}(R_B' - 1) \qquad (15.23)$$

where a_{AB} is the influence coefficient of element B for element A; a_{BA} is the influence coefficient of element A for element B; W is weight fraction; and R' is defined by

$$R_A' = I_{AA}/I_{AS}; \qquad R_B' = I_{BB}/I_{BS} \qquad (15.24)$$

where I_{AA} and I_{AS} are A-line intensity from pure A and from the specimen, and I_{BB} and I_{BS} have similar significance for B-line intensity.

The quantity a_{ij} (some workers use the symbol α_{ij}) is termed the absorption-enhancement *influence coefficient*, *influence factor*, *interaction coef-*

ficient, or *interaction factor* of matrix element j on analyte i. The influence coefficients may be calculated from certain known parameters, but are usually—and preferably—evaluated experimentally with data derived from standards and applied to sets of linear equations. Usually the coefficients are evaluated by inserting in the equations *concentrations* of matrix elements in the standards, but sometimes by inserting *intensities* of matrix-element lines measured from the standards. The intensity methods are applicable to wider matrix composition ranges and tend to reduce third-element effects, but require larger numbers of standards. The concentration methods are applicable to relatively more narrow matrix composition ranges.

Incidentally, in studying the literature on mathematical methods, one must take great care to note the authors' definitions of symbols, especially of a_{ij} and R. Some workers, especially outside the United States, use a_{ij} (or α_{ij}) to represent the influence of *matrix* element i on *analyte* j, the reverse of the definition used in this book. Most authors define R as the ratio of analyte-line intensities from specimen and pure analyte (or other reference standard) (Section 1.4.4), which is the reciprocal of Equation (15.24). The former definition, being more logical, is preferable and used throughout this book. However, in discussions of mathematical correction methods, the equations are presented in the form used by the original authors. Where these authors use the reciprocal form, the symbol R' $(=1/R)$ is used here. However, a_{ij} always signifies the effect of matrix element j on analyte element i, regardless of the original authors' preference.

Many workers have applied, modified, refined, and extended Beattie's and Brissey's basic approach. The following sections describe some of this work. Liebhafsky and his coauthors (*27*) emphasize that all these regression methods are "an extension of the comparative method of analysis (unknown *versus* standards), but by mathematical means. The principal weakness of regression methods is that they are valid only for limited composition ranges, the extent of which cannot be predicted beforehand."

15.5.2. Derivation of the Basic Equations

15.5.2.1. Birks' Derivation [pp. 58–60 in ed. 1 of (*8*)]

The derivation is based on the assumptions that: (1) the specimen is homogeneous, infinitely thick (in the sense of Section 14.7.2), and flat; (2) the primary x-radiation is monochromatic, or, alternatively, that a monochromatic beam has the same effect as the actual polychromatic primary spectrum for all specimen compositions of interest; and (3) en-

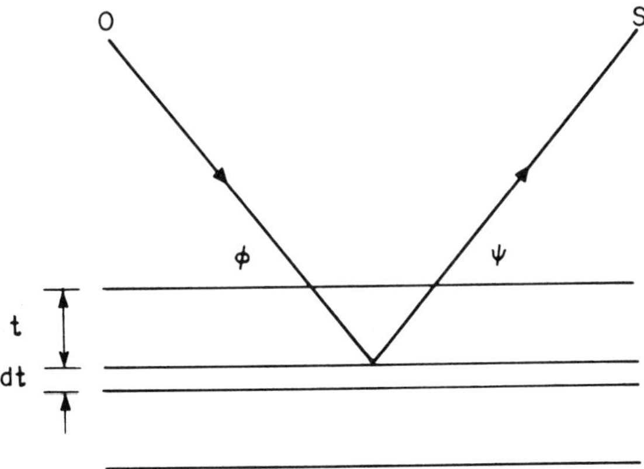

FIG. 15.8. Geometry of secondary excitation.

hancement by interelement secondary excitation within the specimen has the same effect as low matrix absorption—that is, enhancement is regarded as negative absorption.

In Figure 15.8, consider an incremental layer of thickness *dt* at depth *t* in a specimen of density ϱ. The incident angle of the primary beam *O* is ϕ, and the takeoff angle of the secondary beam *S* is ψ. Now consider a single wavelength λ_{pri} having intensity $I_{0,\lambda_{\mathrm{pri}}}$ in the incident primary beam. The intensity arriving at layer *dt* is

$$I_{t,\lambda_{\mathrm{pri}}} = I_{0,\lambda_{\mathrm{pri}}} \exp -[(\mu/\varrho)_{\lambda_{\mathrm{pri}}} \varrho t \csc \phi] \tag{15.25}$$

where $(\mu/\varrho)_{\lambda_{\mathrm{pri}}}$ is the mass-absorption coefficient of the specimen for λ_{pri}. $I_{t,\lambda_{\mathrm{pri}}}$ excites element *i* in layer *dt* to emit its characteristic spectrum, and the intensity of the measured line λ_L at layer *dt* is

$$dI_{t,\lambda_L} = T_i \varrho_i'' I_{t,\lambda_{\mathrm{pri}}}\, dt \tag{15.26}$$

where T_i is an excitation constant relating incident primary intensity and emitted spectral-line intensity, and ϱ_i'' is the density of element *i* in layer *dt*, that is, mass of element *i* per cubic centimeter *of the layer*. The intensity of the spectral line emerging from the specimen is

$$dI_{S,\lambda_L} = dI_{t,\lambda_L} \exp -[(\mu/\varrho)_{\lambda_L} \varrho t \csc \psi] \tag{15.27}$$

where $(\mu/\varrho)_{\lambda_L}$ is the mass-absorption coefficient of the specimen for λ_L.

Substitution of Equations (15.25) and (15.26) into (15.27) gives

$$dI_{S,\lambda_L} = T_i\varrho_i''I_{0,\lambda_{pri}}\{\exp -[(\mu/\varrho)_{\lambda_{pri}}\varrho t \csc \phi]\}\, dt \exp -[(\mu/\varrho)_{\lambda_L}\varrho t \csc \psi] \tag{15.28}$$

$$= T_i\varrho_i''I_{0,\lambda_{pri}} \exp -\{\varrho t[(\mu/\varrho)_{\lambda_{pri}} \csc \phi + (\mu/\varrho)_{\lambda_L} \csc \psi]\}\, dt \tag{15.29}$$

Since ϕ and ψ are constant for a given x-ray spectrometer, and $(\mu/\varrho)_{\lambda_{pri}}$ and $(\mu/\varrho)_{\lambda_L}$ are constant for a given specimen, the term in square brackets in Equation (15.29) may be replaced with a single constant $\overline{(\mu/\varrho)}$, (Section 2.1.1),

$$\overline{(\mu/\varrho)} = (\mu/\varrho)_{\lambda_{pri}} \csc \phi + (\mu/\varrho)_{\lambda_L} \csc \psi \tag{15.30}$$

The constant $\overline{(\mu/\varrho)}$ includes both enhancement and absorption effects. Equation (15.29) then becomes

$$dI_{S,\lambda_L} = T_i\varrho_i''I_{0,\lambda_{pri}}\{\exp[-\overline{(\mu/\varrho)}\varrho t]\}\, dt \tag{15.31}$$

Integration of Equation (15.31) from $t = 0$ to t gives

$$I_{S,\lambda_L} = \frac{T_i\varrho_i''I_{0,\lambda_{pri}}}{\overline{(\mu/\varrho)}\varrho}\{1 - \exp[-\overline{(\mu/\varrho)}\varrho t]\} \tag{15.32}$$

As t approaches infinity, Equation (15.32) becomes

$$I_{S,\lambda_L} = (T_i\varrho_i''I_{0,\lambda_{pri}})/[\overline{(\mu/\varrho)}\varrho] \tag{15.33}$$

Equation (15.33) may now be rewritten by: (1) expressing $\overline{(\mu/\varrho)}$ in terms of new coefficients a_{ij} representing the overall absorption-enhancement effect of element j on element i, so that

$$\overline{(\mu/\varrho)} = \sum a_{ij} \tag{15.34}$$

(2) taking into account all wavelengths that can excite element i; and (3) noting that

$$\varrho = \sum \varrho_i' = \sum \varrho_i'' \tag{15.35}$$

where ϱ_i' is the density of element i in the specimen, that is, mass of element i per cubic centimeter *of the specimen*. Then, from Equation (15.33),

$$\begin{aligned} T_i\varrho_i'I_0/I_i &= \overline{(\mu/\varrho)}\varrho = \sum a_{ij}\varrho_j' \\ &= a_{i1}\varrho_1' + a_{i2}\varrho_2' + \cdots + a_{ii}\varrho_i' + \cdots + a_{in}\varrho_n' \end{aligned} \tag{15.36}$$

where, for simplicity, I_0 is used for $I_{0,\lambda_{pri}}$, and I_i is used for I_{S,λ_L}, the measured analyte-line intensity for element i. The term a_{ii} is always a true constant, and, in fact, has the value 1. This is readily shown by solving Equation (15.36) for the case of pure (100%) element i (100i): all the ϱ_j' terms become zero, except that ϱ_i' becomes ϱ, the specimen density; I_i becomes I_{100i}; and Equation (15.36) becomes

$$T_i \varrho I_0 / I_{100i} = a_{ii} \varrho \tag{15.37}$$

The ϱ's cancel, and from the definition of T_i above, $T_i I_0 = I_{100i}$. Equation (15.37) then reduces to $a_{ii} = 1$. Equation (15.36) may now be rewritten by substitution of these values for $T_i I_0$ and a_{ii} and gathering terms:

$$a_{i1}\varrho_1' + a_{i2}\varrho_2' + \cdots + [1 - (I_{100i}/I_i)]\varrho_i' + \cdots + a_{in}\varrho_n' = 0 \tag{15.38}$$

where I_i and I_{100i} represent analyte-line intensity from element i from the specimen and from pure i, respectively. If Equation (15.38) is divided through by ϱ, and it is noted that ϱ_i'/ϱ is weight fraction W_i, one gets

$$a_{i1}W_1 + a_{i2}W_2 + \cdots + [1 - (I_{100i}/I_i)]W_i + \cdots + a_{in}W_n = 0 \tag{15.39}$$

A similar expression can be written for each element in the specimen, and, with

$$I_{100i}/I_i = R_i' \tag{15.40}$$

the following set of simultaneous equations is obtained:

$$(1 - R_1')W_1 + a_{12}W_2 + a_{13}W_3 + \cdots + a_{1n}W_n = 0 \quad (15.41\text{-}1)$$

$$a_{21}W_1 + (1 - R_2')W_2 + a_{23}W_3 + \cdots + a_{2n}W_n = 0 \quad (15.41\text{-}2)$$

.....

$$a_{n1}W_1 + a_{n2}W_2 + a_{n3}W_3 + \cdots + (1 - R_n')W_n = 0 \quad (15.41\text{-}n)$$

$$W_1 + W_2 + W_3 + \cdots + W_n = 1 \quad (15.42)$$

Each individual member of the set of Equations (15.41) corresponds to one analyte i and contains expressions of the interactions of all the other constituents in the system—that is, of all the matrix elements j—with that analyte. Each analyte is seen to constitute a matrix element for every other analyte. Equation (15.42) is required because the n equations (15.41) are not mutually independent. In practice, if minor constituents are known to be present, but are of no interest or significance, Equation (15.42) may be set equal to the sum of the concentrations of the major and other significant constituents—for example, 0.98. These equations are used both for calibration, that is, evaluation of the a_{ij} coefficients from measurements on suitable standards (Sections 15.5.3.2–15.5.3.5) and for determination, that is, evaluation of the analytical concentrations by use of the a_{ij} values.

15.5.2.2. Müller's Derivation; Regression Coefficients

A similar set of equations is derived in a more direct way by Müller [p. 171 in (*29*)] by use of the regression equation. The derivation is based on the assumptions that: (1) the analyte-line x-rays from the sample and pure analyte are excited by substantially the same wavelength region of the primary spectrum so that the effective primary wavelength is the same for both; and (2) analyte-line intensity is affected only by differences in absorption by the analyte and matrix, and true enhancement by spectral lines of matrix elements can be disregarded.

The relative analyte-line intensity of component 1 in a multicomponent system is approximated by

$$\frac{I_1}{I_{100,1}} \approx \frac{\overline{(\mu/\varrho)}_{11}C_1}{\sum[\overline{(\mu/\varrho)}_{1j}C_j]}$$

$$\approx \frac{\overline{(\mu/\varrho)}_{11}C_1}{\overline{(\mu/\varrho)}_{11}C_1 + \overline{(\mu/\varrho)}_{12}C_2 + \cdots + \overline{(\mu/\varrho)}_{1n}C_n} \qquad (15.43)$$

where I_1 and $I_{100,1}$ are analyte-line intensities of element 1 from the sample and pure element, respectively; $\overline{(\mu/\varrho)}_{11}$ and $\overline{(\mu/\varrho)}_{1j}$ are the total mass-absorption coefficients of elements 1 and j, respectively, for both the primary x-rays and the analyte line of element 1 [Equation (15.30)]; and C is concentration (wt fraction).

From Equation (15.43),

$$C_1 \approx \frac{I_1}{I_{100,1} - I_1} \sum \left[\frac{\overline{(\mu/\varrho)}_{1j}}{\overline{(\mu/\varrho)}_{11}} C_j \right] \qquad (j \neq 1) \qquad (15.44)$$

$$\approx \frac{I_1}{I_{100,1} - I_1} \left[\frac{\overline{(\mu/\varrho)}_{12}}{\overline{(\mu/\varrho)}_{11}} C_2 + \cdots + \frac{\overline{(\mu/\varrho)}_{1n}}{\overline{(\mu/\varrho)}_{11}} C_n \right] \qquad (15.45)$$

In Equation (15.45), the term on the right outside the brackets gives the relationship between C_1 and analyte-line intensity. The terms in the brackets give the interactions of the other components j on the analyte-line intensity and consist of the concentrations of these components and their interaction coefficients with component 1, $\overline{(\mu/\varrho)_{1j}}/\overline{(\mu/\varrho)_{11}}$. Müller terms these factors *regression coefficients* r_{ij}, that is,

$$r_{ij} = \overline{(\mu/\varrho)_{ij}}/\overline{(\mu/\varrho)_{ii}} \tag{15.46}$$

Equations (15.44) and (15.45) now become

$$C_1 \approx \frac{I_1}{I_{100,1} - I_1} \sum (r_{1j}C_j) \qquad (j \neq 1) \tag{15.47}$$

$$\approx \frac{I_1}{I_{100,1} - I_1} (r_{12}C_2 + \cdots + r_{1n}C_n) \tag{15.48}$$

Similar equations can be written for all components of the n-component system. If now each of these n equations is rearranged with all terms on one side of the equality sign and multiplied by $(I_{100i} - I_i)/I_i$, one gets

$$-\frac{I_{100,1} - I_1}{I_1} C_1 + r_{12}C_2 + r_{13}C_3 + \cdots + r_{1n}C_n = 0 \tag{15.49-1}$$

$$r_{21}C_1 - \frac{I_{100,2} - I_2}{I_2} C_2 + r_{23}C_3 + \cdots + r_{2n}C_n = 0 \tag{15.49-2}$$

.....

$$r_{n1}C_1 + r_{n2}C_2 + r_{n3}C_3 + \cdots - \frac{I_{100n} - I_n}{I_n} C_n = 0 \tag{15.49-n}$$

If a_{ij} is substituted for r_{ij} throughout Equations (15.49) (see below), they assume the form of Equations (15.41), since

$$1 - R_i' = 1 - \frac{I_{100i}}{I_i} = -\frac{I_{100i} - I_i}{I_i} \tag{15.50}$$

Equations (15.41) Equations (15.49)

Müller's regression coefficients are evaluated not only from Equation (15.46), but from analyte-line intensities I_i and I_{100i} measured from binary standards having composition C_i–C_j and pure analyte, respectively:

$$r_{ij} = \frac{C_i}{C_j} \frac{I_{100i} - I_i}{I_i} = \frac{C_i}{1 - C_i} \frac{I_{100i} - I_i}{I_i} \tag{15.51}$$

If the value of I_{100i} cannot be measured or estimated, r_{ij} can be evaluated from intensities measured from two binary standards having analyte concentrations C_{i1} and C_{i2}, respectively:

$$r_{ij} = \frac{I_{i2} - I_{i1}}{I_{i2} - I_{i1} + (I_{i1}/C_{i1}) - (I_{i2}/C_{i2})} \tag{15.52}$$

The value of r_{ij} is constant for the entire range of concentrations in the i–j system if the assumptions on which Müller's derivation is based are valid for the i–j system. In fact, comparison of values of r_{ij} from several binary standards representing a wide concentration range permits testing of the validity of these assumptions for a specified i–j system.

Another application of regression coefficients is to predict the curvature of the calibration curve. In the absence of absorption-enhancement effects, analyte-line intensity is given by

$$I_i = C_i I_{100i} \tag{15.53}$$

In a system for which the regression coefficients are valid,

$$I_i = \frac{C_i}{C_i + (1 - C_i)r_{ij}} I_{100i} \tag{15.54}$$

For $r_{ij} < 1$, $I_i > C_i I_{100i}$, and the calibration curve is convex upward (Figure 14.8A, curve A–C); for $r_{ij} = 1$, $I_i = C_i I_{100i}$, and the curve is linear; for $r_{ij} > 1$, $I_i < C_i I_{100i}$, and the curve is concave upward (Figure 14.8A, curve A–B).

15.5.2.3. Influence Coefficient Symbols

Müller's regression coefficients r_{ij} and Beattie's and Brissey's influence coefficients a_{ij} are equivalent. Either may be calculated approximately from mass-absorption coefficients [Equation (15.46)] or evaluated experimentally from binary i–j standards having analyte concentration C_i, that is,

$$\frac{\overline{(\mu/\varrho)_{ij}}}{\overline{(\mu/\varrho)_{ii}}} \approx (a_{ij} \Leftrightarrow r_{ij}) = \frac{C_i}{1 - C_i} \frac{I_{100i} - I_i}{I_i} \tag{15.55}$$

where I_i and I_{100i} are analyte-line intensities from the binary standard and pure i, respectively. The term following the equality sign is obtained from Equation (15.48) for a binary system by setting $C_2 = 1 - C_1$ and solving for r_{12} (or a_{12}).

However, in this chapter, a distinction is drawn between r_{ij} and a_{ij} on the basis of their method of determination. The symbol r_{ij} is reserved for coefficients *calculated* from known or measured mass-absorption coefficients of analyte i and matrix j elements for the primary x-rays (or their effective wavelength) and for the measured analyte line of element i. These coefficients correct only for absorption. The symbol a_{ij} is reserved for coefficients *derived experimentally* from measurements on standards of known composition and that correct for both absorption and true enhancement, which appears as "negative absorption."

An advantage of a_{ij}-type coefficients relative to r_{ij}-type coefficients is that prior knowledge of primary and secondary mass-absorption coefficients is not required. A serious disadvantage of the a_{ij} coefficients is that standards of accurately known composition are required for their evaluation, and the larger the number of analytes, the larger is the number of standards required. Perhaps the most serious limitation is that the coefficients are derived not from true mathematical relationships, but from measured x-ray intensities and "known" standard concentrations that are themselves subject to significant random and systematic errors.

A third symbol, a'_{ij}, is used to indicate experimentally derived influence coefficients not identical with those of Beattie and Brissey. The coefficients of Sherman, Burnham *et al.*, Traill and Lachance, and Lucas-Tooth and Pyne (see below) are of this type.

15.5.3. Solution of the Basic Equations

15.5.3.1. Simplified Solutions

The sets of Equations (15.41) and (15.49) may be solved in many ways with various degrees of convenience and accuracy.

Many multicomponent systems consist of two or three major constituents and numerous minor and trace constituents. In such systems, the analyte-line intensities of the major constituents are influenced largely by absorption-enhancement effects of other major constituents, and the intensities of the lines of the minor constituents are also influenced largely by the major constituents. Thus, to a first approximation, the effect of the minor constituents may be disregarded, and the concentrations of minor constituents as matrix elements j may be disregarded. For such systems, the set of equations may be divided into two subsets—those for the major and minor constituents, respectively. The concentrations of the major constituents are now determined algebraically or by use of binary calibration standards involving pairs of major constituents [Equation (15.55) and Section 15.5.3.2]. Once these concentrations are determined, the concentrations of the minor constituents can be determined, for example, by use of binary calibration standards involving major–minor component pairs.

Another approach is to set

$$a_{ij} \approx r_{ij} \approx \overline{(\mu/\varrho)}_{ij}/\overline{(\mu/\varrho)}_{ii} \approx (\mu/\varrho)_{j,\lambda_i}/(\mu/\varrho)_{i,\lambda_i} \tag{15.56}$$

where $\overline{(\mu/\varrho)}_{ii}$ and $\overline{(\mu/\varrho)}_{ij}$ are the total mass-absorption coefficients of, respectively, elements i and j for both the primary x-rays and the analyte line of element i; and $(\mu/\varrho)_{i,\lambda_i}$ and $(\mu/\varrho)_{j,\lambda_i}$ are the mass-absorption coefficients of, respectively, elements i and j for only the analyte line of element i. Of course, the first ratio gives greater accuracy than the second, but both disregard true enhancement by spectral lines of matrix elements. The first ratio is more difficult to calculate, but can be approximated by assuming an effective wavelength for the primary x-rays.

A still more rapid—and less accurate—approach is to let

$$a_{ij} = 1 \tag{15.57}$$

Equations (15.41) then reduce to

$$(1 - R_1')W_1 + W_2 + W_3 + \cdots + W_n = 0 \tag{15.58-1}$$

$$W_1 + (1 - R_2')W_2 + W_3 + \cdots + W_n = 0 \tag{15.58-2}$$

. . . . .

$$W_1 + W_2 + W_3 + \cdots + (1 - R_n')W_n = 0 \tag{15.58-n}$$

The individual concentrations are then simply

$$W_i = \frac{I_i}{I_{100i}} = \frac{1}{R_i'} \tag{15.59}$$

where I_i and I_{100i} are analyte-line intensities of element i from the sample and pure i, respectively. Insertion of these concentrations in Equations (15.41) permits calculation of the a_{ij}'s to a first approximation. These approximate a_{ij}'s can then be used to determine whether the true values of W_i are larger than, equal to, or smaller than the values estimated by Equation (15.59). For example, if the influence coefficients in Equation (15.41-n) have *average* value >1, then

$$\sum W_j a_{nj} > \sum W_j \qquad (j \neq n) \tag{15.60}$$

and since

$$\sum W_j = 1 - W_n \qquad (j \neq n) \tag{15.61}$$

then

$$(1 - R_n')W_n > 1 - W_n \tag{15.62}$$

and the true concentration of element n is

$$W_n > I_n/I_{100n} \tag{15.63}$$

Similarly, when the values of a_{ij} average <1, the true concentration is smaller than the estimated one, and where the values of a_{ij} are ~ 1, the true concentration is approximately the estimated one. A similar line of reasoning is applicable to Equations (15.49).

15.5.3.2. Evaluation of a_{ij} Coefficients to a First Approximation

Because of the assumptions on which the derivation is based, Equations (15.41) are stated in terms of binary influence coefficients a_{ij}. Each of these coefficients represents the effect of only one element on another, disregarding all other elements present, and is therefore assumed to be constant. Unfortunately, the a_{ij} coefficients are not true constants, but are themselves functions of all other elements present. Sherman (*S28*) has analyzed this complex situation, but his equations can be solved only for specimens having only a few components. However, there are many systems for which acceptable accuracy is realized by considering the a_{ij} coefficients to be constant to a first-order or second-order approximation. The method has been applied especially successfully in metallurgical analyses where reliable standards are available for use in evaluating the coefficients.

To a first approximation, each a_{ij} coefficient may be assumed to depend only on elements i and j, and its value may be derived from the intensities of the spectral line of i from pure i and from a binary standard of i–j having known composition. The i/j concentration ratio in this standard may be 1/1 or, preferably, approximately the ratio of i and j in the samples to be analyzed. In effect, this approach constitutes a division of Equations (15.41) into independent subequations of lower degree. If the samples contain n elements, it is necessary to establish a set of n simultaneous equations of the form of Equations (15.41). These equations contain $n(n - 1)$ coefficients of the form a_{ij} ($j \neq i$). Evaluation of these coefficients requires the n pure elements and their $(n/2)(n - 1)$ binary combinations.

To illustrate the procedure [pp. 60–61 in ed. 1 of (*8*)], suppose alloys of Cr–Fe–Ni–Mo are to be analyzed. Let i (or j) $= 1, 2, 3, 4$, respectively, for these elements. The designation i or j depends on whether the element is being considered as an analyte or matrix element, respectively. To derive a_{CrMo}, the effect of Mo ($j = 4$) on Cr ($i = 1$), Equation (15.41-1) is used, but for a Cr–Mo binary system, all terms drop out except those containing W_{Cr} and W_{Mo}, leaving

$$(1 - R'_{Cr})W_{Cr} + a_{CrMo}W_{Mo} = 0 \tag{15.64}$$

from which

$$a_{CrMo} = (R'_{Cr} - 1)(W_{Cr}/W_{Mo}) \tag{15.65}$$

R'_{Cr}, that is, I_{100Cr}/I_{Cr}, is obtained from the Cr $K\alpha$ intensities from pure chromium and the chromium–molybdenum binary alloy standard. W_{Cr} and W_{Mo} are the known concentrations in the standard. Similarly, a_{MoCr}, the effect of chromium on molybdenum, is derived from Equation (15.41-4) and Mo $K\alpha$ intensities from pure molybdenum and the chromium–molybdenum standard:

$$a_{MoCr}W_{Cr} + (1 - R'_{Mo})W_{Mo} = 0 \tag{15.66}$$

$$a_{MoCr} = (R'_{Mo} - 1)(W_{Mo}/W_{Cr}) \tag{15.67}$$

In general,

$$a_{ij} = (R_i' - 1)W_i/W_j \tag{15.68}$$

Once all the a_{ij} coefficients are evaluated and substituted in Equations (15.41), these equations serve to calculate the composition of any quaternary alloy of these elements to a first approximation. The analysis of each sample consists in: (1) evaluation of R_i' for each of the four elements by measurement of the i spectral-line intensity from pure i and from the sample; (2) substitution of the R_i' values in Equations (15.41), which already contain the a_{ij} coefficients; and (3) solution of the simultaneous equations for the weight fractions W, usually by the method of determinants or with a computer.

The method of first approximations was used by Beattie and Brissey (*B20*) for analysis of chromium–iron–nickel–molybdenum alloys with relative errors of 0.5–6.5%, and by Hakkila and Waterbury (*H4*) for analysis of solutions of chromium–iron–nickel alloys. These analyses provide excellent practical examples of the application of the first-approximation method and are discussed in detail below.

The procedure of Hakkila and Waterbury (*H4*) is as follows:

1. Six solutions were prepared containing 200 mg of, respectively, chromium, iron, nickel, Cr–Fe (1:1), Cr–Ni (1:1), and Fe–Ni (1:1).

2. Cr $K\alpha$, Fe $K\alpha$, and Ni $K\alpha$ net intensities were measured from all six solutions; Cr $K\alpha$ background was obtained by averaging the intensities at 2θ for Cr $K\alpha$ from the pure iron and nickel solutions; Fe $K\alpha$ and Ni $K\alpha$ backgrounds were derived in a similar way.

3. Six net-intensity ratios were calculated: R_{CrFe}, R_{CrNi}, R_{FeCr}, R_{FeNi}, R_{NiCr}, and R_{NiFe}. These ratios had the form

$$R_{\mathrm{CrFe}} = \frac{\text{Cr } K\alpha \text{ intensity from the (1:1) Cr–Fe solution}}{\text{Cr } K\alpha \text{ intensity from the pure Cr solution}} \tag{15.69}$$

4. Six interelement influence coefficients were then calculated: a_{CrFe}, a_{CrNi}, a_{FeCr}, a_{FeNi}, a_{NiCr}, and a_{NiFe}; for example, a_{CrFe} is the influence factor for the effect of iron on Cr $K\alpha$ intensity and is calculated from

$$\frac{1 - R_{\mathrm{CrFe}}}{R_{\mathrm{CrFe}}} = a_{\mathrm{CrFe}} \frac{1 - C_{\mathrm{Cr}}}{C_{\mathrm{Cr}}} \tag{15.70}$$

where C is concentration; since C_{Cr} in the (1:1) Cr–Fe solution is 0.5, the fractional term on the right side of the equation becomes 1, thereby permitting calculation of a_{CrFe}. Once the coefficients were evaluated, the analysis procedure was as follows.

5. The Cr $K\alpha$, Fe $K\alpha$, and Ni $K\alpha$ net intensities were measured from all sample solutions and from the pure chromium, iron, and nickel solutions.

6. Three intensity ratios were calculated for each sample: R_{Cr}, R_{Fe}, and R_{Ni}. These ratios had the form

$$R_{\mathrm{Cr}} = \frac{\text{Cr } K\alpha \text{ intensity from the sample}}{\text{Cr } K\alpha \text{ intensity from the pure Cr solution}} \tag{15.71}$$

7. Chromium concentration was then calculated from the equation

$$\frac{1 - R_{\mathrm{Cr}}}{R_{\mathrm{Cr}}} = \left(\frac{a_{\mathrm{CrFe}} R_{\mathrm{Fe}} + a_{\mathrm{CrNi}} R_{\mathrm{Ni}}}{R_{\mathrm{Fe}} + R_{\mathrm{Ni}}}\right)\left(\frac{1 - C_{\mathrm{Cr}}}{C_{\mathrm{Cr}}}\right) \tag{15.72}$$

Nickel concentration was calculated from a similar equation.

8. Using these calculated values for C_{Cr} and C_{Ni}, iron concentration was calculated from the equation

$$\frac{1 - R_{\mathrm{Fe}}}{R_{\mathrm{Fe}}} = \left(\frac{a_{\mathrm{FeCr}} C_{\mathrm{Cr}} + a_{\mathrm{FeNi}} C_{\mathrm{Ni}}}{C_{\mathrm{Cr}} + C_{\mathrm{Ni}}}\right)\left(\frac{1 - C_{\mathrm{Fe}}}{C_{\mathrm{Fe}}}\right) \tag{15.73}$$

Table 15.5 gives the data of Beattie and Brissey (*B20*) for analysis of chromium–iron–nickel–molybdenum alloys to a first approximation. Part *A* of the table gives the compositions of the 12 binary standards in weight percent *W*. The influence coefficients were evaluated as follows. Cr $K\alpha$ net intensity was measured from the three chromium binaries and from pure chromium; Fe $K\alpha$ net intensity was measured from the iron binaries and

TABLE 15.5. Data for Analysis of Cr–Fe–Ni–Mo Alloys to a First Approximation by the Influence-Coefficient Method[a]

Element i	Element j			
	Cr	Fe	Ni	Mo
A. Composition of binary standards W, wt %				
Cr	—	48.24	48.07	74.28
Fe	50.83	—	51.53	65.33
Ni	48.19	46.65	—	53.70
Mo	23.53	34.44	46.27	—
B. Intensity ratios for binary standards, $R_i' = I_{100i}/I_i$				
Cr	1.000	1.760	1.815	1.841
Fe	3.360	1.000	1.613	1.835
Ni	2.860	3.670	1.000	1.954
Mo	3.770	2.787	2.461	1.000
C. Influence coefficients a_{ij} from Equations (15.41)				
Cr	1.000	0.721	0.813	2.660
Fe	2.482	1.000	0.676	1.582
Ni	1.863	2.420	1.000	1.108
Mo	0.877	0.944	1.260	1.000
D. Intensity ratio for an analytical sample, $R_i' = I_{100i}/I_i$				
—	R'_{Cr}	R'_{Fe}	R'_{Ni}	R'_{Mo}
	5.04	2.346	6.640	15.34

[a] From Beattie and Brissey (*B20*).

from iron; etc. The values of R_i' in part *B* of the table were calculated from these intensities. The influence coefficients in part *C* were obtained by insertion of the R_i' values into a set of four simultaneous equations having the form of Equations (15.41). Then the simultaneous equations for analysis of chromium–iron–nickel–molybdenum alloys are

$$\begin{aligned}
(1 - R'_{\mathrm{Cr}})W_{\mathrm{Cr}} + 0.721W_{\mathrm{Fe}} + 0.813W_{\mathrm{Ni}} + 2.660W_{\mathrm{Mo}} &= 0\\
2.482W_{\mathrm{Cr}} + (1 - R'_{\mathrm{Fe}})W_{\mathrm{Fe}} + 0.676W_{\mathrm{Ni}} + 1.582W_{\mathrm{Mo}} &= 0\\
1.863W_{\mathrm{Cr}} + 2.420W_{\mathrm{Fe}} + (1 - R'_{\mathrm{Ni}})W_{\mathrm{Ni}} + 1.108W_{\mathrm{Mo}} &= 0\\
0.877W_{\mathrm{Cr}} + 0.944W_{\mathrm{Fe}} + 1.260W_{\mathrm{Ni}} + (1 - R'_{\mathrm{Mo}})W_{\mathrm{Mo}} &= 0
\end{aligned} \tag{15.74}$$

An analysis of an alloy of this quaternary system consists in measurement of the Cr $K\alpha$, Fe $K\alpha$, Ni $K\alpha$, and Mo $K\alpha$ net intensities from the sample and from the four respective elements. The four intensity ratios R_i' are then calculated; ratios for a typical alloy are given in Part *D* of Table 15.5. These ratios inserted in Equations (15.74) permit calculation of composition of the sample.

However, it is evident that because there are no constant terms in the simultaneous equations, only a ratio among the four unknowns can be obtained. Such a ratio can be obtained by dividing through the four equations by any of the four unknowns. In this way, one of the four equations —that corresponding to the divisor—can be omitted, leaving three equations to be solved for three unknown ratios. Thus, there are 16 independent ways to solve the four equations. These 16 solutions yield 16 composition ratios, which may differ somewhat because the equations are based on assumptions that are not wholly valid, and the influence coefficients are empirical and evaluated only approximately. Beattie and Brissey give a procedure for preselection of the solution likely to minimize the errors introduced by the assumptions and the coefficients.

15.5.3.3. Evaluation of a_{ij} Coefficients to a Second Approximation

To a second approximation, each a_{ij} coefficient may be evaluated from the intensities of the spectral line of i from pure i, two binary alloys, and a ternary alloy. The simplest case to which this approximation is applicable with high accuracy is that in which one major constituent has approximately the same concentration in all samples. To illustrate the procedure [pp. 61–62 in ed. 1 of (*8*)], suppose that samples of a five-component alloy are to be analyzed in which the concentration of one major constituent is always ~65–75%. Let this element be designated i (or j) = 1. The coefficients for elements, say, 3 and 5 are then evaluated as follows:

1. A first approximation of $a_{31(1)}$ is made from the intensities of the spectral line of element 3 from pure 3 and from a 70:30 1–3 binary standard. From Equations (15.41-3) and (15.68),

$$a_{31(1)} = (R_3' - 1)W_3/W_1 \tag{15.75}$$

2. A first approximation of $a_{51(1)}$ is made from the intensities of the spectral line of element 5 from pure 5 and from a 70:30 1–5 binary standard. Again, from Equations (15.41-5) and (15.68),

$$a_{51(1)} = (R_5' - 1)W_5/W_1 \tag{15.76}$$

3. A second approximation of $a_{35(2)}$ is then made from $a_{31(1)}$ and from the intensity of the spectral line of element 3 from a 70:15:15 1–3–5 ternary standard. From Equation (15.41-3),

$$a_{31(1)}W_1 + (R_3' - 1)W_3 + a_{35(2)}W_5 = 0 \tag{15.77}$$

Similarly, from $a_{51(1)}$, the intensity of the spectral line of element 5 from the 1–3–5 standard, and Equation (15.41-5),

$$a_{51(1)}W_1 + a_{53(2)}W_3 + (R_5' - 1)W_5 = 0 \tag{15.78}$$

All terms in Equations (15.77) and (15.78) are known except the two second approximations $a_{35(2)}$ and $a_{53(2)}$.

For a system of n elements, the following standards are required: the n pure elements, the $(n - 1)$ binaries of element 1 with each of the other elements, and the $[(n - 1)/2](n - 2)$ ternaries of element 1 with all pairs of the other elements.

15.5.3.4. Simultaneous Evaluation of All a_{ij} Coefficients (*C65*)

Equations (15.41) may be written in the somewhat more general form

$$R_i = C_i \Big/ \sum_{j=1}^{j=n} a_{ij}C_j \tag{15.79}$$

where i is the analyte and j the interferant; C is concentration; and R_i is represented symbolically by

$$R_i = f(I_i) = f(C_1, C_2, \ldots, C_n) \tag{15.80}$$

That is, R_i is a function of analyte-line intensity and of the mass concentrations of all elements present; it may be simply analyte-line intensity measured from the specimen I_X, or it may be that intensity expressed as a ratio to the analyte-line intensity from a reference standard I_S or from the pure element I_{100i}, that is,

$$R_i = I_X \quad \text{or} \quad I_X/I_S \quad \text{or} \quad I_i/I_{100i} \tag{15.81}$$

Evaluation of the a_{ij} coefficients by first- and second-order approximations has been considered in the two preceding sections. However, it is best to determine all the coefficients simultaneously from multicomponent standards containing known concentrations of all elements of interest. This approach minimizes the shortcomings of the assumptions in the derivation of Equations (15.41) because secondary absorption effects are represented in the measured intensities.

The coefficients are evaluated by solution of sets of simultaneous equations. For example, in a three-component system, there are nine coefficients (if j may be i), and they are evaluated three at a time. Let 1, 2, and 3 be the three elements and A, B, and C the three required standards, which should cover the full range of composition expected in the samples. First, to evaluate coefficients a_{11}, a_{12}, a_{13}, the intensity of the spectral line of element 1 is measured from standards A, B, and C, and perhaps from a reference standard—which may be a specimen of pure element 1. The value of R_i is calculated for each standard from Equation (15.81) and used in the following set of simultaneous equations:

$$\begin{aligned} C_{1A}/R_{1A} &= a_{11}C_{1A} + a_{12}C_{2A} + a_{13}C_{3A} \\ C_{1B}/R_{1B} &= a_{11}C_{1B} + a_{12}C_{2B} + a_{13}C_{3B} \\ C_{1C}/R_{1C} &= a_{11}C_{1C} + a_{12}C_{2C} + a_{13}C_{3C} \end{aligned} \tag{15.82}$$

where C_{1A} indicates concentration of element 1 in standard A, etc.; R_{1A} indicates intensity function of the spectral line of element 1 from standard A [Equation (15.81)], etc.; and a is the influence coefficient. Since all the C's are known and all the R's are measured, the a's can be calculated.

Coefficients a_{21}, a_{22}, and a_{23} are evaluated in the same way, by measurement of the line intensity of element 2 from standards A, B, and C, and from pure element 2, and application of the data to a set of equations similar to Equations (15.81); a_{31}, a_{32}, and a_{33} are evaluated from the line intensities of element 3 and a third set of equations. Somewhat better values of a_{ij} may be obtained by making measurements on more than the required number of standards, increasing the number of simultaneous equations in the three sets, and calculating a least-squares best-fit set of a_{ij} coefficients. Once derived, the a_{ij} coefficients are used in Equations (15.41) as before.

15.5.3.5. Other Methods for Evaluation of a_{ij} Coefficients

Jenkins and Campbell-Whitelaw (*J14*) show that

$$C_{iX} = (I_{iX}/I_{iS})C_{iS}[1 + a_{ij}(C_{jX} - C_{jS})] \qquad (j \neq 1) \tag{15.83}$$

where C, I, S, and X are concentration, intensity, the standard, and the sample, respectively, and i and j are the analyte and a matrix element, respectively. From this equation,

$$\frac{C_{i\text{X}}}{I_{i\text{X}}}\,\frac{I_{i\text{S}}}{C_{i\text{S}}} = a_{ij}(C_{j\text{X}} - C_{j\text{S}}) + 1 \qquad (j \neq i) \tag{15.84}$$

and

$$a_{ij} = \frac{[(C_{i\text{X}}/I_{i\text{X}})(I_{i\text{S}}/C_{i\text{S}})] - 1}{C_{j\text{X}} - C_{j\text{S}}} \qquad (j \neq i) \tag{15.85}$$

Equation (15.85) permits calculation of a_{ij} from two binary i–j standards (S and X) bracketing a relatively narrow analytical concentration region. Equation (15.84) has the form $y = mx + b$, so that a plot of $(C_{i\text{X}}/I_{i\text{X}})(I_{i\text{S}}/C_{i\text{S}})$ *versus* $C_{j\text{X}} - C_{j\text{S}}$ is linear, crosses the ordinate at 1, and has slope a_{ij}. This equation permits graphic determination of a_{ij} from a large number of binary i–j standards covering a wide composition region. One of the standards may be designated S, the others as X, one at a time; alternatively, a_{ij} can be calculated from all possible pairs of the standards. To a first approximation, the a_{ij} coefficients for a multielement system may be evaluated from individual binary i–j standards, either from Equation (15.85) or graphically from Equation (15.84).

Equation (15.85) is based on the Lachance–Traill correction (Section 15.5.4.4) in that it uses concentration differences in the denominator. A similar equation can be based on the Lucas-Tooth and Pyne correction (Section 15.5.4.5) by use of intensity differences:

$$a_{ij} = \frac{[(C_{i\text{X}}/I_{i\text{X}})(I_{i\text{S}}/C_{i\text{S}})] - 1}{I_{j\text{X}} - I_{j\text{S}}} \qquad (j \neq i) \tag{15.86}$$

A third equation can be based on the "peak-ratio" correction of Shenberg, Haim, and Amiel (*S25*) by replacing the intensity terms with the ratios R of the analyte i or matrix j line and a scattered x-ray tube or radioisotope-source target line (Section 14.5.3):

$$a_{ij} = \frac{[(C_{i\text{X}}/R_{i\text{X}})(R_{i\text{S}}/C_{i\text{S}})] - 1}{C_{j\text{X}} - C_{j\text{S}}} \qquad (j \neq i) \tag{15.87}$$

Equations (15.86) and (15.87), like Equation (15.85), can be used to calculate a_{ij} from two standards for a relatively narrow analytical region, or to determine a_{ij} graphically from many standards for a relatively wide analytical region. Shenberg and Amiel (*S24*) have compared the effectiveness of Equations (15.85)–(15.87) and found the last to be least affected by matrix and analyte concentration, that is, to give the most nearly constant a_{ij} values for wide variation in i and j concentration.

In the influence-coefficient method for x-ray spectrometric analysis, the linear regression equations for the solution of weight fractions are homogeneous. Therefore, regardless of the method of solution, they cannot be solved independently for individual weight fractions since one equation must always be replaced by a nonhomogeneous equation, such as one that defines the sum of the weight fractions [Equation (15.42)]. A difficulty arises with this method—the solutions of one subset of equations differ from those of another subset, that is, the equations are inconsistent. The origin of this difficulty seems to be experimental error in the measurement of the x-ray intensities. Fatemi and Birks (*F11*) give a test for consistency of the regression equations and a method for generation of consistent equations by adjustment of the measured intensities.

15.5.4. Variations of the Influence-Coefficient Method

Sections 15.5.2 and 15.5.3 are based largely on the equations of Beattie and Brissey. The following sections briefly summarize several variations of the influence-coefficient method and demonstrate their essential similarity to Equations (15.41) and (15.49) [pp. 171–80 in (*29*)]. These and several other variations are also compared in Reference (*R6*).

15.5.4.1. Method of Sherman

Apparently Sherman (*S27*) was the first to present a system of linear equations for determination of analyte concentrations in a multicomponent system:

$$(a'_{11} - T_1)C_1 + a'_{12}C_2 + a'_{13}C_3 + \cdots + a'_{1n}C_n = 0 \quad (15.88\text{-}1)$$

$$a'_{21}C_1 + (a'_{22} - T_2)C_2 + a'_{23}C_3 + \cdots + a'_{2n}C_n = 0 \quad (15.88\text{-}2)$$

$$\cdots\cdots$$

$$a'_{n1}C_1 + a'_{n2}C_2 + a'_{n3}C_3 + \cdots + (a'_{nn} - T_n)C_n = 0 \quad (15.88\text{-}n)$$

where T_i is the counting time to accumulate a preset count for the line of analyte i, and the other notation has the significance already defined. The influence coefficients are primed to indicate that they are not identical with those in Equations (15.41). These equations are put in the form of Equations (15.49)—and thereby of Equations (15.41)—by the following procedure.

1. Let $T_{100,1}, T_{100,2}, \ldots, T_{100n}$ represent the counting times from the pure components 1, 2, ..., n.

2. Let $T_1, T_2, \ldots, T_n$ represent the counting times from the multicomponent sample.

3. Let $T_{100i} = a'_{ii}$.

4. Let $r_{ij}T_{100i} = a'_{ij}$.

5. Rewrite Equations (15.88) making these substitutions; for example, Equation (15.88-1) becomes

$$(T_{100,1} - T_1)C_1 + (r_{12}T_{100,1})C_2 + (r_{13}T_{100,1})C_3 + \cdots + (r_{1n}T_{100,1})C_n = 0 \tag{15.89}$$

6. Divide equations 1, 2, ..., n in this set by $T_{100,1}, T_{100,2}, \ldots, T_{100n}$, respectively.

7. Replace counting times T by intensities I, noting that

$$T_i \propto \frac{1}{I_i} \quad \text{and} \quad \frac{T_{100i} - T_i}{T_{100i}} = -\frac{I_{100i} - I_i}{I_i} \tag{15.90}$$

For example, Equation (15.89) now becomes

$$-\frac{I_{100,1} - I_1}{I_1}C_1 + r_{12}C_2 + r_{13}C_3 + \cdots + r_{1n}C_n = 0 \tag{15.91}$$

which has the same form as Equations (15.49) and (15.41).

Note that Sherman and Beattie-Brissey influence coefficients are related as follows:

$$a'_{ij} = a_{ij}T_{100i} \approx r_{ij}T_{100i} \tag{15.92}$$

15.5.4.2. Method of Burnham, Hower, and Jones

These workers (*B94*) have refined and extended Sherman's equations for a ternary system:

$$(a'_{11} - T_1)C_1 + a'_{12}C_2 + a'_{13}C_3 = 0 \tag{15.93-1}$$

$$a'_{21}C_1 + (a'_{22} - T_2)C_2 + a'_{23}C_3 = 0 \tag{15.93-2}$$

$$a'_{31}C_1 + a'_{32}C_2 + (a'_{33} - T_3)C_3 = 0 \tag{15.93-3}$$

where C is concentration; T_i is the time for a preset count for the analyte line of element i from the sample or standard; a'_{ii} is the influence coefficient for pure i, but is taken as the counting time for pure i; and a'_{ij} is the influence coefficient of j on i. These coefficients are the same as Sherman's [Equation (15.92)].

The six a'_{ij} coefficients are evaluated from measurements on the three pure analytes and two ternary standards. If other elements of no analytical interest are present, the known concentrations of elements 1, 2, and 3 in the standards are normalized to 100%.

These equations give good results over relatively limited concentration ranges. The applicability of the method is extended by determination of the a'_{ii} and a'_{ij} coefficients from many standards. For example, the three coefficients in Equation (15.93-1) are derived from the following set of equations:

$$a'_{11} \sum C_1^2 + a'_{12} \sum C_1C_2 + a'_{13} \sum C_1C_3 = \sum T_1C_1^2 \quad (15.94\text{-}1)$$

$$a'_{11} \sum C_1C_2 + a'_{12} \sum C_2^2 + a'_{13} \sum C_2C_3 = \sum T_1C_1C_2 \quad (15.94\text{-}2)$$

$$a'_{11} \sum C_1C_3 + a'_{12} \sum C_2C_3 + a'_{13} \sum C_3^2 = \sum T_1C_1C_3 \quad (15.94\text{-}3)$$

where the summations apply to the concentrations in the standards used. Corresponding sets of equations based on Equations (15.93-2) and (15.93-3) permit evaluation of the other six coefficients.

An outstanding feature of the method of Burnham *et al.* is that it was extended to apply to various sample forms—coupons, turnings, chips, filings, wire—with the same influence coefficients. It is assumed that a change in sample form affects the emitted analyte-line intensity to the same degree for each analyte. An additional parameter S is introduced such that if T_i is the counting time for element i from, say, briqueted chips, ST_i is the counting time from a polished coupon. Equations (15.93) then become

$$(a'_{11} - ST_1)C_1 + a'_{12}C_2 + a'_{13}C_3 = 0 \quad (15.95\text{-}1)$$

$$a'_{21}C_1 + (a'_{22} - ST_2)C_2 + a'_{23}C_3 = 0 \quad (15.95\text{-}2)$$

$$a'_{31}C_1 + a'_{32}C_2 + (a'_{33} - ST_3)C_3 = 0 \quad (15.95\text{-}3)$$

These equations are not solved as readily as the others because they are no longer linear but contain cross-products SC_i. Burnham *et al.* give methods for solution of these equations to give the influence coefficients and the specimen form factor S. A different form factor is required for each specimen form to be analyzed, and each form factor requires a set of Equations (15.95).

Once the influence coefficients and specimen form factors are evaluated, sample concentrations are obtained simply by inserting the measured counting times T_i into Equations (15.95).

These workers have also devised a graphic method that precludes the need to solve Equations (15.95). Figure 15.9A shows a portion of the Cr–

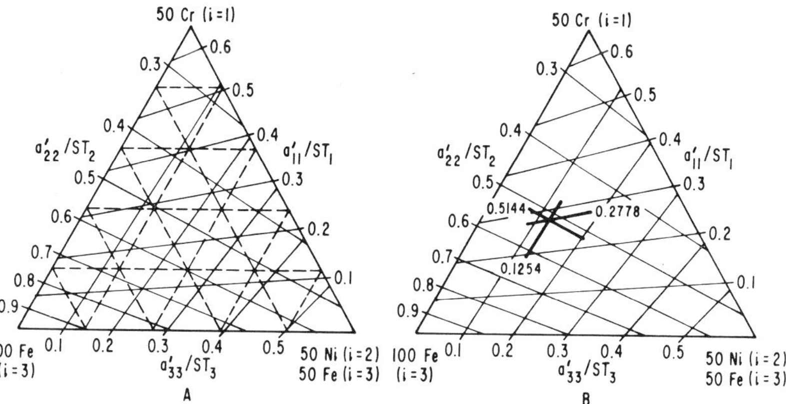

FIG. 15.9. Graphic method of Burnham, Hower, and Jones (*B94*). A. Portion of ternary Cr–Fe–Ni diagram having superimposed grids of a_{ii}/ST_i. B. Overlay for A, to permit analysis of a specific sample.

Ni–Fe composition diagram—the portion 0–50 Cr, 0–50 Ni, 50–100 Fe. In the diagram, Cr, Ni, and Fe are designated $i = 1$, 2, and 3, respectively. Superimposed on the diagram are three grids representing constant values of a'_{11}/ST_1, a'_{22}/ST_2, and a'_{33}/ST_3, respectively. The quantity a'_{ii}/ST_i may be termed relative intensity in the sense that a sample for which, for example, $a'_{11}/ST_1 = 0.5$ would have an analyte-line intensity for element 1 half that from the standard.

Figure 15.9B shows an example of the application of the graphic method to an analysis. Counting times T_1, T_2, and T_3 are measured from the samples; a'_{11}, a'_{22}, a'_{33}, and S are known from previous evaluation. From these data, values of a'_{ii}/ST_i are calculated. The values for the example in Figure 15.9B are given in Table 15.6. Lines representing these

TABLE 15.6. Data for Analysis by the Graphic Method[a]

i	T_i [b]	a'_{ii}	S	a'_{ii}/ST_i
1, Cr	260	27.95	0.3872	0.2778
2, Ni	46	8.13	”	0.5144
3, Fe	278	13.49	”	0.1254

[a] From Burnham, Hower, and Jones (*B94*).
[b] Counting time (s) to accumulate 25,600 counts for the $K\alpha$ line.

values of a'_{ii}/ST_i are plotted on Figure 15.9B, which is overlaid on Figure 15.9A to permit direct reading of C_i.

Another graphic method was devised to permit direct conversion of counting times to concentrations. However, this method is less simple, and the interested reader is referred to the original paper (*B94*).

15.5.4.3. Method of Marti

Marti (*M13*) also modified Sherman's equations in order to make x-ray spectrometric analyses of steels:

$$I_{\mathrm{Cr,cor}} = I_{\mathrm{Cr,measd}}(a_{\mathrm{CrCr}}C_{\mathrm{Cr}} + a_{\mathrm{CrNi}}C_{\mathrm{Ni}} + a_{\mathrm{CrMo}}C_{\mathrm{Mo}} + a_{\mathrm{CrFe}}C_{\mathrm{Fe}} + \cdots) \tag{15.96-1}$$

$$I_{\mathrm{Ni,cor}} = I_{\mathrm{Ni,measd}}(a_{\mathrm{NiCr}}C_{\mathrm{Cr}} + a_{\mathrm{NiNi}}C_{\mathrm{Ni}} + a_{\mathrm{NiMo}}C_{\mathrm{Mo}} + a_{\mathrm{NiFe}}C_{\mathrm{Fe}} + \cdots) \tag{15.96-2}$$

$$I_{\mathrm{Mo,cor}} = I_{\mathrm{Mo,measd}}(a_{\mathrm{MoCr}}C_{\mathrm{Cr}} + a_{\mathrm{MoNi}}C_{\mathrm{Ni}} + a_{\mathrm{MoMo}}C_{\mathrm{Mo}} + a_{\mathrm{MoFe}}C_{\mathrm{Fe}} + \cdots) \tag{15.96-3}$$

where I_{cor} and I_{measd} are corrected and measured intensities, respectively; C is concentration; and a_{ij} is the influence coefficient. In Marti's original notation, the $12 a_{ij}$ coefficients were, in order, α_{Cr}, α_{Ni}, α_{Mo}, α_{Fe}; β_{Cr}, β_{Ni}, β_{Mo}, β_{Fe}; and γ_{Cr}, γ_{Ni}, γ_{Mo}, γ_{Fe}. These equations are put in the form of Equations (15.49)—and thereby of Equations (15.41)—by the following procedure.

1. Since $I_{i,\mathrm{cor}}$, having been corrected for all absorption-enhancement effects, represents strict proportionality between C_i and I_i, $C_i I_{100i}$ is substituted for $I_{i,\mathrm{cor}}$, to which it is equal.
2. Let I_i represent $I_{i,\mathrm{measd}}$.
3. Let $r_{ij} = a_{ij}$.
4. Let a_{ii} (a_{CrCr}, a_{NiNi}, etc.) $= 1$.
5. Rewrite equations (15.96) making these substitutions; for example, Equation (15.96-1) becomes

$$C_{\mathrm{Cr}} I_{100\mathrm{Cr}} = I_{\mathrm{Cr}}(C_{\mathrm{Cr}} + r_{\mathrm{CrNi}}C_{\mathrm{Ni}} + r_{\mathrm{CrMo}}C_{\mathrm{Mo}} + r_{\mathrm{CrFe}}C_{\mathrm{Fe}} + \cdots) \tag{15.97}$$

6. Divide both sides of each equation by I_i.
7. C_i appears on both sides of each equation; transpose the left side of each equation to the right, making each equation equal to zero.

8. Combine the C_i terms, noting that

$$1 - \frac{I_{100i}}{I_i} = -\frac{I_{100i} - I_i}{I_i} \tag{15.98}$$

For example, Equation (15.97) now becomes

$$-\frac{I_{100\mathrm{Cr}} - I_{\mathrm{Cr}}}{I_{\mathrm{Cr}}} C_{\mathrm{Cr}} + r_{\mathrm{CrNi}} C_{\mathrm{Ni}} + r_{\mathrm{CrMo}} C_{\mathrm{Mo}} + r_{\mathrm{CrFe}} C_{\mathrm{Fe}} + \cdots = 0 \tag{15.99}$$

which has the same form as Equations (15.49) and (15.41).

Marti applied his equations to analysis of steels and further simplified them by postulating that

$$a_{\mathrm{CrFe}} = a_{\mathrm{NiFe}} = a_{\mathrm{MoFe}} = 1 \tag{15.100}$$

However, these coefficients are not necessarily equal, let alone unity, so that this simplification is risky at best.

15.5.4.4. Method of Traill and Lachance

The basic equation for this method (*L1*, *T17*, *T18*) is

$$\frac{I_i}{I_{100i}} = \frac{C_i}{1 + \sum (a'_{ij} C_j)} \qquad (j \neq i) \tag{15.101}$$

where I_i and I_{100i} are analyte-line intensities of element i from the specimen and pure i, respectively. The influence coefficients are primed to indicate that they are not identical with those in Equations (15.41). For any system of n components, a series of n simultaneous linear equations like Equation (15.101) can be written and solved for the a'_{ij} coefficients, mass concentrations, or even intensities, depending on the data available. The method has been applied with great success to systems having up to nine elements (Ti, Cr, Mn, Fe, Co, Ni, Nb, Mo, and W).

For a ternary system of elements 1, 2, and 3, the equations are

$$C_1 = (I_1/I_{100,1})(1 + a'_{12} C_2 + a'_{13} C_3) \tag{15.102-1}$$

$$C_2 = (I_2/I_{100,2})(1 + a'_{21} C_1 + a'_{23} C_3) \tag{15.102-2}$$

$$C_3 = (I_3/I_{100,3})(1 + a'_{31} C_1 + a'_{32} C_2) \tag{15.102-3}$$

The a'_{ij} coefficients can be evaluated from these equations and intensities measured from the three pure elements and two ternary standards. If it is not feasible to make measurements on the pure elements, the a'_{ij}'s as well as the net intensities from the pure elements can be calculated from intensities measured from three ternary standards A, B, and C and three sets of equations of the form

$$\begin{aligned} C_{1A}I_{100,1} &= I_{1A}(1 + a'_{12}C_{2A} + a'_{13}C_{3A}) \\ C_{1B}I_{100,1} &= I_{1B}(1 + a'_{12}C_{2B} + a'_{13}C_{3B}) \\ C_{1C}I_{100,1} &= I_{1C}(1 + a'_{12}C_{2C} + a'_{13}C_{3C}) \end{aligned} \tag{15.103}$$

where C_{1A} indicates concentration of element 1 in standard A, etc.; I_{1A} indicates analyte-line intensity for element 1 in standard A, etc.; and $I_{100,1}$ is analyte-line intensity from pure element 1. These equations permit calculation of a'_{12}, a'_{13}, and, if necessary, $I_{100,1}$. Two similar sets of equations permit calculation of a'_{21}, a'_{23}, and $I_{100,2}$ and a'_{31}, a'_{32}, and $I_{100,3}$, respectively.

Even if no ternary standards are available, the influence coefficients can be calculated approximately by equations of the form

$$a'_{12} = \frac{\overline{(\mu/\varrho)_{12}}}{\overline{(\mu/\varrho)_{11}}} - 1 = \frac{(\mu/\varrho)_{2,\lambda_{\text{pri}}} \csc\phi + (\mu/\varrho)_{2,\lambda_1} \csc\psi}{(\mu/\varrho)_{1,\lambda_{\text{pri}}} \csc\phi + (\mu/\varrho)_{1,\lambda_1} \csc\psi} - 1 \tag{15.104}$$

where $(\mu/\varrho)_{1,\lambda_{\text{pri}}}$ and $(\mu/\varrho)_{1,\lambda_1}$ are mass-absorption coefficients of element 1 for the primary λ_{pri} and analyte-line λ_1 radiation, respectively; and ϕ and ψ are the incident and takeoff angles, respectively (Figure 15.8).

Once the a'_{ij} coefficients are evaluated, Equations (15.102) can be used to calculate analyte concentrations from intensities measured from the samples and pure analytes, where feasible.

The relationship of the Traill–Lachance coefficients to those in Equations (15.41) and (15.49) is found as follows. Equation (15.102-1) may be rearranged as follows:

$$-\frac{I_{100,1}}{I_1} C_1 + a'_{12}C_2 + a'_{13}C_3 = -1 \tag{15.105}$$

Addition of +1 to both sides and rearrangement give

$$-\left(\frac{I_{100,1}}{I_1} - 1\right)C_1 + (a'_{12} + 1)C_2 + (a'_{13} + 1)C_3 = 0 \tag{15.106}$$

Thus Traill–Lachance and Beattie–Brissey influence coefficients are related as follows:

$$a'_{ij} = (a_{ij} - 1) \approx (r_{ij} - 1) \tag{15.107}$$

15.5.4.5. Method of Lucas-Tooth and Pyne

Lucas-Tooth and Pyne and others have applied the influence-coefficient method to multicomponent systems having up to 12 elements (*E14*, *L46*, *L47*). Their basic equation is derived from the same assumptions as Equations (15.41) and the following additional one: Over the concentration range of interest, the absorption of the spectral line of element A by element C is assumed to be proportional to the weight fraction of C; that is, if C is added to binary A–B in concentration W_C, A-line intensity will depart from its value in pure A–B, I_A, by $W_C I_A$.

The basic equation is

$$C_{iS} = k_i + I_{iS}\left[a_i + a'_{ii}I_{iS} + \sum_{j=1}^{j=n}(a'_{ij}I_{jS})\right] \qquad (j \neq i) \tag{15.108}$$

$$= k_i + I_{iS}(a_i + a'_{ii}I_{iS} + a'_{i1}I_{1S} + a'_{i2}I_{2S} + \cdots + a'_{in}I_{nS}) \qquad (j \neq i) \tag{15.109}$$

where C_{iS} is concentration of analyte i in specimen S; I_{iS} is analyte-line intensity from specimen S; I_{jS} is line intensity of matrix element j from specimen S; a'_{ij} is the influence coefficient of element j on element i; and a_i and k_i are constants. The constant a_i may contain corrections for nonlinearity of the detector for element i. It is evident that, unlike the influence coefficients in the methods in Sections 15.5.2, 15.5.3, and 15.5.4.1–15.5.4.4, those in the method of Lucas-Tooth and Pyne are evaluated from *intensities*, rather than concentrations, derived from the standards. This tends to reduce third-element effects and permits application of the method to a wider concentration range. However, in general, the intensity method requires more standards.

If there are j elements (including i) in the specimen, there are $j + 2$ constants—k_i, a_i, a'_{ii}, and $(j - 1)$ a'_{ij}'s. Thus, in principle, evaluation of the constants requires a set of $j + 2$ simultaneous equations capable of rigorous solution, each for a different standard of accurately known composition. In practice, a larger number of equations—and standards—should be used and the constants evaluated by a least-squares fit. A different set of $j + 2$ constants, and therefore a different set of equations, is required for each analyte.

The constants can be evaluated by measurement of all intensities I_{iS} and I_{jS} from all $j + 2$ or more standards and solution of the $j + 2$ or more equations for the $j + 2$ constants. The evaluation may be simplified on the basis that, over reasonable concentration intervals, the line intensity of each element is approximately proportional to its concentration. A concept, *apparent concentration*, is then defined as follows:

$$C_{iS(ap)} = k' I_{iS} \approx C_{iS(chem)}; \qquad C_{jS(ap)} = k' I_{jS} \approx C_{jS(chem)} \qquad (15.110)$$

where $C_{iS(chem)}$ is the known concentration of a standard as determined by chemical methods. The apparent concentration approximates the chemical, and by adjustment of k', the two concentrations may actually be made equal at any time on any instrument. In fact, the principal reason for the introduction of the concept of apparent concentration is that measured analyte-line intensity is different from the same specimen on different spectrometers, and even on the same spectrometer, may vary with instrument drift.

Substitution of Equation (15.110) into (15.108) gives

$$C_{iS} = k_i + C_{iS(ap)}\left[a_i + a'_{ii}C_{iS(ap)} + \sum_{j=1}^{j=n} (a'_{ij}C_{jS(ap)})\right] \qquad (j \neq i) \qquad (15.111)$$

The constants are evaluated by solution of $j + 2$ or more equations for $j + 2$ or more standards, by use of the *known* concentrations for $C_{iS(ap)}$ and $C_{jS(ap)}$. If $C_{(ap)}$ is in weight percent, an additional equation may be added to the set:

$$C_{iS(ap)} + \sum_{j=1}^{i=n} C_{jS(ap)} = 100 \qquad (j \neq i) \qquad (15.112)$$

Once the constants are known, determination of analyte i in sample X consists in measurement of intensity I_{iX} and all intensities I_{jX} from the sample X, then solution of a single equation of the form of Equation (15.108) for C_{iX}:

$$C_{iX} = k_i + I_{iX}\left[a_i + a_{ii}I_{iX} + \sum_{j=1}^{j=n} (a_{ij}I_{jX})\right] \qquad (j \neq i) \qquad (15.113)$$

An example of the application of the Lucas-Tooth method is provided by the analysis by Eick *et al.* (*E14*) of gold dental alloys containing, in order of decreasing average concentration in weight percent, 68–92 Au, 6–14 Ag, 2–12 Cu, 0–6 Pt, 0–3 Pd, and 0–2 Zn. The determination of gold will be used for illustration. The method as represented by Equations (15.108) and (15.109) was used, and the equation for gold is

$$C_{AuS} = k_{Au} + I_{AuS}\left[a_{Au} + a'_{AuAu}I_{AuS} + \sum_{j=1}^{j=n} (a'_{Auj}I_{jS})\right] \qquad (j \neq \text{Au}) \qquad (15.114)$$

To establish the method and evaluate the constants in Equation (15.114), the following 25 standards were used: (1) standards 1–5, gold dental-alloy standards analyzed chemically; (2) alloys 1–3, commercial dental alloys having composition determined by careful weighing of the individual metals prior to alloying; and (3) binary alloys: 97, 94, 91, and 88 wt % gold alloyed with copper, platinum, and paladium, and 97, 94, 91, 88, and 75 wt % gold alloyed with silver.

Analyte-line intensity was measured for all six analytes from all 25 standards. The intensities for standards 1–5, alloys 1–3, and the five gold–silver binary alloys are given in Table 15.7. Although the authors do not explicitly say so, these intensities appear to be peak, not net, intensities, as indicated by the values in the Cu $K\alpha$, Pt $L\alpha_1$, Pd $K\alpha$, and Zn $K\alpha$ columns for the gold–silver binary alloys.

The constants k_{Au}, a_{Au}, a'_{AuAu}, and the five $a'_{\mathrm{Au}j}$'s are determined by a least-squares fit for the data in Table 15.7 substituted into a set of equa-

TABLE 15.7. Intensity Data for Analysis of Gold Dental Alloys by the Method of Lucas-Tooth and Pyne[a]

Standard	Au. concn. wt %	Intensity, counts/s					
		Au $L\beta_1$	Ag $K\alpha$	Cu $K\alpha$	Pt $L\alpha_1$	Pd Kα	Zn $K\alpha$
Standard 1	68.3	6251	488	7898	2658	168	2833
” 2	72.0	6545	446	6626	1750	760	2191
” 3	78.1	7044	361	5415	1127	1014	1505
” 4	79.1	7151	555	4091	246	487	250
” 5	91.6	8174	263	1820	”	168	”
Alloy 1	92.5	8154	220	2071	”	300	”
” 2	80.7	7534	458	2372	”	1312	672
” 3	75.0	7046	550	4648	”	1055	1133
Au–Ag binary alloy 1	97.0	8519	170	400	”	168	250
” ” ” 2	94.0	8418	264	”	”	”	”
” ” ” 3	91.0	8257	357	”	”	”	”
” ” ” 4	88.0	8084	452	”	”	”	”
” ” ” 5	75.0	7554	895	”	”	”	”

[a] From Eick, Caul, Smith, and Rasberry (*E14*).

TABLE 15.8. Constants for Determination of Gold in Dental Alloys by the Method of Lucas-Tooth and Pyne[a]

Constant	Value
k_{Au}	-2.685
a_{Au}	2.523×10^{-2}
a'_{AuAu}	-1.484×10^{-6}
a'_{AuAg}	-3.914×10^{-6}
a'_{AuCu}	-1.792×10^{-7}
a'_{AuPt}	-1.262×10^{-7}
a'_{AuPd}	-3.884×10^{-7}
a'_{AuZn}	-2.920×10^{-7}

[a] From Eick, Caul, Smith, and Rasberry (*E14*).

tions of the form of Equation (15.114), each equation for a different standard. The values of the constants for the gold determinations are given in Table 15.8.

Gold concentrations corrected for absorption-enhancement effects are then determined in unknown samples X by use of Equation (15.114) with the constants in Table 15.8:

$$C_{\mathrm{AuX}} = k_{\mathrm{Au}} + I_{\mathrm{AuX}}\left[a_{\mathrm{Au}} + a'_{\mathrm{AuAu}}I_{\mathrm{AuX}} + \sum_{j=1}^{j=n}(a'_{\mathrm{Au}j}I_{j\mathrm{X}})\right] \tag{15.115}$$

$$= k_{\mathrm{Au}} + I_{\mathrm{AuX}}(a_{\mathrm{Au}} + a'_{\mathrm{AuAu}}\,I_{\mathrm{AuX}} + a'_{\mathrm{AuAg}}I_{\mathrm{AgX}} + a'_{\mathrm{AuCu}}I_{\mathrm{CuX}} + a'_{\mathrm{AuPt}}I_{\mathrm{PtX}} + a'_{\mathrm{AuPd}}I_{\mathrm{PdX}} + a'_{\mathrm{AuZn}}I_{\mathrm{ZnX}}) \tag{15.116}$$

$$= -2.685 + I_{\mathrm{AuX}}[0.02523 + (-1.484\times10^{-6})I_{\mathrm{AuX}} + (-3.914\times10^{-6})I_{\mathrm{AgX}} + (-1.792\times10^{-7})I_{\mathrm{CuX}} + (-1.262\times10^{-7})I_{\mathrm{PtX}} + (-3.884\times10^{-7})I_{\mathrm{PdX}} + (-2.920\times10^{-7})I_{\mathrm{ZnX}}] \tag{15.117}$$

Analyte-line intensities of all six elements, I_{AuX} and the five $I_{j\mathrm{X}}$'s, are measured from the sample X and substituted in Equation (15.117). For example, regarding alloy 2 (Table 15.7) as an unknown sample, the corrected gold concentration would be given by the following computation:

$$\begin{aligned} C_{\mathrm{Au,alloy2}} &= -2.685 + 7534[0.02523 - (1.484\times 10^{-6})(7534) \\ &\quad -(3.914\times 10^{-6})(458) - (1.792\times 10^{-7})(2372) \\ &\quad -(1.262\times 10^{-7})(246) - (3.884\times 10^{-7})(1312) \\ &\quad -(2.920\times 10^{-7})(672)] \\ &= 80.9 \text{ wt \% Au} \end{aligned}$$

Equations similar to Equations (15.114) and (15.115) are written and solved in the same way for each of the other five elements.

15.5.4.6. Method of Rasberry and Heinrich

Most absorption-enhancement influence-coefficient procedures treat true enhancement as "negative absorption." However, in Section 12.2.2 and in curve *D* of Figure 12.1A, it is shown that whereas simple absorption effects result in hyperbolic calibration curves, a curve affected by true enhancement is not hyperbolic. Rasberry and Heinrich (*R6*) point out that if a calibration equation is to permit accurate determination of, say, iron in the presence of chromium and nickel, it must also apply in the absence of either chromium or nickel, that is, it must fit both binary curves. Hence, it must take into account the nonhyperbolic nature of the calibration curve for iron in nickel, where nickel gives true enhancement of Fe $K\alpha$. These writers propose the following equation

$$\left(\frac{C}{R}\right)_i = 1 + \sum_{j=1}^{j=n} a_{ij}C_j + \sum_{j=1}^{j=n} \frac{e_{ij}C_j}{1+C_i} \qquad (j \neq i) \tag{15.118}$$

where C is concentration (mass fraction); R is relative intensity (net analyte-line intensity ratio from the specimen and pure analyte); i and j are analyte and matrix elements, respectively; a_{ij} is the *absorption* influence coefficient used when the significant effect of element j on i is absorption; and e_{ij} is the *enhancement* influence coefficient used when the significant effect of element j on i is true enhancement, that is, when j lines excite the i line.

These workers state that there may be cases where both a_{ij} and e_{ij} coefficients may be required for the *same* matrix element j in the *same* equation, that is, for a particular analyte i. However, they further state that they have never observed such a case, so that usually for a specific i–j pair, when the a_{ij} coefficient is indicated, $e_{ij} = 0$, and *vice versa*. Usually the analyst can predict which type of coefficient is required by examination of tables of wavelengths of spectral lines and absorption edges and tables of mass-absorption coefficients, or by consideration of graphs of $(C/R)_i$ *versus* C_i for various matrix elements (Figure 12.1B).

From Equation (15.118), $n - 1$ coefficients are required for determination of each analyte in an n-component system, so that a total of $n(n - 1)$ coefficients are required. These may be determined in either of two ways: The intercept may be evaluated at the limit $C_j \rightarrow 1$ in curves of $(C/R)_i$ *versus* C_i (Figure 12.1B). Alternatively, a set of simultaneous equations of the form of Equation (15.118) may be solved for each element i.

Once the coefficients are evaluated for a specific system, Equation (15.118) may be used for determination of each analyte in the system. However, sets of such equations, one for each analyte, usually do not provide a simple, explicit solution for all the C_i's, so Rasberry and Heinrich use an iteration procedure described in their paper (*R6*). The method was applied with great success to the chromium–iron–nickel system, where absorption-enhancement effects are particularly severe.

15.6. FUNDAMENTAL-PARAMETERS METHOD

The fundamental-parameters method permits the calculation of analytical composition from the measured analyte-line intensity and the tabulated values of three fundamental parameters—primary spectral distribution, absorption coefficient, and fluorescent yield. Tables of these parameters are stored in the computer required to make the extremely complex calculations involved in the application of this method. No calibration standards are required. Several workers have developed fundamental-parameters methods, but the treatment in this section is taken largely from a paper by Criss and Birks [pp. 89–93 in (*8*)] (*C65*). This method, like the influence-coefficient method, assumes that the specimen is homogeneous and infinitely thick and has a plane surface. However, instead of assuming that the incident primary spectrum can be described by a single average effective wavelength, the method uses the actual spectral distribution of the primary beam. This distribution can be estimated theoretically, or it may be measured for a given x-ray tube target, potential, and type of power supply (full-wave or constant-potential). Matrix absorption and analyte-line excitation by matrix elements enter the equation explicitly for each specimen. Analyte concentration is calculated by iteration. The advantages of the fundamental-parameters method are that it avoids, at least in principle, the limited composition ranges of regression methods and that no intermediate standards or empirical coefficients are required for any matrix. The principal limitation is the present uncertainty in mass-absorption coefficient and fluorescent-yield data. However, this shortcoming is being overcome as more laboratories engage in measurement of these parameters. Another limitation is that the spectral distribution of the primary x-ray beam, although determinable experimentally (see below), may vary with the age of the x-ray tube. The calculations are very complex and must be done on a computer.

The fundamental-parameters equation is derived from the geometry of Figure 15.8 and the basic excitation equations, such as Equations (4.7), (4.17), and (15.33). The equation used by Criss and Birks and given here is substantially similar to that derived by other workers, except that it uses a summation over wavelength instead of the more common integration:

$$
\begin{aligned}
[(I_{\lambda_L})_{\text{pri}} + (I_{\lambda_L})_{\text{matrix}}]_X = g_i C_i \sum_{\lambda_{\text{pri}}} \Bigg(& \frac{D_{i,\lambda_{\text{pri}}}(\mu/\varrho)_{i,\lambda_{\text{pri}}}(I_{\lambda_{\text{pri}}}\,\Delta\lambda_{\text{pri}})}{(\mu/\varrho)_{M,\lambda_{\text{pri}}}\csc\phi + (\mu/\varrho)_{M,\lambda_L}\csc\psi} \\
& \times \Bigg\{1 + \frac{1}{2(\mu/\varrho)_{i,\lambda_{\text{pri}}}} \sum_{\lambda_J} [D_{j,\lambda_{\text{pri}}} C_j K_j (\mu/\varrho)_{i,\lambda_J} (\mu/\varrho)_{j,\lambda_{\text{pri}}}] \\
& \times \Bigg[\frac{1}{(\mu/\varrho)_{M,\lambda_{\text{pri}}}\csc\phi} \log_e\!\left(1 + \frac{(\mu/\varrho)_{M,\lambda_{\text{pri}}}\csc\phi}{(\mu/\varrho)_{M,\lambda_J}}\right) \\
& + \frac{1}{(\mu/\varrho)_{M,\lambda_L}\csc\psi} \log_e\!\left(1 + \frac{(\mu/\varrho)_{M,\lambda_L}\csc\psi}{(\mu/\varrho)_{M,\lambda_J}}\right)\Bigg]\Bigg\}\Bigg)
\end{aligned}
\tag{15.119}
$$

where:

C_i	is concentration of analyte element i (wt fraction).
C_j	is concentration of matrix element j (wt fraction).
D_i	has the value 1 if λ_{pri} is short enough to excite λ_L; it has the value 0 for longer wavelengths.
D_j	has the value 1 if λ_{pri} is short enough to excite λ_J; it has the value 0 for longer wavelengths.
g_i	is a function of the absolute intensity of λ_L from element i; g cancels if a relative intensity function is used (see below).
K_j	is $[1 - (1/r)]\omega$ for the particular spectral line of element j that excites λ_L of element i. (See below for r and ω.)
I_{λ_L}	is intensity of analyte line of element i.
$(I_{\lambda_L})_{\text{pri}}$	is intensity of analyte line excited by the primary spectrum.
$(I_{\lambda_L})_{\text{matrix}}$	is intensity of analyte line excited by spectral lines of matrix elements j in the specimen.

$I_{\lambda_{\mathrm{pri}}}$	is intensity of the primary beam in the wavelength interval $\Delta\lambda_{\mathrm{pri}}$.
i	is the analyte element.
j	is a matrix element.
M	is the specimen matrix, including the analyte i.
r	is absorption edge jump ratio.
$\Delta\lambda_{\mathrm{pri}}$	is a wavelength interval into which the primary spectrum is arbitrarily divided for summation purposes.
λ_J	is the wavelength of a spectral line of element j capable of exciting λ_L.
λ_L	is the wavelength of the analyte line of element i.
λ_{pri}	is a wavelength in the primary beam.
$(\mu/\varrho)_i$, $(\mu/\varrho)_j$, $(\mu/\varrho)_{\mathrm{M}}$	are mass-absorption coefficients of analyte i, matrix element j, and matrix M, respectively.
$(\mu/\varrho)_{\mathrm{M},\lambda_J}$, $(\mu/\varrho)_{\mathrm{M},\lambda_L}$, $(\mu/\varrho)_{\mathrm{M},\lambda_{\mathrm{pri}}}$	are mass-absorption coefficients of matrix M for λ_J, λ_L, and λ_{pri}, respectively; similar notations for $(\mu/\varrho)_{i,\lambda}$ and $(\mu/\varrho)_{j,\lambda}$ have analogous significance.
ϕ	is the angle between the central ray of the primary cone and the specimen.
ψ	is the angle between the central ray of the secondary cone and the specimen, the takeoff angle.
ω	is fluorescent yield.

The primary spectrum is divided into wavelength intervals $\Delta\lambda_{\mathrm{pri}}$ each having intensity $I_{\lambda_{\mathrm{pri}}}$. The product of g_iC_i and the first summation over λ_{pri} represents $(I_{\lambda_L})_{\mathrm{pri}}$, the analyte-line intensity excited only by the primary x-rays. The remainder of the equation represents $(I_{\lambda_L})_{\mathrm{matrix}}$, the analyte-line intensity excited by spectral lines of matrix elements j; this interelement excitation contributes as much as 30% to the total measured analyte-line intensity.

A typical spectral distribution for a tungsten-target x-ray tube having a 1-mm (0.040-inch) beryllium window and operating at 50 kV(P) is given in Table 15.9. Birks [pp. 121–8 in (*8*)] (*G13*) describes the experimental method for measuring the distributions and gives tables of distributions for tungsten, molybdenum, copper, and chromium targets operating at 45kV-(CP). Gilfrich *et al.* (*G15*) give the distribution for a rhodium-target tube. Any x-ray tubes having the same target element, window material and thickness, operating potential and wave form (full-wave or constant-potential),

TABLE 15.9. Integrated Intensity Distribution in the Spectrum of a Tungsten-Target Tube at 50 kV(P)[a]

Continuum[b]

λ, Å	$I_\lambda \Delta\lambda$	λ, Å	$I_\lambda \Delta\lambda$	λ, Å	$I_\lambda \Delta\lambda$	λ, Å	$I_\lambda \Delta\lambda$
0.27	30.0	0.87	118	1.47	83.7	2.07	31.0
0.29	53.2	0.89	114	1.49	83.1	2.09	29.7
0.31	76.8	0.91	111	1.51	82.3	2.11	28.5
0.33	104	0.93	108	1.53	81.3	2.13	27.4
0.35	124	0.95	105	1.55	80.2	2.15	26.3
0.37	148	0.97	103	1.57	78.9	2.17	25.3
0.39	169	0.99	100	1.59	77.4	2.19	24.3
0.41	184	1.01	97.9	1.61	75.9	2.21	23.4
0.43	196	1.03	95.8	1.63	74.0	2.23	22.6
0.45	202	1.05	93.8	1.65	72.2	2.25	21.8
0.47	205	1.07	91.8	1.67	70.2	2.27	21.1
0.49	206	1.09	90.0	1.69	68.4	2.29	20.4
0.51	205	1.11	88.4	1.71	66.6	2.31	19.6
0.53	202	1.13	87.0	1.73	64.6	2.33	18.8
0.55	195	1.15	85.8	1.75	62.6	2.35	18.0
0.57	187	1.17	84.6	1.77	60.6	2.37	17.3
0.59	180	1.19	83.5	1.79	58.6	2.39	16.8
0.61	173	1.21[c]	84.2	1.81	56.6	2.41	16.3
0.63	167	1.23	90.2	1.83	54.6	2.43	15.7
0.65	163	1.25	89.8	1.85	52.4	2.45	15.2
0.67	159	1.27	89.3	1.87	50.0	2.47	14.9
0.69	154	1.29	88.8	1.89	47.6	2.49	14.7
0.71	150	1.31	88.3	1.91	45.4	2.51	14.5
0.73	146	1.33	87.8	1.93	43.3	2.53	14.3
0.75	141	1.35	87.4	1.95	41.2	2.55	14.1
0.77	137	1.37	86.9	1.97	39.2	2.57	13.8
0.79	133	1.39	86.4	1.99	37.3	2.59	13.4
0.81	129	1.41	85.8	2.01	35.6		
0.83	125	1.43	85.1	2.03	34.0		
0.85	121	1.45	84.4	2.05	32.4		

TABLE 15.9. *(continued)*

Characteristic lines[d]

Line	λ, Å	$I_\lambda \, \Delta\lambda$
$L\gamma_{2,3}$	1.06460	68.8
$L\gamma_1$	1.09855	155
$L\beta_2$	1.24460	532
$L\beta_1$	1.281809	1464
$L\alpha_1$	1.47639	2072
Ll	1.6782	68.8

[a] From Birks [p. 123 in *(8)*].
[b] λ for the continuum is the middle of the $\Delta\lambda$ interval.
[c] The W LIII absorption edge occurs at 1.216 Å; $I_\lambda \, \Delta\lambda$ is 66.1 for 1.200–1.216 Å, 18.1 for 1.216–1.220 Å.
[d] Wavelengths for spectral lines are taken from Bearden *(59)*; $\Delta\lambda$ for lines is the natural line breadth from Blokhin *(10)*.

and takeoff angle should have the same spectral distribution. Differences in window material and thickness are easily corrected by use of the absorption equation. Differences in takeoff angle are not easy to correct, but differences within $\pm 5°$ usually may be disregarded. Of course, x-ray tube current (mA) has no effect on the spectral distribution.

The fundamental-parameters equation can be based on analyte-line intensity measured from the sample X, as in Equation (15.119). More often, the equation is based on the ratio R of this intensity with that measured from a reference standard S similar to X or from pure analyte $100i$. In the remainder of this section, the symbol R and the term *intensity function* refer to any of these three alternatives, that is,

$$R = I_{\mathrm{X}} \quad \text{or} \quad I_{\mathrm{X}}/I_{\mathrm{S}} \quad \text{or} \quad I_{\mathrm{X}}/I_{100i} \qquad [(15.81)]$$

where I is $(I_{\lambda_L})_{\mathrm{pri}} + (I_{\lambda_L})_{\mathrm{matrix}}$.

The denominator of the second and third forms of Equation (15.81) is represented by another equation identical with Equation (15.119), except that C_i, C_j, $(\mu/\varrho)_{\mathrm{M},\lambda_{\mathrm{pri}}}$, $(\mu/\varrho)_{\mathrm{M},\lambda_L}$, and $(\mu/\varrho)_{\mathrm{M},\lambda_J}$ refer to the standard or pure analyte. If the standard is pure analyte, $(I_{\lambda_L})_{\mathrm{matrix}}$ in the denominator becomes 0. Analyte-line intensity ratios from the sample X and a multicomponent standard S are readily converted to ratios relative to pure analyte as follows:

$$R = \frac{I_X}{I_{100i}} = \left(\frac{I_X}{I_S}\right)_{\text{measd}} \left(\frac{I_S}{I_{100i}}\right)_{\text{calcd}} \qquad (15.120)$$

where I has the same significance as in Equation (15.81).

Equation (15.119) expresses intensity function in terms of specimen composition. However, it is intensity functions that are measured, whereas it is concentrations that are required. Unfortunately, algebraic operation cannot produce explicit equations for mass functions C_1, C_2, etc. Consequently, mass fractions are found by an iteration procedure that makes successively closer estimates of mass fractions until the intensity functions calculated from the estimated concentrations agree satisfactorily with the measured intensity functions. Obviously, the iteration procedure must be performed on a computer. Following is a simple iteration equation:

$$C_i' = \frac{(R_i)_{\text{measd}}}{R_i} C_i \qquad (15.121)$$

where C_i is the present estimate of concentration of element i; C_i' is the next estimate of i concentration in the iteration; $(R_i)_{\text{measd}}$ is the measured intensity function for the spectral line of i; and R_i is the intensity function calculated for C_i.

The iteration procedure is as follows: (1) Initial estimates are made of all analyte concentrations C_i by normalizing their measured intensity functions $(R_i)_{\text{measd}}$ to add to 1, or by some other method. (2) The corresponding intensity functions R_i are calculated. (3) The next estimate C_i' is made by application of the iteration equation to each element i. (4) The C_i' values are scaled to add to 1. (5) The corresponding expected intensity functions R_i' are calculated. (6) A decision is made whether further iteration is indicated. This decision is based on a comparison of values of R_i' and $(R_i)_{\text{measd}}$ or of C_i' and C_i, that is, the latest and the immediately preceding iterations. For example, it may be decided to stop the iteration when the value of C_i' for every element differs from C_i—the value at the preceding iteration—by less than some arbitrary limit, say 0.1%. (7) If further iteration is indicated, the sequence is repeated starting with step 3 and using C_i' and R_i' as the "present" estimate and its calculated intensity function. Table 15.10 gives a series of seven successive iterations of an analysis of a five-component alloy.

The iteration procedure may be varied to meet special requirements. In an analysis of mixed sulfides of iron, cobalt, nickel, and copper, the

TABLE 15.10. Composition of a Five-Component Alloy Calculated in Successive Iterations by the Fundamental-Parameters Method[a]

Iteration	Concentration, wt %				
	Cr	Fe	Co	Ni	Mo
1	26.81	18.77	1.41	39.68	13.33
2	22.90	19.60	1.47	46.07	9.96
3	22.11	18.59	1.44	47.55	10.31
4	22.20	18.38	1.43	47.58	10.41
5	22.23	18.39	1.43	47.53	10.42
6	22.23	18.40	1.43	47.53	10.41
7	22.23	18.40	1.43	47.53	10.41

[a] From Criss and Birks (*C65*).

S $K\alpha$ line was not measured, and the computer calculated sulfur concentration by difference during the iteration (*B11*). The iteration process may be used to evaluate the effect of trace constituents on the minor and major constituents; the iteration is done repetitively, each time disregarding one of the trace elements, and the results compared.

15.7. MULTIPLE-REGRESSION METHOD

Multiple regression (*A15*, *M42*, *S60*) is a statistical method used to study correlation among observed data and to form equations relating one dependent variable with several independent variables. In many x-ray analytical situations, regression methods can be applied to correlate analyte-line intensity I (the independent variable) with concentration C (the dependent variable). Mitchell and Hopper (*M42*) have developed this technique, and the following discussion is derived largely from their paper.

When plotted data obviously lie on a straight line, the dependent variable C may be predicted by use of the single independent variable I in a linear equation such as the following:

$$C = a_0 + a_1 I \tag{15.122}$$

When the plotted curve is nonlinear, other functional relationships must be assumed and the best fit determined, such as a three-term polynomial (parabolic or quadratic) curve:

$$C = a_0 + a_1 I + a_2 I^2 \tag{15.123}$$

In a multicomponent system, if the dependent variable C_1, the concentration of one of the elements, is a linear function of each of the independent variables I_1, I_2, I_3, ..., the line intensities of each of the component elements, a multiple linear equation may be estimated by the least-squares method:

$$C_1 = a_0 + a_1 I_1 + a_2 I_2 + a_3 I_3 + \cdots \tag{15.124}$$

However, if the I–C relationship is nonlinear, a higher-order polynomial may be required. In most cases, x-ray data fit an equation of the form

$$C_1 = a_0 + a_1 I_1 + a_2 I_2 + a_3 I_3 + a_4 I_1^2 + a_5 I_2^2 + a_6 I_3^2 + a_7 I_1 I_2 + a_8 I_1 I_3 + a_9 I_2 I_3 \tag{15.125}$$

The constants are evaluated from intensities measured from numbers of multicomponent standards having accurately known compositions. The calculations can be made on a desk calculator for two- and three-term equations and limited sets of I–C data. However, for more complex equations and large sets of data, a regression program using a digital computer is required. An incidental advantage of the multiple-regression method, compared with the influence-coefficient method, is that it is more easily programmed for digital computers and has smaller storage requirements. The result of the multiple regression procedure is an equation in which the constants a have numerical values giving the best fit to the measured intensity data and the known composition of the standards. Thereafter, the equation is used to provide analytical composition data from intensities measured from samples similar to the standards used to evaluate the constants.

Part VI

Specimen Preparation and Presentation

Chapter 16

Specimen Preparation and Presentation—General; Solids, Powders, Briquets, Fusion Products

16.1. GENERAL CONSIDERATIONS

16.1.1. Introduction

A modern, automatic, sequential x-ray spectrometer collects count data at a rate of 30–100 s per element; semiautomatic and manual instruments are somewhat slower because the settings cannot be made as quickly. A modern, automatic, simultaneous spectrometer collects count data for up to 30 elements in 30–200 s; it has been remarked that simultaneous instruments collect data so fast that the readout time may become significant! An energy-dispersive spectrometer with a multichannel analyzer collects data for up to ~20 elements in 30 s to a few minutes. Clearly, specimen preparation is usually the time-limiting step that determines the sample throughput rate, and when the samples can be measured as received—or substantially so—x-ray spectrochemical analysis is very rapid indeed.

X-ray secondary-emission spectrometry is probably the most versatile of all instrumental analytical chemical methods with respect to the variety of specimen forms to which it is readily applicable nondestructively. Many forms of specimen can be analyzed in the x-ray spectrometer without treatment and without any special mounting arrangement. Other specimens require treatment and/or mounting cells or supports.

In general, specimen considerations, requirements, preparation, and presentation are substantially the same for wavelength- and energy-dispersive and nondispersive x-ray spectrometry, provided that the excitation is by x-radiation. However, if the excitation is by electrons, β-radioisotopes, protons, or other ions, the exciting radiation does not penetrate as deeply into the specimen surface. In these cases, surface-texture and particle-size effects become more severe, and the specimen preparation must be directed more toward dealing with these effects. Also, whereas the entire gamut of specimen forms is commonly used in wavelength-dispersive spectrometry, supported specimens of one kind or another are by far the most commonly used form in energy-dispersive spectrometry.

In this chapter and the next, the term specimen *presentation* is used to indicate the physical form in which the specimen is placed in its compartment, that is, in which it is "presented" to the x-ray spectrometer. The term *preparation* indicates any treatment—other than simply cutting to appropriate size—to which the as-received sample is subjected prior to its presentation to the spectrometer. In this regard, Liebhafsky and his coauthors refer to *original* and *prepared* samples. Specimen preparation may be physical, chemical, or both, and it may involve treatment of the specimen in its present form or conversion to another form. The treatment must be applicable to both samples and standards. The terms *analyte*, *specimen*, *sample*, and *standard*, and *major*, *minor*, and *trace constituent* are defined in Section 3.1, and the term *nondestructive*, as applied to x-ray spectrometric specimens, is discussed in Section 3.3.1.5.

It is of interest and value to classify x-ray spectrometry specimens in several ways as in Table 16.1, which is an expansion of a table compiled by Birks [p. 96 in (*8*)]. The specimen preparation technique, instrumentation, and accuracy of the analysis depend largely on parts *A*, *B*, and *C*. For example, determination of magnesium in low concentration in a solid specimen requires about the same surface preparation, instrument components, and counting strategy regardless of whether the specimen is a metal, mineral, ceramic, wood product, or plastic.

The specimen presentation forms may be classified in seven categories as follows:

1. *Solids*, including (a) unfinished ingot, mineral, etc.; (b) finished bar, rod, sheet, foil, ceramic, glass, plastic, etc.; and even (c) rubber, wood,

TABLE 16.1 Classification of Specimens for X-Ray Spectrometric Analysis

A. By Atomic Number Z of Analyte or Wavelength λ of Analyte Line (Section 16.1.3)

1. $Z > 22$ (Ti), $\lambda < 3$ Å
2. Z 12–22 (Mg–Ti), λ 2–10 Å
3. Z 5–12 (B–Mg), $\lambda > 10$ Å

B. By Analyte Concentration Range

1. 1–100 wt%
2. 0.01–1 wt%
3. 0.0001–0.01 wt% (1–100 ppm)
4. < 0.0001 wt%

C. By Physical Form of Specimen as Presented to the Spectrometer (Section 16.1.1)

1. Solids (Section 16.2)
2. Small fabricated forms and parts (Section 16.2.2.2)
3. Powders (Section 16.3)
4. Briquets (Section 16.3.4.2)
5. Fusion products (Section 16.4)
6. Liquids (Section 17.1)
7. Supported specimens (Section 17.2)

D. By Application (Section 16.1.2)

1. Metals and alloys
2. Minerals and ores
3. Ceramics and glasses
4. Plastic, rubber, wood, textile and paper products, etc.
5. Small fabricated forms and parts
6. Powders
7. Platings, coatings, and films
8. Liquids and solutions
9. Slurries
10. Gases and vapors
11. Inhomogeneous materials—gross analysis
12. Inhomogeneous materials—analysis of individual phases
13. Very low concentrations (trace analysis)
14. Limited total amount (micro analysis)
15. Lighter elements (Z 12–22, Mg–Ti)
16. Lightest elements (Z 5–12, B–Mg)
17. Radioactive materials
18. Dynamic systems
19. Continuous on-stream analysis and process control

paper, and textile products. Bulk (large) specimens may also be included in this group.

2. *Small fabricated forms and parts*, including rod, wire, shot, screws, nuts, washers, lugs, gears, electron-tube and transistor components, and other parts of regular or irregular shape.

3. *Powders.*

4. *Briquets.*

5. *Fusion products* resulting from fusion of the sample material with fluxes, such as potassium pyrosulfate, or with glass-forming fluxes, such as lithium or sodium tetraborate.

6. *Liquids and solutions.*

7. *Supported specimens*, including those distributed on or in filter paper, Millipore filters, Fiberglas, Mylar, Scotch tape, ion-exchange membrane, etc., or on fine quartz or plastic fibers. Thin films and platings may also be included in this group.

In the preceding list, each form can be converted to a form listed below it, and some forms can also be converted to forms listed above them. For example, solutions may be converted to powders by precipitation, evaporation, or absorption in cellulose, and briquets and fusion products are converted to powder by grinding; some powders can be converted to solids by melting or sintering, or by reduction to metal followed by melting or compacting.

Usually, nondestructive x-ray spectrometric analyses of samples in their as-received forms are more rapid and convenient than analyses by any other method. And since the x-ray spectrometer can accommodate almost any specimen form, the analyst is well advised to strive to analyze samples in their original form whenever practical. In fact, one might wonder why it should ever be necessary to alter the specimen form at all—that is, why any preparation other than simple physical treatment of the specimen in its present form should ever be required.

Actually, for qualitative x-ray spectrometric analysis, usually the only requirement is that the specimen fit its compartment, and preparation other than, say, provision of a flat surface on a solid may well be unnecessary. However, for quantitative analysis, elaborate preparation and conversion of form may be required to deal with one or more of a large number of conditions, which may be classified in the six groups listed below. When such elaborate preparation is required, much of the speed and convenience of the x-ray spectrometric method is lost.

1. *Specimen errors must be reduced* when: (a) samples are very inhomogeneous; (b) the surface layer is not representative of the bulk sample; (c) the average particle size or the particle-size distribution varies among the samples; (d) the amount of porosity or pore size varies among the samples; and/or (e) surface texture is unsuitable or varies among the samples.

2. *Additions must be made* to the samples when: (a) absorption-enhancement effects are severe and an internal standard must be added; (b)—or a matrix-masking agent must be added; (c)—or the sample must be diluted; (d) an internal control standard or intensity-reference standard must be added; or (e) a standard-addition or standard-dilution method is to be used. These addition methods are usually not feasible with specimens in the form of solids or fabricated parts.

3. *Chemical separation or other treatment is required* when: (a) the analyte(s) must be preconcentrated or separated; (b) interferants must be separated; or (c) the analyte(s) must be determined by an indirect or association method (Section 14.8.3).

4. *Standards must be matched.* Sometimes, it is necessary to change the as-received specimen form even though it is entirely suitable for analysis. This is the case when: (a) the sample shape is completely irregular and varies among the samples; (b) standards are not available or readily prepared in the as-received form; or (c) the same substance is routinely analyzed in a number of forms—such as bar, drillings, shot, parts, wire, powder—and conversion to a common form—such as solution or fusion product—permits all analyses to be made with a single set of standards.

5. The x-ray spectrometric method is to complement some other analytical method (Section 14.8.4).

6. The sample may be unsuitable because of chemical instability, hygroscopicity, sensitivity to heating or radiolysis by the x-ray beam, etc.

Great ingenuity has been shown in the development of x-ray spectrometric preparation and presentation techniques. The number of such techniques reported in the literature is very large, and there are few new papers on x-ray spectrometry that do not report at least one more. Several comprehensive reviews of x-ray spectrometric preparation and presentation techniques have been published (*25f*, *B43*, *T11*).

The remainder of Section 16.1 gives some general considerations relative to all x-ray spectrometric specimen preparation and presentation techniques. Sections 16.2, 16.3, and 16.4 consider solids, powders and briquets, and fusion products, respectively. Chapter 17 deals with liquids, supported specimens, and micro and trace analysis. For each specimen type, there is a consideration of sources of samples, alternative methods of presentation, favorable and unfavorable features, treatment, precautions and considerations, and, perhaps, some detailed descriptions of certain selected techniques. However, the emphasis is on principles and techniques, rather than specific applications and materials.

16.1.2. Classification of Applications

Birks [p. 71 in ed. 1 of (*8*)] classified x-ray spectrometric applications in 12 groups on the basis of similarities of physical form, nature and concentration range of analyte(s), wavelength region of analyte lines, general problems encountered, and applicable techniques. The following is an expansion of his list:

1. *Metals and alloys* (Section 16.2).

2. *Minerals and ores*. Carl and Campbell (*C25*) further classify minerals and ores in four groups as follows: (a) those in which one analyte varies in concentration in an otherwise substantially constant matrix; (b) those in which two elements vary in all proportions; (c) those in which two elements vary in an otherwise substantially constant matrix; and (d) complex mineral systems. Minerals and ores are characterized by a very wide range of composition, great inhomogeneity, and severe particle-size and absorption-enhancement effects. The analyte may be present in low concentration in a largely unknown and variable matrix. Severe spectral-line interference may result from the presence of many elements in complex minerals. Minerals and ores are usually analyzed as powders, briquets, or fusion products because of their great inhomogeneity. Standards are usually difficult to obtain.

3. *Ceramics and glasses* usually consist of minor and trace elements of relatively high atomic number in a low-Z matrix (*L33*, *R37*). Slags fall in this category rather than with minerals and ores because their molten origin gives them high homogeneity.

4. *Plastic, rubber, wood, textile, and paper products* usually consist of minor and trace elements of low and high atomic number in a carbon–hydrogen–oxygen matrix.

5. *Small fabricated forms and parts* may be analyzed as such or reduced to powder, briquet, fusion product, or solution.

6. *Powders* may be analyzed loose or briqueted, with or without diluents or binders (Section 16.3).

7. *Platings, coatings, and films* (Section 14.7, Chapter 18).

8. *Liquids and solutions* (Chapter 17).

9. *Slurries* (Section 17.1.4.7).

10. *Gases and vapors* containing x-ray spectrometrically determinable elements may be taken up in solution or in a solid absorber. It may be feasible to determine gases not containing such an element by indirect or association analysis (Section 14.8.3). Even the group 0 ("inert") gases (helium, neon, argon, krypton, xenon, and radon) may form stoichiometric association and clathrate compounds with certain substances. For helium and neon, an associated substance containing a determinable element is required. It might be possible to adsorb argon, krypton, and xenon in activated charcoal, which might then be placed in the spectrometer. Provision would be required to cool the charcoal to liquid-nitrogen temperature.

As an example, occluded argon in concentration 0.05–7 wt % is readily determined in argon-sputtered silicon dioxide (SiO_2) films 100 Å to 5 μm thick (*H56, L31*). The relationship between Ar $K\alpha$ intensity and argon concentration is established in either of two ways: Ar $K\alpha$ intensity can be measured from weighed specimens, then argon determined by weight loss on heating the specimens in helium at 600°C for several hours. Alternatively, Ar $K\alpha$ intensity from samples can be applied to curves of Cl $K\alpha$ and K $K\alpha$ intensity measured from potassium chloride (KCl) films of known mass and thickness. Chlorine, argon, and potassium have consecutive atomic numbers, 17–19.

Hudgens and Pish (*H69*) analyzed mixtures of iodomethane (CH_3I, b.p. ~42°C) and bromoethane (CH_3CH_2Br, b.p. ~38°C) by direct measurements of I $K\alpha$ and Br $K\alpha$ from their vapors. The liquid samples were placed in a rubber balloon attached to the side tube of a liquid-specimen cell having a glass body and a 50-μm (0.002-inch) Mylar window. The measurements were made on the emission from the vapor in the cell. The I $K\alpha$/Br $K\alpha$ intensity ratios were calculated and applied to a calibration curve established from standard mixtures.

The $K\alpha$ lines of sulfur and halogens have been measured directly from gas-chromatograph effluents passing through thin-wall plastic tubes in the x-ray spectrometer specimen chamber (Section 14.8.4). Gases and vapors can also be measured by x-ray absorption methods (Section 20.1).

11. *Inhomogeneous materials—gross analysis.* The samples are rotated during measurement or homogenized by pulverization, fusion, or dissolution.

12. *Inhomogeneous materials—analysis of individual phases.* The individual phases may be analyzed after physical separation, or in place by use of selected-area techniques on the x-ray spectrometer (Chapter 19) or electron-probe microanalyzer (Chapter 21).

13. *Very low concentrations—trace analysis* (Section 17.3).

14. *Limited total amount—microanalysis* (Section 17.3). This category includes isolated particles and fibers; micro quantities of analyte separated from a matrix or recovered from solution; and individual grains from a heterogeneous material (class 11 above) after physical separation or in place when analyzed by selected-area techniques.

15. *The lighter elements,* atomic number 9–22 (fluorine to titanium), having spectral lines in the long-wavelength spectral region (Section 16.1.3).

16. *The lightest elements,* atomic number 4–11 (beryllium to sodium), having spectral lines in the ultralong-wavelength spectral region (Section 9.4.3).

17. *Radioactive materials* (*M58*). The problem in this class, aside from personal safety, is to prevent radiation of radioactive origin from interfering with the x-rays to be measured (Section 17.4).

18. *Dynamic systems.* If special specimen-presentation systems are provided, x-ray spectrometry can be applied to the study of dynamic processes such as the following: (a) dissolution of a solid in a liquid; (b) precipitation of a solid from a liquid; (c) admixture of partially miscible liquids; (d) diffusion of solid in solid or liquid in liquid; (e) diffusion of an element to the surface; (f) diffusion of an element from the surface into the bulk; (g) sublimation or desorption; (h) deposition or adsorption; (i) preferential corrosion of a surface; (j) rate of electrodeposition (Section 18.3.8); and (k) continuous paper and gas chromatography (Section 14.8.4). For examples of (a) above, Cu $K\alpha$ intensity has been measured from a cell containing water standing undisturbed over copper(II) sulfate pentahydrate ($CuSO_4 \cdot 5H_2O$) crystals or from a window in a sidearm through which the supernatant solution was circulated from a vessel containing such crystals.

19. *Continuous on-stream analysis and process control* (Section 9.3.1).

16.1.3. Problems Specific to the Light Elements

In x-ray spectrometric analysis, the long-wavelength region extends from ~2.75 to 20 Å and contains the $K\alpha$ lines of elements 22 (titanium) to 9 (fluorine) and the $L\alpha_1$ lines of elements 56 (barium) to 25 (manganese).

The long-wavelength region is that in which a thin-window gas-flow proportional counter and helium or vacuum path may be regarded as essential, although $K\alpha$ lines can be measured in air path from relatively low concentrations of elements down to 19 (potassium). The ultralong-wavelength region is discussed in Section 9.4.3.

Standard commercial x-ray spectrometers are substantially useless for elements having atomic number less than 9 (fluorine), and they are subject to increasing difficulty and inconvenience for elements having atomic number 22 (titanium) to 9. In Section 13.1.3 and Table 13.2, it is shown that even for a very favorable analyte (copper), the x-ray spectrometer is extremely inefficient: For 10^{16} primary x-ray photons emitted by the x-ray tube target and capable of exciting the copper K spectrum, only 4×10^8 Cu $K\alpha$ photons are emitted by a monolayer of copper atoms having area 1 cm^2; and, of these, only $\sim$250 are finally counted by the detector. For a "light element," the efficiency is much lower, and every component and process in the instrument contributes to this inefficiency. In fact, when one considers these individual contributions, one marvels that the light elements can be determined at all!

a. X-ray Tube. The x-ray tube target generates inefficiently the long wavelengths most effective in exciting spectra of the low-Z elements. Then the x-ray tube window has very low transmittance for these long wavelengths; for example, 1-mm (0.040-inch) beryllium transmits only 10, 1, 0.1, and 0.01% of the incident x-ray photons at 3.5, 4.5, 5, and 6 Å, respectively; these wavelengths correspond approximately to the K-absorption edges of potassium, chlorine, sulfur, and phosphorus, respectively; 0.25-mm (0.010-inch) beryllium transmits only 10 and 0.1% at 5 and 8 Å (sulfur and aluminum K edges), respectively, and even 25-μm (0.001-inch) beryllium transmits only 20% at 10 Å (magnesium K edge).

b. Specimen. As the atomic number of the analyte decreases, the wavelength of the analyte line increases, and: (1) infinite thickness decreases; (2) fluorescent yield decreases; (3) matrix absorption increases; (4) absorption in cell windows increases; and (5) displacement of wavelength with chemical state increases. Moreover, as the effective atomic number of the total specimen decreases, both coherent and incoherent scatter increase.

Because of the reduced infinite thickness, only a very thin surface layer contributes to the measured analyte-line radiation, and oxidation, corrosion, surface texture, and particle size have marked effect. Liebhafsky and his colleagues [p. 449 in (*27*)] cite a striking example of the effect of surface condition: a piece of transformer steel containing 3.3% silicon gave a Si $K\alpha$ intensity of 263 counts/s after polishing, 53 counts/s after oxidation, then 725 counts/s after a brief etch in nitric acid.

The fluorescent yield (Section 2.5.2, Figure 2.18) is only 15% for Ca $K\alpha$ (Z 20), only 1% for Na $K\alpha$ (Z 11). The absorption of the analyte line is high even by the analyte itself and by adjacent light elements. Heavier elements, even as minor constituents, cause high absorption. Although a commercial spectrometer may not resolve or permit accurate measurement of wavelength shifts due to chemical effects, the measured intensity may vary with these shifts, providing a source of error. The higher scatter increases the background and thereby the minimum detectable amount. Also, the intensity of the scattered target spectrum increases and with it the possibility of spectral interference. The increased incoherent scatter further complicates the spectrum.

Dilution methods (dry, fusion, solution) are difficult to apply to low-Z elements because analyte-line intensity, already relatively low, is reduced still further, both by dilution and by absorption of the analyte line by the diluent. Finally, internal standardization is less effective and more difficult to apply to low-Z elements.

c. Radiation Path. Absorption of the analyte line in the radiation path increases as the atomic number of the analyte decreases. Helium transmits only 80% at 10 Å (Mg $K\alpha$), and helium enclosures are frequently difficult to work with at the large 2θ angles required by some crystals for low-Z elements. Vacuum enclosures may have a limited 2θ range and force selection of a less favorable crystal.

d. Crystal. The crystals used at long wavelengths are frequently less "reflective" than those applicable at shorter wavelengths (*J8*). Higher background and energy interference may result from the emission of secondary spectra of the elements in the crystal. The temperature coefficient of expansion of crystals is highest at long wavelengths, and PET has the highest coefficient of any commonly used crystal (Figure 5.10).

e. Detector. Detector efficiency decreases at the long wavelengths (*J8*). The transmittance at 10 Å (Mg $K\alpha$) of 6-μm (0.00025-inch) Mylar and 25-μm (0.001-inch) beryllium is only 26.3% and 19.9%, respectively.

f. Spectrum. The increased possibility of spectral interference from scattered target lines and crystal emission has been mentioned already. The spectra of the low-Z elements are also highly prone to interference from higher-order lines of elements of higher atomic number.

16.1.4. Standards

16.1.4.1. General

It has already been mentioned that x-ray spectrometric analysis is a comparison method. Consequently, for a given analyte–matrix system, one, two, three, or more standards may be required, depending on the linearity of the calibration function, the analyte concentration range, and the severity of the matrix effects. Even "absolute" methods may require at least one standard if the pure analyte is unsuitable. It has also been mentioned (Section 3.3.2.6) that standards for x-ray spectrometry must be as similar as possible to the samples, not only in analyte concentration, but in matrix composition, physical form, surface texture, particle size, etc., and in some cases, even in the chemical state of the analyte. Frequently, a rapid, convenient, nondestructive analysis of a metal, ceramic, or small part is precluded because of unavailability of standards in the required form.

Standards may be classified as *primary*, *secondary*, or *synthetic*. A primary standard is one that has been subjected to exhaustive—often round-robin—chemical analysis by independent, highly reliable methods. A secondary standard is one that has been analyzed carefully with respect to a primary standard. A synthetic standard is one that has been prepared by combination of the required constituents in the required proportions.

16.1.4.2. Permanence of Standards

In general, because of the substantially nondestructive nature of x-ray spectrometric analysis, once a set of standards has been obtained, they last more or less indefinitely, if carefully handled and stored. *In principle*, solution standards constitute the only exception; even if they are stable on long storage, the portions actually used are usually discarded. However, in practice there are many exceptions to this, and standards may deteriorate from any of four principal causes:

Radiation damage may occur on repeated prolonged exposure to the intense primary x-ray beam. Solid standards may suffer surface damage with resultant pulverization and sloughing-off of material. Plastic and glass-disk standards may actually develop deep cracks, voids, and fissures. The filter-paper, Mylar, and other substrates of supported standards eventually embrittle and perforate.

Superficial oxidation may occur on repeated exposure to the high ozone (O_3) concentration in the specimen chamber; ozone is also responsible for the characteristic sharp odor noticeable in the chamber and in the vicinity of high-potential electric apparatus. Superficial corrosion also results from perspiration on repeated handling, although this is easily avoided by handling standards only by the edges. Chemical change also may occur in storage. This is especially true of solution standards, which may also evaporate. Deterioration of solutions is accelerated if they are not kept tightly stoppered. Incidentally, in many cases, solution standards must be prepared just prior to use. Powder, briquet, and solid standards may undergo chemical change due to slow reaction with moisture, carbon dioxide, sulfur dioxide, and oxygen in the air. For example, metal standards used for determination of sulfur may increase in apparent sulfur concentration on standing in air. Such processes are particularly serious with plated standards and those having several phases that might deteriorate preferentially. Chemical deterioration is retarded by storing the standards in plastic envelopes or boxes, or in desiccators.

Mechanical damage may occur on repeated handling and in storage. Solid standards may become scratched, abraded, or smeared. Briquets may powder, crumble, swell, or crack, but this damage is minimized by use of binders, backings, rings, or metal caps, or by spraying the edges and faces with clear Krylon or other plastic coating.

Finally, deterioration may occur simply with time—*aging*—partly by chemical change (see above) and partly by diffusion and other physical change. This is especially true of solutions. Powder and briquet standards containing hygroscopic constituents may absorb moisture. Incidentally, this can happen to standards containing lanthanum oxide (La_2O_3) high-absorption diluent (*F1*). Glass disks may slowly devitrify, or craze, or crack from stress. Even stable disks may gradually lose their glassy surface and become dull. Aging may be particularly serious for low-melting alloys. For example, specimens of 10–90 and 63–37 tin–lead solders showed an increase in Sn $K\alpha$ intensity of 0.5% and 1.05%, respectively, after 240 days (*G18*).

As a result of these effects, it may be necessary to refinish or reprocess standards occasionally. Solid and glass-disk standards may be repolished or, in more severe cases, recast. Briquets may be resurfaced or pulverized and recompacted.

16.1.4.3. Sources of Standards

In general, x-ray spectrometric standards are obtained in one of four ways. They may be purchased (*M32*), obtained by analysis of samples, prepared by concentration or dilution of a basic material of known composition, or synthesized.

Standards may be purchased from: (1) the National Bureau of Standards (*N10–12*) and from the corresponding agencies of other countries; (2) optical emission spectrographic supply firms; (3) manufacturers of the type of material to be analyzed; for example, standards for aluminum-base alloys are available from Aluminum Company of America; (4) professional technical societies and institutes concerned with the type of material to be analyzed; and (5) other analytical laboratories engaged in analysis of the same material. A compilation of such sources is available (*M32*).

Samples analyzed very carefully by independent methods may be used as standards. The "independent method" may be an x-ray spectrometric solution or fusion method using synthetic standards. The homogeneity of the prospective standards should be established; for example, if the samples are metal coupons, both sides should be finished, and x-ray spectrometric measurements of analyte-line intensity should be made from both sides with the coupon in various orientations, or with and without rotation. If the coupons are thus shown to be homogeneous, turnings or drillings are taken from one side for analysis. Alternatively, two or three consecutive adjacent slices may be cut from the sample and all sides finished and tested for uniformity in the same way (*F17*). Then one slice is retained, and the adjacent one(s) sacrificed for analysis. Standards for nondestructive analysis of small fabricated forms and parts are obtained in a similar way. Several parts from a lot are tested for uniformity of composition by x-ray spectrometric measurement of analyte-line intensities from each. If the parts all give the same intensities, several are sacrificed for analysis, the others retained as standards. Alternatively, standard parts may be fabricated from material of known composition.

To avoid the necessity for periodic resurfacing or recasting of metal-alloy standards to correct for the aging effect (see above), Lumb (*L53*) prepared standards by suspending the alloy powders in an epoxy resin. The resin is cast into a disk, one surface of which is machined plane and smooth. Analyte-line intensities from this surface and from a freshly surfaced disk of solid alloy of the same composition are compared repeatedly in the x-ray spectrometer. After each comparison, the unsurfaced side of the resin disk is machined to reduce its thickness decrementally. This alternate comparison and machining is continued until the analyte-line intensities from the two disks are identical. Thereafter, the resin disk is used as the standard. No change was noted in the intensities from such resin standards after six months of use.

If a single powder or liquid of the type to be analyzed and of accurately known composition is available, several standards may be prepared from it by quantitative addition of analyte(s) or of analyte-free matrix material or solvent to form more and less concentrated standards, respectively. Analyte(s) may be added to the powder by: (1) quantitative admixture of powder having accurately known analyte concentration and substantially the same particle size as the base powder; or (2) volumetric addition of a standard analyte solution, followed by homogenizing, drying, and grinding (*D43*).

Synthesis of standards is most convenient for powders, briquets, fusion products, and solutions. However, small melts of alloys—especially low-melting alloys—or small firings of ceramics can be made. For metals, small electric vacuum or inert-atmosphere furnaces, or small arc-furnaces having argon atmosphere, may be used to prevent oxidation (*F9*). In arc-melting composite metal powders, dispersion of powder on striking the arc is avoided and homogeneity of the button is improved by briqueting the powder charge at 3500 kg/cm^2 (50,000 lb/inch^2), melting for ~3 min, turning over the button, and remelting for ~3 min (*B77*). Low-melting alloys can be prepared by melting mixtures of the pure elements in proper concentrations under wax and casting in a water-cooled mold. To prevent the segregation that may occur on slow cooling, melts may also be chill-cast by pouring into plaster-ring molds on a cooled metal plate; the chilled surface is the analytical surface. In some cases, metal standards can be prepared from synthetic solutions containing all major, minor, and trace elements in proper relative concentrations by evaporation to dryness, firing and reducing to metal sponge, briqueting, and sintering. This technique has been applied to preparation of nickel alloys (*L23*).

Mulligan *et al.* (*M60*) describe a method of investment or "lost-wax" casting to prepare very thin disks and thereby conserve precious metals or small samples of common metals. A disk 0.5–3 mm thick and up to 2.5 cm in diameter is cut from a sheet of stiff wax and mounted vertically on edge on the tip of a small brass cone. This assembly is placed in an open-top cylindrical cup having inside diameter somewhat larger than that of the disk and height just to its upper edge. The cup is now covered with a cap, the under side of which is tangent to the top of the disk, and having a filling port communicating with the disk. The cup is now filled with a slurry of investment material consisting of equal parts of cristobalite (SiO_2) and gypsum ($CaSO_4 \cdot 2H_2O$), which hardens in a few minutes, encasing the disk. The filled cylinder is placed in a furnace and heated to ~700°C, whereupon the wax volatilizes, leaving a mold cavity having the dimensions of the wax disk and communicating with the filling port. The molten specimen charge, in a small crucible, is forced into this mold in a centrifugal casting machine or by some other means. When being filled, the mold may be preheated to avoid chill-casting and segregation. After cooling, the investment is broken to permit recovery of the specimen disk, which is then trimmed of the waste product at the filling port ("sprue"), mounted in a plastic disk for ease of handling, and surfaced in the usual manner.

Powder standards may be prepared by admixture of individual powder components or from solution by precipitation, evaporation, or absorption in cellulose and firing. Briquet standards may be prepared from the powders. Briquets pressed in metal cups are more rugged and easier to store, and unless low-Z elements are to be determined, the surface should be lightly sprayed with clear Krylon. Powder and briquet standards are discussed further in Section 16.3.2.

Lucite and other plastic standards are made by adding analyte(s) in the form of their organic compounds to solutions of the plastic, which are then evaporated, or to the monomers, which are then polymerized. Polyethylene standards have been made by evaporation of slurries of high-density polyethylene powder in standard solutions of the analyte(s) (*H24*, *H25*) and by impregnation of polyethylene powder with standard solutions of the analyte(s) followed by drying, homogenizing, and briqueting (*F8*).

Fusion products and solutions are most readily synthesized because all inhomogeneities and particle-size effects disappear completely. The fusion-product standards may be in the form of glass disks, or they may be ground and briqueted. Solution standards are usually the easiest to synthesize. Metallic elements are added to organic liquids as organometallic compounds, such as napthenates, or as salts of organic acids. Silicon, phosphorus, and sulfur, and fluorine, chlorine, bromine, and iodine are added as their organic compounds (*D40*, *S17*, *Y3*).

Standards for analysis of films and platings are prepared in several ways. An analyte solution may be sprayed from an atomizer on a blank substrate and area concentration calculated from gain in weight and composition of the sprayed chemical species. The analyte may be vacuum-evaporated on a blank substrate, and area concentration derived from interference colors or fringes, optical transmission, or gain in weight. Alternatively, several substrate blanks may be exposed at geometrically equivalent positions, the uniformity of the deposits verified by x-ray spectrometric measurement, and one or more of the blanks sacrificed for chemical analysis. A weighed portion of analyte may be evaporated completely and area concentration calculated from the geometry, that is, from the solid angle intercepted by the substrate. Finally, metal thin-film standards have been prepared by applying a mixture of organometallic compounds to the substrate and firing in a reducing atmosphere. The method has been applied to preparation of nickel–iron films (*S23*). Thin-film standards can be protected by covering them with 6-μm (0.00025-inch) Mylar film (*G29*).

A measured volume of a slurry having a known weight/volume analyte concentration can be applied to a large, optically flat glass plate and spread into a thin film by means of a casting knife ("doctor blade") (*B39, F14*). The area of the film is measured with a planimeter or by tracing the outline of the film on graph paper and counting squares. The area concentration is then calculated. Blanks can be removed from the film by scoring, flooding with water, and "coaxing" squares of the film to float off. Films having various area concentrations are made by varying either the analyte concentration in the slurry or the film thickness, or by stacking various numbers of layers of film. A film may be placed on a substrate and moistened with solvent to affix it. The substrate may be plane, cylindrical, or of other form. Films may also be vacuum-evaporated on Scotch tape and affixed to substrates of various materials and forms (*S12*).

Surface films of uranium on cylindrical surfaces have been analyzed with flat standards by previously establishing the analyte-line intensity ratio from cylindrical and flat specimens having the same uranium surface concentration (*S12*). This ratio was independent of uranium concentration.

Plated standards are prepared by electroplating a relatively large area of substrate having its back, edges, and a 1-cm border on its front surface masked with insulating tape. These precautions avoid useless plating of the back and nonuniform plating of the edge (edge effect). Calculations based on the electrolysis laws and the exposed area permit plating to approximately the required thickness. Several blanks are cut from the plated area. The uniformity of the plating is verified by x-ray spectrometric measurement of the intensity of a plate-metal line. Then one or more blanks are sacrificed for chemical analysis.

Standards for analysis of samples supported on filter paper, Millipore and Nuclepore filters, Fiberglas, Mylar, ion-exchange membrane, etc. are prepared by volumetric addition of standard analyte solutions to blanks of the support material.

More uniform standards on filter materials are prepared by pipeting the standard solutions onto filter disks of diameter large enough so that the liquid does not spread to the edge, allowing the disk to dry slowly, then measuring the area of spread. At longer wavelengths, x-ray absorption by the filter becomes significant for analyte distributed throughout the thickness of the filter, and correction may be required. Particulate material filtered from gases or liquids is usually retained mostly on the filter surface, and longer-wavelength radiation is absorbed less than for disks prepared as described above. Standards for such samples may be prepared by filtering dilute suspensions of insoluble analyte compounds onto the filters (*G14*). Standards for aerosol specimens are also prepared by bubbling clean air through solutions of analyte salts for various times and passing the spray-laden air through weighed filter disks, which are then dried and reweighed (*C43*).

Addition of wetting agent to solutions ensures more uniform distribution, especially on Mylar and other plastic films. Use of confined-spot filter paper and ring ovens is mentioned in Section 17.2.2.

16.1.5. Effective Layer Thickness

In general, x-ray spectrometric analysis involves a relatively thin surface layer, and this layer is thinner the longer the analyte wavelength and the higher the absorption coefficient of the specimen for that wavelength. However, short-wavelength analyte lines in light matrixes, such as water or lithium tetraborate glass, may have much greater effective layers.

The primary beam from an x-ray tube operating at 50 kV contains wavelengths as short as 0.25 Å [Equation (1.25)]. Also, the angle between the primary beam and specimen surface is usually about twice the takeoff angle (Figure 15.8). These conditions result in primary-beam penetration —and analyte-line excitation—in a *relatively* thick layer on the specimen surface. However, only analyte-line x-rays actually leaving the specimen can be measured. Thus, effective layer thickness is determined not by the depth *to* which *primary* x-rays can *penetrate*, but by the depth *from* which *analyte-line* x-rays can *emerge*. This depth is typically of the order 10-100 μm, so that for specimen area of the order 10 cm², the effective specimen volume is of the order 0.01-0.1 cm³. It is essential that this volume be representative of the bulk specimen and free from surface contamination.

Togel (*T11*) has shown that the depth of the effective specimen layer, that is, the layer in which, say, 99.9% of the measured analyte-line intensity originates, $t_{99.9}$, is given by

$$t_{99.9} = \frac{6.91}{(\mu_{\lambda_{\text{pri}}}/\sin\phi) + (\mu_{\lambda_A}/\sin\psi)} \text{ cm} \tag{16.1}$$

where $\mu_{\lambda_{\text{pri}}}$ and μ_{λ_A} are the linear-absorption coefficients of the specimen for the primary x-rays and analyte line, respectively; and ϕ and ψ are the

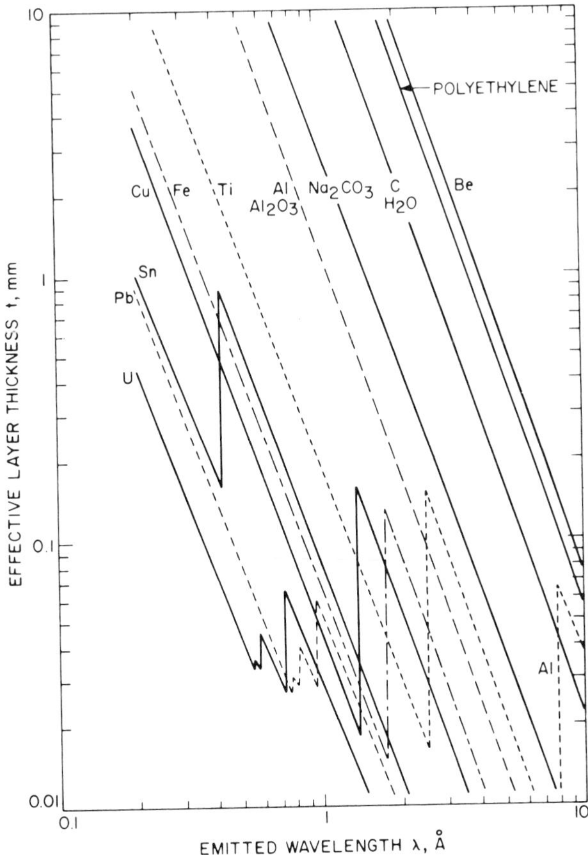

FIG. 16.1. Effective layer thickness (layer that emits 99.9% of the measured analyte line intensity) as a function of wavelength for various matrixes. [K. Togel, in "Zerstörungsfreie Materialprüfung," R. Oldenbourg, Munich, Germany (1961); courtesy of the author and publisher.]

angles made with the specimen plane by the central rays of the primary and secondary beams, respectively. If $\phi = \psi = 45°$,

$$t_{99.9} = \frac{6.91 \sin 45°}{\mu_{\lambda_{\rm pri}} + \mu_{\lambda_{\rm A}}} = \frac{4.9}{\mu_{\lambda_{\rm pri}} + \mu_{\lambda_{\rm A}}} \text{ cm} \tag{16.2}$$

Three cases may be defined on the basis of the relationship of $\lambda_{\rm pri}$, the wavelength of a monochromatic primary beam or the effective wavelength of a polychromatic primary beam; $\lambda_{\rm A}$, the wavelength of the analyte line; and $\lambda_{\rm M_{ab}}$, the absorption edge of the principal matrix constituent:

Case 1. $\lambda_{pri} \approx \lambda_A$, so that both are either $> \lambda_{M_{ab}}$ or $< \lambda_{M_{ab}}$, and $\mu_{\lambda_{pri}} \approx \mu_{\lambda_A}$; then,

$$t_{99.9} \approx 4.9/2\mu_{\lambda_A} \text{ cm} \tag{16.3}$$

Case 2. $\lambda_{pri} < \lambda_{M_{ab}} < \lambda_A$; then, if r is the absorption-edge jump ratio of $\lambda_{M_{ab}}$,

$$\mu_{\lambda_{pri}} \approx r\mu_{\lambda_A} \tag{16.4}$$

and

$$t_{99.9} \approx 4.9/(r\mu_{\lambda_A}) \text{ cm} \tag{16.5}$$

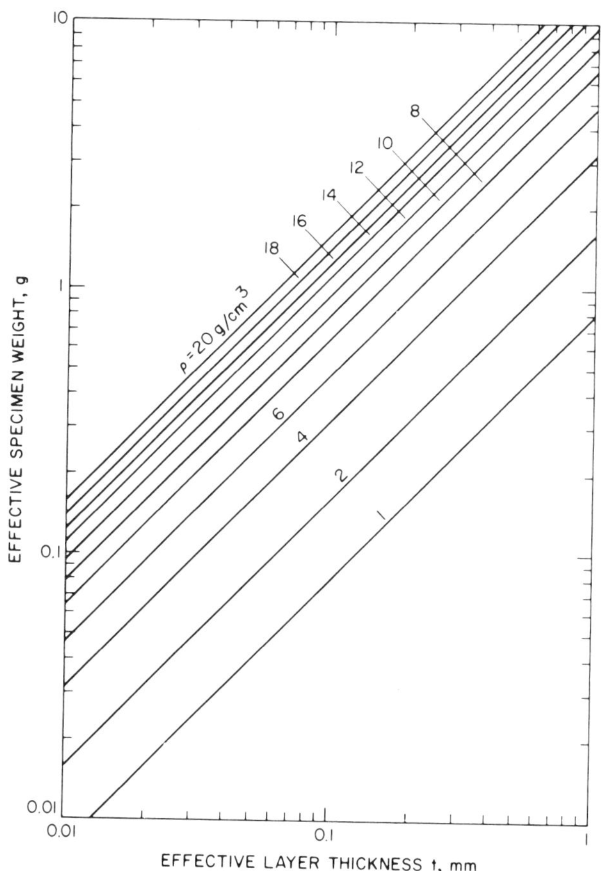

FIG. 16.2. Effective specimen mass as a function of effective layer thickness (Figure 16.1) for various matrix densities. The curves are based on a specimen area 7.84 cm². [K. Togel, in "Zerstörungsfreie Materialprüfung," R. Oldenbourg, Munich, Germany (1961); courtesy of the author and publisher.]

Case 3. $\lambda_{\text{pri}} \ll \lambda_{\text{A}}$, but both are either $>\lambda_{\text{M}_{\text{ab}}}$ or $<\lambda_{\text{M}_{\text{ab}}}$, and $\mu_{\lambda_{\text{pri}}} \ll \mu_{\lambda_{\text{A}}}$; then,

$$t_{99.9} \approx 4.9/\mu_{\lambda_{\text{A}}} \text{ cm} \tag{16.6}$$

Case 3 gives the greatest effective thickness. Figure 16.1 shows effective specimen layer thickness, calculated from Equation (16.6), as a function of analyte-line wavelength λ_{A} for several matrix materials. Figure 16.2 shows effective specimen mass, that is, mass of material in the effective specimen layer, as a function of effective thickness for various densities. The calculations are based on an effective specimen area 2.8×2.8 cm (7.84 cm^2). The data in Figures 16.1 and 16.2 must be used carefully. For example, the first figure shows that the effective layer for Cu $K\alpha$ (1.54 Å) in water is ~1 cm. However, the effective layer in a water *solution* of heavy solutes would be much less.

16.1.6. The Specimen-Preparation Laboratory

Many x-ray spectrometric laboratories deal with only a limited number of sample types. For example, a laboratory in a brass foundry may be required only to cut and finish coupons of copper-base alloys, a laboratory in a cement plant only to grind and briquet cement powders. Such laboratories need only limited specimen-preparation equipment. However, as the diversity of specimen types increases, so does the diversity of equipment required. A laboratory that must deal with the full range of x-ray spectrometric specimens—say, the customer application laboratory of a manufacturer of x-ray spectrometers—requires a very well-equipped specimen-preparation laboratory. Following is a list of most of the equipment required by such a laboratory, in addition to the basic chemicals and apparatus of the analytical chemistry laboratory:

Water wheel.
Diamond wheel.
Belt grinder.
Paper-strip grinder.
Disk grinder.
Rotating polishing wheel, with various abrasive papers and cloths, and abrasive powders of various materials and grits.
Rotating polishing wheel, vibratory type.
Lapping wheel.
Spiral-grooved, lead-covered polishing disk, for use with diamond dust.
Small end mill.

Small lathe.
Small drill press.
Agate, tungsten carbide, silicon carbide, and boron carbide mortars and pestles (in order of increasing hardness).
Mixer/Mill (Spex Industries, Inc., Metuchen, N. J.) or equivalent, with plastic, tungsten carbide, and alundum vials and balls.
Wig-L-Bug mixer.
Freezer/Mill (Spex).
Mechanical grinder, preferably of the type that grinds to uniform particle size.
Wire-cloth sieves.
Briqueting press for pressures up to 7000 kg/cm^2 (100,000 lb/inch^2), with high-temperature and vacuum dies of various sizes.
Drying oven.
Muffle furnace.
Crucible furnace.
Vacuum and inert-atmosphere furnace.
Casting mold.
Book mold.
Roller ball mill.
Film-casting knives ("doctor blades").
Hot plates.
Electroplating apparatus for preparation of plated standards.
Bell-jar vacuum system for evaporation of films.
Ion-exchange columns.
Binocular microscope, 1–100×.
Balance.
Filter flasks and small Buchner funnels, 2.5- and 1.25-cm (1- and 0.5-inch) diameters.

In addition, the preparation laboratory often uses techniques and accessories of related disciplines, including microchemistry, optical emission spectrography, radiochemistry, electron microscopy, and metallography. Finally, the laboratory requires a supply of specimen cells and mounting devices of various sizes and for all types of specimen.

Several special "machines" are available for rapid preparation of uniform briquet and glass-disk specimens (Section 16.4); some of these are described in the following paragraphs. See also Reference (*H21*).

In the Claisse Fluxer (Spex Industries, Inc., Metuchen, N. J.), sample charges are placed in six special platinum crucibles held over six Meker burners mounted on a common gyroscopic joint motor-driven to provide agitation. When fusion is complete, usually ~5 min, turning a single knob inverts all six crucibles at once so that the melts flow into their shallow, cylindrical, flat-bottomed lids, which serve as molds.

Philips provides several commercial specimen-preparation machines that automate or semiautomate, standardize, and greatly expedite the preparation of briquet or borax-glass disk specimens from powders and slurries.

The automatic briquet machine receives ground powders or dried and ground slurries, with or without binder, and, for each, measures out the charge, places it in the press, compacts it at preset pressure and time conditions in a shallow aluminum cap, and conveys the briquet to and properly seats it in the Philips simultaneous automated spectrometer.

The automatic glass-disk machine receives powders or wet slurries and for each dries, pulverizes, and sieves a sample, weighs out the charge, places it in a platinum crucible, heats it at 1200°C to expel water and carbon dioxide, adds a weighed amount of sodium (or lithium) tetraborate, and fuses the charge at 1050°C for a preset time. The machine then applies a jet of air to the melt to cool and anneal it and a high electric potential to free it from the crucible. It then actually tests the bead for mechanical strength, signals if the disk breaks, and conveys a good disk to and properly seats it in the Philips automatic spectrometer.

The fully automatic glass-disk machine above is designed specifically for the cement industry—or at least for laboratories where the samples do not vary widely in nature and composition. For a wider variety of specimens, the semiautomatic glass-disk machine is available, which performs substantially the same functions as the automatic machine, but under control of the operator.

The briquet and glass-disk machines process samples at the rate of one each minute and 5–7 min, respectively. These are about the measurement times for simultaneous and sequential automated spectrometers, respectively.

16.2. SOLIDS

16.2.1. Scope, Advantages, Limitations

The term *solid* is used here to indicate bulk solid, as distinguished from turnings, drillings, chips, small pellets, shot, granules, and powders. The solid may be of metal, rock, mineral, ore, slag, ceramic, glass, plastic, etc.; or it may be of any material that can be formed into a "solid" specimen, such as rubber, wood, textile, and paper. Briquets are not included. The solid may have the form of unfinished ingot, rock, etc.; finished plate, bar, rod, tube, sheet, or even wire; or fabricated parts. It is stated in Section 16.1.1 that solids may be presented to the instrument as such, or they may be reduced to powders, briquets, solutions, or supported specimens. This section is limited to presentation as solid.

If the material can be analyzed in solid form, the analyses are usually very rapid and convenient. Small fabricated parts may require no preparation. Finished materials may require only cutting to appropriate size. Unfinished bulk materials may require only the cutting of a coupon and finishing of one of its surfaces.

The principal disadvantage of the solid form is that any analytical method involving an addition is likely not to be feasible—standard addition and dilution, low- and high-absorption dilution, internal standardization, or internal intensity-reference standardization. The last two methods may be feasible if a suitable element is already present in all specimens in appropriate constant concentration. Chemical concentration or separation is also not feasible. Surface texture and composition are sometimes difficult to reproduce. Standards may not be available and are usually difficult to synthesize.

16.2.2. Presentation

For purely qualitative analysis, it is usually sufficient that the sample fit in the specimen chamber or in a large-specimen accessory. Little attention need be given surface treatment other than, perhaps, to provide a roughly flat surface. Irregular-shaped pieces may be pressed into a Lucite block in a metallographic press for convenience in mounting in the instrument. However, for quantitative and even semiquantitative analysis, it is essential that for all samples and standards in a given analysis, the effective specimen area (Section 4.5) be identical in all respects—area, shape, dimensions, orientation, inclination (ϕ, ψ), position in the plane defined by the mask, position above or below this plane, and surface contour or topography.

16.2.2.1. Flat Specimens

Most solid x-ray spectrometric specimens are objects already having a flat surface, or rectangular or circular coupons cut from bulk solids and finished on one side. Some specimens of this form are shown in *A–G* in Figure 16.3. For such specimens that fit in the specimen chamber, all the

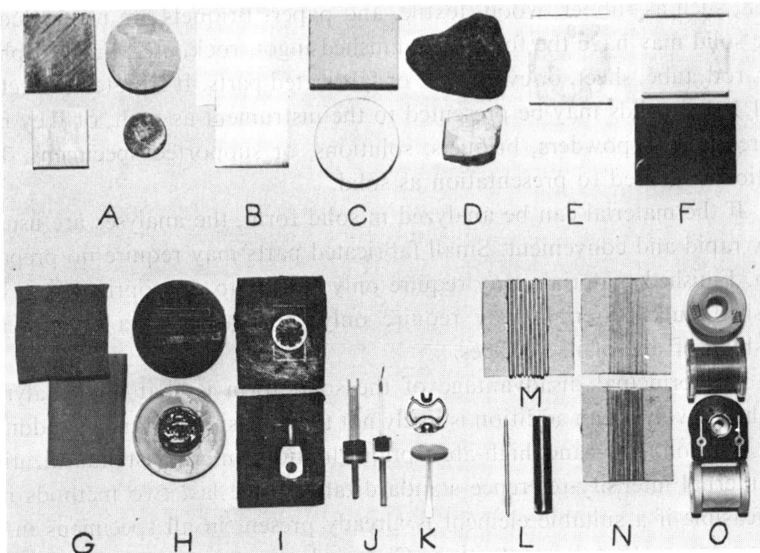

FIG. 16.3. Typical solid-specimen forms. [E. P. Bertin and R. J. Longobucco, *Norelco Reporter* **9**, 31 (1962); courtesy of Philips Electronic Instruments, Inc.] *A*. Metal coupons and stock. *B*. Metal sheet and foil. *C*. Ceramic stock. *D*. Rock, mineral, ore, etc. *E*. Glass stock. *F*. Plastic stock. *G*. Rubber stock. *H*. Metallographic sections. *I*. Small parts held in wax-impression molds. *J*. Small parts held in wax-filled plastic cups (rotable). *K*. Small parts held on plastic pedestals (rotable). *L*. Small rod. *M*. Small rod or stiff wire cemented on card. *N*. Wire wound on cards. *O*. Wire wound on small spools (rotable).

conditions stated above are met by the simple means of placement behind the mask provided in the chamber. Even if the specimen does not fit the chamber, and it is not permissible to cut it, the conditions are met by use of a large-area specimen accessory. For specimens of small area—say 1–50 mm^2—or for small areas of interest on larger specimens, a collimated specimen mask may be used (Section 4.5). For even smaller areas, selected-area techniques may be used, in which a pinhole or slit aperture is placed in the primary or secondary beam (Chapter 19).

16.2.2.2. Fabricated Forms and Parts

For fabricated forms and parts that are not flat and cannot provide flat coupons, other measures than masking must be taken to ensure reproducibility of specimen area, position, and topography.

If the part does have a flat portion, it may be presented behind a mask, or with a selected-area aperture. A single very small part can be analyzed by use of a pinhole aperture. However, higher analyte-line intensity—and therefore greater statistical precision—are realized by cementing a number of parts in a row on a small card and using a slit aperture. In this way, the writer has analyzed lead–tin solders in small (~2 mm square) solder-foil preforms and on small, irregular-shaped solder-coated nickel and phosphor-bronze components for transistors. Because of the inevitable variation of effective area of such specimens, both Sn $K\alpha$ and Pb $L\alpha_1$ lines are measured and their ratio taken. Strips of solder foil or of solder-coated nickel or phosphor-bronze are used as standards.

Wax-impression molds (*Z15, Z17*) (*I* in Figure 16.3) are made by cutting or casting rectangular blocks of sealing wax, warming them on a hot plate to the softening point, and pressing a part of the type to be analyzed into the soft wax. After the wax has cooled and hardened, the part is removed, leaving its impression. The mold is mounted in the spectrometer drawer with the impression centered in the mask window. The mold permits a series of sample and standard parts to be placed reproducibly. The method has been used for determination of thickness of plating or solder coating on small plated and "tinned" parts. Similarly, Glade (*G17*) used a metal block having a machined depression to mount reed-switch reeds.

Components for small electron tubes and semiconductor devices and other small parts may be analyzed in a rotating-specimen drawer of the type shown in Figure 16.4 (*B38*). The drawer has two rotors. Small filamentary or cylindrical parts are implanted in plastic cups (*J* in Figure 16.3) filled with stiff wax. The cups are placed in the upper rotor in Figure 16.4 and rotated about their own axes normal to the primary beam during measurement. Small, more or less flat parts are affixed to plastic pedestals (*K* in Figure 16.3). The pedestals are placed in the lower rotor and rotated in their own planes. In a more ingenious rotating-specimen drawer (*Q1*), both these modes of rotation are effected by a single rotor fitted with a contra-angle U-gear coupling of the type used for dental drills. The coupling can be turned to permit rotation of flat specimens, affixed to suitable pedestals, in their own plane, or of filamentary or cylindrical specimens about their own axes normal to the primary beam.

Some parts may be flattened in a press or vise and treated as ordinary disk specimens. Some parts may be melted to small buttons in a small vacuum electric furnace (*F9*). Of course, there is risk of oxidation, or sublimation of volatile constituents.

Stiff wire and rod of small diameter (*L* in Figure 16.3) may be presented by the wax-impression mold or rotating-spindle techniques described above. Alternatively, V-grooves may be machined on the under side of the specimen

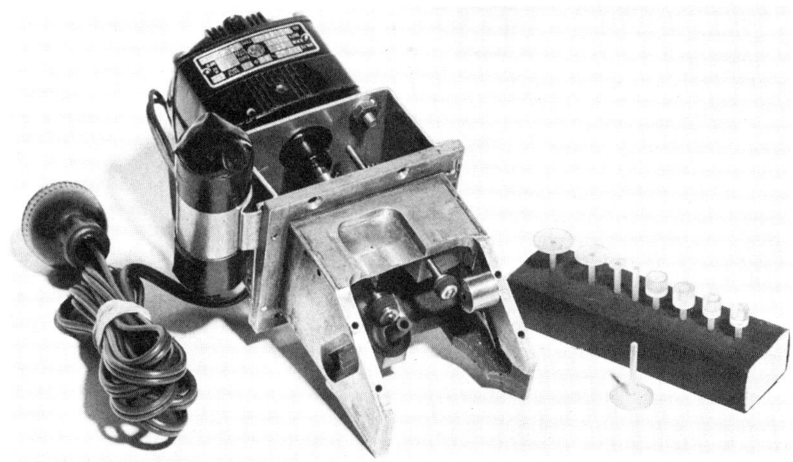

FIG. 16.4. Rotating-specimen drawer and specimen-mounting spindles. Plane specimens are affixed to pedestals (shown), inserted in the center rotor, and rotated in their own plane. Cylindrical or filamentary specimens are supported in cup-type spindles (shown) filled with stiff wax, inserted in the side rotor, and rotated about their axes.

mask. If greater intensity is required, a number of rods may be cemented side by side on a stiff card (*M* in Figure 16.3). Individual wires or rods may also be analyzed with a selected-area slit aperture (Chapter 19), especially if a curved analyzer crystal is used.

A specified number of turns of fine wire may be wound on a stiff card (*N* in Figure 16.3), open frame, or small spool (*O* in Figure 16.3) (*B40*). Provision may be made to rotate the spools during measurement. If the wire is always received on identical spools, these may be placed in the instrument specimen chamber or on a large-area accessory, depending on the spool size. In this way, the need for taking samples is avoided. The effective areas and the surfaces presented by all these wire specimens may vary somewhat among samples and standards, and it is preferable, if not essential, to use some kind of a ratio method in their analysis. For example, if plate thickness is to be determined, the intensity ratio of plate and substrate lines should be used; or if molybdenum–tungsten alloy wires are to be analyzed, the intensity ratio of a line of each constituent should be used. Another way to deal with wire is to gather it into a wad and briquet it.

Metallographic potting and sectioning techniques may be applicable in preparation of small parts. For example, a cross section of an electron tube permits identification of the material of each component by use of selected-area techniques (*H* in Figure 16.3).

16.2.3. Preparation (*M33*)

The specimen may be taken from unfinished or finished bulk material or stock, including "pig," ingot, plate, bar, rod, and sheet. Alternatively, specimens may be taken from furnace melts in small casting molds. Chill-cast molds are preferable to avoid the segregation that occurs on slower cooling. Procedures for casting furnace-melt samples have been reported for steels (*B73*), copper alloys (*M8*), aluminum alloys (*K21*), solders (*Z1*), and gold- and platinum-group alloys (*E14*). The coupon is cut from the stock or cast billet with a hack saw or, preferably, a high-speed abrasive or diamond cut-off wheel.

The finished-stock or wheel-cut surfaces may be satisfactory, but the latter usually require finishing to reduce surface roughness and/or remove work-damaged or nonrepresentative surface layers. The finishing may be done in many ways, including the following: (1) belt grinding followed by polishing with progressively finer wet or dry abrasive papers on a strip polisher (*H10*, *Z1*); (2) steel wool (*H10*); (3) polishing wheel using appropriate papers or slurries (*K11*); the wheel may be of the manual or vibratory type; the abrasives may be, in order of increasing hardness, aluminum oxide, silicon carbide, boron carbide, or diamond dust, depending on the hardness of the specimen; (4) "sandblasting" with a high-pressure jet of aluminum oxide (*C30*); (5) machining on a lathe, end mill, or planer (*M8*, *S9*); (6) electropolishing (*C29*, *J5*), in which the specimen, previously polished with a fine abrasive, is made the anode in an electrolytic cell having a fixed, flat, stainless-steel cathode; and (7) chemical etching. In general, electropolishing and etching are avoided because of possible preferential action on certain constituents.

More recently, the method of *spark planing* or *spark erosion* (*C54*, *J16*) has been used to impart a final high finish to metal specimens, especially when there is possibility of smearing. The surface to be finished and another plane, polished, metal plate are precisely mounted close and parallel and immersed in a dielectric liquid, such as kerosene. The plates are connected across a capacitor, which is charged by a dc power supply through a resistor. The capacitor charges at a rate determined by the resistance until the dielectric breakdown potential is exceeded, whereupon a spark discharges across the point of closest approach of the smooth and specimen plates, that is, at the highest microscopic protrusion on the specimen. The process repeats at a relatively high frequency, each discharge eroding a microprotrusion on the specimen. A 2–10-μm finish can be realized in 5–30 min. F. Breck (unpublished work) reports that improved surfaces are obtained if the

TABLE 16.2. Approximate Particle Size of Abrasives[a]

Nominal grit or mesh	Approximate particle size of abrasive, μm	
	Grit	Mesh
500	20	32
400	25	40
300	35	50
200	60	75
100	150	150
80	200	—
50	350	—
40	450	—
30	650	—
20	1000	—

[a] Buehler, Ltd., Evanston, Ill., "The Metal Analyst," catalog of metallurgical and mineralogical preparation equipment (1962).

polished plate has the form of a rotating disk and if a servomechanism is used to maintain constant spacing between the specimen surface and disk. Jenkins and Hurley (*J16*) have reviewed the applications of spark planing in x-ray spectrometric specimen preparation. They recommend previous polishing of the specimen to a 50-μm finish to reduce the time for spark planing.

A final finish of ~100 μm is usually adequate for measurement of analyte lines of shorter wavelength, 20–50 μm for longer wavelengths down to Al $K\alpha$ or Mg $K\alpha$. Table 16.2 gives approximate abrasive particle sizes for various nominal "grit" or "mesh" designations. In selection of the final abrasive, it is preferable to use a finer abrasive than the required finish. For example, if a 100-μm finish is required, a 200-grit (60-μm) or 200-mesh (75-μm) abrasive should be used.

If the specimen is a multiphase system, smearing of soft constituents may occur. This phenomenon and methods of dealing with it are discussed in Section 16.2.4.

16.2.4. Precautions and Considerations

When the specimen mask is used to define the effective specimen area, the mask window must be filled by all samples and standards, and the finished surface must lie flat against the under side of the mask. No burrs or other protrusions are permissible, because these might tilt the specimen slightly, thereby changing ϕ and ψ.

With thin sheet metals and foils, care must be taken to ensure a plane surface free of warping, wrinkles, and creases. Particular care is necessary to avoid warping on heating under prolonged irradiation. Sheet and foil —as well as thin rubber, paper, or textile materials—must be backed by a rigid support, possibly even cemented to it; however, see the last paragraph of Section 17.2.1.

For low-Z specimens, care must be taken that the critical thickness is exceeded or that all samples and standards have the same thickness. For very low-Z specimens—such as plastics—care must be taken that spectral emission from the specimen retaining clip or other structures under the specimen do not emerge through the specimen and enter the optical path.

Coarse-grained, inhomogeneous, and rough-finished specimens should be rotated during measurement, and no harm is done by rotating any flat specimen. If rotation is provided, care must be taken to ensure that the irradiated surfaces of all samples and standards lie in the same plane, and that the rotation mechanism does not impart a wobble to the plane. The "spinner" must rotate the specimen at least once during the measurement period, preferably much faster. The frequency of rotation must be out of synchronism with, that is, must not be a submultiple of, the power-line frequency. Otherwise, the specimen would tend always to be in the same orientation at a given phase of the wave form of the x-ray tube potential, and the purpose of the rotation would be defeated. In general, this consideration applies only to full-wave x-ray generators, not constant-potential ones.

If rotation is not provided, the specimen should always be loaded in its chamber with linear parallel grind marks oriented the same way, preferably parallel to the plane defined by the directions of the primary and secondary beams (Figure 16.5). This precaution is more necessary the longer is the wavelength of the analyte line to be measured. Figure 16.6 shows the effect of orientation and roughness of grind marks for lines of short (Sn $K\alpha$) and long (Si $K\alpha$) wavelength. If the grind marks are circular or random, orientation may not matter, but rotation is still beneficial.

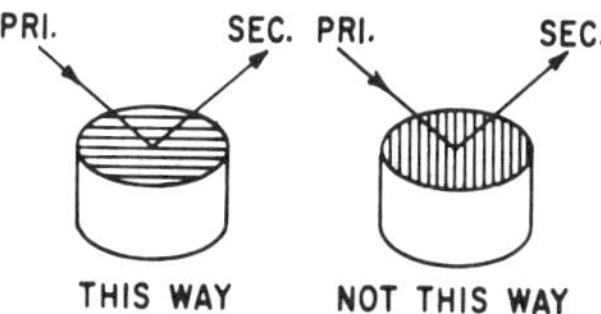

FIG. 16.5. Orientation of specimen grind marks.

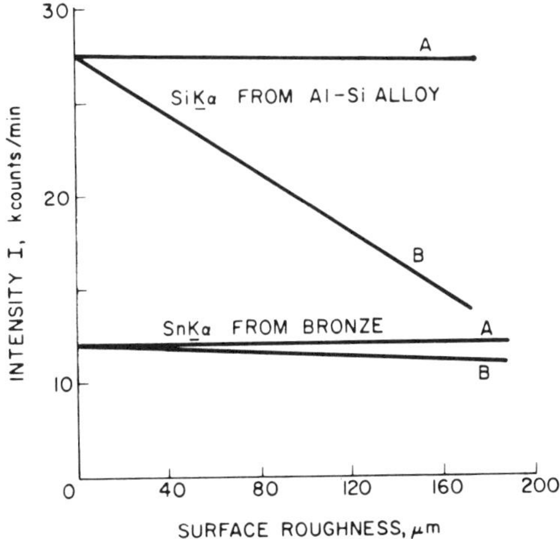

FIG. 16.6. Effect of roughness and direction of parallel grind marks on intensity of analyte lines of long (Si $K\alpha$) and short (Sn $K\alpha$) wavelengths measured from plane solid surfaces. Grind marks for curves *A* were parallel, for curves *B* perpendicular to the plane defined by the central rays of the primary and secondary x-ray beams. [K. Togel, in "Zerstörungsfreie Materialprüfung," R. Oldenbourg, Munich, Germany (1961); courtesy of the author and publisher.]

Although the specimen surface need not be given a high polish, it must be plane and smooth, and the roughness should be reduced to a certain maximum value (Section 12.3). In general, the depth of surface roughness or grind marks should be the same for all samples and standards in a given analysis, and less than the critical thickness of the specimen for the analyte line of longest wavelength. Figure 16.7 shows the effect of surface roughness and absorption coefficient on analyte-line intensity for several wavelengths. The figure shows that a 30–50-μm finish is likely to be adequate for the *K* lines of elements down to atomic number 13 (aluminum). Curves of this type usually apply except when smearing occurs (see below).

The surface of a solid may not be representative of the bulk composition, and in such cases, it must be decided whether analysis is required of the surface, the bulk material, or both. Moreover, surface texture and composition may be difficult to reproduce from specimen to specimen. The remainder of this section is devoted to this problem.

Removal of inclusions may occur during the surface treatment, causing low analytical results for the removed constituent(s). Pickup of surface contamination or abrasive may also occur, and all traces of abrasive, lubricant, and cutting fluid must be removed. Use of aluminum oxide, silicon carbide, and cerium oxide abrasives may result in contamination by these elements, and if the abrasive is supported on a lead- or tin-coated polishing wheel, these elements may be picked up. Electropolishing may result in selective removal of certain constituents and/or in contamination by constituents in the electrolyte. Chemical etches are usually avoided for the same reasons.

Oxidation and other corrosion must be removed. This oxidation and corrosion may be preferential, so that their removal may result in a nonrepresentative surface. Pores and other cavities usually decrease analyte-line intensity in proportion to the surface area they represent. However, cavities may increase analyte-line intensity if the walls of the cavities happen to be enriched in the analyte. Cast and rolled surfaces are very poor for x-ray spectrometric analysis. They are very inhomogeneous; they contain oxide layers, impurities, and inclusions; and their texture may be very coarse-grained and porous. When metal specimens are heated, preferential diffusion of one or more constituents to the surface and/or preferential oxidation may occur. If the metal is heated strongly, sublimation of volatile constituents may occur. Finally, if small buttons of alloys are prepared as standards, their surfaces cool and solidify first and may not be representative of the bulk. Consequently, it is essential to remove sufficient surface material to ensure that the surface is representative of the material to be analyzed.

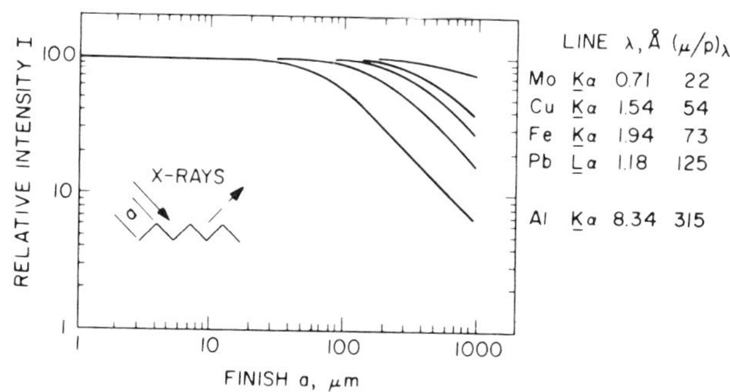

FIG. 16.7. Effect of roughness of parallel grind marks on intensity for analyte lines of various wavelengths measured from plane solid surfaces. The grind marks were perpendicular to the plane defined by the central rays of the primary and secondary x-ray beams as shown. [R. O. Müller, "Spektrochemische Analysen mit Roentgenfluoreszenz," R. Oldenbourg, Munich, Germany (1961); courtesy of the author and publisher.]

A very common surface problem arises when the specimen is a multiphase system having one or more soft phases. Any surface preparation method involving movement of a cutting tool or abrasive powder across the surface may result in smearing of the soft constituent(s). This causes high analytical results for the smeared constituents and low results for the covered constituents. The latter effect is more severe the higher the absorption coefficient of the smeared material and the longer the analyte-line wavelength of the covered material. Surface smearing effects are also likely to be more severe with a chromium-target x-ray tube than with molybdenum, rhodium, silver, tungsten, platinum, or gold tubes.

Etching, electropolishing, and spark-planing (Section 16.2.3) do not involve cutting or abrasion and so do not cause smearing. However, etching and electropolishing may act preferentially on certain constituents. Smearing is less likely with machining than with polishing. When it does occur, it may be removed by final light polishes with 6-μm, then 0.25-μm diamond dust with kerosene lubricant on a cloth polishing wheel (*K11*, *M33*). Alternatively, the smear may be removed by a light chemical etch specific for the smeared constituent or by electropolishing or spark-planing.

Reduction of surface smearing of soft alloys, such as tin–lead–antimony alloys, has also been effected by cooling the blanks to liquid-nitrogen temperature during surface finishing on a lathe, and by pressing the finished disks in vacuum against a mirror-smooth surface (*L56*).

A good test for such superficial smearing is to measure, for a covered element, the $L\alpha_1/K\alpha$ intensity ratio for elements up to atomic number ~60 and the $M\alpha/L\alpha_1$ ratio for the heavier elements. This is done for the possibly smeared specimen and for another specimen of the same alloy known not to be smeared, or even from the pure element. The intensity of the longer-wavelength line ($L\alpha_1$ or $M\alpha$)—and thereby the ratio—is decreased by any overlying smeared constituent. It follows that the technique also serves as a test for removal of the smear.

Another difficulty with soft constituents is the possibility that fine particles of abrasive may become imbedded and packed in soft surface grains, thereby "plugging" them.

16.3. POWDERS AND BRIQUETS

16.3.1. Scope, Advantages, Limitations

Powder samples are derived from: (1) powders; (2) bulk solids—especially minerals, ores, slags, etc.—pulverized because they cannot be analyzed in solid form due to gross inhomogeneity, the need to add an internal standard, or some other reason; (3) metal filings, chips, drillings, and turnings; (4) metals converted to oxide (*A24, D6*); (5) metals put into solution, then precipitated with a suitable reagent (*F6*); (6) precipitates and residues from solutions (*K8*); (7) ground-up fusion products; and (8) ashed, oven-dried, or freeze-dried (lyophilized) organic or biological matter, such as blood, tissue, leaf, and seed.

Powders may be presented to the instrument as: (1) loose powder packed in cells (Section 16.3.4.1); (2) briquets (Section 16.3.4.2); or (3) thin layers (Section 16.3.4.3). These three specimen forms are discussed together in this section because they have in common the feature that the powder particles retain their individual identities. Fusion products are usually prepared from powders, but because the particles disappear on fusion, these melts constitute a basically different type and are considered separately in Section 16.4. It may sometimes be preferable to prepare composite metal powders by arc melting in an inert atmosphere; some relevant considerations are given in Section 16.1.4.3.

Wilgallis and Schneider (*W13*) have compared four methods of specimen preparation and presentation for silicate minerals: (1) loose powder; (2) briqueted powder; (3) ground and briqueted lithium tetraborate ($Li_2B_4O_7$) glass fusion product; and (4) lithium tetraborate glass disk. Best reproducibility was obtained with the briqueted powder, but difficulties were encountered with certain elements. In general, the briqueted glass with 1:1 to 1:10 dilution was recommended.

The principal disadvantage of the solid specimen form—the inapplicability of addition methods—does not apply to powders. Standard addition and dilution, low- and high-absorption dilution, internal standardization, and internal intensity-reference standardization are all readily applied to powders. The applicability of these methods permits great flexibility in dealing with absorption-enhancement effects. Standards are prepared relatively easily (Section 16.3.2). Wherever applicable, powder methods are usually rapid and convenient.

The principal disadvantages of powder specimens are the following: Trace impurities may be introduced by the grinding and briqueting operations, especially when the specimen powder is abrasive. The surface texture of loose-packed powders may be difficult to reproduce, but the difficulty is surmounted by use of back-loading cells (see below) and substantially eliminated by briqueting. Some powders are hygroscopic or react with oxygen or carbon dioxide in the air; such powders are best handled in Mylar-covered cells. Other powders have low cohesion and flow when the specimen plane is inclined; such powders are placed in Mylar-covered cells, mixed with binder and packed in open cells, or briqueted with binder. Powders may outgas in vacuum spectrometers, sometimes ruining the specimen surface and spattering into the chamber; this difficulty is dealt with by use of helium path, Mylar-covered cells, cells covered with microporous film (Section 16.3.4.1), previous outgassing, or vacuum briqueting.

However, the most serious difficulties with powders arise from particle-size effects. The spectral-line intensity of a specified element in a powder depends not only on the concentration of that element, but also on the particle size. The intensity is constant for extremely small particles up to a certain size, then decreases with increasing particle size. The size above which intensity begins to decrease varies with wavelength and is larger the shorter the wavelength. Incidentally, a method has been devised for x-ray *spectrometric* measurement of particle size based on this relationship between intensity and particle size (*H55*); x-ray measurement of particle size is done most often by diffraction methods. If the powder is not homogeneous, particles of different size, shape, density, or composition tend to segregate. The effect is discussed further in Section 16.3.5.1.

16.3.2. Powder and Briquet Standards

Powder standards loosely packed in open cells can be stored, but must be handled very carefully. Powders packed in Mylar-covered cells are more nearly permanent, but the Mylar is likely to require occasional replacement. In either case, if the loose powder consists of particles of different size, density, and/or even shape, occasional repacking may be required due to settling and segregation. Briquets packed in metal cups are substantially permanent, and the absence of a Mylar window is beneficial in the long-wavelength region.

General methods for preparation of standards for various specimen forms are given in Section 16.1.4. However, a more detailed discussion of preparation of loose powder and briquet standards is in order here. It is essential that in any analysis, all powder or briquet samples and standards have the same average particle size and particle-size distribution. This condition is achieved in one of three ways: The standards and samples may be prepared in precisely the same way, or they may be ground to the same particle size or to a particle size so small that differences are no longer significant—say, <30–$50\ \mu m$. Powder and briquet standards are prepared in any of the following ways.

a. Dry Synthesis. The standards may be wholly synthesized from individual component powders (*M56*, *S67*). When this is done, both the standards and samples should be ground to the same particle size or to a size at which particle-size effects disappear.

b. From a Single Standard. A single standard powder of the type to be analyzed may be varied quantitatively in concentration. Standards having higher analyte concentrations than the base powder are prepared by quantitative addition of constituents of analytical interest. Standards having lower analyte concentrations than the base powder are prepared by quantitative addition of matrix constituents to dilute the analyte(s).

c. From Bulk Solids. A bulk solid of known composition may be ground to powder. Samples analyzed by independent methods may be used. This method is particularly advantageous when the samples to be analyzed are also obtained by pulverizing solids.

d. From Solutions. Powder standards may be derived from solutions by precipitation, evaporation, or absorption in cellulose. These methods are useful when the samples are obtained the same way. The standard solutions may be prepared synthetically by dissolution of individual constituents in proper relative concentrations, or they may be prepared by dissolution of solids of known composition, including analyzed samples. If only one such standard material is available, the addition technique described above may be applied.

The powder may be recovered by precipitation or by evaporation to dryness. In either case, the rccovered solid is dried and fired to convert it to oxides. In the evaporation technique, the solids come out of solution in order of increasing solubility. A much more homogeneous residue is obtained by evaporating the solution to near saturation, then absorbing it in cellulose (filter-paper or chromatographic paper pulp). The cellulose is then dried, heated to expel solvent and acids, then fired to burn off the cellulose and convert the salts to oxides (*C16*).

Analytes can be added to powders by: (1) quantitative admixture of powder having accurately known analyte concentration and substantially the same particle size and density as the base powder; or (2) addition of a standard analyte solution, followed by homogenizing, drying, and grinding.

e. From Fusion Products. If the powder samples are derived from fusion products, the standards may be prepared the same way (*M56*). The fusion product may be prepared by fusion of standard materials, analyzed samples, or synthetic mixtures of individual constituents in proper proportion. If only a single standard material is available, the addition technique may be applied prior to fusion.

Methods for mixing, grinding, and making additions to powders are given in the next section.

16.3.3. Preparation of Powders

In the preparation of powder and briquet specimens, powders must be subjected to one or more of the following treatments: drying, calcining, addition, mixing, grinding, casting, dispersion to thin layers, and briqueting.

a. Drying. Prior to any treatment, powders must be dried at a temperature to expel residual solvent from precipitation, etc. Inorganic powders are usually dried at ~120°C for several hours or overnight to expel residual, absorbed, and adsorbed moisture. Plant materials can be dried in this way, then ground and treated as powders (*C42*, *L13*). For such applications, where substantial amounts of water vapor are produced, forced hot-air ovens may be used.

b. Calcining. Powders may be heated at temperatures of several hundred degrees to expel combined water, convert to oxide, and break down crystal structure.

c. Addition. The admixture of powders in the preparation of synthetic standards has already been mentioned. It may also be necessary to add one or another type of material to a powder sample for some required effect. Additives of materials already present in the specimen include analyte(s) for application of standard-addition methods, pure matrix for application of standard-dilution methods, and a specific, interfering, variable minor matrix element to fix its concentration. Diverse additives include internal standards, low-absorption diluents, high-absorption diluents, abrasives, and binders.

An internal standard may be added to compensate absorption-enhancement effects. With powders and briquets, internal standards may perform the additional function of compensating packing density effects and may even preclude the need to briquet. If the internal standard is added as powder, the particle size of the standard and sample materials must be the same, or they must be ground together until they are. The internal standard is best added as a solution that wets the sample powder. In one of the earliest applications of internal standardization to x-ray spectrometry, the sample, internal standard, aluminum binder, and silicon carbide abrasive were mixed, repeatedly mulled to dryness under diethyl ether, then briqueted (*A8*).

A low-absorption diluent (Section 14.6.1,2), such as lithium carbonate (Li_2CO_3), boric acid (H_3BO_3), carbon (*D6*), aluminum (*A7*), starch (*M56*), powdered sugar, and filter-paper or chromatographic-paper pulp (*F3*), may be added to reduce absorption-enhancement effects. A high-absorption buffer (Section 14.6.1,2), such as lanthanum oxide (La_2O_3) (*R24*) or tungstic acid (H_2WO_4) (*C45*), may be added to powders having a high-Z analyte in a low-Z matrix to make the calibration function more nearly linear. Powdered inert abrasives, such as aluminum oxide and silicon carbide, may be added to facilitate grinding (*A8*). This is done most frequently when the powder is to be ground in a mortar. A binder may be required to permit powders having low cohesion to be packed into cells, especially for instruments having inclined specimen planes, and to permit briqueting. The amount of binder used is as small as possible, usually 5–20% of the total specimen weight. Among the binders used are the following: lithium carbonate (Li_2CO_3) (*G36*), boric acid (H_3BO_3), aluminum, stearic acid ($C_{17}H_{35}COOH$), starch (*M56*), nitrocellulose, ethyl cellulose, Lucite (*D6*), carnauba wax, polyvinyl alcohol (PVA), commercial soap or detergent powders, and filter-paper or chromatographic-paper pulp (*R25*). Some additives may serve more than one purpose, as is evident from the duplication in the lists of low-absorption diluents and binders.

Low-absorption diluents, abrasives, and binders are always added in powder form. However, individual constituents of synthetic standards, internal standards, and high-absorption diluents may be added in powder or solution form. If the powders to be mixed all have small particle size (say, −325 mesh or smaller), or substantially the same particle size, shape, and density, they may simply be mixed in dry form. If they are coarse, or differ in particle size or shape, they must be ground separately before admixture, or, preferably, mixed and ground together. If they differ substantially in density, a standard solution of the heavier constituent(s) may be added volumetrically to a weighed portion of the lighter constituent in powder form. Alternatively, standard solutions of the minor and trace constituents may be added to the major constituent(s) in powder form. In either case, the wet powder is homogenized and dried, then ground and mixed.

"X-ray mixes" are available commercially (Chemplex industries, Inc., Scarsdale, N. Y.; Somar Laboratories, Inc., New York, N. Y.) in 44-μm powder and 0.25- and 0.5-g tablets. The mix is said to serve as a grinding aid, binder, low-Z diluent, and backing material. In grinding, the mix reduces caking and adhesion to the vessel walls, provides an abrasive component, and tends to produce uniform particle size. It forms hard, plasticized briquets that are resistant to flaking and breakage on handling and in vacuum, and to deterioration on prolonged irradiation. The mix is chemically inert and produces no spectral lines in the usual x-ray spectral region. The preweighed tablets eliminate the need for weighing the additive charge.

d. Mixing. After the additions have been made, the powder may be mixed or homogenized by spatulation or shaking in a bottle, but best by a mechanical mixer, such as the Spex Mixer/Mill using a plastic vial and ball. Small volumes of powder may be mixed in a dental mixer, such as the Wig-L-Bug, again using a plastic vial and ball.

e. Grinding. Prior to being packed into cells or briqueted, the powder should be ground to −325 or −400 mesh. Coarser powders may be acceptable if only short-wavelength analyte lines are to be measured. If any of the additives mentioned above are to be used, they should be added prior to the grinding operation so that the base powder and additives can be ground to substantially the same particle size.

Grinding may be done manually with an agate, silicon carbide, or boron carbide mortar and pestle. The powder may be ground dry, or it may be ground under ethanol or diethyl ether to dryness several times. However, grinding is preferably done mechanically in a Spex Mixer/Mill using a plastic, alundum, or tungsten carbide vial and ball, depending on the hardness of the material. The Spex Freezer/Mill, which grinds under liquid nitrogen, may be used for soft and resilient materials. In some cases, a Pitchford or equivalent mill may be used that grinds to a substantially uniform particle size. Metals and other reactive materials may be ground in inert atmospheres (carbon dioxide, nitrogen, argon) to prevent oxidation. For extremely small particle size, 1 μm or less, the powder is slurried in an organic vehicle and rolled in a ball mill overnight or longer.

It may not be possible to grind multiphase powders to a uniform small particle size if the phases have very different hardness. In such cases, it is important to use identical grinding technique for all analytically related specimens. This means that all such specimens should be ground either dry or wet, and if wet, the same liquid and weight/volume ratio of powder and liquid should be used. Also, all grinding should be done with the same model of grinder, type of vial and balls, weight of charge, and grinding time.

Casting, briqueting, and fusion are discussed in subsequent sections.

16.3.4. Presentation

16.3.4.1. Loose Powders

For upright and inclined specimen planes, loose powders are usually simply placed in shallow, metal or plastic, circular or rectangular dishlike cells. Most manufacturers of x-ray spectrometers supply powder cells for use with their instruments. Thin-walled aluminum cups for briqueting (Somar Laboratories, New York), Caplugs (Protective Closures Co., Buffalo), and various stainless steel, glass, and plastic "planchets" used in radiochemistry all make excellent disposable powder cells.

The cell may be slightly overfilled, then leveled by drawing the *edge* of a stainless steel spatula across its rim, or it may be made heaping full, then packed and surfaced with the *flat* side of a wide spatula. If care is taken always to load the same measured weight or volume of powder and to avoid any loss, reasonably uniform packing density among the specimens is achieved. If specimen rotation is desired, circular cells are used and placed in rimmed pedestals in a specimen rotator (*B38*). The pedestals are similar to those shown in Figure 16.4, but larger in diameter to closely fit the powder cell.

For inverted specimen planes, the filled circular cells just described may be covered with 6-μm (0.00025-inch) Mylar secured with an *O*-ring or other retaining ring. More conveniently, a liquid-specimen cup is used. The liquid-specimen cells supplied with inverted-geometry instruments consist essentially of open cylinders covered on the bottom with 6-μm Mylar secured with a suitable ring or collar. In fact, a short length of plastic tubing fitted with a Mylar bottom may be used. Enough powder is poured into the cell to exceed infinite thickness, and the powder is settled by tapping the cell sharply on the table. Alternatively, disposable plastic liquid-specimen *cups* (Spex Industries, Metuchen, N. J.) may be used. The powder is placed in the cup, the Mylar window applied and secured, and the cup inverted (Mylar down) and tapped to settle the powder. Most modern instruments having inverted geometry provide for specimen rotation.

A smoother, more uniform surface is obtained by backloading powder cells (*B38*, *L30*). For this purpose, the cells have the form of cylindrical rings or rectangular frames open on both sides. They are loaded by laying them on small glass plates and filling them with powder. The powder is packed firmly in the frames with a stainless steel spatula, and all powder is removed from the rims. Measured weights or volumes of powder may be used to realize uniform packing density. Lucite back covers 3-mm thick and of the same outside diameter or length and width as the cells are now placed on the filled cells. The glass–cell–cover assemblies are then inverted with the glass plates up. The glass plates are removed, exposing the useful surface, and the cells with their backing plates are placed in the specimen drawer with the smooth powder surface up. Sometimes, when the glass plate is removed, particles of nonmetallic powders are attracted from the surface by electrostatic charge. This effect is avoided by use of a metal plate or by sliding the glass plate off sideways, rather than lifting it off. If the backloaded cell is to be rotated, the cylindrical ring is used and loaded as before, but instead of a Lucite plate, a rimmed Lucite pedestal is placed on the cell (*B38*). The glass–cell–pedestal assembly is then inverted, the glass plate removed, and the pedestal shaft inserted in the specimen rotator.

For powders having low cohesion, the tendency to flow in an inclined specimen plane is avoided by use of binders and/or Mylar-covered cells similar to the liquid-specimen cell shown in Figure 17.3. These cells consist of rectangular Lucite dishes having cemented-on 6-μm Mylar windows and filling ports and plugs. Thicker Mylar may be used if only short-wavelength lines are to be measured. When the cell is being filled, a small funnel is fitted to the filling port and a protective Lucite plate is held on the window by a small C-clamp to prevent bulging. The powder is poured through the funnel and settled by tapping the cell on the table top. Again, measured weights or volumes of powder may be used. In general, the plugs are required only for storing standard cells. These covered cells are also good for general-purpose use—upright, inclined, or inverted geometry, and vacuum path. They are much easier to handle, and standards are stored more conveniently. The Mylar cover does substantially attenuate the *K* lines of fluorine, sodium, magnesium, and aluminum. Instructions for preparation of the cells are given in Section 17.1.4.2.

The Chemplex liquid cell (Section 17.1.4.4), having a microporous film cover for use in vacuum, is also applicable to powders in vacuum. The microporous film effects equalization of pressure inside and outside the cell by allowing permeation by gas and vapor, but retains the powder.

A cell for small amounts of loose powder is described in Section 17.1.4.9.

For special applications, loose powders may be presented in other ways. Powders in which only two elements vary in concentration may simply be placed in small cellophane envelopes (*D3*). The variable nature of the specimen is compensated by use of the intensity ratio of the analyte lines of the two constituents. The technique has been used for the 8-quinolinolates of two coprecipitated elements (*D4*). Powders that are unstable in air may be placed in small polyethylene envelopes or in a small, folded sheet of polyethylene and sealed in by thermal "welding" of the open edges (*T11*). Again, unless only qualitative analysis is required, some kind of ratio method must be used. A cell designed to maintain x-ray diffractometer specimens in special atmospheres may also have application in x-ray spectrometry (*M46*). The cell is sealed and has a tabulation to attach to a small cylinder or other source of the required gas.

16.3.4.2. Briquets

The prepared powder may be formed into a self-supporting pellet or disk without loss of identity of the particles by admixture with wax, by suspension in a plastic monomer (*B85*) in a cylindrical cup, followed by polymerization, and by pelletizing with a thermosetting resin, such as methyl methacrylate. However, the most common and most convenient way is by briqueting at high pressure in a press. The briqueting technique is also applicable to metal granules, chips, shot, drillings, turnings (*C69*), and wire, and to textiles (*P19*). Usually a binder is admixed with the powder, but cohesive powders, including metals, may be briqueted without binder.

The powder, with or without binder, is placed in a cylindrical die in a hydraulic press capable of developing up to 7000 kg/cm^2 (100,000 lb/inch^2). The die usually consists of a heavy outer retaining shell surrounding a cylinder or piston chamber in which slides a piston. When the powder is loaded in the die, the piston is withdrawn, and a loading cylinder of the same outside diameter is inserted in its place. The powder is introduced through the loading cylinder to avoid getting powder on the inside wall of the piston chamber, where it would soon score the chamber and piston. The loading cylinder is then withdrawn and the piston replaced. Dies of several diameters should be available. It is preferable that provision be made to heat and evacuate the die for certain applications. For any given series of samples and standards, uniform packing density is achieved by compacting the same weight of powder at the same pressure for each specimen.

Briquets may be formed, in several ways as follows:

1. *Without backing or other support*. The powder is simply compacted as is.

2. *With a backing of pure binder*. A layer of pure binder (*R24*), cellulose (*A18*), or boric acid (H_3BO_3) [p. 125 in (*1*)] is placed in the die and com-

pacted lightly with a manually inserted piston or at full pressure in the press. Then the powder is loaded and the two-layer briquet compacted at full pressure. Alternatively, a supply of binder or cellulose briquets may be made in advance. One of these disks is placed in the die prior to loading each powder charge.

3. *With a backing and jacket of pure binder*. A special die arrangement permits the analytical powder to be ringed as well as backed with pure binder or cellulose (*B17*).

4. *With a metal or plastic jacket*. A thin-walled cylindrical metal (*M40*) or Bakelite (*B2*) ring having outside diameter just less than the inside diameter of the piston chamber is placed in the die prior to loading of the powder.

5. *With a metal jacket and binder backing*. This technique is a combination of 2 and 4 above.

6. *In a metal cup*. The technique is the same as 4 above, except that a thin-walled cylindrical metal cup (Somar Laboratories, New York) is used (*M18*). The die may be operated so as to slightly crimp over the top rim of the cup.

If a glass disk is placed in the die prior to loading of the powder, the briquet surface that was in contact with the glass is exceptionally smooth (*V13*).

Fabbi (*F2*) describes a die and briqueting procedure that produce durable pellets having smooth, reproducible surfaces and a hard protective outer edge and backing. A glass disk is placed on the bottom of a cylindrical outer chamber. A cylindrical die sleeve is placed in the chamber, resting on the glass. A cylindrical casting sleeve is placed inside the die sleeve, also resting on the glass. The sample powder is poured into the casting sleeve. A cylindrical casting piston is now inserted in the casting sleeve and pressed manually to form a semirigid sample briquet. The casting sleeve and die are now withdrawn, leaving the sample disk and die sleeve in place. A 20–30-ml portion of methyl cellulose is poured over the sample disk, covering it and surrounding it in the annular space left on withdrawal of the casting sleeve. A cylindrical die piston is now inserted in the die sleeve, and the entire assembly is placed in a press, where the sample is briqueted at 2100 kg/cm^2 (30,000 lb/inch^2). The die piston and sleeve, with the sample briquet adhering, are removed from the outer chamber and replaced in the press with the briquet on top. A pellet receiver is placed over the briquet, and minimal pressure is applied to extrude the briquet into the receiver. All components of the die are made of carbon tool steel. The glass disk, actually a flashlight "lens," is cleaned thoroughly before each use by immersion in chromium (VI) oxide–sulfuric acid cleaning solution (CrO_3–H_2SO_4), rinsing in water, immersion in dilute hydrochloric acid containing a little hydrogen peroxide (HCl–H_2O_2), rinsing in distilled water, and drying between filter papers.

Oswin (*O4*) noted that briquets of powdered blast-furnace slag and borate-fusion glass tended to fail by compressive shear caused by the compressive force of the mold when the pressure is removed from the die. Very stable, strain-free disks were obtained by briqueting these powders in a cylindrical lead ring (4 cm I.D., 0.3 cm wall, 0.6 cm high) between two lapped, hardened steel plates at 2800–8400 kg/cm^2 (20–60 tons/inch^2).

All the foregoing techniques are for conversion of a powder to a briquet. The solution–briquet technique (*R25*), in effect, permits substantially direct conversion of a solution to a briquet. The prepared solution is evaporated to remove some solvent, then taken up in 500 mg of chromatographic-paper pulp. The pulp is then homogenized, dried, ground in a boron carbide mortar, and pressed into a double-layer disk against a backing of previously pelletized chromatographic paper.

Another example of a solution technique is the early work of Fagel, Balis, and Bronk (*F6*). They coprecipitated low concentrations of chromium, manganese, iron, nickel, and copper and a high concentration of aluminum, added to serve as a carrier and to provide bulk, with 8-quinolinol. The precipitate was filtered, dried, ground, and briqueted. Standards were prepared the same way from solutions having known concentrations of the several analytes.

Briquet surfaces and edges may be sprayed with clear Krylon or 10% wt/vol collodion in acetone to prevent flaking and crumbling (*A18*). However, some attenuation of wavelengths greater than 6 or 7 Å results from spraying the measured surface.

Some considerations relevant to briqueting are given in Section 16.3.5.3. Thin—less than infinitely thick—briquets are mentioned briefly in Section 16.3.4.3. An automatic machine that presses briquets at a rate of about one per minute is described briefly in Section 16.1.6.

16.3.4.3. Thin Layers

The effects of particle size and inhomogeneity and absorption-enhancement effects disappear if the powder is ground to extremely fine particle size and distributed in a thin layer. Under these conditions, the powder assumes, to a certain extent, the nature of a thin film. Another

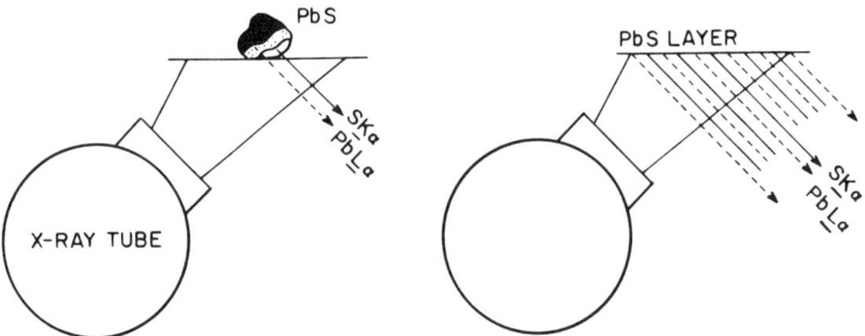

FIG. 16.8. Effect of specimen presentation on S $K\alpha$ (5.37 Å) and Pb $L\alpha_1$ (1.18 Å) line intensity measured from lead sulfide. For the particulate specimen, only the unshaded region contributes to the measured S $K\alpha$ intensity, the unshaded and dotted regions to the Pb $L\alpha_1$ intensity. If the specimen is dispersed as a film, the entire mass contributes to the measured intensity of both lines. [R. Jenkins and J. L. de Vries, "Practical X-Ray Spectrometry," Springer-Verlag N. Y. (1967); courtesy of the authors and publisher.]

advantage of the thin-layer presentation is realized when only a small piece of material is available. The measured analyte-line intensity is greater if the small piece is ground to fine powder and spread out, as shown in Figure 16.8. Powders may be spread into thin layers dry or as slurries.

In the dry technique, the powder is lightly dusted on the adhesive surface of a piece of Scotch tape (*A3*) or Mylar film (*G29*, *W16*); excess powder is gently blown off the Scotch tape. Ordinary Scotch tape withstands the intense irradiation by the primary beam quite well, but Scotch "Magic" tape deteriorates rapidly. Loose powder can also be supported on thin Mylar film coated lightly with Scotch tape adhesive (Section 17.2.1). A better technique is as follows (*S4*, *S5*): A 6-μm (0.00025-inch) Mylar film is tightly stretched, in the manner of a drum head, across an empty plastic liquid-specimen cup or a short length of plastic tubing 2–3 cm in diameter. The Mylar is secured with an *O*-ring or other suitable retainer. A measured weight or volume of fine powder is placed on the Mylar and distributed as uniformly as possible. A second Mylar film is now stretched over the cup and secured, enclosing the powder between the two tightly stretched films. After assembly, the powder may be distributed further by rubbing a strip of Mylar film over the "sandwich," more or less in the manner of a shoeshine cloth, and by tugging at the edges of the films extending outside the retaining rings. Standards and samples are prepared the same way. When this technique is used with liquid-specimen cups in vacuum spectrometers, evacuation holes must be punched in the cups.

A better way to collect fine powders on filter disks is given by use of the simple apparatus shown in Figure 16.9 (*G9*). The specimen, pulverized to −325 mesh (44 μm) or finer, is placed in the jar, and an 0.8-μm Millipore filter 2.5 cm in diameter is affixed to the glass frit. A constant vacuum is maintained on the filter tube while short blasts of air are admitted to the vessel by rapid manipulation of the stopcock. The air disperses the powder as a fine dust, which is collected on and imbedded in the Millipore disk. This technique may be unsatisfactory in some cases where more than one phase is present and fine particles of one phase selectively adhere to the wall so that a nonrepresentative specimen is collected. This may be prevented by fusing the specimen with lithium tetraborate ($Li_2B_4O_7$) to form a homogeneous material, which is then pulverized and prepared as above.

In the slurry technique, a small amount of fine powder is slurried with a solution of 2–5% (wt/vol) nitrocellulose in amyl acetate and spread over a scrupulously clean microscope slide (*D6*). The film may be presented to the spectrometer thus supported on glass, or it may be scored around the edge and floated off on water, then scooped up and supported some other way.

A more versatile slurry technique is that of Finnegan (*F14*). A weighed portion of powder is mixed with a weighed portion of ethylcellulose or

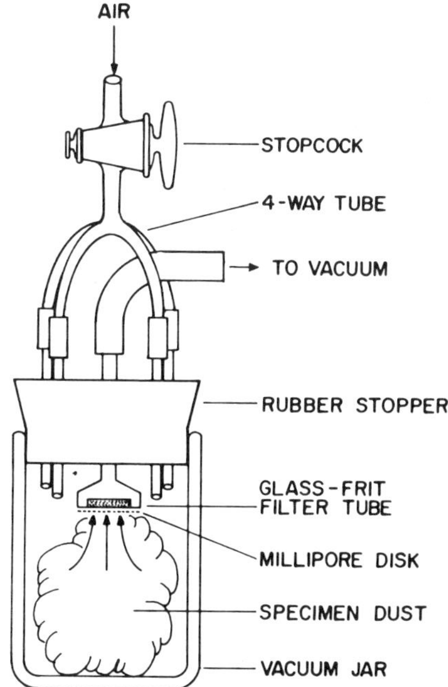

FIG. 16.9. Apparatus for collection of thin powder layers on filter disks by the "puff technique."

nitrocellulose binder and added to a measured volume of amyl or butyl acetate. The slurry is best mixed by rolling on a roller mill, and the powder may be ground further to very small particle size if the slurry is rolled with grinding balls. A measured volume of the slurry is placed on an optically flat glass plate and spread in a uniform thin film with a casting knife (doctor blade). Blanks are removed from the film by scoring, flooding with water, and coaxing squares of the film to float off. Preparation of standards for analysis of such films is described in Section 16.1.4. A similar method has been used to analyze paints (*M24*) and sublimates in electron tubes (*B39*, *R1*).

In the filter technique (*E13*), the powder is ground to extremely fine particle size, mixed with 10 times its weight of sodium chloride (NaCl), then ground some more to prevent agglomeration. A 10-mg portion of this mix is suspended in ~30 ml of water. This solution-slurry is placed in the funnel of a suction filter and the solid collected on a fritted-glass filter disk. During filtration, the slurry is continuously stirred and the walls of the funnel continuously washed down. The result is a very uniform film of fine powder on the filter disk, which is dried in an oven and used as the sample. The film is only 1000–2000 Å thick.

Powders may be dispersed in thin wax films as follows (*C27*). A 0.33-g portion of powder, 350 mesh or finer, is mixed thoroughly with 0.67 g of powdered paraffin (melting point 65–71°C), then ground with ~5 ml of benzene (C_6H_6) in an agate mortar to dryness. The charge is placed on a plane, polished steel plate covered with a double layer of 6-μm (0.00025-inch) Mylar. A 0.05-cm-thick spacer ring is placed on the plate to limit the final film thickness, then another similar Mylar-covered plate. The sandwich is then pressed at ~10^4 kg (~10 tons). The plates are then removed, the Mylar films peeled off the wax film, and circles of suitable diameter cut from the film.

For layer thickness $\lesssim$0.25 mm, excellent uniformity is realized by mixing equal volumes of powder specimen and vaseline and pressing the paste between spaced 25-μm (0.001-inch) Mylar films (*S4*).

Thin—less than infinitely thick—briquets ~2.5 cm in diameter and <1 mm thick may be prepared, especially of dried biological tissue. These may be self-supporting if briqueted with an imbedded fine Nylon mesh of the type used to support detector windows, or if backed by disks of chromatographic paper or pulp.

Powder specimens prepared as thin layers actually constitute "supported specimens." Some means of reducing background with such specimens are given in Section 17.2.1.

16.3.5. Precautions and Considerations

16.3.5.1. Particle-Size Effects

With regard to surface roughness, the particle-size effect in powders and briquets is analogous to the surface roughness effect in solids (Section 16.2.4, Figure 16.7). Just as the surface roughness of solids should be less than the critical thickness of the solid for the longest analyte line to be measured, so should the roughness—that is, the particle size—of powders and briquets. In general, particle-size effects substantially disappear, even for Mg $K\alpha$, at about −400 mesh. Coarser particles may be satisfactory for short wavelengths, and finer particles are required for Na $K\alpha$ and F $K\alpha$. If the particle size is such that it still affects the analyte-line intensity, then all samples and standards in a given analysis must have the same average particle size and particle-size distribution, and measures should be taken to avoid segregation (*A3*, *H2*). (See Section 12.3.)

In powders having nonuniform particle size, segregation may occur by size, and, in nonhomogeneous powders, segregation may occur by density. In either case, sifting is likely to accentuate the separation.

Particle-size effects are minimized by one or more of the following methods: (1) grinding so fine that particle-size effects substantially disappear; (2) grinding of all samples and standards to substantially the same particle size or distribution by use of a standardized grinding procedure; (3) dry dilution, preferably with a powder having mass-absorption coefficient for primary and analyte-line x-rays similar to that of the particles containing the analyte; (4) briqueting at high pressure; (5) use of an x-ray tube target having short-wavelength spectral lines to reduce the primary component of the specimen mass-absorption coefficient; and (6) mathematical methods (Section 12.3).

In filling cells for instruments having inverted geometry, the powder is usually poured onto the cell window. The small particles tend to flow through gaps between the large particles and collect on the window, forming a layer behind which lie the large particles. In filling cells for instruments having upright geometry, the powder is placed in the cell and packed by exertion of a small manual pressure. The larger particles are pressed toward the bottom, leaving a layer of small particles on top, which is the analytical surface. Thus, in either case, the specimen surface is likely to consist of an abnormally high proportion of small particles. However, in filling back-loaded cells (Section 16.3.4.1), these effects are not as marked. When the powder is poured into the cell, the first effect occurs, forming a layer of small particles on the bottom. Then, when the powder is packed and leveled, the second effect occurs, tending to push large particles into the layer of small particles on the bottom, which is inverted and becomes the specimen surface. These effects are substantially eliminated by briqueting or by grinding to -325 mesh or smaller.

Consider a heterogeneous powder consisting of large A-rich particles and small B-rich particles. If the effective specimen depth contributing to the measured A-line and B-line intensities is less than the average size of the large A-rich particles, B-line intensity will be abnormally high, A-line intensity abnormally low. This condition is minimized by further grinding to reduce the A-rich particle size.

A good test for a nonrepresentative analytical surface caused by any of the effects described above is to measure the $L\alpha_1/K\alpha$ intensity ratio for analytes up to atomic number ~60 and the $M\alpha/L\alpha_1$ ratio for heavier elements. The intensity of the longer-wavelength line—and thereby the ratio—is strongly increased by abnormally high analyte concentration on the irradiated surface. Some workers adopt the policy of grinding powders to constant $L\alpha_1/K\alpha$ or $M\alpha/L\alpha_1$ ratio. This technique is directly applicable down to perhaps atomic number 25 (Mn $L\alpha_1 \approx 19.5$ Å). However, it is in the analysis of minerals, cements, ceramics, slags, etc. that powder

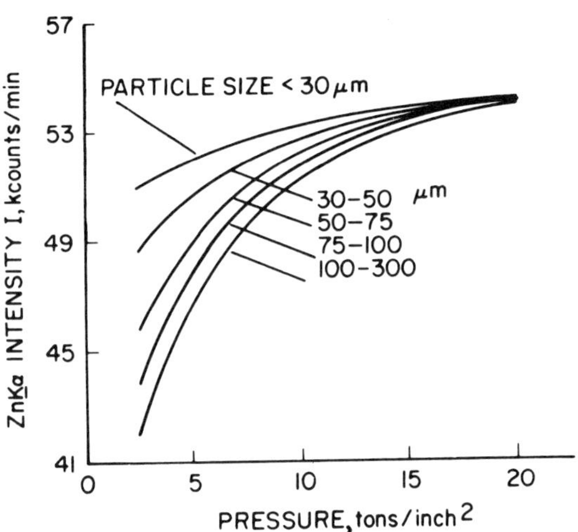

FIG. 16.10. Effect of briqueting pressure on Zn $K\alpha$ intensity measured from briquets of zinc powder having various particle sizes. (1 ton/inch² = 140 kg/cm²). [K. Togel in "Zerstörungsfreie Materialprüfung," R. Oldenbourg, Munich, Germany (1961); courtesy of the author and publisher.]

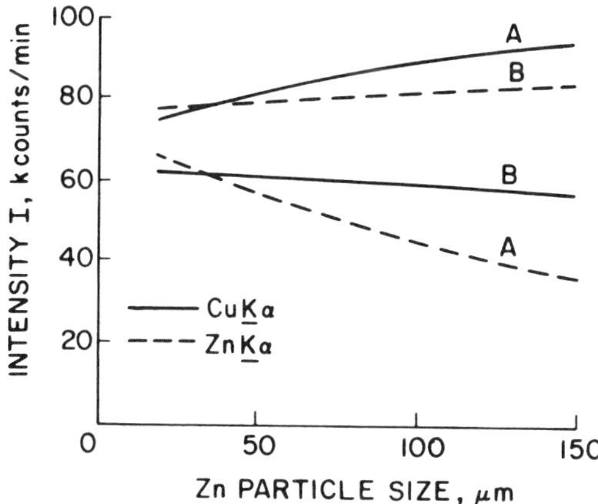

FIG. 16.11. Effect of particle size and surface facing on analyte-line intensity measured from briqueted 1:1 mixtures of copper and zinc powders. For all briquets, copper particle size was constant at a few micrometers, and briqueting pressure was 560 kg/cm² (4 tons/inch²). Briquets were measured as-pressed (curves *A*) and after removal of the surface to a depth of 0.1 mm (curves *B*). [K. Togel, in "Zerstörungsfreie Materialprüfung," R. Oldenbourg, Munich, Germany, (1961); courtesy of the author and publisher.]

methods are most widely used, and these substances are composed mostly of the lighter elements. Nevertheless, even with such specimens, a trace or minor constituent of higher atomic number can be used to evaluate the specimen surface by the line-ratio method.

In a completely homogeneous powder, the measured analyte-line intensity decreases as the path length of the measured wavelength in the specimen approaches the particle size. This phenomenon is due to mutual shielding of particles and is partially dealt with by dilution.

For a given briqueting pressure, the smaller the particle size, the higher is the analyte-line intensity. For a given particle size, the higher the briqueting pressure, the higher is the analyte-line intensity. These relationships are shown in Figure 16.10. The use of a binder or diluent accentuates the decreased intensity with increased particle size, and reduces the effect of pressure in overcoming this decrease.

On briqueting, ductile particles tend to flow around less ductile ones. This effect is most marked with metals, and, for this reason, briquets of metal filings, granules, chips, drillings, turnings, or wire should be surfaced to a depth of 0.1–0.5 mm. Figure 16.11 illustrates this effect.

16.3.5.2. Additives

Binders have the following advantages: (1) Stable briquets are formed even with powders having low cohesion; (2) better homogeneity is realized in packed and briqueted powders of nonuniform particle size or density; (3) higher packing densities and smoother surfaces are obtained; and (4) absorption-enhancement effects are reduced by the dilution. However, binders also have disadvantages. A few percent of binder may not significantly affect the absorption of a medium-Z matrix, but it may reduce the effective atomic number enough to increase primary-beam scatter and thereby the background. With low-Z analytes, binders reduce analyte-line intensity, both by dilution and by absorption. Figure 16.12 shows the effect of binder concentration on analyte-line intensity in the long-wavelength region. Other adverse effects of binders are cited above. In general, it is advisable to mix the minimum effective amount of binder with the specimen, then briquet the mixed powder against a backing of pure binder.

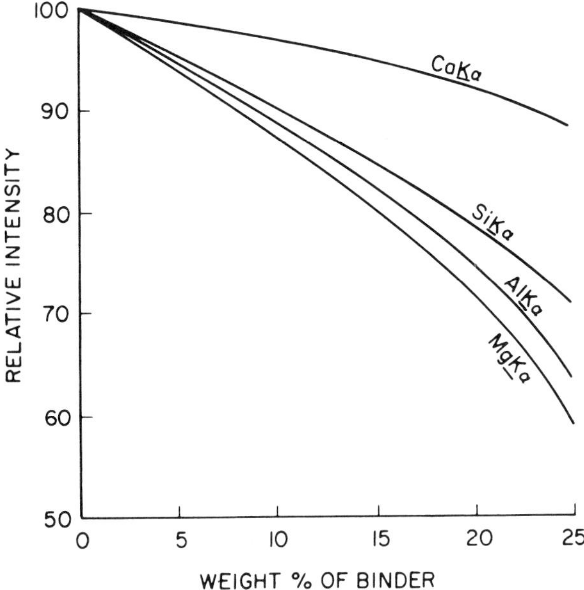

FIG. 16.12. Effect of concentration of binder or diluent on measured intensity for analyte lines of various wavelengths. The specimen material was a slag containing 6.5% MgO, 16.8% Al_2O_3, 39.7% SiO_2, and 29.5% CaO. The binder was starch. [K. Togel, in "Zerstörungsfreie Materialprüfung," R. Oldenbourg, Munich, Germany (1961); courtesy of the author and publisher.]

16.3.5.3. Briqueting

Figure 16.10 shows that the smaller the particle size, the lower is the briqueting pressure required to realize a specified analyte-line intensity. However, even for very small particle size, pressures of 2800 kg/cm^2 (20 tons/inch^2) are required to attain maximum Zn $K\alpha$ intensity. Even higher pressures may be required in the long-wavelength region, and lower pressures may be satisfactory for short wavelengths.

The briqueting die should be evacuated before and after pressing so that occluded air does not effuse and spoil the surface when pressure is released or later when the briquet is used in a vacuum spectrometer. Maximum pressure should be maintained for a minute or so to overcome any elasticity of the powder. If powders containing a high proportion of relatively large, hard powders are briqueted at high pressure, these particles may become highly strained. When the briquets are removed from the press and this strain is relieved, the briquet surface may become damaged. Powders should be briqueted in a thin-walled metal cup, or at least in a metal or plastic cylindrical jacket with a backing of cellulose or pure binder.

16.3.5.4. Other Considerations

Variations in packing density may cause variations in analyte-line intensity unless an internal standard is used, or unless the same weight of each specimen in the analysis is briqueted at the same pressure. Mixing, grinding, and briqueting may introduce trace impurities, especially when the specimen has an abrasive nature. Powders and briquets may outgas or even mildly explode in vacuum spectrometers unless previously outgassed in a vacuum chamber or vacuum oven, or unless briqueted in a vacuum die. Rotation of powder cells and briquets is recommended. If rotation is provided, care must be taken to ensure that the irradiated surfaces of all samples and standards lie in the same plane and that the rotation mechanism does not impart a wobble to the specimen.

Campbell and Thatcher (*C19*) have compared briqueting, fusion, and solution techniques for determination of calcium in wolframite concentrates.

16.4. FUSION PRODUCTS

16.4.1. Scope, Advantages, Limitations

The material to be analyzed may be fused with a flux for one of three reasons: (1) to render an insoluble or difficultly soluble material readily soluble prior to analysis as a solution; (2) to homogenize a material with respect to composition and density and to eliminate its particulate structure so that it may be ground to a homogeneous powder that may be used as such or as a briquet; or (3) to provide a homogeneous solid or glass pellet suitable for direct use as a specimen. The first application is the concern of Chapter 17; the other two are the subject of this section.

Fusion methods have many advantages regardless of whether the "melt" is used directly or powdered or briqueted. All inhomogeneity of composition, density, and particle size is eliminated. Standard addition and dilution methods and internal standardization are readily feasible. Absorption-enhancement effects are reduced or even eliminated by the leveling effect of the flux. A high-absorption diluent can be added, or a specific, interfering, variable minor matrix element can be added to fix its concentration. Standards are easily prepared—synthetically, if necessary. If the melt is ground and briqueted, it need be ground only to −200 mesh or so because of the homogeneity. If the solidified melt is used as the specimen, additional advantages are realized. The surface texture is smooth and uniform, although some polishing may be required, and standards are very conveniently stored. Properly prepared disks have been known to undergo no noticeable change after 20 h of irradiation in the primary beam.

The most serious disadvantages are that analyte-line intensities from low-Z elements are reduced by dilution and absorption, and that trace and minor constituents are highly diluted. The fusion requires considerable time and effort and, if glass disks are to be prepared, skill. Glass-disk standards may break in storage due to stress or devitrification. The pieces can be remelted and recast if a crucible is used from which the glass pours cleanly without adherence.

Automatic and semiautomatic machines that produce glass disks at a rate of one per 1–7 min are described briefly in Section 16.1.6.

16.4.2. Materials

For fusions other than glass-forming fusions, by far the most commonly used flux is potassium pyrosulfate ($K_2S_2O_7$). These fusions are done in Vycor or even Pyrex crucibles.

The principal fluxes for preparation of specimens in the form of loose or briqueted glass powders and glass disks are sodium tetraborate ($Na_2B_4O_7$) and lithium tetraborate ($Li_2B_4O_7$). The lithium salt, having the lower mass-absorption coefficient, is preferable to the sodium salt for light-element specimens. $Na_2B_4O_7$, being the salt of a strong base (sodium hydroxide,

NaOH) and very weak acid (boric acid, H_3BO_3), is relatively unreactive to strongly basic specimens. $Li_2B_4O_7$, being the salt of a weaker base (lithium hydroxide, LiOH) and very weak acid, is more reactive to basic specimens. In place of lithium tetraborate, a mixture of lithium carbonate (Li_2CO_3) and boron oxide (B_2O_3) in 1:2 mole ratio can be used.

If a lighter matrix than sodium tetraborate is required, but also a lower fusion temperature than lithium tetraborate (melting point 930°C, compared with 741°C for $Na_2B_4O_7$), the eutectic formed by fusion of lithium tetraborate and lithium carbonate in weight ratio 6:1 can be used. This eutectic actually has a lower melting point than sodium tetraborate. Unfortunately, the fused eutectic is hygroscopic, and so are the glass specimen disks prepared from it.

A supply of each flux that is to be used—including potassium pyrosulfate—should be fused in a large crucible to expel moisture, then pulverized and stored in a tightly stoppered bottle or desiccator.

For various purposes, other materials may be added to the charge prior to fusion: Lithium carbonate (Li_2CO_3) may be added to lithium tetraborate charges to reduce the fusion temperature, as mentioned above. Lithium carbonate and lithium fluoride (LiF) may be added to lithium tetraborate to render acidic and basic samples, respectively, more soluble. Sodium carbonate and sodium fluoride may be added to sodium tetraborate for the same purposes. Also, during fusion, the carbonates decompose with evolution of carbon dioxide, which may aid in agitation and mixing of the melt, but which may also form bubbles in the glass. High-absorption diluents (absorption buffers) include lanthanum oxide (La_2O_3), barium oxide (BaO), barium sulfate ($BaSO_4$), and potassium pyrosulfate ($K_2S_2O_7$). Oxidants, such as sodium nitrate ($NaNO_3$), potassium chlorate ($KClO_3$), barium peroxide (BaO_2), and cerium oxide (CeO_2), may be added to sulfides and other specimens for oxidation. Barium peroxide and cerium oxide serve as both oxidants and high-absorption diluents. The amount to be added may be calculated from the amount of oxygen required for the oxidation. A small amount of manganese dioxide (MnO_2) may be added to catalyze the oxidation. For silicate minerals and ceramics, a few grams of potassium fluoride (KF) or sodium fluoride (NaF) may be added to volatilize silicon as silicon tetrafluoride (SiF_4). Small amounts of these fluorides may be added to other specimens to ensure production of clear glass disks. It is claimed that a little sodium sulfate (Na_2SO_4) improves the resistance of disks to cracking (*H54*). Aluminum oxide (Al_2O_3), silicon dioxide (SiO_2), or germanium oxide (GeO_2) may be added to alkaline earth samples, which tend to produce glass disks that crack easily. One or another of these oxides is chosen to avoid spectral-line interference or on the basis of absorption-coefficient. Of course, if an internal standard or standard additive is to be used, it is added quantitatively to the charge prior to fusion.

Flux mixtures may be prepared, ground, dried, and stored. One such mixture is available commercially (Johnson Matthey Chemicals, Ltd., London) and contains $Li_2B_4O_7$, Li_2CO_3, and La_2O_3 in weight proportions 47.03, 36.63, and 16.34, respectively; $NaNO_3$ oxidant may also be added in proportion 13.33 g per kilogram of the mixed flux (*H21*).

The borate glasses wet quartz, nickel, and even platinum crucibles, and these metals are unsatisfactory for the fusions. Graphite is satisfactory, but may reduce some elements to the metallic state. Vitreous carbon, Palau (80Au–20Pd), platinum–gold alloys (97Pt–3Au and 95Pt–5Au), platinum–rhodium alloy (80Pt–20Rh), and gold-plated platinum–iridium alloy (90Pt–10Ir) are not wet by the glasses, which pour cleanly from crucibles of these materials. Incidentally, if the glass is to be pulverized rather than cast into disks, platinum crucibles can be used: As soon as the melt has solidified, but before it has cooled, the crucible is set in a shallow pan of water to quench the melt, shatter it, and separate it from the crucible. The fractured glass is then dumped out of the crucible, leaving it substantially clean. Expendable graphite crucibles are available for preparation of glass specimens.

The glass melt may be poured on or into a preheated plate or mold of graphite, aluminum, copper, duralumin, or gold-plated stainless steel.

16.4.3. Specific Fusion Procedures

In the *potassium pyrosulfate fusion* (*C68*), the charge typically consists of 200 mg of sample and 10 g of potassium pyrosulfate. If graphite is present, 200 mg of potassium persulfate ($K_2S_2O_8$) is added. For silicate minerals, a few grams of potassium fluoride (KF) or sodium fluoride (NaF) is added to volatilize silicon as silicon tetrafluoride (*C71*). The charge is heated in a Pyrex or Vycor crucible at $<700°C$ for ~ 3 min. The melt is allowed to cool, ground to -200 mesh, and briqueted at >840 kg/cm^2 (12,000 lb/inch^2). If the melt is allowed to cool very slowly—usually overnight—a very homogeneous bead results that can be surfaced with fine abrasive paper and used as a specimen (*M29*).

In the first reported application of the *borate-glass fusion method* (*A21*), the charge consisted of 300 mg of sample, 1.5 g of lithium carbonate, and 1.5 g of boron oxide (or about 3 g of lithium tetraborate). The charge was heated in a graphite crucible at 1150°C for ~ 20 min, then poured into a brass ring 1.6 cm ($\frac{5}{8}$ inch) in diameter on a copper-surfaced hot plate. Immediately on solidifying, the pellet was imbedded in asbestos for ~5 min to anneal. It was then removed from the asbestos and allowed to cool. The flat surface was polished, if necessary, on a belt grinder.

In the *Claisse sodium tetraborate fusion* (*C47*, *M7*), the charge typically consists of 100 mg of sample and 10 g of sodium tetraborate. If a high-absorption diluent is required, 100 mg of lanthanum oxide (La_2O_3) may be added. Claisse used platinum crucibles, but other materials are preferable (Section 16.4.2). The charge is fused over a Meker burner or in a muffle furnace at 800–1000°C for 5–15 min. By means of platinum-tipped tongs, the crucible is tipped and agitated occasionally to ensure homogeneity and to aid the expulsion of gas bubbles. The melt is poured on a polished aluminum plate heated at 450°C on a hot plate. If the melt is very fluid, an aluminum ring of ~3 cm ($1\frac{1}{4}$ inch) inside diameter and at room temperature is placed on the aluminum plate and the melt poured into it. The pellet is kept on the hot plate for ~2 min, then pushed with a Transite rod onto a Transite plate previously heated to 450°C. The pellet is allowed to cool with the Transite to room temperature. The ring is removed. The bottom surface of the glass pellet may be satisfactory as is; if not, it may be polished.

Strain-free glass disks are prepared as follows (*D27*). The charge consists of 3 g of sample and 9 g of fused lithium tetraborate, or, if a high-absorption diluent is required, 1 g of sample, 1 g of lanthanum oxide, and 8 g of lithium tetraborate. The charge is fused in a 97Pt–3Au alloy crucible at 1250°C for ~10 min. The crucible is kept covered, but is stirred occasionally with a rod of the same alloy. The crucible and charge are allowed to cool to room temperature. The crucible is then inverted, and the glass button drops out. The button is ground to powder, which is returned to the crucible, remelted at 1250°C for 5 min, then allowed to cool. The button is recovered as before, replaced in the crucible upside down, and again remelted at 1250°C for 5 min. The button, still in the crucible, is annealed at 650°C for 30 min, then allowed to cool with the furnace. The flat side of the button is polished to a final 600-grit surface with silicon carbide.

Another method for preparation of strain-free glass disks has several other noteworthy features (*L10*). The flux is a glass made by fusing boric oxide (B_2O_3), sodium carbonate, and lithium tetraborate in weight ratio 6:2:1. A supply of this glass is prepared and ground to −100 mesh. The charge consists of 0.8 g of sample and 7.2 g of glass. The charge is heated in a 30-ml platinum crucible at 1050–1100°C for 10–15 min with occasional swirling. The melt is poured in a graphite mold consisting of a cylindrical depression 2.5 cm in diameter and 3 mm deep in a graphite block, then pressed with a flat graphite block. While still hot, the disk is transferred to a Pyrex petri dish on a hotplate at ~200°C, then annealed by simply turning off the hotplate and allowing it to cool or, after a few minutes, sandwiching the disk between pieces of asbestos cloth. Errors arising from surface texture are compensated by introducing suitable internal standards in the melt, thereby precluding the need for polishing.

In the *minimum-flux technique* (*A18*), 1 g of sample and 1 g of lithium tetraborate are mixed and fused in a graphite crucible at 1400°C for 1.5 min, then poured on a copper block and allowed to cool. This glass is unsuitable for use as a disk. It is ground and briqueted without binder, but with a cellulose backing. The low dilution ratio makes this technique favorable for low-Z analytes and minor and trace constituents.

In the *high-concentration, glass-disk method*, Longobucco (*L38*) was able to achieve high sample-to-flux ratio and still use the melt in the form of glass disks. She applied the method to determination of magnesium, aluminum, silicon, potassium, and calcium in clay minerals. A charge consisting of 5 g of sample, 5.5 g of sodium tetraborate, and 2 g of lithium carbonate is thoroughly mixed, placed in a platinum crucible, and heated slowly over a Meker burner until fusion begins. The temperature is then gradually increased until the crucible and charge are at a bright red heat. The crucible is heated at this temperature with constant swirling until fusion is complete, then placed in a muffle furnace at 1350°C for 3 min to eliminate gas bubbles. The melt is then poured on a polished aluminum plate maintained at ~260°C on a hot plate and allowed to cool slowly. Finally, the flat side of the disk is polished on a wet abrasive paper.

Great care is required in preparation of the disks. If the aluminum plate is too hot, the disk will wet and adhere to it; if it is not hot enough, the bottom of the disk may be rippled. If the hot disk comes in contact with a cold (room temperature) object, it will crack. If it cools too quickly, it will be so strained that it may "explode" like a Prince Rupert drop when handled. Nevertheless, standard disks prepared in this way have withstood years of use and storage. Longobucco later applied the same method to preparation of lithium tetraborate glass disks.

In the *heavy-absorber, fusion-briquet method* (*R24*), 125 mg of sample, 125 mg of lanthanum oxide, and 1.000 g of lithium tetraborate are mixed in a boron carbide mortar, transferred to a graphite crucible, and fused at 1100°C for ~10 min. The crucible is then removed from the furnace, and the melt is allowed to cool in the crucible. This glass is also unsuitable for use as a disk. The pellet is removed and weighed. Enough boric acid (H_3BO_3)

is added to bring the total weight to 1.300 g to level the final sample weight and compensate ignition weight loss. The bead and boric acid are placed in a tungsten carbide Mixer/Mill grinding vial, and the bead is broken up. The vial is then placed on the Mixer/Mill and ground for ~10 min to about −300 mesh. The ground powder is briqueted at 3500 kg/cm² (50,000 lb/inch²) against a backing of a previously briqueted boric acid disk. Here again, the low dilution factor makes this technique favorable for low-Z and trace analytes. The lanthanum oxide reduces absorption-enhancement effects. Norrish and Hutton (*N20*) describe a somewhat similar method using a mixture of lithium tetraborate, lithium carbonate, and lanthanum oxide. Both these methods are widely used for geological specimens.

In the *chemical-separation, borax-disk method* (*L48*), the glass-disk method is applied to trace and micro analysis. Net intensities of 1–5 counts/s per μg are obtained for the $K\alpha$ lines of elements 20–42 (calcium to molybdenum). In order to avoid the need for internal standardization, the analytes are separated by chemical means prior to fusion. Each separated analyte, as salt or oxide, is placed in a separate 8-ml platinum crucible, and 2 g of $Na_2B_4O_7$ is added and fused. An aluminum plate 0.6 cm ($\frac{1}{4}$ inch) thick having several plane, polished, circular depressions 3.8 cm (1.5 inch) in diameter and 0.2 cm ($\frac{1}{16}$ inch) deep is heated on a hot plate to 225°C. The melt for each analyte is poured into a depression and immediately flattened with a hot $5\times5\times0.6$ cm ($2\times2\times\frac{1}{4}$ inch) aluminum plate having a wood handle. Thus, all disks have uniform 0.2-cm ($\frac{1}{16}$-inch) thickness. When a disk has been cast in each depression, the aluminum plate is removed from the hot plate and allowed to cool with the disks in place. Standard disks are prepared the same way from known weights of analyte salts or oxides. In another study (*S14*), as little as 0.5 wt % of a favorable element was determined in a 2-mg sample fused with 1 g of sodium tetraborate.

In the *pressed-disk method* (*H21*), 0.3700 g of sample is mixed with 2.000 g of $Li_2B_4O_7$–Li_2CO_3–La_2O_3 flux mixture (47.03–36.63–16.34 wt %), placed in a 95Pt–5Au crucible, and fused at 1000°C. The melt is poured in a heated, shallow, cylindrical, duralumin mold and pressed with a heated duralumin plunger. The disk is then placed in a duralumin block having a shallow cylindrical depression and covered with another similar block, both blocks heated at 200°C. After 10 min, the disk is removed and allowed to cool on an asbestos surface. Reference (*H21*) describes equipment and gives the procedure for preparation of about 100 disks a day.

The *thin-layer, coated-metal-disk method* (*S54*) combines features of the glass-disk and thin-layer methods. The $Na_2B_4O_7$-glass specimen melt is poured over a preheated metal disk and spreads over the disk, forming a thin, uniform, bonded layer, which is immediately cooled by a stream of compressed air. Nickel disks may be used, and, in order to preclude excessive oxidation, are brought to temperature just before the melts are ready to pour. Temperature matching between melt and disk is extremely important: If the disk is too hot, the molten glass reacts with and is contaminated by the nickel; if the disk is not hot enough, the glass does not flow evenly. The method was applied to blast-furnace slags.

Strasheim and Brandt (*S65*) derived absorption corrections for mineral analyses by fusing two portions of each sample in $Na_2B_4O_7$ and $Li_2B_4O_7$, respectively. The melts were cast in graphite molds at 600–700°C for 15 min, then tipped out onto asbestos to cool to room temperature. Measurements on the two disks permitted calculation of the mass-absorption coefficients affecting the results and correction of the concentrations found for the individual elements.

The reader is referred to the bibliography for further details on applications of the glass-disk method to various sample types as follows: (1) ferrous and nonferrous alloys (*B86, S14*); (2) Alnico alloys (*B69*); (3) ferrous and sulfide ores (*C47*); (4) rare-earth materials (*M7*); (5) cements and cement materials (*A18, R18*); (6) ceramics (*L38*); (7) glasses, glass-batch materials, and refractories (*D27*); (8) rocks, minerals, ores, and slags (*A21, H62, R24, S65, W3, W6*); (9) metallic elements after chemical separation (*L48*); and (10) microchemical applications (*L48, S14*).

16.4.4. Considerations

If the borax fusion product is to be pulverized and presented to the spectrometer as a loose or briqueted powder, the fusion process and recovery of the glass are not particularly critical. However, admittedly, successful preparation of glass disks is an art. Each worker has his own little "tricks," and probably no two workers prepare disks the same way. Many of these individual techniques are cited in Section 16.4.3 in the discussions of specific fusion procedures. Others follow.

Sometimes it is advantageous to pretreat the specimens prior to fusion with the glass-forming flux. For example, sulfides may be fused with about 10 times their weight of potassium pyrosulfate, then subjected to the glass fusion. Metals may be converted to sulfides by heating with sulfur, then treated in turn by the pyrosulfate and tetraborate fusions (*S10*). Samples containing organic material—soils, for example—are heated for several hours at ~450°C.

Weight proportions of sample to flux range from 1:2 to 1:100; common practice ranges from 1:4 to 1:10. Use of additives to reduce fusion temperature, render refractory samples soluble, clarify the glass, increase the matrix absorption, etc., and use of crucibles not wet by the melt are discussed in Section 16.4.2. The sample should be ground to relatively small particle size, <0.1 mm, prior to admixture with the flux, and all components of the charge should be mixed thoroughly prior to fusion.

The purpose of the fusion is to produce a clear melt free of bubbles. If the melt is not clear, incomplete fusion, decomposition, or reaction is indicated. If the melt contains bubbles, inadequate measures were taken to permit their escape, or perhaps the fusion temperature is too low or the melt too viscous. It may be necessary to heat the melt to temperatures higher than those indicated in Section 16.4.3 to effect complete fusion—up to 1400°C for $Li_2B_4O_7$ melts. Use of additives to effect complete dissolution of refractory materials is discussed in Section 16.4.2. Heating should be continued until bubble formation ceases, indicating completion of the reaction. To permit bubbles to escape, the crucible is tipped, rotated, and/or stirred with a rod of alloy not wet by the melt. Also, heating over a Meker burner helps by initiating convection currents in the melt. When the melt becomes clear and free of bubbles, it is recommended that the crucible be placed in a muffle furnace to permit the charge to equilibrate at the pouring temperature, usually ~950°C. Radiofrequency induction heating has also been used to fuse the charges.

The disks are cast on a metal plate which they will not stick to or be contaminated by. Although aluminum was used by Claisse and most other workers, graphite, copper, duralumin, and gold-plated stainless steel are also used. If the plate is too hot, the disk will bond to it; if it is not hot enough, the bottom of the disk will be irregular and require polishing.

If it is required that all disks have the same diameter, any of several procedures may be used: The techniques of pouring the melt into a brass ring on a heated copper plate or into an aluminum ring on a heated aluminum plate are mentioned in Section 16.4.3. Bruch (*B86*) uses a similar technique: Aluminum and Transite sheets 5 mm thick are placed side by side on a hotplate at 300°C and allowed to come to temperature. A cylindrical ring of steel containing ~13% chromium is placed on the aluminum sheet. The ring is 30 mm I.D., 50 mm O.D., and 10 mm high. The melt, at ~950°C, is poured into the ring immediately. After ~5 min, the ring with the solidified melt is pushed onto the adjacent Transite sheet, and the ring is removed. After another 3 min, the Transite sheet with the disk is removed from the hotplate and allowed to cool to room temperature.

In any of these ring techniques, if the ring is allowed to become too hot before pouring, the disk may become irretrievably bonded to it.

Glass disks may be cast to any required diameter also by pouring the melts into suitable graphite molds at 210–400°C. After 1–2 min, the molds are tipped to dump the disks onto a heated asbestos plate, where they are allowed to cool (*R16*). The bottom surfaces are smooth and flat. Alternatively, the charges can be fused and cooled in special graphite crucibles having cylindrical chambers with smooth, flat bottoms. The cool glass buttons can be jarred from the crucibles or the (expendable) crucibles can be broken away. In still another technique (*K22*), the initial melts are quenched, removed from their crucibles, and pulverized. The powders are packed in graphite molds and heated at 1000°C to remelt the powders and form disks.

If it is required that all disks have the same thickness and/or parallel top and bottom surfaces, any of several techniques may be used: The melt may be poured onto a preheated metal plate and immediately pressed by a heated aluminum block having a shallow cylindrical depression as deep as the required thickness and having a smooth, flat bottom. The melt may be poured into a heated, shallow, flat-bottomed mold and pressed with a close-fitting heated cylindrical plunger (*H21*). Some workers abandon the pouring phase altogether. Stephenson (*S59*) fused the charge in a vitreous carbon crucible, then inserted a loose-fitting cylindrical carbon rod having a plane, polished bottom, in the manner of a piston, to rest on the melt and impart a flat top surface on the glass pellet. After the system has cooled to room temperature, the piston is removed and the crucible inverted and tapped on the table to jar the glass disk loose. Other workers melt small (1–3-g) charges in shallow dishes (~2.5 cm diameter, 1–5 mm deep) of a metal not wet by the glass. After cooling, the disks are tipped out ready to use. Still other techniques are described in Section 16.4.3 in the *strain-free disk* and *chemical-separation*, *borax-disk* methods.

While still hot, the disks may be handled, transferred, or prodded only with preheated Transite or asbestos rods or with tongs tipped with these materials. Otherwise, they are sure to crack or develop strains. If the disks strongly bond to the metal plate, mold, or ring on or in which they are poured, little can be done about it. Dijksterhuis (*D17*) describes electrical methods for freeing tightly bound disks. A high-potential, high-frequency ac or 30-kV dc potential is applied between an electrode mounted above—but not in contact with—the glass and the electrically grounded crucible, mold, plate, or ring; for the dc, the electrode is positive. A spark discharge occurs between the electrode and disk, and the electric current reduces the oxides at the interface between glass and metal, thereby weakening the bond and freeing the bead.

After solidification, the disks may be placed on edge in slots in a fire brick, which is then placed in an annealing oven. Correct annealing temperature may be based on experience or trial and error. Or a differential thermal analysis (DTA) curve may be recorded from a sample of powdered disk. The curve is a plot of heat evolution or absorption as a function of temperature and/or as a result of physical or chemical change in the sample on heating or cooling. The annealing temperature should be ~50°C below the temperature of the differential thermal analysis peak corresponding to stress relief. A typical annealing temperature for lithium tetraborate glass is ~525°C. Minimum annealing time for disks ~3 mm thick is ~1/2 h.

After annealing, or even if not annealed, the disks are allowed to cool very slowly to room temperature. Some workers cover the hot disks with asbestos powder to ensure slow cooling. The effectiveness of the anneal is best evaluated by viewing the disks through crossed Polaroid sheets. If properly annealed, the disk is isotropic and does not rotate the plane of polarized light, so the crossed Polaroid sheets remain uniformly dark (*O4*). The disks may be tested further by placing them in an x-ray diffractometer and observing their Debye–Scherrer diffraction patterns. The appearance of diffraction peaks indicates the presence of residual crystalline material.

After the disks have cooled, their bottoms may be polished, if necessary, on a belt grinder, polishing wheel, or abrasive paper. Silicon carbide or aluminum oxide may be used, wet or dry. The final polish should be ~300 grit. Another way to deal with surface texture is by internal standardization (*L10*).

Disks that may be hygroscopic—such as those prepared from lithium tetraborate–carbonate eutectic (Section 16.4.2)—should be stored in a desiccator.

The problem of weight loss has been concisely discussed by Harvey *et al.* (*H21*), who point out that the total loss is the sum of contributions from the flux Δw_F and sample powder Δw_X and give four ways to deal with it: (1) When Δw_F and Δw_X are both of about the same magnitude *or* very small, they may be disregarded. (2) If a supply of flux has been previously dried at 400°C, or fused and ground, then stored in a desiccator, Δw_F may be assumed to be zero; Δw_X may then be obtained by weighing the flux-sample charge before and after fusion or by measuring weight loss on ignition of a separate portion of powder. (3) Weighed portions of flux may be fused along with the flux-sample charges to obtain Δw_F; Δw_X is then determined as in (2) above. (4) The weighed portions of flux are placed in their crucibles, fused, cooled, and weighed to obtain Δw_F. The weighed sample powders are then added and the charges fused, cooled, and weighed again to obtain Δw_X. In this way, both Δw_F and Δw_X are determined on each sample itself. These workers (*H21*) claim that this method gives highest accuracy.

Special "machines" for preparation of glass disks are described in Section 16.1.6.

Chapter 17

Specimen Preparation and Presentation—Liquids; Supported Specimens

17.1. LIQUID SPECIMENS

17.1.1. Introduction

In Chapter 16, it was pointed out that usually x-ray spectrometric analyses of samples in their as-received forms are more rapid and convenient than analyses by any other method. Thus, the analyst strives to analyze samples in their original forms whenever this is practical. However, it is often necessary to change the original form—especially of solid samples and parts—when: (1) specimen errors arising from inhomogeneity, particle size, etc. must be eliminated; (2) a method involving addition is to be used; (3) chemical separation or other treatment is required; (4) the samples have form different from that of the available standards, or the same substance is received in many forms, and it is desired to use a single set of standards for all; (5) the x-ray spectrometric method is to complement some other analytical method; or (6) the samples are chemically or radiolytically unstable. In such cases, all specimens are usually reduced to powder, fusion product, or solution. If the decision is made to put the specimens into solution, many alternative possibilities for subsequent preparation and treatment, presentation, and analysis become feasible that would not be so for other forms. Many general papers dealing with x-ray spectrometric solution techniques have appeared (*25f, B34, B37, C10, C20, C22, C73, C74, G35, L5, T11*).

Of course, many samples are received in liquid form, including the following: (1) liquids and solutions, aqueous and nonaqueous, including plating solutions and hydrocarbons; (2) dissolved solids, powders, and parts; (3) dissolved fusion products; (4) leach (*L58*) and wash liquids; (5) brines; (6) eluates from ion-exchange columns; (7) liquid–liquid and liquid–solid extractions; (8) biological fluids; (9) emulsions and gels; (10) sols (*G34*); (11) slurries (*H70*); (12) greases, waxes, etc.; (13) plastic monomers; (14) liquid ion-exchange resins; and (15) melts.

A specimen in liquid or solution form may be readily subjected to various treatments (*D8, H68, M47*), including the following: (1) dilution; (2) evaporation (concentration); (3) absorption in cellulose or other material and firing to powder (*C16*); (4) evaporation to dryness; (5) precipitation of analyte or matrix constituents, selectively or collectively; (6) selective dissolution; (7) liquid–liquid extraction (*B67, B68, K5, M55, W20*); (8) ion exchange with granular resin by batch or column technique, or with ion-exchange liquid, membrane, or paper (*C23, V1, V2*); (9) paper chromatography (*J1, J2*); (10) gas chromatography (*S15*); (11) paper electrophoresis (*W7*); (12) electrodeposition on solid (*L39*) or mercury cathode (*C35*); (13) distillation, including steam distillation; (14) dry or wet ashing (*D9, L59, R30, S34*); (15) gas formation—for example, dissolved sulfur may be converted to hydrogen sulfide and absorbed in paper impregnated with lead acetate, and arsenic may be converted to arsine and absorbed in mercury(II) bromide paper (*J3*); (16) quantitative addition; and (17) centrifugation.

Quantitative additives include: (1) analyte (standard-addition method); (2) solvent (standard-dilution method); (3) a specific, interfering, variable minor matrix element (to fix its concentration); (4) a metal salt or a matrix constituent of no analytical interest (matrix-masking agent); (5) low- or high-absorption diluent; (6) internal standard; (7) internal control or intensity-reference standard; or (8) a relatively heavy element to combine stoichiometrically with a low-Z element, organic compound, or non-determinable substance to make its determination feasible or more convenient (indirect or association method).

Solutions may be converted to powders by precipitation, evaporation, or absorption in cellulose; and the powders may be converted to briquets or fusion products. Solutions may also be presented as supported specimens of various types (Section 17.2). Solutions are perhaps more readily applicable than other forms to x-ray analytical methods other than x-ray secondary-emission spectrometry, including polychromatic and monochromatic absorptiometry, absorption-edge spectrometry, and methods based on x-rays scattered from the specimen. Finally, when x-ray spectrometry is to supplement another, wholly independent analytical method, the samples must usually be in solution form.

The remainder of Section 17.1 deals with the presentation of specimens in liquid form, Section 17.2 with supported specimens of various types.

17.1.2. Advantages

The principal advantages of specimen presentation in liquid form are the following:

1. Solution techniques are applicable to almost any type of sample. They are directly applicable to liquids, solutions, slurries, and sols; and solids, fabricated parts, powders, and fusion products may be put into solution.

2. Solution techniques are particularly advantageous when the same material is received in various forms, such as bar, foil, drillings, powder, wire, and parts. A single specimen procedure and set of standards can deal with all sample forms.

3. For samples already in liquid, sol, or slurry form, often little or no preparation is required; what preparation is required is usually simple and convenient. For other sample forms, preparation is also usually simple and convenient.

4. Samples and standards are perfectly homogeneous.

5. The analytical results are representative of the bulk sample, not just the surface; the effects of nonrepresentative surface composition, surface texture, and particle size are eliminated.

6. Standards are usually prepared easily by dissolution of the individual constituents in proper concentrations, dissolution of standard materials, or quantitative variation of the composition of a single dissolved standard or of a standard solution.

7. Blanks for measurement of background and for compensation of contamination are usually prepared easily.

8. Standard addition and dilution techniques are applicable which preclude the need for standards; such analyses are particularly useful for one-of-a-kind analyses.

9. The generally low effective atomic number of liquid matrixes results in relatively high scattered background. This higher background is usually

disadvantageous, but may be beneficial in facilitating the use of methods for compensation of absorption-enhancement effects and analytical methods based on scattered x-rays.

10. Chemical separations, if required, are usually made conveniently (*D8*, *H68*, *M47*).

11. Concentration or dilution to appropriate analyte concentrations is usually feasible.

12. The absorption-enhancement effects of the total solute are reduced or eliminated by the dilution and leveling effect of the solvent.

13. The matrix (solvent) usually consists of elements which have low atomic number and which contribute little absorption and no enhancement of the analyte lines.

14. When residual absorption-enhancement effects from the original sample persist, a wide choice of techniques is available to deal with them (*C22*, *C73*, *G40*). Some of these techniques are applicable only to solutions; others are more convenient with solutions. These techniques are as follows: (a) separation of analyte or interferant; (b) use of sets of calibration curves; (c) dilution with low or high absorber; (d) internal standardization; (e) internal control or intensity-reference standardization; (f) background-ratio method; (g) coherently or incoherently scattered target-line ratio methods; (h) Reynolds–Ryland scatter method; (i) standard addition and dilution methods; (j) matrix masking; (k) fixing of concentration of a specific interfering variable minor matrix element; (l) special experimental methods, such as those described in Section 18.2.2; and (m) mathematical methods (*D39*, *K4*).

15. The low absorption by the matrix leads to high penetration by the primary beam and low absorption of the emitted secondary radiation, and, consequently, high absolute sensitivity. Unfortunately, this advantage may be canceled by the dilution of the analyte and by the increased background and consequent decreased peak-to-background ratio.

16. On dilution of a solution, analyte-line intensity decreases at a much slower rate than analyte concentration (Table 14.3).

17. The concentrations of heavy elements in their compounds with low-Z elements can be determined in solution. For example, the Fe $K\alpha$ intensity from FeO, Fe_2O_3, and Fe_3O_4 is about the same for the solids due to low contrast or slope in the calibration function (Table 12.2); if equal weights of these substances are put into equal volumes of solutions, the Fe $K\alpha$ intensities have much greater spread. Differences as small as 0.1 in the uranium:oxygen ratio of various samples of uranium(IV) oxide (UO_2) can be determined in nitric acid solution (*H65*).

18. Solution and slurry techniques are particularly applicable to dynamic systems and to on-line process control.

19. Dissolution of solid and powder samples opens up many alternative possibilities for subsequent preparation, presentation, and analysis (Section 17.1.1).

20. Even when the samples are to be analyzed in another form, a solution method may be used to provide standards. Samples of the required form are selected as prospective standards and analyzed by a solution method using synthetic solution standards.

17.1.3. Disadvantages

The principal disadvantages of specimen presentation in liquid form are the following:

1. Except for samples already in the form of liquids, solutions, sols, and slurries, liquid-specimen methods are destructive.

2. For solid and powder samples, a solution technique is usually less rapid and convenient than a technique that uses the sample in its original form.

3. Some elements are very difficult to put into and/or keep in solution. Fusion may be required.

4. The portions of liquid standards actually used for the analysis are usually discarded. When large numbers of samples are to be analyzed and the analysis requires many hours or all day, standards must be measured at intervals throughout the analysis. In such cases, sometimes a cell of standard can be measured only a few times because of deterioration on repeated prolonged exposure to the primary beam. Thus, relatively large volumes of standards must be prepared and stored, or else standards must be prepared frequently. Moreover, liquid standards may not be stable in long storage. Conversely, solid, powder, briquet, and glass-disk standards are substantially permanent and require little storage space. It is sometimes feasible to analyze liquid samples against solid—especially plastic—standards (*C5*).

5. Minor and trace constituents and micro samples may be more difficult to analyze because of the high dilution. This applies especially for micro samples diluted to sufficient volume for infinite thickness.

6. The intensities from analytes of low atomic number are seriously reduced by dilution and by absorption in the liquid matrix and cell windows.

7. Selected-area analyses are not feasible.

8. Because of the low effective atomic number of the solvents and cell windows, the scattered primary background is relatively high. This increases

the minimum detectable concentration, and the coherently and incoherently scattered target lines increase the possibility of spectral-line interference.

9. Evaporation of solvent may occur during irradiation, especially with volatile solvents and in vacuum spectrometers, with consequent change in concentration and, possibly, in specimen plane.

10. The specimen compartment, drawer, mask, and, worse, the beryllium window of the x-ray tube may be subjected to leakage and corrosive vapors. With inverted specimen planes, perforation or rupture of the cell window may allow specimen liquid in quantity to flood these components.

11. Liquids may outgas vigorously, spatter, or rupture their cell windows when used in vacuum spectrometers (see below).

12. Liquids may heat under prolonged irradiation, causing expansion and, in closed cells, distension of the cell window.

13. Because of the outgassing, heating, and expansion, the specimen plane may vary during the irradiation.

14. Some liquid cells are difficult to fill properly or are leaky and messy.

15. Bubbles often form in the liquid during irradiation, adhere to the cell window, and reduce the measured analyte-line intensity. The effect is reduced, but by no means eliminated, by use of inverted specimen plane.

16. Some solutions undergo x-ray-induced radiolysis, with consequent precipitation. In inverted geometry, the precipitate settles on the window, which is the specimen surface. In upright geometry, the precipitate settles to the bottom of the cell, away from the measured surface. Thus, assuming that the precipitate is analyte-rich, the measured analyte-line intensity is increased in inverted geometry, decreased in upright geometry. In general, colloidal precipitation that does not flocculate is much less serious than precipitation that does settle out. However, any precipitation must be regarded as undesirable.

17. Müller [p. 184 in (*29*)] warns of possible difficulty even with liquids and solutions that produce schlieren (regions of different index of refraction arising from differences in density).

When liquid specimens are used in vacuum-path spectrometers, some of the problems attending their use in air or helium path are accentuated and some new problems arise. In closed cells, the thermal expansion, vapor pressure, and outgassing of the liquid cause the cell window to distend much more than in air or helium path, displacing the specimen plane and increasing the possibility of perforation or rupture of the window. The effect is more severe the more volatile the liquid and the thinner the window. In open cells, the liquid outgasses and evaporates more rapidly than in air path, and may actually boil at the reduced pressure. The evaporation and ebullition are more severe the more volatile the liquid. In inverted geometry, window distension from the weight of the liquid is more severe in vacuum than in air path. The effect is more severe the thinner the window and the greater the density and volume of the liquid. If only short-wavelength spectral lines are to be measured from nonvolatile liquids, relatively thick (>50-μm, 0.002-inch) Mylar or even 0.5-mm polystyrene windows can be used, and the problems may be negligible. However, long-wavelength lines require thin windows, and volatile liquids accentuate vapor-pressure and boiling problems.

Many of these unfavorable features may be eliminated or minimized by suitable precautions (Section 17.1.5) or treatment (Section 17.1.1), or by use of alternative specimen-presentation techniques (Section 17.2).

17.1.4. Liquid-Specimen Cells

17.1.4.1. Forms and Materials

The early Philips x-ray spectrometers had flat specimen drawers and a horizontal, upright specimen plane. Several liquid-specimen drawers having a liquid chamber as an integral part of the drawer were made for these instruments. Some of these drawers were water-cooled (*D24*). However, present-day instruments all use a liquid-specimen cell having a body consisting of a cup, dish, or open cylinder and a thin window attached semipermanently or with some form of retaining ring or cap.

Each x-ray spectrometer manufacturer provides cells for its own instruments (Figure 17.1), and disposable cells that fit all commercial instruments are available from Spex Industries (Figure 17.2A). However, great ingenuity has always been shown in the design and construction of cells in the laboratory. Cells for upright, horizontal, or inclined specimen planes must have the form of cups or dishes. Cells for inverted geometry may have the same form, or they may be simple open cylinders.

Cell bodies have been made of aluminum (*D40*), stainless steel or gold-plated steel (*D34*), glass (*K5*, *P21*), and plastic (Lucite, polyethylene, polypropylene, nylon, and Teflon). Cylindrical cells have been made from tubing of these materials cut to appropriate length (Figure 17.2C). Cup cells have been made of polyethylene vial caps of various sizes and of polyethylene or polystyrene vials of various diameters cut off at appropriate depths. Caplugs (Protective Closures Co., Buffalo, N. Y.) provide excellent

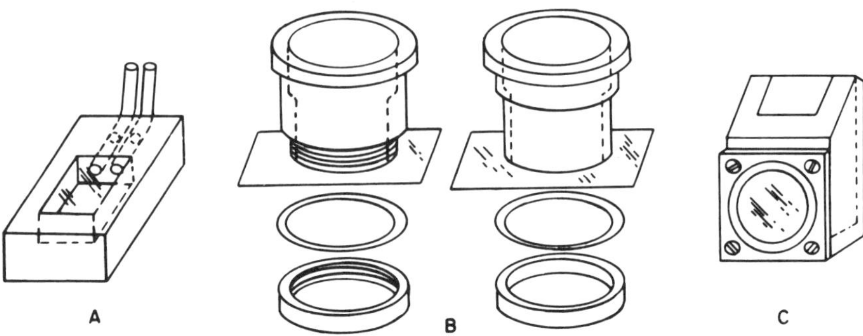

FIG. 17.1. Liquid-specimen cells for some commercial x-ray spectrometers. A. Early Philips spectrometer; the specimen plane is horizontal; the window is cemented on semipermanently. B. Philips Inverted Three-Position and Universal-Vacuum spectrometers; the rings are gaskets; the retaining collars may screw (left) or push (right) on; the drawings are schematic only. C. Early Siemens spectrometer; the window is held in place by the mask, which, in turn, is secured by the four screws.

disposable cups of various sizes (Figure 17.2B). Only white Caplugs should be used because the pigments in colored plugs contain traces of heavy elements (*B66*). Sometimes metal cells can be used with corrosive liquids if their inside surfaces are coated with paraffin (*H4*).

Rigid cell windows have been made of thin mica, glass microscope-slide covers, polystyrene, methyl methacrylate, and beryllium sheet (*D40*). Such windows are usually permanently or semipermanently cemented to the cell body (*K5*) or sealed to it with a gasket. However, usually windows are made by stretching thin plastic film over the rim of the cell body. Formerly, cellophane (*B64*) and even Scotch tape were used. However, today, the most commonly used films are 3–125-μm (0.00012–0.005-inch) Mylar, 25–125-μm (0.001–0.005-inch) polyethylene and polypropylene, and micro-

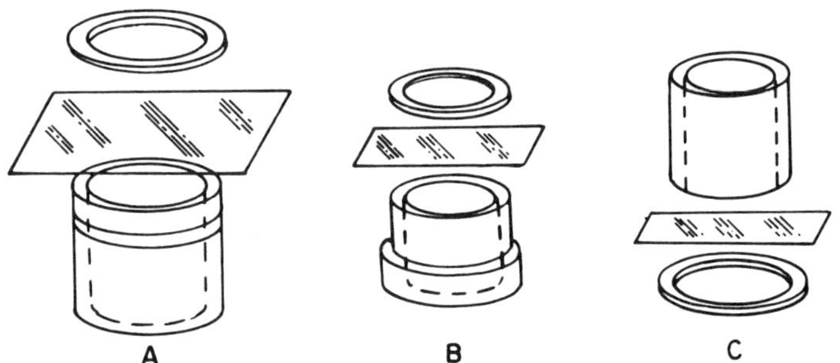

FIG. 17.2. Disposable liquid-specimen cells made from (A) Spex cups, (B) Caplugs, and (C) cylinders cut from plastic tubing. The rings slip over the outside of the cylindrical bodies.

porous film. Saran must be avoided because its chlorine content results in relatively high absorption and, of course, emission of chlorine K lines. Ordinary Scotch tape withstands the intense irradiation of the primary beam fairly well, but Scotch "Magic" tape deteriorates rapidly. Mylar is remarkably stable in the presence of most liquids, including organic solvents, but deteriorates rapidly in contact with strongly alkaline liquids ($p\text{H} > 10$). The longer the wavelengths to be measured, the thinner the window must be; 6-μm (0.00025-inch) Mylar has only 40% transmission for Al $K\alpha$, only 20% for Mg $K\alpha$. However, when only short-wavelength lines are to be measured, thicker films may be used. There is no advantage to thicker films for windows replaced after each specimen, but thick films have longer life for semipermanently attached windows.

Mylar is the most widely used film for cell windows, and, in general, it is excellent for this purpose. However, when compared with polypropylene, Mylar has some disadvantages. Mylar having thickness 6 μm (0.00025 inch) is most commonly used. Mylar having thickness 3 μm (0.00012 inch) is available and would be advantageous for the $K\alpha$ lines of elements 9–14 (fluorine to silicon). However, the thickness of these thin Mylar films is not as uniform as it is for thicker films (*A14*), and pinholes are likely to be present in the 3-μm film. Mylar has a higher content of high-Z trace impurities than polypropylene; the ash contents of the two films are ~0.2% and <0.005%, respectively. Moreover, the trace impurities in Mylar are not evenly distributed, while those in polypropylene are. Mylar is polyethyleneterephthalate, a polymer consisting of carbon, hydrogen, and oxygen, whereas polypropylene is a hydrocarbon polymer, and has effective atomic number ~1/2 and mass absorption coefficient ~1/5 those of Mylar. A greater thickness of polypropylene is required to prevent sag under the weight of the specimen, but, due to its lower absorption 12-μm (0.0005-inch) polypropylene has less than half the attenuation of 6-μm (0.00025-inch) Mylar. For maximum strength, only fully oriented polypropylene film should be used. Both these plastics, as well as most others, have the property of adsorbing metal ions and, under intense x-radiation, actually incorporating them into their polymer structures. This applies to film used for windows and to the heavier stock used for cell bodies.

Thin film windows are usually stretched over the rims of cylinder or cup cells in the manner of drum heads and secured with plastic curtain rings, O-rings, or rings of nylon, Teflon, or metal. Windows have also been affixed to cells with grease, and semipermanently cemented on with Pliobond rubber cement (Goodyear Tire and Rubber Co.). Some liquid-specimen cells require gaskets; these are usually made of polyethylene, nylon, or Teflon. In practice, the cell window is usually replaced after use with each specimen, and some cells are so inexpensive (Spex cells, Caplugs) that the entire cell may be discarded after one use.

The following sections describe some liquid-specimen cells worthy of special note.

17.1.4.2. General-Purpose Cell

A completely clean, leak-free cell especially useful for instruments having upright, inclined specimen planes is shown in Figure 17.3 (*B38*, *B41*). The cell body *A* is 1 cm ($\frac{3}{8}$ inch) deep and may be of Lucite, polyethylene, or Teflon. The depression *B* is 0.6 cm ($\frac{1}{4}$ inch) deep and holds 4–5 ml. The Mylar cover *C* may be 6–125 μm (0.00025–0.005 inch) thick, depending on the wavelength to be measured, and is cemented to the cell with Pliobond rubber cement. When the cell is in place in the specimen drawer, the triangular space at the neck of the cell is upward and covered by the specimen mask. This space serves to conceal the air volume at the tapered filling port *D* and provides an expansion chamber for the liquid. The plug *E* is required only for volatile or corrosive liquids. Alternatively, when the cell is in use, a short length of Tygon tubing may be attached to the filling port to conduct corrosive vapors out of the specimen compartment. The cells cannot be

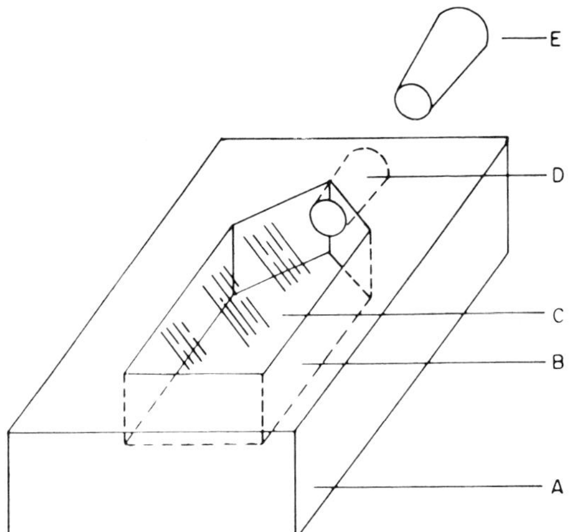

FIG. 17.3. Leak-free liquid-specimen cell for upright or inverted, horizontal or inclined specimen planes (Bertin). *A*. Cell body. *B*. Depression. *C*. Mylar cover. *D*. Tapered filling port. *E*. Tapered plug.

made of any readily available component and must be made in a machine shop.

Five to ten cells may be covered at a time with Mylar 6–125 μm thick as follows. A strip of Mylar, somewhat wider than the cell length and long enough to accommodate five to ten cells side by side, is stretched extremely tightly on a smooth plane surface and secured at both ends. Each cell is coated on its top face with a thin layer of Pliobond and laid face down on the Mylar. A heavy weight—2 kg or so—is placed on each cell and left undisturbed for an hour or so. The Mylar is then cut between the cells, and any wrinkles in the window are stretched out. The cells are then replaced face down on the flat surface, weighted again, and left undisturbed overnight. Thicker windows are applied the same way, but less care is required to avoid wrinkling. Windows 250 μm thick never do sag or wrinkle, and windows 6 μm thick have been used up to 100 times at 50 kV, 50 mA before they finally sagged or perforated. When this happens, the window is peeled off, the cell resurfaced on a 400-grit wet paper, and a new window applied.

Filling pipets are made from 5-ml pipets by drawing down the tips and blowing spherical chambers in the upper part of the enlarged chambers. One of these pipets is shown in Figure 20.3. After each use, the cell is emptied, rinsed with one or two portions of the next specimen, then filled with the next specimen.

The principal disadvantage of the cells is that some specimens tend to form adherent deposits on the window. However, often, complexing agents can be added to the specimen to prevent this, or a suitable rinse can be used to remove the deposits between fillings.

17.1.4.3. Somar "Spectro-Cup"

The Somar "Spectro-Cup" (Somar Laboratories, Inc., New York) is a versatile cell for all liquids and slurries in air or helium path and for non-volatile liquids and slurries in vacuum. The cell is also useful for powders in any path, either loosely packed or spread in thin layers between Mylar films (Section 16.3.4.3). Provision is made for equalization of internal and external pressures. The cells are filled rapidly and conveniently, inexpensive enough to be disposable, yet suitable for reuse and for storage of samples and standards for future use. However, the cell is applicable only to instruments having inverted geometry.

The cell, shown in Figure 17.4, consists of a cylindrical body *A*, retaining ring *B*, and snap-on cap *C* with handle *D*, all of polyethylene. The laby-

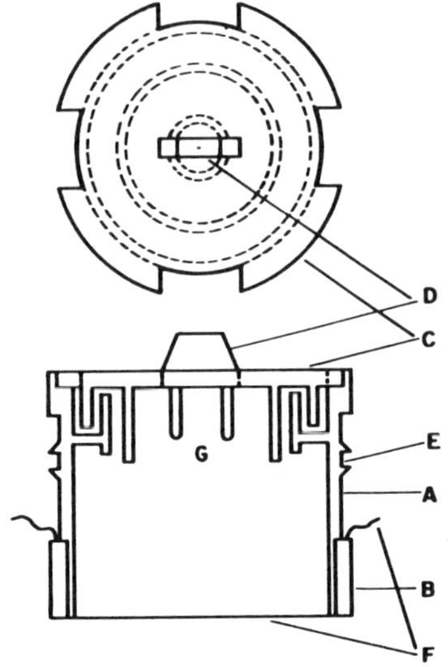

FIG. 17.4. Somar Laboratories liquid- (and powder-) specimen cell. All components are of polyethylene except the Mylar film. *A*. Cylindrical cell body. *B*. Retaining ring for Mylar film *F*. *C*. Snap-on cap. *D*. Handle for *C*. *E*. Groove to receive retaining ring (not shown) for second Mylar film for thin-film powder method (Section 16.3.4.3). *F*. Mylar film. *G*. Female cylindrical receptacle for male cylindrical internal intensity reference standard (Section 17.1.4.9, Figure 17.10).

rinthine structure between the cell body and cap provides a liquid anticreep undercut and prevents escape of specimen material into the specimen chamber. Equalization of internal and external pressures also occurs through the labyrinth, which opens very slightly into the four slots in the cap. The ring groove *E* receives a flat polyethylene ring (supplied, but not shown), as explained below.

For use with liquids, slurries, and loose powders, the cup is inverted, and a Mylar film *F* (6 or 3 μm, 0.00025 or 0.00012 inch) is stretched over the end and secured with the ring *B*. The cup is then set upright as shown, filled, and covered with the cap *C*.

For use with thin-layer sandwiched powder specimens, the cup is inverted, and a Mylar film is stretched over the end and secured, this time with the polyethylene ring in the groove *E*. The powder is spread uniformly over the film. Then a second film *F* is stretched over the powder and secured with ring *B*.

17.1.4.4. Cells for Use in Vacuum

When liquid specimens are used in vacuum-path spectrometers, some of the problems attending their use in air or helium path are accentuated and some new problems arise (Section 17.1.3). One way to equalize internal and external pressures and prevent window distension in closed cells is by provision of "breather" ports. However, this approach has had only limited success. A small hole may not permit vapor or gas to effuse fast enough, and a large hole may permit excessive outgassing, evaporation, or even boiling. When such cells are used with highly concentrated solutions, the ports may become plugged during measurement with residue from evaporated solution. Possibly the best way to equalize pressures in closed cells in vacuum path is by use of microporous film (see the last paragraph of this section). Window distension is substantially eliminated by supporting the window on the outside with a grid of nylon threads.

General Electric supplied a liquid-specimen cell capable of presenting a reproducible surface in the vacuum x-ray spectrometer. The cell, shown in Figure 17.5, consists of a cylindrical aluminum housing *A*, aluminum or copper window grid *B*, specimen chamber *C*, lower chamber *D*, and plug *E*.

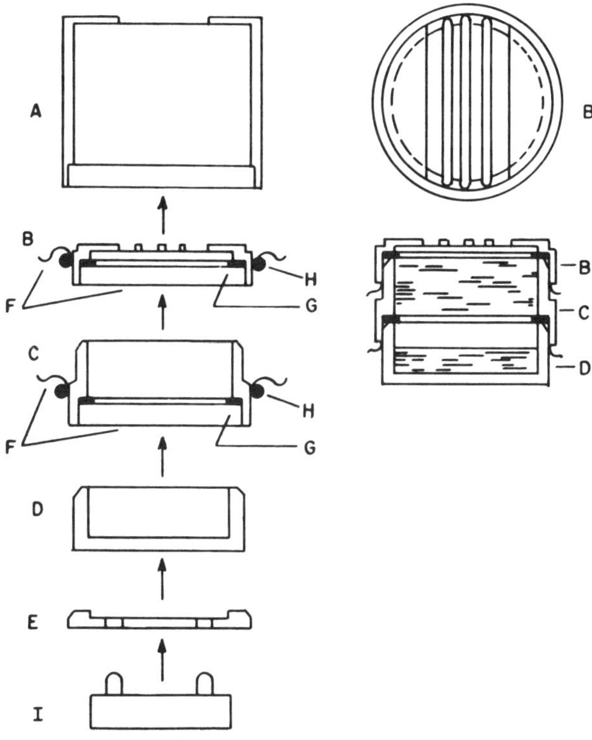

FIG. 17.5. Liquid-specimen cell for use in vacuum spectrometers. [Courtesy of General Electric Co., X-Ray Dept.] *A*. Aluminum housing. *B*. Window-retaining grid (aluminum or copper). *C*. Specimen chamber. *D*. Lower chamber. *E*. Screw plug. *F*. 6-μm (0.00025-inch) Mylar film. *G*. Gasket. *H*. O-ring used temporarily during filling and assembly of the cell. *I*. Key for screw plug *E*.

The assembled cell also has two 6-μm (0.00025-inch) Mylar windows *F*, and two flat gaskets *G*. During the assembly and filling of the cell, temporary use is made of two retaining rings *H* and a two-pronged key *I* for inserting and tightening the plug *E*, which has a threaded edge, into the bottom of the housing, which has an inside thread. The cells are assembled and filled as follows.

Flat gaskets *G* are placed in the grid *B* and specimen cell *C*, which are then fitted with tightly stretched Mylar covers *F*, temporarily secured with rings *H*, as shown. The lower chamber *D* provides a sort of expansion chamber or cushion for the specimen. It may be empty for specimens having low vapor pressure, or it may be filled ~1/3 full of specimen liquid or ethanol for specimens having higher vapor pressure. The specimen chamber *C* is pressed over the lower chamber *D* until the rim of *D* deforms the Mylar film on the bottom of *C* and seats against its gasket. The specimen chamber is filled with specimen (~10 ml) until a convex meniscus protrudes just slightly above the rim. The grid is now pressed over the specimen chamber *C* until the rim of *C* deforms the Mylar film on the bottom of the grid and seats against its gasket. Any wrinkles in either Mylar film are smoothed by tugging at the excess protruding around the edge. The retaining rings *H* are now removed, the excess Mylar is trimmed away, and any spilled liquid is blotted. The assembly, appearing as shown at the right in Figure 17.5, is inserted in the housing *A* and secured with the plug *E*, which is tightened with the key *I*.

The filling procedure is obviously complex, but, by application of the gaskets and Mylar covers to a number of specimen chambers and grids at a time, much time is saved. It is claimed that four cells can be filled and assembled in 10 min.

In the vacuum spectrometer, the top Mylar window bulges through the openings in the grid, but no air space is formed, because the air or vapor in the lower chamber expands against the Mylar on the bottom of the specimen cell. Apparently, the bulging is uniform from cell to cell, because excellent precision is realized with the cells.

Hughes and Davey (*H71*) describe another liquid-specimen cell for use with volatile liquids in vacuum path. Their cell has a double-cell construction, consisting of an open inner cell, containing the liquid, within a sealed outer cell. A cross section of the assembly is shown in Figure 17.6. A stainless-steel inner cell *A* has a thin Mylar window *B* secured with a push-fit retaining ring *C* and is provided with two exhaust holes *D* above the liquid level. The stainless-steel outer cell *E* has a thin Mylar window *F* held by an aluminum flange *G* secured with a screw-on stainless steel

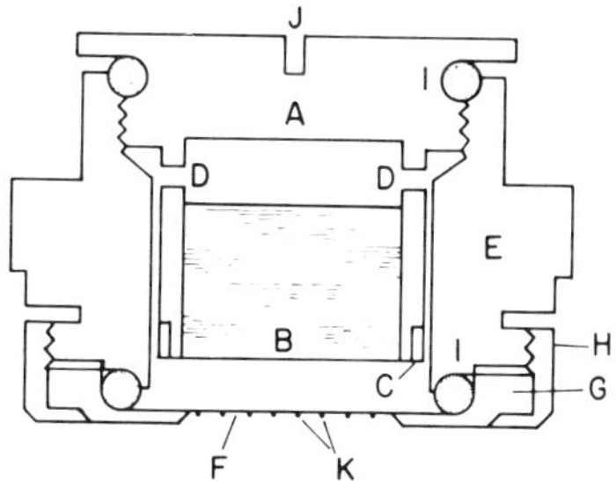

FIG. 17.6. Liquid-specimen cell for use in vacuum spectrometers. *A*. Stainless steel inner cell. *B*. Mylar window of inner cell. *C*. Retaining ring. *D*. Exhaust holes. *E*. Stainless steel outer cell. *F*. Mylar window of outer cell. *G*. Aluminum flange. *H*. Stainless steel retaining ring. *I*. O-ring. *J*. Slot. *K*. Nylon mesh.

retaining ring *H*. The seal between the flange and outer cell is made vacuum-tight by an *O*-ring *I*. Before insertion in the outer cell, the inner cell is filled through one of its exhaust holes by means of a hypodermic syringe, then threaded into the outer cell by use of a screw driver or coin in the slot *J*. A second *O*-ring *I* makes the entire assembly vacuum-tight.

This arrangement provides equal pressures on both sides of the inner cell window. Only the outer window is likely to distend, but even this is undesirable because it would provide a vapor layer of variable thickness between the two windows. Therefore, the outer window is supported by a nylon mesh *K*.

A novel approach is taken in the liquid cell supplied by Chemplex Industries, Inc., Scarsdale, N. Y. The cell consists of a slightly nonrigid, clear, plastic cylinder, open and lipped at both ends and covered at both ends with thin microporous plastic film secured with *O*-rings. The cell is covered on the bottom, filled, then covered on top. The microporous film effects equalization of pressure inside and outside the cell by allowing permeation by gas and vapor, but not by liquid.

17.1.4.5. Uncovered Cell

Mylar cell covers trap bubbles, scatter the primary beam, and absorb the long-wavelength spectral lines of elements of low atomic number. Elimination of the cover should improve precision, peak-to-background ratio, and detection limit, or permit reduced counting times for a given precision, with consequent reduction of thermal or radiolytic effects. Beard and Proctor (*B18*) describe an uncovered cell for use with the Picker spectrometer, in which an end-window x-ray tube and the spectrogoniometer constitute an integral unit, which may be mounted in any position. Ordinarily, the instrument is mounted with the x-ray tube axis vertical; the specimen plane is then inclined to the horizontal at 25°. Beard and Proctor tipped the instrument so that the specimen plane is horizontal.

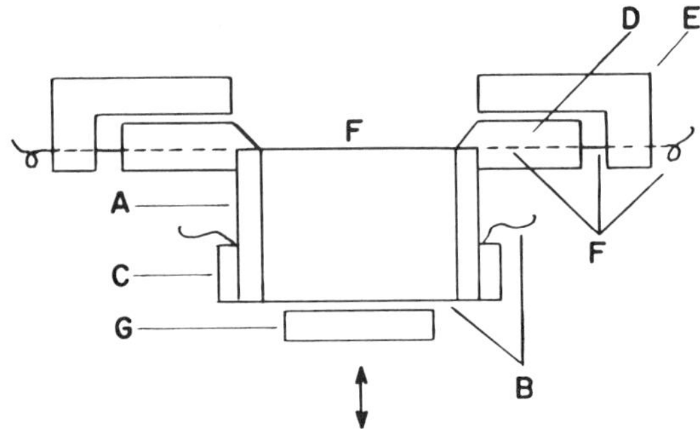

FIG. 17.7. Uncovered liquid-specimen cell (Beard and Proctor). *A*. Cylindrical Plexiglas cell body. *B*. Polyethylene film. *C*. Cylindrical Plexiglas retaining ring. *D*. Micarta specimen mask. *E*. Removable frame. *F*. Nylon threads stretched on *E* and, when *E* is in place on the mask *D*, lying in grooves in *D*. *G*. Movable piston for adjustment of liquid level by deformation of *B*.

The cell, shown in Figure 17.7, consists of a cylindrical Plexiglas body *A* having a polyethylene film *B* stretched across its bottom and secured with a retaining ring *C*, and is mounted on a Micarta specimen mask *D*. The cell is filled nearly to the brim, and a frame *E* fitted with two stretched parallel nylon threads *F* is placed over the mask temporarily. The mask has grooves to receive the threads, which then lie in the plane of the rim of the cell. The movable disk *G* presses against the polyethylene cell bottom *B* and is adjusted to bring the liquid surface into contact with the threads. Then *E* is removed, and the cell is ready for x-ray measurements.

Comparison of results for solutions in the uncovered cells and in the same cells with 6-μm (0.00025-inch) Mylar covers showed net intensity gains for Al $K\alpha$, Si $K\alpha$, and S $K\alpha$ by factors of 6, 3.3, and 2, respectively.

17.1.4.6. Frozen-Specimen Cell

Freezing eliminates many of the disadvantages of liquid specimens, including bubble formation, expansion, radiolysis, inconvenience with inclined specimen planes and vacuum spectrometers, and scatter from and absorption in Mylar cell covers. Birks [p. 79 in ed. 1 of (*8*)] was probably the first to suggest that certain liquids could be analyzed in the x-ray spectrometer in frozen form. He suggested an arrangement in which nitrogen gas is circulated through a coil immersed in liquid nitrogen, then through a channel in a metal block having a shallow depression to hold the specimen.

He also suggested that the frozen specimen might be more homogeneous if the specimen liquid were sprayed, rather than pipeted, onto the cold block, and that such frozen specimens would be particularly advantageous for low-Z analytes. For example, at room temperature, liquid water has a vapor pressure of ~17.5 torrs (mm Hg) and absorbs ~58% of Mg $K\alpha$ intensity in a 20-cm path. However, ice at −20°C has a vapor pressure of <1 torr and absorbs only ~4% of the Mg $K\alpha$. Frozen specimens are to be distinguished from freeze-dried or lyophilized specimens.

Figure 17.8 shows a frozen-specimen cell built by Chan (*C39*). A solid aluminum cylinder *B*, previously cooled in liquid nitrogen or a freezing mixture, is placed in a thermal insulating cup consisting of a Plexiglas inner cup *C*, aluminum reflecting cup *D*, and Plexiglas outer cup *E*. The frozen specimen *A* (see below) is placed on *B*, followed by an insulating cork disk *F* and metal mask *G*. The assembly is secured with the screws *H*. *F* and *G* have rectangular windows *I*.

Three types of specimen are described: (1) A metal disk 33 mm in diameter and 3 mm thick has a rectangular depression 20×12×2 mm. Liquids requiring no Mylar cover (low vapor pressure, noncorrosive) are placed in the depression. The disk is then placed between cold metal blocks to freeze the specimen. (2) A shallow metal planchet 33 mm in diameter and 0.5 mm deep holds a thick filter-paper disk impregnated with ~0.5 ml of solution. A 6-μm (0.00025-inch) Mylar film may be placed on the planchet and secured with a retaining ring. The specimen is frozen as before. (3) A shallow metal planchet 33 mm in diameter and 3 mm deep is filled with aqueous solution thickened with gelatin or with organic solution thickened with alkali metal stearate, covered with Mylar, and frozen.

All specimen preparation is done in a dry-box fitted with rubber gloves and provided with dewar vessels containing the supply of aluminum cylinders and freezing blocks. A continuous stream of thoroughly dried nitrogen is passed through the box. The box communicates directly with

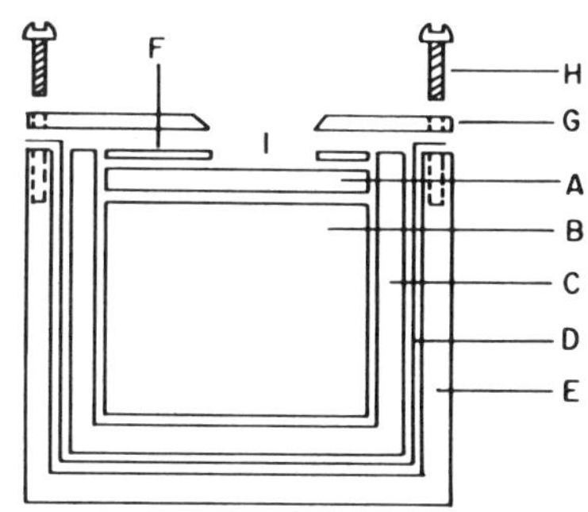

FIG. 17.8. Frozen liquid-specimen cell (Chan). *A*. Frozen specimen (see text). *B*. Precooled aluminum cylindrical block. *C*. Plexiglas inner thermal-insulation cup. *D*. Aluminum thermal-reflection cup. *E*. Plexiglas outer thermal-insulation cup. *F*. Cork disk. *G*. Metal specimen-mask disk. *H*. Retaining screws. *I*. Rectangular specimen window in *F* and *G*.

the spectrometer specimen chamber through a flexible plastic tunnel. The helium and air inlets to the spectrometer are provided with drying trains.

17.1.4.7. Cell for Slurries

X-ray spectrometric analyses of powders and briquets are limited to thin surface layers, and may be influenced by inhomogeneity and particle size. In such cases, especially if the powder is difficult to dissolve or fuse, it may be advantageous to analyze it as a slurry. Hudgens and Pish (*H70*) describe a slurry cell consisting of a sealed body having 5–6-ml capacity, 75-μm (0.003-inch) Mylar cover, externally powered stirrer mounted asymmetrically for maximum turbulence, and a bleeder valve to allow escape of gases produced by radiolysis and heating. The sample powder and internal standard are placed in the disassembled cell, covered with liquid vehicle, and stirred with a toothpick to remove occluded air. The cell is then assembled and filled with vehicle through the bleeder hole by use of a hypodermic syringe. The cell was evaluated for determination of the thorium and uranium in mixtures of their oxides using strontium sulfate ($SrSO_4$) as internal standard.

17.1.4.8. High-Temperature Liquid-Specimen Cell

Figure 17.9 shows a high-temperature liquid-specimen cell built by Campbell and Leon (*C18*). A ceramic or metal crucible *A* rests in a depression in an Inconel block *B* surrounded by a Nichrome heating element *C* wound on a refractory ceramic cylinder *D* in a brass housing *F* provided with water-cooling coils *G* and packed with high-temperature insulation *E*. A thermocouple well *H* and an inert gas inlet (not shown) are provided.

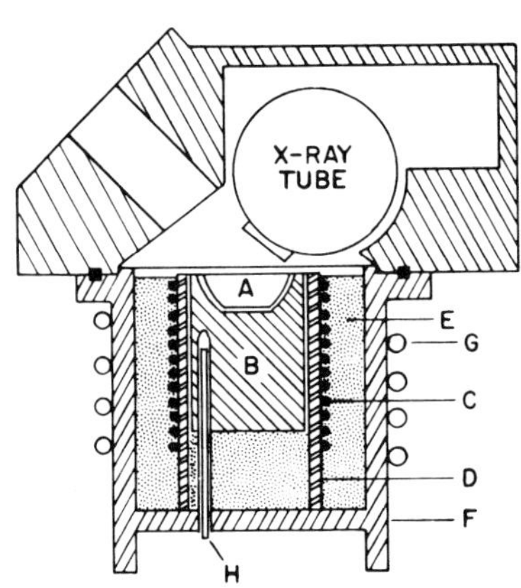

FIG. 17.9. High-temperature liquid-specimen cell (Campbell and Leon). *A*. Ceramic or metal crucible. *B*. Inconel cylinder with depression for *A*. *C*. Nichrome heater coil. *D*. Refractory cylinder, support for *C*. *E*. Thermal insulation. *F*. Brass shell. *G*. Water cooling coils. *H*. Thermocouple tube.

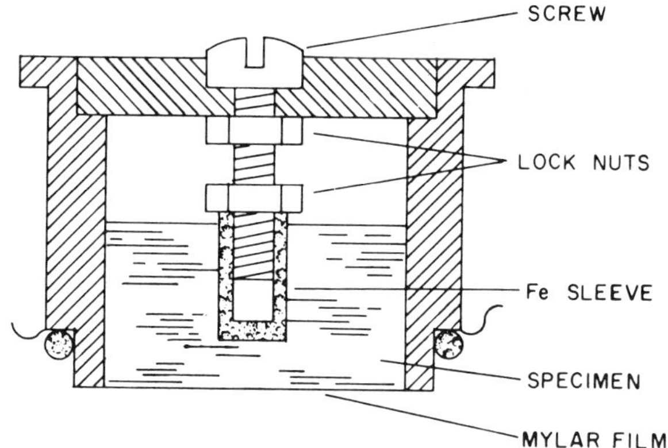

FIG. 17.10. Liquid-specimen cell with built-in iron intensity-reference standard (Jones).

With this arrangement, specimen temperatures up to 900°C are attainable, but Campbell and Leon suggest that with a platinum or alumina crucible, platinum block, platinum–rhodium alloy heater coil, and alumina support, temperatures up to 1400°C should be attainable. The x-ray tube window was 12 mm away from the specimen surface and was protected by a thin sheet of mica. No adverse effect on the x-ray tube was observed after 50 runs of up to 5 h at up to 800°C. The cell was used to make dynamic studies of the selective oxidation rates of traces of arsenic and antimony in molten lead. The cell also provides a novel way of preconcentration of the arsenic and antimony trace impurities because they oxidize preferentially and collect on the surface, where they are readily determined.

17.1.4.9. Other Liquid-Specimen Cells

For determination of traces of manganese in gasoline, Jones (*J25*) used the cell shown in Figure 17.10. The screw in the cell cover has an adjustable iron cap or sleeve, which serves as an internal intensity-reference standard. The analysis consists in measurement of Mn $K\alpha$ intensity from the analyte and Fe $K\alpha$ intensity from the cap. The Mn $K\alpha$/Fe $K\alpha$ intensity ratio is applied to a calibration curve prepared from standards measured in the same cell. Later, Gunn (*G39*) described a similar cell having an internal intensity-reference standard in the form of a platinum disk. The cell was used for determination of lead in hydrocarbon oils from the Pb $L\alpha_1$/Pt $L\alpha_1$ intensity ratio. The position of the intensity-reference standard is adjusted to give reference intensity about the same as the analyte-

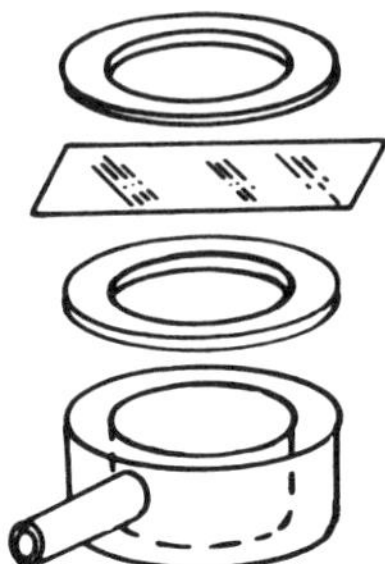

FIG. 17.11. All-glass liquid-specimen cell for upright, inclined specimen plane. The two rings are gaskets. The window may be thin borosilicate glass or Mylar film. The assembled cell is tightly clamped to the under side of the specimen mask and is filled through the side tube with a hypodermic syringe. Alternatively, a thin glass window may be cemented to the cell rim.

line intensity from the samples. When using these metal internal intensity-reference standards for highly corrosive liquids, the stud or disk may be protected by thin Mylar or polyethylene film (*W21*).

An all-glass cell for upright, inclined specimen planes is shown in Figure 17.11 (*K5, P21*). An all-polyethylene cell for use in the same geometry with hydrofluoric acid solutions is shown in Figure 17.12 (*W1*). A cell through which liquid can be circulated continuously is shown in Figure 17.13 (*C11*).

The liquid cells supplied with most commercial inverted-geometry spectrometers have ~10 ml capacity. Inserts for such cells have been described that permit use of specimen volumes ⪅1 ml (*W14*). One such insert is the following. A microspecimen cell that permits analysis of ⪅1 ml of liquid or ⪅1 g of powder in inverted-geometry instruments (*N17*) consists of a glass tube having a thin Mylar window and inserted in a cylindrical Teflon base. The base has diameter the same as that of standard liquid cells (for a specified instrument) and height 2.5 cm (1 inch), and has

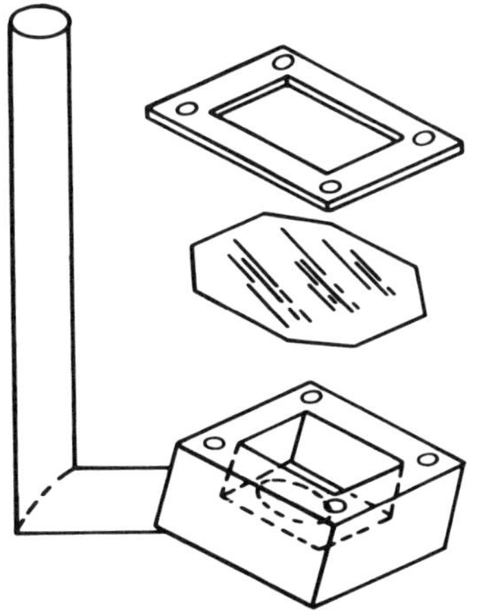

FIG. 17.12. All-polyethylene liquid-specimen cell for use in specimen drawers having upright, inclined specimen planes and for use with hydrofluoric acid solutions (Waehner). The assembled cell is affixed to the under side of the specimen mask by screws in the four holes shown in the gasket and cell body. The horizontal portion of the reservoir tube passes through a hole in the specimen drawer so that the upright tube is outside the drawer.

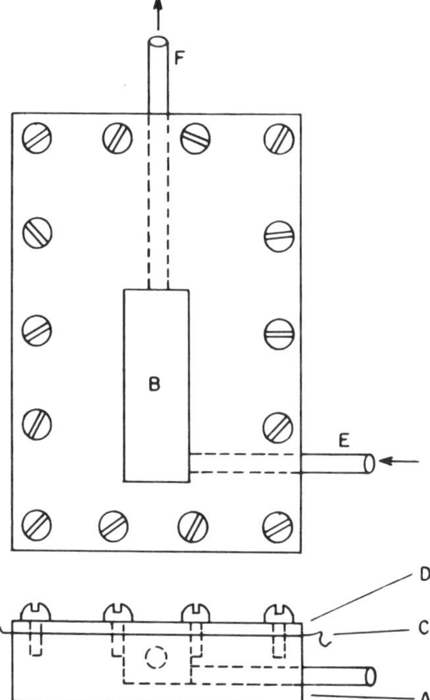

FIG. 17.13. Continuous-flow liquid-specimen cell (Campbell). *A*. Plexiglas body. *B*. Depression. *C*. Mylar window. *D*. Aluminum cover plate. *E*. Specimen inlet. *F*. Outlet.

an axial cylindrical hole 1 cm ($\frac{3}{8}$ inch) in diameter. The insert consists of a 3.5-cm ($1\frac{3}{8}$-inch) length of 9-mm O.D. glass tubing firepolished at both ends. The cell is prepared by laying a 6-cm (2.5-inch) square of 6-μm (0.00025-inch) Mylar film on the base and pushing the film through the hole with the glass tube until the tube is flush with the bottom of the base. The film forms a tightly stretched cell window and effects a snug fit of the tube in the base. The 1 cm of glass protruding from the top of the base serves as a handle for inserting and removing the cell. Teflon gives the lowest scattered background of any practical material, including aluminum, Bakelite, Lucite, and nylon. The cell should be rotated during measurement to compensate variations in cell geometry.

Cells for liquid micro specimens (*T11*) are also made by pressing concave depressions into polyethylene disks warmed on a hot plate. The drop of specimen is placed in the depression, and a 25-μm (0.001-inch) polyethylene film is hotwelded to the rim to enclose the drop. The welding may be done with a small pencil soldering iron or with a small heated cylinder.

Liquid specimens may be cooled by provision of a water-cooled jacket for the cell (*L4*), by water-cooling the specimen drawer or chamber (*D41*), or by directing on the cell a stream of gas cooled by passage through a coil immersed in liquid nitrogen (*M48*).

Thick syrups, oils, greases, silicones, waxes, and semisolids may be placed in liquid-specimen cells or shallow cylindrical cups or planchets, then covered with Mylar film. For example, Viscose has been analyzed this way (*P17*). Hydrocarbon liquids may be thickened to self-supporting semisolids by addition of heavy grease or silicone, wax, or lithium or sodium stearate. Inorganic solutions may be thickened by addition of gelatin. The thickened liquids may be placed in shallow cylindrical cups and covered with Mylar. The gelatin preparations can be dried to form small disks (*R28*). Waxes can be melted and poured into plastic cups (Spex cups or Caplugs), which are then covered with Mylar and inverted to permit the wax to set with a smooth, flat surface.

17.1.5. Precautions and Considerations

17.1.5.1. Interaction of the Primary Beam with Liquid Specimens

In general, only liquid specimens interact with the primary beam in such a way as to significantly affect the measured analyte-line intensity. The primary beam causes heating and radiolysis with consequent expansion, bubble formation, precipitation, and displacement of the specimen plane. These interactions may be reduced by freezing the specimen and by reduction of excitation power if the intensity is not reduced below values that give acceptable statistical precision. In absorption-edge spectrometry, another means of reducing incident flux is to derive the monochromatic bracketing wavelengths from secondary targets. Other ways of dealing with interaction effects are given by Cullen (*C73*), and the following discussion is based largely on his paper.

a. Specimen Heating. Cullen reports that water in a liquid cell was warmed only 4°C in 10 min by a Machlett OEG-50 tungsten-target tube operating at 50 kV, 45 mA in air path. The corresponding expansion is only ~0.02%, and the effect on analytical results would be less than the statistical counting error. However, specimen heating increases with concentration and absorption coefficient of solute, but may be reduced by use of: (1) modern, cooler-operating x-ray tubes; (2) water-cooled jacket on the x-ray tube head; (3) water-cooling of the specimen drawer and/or compartment; (4) water-cooled jacket for the specimen cell; (5) thermal insulation of the specimen by use of plastic cells and specimen masks to reduce heat conduction from the drawer or compartment; and (6) continuous helium flow.

b. Bubble Formation. Bubbles usually form in liquid specimens during irradiation. The bubbles originate from three causes—expulsion of dissolved gases, volatilization of the solvent, and/or x-ray-induced radiolysis of solvent or solute. Measures to reduce the invariably adverse effects of bubbles regardless of origin are of two basic types—preventive and remedial.

The preventive measures include: (1) reduction of specimen heating by the means listed above; (2) use of previously boiled or otherwise outgassed water and acids for specimen dilution and preparation; (3) avoidance of gas-forming reagents insofar as is feasible, such as peroxides, ammonium hydroxide (NH_4OH), and ammonium persulfate [$(NH_4)_2S_2O_8$]; (4) use of gas inhibitors—for example, perchlorates reduce gas formation in iron(III) chloride ($FeCl_3$) solutions; and (5) outgassing of specimen solutions by heating or exposure to vacuum in a vacuum desiccator prior to loading them in cells; the filled cells may also be exposed to vacuum, but the bubbles must then be dislodged by sharply jarring the cells.

The remedial measures include: (1) use of inverted optics; (2) rotation of cells in inverted geometry to possibly dislodge some of the bubbles from the cell window; (3) in upright geometry, use of cells having a space covered by the specimen mask and into which bubbles may rise (Figure 17.3); (4) use of antiwetting agent (such as Desicote) on cell windows to reduce bubble formation and adherence to the window (*H4*); (5) preirradiation for a period of $\sim$60 s after the cell is placed in the primary beam, to permit the bubbles to reach equilibrium before the measurements are taken; (6) when two or more analytes and, perhaps, their backgrounds are to be measured from each specimen, they should always be measured in the same sequence from each specimen; and (7) the measuring time should be as short as possible consistent with the accumulation of enough counts for the required statistical precision.

It is often assumed that the adverse affects of bubble formation disappear with inverted geometry. This is not necessarily true. Very small bubbles formed on the cell window may be very adherent and may not dislodge and rise to the liquid surface until they have grown relatively large. However, bubbles are always less troublesome with inverted geometry.

c. Precipitation. Avoidance of high concentrations of radiolytically labile constituents and use of complexing agents reduce precipitation. For example, cobalt, nickel, indium, and gold plating solutions are more stable in the presence of potassium citrate than with potassium cyanide. When it does occur, precipitation may equilibrate after ~60 s, and a preirradiation period like that recommended above may be beneficial. When there is any possibility of even minimal adherence of precipitate on the cell window, cells must be used in which the window is replaced for each specimen, or the cell must be rinsed with a reagent that removes any precipitate from the cell window.

d. Specimen Plane. Variation in specimen plane occurs with inverted geometry in open or closed cells from sagging of the specimen window under the weight of the specimen, and in closed cells from thermal expansion or outgassing of the specimen. Clearly, specimen-plane variation is minimized by use of open cells, by masking of the specimen window to support part of the specimen, and by use of a nylon or other grid.

Variation in specimen plane occurs with upright geometry in closed cells from expansion and outgassing. Use of cells having open side-arms, filling ports, or breather holes is indicated. There is no problem of sagging under specimen weight with upright geometry. However, if the volume of liquid in a closed cell is slightly smaller than the volume of the cell, the window may be drawn inward slightly.

17.1.5.2. Composition

Concentrations of all analytes should be as high as possible to give high intensities. Solutions must be stable and not prone to decomposition during irradiation or, for standards, in storage. Complexing agents should be used to improve the stability of solutions. Volatile and corrosive solvents are avoided if possible.

The nature and concentration of the acid used in preparation of the solutions must be considered from two aspects. The higher the mass-absorption coefficient and concentration, the lower is the measured analyte-line intensity, especially for long-wavelength lines; in general, the decreasing order of preference in this respect is nitric (HNO_3), hydrofluoric (HF), phosphoric (H_3PO_4), sulfuric (H_2SO_4), perchloric ($HClO_4$), and hydrochloric (HCl). Whatever acid is used, it is necessary that the concentration be the same in all samples and standards. The magnitude of the effect of acid and concentration is illustrated by the following data: Cr $K\alpha$ intensity from chromium in 3 N acid is decreased 1% by addition of 40 ml of concentrated acid per liter of sample for nitric acid, by 4 ml per liter for hydrochloric acid. The higher the effective atomic number and concentration

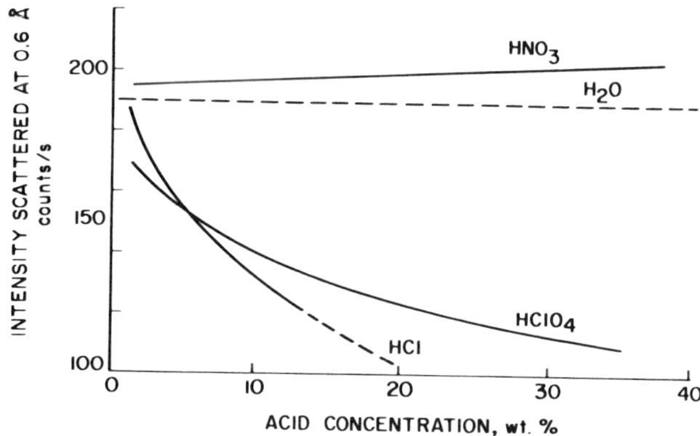

FIG. 17.14. Effect of acid concentration on intensity of scattered 0.6-Å primary x-rays for various acids and concentrations. The intensity scattered from water is shown for comparison. [K. Togel, in "Zerstörungsfreie Materialprüfung," R. Oldenbourg, Munich, Germany (1961); courtesy of the author and publisher.]

of the acid, the lower is the scattered primary x-ray intensity, as shown in Figure 17.14.

In preparation of standards for analysis of hydrocarbon liquids, the carbon:hydrogen ratio of the solvent must be substantially the same as that in the samples. For any kind of solution, the matrix must be substantially the same in all samples and standards. However, substantially greater latitude is permissible if an internal standard is added.

Emulsified liquid and suspended solid must be removed by centrifugation, coagulation, or other means. If the emulsified liquid is volatile, such as residual organic liquid in an aqueous solution, it may be distilled off.

17.1.5.3. Miscellany

With inverted specimen planes, there is no problem with air bubbles entrapped during filling and covering of the cells, because the bubbles rise to the top surface of the liquid, where they do no harm. With upright, horizontal specimen planes, care must be taken to avoid entrapping any air. This is done by filling the cell until the meniscus rises just above the rim, then carefully applying the Mylar window. With upright, inclined specimen planes, an air bubble is permissible if the specimen surface is masked in such a way that the bubble is hidden.

Care is taken when liquid specimens are used in vacuum spectrometers to avoid rupture of the thin cell windows. Special cells may be used (Section 17.1.4.4) or nylon grids may be used on the specimen mask.

Infinite thickness is much greater for liquids than for solids and varies greatly with the wavelength of the measured line, and with the nature and concentration of the analyte and matrix. Consequently, it is desirable that the depth of the liquid cells be infinite for the shortest wavelength to be measured. Otherwise, the specimen depth must be constant for all specimens. Lambert used an immiscible, heavy organic liquid, such as carbon tetrachloride, to raise the specimen level in cells for upright horizontal geometry.

For inverted geometry, it may be prudent to provide a protective cover for the x-ray tube window in case of rupture of a cell window. Campbell states that Mylar covers are too short-lived because of embrittlement by the extremely high intensities at the x-ray tube window; he recommends a thin, removable beryllium cover.

17.2. SUPPORTED SPECIMENS

17.2.1. General

The term *supported specimen* as used here indicates a specimen in which a relatively small amount of analytical material—less than enough to provide an infinitely thick layer—is supported on some type of substrate. Usually, the analytical material is derived from a solution or some other material by a preconcentration technique, and supported specimens are the most commonly used forms in trace and micro analysis and in energy-dispersive analysis. Davis *et al.* (*D8*), Huberman *et al.* (*H68*), and Mizuike (*M47*) have reviewed these preconcentration techniques.

Some materials inherently constitute supported specimens in that small amounts of analytes are already distributed on or in a substrate, preferably a radiolucent one. In this class are coatings, impregnants, fillers, sizings, pigments, dyes, inks, fire retardants, trace impurities, etc. on or in papers, fabrics, thin plastic or rubber sheet, etc. Interesting examples are the distinguishing of genuine and counterfeit U. S. currency by differences in the x-ray spectra of their ink and paper [pp. 309–12 in (*27*), p. 3.16 in (*39*)] and determination of silver, chlorine, bromine, and iodine in photographic emulsions on paper and film bases (Section 14.7.3). Perhaps platings, paints, corrosion layers, thin films, etc. may also be regarded in this way (Chapter 18).

Other specimens fall into the supported category only as a result of the preparation technique. Such specimens are considered in Sections 17.2.2–17.2.6.

Preparation and presentation criteria for supported specimens are as follows (*G9*): (1) Critical thicknesses for the analyte lines are not exceeded. (2) Absorption-enhancement effects for the analyte lines remain negligible or small. (3) Intensities do not exceed the capabilities of the detection–readout system; this is especially important in energy-dispersive spectrometers having semiconductor detectors. (4) Intensities are sufficiently high so that acceptable statistical precision is obtained in reasonable counting times. (5) Finally, specimen geometry may not vary significantly among samples and standards.

The ideal specimen support should be: (1) as thin as possible to minimize x-ray absorption and scatter; (2) free of all elements of atomic number $\lesssim 9$ (fluorine), even as traces; (3) retentive to particulate matter, such as particles filtered from liquids or gases, and residues from evaporated solutions; (4) wettable by liquids, if the specimen is to be applied in liquid form; (5) weighable accurately; and (6) reasonably strong.

Birks [Reference (*53*) for 1974] gives the following equation for line-to-background ratio of specimens supported on thin substrates:

$$I_L/I_B \propto (C\tau\omega)_i/[\sum (C\sigma)_j + (C\sigma)_s] \qquad (17.1)$$

where C is concentration; ω is fluorescent yield; τ is linear photoelectric absorption coefficient; σ is linear scatter coefficient; i is the analyte; j refers to individual matrix elements, including i; and s refers to the substrate.

Commonly used support materials include the following: (1) filter paper, (2) black filter paper (for better visibility of small white precipitates), (3) chromatographic paper, (4) Fiberglas or other glass-fiber filters, (5) Millipore and Nuclepore filters, (6) thin Mylar and polyethylene film, (7) ultrathin collodion or Formvar film, (8) microscope-slide cover glasses, (9) ion-exchange resin granules, (10) ion-exchange resin membrane, (11) ion-exchange resin-impregnated filter paper, (12) abrasive paper and cloth, (13) Scotch tape, and (14) glass, quartz, and plastic capillaries and fibers. Filter-paper, glass-fiber, Millipore and Nuclepore filters, and nylon-reinforced Nuclepore filters are available in disks 2.1 cm in diameter, and disks of any diameter can be cut from the sheet stock. Confined spot-test papers consisting of small filter-paper squares having impregnated wax rings ~1 cm in inside diameter are available in several thicknesses. Bioassay disks of filter paper ~0.5 mm (0.020 inch) thick and 12½ or 6 mm (½ or ¼ inch) in diameter are available. Certain chromatographic papers have the advantage that they absorb a greater volume of liquid per unit weight and do not warp when dry.

The x-ray scattering power of a substrate, and therefore background and detection limit, decrease linearly with area density (mass thickness). The common substrates follow, in order of decreasing area density in milligrams per square centimeter: filter paper (~10), Fiberglas (~10), Millipore (~5), Nuclepore (~1), 6-μm (0.00025-inch) Mylar (~1), 3-μm (0.00012-inch) Mylar (~0.5), 1-μm polypropylene (~0.2), and 1000-Å collodion or Formvar (~0.02) (Section 6.2.1.2). With regard to absence of trace impurities, Dams *et al.* (*D1*) evaluated 10 commercial filter materials and found Whatman No. 41 filter paper to be best and Millipore filters also very good.

A very thin, low-background adhesive support is made by dissolving the adhesive from Scotch tape in acetone or toluene (~10 ml of solvent for each 10 cm^2 of tape—10 cm of $\frac{1}{2}$-inch tape), applying the solution to 6- or 3-μm Mylar film, and allowing it to dry.

Micro Buchner funnels and suction flasks are useful in collecting specimens on small filter disks.

The nature of supported specimens leads to several problems peculiar to this specimen form. Such specimens are not as uniform or reproducible as other specimen forms. Consequently, it is usually necessary to provide internal standardization and/or, when feasible, to rotate the specimen in its own plane during measurement. When the specimen is supported on an x-ray-transparent support, such as filter paper or Mylar film, the side containing the material, or the side to which it was pipeted or on which it was collected, should face the primary x-ray beam. Supported specimens tend to be loose and fluffy, and loss of material can occur. Specimens supported on thin Mylar or filter disks may be sprayed lightly with clear Krylon or 10% wt/vol collodion in acetone, or covered with or sandwiched between 6- or 3-μm Mylar films. Such sandwiched specimens may be stretched across the end of a plastic cylinder or the rim of a plastic cup and secured with an *O*-ring. Powders have also been suspended in solutions of 1% glue in methanol (CH_3OH) or 3% gelatin in water, then filtered and collected on filter-paper disks (*B9*). The glue or gelatin secures the powder to the paper.

Specimens having substrates of thin filter paper, Mylar, etc. must be supported in a way that presents them to the x-ray spectrometer as plane surfaces. A simple way to do this is to back them with Lucite or other rigid plates. However, most thin substrates are highly transparent to the primary x-ray beam, and, if such backings are used, much primary radiation will be scattered back up through the specimen, thereby increasing the background. This is especially true with 6- or 3-μm Mylar, which, by itself, gives very low scattered background. Consequently, such substrates are preferably mounted by stretching them on open, rigid frames, or over the ends of plastic, glass, or metal cylindrical tubes (~3 cm diameter, 0.6–1 cm high), or over the open ends of (empty) liquid-specimen cups. If the specimens are supported over such tubes or cups, background, and therefore detection limit, can be reduced by taking care to prevent primary-beam irradiation of the inside walls of these supports and thereby to avoid scatter from the walls. This may be done by use of primary beam apertures, large-diameter support cells, and/or support cells having truncated conic form with increasing diameter away from the x-ray source (*B58*). A primary x-ray beam trap may be placed to intercept the x-rays transmitted by the radiolucent specimen. The trap may have the form of a short collimator having thin-walled tubes or foils (Figure 5.2) and resting on a high-Z base, such as lead. Only primary x-rays directed parallel to the collimator reach the base, which has low scattering power. Of the x-rays scattered and emitted by the base, only those directed back through the collimator channels avoid striking its tubes or blades, and this directly backward radiation does not enter the spectrometer. Foil-type ("Soller") traps must be oriented with the foils perpendicular to the direction toward the spectrometer.

17.2.2. Specimens Derived from Liquids and Gases

Small volumes of solutions may be pipeted on filter-paper disks, glass-fiber filters, Millipore or Nuclepore filters, or nylon-reinforced Millipore filters and evaporated to dryness (*L54*, *M6*, *M16*). Glass-fiber filters contain relatively high concentrations of heavy elements and should be avoided. The dried disks constitute the specimens. Natelson and his co-workers (*N2*, *N3*) have developed highly refined filter-paper techniques for analysis of biological fluids, and the techniques are readily applicable to other solution samples: (1) Filter paper is treated with hydrochloric acid to remove trace impurities, and is imprinted with wax rings by use of cork borers dipped in melted wax or ballpoint pens filled with a hexanone solution of wax (*N6*, *N9*). Other workers have imprinted wax rings by means of heated cylindrical stamps. If the analytical volume as well as area is to be uniform for all specimens, it is necessary that the wax penetrate to the bottom of the paper to preclude diffusion of liquid under the ring. Samples are micropipeted inside the rings. Similar confined-spot papers are available commercially (*J21*). (2) It is well known that when drops of solution are placed on filter paper, each drop tends to dissolve the residue from the preceding drops and diffuse outward, forming concentric rings of residues of various solubilities around a relatively residue-free center. More uniform deposits are obtained by pipeting the sample solutions onto the filter paper in an apparatus combining a ring oven and aspirator (*N4*, *N8*). (3) The analyses are automated by use of rolls of filter-paper ribbon imprinted with wax rings and driven mechanically through the spectrometer sample compartment by a mechanical accessory (*N5*). (4) Even greater speed is realized by incorporation of a modified 35-mm photographic slide changer into the sample compartment (*N7*). Individual filter-paper squares are placed in the slide changer and indexed automatically or manually through the sample compartment. It seems likely that much lower background would be realized by this technique by stretching 3-μm (0.00012-inch) Mylar film on cardboard 35-mm slide frames and applying the specimens on these supports. In both automated techniques, sample concentration is derived from peak heights on ratemeter–recorder charts.

Possibly, very small volumes of solutions could be absorbed in a textile fiber, or evaporated on a quartz fiber, or taken up in a plastic, quartz, or glass capillary and used as a line source in a curved-crystal spectrometer. In a similar technique, metallic analytes are electrodeposited on the tip or along the length of a platinum or other metal wire or carbonized quartz fiber and used in a curved-crystal spectrometer. The solution may be subjected to paper-chromatographic techniques to separate the analytes (Section 14.8.4). The paper chromatogram constitutes the specimen.

Small volumes of solutions may be placed in small dishes or planchets and evaporated to dryness (*M15*). The dishes or planchets constitute the specimens. If this is done dropwise or by pipeting, the residue is likely to be unevenly distributed, with the smallest accumulation at the center, much like what happens with filter-paper disks (see above). This can be avoided by spraying the liquid into the planchet, or by using planchets having slightly concave, rather than flat, bottoms. Birks (*B54*) evaporated 1-ml portions of solutions drop by drop into small depressions in electrically heated aluminum-foil strips. Small volumes of solutions may be sprayed or distributed on microscope-slide cover glasses or thin Mylar (*G37*, *H8*, *M5*) or even ultrathin collodion or Formvar films (*M33*). The plastic films have very low background compared with filter paper. Desicote or other wetting agents effect more uniform distribution on metal, glass, and plastic. If the solvent is organic, polystyrene or some other plastic binder may be added to the solution to leave a protective binder and coating on the dry residue (*K33*).

Oils may be spread in thin films on metal substrates (*S17*).

The analytes may be precipitated, with or without gathering agent, and collected on small filter disks in a micro Buchner funnel. The dried disks constitute the specimens. Alternatively, the powders may be transferred to small plastic envelopes or distributed on Scotch tape, or used in the micro borax-glass disk fusion technique (Section 16.4.3). The technique of homogeneous precipitation may be used: Chan (*C37*) allowed homogeneously precipitated selenium to settle for 1–2 days onto filter-paper disks secured to the bottoms of Pyrex reaction tubes 18 cm high. The disks were dried and mounted between protective 6-μm (0.00025-inch) Mylar films.

Organic reagents such as 8-quinolinol, cupferron (ammonium nitrosophenylhydroxylamine), dithizone (diphenylthiocarbazone), and tannin may be used as general, group, or specific precipitants (*D4, D8, H47, H49, H68, L50, R35, T13*). Individual analytes may be precipitated with specific reagents. Several organic or inorganic reagents may be added to the same solution, each under appropriate conditions, to precipitate various analytes. The precipitates may be collected separately or all together. Complexing agents, such as EDTA (ethylenediaminetetraacetic acid), may be used to prevent precipitation of selected analyte or matrix constituents. Indirect or association analysis may be effected by stoichiometric precipitation of a low-Z element or organic compound with an easily determined heavier element.

Extremely dilute solutions may be reduced in volume and used as a liquid, or the analyte(s) may be recovered on a support (*P16*). Probably the best technique is to take up the analyte(s) by prolonged agitation with a piece of ion-exchange membrane (*G33*). Suspended solids in liquids are collected on filter disks (*H53*).

Gas-borne dusts may also be collected by filtration or by impingement, that is, by drawing the gas through or directing it on, filter-paper, Millipore, or other filter disks (*C44, H53, S32*). If the filter is too porous, the particles may imbed deep in the filter and long-wavelength lines may be severely attenuated.

Luke (*L50, M45*) has highly refined the filtration technique in his "coprex" (*copre*cipitation-*x*-ray) method for trace analysis: ~0.5-g solid samples are dissolved in several milliliters of water, acid, or other solvent; organic or other matrix material and interfering elements are separated as required; 50–200 μg of a suitable element is added as a coprecipitant; and the trace analytes and coprecipitant are precipitated with a suitable reagent. The reagent may be specific for one analyte or applicable to a group of analytes. The precipitate is collected on a filter-paper, Millipore, or other filter disk in the special filter apparatus shown in Figure 17.15. Trace analytes in liquids and solutions are treated the same way. Solid material suspended in liquids may be filtered without treatment, or perhaps with a gathering agent. There is no reason why gases cannot be drawn through a similar

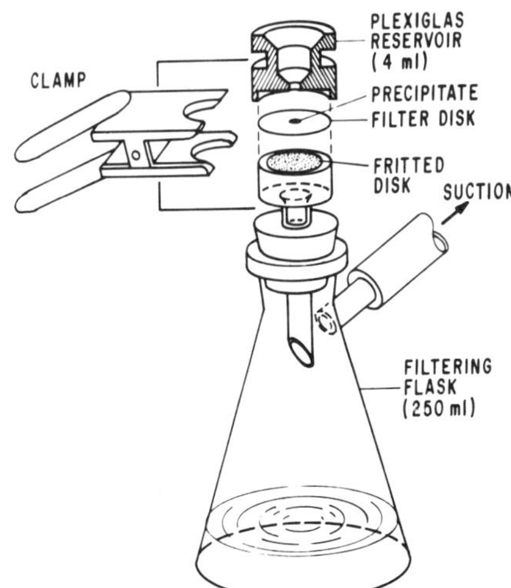

FIG. 17.15. Coprex filtration apparatus (*M45*). The diameter of the precipitate spot may be 1 mm to 2 cm and is determined by the diameter of the opening in the bottom of the reservoir.

arrangement to collect suspended solid material. The filter disk constitutes the specimen. Calibration curves are established from standard disks prepared the same way from known amounts of analyte. Reagent-blank disks for background measurement are prepared by filtration of solutions processed the same way but without analyte.

The opening in the bottom of the reservoir in the filter apparatus (Figure 17.15) determines the diameter of the residue spot on the filter and may be 2 cm to 1 mm. The larger areas may be measured in conventional flat-crystal spectrometers, the smallest in focusing, curved-crystal spectrometers (Figure 19.9). Any of the sizes is suitable for energy-dispersive instruments.

The great success of the method is attributable to three features: (1) elimination of matrix and possible interfering elements by preliminary separation; (2) substantially quantitative recovery of trace analytes by use of a coprecipitant; and (3) substantial elimination of absorption-enhancement effects by dispersion of the precipitate in a uniform, thin layer on a low-Z substrate.

Sensitivities of 10 ng to 1 μg are realized with large-area specimens in flat-crystal spectrometers. Sensitivities in the picogram range are attained with "microdot" specimens in full-focusing spectrometers. The calibration curves are linear in the region 0–100 ng or more.

Once separated, the submicrogram amounts of analyte can be taken up in 3-mm ($\frac{1}{8}$-inch) disks of ion-exchange membrane and measured in a curved-crystal spectrometer (Section 17.2.4).

Vassos and his co-workers have developed methods for electrodeposition of analytes on pyrolytic graphite. One method is for preparation of specimens supported on thin pyrolytic graphite disks (*V6*). Cylinders 1 cm in diameter and 1 cm high are cut from pyrolytic graphite plate stock 1 cm thick. The crystallographic *c* axis is normal to the surface of the stock so that the *c* axis coincides with and the cleavage plane is normal to the axes of the cut cylinders. The cylinder sides are fitted with heat-shrunk plastic tubes to render them water-tight and prevent seepage of electrolyte along the cylinder sides. The analytes are then electroplated on one end of the cylinder, and the plated end is dried and sprayed with clear, colorless Krylon, then cleaved off with a razor blade at a thickness of 0.2–0.5 mm. The disks are durable and easy to store, and because of their thinness and high purity, give minimal background and spectral interference. Another paper describes the electrolytic preparation of standards on pyrolytic graphite disks (*V9*).

The method of *double deposition* (*V7*) permits preconcentration of dissolved constituents in extremely dilute solutions in a manner somewhat analogous to ion exchange. A large volume of the solution is passed through an electrolytic cell having the form of a column packed with granular pyrolytic graphite. The cationic solutes plate out on the graphite. Then a small volume of a stripping electrolyte is passed through the column to "elute" —that is, deplate—the stored electroplated elements. This solution can then be analyzed as a liquid specimen or the solute can be recovered and presented some other way.

17.2.3. Specimens Derived from Solids

Surface deposits can be analyzed without removal from the substrate if the substrate does not contain elements present in the deposit. Otherwise, the deposit can be removed in any of several ways. It may be gently abraded off with a disk of fine abrasive paper, stripped off with Scotch tape, or wiped off with a filter-paper disk moistened with a suitable reagent. Stainless steel in the form of bulk metal, chips, wire, filings, etc. can be rubbed on or between thin disks of aluminum oxide, silicon carbide, or other low-Z abrasive so as to deposit thin layers. Uniformity of the layers can be improved by rubbing two such sample-coated disks together. Background intensity is measured from a clean disk (*F18*). Another stripping technique is to coat the surface with a solution of collodion in amyl acetate, allow the plastic to dry, then strip it off with the adherent deposit, perhaps by use of Scotch tape. Some coatings may be removed by scoring them with a pattern of crossed lines ~5 mm apart, then immersing the specimen in a reagent that dissolves the substrate; squares of the coating float off. Finally, a drop of reagent can be spread on the surface to dissolve the deposit, then absorbed in a small filter-paper disk or even taken up in a micropipet and delivered to some other substrate.

Segregations can be analyzed without removal from the specimen surface by selected-area techniques. Segregations may be dislodged under a microscope by prying with a stainless steel or platinum needle, or by drilling with a small standard drill or with a microdrill. A standard number 80 drill is 0.0134 inch (0.34 mm) in diameter. Microdrills 0.005–0.001 inch (0.125–0.025 mm) in diameter are available. This technique has also been used to remove micro samples from priceless antiquities and other art objects (*S2*). Alternatively, selective etching techniques may be used to etch the matrix away and leave the segregations either loose or in relief. The protruding segregations may be dislodged with an ultrasonic vibrator. The loose segregations may be picked up on a glass, quartz, or plastic fiber coated with vaseline and used in a curved-crystal spectrometer. The loose particles can also be affixed to thin Mylar film coated lightly with Scotch tape adhesive (Section 17.2.1). Alternatively, a "micro vacuum cleaner" may be made by placing a hypodermic needle in the end of a rubber tube and applying gentle suction. A small piece of filter paper or glass-fiber paper is held at the end of the needle, and the particles are drawn into the paper. Finally, the extraction-replica technique may be used (*B74*). The segregations are etched into relief as before, but the surface is then coated with collodion or Formvar to imbed the protruding particles. The plastic is then backed with Scotch tape and stripped off with the particles.

The applicability of the laser to x-ray spectrometric specimen preparation has been evaluated (*B3*). The Jarrell-Ash "Laser-Probe," designed for use in optical emission spectrography, was used. The instrument consists of a laser, microscope, and specimen stage. The specimen, on its stage, is placed with the point to be sampled in the microscope cross hairs, which locate the point of incidence of the laser beam. A square of 6-μm (0.00025-inch) Mylar is sandwiched between two rigid cardboard or plastic squares ~0.5 mm thick and having circular openings ~1 cm in diameter. This assembly is placed directly on the specimen surface on the microscope stage, so the Mylar collecting film is just 0.5 mm off the specimen. The laser beam is directed through the Mylar onto the specimen. The Mylar, being completely transparent to the laser light, is not affected, but the specimen is vaporized at the point of incidence, and the vapors condense as a thin film on the Mylar. The sandwich assembly constitutes the x-ray spectrometer specimen. The laser crater is typically 50–100 μm in diameter and 10–20 μm deep. The technique makes selected-area analysis feasible without special accessories for the spectrometer.

Many of the above techniques are applied for reasons other than for removal of surface layers and segregations. For example, they may be used to take samples from surfaces from which it is not possible or permissible to take samples by cutting, drilling, chipping, scraping, etc. They may also be used to obtain samples having minimal absorption-enhancement effects.

17.2.4. Ion-Exchange Techniques

17.2.4.1. Principles

Ion exchange may be defined as a reversible exchange of ions between a resin and a liquid without substantial change in the structure of the resin. The *ion-exchange resin* consists of a three-dimensional hydrocarbon polymer to which are attached a very large number of ionizable—and therefore exchangeable—chemical groups. Physically, the resin may have the form of granules, liquid, membrane, or sheet, or impregnated filter paper. Chemically, the resin may have cation-exchange (acidic) or anion-exchange (basic) form, and the ions to be absorbed must be in the same form as the resin.

Prior to use, cation-exchange resins are usually saturated with H^+ or Na^+—that is, all the ionizable sites are occupied by H^+ or Na^+. Similarly, anion-exchange resins are usually saturated with OH^- or Cl^-. The ion-exchange reaction may be represented by simple equations of the following forms, where R represents the hydrocarbon polymer matrix:

Cation exchange:

$$(RH) + Ag^+ \rightleftarrows (RAg) + H^+$$

$$(RNa) + Fe^{3+} \rightleftarrows (RFe) + 3Na^+$$

Anion exchange:

$$(ROH) + Br^- \rightleftarrows (RBr) + OH^-$$

$$(RCl) + SO_4^{2-} \rightleftarrows (RSO_4) + 2Cl^-$$

The *affinity* or *retentivity* of the resin varies widely for different ions and is affected by many factors, such as the chemical nature of the resin and solution, pH, concentration, and temperature. However, in general, the affinity of a resin for ions increases as the ionic radius of the hydrated ion increases, that is, as its valence and atomic number increase. However, this "rule" is by no means rigorously applicable, as is evident from the following *typical* decreasing retentivity series for cation and anion exchange:

$$La^{3+} > Nd^{3+} > Eu^{3+} > Y^{3+} > Fe^{3+} > Al^{3+} > Ba^{2+} \sim Ag^{+} > Pb^{2+}$$
$$> Hg^{2+} > Sr^{2+} > Ca^{2+} > Cu^{+} > Ni^{2+} > Cd^{2+} > Cu^{2+} \sim Zn^{2+}$$
$$> Fe^{2+} \sim Mg^{2+} > Cs^{+} > Rb^{+} > K^{+} > Na^{+} > H^{+} > Li^{+}$$
$$I^{-} > HSO_4^{-} > Br^{-} > Cl^{-} > F^{-} \sim OH^{-}$$

17.2.4.2. Techniques

Analytes in solution can be taken up in ion-exchange resin by any of three processes (*V1, V2*):

1. In the *batch technique*, the solution is allowed to stand or is shaken with a portion of ion-exchange granules or liquid, or with a square or circle of membrane or resin-impregnated paper. The technique of allowing a piece of membrane or paper to stand in contact with a solution has been termed "danglation" (*G30*). The granules are separated and washed by filtration, then placed in a powder cell wet or dry, or briqueted (*C55*). The liquid resin is separated and washed by decantation and used in a liquid cell (*V1*). The membrane or paper is scooped out, rinsed, dried, and mounted as is (*C10, G33, Z10*).
2. In the *filtration technique*, the solution is filtered several times through an ion-exchange filter paper on a Buchner or other filter funnel in a suction flask until quantitative recovery of the analytes is realized (*C23*).
3. In the *column technique* (*V1, V2*), the solution is passed through a column of ion-exchange granules packed in a glass apparatus resembling a short buret. The analytes are retained on the column and removed collectively, or individually in order of increasing retentivity by elution with a solution of complexing agent. The eluate constitutes a liquid specimen.

In the batch and filter techniques, it is wise to establish the agitation time or number of filtrations required to effect complete equilibrium or recovery prior to undertaking the actual analysis. This is done by plotting analyte-line intensity *versus* time or number of filtrations. The point is noted beyond which no further increase in intensity occurs. Equilibrium is established more rapidly with liquid resins than with granules, but sensitivity is lower with liquids. Equilibrium with granules is expedited by presoaking. Care must be taken not to exceed the capacity of the resin, which is usually stated in milliequivalents per gram. In the filtration technique, completeness of recovery can be tested by passing the filtrate through a second, fresh ion-exchange filter disk, which is then measured on the x-ray spectrometer. For use with very dilute analyte solutions, it may be necessary to purify the ion-exchange resin by repeated washing or elution with pure eluant.

Ion-exchange filters may have several forms:

1. Filter paper chemically treated to impart ion-exchange properties.
2. Filter paper impregnated with cation- or anion-exchange resin having low selectivity, that is, useful for a wide range of cations or anions (*L12*). The weight ratio of paper to resin is usually about one to one and the capacity ~1.5–2 meq/g. For a 3.5-cm disk, this comes to ~0.2 meq—~4 mg of Fe^{3+}, for example.
3. Filter paper impregnated with cation- or anion-exchange resin having high specificity for a certain element or group of elements. Such resins are termed *selective* or *chelating* resins (*B71, L25*). For example, Dowex A-1 resin has relative retentivities for Hg^{2+}, Cu^{2+}, Ni^{2+}, Zn^{2+}, and Mg^{2+} of about 1060, 126, 4.4, 1, and 0.01, respectively. The nonspecific ion-exchange papers can be made selective by appropriate regulation of *p*H and by use of chemical masking and complexing reagents. Filter paper impregnated with chemical reagents specific for one or more elements, such as lead acetate paper for sulfides, performs the same function in a different way.

Campbell and Thatcher (*C10, C20*) and Green *et al.* (*G30*) discuss in detail the use of ion-exchange membranes and filter papers, comparing filtration and immersion and mechanical and ultrasonic agitation, describing filtering apparatus and specimen holders, and giving mass-absorption coefficients.

Internal standards may be added to ion-exchange resins in three ways: An internal standard may be added to the sample solution prior to treatment with the resin (*C55*); this technique is not recommended, because the internal standard may not coabsorb reproducibly under conditions optimal for the analyte. A solution of internal standard in exchangeable form may be added volumetrically to the dry or moist analyte-loaded resin in the specimen cell. Probably the best way is to quantitatively add to the dry analyte-loaded resin in the cell a solution of internal standard in *non-*exchangeable form. The solution fills the voids between the granules. The specimen cell is then covered with Mylar and measured in the wet form.

Ion-exchange techniques have been applied to trace and micro analysis. Grubb and Zemany (*G33*) determined 1 μg of cobalt in 1 liter of water by collecting it on 1 cm^2 of ion-exchange membrane stirred in the water for ~48 h. Zemany *et al.* (*Z10*) determined microgram quantities of potassium by a similar technique. Campbell *et al.* (*C23*) determined 11 trace elements in tungsten by dissolution of the samples in tartrate medium, passage of the solutions through an ion-exchange column to retain the trace analytes, elution of the analytes, and "filtration" of the eluate through disks of ion-exchange paper. Seven passes through the disks gave >99% retention of all trace analytes. Luke (*L49*) determined analytes in quantities as small as 0.01 μg by isolating them chemically, taking them up in 1 ml of acid or water, then collecting them in individual disks of ion-exchange membrane 3 mm ($\frac{1}{8}$ inch) in diameter. The disks were measured in a curved-crystal spectrometer.

17.2.5. Ashing Techniques

Metallic analytes in organic and biological materials in concentration range 0.01–10 ppm may be preconcentrated by ashing. One of four essentially distinct processes may be used.

a. Dry Ashing or Combustion. This process is applicable only to combustible samples, especially hydrocarbons. The hydrocarbon is actually ignited and burns to a carbonaceous mass, which may be collected for analysis or ignited further on a burner or in a muffle furnace to a clean ash. This method is the most prone of the four to loss of volatile analytes. However, the soot may be collected on a cold surface and analyzed by itself or combined with the carbonaceous residue.

b. Wet Ashing (*D9, L59*). The sample is heated with a mixture of a strong dehydrating acid and a strong oxidizing acid, usually concentrated sulfuric and nitric acids, respectively. In some applications, concentrated perchloric acid or fuming nitric acid may perform both functions. It may be necessary to heat the sample repeatedly with successive portions of acid until, finally, a clean ash remains after all the acid has been fumed off. This method results in no loss of analytes if carefully conducted, but is very time-consuming.

c. Dehydration–Ignition. The sample is reduced to a carbonaceous residue by heating with a dehydrating acid alone—usually sulfuric. The carbon residue is then ignited on a burner or in a muffle furnace to burn off the carbon and form a clean ash. This method also results in no loss of analytes and is much more rapid than wet ashing.

d. Wet Ashing and Combustion (*R30*, *S34*). This process, like dry ashing, is also applicable only to combustible samples. It has been shown that when hydrocarbons containing trace metallic impurities are heated with benzenesulfonic acid ($C_6H_5SO_3H$) or a homologous compound—such as xylenesulfonic acid [$(CH_3)_2C_6H_3SO_3H$]—the metal–organic complexes are destroyed, leaving the metallic elements in a nonvolatile condition. It is then feasible to burn the sample without loss of analytes. The procedure consists in heating the sample with benzenesulfonic acid, then burning it to a carbonaceous residue, as in dry ashing. The residue may be collected for analysis or ignited on a burner or in a muffle furnace to a clean ash.

Prior to ashing, the sample may be treated with additives. If the ash is to be collected and analyzed as is—rather than after dissolution—a gathering agent may be added to impart some bulk and facilitate the collection. Magnesium or aluminum, added as naphthenate, salt of an organic acid, or other organic compound, is usually used because it provides a light matrix. Because of the generally nonreproducible nature of the applicable specimen-presentation forms (see below), an internal standard or internal intensity-reference standard is usually added, again a suitable metal as an organic compound.

The ash may be presented in several ways. The dry ash may be transferred to a Mylar film (*R30*) or to a paper, glass-fiber, or Millipore filter disk, or it may be washed with water or a hydrocarbon liquid onto a filter disk in a Buchner funnel in a suction flask (*S34*). The Mylar or filter disk may be used as is, or sprayed with Krylon, or sandwiched between Mylar films. Alternatively, the ash may be dissolved. The solution may be absorbed in a filter disk and dried, or evaporated on an aluminum plate or microscope-slide cover glass (*D9*), or evaporated to dryness and used as a powder. The analytes may also be taken up in ion-exchange resin or membrane, or precipitated individually or collectively and filtered (*L59*).

17.2.6. Bomb Techniques

Two other ways to destroy organic matrixes prior to x-ray spectrometric analysis are the following.

In the *Parr peroxide bomb* technique, the sample is mixed with sodium peroxide (Na_2O_2) in a crucible, which is then placed in a sturdy high-pressure vessel ("bomb"). The oxidation reaction is initiated electrically or by external localized application of a small, hot flame. A little potassium chlorate ($KClO_3$) or some other oxidant may be added to the charge to initiate the reaction at a lower temperature.

Organic (hydrocarbon) matter reacts with the sodium peroxide to produce carbon dioxide (CO_2), steam (H_2O), and sodium oxide (Na_2O), which react further to produce sodium carbonate (Na_2CO_3) and sodium hydroxide (NaOH), thereby consuming the gaseous products. Carbon, silicon, phosphorus, sulfur, arsenic, and the halogens (fluorine, chlorine, bromine, and iodine) react to form water-soluble sodium salts. Most metallic elements form water-insoluble oxides. On completion of the reaction, the soluble and insoluble compounds are recovered by dissolution in water and filtration, respectively.

In the *oxygen bomb* technique, the sample is mixed with an inert refractory powder, such as aluminum oxide (Al_2O_3), to act as a retaining matrix. The charge, as loose powder or briquet, is placed in a crucible in a high-pressure vessel, and a stream of oxygen and inert carrier gas is passed through. The combustion is initiated, and the sample material burns. Carbon dioxide and water vapor, as well as gaseous products of analytical interest, such as sulfur dioxide (SO_2), are swept out by the carrier gas and may be recovered by passage through suitable absorbers external to the bomb. Metallic oxides remain in the matrix.

Semimicro and micro Parr and oxygen bombs are also available.

17.3. TRACE AND MICROANALYSIS

Trace analysis is the detection, estimation, or determination of analytes having very low concentrations ($\leq$0.1 wt %)—trace constituents— in relatively large samples. *Microanalysis* is the detection, estimation, or determination of analytes having relatively high concentrations—major and perhaps minor constituents—in very small samples. Detection, estimation, and determination constitute, respectively, qualitative, semiquantitative, and quantitative analysis. In favorable cases, x-ray fluorescence spectrometry, applied on standard, commercial, flat-crystal instruments, can detect concentrations down to ~0.0001 wt % in solids and ~1 μg/ml in solutions, isolated masses down to ~1 ng, and films as thin as ~0.01 monolayer. Trace and microanalyses approach these levels and constitute four of the categories of applications listed in Section 16.1.2 and Table 16.1: (1) very low concentrations, with or without preconcentration; (2) limited total amount; (3) analysis of individual phases in inhomogeneous materials, either after physical separation from the matrix or in place by selected-area techniques (Chapter 19); and (4) thin platings, coatings, and films. Preconcentration of trace analytes and physical separation of an individual grain from a heterogeneous specimen usually result in a specimen in the limited-total-amount class.

In x-ray spectrometric trace and micro analysis, all the various sources of error assume increased significance, and errors that may be disregarded in the determination of major and even minor constituents may limit the precision in the determination of traces. Details and refinements of technique unnecessary at higher concentrations must be applied for traces. Thus, x-ray spectrometric trace and micro analyses involve no basically special methods, but rather a refinement of methods used for higher concentrations or larger specimens. This section is intended as a guide to these refinements.

Trace and micro analyses are characterized by low analyte-line intensity, that is, analyte peak intensity approaching background intensity. For 95% (2σ) confidence level, the lower limit of detectability is given by Equation (13.4):

$$C_{\mathrm{DL}} = (3/m)(I_B/T_B)^{1/2}$$

where m is the slope of the calibration curve (I_A *versus* C_A); I_B is background intensity; and T_B is background counting time, which, in this case, is of the same order as analyte-line counting time. Consequently, in trace and micro analysis, background intensity must be measured with the same statistical precision as analyte-peak intensity and the method for measuring background must be chosen carefully (Section 7.3.3). The lower limit of determination for quantitative analysis is defined as three times the lower limit of detectability (Section 13.1.1). It is evident from the equation above that there are three ways to reduce the lower limit of detection and determination: increase analyte-line intensity and thereby m, reduce background intensity, and increase counting time.

The third alternative—increased counting time—is practical only to the point where instrumental instability during the long counting time becomes the factor limiting the precision. It follows that instrumental stability must be high. Methods for increasing the analyte-line intensity are given in Section 13.1.2. However, of the methods given there, it is necessary to choose those that do not also substantially increase background. Methods for decreasing background are given in Section 7.3.4. Here, again, it is necessary to choose methods that do not also substantially decrease analyte-line intensity. Methods for increasing analyte-line intensity and decreasing background fall into two groups: those involving the instrument and its operating conditions and those involving specimen preparation and presentation. If meaningful comparisons of analyte-line intensities are to be made from different samples or from samples and standards, operational and specimen errors (Sections 11.3.2.3 and 11.3.2.4) must be reduced to a minimum.

The instrumental methods most effective in increasing line intensity for trace analytes include use of: (1) a constant-potential generator for lines having high excitation potential; (2) an x-ray tube target having strong lines close to the short-wavelength side of the analyte absorption edge; (3) x-ray tube operating potential such that the primary continuum hump lies just on the short-wavelength side of the analyte absorption edge; (4) tapered collimation; (5) an analyzer crystal having high "reflectivity" and low emission; and (6) a detector having high quantum efficiency for the analyte line.

The instrumental methods most effective in reducing background intensity for trace analytes include use of: (1) filters to reduce continuum intensity in the region of the analyte line; (2) an analyte line at relatively high 2θ where background is low; (3) an evacuated or helium-flushed specimen chamber to eliminate air-scatter of the primary beam; (4) baffles, collimators, and/or apertures to confine the primary beam and intercept background; (5) spectrometer geometry such that the crystal axis and specimen plane are not parallel (Section 7.3.4); and (6) pulse-height selection.

In the determination of traces homogeneously distributed in a large specimen, invariably the detection limit is reduced by preconcentration by chemical or physical means (*D8*, *H68*, *M47*). The chemical methods include precipitation (selective or collective), selective dissolution, ion-exchange (on granules, membrane, paper, or liquid), liquid–liquid extraction, paper chromatography, paper electrophoresis, electrodeposition, wet or dry ashing, and gas-formation. The physical methods include distillation, sublimation (including use of laser beams), prying with needles, microdrilling, and extraction replication. The last four techniques are discussed further in Section 17.2.3.

In the determination of homogeneous traces in large specimens without preconcentration, it is often preferable to measure a small area with a pinhole or slit aperture and curved-crystal optics, rather than the large area with flat-crystal optics.

For a trace analyte, emitted spectral-line intensity is directly proportional to concentration, and the calibration curve is linear. The matrix determines only the slope of the curve; the lower the mass-absorption coefficient of the matrix for the analyte line, the steeper is the calibration curve. Trace analytes, unlike those at higher concentrations, can often be determined by use of standards having a matrix entirely different from that of the samples. From Equation (12.6), analyte-line intensities from a sample I_X and standard I_S are related by

$$\frac{I_X}{I_S} = \frac{\overline{(\mu/\varrho)_S}}{\overline{(\mu/\varrho)_X}} \approx \frac{(\mu/\varrho)_S}{(\mu/\varrho)_X} \tag{17.2}$$

where $\overline{(\mu/\varrho)}$ and (μ/ϱ) are the mass-absorption coefficients for, respectively, combined primary and analyte-line x-rays [Equation (2.14)] and analyte-line x-rays alone. From this, trace-analyte concentration in a sample C_X is obtained from a standard having analyte-line concentration C_S in a different matrix from

$$C_X = \frac{I_X\overline{(\mu/\varrho)_X}}{I_S\overline{(\mu/\varrho)_S}}\, C_S \tag{17.3}$$

The limited-total-amount category includes isolated particles and fibers, microquantities of analyte separated from a matrix or recovered from a solution or gas, and individual phases in heterogeneous specimens. Such specimens present two special problems: (1) They must be supported—and in such a way as to give maximum excitation of the analyte and minimum background, and (2) the x-ray optical system must collect as much of the emitted analyte-line radiation as possible.

In supporting microquantities of analyte, scattered background is minimal when the analyte is suspended as nearly as possible in "free space." Although the glass-disk method has been applied successfully to preconcentrated trace analytes (Section 16.4.3), the support should preferably be as thin and of as low atomic number as possible to reduce primary beam scatter. Thin filter paper is preferable to thick filter or chromatographic paper, 6- or 3-μm Mylar is preferable to filter paper, and stretched (1-μm) polypropylene (Section 6.2.1.2) or ultrathin Formvar are even better. Millipore filter disks and ion-exchange membrane or impregnated filter paper are also useful. All such x-ray-transparent supports must be stretched over open frames or across the open ends of short cylindrical tubes or liquid-specimen cups. Otherwise, the purpose of the thin substrates is defeated by primary-beam scatter from underlying material (Section 17.2.1). However, enhancement radiators (Section 4.4.3) may be used behind such supports to excite additional analyte-line radiation.

Many successful methods have been based on large-area supported specimens used with flat-crystal optics, and conditions favorable to efficient collection of analyte-line radiation in flat-crystal spectrometers are given in Sections 13.1.2.3 and 13.1.2.4. However, it is almost invariably preferable to confine the preconcentrated analyte(s) to a small area and use selected-area apertures and curved-crystal optics. The small area may be circular (1–4 mm in diameter), as in the "Coprex" (Section 17.2.2) and ion-exchange disk techniques (Section 17.2.4). Or the analyte may be electrodeposited on a pure metal wire or carbonized quartz fiber, held in a thin-walled plastic capillary, absorbed in a cotton fiber, or affixed to a cotton, plastic, or quartz fiber with Vaseline. Of course, particulate or fibrous micro specimens should be placed in the most intense region of the primary beam (Section 19.2.1). For large and small specimens, the energy-dispersive mode with the detector close to the specimen may be advantageous.

17.4. RADIOACTIVE SPECIMENS

The problems peculiar to radioactive specimens are personal safety and prevention of interference with the analyte-line x-rays by radiation of direct or indirect radioactive origin.

Personal safety involves prevention of: (1) ingestion of radioactive material; (2) irradiation of personnel, at least above permissible dosage (Section 9.5); and (3) radioactive contamination of the laboratory environment, including the x-ray spectrometer. Ingestion is avoided by caution and orderly work practices, use of respirators, and encapsulation of the radioactive specimens. Irradiation is minimized by use of shielding and large working distances and by limitation of work time with the radioactive materials. Contamination is avoided by caution and orderly work practices, use of glove boxes, provision of restricted areas for specimen preparation, and encapsulation of specimens.

The precautions required to preclude interference with x-ray measurements depend on the type (α, β, γ, or x), energy (keV), and activity (Ci) of the radioactivity. In general, γ-rays are by far the most penetrating, α-rays the least. Usually α- and β-rays are adequately confined by encapsulation, but they may generate within the specimen x-rays that can penetrate the encapsulation. In general, γ-rays are not confined by encapsulation and may reach the detector by penetration of the specimen chamber or by passing through the source collimator and scattering from the crystal; γ-rays also generate x-rays in the specimen. Some radioactive isotopes emit x-rays directly as a result of their decay process (Section 1.6.3). X-rays originating directly or indirectly from the radioactivity contribute to the x-ray background and reduce the analyte peak-to-background ratio.

The principal techniques used in x-ray spectrometric analysis of radioactive specimens are the following (*M58*).

1. *Encapsulation.* Preferably, all radioactive specimens are *doubly* encapsulated. Solid specimens can be sealed into plastic envelopes, coated with an adherent plastic layer, or molded in plastic, then sealed in outer plastic envelopes. Liquids can be presented in double cells having Mylar, polystyrene, mica, or glass windows. An inner cell is filled and sealed, then totally enclosed in a larger outer cell or even a thin Mylar or polyethylene envelope. Glass disks are already encapsulated, in effect, by the glass and may be sealed in plastic envelopes. Supported specimens on filter paper or other substrates, or in metal, glass, or plastic planchets, can be sealed in plastic envelopes. Of course, care must be taken that all samples and standards are encapsulated the same way so that the attenuation of the emitted analyte-line and radioactive radiation and scatter of the primary x-rays by the plastic are the same for all specimens.

2. *Shielding* between the specimen chamber and detector may be provided to minimize direct γ-irradiation of the detector.

3. *Pulse-height selection* (*K25*). If a scintillation counter is used and if the γ-rays or α-, β-, or γ-excited x-rays have high energy, some discrimination may occur in the scintillation crystal itself. The crystal has high quantum efficiency for the relatively low-energy analyte-line x-rays, but low quantum efficiency for the more penetrating high-energy γ- or x-rays. The pulse distributions arising from the γ- or x-rays of radioactive origin are likely to be discriminated by the pulse-height selector in the usual way.

4. *"Phoswich" detection and pulse-shape selection* (Section 7.1.4.3). X- and γ-rays can be discriminated by a scintillation counter having a composite phosphor consisting of a thin, front phosphor that absorbs x-rays but transmits high-energy γ-rays and a thick, back phosphor that absorbs the γ-rays. The two groups of detector-output pulses are separated on the basis of their shapes by a pulse-shape discriminator.

5. *Radiation blank.* The intensity at the 2θ angle of the analtye line can be measured from the radioactive specimen in the x-ray spectrometer operating normally, then operating without power applied to the x-ray tube. The latter measurement is the contribution to background by γ-rays and x-rays of radioactive origin, and is subtracted from the intensity measured with x-ray tube power.

Part VII

Unconventional Modes of Operation; Related X-Ray Methods of Analysis

Chapter 18

Measurement of Thickness of Films and Platings

18.1. PRINCIPLES AND BASIC METHODS

Actually, the x-ray spectrometer directly provides only two types of information about films and platings—composition (Section 14.7) and area density (mass thickness). Thickness is derived from these quantities indirectly and only after assumptions are made about the uniformity of the film thickness and about the actual densities of the film and the constituent elements in the film, as distinguished from bulk densities. Section 14.7 discusses x-ray spectrometric determination of film composition, particularly as a means of dealing with absorption-enhancement effects. This chapter discusses determination of film thickness (*25j*, *B42*, *L27*).

Incidentally, x-ray spectrometric thin-film studies may provide other types of information. For example, Spielberg and Abowitz (*S48*) noted that in vacuum evaporation of various metals, under identical conditions, different amounts of metal deposited on beryllium, glass, and quartz substrates. Thus x-ray spectrometry provides a means of measuring "sticking coefficients" of relatively nonvolatile substances.

The following notation is used throughout this chapter (Figure 18.1):

A is the measured analytical secondary x-ray beam.
O is the incident primary x-ray beam.
p refers to the plating or film.
s refers to the substrate or base metal.

I_0	is the intensity of the incident primary beam.
$I_{p,t}$	is the intensity of the plate-element line from a plating of thickness t.
$I_{p,\infty}$	is the intensity of the plate-element line from infinitely thick plate metal.
$I_{p,s}$	is the intensity of the plate-element line from plate element in the substrate after emerging through the overlying plating.
$I_{p,s,0}$	is the intensity of the plate-element line from plate element in the substrate before emerging through the overlying plating.
I_s	is the intensity of the substrate-element line from unplated substrate.
$I_{s,p}$	is the intensity of the substrate-element line after emerging through the overlying plate.
$I_{K\alpha}$, $I_{K\beta}$, $I_{L\alpha}$, $I_{L\beta}$	are the intensities of the indicated specific lines.
t	is thickness of the film or plate.
$\overline{(\mu/\varrho)}$	is defined in Equation (14.49).

The film may be presented to the instrument in any of three ways: (1) after removal from the substrate without regard to preservation of film integrity; (2) after removal from the substrate intact; and (3) nondestructively, that is, in place on the substrate. Unsupported films and films on thin, x-ray-transparent substrates belong by their nature in the second group. The first technique is not considered here. The work of Rhodin (*R14*) is an outstanding example of the second technique. He studied films prepared in three ways: (1) Chromium, iron, nickel, and stainless steel films up to 300 Å thick were vacuum-evaporated on 6-μm (0.00025-inch) Mylar. (2) Oxide films were prepared by evaporation of chromium, iron, nickel, and stainless steel on Formvar and floating the Formvar-backed films on a solution of potassium dichromate in nitric acid for various times to form oxide films of various thicknesses. Unoxidized metal was removed in a bromine–methanol solution in an inert atmosphere, and the Formvar was removed in ethylene chloride. The oxide films were scooped up on Mylar film. (3) Oxide films were removed from bulk metals by chemical stripping in bromine–methanol solution, then scooped up on Mylar. Rhodin obtained sensitivities for chromium, iron, and nickel of 8.5, 16.5, and 27 counts/s per $\mu g/cm^2$, respectively.

The remainder of this section is devoted to consideration of x-ray spectrometric determination of thickness of films and platings in place on the substrate. Such methods may be classified in seven categories as follows:

1. The *substrate-line attenuation method* (*B23, L15, L16, Z8*), shown in Figure 18.1A, consists in measurement of the attenuation of a substrate line on passing through the overlying plating. From Equation (2.4),

$$\log_e(I_s/I_{s,p}) = \overline{(\mu/\varrho)}\varrho t \tag{18.1}$$

where ϱ is film density (g/cm³). A plot of $I_{s,p}$ or of $I_s/I_{s,p}$ *versus* thickness may be prepared from standards having known plate thickness.

The absorption method becomes more sensitive to thickness of very thin layers as the takeoff angle is decreased because thereby the path length

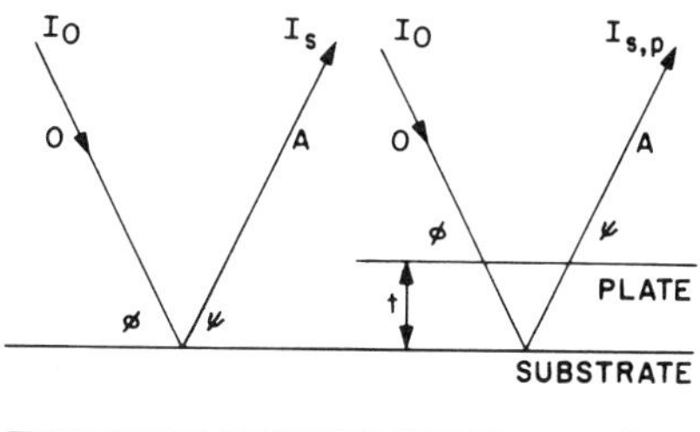

A

FIG. 18.1. Basic methods for x-ray spectrometric determination of film and plating thickness. A. Attenuation of the line of a substrate element absent in the plate. B. Emission of the line of a plate element absent in the substrate. C. Emission of the line of a plate element also present in the substrate.

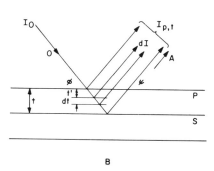

B

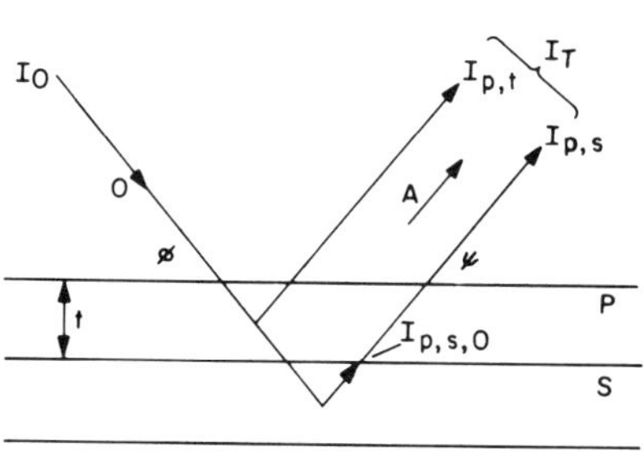

C

in the layer is increased. However, surface texture may introduce serious error at low angles.

2. The *plate-line emission method* (*K24*), for systems in which the plate element is absent from the substrate, is shown in Figure 18.1B, and consists in measurement of the intensity of the plate line. From Equation (14.60),

$$\log_e[1 - (I_{p,t}/I_{p,\infty})] = -\overline{(\mu/\varrho)}\varrho t \tag{18.2}$$

A plot of $I_{p,t}$ or of $1 - (I_{p,t}/I_{p,\infty})$ *versus* thickness may be prepared from standards.

Cameron and Rhodes (*C8*) define the maximum thickness that can be determined by this method in terms of the mass-absorption coefficients of the plating for the primary and analyte-line x-rays [Equation (2.13)] as

$$t_{\max} = \frac{3}{(\mu/\varrho)_{\lambda_{\text{pri}}} + (\mu/\varrho)_{\lambda_A}} = \frac{3}{(\mu/\varrho)'} \tag{18.3}$$

The minimum thickness is determined by the background (Section 13.1.1).

3. The *plate-line emission method* (*K1*), for systems in which the plate element is present in the substrate, is shown in Figure 18.1C. If I_T is the total measured plate-element line intensity,

$$I_T = I_{p,s} + I_{p,t} \tag{18.4}$$

in which

$$I_{p,s} = I_{p,s,0} \exp -\overline{(\mu/\varrho)}\varrho t \tag{18.5}$$

$$I_{p,t} = I_{p,\infty} - I_{p,\infty} \exp -\overline{(\mu/\varrho)}\varrho t \tag{18.6}$$

Then,

$$I_T = I_{p,s,0} \exp[-\overline{(\mu/\varrho)}\varrho t] + I_{p,\infty} - I_{p,\infty} \exp -\overline{(\mu/\varrho)}\varrho t \tag{18.7}$$

$$I_T - I_{p,\infty} = (I_{p,s,0} - I_{p,\infty}) \exp -\overline{(\mu/\varrho)}\varrho t \tag{18.8}$$

$$\log_e \frac{I_T - I_{p,\infty}}{I_{p,s,0} - I_{p,\infty}} = -\overline{(\mu/\varrho)}\varrho t \tag{18.9}$$

A plot of $(I_T - I_{p,\infty})/(I_{p,s,0} - I_{p,\infty})$ *versus* thickness may be prepared from standards.

4. Many ratio methods have been used for determination of plate thickness. Calibration curves or mathematical relationships are established from standards. Any of the following quantities may be plotted as a function of thickness (see paragraph 3 of this section for notation):

$I_{p,t}/I_B$	where *B* refers to the background at or adjacent to the peak or in the continuum hump (*G23*, *K7*).
$I_{p,t}/I_{p,\infty}$	(*H30*).
$(I_{p,\infty} - I_{p,t})/I_{p,\infty}$	(*Z15*, *Z17*).
$I_{p,\infty}/(I_{p,\infty} - I_{p,t})$	(*S21*).
$1 - (I_{p,t}/I_{p,\infty})$	(*H51*).
$(I_{K\alpha}/I_{K\beta})_p$	(for low-*Z* platings) (*Z15*, *Z17*).
$(I_{L\alpha}/I_{L\beta})_p$	(for high-*Z* platings) (*Z15*, *Z17*).
$I_{p,t}/I_{s,p}$	(*B40*).
$I_{s,p}/I_s$	(*Z15*, *Z17*).
$(I_{K\alpha}/I_{K\beta})_s$	(for low-*Z* substrates) (*Z15*, *Z17*).
$(I_{L\alpha}/I_{L\beta})_s$	(for high-*Z* substrates) (*Z15*, *Z17*).
$(I_1/I_2)_s$	where 1 and 2 represent lines of different substrate elements.

The intensities may be peak or net. For a single plating on a substrate, the functions $I_{p,t}$, $I_{s,p}/I_s$, or even $I_{p,t}/I_{s,p}$ are of comparable effectiveness. Multiple platings are considered below. The methods involving $(I_\alpha/I_\beta)_s$ and $(I_1/I_2)_s$ are more effective the greater the difference in the absorption coefficient of the plating for the two wavelengths. For example, the ideal case would be the fortuitous occurrence of an absorption edge of the plate metal between the α and β substrate lines; another favorable case would be two widely separated lines λ_1 and λ_2 from different elements in an alloy substrate. In general, the α–β methods usually give calibration curves having relatively shallow slopes. However, two-wavelength methods have the advantage, compared with one-wavelength methods, that they are less prone to error due to pinholes, cracks, scratches, etc. in the plating. They are also applicable to irregular-shaped plated objects (Section 18.3.1).

5. The *transmission–absorption method* may be used if the substrate is transparent to x-rays. The specimen is placed between the analyzer crystal and the detector, and a monochromatic beam diffracted by the crystal is passed through the specimen. Transmitted intensity is given by

$$I = I_0 \exp[-(\mu/\varrho)\varrho t] \tag{18.10}$$

where μ/ϱ is the mass-absorption coefficient (cm^2/g) of the film for the monochromatic wavelength; ϱ is the density (g/cm^3) of the film; and I_0 is the intensity transmitted by uncoated substrate. The monochromatic wavelength may be derived from the direct primary beam, from primary x-rays scattered from a low-*Z* scatter target in the specimen compartment, or from the secondary-emission spectrum of an appropriate element in the compartment.

6. The *method of variable takeoff angle*, developed by Ebel, permits determination of thickness (area density) of single-element and multielement films without thickness standards. This method is discussed in detail in Section 14.8.2.

7. If the plating is an alloy of elements A, B, C, . . . on substrate *s*, a line of each plate element and the principal substrate element may be measured and the following ratio used: $(\sum I_i)_p/I_s$ or $(I_A + I_B + I_C + \cdots)/I_s$ *versus* plate thickness. This function is substantially linear over a wide thickness range, especially if the substrate radiation does not enhance the line of any major plate constituent. The method is particularly useful for platings of alloys of elements of consecutive atomic number, or nearly so.

The relative effectiveness of the substrate-absorption and plate-emission methods (1 and 2 above) depends largely on the specific system, and it is difficult to draw generalizations. For thick layers, the emission method is limited to somewhat less than the thickness through which the plate line can emerge, the absorption method to somewhat less than the thickness through which the substrate line can emerge. (It is assumed that neither method is limited by penetration of the primary beam.) For very thin layers, if the substrate has a short-wavelength line and the layer low absorption —aluminum-clad molybdenum, for example—the emission method is preferable. If the substrate has a long-wavelength line and the layer high absorption—tin-coated steel, for example—the absorption method is preferable. In general, regardless of which of the seven methods is used, relatively long-wavelength plate or substrate lines are preferable for thin, low-Z layers, and short-wavelength lines are preferable for thick, high-Z layers.

18.2. MULTIPLE-LAYER AND ALLOY PLATINGS

18.2.1. Multiple-Layer Platings

Birks *et al.* (*B62*) determined the thickness of both platings in nickel/copper-plated steel. Nickel thickness was obtained from Ni $K\alpha$ intensity. Copper thickness was then obtained from a set of curves of Fe $K\alpha$ intensity *versus* copper thickness, each curve for a different overlying nickel thickness. Keesaer (*K7*) determined the thickness of chromium and nickel in "triplate" (chromium/nickel/copper-plated steel). Chromium thickness was obtained from Cr $K\alpha$ intensity, nickel thickness from Ni $K\alpha$ intensity corrected for absorption in the overlying chromium plate. Gerold (*G7*) combined x-ray diffraction and spectrometry to determine chromium and nickel in chromium/nickel-plated brass. Chromium was determined by diffractometry, nickel by spectrometry of the Ni $K\alpha$ line. Glade (*G17*) determined rhodium and gold thickness on small nickel–iron reed-switch reeds plated successively with gold and rhodium. However, the gold plating was measured before the rhodium was plated on, either by attenuation of Ni $K\alpha$ from the substrate or by Au $L\alpha_1$ emission. The rhodium was then determined from Rh $K\alpha$ intensity. Under certain conditions, Ebel's variable-takeoff-angle method (Section 14.8.2) is useful for multiple layers and was applied to layers of ~5570 Å of tin on ~940 Å of copper.

18.2.2. Alloy Platings—Composition and Thickness

Niobium and tin have been determined in cryogenic Nb_3Sn coatings on metal and ceramic substrates (*B36*) by application of the binary-ratio method (Section 14.3.2.3). Most of the coatings were infinitely thick for both Nb $K\alpha$ and Sn $K\alpha$; those that were not had constant thickness. For such films, a log-log plot of Nb $K\alpha$/Sn $K\alpha$ intensity ratio *versus* Nb/Sn concentration ratio was established from standards having coatings of known composition. If the thickness was less than infinite for the K lines and variable, the L lines were used, so that infinite thickness was smaller.

Many methods have been reported for simultaneous determination of both composition and thickness of films. Most of these methods were developed for analysis of Ni–Fe (Permalloy) films.

Schindler *et al.* (*S13*) report solution and nondestructive methods for determination of composition and thickness of nickel–iron films on gold in the ranges 0–40% iron, 10^2–10^5 Å. Five stock solutions were prepared having concentration 10 mg Ni + Fe per ml and Ni:Fe weight ratios 10:0, 9:1, 8:2, 7:3, and 6:4. Aliquots of each of these five solutions were diluted to give five solutions having concentrations 1, 2, 3, 4, and 5 mg Ni + Fe per ml for each Ni:Fe ratio. Ni $K\alpha$ and Fe $K\alpha$ intensities were measured from all 25 solutions, and the data were plotted on a graph of Fe $K\alpha$ intensity *versus* Ni $K\alpha$ intensity. A grid consisting of two sets of lines was drawn on the plot. One set represented the same iron concentration for different total weights of Ni–Fe. The other set represented the same total weight of Ni–Fe for different iron concentrations.

Next, Ni $K\alpha$ and Fe $K\alpha$ intensities were measured on 150 Ni–Fe films having various compositions and thicknesses on gold disks. The films were then dissolved quantitatively, and Ni $K\alpha$ and Fe $K\alpha$ intensities measured on the solutions. Composition (Ni:Fe) of each film was derived from the grid. Thickness was derived from the film weight, area, and composition, and from the densities of nickel and iron. The film weight was obtained from the weights of the disks before and after dissolution of the film. A plot of Fe $K\alpha$ *versus* Ni $K\alpha$ intensity was made of the data from the *films*. A grid consisting of two sets of lines was drawn on the plot. One set represented the same iron concentration for different thicknesses. The other set represented the same thickness for different iron concentrations. Finally, the grid on the plot of *solution* data was redrawn in terms of thickness rather than total weight of Ni–Fe, so that thickness could be derived directly either nondestructively or by the solution technique. The form of the calibration grid is shown in Figure 18.2.

Verderber (*V10*) developed a method for determination of composition and thickness of Ni–Fe films on magnetic tapes in the ranges ~81Ni–19Fe and 1000–4000 Å. Three stock solutions were prepared having concentration 3 mg Ni + Fe per ml and Ni:Fe weight ratios 85:15, 80:20, and 75:25. Aliquots of each of these three solutions were diluted to give seven solutions of different concentration for each Ni:Fe ratio, the most dilute having

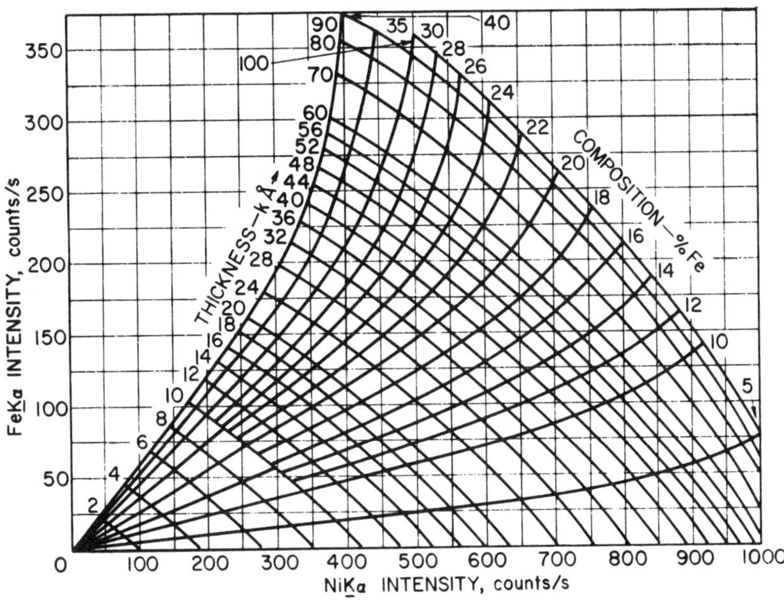

FIG. 18.2. Nondestructive determination of composition and thickness of nickel–iron films.

0.4 mg Ni + Fe per ml. Ni $K\alpha$ and Fe $K\alpha$ intensities were measured from all 21 solutions, and two calibration curves were established: Ni $K\alpha$ intensity *versus* mg Ni/ml and Fe $K\alpha$ intensity *versus* mg Fe/ml.

Next, 21 evaporated Ni–Fe films were selected, all having nominal 81Ni–19Fe composition, but varying in thickness from 1000 to 4000 Å. About 100 cm^2 of each film was dissolved quantitatively, and Ni $K\alpha$ and Fe $K\alpha$ intensities were measured from each solution. Nickel and iron concentrations were derived from the calibration curves and used to calculate film composition.

On an unused portion of each film, thickness was measured by a physical method, and Ni $K\alpha$ and Fe $K\alpha$ intensities were measured. Two new calibration curves were now established from the data from the films: Ni $K\alpha$ intensity *versus* (thickness × %Ni) and Fe $K\alpha$ intensity *versus* (thickness × %Fe). Thereafter, these curves were used for nondestructive analysis of films.

Weyl (*W9*) and Pluchery (*P23*) developed absolute mathematical methods for determining thickness and composition of thin alloy films without calibration standards other than the pure constituent elements. Their equations correct for spectral intensity distribution in the primary beam, fluorescent yields of the constituent elements, secondary excitation of low-atomic-number elements by high-atomic-number elements, and attenuation of the spectral emission on its way to the detector. Pluchery used filters to remove target lines from the primary beam, and corrected for scattered continuous background. Weyl and Pluchery evaluated their equations for simple Ni–Fe films.

Hirokawa *et al.* (*H51*) prepared a series of nickel, iron, and 80Ni–20Fe films having thickness up to ~3 μm. Ni $K\alpha$ and Fe $K\alpha$ intensities were measured from all the standards ($I_{p,t}$) and from infinitely thick nickel and iron ($I_{p,\infty}$). Four curves were plotted on a graph of the function $\log[1 - (I_{p,t}/I_{p,\infty})]$ *versus* thickness: Ni $K\alpha$ intensity from nickel and from Ni–Fe, and Fe $K\alpha$ intensity from iron and from nickel–iron. All the curves were linear. From these curves, two pairs of simultaneous equations were derived for determination of composition and thickness near the composition 80Ni–20Fe. The two pairs of equations were for thickness >0.1 μm and <0.1 μm, respectively.

Gianelos (*G8*) determined composition and thickness (plating weight) of 0.25–0.75-μm brass (~70Cu–30Zn) platings on steel tire-cord wires 0.1–0.4 mm in diameter. About 40 4-cm (1.5-inch) lengths of wire were affixed side by side to a Plexiglas square with two-sided Scotch tape. The number of wires is not critical, but their ends, where steel is exposed, must be covered by the specimen mask. Each sample may consist of single or 2–8-stranded wires or mixtures of these. Composition was determined from plots of Cu $K\alpha$/Zn $K\alpha$ net-intensity ratio *versus* %Cu and Zn $K\alpha$/Cu $K\alpha$ net-intensity ratio *versus* %Zn; alternatively, either element may be determined by difference. Plating weight (g plate/kg plated wire) was determined from a plot of (Cu $K\alpha$ + Zn $K\alpha$)/Fe $K\alpha$ net-intensity ratio *versus* plating weight. The Cu $K\alpha$ and Zn $K\alpha$ sum is very effective for the same lines ($K\alpha$) of adjacent elements. The same set of standards can be used for composition of all wire sizes, unstranded or stranded. The same set of, say, 0.25-mm standards can be used for plating weight of all 0.25-mm wire samples, unstranded or stranded, but a different set of standards is required for each wire size.

Silver and Chow (*S36*) compared optical interferometric, stylus, and x-ray spectrometric methods for determination of thickness of Ni–Fe films having composition 81–92% nickel and thickness 400–3500 Å. Thicknesses of several standard films were established by chemical determination of composition and physical determination of area density by measurement of mass and area:

$$t_S = \frac{100(m/a)}{(W\varrho)_{Ni} + (W\varrho)_{Fe}} \tag{18.11}$$

where t_S is thickness of the standard film (Å); m is mass (μg); a is area (cm^2); W is weight fraction; and ϱ is bulk density (g/cm^3). The quantity (m/a) is area density (μg/cm^2). An x-ray spectrometric thickness measurement consisted in measurement of the Ni $K\alpha$ and Fe $K\alpha$ intensities from the samples and from one of the standards. Then.

$$t_X = t_S \frac{(kI_{Ni} + I_{Fe})_X}{(kI_{Ni} + I_{Fe})_S} \tag{18.12}$$

where t is thickness (Å); I is net intensity (counts/s); X and S refer to the sample and standard, respectively; and k is calculated from the data measured from the standard:

$$k = \frac{I_{Fe}/I_{Ni}}{C_{Ni}/C_{Fe}} \tag{18.13}$$

where C is concentration (wt %).

Chan (*C38*) analyzed films of semiconductive II–VI compounds, that is, compounds of zinc, cadmium, and/or mercury with sulfur, selenium, and/or tellurium. He also compared the x-ray spectrometric results with those from analyses by substrate weight gain, neutron activation, and optical emission spectrography with laser-probe excitation.

18.3. SPECIAL TECHNIQUES

18.3.1. Special Specimen Forms

Zimmerman (*Z15, Z17*) applied three techniques to determination of plate thickness on a wide variety of small, irregular-shaped, plated parts, such as washers, solder lugs, and connectors: selected-area apertures, masks, and impression molds in sealing wax to permit reproducible positioning of irregular-shaped parts in the spectrometer specimen drawer (*I* in Figure 16.3). Selected-area methods have also been used to evaluate the uniformity of plate thickness around the edges of small disks and cylindrical rods (Section 19.5).

Plating uniformity has been evaluated on metal disks up to 14 inches in diameter, both radially from center to edge and around the disk at various distances from the center. A special motor-driven holder allows the disk to be translated radially or rotated in the primary beam while the intensity of a plate-metal line is recorded. The measured area may be the normal specimen area, or primary or secondary limiting apertures may be used (*P1*).

Bertin and Longobucco (*B40*) developed two methods for determining plate thickness on wires having diameter 0.01–0.25 mm (0.0004–0.010 inch). One method is applicable to large numbers of specimens consisting of infinitely thick layers of wire ($\geq$3–5 layers) wound on identical spools that fit the spectrometer specimen compartment. With the instrument set at a plate-metal line, each spool is placed in turn in the specimen drawer and the ratemeter allowed to trace plate-metal intensity for 15 s or more. A set of spools wound with standard wires is treated the same way. As many as four specimens can be analyzed each minute. Figure 18.3 shows a typical ratemeter chart.

For more accurate analysis, a few meters of wire is wound on a small open frame, card, or spool. The intensities of a plate and substrate line are measured and their ratio calculated. A calibration curve of plate/substrate intensity ratio *versus* plate thickness is established from wires of known plate thickness on the same type of support. The ratio method corrects for variations in number, spacing, and orientation of the turns of wire in the specimen. A set of standards is required for each combination of plate metal and substrate, wire diameter, and specimen mount (card, spool diameter, etc.).

Determination of composition and plating weight of brass plating on steel wire is described in Section 18.2.2. Evaluation of uniformity of platings or coatings on continuously moving or indexed wires and tapes is described in Section 18.3.8.

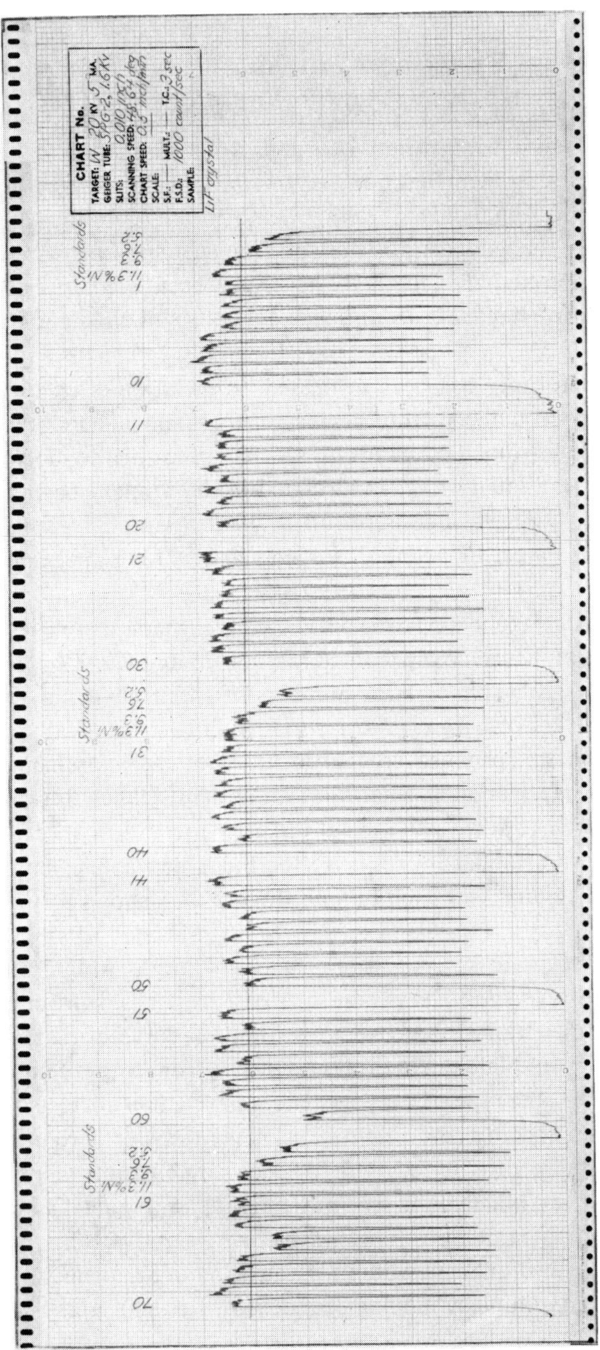

FIG. 18.3. Determination of plate metal on plated wire by the ratemeter method. [E. P. Bertin and R. J. Longobucco, *Analytical Chemistry* **34**, 804 (1962); courtesy of the American Chemical Society.]

Plate thickness on irregular-shaped objects is determined most accurately by measurement of the intensity ratio of the $K\alpha$ and $K\beta$ or $L\alpha_1$ and $L\beta_1$ lines of a substrate element, or of two lines from different elements in an alloy substrate [pp. 110–12 in *(23)*]. For example, suppose it is required to determine the chromium plate thickness on small irregular-shaped chromium-plated steel parts. The measured Fe $K\alpha$/Fe $K\beta$ net-intensity ratio for the substrate is given by:

$$\left(\frac{I_{\mathrm{Fe}K\alpha}}{I_{\mathrm{Fe}K\beta}}\right)_{s,p} = \left(\frac{I_{\mathrm{Fe}K\alpha}}{I_{\mathrm{Fe}K\beta}}\right)_{s} \frac{\exp -[(\mu/\varrho)_{\mathrm{Cr,Fe}K\alpha}\varrho_{\mathrm{Cr}}t]}{\exp -[(\mu/\varrho)_{\mathrm{Cr,Fe}K\beta}\varrho_{\mathrm{Cr}}t]} \tag{18.14}$$

where $(I)_{s,p}$ is measured substrate-line intensity after emerging through the plating; $(I)_s$ is substrate-line intensity before emerging through the plating; ϱ_{Cr} is the density of chromium; and t is plate thickness. $(I_{\mathrm{Fe}K\alpha}/I_{\mathrm{Fe}K\beta})_{s,p}$ is measured from plated samples; $(I_{\mathrm{Fe}K\alpha}/I_{\mathrm{Fe}K\beta})_s$ is measured from pure, unplated iron, or, preferably, from an unplated specimen; $(\mu/\varrho)_{\mathrm{Cr}}$ for Fe $K\alpha$ and Fe $K\beta$ are known. Thus, all quantities in the equation are known except t.

18.3.2. Selected-Area Analysis

By use of selected-area techniques (Chapter 19), analysis of small selected areas and point-by-point measurement of uniformity of composition and/or thickness of films become feasible. With the spectrometer set at the spectral line of an element in the film, the specimen may be translated continuously with a motor drive while intensity is recorded as a function of distance in a line across the specimen. Alternatively, the specimen may be moved manually in small increments and intensity measured at each point with the scaler. Successive scans can be made at different places and/or at the same place for different elements. Using such techniques and a 0.5-mm secondary-beam pinhole aperture, Sloan *(S39)* evaluated point-by-point variation of composition and thickness of Nichrome (nickel–chromium) films ~1500 Å thick. Ni $K\alpha$ and Cr $K\alpha$ intensities were measured at 1-mm intervals over the film. The individual intensities indicated film thickness; their ratio indicated composition. Zimmerman *(Z15, Z16)* measured variation in plate thickness over plated surfaces. He also investigated wear of gold-plated nickel by scanning a wear track and recording Ni $K\alpha$ intensity along or across the track. Bertin and Longobucco *(B39)* analyzed and mapped the distribution of sublimates on bulbs, stems, micas, grids, and plates of electron tubes and of platings around the rims of small disks and cylindrical rods (Section 19.5).

18.3.3. Selective Excitation

Beeghley (*B23*) measured tin-plate thickness by operation of the x-ray tube below the *K*-excitation potential of tin (29 kV) so as to excite only Fe $K\alpha$ (excitation potential 7 kV) from the base metal. The Sn *L* lines (excitation potential ~4 kV) are absorbed by the air and do not reach the detector. The selective-excitation method has since been applied by others (*B62*, *L27*).

18.3.4. Enhancement

If the specimen is an unsupported film or a film on a highly x-ray-transparent substrate, most of the primary radiation passes through the specimen and is lost. Hirokawa *et al.* (*H52*) placed under such specimens a secondary radiator having a strong spectral line of wavelength just shorter than that of the absorption edge of the element to be excited in the film. The primary x-rays transmitted by the film excite the radiator spectrum, which in turn excites additional emission from the film. Enhancements up to 50% were obtained for chromium and nickel films. Even a simple low-*Z* scatter target to direct the transmitted primary x-rays back through the film is very beneficial. These techniques should also apply to microgram quantities of analytical material in specimens supported on filter paper, Mylar, Millipore and Nuclepore filters, ion-exchange membrane, etc.

Pollai *et al.* (*P26*) calculated from theoretical considerations the enhancement of lines of elements in thin films by various phenomena. Enhancement of film lines by substrate lines may be as high as 25%, of substrate lines by film lines as high as 5%; in alloy films, excitation of the lines of one element by those of another is relatively small, only ~2% maximum in 1000-Å films. Excitation of film lines by scattered continuum and target-line x-rays may be as much as 30%.

Alternatively, for x-ray-opaque film-substrate systems, the radiator may be placed in the primary beam (*G38*). However, this technique reduces the total primary intensity incident on the specimen and therefore reduces the measured intensity by ~50%. The only advantage is improved peak-to-background ratio. This technique is discussed in more detail in Section 4.4.2.

18.3.5. Excitation by Radioactive Sources

Cook *et al.* (*C57*) used radioactive isotopes to excite, and an energy-dispersive spectrometer to measure, plate-metal emission. Excitation of base-metal emission by radioactive isotopes is described by the same workers, by Cameron and Rhodes (*C7*), and by Zemany (*Z6*).

Cameron and Rhodes (*C7*) report the use of tritium sources for excitation of Fe *K* radiation in galvanized (zinc-coated) steel and in copper-plated steel, Ti *K* radiation in platinum-clad titanium, and Cu *K* radiation in silver-plated copper wire. Their work was done on an energy-dispersive instrument.

Zemany's work (*Z6*) is particularly noteworthy because it combines the techniques of excitation by radioactive sources, selective excitation, and energy-dispersive detection. He used ^{55}Fe, a *K*-capture isotope emitting only Mn *K* x-rays, which can excite only *K* spectra of elements *below* atomic number 25 (manganese), and *L* spectra of elements *below* 61 (promethium). If the measurements are made in air, absorption eliminates the *K* spectra of elements below 19 (potassium) and *L* spectra of elements below 47 (silver). If the method is to be applicable to one element plated on another, it is necessary that the plate be among the elements potassium through chromium, or silver through neodymium, but that the base metal not be among these elements. Zemany applied the method to titanium-plated Kovar (Fe–Co–Ni), using a gas-flow proportional counter as detector. He suggests that other *K*-capture isotopes might extend the utility of the technique.

18.3.6. Energy-Dispersive Operation

Use of specially built energy-dispersive spectrometers has been mentioned above in connection with radioactive sources. Standard commercial x-ray spectrometers can be operated in energy-dispersive mode by setting the goniometer at 0° 2θ, removing the crystal, and using a pulse-height selector. Elimination of the crystal may increase the measured intensity by as much as two orders. Further increase in sensitivity may be realized by moving the detector closer to the specimen compartment. Achey and Serfass (*A1*) used balanced (Ross) filters (*K16*, *R27*) to isolate the spectral line to be measured. The disadvantages are that the x-rays from the specimen may contain secondary emission from both plate and substrate, and/or primary x-rays diffracted from crystalline plate or substrate. Selective excitation may eliminate the emission from either plate or substrate, depending on which has the higher excitation potential. Balanced filters may isolate the line to be measured from either emitted or diffracted radiation.

18.3.7. Decoration

Lundquist (*L55*) evaluated porosity of nickel electroplatings on uranium as follows. The nickel-plated uranium was anodized in a solution containing potassium ferrocyanide [$K_4Fe(CN)_6$] under conditions that minimize attack of nickel, but precipitate uranyl ferrocyanide $[(UO_2)_2Fe(CN)_6]$ wherever uranium is exposed. Fe $K\alpha$ intensity is then measured. This technique should be valuable for investigating porosity and discontinuity of other kinds of films and platings.

18.3.8. Dynamic Studies

A special electroplating cell designed by Burbank (*B88*) for dynamic x-ray diffractometric studies of the plating process is also applicable in principle to dynamic x-ray spectrometric studies. A rotating-disk cathode turns half-immersed in the cell containing the electrolyte and anode. The unimmersed surface constitutes the diffractometer specimen surface. Only crystalline material diffracts x-rays, so the thin layer of plating solution on the inspected surface is not detrimental. However, during the early stage of the plating, when the plate is still very thin, the intensity of the plate-metal line from the solution would exceed that from the plating. Provision would be required to remove the plating solution continuously. The cell could be modified so that a continuous-belt cathode would pass in turn through the plating cell, a rinse, and the spectrometer specimen chamber.

Uniformity of platings or coatings can be evaluated along the lengths of thin wires and plastic, textile, paper, and thin metal tapes—for example, tapes for audio or video recording and computer memory storage. The wire or tape can be passed through the specimen chamber continuously and the intensity of a plate or substrate line recorded continuously or scaled repetitively. Alternatively, the wire or tape can be indexed and stopped at appropriate intervals and the intensity recorded or scaled. The specimen holder is fitted with feed and takeup spools or reels and wire or tape guides arranged so that the length of wire in the analytical position always occupies precisely the same position and is free of curvature and vibration. The takeup reel is motor-driven. Such work can be done on flat-crystal spectrometers, but the measured intensities are much greater on curved-crystal instruments—even semifocusing ones. Wire and narrow ribbon constitute ideal specimens for curved-crystal spectrometers; wider tapes require a slit.

18.4. CONSIDERATIONS

X-ray spectrometric methods for determination of composition and thickness of films and platings are prone to several sources of error. The methods are impractical when there are more than two or three layers to be determined, and are more difficult when the plate element is also present in the substrate. Intensities from thin films are very low, and those from thick films are affected by surface texture—rough, matte, polished, porous, etc. Thickness and composition may not be uniform. Films and platings may be single-crystalline or highly oriented and thus produce diffraction effects.

In the measurement of plate thickness, serious errors arise from the presence of very thin platings ("strikes" or "flashes") and/or interfacial alloying between plate and substrate. The alloying occurs on annealing or other heat treatment. The effects of interfacial strikes and alloying are discussed below in terms of nickel-plated molybdenum wire as an example (*B40*). However, the principles apply as well to any plated specimen.

Figure 18.4A shows the simple case of a plate on a base wire with no strike or alloy formation at the interface. If such a wire is placed in the x-ray spectrometer, the primary x-radiation in turn excites the plate metal emission, undergoes some attenuation as it penetrates the plate, and excites the base-metal emission. The base-metal emission also undergoes some attenuation as it emerges through the plate, and possibly enhances the plate emission. Plate or substrate-line intensity or their ratio may be used as the calibration function.

Suppose now that there is a strike between the plate and base wire, as shown in Figure 18.4B. The intensity of the plate emission remains unchanged, unless the strike consists of an element capable of enhancing the plate emission, or unless the plate emission is enhanced by the base emission, and the strike attenuates the base emission. Thus, calibration functions involving only the plate-element line intensity are usually not substantially affected by strikes. However, both the base-metal emission and the primary

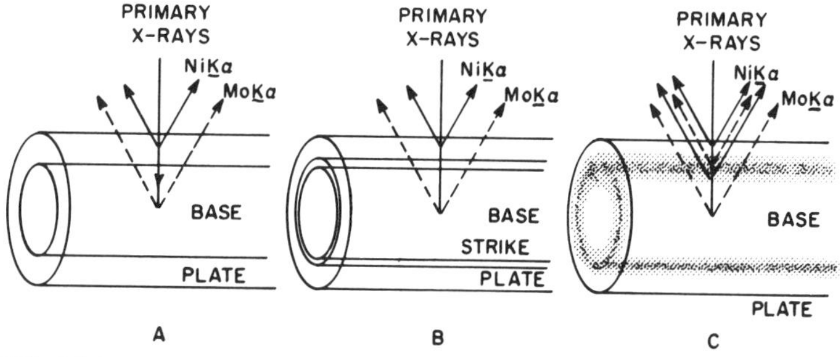

FIG. 18.4. Effect of interfacial strike and alloying on determination of plate thickness. [E. P. Bertin and R. J. Longobucco, *Analytical Chemistry* **34**, 804 (1962); courtesy of the American Chemical Society.] A. Absence of strike and alloying. B. Presence of strike. C. Presence of alloying.

radiation which excites it must pass through the strike. If the strike consists of an element of low atomic number (*e.g.*, chromium), and the base-metal emission is of relatively short wavelength (*e.g.*, Mo $K\alpha$, 0.7 Å), the base emission undergoes little reduction in intensity. Conversely, if the strike consists of an element of high atomic number (*e.g.*, silver or gold), and/or the base-metal emission is of relatively long wavelength (*e.g.*, Ni $K\alpha$, 1.66 Å), the base emission may undergo substantial reduction in intensity. This reduction increases with increasing wavelength of the base-metal emission, and with increasing absorption coefficient and/or thickness of the strike. Of course, reduction in base-metal intensity increases the intensity ratio.

Because of these effects, calibration functions involving substrate lines should be applied very cautiously to platings having strikes. In some cases, the effect of the strike can be compensated by use of standards having similar strikes or by determination of the strike thickness and correcting the substrate line intensity.

Now suppose that alloying has occurred between the plate and base wire, as shown in Figure 18.4C. Consider what happens, using nickel-plated molybdenum as an example. That part of the Ni $K\alpha$ emission originating in the unalloyed part of the nickel layer is excited in the same manner as in unalloyed wire. That part of the Ni $K\alpha$ emission originating in the alloyed layer and the primary radiation which excites it must pass through a certain amount of molybdenum. In doing so, they both undergo some absorption, with resultant reduction in Ni $K\alpha$ intensity. This phenomenon becomes more marked the deeper the nickel has diffused into the molybdenum. That part of the Mo $K\alpha$ emission originating in the unalloyed "core" of the base wire is excited in the same manner as before, that is, by primary radiation that has passed through all the nickel. Similarly, the Mo $K\alpha$ must emerge through all the nickel. Finally, that part of the Mo $K\alpha$ radiation originating in the alloyed layer and the primary radiation which excites it pass through only a part of the nickel layer. Consequently, they undergo less absorption than in unalloyed wire, with resultant increase in Mo $K\alpha$ intensity.

For a given percent nickel, the Ni $K\alpha$ intensity is lower and the Mo $K\alpha$ intensity higher for alloyed wire than for unalloyed wire. In the intensity-ratio method, the Ni $K\alpha$/Mo $K\alpha$ intensity ratio is smaller. Thus, for nickel-plated molybdenum, all three analytical methods give low results. If the substrate were a light element having relatively low absorption for Ni $K\alpha$, a similar consideration would show that for a given percent nickel, both Ni $K\alpha$ and substrate-line intensity are higher. The change in the ratio would depend on the relative increases in the individual intensities.

Chapter 19

Selected-Area Analysis

19.1. PRINCIPLE AND SCOPE

19.1.1. Principle

If the conventional aluminum-specimen-mask technique is applied to very small specimens or to small areas on large specimens, analyte peak-to-background ratio may decrease prohibitively. As analyte-line intensity decreases with specimen area, primary scatter from the increasing mask area increases, and primary scatter from air in the specimen compartment assumes greater significance. Scatter from the mask is reduced by use of a high-Z metal, but then the spectrum of this element is strongly excited, increasing the probability of spectral-line interference. Small specimens can be supported on thin Mylar or fine fibers, but air scatter remains, unless helium or vacuum is used. A better approach is to confine the primary beam to the area of interest, or to exclude all secondary radiation from the optical path except that originating in the area of interest.

If a small aperture is placed in the primary beam, only that incremental specimen area lying in the projection of the aperture receives primary radiation. Alternatively, if the aperture is placed in the secondary beam, the entire specimen is irradiated, but only secondary radiation originating in the projection of the aperture enters the optical path. Such apertures provide the basis for selected-area analysis. This chapter is devoted to analysis of small specimens and areas by x-ray secondary-emission spectrometry. Emphasis is placed on techniques that are applicable to commercial instruments, particularly with relatively simple accessories and minimal modification. The subject has been reviewed by Adler [pp. 101–8 in (*1*); and (*A5*)], Duncumb (*D32*), and Loomis and Vincent (*25n*).

An x-ray spectrometer arranged for selected-area analysis is often termed an x-ray "macroprobe" or "milliprobe" (compare with electron "microprobe") because: (1) excitation of the "probed" (selected) area is by x-rays rather than electrons, and (2) the probed area has dimensions of the order of millimeters or mils rather than micrometers.

Four other x-ray methods for evaluating point-to-point composition in place on the specimen are discussed in other sections: the Von Hamos image spectrograph (Section 5.5.3.4), x-ray contact microradiography (Section 20.1.5), x-ray absorptiometry (Section 20.1.3), and the electron-probe microanalyzer (Chapter 21).

19.1.2. Applications

Selected-area techniques are applicable to two classes of specimens: very small specimens, which may be regarded as "preselected" areas, and small selected areas on extensive specimens.

The first group includes the following: (1) particles, flakes, very small chips, etc.; (2) fibers, filaments, fine wires, etc.; (3) submilligram weights of powders on fibers or in fine plastic capillary tubes (these are the types of specimen used in x-ray powder diffraction analysis, and selected-area techniques may serve as a preliminary to a powder analysis); (4) metallic trace analytes electroplated on fine wire or on carbonized quartz fibers (*L39*); (5) flecks of purified trace analyte after microchemical separation from the matrix; (6) very small disks of filter paper or ion-exchange membrane supporting microgram amounts of analyte (*L49*); (7) small spots of analyte residue formed by evaporating droplets of solution on thin Mylar film; and (8) segregations, grains, and inclusions extracted from materials by micro drilling, extraction replication (*B74*), laser volatilization (Section 17.2.3), or preferential dissolution or etching away of the matrix.

The second group includes samples submitted for the following types of work:

1. Mapping of composition distribution of analyte(s) and evaluation of homogeneity of composition (*C40*).

2. Analysis of individual grains of various phases in heterogeneous alloys, minerals, ceramics, etc. (*A6*, *A9*).

3. Analysis of inclusions, segregations, etc. in the same materials (*J22*, *J23*).

4. Analysis of particles filtered from gases or liquids (*L24*).

5. Evaluation of composition, thickness, and uniformity of thin films and platings (*B39*, *R1*, *S39*, *Z15–18*).

6. Measurements of extent of diffusion in diffusion couples (*R17*).

7. Identification of layers and measurement of extent of diffusion in sections of metal-to-metal and metal-to-ceramic interfaces (*B32*, *B35*).

8. Measurement of extent of diffusion from liquid to solid phase (*E16*).

9. Investigation of sections of plated, coated, or corroded surfaces (*Z15–18*).

10. Analysis of parts of electronic devices or other small parts (*G17*).

11. Determination and distribution of natural and foreign elements in biological materials (*H12*, *H13*, *L37*).

12. Determination of minor and trace constituents in bulk specimens. Sometimes, greater sensitivity is realized by analyzing a small area of specimen with a curved crystal than by analyzing a large area with a flat crystal (*R26*). For example, in the determination of manganese and iron in aluminum alloys, Mn $K\alpha$ and Fe $K\alpha$ intensity, resolution, and peak-to-background ratio are better from a 1-mg sample (containing <10 μg of manganese and iron) in a curved-crystal spectrometer than from a 10-g sample in a flat-crystal spectrometer (*B55*).

Selected-area apertures may serve in place of constant-area masks for small and irregular-shaped specimens. For example, the writer inspected samplings from all incoming lots of about 20 different parts for an electron device to ascertain that they were of Invar (64Fe–36Ni). The parts ranged in size from a bent wire 0.5 mm in diameter and 5 mm long to a tube 2 cm in diameter and 20 cm long. A secondary-beam aperture was selected having a projection lying entirely on the smallest parts—the wire after flattening, and a small hexagonal nut. Ni $K\alpha$ intensity was observed on the panel ratemeter for each sample and from pieces of Invar. A sample could be verified in less than 2 min, most of the time being spent in aligning it with the pinhole.

Specimen areas somewhat greater than $\sim$1 mm^2 and substantially smaller than $\sim$1 cm^2 represent a case intermediate between conventional and selected-area techniques and may be measured in a *collimated specimen drawer* (*C40*, *G17*). The primary beam is conducted to the specimen and the secondary beam to the optical path in a continuous L-shaped tunnel in a block on the mask plate. This arrangement reduces primary scatter from air and mask.

19.1.3. Advantages and Limitations

The technique of selected-area analysis on the x-ray secondary-emission spectrometer has not assumed the degree of application that its capabilities and advantages warrant. The principal reason is the ready availability of many models of electron-probe microanalyzers. However, the two techniques should be complementary rather than competitive, and, for certain applications, the x-ray probe is actually preferable.

The advantages of the x-ray probe over the electron probe include the following:

1. The x-ray probe is simple and convenient to use; the electron-probe is not (Section 21.9).

2. There are applications where the extremely high resolution of the electron probe is actually detrimental. If the average composition of an area, say, 0.5 mm in diameter or of a narrow layer ~1 cm long is required, a single x-ray measurement for each analyte is sufficient. However, if the area is inhomogeneous on a 1–100-μm scale, the average of several electron-probe measurements for each analyte is required, or the electron beam must be defocused to cover the larger area.

3. Specimen translations up to 2 cm are usually possible in x-ray probes; some electron probes permit translations of only ~0.5 cm or less. Moreover, it is practical to map concentration profiles over distances of 1–2 cm with the x-ray probe. It may be difficult to keep the electron probe in focus over such distances.

4. The equipment is much simpler and less expensive. Electron-optical and vacuum equipment are not required. Many techniques are applicable to standard commercial x-ray spectrometers with simple accessories.

5. Specimen preparation, manipulation, and measurement are more rapid and convenient. Electrically nonconductive specimens require no provision of conductive film or paint.

6. Since primary x-rays penetrate deeper into the specimen than electrons, surface condition and takeoff angle are not as critical, and analyses are more representative. The relatively high takeoff angle (30–60°) of commercial x-ray spectrometers is favorable.

7. Secondary excitation gives lower background than primary.

8. Biological and volatile specimens are usually not damaged because the specimen is not in vacuum and is not bombarded by electrons.

9. The instrument can be arranged to accommodate larger specimens than the electron probe, and so can deal with objects that may not be damaged or marred (*B5*, *H11*, *Y5*). One instrument has been designed with

the specimen plane vertical at the back of the instrument to permit paintings, art objects, and antiquities to be studied (*Y5*).

10. The instrument is applicable to routine analysis of large numbers of samples and has been applied to incoming inspection—an example of which is cited above—and even to production control of plate thickness on very small critical parts (*G17*, *Z15–18*).

The disadvantages of the x-ray probe include the following:

1. Secondary excitation is two to three orders less intense than primary. As a result, second- and higher-order lines are useful only for relatively high concentrations, and, when *L* lines must be used, analyte sensitivity decreases sharply. Sensitivity for the low-*Z* elements (11–18) is very poor.
2. Spacial resolution is poorer by two to three orders (10–100 μm compared with ~1 μm).
3. The x-ray probe has only limited application to thin layers because the primary x-rays undergo little absorption in the thin layers and penetrate to the substrate, whereas electrons are likely to be absorbed in the analytical layer. This phenomenon adversely affects the x-ray probe in three ways: The composition of a thin layer is difficult to determine because spectral lines of the substrate "show through." In obliquely sectioned layer structures, the leading edges of the exposed layers are difficult to locate for the same reason. Finally, the analyte-line intensity from the layer may be low because the primary x-rays are not substantially absorbed therein.
4. The electron probe is useful for elements down to beryllium.
5. The electron probe can be used for all elements without change of accessories. Two simultaneously operable spectrometers are provided, each of which may be arranged for a different spectral region. Because of the high primary excitation efficiency, *M*, as well as *K* and *L* lines, are useful.
6. The electron probe may be operated in many modes based on phenomena other than x-ray emission, the two principal ones being scattered primary electrons and specimen current. Each of these phenomena gives a particular kind of information.
7. In the electron probe, analyte-line intensity and scattered electron and specimen current distribution along a line or over an area can be displayed on a cathode-ray oscilloscope.

Compared with conventional secondary-emission spectrometry, selected-area techniques give improved spectral resolution, reduced background, and, when a curved crystal is used, increased intensity and sensitivity.

19.2. SELECTED-AREA ANALYSIS ON STANDARD COMMERCIAL X-RAY SPECTROMETERS

19.2.1. Specimen Irradiation

The primary radiation intensity at a point *i* on the specimen is given by

$$I_i = (J_{\mathrm{pri}} \sin^3 \phi_i)/L^2 \tag{19.1}$$

where I_i is primary intensity at point i; J_{pri} is x-ray flux emitted by the x-ray tube per unit solid angle; ϕ_i is the angle between the specimen plane and the line joining i and the center of the x-ray tube focal spot; and L is the perpendicular distance from the specimen surface to the center of the focal spot.

For a given x-ray optical system, measured analyte-line intensity from a very small specimen varies with position in the specimen plane, position above or below the specimen plane, and goniometer angle 2θ. Primary intensity distribution over the specimen plane is nonuniform because the x-ray tube focal spot distance is different to different parts of the plane.

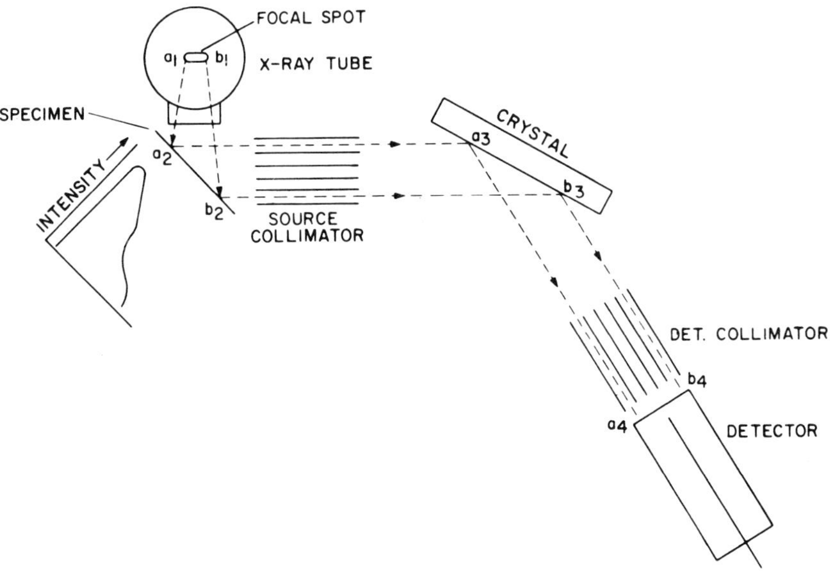

FIG. 19.1. Effect of specimen position and goniometer angle (2θ) on measured analyte-line intensity from small specimens. [R. Jenkins and J. L. de Vries, "Practical X-Ray Spectrometry," Springer-Verlag New York, Inc., p. 153 (1967); courtesy of the authors and publisher.]

This is shown in Figure 19.1, where primary intensity at the specimen plane decreases from a_2 to b_2 because the focal spot distance a_1a_2 is smaller than b_1b_2. When the angle between the primary beam and specimen plane is nominally, say, 60°, the actual angle over the specimen plane varies from this by $\pm 8°$ or so. For a typical target-specimen geometry, this may represent a variation in primary intensity of ~10%. Moreover, the variation in incident angle causes a variation in depth of penetration and critical thickness. Even if the specimen plane were uniformly irradiated, the measured intensity from a small specimen would vary with position. Different parts of the analyzer crystal diffract radiation from different parts of the specimen plane, and the crystal surface may not be uniformly "reflective." Finally, secondary x-rays originating at a_2 follow path $a_2a_3a_4$ to the detector, while rays originating at b_2 follow path $b_2b_3b_4$. Movement of a small specimen from a_2 to b_2 has the effect of decreasing the goniometer setting (2θ) at which maximum measured intensity occurs. This effect is minimized by use of a detector with a large useful solid angle, but may assume values as high as 0.5° 2θ. All these factors apply as well to variations in specimen position above and below the specimen plane.

A similar phenomenon arises from the fact that different parts of the crystal may present slightly different angles to the incident secondary beam from the source collimator. Suppose that a ratio method—analyte/internal standard, minor analyte/major matrix element, analyte/scattered target line, etc.—is to be used for analysis of a micro specimen. If the two lines of the ratio are not fairly close in 2θ, they may follow slightly different paths to, and diffract from different areas on, the crystal. In such cases, small differences in size, shape, and position of micro samples can affect the ratio. However, the difficulty does not arise if the lines are close in 2θ.

Because primary intensity varies over the specimen plane, small specimens and areas must be placed so as to receive maximum primary intensity. Reproducibility of this position is particularly important when a series of specimens or of standards and samples are to be compared quantitatively. The use of glass microscope slides and unexposed x-ray film to delineate the area of primary irradiation is described in Section 4.2.4.4. However, other techniques are required to establish intensity distribution within this area. Figure 19.2A shows a typical primary-beam distribution pattern on the specimen plane in the form of "isointensity" contour lines. The smallest (central) contour line defines the area of greatest primary intensity and therefore the optimal position for a small specimen or selected area. Successive contour lines represent progressively lower intensities. The pattern is ob-

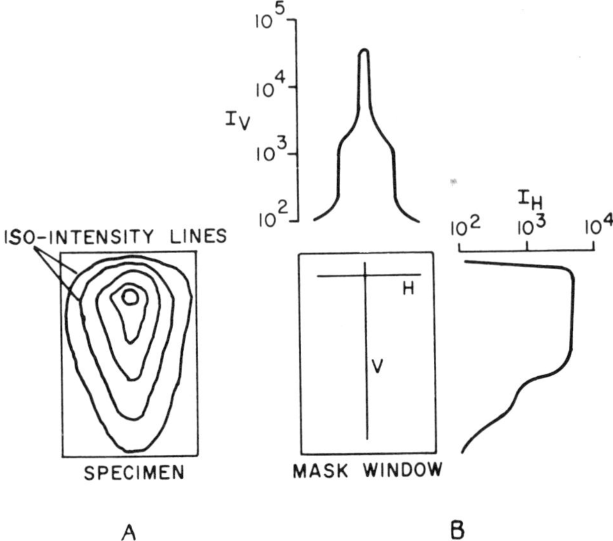

FIG. 19.2. (A) Variation of primary intensity over the specimen plane and (B) effect of specimen position on measured analyte-line intensity from filamentary specimens.

tained by mounting a piece of graph paper (10 × 10 divisions per inch) in the specimen window and scaling Ni $K\alpha$ intensity from a very small square of nickel foil placed successively in each square. The intensities are written in their relative positions, and contour lines are drawn through positions having about the same intensity.

Figure 19.2B shows the effect of position on intensity measured from linear specimens by use of curved crystals. A nickel wire 0.06 mm (0.0025 inch) in diameter is oriented parallel to the top edge of the specimen window and translated downward, then oriented parallel to a side edge and translated laterally. Ni $K\alpha$ intensity is recorded during the translations. Figure 19.2B shows the Ni $K\alpha$ intensity profiles and the optimal specimen positions for horizontally H and vertically V oriented filamentary specimens or linear areas. The strong intensity maximum for vertically oriented specimens does not necessarily lie at the center of the window, but may be placed there by slight rotation of the x-ray tube about its axis. The data for Figure 19.2 were obtained on an instrument having the geometry of the bottom row of Figure 9.2; the long edge of the window (Figure 19.2) is the inclined edge. No aperture was used in making any of the measurements. When selected-area work is to be done, intensity profiles of these types should be plotted each time the x-ray tube is changed.

19.2.2. Selected-Area Apertures

19.2.2.1. General

Possibly the simplest selected-area technique is to place a metal foil having an aperture on the specimen with the aperture over the area of interest (*A6*), but this is simply the masking technique. This section is devoted to more versatile arrangements in which provision is made to translate the specimen to place the area of interest under a fixed aperture.

The aperture can be mounted on the specimen compartment or, preferably, the specimen drawer, which can be removed to permit convenient alignment of the selected area with the aperture. The aperture can be mounted in the primary (*A9, L39, L40, T10*) or secondary (*H28, M35*) beam, or both (*B33, P13, R17*). With a primary aperture, only the selected area receives primary radiation. With the secondary aperture, only secondary x-rays originating in the selected area enter the optical path. The aperture can have the form of a pinhole, slit, or tunnel. Figure 19.3 shows some typical apertures. The collimated specimen drawer (Section 19.1.2) and the slit aperture shown in Figure 19.8 are examples of primary-and-secondary aperture systems.

19.2.2.2. Pinholes

Pinhole apertures have been made from drilled metal disks (*A6*); heavy-wall glass capillary, thermometer, and barometer tubing (*H28*); hypodermic needles (*H28*); and Leroy lettering pen points (*S39, W17, W18*). Square-bore "pinholes" (*C40*) have been used for evaluation of homogeneity to permit complete point-by-point coverage of the specimen surface with minimum overlap from adjacent incremental areas. The optimal pinhole

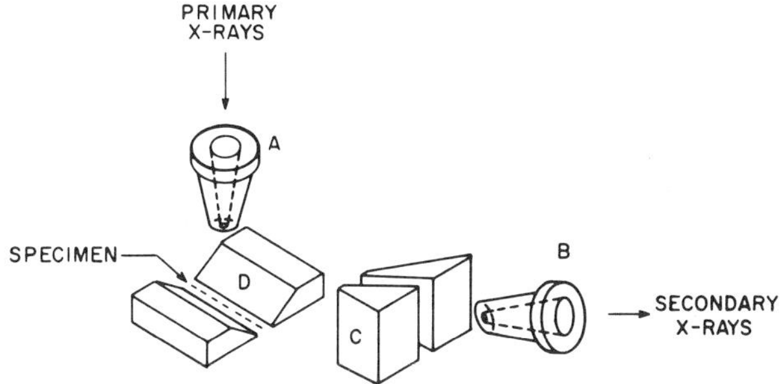

FIG. 19.3. Forms and positions of selected-area apertures. *A*. Primary pinhole. *B*. Secondary pinhole. *C*. Secondary slit. *D*. Primary–secondary slit or edge.

aperture has a conically tapered bore, as shown in *A* and *B* of Figure 19.3 (*25n*, *A6*, *A9*, *L39*, *L40*). In the primary beam, the tapering allows the selected area on the specimen to receive primary radiation from the entire x-ray tube focal spot. It is possible to irradiate the area with a demagnified pinhole image of the x-ray tube focal spot, giving optimal use of primary energy. In the secondary beam, the tapering allows the specimen to emit analyte-line radiation into the spectrometer through a relatively large solid angle.

The end of a secondary pinhole toward the specimen should be inclined as shown in *B* in Figure 19.3 to shield the pinhole itself from the primary beam (*H28*). Otherwise, the characteristic radiation of the aperture will be excited, and some of it will pass into the spectrometer. For maximum resolution, a secondary pinhole must approach the specimen plane as closely as possible without blocking the primary beam; a primary aperture must approach as closely as possible without blocking the secondary beam.

Figure 19.4 shows a secondary pinhole accessory available commercially for the General Electric x-ray spectrometer (*H28*). A brass block on the mask plate has a bore to receive pinholes having diameters 1, 0.1, and 0.05 mm. Under the mask is a specimen stage that can be translated in two

FIG. 19.4. Specimen drawer fitted with secondary-beam pinhole. [Courtesy of General Electric Co., X-Ray Dept.]

FIG. 19.5. Philips Macro-Probe Attachment in place on the Philips Universal Vacuum spectrometer. Specimen-translation micrometers, specimen illuminator, and viewing microscope are visible on the principal unit on the spectrometer specimen platform. The crystal-curving mechanism and curved crystal are mounted in the spectrometer chamber. [Courtesy of Philips Electronic Instruments, Inc.]

mutually perpendicular directions by the micrometer drives shown. A simple slit to replace the detector collimator and a specimen motor-drive are also supplied. The motor-drive couples to a micrometer drive by means of a flexible shaft. A collimated specimen drawer is also available for the GE instrument (*C40*, *G17*). A somewhat more sophisticated secondary pinhole accessory was available commercially for the Picker spectrometer. The device provided for specimen alignment in place in its compartment while it was observed through a 32× microscope. These accessories are used with flat of semifocusing curved crystals. A full-focusing primary pinhole accessory is available commercially for the Philips Universal Vacuum spectrometer. Features of the device include 0.5- and 0.05-mm pinholes, 70× microscope for observation and alignment of the specimen in its compartment, and continuous variation of curved-crystal radius with 2θ. Lithium fluoride and mica crystals are supplied. The accessory is shown in place on the spectrometer in Figure 19.5.

19.2.2.3. Slits

Slits are preferable to pinholes for some specimens, such as sections of long, linear inclusions; diffusion couples; metal-to-metal and metal-to-ceramic interfaces; plated, coated, or corroded surfaces; and other layered structures where composition varies in the direction across, but not along, the layers (*B32*, *B33*, *B35*, *L8*). Figure 19.6 shows a section of a typical ceramic-to-metal system. The ceramic has been metallized, plated, and brazed to the metal. The metallizer may diffuse into the ceramic, the braze into the metal, and the metallizer, plating, and braze into each other. Inclusions or segregations occur in the ceramic and metal. Selected-area techniques can be applied to identify the metallizer, plating, braze, and inclusions, and to evaluate the extent of diffusion. The figure shows the projections of pinhole and slit apertures on the specimen.

The roughly circular inclusions are identified by use of the pinhole. However, if a pinhole is used to identify the linear inclusions or interfacial layers, or to evaluate the extent of diffusion, only that incremental specimen layer under the pinhole contributes to the measured analyte-line intensity. A slit is advantageous for these applications because it passes secondary radiation from the entire length of the incremental layer. Increased intensity, and therefore sensitivity and statistical precision, are realized without loss of resolution by replacing a pinhole of given diameter with a slit of the same width. Increased resolution is realized without loss of intensity by replacing the pinhole with a slit of smaller width. Moreover, a more representative sampling of a layer is obtained with a slit than with a pinhole. For example, the writer, using the binary-ratio technique (Section 14.3.2.3) determined the lead/tin ratio for a wide variety of very small solder-coated parts and solder-foil preforms used in making transistors. It would have been feasible to use a pinhole and make measurements on in-

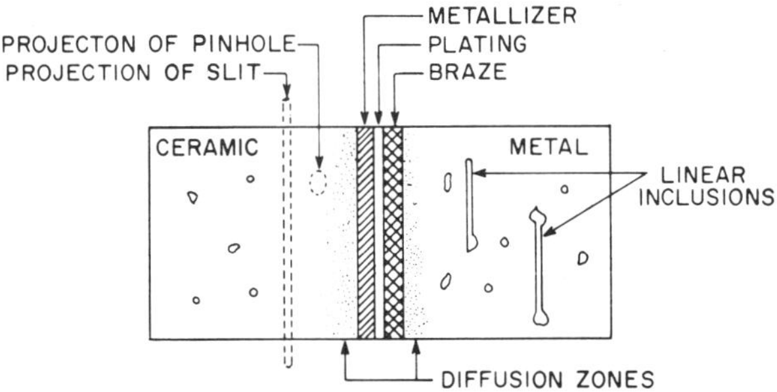

FIG. 19.6. Applications of pinhole and slit apertures.

FIG. 19.7. Specimen drawer in Figure 19.4 fitted with secondary-beam slit aperture.

dividual parts. However, higher intensities, improved precision, and more representative sampling were realized by using a slit and making measurements on a row of parts mounted on two-sided masking tape.

Figure 19.7 shows a slit accessory consisting of the drawer shown in Figure 19.4 in which the pinhole mounting block has been replaced with a secondary-beam slit (*B32*). Figure 19.8 shows the mask plate of the same drawer fitted with a slit lying in both the primary and secondary beams (*B33*).

A slit of the type shown in *C* of Figure 19.3 and in Figure 19.7 has many limitations that do not apply to pinholes (*B32*):

1. Its applicability is limited to the types of specimen listed above.

2. It is extremely wasteful of secondary intensity. The jaws may not be tapered in a way analogous to the tapering of a pinhole because the jaws cannot approach the specimen surface without blocking the primary beam. Consequently, tapered jaws would accept secondary radiation from a relatively wide specimen area, and resolution would be lost.

3. Effective slit width (see below) and resolution vary from top to bottom of the slit because distance from slit to specimen plane varies.

4. Analyte-line intensity measured from a small specimen decreases as the specimen is moved from opposite the top to opposite the bottom of the

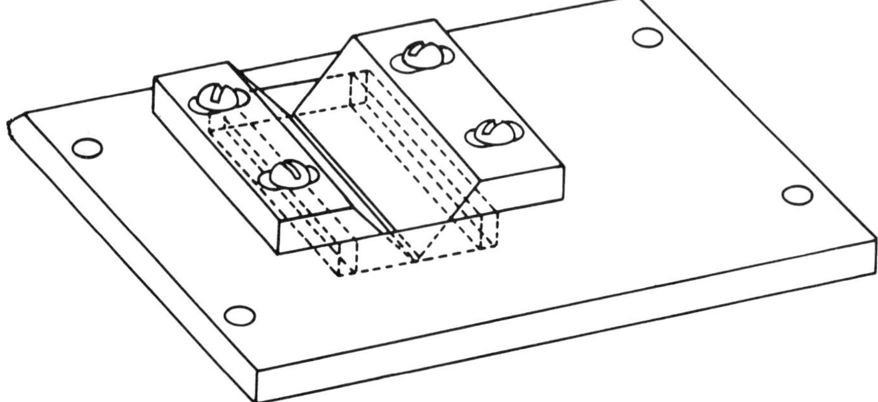

FIG. 19.8. Specimen mask plate fitted with slit aperture lying in both primary and secondary beams.

slit, because distance between x-ray tube focal spot and sample plane increases.

5. It follows from the preceding two considerations that if several specimens, or samples and standards, are to be compared quantitatively, they must have the same height (in the direction parallel to the projection of the slit) and the same position with respect to the slit.

6. The lateral position of the slit is very critical if full advantage is to be taken of the intense central "hot spot" (*V* in Figure 19.2).

7. Finally, the linear specimen area must be precisely aligned with the slit. Nevertheless, a simple slit accessory used with a semifocusing curved crystal on a commercial spectrometer permits much sophisticated work to be done that would require an elaborate full-focusing curved-crystal accessory with a pinhole.

The two-jaw primary–secondary slit (*B33*) shown in *D* of Figure 19.3 and in Figure 19.8 gives analyte-line intensity about an order higher than the slit discussed above. The beveled jaws permit the selected area to receive full primary irradiation and to emit secondary radiation into the optical system through a wide solid angle. Resolution is excellent because the slit jaws lie nearly in contact with the specimen surface. However, this arrangement constitutes little more than a mask. Background is high because of primary scatter and secondary emission from the jaws. With narrow slit widths, one of the jaws must be withdrawn to permit observation of the specimen for alignment; this is no problem with an individual specimen.

However, when a number of specimens are to be compared quantitatively, the slit must be opened and reset for each, and the analytical results are impaired by slight nonreproducibility of slit width.

Slit width is set with shims. Mylar film and sheet are available in a range of thicknesses from 0.003 mm (0.00012 inch) up and make excellent shims that do not wear the slit jaws as much as metal shims. When setting slits to very narrow width—0.125 mm (0.005 inch) or less—the Mylar must be free of wrinkles and dust.

Rizzo (*R17*) studied diffusion of electroplated metal into its substrate with a similar slit having only one jaw. The sectioned specimen was mounted with the plate–substrate interface parallel to the edge, then translated incrementally as analyte-line intensity was scaled or recorded.

Proctor (*P31*) has shown that a single pinhole in a thin mask acts as a slit when mounted perpendicular to the central ray of the secondary beam between specimen and crystal; this is because the crystal diffracts rays from a row of points on the specimen surface, the row being coplanar with the crystal's axis of rotation. This arrangement is extremely simple and convenient, but gives lower intensity than a true slit because the pinhole passes only one ray from each point on the specimen to the crystal, whereas the slit passes a V-shaped shaft of rays from each point. Two such pinholes in register a short distance apart act as a pinhole.

19.2.2.4. Resolution

The resolution attainable with a pinhole or slit increases as the pinhole diameter or slit width decreases. The *effective* aperture width (pinhole diameter or slit width) is measured by translating a very sharp two-element interface past the aperture while recording intensity of a spectral line of either element. The effective width is defined as the distance between the two points on the specimen where response is just distinguishable above background and where it attains its maximum value. A test specimen is made by bolting highly polished flat pieces of copper and nickel together, sectioning the couple, and polishing the section. Alternatively, a copper or nickel wire can be imbedded in a lead disk and polished flat. Thermally bonded couples are unsuitable because diffusion may blur the sharpness of the interface. The effective width is always greater than the true width, but the closer the aperture approaches the specimen, the more nearly equal are the two widths.

The projection of a pinhole on an inclined specimen plane (Figure 19.3) is an ellipse having its major axis lying in the direction of the inclination. Thus, concentration profiles should be scanned in the other direction, where effective pinhole diameter is narrower and resolution better.

Slits of the type shown in *C* of Figure 19.3 and in Figure 19.7 cannot approach the specimen plane as close as a pinhole without blocking the primary beam. Therefore, effective slit width is always substantially greater than the true width. Fortunately, this condition can be compensated by use of narrower slit widths. The distance from the slit jaws to the specimen plane increases from bottom to top of the slit, and so does effective slit width.

19.2.3. Dispersion

An incremental area on a specimen emits secondary x-radiation in all directions. A flat crystal diffracts only that narrow V-shaped shaft of these x-rays that is incident on the crystal at the Bragg angle θ. With an extensive specimen, the Soller collimator permits diffraction of such a V-shaped shaft of x-rays from each incremental area, and high intensity reaches the detector. But when the specimen area is very small, only one such shaft is effective, and all other secondary radiation is wasted. However, with a curved crystal of appropriate radius, all divergent secondary x-rays intercepted by the crystal strike the tangent to its surface at the point of intersection at the Bragg angle and are diffracted. The diffracted rays converge to a "focus" at the focusing circle, where the detector slit is placed.

Selected-area analysis is feasible with flat crystals. The collimators are removed to avoid possible blocking of the small effective secondary beam by a Soller foil. A simple slit may be placed before the detector window to reduce background and improve resolution. However, flat crystals are satisfactory only for pinholes 0.05–1 mm in diameter, and the smaller diameters are useful only for relatively high concentrations of elements having intense lines. In general, sensitivity falls off sharply when *L* lines or crystals having "reflectivity" substantially lower than lithium fluoride must be used. For linear specimens or areas, slits may permit measurements to be made where pinholes would be useless. Slit widths as narrow as 0.003 mm can be used in favorable cases.

Measured intensity from pinholes and slits can be increased as much as an order by use of semifocusing curved-crystal optics, that is, by simply replacing the flat crystal with a fixed-radius curved crystal. The collimators are removed, but for a different reason—because the parallel foils would intercept the divergent rays. Again, a slit before the detector window may be beneficial. One crystal having radius optimal for the middle of the 2θ range may suffice if some line broadening at high and low 2θ can be tolerated. For high resolution over the full 2θ range, two or three crystals having different radii are required. A set of such crystals can be mounted on a multiple-crystal changer.

Selected-area analysis on a commercial spectrometer is best accomplished by fitting it with a full-focusing curved-crystal accessory. Such accessories have been designed for General Electric (*25n*, *L39*, *L40*), Philips (*D35*, *L24*, *M35*, *P13*), and Siemens (*J22*, *J23*) instruments, and an accessory for the Philips Universal vacuum x-ray spectrometer is available commercially (*H42*). For selected-area analysis, a full-focusing x-ray optical system must "view" the same point or line on the specimen at all values of 2θ, that is, either (1) the instrument must have a constant takeoff angle, or (2) the specimen point or line must lie on the focusing circle. The first condition is met by instruments having the ARL design or provision for continuous variation of crystal radius with 2θ; these instruments are shown in Figures 5.23 and 5.21, respectively. The second condition is met by instruments having the NRL design shown in Figure 5.22 with the x-ray tube and specimen in the alternate positions indicated.

Figure 19.9 (*25n*) shows an excellent accessory having the modified NRL design for use on the basic General Electric spectrogoniometer. A 75-kV x-ray tube irradiates the selected area on the specimen through a primary-beam pinhole having a tapered bore 0.02–1.6 mm in minimum diameter. Secondary radiation emerges from the specimen chamber through a rectangular exit port that intercepts radiation that would miss the crystal, but permits maximum coverage of the crystal. The crystals (LiF and EDDT) are curved to a 35-cm (14-inch) radius, and, for optimal resolution, should also be ground to a 17.5-cm (7-inch) radius. The diffracted rays converge on an adjustable detector slit behind which is the conventional detector. As 2θ is varied, the entire specimen chamber rotates about a vertical axis concentric with the primary pinhole, and a system of rods and couplings maintains the selected area, rectangular exit port, crystal, and detector in the proper geometric relationship. The specimen-stage assembly, including the translation mechanism and aperture, can be removed for loading the specimen and for aligning it under a microscope. Bellowed plastic tunnels can be placed between the specimen and crystal chamber and the crystal chamber and detector to permit use of helium path.

In general, curved-crystal transmission instruments are avoided for selected-area work because of the loss of intensity due to absorption in the crystal.

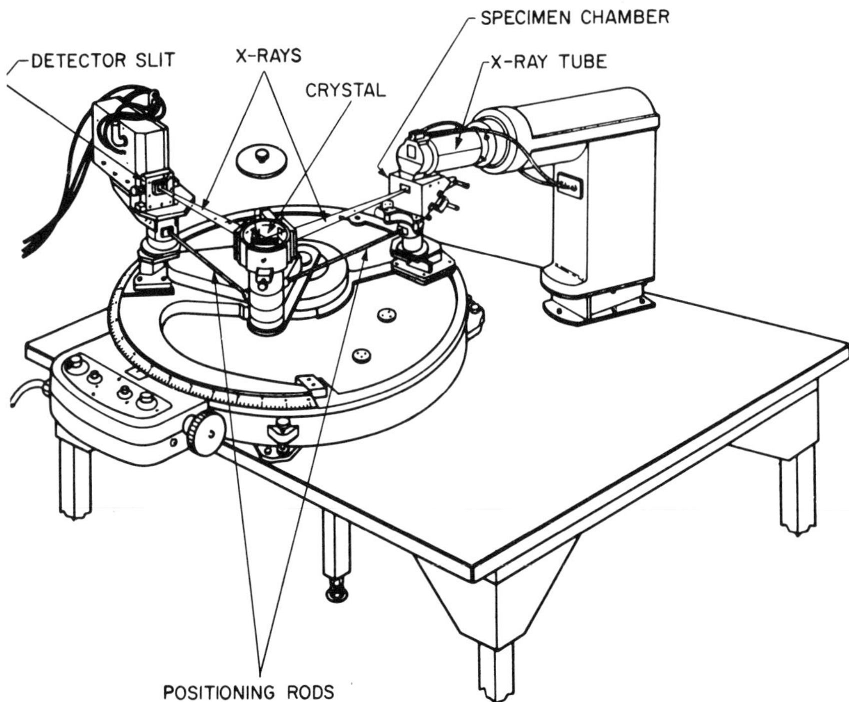

FIG. 19.9. Full-focusing curved-crystal accessory for use with General Electric x-ray spectrogoniometer. [T. C. Loomis and S. M. Vincent, in "Handbook of X-Rays," E. F. Kaelble, ed.; McGraw-Hill Book Co., p. **37**-12 (1967); courtesy of the authors and publisher.]

Incidentally, when the specimen is homogeneous and selected-area technique is used simply to take advantage of curved-crystal optics to realize increased sensitivity, the instrument need not necessarily meet the two conditions cited above.

19.3. INSTRUMENTS FOR SELECTED-AREA X-RAY SPECTROMETRIC ANALYSIS

The instruments that have been used for selected-area x-ray spectrometric analysis may be classified as follows:

1. Electron-probe x-ray primary-emission spectrometers (Chapter 21).

2. Microfocus x-ray tubes (*14, C61, D15, L35–37, O2, R29, Z2, Z4*). An electron-optical system forms a fine-focus electron image of its filament on a thin metal foil serving as both target and x-ray window. Thin specimens

are placed just outside this window, and the emitted characteristic emission is measured in a curved-crystal or energy-dispersive spectrometer. Alternatively, a standard commercial microfocus tube can be used of the type applied in x-ray topography and microdiffraction equipment.

3. Laboratory-built, full-focusing, curved-crystal spectrometers (*A6, A9, A12, B5, B10, B55, N1, S30, T9, Y5*).

4. Standard commercial x-ray spectrometers fitted with apertures and full-focusing curved-crystal accessories (*25n, D35, J22, J23, L39, L40, M35, P13*). The instrument shown in Figure 19.9 (*25n*) is of this type.

5. Standard commercial x-ray spectrometers fitted with apertures and semifocusing curved-crystal accessories (*B32, B33, B35*).

6. Standard commercial x-ray spectrometers fitted with apertures and flat-crystal accessories (*B38, C40, H28, L57, M35, R17, S39, T10, W17, W18, Z15–18*).

7. Energy-dispersive spectrometers having x-ray sources (*D15, H12, H13*).

8. Nondispersive spectrometers having radioisotope sources.

9. Instruments based on special principles. Two examples follow (*D15*). In the method of *selective excitation*, a single wavelength on the short side of the analyte absorption edge is selected from the primary beam, and diffracted and focused by a doubly curved aluminum-crystal monochromator on a selected area ~1 mm² on the specimen. The analyte-line emission is measured in a curved-crystal spectrometer in the usual way. Unless a target line can be used for the selected wavelength, the technique is very wasteful of primary intensity, and gives low intensity. In the method of *differential excitation*, the primary beam is dispersed into a spectrum by a cylindrically curved aluminum crystal. A mask plate having a small aperture can be placed at any position—and therefore any wavelength—around the focusing circle. The specimen is placed behind the mask with the aperture over the area of interest. The detector window is situated to receive x-rays emitted from the selected area. The mask–specimen–detector assembly is placed to allow the selected area to be irradiated in turn by two wavelengths λ_S and λ_L on, respectively, the short- and long-wavelength sides of the analyte absorption edge. The x-rays emitted and/or scattered by the specimen pass directly into the detector. The detector receives analyte-line emission at λ_S, but not at λ_L. The difference in the intensities at the two wavelengths gives analyte-line intensity—*without pulse-height analysis*. The technique has been applied to areas of ~0.5 mm². The disadvantage cited above for the selective excitation technique applies here as well, but is partially compensated by the nondispersive arrangement.

19.4. SPECIMEN TECHNIQUES

19.4.1. Preparation

Very small specimens may require no preparation and are usually supported on Scotch tape, thin Mylar, or fine fibers. Small volumes of solution can be evaporated on the flat ends of small (~2-mm diameter) rods, or solutes can be taken up in ion-exchange membrane disks of similar diameter. If a curved crystal is used, an aperture, although preferable, may not be required because the specimen itself constitutes a point or line source. Linear (filamentary) specimens must be oriented in the plane defined by the directions of the central rays of the primary and secondary beam cones, even when a flat crystal is used.

Extensive specimens are supported in the specimen compartment, and an aperture must be used. The projection of the aperture on the specimen defines the selected area and constitutes the source of measured secondary radiation. Such specimens usually require some kind of preparation, usually the metallographic techniques of potting, sectioning, and polishing. Oblique sectioning increases layer width in layered specimens; however, if the cut is too oblique, x-rays from lower layers may penetrate the thin edges of overlying layers and give erroneous layer widths and diffusion depths. A dull matte finish (~600 grit) is preferable to a high metallographic polish because, with the latter, it is difficult to see the light beam when aligning the specimen with the aperture (see below). Specimens should be etched only to eliminate surface smear or contamination, and preferential etching must be avoided. If layered specimens are to be measured with a slit, all layers must have the same length, or else be masked, and the layers must be oriented in the same way as filamentary specimens (see above).

19.4.2. Alignment of Selected Area and Aperture

In principle, it is preferable to align the specimen with the aperture in place in the spectrometer. However, the alignment is much more convenient and nearly as accurate if the aperture is mounted on a specimen drawer that can be removed and placed on a microscope stage. Final adjustment can then be made with the drawer in the instrument. If the major constituent of the selected area is known, the goniometer can be set at its line and the specimen micrometers adjusted slightly for maximum response. If the composition is unknown, a 2θ scan can be made and alignment refined later.

Five alignment techniques are commonly used:

1. Sighting by eye. If the aperture is large and the specimen area can be illuminated, the specimen can be aligned while being observed through the aperture with the unaided eye.

2. Light-beam technique. A light beam is directed through the aperture onto the specimen as the latter is viewed through a microscope. The specimen is translated to bring the selected area into the light beam. This technique is most convenient with drawer-type accessories of the types shown in Figures 19.4 and 19.7, which can be removed and placed on a microscope stage. However, a microscope and light source can also be provided on accessories in which the aperture is mounted on the specimen compartment. With very small pinholes and slits, the light beam is extremely difficult to see.

3. Fine, stiff wires and thin, stiff shims may be passed through, respectively, pinholes and slits and advanced to contact with the specimen, which is then translated to bring the selected area into alignment with the wire or shim.

4. Fluorescent technique. A small grain of fluorescent material is placed directly over the selected area. With the specimen in place in the x-ray spectrometer, x-rays are turned on and the specimen translated to give maximum fluorescence. The particle is then removed. This technique is applicable only to primary apertures.

5. A microscope having cross-hairs in the eyepiece can be adjusted so that the cross-hairs mark the projection of a pinhole or so that one of the hairs lies parallel to the projection of a slit. Thereafter, it is simply necessary to bring the selected area into alignment with the cross-hairs.

19.5. ANALYTICAL TECHNIQUES

The small specimen or area can be placed in the projection of the pinhole or slit and a qualitative analysis made by recording a 2θ scan on the ratemeter–recorder, or a quantitative analysis made by scaling the analyte-line intensities. The 2θ scan charts are similar to those shown in Figure 10.1. The procedure is repeated at other points of interest.

The concentration distribution profile of a selected analyte along a line across the specimen can be mapped by setting the spectrometer at the analyte line, moving the specimen manually in small increments, and scaling the intensity at each point. For layered specimens, a slit is preferable; otherwise, a pinhole must be used. The data are plotted manually. The same distribution profile can be mapped by translating the specimen with a motor-drive while recording analyte-line intensity on the recorder. The two techniques are compared below. Successive scans can be made for different analytes, and, with pinholes, scans can be made along different lines for the same analyte.

The motor-drive should be slow enough to allow the recorder to follow small irregularities in concentration. In the accessory shown in Figures 19.4 and 19.7, two turns of the specimen-translation micrometers displace the specimen 1 mm. A motor speed of 20 revolutions per hour moves the specimen at a rate 1 cm/h. The writer uses a motor having speeds 1, 2, 4, 12, 30, and 60 revolutions per hour.

The writer has evaluated the uniformity of molybdenum metallizing and nickel plating on the edges of small ceramic disks as follows. The disks were mounted on suitable spindles in the *side* rotor of the rotating specimen drawer shown in Figure 16.4 so that the plane of the disks coincided with the plane defined by the central rays of the primary and secondary x-ray beams. A secondary pinhole was mounted on the drawer in line with the edge of the disk. The intensity of the Mo $K\alpha$ or Ni $K\alpha$ line was recorded as the disk slowly rotated. In a similar manner, Lublin (*L43*) used a primary slit aperture to determine uniformity of cladding thickness around the circumference of small rotating rods.

Campbell *et al.* (*C9*) describe an accessory for photography of a color mosaic picture of the analyte-line intensity distribution over the specimen surface. By means of *X* and *Y* stepping motors, the specimen stage is translated in mutually perpendicular directions in such a way that, in effect, a primary-beam aperture scans a raster on the specimen surface. The *X* motor advances the specimen for a predetermined number of 0.3- or 1-mm steps; then the *Y* motor moves one step. The *X* motor reverses and scans the second line in the opposite direction; then the *Y* motor moves another step. The *X* motor scans the third line in the forward direction, etc. until the raster is scanned. Simultaneously, a Polaroid color-film holder is driven by two other *X* and *Y* stepping motors in synchronism with the specimen drive motors.

At each point on the specimen scan, the corresponding point on the color film is illuminated by a beam of light passing through a mechanical shutter and a 10-sector color disk, each sector of which is a different color. The disk is rotated by a repeater motor driven by the ratemeter output. Thus, at each point, the position of the disk and the color of the light beam are determined by the ratemeter, and thereby the analyte-line intensity. In this way, a color mosaic photograph of the analyte distribution is built up, the colors representing a 10-step gradation of the intensity range.

By alteration of the ratio of the specimen and film scanning linkages, magnification or reduction of the photograph may be realized.

A color photograph made with this apparatus is shown in Figure 19.10. The test specimen was an aluminum plate 25 mm square having shallow circular depressions packed with nickel oxide powder. The specimen was translated in 0.4-mm increments under a 1-mm x-ray "probe." The *apparent* change in nickel concentration around each circle is due mostly to partial coverage of the 1-mm x-ray probe by nickel oxide as the probe moves on or off the nickel oxide area.

Secondary-emission selected-area techniques are best suited to qualitative and semiquantitative analysis, and to mapping of concentration profiles. Standards are usually not required for such work. If the identity of the selected area is suspected or at least narrowed to a few possibilities, it may be identified by comparison with specimens of the likely substances. The presence or absence of some distinguishing element or combination of elements, or the concentration level of some key element, may also permit identification. Sometimes, the intensities of several elements in the sample can be measured, ratioed with the highest intensity, and compared with similar ratios from specimens of likely substances.

Quantitative analysis is also feasible, and high accuracy is attainable when an aperture and curved crystal are used to realize higher intensity from a trace element in a homogeneous sample. Also, quantitative analysis may be simplified for very small isolated quantities and thin films, for which matrix effects may be disregarded. However, when selected-area techniques are applied to quantitative analysis of heterogeneous samples, difficulties arise that are not encountered in conventional x-ray spectrometric analysis. The selected area under the aperture is not likely to be uniform in composition. The aperture may cover matrix or layers adjacent to the spot or layer of interest. A diffusion zone has a range of compositions, and a layer may contain grains of different phases. Intensity may be low, especially for minor constituents, with consequent low statistical precision. Long counting times may be risky because prolonged exposure to the primary beam may cause the specimen to heat and shift slightly. Standards must be homogeneous on a microscopic scale. Nevertheless, accurate quantitative analyses can often be made when suitable calibration standards are available.

Sommer (*S44*) has used selected-area techniques to obtain x-ray spectrometric analyses and x-ray-excited optical fluorescence spectral distribution curves from small selected areas, such as individual grains and phases (Section 20.5).

An oustanding example of application of x-ray spectrometric selected-area analysis on a larger scale is the Kevex-Scan thyroid scanner (Kevex Corp., Burlingame, Calif.). This instrument permits mapping the iodine distribution in the human thyroid gland *in vivo* without injecting the patient with radioactive iodine and thereby at greatly reduced exposure of the patient to radiation. The iodine concentration in normal human thyroid is ~0.04 wt %. A 5-Ci ^{241}Am radioisotope disk source is mounted behind a focused collimator (Section 5.6.5, Figure 5.24), which "focuses" the γ-rays from the source to a small area on the patient's neck, exciting the I $K\alpha$ x-rays. The source–collimator system is then mechanically scanned over the thyroid area on the patient in synchronism with the electron beam over the display screen in a cathode-ray tube in a storage oscilloscope. The scan requires 15–20 min. The I $K\alpha$ x-rays are detected, and their pulses amplified, passed through a single-channel pulse-height selector, and used to modulate the instantaneous brightness of the display. In this way, the iodine distribution in the thyroid gland is mapped. The I $K\alpha$ intensity at any or every point can also be scaled for semiquantitative analytical purposes or for printout in the form of Figure 21.5.

19.6. PERFORMANCE

The sensitivity attainable by selected-area techniques varies widely with the particular system. However, Table 19.1 gives some idea of the performance for several typical arrangements. The data in part *C* of the table for a full-focusing instrument probably represent maximum attainable performance from an accessory for a commercial spectrometer. This accessory is capable of producing measurable intensities from areas approaching those involved in electron-probe microanalysis.

The effectiveness of a selected-area analysis in excluding surrounding matrix is shown in Figure 19.11 (*L40*) for a nickel wire 0.04 mm (0.0015 inch) in diameter imbedded in Wood's metal (Bi–Pb–Cd–Sn). Ratemeter–recorder 2θ scans are shown in the Ni $K\alpha$ spectral region for conventional flat-crystal operation and for selected-area operation with a curved-crystal and three pinhole apertures having different diameters. Note the progressive decrease in intensity of matrix lines and background without loss of Ni $K\alpha$ intensity.

Typical performance of a secondary slit aperture (Figure 19.7) with a semifocusing curved crystal is shown in Figures 19.12 and 19.13 (*B32*). The slit probe was applied to an investigation of a disk-shaped system

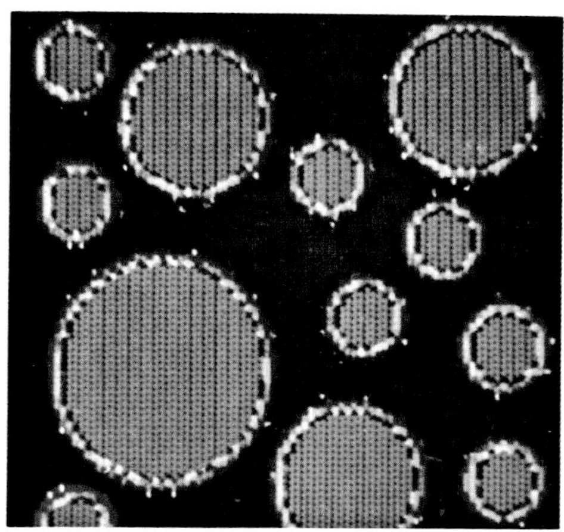

FIG. 19.10. Color-mosaic photograph of Ni $K\alpha$ intensity distribution over the specimen surface. [J. T. Campbell, F. W. J. Garton, and J. D. Wilson, *Talanta* **15**, 1205 (1968); courtesy of the authors and Pergamon Press.]

TABLE 19.1. Performance of X-Ray Secondary-Emission Spectrometers for Small Selected Areas

Z-line[a]	Target, kV, mA	Path	Detector	Aperture[b]	Crystal[c]	Intensity, counts/s	
						Line	Background
A. Cylindrical secondary pinhole aperture, flat crystal [Heinrich (*H28*)]							
$_{14}$Si $K\alpha$	W, 50, 50	He	Flow	0.8-mm	EDDT	18	0
$_{16}$S $K\alpha$	”	”	”	”	NaCl	360	2
$_{22}$Ti $K\alpha$	”	”	”	”	LiF	3,000	10
”	”	”	”	0.1-mm	”	40	0
$_{26}$Fe $K\alpha$	”	Air	Scint.	0.8-mm	”	6,250	7
”	”	”	”	0.1-mm	”	125	0
$_{29}$Cu $K\alpha$	”	”	”	0.8-mm	”	11,000	12
”	”	”	”	0.1-mm	”	160	0
$_{40}$Zr $K\alpha$	”	”	”	0.8-mm	”	4,700	12
”	”	”	”	0.1-mm	”	80	0
$_{47}$Ag $K\alpha$	”	”	”	0.8-mm	”	2,300	15
”	”	”	”	0.1-mm	”	43	0
$_{50}$Sn $K\alpha$	”	”	”	0.8-mm	”	1,500	12
”	”	”	”	0.1-mm	”	25	0
$_{73}$Ta $L\alpha_1$	”	He	Flow	0.8-mm	”	1,350	3
”	”	”	”	0.1-mm	”	21	0
$_{82}$Pb $L\alpha_1$	”	Air	Scint.	0.8-mm	”	1,150	3
”	”	”	”	0.1-mm	”	15	1

TABLE 19.1. *(continued)*

Z-line[a]	Target, kV, mA	Path	Detector	Aperture[b]	Crystal[c]	Intensity, counts/s	
						Line	Background
B. Tapered-bore secondary pinhole aperture, semifocusing curved crystal [Loomis (*L39*)]							
$_{12}$Mg $K\alpha$	Cr, 50, 40	He	Flow	0.030-inch	ADP, 8 inch	253	—
$_{13}$Al $K\alpha$	,,	,,	,,	,,	EDDT, 8 inch	1,418	—
$_{14}$Si $K\alpha$	,,	,,	,,	,,	,,	2,386	—
$_{16}$S $K\alpha$[a]	,,	,,	,,	,,	NaCl, 7 inch	9,785	—
$_{17}$Cl $K\alpha$[a]	,,	,,	,,	,,	,,	14,098	—
$_{19}$K $K\alpha$[a]	,,	,,	,,	,,	LiF, 8 inch	7,748	—
$_{20}$Ca $K\alpha$[a]	,,	,,	,,	,,	,,	29,290	—
$_{22}$Ti $K\alpha$	,,	,,	,,	,,	,,	124,500[d]	—
$_{24}$Cr $K\alpha$	,,	,,	,,	,,	,,	6,827	—
$_{28}$Ni $K\alpha$	,,	,,	,,	,,	LiF, 14 inch	14,610	—
$_{30}$Zn $K\alpha$	,,	Air	Scint.	,,	,,	14,390	—
$_{39}$Y $K\alpha$	,,	,,	,,	,,	LiF, 24 inch	32,450	—
$_{42}$Mo $K\alpha$	,,	,,	,,	,,	,,	13,420	—
C. Tapered-bore primary pinhole aperture, full-focusing curved crystal [Loomis (*L39*)]							
$_{12}$Mg $K\alpha$	Cr, 50, 40	Vacuum	Flow	0.030-inch	ADP, 8 inch/4 inch	320	—
$_{13}$Al $K\alpha$	,,	,,	,,	,,	EDDT, 8 inch/4 inch	1,560	—
$_{14}$Si $K\alpha$	,,	,,	,,	,,	,,	2,040	—
$_{15}$P $K\alpha$[a]	,,	,,	,,	,,	,,	640	—
$_{16}$S $K\alpha$[a]	,,	,,	,,	,,	NaCl, 8 inch/4 inch	9,600	—
$_{17}$Cl $K\alpha$[a]	,,	,,	,,	,,	,,	10.960	—
$_{19}$K $K\alpha$[a]	,,	,,	,,	,,	LiF, 8 inch/4 inch	28,600	—
$_{20}$Ca $K\alpha$[a]	,,	,,	,,	,,	,,	154,000[d]	—
$_{22}$Ti $K\alpha$	,,	,,	,,	,,	,,	303,600[d]	—

$_{24}$Cr $K\alpha$	W, 50, 40	”	”	”	”	152,000[d]	—
$_{26}$Fe $K\alpha$	”	”	”	”	”	216,400[d]	—
$_{28}$Ni $K\alpha$	”	”	”	”	”	208,800[d]	—
$_{47}$Ag $L\alpha_1$	Cr, 50, 40	”	”	”	”	4,640	—
D. Secondary slit aperture, semifocusing curved crystal [Bertin (*B32*)]							
$_{24}$Cr $K\alpha$	W, 50, 45	Air	Flow	0.00012-inch	LiF, 14 inch	71	0
$_{28}$Ni $K\alpha$	”	”	Kr prop.	”	”	186	1
$_{30}$Zn $K\alpha$	”	”	”	”	”	223	1
$_{32}$Ge $K\alpha$	”	”	”	”	”	213	1
$_{73}$Ta $L\alpha_1$	”	”	”	”	”	26	1
$_{78}$Pt $L\alpha_1$	”	”	”	”	”	33	1
$_{82}$Pb $L\alpha_1$	”	”	”	”	”	35	1
E. Comparison of secondary slit and pinhole apertures, flat crystal [Bertin (*B35*)]							
$_{29}$Cu $K\alpha$	W, 50, 45	Air	Kr prop.	0.1-mm pinhole	LiF	898	11
”	”	”	”	0.05-mm pinhole	”	183	3
”	”	”	”	0.025-mm pinhole	”	30	3
”	”	”	”	0.025-mm slit	”	3,152	40
$_{82}$Pb $L\alpha_1$	”	”	”	0.1-mm pinhole	”	86	1
”	”	”	”	0.05-mm pinhole	”	19	1
”	”	”	”	0.025-mm pinhole	”	5	1
”	”	”	”	0.025-mm slit	”	406	6

[a] All secondary targets were pure element except the following: P—ADP crystal (27% P). S—Rubber (40% S). Cl—Polyvinyl chloride (35% Cl). K—Fused $K_2S_2O_7$ (30% K). Ca—Calcite (40% Ca).

[b] 0.030 and 0.00012 inch are 0.75 and 0.003 mm, respectively.

[c] Numbers are crystal radii; for example, 8-inch indicates that the crystal is curved to 8-inch radius, 8-inch/4-inch indicates that the crystal is curved to 8-inch radius, ground to 4-inch radius. The inch-to-cm conversions are as follows: 4 inches = 10 cm, 7 = 17.5, 8 = 20, 14 = 35, 24 = 60.

[d] Loomis (*L39*) reports the data in terms of counts/s per mA; the data here are all reported in terms of 40 mA to allow direct comparison with the other parts of the table.

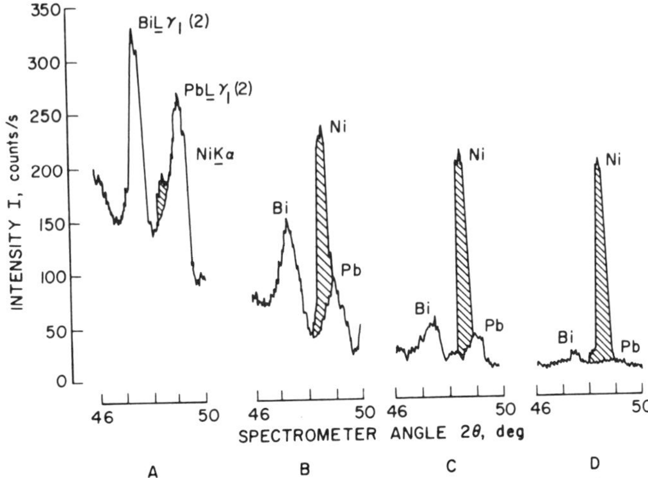

FIG. 19.11. Partial spectra from a 0.04-mm nickel wire imbedded in Wood's metal (Bi–Pb–Cd–Sn). [T. C. Loomis and K. H. Storks, *Bell Laboratories Record* **45**, 2 (1967); courtesy of the authors and the Bell Telephone Laboratories.] A. Conventional flat-crystal x-ray spectrometer. B. Full-focusing curved-crystal x-ray probe (Figure 19.9) with 1.6-mm pinhole. C. Same as B with 0.75-mm pinhole. D. Same as B with 0.35-mm pinhole.

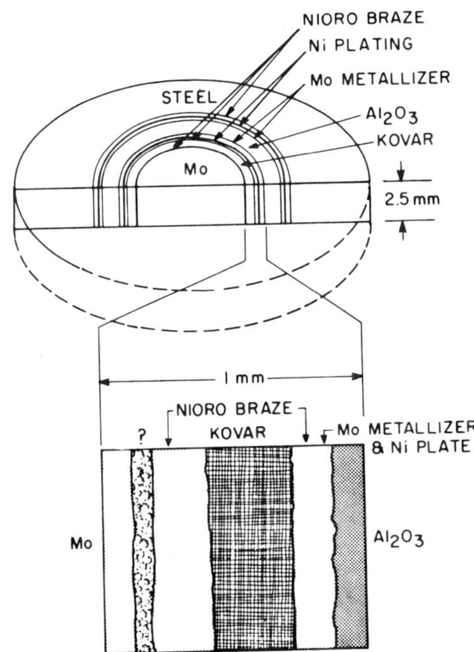

FIG. 19.12. Diagram and photomicrograph of circular metal-ceramic bonded system, which was sectioned as shown and subjected to selected-area analysis over the 1-mm width indicated (Figure 19.13). On the diagram, thicknesses of metallizer, plating, and braze layers are exaggerated.

involving both metal-to-metal and metal-to-ceramic bonds. Figure 19.12 shows the various concentric layers in the disk as established by metallography and x-ray-probe spectrometry, and a photomicrograph of the region of specific interest in the x-ray study. Particular curiosity was aroused by the layer indicated by the question mark on the photomicrograph; only a braze was expected between the Kovar and the molybdenum. The disk was sectioned on a chord just intersecting the molybdenum center area, rather than on the diameter as shown in Figure 19.12; this section results in somewhat wider layers in the zone of interest and facilitates the x-ray

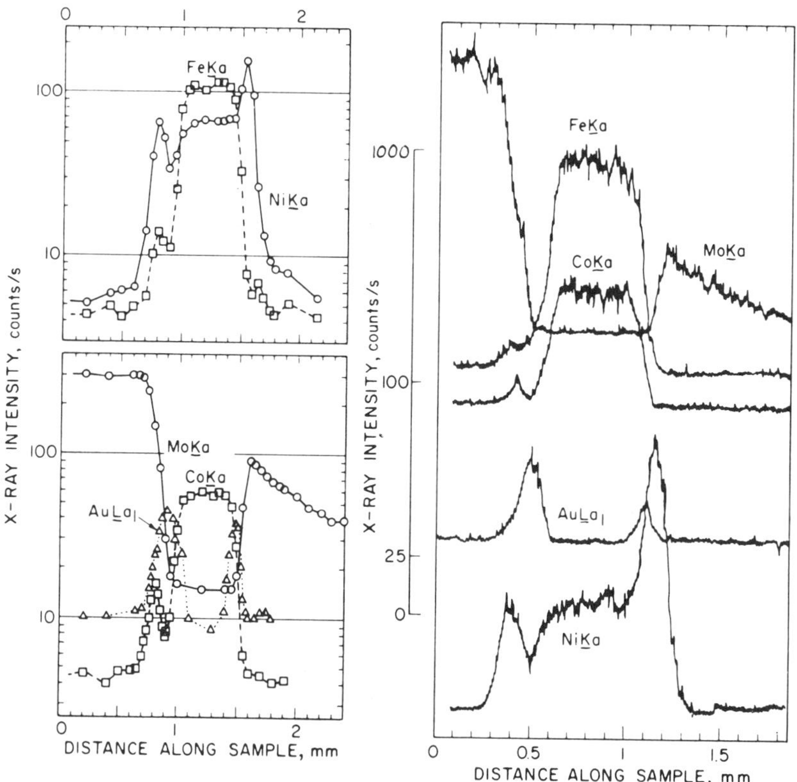

FIG. 19.13. Intensity profiles across the indicated portion of the system shown in Figure 19.12 made with manual incremental sample translation and the scaler–timer (left) and with motor-driven sample translation and the ratemeter–recorder (right). The selected-area accessory was that shown in Figure 19.7 set at 0.006-mm (0.00025-inch) slit width. To avoid confusion, the curves at right were traced from the ratemeter charts and displaced vertically with respect to one another. The low-intensity portion of each curve corresponds to 0 on the intensity scale. The scale is linear up to ~25 counts/s, logarithmic above.

study. The disk was ~2.5 mm thick, covering substantially less than the full useful height of the slit. The specimen was scanned with a 0.006-mm (0.00025-inch) slit for Fe $K\alpha$, Co $K\alpha$, Ni $K\alpha$, Mo $K\alpha$, and Au $L\alpha_1$ both manually and with the motor drive. The results of both series of scans are shown in Figure 19.13, where the following features are evident: (1) The ceramic was metallized with molybdenum, which diffused a considerable distance into the ceramic; (2) the molybdenum was nickel-plated to improve bonding; (3) the brazing metal was a gold–nickel alloy, probably Nioro (82Au–18Ni); (4) separation of gold and nickel in the brazing alloy has occurred; and (5) the well-defined phase indicated by the question mark on the photomicrograph in Figure 19.12 is seen to contain cobalt and iron, which appear to have been dissolved from the Kovar by the Nioro during brazing and deposited near the molybdenum.

Figure 19.13 permits comparison of the manual and motor-drive methods. It is evident that the information available from the two methods is substantially the same. However, whereas the manual study required nearly half a work day, each of the motor-drive scans required ~18 min, the entire study ~1.5 h.

More recently, the writer has identified eight distinct layers involving six elements in a distance of ~0.25 mm using a 0.003-mm (0.00012-inch) secondary slit (Figure 19.7) and a specimen translation speed of 0.5 mm/h. Comparable performance is to be expected with pinholes and full-focusing curved-crystal accessories.

Chapter 20

Other Analytical Methods Based on Emission, Absorption, and Scatter of X-Rays; Other Spectrometric Methods Involving X-Rays

The x-ray secondary-emission spectrometer is an extremely versatile instrument and need not necessarily be operated in the mode for which it was designed. Use of the instrument for energy-dispersive (Section 8.2), nondispersive (Section 8.3), and selected-area (Chapter 19) analysis has already been described, and this chapter is devoted principally to still other modes of operation. Brief consideration is also given to some other spectrometric methods involving x-rays, but for which special instrumentation is required. Each of the methods summarized in this chapter would require a book of its own for adequate treatment. This chapter is intended only to give the x-ray spectrographer an elementary introduction to some analytical methods related to his field.

20.1. X-RAY ABSORPTION METHODS

Whereas x-ray emission spectrometry is based on measurement of the characteristic x-rays *emitted* by the specimen, x-ray absorption methods are based on measurement of x-rays *transmitted* by the specimen. We have already considered emission–absorption methods (Section 14.8.1). In this section, we consider methods involving only absorption (*L26*). Simple polychromatic and monochromatic x-ray absorptiometers can be made consisting of a radioisotope source, specimen cell, and detector–readout system. This section is limited to application of standard, commercial, flat-crystal x-ray spectrometers to x-ray absorptiometry. However, the principles are the same for both types of instrument. Specimen preparation for all x-ray absorption methods is discussed in Section 20.1.7.

20.1.1. Polychromatic X-Ray Absorptiometry

20.1.1.1. Principles and Instrumentation

In polychromatic x-ray absorptiometry (*25m*), which is analogous to photometry with white light, the primary x-ray beam is directed through samples and standards of equal thickness and similar composition. Transmitted intensity, which varies inversely with analyte concentration, is measured. Calibration curves or mathematical relationships are established from the standard data and permit analysis of the samples. Sensitivity is proportional to the difference in mass-absorption coefficient of the analyte across its absorption edge and to the difference in mass-absorption coefficients of analyte and matrix on the short side of the edge. From Equations (2.10) and (2.12),

$$I = I_0 \exp -\{[C_A(\mu/\varrho)_A + \textstyle\sum C_i(\mu/\varrho)_i]t\} \tag{20.1}$$

$$I = I_0 \exp -\{[C_A(\mu/\varrho)_A + (1 - C_A)(\mu/\varrho)_M]t\} \tag{20.2}$$

where I_0 and I are incident and transmitted intensity, respectively; μ/ϱ is mass-absorption coefficient (cm²/g) for the polychromatic x-rays; C is mass concentration (g/cm³); t is specimen thickness (cm); i refers to individual matrix elements; A and M refer to analyte and matrix, respectively; and $(1 - C_A)$ is matrix concentration.

The instrument arrangements are shown in Figures 9.2 and 9.3 for goniometers having horizontal and vertical planes, respectively. The x-ray tube is rotated on its axis so the window is in position *W2*, where it faces directly into the source collimator. The absorption specimen is placed in position 7 or 8. Selected-area apertures *PA* and *SA* are omitted, but a tunnel of appropriate area and shape may be used to confine the x-ray beam to the specimen area. The goniometer is set at 0° 2θ so the x-ray tube window, specimen, collimator, and detector are in line, and the crystal is removed so it will not partially block the x-ray beam.

Polychromatic absorptiometry is most conveniently practiced on a diffractometer, which already has the x-ray beam path described above. Alternatively, a spectrogoniometer may be arranged for polychromatic absorptiometry in one of three ways: (1) It may be fitted with diffractometer accessories; (2) the collimators and detector may be elevated on spacers so they are in line with the x-ray tube window after rotation to position $W2$; (3) the spectrogoniometer may be arranged for conventional x-ray secondary-emission spectrometry with the x-ray tube window in position $W1$, but with the crystal removed and the goniometer set at $0° \, 2\theta$. A very low-Z scatter target is placed in the specimen compartment in position 2, the same position as an emission specimen. The absorption specimen is placed in position 5 or 6. The scatter target scatters primary x-rays into the x-ray optical path, but at greatly reduced intensity.

In order to minimize the effects of minor variations in the specimen matrix, the x-ray tube is operated at a potential selected to make $C_A(\mu/\varrho)_A - \sum C_i(\mu/\varrho)_i$ [Equation (20.1)] as large as possible. In practice, this means a potential such that the peak of the continuum lies on the short-wavelength side of the analyte absorption edge, where analyte absorption is maximum and transmitted intensity most sensitive to analyte concentration. Transmitted intensity may be measured without pulse-height selection. However, some improvement is realized by using the pulse-height selector with the window set to pass a relatively wide band of wavelengths on the short side of the analyte absorption edge. It may be feasible to determine two or three analytes having widely separated absorption edges by using a relatively narrow window set on the short side of each edge in turn.

Polychromatic absorption measurements are also made on specially built x-ray photometers and on small, compact instruments consisting of a radioisotope source, detector, and electronic system. A novel method of polychromatic x-ray absorptiometry involving modulated excitation is described in Section 8.3.4.

20.1.1.2. Advantages and Limitations

The method is simple and convenient. When some accuracy can be sacrificed, transmitted intensity can be read on the panel ratemeter or on the recorder; otherwise, the scaler is used. The high primary-beam intensity gives relatively high transmitted intensity, so good statistical precision is attained, or else, some precision may be sacrificed so that specimens may be thicker or have higher effective atomic number. If very long path length is feasible, gas and vapor specimens can be analyzed. The method is readily adaptable to on-stream analysis and process control, of lead content in gasoline, for example.

However, the method has serious limitations:

1. It is substantially nonspecific and limited to a single analyte having relatively high absorption with respect to the matrix, or at best to a very few such analytes having widely separated absorption edges.

2. The method is useless for qualitative analysis and for quantitative analysis of samples having substantial concentrations of unknown elements or variable matrix composition.

3. Applicability to low-Z elements is very limited. Absorption edges of these elements lie at long wavelengths, where primary intensity is low and matrix absorption high. Consequently, the transmitted beam is mostly shorter-wavelength radiation largely unaffected by the analyte absorption edge.

4. The high primary-beam intensity, although advantageous for statistical precision, causes serious difficulty with liquid specimens.

5. Because of unequal absorption of long and short wavelengths, the effective wavelength of the polychromatic beam changes continuously with distance through the specimen and with analyte concentration. Therefore, calculations of corrections for matrix absorption and other effects are complex and inaccurate, and such corrections are best derived empirically.

6. A very serious limitation in polychromatic x-ray absorptiometry is the severe dependence of detector output on fluctuations in x-ray tube potential, detector output varying by as much as the 24th power of x-ray tube potential. For example, in one study (*Z11*), scintillation counter output increased from 150 to 1000 μA for an increase in x-ray tube potential from 14.0 to 15.2 kV. This condition is dealt with by regulation of the x-ray tube potential to $\pm 0.01\%$, simultaneous measurement of I_0 and I, or alternate repetitive measurement of I_0 and I. These last two techniques are discussed in Section 20.1.4.3.

20.1.2. Monochromatic X-Ray Absorptiometry

20.1.2.1. Principles and Instrumentation

The principle of monochromatic x-ray absorptiometry is the same as that of polychromatic absorptiometry except that monochromatic radiation is used having wavelength on the short side of the analyte absorption edge

(*25g*, *L21*). The monochromatic radiation is usually obtained in one of three ways: (1) The primary beam can be directed through the specimen as in polychromatic absorptiometry, but a filter is placed in the beam to pass the principal target line and attenuate other lines and continuum; (2) the primary beam can be directed through the specimen as before, but the goniometer is set to diffract and detect only the selected wavelength; (3) the primary beam can be directed on a secondary target (radiator or "fluorescer") having a strong spectral line at the appropriate wavelength.

In order to realize sensitivity and specificity, one would intuitively select a wavelength just short of the analyte absorption edge, and Equation (20.1) largely confirms this choice: For highest sensitivity in measurement of C_A, $(\mu/\varrho)_A t$ should be as large as possible. However, there are other considerations: It is desirable that the mass-absorption coefficients for the matrix elements be as small and as nearly the same as practical so that $\sum C_i(\mu/\varrho)_i$ is small and nearly constant. Also, $(\mu/\varrho)_A t$ may not be too large, or the transmitted intensity will be too low, necessitating prohibitive counting times.

The instrument arrangements are shown in Figures 9.2 and 9.3 for goniometers having horizontal and vertical planes, respectively. If the monochromatic wavelength is to be derived from the primary beam, the instrument arrangement is the same as for polychromatic absorptiometry, with the x-ray tube window in position *W*2 and the absorption specimen in position 7 or 8. For filtered primary beams, the crystal is removed and the goniometer set at 0° 2θ; this technique avoids the severe intensity loss incurred by crystal monochromatization, but is generally limited to the target lines of commercially available tubes. For unfiltered beams, the crystal is replaced and the goniometer set to diffract the selected wavelength, which may be a target line or a wavelength in the continuum. If the monochromatic wavelength is to be derived from a secondary target, the instrument is arranged for conventional x-ray spectrometry, with the x-ray tube window in position *W*1. The secondary target is placed in the specimen compartment in position 2, the same position as an emission specimen. The absorption specimen is placed in position 5 or 6. The goniometer may be set at 0° 2θ with the crystal removed or at 2θ for a specific secondary target line with the crystal in place. The former technique (0° 2θ) gives higher intensity, but all secondary target lines are measured, and so all but the weakest ones must have wavelengths shorter than the analyte absorption edge. The latter technique gives strictly monochromatic radiation, but at greatly reduced intensity. In any of these arrangements, a tunnel should be provided to confine the incident x-rays. For monochromatic wavelengths derived directly from the primary x-ray beam, Section 20.1.4.1 briefly discusses the relative merits of placing the specimen cell between the x-ray source and crystal and between the crystal and detector.

Monochromatic absorption measurements are also made on special x-ray photometers and on instruments having radioactive sources (*G31*, *H72*).

20.1.2.2. Advantages and Limitations

Monochromatic absorptiometry eliminates or minimizes most of the limitations of polychromatic absorptiometry:

1. It is more specific, and several analytes can be determined if their absorption edges are relatively widely spaced and there are no interfering edges of matrix elements.

2. When the incident wavelength is derived from a secondary target, the method is better than polychromatic absorptiometry for low-Z elements. The secondary emission line is likely to be more intense than the primary beam in the long-wavelength region. Nevertheless, high matrix absorption for long wavelengths still limits the utility of the method for low-Z elements.

3. When the incident wavelength is derived from a secondary target, the specimen is not subjected to the direct primary beam, and the reduced incident intensity minimizes heating and radiolysis effects in liquid specimens.

4. The "effective wavelength" does not change with distance through the specimen or with analyte concentration, so calculations of matrix absorption coefficient and other corrections are simpler and analytical results more accurate.

The principal disadvantage, compared with polychromatic absorptiometry, is the much lower incident intensity at medium and short wavelengths. In polychromatic absorptiometry, a wide band of the primary continuum is effective, whereas in monochromatic absorptiometry, only a single wavelength is effective—and even this may be derived by secondary excitation. Another disadvantage is that, unless radioisotope sources are used, more elaborate equipment is required.

20.1.3. Applications of X-Ray Absorptiometry

A typical x-ray absorptiometric analysis consists in the measurement of incident I_0 and transmitted I intensities [Equation (20.1)] from samples and standards. The I_0 may be measured through an empty specimen cell or support, or from a specimen of analyte-free matrix or solvent. Analytical data may be derived from a curve of I/I_0 *versus* C_A.

X-ray absorptiometry is well suited to determination of a relatively high-Z analyte in solutions having relatively constant low-Z matrix compositions; for example, phosphorus, sulfur, chlorine, lead tetraethyl, ethylene bromide, or heavy-metal additives in gasoline and other hydrocarbon liquids; plate metals in plating solutions; and heavy elements in mineral processing effluents.

Porosity of solid and compacted materials can be evaluated (*C51*):

$$P = \frac{v_{\text{ap}} - v_{\text{tr}}}{v_{\text{ap}}} = 1 - \frac{v_{\text{tr}}}{v_{\text{ap}}} = 1 - \frac{(\varrho t)_{\text{tr}}}{(\varrho t)_{\text{ap}}} \tag{20.3}$$

Since $I = I_0 \exp[-(\mu/\varrho)\varrho t]$ [Equation (2.10)],

$$2.303 \log_{10}(I_0/I) = (\mu/\varrho)\varrho t \tag{20.4}$$

Therefore

$$P = 1 - \frac{2.303 \log_{10}(I_0/I)}{(\mu/\varrho)(\varrho t)_{\text{ap}}} \tag{20.5}$$

where P is porosity; v is volume (cm^3); I_0 and I are incident and transmitted intensities, respectively; μ/ϱ is mass-absorption coefficient (cm^2/g); ϱ is density (g/cm^3); t is thickness (cm); ϱt is mass thickness (g/cm^2); and subscripts "ap" and "tr" indicate apparent and true values, respectively. Thus, if the mass-absorption coefficient of the material is known, porosity is derived from the measured values of I_0, I, and $(\varrho t)_{\text{ap}}$. Even if mass-absorption coefficient is not known, porosity is readily compared for specimens having the same composition and thickness. The technique is especially applicable to paper and wood products, asbestos products, ceramics and glasses, plastics, filled materials, etc.

Gases and vapors are readily analyzed by even polychromatic absorptiometry, and analysis of gases containing a relatively heavy element ($Z >$ 10, neon) is particularly simple. However, the following precautions should be taken (*W15*): (1) long specimen cells having highly transparent windows should be used; (2) long x-ray wavelengths are preferable; (3) if polychromatic x-rays are used, changes in effective wavelength must be taken into account, especially at long wavelengths; and (4) the specimen mass or density is preferably calculated from the gas laws, in which case pressure and temperature must be known.

Absorptiometry is also convenient for: (1) measurement of thickness of relatively high-Z coatings on x-ray-transparent substrates, such as paper and wood products, plastics, and ceramics (*S48*); (2) determination of relatively high-Z impurities and fillers in similar products and in textiles; (3) identification and evaluation of purity of organic chemicals; (4) identification of plastics and ceramics by comparison of transmittance of specimens of equal thickness; (5) determination of chlorine in plastics; (6) determination of ratio of inorganic and organic elements in plant and animal tissue; (7) selected-area analysis; and (8) dynamic process control. Because of the high primary-beam intensities in polychromatic absorptiometry, useful measurements may be obtained with very small beams, and selected-area and point-by-point analysis is feasible. The uniformity of cadmium-, barium-, and lead-salt impregnation of carbon used for making electric motor brushes has been evaluated in this way.

20.1.4. X-Ray Absorption-Edge Spectrometry (Differential X-Ray Absorptiometry)

20.1.4.1. Principles and Instrumentation

Referring to Figure 2.2, suppose that monochromatic x-ray beams of wavelength 0.40 Å (Cs $K\alpha$) and 0.45 Å (Te $K\alpha$) are passed in turn through a thin foil of ship-nail brass (64Cu–25Zn–8.5Pb–2.5Sn). Only tin, whose K-absorption edge (0.42 Å) is bracketed by these wavelengths, undergoes a marked change in absorption coefficient, and therefore in transmittance. Similarly, if beams of wavelength 1.34 Å (Ga $K\alpha$) and 1.54 Å (Cu $K\alpha$) were passed through the foil, only copper ($\lambda_{K_{\mathrm{ab}}}$ 1.38 Å) would undergo a marked change in transmittance. X-ray absorption-edge spectrometry (or differential x-ray absorptiometry), first applied by Glocker and Frohnmeyer (*G19*, *G20*), is based on this phenomenon (*25g*, *D20*). One or more wavelengths on each side of the analyte absorption edge are passed in turn through each sample and through an empty cell or blank. Analyte concentration is calculated from these measured intensities, the absorption coefficient of the analyte at the bracketing wavelengths, and the specimen thickness. In general, standards are not required.

An expression having the form of Equation (20.1) can be written for each of the two wavelengths λ_S and λ_L bracketing the analyte absorption edge on the short and long sides, respectively, as shown in Figure 20.1. Incidentally, throughout Section 20.1.4 and in Figure 20.1, subscripts L and S refer to the long- and short-wavelength sides, respectively, of the

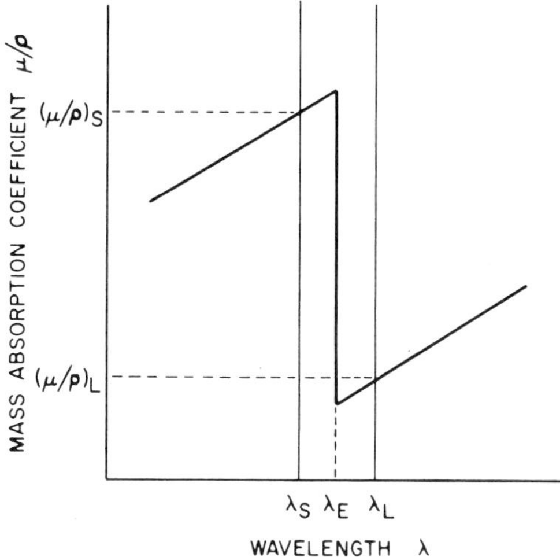

FIG. 20.1. Principle of absorption-edge spectrometry.

analyte absorption edge, rather than analyte line and standard, their usual significance elsewhere in this book. If the expression for λ_S is divided by that for λ_L, and if we let

$$\Delta(\mu/\varrho) = (\mu/\varrho)_S - (\mu/\varrho)_L \tag{20.6}$$

the following equation results:

$$I_S/I_L = (I_S/I_L)_0 \exp\{-[C_A\,\Delta(\mu/\varrho)_A + (1 - C_A)\,\Delta(\mu/\varrho)_M]t\} \tag{20.7}$$

where the notation is as defined in Equations (20.1) and (20.6). The exponential term may be separated as follows:

$$I_S/I_L = (I_S/I_L)_0 \exp[-(1 - C_A)\,\Delta(\mu/\varrho)_M t] \exp[-C_A\,\Delta(\mu/\varrho)_A t] \tag{20.8}$$

The matrix undergoes little change in absorption coefficient between the bracketing wavelengths, that is, $\Delta(\mu/\varrho)_M \approx 0$, so the first exponential term becomes exp($\sim$0), or $\sim$1. Also, for a given series of measurements, the incident intensities at the bracketing wavelengths, and therefore $(I_S/I_L)_0$, are constant. Then,

$$I_S/I_L = K \exp[-C_A\,\Delta(\mu/\varrho)_A t] \tag{20.9}$$

It is evident that analytical sensitivity—the slope of the curve of Equation (20.9)—increases with path length t and with $\Delta(\mu/\varrho)_A$, which, in turn, increases with the analyte absorption-edge jump difference (Figure 2.4) and with proximity of the bracketing lines to the analyte edge. The bracketing wavelengths should also be as close together—and therefore as close to the edge—as possible to minimize difference in matrix absorption and to avoid bracketing an absorption edge or spectral line of a matrix element. Referring to Equation (20.9), in principle, the more nearly λ_L approaches λ_S, that is, the more nearly $\lambda_L - \lambda_S$ approaches zero, the more nearly $I_{L,0}$ for the matrix approaches $I_{S,0}$ and $(\mu/\varrho)_{M,L}$ approaches $(\mu/\varrho)_{M,S}$. However, the two wavelengths should be far enough from the edge, especially on the short-wavelength side, to avoid absorption-edge fine structure (see below).

In general, only *K* and *L*III absorption edges have jump ratios $(\mu/\varrho)_S/(\mu/\varrho)_L$ large enough to be useful for this application, and *K*-edge jumps are larger than *L*III-edge jumps for all elements (Figure 2.4). However, it is the absorption-edge *difference* $(\mu/\varrho)_S - (\mu/\varrho)_L$ that enters into the equations for absorption-edge spectrometry, and for the heavier elements, *L*III edges have higher jump differences than *K* edges (Figure 2.4). Consequently, the *L*III edge may be more favorable than the *K* edge for these elements; then again, it may not, because the *L*III edge of an element lies at longer wavelength than the *K* edge, so the overall absorption will be higher and the transmitted intensities lower.

When measurement is made at only one wavelength on each side of the absorption edge, a correction for change in matrix absorption $\Delta(\mu/\varrho)_M$ may be required. This may be done by making measurements at λ_S and λ_L on analyte-free matrix or other material. Some techniques for making this correction are given in Section 20.1.4.3. Alternatively, measurements may be made at several wavelengths on each side of the edge and extrapolated to the edge to get $\Delta(\mu/\varrho)$. When this is done, no correction is required. This extrapolation is preferable to actually measuring λ_S and λ_L extremely close to the edge because of the presence of fine structure near the edge (Section 20.1.6).

The instrument arrangements are identical with those for monochromatic absorptiometry with the crystal in place. The bracketing wavelengths may be obtained from the primary beam or from secondary targets in the specimen compartment. In either case, the goniometer is set at 2θ for each of the wavelengths in turn. The secondary wavelengths may be derived from separate targets or from the same target. Often, an element can be found that has $K\alpha$ and $K\beta$ or $L\alpha_1$ and $L\beta_1$ lines bracketing the analyte absorption edge. Otherwise, an alloy target or a briquet of mixed element, oxide, or other compound powders may be used.

In principle, identical results are obtained with the specimen cell between the x-ray source and crystal and between the crystal and detector.

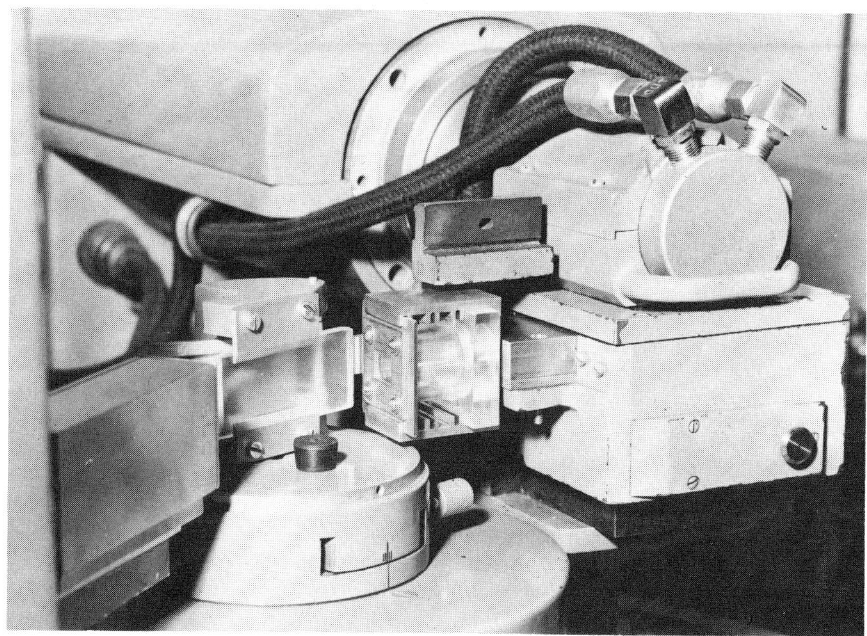

FIG. 20.2. Accessory for x-ray absorptiometry and absorption-edge spectrometry on the secondary-beam exit port of an x-ray spectrometer. [E. P. Bertin, R. J. Longobucco, and R. J. Carver, *Analytical Chemistry* **36**, 641 (1964); courtesy of the American Chemical Society.]

However, when the bracketing wavelengths are obtained from the primary x-ray beam, the former position has disadvantages: Excitation of unwanted characteristic x-rays of matrix elements may occur, and liquid and plastic specimens may heat, expand, outgas, and undergo radiolysis. Consequently, with primary x-ray sources, this specimen position is avoided, except for radioactive specimens and in cases where the intensity of scattered x-rays reaching the detector must be minimized.

Figure 20.2 shows an accessory for absorption-edge spectrometry—or for polychromatic or monochromatic absorptiometry—on the secondary-beam exit port of the specimen compartment of a commercial x-ray spectrometer (*B45*). Figure 20.3 shows a series of liquid-specimen absorption cells having path length 0.5–5 cm (*B45*). The construction of these cells is shown in Figure 20.4. All the cells are made from Lucite blocks *A* 4 cm square and 1 cm thick: 0.5-cm cells are made by machining cylindrical holes 2 cm in diameter through the blocks and concentric circular recesses *C* 3 cm in diameter and 0.25 cm deep on both sides of the blocks; 1-cm cells are made the same way without the recesses; cells having longer path length

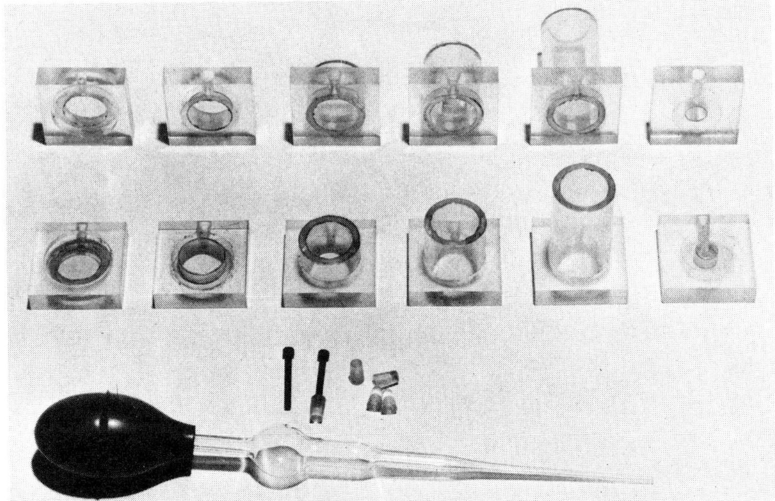

FIG. 20.3. Liquid-specimen cells having various path lengths for x-ray absorptiometry and absorption-edge spectrometry, with plugs and filling pipet. [E. P. Bertin, R. J. Longobucco, and R. J. Carver, *Analytical Chemistry* **36**, 641 (1964); courtesy of the American Chemical Society.]

are made by machining through the blocks cylindrical holes 2.5 cm in diameter in which are cemented cylindrical Lucite inserts *B* having 2.5 cm outside diameter, 2 cm inside diameter, and 2, 3, or 5 cm length. Circular Mylar windows *D* 0.125 mm (0.005 inch) thick are cemented on both sides of all cells with Pliobond rubber cement. All the cells have tapered filling

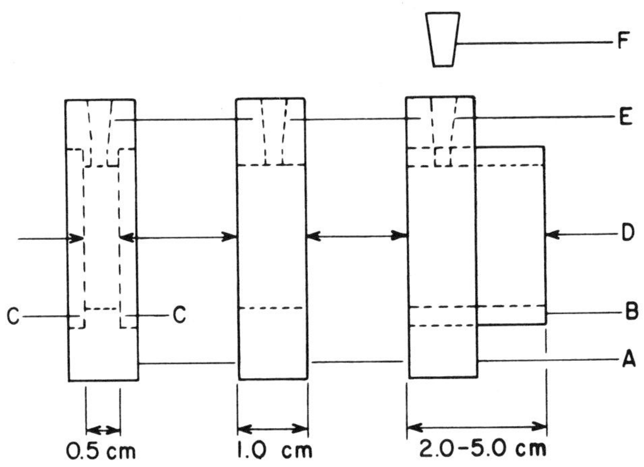

FIG. 20.4. Construction of the cells shown in Figure 20.3.

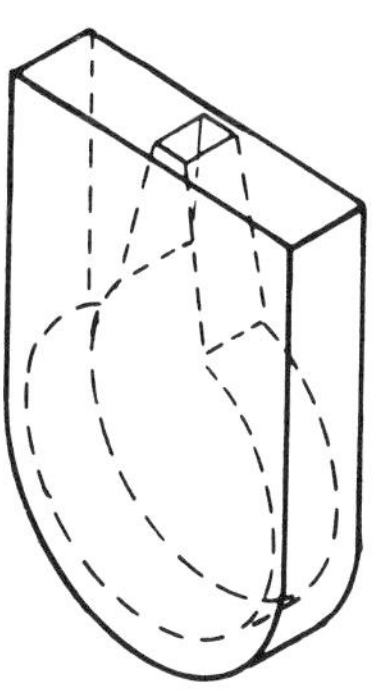

FIG. 20.5. All-polystyrene liquid-specimen cell for x-ray absorptiometry and absorption-edge spectrometry. [E. A. Hakkila and G. R. Waterbury, *Developments in Applied Spectroscopy* **2**, 297 (1963); courtesy of the authors and Plenum Press.]

ports *E* and plugs *F*. The cells may also be made of polyethylene or Teflon and the windows of polystyrene or even thin glass or quartz. Figure 20.5 shows an all-polystyrene cell having windows 0.8 mm ($\frac{1}{32}$ inch) thick (*H5*). Supports for other specimen forms are described in (*B45*).

Multichannel x-ray spectrometers may be used to determine more than one element simultaneously. For example, lead and bromine can be determined in gasoline by placing a sample cell in each of three channels set for Sr $K\alpha$ (0.88 Å), Rb $K\alpha$ (0.93 Å), and Br $K\alpha$ (1.04 Å), respectively. The first two wavelengths bracket the bromine *K* edge (0.92 Å), the last two the lead *L*III edge (0.95 Å). The lines are derived from a briquet of compounds of the three elements.

For maximum accuracy, the x-ray tube power should be highly filtered constant potential, and the operating potential should be well above the potential corresponding to the analyte absorption edge plus the ripple. High-order diffraction from the crystal at the 2θ angles of the bracketing wavelengths must be avoided by: (1) use of secondary targets; (2) operation of the x-ray tube at a potential too low to generate the interfering wavelengths; (3) use of a crystal not having the interfering order; or (4) use of pulse-height analysis. The elements in the crystal, detector window, and scintillation crystal or detector gas must have no absorption edge between the bracketing wavelengths. Detector escape peaks must be avoided or excluded unless they arise from the wavelength being measured, in which case they must be counted.

20.1.4.2. Advantages and Limitations

In view of the many advantages of absorption-edge spectrometry and its ready applicability to commercial x-ray spectrometers, it is difficult to understand why the method is not more widely used:

1. The preparation of standards and calibration curves for x-ray fluorescence spectrometric analysis of a multicomponent system is warranted if there are many samples to be analyzed from time to time, but usually not for one or a few samples on a single occasion. Absorption-edge spectrometry eliminates the need for preparation and storage of standards and for their measurement each time an analysis is made. Consequently, the method is particularly well suited to one-of-a-kind analyses.

2. The method is applicable to liquids, solids, briquets, powders, foils, films, and even gases.

3. Liquid volumes as small as 0.2 ml and very small solid samples may be sufficient.

4. Matrix effects are minimized because only the analyte undergoes a marked change in absorption coefficient at the bracketing wavelengths.

5. The method is highly specific; as many analytes can be determined in each sample by absorption-edge spectrometry as by secondary-emission spectrometry.

6. Because the calculations involve ratios of the intensities transmitted by the empty and sample-filled cells, the method has many of the advantages of ratio methods in general.

7. The disproportionate intensity–concentration relationship for an element having short-wavelength spectral lines in a low-Z matrix (Section 12.2.7) does not occur in absorption-edge spectrometry.

The technique of deriving the bracketing wavelengths from secondary targets brings additional advantages:

8. Most primary-beam absorption-edge procedures are carried out on diffractometers. However, determination of chemical elements is usually the concern of x-ray spectrometric laboratories, many of which are not equipped for diffractometry, or at least cannot afford the time required to rearrange their instruments from time to time for diffractometric operation. The secondary target method is readily applied on x-ray spectrometers with little modification.

9. The total x-ray intensity incident on the sample is only a fraction of that in the primary-beam method; this reduces heating and radiolysis of liquid samples, destruction of thin-film samples, and secondary emission and scatter from the sample.

10. If the bracketing lines are derived from different secondary targets, they are not coexistent in the incident radiation and may be very close without exceeding the resolution capability of the instrument.

11. In the primary-beam method, it may be necessary to operate the x-ray tube at a potential below that at which harmonics of the analyte absorption edge appear (unless pulse-height selection or a suitable crystal is used). There is no such restriction on x-ray tube potential in the secondary-target method.

12. Finally, a relatively high x-ray intensity is available at all wavelengths required for bracketing the analyte absorption edge; this is not necessarily true for the primary-beam method, particularly when the tube is operated at reduced potentials to avoid generating harmonics of the absorption edge, as mentioned above.

The principal limitations of the method are the following:

1. Compared with x-ray fluorescence spectrometry, the method is not as readily applicable to elements of low atomic number (fluorine through titanium), because of high absorption of the long wavelengths in the matrix and cell windows.

2. It is not readily made as sensitive to low concentrations as x-ray emission.

3. Sensitivity is reduced by the presence in the matrix of elements having high atomic number, and the reduction is more severe the higher the atomic number and concentration of the interfering element.

4. The principal source of specific interference is the relatively infrequent occurrence of absorption edges of matrix elements between the bracketing wavelengths.

5. The transmitted intensities may be very low and require long counting times for acceptable statistical precision.

6. Many of the routine procedures of wide general applicability require more or less elaborate preliminary experimental evaluation of mathematical constants whenever the procedure is applied to a new analyte for the first time. However, the increasing availability of tables of accurate absorption coefficients is eliminating the need for much of this work.

7. Finally, the calculations are complex—but no more so than elaborate matrix corrections in x-ray emission spectrometry.

20.1.4.3. Applications

Many absorption-edge spectrometric procedures have been reported, and some of these are routine methods of wide general applicability. A few of these are considered as examples (*B6*, *B45*, *D34*, *K20*, *S63*).

Dunn's procedure (*D34*) is for analysis of solutions and uses a diffractometer, deriving the bracketing wavelengths from the primary beam. The analytical equation is

$$C_{\mathrm{A}} = \frac{2303}{\Delta(\mu/\varrho)_E t}(m_S \log_{10} I_S - m_L \log_{10} I_L - a) \tag{20.10}$$

where C_{A} is analyte concentration (mg/ml); $\Delta(\mu/\varrho)_E$ is analyte absorption-edge jump difference (cm^2/g); t is cell length (cm); I is transmitted intensity; S and L refer to the bracketing wavelengths on the short and long sides of the analyte absorption edge, respectively; and m_S, m_L, and a are experimentally evaluated constants. The method gives good results, but the preliminary work required to evaluate the constants may be time-consuming.

The writer (*B45*) developed a simple routine procedure for solutions, thin solids, foils, briquets, powders, and films. The measurements are made on a standard commercial spectrometer with only minor modification and simple accessories, some of which are shown in Figures 20.2–20.4. The bracketing wavelengths are derived from secondary targets.

Putting Equation (20.2) in common logarithmic form gives

$$2.303 \log_{10}(I_0/I) = C_{\mathrm{A}}(\mu/\varrho)_{\mathrm{A}} t + (1 - C_{\mathrm{A}})(\mu/\varrho)_{\mathrm{M}} t \tag{20.11}$$

where I_0 and I are incident and transmitted intensities, respectively; C_{A} and $1 - C_{\mathrm{A}}$ are analyte and matrix concentrations, respectively (g/cm^3); μ/ϱ is mass-absorption coefficient (cm^2/g); t is sample thickness (cm); and A and M refer to analyte and matrix, respectively. If now an equation of this form is written for each of the bracketing wavelengths λ_S and λ_L, the equation for λ_L is subtracted from that for λ_S, it is assumed that $(\mu/\varrho)_{\mathrm{M}}$ is the same at λ_S and λ_L, and the resulting expression is solved for C_{A}, one gets

$$C_{\mathrm{A}} = \frac{2.303}{[(\mu/\varrho)_S - (\mu/\varrho)_L]_{\mathrm{A}} t}\left[\log_{10}\left(\frac{I_0}{I}\right)_S - \log_{10}\left(\frac{I_0}{I}\right)_L\right] \tag{20.12}$$

where S and L refer to λ_S and λ_L, respectively.

In practice, matrix absorption is somewhat different at λ_S and λ_L, and if Equation (20.12) is to give high accuracy, a correction must be made for this difference. For this purpose, a correction method devised by Barieau (*B6*) was adopted. In a matrix having components 1, 2, . . . , n, the ratios of mass-absorption coefficients $(\mu/\varrho)_S/(\mu/\varrho)_L$ at λ_S and λ_L are related as follows:

$$\left[\frac{(\mu/\varrho)_S}{(\mu/\varrho)_L}\right]_1 = \left[\frac{(\mu/\varrho)_S}{(\mu/\varrho)_L}\right]_2 = \cdots = \left[\frac{(\mu/\varrho)_S}{(\mu/\varrho)_L}\right]_n = \left(\frac{\lambda_S}{\lambda_L}\right)^a = k \tag{20.13}$$

where k and a are experimentally evaluated constants which are different for each λ_S–λ_L line pair. The exponent a is determined largely by the matrix elements, but not their concentrations, and is in fact a measure of the matrix effect; it is minimized by selection of the bracketing wavelengths to be as close as possible. Equation (20.13) assumes that no matrix element has an absorption edge between λ_S and λ_L. Incorporation of the correction constant into Equation (20.12) and conversion of C_A to mg/cm³ gives

$$C_A = \frac{2303}{[(\mu/\varrho)_S - k(\mu/\varrho)_L]_A t}\left[\log_{10}\left(\frac{I_0}{I}\right)_S - k\log_{10}\left(\frac{I_0}{I}\right)_L\right] \tag{20.14}$$

where C_A is in mg/cm³ (or mg/ml). For foils and films, the equation may be put in the form

$$C_A t = \frac{2303}{[(\mu/\varrho)_S - k(\mu/\varrho)_L]_A}\left[\log_{10}\left(\frac{I_0}{I}\right)_S - k\log_{10}\left(\frac{I_0}{I}\right)_L\right] \tag{20.15}$$

where $C_A t$ is area concentration in mg/cm². The first term on the right in Equations (20.14) and (20.15) is constant for each combination of analyte, path length t, and λ_S–λ_L pair.

The correction factor k is evaluated from the intensities I_{cor} transmitted at λ_S and λ_L by an analyte-free correction standard. The author used an aluminum plate for all analyses, but greater accuracy would result if the correction standard were similar to the matrix in composition. Since the first term on the right in Equations (20.14) and (20.15) is finite, it follows that if $C_A = 0$,

$$\log_{10}(I_0/I_{cor})_S - k\log_{10}(I_0/I_{cor})_L = 0 \tag{20.16}$$

from which

$$k = \frac{\log_{10}(I_0/I_{cor})_S}{\log_{10}(I_0/I_{cor})_L} \tag{20.17}$$

An analysis consists in the measurement of the intensities transmitted by the sample-filled cell I, empty cell I_0, and correction standard I_{cor} at λ_S and λ_L. The I_0 and I_{cor} data are used in Equation (20.17) to calculate k; the I_0 and I data and k are used in Equations (20.14) and (20.15) to calculate the analytical result. The absorption coefficients are obtained from the literature directly or by interpolation.

Hakkila *et al.* (*H7*) developed a three-wavelength method for determination of plutonium. Intensities transmitted by the absorption cell filled in turn with water and a standard solution of known plutonium concentration are measured at three wavelengths: Nb $K\beta$ (λ_1) and Mo $K\alpha$ (λ_2), which bracket the Pu *L*III absorption edge, and Nb $K\alpha$ (λ_3). Then the intensities transmitted by the same cell filled with the sample(s) are measured at the two bracketing wavelengths (λ_1 and λ_2). The measurements on the water and standard are used to calculate k_1/k_2:

$$\frac{k_1}{k_2} = \left[\frac{\log_e(I_0/I)_3}{\log_e(I_0/I)_2}\right]^{\log_e(\lambda_2/\lambda_1)/\log_e(\lambda_3/\lambda_2)} \tag{20.18}$$

Then,

$$C = k_1 \log_e(I_0/I)_1 - k_2 \log_e(I_0/I)_2 \tag{20.19}$$

I_0 is the intensity transmitted by the water-filled cell, and I is the intensity transmitted by the cell filled with standard [Equation (20.18)] or sample [Equation (20.19)].

Knapp *et al.* (*K20*) developed an absorption-edge spectrometric method for standard commercial x-ray spectrometers. The bracketing wavelengths are derived from secondary targets, usually the $K\alpha$–$K\beta$ lines of a single element. The transmitted intensity is measured at these two wavelengths from a series of standards in a certain solvent, and a calibration curve is established having the equation

$$\log_e(I_L/I_S) = m_M C_A + b_M \tag{20.20}$$

where I is transmitted intensity; C_A is analyte concentration; m_M is the slope of the curve; b_M is the intercept; L and S refer to the bracketing wavelengths; and M refers to the matrix, that is, the solvent. Calibration curves for other solvents M′ are calculated simply from the absorption coefficients of the pure solvents M and M′ at λ_S and λ_L.

Cullen (*C72*) describes three novel *internal differential absorption-edge* spectrometric methods in which the wavelengths bracketing the analyte absorption edge are either generated within or scattered by the specimen itself: (1) An element having bracketing spectral lines may already be present in the sample; for example, nickel can be determined in nickel–copper alloys from the intensities of the Cu $K\alpha$ and Cu $K\beta$ lines. (2) The element having the bracketing lines may be added to the sample; for example, selenium can be determined in solutions by adding bromine and measuring Br $K\alpha$ and Br $K\beta$ intensities. (3) The two bracketing wavelengths can be derived from the continuous spectrum originating in the x-ray tube and scattered by the sample; for example, silver ($\lambda_{K_{ab}}$ 0.49 Å) can be determined from the intensities scattered at 0.46 and 0.63 Å. Scattered characteristic target lines may be used if they have suitable wavelengths. In all three methods, calibration curves of the function (I_L/I_S) *versus* C_A are established from measurements on standards. These methods, like that of Knapp *et al.* above, lack one of the advantages of most absorption-edge methods, in that they do require standards.

If the two bracketing wavelengths are coexistent—as when both are generated by the same secondary target—some provision must be made to permit their measurement individually. The most convenient means is crystal dispersion. Pulse-height selection for two such close wavelengths is feasible only with Si(Li) detectors, and then only at wavelengths $<\sim 2$ Å. An alternative means is reported by Roy *et al.* (*R31*) in a "nondispersive" absorption-edge method for determination of iodine in solution. The primary x-ray beam excites a lanthanum secondary target to emit its L spectrum, two lines of which closely bracket the I LIII absorption edge (2.719 Å): La $L\alpha_1$ (2.665 Å) and La $L\eta$ (2.740 Å). The lanthanum L radiation passes in turn through the iodine specimen solution and a two-position filter to a scintillation counter. The two filters consist of solutions of barium and iodine salts, respectively. The barium filter solution passes both the bracketing La L lines because both have wavelength longer than the Ba LIII absorption edge (2.363 Å). However, the iodine filter solution strongly absorbs the La $L\alpha_1$ line. The intensity passing through the iodine *specimen* solution is measured in turn through the barium and iodine *filter* solutions. The ratio of the two intensities for standard iodine solutions is plotted as a function of iodine concentration. The calibration curve is then used to analyze iodine sample solutions.

In x-ray absorption-edge spectrometry—as well as x-ray absorptiometry—accuracy is greatly improved by simultaneous measurement of the incident I_0 and transmitted I intensities. This is effected by use of two two-side-window detectors in tandem with the absorption specimen between them (*H14, H74, P2*). The first (front) detector permits measurement of I_0 and also acts as a monitor. The second (back) detector permits measurement of I. Each detector has its own preamplifier, amplifier, pulse-height selector, and scaler. The scaler for the monitor (front) detector is set for a preset count N_1, and on accumulation of this count in scaler 1, both scalers stop simultaneously. At each wavelength measured, the relationship between the count rates from the two detectors is established without the absorption specimen in place. Then the specimen is inserted between the detectors and another pair of measurements is made. If monochromatic x-rays are used, backgrounds can also be measured. The product of linear absorption coefficient μ and absorber thickness t is given by either of the following equations:

$$\mu t = \log_e(N_1/N_2) + \log_e(N_{2,0}/N_1) \tag{20.21}$$

$$= \log_e \frac{I_0}{I} = \log_e \frac{[(N_{2,0}/N_1)/T]N_1}{N_2/T} \tag{20.22}$$

where N_1 is the preset count for the monitor scaler (detector 1); N_2 and $N_{2,0}$ are the counts accumulated from detector 2 with the absorption specimen in and out, respectively; and T is the counting time. Of course, N_2 and $N_{2,0}$ are accumulated in scaler 2 during the time T during which N_1 is accumulated in scaler 1. The ratio $N_{2,0}/N_1$ represents the count-rate correlation of the two detectors. All N's may be corrected for background.

Alternatively, the total counting times for I_0 and I may be divided into a number of interspersed intervals, I being measured with the specimen cell in the x-ray beam, I_0 with it out. The cell may be placed in and out manually (*V19*) or by an automatic mechanical oscillating arrangement (*H14*).

20.1.5. X-Ray Contact Microradiography

If a uniformly thin piece of metal, ceramic, mineral, plastic, wood, paper or other material is placed in contact with the emulsion of a fine-grain photographic plate and irradiated with x-rays, the emulsion records the distribution pattern of the x-rays transmitted by the specimen. Such contact microradiographs may be magnified up to 200× and reveal microscopic voids, cracks, inclusions, segregations, inhomogeneities, and other irregularities in the specimen, or they may reveal the distribution of selected elements over the specimen. An x-ray spectrometer constitutes a convenient instrument for contact microradiography (*25b*, *T19*).

In polychromatic contact microradiography, the primary x-ray beam is used to expose the microradiograph. A small microradiography camera may be placed in the specimen compartment in the position usually occupied by the emission specimen—position 2 in Figures 9.2 and 9.3. The inclination of the camera to the primary beam may prove advantageous, in that the target-to-film distance varies over the specimen, giving a range of exposure conditions. Alternatively, a specimen drawer can be made in which the camera is perpendicular to the x-ray beam—position 4. Provision may be made to incline the camera for stereomicroradiography.

In monochromatic contact microradiography (*S53*), the specimen is photographed with one or more single wavelengths. In Figure 2.2, suppose that 0.4 Å (Cs $K\alpha$) x-rays are passed through a thin foil of Cu–Zn–Pb–Sn alloy in contact with a photographic plate; then suppose a new plate is provided and 0.45-Å (Te $K\alpha$) x-rays are passed through the foil. Only tin, whose K absorption edge lies between these wavelengths, would appear different on the two photographs; if the alloy were inhomogeneous, tin-rich areas would appear lighter (relatively unexposed) in the 0.4-Å photograph where tin absorption is high, darker in the 0.45-Å photograph. The distribution of the other elements can be revealed by photographing the alloy with other appropriate pairs of wavelengths, as indicated in Figure 2.2. Tables of such bracketing spectral lines have been compiled for the K and LIII edges of most of the elements (*25b*, *B45*).

For monochromatic microradiography, the primary x-ray beam is used to excite various secondary targets placed in the specimen drawer (position 2 in Figures 9.2 and 9.3) in the same manner as specimens for secondary-emission spectrometry. These targets are chosen to have strong spectral lines at the wavelengths required for the microradiograph. A small microradiography camera can be mounted directly outside the secondary-beam exit window of the specimen compartment, or in position 5. In instruments having specimen drawers, a special drawer can be made in which the secondary target is inclined in the opposite direction, away from the crystal, toward the front of the drawer (position 3). A window may be cut in the front of the drawer and the camera mounted at this window in position 1. Commercial direct or stereo cameras may be used in either case.

For strictly monochromatic microradiography, the camera may be mounted in position 6 or 8, beyond the crystal, which is set to diffract the required wavelength from a secondary target in the specimen drawer or from the primary beam. Of course, the diffracted intensities are relatively low, so that very long exposure times are required. Also, any pattern of nonuniform diffracted intensity over the crystal surface is superposed on the microradiograph.

20.1.6. X-Ray Absorption-Edge Fine Structure

Throughout this book, absorption edges have been represented as simple abrupt discontinuities in absorption coefficient at a photon energy just sufficient to expel electrons from a specific inner level in the atom. Fine structure (*25q*) consists of deviations from this simple step function and includes deviations both in the abrupt rise and in the region on the high-energy side of the rise. In general, fine structure is confined to within ~200 eV of the edge. Figure 20.6 is a composite absorption-edge fine-

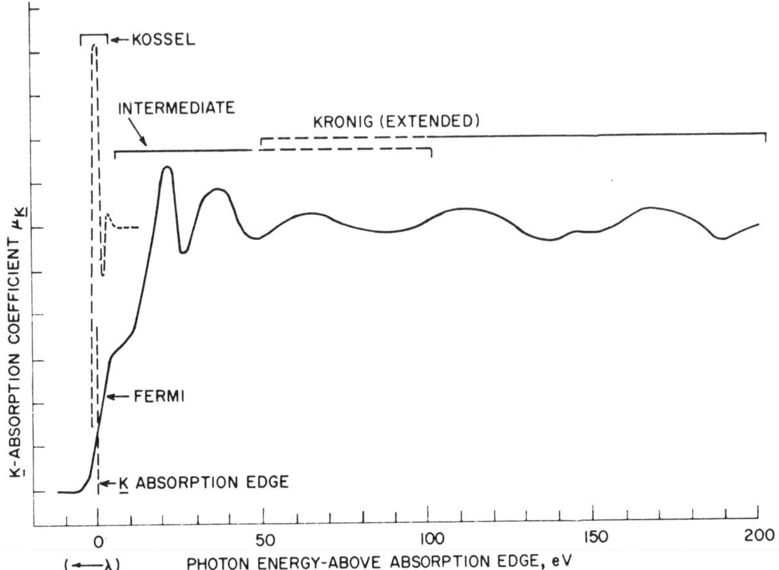

FIG. 20.6. Composite absorption-edge fine-structure curve showing four principal types of fine structure.

structure curve showing the four principal recognized types. All these types do not occur simultaneously.

1. *Fermi-level* fine structure is the first rise in absorption at the K edge of a metal, and is attributed to elevation of a K electron to the Fermi level in the conduction band.
2. *Kossel* fine structure occurs within ~3 eV of the high-energy (short-wavelength) side of the edge, and is attributable to electron transitions to bound atomic states in accordance with the selection rules. This type of fine structure has been observed only in monatomic gases.
3. *Intermediate* fine structure occurs in the region 5–100 eV on the high-energy side of the edge, and is attributable to short-range order, valence state, and neighboring atoms.
4. *Kronig* (or *extended*) fine structure consists of low-amplitude fluctuations in absorption coefficient in the region 50 to ~200 eV on the high-energy side of the edge, and is related to lattice parameter and temperature.

In the abscissa of Figure 20.6, energy increases to the right, wavelength to the left. This sense is opposite to that used throughout this book, but is conventional for fine-structure work. The term *K-absorption coefficient* μ_K is defined below.

Fine-structure measurements may be made on diffractometers, flat-crystal spectrometers, or full-focusing curved-crystal spectrometers; quartz or silicon crystals are used because of their narrow half-width and consequent high resolution. The most precise measurements are made with two-crystal spectrometers.

Incident I_0 and transmitted I intensities are measured at small increments of photon energy (or wavelength) from 10–20 eV below the K edge of the element of interest to 100–200 eV above; $\log_e(I_0/I)$ is calculated for each increment. The K-absorption coefficient μ_K is derived from $\log_e(I_0/I)$ by (1) correcting it for all absorption other than that responsible for expelling K electrons from the element of interest, and then (2) normalizing the corrected values relative to unity at an energy beyond the fine-structure region. Matrix absorption is substantially constant in the region of the absorption edge (if no matrix element has an edge in that region), and no K excitation occurs below the edge. Consequently, $\log_e(I_0/I)$ is relatively low and constant below the edge. This constant value—$\log_e(I_0/I)_{\text{low}}$—is used to correct for absorption from all sources other than K excitation of the element of interest. Somewhere beyond 50 to ~200 eV above the edge, $\log_e(I_0/I)$ becomes substantially constant at some relatively high value. It is this value—$\log_e(I_0/I)_{\text{high}}$—to which the corrected values are normalized. At each incremental photon energy i, K-absorption coefficient is calculated from the following equation:

$$(\mu_K)_i = \frac{\log_e(I_0/I)_i - \log_e(I_0/I)_{\text{low}}}{\log_e(I_0/I)_{\text{high}}} \tag{20.23}$$

The applications of absorption-edge fine structure are based principally on its sensitivity to chemical state. The method is used to establish the nature of chemical bonds; the electronic band structure in solids; identity or class of compounds, especially when the absorbing element is present in more than one oxidation state; oxidation state(s) of the absorbing element; and the nature of catalysts on x-ray-transparent substrates.

20.1.7. Specimen Preparation for X-Ray Absorption

Specimen preparation for x-ray absorption is somewhat more critical than for emission because the x-ray beam passes through the specimen. For quantitative analysis by absorptiometry and absorption-edge spectrometry, the specimen must be thin enough to transmit x-radiation sufficiently intense to be measured with acceptable precision in a reasonable counting time. For microradiography, it must be thin enough to permit photography of the image in a reasonable exposure time. The thickness must be uniform, and opposite surfaces must be plane and parallel. All specimens in a given analysis must have the same thickness. For absorption-edge fine structure, the requirement for uniformity of thickness may be relaxed somewhat.

Films on x-ray-transparent substrates, foils, and uniformly thin sheets of metal, ceramic, plastic, rubber, paper, etc. may usually be used without preparation. Metals, ceramics, minerals, etc. may be reduced to uniformly thin plates by combinations of sawing, rolling, abrading, etching, and electropolishing. Care must be taken not to etch certain elements preferentially. Powders should have small, uniform particle size to minimize packing voids. High-Z powders may be diluted with carbon, lithium carbonate, starch, etc. In a given analysis, the same accurately weighed amount of all specimens may be packed into open, windowless frames or windowed cells having the same volume, or pressed into briquets of the same diameter and thickness. Weighed amounts of powder may be slurried and cast into films of uniform thickness with a casting blade. These films may be stripped and mounted in open frames, or they may be cast on x-ray-transparent substrates. For fine-structure studies, powders may be dusted on adhesive cellulose tape. Liquids are held in cells having windows rigid enough not to bow on filling, yet transparent to x-rays, or small, constant volumes of liquid may be absorbed in filter-paper disks. The lower the analyte concentration, the longer the cell length must be. Gases, having very low absorption coefficients, require very long cell lengths—up to 100 cm—and it may be necessary to calculate the mass of gas from temperature and pressure and the gas laws.

Absorption specimens likely to be nonuniform may be rotated (*B6*); such specimens include packed and briqueted powders, cast films, and specimens supported on filter paper or other material.

20.2. X-RAY SCATTER METHODS

The occurrence of x-ray scatter in spectrometers was formerly regarded only as a nuisance (*J20*), and, in fact, scatter is the principal source of background, limits analytical sensitivity, and may cause spectral-line interference. However, more recently, scattered x-rays have proved to be of great value in x-ray spectrometry. Their use in correction of absorption-enhancement and other errors has already been discussed (Section 14.5). This section describes three analytical methods applicable to commercial x-ray spectrometers and based on efficiency of coherent and/or incoherent scatter.

20.2.1. Coherent Scatter

Ziegler and his co-workers (*M22, Z13*) have developed a method for determination of heavy elements in light matrixes based on coherent (Rayleigh) scatter. The technique may be regarded as x-ray nephelometry. A high-energy (80–150-kV) polychromatic x-ray beam irradiates the specimen solution, but, rather than an analyte spectral line, coherently scattered radiation is measured at an appropriate angle. Maximum differentiation of coherent and incoherent (Compton) scatter is realized by filtering the long-wavelength components from the primary beam, measuring the scatter at a large angle ($\sim 140°\ 2\theta$), and using a discriminator to pass only detector output pulses at or near the maximum primary energy. Since coherent and incoherent scatter are proportional, respectively, to Z^3 and Z, the scattered intensity is highly dependent on concentration of high-Z elements.

The method was used to determine uranium in concentrations 1–10 g/l in solutions having high concentrations of aluminum and stainless steel. Scattered intensity was measured from the sample(s) and from two standards having known uranium concentrations bracketing that of the samples. The analytical equation is

$$C_X = C_1 + (C_2 - C_1)\frac{I_X - I_1}{I_2 - I_1} \tag{20.24}$$

where C is concentration; I is scattered intensity at the selected angle; subscripts 1 and 2 refer to the standards; and X refers to the sample. The work apparently was done on a specially constructed instrument using a full-wave x-ray radiography unit as the x-ray source. However, the technique should be feasible on a 75- or 100-kV x-ray spectrometer, particularly one having a constant-potential generator.

20.2.2. Coherent/Incoherent Scatter Ratio

Dwiggins (*D38, D39*) has made an extremely significant contribution with his method of determining carbon in hydrocarbon liquids using an unmodified conventional x-ray spectrometer. The intensity ratio R of the coherently and incoherently scattered W $L\alpha_1$ line from the x-ray tube is measured from the sample(s) and from a series of pure hydrocarbon liquid compounds having various known carbon/hydrogen ratios. A calibration curve of R *versus* %C is established from the known liquids; analytical concentrations are derived from this curve.

C—H bond type has no effect. If a few percent of nitrogen and/or oxygen is present, the method gives %(C + N + O), and independent determinations of nitrogen and oxygen are required to obtain carbon concentration. In the presence of sulfur in concentrations up to ~10%, if the C/H ratio is constant, *R* is proportional to sulfur concentration; if sulfur concentration is constant, *R* is proportional to C/H ratio. Dwiggins describes ways to correct for sulfur. An uncovered water-cooled liquid-specimen drawer was used in the original work (*D41*), but later, inverted geometry and cells covered with 6-μm (0.00025-inch) Mylar were used. Toussaint and Vos (*T16*) have extended the method to solid hydrocarbons.

20.2.3. Determination of Dry Mass

The intensities of scattered target lines have also been used to determine dry mass of biological and medical specimens (*Z3*). The x-ray tube irradiates the specimen supported on thin (900–1300 Å) nylon film in a vacuum chamber. The vacuum is required to eliminate target-line scatter from air, which would have intensity comparable with that from thin biological specimens. A full-focusing curved-crystal spectrometer and a proportional-counter energy-dispersive spectrometer "view" the specimen at takeoff angles of 45°. The crystal spectrometer measures analyte-line intensities in the usual way. The energy-dispersive spectrometer measures the scattered target lines. A calibration curve of scattered intensity *versus* dry mass is established from different amounts of sucrose ($C_{12}H_{22}O_{11}$) and is linear from <10 to >1000 μg.

The scattered intensity per gram of material is given by

$$I_{sc} = \sum_{i=1}^{i=n} [I_{at,i} C_i (N/A_i)] = N \sum_{i=1}^{i=n} [I_{at,i} (C_i/A_i)] \tag{20.25}$$

in which,

$$I_{at} = I_e \left[\left(\sum_{i=1}^{i=Z} f_i \right) + Z - \sum_{i=1}^{i=Z} f_i^2 \right] \tag{20.26}$$

where I_{sc}, I_{at}, and I_e are scattered intensity per gram, per atom, and per electron, respectively; A is atomic weight; C is concentration (wt fraction); N is Avogadro's number (6.02×10^{23} atoms/mol); Z is atomic number; and f_i is the structure factor of the ith electron in the atom. In Equation (20.25), the summation is for each of the n elements present in the specimen, and i refers to individual elements. In Equation (20.26), the summations are for each of the Z electrons in the atom, and i refers to individual electrons.

In another method for determination of dry mass of biological and medical specimens (*O3*), known amounts of two elements are added to and thoroughly mixed with the samples, one element having a short-wavelength $K\alpha$ line, the other a long-wavelength $K\alpha$ line. The ratio of the intensities of these lines in the absence of absorption is a readily determined constant depending only on the relative amounts of the two elements and the wavelength of the excitation radiation. Monochromatic excitation is used. The absolute intensities of the additive lines serve as a reference for measurement of the excitation radiation; their ratio permits calculation of mass-absorption coefficients of the specimen at the two wavelengths; and Compton-scattered primary intensity permits determination of dry mass.

Still another scattered-intensity method for determination of dry mass is the slurry-density or pulp-density method described in Section 14.5.6.

20.3. SCANNING X-RAY MICROSCOPY

20.3.1. Introduction

There are six basic techniques of x-ray microscopy, that is, of producing x-ray images, usually magnified, of small specimens (*14*, *C61*, *E17*):

1. *Contact microradiography.* A contact microradiograph, made on a fine-grained photographic plate, is enlarged optically; there is no x-ray magnification.

2. *Point-projection* or *shadow microscopy.* A point source of divergent x-rays irradiates the specimen, which is separated from a photographic plate or fluorescent screen and therefore casts an enlarged shadow image.

3. *Von Hamos image spectrograph.* A series of true monochromatic x-ray images of the specimen surface, formed by a curved crystal on a photographic plate, is enlarged optically; there is no x-ray magnification. Each monochromatic image represents the intensity distribution of a different spectral line—and thereby the concentration distribution of the corresponding element.

4. *Reflection microscopy.* A true magnified x-ray optical image is formed by total reflection at a glancing angle by two crossed cylindrical mirrors in series.

5. *Scanning x-ray microscopy with electron beams.*

6. *Scanning x-ray microscopy with x-ray beams.*

These last two techniques are discussed in the next two sections. Contact microradiography and the Von Hamos technique are discussed in Sections 20.1.5 and 5.5.3.4, respectively. Point projection and reflection microscopy are not really relevant to x-ray spectrometry and are not considered further.

Contact microradiography and projection microscopy are based on absorption of x-rays in thin radiolucent specimens. The distribution of an element can be mapped if successive micrographs are made at each of two wavelengths bracketing its absorption edge (Section 20.1.5). Image spectrography, reflection microscopy, and scanning microscopy with electron beams are based on x-ray emission. Scanning microscopy with x-ray beams is applicable to either emission or absorption.

20.3.2. Scanning X-Ray Emission Microscopy

In this technique, the specimen is scanned by a very small—of the order 1 μm—electron or x-ray beam in synchronism with the electron beam in a cathode-ray tube. The instantaneous x-ray emission excited at each incremental area on the specimen is dispersed and detected (or *vice versa*) in a wavelength- or energy-dispersive spectrometer. The instantaneous output intensity of the selected x-ray line is used to modulate the instantaneous brightness of the cathode-ray tube display. Thus the brightness of each incremental area on the display is proportional to the intensity of the selected x-ray line from the corresponding incremental area on the specimen. In this way, the distribution of any selected element can be displayed.

The emission can be measured from the scanned side of the specimen or from the opposite side of thin radiolucent specimens. The magnification is simply the ratio of the scanned areas of the display and specimen and may be of the order 10^3–10^4. Submicrometer resolution can be realized. Stereo x-ray micrographs can be made by photographing two displays with the specimen appropriately tilted for each. If the scanning is done incrementally instead of continuously, the intensity can be scaled at each point and used for other types of display (for example, Figures 21.5 and 21.6), image enhancement, quantitative analysis, etc.

In the electron-probe microanalyzer, the x-ray emission is excited by a scanning electron beam. This instrument is described in detail in Chapter 21.

There are three principal ways to produce scanning x-ray "probes": (1) "focusing" a divergent x-ray source with curved crystals or totally reflecting mirrors; (2) pinhole collimation of a very high x-ray flux; and (3) scanning x-ray tubes, that is, tubes in which the electron beam is focused to a very small spot and scans a rectangular raster on the target so that the x-ray source is a "flying spot," rather than a fixed spot.

With sources 1 and 2 above, the specimen is scanned mechanically under the fixed x-ray beam spot in synchronism with the electron beam in the cathode-ray display tube. Horowitz and Howell (*H63*) built such a scanning x-ray microscope using pinhole collimation of the high x-ray flux produced by an electron synchrotron.

Scanning x-ray tubes have been described by Pattee (*P14*) and Moon (*M51*). In Pattee's tube, the electron beam scans a thin-foil target which also serves as the x-ray window. The specimen, which must be thin and radiolucent, is placed outside but in direct contact with the window and is therefore scanned by the moving x-ray point source generated by the scanning electron beam on the inside of the thin target. In Moon's tube, the useful x-rays are taken from the same side of the target as is scanned by the electron beam. If the tube is to be operated at low potential, $\lesssim 50$ kV, an x-ray window must be provided. A lead shield having a small pinhole is placed just outside this window between the x-ray tube and specimen. Although x-ray emission occurs in all directions from the point of impact of the scanning electron beam on the target, only a linear ray defined by the instantaneous electron-beam focal spot and the pinhole arrives at the specimen. Thus, as the electron beam scans the target line-by-line, say, from left to right, top to bottom, the x-ray beam scans the specimen from right to left, bottom to top.

20.3.3. Scanning X-Ray Absorption Microscopy

When a scanning x-ray microscope having an x-ray probe is used with a thin radiolucent specimen, the instantaneous x-ray intensity *transmitted* by each incremental area on the specimen can be detected and used to modulate the display-tube beam. If no pulse selection is used, the display is simply a type of shadow image showing the distribution of density or mean atomic number. Incidentally, Moon's scanning x-ray tube was designed for precisely this application, but on a vastly larger scale—medical fluoroscopy. The scanning x-ray beam covered the area of interest on the patient. The transmitted x-rays fell on a large, truncated, conic, fluorite (CaF_2) scintillator having large and small diameters 14 and 2.5 cm and thickness 6 cm, and affixed at the small end to a multiplier phototube. Moon also demonstrated the applicability of his system to small objects.

However, the distribution of a specific element can be mapped by making two x-ray micrographs of the same specimen area, each with one of two monochromatic x-ray beams having wavelengths bracketing its absorption edge. Any difference in the two displays is due to the selected element, which appears brighter and darker on the cathode-ray display for the longer and shorter wavelength, respectively. This can be done with monochromatic scanning x-ray beams or with a polychromatic beam and pulse-height selection. The latter method requires high energy resolution if the pulse distributions of two wavelengths closely bracketing the absorption edge are to be separated.

20.4. X-RAY PHOTOELECTRON AND AUGER-ELECTRON SPECTROMETRY

20.4.1. Introduction

When matter is irradiated with photons or bombarded with electrons, among the several phenomena that may occur is the ejection of bound electrons from atoms in the matter. These ejected electrons have energies depending on the atoms of their origin, and investigation of electron-energy spectra gives chemical information. The discipline concerned with study of these electron spectra is *electron spectrometry for chemical analysis* (ESCA). The method has been known since the 1920's, but has only recently undergone rapid development and increased application.

Irradiation by photons excites photoelectron spectra (PES); bombardment by electrons excites secondary-electron spectra (SES). If the incident photons or electrons have low energy, only outer-shell electrons are ejected, but if they have high energy, inner-shell (K, L, M) electrons may be ejected as well. With photon excitation, the incident photons are totally absorbed and disappear, but with electron excitation, the incident electrons scatter, and the scattered electrons appear in the secondary-electron spectra.

Photoelectron and secondary-electron spectra arise from one-step processes—photoionization and ionization by collision, respectively. Another type of electron spectrum arises from the Auger effect (Section 2.5), which is a two-step process—inner-shell ionization (by photon or electron) followed by radiationless atomic rearrangement and ejection of an Auger electron.

This section is intended to introduce the reader to two branches of electron spectrometry that involve x-rays: x-ray-excited photoelectron spectrometry and Auger-electron spectrometry (AES).

20.4.2. X-Ray Photoelectron Spectrometry (*H58*)

In secondary excitation of *K* x-ray spectra, the energy of each incident primary photon is expended as follows:

$$E_x = E_K + \phi + E_{pe} \tag{20.27}$$

where E_x is energy of the incident photon; E_K is the energy required to expel a *K* electron (*K*-excitation potential); ϕ is the energy required to remove an electron from the specimen surface (work function); and E_{pe} is the kinetic energy of the free photoelectron. All energies are in electron volts. Since ϕ is only a few electron volts,

$$E_{pe} \sim E_x - E_K \tag{20.28}$$

Similar expressions apply to photoelectron emission from *L*I, *L*II, *L*III, and the five *M* shells. Thus, for monoenergetic primary radiation, photoelectron energy should be characteristic of the secondary emitter. However, only photoelectrons emitted from atoms on the surface have the energy indicated by Equation (20.28). Those emitted from atoms beneath the surface undergo stepwise deceleration and emerge with less than their initial energy, and only those emitted in an extremely thin surface layer can emerge from the surface at all.

Thus, a monochromatically excited x-ray photoelectron energy spectrum should show a series of peaks, one for each atom energy level (*K*, *L*I, *L*II, etc.). The peaks should have vertical edges on the high-energy side, as given by Equation (20.28), and fall off exponentially on the low-energy side. The lower the excitation potential of the shell, the smaller is the fraction of the primary energy expended in ionizing it, and the greater is the balance remaining to impart kinetic energy to the photoelectron. Figure 20.7 (*S56*) shows the photoelectron energy spectra of copper and silver excited by Mo $K\alpha$ x-rays. The spectra have essentially the predicted form, but may be complicated by several phenomena. Absorption and scatter of photoelectrons in the specimen may broaden the peaks. The incident x-rays are usually only filtered, rather than truly monochromatic, to avoid the prohibitive loss of intensity resulting from crystal monochromatization. Although the photoelectrons originate in inner atomic orbitals and are therefore relatively independent of chemical state, they are extremely sensitive to surface conditions, which may broaden and displace the peaks and reduce their intensity. Finally, Auger electrons are also produced, and their spectral distribution is superposed on that of the photoelectrons. For

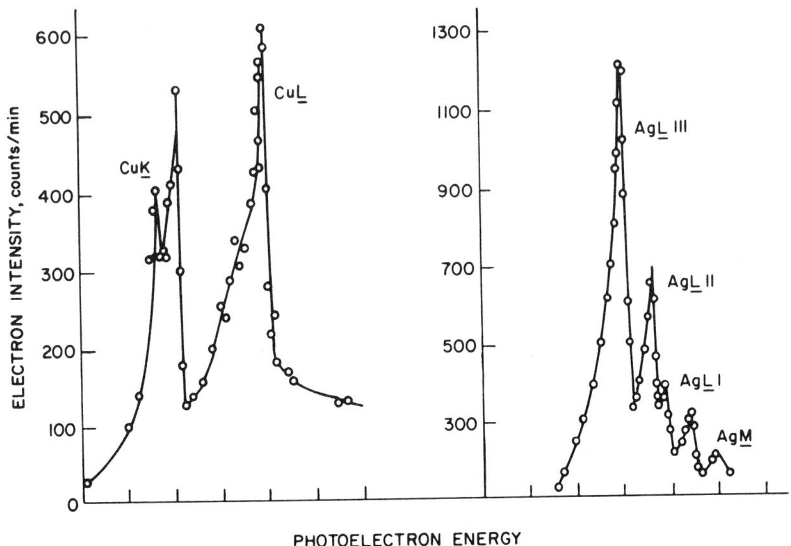

FIG. 20.7. X-ray photoelectron energy spectra of copper and silver excited by Mo $K\alpha$ x-rays. [R. G. Steinhardt, Jr., and E. J. Serfass, *Analytical Chemistry* **23**, 1585 (1951); courtesy of the authors and the American Chemical Society.]

a given intensity and spectral distribution of primary x-rays, photoelectron intensity is approximately proportional to Z^5.

Krause (*K30*) describes a method of x-ray spectrometric analysis in which, rather than measuring the x-ray energies or wavelengths in an x-ray spectrometer, one allows the x-rays to strike a target and eject photoelectrons, whose energies are then measured in an electron spectrometer. He terms the method *photoelectron spectrometry for analysis of x-rays* (PAX).

Little application has been made of x-ray photoelectron spectrometry (*B47*, *F4*, *S56–58*), but perhaps interest in the method may be renewed by recent developments in Auger-electron spectrometry (see below). Many of the hoped-for applications of Auger spectrometry would be feasible for photoelectron spectrometry as well.

20.4.3. Auger-Electron Spectrometry

Energies of Auger electrons (Section 2.5), like those of photoelectrons, are characteristic of the emitting element, but Auger-electron spectra are less intense and less sharply defined than photoelectron spectra.

Assuming monochromatic excitation, a photoelectron originating in, say, the *L*III shell of an atom can have only one *initial* energy—the difference between incident photon energy and *L*III excitation energy. An Auger electron originating in the same shell may have any of a large number of energies. A vacancy in, say, the *K* shell can be filled by an electron transition from the *L*II, *L*III, *M*II–*M*V, or *N*II–*N*VII shells. Each such transition produces a photon of different energy, each of which can undergo internal conversion in the *L*III shell, producing an Auger electron. Moreover, the *L*III Auger electron may be ejected by internal conversion of photons originating from electron transitions to the *L*I or *L*II shells. In each case, the initial energy of the Auger electron is different. In practice, the most common Auger energies correspond to the most probable electron transitions—those giving rise to the strongest x-ray spectral lines. The Auger electrons involved in spectrometry have energies of only ~50–1000 eV, somewhat less energy than x-ray photoelectrons. Thus, Auger electrons undergo more severe absorption in emerging from the emitter, and probably only the top several atom layers contribute to the observed Auger spectrum. For example, Figure 20.8 shows the Auger-electron spectrum of beryllium metal; it is evident that the oxygen peak from the surface oxide layer is more prominent than the beryllium peak.

The observed Auger-electron energy spectrum consists of relatively weak, broad peaks slightly below the corresponding transition energies of the emitting atoms. The peaks are superposed on photoelectron spectra when excited by x-rays, and on secondary and scattered electron spectra

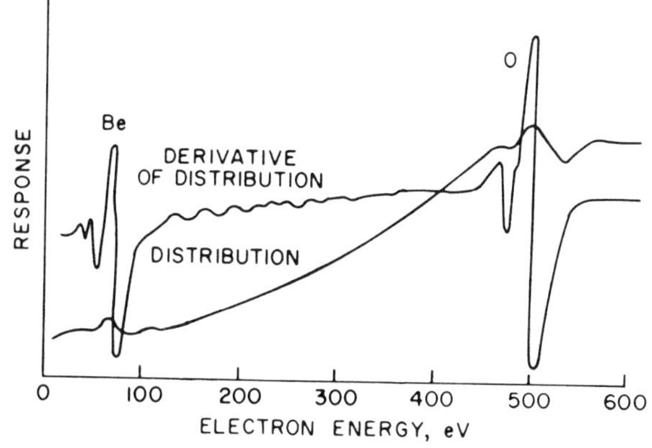

FIG. 20.8. Secondary-electron energy spectra of beryllium, showing Auger-electron spectra of beryllium and oxygen. [L. A. Harris, *Industrial Research* **10**(2), 52 (1968); courtesy of the author and publisher.]

when excited by electrons. Auger peaks, like photoelectron peaks, taper off toward low energies. The Auger-electron spectrum of beryllium is shown in Figure 20.8. Fortunately, a way has been found to accentuate this poorly defined Auger spectrum (see below). The Auger spectra of low-Z elements are simple and have well-separated and easily identified peaks. The observed Auger spectra of high-Z elements are much more complex, and overlap. The high-Z elements also have simple Auger spectra, but they lie outside the energy range convenient for observation.

Auger-electron spectrometry promises to be particularly advantageous for elements of low atomic number. It has been shown (Section 2.5) that the utility of x-ray emission spectrometry decreases with Z because of decreasing fluorescent yield. The reason for the low fluorescent yield is internal conversion of characteristic x-ray photons, with consequent production of Auger electrons. Thus, as Z decreases, x-ray spectral line intensity decreases and Auger electron intensity increases.

Auger yield (Section 2.5) is greater than fluorescent yield for atomic numbers less than $\sim$32 (germanium). Moreover, x-ray spectrometric analysis of the light elements is beset by other problems (Section 16.1.3), such as high absorption of their long-wavelength spectra, and crystals of relatively low reflectivity and resolution. Conversely, Auger electrons are more numerous than x-ray spectral-line photons for low-Z elements, and they are readily resolved by present-day electron-velocity analyzers and efficiently detected by scintillation counters.

Because of the low electron energies involved, Auger-electron spectrometry (*H17*, *H18*, *H39*) is a very sensitive method for chemical analysis of extremely thin surface layers, $<$100 Å at most. The entire observed Auger-electron spectrum probably arises in the top 5–10 atom layers, and as little as 1/50 atom layer of cesium has been detected and identified on a silicon surface. The method is being applied, or at least has promise of application, in such areas as the following: (1) analysis of surfaces; (2) analysis of extremely thin films; (3) surface contamination, oxidation, and corrosion; (4) absorption and desorption; (5) material transfer by friction, evaporation, etc.; (6) diffusion to the surface; (7) migration over and/or segregation on surfaces; (8) segregation at grain boundaries; (9) degree of surface coverage; (10) evaluation of superclean surfaces; (11) analysis of residues from wash solutions; (12) determination of dopants and identification of contaminants on semiconductor and other electronically active surfaces; (13) analysis of catalytic surfaces; and (14) bulk analyses for certain elements with previously unattainable sensitivity when the surface and bulk compositions are the same; for example, carbon in concentration

<40 μg/g (ppm) in iron and sulfur in concentration <10 μg/g in nickel have been determined, and the method is also applicable to oxygen.

If the Auger electrons are excited by a fine-focus electron beam having diameter of the order 1 μm, Auger-electron analysis is feasible at a point, along a line, and over a raster (Sections 21.4.2–21.4.4); this last method is known as *scanning Auger microscopy* (SAM) (*H23*, *M3*). If an ion-etching gun is provided, successively deeper layers can be analyzed after repetitive removal of thin layers, and if an evaporation source is provided, successive freshly evaporated layers can be studied. If provision is made to permit slow leakage of gas from a gas-sample cell into the vacuum specimen chamber, gases can be studied for composition, oxidation states, chemical bonding, etc.

20.4.4. Photo- and Auger-Electron Spectrometer

The instrument is essentially the same for both photoelectron and Auger-electron spectrometry, except for the excitation. A typical arrangement is shown in Figure 20.9 and consists of four parts—exciter, specimen, electron-energy analyzer, and measurement system. The drawing shows both x-ray and electron excitation for convenience in explaining the function of the instrument. The use of both in the same instrument is unlikely.

Photoelectrons are produced only during secondary excitation of x-ray spectra, so, for photoelectron spectrometry, the exciter is an x-ray tube. The beam is filtered to provide substantially monochromatic target-line radiation. Of course, some Auger electrons are produced along with the photoelectrons. Auger electrons are produced during both primary and secondary excitation of x-ray spectra. However, primary excitation is more efficient by two or three orders, and the greater the number of orbital electron vacancies created, the greater is the probability of Auger-electron production. Thus, for Auger-electron spectrometry, the exciter is an electron gun. Many scattered and secondary electrons are produced along with the Auger electrons, and instruments having electron-gun excitation are often referred to as secondary-electron spectrometers.

The specimen is usually a solid having a plane, relatively smooth surface. A rotatable wheel holding many specimens is provided to permit changing specimens without breaking the vacuum. Provision may be made for heating the specimen.

The photo, Auger, secondary, and scattered electrons ejected from the specimen by the x-ray tube or electron gun enter the electron analyzer through an entrance slit. The analyzer shown in the figure is a 127° electro-

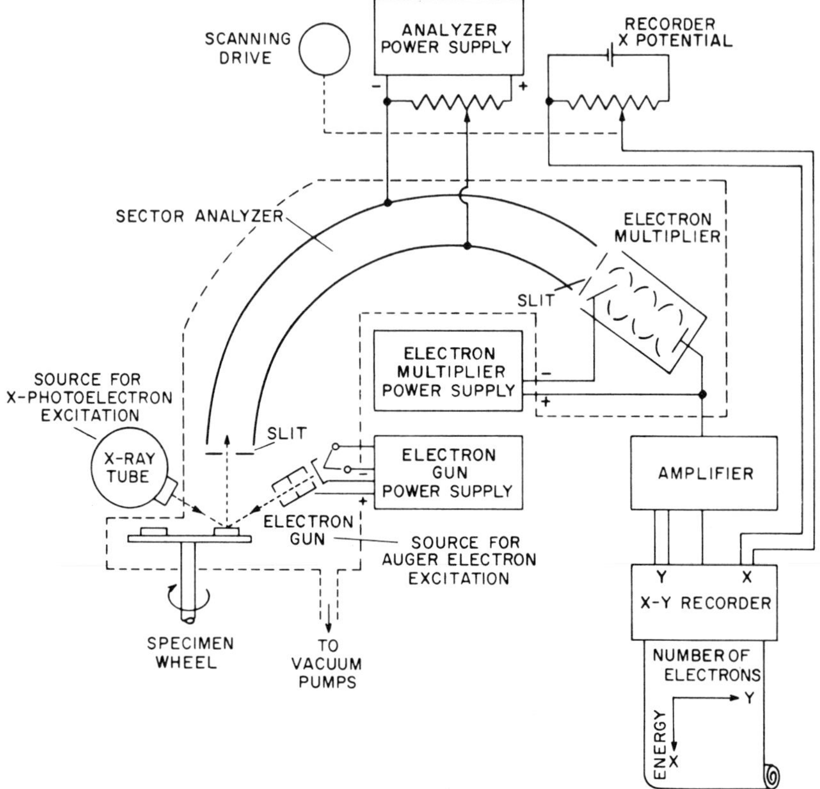

FIG. 20.9. Instrument for x-ray photoelectron spectrometry and Auger-electron spectrometry. The instrument is shown with both x-ray and electron excitation for illustration only.

static electron analyzer in which electrons are sorted according to energy by passing them between two charged, parallel, curved plates. With a given potential across the plates, only electrons having a certain velocity pass through the analyzer; all others strike the wall of the positive plate. The potential applied to the plates is varied continuously over a specified range by a motor-driven potentiometer, thereby permitting electrons having a range of energies to pass in sequence. The energy scan requires ~5 min.

The electrons that pass through the analyzer enter the electron multiplier, where each produces an output pulse of 10^6 or more electrons. The electron multiplier functions the same way as that in a multiplier phototube. The multiplier output is amplified, electronically differentiated, and passed to the Y input of the X–Y recorder. The X input comes from a potentiometer coupled to the electron-energy motor drive as shown.

The entire electron path, including the electron gun, is enclosed in a vacuum housing. Although some work is done with oil-diffusion pumps and rubber or other greased gaskets, Auger electrons are so extremely sensitive to surface contamination that metal gaskets, ion pumps, and pressures down to 10^{-8} torr (mm Hg) or less are preferable.

The recorder may plot a simple electron-energy distribution curve—electron current *versus* energy. Such curves are satisfactory for photoelectron spectrometry because secondary excitation produces only photo and relatively few Auger electrons. However, in Auger-electron spectrometry, the Auger peaks are small, poorly defined, and superposed on a relatively high background of scattered and secondary electrons. The Auger peaks are accentuated by displaying the derivative of the energy distribution curve with respect to energy—in effect, a plot of the instantaneous slope of the energy distribution curve. The electronic differentiation mentioned above produces this plot, and the great improvement is evident in Figure 20.8. Alternatively, the energy spectrum may be scanned manually in small increments and the amplified electron multiplier output pulses counted by a scaler-timer. The photoelectron spectra in Figure 20.7 were plotted from data obtained in this way.

Several Auger-electron spectrometers are available commercially.

20.5. X-RAY-EXCITED OPTICAL-FLUORESCENCE SPECTROMETRY

X-rays excite optical fluorescence in many solids and gases (*D12*). The phenomenon is especially prominent among trivalent rare-earth ions. X-rays excite most rare-earth impurities in Y_2O_3, CeO_2, La_2O_3, Gd_2O_3, and Lu_2O_3 to emit intense, highly characteristic optical line fluorescence in the 3–10-kÅ region. The excitation mechanism is not fully understood, but it is believed that x-ray-excited photoelectrons preferentially transfer their energy to the impurity atoms by inelastic scattering. Minimum detectable concentrations in any of these five matrixes are of the order 1 ng/g (ppb) for neodymium, terbium, and dysprosium; 10 ng/g for samarium, europium, and thulium; and 100 ng/g for praseodymium, gadolinium, holmium, erbium, and ytterbium. Yttrium, lanthanum, and lutetium emit no such spectra; the cerium spectrum is farther in the infrared region; and the gadolinium spectrum is relatively simple. In rare-earth oxide matrixes other than these five, the impurity spectra are severely quenched by the host oxide. This difficulty is overcome by 1000-fold dilution with Y_2O_3, but the sensitivity is decreased 1000-fold by the dilution.

The instrument is shown in Figure 20.10 (*B90*, *L29*). X-rays from an x-ray spectrometer excitation tube operating at 50 kV, 50 mA irradiate the specimen packed in a shallow planchet. The optical fluorescence is reflected by a series of mirrors through a rotating chopper into an optical grating monochromator. The monochromator disperses and scans the optical spectrum, permitting the multiplier phototube to measure the intensity at each wavelength in turn. The detector output is amplified and plotted on the recorder as a graph of intensity *versus* wavelength. The chopper interrupts the fluorescent beam at a rate ~13 times per second, so that the detector output can be amplified by an ac, rather than dc, amplifier.

X-ray excitation has the advantage that the rare-earth ions are excited to higher energy states than would be the case with ultraviolet excitation. Thus, all of the allowed optical transitions can be observed. Another advantage is that there is no interference of excitation and fluorescence radiation, because the x-rays are not transmitted by the optical system. At present, the method is applied principally to analysis of rare-earth oxides and to investigation of energy levels in solids.

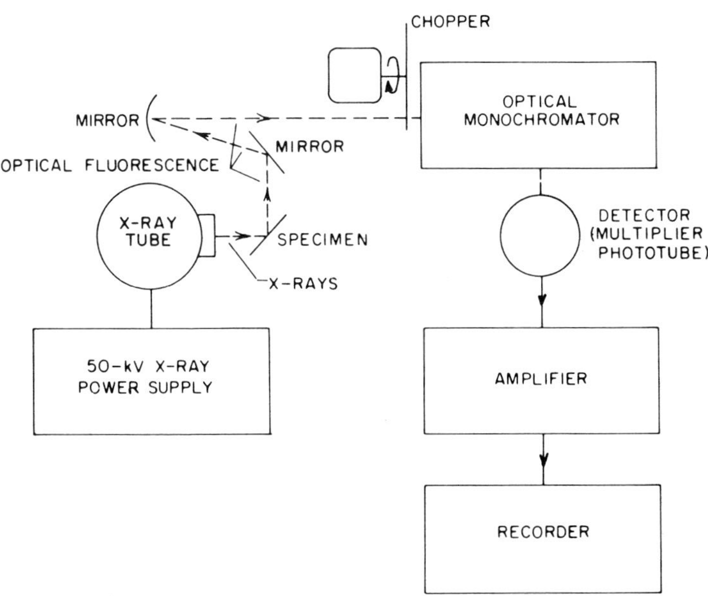

FIG. 20.10. X-ray-excited optical fluorescence spectrometer.

Sommer (*S44*) has used selected-area techniques (Section 19.5) to obtain x-ray spectrometric analyses and x-ray-excited optical fluorescence spectral distribution curves from small selected areas, such as individual grains and phases. He used the General Electric secondary-pinhole drawer accessory (Section 19.2.2.2, Figure 19.4). The area of interest is selected by the light-beam technique (Section 19.4.2) and the x-ray spectrum recorded in the usual way. Only x-ray spectral-line emission originating in the projection of the secondary aperture on the specimen surface passes to the analyzer crystal and x-ray detector. The specimen drawer is then reversed end-for-end in its compartment so that the specimen and secondary aperture now face away from the crystal and x-ray detector. A Lucite light pipe is now placed at the end of the secondary-beam aperture to collect and direct optical luminescence to an optical spectrometer, which records the optical spectral distribution. Only x-ray-excited optical fluorescence originating in the projection of the secondary aperture passes to the light pipe. Alternatively, the optical spectrometer may be set at a specified wavelength and the specimen translated past the secondary-beam aperture as the emission profile of that wavelength across the specimen is plotted. The technique is applicable to studies of minerals, rare-earth materials, phosphors, etc.

20.6. X-RAY LASERS

A *laser* (from "light amplification by stimulated emission of radiation") is a device that produces a highly coherent, collinear, monochromatic, and usually intense beam of light or other electromagnetic radiation by the process of stimulated emission from an atomic inversion population (see below). Here the term *coherent* means that all waves are in phase, as shown in Figure 2.12B. Although only certain substances can be made to "lase," these substances may have the form of single crystals, semiconductors, glasses, plastics, liquids, gases, and plasmas.

Briefly, the laser principle is as follows. When all the electrons in an atom are in their lowest possible orbitals, the atom is said to be in the ground state; this may be regarded as the normal state of the atom. Each electron in the ground state can be elevated to higher orbitals if the atom absorbs a discrete amount of energy—that is, if it absorbs a photon of wavelength λ—corresponding to the specified transition. Ordinarily, such excited electrons spontaneously and almost instantly return to the ground-state level, whereupon the atom emits the energy it absorbed as a photon of the same wavelength λ. However, the excited electron may fall instead into some intermediate level from which spontaneous transition to the ground-state level is relatively improbable and slow, and become, in a sense, trapped there. If more atoms are in this metastable state than in the ground state, the exciting radiation λ is said to have *pumped* most of the specified electrons into the metastable state, and the material is said to have undergone a *population inversion* of the specified electrons. It must be emphasized that only one specific electron in each atom undergoes this transition. The transition from metastable to ground state can be triggered or *stimulated* by radiation of wavelength $\lambda' > \lambda$ corresponding to the energy of the transition. This radiation is not absorbed or modified by the stimulation process. In other words, when a metastable atom reverts to the ground state, a quantum of energy is produced which promptly stimulates other metastable atoms to decay. Thus, as the λ' radiation proceeds through the inverted atoms, it is amplified in intensity.

One form of laser consists of a ruby rod [single-crystalline sapphire (aluminum oxide, Al_2O_3) with a trace chromium impurity] having highly plane polished ends. Both ends are silvered, one end with a light-opaque layer, the other having ~90% transmittance. The rod is placed along the axis of a helical, electronic pulsed flash lamp.

When the lamp flashes, most of the chromium atoms absorb a certain wavelength λ from its intense white light and are pumped into a metastable state; that is, the chromium atoms undergo a population inversion. In a few atoms, the excited electrons spontaneously return to the ground level with the emission of photons of wavelength λ'. These photons stimulate other such transitions with emission of more photons, etc. Some of these photons reach the ends of the rod and are reflected back to continue the process. Others emerge from the partially silvered end as an intense, collinear, coherent, monochromatic (λ') laser-light beam. The process continues until all the chromium atoms have returned to the ground state, whereupon the second flash from the lamp reinitiates the pumping, etc.

Lasers have been made to operate at wavelengths from ~21 cm in the microwave region ("masers") down to ~1500 Å in the ultraviolet. In principle, there is no reason why an x-ray laser should not be feasible, but in practice the obstacles are very great indeed. A substance is required in which it is possible to create a population inversion at a level above ground state corresponding to an energy in the x-ray region. This substance must have low attenuation for the laser x-ray wavelength. As the energy of the metastable-to-ground interval increases, the lifetime of the state decreases, so that for x-ray energies, it becomes increasingly difficult to maintain the population inversion. Because of the high energy of transitions corresponding to the x-ray region (say, 1–100 keV) compared with those corresponding to the optical region (<10 eV), high power is required to pump the x-ray laser, and the laser itself must withstand the thermal and radiation damage. Finally, provision must be made for "reflection" of the laser x-rays from the ends of the laser to sustain the stimulation process.

Kepros, Eyring, and Cagle (*K10*) report evidence of x-ray laser production. A gel was prepared consisting of 0.444 g of unflavored Knox gelatin in 28 ml of 1–10 m*M* (millimolar) copper (II) sulfate ($CuSO_4$) solution. A layer of this gel $\lesseqgtr$0.1 mm thick was sandwiched between two microscope-slide cover glasses 0.13 mm thick. Light (λ 1.06 μm) from a neodymium-glass laser was used to pump the x-ray laser sandwich. The sandwich was mounted parallel to and 4 cm from a cylindrical condensing lens, and the lens-sandwich assembly was mounted at $\sim 25°$ to the laser beam. In this way, the laser-illuminated area on the sandwich was a 1×0.01-cm line. A piece of double-emulsion (emulsion on both sides) Kodak No-Screen Medical x-ray film was wrapped in four layers of black photographic paper, then in four layers of 12-μm (0.0005-inch) aluminum foil to exclude soft x-rays. The film packet was placed 10–110 cm away from the edge of the sandwich, in line with and perpendicular to the laser-illuminated line.

In about one trial in ten, a small exposed spot was produced on the developed x-ray film. In the most successful and significant experiments, with 1-m*M* (millimolar) copper (II) sulfate gel and 30-cm sandwich-to-film distance, a dot 0.2 mm in diameter was produced, and it did not increase in size on moving the film to 110 cm, indicating high collinearity of the beam. X-ray wavelength was not measured accurately, but the beam penetrates 110 cm of air, 50 μm of aluminum, and four layers of paper, and is attenuated by iron and nickel foils to the degree predicted for Cu $K\alpha$.

X-ray lasers would make feasible scanning-spot or holographic x-ray microscopes having 1-Å resolution, permitting observation of the atomic structure of matter.

20-7. X-RAY APPEARANCE-POTENTIAL SPECTROMETRY

X-ray appearance-potential spectrometry (APS) is an energy-dispersive x-ray spectrometric method, introduced by Park, Houston, and Schreiner (*P6*), for determination of binding energies of core electrons of atoms of a solid surface (*P5, V11*).

The method is based on the fact that total electron-excited x-ray emission intensity, plotted as a function of bombarding electron potential, shows abrupt discontinuities at the "appearance potentials," that is, excitation potentials, of characteristic x-rays. The intensity of the characteristic-line component of this emission depends on both the surface concentrations of the emitting elements and the densities of unfilled electron states.

The instrument is shown schematically in Figure 20.11. The specimen target, a thermionic electron-emissive tungsten filament, and an x-ray detector are housed in a vacuum chamber. The target–detector distance is ~15 cm. The detector may be a windowless multiplier phototube or a thin-window flow-proportional counter. The latter has the advantage of excluding electrons, ions, and low-energy continuous x-rays, the disadvantages of gas leakage into the vacuum chamber, x-ray absorption in the window, and relatively low maximum linear count rate ($\sim 10^4$/s) with resultant long measurement times. The windowless multiplier phototube must be optically shielded from the filament. The vacuum is provided by a sublimation–getter-ion pump system and is of the order 10^{-9} and 10^{-7} torr for multiplier and flow detectors, respectively. The target may be baked at temperatures up to 1000°C by filament-electron bombardment or by ohmic heating.

A dc potential varying linearly between, say, 0 and 2 kV is provided by a power supply swept by a linear "ramp" generator and applied to the target. The filament is heated by a current from the filament supply and

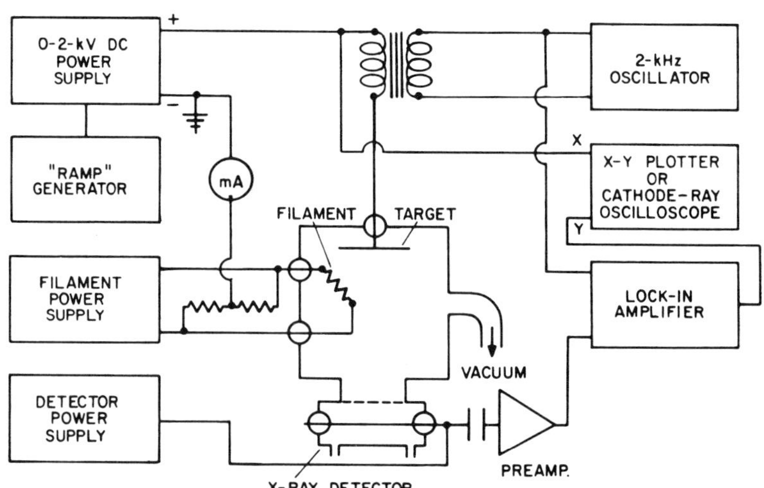

FIG. 20.11. X-ray appearance-potential spectrometer.

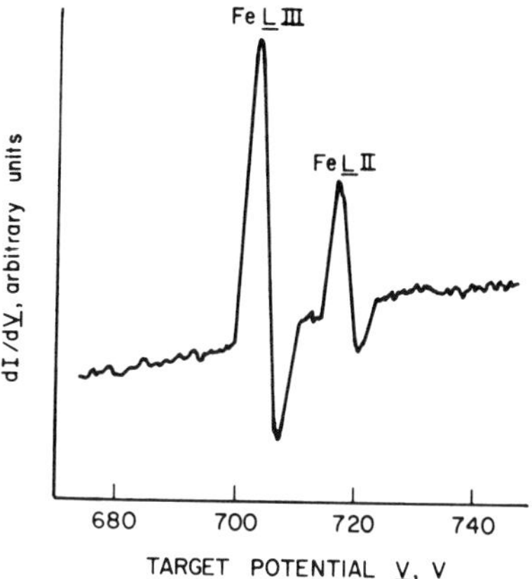

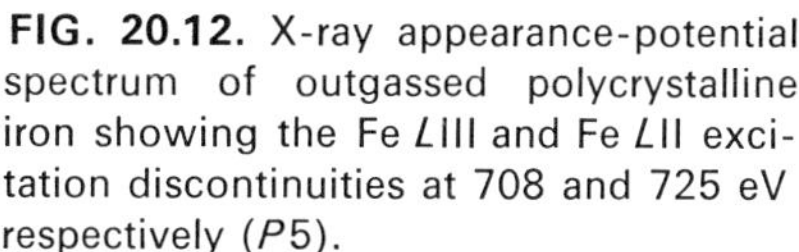
FIG. 20.12. X-ray appearance-potential spectrum of outgassed polycrystalline iron showing the Fe *L*III and Fe *L*II excitation discontinuities at 708 and 725 eV respectively (*P*5).

emits electrons, which are accelerated to the target and excite x-ray emission. A portion of these x-rays is intercepted by the detector.

The discontinuities that appear in the x-ray emission as target potential is swept are accentuated with respect to the background by electronic differentiation. The first derivative of the measured intensity with respect to target potential is obtained by superposition on the dc target potential of a modulating ac potential of, say, 5 V supplied by an oscillator of frequency, say, ~50 Hz to 2 kHz. The resulting fluctuation in intensity is measured synchronously with a lock-in amplifier tuned to the modulating frequency.

The first derivative of the intensity with respect to electron acceleration potential dI/dV is plotted on an X–Y recorder or displayed on a cathode-ray oscilloscope. The target-potential sweep may be incremental or continuous; the latter mode may require from 0.5 to 30 min, depending on the resolution required. Resolution as high as 0.5 eV has been attained. Figure 20.12 shows an x-ray appearance-potential spectrum of iron in the region of its *L*III and *L*II excitation potentials.

Because of the low penetration of 2-keV electrons, the effective surface layer contributing the measured x-rays is only 10–20 atom layers. Consequently, appearance-potential spectrometry is an excellent method for studying surfaces, including composition, core-electron binding energies and chemical shifts thereof, band structures, and effect of contamination.

Chapter 21

Electron-Probe Microanalysis

21.1. INTRODUCTION

Electron-probe microanalysis (EPMA), or electron-microprobe analysis, is essentially a nondestructive instrumental method of qualitative and quantitative analysis for chemical elements based on measurement of the wavelengths and intensities of their characteristic x-ray spectra excited by an electron beam having diameter of the order 1 μm; it permits in-place determination of composition and spacial variation of composition on a microscopic scale.

The electron-probe microanalyzer is basically an electron-probe x-ray primary-emission *spectrometer*. A fine-focus electron beam excites the x-ray spectra of the elements in a specimen region $\sim$1 μm in diameter and $\sim$1 μm^3 in volume. The x-radiation is dispersed, detected, and read out in one or more curved-crystal x-ray spectrometers and/or a Si(Li) energy-dispersive spectrometer. However, the interaction of the electron beam and specimen results in many other phenomena—principally backscattered, secondary, and absorbed electrons and cathodoluminescence—that may provide other types of information. Thus, the more general term, electron-probe *microanalyzer*, is more appropriate.

For *qualitative analysis*, in favorable cases, the method is applicable to all chemical elements down to atomic number 4 (beryllium); concentrations down to $\sim$0.01 wt %; isolated masses down to $\sim$1 fg (10^{-15} g); films as thin as a fraction of a monolayer; areas 1–100 μm in diameter; depths down to several micrometers; and specimens in the form of bulk material, film, fiber, powder, small fabricated parts and devices, and metallographic sections, all without serious limitations on surface finish.

For *quantitative analysis*, the minimum determinable concentration is likely to be greater than for qualitative analysis, and there are rigid limitations on specimen form and finish. In favorable cases, precision may be of the order ± 0.5–2% of the amount present for major and minor constituents, but may be much poorer as concentration decreases. Calibration standards are preferably specimens of known composition similar to the samples; analytical concentrations are obtained by applying intensity data measured from the samples to a calibration curve established from intensity data measured from the standards. Otherwise, the pure constituent elements or suitable alloys or compounds different from the samples, but of known composition, may serve as standards; analytical concentrations are obtained from intensity data measured from samples and standards by means of complex mathematical equations. In either case, samples and standards must have identical surface finishes, and standards must be homogeneous on a micrometer scale.

Spacial variation of composition (or of electron backscattering, secondary electron emission, specimen current, or cathodoluminescence) along a line or over an area can be mapped by moving the specimen under a fixed electron beam, by scanning the electron beam over a fixed specimen, or by a combination of both (Sections 21.4.3 and 21.4.4).

This chapter, like Chapter 20, is intended only to give the x-ray fluorescence spectrographer an elementary introduction to a related discipline. Good introductions to the method and instrumentation of electron-probe microanalysis appear in references (*1*, *7*, *8*, *25c*, *A20*, *B16*).

21.2. INSTRUMENTATION

21.2.1. General

Two types of instrument for electron-probe microanalysis must be distinguished: electron-probe microanalyzers designed specifically for this application, and scanning electron microscopes (SEM's) fitted with energy-dispersive x-ray spectrometers. These instruments are essentially very similar.

Electron-probe microanalyzers are intended primarily for x-ray spectrometric analysis and so are designed to operate at relatively high electron-beam currents (usually 0.01–10 μA) at the expense of high spacial resolution ($\lesssim$0.5 μm). Cathode-ray tube displays of backscattered and secondary electrons are of secondary importance and are used principally to examine the specimen surface as an aid to the x-ray spectrometric analysis. Magnifications greater than a few thousand are of poor quality. The high electron-beam current gives analyte-line intensities adequate for wavelength-dispersive (crystal) spectrometers.

Alternatively, scanning electron microscopes are intended primarily to give electron images of the specimen surface at the highest possible magnification ($\lesssim 50{,}000\times$) and resolution ($\lesssim 100$ Å) and so are designed to operate at low electron-beam currents (0.01–1 nA). X-ray spectrometric analysis, when required at all, is of secondary importance. The low electron-beam current gives analyte-line intensities too low for wavelength dispersion, but energy-dispersive spectrometers using Si(Li) detectors and multichannel analyzers are well suited to this application.

A schematic diagram of an electron-probe microanalyzer is shown in Figure 21.1. The instrument consists of six systems: electron optics, specimen stage, light optics, x-ray analysis, electron analysis, and vacuum. These components are housed in or near an evacuated electron-optical column similar to that of an electron microscope, one or more attached x-ray spectrometers, and an electronics console. These instrument systems are described in Sections 21.2.2 and 21.2.3.

21.2.2. Electron-Optical Systems

The electron-optical column always contains the first three, and usually all, of the following components: (1) an electron gun consisting of a V-shaped tungsten filament, grid cup (Wehnelt cylinder), and anode; (2) two electromagnetic reducing electron lenses; (3) two beam-limiting apertures; (4) a fluorescent viewing screen; (5) an electron-beam scanning system; (6) an astigmatism compensator; and (7) an electron-beam current regulator. Electrons emitted by the incandescent filament are formed into a thin beam and accelerated to high energy by the electron gun operated at up to 50 kV dc. The two reducing lenses demagnify and focus the beam to a final crossover $\sim$1 μm in diameter at the specimen plane, where x-ray excitation occurs. The column may be regarded as a highly sophisticated x-ray tube with the specimen as target. The lenses consist of soft iron cylinders having cross-sectional form similar to that shown in Figure 21.1. The lower portions of the cylinders are formed into pole pieces having apertures to limit the electron-beam diameter. The cylinders house electromagnetic coils through which pass the lens currents.

The electron beam can be made to scan the specimen surface by means of two horizontal, mutually perpendicular deflection coils connected to

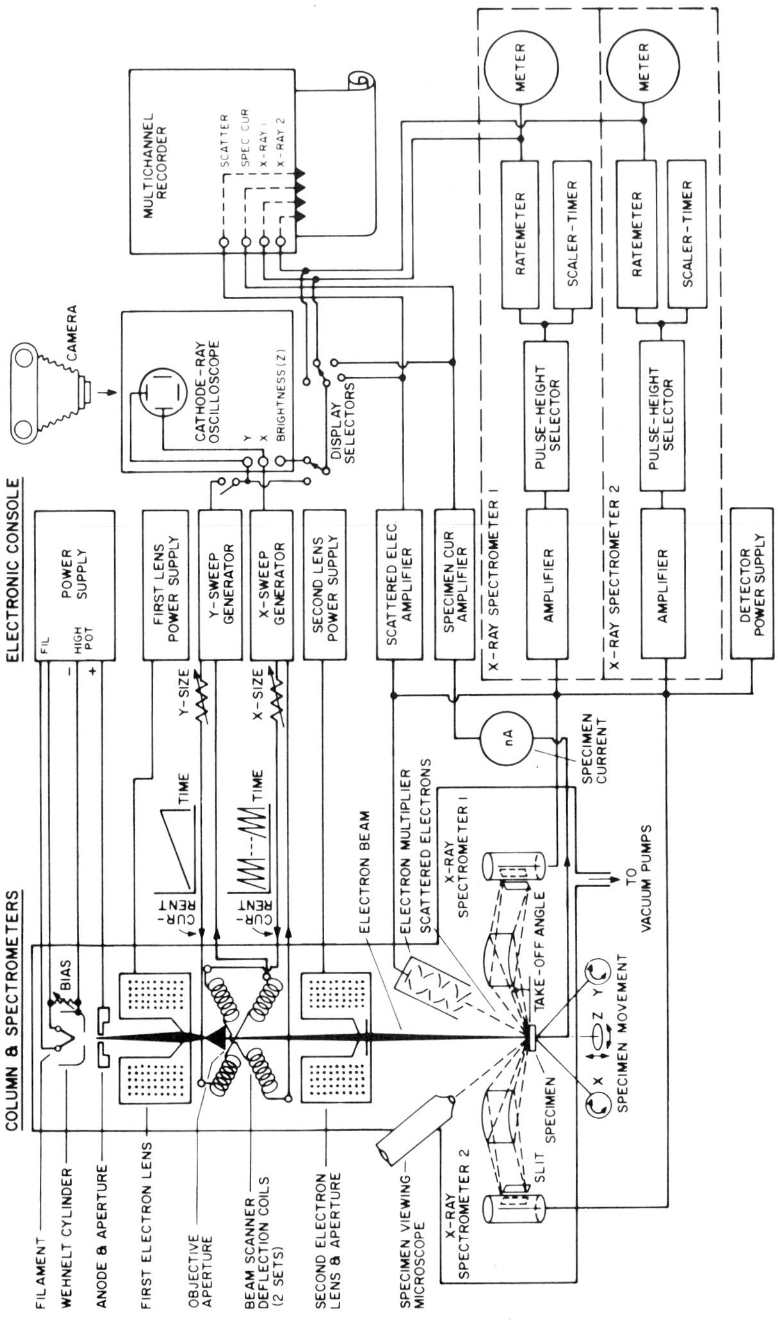

FIG. 21.1. Electron-probe microanalyzer.

sawtooth sweep (time-base) generators, as shown. The X-sweep coils deflect the beam forth and back in the X direction, say, 200 times a second, while the Y-sweep coils deflect it in the Y direction once each second. Actually, two sets of deflection coils like those shown are required, one set above the other. If only one set were used as shown, the electron beam would be deflected away from the second aperture. The lower coils deflect the beam back, that is, in the direction opposite that imparted to it by the upper coils, so that it passes through the aperture, but at a small angle to the column axis. Alternatively, a single set of coils like the one shown is sufficient if it is placed below the second lens aperture. By appropriate use of the deflection system, the specimen area covered by the electron beam can be made to assume any of three forms—point, line, or raster (rectangular area):

1. With neither X or Y deflection, the intersection of the electron beam with the specimen surface is a fixed point.

2. With only X (or Y) deflection, the electron beam scans a line across the specimen in the X (or Y) direction. By variation of the X size control shown, which varies the amplitude of the X-sweep lens current, the length of the scanned line can be varied from a point to perhaps 500 μm. The line scan rate can be varied; typical fast and slow limits are 100/s to once every 6 s.

3. With both X and Y deflection, the electron beam scans a rectangular, usually square, line raster similar to that on a television picture tube. The number of lines in the raster is varied by changing the frequency of the X-sweep current, typically from 100 to 500 lines per frame. The frame scan rate can be varied by changing the *frequency* of the Y-sweep current, typically from 1 to 150 s per frame. The X and Y size of the raster on the specimen can be varied by changing the *amplitudes* of the X and Y sweep currents with the X and Y size controls, typically from a point to ~500 μm square.

The X- and Y-sweep generators also provide deflection for a cathode-ray display oscilloscope, as shown in Figure 21.1. Thus the electron beam in the cathode-ray tube scans a line or line raster on its fluorescent display screen in synchronism with the electron beam on the specimen surface. The cathode-ray display is typically 7–10 cm square.

Provision may be made to introduce a fluorescent screen at an appropriate place in the path of the electron beam to aid in column alignment. A fluorescent screen in the specimen plane is always provided to aid in aligning, focusing, and positioning the beam. The astigmatism compensator permits adjustment of the beam spot on the specimen plane to circular shape.

The electron-beam current regulator feeds back a portion of the current intercepted by one of the beam apertures so as to vary the first lens current and regulate fluctuations in the beam current.

21.2.3. Other Instrument Systems

The *specimen stage* accommodates one or more specimen mounts having diameter up to 3 cm ($1\frac{1}{4}$ inches) and provides for coarse and fine linear specimen translation in two mutually perpendicular directions (X, Y) in its own plane, and normal to this plane (Z). Provision is also made to rotate the specimen in its own plane to a specified orientation. The X and Y movements may be manual or motor-driven in continuous or stepwise modes. Each specimen *mount* may contain a large number of individual specimens. The specimen stage must also accommodate various special accessories, such as hot stages and cameras for Kikuchi and micro electron diffraction or x-ray projection microradiography.

The *light microscope* permits visual observation of the specimen at magnification up to 400× for positioning the specimen and beam spot. Moreover, the specimen plane is defined by placing it in focus in the light microscope at highest magnification.

The *x-ray analysis system* consists of one to four conventional, full-focusing, curved-crystal, wavelength-dispersive x-ray spectrometers, each fitted with its own slits, crystal, detector, preamplifier, and manual and/or motor-driven wavelength drive. Gas-filled detectors are used almost exclusively for this application. Incidentally, the x-ray spectrometer planes are usually arrayed vertically (axes horizontally) instead of horizontally as shown for convenience in Figure 21.1. Also, probably the type of spectrometer most commonly used is that described in Section 5.6.4 and shown in Figure 5.23, rather than that shown in Figure 21.1. Each such spectrometer has a complete readout chain consisting of an amplifier, pulse-height selector, ratemeter, and scaler–timer. An energy-dispersive spectrometer may also be provided having a cryogenic lithium-drifted silicon [Si(Li)] detector (Section 6.4) with a multichannel analyzer and oscilloscope and/or X–Y-recorder readout (Section 8.1.2.3), or a complete energy-dispersive multichannel x-ray spectrometer system (Section 8.2.2.4).

The *electron-analysis system* consists of detectors for the backscattered and secondary electrons, and amplifiers and readout and display units for these and the specimen current (see below).

The *vacuum system* consists of a rotary mechanical pump providing rough vacuum for the column and fore vacuum for an oil-diffusion pump, an automatic valving arrangement, and protective devices against failure of power, diffusion-pump cooling water, and fore vacuum.

The *electronics console* houses the following components: (1) filament and 2–50-kV dc supplies for the electron gun; these supplies must be extremely well regulated ($\pm 0.01\%$ or less), not so much for constancy of x-ray emission, but for constancy of electron spot diameter and position; (2) electron lens current supplies; (3) X- and Y-sweep generators; (4) backscattered electron current amplifier; (5) secondary electron current amplifier; (6) specimen current amplifier; (7) a complete electronic readout system for each x-ray spectrometer, including amplifier, pulse-height selector, ratemeter, and scaler–timer; (8) power supplies for x-ray and electron detectors; (9) multichannel recorder, having a channel and pen for each of the x-ray ratemeters, the specimen current amplifier, and perhaps the backscattered electron current amplifier; and (10) cathode-ray display oscilloscope. The display tube phosphor must have relatively high persistence because of the relatively slow scanning rates, and highly actinic fluorescence to permit photography of the displays in a single line-scan or raster-scan time. If an energy-dispersive x-ray spectrometer is provided, the console also houses the multichannel analyzer and its oscilloscope and/or X–Y-recorder readout components.

The functions of positioning the specimen stage, setting the x-ray spectrometers, scaling and reading out the x-ray intensities and specimen current, etc., can all be automated. Instruments have been built and are available commercially that provide equipment for operation of an electron-probe microanalyzer for many hours substantially unattended. The intensity data are printed out on a typewriter or continuous-roll printer. Moreover, calculation of analytical concentrations from the measured intensity data from samples and standards can be done and the results printed out by a computer. Even the complex intensity corrections (Section 21.6.2) can be made. A "dedicated" computer can be provided for the instrument, or the intensity data can be transmitted to a central, time-shared computer.

21.3. INTERACTION OF THE ELECTRON BEAM AND SPECIMEN

21.3.1. Interaction Phenomena

At the point of intersection of the electron beam and specimen, many phenomena occur that can be used to provide information about the specimen.

21.3.1.1. Electron Phenomena

a. Backscattered Electrons. Some of the incident electrons are scattered back in the general direction from which they arrived. The fraction so scattered varies with the effective atomic number of the incremental area: about half are scattered by the heaviest elements, very few by the lightest. Backscatter is also strongly influenced by topography. The backscattered electrons have relatively high energy, ranging up to that of the incident electrons, and their paths have high angles with respect to the specimen surface.

b. Secondary Electrons. The incident electrons eject some true secondary electrons from the specimen surface. The number of secondary electrons has no simple relation to atomic number, and their distribution is strongly dependent on surface topography. The secondary electrons have low energy ($<$50 eV), and their paths have low angles with respect to the specimen surface.

c. Absorbed Electrons. Some of the incident electrons are collected by the specimen and constitute what is variously termed the *absorbed-electron*, *target*, or *specimen current*, analogous to x-ray tube current. The fraction so collected varies inversely with the effective atomic number: Nearly all are collected by beryllium, about half by uranium. The backscattered and absorbed electrons are essentially complementary, and the absorbed current is approximately equal to the beam current minus the backscattered electron current. If the electron beam impinges on a Faraday cup, the specimen current is, in fact, the beam current. A suitable Faraday cup is a cylindrical hole 2–4 mm in diameter and $\sim$5 mm deep, drilled part way through a metal plate, and covered by a metal foil having an aperture of diameter sufficient to admit the entire electron beam. The cup prevents escape of any backscattered or secondary electrons.

d. Transmitted Electrons. If the specimen is extremely thin, electrons may pass through it in numbers inversely proportional to its thickness and atomic number.

e. Diffracted Electrons. If the specimen is crystalline, electron microdiffraction occurs. For specimens infinitely thick to the electrons, only back-reflection diffraction is feasible, but for thin specimens or thin edges or protrusions on thick specimens, transmission diffraction is feasible. High-energy electron diffraction (HEED) and low-energy electron diffraction (LEED) are feasible, depending on the electron acceleration potential.

f. Kikuchi-Diffracted Electrons. The point of incidence of the electron beam on the specimen surface acts as a point source of scattered, spherically divergent, monoenergetic electrons. If the specimen is thin enough, a solid cone of electrons is transmitted through the specimen. In the absence of diffraction, photographic plates placed above and below the specimen would be uniformly exposed by the scattered and transmitted electrons, respectively. However, if the specimen is a thin single crystal, certain rays of the divergent electron cone will strike certain crystallographic planes at the correct angle for diffraction. The electron density will be increased in the directions of the diffracted rays (reflection lines), and decreased in the directions of the corresponding incident rays (extinction lines). In this way, a pattern of dark and light circles, ellipses, parabolas, and hyperbolas is formed on the photographic plates. These *Kikuchi lines* convey information about crystallographic perfection, orientation, lattice parameter, stress, etc.

g. Auger Electrons. When an incident electron expels an electron from an inner orbit of an atom and an electron from an orbit farther out falls into the vacancy, a characteristic x-ray photon is produced. There is a certain probability that this photon will not emerge from the atom of its origin, but rather will be absorbed there with consequent expulsion of another orbital electron from the atom. This phenomenon is the *Auger effect* (Section 2.5), and its probability increases as atomic number decreases, approaching unity for the lightest elements. The Auger electrons, like characteristic x-ray spectra, have energies characteristic of the atoms of their origin and are useful for analysis of extremely thin surface layers. The Auger electrons are actually one component of the secondary electrons mentioned above.

21.3.1.2. X-Ray Phenomena

a. Continuous X-Ray Spectrum (Section 1.5). The incident electrons undergo incremental deceleration in the specimen, thereby generating a continuous x-ray spectrum consisting of a continuous band of wavelengths, analogous to white light, having an abrupt minimum wavelength, rising to a peak of maximum intensity, then falling off in intensity at longer wavelengths. The integrated intensity of this continuum is proportional to the specimen current, the square of the acceleration potential, and the effective atomic number of the specimen at the point of impact. The continuum is significant as the principal source of x-ray background.

b. Characteristic X-Ray Spectrum (Section 1.6). The electron beam excites each element at the point of incidence to emit its characteristic x-ray spectrum consisting of discrete spectral lines having wavelengths characteristic of that element and intensities related to its concentration. The incident electrons expel electrons from the inner orbitals of the atoms in the specimen; electrons from orbitals farther out fall into the vacancies with resultant emission of the characteristic x-ray spectrum.

c. *Kossel-Diffracted X-Rays.* Kossel diffraction is the x-ray analog of Kikuchi diffraction of electrons (Section 21.3.1.1f). The point of incidence of the electron beam on the specimen surface generates a point source of spherically divergent x-rays. The most intense wavelength present is likely to be the strongest spectral line of the major element present. The divergent x-rays act in a manner analogous to electrons in Kikuchi diffraction to produce transmission and back-reflection Kossel patterns of dark and light circles, ellipses, parabolas, and hyperbolas on photographic plates. Alternatively, a thin film of a suitable element can be deposited on, or supported just above, the specimen surface to act as an emitter of x-rays of suitable wavelength. The resulting patterns are known as *pseudo-Kossel* or *Lonsdale* patterns and are substantially identical with true Kossel patterns.

d. *Projection X-Ray Microradiography.* The point sources of spherically divergent x-rays (above) may also be used for projection x-ray microradiography. The thin radiolucent specimen lies between the point source and a fine-grain photographic plate. The point source may be generated by impact of the electron beam on the surface of the specimen itself, on a thin film deposited thereon, or on a thin film supported some distance above it. The magnification is the ratio of the distances from the source to the specimen and plate. The photographic image reveals composition distribution (Section 20.1.5) and microstructural features and defects.

21.3.1.3. Cathodoluminescence

The electron beam may excite luminescence in the visible and ultraviolet regions, particularly in mineral and ceramic specimens, and in certain semiconductor specimens.

21.3.1.4. Potential Distribution Pattern

If the specimen is a semiconductor, the incident electrons may create a charge pattern over its surface. This pattern is influenced by composition; dopant; oxide and diffusion layers; semiconductor type (n, p, i); electrical connections; etc. The backscattered, secondary, and absorbed electron line and raster displays may reveal these features, even when compositional differences between them are too small or dopant concentrations too low to be revealed by x-ray methods. If the specimen is a semiconductor device, provision may be made to observe it in operation, in which case the observed potential distribution is also affected by applied potentials.

21.3.2. Detection

Appropriate means can be provided to measure any of the phenomena listed in Section 21.3.1.

21.3.2.1. Electron Detection

Backscattered electrons can be detected by an electron detector placed at a relatively large angle with respect to the specimen surface as shown in Figure 21.1. The detector may be an open-window electron multiplier, a scintillation counter consisting of a plastic scintillator and multiplier phototube, or a solid-state electron detector. The detector output is proportional to the effective atomic number of the specimen spot under the electron beam, but is also highly affected by topography. A single electron detector gives scattered-electron raster images (Section 21.4.4) having a surprising degree of three-dimensional quality. However, much more striking three-dimensional images are obtained by the following arrangement. Two small solid-state electron detectors are mounted near the specimen surface, facing each other on opposite sides of the electron beam and inclined at the same angle to the specimen surface; this is the arrangement for stereo viewing. The *sum* of the outputs from the two detectors gives the same information as a single detector—signal current approximately proportional to atomic number with some three-dimensionality. The *difference* signal from the two detectors provides striking stereoscopic displays of surface topography (Section 21.4.4).

Secondary electrons are detected by a scintillation detector at a small angle with respect to the specimen surface, usually with provision of a positive acceleration potential. A commonly used detector consists of an aluminized hemispherical plastic scintillator attached to the end of a Lucite light pipe, the other end of which is attached to the window of a multiplier phototube. The aluminized coating is maintained at a positive potential of up to a few hundred volts and serves both to exclude stray light and accelerate the low-energy secondary electrons to the underlying scintillator.

Absorbed electrons constitute a simple direct current and require no special detector. *Transmitted electrons* are collected in a Faraday cup below the specimen. *Diffracted*, including *Kikuchi*, *electrons* are recorded on photographic plates in a small camera, which usually contains the specimen itself.

21.3.2.2. X-Ray, Luminescence, and Potential Detection

Kossel patterns and *projection x-ray microradiographs* are recorded on photographic plates in cameras in the same way as Kikuchi electrons (above).

Characteristic x-ray spectra are detected and measured by one to four x-ray spectrometers arrayed about the lower part of the column as shown in Figure 21.1, each complete with electronic readout system. The point source of x-rays is ideal for use with curved-crystal spectrometers. The entire path from specimen to detector is evacuated, and, by use of appropriate crystals and detectors, K spectra of elements down to beryllium can be measured. The favorable combination of primary excitation, curved-crystal optics, and vacuum path results in high excitation and detection efficiency. As a result, L spectra of elements of intermediate atomic number and M spectra of elements of high atomic number are readily measured. Alternatively, the spectrometers can be operated in air path if a beryllium or Mylar window is placed between the spectrometer and column. If the spectrometers can be set at $0° \ 2\theta$, they may be operated in energy-dispersive mode. However, it is preferable to provide a separate instrument for energy-dispersive operation, consisting of a lithium-drifted silicon detector, multichannel analyzer, and oscilloscope or X–Y recorder.

Cathodoluminescence is detected by conveying the luminescence from the specimen surface to a multiplier phototube, either through a Lucite light pipe or by means of a lens–mirror system.

Potential distribution patterns reveal themselves on backscattered, secondary, and absorbed electron patterns, or may be detected by special potential sensors.

21.4. MODES OF MEASUREMENT AND DISPLAY

21.4.1. General

An electron-probe microanalyzer can be fitted to detect, measure, and display any or all of the phenomena listed above. However, the following discussion is limited largely to the four most commonly applied phenomena

—characteristic x-rays and backscattered, secondary, and absorbed electrons.

The electron-probe microanalyzer is capable of great versatility, not only in the number of phenomena it can measure (Section 21.3), but in the ways in which these phenomena can be measured, then read out and displayed. This section deals with the various ways in which the measurements can be made on the specimen and read out and/or displayed on panel meters, digital registers, recorders, printers, cathode-ray oscilloscopes, and photographic cameras. These modes of readout and display are summarized in Table 21.1.

21.4.2. Measurement at a Point

With neither X nor Y deflection, the electron-beam intersects the specimen at a fixed point. The specimen is placed so that the beam falls at a point of interest. A qualitative x-ray spectrometric analysis is made by recording the full x-ray spectrum with the x-ray spectrometer(s). If, say, three spectrometers are available, they may be fitted with crystals and detectors for the short-, intermediate-, and long-wavelength regions, respectively. Thus these three regions can be recorded simultaneously on a multipen recorder. The recorded spectrum consists of a series of peaks on a strip chart just like the spectrum from an x-ray fluorescence spectrometer shown in Figure 10.1. The elements present at the point of impact of the electron beam on the specimen are identified on the chart by the peak positions—that is, wavelengths—and their concentrations may be estimated by the peak heights—that is, intensities. Special wavelength tables for use in electron-probe microanalysis are described in Section 10.4.1. A quantitative x-ray analysis is made by measuring the intensities of the x-ray lines of the analyte elements from samples and standards or pure elements. The intensities are displayed on the digital count registers of the scalers or printed out by a teletypewriter or printer.

21.4.3. Measurement along a Line (One-Dimensional Analysis)

The electron beam can be made to scan a line across the specimen in any of several ways: The *specimen* can be advanced manually in small increments, or motor-driven incrementally or continuously, or the *electron beam* can be deflected forth and back in the X (or Y) direction by the scanning coils. The manual and motor-driven translations can be made for

TABLE 21.1. Classification of Modes of Operation of the Electron-Probe Microanalyzer

Phenomena (principal)	*Readout modes*
Characteristic x-ray spectra	Visual (cathodoluminescence)
Electrons	Meter (x-ray intensity and specimen current)
Backscattered	Recorder (x-ray intensity and specimen current)
Secondary	Digital scaler (x-ray intensity and specimen current)
Absorbed (specimen current)	Printer (x-ray intensity and specimen current)
Cathodoluminescence	Cathode-ray oscilloscope (x-ray intensity, backscattered and secondary electrons, and specimen current)
	Photographic camera (images on cathode-ray oscilloscope and in light microscope)

Spacial distribution

Extent		Scan	Movement	Distance	Principal applicable modes of readout and display	Figure
Point	—	—	—	—	Ratemeter, scaler	—
Line	Step	....	Manual	1 cm	Scaler	19.13 (left)
			Mechanical	1 cm	,,	
			Electronic	0.5 mm	,,	—
	Continuous	———	Manual	1 cm	Ratemeter	19.13 (right)
			Mechanical	1 cm	,,	
			Electronic	0.5 mm	Oscilloscope	21.2, 21.3B
Area	Step	⋮⋮⋮ ↓	Manual	(1 cm)^2	Scaler	21.5
			Mechanical	(1 cm)^2	,,	
			Electronic	(0.5 mm)^2	,,	21.5, 21.6
	Continuous	☰ ↓	Mechanical	(1 cm)^2	Ratemeter	—
			Electronic	(0.5 mm)^2	Oscilloscope	21.2, 21.3A,C, 21.4, 21.8, 21.9

distances up to the order of 1 cm in increments as small as 0.1 μm or for continuous rates as slow as 0.1 μm/s, and read out on the ratemeter or scaler. A motor-driven step-scanner automatically translates the specimen in preselected increments of 0.1, 0.5, 1, 5, 10 or 50 μm and stops at each point to scale and print out analyte-line intensities and specimen current. Electron-beam scanning covers a distance up to ~500 μm in 0.1–10 s and must be displayed on the cathode-ray oscilloscope.

The concentration distribution profile of a selected analyte along a line across the specimen is mapped by setting an x-ray spectrometer at the analyte line, moving the specimen manually or with the step-scanner in small increments, and scaling the intensity at each point. Profiles of two to four different analytes can be scanned simultaneously if two to four spectrometers are available. Subsequent line scans can be made along the same line for different elements or along different lines for the same elements. The intensity data are plotted manually on coordinates of intensity *versus* distance. Such plots are very similar to those shown at the left in Figure 19.13, which were obtained by selected-area x-ray fluorescence spectrometry. However, in electron-probe microanalysis, the total and incremental distances may be smaller by one to three orders.

The same profiles are mapped much more rapidly and conveniently, but less precisely, by translating the specimen continuously with the motor-drive while recording analyte-line intensity on the ratemeter–recorder. Profiles of two to four elements can be recorded simultaneously with a multipen recorder. Such profiles are very similar, except for the distances, to those shown at the right in Figure 19.13, which were obtained by selected-area x-ray fluorescence spectrometry.

Finally, the concentration profile of a selected analyte along a line can be displayed almost instantaneously by setting one of the x-ray spectrometers at the analyte line, allowing the electron beam to line-scan the specimen, and displaying analyte-line intensity on the cathode-ray oscilloscope. The *X*-sweep coil deflects the electron beam in the cathode-ray tube across the display screen in synchronism with the electron beam across the specimen surface. The x-ray spectrometer ratemeter output then provides the *Y* deflection of the cathode-ray tube beam, that is, deflects the beam vertically in proportion to instantaneous analyte-line intensity. The result is a bright-line cathode-ray trace of intensity *versus* distance, analogous to the ratemeter–recorder charts or scaler plots described above. Such line scans are shown in Figures 21.2, 21.3B, and 21.4.

In any of the foregoing modes, it is customary to scale, record, or display specimen or backscattered electron current along with the x-ray

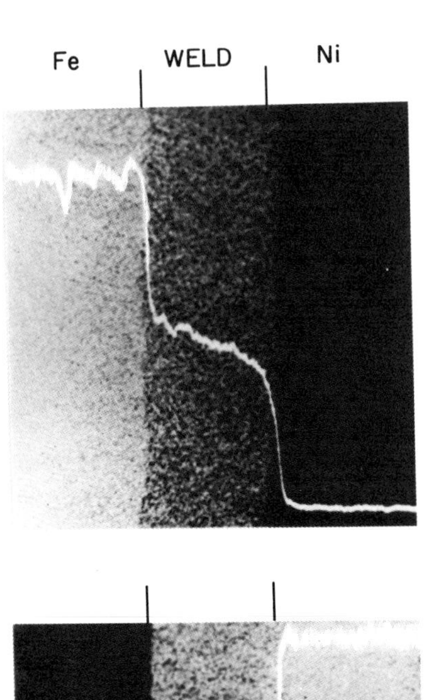

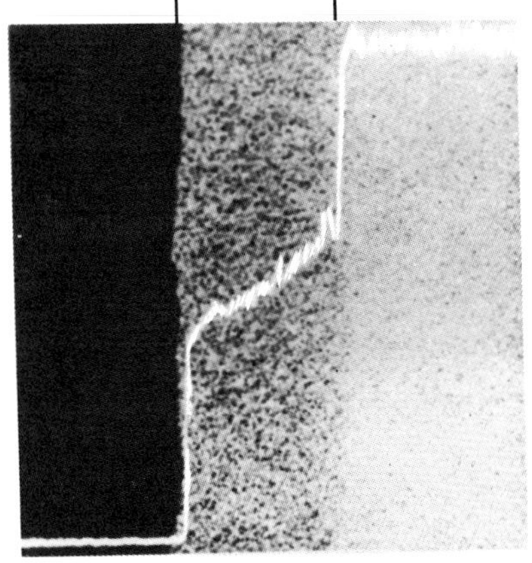

FIG. 21.2. Electron-probe microanalyzer oscilloscope displays: Fe $K\alpha$ and Ni $K\alpha$ x-ray raster images (intensity modulation) with superimposed line scans (deflection modulation) from a section of an iron-to-nickel weld. [Courtesy of JEOLCo (USA), Inc.]

intensities. These electron profiles reveal variations in effective atomic number and surface topography that may be correlated with the x-ray intensity profiles.

21.4.4. Measurement over a Raster (Two-Dimensional Analysis)

The electron beam can be made to scan a raster on the specimen in either of two ways—a *point* raster by incremental mechanical specimen translation, and a *line* raster by continuous mechanical translation or electron-beam scanning. A raster can also be scanned by *manual* incremental

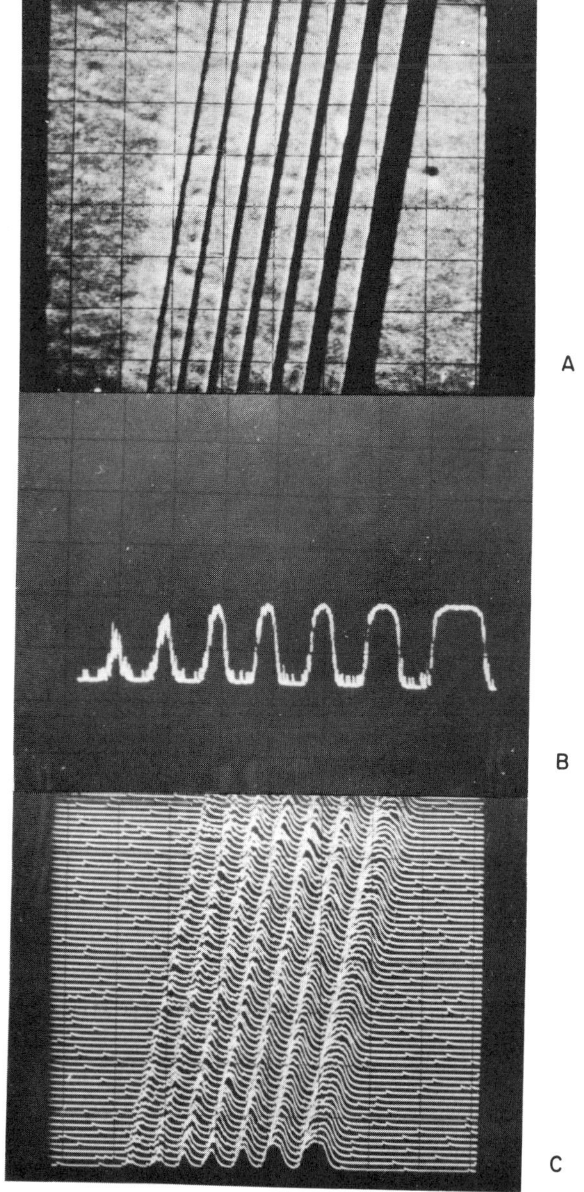

FIG. 21.3. Electron-probe microanalyzer oscilloscope displays of a section of a layered gold–copper specimen in which the gold layers have thicknesses 0.2, 0.3, 0.6, 0.8, 1, 2, and 4 μm, respectively. [Courtesy of Materials Analysis Co.] A. Specimen current raster image (intensity modulation); gold (Z 79) backscatters more electrons than copper (Z 29) and therefore appears dark. B. Au $M\alpha_1$ x-ray line scan (deflection modulation). C. Au $M\alpha_1$ x-ray raster image (deflection modulation).

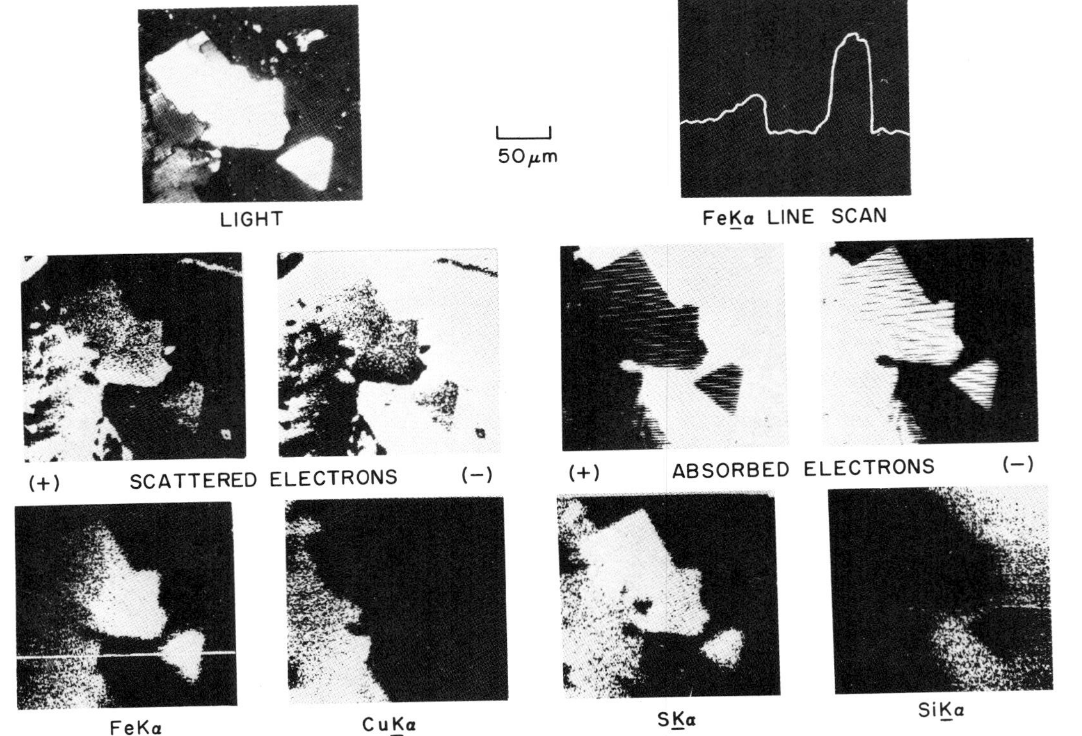

FIG. 21.4. Electron-probe microanalyzer oscilloscope displays of the same area of a heterogeneous mineral consisting mostly of pyrite (FeS_2) and chalcopyrite [$(Fe,Cu)S_2$] with some siliceous material. The top left image is a photomicrograph taken through the light microscope. The Fe $K\alpha$ x-ray line scan (deflection modulation) was made at the position indicated by the bright-line marker on the Fe $K\alpha$ x-ray raster image. All the other displays are intensity-modulation electron or x-ray raster images. The scattered-electron and absorbed-electron (specimen-current) images marked + are the normal images—higher atomic number brighter in the scattered-electron image, darker in the absorbed-electron image; the images marked − are merely electronically reversed. [Courtesy of JEOLCo

specimen translation, but this procedure is extremely time-consuming and tedious.

In the first method, a line is scanned across the specimen in the X direction by step-scanned incremental translation, as described above. The specimen is then translated one increment in the Y direction and another X line scanned. This procedure is continued until a point raster has been covered. At each point, x-ray intensity is read out on a digital scaler. Rasters 1 cm square or more can be scanned in this way. The scaler data are then entered manually on a two-dimensional grid. "Isointensity" lines may be drawn joining points of substantially the same intensity. Step-scanners are available to perform such a raster scan automatically, scaling the intensity at each point and printing it out.

The multichannel analyzer, operated in multichannel-scaler mode (Section 8.1.2.3), is extremely valuable for storing and displaying point-raster data. The accumulated analyte-line counts from successive raster points are stored in successive channels in the multichannel scaler. Then, at the end of the scan, the data may be read out and displayed in any of several ways. The accumulated counts in the individual channels may simply be printed out digitally in columns for manual plotting, as above. Or they may be printed out in a two-dimensional array, each count in its correct relative position in the raster; Figure 21.5 [p. 81 in (*7*)] shows such a two-dimensional printout, point-raster, intensity topograph. Alternatively, the data may be displayed with each channel—that is, each raster point represented by a dot. However, instead of each dot occupying its position in the rectilinear raster, each dot is displaced vertically from this position. The vertical displacement is proportional to the number of analyte-line counts accumulated in that channel, and therefore to the number of analyte-line photons emitted from the corresponding point on the specimen. Figure 21.6 [p. 100 in (*7*)] shows such a topograph, which may be displayed on an X–Y recorder or cathode-ray oscilloscope. The displaced-dot array has the appearance of a topographic map viewed in perspective and very effectively portrays the analyte-element distribution over the specimen surface.

In the electron-beam scanning method, the beam scans a raster line-by-line on the specimen surface. The dimensions can be varied from nearly a point to ~500 μm square by use of the X- and Y-size controls. The concentration distribution of a selected analyte over the raster area is displayed on the oscilloscope. The X- and Y-sweep coils deflect the electron beam in the cathode-ray tube in a line raster on the display screen in synchronism with the electron beam on the specimen surface. The x-ray spectrometer pulse-height selector or ratemeter output then modulates the *intensity*

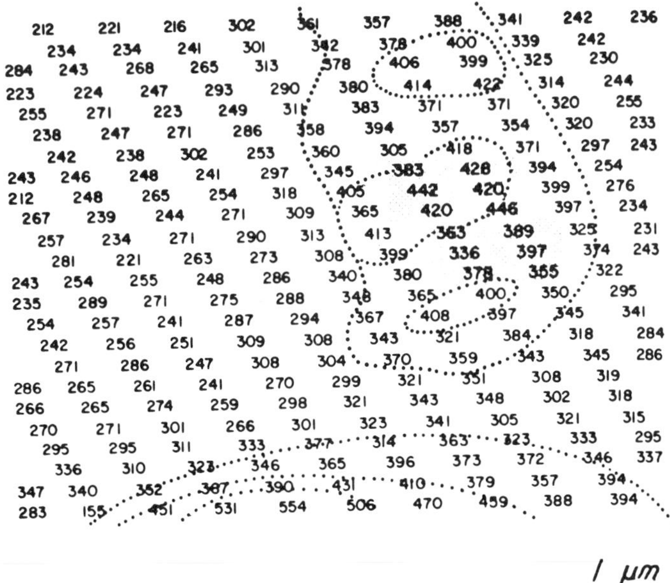

FIG. 21.5. Electron-probe microanalyzer multichannel-scaler printout intensity topograph (point raster) of iron-rich inclusions in an aluminum alloy. The numbers are Fe $K\alpha$ counts accumulated in the multichannel scaler channels. The isointensity lines were drawn in manually by joining points of substantially equal intensity and delineate the iron-rich regions. [L. S. Birks, "Electron-Probe Microanalysis," ed. 2, Wiley-Interscience, N. Y., p. 81 (1971); courtesy of the author and publisher.]

(Z axis) of the cathode-ray beam in proportion to instantaneous analyte-line intensity, so the brightness of each incremental area on the display is proportional to analyte-line intensity—and concentration—at the corresponding incremental area on the specimen.

Both of the above raster scanning techniques are used in the combined electron-beam specimen-translation method. Electron-beam scanning is used in the direction perpendicular to the line joining the specimen and x-ray spectrometer because in this direction the spectrometer is relatively insensitive to the position of the beam spot. Mechanical specimen displacement is used in the other direction.

Absorbed, backscattered, and secondary electron currents are displayed the same way as the x-rays. The higher the effective atomic number of the specimen area under the electron beam, the brighter is the corresponding point on the backscatter display pattern, and the darker is the point on the absorbed current pattern. When the electron-probe microanalyzer is operated with electron displays in this way, it is, in effect, a scanning electron microscope (SEM). However, it is incapable of magnification, resolution, or depth of electron focus as high as an instrument designed specifically for scanning electron microscopy.

Figures 21.2, 21.3A, and 21.4 show intensity-modulated x-ray and electron oscilloscope raster displays.

The electron displays—especially the secondary-electron display—are influenced by surface topography and often show a marked degree of three-dimensionality. If two scattered-electron detectors are provided, as described in Section 21.3.2.1, and if their difference signal is used to modulate the brightness of the oscilloscope display, the electron display has the appearance of a stereoscopic map of surface topography in which the relief may be very striking. The sum signal of the two detectors gives the same display as a single detector, where brightness is proportional to backscatter

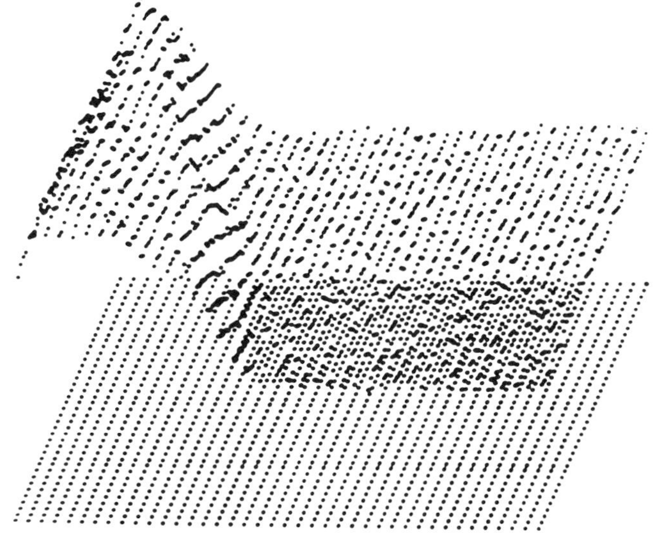

FIG. 21.6. Electron-probe microanalyzer multichannel-scaler print-out intensity topograph (point raster) of the edge of a diffusion zone. The regular rectilinear dot array below shows the raster position corresponding to each channel. In the dot topograph above, each of these points is displaced above its raster position in proportion to the accumulated analyte-line count in that channel. The regular array below would represent zero analyte-line intensity of all raster points. [L. S. Birks, "Electron-Probe Microanalysis," ed. 2, Wiley-Interscience, N. Y., p. 100 (1971); courtesy of the author and publisher, and of K. F. J. Heinrich, National Bureau of Standards.]

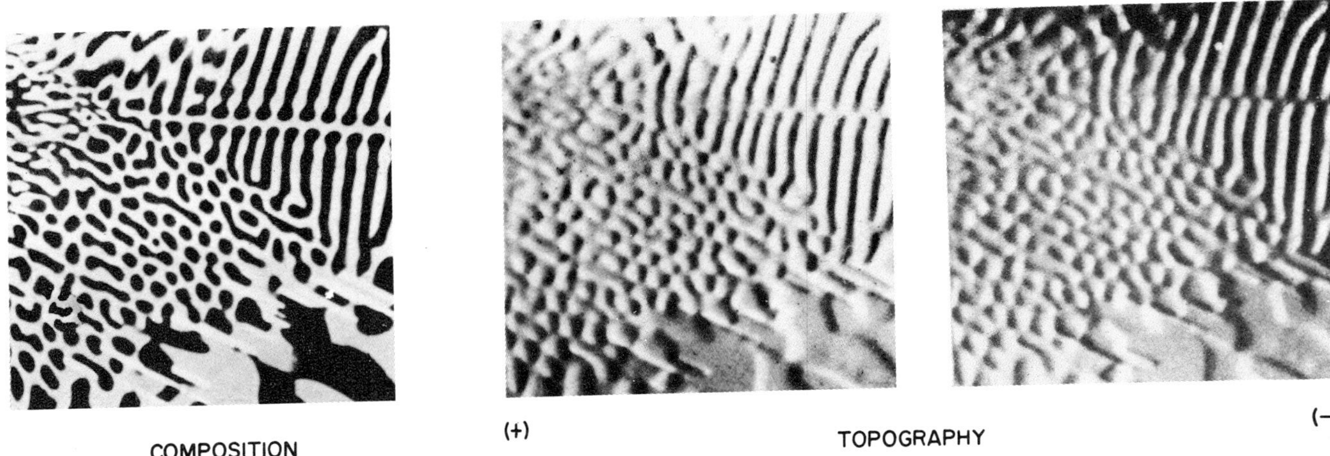

FIG. 21.7. Electron-probe microanalyzer oscilloscope displays of 67Al–33Cu alloy: backscattered-electron raster images (intensity modulation) from two stereo solid-state detectors. The composition display is the sum of the two detector outputs, and the brightness is proportional to atomic number. The topography displays are the difference of the detector outputs and show surface relief. [Courtesy of JEOLCo (USA), Inc.]

efficiency, that is, to atomic number. Figure 21.7 shows intensity-modulated backscattered-electron oscilloscope raster displays of the output of such a dual detector.

Another type of oscilloscope line-raster display is provided by combination of the line and raster modes. The method is the same as for the electron-beam scanning raster method already described, except that each line in the raster is *deflection*-modulated, as in a simple line scan—rather than *intensity*-modulated—by the x-ray analyte-line intensity or the absorbed or scattered electron profile along that line. Figure 21.3C shows a deflection-modulated oscilloscope raster display. Such displays can also be obtained on *X–Y* recorders, as shown in Figure 21.8 (*H29*).

In all the intensity-modulation x-ray raster images described so far, the brightness of each incremental area of the cathode-ray tube display is directly proportional to the number of x-ray photons emitted by the corresponding incremental area on the specimen. This type of display is sometimes termed *density-modulation*. However, the human eye has logarithmic sensitivity to this linear variation of brightness, so analyte-line intensity distribution—and thereby analyte concentration distribution—is difficult to evaluate from the display. This problem is solved in a remark-

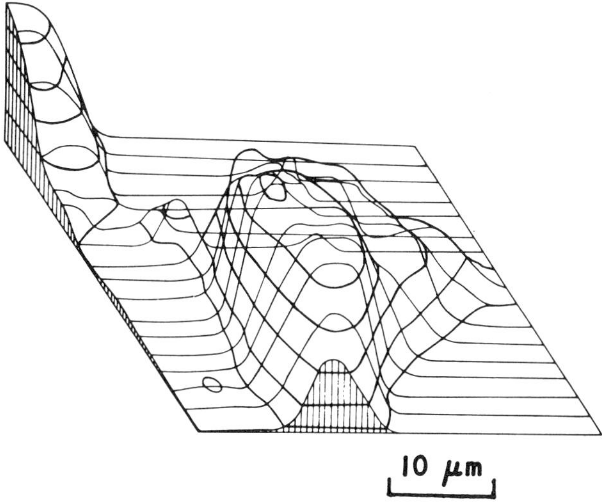

FIG. 21.8. Electron-probe microanalyzer concentration map of lead inclusions in brass made by successive ratemeter line-scans of the Pb $L\alpha_1$ line on an *X–Y* recorder. The starting points of successive lines are displaced along an oblique line with respect to the distance (or time) axis. [K. F. J. Heinrich, *Advances in X-Ray Analysis* **7**, 382 (1964); courtesy of the author and Plenum Press.]

ably ingenious technique of semiquantitative electron-probe microanalysis known as *content mapping* (*T14*).

In content mapping, the raster area is divided into 2500 (50×50) square *mapping elements*. On the cathode-ray tube display, each of these elements has one of, say, 10 *discrete* logarithmic brightness levels determined by the linear intensity of the x-ray emission at the corresponding element on the specimen. The electron beams on the specimen and cathode-ray tube screen automatically and synchronously step-scan the mapping elements in the following sequence. The top line of 50 elements is step-scanned from left to right. The beams then return substantially instantly to the left edge, index one picture element downward, step-scan the second line of 50 elements, and so on until all 50 lines—all 2500 elements—have been covered. Typical scanning time is ~10 min. During the scan, the output of the pulse-height analyzer goes to an electronic content-mapping unit which divides the signal into, say, 10 intensity ranges or steps of,

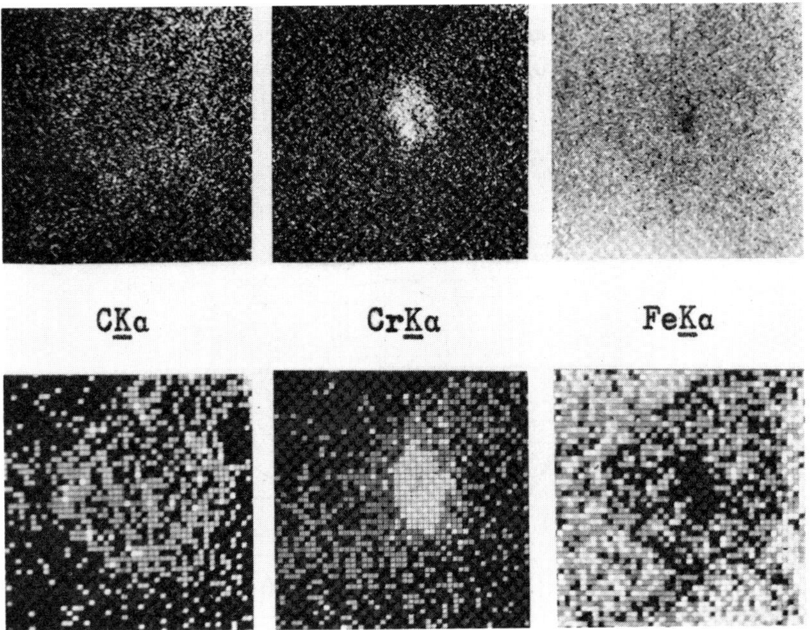

FIG. 21.9. Comparison of conventional density-modulation (above) and content-mapping (below) raster displays of C *Kα*, Cr *Kα*, and Fe *Kα* x-rays from a chromium-rich inclusion in a steel specimen. The content-mapping displays have five intensity steps. The images represent specimen areas 50 μm square. [T. Tomura, H. Okano, K. Hara, and T. Watanabe, *Advances in X-Ray Analysis* **11**, 316 (1968); courtesy of the authors, Hitachi, Ltd., and Plenum Press.]

say, 50 counts/s each. That is, the instantaneous pulse-height analyzer output is "classified" into 10 intensity steps of 0–50, 50–100, 100–150, . . . , 450–500 counts/s, respectively. A pulser unit uses these intensity steps to modulate the instantaneous brightness of the cathode-ray tube on a logarithmic scale. Thus, the 2500 display elements have one of 10 discrete logarithmic brightness levels corresponding to the 10 linear intensity steps. Alternatively, a typewriter can print the number of the intensity step for each mapping element in its proper position on the paper. Such printout topographs are more striking if each digit is printed in a different color. A photograph of such a color display is given in Reference (*T14*). In practice, the operator has a selection of, say, 2, 3, 5, 7, or 10 steps of 50, 100, 200, or 500 counts/s per step, thus giving him 20 different step–intensity combinations from 2 steps of 50 counts/s each to 10 steps of 500 counts/s each. Figure 21.9 permits a comparison to be made of conventional density modulation and content-mapping raster displays of C $K\alpha$, Cr $K\alpha$, and Fe $K\alpha$ x-rays from a steel specimen.

21.4.5. Measurement Perpendicular to the Specimen Surface (Three-Dimensional Analysis)

The dependence of electron penetration on acceleration potential provides the basis for measurements along a line perpendicular to specimen surfaces that consist of a series of thin (10–10^4-Å) layers of different composition. Analyte-line intensities may be measured at a series of acceleration potentials from 50 to 2 kV or less in, say, 2.5-kV decrements. A plot of analyte-line intensity *versus* potential may reveal the composition and approximate thickness of the various layers. The technique also permits analysis of a surface layer to the exclusion of interference from underlying layers. Of course, the method is seriously complicated and limited by excitation potentials and x-ray absorption coefficients of the various elements in the layers. A much better way to obtain such data is to probe the specimen after sectioning or angle-lapping to reveal the layers.

21.4.6. Other Methods of Readout and Display

Of course, any of the line and raster displays on the oscilloscope, as well as the images in the light microscope, can be photographed. Light photomicrographs are often useful for correlation with x-ray raster images of concentration distributions of different elements, although usually backscattered electron images are as good or even better for this purpose.

Each of the x-ray ratemeters has its own panel meter indicating x-ray intensity in counts per second, and the specimen current is indicated on a 0–10 microammeter having scales down to 10 nA. As already mentioned, if the "specimen current" is measured in a Faraday cup, it becomes the beam current.

Absorbed, scattered, and secondary electron currents may also be read out digitally: The dc electron currents are applied to the input of a *digitizer* where they vary the frequency of an oscillator—the higher the current, the higher the frequency. The output frequency is then rectified and the remaining alternate pulses counted in a scaler and read out on a count register in the usual manner.

The outputs of the x-ray scalers can be applied directly to a computer, which can be programmed to correct the intensity data (Section 21.6.2) and compute analyte concentrations directly.

21.4.7. Color Displays

Color photography can be used to aid in the visualization and interpretation of electron-probe raster displays. Different colors may be used to represent different chemical elements or different concentrations of the same element.

The first application can be realized in several ways. Successive raster images of the same specimen area can be displayed on the oscilloscope from, say, three x-ray spectrometers set for x-ray spectral lines of three different elements. Each image is photographed on the same color film, but through a filter of a different color. For example, the images of the shortest, intermediate, and longest wavelengths can be photographed through blue, green, and red filters—that is, in blue, green, and red light—respectively. In this way, a color picture of the specimen area is obtained in which different colors represent different elements, and variations in color density over the image represent variations in concentration of the element represented by that color. This technique may be termed the *direct method* (*H29*, *I1*, *J24*). Figure 21.10 (*I1*) shows such a color photograph.

Alternatively, the several oscilloscope images can be photographed on separate black-and-white positive films, then each film exposed in turn through a different color filter onto the same color film (*J24*, *Y2*). A third technique is to photograph the oscilloscope images on separate black-and-white negative films, each of which is then used to expose a different color diazochrome film to produce a positive color transparency. These color transparencies are then superimposed in register, and the composite exposed

FIG. 21.10. Direct-method composite color photograph of electron-probe microanalyzer displays of a section of an iron rod plated with successive layers of copper, nickel, copper, and chromium. Oscilloscope displays of Fe $K\alpha$, Cu $K\alpha$, Ni $K\alpha$, and Cr $K\alpha$ x-ray raster images (intensity modulation) were photographed successively on the same color film through orange-red, yellow, blue, and green color filters, respectively. [R. M. Ingersoll and D. H. Derouin, *Review of Scientific Instruments* **40**, 637 (1969); courtesy of the authors and the American Institute of Physics.]

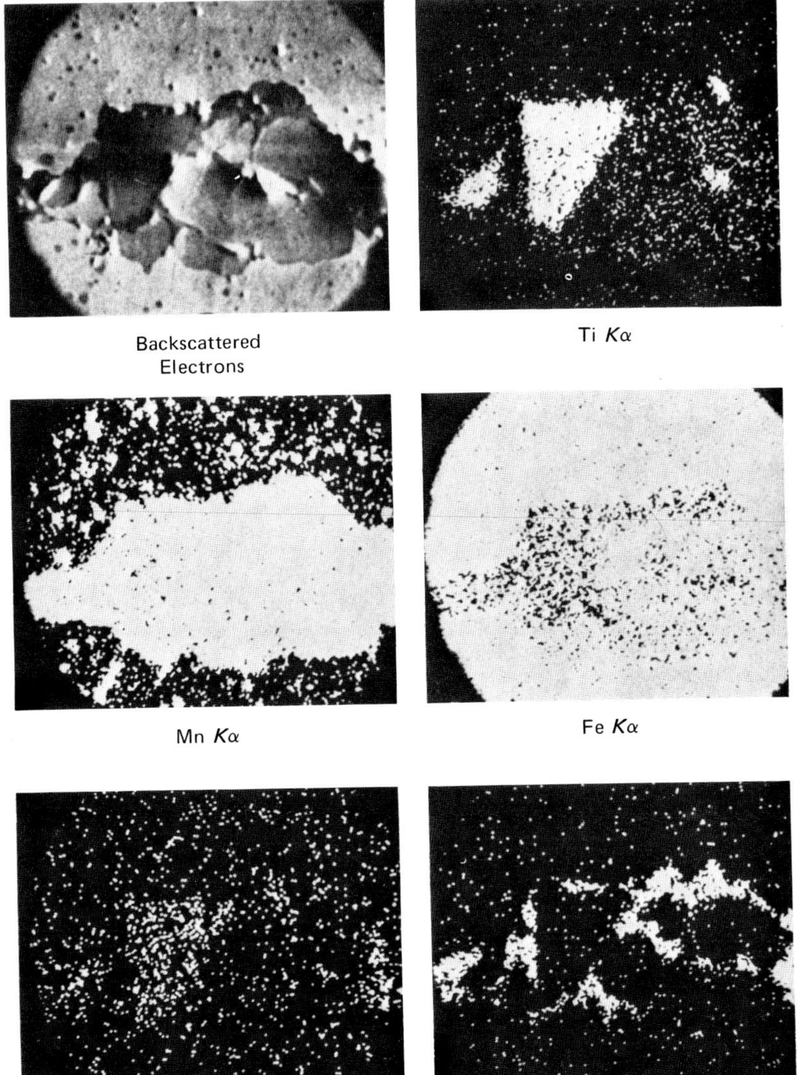
Backscattered
Electrons
Ti $K\alpha$
Mn $K\alpha$
Fe $K\alpha$
Al $K\alpha$
Si $K\alpha$

A

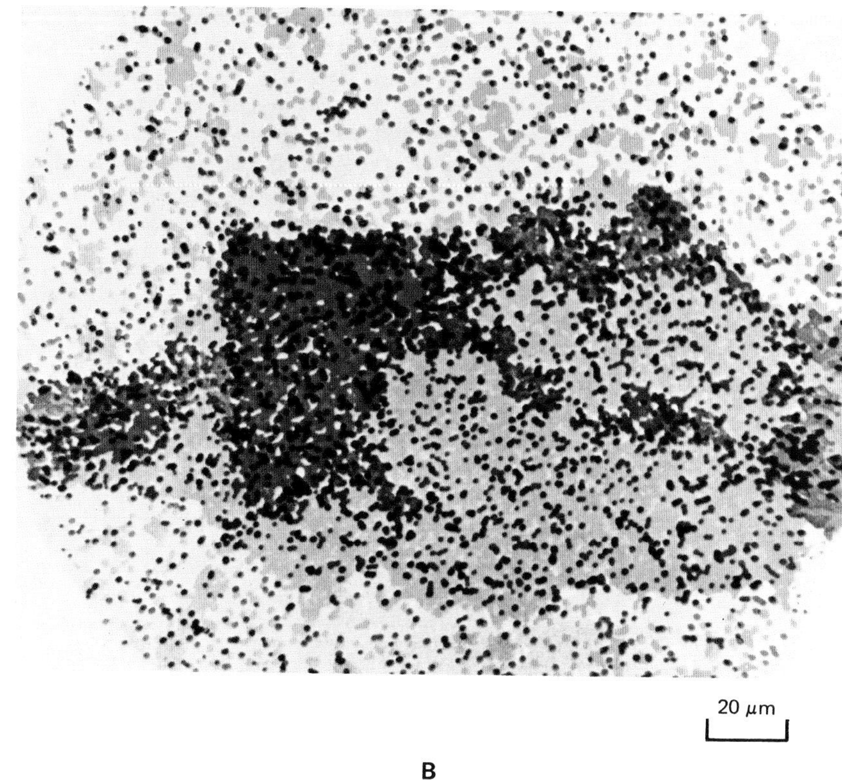

B

FIG. 21.11. Electron-probe microanalyzer displays of the same area of an inclusion in iron. [L. P. Salvage, *Metallography* **2**, 101 (1969); courtesy of the author, Elsevier Publishing Co., and the International Microstructural Analysis Society.] A. Oscilloscope raster images (intensity modulation) of backscattered electrons and $K\alpha$ x-rays of Ti, Mn, Fe, Al, and Si. B. Indirect-method composite color photograph of all the x-ray images in A. The oscilloscope raster images in A were photographed on separate black-and-white negative films, each of which was then used to expose a different color diazochrome film to produce a color transparency. These color transparencies were then superimposed in register and the composite exposed onto a color film. The color code is as follows: red—Ti $K\alpha$; yellow—Mn $K\alpha$; light blue—Fe $K\alpha$; dark blue—Al $K\alpha$; and green—Si $K\alpha$. The dots represent individual x-ray photons; the Fe $K\alpha$ (light blue) and Mn $K\alpha$ (yellow) images appear continuous because the intensities of these lines are high.

onto a color *film* (*S8*). Figure 21.11A shows the several black-and-white oscilloscope x-ray displays from an inclusion in iron, and Figure 21.11B shows the resulting color composite prepared by this technique (*S8*). The two techniques described in this paragraph may be termed *indirect methods*.

Incidentally, the cathode-ray tube used for the direct method must have a phosphor having white—rather than yellow, green, or blue—luminescence; this does not necessarily apply to tubes used for the indirect methods.

Yakowitz and Heinrich (*Y2*) give detailed considerations for preparation and comparison of photographic color composites, specifically color mixing, choice of filters to match film characteristics, exposure criteria, and choice of color for each partial image.

Finally, the same type of color display can be obtained simultaneously by using the outputs of the three x-ray spectrometers to modulate the brightness of the red, green, and blue electron guns, respectively, of a color-television picture tube. The advantages of this method are that the picture is seen directly and can be adjusted to the operator's satisfaction prior to photographing it, and that errors arising from slight differences in size and orientation of the raster images and slight misalignments of superimposed images do not occur.

An example of the use of color to represent different concentrations of the same element follows. By means of X and Y stepping motors, the specimen is translated in mutually perpendicular directions so as to scan a point raster on its surface. The X motor advances the specimen for a preset number of increments of preset size. Then the Y motor moves one step, and the X motor reverses and scans the second line in the opposite direction. Then the Y motor moves another step, and the X motor scans the third line in the forward direction, etc., until the raster is scanned. Simultaneously, a Polaroid color-film holder is driven by two other X and Y stepping motors in synchronism with the specimen-drive motors.

At each point on the specimen scan, the corresponding point on the color film is illuminated by a beam of light passing through a shutter and a 10-sector color disk, each sector of which is a different color. The disk is rotated by a repeater motor driven by the x-ray spectrometer ratemeter output. Thus, at each point, the position of the disk—and therefore the color of the light beam—is determined by the output of the x-ray ratemeter, and thereby by the analyte-line intensity. In this way, a color mosaic photograph of the analyte distribution is built up, the colors representing a 10-step gradation of the intensity range. Figure 19.10 shows such a color display obtained by selected-area x-ray spectrometry (*C9*). By alteration of the ratio of the specimen and film scanning linkages, magnification or reduction of the photograph is effected.

Another type of color display is described in Section 21.4.4. Color-coding of x-ray energy spectra is shown in Figure 8.23.

21.4.8. Considerations

In most instruments, the output of an x-ray spectrometer can be displayed on the oscilloscope simultaneously with absorbed or scattered electron current. Both the simultaneous signals may be displayed as rasters, or one as a raster, the other as a line. Figure 21.2 shows oscilloscope displays with superposed x-ray raster and line displays. Provision is usually made to permit: (1) selecting a line of particular interest on a scattered or absorbed electron raster display; (2) marking it electronically; and (3) superimposing on the raster display the profile of a specified analyte x-ray line or of the absorbed or scattered electrons along the selected line. Such superpositions facilitate correlation of composition and topography.

Provision may also be made to reverse the positive–negative polarity of the raster displays (bright to dark, or upward to downward deflection) to aid in their visual conception and interpretation. This feature is particularly useful in topographic scattered and secondary electron displays. The negative rasters also permit direct photography of *positive* transparencies. In Figure 21.4, the backscattered and absorbed electron displays are shown in both positive (normal) and negative polarity.

The magnification of the line and area displays is given by the ratio of the distance between corresponding points on the cathode-ray tube display and on the specimen itself. The magnification is varied by changing the length or dimensions of the scan on the specimen while leaving the line or raster on the cathode-ray tube unchanged. A typical magnification range is 100–10,000×.

In an electron-probe microanalyzer, the cross section of the electron beam can be made to cover a specimen area having diameter of the order 1–100 μm. When larger incremental areas—0.01–1 mm in diameter—are to be analyzed, it is often more convenient to use an x-ray secondary-emission (fluorescence) spectrometer fitted with selected-area accessories (Chapter 19). This instrument can be operated in the same *manual* and *motor-driven* modes as the electron-probe microanalyzer: qualitative and quantitative analysis at a point, and intensity profiles along a line or over a point raster by manual or motor-driven incremental or continuous specimen translation. However, of course, electron-beam scanning does not apply. The "x-ray probe" provides the same type of information in the same form as the electron probe, but from an incremental area larger by two or three orders.

21.5. SPECIMEN CONSIDERATIONS

For accurate quantitative electron-probe microanalysis, or even comparison of intensities, all samples, standards, and blanks should be potted in the same Araldite mold or cemented to the same mount, then polished to an identical surface plane and 0.25-μm finish. Any less careful preparation is likely to reduce whatever accuracy is attainable by the electron-probe method for the particular specimen type. Some specimens by their very nature cannot be so treated, but less accuracy is to be expected in such cases. For qualitative analysis, preparation requirements are much less rigid.

In x-ray secondary-emission (fluorescence) spectrometry, the depth of the surface layer contributing the measured x-rays is limited by the depth *from* which the characteristic x-rays can emerge and is of the order 100 μm. In primary-emission spectrometry, the limit is imposed by the depth *to* which the electrons penetrate (with sufficient energy to exceed the analyte excitation potential), which is of the order 1 μm. As a result, absorption-enhancement (matrix) effects are *less* serious and surface-texture effects *more* serious in electron-probe microanalysis than in fluorescence spectrometry. Microscopic topography can be detrimental in three ways: incident electrons may be shielded from surface features, and x-rays originating in micro voids or depressions may be shielded from the x-ray spectrometers. Also, microscopic slopes present a takeoff angle (see below) different from that of the specimen plane, and takeoff angle affects the x-ray intensity in the direction of the x-ray spectrometer.

The implications of the reduced matrix effects are considered in Section 21.6.1. The more severe surface-texture effects necessitate highly plane and polished specimen surfaces. The final polish should be made with 0.25-μm diamond dust on a cloth lap, but a finer finish may be preferable. Other abrasives—silicon carbide, aluminum oxide, cerium oxide, etc.—and lead- or tin-coated polishing wheels are avoided to prevent contamination of the specimen surface. Etches are avoided because they alter composition and create topography. When multiple thin-layer specimens are sectioned to permit analysis of the individual layers, oblique sectioning (angle-lapping) may be preferable to normal sectioning and lapping to widen the layers. It is necessary to evaporate a carbon film ~100 Å thick on electrically nonconductive specimens to prevent buildup of electric charge.

It is usually difficult or impossible to observe features on highly polished surfaces under the light microscope. In such cases, the feature may be located as follows. Spacial variations in composition may be satisfactorily revealed in the scattered or absorbed electron raster displays. Alternatively, the specimen may be etched to reveal the features, then, under observation in a light microscope, scored, or pitted with a micro hardness tester to locate the features of interest. The specimen is then polished to remove the etch topography, but not the scoring or microhardness marks.

Optimal performance requires that the point to be analyzed on the specimen surface also be the common point of intersection of the electron-optical column axis, x-ray beam paths to the spectrometers, light-microscope axis, and electron beam. The x-ray beam path is the straight line joining the beam spot and the centers of the x-ray spectrometer entrance apertures and crystals. The angle between this line and the specimen plane is the x-ray *takeoff angle*. In general, a relatively large takeoff angle ($\gtrsim 30°$) is advantageous for two reasons: The intensity–concentration function is more linear, especially for long-wavelength spectral lines, because x-ray beam path length in the specimen is shorter. Also, the measured intensity is influenced less by shadowing effects of surface topography.

For any mode of measurement (Section 21.4), the region of interest is selected by use of the light microscope and specimen translators. If observation at higher magnification is required (up to 10,000×), the absorbed, scattered, or (preferably) secondary electron oscilloscope raster display may be viewed. The exact spot, line, or area to be studied may be placed by use of the light-microscope cross-hairs or, more precisely, the oscilloscope displays.

After the electron beam has bombarded the selected region for a while, a visible dark spot, line, or square appears on the specimen viewed through the light microscope. The darkening results from carbonization, by the electron beam, of adsorbed residual oil vapors from the diffusion pump. These deposits may interfere with the x-ray measurements in one or more of three ways: (1) They reduce the energy of the incident electrons; (2) they absorb the emergent x-rays, although this effect is serious only for very long wavelengths; and (3) they contribute analyte-line intensity when carbon or other elements present in the deposit are to be determined. Otherwise, the deposits are beneficial in that they mark precisely the spot, line, or area investigated.

In electron-probe microanalyzers, the electron-beam spot diameter at the specimen plane is typically ~1 μm. However, for highest resolution, it is possible to reduce the spot diameter to ~0.5 μm. The smaller the beam spot, the higher the resolution, but the smaller is the maximum attainable beam current and therefore the smaller is the maximum attainable x-ray intensity. There are also reasons for using larger spot diameters. An increase in beam spot diameter from, say, 1 to 3–5 μm increases the maximum attainable beam current from, say, 1 to 10 μA and increases the net analyte-line intensity proportionally. However, background also increases, so sensitivity remains substantially unimproved. Larger beam spots—up to ~50 μm—are also used when the specimen surface finish is not as smooth or homogeneous as desired and to reduce the rate of contamination buildup. In scanning electron microscopes fitted with energy-dispersive spectrometers, the electron beam spot is typically ~0.1 μm in diameter and may be made as small as ~200 Å.

21.6. QUANTITATIVE ANALYSIS

21.6.1. Principles

In principle, in either primary or secondary x-ray emission spectrometry, one would expect the concentration $C_{A,X}$ of analyte A in sample X to be given by

$$C_{A,X} = I_{A,X}/I_{A,A} = R_A \tag{21.1}$$

where $I_{A,X}$ and $I_{A,A}$ are analyte-line intensities from the sample and pure analyte, respectively. Although this equation may give high accuracy in favorable cases in both methods, it is not generally satisfactory in either. The reason is that the incident electron (or x-ray) beam and the emergent analyte-line x-rays are affected not only by the analyte, but also by the specimen matrix. The equation is usually more accurate in electron-probe microanalysis than in fluorescence spectrometry because the measured x-rays originate in a very thin surface layer, so the effect of the matrix is reduced.

When the equation does not give acceptable accuracy, the matrix effect must be dealt with in one of two ways, with calibration standards or by mathematical calculation.

The first of these methods is the more convenient and satisfactory. Analyte-line intensity is measured from the samples and from two or more standards of the same substance of known composition. The standards must be homogeneous on a submicrometer scale, and standards and samples must have identical surface finishes. An intensity–concentration relationship is established from the standard data and used to derive analyte concentrations for the samples.

When suitable standards are not available, resort is made to the mathematical methods. Essentially, the procedure is as follows.

1. Analyte-line intensity is measured from the samples and from a reference standard. Usually the standard is the pure analyte, but when its use is not feasible (gaseous elements, sodium, sulfur, bromine, etc.), a suitable compound or alloy having known composition is used.

2. The measured intensities are corrected for dead time, background, instrument drift, etc. (Section 21.6.2).

3. The *relative intensity*—ratio of analyte-line intensities from the sample and reference standard—is calculated for each sample.

4. The relative intensity ratio is corrected for the atomic number (Z), absorption (A), and fluorescence (F) effects to give analyte concentration, as indicated in Equation (21.2) below.

21.6.2. Intensity Corrections

The corrections referred to in Section 21.6.1 are of two kinds: those required to increase the accuracy of the measured intensity, and those required to make the measured intensity more accurately representative of the intensity actually generated in the specimen.

The first group comprises corrections for one or more of the following.

1. *Wavelength setting*—differences among the specimens in peak analyte-line wavelength due to chemical effects. Of course, the need for this correction is precluded by "peaking" the spectrometer for each specimen.

2. *Contamination*—differences among the specimens in the rate of buildup of the contamination spot (Section 21.5). This correction is most important at low acceleration potentials and ultralong x-ray wavelengths.

3. *Instrument drift*—the slow, continuous drift of electron-beam current due to gradual displacement of the filament with respect to the electron gun.

4. *Dead time*—coincidence loss in the detector–readout system. This correction is usually negligible at analyte-line intensities below ~20,000 counts/s for crystal spectrometers using proportional counters.

5. *Background*—the intensity at the analyte-line wavelength in the absence of analyte. This may be measured adjacent to the analyte-line peak from the specimens themselves, or at the analyte-line peak from a "blank" having the same effective atomic number as the specimens but no analyte.

The second group of corrections consists of the following.

1. *Atomic number correction.* As the average atomic number of the matrix decreases, the beam electrons penetrating the specimen lose energy more rapidly because light elements are more readily ionized; thus the matrix absorbs more electrons, tending to reduce the number available for analyte-line excitation. If this were the only atomic number effect, a specified concentration of a specified analyte would give a lower analyte-line intensity from a low-Z matrix than from a high-Z matrix. However, as the average atomic number of the matrix decreases, the fraction of beam electrons that backscatter also decreases, making more electrons available for ionization—that is, for excitation—and tending to give a higher analyte-line intensity than in a high-Z matrix. These penetration and backscatter phenomena, which are also functions of beam acceleration potential, tend to compensate one another. A third atomic number effect is the ionization effect. The number of x-ray-producing ionizations excited by the electrons in unit mass of analyte is given by the ratio of the number of orbital electrons and mass of atoms of that element, that is, by the quotient of atomic number and weight, Z/A.

2. *Absorption correction.* Analyte-line x-rays excited beneath the specimen surface undergo absorption in emerging from the specimen.

3. *Fluorescence correction.* Analyte-line x-rays arise not only by primary excitation by the electron beam, but also by secondary excitation by spectral-line x-rays of other elements in the specimen and by the continuous x-rays. Such secondary excitation is efficient only for relatively intense spectral lines and continuum having wavelength shorter than, and relatively near, the analyte-line absorption edge.

In general, analyte concentration C_A (wt fraction) is given by

$$C_A = R_A(ZAF) \tag{21.2}$$

where R_A is the ratio of analyte-line intensities from the sample X and pure analyte A, $I_{A,X}/I_{A,A}$; and (ZAF) is the total correction for atomic number (Z), absorption (A), and fluorescence (F) effects. The atomic number correction contains factors for both electron backscatter and ionization-penetration. The fluorescence correction is usually limited to enhancement by lines of matrix elements, but may include enhancement by continuum. Incidentally, in the literature of electron-probe microanalysis, the symbol k is usually used for relative intensity, rather than R.

A comprehensive treatment of intensity corrections for electron-probe microanalysis is beyond the scope of this introductory chapter, but the chapter would not be complete without consideration of a typical correction procedure as an example. The original version of Colby's MAGIC (Microprobe Analysis Generalized Intensity Corrections) procedure (*C53*) provides an excellent example. In the following discussion, subscripts A and X, when used on r, s, $f(\chi)$, and (μ/ϱ), refer to pure analyte and sample, respectively.

Equation (21.2) may be expanded as follows:

$$C_{\mathrm{A}} = \frac{I_{\mathrm{A,X}}}{I_{\mathrm{A,A}}}\left[\frac{r_{\mathrm{A}}}{r_{\mathrm{X}}}\ \frac{s_{\mathrm{X}}}{s_{\mathrm{A}}}\right]\left[\frac{f(\chi)_{\mathrm{A}}}{f(\chi)_{\mathrm{X}}}\right]\left[\frac{1}{1+\sum(I_j/I_e)+(I_{\mathrm{cont}}/I_e)}\right] \qquad (j \neq \mathrm{A}) \tag{21.3}$$

where the three bracketed terms are the atomic number, absorption, and fluorescence corrections, respectively, and are expanded below.

In the *atomic number correction*, r is the electron-backscatter coefficient and is the ratio of ionization produced to that which would be produced in the absence of backscatter; and s is the electron stopping power,

$$s_i = \frac{Z}{A}\log_{\mathrm{e}}\left[1.17\,\frac{(V+V_i)/2}{\bar{V}_i}\right] \tag{21.4}$$

where Z and A are atomic number and weight, respectively; V and V_i are electron acceleration potential and excitation potential, respectively; and $\bar{V}_i$ is the mean ionization potential [p. 114 in (*7*)].

In the *absorption correction*, $f(\chi)$ ("f of chi") is the fraction of the generated analyte-line photons that are actually emitted:

$$f(\chi)_i = \frac{1+h}{[1+(\chi/\sigma)]\{1+h[1+(\chi/\sigma)]\}} \tag{21.5}$$

where χ is the attenuation of the emergent analyte-line x-rays, h is an electron penetration term, and σ is the effective Lenard electron absorption coefficient as modified by Heinrich:

$$\chi_i = (\mu/\varrho)_{i,\lambda_{\mathrm{A}}} \csc\psi \tag{21.6}$$

$$h_i = 1.2(A/Z^2)_i \tag{21.7}$$

$$\sigma_i = (4.5\times10^5)/(V^{1.67} - V_i^{1.67}) \tag{21.8}$$

where $(\mu/\varrho)_{i,\lambda_A}$ is the mass-absorption coefficient of element i for the analyte line λ_A; ψ is the x-ray emergence angle; and A, Z, V, and V_i are defined above.

In the *fluorescence correction*, the enhancement of the analyte line λ_A by spectral lines of matrix elements j is given by the ratio of analyte-line intensities excited by elements j, I_j, and directly by the electron beam, I_e, summed for all matrix lines that can significantly enhance λ_A:

$$\left(\frac{I_j}{I_e}\right)_{\lambda_A} = 0.5P_{Aj}C_j \frac{r_A - 1}{r_A} \omega_j \frac{A_A}{A_j} \left(\frac{U_j - 1}{U_A - 1}\right)^{1.67} \frac{(\mu/\varrho)_{A,\lambda j}}{(\mu/\varrho)_{X,\lambda_j}} \times \left[\frac{\log_e(1+u)}{u} + \frac{\log_e(1+v)}{v}\right] \quad (j \neq A) \qquad (21.9)$$

in which

$$U_A = V/V_A \qquad \text{and} \qquad U_j = V/V_j \qquad (21.10)$$

$$u = [(\mu/\varrho)_{X,\lambda_A} \csc \psi]/(\mu/\varrho)_{X,\lambda_j} \qquad (21.11)$$

$$v = \sigma/(\mu/\varrho)_{X,\lambda_j} \qquad (21.12)$$

and where C_j is concentration of enhancing element j; r_A is analyte absorption-edge jump ratio; ω is fluorescent yield; and A, Z, V, V_j, σ, χ, and ψ have the significance given above. The constant P_{Aj} has the value 1 for a j K line enhancing an A K line or a j L line enhancing an A L line, 4 for a K line enhancing an L line, and 0.25 for an L line enhancing a K line. The enhancement of λ_A by continuum $(I_{cont}/I_e)_{\lambda_A}$ may often be disregarded and is not considered here.

In all the foregoing equations, parameters for the sample X are calculated as follows:

$$r_X = \sum (C_i r_i) \qquad (21.13)$$

$$s_X = \sum (C_i s_i) \qquad (21.14)$$

$$f(\chi)_X = \sum [C_i(\mu/\varrho)_{i,\lambda_A}] \csc \psi \qquad (21.15)$$

$$(\mu/\varrho)_{X,\lambda} = \sum [C_i(\mu/\varrho)_{i,\lambda}] \qquad (21.16)$$

in which i represents the individual constituent elements—including the analyte—in the sample, and C_i is in weight fraction.

There are simplified practical versions of some intensity correction systems, and tables of constants and other data for their solution are available. In many such cases, corrections for two-, three-, and possibly even four-component systems can be calculated manually on a desk calculator [pp. 201–20 in (*1*), pp. 112–19 in (*7*)]. However, in general, a computer is required for these calculations. An excellent comprehensive review of 40 computer programs for making intensity corrections in electron-probe microanalysis has been compiled by Beaman and Isasi (*B15*).

21.7. PERFORMANCE

Sensitivity performance of a modern commercial electron-probe microanalyzer is summarized in Table 21.2 for the $K\alpha$ lines of selected chemical elements from atomic number 4 to 29 (beryllium to copper). However, the

TABLE 21.2. Performance of a Modern Electron-Probe Microanalyzer[a]

Element and atomic number	Specimen[b]	Crystal[c]	Line intensity[d] I_L, counts/s / μA	Peak-to-background ratio I_P/I_B	Limit of detectability,[e] wt%
$_4$Be	Be	PbLig	6.0×10^2	50	1.7
$_5$B	B	PbSt	1.5×10^4	50	0.35
$_6$C	C	"	4.0×10^4	70	0.18
$_8$O	SiO_2[b]	"	3.0×10^4	20	0.20
$_9$F	LiF[b]	KHP	1.1×10^4	280	0.12
$_{11}$Na	NaCl[b]	"	3.2×10^4	550	0.03
$_{12}$Mg	Mg	"	4.0×10^5	1500	0.01
$_{13}$Al	Al	"	8.0×10^5	1300	0.01
$_{14}$Si	Si	"	1.2×10^6	800	0.01
$_{22}$Ti	Ti	PET	1.8×10^6	700	0.01
$_{26}$Fe	Fe	LiF	2.5×10^5	460	0.03
$_{29}$Cu	Cu	"	1.5×10^5	230	0.05

[a] Courtesy of JEOLCo (USA), Inc.

[b] SiO_2 is 53% O; LiF is 73% F; NaCl is 39% Na.

[c] The symbols indicate, in order, lead lignocerate, lead stearate, potassium hydrogen phthalate, pentaerythritol, and lithium fluoride.

[d] All $K\alpha$ lines; acceleration potential 25 kV.

[e] Defined as $(3\sqrt{I_B/I_L})C$, where I_L and I_B are analyte-line and background intensities, respectively, and C is analyte concentration (wt%); $C = 100\%$, except as specified in note *b* above.

detection limits given in the table apply only to specimens having effective atomic number similar to that of the analyte itself. The sensitivity may be much less than that given for a heavy matrix, and much greater for a light matrix. The data probably represent optimal performance. Background is relatively high because of the electron excitation, but peak-to-background ratios are reasonably high.

The electron-beam spot size at the specimen surface is typically ~1 μm, although it can be made as small as 0.5 μm (5000 Å), and even smaller in scanning electron microscopes. However, when the specimen surface is heterogeneous on a micrometer scale or does not have an optically fine finish (0.25 μm or less), spot sizes as large as 50 μm may be used.

If the beam diameter and penetration are both ~1 μm, the effective specimen volume is ~1 μm^3 or 10^{-12} cm^3. For a density of 10 g/cm^3, the effective specimen mass is 10^{-11} g, and, at the detection limit—say, 0.01 wt %—the contributing analyte mass is only 1 fg (10^{-15} g), compared with 1 ng (10^{-9} g) for x-ray fluorescence spectrometry. However, direct comparison of the detection limits of electron-probe microanalysis and x-ray fluorescence spectrometry is not valid. A local concentration of 0.01 wt % (100 ppm) in the microanalyzer may correspond to an average concentration in the bulk sample of 1–0.001 ppm (1 ppm to 1 ppb).

The depth of electron penetration as a function of acceleration potential and atomic number is shown in Figure 21.12 (*17*). However, of more practical significance than electron-beam diameter and penetration are the areas and depths in which x-rays of specified wavelength originate. The depth of origin of $K\alpha$, $L\alpha_1$, and $M\alpha_1$ x-ray lines of the elements is given in Figure 21.13 (*A20*) as a function of acceleration potential and specimen

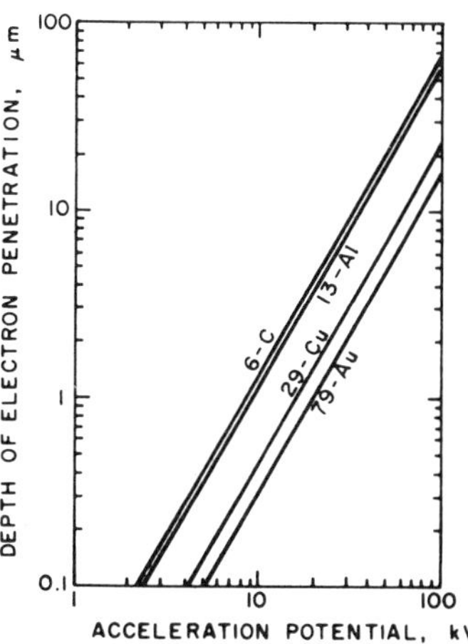

FIG. 21.12. Electron penetration as a function of acceleration potential for several chemical elements. [H. A. Elion, "Instrument and Chemical Analysis Aspects of Electron Microanalysis and Macroanalysis," Pergamon Press, N. Y., p. 169 (1966); courtesy of the author and publisher.]

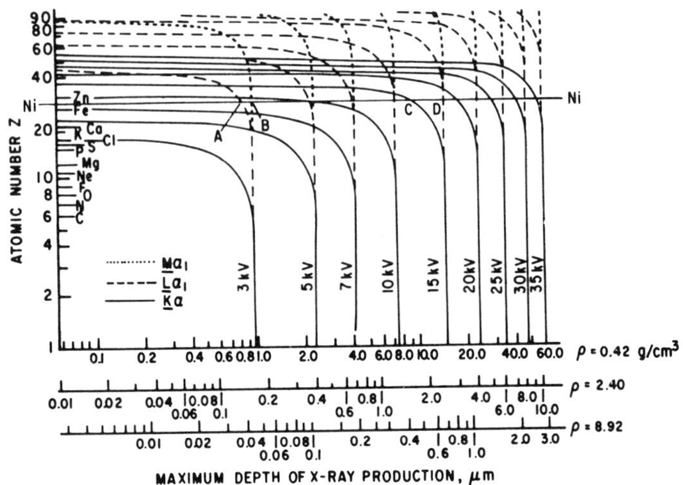

FIG. 21.13. Maximum depth of x-ray production for $K\alpha$, $L\alpha_1$, and $M\alpha_1$ x-ray lines of the chemical elements for various acceleration potentials and for specimens of various densities. [C. A. Andersen, in "Methods of Biochemical Analysis," Vol. 15, Interscience Publishers, Inc., N. Y., p. 147 (1967); courtesy of the author and publisher.]

density. Use of the diagram is illustrated by the following example, assuming nickel to be the analyte. Ni (Z 28) is located on the ordinate midway between Fe (Z 26) and Zn (Z 30). Following the Ni—Ni line parallel to the abscissa, one derives the maximum depths of x-ray production of the Ni $K\alpha$ and Ni $L\alpha_1$ lines for various acceleration potentials and specimen densities as follows. At 3 kV: The Ni—Ni line intersects the $L\alpha_1$ curve at A; following down from A parallel to the ordinate, the maximum depths for Ni $L\alpha_1$ in specimens having densities 0.42, 2.40, and 8.92 g/cm³ are 0.85, 0.15, and 0.04 μm, respectively. The Ni line intersects the $M\alpha_1$ curve at B; following down from B, the maximum depths for Ni $M\alpha_1$, if there were such a line, at the three densities would be 0.98, 0.16, and 0.046 μm, respectively. At 15 kV: The Ni line intersects the $K\alpha$ curve at C, and the respective maximum depths for Ni $K\alpha$ are 9.5, 1.75, and 0.45 μm. The Ni line intersects the $L\alpha_1$ curve at D, and the respective depths for Ni $L\alpha_1$ are 15, 2.6, and 0.7 μm. Maximum depths at intermediate densities are estimated by interpolation.

Figure 21.14 shows cross sections of the specimen surface showing diffusion zones of electrons as functions of acceleration potential and atomic number (*D33*). Figure 21.14A shows that: (1) the electron diffusion zone deepens as atomic number decreases and acceleration potential increases;

(2) the zone widens—that is, more electrons scatter laterally in the specimen—as atomic number and potential increase; (3) the zone from which backscattered electrons (solid arrows directed upward from the specimen surface) arise is wider than the zone actually bombarded by the incident electron beam, and the higher the atomic number, the wider it is; and (4) the zone from which secondary electrons (dashed arrows directed upward) arise is the same as the bombarded zone.

Figure 21.14B shows that *K*, *L*, and *M* lines of the same element arise from successively deeper zones. As the electrons diffuse into the specimen surface, their initial energy is decreased incrementally and successively falls below the *K*, *L*, and *M* excitation potentials. Figure 21.14C shows that the *K* (or *L* or *M*) lines of elements of successively lower atomic number arise from successively deeper zones for the same reason. The difference in this respect between electron and x-ray excitation is discussed in Section 1.7.2.

Although most instruments are capable of operation up to 50 kV, 25 kV is a typical practical maximum operating potential. Lower potentials are used to: (1) limit the effective specimen thickness to that of the thinnest

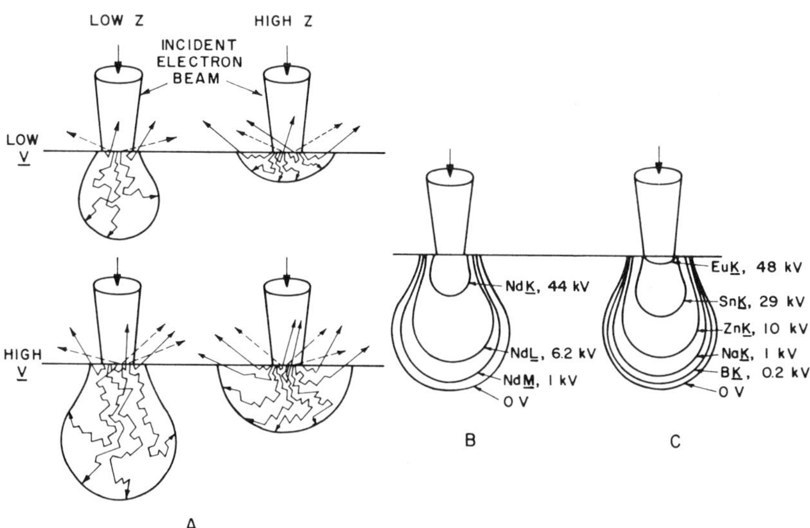

FIG. 21.14. Sections through the specimen surface showing variation of electron penetration, electron scattering, and depth of origin of x-ray spectral lines as functions of atomic number and acceleration potential. A. Electron penetration and scatter (*D*33). B. Depth of origin of x-ray lines of different series from the same element. C. Depth of origin of x-ray lines of the same series from different elements. In B and C, the numbers are excitation potentials.

film in the group to be analyzed; (2) avoid excitation of underlying layers; (3) avoid excitation of interfering spectral lines; (4) improve spacial resolution by reduction of lateral electron scatter (Figure 21.14); and (5) reduce the magnitude of the atomic number and absorption intensity corrections (Section 21.6.2).

21.8. APPLICATIONS

The electron-probe x-ray microanalyzer fills the great need for in-place nondestructive determination of composition and spacial variation of composition on a microscopic scale in analytical chemistry, metallurgy, ceramics, mineralogy, geology, and biology and medicine. The instrument is extremely valuable in materials research because the properties of materials depend not only on average composition, but also on local concentrations of elements in individual grains, at grain boundaries, etc. Among the many classes of applications are the following: (1) analysis of individual crystallites or grains of phases in heterogeneous alloys, ceramics, and minerals; (2) analysis of microscopic precipitates, inclusions, and segregations in the same materials; (3) analysis of grain boundaries; (4) analysis of individual layers in sections of coated, plated, or corroded surfaces or of bonded surfaces; (5) mapping concentration gradients at metal-to-metal or metal-to-ceramic interfaces, diffusion couples, diffusion zones, heterogeneous grain boundaries, etc.; (6) evaluation of microscale homogeneity; (7) mapping distribution of elements over the surfaces of heterogeneous alloys, ceramics, minerals, etc.; (8) analysis of very small particles filtered from gases and liquids; (9) determination and distribution of natural and foreign elements in biological materials; (10) analysis of specific areas on small semiconductor electron devices and other very small parts; and (11) determination of composition and thickness of thin films.

21.9. COMPARISON WITH X-RAY FLUORESCENCE SPECTROMETRY

There are probably several thousand x-ray fluorescence spectrometers in use in laboratories and factories in the United States alone, while there are probably not more than a few hundred electron-probe microanalyzers in the whole world. Because, as already mentioned, the electron-probe microanalyzer is, in effect, an x-ray primary-emission spectrometer, operators of x-ray fluorescence spectrometers may conclude that the microanalyzer is as simple, convenient, rapid, and reliable as the fluorescence spectrometer. The writer feels almost compelled to append this section to set this misconception straight.

The electron-probe microanalyzer is much more complex than the x-ray fluorescence spectrometer. A highly sophisticated and "temperamental" electron-optical column replaces the simple, stable, sealed-off x-ray tube. An automatically valved high-vacuum system replaces the simple rotary pump of the vacuum fluorescence spectrometer. Even the specimen stage is complex, with its several micrometer-driven movements and its requirements for highly reproducible positioning and freedom from backlash. Focusing curved-crystal spectrometers replace the nonfocusing flat-crystal spectrometers.

In the x-ray fluorescence spectrometer, there is rarely any problem of constancy of excitation conditions. When there is, the x-ray tube potential and current are observed on their respective meters and may be reset as required with their respective controls. In the electron-probe microanalyzer, the specimen current may vary with time, and it is difficult to monitor this current because it varies among specimens and from point to point on the same specimen if the average atomic number varies. Moreover, it is not permissible to reset the specimen current for each measurement if mathematical atomic number correction is to be applied; in order to reset the specimen current, the beam spot must be returned to a Faraday cup or to the same reference material—usually pure analyte.

In the x-ray fluorescence spectrometer, the simple act of placing a plane specimen in its compartment correctly locates the specimen plane. Each analyte line and background is measured by simply setting the goniometer to the required 2θ and scaling the intensity. In the electron-probe microanalyzer, the specimen plane is set by bringing the specimen surface into focus at the highest magnification of the light microscope. With the specimen so placed, the electron beam, the axes of the electron-optical column and light microscope, and the x-ray paths to the several spectrometers must meet at a common point—the analytical point—on this plane within very close tolerance. It is well known that the focusing of a light microscope can be highly subjective, differing among operators and even with repeated trials by the same operator. Such slight differences in focus may cause substantial differences in the measured analyte-line intensity. If the specimen surface is not highly polished, measured intensity is likely to be different if high, low, or intermediate relief is focused. A highly polished, very clean surface may present no feature at all on which to focus, and other means are required to set the specimen plane. Even when the surface is favorable for focusing, it may be difficult to find the particular microscopic feature to be studied.

In the electron-probe microanalyzer, the full-focusing spectrometers give extremely sharp peaks. These are more difficult to "peak" than the nonfocusing x-ray fluorescence spectrometers, and slight departures from the peak cause much greater decrease in measured intensity. Thermal effects on the analyzer crystal are also more severe. Once an x-ray fluorescence spectrometer has been peaked for a specified spectral line, the 2θ setting is noted and may be reset any number of times; repeaking is required only if the crystal or collimator is changed. This is usually true even if the instrument is peaked with pure analyte, then used to measure specimens having the analyte in a different chemical state. In the electron-probe microanalyzer, the peak setting is affected by—or, at a fixed peak setting, the measured intensity is affected by—(1) the focusing of the specimen plane in the light microscope; (2) differences in the microfinish of the specimen surface; (3) slight differences in the tilt of the specimen surface—overall or locally; (4) differences among the specimens in chemical state of the analyte; (5) slight movement and distortion of the filament; and (6) drift in operating conditions of the electron-optical column. The result is that all of the one to four x-ray spectrometers may require peaking for *every* individual measurement. This may be very difficult for low-intensity peaks because of the "noise" in the panel ratemeter.

As a result of the features discussed in the preceding two paragraphs, the process of locating an analytical point, focusing it, peaking the several spectrometers, and scaling the analyte-line intensities may be very time-consuming and tedious.

The components of the x-ray spectrometers in the microanalyzer are usually inaccessible, so that changing of crystals, slits, and detectors is inconvenient.

Because of the small size of the electron-beam spot, problems of microhomogeneity and microtopography are severe. For x-ray fluorescence spectrometry, 30–100-μm finishes are usually satisfactory. For electron-probe microanalysis, 0.25-μm finishes or finer may be required, and there is little value in even semiquantitative comparison of intensities from specimens having substantially different finishes.

Background is higher and much more difficult to measure accurately in electron-probe microanalysis.

Chemical effects on wavelength are the exception in x-ray fluorescence spectrometry, the rule in electron-probe microanalysis when the chemical state of the analyte varies among the specimens or when pure analyte is used to peak the spectrometer and the specimens are compounds.

The buildup of carbon at the beam spot can result in decrease in measured intensity during the actual measurement. It is almost always necessary to make replicate measurements at different spots, and often to translate the specimen to continuously expose clean surface during the measurement; this latter procedure is frequently necessary when C $K\alpha$ is measured.

Suitable standards are much more difficult to obtain for electron-probe microanalysis because microhomogeneity is required and micro-topography of standards and samples must be the same.

Nevertheless, in fairness it must be remembered that the electron-probe microanalyzer is one of the very few instruments that can determine composition and spacial variation of composition at the micrometer level, and one must expect and be willing to assume some inconvenience to obtain such information.

Part VIII

Appendixes, Bibliography

Appendixes

The tables in these appendixes do not necessarily give the most accurate, up-to-date values available, nor are they intended to. However, they do give good working values suitable for all but the most precise practical applications of x-ray spectrochemical analysis. For more accurate values, refer to the section on *Tables of Wavelengths, 2θ Angles, and Mass-Absorption Coefficients* in the Bibliography.

1. Wavelengths of the Principal X-Ray Spectral Lines of the Chemical Elements—*K* Series.
2. Wavelengths of the Principal X-Ray Spectral Lines of the Chemical Elements—*L* Series.
3. Wavelengths of the Principal X-Ray Spectral Lines of the Chemical Elements—*M* Series.
4. Photon Energies of the Principal *K* and *L* X-Ray Spectral Lines of the Chemical Elements.
5. Wavelengths of the *K*, *L*, and *M* X-Ray Absorption Edges of the Chemical Elements.
6. *K*, *L*, and *M* X-Ray Excitation Potentials of the Chemical Elements.

7A. X-Ray Mass-Absorption Coefficients of the Chemical Elements at 0.1–30 Å.

7B. X-Ray Mass-Absorption Coefficients of Elements 2–11 (He–Na) at 40–100 Å.

8. Values of the *K* and *L*III Absorption-Edge Jump Ratios r and $(r - 1)/r$ of the Chemical Elements.

9. Average Values of the K, L, and M Fluorescent Yields of the Chemical Elements.
10. X-Ray Spectrometer Analyzer Crystals and Multilayer Films.
11A. Glossary of Frequently Used Notation.
11B. Prefixes for Physical Units.
12. Periodic Table of the Chemical Elements.

APPENDIX 1. Wavelengths of the Principal X-Ray Spectral Lines of the Chemical Elements—*K* Series (*89, 87*)

Wavelength, Å

Line	α	α_1	α_2	β_1	β_3	β_2
Electron transition $K \leftarrow$		*L*III	*L*II	*M*III	*M*II	*N*II, *N*III
Approximate relative intensity	150	100	50	20		5
3-Li	240	—	—	—	—	—
4-Be	113	—	—	—	—	—
5-B	67	—	—	—	—	—
6-C	44	—	—	—	—	—
7-N	31.603	—	—	—	—	—
8-O	23.707	—	—	—	—	—
9-F	18.307	—	—	—	—	—
10-Ne	14.615	—	—	14.460	—	—
11-Na	11.909	—	—	11.617	—	—
12-Mg	9.889	—	—	9.558	—	—
13-Al	8.339	8.338	8.341	7.981	—	—
14-Si	7.126	7.125	7.127	6.769	—	—
15-P	6.155	6.154	6.157	5.804	—	—
16-S	5.373	5.372	5.375	5.032	—	—
17-Cl	4.729	4.728	4.731	4.403	—	—
18-Ar	4.192	4.191	4.194	3.886	—	—
19-K	3.744	3.742	3.745	3.454	—	—
20-Ca	3.360	3.359	3.362	3.090	—	—
21-Sc	3.032	3.031	3.034	2.780	—	—
22-Ti	2.750	2.749	2.753	2.514		—
23-V	2.505	2.503	2.507	2.285		—
24-Cr	2.291	2.290	2.294	2.085		—
25-Mn	2.103	2.102	2.105	1.910	—	—
26-Fe	1.937	1.936	1.940	1.757	—	—
27-Co	1.791	1.789	1.793	1.621	—	—
28-Ni	1.659	1.658	1.661	1.500	—	1.489
29-Cu	1.542	1.540	1.544	1.392	1.393	1.381
30-Zn	1.437	1.435	1.439	1.296	—	1.284
31-Ga	1.341	1.340	1.344	1.207	1.208	1.196
32-Ge	1.256	1.255	1.258	1.129	1.129	1.117
33-As	1.177	1.175	1.179	1.057	1.058	1.045

APPENDIX 1 *(continued)* Wavelength, Å

Line	α	α_1	α_2	β_1	β_3	β_2
Electron transition $K \leftarrow$		LIII	LII	MIII	MII	NII, NIII
Approximate relative intensity	150	100	50	20		5
34-Se	1.106	1.105	1.109	0.992	0.993	0.980
35-Br	1.041	1.040	1.044	0.933	0.933	0.921
36-Kr	0.981	0.980	0.984	0.879	0.879	0.866
37-Rb	0.927	0.926	0.930	0.829	0.830	0.817
38-Sr	0.877	0.875	0.880	0.783	0.784	0.771
39-Y	0.831	0.829	0.833	0.740	0.741	0.728
40-Zr	0.788	0.786	0.791	0.701	0.702	0.690
41-Nb	0.748	0.747	0.751	0.665	0.666	0.654
42-Mo	0.710	0.709	0.713	0.632	0.633	0.621
43-Tc	0.674	0.673	0.676	0.601	0.602	0.590
44-Ru	0.644	0.643	0.647	0.572	0.573	0.562
45-Rh	0.614	0.613	0.617	0.546	0.546	0.535
46-Pd	0.587	0.585	0.590	0.521	0.521	0.510
47-Ag	0.561	0.559	0.564	0.497	0.498	0.487
48-Cd	0.536	0.535	0.539	0.475	0.476	0.465
49-In	0.514	0.512	0.517	0.455	0.455	0.445
50-Sn	0.492	0.491	0.495	0.435	0.436	0.426
51-Sb	0.472	0.470	0.475	0.417	0.418	0.408
52-Te	0.453	0.451	0.456	0.400	0.401	0.391
53-I	0.435	0.433	0.438	0.384	0.385	0.376
54-Xe	0.418	0.416	0.421	0.369	—	0.360
55-Cs	0.402	0.401	0.405	0.355	0.355	0.346
56-Ba	0.387	0.385	0.390	0.341	0.342	0.333
57-La	0.373	0.371	0.376	0.328	0.329	0.320
58-Ce	0.359	0.357	0.362	0.316	0.317	0.309
59-Pr	0.346	0.344	0.349	0.305	0.305	0.297
60-Nd	0.334	0.332	0.337	0.294	0.294	0.287
61-Pm	0.322	0.321	0.325	0.283	0.284	0.276
62-Sm	0.311	0.309	0.314	0.274	0.274	0.267
63-Eu	0.301	0.299	0.304	0.264	0.265	0.258
64-Gd	0.291	0.289	0.294	0.255	0.256	0.249
65-Tb	0.281	0.279	0.284	0.246	0.246	0.239
66-Dy	0.272	0.270	0.275	0.237	0.238	0.231
67-Ho	0.263	0.261	0.266	0.230	0.231	0.224
68-Er	0.255	0.253	0.258	0.222	0.223	0.217
69-Tm	0.246	0.244	0.250	0.215	0.216	0.210

APPENDIX 1 (*continued*) Wavelength, Å

Line	α	α_1	α_2	β_1	β_3	β_2
Electron transition $K \leftarrow$		LIII	LII	MIII	MII	NII, NIII
Approximate relative intensity	150	100	50	20		5
70-Yb	0.238	0.236	0.241	0.208	0.209	0.203
71-Lu	0.231	0.229	0.234	0.202	0.203	0.197
72-Hf	0.224	0.222	0.227	0.195	0.196	0.190
73-Ta	0.217	0.215	0.220	0.190	0.191	0.185
74-W	0.211	0.209	0.213	0.184	0.185	0.179
75-Re	0.204	0.202	0.207	0.179	0.179	0.174
76-Os	0.198	0.196	0.201	0.173	0.174	0.169
77-Ir	0.193	0.191	0.196	0.168	0.169	0.164
78-Pt	0.187	0.185	0.190	0.163	0.164	0.159
79-Au	0.182	0.180	0.185	0.159	0.160	0.155
80-Hg	0.177	0.175	0.180	0.154	0.155	0.150
81-Tl	0.172	0.170	0.175	0.150	0.151	0.146
82-Pb	0.167	0.165	0.170	0.146	0.147	0.142
83-Bi	0.162	0.161	0.165	0.142	0.143	0.138
84-Po	0.158	0.156	0.161	0.138	0.139	0.134
85-At	0.154	0.152	0.157	0.134	0.135	0.131
86-Rn	0.150	0.148	0.153	0.131	0.132	0.127
87-Fr	0.146	0.144	0.149	0.127	0.128	0.124
88-Ra	0.142	0.140	0.145	0.124	0.125	0.120
89-Ac	0.138	0.136	0.141	0.121	0.121	0.117
90-Th	0.135	0.133	0.138	0.117	0.118	0.114
91-Pa	0.131	0.129	0.134	0.114	0.115	0.111
92-U	0.128	0.126	0.131	0.111	0.112	0.108
93-Np	0.125	0.123	0.128	0.109	—	0.105
94-Pu	0.122	0.120	0.125	0.106	—	0.103
95-Am	0.119	0.117	0.122	0.103	—	0.100
96-Cm	0.116	0.114	0.119	0.101	—	0.098
97-Bk	0.113	0.111	0.116	0.098	—	0.095
98-Cf	0.110	0.108	0.113	0.097	—	0.093

APPENDIX 2. Wavelengths of the Principal X-Ray Spectral Lines of the Chemical Elements—*L* Series (*64*, *87*, *89*)

Wavelength, Å

Line	α_1	α_2	β_1	β_2	β_3	β_4	β_5	γ_1	γ_2	γ_3	γ_4	γ_6	l	η
Electron transition														
*L*III ←	*M*V	*M*IV		*N*V			*O*IV, *O*V						*M*I	
*L*II ←			*M*IV					*N*IV				*O*IV		*M*I
*L*I ←					*M*III	*M*II			*N*II	*N*III	*O*III			
Approximate relative intensity	100	10	80	30	3	2	2	10	1	2	1	1	5	2
16-S	—	—	—	—	—	—	—	—	—	—	—	—	83.400	
17-Cl	—	—	—	—	—	—	—	—	—	—	—	—	67.840	67.250
18-Ar	—	—	—	—	—	—	—	—	—	—	—	—	56.212	56.813
19-K	—	—	—	—	—	—	—	—	—	—	—	—	47.835	47.325
20-Ca	36.393		36.022	—	—	—	—	—	—	—	—	—	41.042	40.542
21-Sc	31.393		31.072	—	—	—	—	—	—	—	—	—	35.671	35.200
22-Ti	27.445		27.074	—	—	—	—	—	—	—	—	—	31.423	30.942
23-V	24.309		23.898	—	21.890		—	—	—	—	—	—	27.826	27.375
24-Cr	21.713		21.323	—	19.429		—	—	—	—	—	—	24.840	24.339
25-Mn	19.489		19.158	—	17.575		—	—	—	—	—	—	22.315	21.864

26-Fe	17.602		17.290	—	15.710		—	—	—	—	—	—	20.201	19.730
27-Co	16.000		15.698	—	14.240		—	—	—	—	—	—	18.358	17.860
28-Ni	14.595		14.308	—	13.146		—	—	—	—	—	—	16.693	16.304
29-Cu	13.357		13.079	—	12.115		—	—	—	—	—	—	15.297	14.940
30-Zn	12.282		12.009	—	11.185		—	—	—	—	—	—	14.081	13.719
31-Ga	11.313		11.045	—	10.365		—	—	—	—	—	—	12.976	12.620
32-Ge	10.456		10.194	—	9.580	9.640	—	—	—	—	—	—	11.944	11.608
33-As	9.671		9.414	—	8.930		—	—	—	—	—	—	11.069	10.732
34-Se	8.990		8.735	—	8.321		—	—	—	—	—	—	10.293	9.959
35-Br	8.375		8.126	—	7.767		—	—	—	—	—	—	9.583	9.253
36-Kr	7.817		7.576	—	7.264	7.304	—	—	—	—	—	—	8.946	8.626
37-Rb	7.318	7.325	7.075	—	6.788	6.821	—	—	6.045		—	—	8.363	8.042
38-Sr	6.863	6.870	6.623	—	6.367	6.403	—	—	5.644		—	—	7.836	7.517
39-Y	6.449	6.456	6.211	—	5.983	6.018	—	—	5.283		—	—	7.356	7.040
40-Zr	6.070	6.077	5.836	5.586	5.632	5.668	—	5.384	4.953		—	—	6.918	6.606
41-Nb	5.725	5.732	5.492	5.238	5.310	5.346	—	5.036	4.654		—	—	6.517	6.210
42-Mo	5.406	5.414	5.176	4.923	5.013	5.048	—	4.726	4.380		—	—	6.150	5.847
43-Tc	5.114	5.123	4.887	4.636	4.737	4.773	—	4.440	4.138		—	—	5.819	5.518
44-Ru	4.846	4.854	4.620	4.372	4.487	4.523	—	4.182	3.897		—	—	5.503	5.204
45-Rh	4.597	4.605	4.374	4.130	4.253	4.289	—	3.944	3.685		—	—	5.217	4.922
46-Pd	4.368	4.376	4.146	3.909	4.034	4.071	—	3.725	3.489		—	—	4.952	4.660
47-Ag	4.154	4.162	3.935	3.703	3.834	3.870	—	3.523	3.307		—	—	4.707	4.418
48-Cd	3.956	3.965	3.739	3.514	3.644	3.681	—	3.336	3.137		—	—	4.480	4.193
49-In	3.752	3.781	3.555	3.339	3.470	3.507	—	3.162	2.980		2.926	—	4.269	3.983
50-Sn	3.600	3.609	3.385	3.175	3.306	3.344	—	3.001	2.835		2.778	—	4.071	3.789
51-Sb	3.439	3.448	3.226	3.023	3.152	3.190	—	2.852	2.695		2.639	—	3.888	3.607
52-Te	3.290	3.299	3.077	2.882	3.009	3.046	—	2.712	2.567		2.511	—	3.716	3.438
53-I	3.148	3.157	2.937	2.751	2.874	2.912	—	2.582	2.447		2.391	—	3.557	3.280

APPENDIX 2 *(continued)*

Wavelength, Å

Line	α_1	α_2	β_1	β_2	β_3	β_4	β_5	γ_1	γ_2	γ_3	γ_4	γ_6	l	η
Electron transition LIII ←	MV	MIV		NV			OIV, OV						MI	
LII ←			MIV					NIV				OIV		MI
LI ←					MIII	MII			NII	NIII	OIII			
Approximate relative intensity	100	10	80	30	3	2	2	10	1	2	1	1	5	2
54-Xe	3.015	3.025	2.803	2.626	2.745	2.784	—	2.462	2.338	2.331	—	—	3.421	3.143
55-Cs	2.892	2.902	2.683	2.511	2.628	2.666	—	2.348	2.237	2.233	2.174	—	3.267	2.994
56-Ba	2.776	2.785	2.567	2.404	2.516	2.555	—	2.242	2.138	2.134	2.075	—	3.135	2.862
57-La	2.665	2.674	2.458	2.303	2.410	2.449	—	2.141	2.046	2.041	1.983	—	3.006	2.740
58-Ce	2.561	2.570	2.356	2.208	2.311	2.349	—	2.048	1.960	1.955	1.899	—	2.892	2.620
59-Pr	2.463	2.473	2.259	2.119	2.216	2.255	—	1.961	1.879	1.874	1.819	—	2.784	2.512
60-Nd	2.370	2.382	2.166	2.035	2.126	2.166	—	1.878	1.801	1.797	1.745	1.855	2.675	2.409
61-Pm	2.283	2.292	2.081	1.956	2.042	2.081	—	1.799	1.729	1.724	—	—	2.591	2.322
62-Sm	2.199	2.210	1.998	1.882	1.962	2.000	1.779	1.726	1.659	1.655	1.606	—	2.482	2.218
63-Eu	2.120	2.131	1.920	1.812	1.887	1.926	—	1.657	1.597	1.591	1.544	—	2.395	2.131
64-Gd	2.046	2.057	1.847	1.746	1.815	1.853	—	1.592	1.534	1.529	1.485	—	2.312	2.049
65-Tb	1.976	1.986	1.777	1.682	1.747	1.785	1.577	1.530	1.477	1.471	1.427	—	2.234	1.973
66-Dy	1.909	1.920	1.710	1.623	1.681	1.720	—	1.473	1.423	1.417	1.374	—	2.158	1.898
67-Ho	1.845	1.856	1.647	1.567	1.619	1.658	—	1.417	1.371	1.364	1.323	—	2.086	1.826
68-Er	1.785	1.796	1.587	1.514	1.561	1.601	—	1.364	1.321	1.315	1.276	—	2.019	1.757
69-Tm	1.726	1.738	1.530	1.463	1.505	1.544	—	1.316	1.274	1.268	1.229	—	1.955	1.695
70-Yb	1.672	1.682	1.476	1.416	1.452	1.491	1.387	1.268	1.228	1.222	1.185	1.243	1.894	1.635

71-Lu	1.619	1.630	1.424	1.370	1.402	1.441	1.342	1.222	1.185	1.179	1.143	1.198	1.836	1.478
72-Hf	1.569	1.580	1.374	1.327	1.353	1.392	1.298	1.179	1.144	1.138	1.103	1.155	1.782	1.523
73-Ta	1.522	1.533	1.327	1.285	1.307	1.346	1.256	1.138	1.105	1.099	1.065	1.114	1.728	1.471
74-W	1.476	1.487	1.282	1.245	1.263	1.302	1.215	1.098	1.068	1.062	1.028	1.074	1.678	1.421
75-Re	1.433	1.444	1.238	1.206	1.220	1.260	1.177	1.061	1.032	1.026	0.993	1.037	1.630	1.374
76-Os	1.391	1.402	1.197	1.169	1.179	1.218	1.140	1.025	0.998	0.992	0.959	1.001	1.585	1.328
77-Ir	1.352	1.363	1.158	1.135	1.141	1.179	1.106	0.991	0.966	0.959	0.928	0.967	1.541	1.285
78-Pt	1.313	1.325	1.120	1.102	1.104	1.142	1.072	0.958	0.934	0.928	0.897	0.934	1.499	1.243
79-Au	1.277	1.288	1.083	1.070	1.068	1.106	1.040	0.927	0.905	0.898	0.867	0.903	1.460	1.202
80-Hg	1.242	1.253	1.049	1.040	1.034	1.072	1.010	0.897	0.876	0.869	0.839	0.873	1.422	1.164
81-Tl	1.207	1.218	1.015	1.010	1.001	1.039	0.981	0.868	0.848	0.842	0.812	0.845	1.385	1.127
82-Pb	1.175	1.186	0.982	0.983	0.969	1.007	0.953	0.840	0.822	0.815	0.786	0.817	1.350	1.092
83-Bi	1.144	1.155	0.952	0.955	0.939	0.977	0.926	0.814	0.796	0.790	0.761	0.791	1.317	1.058
84-Po	1.114	1.125	0.922	0.929	0.909	0.948	0.900	0.788	0.772	0.765	0.736	0.765	1.283	1.024
85-At	1.085	1.097	0.894	0.905	0.881	0.919	0.875	0.763	0.747	0.740	0.713	0.741	1.256	0.997
86-Rn	1.057	1.069	0.866	0.881	0.854	0.892	0.852	0.739	0.725	0.718	0.692	0.717	1.223	0.968
87-Fr	1.030	1.042	0.840	0.858	0.828	0.867	0.829	0.716	0.703	0.696	0.670	0.695	1.199	0.938
88-Ra	1.005	1.017	0.814	0.836	0.803	0.841	0.807	0.694	0.682	0.675	0.649	0.673	1.167	0.908
89-Ac	0.980	0.992	0.789	0.814	0.778	0.816	0.786	0.671	0.662	0.655	0.630	0.653	1.144	0.882
90-Th	0.956	0.968	0.766	0.794	0.755	0.793	0.765	0.653	0.642	0.635	0.611	0.632	1.115	0.855
91-Pa	0.933	0.945	0.742	0.774	0.732	0.770	0.746	0.634	0.624	0.617	0.594	0.613	1.091	0.830
92-U	0.911	0.923	0.720	0.755	0.710	0.748	0.726	0.615	0.605	0.598	0.577	0.595	1.067	0.806
93-Np	0.889	0.901	0.698	0.736	0.689	0.727	0.708	0.597	0.587	0.581	0.558	0.577	1.043	0.781
94-Pu	0.869	0.880	0.678	0.720	0.669	0.707	0.691	0.579	0.571	0.564	0.542	0.560	1.023	0.759
95-Am	0.849	0.860	0.658	0.701	0.649	0.686	0.674	0.562	0.554	—	—	0.543	—	—
96-Cm	0.829	0.841	0.639	0.685	—	—	—	0.546	—	—	—	—	—	—
97-Bk	0.810	0.822	0.621	0.669	—	—	—	0.530	—	—	—	—	—	—
98-Cf	0.792	0.804	0.603	0.653	—	—	—	0.515	—	—	—	—	—	—

APPENDIX 3. Wavelengths of the Principal X-Ray Spectral Lines of the Chemical Elements—*M* Series (*59, 64*)

Wavelength, Å

Line	α_1	α_2	β	γ	ζ_1	ζ_2	—
Electron transition							
*M*V ←	*N*VII	*N*VI			*N*III		
*M*IV ←			*N*VI			*N*II	
*M*III ←				*N*V			
*M*II ←							*N*IV
Approximate relative intensity	50	50	80	5	1	1	5
35-Br	—	—	—	—	192.6	191.1	—
36-Kr	—	—	—	—	—	—	—
37-Rb	—	—	—	—	128.7	127.7	—
38-Sr	—	—	—	—	108.7	108.0	—
39-Y	—	—	—	—	93.400		—
40-Zr	—	—	—	38.390	82.100		37.000
41-Nb	—	—	—	34.900	72.190		33.100
42-Mo	—	—	—	32.700	64.380		31.400
43-Tc	—	—	—	30.100	59.500		28.800
44-Ru	—	—	—	26.900	52.340		25.500
45-Rh	—	—	—	25.010	47.670		24.450
46-Pd	—	—	—	23.300	43.600		22.100
47-Ag	—	—	—	21.820	39.770		20.660
48-Cd	—	—	—	20.470	36.800		19.400
49-In	—	—	—	19.210	33.200		18.240
50-Sn	—	—	—	17.940	31.240		16.930
51-Sb	—	—	—	16.920	28.880		15.980
52-Te	—	—	—	15.930	26.720		15.020
53-I	—	—	—	15.010	24.650		14.150
54-Xe	—	—	—	14.180	23.020		13.310
55-Cs	—	—	—	13.420	21.690		12.580
56-Ba	—	—	—	12.750	20.640		11.890
57-La	14.880		14.510	12.080	19.440		11.280
58-Ce	14.040		13.750	11.530	18.350		10.690
59-Pr	13.343		13.060	10.998	17.380		10.180
60-Nd	12.680		12.440	10.505	16.460		9.700
61-Pm	—	—	—	10.050	15.680		9.260
62-Sm	11.470		11.270	9.600	14.910		8.840
63-Eu	10.960		10.750	9.211	14.220		8.450

APPENDIX 3 (*continued*)

Wavelength, Å

<table>
<tr><td>Line</td><td>α_1</td><td>α_2</td><td>β</td><td>γ</td><td>ζ_1</td><td>ζ_2</td><td>—</td></tr>
<tr><td>Electron transition</td><td></td><td></td><td></td><td></td><td></td><td></td><td></td></tr>
<tr><td>MV ←</td><td>NVII</td><td>NVI</td><td></td><td></td><td>NIII</td><td></td><td></td></tr>
<tr><td>MIV ←</td><td></td><td></td><td>NVI</td><td></td><td></td><td>NII</td><td></td></tr>
<tr><td>MIII ←</td><td></td><td></td><td></td><td>NV</td><td></td><td></td><td></td></tr>
<tr><td>MII ←</td><td></td><td></td><td></td><td></td><td></td><td></td><td>NIV</td></tr>
<tr><td>Approximate relative intensity</td><td>50</td><td>50</td><td>80</td><td>5</td><td>1</td><td>1</td><td>5</td></tr>
<tr><td>64-Gd</td><td colspan="2">10.460</td><td>10.254</td><td>8.844</td><td colspan="2">13.570</td><td>8.120</td></tr>
<tr><td>65-Tb</td><td colspan="2">10.000</td><td>9.792</td><td>8.486</td><td colspan="2">12.980</td><td>7.740</td></tr>
<tr><td>66-Dy</td><td colspan="2">9.590</td><td>9.357</td><td>8.144</td><td colspan="2">12.430</td><td>7.460</td></tr>
<tr><td>67-Ho</td><td colspan="2">9.200</td><td>8.965</td><td>7.865</td><td colspan="2">11.860</td><td>7.160</td></tr>
<tr><td>68-Er</td><td colspan="2">8.820</td><td>8.592</td><td>7.546</td><td colspan="2">11.370</td><td>6.860</td></tr>
<tr><td>69-Tm</td><td colspan="2">8.480</td><td>8.249</td><td>7.318</td><td colspan="2">10.920</td><td>6.540</td></tr>
<tr><td>70-Yb</td><td colspan="2">8.149</td><td>7.909</td><td>7.024</td><td colspan="2">10.480</td><td>6.270</td></tr>
<tr><td>71-Lu</td><td colspan="2">7.840</td><td>7.601</td><td>6.768</td><td colspan="2">10.060</td><td>6.020</td></tr>
<tr><td>72-Hf</td><td colspan="2">7.539</td><td>7.303</td><td>6.544</td><td colspan="2">9.686</td><td>5.770</td></tr>
<tr><td>73-Ta</td><td colspan="2">7.252</td><td>7.023</td><td>6.312</td><td>9.316</td><td>9.330</td><td>5.570</td></tr>
<tr><td>74-W</td><td>6.983</td><td>6.992</td><td>6.757</td><td>6.092</td><td>8.962</td><td>8.993</td><td>5.357</td></tr>
<tr><td>75-Re</td><td colspan="2">6.729</td><td>6.504</td><td>5.885</td><td>8.629</td><td>8.664</td><td>5.150</td></tr>
<tr><td>76-Os</td><td colspan="2">6.490</td><td>6.267</td><td>5.682</td><td>8.310</td><td>8.359</td><td>4.955</td></tr>
<tr><td>77-Ir</td><td>6.262</td><td>6.275</td><td>6.038</td><td>5.500</td><td>8.021</td><td>8.065</td><td>4.780</td></tr>
<tr><td>78-Pt</td><td>6.047</td><td>6.058</td><td>5.828</td><td>5.319</td><td>7.738</td><td>7.790</td><td>4.601</td></tr>
<tr><td>79-Au</td><td>5.840</td><td>5.854</td><td>5.624</td><td>5.145</td><td>7.466</td><td>7.523</td><td>4.432</td></tr>
<tr><td>80-Hg</td><td>5.648</td><td>5.677</td><td>5.432</td><td>4.984</td><td>7.232</td><td>7.250</td><td>4.266</td></tr>
<tr><td>81-Tl</td><td>5.460</td><td>5.472</td><td>5.249</td><td>4.823</td><td>6.974</td><td>7.032</td><td>4.116</td></tr>
<tr><td>82-Pb</td><td>5.286</td><td>5.299</td><td>5.076</td><td>4.674</td><td>6.740</td><td>6.802</td><td>3.968</td></tr>
<tr><td>83-Bi</td><td>5.118</td><td>5.130</td><td>4.909</td><td>4.532</td><td>6.521</td><td>6.585</td><td>3.834</td></tr>
<tr><td>84-Po</td><td>4.955</td><td>4.958</td><td>4.736</td><td>4.361</td><td>6.290</td><td>6.349</td><td>3.680</td></tr>
<tr><td>85-At</td><td>4.802</td><td>4.802</td><td>4.581</td><td>4.234</td><td>6.096</td><td>6.156</td><td>3.559</td></tr>
<tr><td>86-Rn</td><td>4.655</td><td>4.657</td><td>4.436</td><td>4.124</td><td>5.911</td><td>5.971</td><td>3.448</td></tr>
<tr><td>87-Fr</td><td>4.515</td><td>4.521</td><td>4.303</td><td>4.008</td><td>5.737</td><td>5.801</td><td>3.322</td></tr>
<tr><td>88-Ra</td><td>4.383</td><td>4.392</td><td>4.178</td><td>3.892</td><td>5.579</td><td>5.642</td><td>3.220</td></tr>
<tr><td>89-Ac</td><td>4.256</td><td>4.270</td><td>4.060</td><td>3.798</td><td>5.389</td><td>5.489</td><td>3.118</td></tr>
<tr><td>90-Th</td><td>4.138</td><td>4.151</td><td>3.941</td><td>3.679</td><td>5.245</td><td>5.340</td><td>3.011</td></tr>
<tr><td>91-Pa</td><td>4.022</td><td>4.035</td><td>3.827</td><td>3.577</td><td>5.092</td><td>5.193</td><td>2.910</td></tr>
<tr><td>92-U</td><td>3.910</td><td>3.924</td><td>3.716</td><td>3.479</td><td>4.946</td><td>5.050</td><td>2.817</td></tr>
</table>

APPENDIX 4. Photon Energies of the Principal *K* and *L* X-Ray Spectral Lines of the Chemical Elements

Photon energy, keV

Element	$K\alpha_1$	$K\alpha_2$	$K\beta_1$	$L\alpha_1$	$L\alpha_2$	$L\beta_1$	$L\beta_2$	$L\gamma_1$
3-Li	0.052		—	—	—	—	—	—
4-Be	0.110		—	—	—	—	—	—
5-B	0.185		—	—	—	—	—	—
6-C	0.282		—	—	—	—	—	—
7-N	0.392		—	—	—	—	—	—
8-O	0.523		—	—	—	—	—	—
9-F	0.677		—	—	—	—	—	—
10-Ne	0.851		—	—	—	—	—	—
11-Na	1.041		1.067	—	—	—	—	—
12-Mg	1.254		1.297	—	—	—	—	—
13-Al	1.487	1.486	1.553	—	—	—	—	—
14-Si	1.740	1.739	1.832	—	—	—	—	—
15-P	2.015	2.014	2.136	—	—	—	—	—
16-S	2.308	2.306	2.464	—	—	—	—	—
17-Cl	2.622	2.621	2.815	—	—	—	—	—
18-Ar	2.957	2.955	3.192	—	—	—	—	—
19-K	3.313	3.310	3.589	—	—	—	—	—
20-Ca	3.691	3.688	4.012	0.341		0.344	—	—
21-Sc	4.090	4.085	4.460	0.395		0.399	—	—
22-Ti	4.510	4.504	4.931	0.452		0.458	—	—
23-V	4.952	4.944	5.427	0.510		0.519	—	—
24-Cr	5.414	5.405	5.946	0.571		0.581	—	—
25-Mn	5.898	5.887	6.490	0.636		0.647	—	—
26-Fe	6.403	6.390	7.057	0.704		0.717	—	—

APPENDIX 4 (*continued*)

Photon energy, keV

Element	$K\alpha_1$	$K\alpha_2$	$K\beta_1$	$L\alpha_1$	$L\alpha_2$	$L\beta_1$	$L\beta_2$	$L\gamma_1$
27-Co	6.930	6.915	7.649	0.775		0.790	—	—
28-Ni	7.477	7.460	8.264	0.849		0.866	—	—
29-Cu	8.047	8.027	8.904	0.928		0.948	—	—
30-Zn	8.638	8.615	9.571	1.009		1.032	—	—
31-Ga	9.251	9.234	10.263	1.096		1.122	—	—
32-Ge	9.885	9.854	10.981	1.186		1.216	—	—
33-As	10.543	10.507	11.725	1.282		1.317	—	—
34-Se	11.221	11.181	12.495	1.379		1.419	—	—
35-Br	11.923	11.877	13.290	1.480		1.526	—	—
36-Kr	12.648	12.597	14.112	1.587		1.638	—	—
37-Rb	13.394	13.335	14.960	1.694	1.692	1.752	—	—
38-Sr	14.164	14.097	15.834	1.806	1.805	1.872	—	—
39-Y	14.957	14.882	16.736	1.922	1.920	1.996	—	—
40-Zr	15.774	15.690	17.666	2.042	2.040	2.124	2.219	2.302
41-Nb	16.614	16.520	18.621	2.166	2.163	2.257	2.367	2.462
42-Mo	17.478	17.373	19.607	2.293	2.290	2.395	2.518	2.623
43-Tc	18.410	18.328	20.585	2.424	2.420	2.538	2.674	2.792
44-Ru	19.278	19.149	21.655	2.558	2.554	2.683	2.836	2.964
45-Rh	20.214	20.072	22.721	2.696	2.692	2.834	3.001	3.144
46-Pd	21.175	21.018	23.816	2.838	2.833	2.990	3.172	3.328
47-Ag	22.162	21.988	24.942	2.984	2.978	3.151	3.348	3.519
48-Cd	23.172	22.982	26.093	3.133	3.127	3.316	3.528	3.716
49-In	24.207	24.000	27.274	3.287	3.279	3.487	3.713	3.920
50-Sn	25.270	25.042	28.483	3.444	3.435	3.662	3.904	4.131

APPENDIX 4 *(continued)* Photon energy, keV

Element	$K\alpha_1$	$K\alpha_2$	$K\beta_1$	$L\alpha_1$	$L\alpha_2$	$L\beta_1$	$L\beta_2$	$L\gamma_1$
51-Sb	26.357	26.109	29.723	3.605	3.595	3.843	4.100	4.347
52-Te	27.471	27.200	30.993	3.769	3.758	4.029	4.301	4.570
53-I	28.610	28.315	32.292	3.937	3.926	4.220	4.507	4.800
54-Xe	29.802	29.485	33.644	4.111	4.098	4.422	4.720	5.036
55-Cs	30.970	30.623	34.984	4.286	4.272	4.620	4.936	5.280
56-Ba	32.191	31.815	36.376	4.467	4.451	4.828	5.156	5.531
57-La	33.440	33.033	37.799	4.651	4.635	5.043	5.384	5.789
58-Ce	34.717	34.276	39.255	4.840	4.823	5.262	5.613	6.052
59-Pr	36.023	35.548	40.746	5.034	5.014	5.489	5.850	6.322
60-Nd	37.359	36.845	42.269	5.230	5.208	5.722	6.090	6.602
61-Pm	38.649	38.160	43.945	5.431	5.408	5.956	6.336	6.891
62-Sm	40.124	39.523	45.400	5.636	5.609	6.206	6.587	7.180
63-Eu	41.529	40.877	47.027	5.846	5.816	6.456	6.842	7.478
64-Gd	42.983	42.280	48.718	6.059	6.027	6.714	7.102	7.788
65-Tb	44.470	43.737	50.391	6.275	6.241	6.979	7.368	8.104
66-Dy	45.985	45.193	52.178	6.495	6.457	7.249	7.638	8.418
67-Ho	47.528	46.686	53.934	6.720	6.680	7.528	7.912	8.748
68-Er	49.099	48.205	55.690	6.948	6.904	7.810	8.188	9.089
69-Tm	50.730	49.762	57.576	7.181	7.135	8.103	8.472	9.424
70-Yb	52.360	51.326	59.352	7.414	7.367	8.401	8.758	9.779
71-Lu	54.063	52.959	61.282	7.654	7.604	8.708	9.048	10.142
72-Hf	55.757	54.579	63.209	7.898	7.843	9.021	9.346	10.514
73-Ta	57.524	56.270	65.210	8.145	8.087	9.341	9.649	10.892
74-W	59.310	57.973	67.233	8.396	8.333	9.670	9.959	11.283
75-Re	61.131	59.707	69.298	8.651	8.584	10.008	10.273	11.684

APPENDIX 4 (*continued*) Photon energy, keV

Element	$K\alpha_1$	$K\alpha_2$	$K\beta_1$	$L\alpha_1$	$L\alpha_2$	$L\beta_1$	$L\beta_2$	$L\gamma_1$
76-Os	62.991	61.477	71.404	8.910	8.840	10.354	10.596	12.094
77-Ir	64.886	63.278	73.549	9.173	9.098	10.706	10.918	12.509
78-Pt	66.820	65.111	75.736	9.441	9.360	11.069	11.249	12.939
79-Au	68.794	66.980	77.968	9.711	9.625	11.439	11.582	13.379
80-Hg	70.821	68.894	80.258	9.987	9.896	11.823	11.923	13.828
81-Tl	72.860	70.820	82.558	10.266	10.170	12.210	12.268	14.288
82-Pb	74.957	72.794	84.922	10.549	10.448	12.611	12.620	14.762
83-Bi	77.097	74.805	87.335	10.836	10.729	13.021	12.977	15.244
84-Po	79.296	76.868	89.809	11.128	11.014	13.441	13.338	15.740
85-At	81.525	78.956	92.319	11.424	11.304	13.873	13.705	16.248
86-Rn	83.800	81.080	94.877	11.724	11.597	14.316	14.077	16.768
87-Fr	86.119	83.243	97.483	12.029	11.894	14.770	14.459	17.301
88-Ra	88.485	85.446	100.136	12.338	12.194	15.233	14.839	17.845
89-Ac	90.894	87.681	102.846	12.650	12.499	15.712	15.227	18.405
90-Th	93.334	89.942	105.592	12.966	12.808	16.200	15.620	18.977
91-Pa	95.851	92.271	108.408	13.291	13.120	16.700	16.022	19.559
92-U	98.428	94.648	111.289	13.613	13.438	17.218	16.425	20.163
93-Np	101.005	97.023	114.181	13.945	13.758	17.740	16.837	20.774
94-Pu	103.653	99.457	117.146	14.279	14.082	18.278	17.254	21.401
95-Am	106.351	101.932	120.163	14.618	14.411	18.829	17.677	22.042
96-Cm	109.098	104.448	123.235	14.961	14.743	19.393	18.106	22.699
97-Bk	111.896	107.023	126.362	15.309	15.079	19.971	18.540	23.370
98-Cf	114.745	109.603	129.544	15.661	15.420	20.562	18.980	24.056
99-Es	117.646	112.244	132.781	16.018	15.764	21.166	19.426	24.758
100-Fm	120.598	114.926	136.075	16.379	16.113	21.785	19.879	25.475

APPENDIX 5. Wavelengths of the *K*, *L*, and *M* X-Ray Absorption Edges of the Chemical Elements (*64*; *26*)

Element	Wavelength, Å								
	K	*L*I	*L*II	*L*III	*M*I	*M*II	*M*III	*M*IV	*M*V
3-Li	226.950	—	—	—	—	—	—	—	—
4-Be	107.200	—	—	—	—	—	—	—	—
5-B	65.604	—	—	—	—	—	—	—	—
6-C	43.887	—	—	—	—	—	—	—	—
7-N	31.220	—	—	—	—	—	—	—	—
8-O	23.233	—	—	—	—	—	—	—	—
9-F	17.897	—	—	—	—	—	—	—	—
10-Ne	14.170	—	—	—	—	—	—	—	—
11-Na	11.475	—	—	—	—	—	—	—	—
12-Mg	9.512	197.300	—	220.534	—	—	—	—	—
13-Al	7.951	142.500	163.022	163.942	—	—	—	—	—
14-Si	6.745	105.000	122.028	122.761	—	—	—	—	—
15-P	5.787	81.000	95.070	95.801	—	—	—	—	—
16-S	5.018	64.100	75.314	75.868	—	—	—	—	—
17-Cl	4.397	52.100	60.739	61.222	591.900	—	—	—	—
18-Ar	3.871	43.200	50.025	50.446	476.587	—	—	—	—
19-K	3.436	36.400	41.754	42.179	373.456	—	—	—	—
20-Ca	3.070	30.700	35.797	36.173	309.146	—	—	—	—
21-Sc	2.757	26.800	30.943	31.296	264.440	—	—	—	—
22-Ti	2.497	23.400	26.936	27.290	228.816	—	—	—	—

23-V	2.269	19.803	23.877	24.229	201.484	—	—	—	—
24-Cr	2.070	17.840	21.294	21.637	171.626	—	—	—	—
25-Mn	1.896	16.138	19.086	19.417	150.739	—	—	—	—
26-Fe	1.743	14.650	17.188	17.504	132.991	—	—	—	—
27-Co	1.608	13.333	15.545	15.843	118.175	—	—	—	—
28-Ni	1.488	12.201	14.104	14.387	103.062	—	—	—	—
29-Cu	1.380	11.172	12.841	13.113	91.685	137.152	—	793.498	—
30-Zn	1.283	10.262	11.725	11.987	81.898	116.000	—	546.224	—
31-Ga	1.196	9.416	10.728	10.930	73.146	98.670	107.500	405.000	—
32-Ge	1.116	8.692	9.840	10.089	65.116	89.857	93.488	299.437	—
33-As	1.045	8.067	9.056	9.298	58.581	79.123	82.205	238.357	—
34-Se	0.980	7.456	8.347	8.584	52.835	69.999	72.578	187.662	—
35-Br	0.920	6.920	7.721	7.952	46.679	62.341	64.705	150.218	—
36-Kr	0.866	6.444	7.158	7.377	42.071	54.882	57.020	129.756	—
37-Rb	0.816	5.995	6.642	6.861	37.696	49.562	51.527	108.466	109.986
38-Sr	0.770	5.591	6.173	6.387	34.553	44.124	45.866	90.985	92.251
39-Y	0.728	5.226	5.752	5.961	31.453	39.733	41.308	77.850	78.973
40-Zr	0.689	4.889	5.378	5.565	28.474	35.564	36.972	66.136	67.184
41-Nb	0.653	4.595	5.031	5.230	26.482	32.708	34.155	59.815	60.608
42-Mo	0.620	4.323	4.716	4.913	24.413	30.083	31.401	53.277	54.201
43-Tc	0.589	4.068	4.431	4.623	22.500	27.578	28.853	47.507	48.140
44-Ru	0.560	3.837	4.169	4.358	20.945	25.461	26.510	42.660	43.038
45-Rh	0.534	3.623	3.928	4.113	19.453	23.341	24.492	38.561	39.286
46-Pd	0.509	3.425	3.706	3.889	18.109	21.603	22.699	34.940	35.494
47-Ag	0.486	3.243	3.501	3.680	16.877	20.011	21.061	31.793	32.299
48-Cd	0.464	3.073	3.312	3.488	15.873	18.602	19.614	29.269	29.507

APPENDIX 5 *(continued)*

Wavelength, Å

Element	*K*	*L*I	*L*II	*L*III	*M*I	*M*II	*M*III	*M*IV	*M*V
49-In	0.444	2.916	3.137	3.311	14.764	17.314	18.285	26.717	27.165
50-Sn	0.425	2.648	2.975	3.146	13.856	16.050	17.200	24.500	24.900
51-Sb	0.407	2.634	2.824	2.994	13.020	15.072	16.014	22.699	23.114
52-Te	0.390	2.508	2.685	2.853	12.274	14.185	15.079	21.124	21.528
53-I	0.374	2.390	2.555	2.721	11.575	13.344	14.193	19.670	20.050
54-Xe	0.358	2.277	2.434	2.597	10.800	12.532	13.381	18.300	18.716
55-Cs	0.345	2.175	2.321	2.482	10.338	11.824	12.638	17.158	17.607
56-Ba	0.331	2.078	2.214	2.374	9.787	11.156	11.957	16.087	16.513
57-La	0.318	1.988	2.115	2.273	9.316	10.560	11.343	15.060	15.470
58-Ce	0.306	1.902	2.022	2.178	8.844	9.997	10.758	14.242	14.560
59-Pr	0.295	1.822	1.934	2.089	8.394	9.498	10.242	13.429	13.792
60-Nd	0.284	1.747	1.852	2.007	8.026	9.033	9.787	12.789	13.092
61-Pm	0.274	1.675	1.775	1.928	7.550	8.610	9.342	12.160	12.461
62-Sm	0.265	1.608	1.703	1.855	7.338	8.199	8.926	11.547	11.828
63-Eu	0.256	1.545	1.634	1.785	7.013	7.827	8.548	10.975	11.257
64-Gd	0.247	1.485	1.569	1.719	6.703	7.496	8.200	10.457	10.747
65-Tb	0.238	1.428	1.507	1.656	6.398	7.135	7.829	9.958	10.233
66-Dy	0.230	1.375	1.449	1.597	6.139	6.865	7.547	9.506	9.791
67-Ho	0.223	1.323	1.393	1.540	5.892	6.571	7.225	9.079	9.342
68-Er	0.216	1.274	1.341	1.487	5.675	6.283	6.950	8.664	8.950
69-Tm	0.209	1.227	1.291	1.436	5.410	5.985	6.661	8.275	8.541
70-Yb	0.202	1.183	1.243	1.387	5.192	5.724	6.361	7.912	8.164
71-Lu	0.196	1.140	1.198	1.341	4.981	5.480	6.119	7.573	7.813

72-Hf	0.190	1.100	1.155	1.297	4.765	5.238	5.876	7.216	7.450
73-Ta	0.184	1.062	1.114	1.255	4.594	5.041	5.675	6.951	7.183
74-W	0.178	1.025	1.074	1.215	4.407	4.831	5.453	6.649	6.875
75-Re	0.173	0.990	1.037	1.177	4.237	4.632	5.250	6.371	6.597
76-Os	0.168	0.956	1.001	1.140	4.067	4.440	5.052	6.097	6.313
77-Ir	0.163	0.924	0.966	1.105	3.909	4.262	4.858	5.848	6.076
78-Pt	0.158	0.893	0.933	1.071	3.761	4.082	4.679	5.622	5.820
79-Au	0.153	0.863	0.902	1.039	3.669	3.936	4.518	5.374	5.584
80-Hg	0.149	0.835	0.872	1.008	3.476	3.771	4.352	5.182	5.409
81-Tl	0.145	0.807	0.843	0.979	3.348	3.626	4.192	4.992	5.187
82-Pb	0.141	0.781	0.815	0.950	3.217	3.484	4.035	4.793	4.989
83-Bi	0.137	0.756	0.788	0.923	3.097	3.342	3.890	4.601	4.793
84-Po	0.133	0.731	0.762	0.897	2.989	3.209	3.742	4.430	4.624
85-At	0.129	0.708	0.738	0.872	2.850	3.092	3.616	4.267	4.458
86-Rn	0.125	0.686	0.715	0.848	2.740	2.982	3.506	4.112	4.301
87-Fr	0.122	0.665	0.692	0.825	2.650	2.866	3.392	3.966	4.151
88-Ra	0.119	0.644	0.671	0.803	2.576	2.769	3.280	3.829	4.008
89-Ac	0.116	0.625	0.650	0.782	2.470	2.673	3.185	3.698	3.871
90-Th	0.113	0.606	0.629	0.761	2.394	2.577	3.076	3.568	3.742
91-Pa	0.110	0.587	0.610	0.741	2.314	2.483	2.977	3.446	3.628
92-U	0.108	0.570	0.592	0.722	2.240	2.394	2.887	3.339	3.509

APPENDIX 6. *K*, *L*, and *M* X-Ray Excitation Potentials of the Chemical Elements (*64*, *66*)

Element	Excitation potential, kV								
	K	*L*I	*L*II	*L*III	*M*I	*M*II	*M*III	*M*IV	*M*V
1-H	0.014	—	—	—	—	—	—	—	—
2-He	0.025	—	—	—	—	—	—	—	—
3-Li	0.055	—	—	—	—	—	—	—	—
4-Be	0.116	—	—	—	—	—	—	—	—
5-B	0.192	—	—	—	—	—	—	—	—
6-C	0.283	—	—	—	—	—	—	—	—
7-N	0.399	—	—	—	—	—	—	—	—
8-O	0.531	—	—	—	—	—	—	—	—
9-F	0.687	—	—	—	—	—	—	—	—
10-Ne	0.874	0.048	0.022	0.022	—	—	—	—	—
11-Na	1.080	0.055	0.034	0.034	—	—	—	—	—
12-Mg	1.303	0.063	0.050	0.049	—	—	—	—	—
13-Al	1.559	0.087	0.073	0.072	—	—	—	—	—
14-Si	1.838	0.118	0.099	0.098	—	—	—	—	—
15-P	2.142	0.153	0.129	0.128	—	—	—	—	—
16-S	2.470	0.193	0.164	0.163	—	—	—	—	—
17-Cl	2.819	0.238	0.203	0.202	0.020	—	—	—	—
18-Ar	3.203	0.287	0.247	0.245	0.026	—	—	—	—
19-K	3.607	0.341	0.297	0.294	0.033	—	—	—	—

20-Ca	4.038	0.399	0.352	0.349	0.040	—	—	—	—
21-Sc	4.496	0.462	0.411	0.406	0.046	—	—	—	—
22-Ti	4.964	0.530	0.460	0.454	0.054	—	—	—	—
23-V	5.463	0.604	0.519	0.512	0.061	—	—	—	—
24-Cr	5.988	0.679	0.583	0.574	0.072	—	—	—	—
25-Mn	6.537	0.762	0.650	0.639	0.082	—	—	—	—
26-Fe	7.111	0.849	0.721	0.708	0.093	—	—	—	—
27-Co	7.709	0.929	0.794	0.779	0.104	—	—	—	—
28-Ni	8.331	1.015	0.871	0.853	0.120	—	—	—	—
29-Cu	8.980	1.100	0.953	0.933	0.135	0.090	—	0.015	—
30-Zn	9.660	1.200	1.045	1.022	0.151	0.106	—	0.022	—
31-Ga	10.368	1.300	1.134	1.117	0.169	0.125	0.115	0.030	—
32-Ge	11.103	1.420	1.248	1.217	0.190	0.137	0.132	0.041	—
33-As	11.863	1.529	1.359	1.323	0.211	0.156	0.150	0.052	—
34-Se	12.652	1.652	1.473	1.434	0.234	0.177	0.170	0.066	—
35-Br	13.475	1.794	1.599	1.552	0.265	0.198	0.191	0.082	—
36-Kr	14.323	1.931	1.727	1.675	0.294	0.225	0.217	0.095	—
37-Rb	15.201	2.067	1.866	1.806	0.328	0.250	0.240	0.114	0.112
38-Sr	16.106	2.221	2.008	1.941	0.358	0.280	0.270	0.136	0.134
39-Y	17.037	2.369	2.154	2.079	0.394	0.312	0.300	0.159	0.156
40-Zr	17.998	2.547	2.305	2.220	0.435	0.348	0.335	0.187	0.184
41-Nb	18.987	2.706	2.467	2.374	0.468	0.379	0.362	0.207	0.204
42-Mo	20.002	2.884	2.627	2.523	0.507	0.412	0.394	0.232	0.228
43-Tc	21.054	3.054	2.795	2.677	0.551	0.449	0.429	0.260	0.257
44-Ru	22.118	3.236	2.966	2.837	0.591	0.486	0.467	0.290	0.288
45-Rh	23.224	3.419	3.145	3.002	0.637	0.531	0.506	0.321	0.315
46-Pd	24.347	3.617	3.329	3.172	0.684	0.573	0.546	0.354	0.349
47-Ag	25.517	3.810	3.528	3.352	0.734	0.619	0.588	0.389	0.383

APPENDIX 6 *(continued)*

Excitation potential, kV

Element	*K*	*L*I	*L*II	*L*III	*M*I	*M*II	*M*III	*M*IV	*M*V
48-Cd	26.712	4.019	3.727	3.538	0.781	0.666	0.632	0.423	0.420
49-In	27.928	4.237	3.939	3.729	0.839	0.716	0.678	0.464	0.456
50-Sn	29.190	4.464	4.157	3.928	0.894	0.772	0.720	0.506	0.497
51-Sb	30.486	4.697	4.381	4.132	0.952	0.822	0.774	0.546	0.536
52-Te	31.809	4.938	4.613	4.341	1.010	0.873	0.822	0.586	0.575
53-I	33.164	5.190	4.856	4.559	1.071	0.929	0.873	0.630	0.618
54-Xe	34.579	5.452	5.104	4.782	1.147	0.989	0.926	0.677	0.662
55-Cs	35.959	5.720	5.358	5.011	1.199	1.048	0.981	0.722	0.704
56-Ba	37.410	5.995	5.623	5.247	1.266	1.111	1.036	0.770	0.750
57-La	38.931	6.283	5.894	5.489	1.330	1.173	1.092	0.823	0.801
58-Ce	40.449	6.561	6.165	5.729	1.401	1.240	1.152	0.870	0.851
59-Pr	41.998	6.846	6.443	5.968	1.476	1.305	1.210	0.923	0.898
60-Nd	43.571	7.144	6.727	6.215	1.544	1.372	1.266	0.969	0.946
61-Pm	45.207	7.448	7.018	6.466	1.642	1.439	1.327	1.019	0.994
62-Sm	46.846	7.754	7.281	6.721	1.689	1.512	1.388	1.073	1.048
63-Eu	48.515	8.069	7.624	6.983	1.767	1.584	1.450	1.129	1.101
64-Gd	50.229	8.393	7.940	7.252	1.849	1.653	1.511	1.185	1.153
65-Tb	51.998	8.724	8.258	7.519	1.937	1.737	1.583	1.245	1.211
66-Dy	53.789	9.083	8.621	7.850	2.019	1.805	1.642	1.304	1.266
67-Ho	55.615	9.411	8.920	8.074	2.104	1.886	1.715	1.365	1.327
68-Er	57.483	9.776	9.263	8.364	2.184	1.973	1.783	1.430	1.385
69-Tm	59.335	10.144	9.628	8.652	2.291	2.071	1.861	1.498	1.451
70-Yb	61.303	10.486	9.977	8.943	2.387	2.165	1.948	1.566	1.518
71-Lu	63.304	10.867	10.345	9.241	2.488	2.262	2.025	1.637	1.586
72-Hf	65.313	11.264	10.734	9.556	2.601	2.366	2.109	1.718	1.664

73-Ta	67.400	11.676	11.130	9.876	2.698	2.459	2.184	1.783	1.725
74-W	69.508	12.090	11.535	10.198	2.812	2.566	2.273	1.864	1.803
75-Re	71.662	12.522	11.955	10.531	2.926	2.676	2.361	1.946	1.879
76-Os	73.860	12.965	12.383	10.869	3.047	2.792	2.453	2.033	1.963
77-Ir	76.097	13.413	12.819	11.211	3.171	2.908	2.551	2.119	2.040
78-Pt	78.379	13.873	13.268	11.559	3.296	3.036	2.649	2.204	2.129
79-Au	80.713	14.353	13.733	11.919	3.379	3.149	2.744	2.307	2.220
80-Hg	83.106	14.841	14.212	12.285	3.566	3.287	2.848	2.392	2.291
81-Tl	85.517	15.346	14.697	12.657	3.702	3.418	2.957	2.483	2.389
82-Pb	88.001	15.870	15.207	13.044	3.853	3.558	3.072	2.586	2.484
83-Bi	90.521	16.393	15.716	13.424	4.003	3.709	3.186	2.694	2.586
84-Po	93.112	16.935	16.244	13.817	4.147	3.863	3.312	2.798	2.681
85-At	95.740	17.490	16.784	14.215	4.350	4.008	3.428	2.905	2.780
86-Rn	98.418	18.058	17.337	14.618	4.524	4.156	3.536	3.014	2.882
87-Fr	101.147	18.638	17.904	15.028	4.678	4.324	3.654	3.125	2.986
88-Ra	103.927	19.233	18.481	15.442	4.811	4.477	3.779	3.237	3.093
89-Ac	106.759	19.842	19.078	15.865	5.019	4.637	3.892	3.352	3.202
90-Th	109.630	20.460	19.688	16.296	5.176	4.810	4.030	3.474	3.313
91-Pa	112.581	21.102	20.311	16.731	5.355	4.993	4.164	3.597	3.416
92-U	115.591	21.753	20.943	17.163	5.532	5.177	4.293	3.712	3.533
93-Np	118.619	22.417	21.596	17.614	—	—	—	—	—
94-Pu	121.720	23.097	22.262	18.066	—	—	—	—	—
95-Am	124.876	23.793	22.944	18.525	—	—	—	—	—
96-Cm	128.088	24.503	23.640	18.990	—	—	—	—	—
97-Bk	131.357	25.230	24.352	19.461	—	—	—	—	—
98-Cf	134.683	25.971	25.080	19.938	—	—	—	—	—
99-Es	138.067	26.729	25.824	20.422	—	—	—	—	—
100-Fm	141.510	27.503	26.584	20.912	—	—	—	—	—

APPENDIX 7A. X-Ray Mass-Absorption Coefficients of the Chemical Elements at 0.1–30 Å[a]

Mass-absorption coefficient μ/ϱ, cm²/g

Z	Element	Wavelength, Å												
		0.1	0.15	0.20	0.25	0.3	0.4	0.5	0.6	0.7	0.8	0.9	1.0	1.5
1	H	0.29	0.32	0.34	0.35	0.37	0.38	0.40	0.42	0.43	0.44	0.45	0.45	0.49
2	He	0.11	0.12	0.12	0.13	0.14	0.14	0.17	0.86	0.20	0.22	0.23	0.25	0.35
3	Li	0.12	0.13	0.13	0.14	0.15	0.15	0.18	0.22	0.25	0.30	0.36	0.42	1.02
4	Be	0.13	0.13	0.14	0.14	0.15	0.16	0.19	0.23	0.28	0.34	0.43	0.53	1.54
5	B	0.13	0.14	0.14	0.15	0.16	0.19	0.24	0.31	0.40	0.54	0.70	0.92	2.87
6	C	0.14	0.14	0.15	0.16	0.17	0.23	0.31	0.42	0.59	0.83	1.14	1.54	4.79
7	N	0.14	0.15	0.16	0.17	0.20	0.28	0.40	0.59	0.88	1.26	1.76	2.37	7.38
8	O	0.14	0.15	0.16	0.18	0.23	0.34	0.53	0.83	1.27	1.84	2.56	3.45	10.74
9	F	0.14	0.15	0.18	0.21	0.26	0.43	0.70	1.14	1.77	2.57	3.57	4.81	14.96
10	Ne	0.14	0.16	0.19	0.24	0.31	0.54	0.93	1.54	2.38	3.45	4.80	6.47	20.11
11	Na	0.15	0.17	0.21	0.27	0.36	0.66	1.21	2.02	3.11	4.51	6.28	8.45	26.29
12	Mg	0.15	0.18	0.23	0.31	0.43	0.83	1.54	2.57	3.97	5.76	8.02	10.79	33.56
13	Al	0.15	0.19	0.26	0.36	0.50	1.03	1.93	3.22	4.96	7.21	10.03	13.50	41.97
14	Si	0.15	0.20	0.29	0.42	0.59	1.27	2.38	3.97	6.11	8.88	12.35	16.63	51.69
15	P	0.16	0.22	0.31	0.47	0.70	1.54	2.88	4.81	7.41	10.77	14.98	20.14	62.68
16	S	0.17	0.23	0.35	0.54	0.84	1.85	3.46	5.77	8.88	12.90	17.95	24.16	75.10
17	Cl	0.17	0.25	0.39	0.61	0.98	2.19	4.10	5.89	10.52	15.29	21.27	28.63	88.98
18	A	0.18	0.27	0.45	0.70	1.15	2.57	4.81	8.02	12.34	17.93	24.95	33.58	104.38
19	K	0.19	0.29	0.50	0.81	1.34	2.99	5.59	9.32	14.36	20.86	29.01	39.05	121.37
20	Ca	0.20	0.32	0.55	0.92	1.54	3.45	6.44	10.74	16.54	24.04	33.43	44.99	139.85
21	Sc	0.21	0.36	0.61	1.06	1.77	3.95	7.37	12.29	18.93	27.50	38.25	51.49	160.03
22	Ti	0.22	0.38	0.69	1.20	2.01	4.49	8.38	13.98	21.53	31.27	43.50	58.55	181.98
23	V	0.23	0.40	0.76	1.36	2.27	5.07	9.47	15.79	24.31	35.31	49.12	66.11	205.47
24	Cr	0.24	0.43	0.83	1.53	2.55	5.70	10.63	17.73	27.30	39.66	55.16	74.25	230.78
25	Mn	0.25	0.46	0.92	1.71	2.85	6.37	11.89	19.82	30.52	44.33	61.66	83.00	257.97
26	Fe	0.26	0.50	1.02	1.90	3.17	7.09	13.23	22.06	33.96	49.34	68.62	92.36	287.08
27	Co	0.28	0.56	1.13	2.11	3.52	7.86	14.66	24.45	37.64	54.68	76.06	102.37	318.19 K
28	Ni	0.30	0.61	1.25	2.33	3.88	8.68	16.19	27.00	41.57	60.38	83.99	113.04	43.84
29	Cu	0.32	0.68	1.38	2.56	4.27	9.55	17.82	29.71	45.74	66.45	92.42	120.48	48.58
30	Zn	0.35	0.72	1.51	2.81	4.69	10.47	19.54	32.58	50.17	72.88	101.37	136.44	53.65
31	Ga	0.37	0.76	1.73	3.08	5.13	11.46	21.39	35.66	54.91	79.76	110.94	149.32	59.31
32	Ge	0.40	0.82	1.81	3.35	5.59	12.49	23.31	38.87	59.84	86.93	120.91	162.75	65.10
33	As	0.42	0.88	1.96	3.65	6.08	13.59	25.36	42.28	65.09	94.56	131.52	177.03	71.02
34	Se	0.47	0.96	2.13	3.96	6.60	14.75	27.51	45.87	70.63	102.61	142.71	25.02	77.85
35	Br	0.50	1.04	2.31	4.29	7.15	15.97	29.78	49.66	76.46	111.07	154.49	27.29	84.93
36	Kr	0.53	1.12	2.49	4.63	7.72	17.25	32.18	53.65	82.60	120.00	22.05	29.69	92.27
37	Rb	0.57	1.21	2.69	4.99	8.32	18.59	34.68	57.82	89.03	129.33	23.99	32.29	100.38
38	Sr	0.61	1.30	2.89	5.37	8.96	20.00	37.32	62.22	95.80	18.71	26.03	35.03	108.89
39	Y	0.65	1.40	3.11	5.77	9.62	21.48	40.08	66.82	102.88	20.24	28.16	37.90	117.80
40	Zr	0.69	1.50	3.33	6.19	10.31	23.03	42.96	71.63	15.06	21.88	30.43	40.96	127.33
41	Nb	0.74	1.60	3.57	6.62	11.03	24.65	45.98	76.66	16.19	23.52	32.72	44.04	136.88
42	Mo	0.79	1.71	3.81	7.07	11.79	26.33	49.12	81.90	17.38	25.25	35.12	47.28	146.95
43	Tc	0.84	1.83	4.07	7.55	12.58	28.09	52.41	12.12	18.66	27.11	37.70	50.75	157.73
44	Ru	0.89	1.95	4.33	8.04	13.40	29.93	55.83	12.97	19.97	29.02	40.36	54.33	168.86
45	Rh	0.94	2.07	4.61	8.55	14.26	31.84	59.39	13.87	21.36	31.03	43.16	58.10	180.58
46	Pd	1.00	2.20	4.90	9.09	15.14	33.82	63.18	14.81	22.80	33.12	46.07	62.01	192.75
47	Ag	1.05	2.34	5.19	9.64	16.07	35.89	9.4	15.7	24.3	35.3	49.1	66.1	205.
48	Cd	1.10	2.48	5.51	10.22	17.03	38.04	10.0	16.8	25.8	37.6	52.2	70.3	218.
49	In	1.15	2.62	5.83	10.82	18.03	40.27	10.7	17.8	27.5	39.9	55.5	74.7	232.
50	Sn	1.20	2.77	6.16	11.44	19.06	42.58	11.9	19.9	30.7	44.7	62.1	79.5	260.
51	Sb	1.25	2.93	6.51	12.08	20.14	44.97	12.0	20.1	30.9	45.0	62.5	84.2	261.
52	Te	1.30	3.09	6.87	12.75	21.25	6.85	12.7	21.3	32.7	47.6	66.2	89.1	277.
							K							

[a] H. A. Liebhafsky, H. G. Pfeiffer, E. H. Winslow, and P. D. Zemany, "X-Rays, Electrons, and Analytical Chemistry," Wiley-Interscience, N. Y., pp. 525–8 (1972); courtesy of the authors and publisher.

APPENDIX 7A (*continued*)

Mass-absorption coefficient μ/ϱ, cm²/g

Wavelength, Å													
2.0	2.5	3.0	4.0	5.0	6.0	7.0	8.0	9.0	10.0	15.0	20.0	25.0	30.0
0.52	0.62	0.75	1.25	2.12	3.28	4.85	7.1	10.0	13.7	32.	69.	127.	208.
0.71	1.04	1.48	3.55	6.90	11.60	18.1	26.6	37.7	51.	107.	268.	540.	970.
2.18	3.98	6.60	15.2	28.80	48.80	76	113.0	157.	213.	402.	970.	1900.	3270.
3.45	6.44	10.74	24.04	44.91	74.84	115.2	167.4	232.	312.	973.	2178.	4068	6778.
6.43	12.01	20.03	44.80	83.69	139.4	214.7	312.0	434.	583.	1814.	4059.	7583.	12633.
10.72	20.01	33.35	74.61	139.3	232.2	357.5	519.6	722.	970.	3020.	6760	12627	21037.
16.51	30.83	51.38	114.9	214.6	357.7	550.7	800.4	1113.	1495.	4652.	10413.	19450.	32407.
24.02	44.85	74.75	167.1	312.3	520.4	801.2	1164.5	1619.	2175.	6768.	15149.	1005.	1675.
33.45	62.45	104.0	232.8	434.8	724.7	1115.6	1621.4	2254.	3029.	9425.	932.	1741.	2900.
44.98	83.98	139.9	313.0	584.8	974.5	1500.3	2180.5	3032.	4073.	659.	1475.	2756.	4593.
58.79	109.75	182.9	409.1	764.2	1273.5	1960.6	2849.5	3962.	5323.	951.	2130.	3979.	6629.
75.05	140.11	233.5	522.3	975.6	1625.9	2503.0	3637.8	5059.	467.	1453.	3252.	6075.	10122.
93.86	175.22	292.0	653.2	1220.1	2033.3	3130.3	336.1	467.	628.	1954.	4373.	8168.	13608.
115.58	215.78	359.6	804.4	1502.5	2503.9	305.3	443.8	617.	829.	2579.	5773.	10784.	17966.
140.16	261.66	436.1	975.4	1822.0	251.1	386.6	561.9	781.	1049.	3266.	7310.	13655.	22750
167.92	313.48	522.4	1168.6	2182.9	310.7	478.4	695.3	966.	1298.	4041.	9044.	16894.	28146.
198.97	371.45	619.0	1384.7	225.1	375.2	577.6	839.5	1167.	1568.	4880.	10921.	20400.	33987
233.39	435.70	726.1	156.2	291.9	486.5	748.9	1092	1510	2033.	6327.	14160.	26450.	44066
271.38	506.63	844.4	190.8	356.4	594.0	914.5	1329.	1848.	2483.	7725.	17289.	32295.	53804
312.71	583.78	972.9	232.7	434.7	724.4	1115.	1620.	2254.	3028.	9422.	21086.	39387.	65624.
357.84	668.03	121.9	272.6	509.3	848.8	1306.	1899.	2641.	3547.	11039.	24705.	46146.	65378.
406.91	85.71	142.7	319.3	596.6	994.2	1530.	2224.	3093.	4155.	12930.	28939.	45874.	8845.
459.45	104.07	173.4	387.9	724.7	1207.	1859.	2702.	3757.	5048.	15707.	28265.	5960.	9930.
516.03	117.54	195.9	438.1	818.5	1364.	2099.	3051.	4244.	5701.	17740.	32301.	6897.	11490.
70.76	132.11	220.1	492.5	919.9	1533.	2360.	3430.	4770.	6408.	19938.	4155.	7761.	12930.
79.21	147.88	246.4	551.3	1029.	1716.	2641.	3839.	5339.	7173.	18597.	4656.	8698.	14490.
88.41	165.04	275.0	615.2	1149.	1915.	2948.	4285.	5959.	8005.	20828.	5184.	9684.	16134.
98.04	183.02	305.0	682.3	1274.	2123.	3269.	4752.	6608.	8877.	2623.	5872.	10968.	18273.
108.64	202.82	338.0	756.1	1412.	2353.	3623.	5266.	7323.	9838.	2918.	6531.	12199.	20325.
119.96	223.95	373.2	834.8	1559.	2598.	4000.	5814.	8086.	10862.	3234.	7237.	13519.	22523.
132.61	247.57	412.6	922.9	1723.	2872.	4422.	6428.	8939.	10315.	3584.	8021.	14982.	24961.
145.58	271.78	452.9	1013.	1892.	3153.	4855.	7056.	8535.	11079.	3984.	8916.	16654.	27746.
158.81	296.47	494.1	1105.	2064.	3440.	5296.	7697.	9389.	1409.	4386.	9816.	18336.	30549.
174.08	324.98	541.6	1211.	2263.	3771.	5805.	7397.	10287.	1548.	4818.	10783.	20141.	33556.
189.90	354.52	590.8	1321.	2468.	4113.	5574.	994.	1383.	1858.	5781.	12938.	24166.	40262.
206.33	385.19	641.9	1435.	2682.	4469.	6090.	1122.	1561.	2088.	6525.	14604.	27279.	45447.
224.45	419.02	698.3	1562.	2917.	4315.	877.	1275.	1774	2383.	7416.	16598.	31004.	51652.
243.49	454.57	757.6	1694.	3165.	4699.	971.	1412.	1963.	2638.	8208.	18371.	34315.	57169.
263.41	491.75	819.5	1833.	3424.	704.	1084.	1575.	2191.	2943.	9159.	20498.	38289.	62686.
284.71	531.50	885.8	1981.	3313.	790.	1217.	1769.	2460.	3305	10286.	23020.	42998.	55280.
306.07	571.37	952.3	2130.	3579.	860.	1324.	1925.	2678.	3597.	11193.	25050.	46791.	60946.
328.59	613.43	1022.	2286.	568.	946.	1456.	2117.	2944.	3955.	12307.	27543.	39976.	66612.
352.68	658.40	1097.	2454.	624.	1040.	1602.	2328.	3238.	4350.	13535.	30293.	44632.	39444.
378.03	704.85	1174.	2385.	678.	1131.	1741.	2531.	3520.	4728.	14713.	32929.	48987.	44717.
403.77	753.78	1256.	2423.	739.	1233.	1898.	2758.	3836.	5154.	16036.	29040.	30192.	50304.
431.00	804.61	1341.	430.	804.	1340.	2063.	2999.	4170.	5603.	17433.	31764.	33872.	56432.
459.	857.	1429.	467.	873.	1455.	2240.	3255.	4527.	6082.	18924.	34727.	37812.	62996.
489.	913.	1522.	502.	937.	1562.	2406.	3497.	4863.	6533.	20327.	22293.	41641.	18612.
519.	970.	1485.	546.	1020.	1700.	2618.	3805.	5291.	7108.	19536.	24793.	46311.	20439.
581.	1086.	1480.	588.	1098.	1831.	2819.	4097.	5736.	7654.	20067.	27429.	12951.	21578.
585.	1093.	282.	632.	1181.	1969.	3031.	4405.	6126.	8230.	21593.	29983.	14367.	23935.
619.	1157.	302.	677.	1265.	2109.	3247.	4719.	6563.	8816.	21580.	32605.	15196.	25317.

K L_I L_{II} L_{III} M_I M_{II} M_{III} M_{IV} M_V

APPENDIX 7A (*continued*)

Mass-absorption coefficient μ/ϱ, cm²/g

Z	Element	0.1	0.15	0.20	0.25	0.3	0.4	0.5	0.6	0.7	0.8	0.9	1.0	1.5
		Wavelength, Å												
53	I	1.36	3.26	7.24	13.44	22.40	7.24	13.5	22.5	34.6	50.4	70.1	94.3	293.
54	Xe	1.42	3.43	7.63	14.16	23.59	7.66	14.2	23.8	36.6	53.3	74.1	99.7	310.
55	Cs	1.48	3.61	8.03	14.90	24.82	8.08	15.0	25.1	38.7	56.2	78.2	105.3	327.
56	Ba	1.53	3.80	8.44	15.66	26.10	8.52	15.9	26.5	40.8	59.3	82.5	111.0	345.
57	La	1.60	3.99	8.86	16.45	27.42	8.98	16.7	27.9	42.9	62.4	86.8	116.9	363.
58	Ce	1.66	4.19	9.30	17.27	28.78	9.45	17.6	29.4	45.2	65.7	91.4	123.1	382.
59	Pr	1.72	4.39	9.76	18.11	4.45	9.94	18.5	30.9	47.5	69.1	96.1	129.4	402.
60	Nd	1.80	4.60	10.23	18.98	4.67	10.4	19.4	32.4	49.9	72.6	101.0	135.9	422.
61	Pm	1.86	4.82	10.71	19.88	4.90	10.9	20.4	34.0	52.4	76.2	106.0	142.7	443.
62	Sm	1.93	5.04	11.21	20.80	5.14	11.4	21.4	35.7	55.0	79.9	111.2	149.7	465.
63	Eu	2.02	5.28	11.73	21.76	5.39	12.0	22.4	37.4	57.7	83.8	116.5	156.9	487.
64	Gd	2.09	5.52	12.26	3.38	5.64	12.6	23.5	39.2	60.3	87.6	122.0	164.2	478.
65	Tb	2.18	5.76	12.80	3.54	5.91	13.2	24.6	41.0	63.2	91.8	127.7	171.7	501.
66	Dy	2.26	6.02	13.37	3.70	6.18	13.8	25.7	42.9	66.0	95.9	133.5	179.6	469.
67	Ho	2.33	6.28	13.95	3.87	6.46	14.4	26.9	44.8	69.0	100.3	139.5	187.8	489.
68	Er	2.42	6.55	14.55	4.05	6.75	15.0	28.1	46.8	72.1	104.8	145.8	196.3	107.
69	Tm	2.50	6.82	15.16	4.23	7.05	15.7	29.3	48.9	75.4	109.5	152.3	205.1	113.
70	Yb	2.58	7.11	15.90	4.42	7.36	16.4	30.6	51.1	78.7	114.4	159.1	214.1	118.
71	Lu	2.66	7.40	2.48	4.61	7.68	17.1	32.0	53.3	82.1	119.3	166.0	223.5	124.
72	Hf	2.75	7.71	2.59	4.80	8.01	17.8	33.3	55.6	85.6	124.4	173.0	232.9	131.
73	Ta	2.82	8.02	2.70	5.01	8.35	18.6	34.7	57.9	89.3	129.7	180.4	242.8	136.
74	W	2.90	8.34	2.81	5.22	8.70	19.4	36.2	60.4	93.0	135.1	188.0	253.0	143.
75	Re	2.96	8.67	2.93	5.44	9.06	20.2	37.7	62.9	96.9	140.8	195.8	249.4	150.
76	Os	3.03	9.00	3.05	5.66	9.44	21.0	39.3	65.5	100.9	146.6	204.0	260.1	157.
77	Ir	3.10	9.35	3.17	5.89	9.82	21.9	40.9	68.2	105.0	152.5	212.2	231.5	165.
78	Pt	3.17	9.71	3.30	6.13	10.2	22.8	42.5	70.9	109.2	158.7	210.1	240.0	172.
79	Au	3.23	10.08	3.43	6.37	10.6	23.7	44.2	73.8	113.6	165.1	218.2	249.0	177.
80	Hg	3.30	1.60	3.57	6.62	11.0	24.6	46.0	76.7	118.1	171.5	191.6	257.9	189.
81	Tl	3.36	1.67	3.71	6.89	11.4	25.6	47.8	79.7	122.8	178.4	198.3	63.5	197.
82	Pb	3.41	1.73	3.86	7.16	11.9	26.6	49.7	82.9	127.7	176.5	205.3	66.6	207.
83	Bi	3.45	1.80	4.01	7.44	12.4	27.7	51.6	86.1	132.6	152.7	212.3	69.6	216.
84	Po	3.52	1.87	4.17	7.73	12.8	28.7	53.7	89.5	137.8	157.8	54.1	72.7	225.
85	At	3.56	1.94	4.33	8.03	13.3	29.8	55.7	92.9	143.1	163.1	57.1	76.8	238.
86	Rn	3.61	2.02	4.49	8.33	13.8	31.0	57.8	96.4	141.6	168.5	59.7	80.4	250.
87	Fr	3.66	2.09	4.65	8.64	14.4	32.1	60.0	100.0	119.8	174.0	62.1	83.6	260.
88	Ra	3.70	2.17	4.83	8.96	14.9	33.3	62.2	103.7	123.6	179.6	64.2	86.4	268.
89	Ac	3.75	2.25	5.00	9.29	15.4	34.5	64.5	107.5	127.6	48.5	67.4	90.8	282.
90	Th	3.81	2.33	5.19	9.63	16.0	35.8	66.8	111.4	131.6	50.2	69.9	94.1	292.
91	Pa	3.86	2.42	5.39	9.98	16.6	37.1	69.3	110.5	135.7	52.3	72.7	97.9	304.
92	U	3.91	2.51	5.58	10.35	17.4	38.5	71.8	90.8	139.9	54.3	75.5	101.5	316.
93	Np	3.95	2.58	5.66	10.7	18	40	83	93	143.0	57	77	104.	328.
94	Pu	4.00	2.66	5.74	11.0	19	41	85	96	39.1	59	79	107	342.
95	Am	4.05	2.74	5.82	11.4	20	43	87	99	41	61	81	110	354.
96	Cm	2.50	2.82	5.90	11.8	21	44	89	101	43	63	84	113	367.
97	Bk	2.56	2.90	5.97	12.3	21	45	91	104	46	65	86	117	382.
98	Cf	2.62	2.96	6.03	12.8	22	47	89	106	48	67	89	121	395.
99	Es	2.68	3.03	6.09	13.3	22	50	71	110	50	69	92	125	412.
100	Fm	2.74	3.10	6.15	13.7	23	51	73	35	52	71	95	128	427.

K; L_I; L_{II}; L_{III} (absorption edges)

APPENDIX 7A (*continued*)

Mass-absorption coefficient μ/ϱ, cm^2/g

	Wavelength, Å													
	2.0	2.5	3.0	4.0	5.0	6.0	7.0	8.0	9.0	10.0	15.0	20.0	25.0	30.0
	655.	1132.	324.	725.	1355.	2258.	3477.	5053.	7027.	9440.	15831.	34719.	16259.	27088.
	693.	1110.	351.	786.	1469.	2448.	3769.	5478.	7619.	10235.	17221.	9323.	17414.	29012.
	731.	222.	370.	827.	1546.	2576.	3966.	5764.	8016.	10769.	18564.	9873.	18442.	30725.
	771.	236.	394.	882.	1648.	2746.	4228.	6144.	8545.	9855.	20012.	10461.	19540.	32554.
	755.	250.	417.	934.	1745.	2908.	4478.	6508.	9051.	10506.	21612.	11061.	20661.	34422.
L_I	765.	266.	443.	992.	1854.	3090.	4758.	6915.	8338.	10195.	5564.	12454.	23262.	38755.
	766.	283.	471.	1055.	1971.	3284.	5056.	7349.	8853.	10887.	5960.	13339.	24916.	41511.
L_{II}	805.	298.	497.	1111.	2076.	3460.	5327.	7743.	9382.	8404.	6263.	14016.	26181.	43619.
	171.	320.	533.	1193.	2230.	3716.	5721.	7134.	9150.	8913.	6631.	14841.	27721.	46184.
L_{III}	177.	331.	551.	1234.	2305.	3842.	5915.	7546.	7048.	9466.	7000.	15667.	29264.	48754.
	186.	349.	581.	1301.	2430.	4050.	6236.	7194.	7478.	10044.	7460.	16695.	31184.	51954.
	197.	367.	613.	1371.	2561.	4269.	5769.	7606.	7910.	10627.	7943.	17778.	33207.	55323.
	208.	388.	647.	1448.	2704.	4507.	6112.	6022.	8374.	10887.	8453.	18918.	35336.	58871.
	218.	407.	679.	1519.	2838.	4730.	5723.	6357.	8840.	2890.	8994.	20130.	37600.	62642.
	229.	427.	712.	1594.	2977.	4369.	6022.	6707.	9327.	3073.	9561.	21399.	39971.	66597.
	239.	446.	744.	1665.	3110.	4603.	4873.	7082.	2433.	3268.	10169.	22759.	42512.	54286.
	252.	472.	787.	1760.	3288.	4300.	5141.	7472.	2585.	3472.	10804.	24179.	45164.	57823.
	265.	495.	825.	1847.	3450.	4537.	5417.	7632.	2745.	3687.	11474.	25679.	47966.	61517.
	278.	520.	866.	1938.	3240.	4747.	5701.	2095.	2914.	3914.	12180.	27258.	39587.	65211.
	293.	547.	912.	2041.	3415.	3917.	6032.	2224.	3093.	4155.	12929.	28936.	42537.	59733.
	306.	571.	952.	2130.	3571.	4092.	6124.	2361.	3284.	4411.	13726.	30721.	45622.	63626.
	321.	599.	999.	2236.	3258.	4309.	1727.	2511.	3492.	4691.	14597.	32670.	48933.	37825.
	336.	628.	1046.	2341.	3406.	4530.	1837.	2670.	3713.	4988.	15520.	34735.	43618.	41124.
	352.	658.	1097.	2455.	3562.	4767.	1951.	2836.	3944.	5299.	16488.	29811.	46504.	44644.
	369.	689.	1149.	2324.	3004.	4786.	2077.	3019.	4199.	5640.	17551.	31973.	49560.	48379.
	386.	721.	1202.	2444.	3144.	1433.	2207.	3208.	4461.	5993.	18647.	34038.	31491.	52468.
	397.	742.	1237.	2172.	3314.	1525.	2348.	3413.	4747.	6377.	19843.	29965.	34034.	56784.
	423.	791.	1318.	2268.	3458.	1623.	2498.	3631.	5050.	6784.	21110.	31981.	36864.	61416.
	442.	826.	1377.	2370.	3455.	1723.	2652.	3855.	5362.	7110.	18501.	33997.	39791.	66048.
	463.	865.	1442.	2027.	1098.	1829.	2817.	4094.	5693.	7648.	19714.	22997.	42956.	21595.
	484.	905.	1508.	2126.	1166.	1943.	2992.	4348.	6047.	8124.	20961.	24795.	46314.	23810.
	505.	943.	1447.	2222.	1238.	2063.	3176.	4616.	6420.	8625.	17953.	26718.	49906.	26221.
	534.	997.	1510.	2322.	1310.	2184.	3362.	4887.	6796.	9130.	19300.	28660.	17312.	28844.
	559.	1043.	1305.	2424.	1386.	2310.	3557.	5170.	7190.	9659.	20448.	30902.	19023.	31694.
	581.	1085.	1356.	2328.	1464.	2440.	3757.	5460.	7593.	10201.	21674.	33017.	20880.	34787.
	600.	1121.	1410.	2497.	1544.	2573.	3961.	5758.	8007.	10757.	15740.	33026.	24647.	41063.
	631.	1005.	1459.	872.	1630.	2716.	4181.	6077.	8452.	9673.	16748.	14827.	27696.	46142.
	654.	1121.	1520.	923.	1724.	2873.	4424.	6429.	8942.	10250.	17797.	16640.	31082.	51783.
	680.	947.	1331.	964.	1801.	3001.	4620.	6715.	9339.	10852.	18837.	17647.	32963.	54918.
	707.	981.	1381.	1019.	1903.	3172.	4883.	7098.	8406.	9091.	19924.	19373.	36188.	60289.
	729.	1000.	1435.	1070	2020.	3360.	5170.	7450.	N_I	N_{II}	N_{III}	N_{IV} N_V		
	755.	1021.	1470.	1120.	2140.	3750.	5460.							
	785.	1042.	1400.	1170.	2260.	3960.	5750.							
	530.	1063.	1450.	1220.	2370.	4180.	6080.							
	420.	910.	510.	1270.	2490.	4420.	6430.							
	435.	930.	525.	1320.	2620.	4620.								
	452.	950.	540.	1370.	2750.	4880.								
	470.	980.	555.	1420.	2880.	5100.								
	M_I M_{II}	M_{III}	M_{IV} M_V											

APPENDIX 7B. X-Ray Mass-Absorption Coefficients of Elements 2–11 (He–Na) at 40–100 Å (27)

Mass-absorption coefficient μ/ϱ, cm^2/g

Element	Wavelength, Å						
	40	50	60	70	80	90	100
2-He	2,360	4,670	8,000	12,600	18,500	25,800	34,500
3-Li	7,500	14,000	23,000	34,400	48,400	65,000	83,000
4-Be	16,500	29,100	45,500	65,000	88,000	114,000	142,000
5-B	28,300	47,900	22,000	3,650	5,100	6,700	8,500
6-C	44,100	3,140	4,890	7,000	9,500	12,400	15,700
7-N	2,960	5,100	7,900	11,400	15,600	20,500	26,000
8-O	4,560	7,900	12,400	17,900	24,400	32,000	40,600
9-F	6,600	11,500	17,900	25,800	34,900	45,100	56,000
10-Ne	10,300	17,800	27,300	38,400	51,000	64,000	78,000
11-Na	14,100	23,900	35,800	49,200	64,000	78,000	93,000

APPENDIX 8. Values of the *K* and *L*III Absorption-Edge Jump Ratios r and $(r-1)/r$ of the Chemical Elements (*73, 78*)

Element	*K* edge		*L*III edge	
	r	$(r-1)/r$	r	$(r-1)/r$
4-Be	35	0.970	—	—
5-B	28.3	0.965	—	—
6-C	24.2	0.959	—	—
7-N	21.4	0.953	—	—
8-O	19.3	0.948	—	—
9-F	17.5	0.943	—	—
10-Ne	15.94	0.937	—	—
11-Na	14.78	0.932	—	—
12-Mg	13.63	0.927	—	—
13-Al	12.68	0.921	—	—
14-Si	11.89	0.916	—	—
15-P	11.18	0.911	—	—
16-S	10.33	0.903	—	—
17-Cl	9.49	0.895	—	—
18-Ar	9.91	0.899	—	—
19-K	8.84	0.887	—	—
20-Ca	9.11	0.890	—	—
21-Sc	8.58	0.883	—	—
22-Ti	8.53	0.883	—	—
23-V	8.77	0.886	—	—
24-Cr	8.78	0.886	—	—
25-Mn	8.61	0.884	—	—
26-Fe	8.22	0.878	—	—
27-Co	8.38	0.881	—	—
28-Ni	7.85	0.873	2.77	0.639
29-Cu	7.96	0.874	2.87	0.652
20-Zn	7.60	0.868	5.68	0.824

APPENDIX 8 (*continued*)

Element	*K* edge		*L*III edge	
	r	$(r-1)/r$	r	$(r-1)/r$
31-Ga	7.40	0.865	5.67	0.824
32-Ge	7.23	0.862	5.70	0.825
33-As	7.19	0.861	4.88	0.795
34-Se	6.88	0.855	4.59	0.782
35-Br	6.97	0.857	4.58	0.782
36-Kr	7.04	0.858	4.17	0.760
37-Rb	6.85	0.854	4.22	0.763
38-Sr	7.06	0.858	3.91	0.744
39-Y	6.85	0.854	4.04	0.752
40-Zr	6.75	0.852	3.98	0.748
41-Nb	7.13	0.860	3.77	0.735
42-Mo	6.97	0.856	3.68	0.728
43-Tc	6.80	0.853	3.59	0.722
44-Ru	6.76	0.852	3.43	0.708
45-Rh	6.53	0.847	3.72	0.731
46-Pd	6.93	0.856	3.40	0.706
47-Ag	6.58	0.848	3.22	0.690
48-Cd	6.50	0.846	3.25	0.692
49-In	6.26	0.840	3.25	0.693
50-Sn	6.47	0.845	3.06	0.673
51-Sb	6.35	0.843	2.94	0.660
52-Te	6.21	0.839	2.98	0.664
53-I	6.16	0.838	2.86	0.650
54-Xe	6.08	0.835	2.88	0.653
55-Cs	5.95	0.832	2.85	0.649
56-Ba	5.80	0.828	2.84	0.648
57-La	6.05	0.835	2.72	0.632
58-Ce	5.90	0.830	2.74	0.635
59-Pr	5.82	0.828	2.70	0.629
60-Nd	5.99	0.833	2.66	0.624

APPENDIX 8 (*continued*)

Element	K edge		LIII edge	
	r	$(r-1)/r$	r	$(r-1)/r$
61-Pm	5.91	0.831	2.70	0.630
62-Sm	5.79	0.827	2.68	0.627
63-Eu	5.69	0.824	2.72	0.633
64-Gd	5.77	0.827	2.70	0.630
65-Tb	5.52	0.819	2.71	0.631
66-Dy	5.48	0.818	2.75	0.636
67-Ho	5.33	0.812	2.86	0.650
68-Er	5.50	0.818	2.93	0.659
69-Tm	5.34	0.813	2.76	0.637
70-Yb	5.19	0.807	2.57	0.611
71-Lu	5.22	0.808	2.62	0.618
72-Hf	5.45	0.816	2.42	0.586
73-Ta	5.02	0.801	2.60	0.615
74-W	5.12	0.805	2.62	0.618
75-Re	4.79	0.791	2.68	0.626
76-Os	5.06	0.803	2.53	0.605
77-Ir	5.18	0.807	2.39	0.581
78-Pt	5.12	0.805	2.63	0.620
79-Au	4.92	0.797	2.44	0.590
80-Hg	5.02	0.801	2.40	0.583
81-Tl	4.88	0.795	2.50	0.600
82-Pb	4.79	0.791	2.47	0.591
83-Bi	4.73	0.788	2.34	0.572
86-Rn	4.72	0.788	2.34	0.573
90-Th	4.39	0.772	2.39	0.581
92-U	4.41	0.773	2.28	0.562
94-Pu	4.53	0.779	2.25	0.556

APPENDIX 9. Average Values of the *K*, *L*, and *M* Fluorescent Yields of the Chemical Elements (*C53*)

Element	*K*	*L*	*M*	Element	*K*	*L*	*M*
6-C	0,001	—	—	54-Xe	0.876	0.110	0.003
7-N	0.002	—	—	55-Cs	0.882	0.118	0.004
8-O	0.003	—	—	56-Ba	0.888	0.126	0.004
9-F	0.005	—	—	57-La	0.893	0.135	0.004
10-Ne	0.008	—	—	58-Ce	0.898	0.143	0.005
11-Na	0.013	—	—	59-Pr	0.902	0.152	0.005
12-Mg	0.019	—	—	60-Nd	0.907	0.161	0.006
13-Al	0.026	—	—	61-Pm	0.911	0.171	0.006
14-Si	0.036	—	—	62-Sm	0.915	0.180	0.007
15-P	0.047	—	—	63-Eu	0.918	0.190	0.007
16-S	0.061	—	—	64-Gd	0.921	0.200	0.008
17-Cl	0.078	—	—	65-Tb	0.924	0.210	0.009
18-Ar	0.097	—	—	66-Dy	0.927	0.220	0.009
19-K	0.118	—	—	67-Ho	0.930	0.231	0.010
20-Ca	0.142	0.001	—	68-Er	0.932	0.240	0.011
21-Sc	0.168	0.001	—	69-Tm	0.934	0.251	0.012
22-Ti	0.197	0.001	—	70-Yb	0.937	0.262	0.013
23-V	0.227	0.002	—	71-Lu	0.939	0.272	0.014
24-Cr	0.258	0.002	—	72-Hf	0.941	0.283	0.015
25-Mn	0.291	0.003	—	73-Ta	0.942	0.293	0.016
26-Fe	0.324	0.003	—	74-W	0.944	0.304	0.018
27-Co	0.358	0.004	—	75-Re	0.945	0.314	0.019
28-Ni	0.392	0.005	—	76-Os	0.947	0.325	0.020
29-Cu	0.425	0.006	—	77-Ir	0.948	0.335	0.022
30-Zn	0.458	0.007	—	78-Pt	0.949	0.345	0.024
31-Ga	0.489	0.009	—	79-Au	0.951	0.356	0.026
32-Ge	0.520	0.010	—	80-Hg	0.952	0.366	0.028
33-As	0.549	0.012	—	81-Tl	0.953	0.376	0.030
34-Se	0.577	0.014	—	82-Pb	0.954	0.386	0.032
35-Br	0.604	0.016	—	83-Bi	0.954	0.396	0.034
36-Kr	0.629	0.019	—	84-Po	0.955	0.405	0.037
37-Rb	0.653	0.021	0.001	85-At	0.956	0.415	0.040
38-Sr	0.675	0.024	0.001	86-Rn	0.957	0.425	0.043
39-Y	0.695	0.027	0.001	87-Fr	0.957	0.434	0.046
40-Zr	0.715	0.031	0.001	88-Ra	0.958	0.443	0.049
41-Nb	0.732	0.035	0.001	89-Ac	0.958	0.452	0.052
42-Mo	0.749	0.039	0.001	90-Th	0.959	0.461	0.056
43-Tc	0.765	0.043	0.001	91-Pa	0.959	0.469	0.060
44-Ru	0.779	0.047	0.001	92-U	0.960	0.478	0.064
45-Rh	0.792	0.052	0.001	93-Np	0.960	0.486	0.068
46-Pd	0.805	0.058	0.001	94-Pu	0.960	0.494	0.073
47-Ag	0.816	0.063	0.002	95-Am	0.960	0.502	0.077
48-Cd	0.827	0.069	0.002	96-Cm	0.961	0.510	0.083
49-In	0.836	0.075	0.002	97-Bk	0.961	0.517	0.088
50-Sn	0.845	0.081	0.002	98-Cf	0.961	0.524	0.093
51-Sb	0.854	0.088	0.002	99-Es	0.961	0.531	0.099
52-Te	0.862	0.095	0.003	100-Fm	0.961	0.538	0.106
53-I	0.869	0.102	0.003				

APPENDIX 10. X-Ray Spectrometer Analyzer Crystals and Multilayer Films

Information is given for 72 crystal and multilayer-film analyzers arranged in order of increasing $2d$ spacing, that is, in order of increasing useful wavelength region. An *Alphabetical Index for Appendix 10*, listing common and chemical names and letter-symbols, and a plot of the useful spectral regions of the 72 analyzers follow the table.

A *crystal* analyzer is properly designated by its name or letter-symbol *and* the Miller indexes (hkl), or perhaps the $2d$ spacing, of the diffracting planes; for example, quartz(502), quartz($2d$ 1.62), LiF(220), Si(111), ADP(101). However, some crystals are used in only one "cut," and the indexes or spacing are commonly omitted; for example, PET, EDDT, KHP (or "KAP"). A multilayer-film analyzer need be designated only by its chemical name or letter-symbol.

The table gives information as follows:

Column 1 gives the serial numbers of the analyzers in the table; these are also the numbers referred to in the *Alphabetical Index for Appendix 10.*

Column 2 gives commonly used letter-symbols (if any), common and chemical names, and chemical formulas; for organic compounds, the formulas are given in a form that indicates the molecular structure.

Column 3 gives three parameters: (1) The Miller indexes (hkl) or ($hkil$) of the diffracting planes parallel to the diffracting surfaces for crystals; (?) indicates that the indexes are not known or are not ascertainable by the compiler; (LBF) indicates that the analyzer is a multilayer Langmuir–Blodgett film. (2) Twice the interplanar spacing for crystals or twice the repeat interval between metal-atom layers for multilayer films, both in angstroms; the $2d$ value is also the longest wavelength that the analyzer can diffract (Section 5.4.2.1). (3) The crystal system; (?) indicates that the crystal system is not known or is not ascertainable by the compiler; of course, crystal system has no significance for multilayer films.

Column 4 (λ, Å:, $K\alpha$:, $L\alpha_1$:) applies to columns 5 and 6.

Column 5 gives the longest wavelength ($2d$ Å) and the elements having the longest-wavelength $K\alpha$ and $L\alpha_1$ lines that each analyzer can diffract.

Column 6 gives the practical useful wavelength region lying in the interval 10–140° 2θ and the $K\alpha$ and $L\alpha_1$ lines lying in this region; this is the spectral region plotted in the accompanying figure; only in special cases should the analyzer be used outside these limits, which are themselves a little generous.

Column 7. No comprehensive compilations of intercomparable relative diffracted intensities or resolving powers of analyzer crystals are known to the compiler. However, some relative intensities and resolving powers are given here and in column 8. Column 7 also gives possible secondary x-ray emission lines from the analyzer ("crystal emission").

Column 8 gives optimal and other applications, advantages and limitations, other practical information, and, for some analyzers, literature references.

The footnotes cited throughout the table are given at the end of the table.

APPENDIX 10 (*continued*)

No.*	Symbol(s)[a,b] (if any) Common name(s) Chemical name(s) Chemical formula(s)	(*hkl*), (*hkil*) 2*d*, Å Crystal system	→	Longest wavelength and lowest atomic no. diffracted	Practical useful region (10–145° 2θ)	Reflectivity Resolution Secondary emission	Optimal application Remarks Reference(s)
1	QTZ(502), QTZ-1.62 α-Quartz Silicon dioxide SiO_2	$(50\bar{5}2)$ 1.624 Hexagonal	λ, Å: *Kα*: *Lα*:	1.624 29-Cu 71-Lu	0.142–1.55 88-Ra to 29-Cu $\lesssim$73-Ta	Very low Very high Si *K*	Shortest 2*d* of any practical crystal Good for high-*Z K* lines excited by 100-kV generators
2	LiF(422), LiF-1.65 Lithium fluoride LiF	(422) 1.652 Cubic	λ, Å: *Kα*: *Lα*:	1.652 29-Cu 71-Lu	0.144–1.58 87-Fr to 29-Cu $\lesssim$72-Hf	High High —	Better than QTZ(502) for the same applications (*J12*)
3	COR(146), COR-1.66 Corundum, sapphire, alumina Aluminum oxide Al_2O_3	$(14\bar{5}6)$ 1.660 Hexagonal	λ, Å: *Kα*: *Lα*:	1.660 29-Cu 71-Lu	0.145–1.584 87-Fr to 29-Cu $\lesssim$72-Hf	— Very high Al *K*	Same applications as QTZ(502)
4	LiF(420), LiF-1.80 Lithium fluoride LiF	(420) 1.80 Cubic	λ, Å: *Kα*: *Lα*:	1.80 27-Co 68-Er	0.157–1.72 84-Pb to 28-Ni $\lesssim$70-Yb	— — —	Similar to LiF(422) Free from "forbidden" reflections (see remarks for LiF(220) below) (*J12*)
5*	— Topaz(303), Topaz-2.71[c] Hydrated aluminum fluorosilicate $Al_2(F,OH)_2SiO_4$	(303) 2.712 Orthorhombic	λ, Å: *Kα*: *Lα*:	2.712 23-V 57-La	0.236–2.59 70-Yb to 23-V $\lesssim$58-Ce	Medium Very high F, Al, Si *K*	Improves dispersion in the region V *K* to Ni *K*, where Z *Kβ* overlaps (Z + 1) *Kα* Improves dispersion of the *L* lines of the lanthanon elements (*Z* 57–71)

							Inferior to LiF(220); diffracted intensity ~0.25–0.5× LiF(220) Diffracted intensity ~0.05–0.1× LiF(200) Individual crystals may vary widely in diffracted intensity May give spurious reflections from planes other than (303)
6	COR(030), COR-2.75 Corundum, sapphire, alumina Aluminum oxide Al_2O_3	$(03\bar{3}0)$ 2.748 Hexagonal	λ, Å: $K\alpha$: $L\alpha$:	2.748 23-V 57-La	0.240–2.622 69-Tm to 23-V $\lesssim$58-Ce	— Very high Al *K*	Diffracted intensity ~2–4× Topaz (303) and QTZ(203) with the same or better resolution (*B26*)
7*	QTZ(203), QTZ-2.75 α-Quartz Silicon dioxide SiO_2	$(20\bar{2}3)$ 2.749 Hexagonal	λ, Å: $K\alpha$: $L\alpha$:	2.749 23-V 57-La	0.240–2.62 59-Tm to 23-V $\lesssim$58-Ce	Low Very high Si *K*	Same applications as Topaz(303), LiF(220) Somewhat sharper lines, but lower intensity than Topaz(303) Inferior to LiF(220)
8	— Topaz(006), Topaz-2.80 Hydrated aluminum fluorosilicate $Al_2(F,OH)_2SiO_4$	(006) 2.795 Orthorhombic	λ, Å: $K\alpha$: $L\alpha$:	2.795 22-Ti 56-Ba	0.244–2.666 69-Tm to 23-V $\lesssim$57-La	— — F, Al, Si *K*	—
9*	LiF(220), LiF-2.85 Lithium fluoride LiF	(220) 2.848 Cubic	λ, Å: $K\alpha$: $L\alpha$:	2.848 22-Ti 56-Ba	0.248–2.72 68-Er to 22-Ti $\lesssim$57-La	High Very high F *K*	Same application as Topaz(303) Diffracted intensity ~2–4× Topaz (303), ~0.4–0.8× LiF(200) Diffracted spectrum may be cluttered by broad "forbidden" (110), (330), (550), etc. reflections (*P15*)

APPENDIX 10 (*continued*)

No.*	Symbol(s)[a,b] (if any) Common name(s) Chemical name(s) Chemical formula(s)	(*hkl*), (*hkil*) 2*d*, Å Crystal system	→	Longest wavelength and lowest atomic no. diffracted	Practical useful region (10–145° 2θ)	Reflectivity Resolution Secondary emission	Optimal application Remarks Reference(s)
10	MICA(33$\bar{1}$), MICA-3.00 Mica, Muscovite $K_2O \cdot 3Al_2O_3 \cdot 6SiO_2 \cdot 2H_2O$	(33$\bar{1}$) 3.00 Monoclinic	λ, Å: $K\alpha$: $L\alpha$:	3.00 22-Ti 55-Cs	0.262–2.86 67-Ho to 22-Ti $\lesssim$56-Ba	— — Al, Si, K *K*	Transmission spectrometers, especially Cauchois and Dumond
11	QTZ(211), QTZ-3.08 α-Quartz Silicon dioxide SiO_2	(21$\bar{3}$1) 3.082 Hexagonal	λ, Å: $K\alpha$: $L\alpha$:	3.082 21-Sc 54-Xe	0.269–2.940 66-Dy to 22-Ti $\lesssim$55-Cs	— — Si *K*	—
12	QTZ(112), QTZ-3.64 α-Quartz Silicon dioxide SiO_2	(11$\bar{2}$2) 3.636 Hexagonal	λ, Å: $K\alpha$: $L\alpha$:	3.636 20-Ca 50-Sn	0.317–3.469 61-Pm to 20-Ca $\lesssim$51-Sb	— — Si *K*	—
13	Si(220), Si-3.84 Silicon Si	(220) 3.84 Cubic	λ, Å: $K\alpha$: $L\alpha$:	3.84 19-K 49-In	0.335–3.66 59-Pr to 20-Ca $\lesssim$50-Sn	— — Si *K*	—
14	CaF_2(220), CaF_2-3.86 Fluorite Calcium fluoride CaF_2	(220) 3.862 Cubic	λ, Å: $K\alpha$: $L\alpha$:	3.862 19-K 49-In	0.337–3.68 59-Pr to 20-Ca $\lesssim$50-Sn	Medium Medium F, Ca *K*	—

15	Ge(220), Ge-4.00	(220)	λ, Å:	4.00	0.349–3.82	—	—
	Germanium	4.00	$K\alpha$:	19-K	58-Ce to 19-K	—	
	Ge	Cubic	$L\alpha$:	48-Cd	$\gtrsim$49-In	Ge K, L	
16*	LiF(200), LiF(4.03)[c]	(200)	λ, Å:	4.027	0.351–3.84	Very high	Best general crystal for K K to Lr L
	Lithium fluoride	4.027	$K\alpha$:	19-K	58-Ce to 19-K	Very high	Highest intensity for largest number of elements of any crystal
	LiF	Cubic	$L\alpha$:	48-Cd	$\gtrsim$49-In	F K	Combines high intensity and high dispersion
17	Al(200), Al-4.05	(200)	λ, Å:	4.048	0.353–3.86	High	Curved, especially doubly curved, optics
	Aluminum	4.048	$K\alpha$:	19-K	58-Ce to 19-K	—	
	Al	Cubic	$L\alpha$:	48-Cd	$\gtrsim$49-In	Al K	
18	QTZ(102), QTZ-4.56	$(10\bar{1}2)$	λ, Å:	4.564	0.398–4.354	—	Used in the prototype Laue spectrometer
	α-Quartz	4.564	$K\alpha$:	18-Ar	55-Cs to 18-Ar	—	
	Silicon dioxide	Hexagonal	$L\alpha$:	46-Pd	$\gtrsim$47-Ag	Si K	
	SiO_2						
19	—	(200)	λ, Å:	4.64	0.405–4.426	—	—
	Topaz(200), Topaz-4.64	4.64	$K\alpha$:	18-Ar	54-Xe to 18-Ar	—	
	Hydrated aluminum fluorosilicate	Orthorhombic	$L\alpha$:	45-Rh	$\gtrsim$46-Pd	F, Al, Si K	
	$Al_2(F,OH)_2SiO_4$						
20	Al(111), Al-4.68	(111)	λ, Å:	4.676	0.408–4.46	Very high	Curved, especially doubly curved, optics
	Aluminum	4.676	$K\alpha$:	18-Ar	54-Xe to 18-Ar	—	
	Al	Cubic	$L\alpha$:	45-Rh	$\gtrsim$46-Pd	Al K	

APPENDIX 10 *(continued)*

No.*	Symbol(s)[a,b] (if any) Common name(s) Chemical name(s) Chemical formula(s)	(*hkl*), (*hkil*) 2*d*, Å Crystal system	→	Longest wavelength and lowest atomic no. diffracted	Practical useful region (10–145° 2θ)	Reflectivity Resolution Secondary emission	Optimal application Remarks Reference(s)
21	GYP(002), GYP-4.99 Gypsum Calcium sulfate dihydrate $CaSO_4 \cdot 2H_2O$	(002) 4.990 Monoclinic	λ, Å: *K*α: *L*α:	4.990 17-Cl 44-Ru	0.435–4.76 53-I to 17-Cl ≲45-Rh	— — S, Ca *K*	—
22*	NaCl Rock salt, halite Sodium chloride NaCl	(200) 5.641 Cubic	λ, Å: *K*α: *L*α:	5.641 16-S 42-Mo	0.492–5.38 50-Sn to 16-S ≲44-Ru	Very high High Na, Cl *K*	S *K*α and Cl *K*α in matrixes lighter than Mg Like LiF(200), good general crystal for S *K*α to Lr *L* May give slightly higher intensity than LiF(200) for Ge *K*α–Ba *K*α Background may be higher than for LiF(200) due to Na and Cl *K*α emission Original standard for the x-unit
23	— Calcite Calcium carbonate $CaCO_3$	(20$\bar{2}$0) 6.071 Hexagonal	λ, Å: *K*α: *L*α:	6.071 16-S 41-Nb	0.529–5.79 48-Cd to 16-S ≲41-Nb	Medium Very high Ca *K*	Very precise measurement of wavelength Extremely high degree of crystal perfection with resultant sharp lines Replaced NaCl(200) as standard for the x-unit

24*	ADP(112), ADP-6.14 Ammonium dihydrogen phosphate $NH_4H_2PO_4$	(112) 6.14 Tetragonal	λ, Å: $K\alpha$: $L\alpha$:	6.14 16-S 40-Zr	0.535–5.86 48-Cd to 16-S ⋝41-Nb	— — P *K*	—
25*	Si(111), Si-6.27[c] Silicon Si	(111) 6.271 Cubic	λ, Å: $K\alpha$: $L\alpha$:	6.271 15-P 40-Zr	0.547–5.98 47-Ag to 16-S ⋝41-Nb	High — Si *K*	Eliminates second-order[d] spectral-line interference in measurement of intermediate- and high-*Z* elements where Si $K\alpha$ emission from the crystal is absorbed by the air path Very rugged and stable (*L42*)
26	KCl Sylvite Potassium chloride KCl	(200) 6.292 Cubic	λ, Å: $K\alpha$: $L\alpha$:	6.292 15-P 40-Zr	0.549–6.00 47-Ag to 16-S ⋝41-Nb	— — Cl, K *K*	—
27	CaF_2(111), CaF_2-6.31 Fluorite Calcium fluoride CaF_2	(111) 6.306 Cubic	λ, Å: $K\alpha$: $L\alpha$:	6.306 15-P 40-Zr	0.550–6.02 47-Ag to 16-S ⋝41-Nb	Medium Medium F, Ca *K*	Very weak second order, strong third order
28*	Ge(111), Ge-6.53[c] Germanium Ge	(111) 6.532 Cubic	λ, Å: $K\alpha$: $L\alpha$:	6.532 15-P 39-Y	0.570–6.23 46-Pd to 15-P ⋝40-Zr	High Very high Ge *K*	Eliminates second-order[d] spectral-line interference in measurement of intermediate- and low-*Z* elements where Ge $K\alpha$ emission from the crystal is minimized by pulse-height selection Very rugged and stable (*L42*)

APPENDIX 10 (*continued*)

No.*	Symbol(s)[a,b] (if any) Common name(s) Chemical name(s) Chemical formula(s)	(*hkl*), (*hkil*) 2*d*, Å Crystal system	→ Longest wavelength and lowest atomic no. diffracted	Practical useful region (10–145° 2θ)	Reflectivity Resolution Secondary emission	Optimal application Remarks Reference(s)
29	KBr Potassium bromide KBr	(200) 6.584 Cubic	λ, Å: 6.584 $K\alpha$: 15-P $L\alpha$: 39-Y	0.574–6.28 46-Pd to 15-P $\lesssim$40-Zr	— — K, Br *K*	—
30*	QTZ(101), QTZ-6.69 α-Quartz Silicon dioxide SiO_2	(10$\bar{1}$1) 6.687 Hexagonal	λ, Å: 6.687 $K\alpha$: 15-P $L\alpha$: 39-Y	0.583–6.38 46-Pd to 15-P $\lesssim$40-Zr	High Very high Si *K*	P $K\alpha$ in low-*Z* matrixes, especially in presence of Ca P $K\alpha$ intensity ~1.8× EDDT, but less than PET Intensity for P $K\alpha$ to K $K\alpha$ greater than EDDT, but less than PET
31*	GRAPH Graphite Carbon C	(0002) 6.708 Hexagonal	λ, Å: 6.708 $K\alpha$: 15-P $L\alpha$: 39-Y	0.585–6.40 46-Pd to 15-P $\lesssim$40-Zr	— — C *K*	P, S, Cl *K* lines Intensity for P $K\alpha$ >5× EDDT Relatively poor resolution Has highest diffracted intensity of any crystal used for x-ray spectrometry—up to 3× LiF(200), up to 7× PET (*G25*, *J12*)
32	— Ammonium hydrogen citrate $(NH_4)_2HC_6H_5O_7$ $C(OH)(COOH)(CH_2COONH_4)_2$	(?) 7.38 (?)	λ, Å: 7.38 $K\alpha$: 14-Si $L\alpha$: 37-Rb	0.644–7.040 44-Ru to 15-P $\lesssim$38-Sr	— — None	(*H38*)

[illegible]	[illegible] (200), ADP 7.50 Ammonium dihydrogen phosphate $NH_4H_2PO_4$	(200) 7.50 Tetragonal	λ, Å: $K\alpha$: $L\alpha$:	7.50 14-Si 37-Rb	0.654–7.16 43-Tc to 14-Si ≳38-Sr	— — P K	Higher intensity than EDDT ($B13$)
34	— Topaz (002), Topaz-8.37 Hydrated aluminum fluorosilicate $Al_2(F,OH)_2SiO_4$	(002) 8.374 Orthorhombic	λ, Å: $K\alpha$: $L\alpha$:	8.374 13-Al 36-Kr	0.730–7.989 41-Nb to 14-Si ≳36-Kr	— — F, Al, Si K	—
35*	QTZ(100), QTZ-8.52[c] α-Quartz Silicon dioxide SiO_2	$(10\bar{1}0)$ 8.52 Hexagonal	λ, Å: $K\alpha$: $L\alpha$:	8.52 13-Al 35-Br	0.742–8.12 41-Nb to 14-Si 100-Fm to 36-Kr	Medium Very high Si K	Same applications as EDDT and PET; higher resolution, but lower intensity
36*	PET Pentaerythritol Tetrakis (hydroxymethyl)-methane[e] $C(CH_2OH)_4$	(002) 8.742 Tetragonal	λ, Å: $K\alpha$: $L\alpha$:	8.742 13-Al 35-Br	0.762–8.34 40-Zr to 13-Al 99-Es to 36-Kr	High Low None	Al, Si, P, S, Cl $K\alpha$, especially P and S $K\alpha$ in intermediate- and high-Z matrixes Good general crystal for Al–Sc $K\alpha$ Al, Si, P, S $K\alpha$ intensities ~1.5–2× EDDT; Al $K\alpha$ intensity ~2.5× KHP Low background due to absence of any element to emit secondary spectral lines Crystal is soft, must be mounted in a rigid support and handled carefully Tends to deteriorate with age and should be kept in a desiccator when not in use

APPENDIX 10 *(continued)*

No.*	Symbol(s)[a,b] (if any) Common name(s) Chemical name(s) Chemical formula(s)	(*hkl*), (*hkil*) 2*d*, Å Crystal system	→	Longest wavelength and lowest atomic no. diffracted		Practical useful region (10–145° 2θ)	Reflectivity Resolution Secondary emission	Optimal application Remarks Reference(s)
								Tends to deteriorate on prolonged exposure to high-intensity x-rays Highest thermal expansion coefficient of all common analyzer crystals
37	— Ammonium tartrate $(NH_4)_2C_4H_4O_6$ $COONH_4$–$(CHOH)_2$–$COONH_4$	(?) 8.80 Monoclinic		λ, Å: $K\alpha$: $L\alpha$:	8.80 13-Al 35-Br	0.767–8.40 40-Zr to 13-Al $\lesssim$35-Br	— — None	(*H38*)
38*	EDDT, EDdT, EDT Ethylenediamine-*d*-tartrate (*d* for *dextro*, not *di*) $C_6H_{14}N_2O_6$ $NH_2 \cdot CH_2 \cdot CH_2 \cdot NH_2$ / $COOH \cdot (CHOH)_2 \cdot COOH$	(020) 8.808 Monoclinic		λ, Å: $K\alpha$: $L\alpha$:	8.808 13-Al 35-Br	0.768–8.40 41-Nb to 13-Al 99-Es to 35-Br	Medium Low None	Al and Si $K\alpha$ in any matrix P, S, Cl $K\alpha$ in intermediate- and high-*Z* matrixes Same remarks as for PET, except: substantially lower intensity than PET; substantially lower thermal expansion coefficient than PET Rugged and stable

No.	Crystal	Plane, 2d, system			Range	Interference	Remarks
39*	ADP(101), ADP–10.64[c] Ammonium dihydrogen phosphate $NH_4H_2PO_4$	(101) 10.640 Tetragonal	λ, Å: $K\alpha$: $L\alpha$:	10.640 12-Mg 32-Ge	0.928–10.15 37-Rb to 12-Mg 91-Pa to 33-As	Low — P K	Mg $K\alpha$ Background may be high due to P $K\alpha$ emission Can be used for Al and Si $K\alpha$ when Mg $K\alpha$ is also to be measured, but intensity is much lower than EDDT and PET (*D14*)
40	— Oxalic acid dihydrate Ethanedioic acid dihydrate[e] $(COOH)_2 \cdot 2H_2O$	(001)[f] 11.92 Monoclinic	λ, Å: $K\alpha$: $L\alpha$:	11.92 11-Na 31-Ga	1.039–11.37 35-Br to 12-Mg 86-Rn to 31-Ga	— — None	—
41	SHA Sorbitol hexaacetate $C_6H_8O_6(COCH_3)_6$ $HO \cdot CH \cdot CO \cdot CH_3$ | $(HO \cdot C \cdot CO \cdot CH_3)_4$ | $HO \cdot CH \cdot CO \cdot CH_3$	(110)[f] 13.98 Monoclinic	λ, Å: $K\alpha$: $L\alpha$:	13.98 11-Na 29-Cu	1.22–13.34 32-Ge to 11-Na 80-Hg to 29-Cu	— — None	Bridges interval between ADP ($2d$ 10.6 Å) and KHP ($2d$ 26.6 Å) May completely replace ADP(101) when consistently good crystals become available Diffracted intensity ~2× GYP (020) for Cu $K\alpha$ Not as good as RHP or GYP for Na $K\alpha$ High resolution Stable in high vacuum Available only in relatively small pieces (*J12*, *J17*, *R32*)

APPENDIX 10 (*continued*)

No.*	Symbol(s)[a,b] (if any) Common name(s) Chemical name(s) Chemical formula(s)	(*hkl*), (*hkil*) 2*d*, Å Crystal system	→	Longest wavelength and lowest atomic no. diffracted	Practical useful region (10–145° 2θ)	Reflectivity Resolution Secondary emission	Optimal application Remarks Reference(s)
42	— Rock sugar (001), -15.12 Sucrose $C_{12}H_{22}O_{11}$	(001)[f] 15.12 Monoclinic	λ, Å: $K\alpha$: $L\alpha$:	15.12 10-Ne 28-Ni	1.318–14.42 31-Ga to 11-Na 77-Ir to 29-Cu	— — None	—
43*	GYP(020), GYP-15.18[c] Gypsum Calcium sulfate dihydrate $CaSO_4 \cdot 2H_2O$	(020) 15.185 Monoclinic	λ, Å: $K\alpha$: $L\alpha$:	15.185 10-Ne 28-Ni	1.32–14.49 31-Ga to 11-Na 77-Ir to 29-Cu	Medium — S, Ca *K*	Na $K\alpha$, but not quite as good as KHP or Pb stearate Useful for Mg–Cl *K* lines when Na is also to be measured, but applicability limited due to presence of S and Ca Background higher than EDDT or PET due to Ca $K\alpha$ and S $K\alpha$ emission from the crystal May effloresce in low humidity or prolonged use in vacuum path
44	— Beryl $3BeO \cdot Al_2O_3 \cdot 6SiO_2$	$(10\bar{1}0)$ 15.954 Hexagonal	λ, Å: $K\alpha$: $L\alpha$:	15.954 10-Ne 28-Ni	1.391–15.22 30-Zn to 10-Ne 76-Os to 28-Ni	— — Al, Si *K*	Large (2.5 × 3.5 cm, 1 × 3-inch) blanks may be difficult to obtain Individual crystals may vary somewhat in 2*d*

No.	Name	Plane, 2d, System		Long λ	Range	Relative intensity / Interfering	Remarks
45	— Bismuth titanate $Bi_2(TiO_3)_3$	(04$\bar{4}$0) 16.40 Orthorhombic	λ, Å: $K\alpha$: $L\alpha$:	16.40 10-Ne 27-Co	1.430–15.65 30-Zn to 10-Ne 75-Re to 28-Ni	— — Ti *K*; Bi *L*, *M*	—
46	— Itaconic acid Methylenebutanedioic acid[e] CH_2:C(COOH)CH_2COOH	(02$\bar{2}$0)[f] 18.50 Orthorhombic	λ, Å: $K\alpha$: $L\alpha$:	18.50 9-F 26-Fe	1.613–17.65 28-Ni to 10-Ne 71-Lu to 26-Fe	— — None	—
47*	MICA(002), MICA-19.84[c] Mica, muscovite $K_2O\cdot 3Al_2O_3\cdot 6SiO_2\cdot 2H_2O$	(002) 19.84 Monoclinic	λ, Å: $K\alpha$: $L\alpha$:	19.84 9-F 25-Mn	1.73–18.93 27-Co to 9-F 69-Tm to 26-Fe	Low — Al, Si, K *K*	Variable-radius curved-crystal spectrometers, especially for selected-area analysis and electron-probe microanalyzers Several higher orders of about equal, relatively high intensity (*H38*)
48	— Silver acetate CH_3COOAg	(001) 20.0 (?)	λ, Å: $K\alpha$: $L\alpha$:	20.0 9-F 25-Mn	1.744–19.08 27-Co to 9-F 68-Er to 26-Fe	Low — Ag *K*, *L*	—
49	— Rock sugar (100), -20.12 Sucrose $C_{12}H_{22}O_{11}$	(100)[f] 20.12 Monoclinic	λ, Å: $K\alpha$: $L\alpha$:	20.12 9-F 25-Mn	1.75–19.19 27-Co to 9-F 68-Er to 26-Fe	Low — None	—
50	THP, TlHP; TAP, TlAP[g,h] Thallium hydrogen phthalate[e] $TlHC_8H_4O_4$, $TlOOC\cdot C_6H_4\cdot COOH$ C_6H_4(COOTl)(COOH)	(10$\bar{1}$0) 25.9 Orthorhombic	λ, Å: $K\alpha$: $L\alpha$:	25.9 8-O 23-V	2.26-24.7 24-Cr to 8-O 61-Pm to 23-V	High — Tl *L*, *M*	Same applications as RHP, KHP Diffracted intensity ~2× RHP (*B72*, *L22*)

APPENDIX 10 (*continued*)

No.*	Symbol(s)[a,b] (if any) Common name(s) Chemical name(s) Chemical formula(s)	(*hkl*), (*hkil*) 2*d*, Å Crystal system	→	Longest wavelength and lowest atomic no. diffracted	Practical useful region (10–145° 2θ)	Reflectivity Resolution Secondary emission	Optimal application Remarks Reference(s)
51*	RHP, RbHP; RAP, RbAP[g,h] Rubidium hydrogen phthalate[e] $RbHC_8H_4O_4$, $RbOOC \cdot C_6H_4 \cdot COOH$ (benzene ring with COORb, COOH)	(10$\bar{1}$0) 26.121 Orthorhombic	λ, Å: *K*α: *L*α:	26.12 8-O 23-V	2.28–24.92 24-Cr to 8-O 61-Pm to 23-V	High — Rb *K*, *L*	Same applications as KHP, which it may replace Diffracted intensity ~3× KHP for Na, Mg, Al *K*α and Cu $L\alpha_1$; ~4× KHP for F *K*α; ~8× KHP for O *K*α Resolution about the same as KHP Peak/background ratio higher than KHP F *K*α pulse distribution from Ar-filled detector free from interference (see Remarks for KHP below) Substantially weaker anomalous peak for O *K*α than KHP (*J12*, *R32*)
52*	KHP; KAP[g,h] Potassium hydrogen phthalate[e] $KHC_8H_4O_4$, $KOOC \cdot C_6H_4 \cdot COOH$ (benzene ring with COOK, COOH)	(10$\bar{1}$0) 26.632 Orthorhombic	λ, Å: *K*α: *L*α:	26.632 8-O 23-V	2.32–25.41 23-V to 8-O 60-Nd to 23-V	Medium — K *K*	Good general crystal for all low-*Z* elements when F and/or Na are also to be determined K *K*α crystal emission may give high background K *K*α escape peak with Ar-filled detector overlaps F *K*α pulse distribution

						Diffracted intensity may increase with age May give anomalous second peak with O $K\alpha$ See remarks for Pb stearate (*B19*, *H38*)
53*	CHLOR Clinochlore Hydrated Mg Al Fe silicate aluminate	(001) 28.392 Monoclinic	λ, Å: 28.39 $K\alpha$: 8-O $L\alpha$: 22-Ti	2.48–27.09 23-V to 8-O 58-Ce to 23-V	— — Mg, Al, Si, Fe K	O–P $K\alpha$ O $K\alpha$ ~4× KHP, but only ~0.2× Pb stearate High mechanical, thermal, and chemical stability Intensity relatively low at wavelengths $<\lambda$ O $K\alpha$ Gives intense higher orders with consequent spectral-line interference for multielement specimens (*B14*, *H38*)
54	PEN Penninite Hydrated Mg Al Fe silicate aluminate (similar to clinochlore)	(001) 28.4 Monoclinic	λ, Å: 28.4 $K\alpha$: 8-O $L\alpha$: 22-Ti	2.48–27.1 23-V to 8-O 58-Ce to 23-V	— — Mg, Al, Si, Fe K	Same applications as clinochlore (*B14*, *H38*)
55	— Potassium hydrogen cyclohexane-1,2-diacetate $KHC_{10}H_{14}O_4$ $KOOC\cdot CH_2\cdot C_6H_{10}\cdot CH_2COOH$ (cyclohexane ring) CH_2COOK, CH_2COOH	(?)[f] 31.2 (?)	λ, Å: 31.2 $K\alpha$: 8-O $L\alpha$: 22-Ti	2.721–29.76 22-Ti to 8-O 56-Ba to 22-Ti	Very high — K K	Good crystallinity

APPENDIX 10 (*continued*)

No.*	Symbol(s)[a,b] (if any) Common name(s) Chemical name(s) Chemical formula(s)	(*hkl*), (*hkil*) 2*d*, Å Crystal system	→	Longest wavelength and lowest atomic no. diffracted	Practical useful region (10–145° 2θ)	Reflectivity Resolution Secondary emission	Optimal application Remarks Reference(s)
56	— Tetradecanoamide $CH_3(CH_2)_{12}CONH_2$	(?)[f] ~54 (?)	λ, Å: $K\alpha$: $L\alpha$:	~54 6-C Any	4.71–51.5 17-Cl to 6-C $\lesssim$44-Ru	Very high — C, O *K*	Ultralong-wavelength region down to C $K\alpha$ Good crystallinity
57	HHM Hexadecyl hydrogen maleate Hexadecyl hydrogen *cis*-butenedioate[e] $CH_3(CH_2)_{15}OOC\cdot CH{:}CH\cdot COOH$	(?)[f] 58.0 (?)	λ, Å: $K\alpha$: $L\alpha$:	58.0 6-C Any	5.06–55.3 16-S to 6-C $\lesssim$43-Tc	— — C, O *K*	Ultralong-wavelength region down to C $K\alpha$ (*R33*)
58	OHM Octadecyl hydrogen maleate Octadecyl hydrogen *cis*-butenedioate[e] $CH_3(CH_2)_{17}OOC\cdot CH{:}CH\cdot COOH$	(?)[f] 63.5 (?)	λ, Å: $K\alpha$: $L\alpha$:	63.5 6-C Any	5.54–60.56 15-P to 6-C $\lesssim$41-Nb	— — C, O *K*	Ultralong-wavelength region down to C $K\alpha$ (*H38*, *R32*, *R33*)
59*	Llaur Lead laurate Lead dodecanoate[e] $[CH_3(CH_2)_{10}COO]_2Pb$	(LBF) ~70	λ, Å: $K\alpha$: $L\alpha$:	~70 5-B Any	6.10–66.8 15-P to 5-B $\lesssim$39-Y	— — C, O *K*; Pb *L*, *M*[i]	Ultralong-wavelength region down to B $K\alpha$ (*H38*)
60	BHM Behenyl hydrogen maleate	(?)[f] ~74	λ, Å: $K\alpha$:	~74 5-B	6.45–70.6 14-Si to 5-B	High —	Ultralong-wavelength region down to B $K\alpha$

	Docosyl hydrogen *cis*-butenedioate[e] $CH_3(CH_2)_{21}OOC \cdot CH{:}CH$ $\cdot COOH$	(?)	$L\alpha$:	Any	≥39-Y	C, O K	(*R33*)
61	LTD, Lmyr Lead myristate Lead tetradecanoate[e] $[CH_3(CH_2)_{12}COO]_2Pb$	(LBF) 80.5	λ, Å: $K\alpha$: $L\alpha$:	80.5 5-B Any	7.02–76.8 14-Si to 5-B ≲37-Rb	Medium — C, O K; Pb L, M[i]	Ultralong-wavelength region down to B $K\alpha$
62	OTO Dioctadecyl terephthalate $C_6H_4[COO(CH_2)_{17}CH_3]_2$ $COO(CH_2)_{17}CH_3$  $COO(CH_2)_{17}CH_3$	(?)[f] ~84 (?)	λ, Å: $K\alpha$: $L\alpha$:	~84 5-B Any	7.32–80.1 13-Al to 5-B ≲37-Rb	High — C, O K	Ultralong-wavelength region down to B $K\alpha$ Excellent crystallinity (*R33*)
63	Lpal Lead palmitate Lead hexadecanoate[e] $[CH_3(CH_2)_{14}COO]_2Pb$	(LBF) ~90	λ, Å: $K\alpha$: $L\alpha$:	~90 5-B Any	7.85-85.9 13-Al to 5-B ≲35-Br	— — C, O K; Pb L, M[i]	Ultralong-wavelength region down to B $K\alpha$
64	OAO Dioctadecyl adipate Dioctadecyl hexanedioate[e] $(CH_2)_4[COO(CH_2)_{17}CH_3]_2$	(?)[f] ~94 (?)	λ, Å: $K\alpha$: $L\alpha$:	~94 5-B Any	8.20–89.7 13-Al to 5-B ≲35-Br	Very high — C, O K	Ultralong-wavelength region down to B $K\alpha$ Excellent crystallinity (*B72*, *R33*)
65	OHS Octadecyl hydrogen succinate Octadecyl hydrogen butanedioate[e] $CH_3(CH_2)_{17}OOC \cdot CH_2$ $\cdot CH_2COOH$	(?)[f] ~97 (?)	λ, Å: $K\alpha$: $L\alpha$:	~97 5-B Any	8.46–92.5 12-Mg to 5-B ≲34-Se	— — C, O K	Ultralong-wavelength region down to B $K\alpha$ (*R33*)

APPENDIX 10 (*continued*)

No.*	Symbol(s)[a,b] (if any) Common name(s) Chemical name(s) Chemical formula(s)	(*hkl*), (*hkil*) 2*d*, Å Crystal system	→	Longest wavelength and lowest atomic no. diffracted	Practical useful region (10–145° 2θ)	Reflectivity Resolution Secondary emission	Optimal application Remarks Reference(s)
66*	LSD,[j] LOD[k] Lead stearate decanoate[l] Lead octadecanoate decanoate[e] $CH_3(CH_2)_{16}COO \cdot Pb \cdot OOC(CH_2)_8CH_3$	(LBF) ~100	λ, Å: $K\alpha$: $L\alpha$:	~100 5-B Any	8.72–95.4 12-Mg to 5-B ≲34-Se	— — C, O *K*; Pb *L*, *M*[i]	Ultralong-wavelength region down to B $K\alpha$ Diffracted intensity ~2× Pb stearate at wavelengths <44 Å
67*	LOD,[k] Lstear Lead stearate[l] Lead octadecanoate[e] $[CH_3(CH_2)_{16}COO]_2Pb$	(LBF) 100.4	λ, Å: $K\alpha$: $L\alpha$:	100.4 5-B Any	8.75–95.75 12-Mg to 5-B ≲34-Se	— — C, O *K*; Pb *L*, *M*[i]	Ultralong-wavelength region down to B $K\alpha$ LOD/KHP intensity ratios: slightly <1 for Al $K\alpha$, ~1 for Mg $K\alpha$, ~1.5 for Na $K\alpha$, ~2.5 for F $K\alpha$ (*E10*, *H38*)
68	Lara Lead arachidate Lead eicosanoate[e] $[CH_3(CH_2)_{18}COO]_2Pb$	(LBF) ~110	λ, Å: $K\alpha$: $L\alpha$:	~110 5-B Any	9.6–105 12-Mg to 5-B ≲33-As	— — C, O *K*; Pb *L*, *M*[i]	Ultralong-wavelength region down to B $K\alpha$
69	Lbeh Lead behenate Lead docosanoate[e] $[CH_3(CH_2)_{20}COO]_2Pb$	(LBF) ~120	λ, Å: $K\alpha$: $L\alpha$:	~120 4-Be Any	10.5–114 11-Na to 4-Be ≲32-Ge	— — C, O *K*; Pb *L*, *M*[i]	Ultralong-wavelength region down to Be $K\alpha$
70*	LTF, Llig	(LBF)	λ, Å:	~126	11.35–124	—	Ultralong-wavelength region down

	Lead lignocerate, lead	~120	$K\alpha$:	4-Be	11-Na to 4-Be	—	to Be $K\alpha$
	carnaubate		$L\alpha$:	Any	$\precsim$31-Ga	C, O K; Pb L, M[i]	($E10$, $H38$)
	Lead tetracosanoate[e]						
	$[CH_3(CH_2)_{22}COO]_2Pb$						
71	Lcer	(LBF)	λ, Å:	~140	12.2–134	—	Ultralong-wavelength region down
	Lead cerotate	~140	$K\alpha$:	4-Be	10-Ne to 4-Be	—	to Be $K\alpha$
	Lead hexacosanoate[e]		$L\alpha$:	Any	$\precsim$30-Zn	C, O K; Pb L, M[i]	
	$[CH_3(CH_2)_{24}COO]_2Pb$						
72*	LTC, Lmel	(LBF)	λ, Å:	~156	14.0–156	—	Ultralong-wavelength region down
	Lead melissate	~156	$K\alpha$:	4-Be	10-Ne to 4-Be	—	to Be $K\alpha$
	Lead triacontanoate[e]		$L\alpha$:	Any	$\precsim$28-Ni	C, O K; Pb L, M[i]	($H38$)
	$[CH_3(CH_2)_{28}COO]_2Pb$						

* Included in ASTM Table DS-37A (*86*).

[a] Many of these symbols are not generally agreed upon, and different manufacturers and authors may use different symbols for the same crystal.

[b] The parenthetic numbers are Miller indexes (hkl). In this column, Miller indexes for hexagonal and rhombohedral crystals ($hkil$) are given as (hkl), since $i = -(h + k)$; full ($hkil$) indexes are given in column 3. The other number is twice the interplanar spacing $2d$ in angstroms.

[c] This crystal is used in more than one "cut"; when reference is made to this crystal without specification of (hkl) or $2d$, it is likely to be this cut that is indicated.

[d] Si(111) and Ge(111) crystals eliminate all even orders such that $n = 4i + 2$, where $i = 0, 1, 2, \ldots$; thus, orders 2, 6, 10, etc. are eliminated, but not 4, 8, 12, etc.

[e] Preferred chemical nomenclature.

[f] A true organic crystal, as distinguished from a multilayer smectic soap film.

[g] For thallium, rubidium, or potassium *acid* phthalate, which is not preferred nomenclature.

[h] Sodium, cesium, and ammonium (NH_4^+) hydrogen phthalates have also been evaluated, but are less efficient than the potassium and rubidium salts. Also, CsHP would be very expensive. All these alkali hydrogen phthalates have substantially the same $2d$ spacing, ~26 Å. All except the ammonium salt may give crystal emission, depending on the wavelength being diffracted; for specific analyte lines, one or another of the crystals may be chosen to avoid this emission.

[i] Also possibly the spectral lines of elements in the substrate, usually mica or glass.

[j] Not to be confused with the hallucinogenic drug, which is *lys*ergic acid *d*iethylamide.

[k] LOD is commonly used for both *l*ead *o*cta*d*ecanoate (stearate) and *l*ead *o*ctadecanoate *d*ecanoate.

[l] The barium and strontium salts have substantially the same $2d$ spacing as the lead salt, but substantially lower diffracted intensity.

Alphabetical Index for Appendix 10

The number following each entry indicates the serial number (column 1) of that crystal in the table.

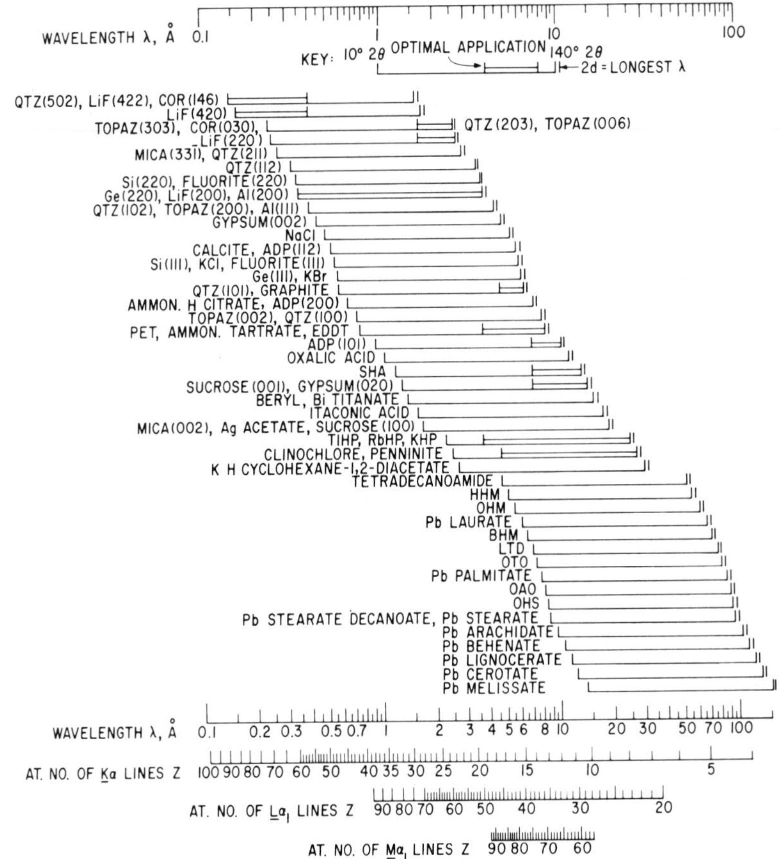

FIG. A10.1. Crystals and multilayer Langmuir–Blodgett films used as analyzers in wavelength-dispersive x-ray spectrometers.

APPENDIX 11A. Glossary of Frequently Used Notation

This glossary lists and defines most of the symbols used in this book, but some symbols used only once or twice are not listed. Of necessity, some symbols listed here, although used in the book generally with the meanings given, may also be used in specific cases with other meanings; of course, in all such cases, the definition is given at the place of use. In general, the notation of original literature is changed as required to be consistent with that used in this book. Insofar as possible, commonly accepted symbols are adopted in this book; notable exceptions are the retention of μ/ϱ for mass-absorption coefficient and a_{ij} for absorption-enhancement influence coefficient, rather than μ and α_{ij}, which are gaining increasing acceptance. Certain styles of notation are avoided; these are listed below with the reasons for their rejection. Also given are a few other guidelines followed in selection of the notation used in this book.

Notation such as $(\mu/\varrho)_Z{}^\lambda$ (mass-absorption coefficient of element Z for wavelength λ) and $C_i{}^S$ (concentration of element i in standard S) is not used; these symbols are written as $(\mu/\varrho)_{Z,\lambda}$ and C_{iS}, respectively. The superscript position following the symbol of a physical quantity (as distinguished from the symbol of a chemical element) should be reserved exclusively for mathematical exponents and never used to modify, specify, or restrict the unit; these functions are properly filled by the *sub*script. Mathematically, the notation $C_i{}^S$ indicates C_i to the power S. For the same reason, notation such as I^* (intensity from the standard) and I^0 (incident intensity) is not used; these symbols are written as I_S and I_0, respectively. However, this policy does not apply to primes (′, ″), which are used in this book.

Notation such as $(\mu/\varrho)_i(\lambda)$ (mass-absorption coefficient of element i for wavelength λ) and $I(\lambda_i)$ (intensity of the line of element i) is not used; these symbols are written as $(\mu/\varrho)_{i,\lambda}$ and I_{λ_i}, respectively. Mathematically, the notation $(\mu/\varrho)_i(\lambda)$ indicates the product of $(\mu/\varrho)_i \times \lambda$. Of course, this policy does not apply to terms of the form $f(x)$ (function of x).

In general, well-established and widely used symbols from other disciplines are retained, unless there is a compelling reason not to do so. Examples of such retained symbols are A (atomic weight), c (velocity of light), e (electronic charge), M (mass number), N (Avogadro number), Z (atomic number), and λ (wavelength). Of course, the international symbols and prefixes for physical units are strictly adopted. Conversely, in the disciplines of electrical engineering and electronics, E, I, and R signify electric potential, current, and resistance, respectively, and ordinarily, one would retain these symbols. However, in this book, E and I are reserved for (photon) energy and (x-ray) intensity, respectively, quantities that are basic to x-ray spectrochemical analysis and therefore used far more frequently than potential and current. Italic V ("voltage") is used for potential (roman V for the volt) and lower-case i for current. R is retained for resistance because this is a term used relatively infrequently in x-ray spectrochemical analysis, so there is little likelihood of its confusion with relative intensity, also designated by R.

In general, the same symbol is usually used for the same physical quantity, and specific applications are distinguished by subscripts. For example, the commonly used symbols T (counting time), $t_{1/2}$ (radioactive half-life), and τ (detector resolving time) are written in this book as T, $T_{1/2}$, and T_R, respectively.

Insofar as possible, symbols are chosen to be suggestive of the quantities they represent, for example, C for concentration, I for intensity, S for standard, and X for (unknown) sample.

Only upper- and lower-case roman, italic, and Greek letters are used. Boldface, sans-serif, script, and other less common type fonts are not used. Of course, ′, ″, %, °, etc. are used.

A comment is in order about the spelling of the word *x-ray*. Except for publications of the American Institute of Physics, which distinguish x ray (noun) and x-ray (adjective), most technical editors hyphenate the word as both noun and adjective. However, technical editors are about evenly divided on use of upper- or lower-case x. In the writer's opinion, the small x is clearly preferable for five reasons: (1) Current editions of *Webster's International* and other authoritative dictionaries list *x-ray* as the correct or preferred form. [Admittedly, this is not a particularly cogent reason, because standard (nontechnical) dictionaries are not necessarily authoritative sources for scientific and technical usage.] (2) The *Style Manual* of the American Institute of Physics specifies lower-case x. (3) The word is, after all, a *common* noun or an adjective derived from a common noun. (4) The x was taken by Roentgen from the mathematical x (unknown), which is usually lower-case. (5) Finally, lower-case x is consistent with the lower-case letters e, n, p, α, β, γ, etc., used for other types of radiation; in fact, there is much to be said for using lower-case x even in titles and as the first word of a sentence—as is done with these other symbols. In this book, the word is spelled "x-ray" as both noun and adjective, except in titles and as the first word of a sentence.

A .... Analyte, the element of analytical interest.
A .... Ampere, unit of electric current; derivatives: mA, μA.
A .... Atomic weight.
A .... Geometric factor, $\sin\phi/\sin\psi$ (Section 4.1.2).
Å .... Angstrom ($\equiv 10^{-10}$ m), unit of wavelength, based on the international meter.
$\overset{*}{\text{A}}$ .... Angstrom ($\sim 10^{-10}$ m), unit of wavelength, based on the W $K\alpha_1$ line (Section 1.4.2).
a_{ij} .... Matrix influence coefficient of element j on element i.

B .... Background.
b .... Intercept of linear calibration curve (I *vs.* C) on the intensity axis.

C .... Concentration (various units specified where used and by subscript, except that concentration in weight fraction or weight percent may be indicated by W).
C_{A} .... Concentration of analyte.
$C_{\text{A,M}}$ .... Concentration of analyte in matrix M.
C_{DL} .... Concentration at the detection limit.
C_{IS} .... Concentration of internal-standard element.
C_{S}, C_{X} .... Concentrations of specified element (usually analyte) in standard and sample, respectively.
C_i, C_j .... Concentrations of elements i and j, respectively.
Ci .... Curie, unit of radioactivity; derivatives: mCi, μCi.
c .... Velocity of light in vacuum (3×10^{10} cm/s).

$d, d_{(hkl)}$ .... Interplanar spacing of the (hkl) crystallographic planes (Å).

E .... Energy (keV, unless other unit indicated by subscript).
E_x .... X-ray photon energy (keV).
e .... Charge on the electron (4.8×10^{-10} esu).
eV .... Electron volt, unit of energy; derivatives: keV, MeV, GeV.

F .... Count fraction (Section 1.4.4).
F .... Fano factor (Section 6.5.1.12).
FWHM .... Full width at half maximum, or half-width, of a 2θ peak or pulse-height distribution; see also $\Delta V_{1/2}$, $W_{1/2}$.
f .... Frequency (of ac current or potential) (Hz, s^{-1}); see also ν.

g .... Gram, unit of mass; derivatives: mg, μg, ng.
g_L .... Probability factor, relative intensity of line L in its series [Equation (4.4)].

Hz................ Hertz, unit of frequency.
h .... Hour, unit of time.
h .... Planck's constant (6.6×10^{-27} erg·s).

I .... Intensity (counts/s).
I_A .... Intensity of analyte line.
I_B, I_L, I_P .... Intensities of the background, line, and peak, respectively: $I_L = I_P - I_B$.
I_{IS} .... Intensity of internal-standard line.
I_0 .... Incident intensity.
I_S, I_X .... Intensities of a line of a specified element (usually analyte) from standard and sample, respectively.
I_i, I_j .... Intensities of lines of elements i and j, respectively.
I_{iM}, I_{iS}, I_{iX} ..... Intensities of a line of element i from matrix, standard, and sample, respectively.
I_{int} .... Integrated intensity.
I_{measd} .... Measured intensity.
I_{rel} .... Relative intensity; not to be confused with R (Section 1.4.4).
I_{true} .... True intensity, corrected for coincidence loss.
I_{100i} .... Intensity of a line of element i from pure (100%) i.
I_λ.............. Intensity at wavelength λ.
i.................. Electric current (A, mA, μA).
i.................. Individual element, usually the analyte.
i.................. Individual value in a series of numbers or measurements.

$J(\lambda_{pri})$ .... Spectral distribution of the primary x-ray beam.
j .... Individual matrix element.

k .... General symbol for constants, defined where used.
kV(CP) .... Kilovolts, constant potential.
kV(P) .... Kilovolts, peak.

L .... X-ray spectral line.
L_{tgt} .... X-ray tube target line.

M .... Matrix.
M .... Atomic mass number.
m .... Meter, unit of length; derivatives: cm, mm, μm.
m .... Slope of linear calibration curve (I *vs.* C) (counts/s per unit concentration).
min .... Minute, unit of time.

N .... Avogadro number (6.02×10^{23} atoms/mol).
N .... Number of counts.
N_A, N_B, N_L,
N_{IS}, N_P, N_S, N_X . Have the same significance as the corresponding intensities.
$\bar{N}$ .... Average number of counts.
n .... Diffraction order [Equation (2.42)].
n .... General symbol for "number of" (specified where used).

P .... Excitation factor [Equation (4.8)].
P .... Peak.
pri .... (Subscript) Primary x-ray beam.
ppm .... Part per million, unit of concentration.

R .... Roentgen, unit of x-ray dose (Section 9.5); derivatives: MR, mR.
R .... Relative intensity (Section 1.4.4); not to be confused with I_{rel}.
R .... Electric resistance (Ω).
R .... Resolution.
R .... Radius of curved crystal.
R' .... Reciprocal of relative intensity, $1/R$.
r .... Absorption-edge jump ratio [Equation (2.27)].

S .... Standard; also specimen, when the specimen can be either a standard or sample.
s .... Second, unit of time.

T .... Counting time (s).
T_A, T_B, T_L,
T_{IS}, T_P, T_S, T_X .. Have the same significance as the corresponding intensities.
T_R .... Resolving time (s, μs).
t .... Thickness (cm, μm).
$t_{1/2}$ .... Half-thickness (cm) (Section 2.1.7).
t_∞ .... Infinite thickness (cm) (Section 14.7.2).
t .... Temperature (°C).

V .... Volt, unit of electric potential; derivative: kV.
V .... Electric potential (V, kV).
V_{ZK}, V_{ZLIII}, etc. . Critical excitation potential of the K, LIII, etc. series of element Z.
V_i .... Ionization potential (eV).
$\Delta V_{1/2}$ .... Half-width of a pulse-height distribution (V); same as FWHM, $W_{1/2}$.
v .... Statistical variance [Equation (11.7)].

W .... Watt, unit of electric power; derivatives mW, kW.
W .... Concentration (wt %, weight fraction).

$W_{1/2}$ .... Half-width of a 2θ peak (deg) or pulse-height distribution (V); same as FWHM, $\Delta V_{1/2}$.

w .... Weight (g, mg, μg).

X................. Sample, unknown.

xu .... X-unit, unit of x-ray wavelength; derivative: kxu.

Z .... Atomic number.

$\bar{Z}$ .... Mean atomic number.

δ .... Absorption-edge jump difference [Equation (2.28)].

ε .... Statistical coefficient of variation or relative standard deviation [Equation (11.8)].

ε_I, ε_N, ε_{FC}, ε_{FT}, ε_{FTO}........ Coefficients of variation of, respectively, an intensity, an accumulated count, and fixed-count, fixed-time, and optimal-fixed-time measurements.

θ .... Bragg angle [Equation (2.42)].

Λ .... Conversion factor, kxu to Å [Equation (1.8)].

Λ^* .... Conversion factor, $\overset{*}{\text{A}}$ to Å [Equation (1.10)].

λ .... Wavelength (Å; other units indicated by subscript).

λ_A .... Wavelength of the analyte line.

$\lambda_{I_{max}}$ .... Wavelength of maximum intensity in the continuum.

λ_L .... Wavelength of a line.

$\lambda_{ZK_{ab}}$, $\lambda_{ZL\text{III}_{ab}}$, etc. Wavelengths of the indicated absorption edges of element Z.

$\lambda_{ZK\alpha}$, $\lambda_{ZL\alpha_1}$, etc. . Wavelengths of the indicated spectral lines of element Z.

λ_{ab} .... Wavelength of an absorption edge.

λ_{eff} .... Effective wavelength.

λ_{min} .... Short-wavelength limit of the continuum.

λ_{pri} .... Wavelength of the primary x-rays.

μ .... Linear absorption coefficient (cm^{-1}) [Equation (2.5)].

μ_a .... Atomic absorption coefficient (cm^2/atom) [Equation (2.7)].

μ_{mol} .... Molar absorption coefficient (cm^2/mol) [Equation (2.8)].

μ/ϱ .... Mass-absorption coefficient (cm^2/g) [Equation (2.6)].

$(\mu/\varrho)_{Z,\lambda}$ .... Mass-absorption coefficient of element Z for wavelength λ.

$(\mu/\varrho)_{A,\lambda}$, $(\mu/\varrho)_{M,\lambda}$, $(\mu/\varrho)_{i,\lambda}$, $(\mu/\varrho)_{j,\lambda}$. Mass-absorption coefficient of, respectively, analyte, matrix, element *i*, and element *j* for wavelength λ.

$(\mu/\varrho)'$ .... Mass-absorption coefficient for both primary and analyte-line radiation [Equation (2.13)].

$\overline{(\mu/\varrho)}$............ Same as $(\mu/\varrho)'$ but with geometric correction [Equations (2.14) and (2.15)].

ν .... Frequency (of electromagnetic radiation) (Hz, s^{-1}); see also *f*.

π, π/ϱ .... Linear and mass pair-production coefficients (cm^{-1}, cm^2/g).

ϱ .... Density (g/cm^3).
ϱ_i' .... Density of element i in the specimen (g/cm^3) [Equation (15.35)].

σ, σ/ϱ .... Linear and mass scatter absorption coefficients (cm^{-1}, cm^2/g).
σ .... X-ray absorption cross section (cm^2/atom) (Section 2.1.7).
σ .... Statistical standard deviation [Equation (11.6)].
σ_N .... Standard counting error for N counts.
σ_{FC}, σ_{FT}, σ_{FTO} .. Standard counting errors for fixed-count, fixed-time, and optimal-fixed-time measurements, respectively.

τ, τ/ϱ .... Linear and mass photoelectric (true) absorption coefficients (cm^{-1}, cm^2/g).

ϕ .... Angle of incidence (Figure 4.1) between primary beam and specimen.

ψ .... Takeoff angle (Figure 4.1) between secondary beam and specimen.

Ω.................. Ohm, unit of electrical resistance.
ω .... Fluorescent yield.
ω_{ZK}, ω_{ZL}, etc....... Fluorescent yield of the indicated spectral-line series of element Z.

APPENDIX 11B. Prefixes for Physical Units

Prefix	Symbol	Factor
tera	T	10^{12}
giga	G	10^{9}
mega	M	10^{6}
kilo	k	10^{3}
centi	c	10^{-2}
milli	m	10^{-3}
micro	μ	10^{-6}
nano	n	10^{-9}
pico	p	10^{-12}
femto	f	10^{-15}
atto	a	10^{-18}

APPENDIX 12. Periodic Table of the Chemical Elements Showing Atomic Number Z and Wavelengths of Selected Emission Lines[a]

	IA	IIA	IIIB	IVB	VB	VIB	VIIB	VIII	VIII	VIII	IB	IIB	IIIA	IVA	VA	VIA	VIIA	O	
	H 1																	He 2	
$K\alpha$	Li 3 230.0	Be 4 113.0											B 5 67.00	C 6 44.00	N 7 31.60	O 8 23.71	F 9 18.31	Ne 10 14.61	$K\alpha$
$K\alpha$ $K\beta_1$	Na 11 11.91 11.62	Mg 12 9.889 9.570											Al 13 8.339 7.982	Si 14 7.126 6.778	P 15 6.155 5.804	S 16 5.373 5.032	Cl 17 4.728 4.403	Ar 18 4.192 3.886	$K\alpha$ $K\beta_1$
$K\alpha$ $K\beta_1$ $L\alpha$	K 19 3.742 3.454	Ca 20 3.360 3.090 36.39	Sc 21 3.032 2.780 31.39	Ti 22 2.750 2.514 27.45	V 23 2.505 2.284 24.31	Cr 24 2.291 2.085 21.71	Mn 25 2.103 1.910 19.49	Fe 26 1.937 1.756 17.60	Co 27 1.790 1.621 16.00	Ni 28 1.659 1.500 14.60	Cu 29 1.542 1.392 13.36	Zn 30 1.436 1.295 12.28	Ga 31 1.341 1.208 11.31	Ge 32 1.255 1.129 10.46	As 33 1.177 1.057 9.671	Se 34 1.106 0.992 8.990	Br 35 1.041 0.933 8.375	Kr 36 0.982 0.878 7.817	$K\alpha$ $K\beta_1$ $L\alpha$
$K\alpha$ $K\beta_1$ $L\alpha_1$	Rb 37 0.927 0.829 7.318	Sr 38 0.876 0.783 6.862	Y 39 0.830 0.741 6.448	Zr 40 0.787 0.702 6.070	Nb 41 0.748 0.666 5.724	Mo 42 0.711 0.632 5.406	Tc 43 0.676 0.601 5.114	Ru 44 0.644 0.572 4.846	Rh 45 0.615 0.546 4.597	Pd 46 0.587 0.520 4.368	Ag 47 0.561 0.497 4.154	Cd 48 0.536 0.475 3.956	In 49 0.514 0.454 3.772	Sn 50 0.492 0.435 3.600	Sb 51 0.472 0.417 3.439	Te 52 0.453 0.400 3.289	I 53 0.435 0.384 3.148	Xe 54 0.417 0.368 3.015	$K\alpha$ $K\beta_1$ $L\alpha_1$
$K\alpha$ $L\alpha_1$ $M\alpha$ N	Cs 55 0.402 2.892	Ba 56 0.386 2.775	La 57 0.372 2.665 14.88	Hf 72 0.224 1.570 7.539	Ta 73 0.217 1.522 7.252 58.20	W 74 0.211 1.476 6.983 55.80	Re 75 0.204 1.433 6.728	Os 76 0.198 1.391 6.490 51.90	Ir 77 0.193 1.351 6.262 50.20	Pt 78 0.187 1.313 6.046 48.10	Au 79 0.182 1.276 5.840 46.80	Hg 80 0.177 1.241 5.666 45.20	Tl 81 0.172 1.207 5.461 46.50	Pb 82 0.167 1.175 5.285 42.30	Bi 83 0.162 1.144 5.118 13.36	Po 84 0.158 1.114	At 85 0.154 1.085	Rn 86 0.150 1.057	$K\alpha$ $L\alpha_1$ $M\alpha$ N
$K\alpha$ $L\alpha_1$ $M\alpha$ N	Fr 87 0.146 1.030	Ra 88 0.142 1.005	Ac 89 0.138 0.980																
				→ $K\alpha$ $L\alpha_1$ $M\alpha$ N	Ce 58 0.359 2.561 14.06	Pr 59 0.346 2.463 13.34	Nd 60 0.340 2.370 12.68	Pm 61 0.322 2.283	Sm 62 0.310 2.199 11.43	Eu 63 0.300 2.120 10.95	Gd 64 0.290 2.046 10.41	Tb 65 0.280 1.975 9.937 86.00	Dy 66 0.271 1.909 9.543 83.00	Ho 67 0.262 1.845 9.161	Er 68 0.254 1.784 8.801 72.70	Tm 69 0.246 1.726 8.460	Yb 70 0.238 1.672 8.146 65.10	Lu 71 0.231 1.619 7.840 63.00	$K\alpha$ $L\alpha_1$ $M\alpha$ N
				→ $K\alpha$ $L\alpha_1$ $M\alpha$ N	Th 90 0.134 0.956 4.138 9.442	Pa 91 0.131 0.933 4.022	U 92 0.128 0.910 3.910 8.808	Np 93 0.125 0.889	Pu 94 0.122 0.869	Am 95 0.119 0.849	Cm 96 0.116 0.829	Bk 97 0.113 0.810	Cf 98 0.110 0.792	Es 99	Fm 100	Md 101	No 102	Lr 103	$K\alpha$ $L\alpha_1$ $M\alpha$ N

[a] After E. W. White, G. V Gibbs, G. G. Johnson, Jr., and G. R. Zechman, Jr., "X-Ray Emission Line Wavelength and Two-Theta Tables," *ASTM Data Series* **DS-37**, American Society for Testing and Materials (1965); courtesy of the authors and ASTM.

Bibliography

BOOKS

(*1*) Adler, I., "X-Ray Emission Spectrography in Geology," Elsevier Publishing Co., N. Y., 258 pp. (1966).

(*2*) American Society for Testing and Materials, "Energy-Dispersion X-Ray Analysis —X-Ray and Electron-Probe Analysis," *Spec. Tech. Publ.* **485**, 285 pp. (1971).

(*3*) American Society for Testing and Materials, "Symposium on Fluorescent X-Ray Spectrographic Analysis," *Spec. Tech. Publ.* **157**, 68 pp. (1954); Xerox copies available from University Microfilms, Inc.

(*4*) American Society for Testing and Materials, "Symposium on X-Ray and Electron-Probe Analysis," *Spec. Tech. Publ.* **349**, 209 pp. (1964).

(*5*) Azaroff, L. V., ed., "X-Ray Spectroscopy," McGraw-Hill Book Co., N. Y., 560 pp. (1974).

(*6*) Bertin, E. P., "Principles and Practice of X-Ray Spectrometric Analysis," ed. 2, Plenum Press, N. Y. 1071 pp. (1975).

(*7*) Birks, L. S., "Electron Probe Microanalysis," ed. 2, Interscience Publishers, N. Y., 190 pp. (1971).

(*8*) Birks, L. S., "X-Ray Spectrochemical Analysis," ed. 2, Interscience Publishers, N. Y., 143 pp. (1969).

(*9*) Blokhin, M. A., (translated from Russian by F. L. Curzon), "Methods of X-Ray Spectroscopic Research," Pergamon Press, N. Y., 448 pp. (1965).

(*10*) Blokhin, M. A., "Physics of X-Rays," State Publishing House of Technical-Theoretical Literature, Moscow (1957); *U. S. At. Energy Comm. Rep.* **AEC-TR-4502,** 429 pp. (1957).

(*11*) Castaing, R., P. Deschamps, and J. Philibert, eds., "Optique des Rayons X et Microanalyse"; Proceedings of the 4th International Symposium on X-Ray Optics and Microanalysis, Orsay, France (1965); Hermann et Cie., Paris, 707 pp. (1966).

(*12*) Clark, G. L., "Applied X-Rays," ed. 4, McGraw-Hill Book Co., N. Y., 843 pp. (1955).

(*13*) Compton, A. H., and S. K. Allison, "X-Rays in Theory and Experiment," D. Van Nostrand Co., N. Y., 828 pp. (1935).

(*14*) Cosslett, V. E., and W. C. Nixon, "X-Ray Microscopy," Cambridge University Press, Cambridge, Eng., 406 pp. (1960).

(*15*) Cosslett, V. E., A. Engström, and H. H. Pattee, eds., "X-Ray Miscroscopy and Microradiography"; Proceedings of the First International Symposium on X-Ray Microscopy and Microradiography, Cambridge, Eng. (1956); Academic Press, Inc., N. Y., 645 pp. (1957).

(*16*) Dewey, R. D., R. S. Mapes, and T. W. Reynolds, "Handbook of X-Ray and Microprobe Data"; Progress in Nuclear Energy; Series 9, Analytical Chemistry; vol. 9; Pergamon Press, N. Y., 353 pp. (1969).

(*17*) Elion, H. A., "Instrument and Chemical Analysis Aspects of Electron Microanalysis and Macroanalysis," Progress in Nuclear Energy; Series 9, Analytical Chemistry; vol. 5; Pergamon Press, N. Y., 256 pp. (1966).

(*18*) Engström, A., V. E. Cosslett, and H. H. Pattee, eds., "X-Ray Microscopy and X-Ray Microanalysis"; Proceedings of the Second International Symposium on X-Ray Microscopy and X-Ray Microanalysis, Stockholm (1960); Elsevier Publishing Co., N. Y., 542 pp. (1960).

(*19*) Glasser, O., "Dr. W. C. Röntgen," ed. 2, Charles C. Thomas, Publisher, Springfield, Ill., 169 pp. (1958).

(*20*) Jaffe, B., "Moseley and the Numbering of the Elements," Doubleday and Co., N. Y., 178 pp. (1971).*

(*21*) Jenkins, R., "Introduction to X-Ray Spectrometry," Heyden and Son, Ltd., London, 163 pp. (1974).

(*22*) Jenkins, R., and J. L. de Vries, "Practical X-Ray Spectrometry," ed. 2, Philips Technical Library, Springer-Verlag New York, N. Y., 190 pp. (1969).

(*23*) Jenkins, R., and J. L. de Vries, "Worked Examples in X-Ray Spectrometry," Philips Technical Library, Springer-Verlag New York, N. Y., 132 pp. (1970).

(*24*) Jenkins, R., R. W. Gould, and D. A. Gedke, "Quantitative X-Ray Spectrometry," Marcel Dekker, Inc., N. Y., in press (1975).

(*25*) Kaelble, E. F., ed., "Handbook of X-Rays," McGraw-Hill Book Co., N. Y. (1967); chapters pertinent to this book:

a) Adler, I., [X-Ray] Optics, ch. 31, 20 pp.

b) Bertin, E. P., and R. L. Samber, Contact Microradiography, ch. 45, 28 pp.

c) Birks, L. S., Electron-Probe Microanalysis, ch. 39, 12 pp.

d) Buhrke, V. E., Detection and Measurement of X-Rays, ch. 3, 46 pp.

e) Cave, W. T., Effects of Chemical State of Specimen, ch. 34, 3 pp.

f) Croke, J. F., and W. R. Kiley, Specimen-Preparation Techniques, ch. 33, 22 pp.

g) Dodd, C. G., Absorptiometry with Monochromatic X-Rays, ch. 42, 14 pp.

h) Gunn, E. L., Quantitative Techniques, ch. 36, 48 pp.

i) Holliday, J. E., Soft X-Ray Emission Spectroscopy in the 10-150-Å Region, ch. 38, 41 pp.

j) Johns, W. D., Measurement of Film Thickness, ch. 44, 12 pp.

k) Kaelble, E. F., Introduction to X-Ray Emission Spectroscopy, ch. 30, 5 pp.

l) Kemp, J. W., High-Speed Automatic Spectrometers and Continuous Analyzers, ch. 40, 8 pp.

* Another biography of Moseley was published while the writer's book was in press: J. L. Heilbron, "H. G. J. Moseley, the Life and Letters of an English Physicist, 1887-1915," University of California Press, Berkeley, 312 pp. (1974).

m) Lambert, M. C., Absorptiometry with Polychromatic X-Rays, ch. 41, 24 pp.
n) Loomis, T. C., and S. M. Vincent, Trace and Microanalysis, ch. 37, 16 pp.
o) Salmon, M. L., Factors which Determine Sensitivity, ch. 32, 8 pp.
p) Salmon, M. L., Qualitative and Semiquantitative Techniques, ch. 35, 18 pp.
q) Van Nordstrand, R. A., Fine Structure in Absorption-Edge Spectra, ch. 43, 18 pp.

(*26*) Liebhafsky, H. A., H. G. Pfeiffer, E. H. Winslow, and P. D. Zemany, "X-Ray Absorption and Emission in Analytical Chemistry," John Wiley and Sons, N. Y., 357 pp. (1960).

(*27*) Liebhafsky, H. A., H. G. Pfeiffer, E. H. Winslow, and P. D. Zemany, "X-Rays, Electrons, and Analytical Chemistry—Spectrochemical Analysis with X-Rays," Wiley-Interscience, N. Y., 566 pp. (1972).

(*28*) Möllenstedt, G., and K. H. Gaukler, eds., "Vth International Congress on X-Ray Optics and Microanalysis," Tübingen, Ger. (1968); Springer-Verlag, Berlin, 612 pp. (1969).

(*29*) Müller, R. O. (translated from German by K. Keil), "Spectrochemical Analysis by X-Ray Fluorescence," Plenum Press, N. Y., 326 pp. (1972).

(*30*) Müller, R. O., "Spektrochemische Analysen mit Röntgenfluoreszens," R. Oldenbourg, Munich, 315 pp. (1967); in German.

(*31*) Parrish, W., ed., "X-Ray Analysis Papers," Centrex Publishing Co., Eindhoven, Netherlands, 310 pp. (1965).

(*32*) Pattee, H. H., V. E., Cosslett, and A. Engström, eds. "X-Ray Optics and X-Ray Microanalysis"; Proceedings of the Third International Symposium on X-Ray Optics and X-Ray Microanalysis, Stanford University, Calif. (1962); Academic Press, N. Y., 622 pp. (1963).

(*33*) Polonio, J. Bermudez, "Teoria y Practica de la Espectroscopia de Rayos X," Editorial Alhambra, S. A., Madrid, 267 pp. (1967); in Spanish.

(*34*) Russ, J. C., "Elemental X-Ray Analysis of Materials EXAM Methods," EDAX Laboratories Div., EDAX International, Inc., Prairie View, Ill., 89 pp. (1972).

(*35*) Shinoda, G., K. Kohra, and T. Ichinokawa, eds., "Proceedings of the Sixth International Conference on X-Ray Optics and Microanalysis," Osaka, Jap. (1971); University of Tokyo Press, 908 pp. (1972).

(*36*) Siegbahn, M., (translated from German by G. A. Lindsay), "The Spectroscopy of X-Rays," Oxford University Press, London, 287 pp. (1925); Xerox copies available from University Microfilms, Inc.

(*37*) Sproull, W. T., "X-Rays in Practice," McGraw-Hill Book Co., N. Y., 615 pp. (1946).

(*38*) Von Hevesy, G., "Chemical Analysis by X-Rays and Its Applications," McGraw-Hill Book Co., N. Y., 333 pp. (1932); Xerox copies available from University Microfilms, Inc.

(*39*) Woldseth, R., "X-Ray Energy Spectrometry," Kevex Corp., Burlingame, Calif., 151 pp. (1973).

(*40*) Ziegler, C. A., ed.; "Applications of Low-Energy X- and Gamma-Rays," Gordon and Breach Science Publishers, N. Y., 467 pp. (1971).

(*41*) "Advances in X-Ray Analysis" (various editors), Proceedings of the Annual Conference on Applications of X-Ray Analysis, Denver; sponsored by Denver Research Institute, University of Denver; Plenum Press, N. Y., vols. 1–17 (1958–74).

(*42*) "Developments in Applied Spectroscopy" (various editors), Proceedings of the Annual Mid-America Spectroscopy Symposium, Chicago; Plenum Press, N. Y., vols. 1–10 (1962–72).

PERIODICALS

Advances in X-Ray Analysis (annual); Proceedings of the Annual Conference on Applications of X-Ray Analysis, Denver; Plenum Press, 227 West 17th St., New York, N. Y. 10011.

The Analyst (monthly); Society for Analytical Chemistry, 9/10 Savile Row, London W1X 1AF, England.

Analytica Chimica Acta (monthly); Elsevier Publishing Co. Inc., 52 Vanderbilt Av., New York, N. Y. 10017.

Analytical Abstracts (monthly); Society for Analytical Chemistry, 9/10 Savile Row, London W1X 1AF, England.

Analytical Chemistry (monthly); American Chemical Society, 1155 16th St. NW, Washington, D. C. 20036.

Applied Spectroscopy (bimonthly); Society for Applied Spectroscopy, 428 East Preston St., Baltimore, Md. 21202.

Canadian Spectroscopy (five issues per year); Spectroscopic Society of Canada; Multiscience Publications, Ltd., 1253 McGill College, Suite 404, Montreal 110, Quebec, Canada.

Chemical Abstracts (semimonthly); Chemical Abstracts Service, Ohio State University, Columbus, Ohio 43210; American Chemical Society, 1155 16th St. NW, Washington, D. C. 20036.

Fresnius' Zeitschrift fuer Analytische Chemie (approx. 10 issues per year); Springer-Verlag, 175 5th Av., New York, N. Y. 10010.

Japan Analyst (*Bunseki Kagaku*) (monthly); Japan Society for Analytical Chemistry, 1-1 Honmachi, Shibuya-ku, Tokyo, Japan.

Methodes Physiques d'Analyse (quarterly); Groupement pour l'Advancement des Methodes Physiques d'Analyse, 10 Rue de Delta, 75 Paris 9e, France.

Norelco Reporter (semiannual); Philips Electronic Instruments, 750 South Fulton Av., Mount Vernon, N. Y. 10550.

Siemens Review, Special Issues, "X-Ray and Electron Microscopy News" (approx. annual); Siemens Aktiengesellschaft; Siemens-America, Inc., 186 Wood Av. South, Iselin, N. J. 08830.

Spectrochimica Acta, Part B. Atomic Spectroscopy (monthly); Pergamon Press, Maxwell House, Fairview Park, Elmsford, N. Y. 10523.

X-Ray Fluorescence Spectrometry Abstracts (quarterly); Science and Technology Agency, 3 Harrington Road, South Kensington, London SW7 3ES, England.

X-Ray Spectrometry (quarterly); Heyden and Son, Ltd., Spectrum House, Alderton Crescent, London NW4 3XX, England.

Zeitschrift fuer Analytische Chemie; see *Fresnius' Zeitschrift fuer Analytische Chemie* above.

GENERAL REVIEWS

(*43*) Adler, I., and H. S. Rose, Jr., "X-Ray Emission Spectrography; in "Trace Analysis, Physical Methods," C. H. Morrison, ed.; Interscience Publishers, N. Y., ch. 8, pp. 271–324 (1965).

(*44*) Campbell, W. J., X-Ray Emission and Absorption; in "Standard Methods of Chemical Analysis," F. J. Welcher, ed.; D. Van Nostrand Co., N. Y., ed. 6, vol. 3, part A, ch. 9, pp. 169–185 (1966).

(*45*) Carr-Brion, K. G., and K. W. Payne, X-Ray Fluorescence Analysis, a Review, *Analyst (London)* **95**, 977–91 (1970).

(*46*) Hakkila, E. A., X-Ray Spectroscopic Methods; in "Modern Analytical Techniques for Metals and Alloys," Part 1; R. F. Bunshah, ed.; Techniques of Metals Research Series, vol. 3, part 1; Interscience Publishers, N. Y., pp. 275–323 (1970).

(*47*) Hall, T., X-Ray Spectroscopy; in "Trace Analysis," J. H. Yoe and H. J. Koch, eds.; John Wiley and Sons, N. Y., ch. 17, pp. 458–92 (1957).

(*48*) Jenkins, R., X-Ray Fluorescence Analysis; in "Analytical Chemistry—Part 2," T. S. West, ed.; vol. 13 of *Physical Chemistry, Series One*; *MTP* [Medical and Technical Publishing Co.] *International Review of Science*; Butterworths, London, and University Park Press, Baltimore; pp. 95–126 (1973).

(*49*) Jenkins, R., and J. L. de Vries, X-Ray Spectrometry [Metallurgical Application], *Met. Rev.* **16**, 125–41 (1971).

(*50*) Narbutt, K. I., and N. F. Losev, Progress in X-Ray Spectral Analysis in the Soviet Union, *Ind. Lab.* **33**, 1432–8 (1967).

(*51*) Sandström, A. E., Experimental Methods of X-Ray Spectroscopy, Ordinary Wavelengths; in "Handbuch der Physik," S. Flügge, ed.; Springer-Verlag, Berlin, vol. 30, pp. 78–245 (1957).

(*52*) Shalgosky, I., Fluorescent X-Ray Spectrography: in "Methods in Geochemistry," A. A. Smales and L. R. Wager, eds.; Interscience Publishers, N. Y., pp. 111–47 (1960).

(*53*) X-Ray Absorption and Emission, Annual (1949–52) and Biennial (1954–) Reviews, *Anal. Chem.*:

H. A. Liebhafsky, E. H. Winslow, and H. G. Pfeiffer, **21**, 17–24 (1949); **22**, 15–16 (1950); **23**, 14–16 (1951); **24**, 16–20 (1952); **26**, 26–31 (1954); **28**, 583–91 (1956); **30**, 580–9 (1958); **32**, 240R–248R (1960); **34**, 282R–294R (1962).

W. J. Campbell, J. D. Brown, J. V. Gilfrich, and J. W. Thatcher, **36**, 312R–328R (1964); **38**, 416R–439R (1966); **40**, 364R–375R (1968); **42**, 248R–268R (1970).

L. S. Birks and J. V. Gilfrinch, **44**, 557R–562R (1972); **46**, 360R–366R (1974).

See also (*3*), (*4*), and (*25*) above and (*L5*) and (*P11*) below.

BIBLIOGRAPHIES

(*54*) American Society for Testing and Materials, "Index to the Literature on X-Ray Spectrographic Analysis. Part 1. 1913–1957," *Spec. Tech. Publ.* **292**, 42 pp. (1961).

(*55*) N. V. Philips Gloeilampenfabrieken, Scientific Instruments Dept., "Review of Literature, X-Ray Spectrometry," ed. 2, 55 pp. (1965).

(*56*) Stoddart, H. A., and W. A. Dowden, "X-Ray Fluorescence Analysis: Part 1. Bibliography. Part 2. Indexes," *At. Energy Res. Estab. Gt. Brit., Rep.* **AERE-Bib 136**, 203 pp. (1966).

(*57*) Yakowitz, H., and J. R. Cuthill, "Annotated Bibliography on Soft X-Ray Spectroscopy," *Nat. Bur. Stand.* (*U.S.*) *Monogr.* **52**, 109 pp. (1962).

See also *X-Ray Fluorescence Spectrometry Abstracts* in Periodicals List above.

TABLES OF WAVELENGTHS, 2θ ANGLES, AND MASS-ABSORPTION COEFFICIENTS

(*58*) Amsbury, W. P., W. W. Lee, J. H. Rowan, and G. E. Walden, "X-Ray Fluorescence Tables," *U. S. At. Energy Comm. Rep.* **Y-1470** (1964); in 9 parts: A. Program Description, 47 pp.; B. Topaz, 116 pp.; C. LiF, 135 pp.; D. EDDT, 171 pp.; E. Gypsum, 187 pp.; F. Mica (2*d* 19.896 Å), 191 pp.; G. Mica (2*d* 19.96 Å), 192 pp.; H. KHP, 196 pp.; I. Ba Stearate, 206 pp.

(*59*) Bearden, J. A., "X-Ray Wavelengths," *U. S. At. Energy Comm. Rep.* **NYO-10586,** 533 pp. (1964).

(*60*) Bearden, J. A., X-Ray Wavelengths, *Rev. Mod. Phys.* **39**, 78–124 (1967); *Nat. Bur. Stand.* (*U. S.*), *Nat. Stand. Ref. Data Ser.* **NSRDS-NBS-14**, 66 pp. (1967).

(*61*) Berman, M., and S. Ergun, Angular Positions of X-Ray Emission Lines of the Elements for Common Analyzing Cıystals, *U. S. Bur. Mines Inf. Circ.* **8400**, 309 pp. (1969): Topaz, LiF, NaCl, Quartz(101), PET, EDDT, ADP, Gypsum, KHP.

(*62*) Bracewell, B. L., and W. J. Veigele, Tables of X-Ray Mass Attenuation Coefficients for 87 Elements at Selected Wavelengths, *Develop. Appl. Spectrosc.* **9**, 357–400 (1971).

(*63*) Cauchois, Y., and H. Hulubei, "Tables of Constants and Numerical Data. I. Selected Constants—Wavelengths of X-Ray Emissions and of X-Ray Absorption Discontinuities," Hermann & Cie., Paris, 200 pp. (1947).

(*64*) Dewey, R. D., R. S. Mapes, and T. W. Reynolds, "Computed X-Ray Wavelengths with Wavelength Tables," Reynolds Metals Co., Metallurgical Research Div., Richmond, Va., 198 pp. (1966).

(*65*) Dewey, R. D., R. S. Mapes, and T. W. Reynolds, "A Study of X-Ray Mass Absorption Coefficients with Tables of Coefficients," Reynolds Metals Co., Metallurgical Research Div., Richmond Va., 51 pp. (1967).

(*66*) Fine, S., and C. F. Hendee, Table of X-Ray *K* and *L* Emission and Critical Absorption Energies for All the Elements, *Nucleonics* **13**(3), 36–7 (1955); *Norelco Rep.* **3**, 113–15 (1956).

(*67*) Frazer, J. Z., Computer Fit to Mass-Absorption Coefficient Data, *Scripps Inst. Oceanogr. Rep.* **67-29**, 38 pp. (1967).

(*68*) General Electric Co., X-Ray Dept., "X-Ray Wavelengths for Spectrometer," Cat. no. A4961DA, ed. 5 (1969): ADP, EDDT, Ge, Gypsum, KHP, LiF(200), LiF(220), NaCl, PET, Quartz (203), (101), Si, Topaz; Pb myristate, Pb stearate.

(*69*) Heinrich, K. F. J., X-Ray Absorption Uncertainty [Including Table of Mass Absorption Coefficients]; in "The Electron Microprobe," E. D. McKinley, K. F. J. Heinrich, and D. B. Wittry, eds.; John Wiley and Sons, N. Y., pp. 296–377 (1966).

(*70*) Heinrich, K. F. J., and M. A. M. Giles, "X-Ray Wavelength Conversion Tables and Graphs for Qualitative Electron-Probe Microanalysis," *Nat. Bur. Stand.* (*U. S.*) *Tech. Note* **406**, 58 pp. (1967).

(*71*) Henke, B. L., and E. S. Ebisu, Low-Energy X-Ray and Electron Absorption within Solids (100–1500-eV Region), *Advan. X-Ray Anal.* **17**, 150–213 (1974).

(*72*) Henke, B. L., and R. L. Elgin, X-Ray Absorption Tables for the 2-to-200-Å Region, *Advan. X-Ray Anal.* **13**, 639–65 (1970).

(*73*) Henke, B. L., R. L. Elgin, R. E. Lent, and R. B. Ledingham, "X-Ray Absorption in the 2 to 200 Å Region," *U. S. Air Force, Office Sci. Res. Rep.* **AFOSR-67-1254,** 35 pp. (1967).

(*74*) Hubbell, J. H., Photon Cross Sections, Attenuation Coefficients, and Energy Absorption Coefficients from 10 keV to 100 GeV, *Nat. Bur. Stand. (U. S.), Nat. Stand. Ref Data Ser.* **NSRDS-NBS-29**, 80 pp. (1969).

(*75*) International Union of Crystallography, Absorption; in "International Tables for X-Ray Crystallography," Kynoch Press, Birmingham, Eng.; vol. 3, ch. 3.2, p. 157–200 (1962).

(*76*) Jenkins, R., Principal Emission Lines of the *M* Series, *X-Ray Spectrom.* **2**, 207–8 (1973).

(*77*) Johnson, G. G., Jr., and E. W. White, X-Ray Emission Wavelengths and keV Tables for Nondiffractive Analysis, *Amer. Soc. Test. Mater. Data Ser.* **DS-46**, 38 pp. (1970).

(*78*) McMaster, W. H., N. K. Del Grande, J. H. Mallett, and J. H. Hubbell, Compilation of X-Ray Cross Sections, Sec. 2, Rev. 1, Univ. of Calif., Livermore, *U. S. At. Energy Comm. Rep.* **UCRL-50174**, 350 pp. (1969).

(*79*) N. V. Philips Gloeilampenfabrieken, Application Lab., "Table of X-Ray Mass Absorption Coefficients," *Norelco Rep.* **9**(3) (1962).

(*80*) Powers, M. C., "X-Ray Fluorescent Spectrometer Conversion Tables," Philips Electronic Instruments, Inc., Mt. Vernon, N. Y. (1957): ADP, EDDT, LiF, NaCl, Topaz.

(*81*) Robinson, J. W., ed., "CRC Handbook of Spectroscopy", Vol. 1, CRC Press, Div. of Chemical Rubber Co., Cleveland, O., 913 pp. (1974).

(*82*) Sagel, K., "Tabellen zur Röntgen Emissions und Absorptions Analyse," Springer-Verlag, Berlin, 135 pp. (1959).

(*83*) Stainer, H. M., "X-Ray Mass Absorption Coefficients—A Literature Survey," *U. S. Bur. Mines Inf. Circ.* **8166**, 124 pp. (1963).

(*84*) Veigele, W. J., E. Briggs, B. Bracewell, and M. Donaldson, X-Ray Cross-Section Compilation, *Kaman. Nucl. Corp. Rep.* **KN-789-69-2(R)** (1969).

(*85*) White, E. W., and G. G. Johnson, Jr., X-Ray Emission and Absorption-Edge Wavelengths and Interchange Settings for LiF-Geared Curved-Crystal Spectrometers, Mineral Industries Exp. Sta., Pa. State Univ., University Park, Pa., *Spec. Publ.* **1-70**, 168 pp. (1970).

(*86*) White, E. W., and G. G. Johnson, Jr., X-Ray Emission and Absorption Wavelengths and Two-Theta Tables, *Amer. Soc. Test. Mater. Data Ser.* **DS-37A**, ed. 2, 293 pp. (1970): ADP(101), (112), (200); Clinochlore; EDDT; Ge; Graphite; Gypsum; KHP; LiF(200), (220); Mica(002); NaCl; Pb laurate, lignocerate, melissate, stearate; PET; Quartz(100), (101), (203); RHP; Si, Topaz.

(*87*) White, E. W., G. V. Gibbs, G. G. Johnson, Jr., and G. R. Zechman Jr., "X-Ray Emission Line Wavelength and Two-Theta Tables," *Amer. Soc. Test. Mater. Data Ser.* **DS-37** (1965): ADP, $CaCO_3$, EDDT, Gypsum, KHP, Lead stearate decanoate, LiF, Mica, NaCl, PET, Quart (101).

(*88*) White, E. W., G. V. Gibbs, G. G. Johnson, Jr., and G. R. Zechman, Jr., "X-Ray Wavelengths and Crystal Interchange Settings for Wavelength-Geared Curved-Crystal Spectrometers," Mineral Industries Exp. Sta., Pa. State Univ., University Park, Pa., *Spec. Publ.* **3-64**, ed. 2 (1965).

(*89*) Zingaro, P. W., X-Ray Spectrographic Data [Emission Lines and Absorption Edges]. *Norelco Rep.*: I. *K*-Series, **1**, 67–8 (1953); II. *L*-Series, **1**, 70, 78–9 (1953); III. X-Ray Emission Lines of the Elements Arranged in Order of Their Wavelengths, **2**, 92–5 (1955).

See also Appendixes 2–5 in (*27*).

PAPERS AND REPORTS

(*A1*) Achey, F. A., and E. J. Serfass, A New Approach to Coating Thickness by Fluorescent X-Ray Absorption, *J. Electrochem. Soc.* **105**, 204–5 (1958).

(*A2*) Adam, J., X-Ray Spectra Calculator, *X-Ray Spectrom.* **1**, 49–50 (1972).

(*A3*) Addink, N. W. H., Note on Analysis of Small Quantities of Material by X-Ray Fluorescence, *Rev. Universelle Mines* **15**, 530–2 (1959).

(*A4*) Addink, N. W. H., H. Kraay, and A. W. Witmer, Putting to Advantage the Absorbing Qualities of Diluting Agents for Obtaining 45-Degree Calibration Curves in X-Ray Fluorescence Analysis, *Colloq. Spectrosc. Int., 9th, Lyons, 1961* **3**, 368–84 (1962).

(*A5*) Adler, I., Intermediate X-Ray Probe; in "Encyclopedia of X-Rays and Gamma Rays," G. L. Clark, ed.; Reinhold Publishing Corp., N. Y., pp. 834–40 (1963).

(*A6*) Adler, I., and J. M. Axelrod, The Curved-Crystal X-Ray Spectrometer—A Mineralogical Tool, *Amer. Mineral.* **41**, 524–7 (1956).

(*A7*) Adler, I., and J. M. Axelrod, Aluminum Powder as a Binder in Sample Preparation for X-Ray Spectrometry, *Anal. Chem.* **26**, 931–2 (1954).

(*A8*) Adler, I., and J. M. Axelrod, Determination of Thorium by Fluorescent X-Ray Spectrometry, *Anal. Chem.* **27**, 1002–3 (1955).

(*A9*) Adler, I., and J. M. Axelrod, The Reflecting Curved-Crystal X-Ray Spectrograph —A Device for the Analysis of Small Mineral Samples, *Econ. Geol.* **52**, 694–701 (1957).

(*A10*) Adler, I., and J. M. Axelrod, A Multi-Wavelength Fluorescence Spectrometer, *J. Opt. Soc. Amer.* **43**, 769–72 (1953).

(*A11*) Adler, I., and J. M. Axelrod, Internal Standards in Fluorescent X-Ray Spectroscopy, *Spectrochim. Acta* **7**, 91–9 (1955).

(*A12*) Adler, I., J. M. Axelrod, and J. R. Branco, Further Application of the Intermediate X-Ray Probe, *Advan. X-Ray Anal.* **2**, 167–73 (1959).

(*A13*) Adler, I., J. I. Trombka, and P. Gorenstein, Remote Chemical Analysis during the Apollo 15 Mission, *Anal. Chem.* **44**(3), 28A–35A (1972).

(*A14*) Alley, B. J., and J. H. Higgins, Empirical Corrections for Variable Absorption of Soft X-Rays by Mylar, *Norelco Rep.* **10**, 77–80 (1963).

(*A15*) Alley, B. J., and R. H. Meyers, Corrections for Matrix Effects in X-Ray Fluorescence Analysis Using Multiple Regression Methods, *Anal. Chem.* **37**, 1685–90 (1965).

(*A16*) American National Standards Institute, American National Standard: Radiation Safety for X-Ray Diffraction and Fluorescence Analysis Equipment, *Nat. Bur. Stand.* (*U. S.*) *Handb.* **111**, 20 pp. (1972).

(*A17*) Anater, T. F., Semiquantitative Analysis of Unknown Samples by X-Ray Fluorescence Spectrography, *U. S. At. Energy Comm. Rep.* **WAPD-321**, 25 pp. (1968).

(*A18*) Andermann, G., and J. D. Allen, X-Ray Emission Analysis of Finished Cements, *Anal. Chem.* **33**, 1695–9 (1961).

(*A19*) Andermann, G., and J. W. Kemp, Scattered X-Rays as Internal Standards in X-Ray Emission Spectroscopy, *Anal. Chem.* **30**, 1306–9 (1958).

(*A20*) Andersen, C. A., An Introduction to the Electron-Probe Microanalyzer and Its Applications to Biochemistry; in "Methods of Biochemical Analysis," vol. 15, D. Glick, ed.; Interscience Publishers, N. Y., pp. 147–270 (1967).

(*A21*) Applied Research Laboratories, Inc., X-Ray Analysis of Metallic Elements in Nonmetallic Material, *ARL Spectrogr. News Lett.* **7**(3), 1–3 (1954).

(*A22*) Arndt, U. W., W. A. Coates, and A. R. Crathorn, Gas-Flow X-Ray Diffraction Counter, *Proc. Phys. Soc. (London)* **67B**, 357–9 (1954).

(*A23*) Ashby, W. D., and E. M. Proctor, Measurement of X-Ray Source Stability, *Picker Analyzer* **4**(1), 1–8 (1965).

(*A24*) Ashley, R. W., and R. W. Jones, Application of X-Ray Fluorescence Methods to the Analysis of Zircaloy, *Anal. Chem.* **31**, 1632–5 (1959).

(*A25*) Auger, P., Compound Photoelectric Effect, *Ann. Phys.* **6**, 183–253 (1926).

(*A26*) Auger, P., Secondary β-Rays Produced in O Gas by X-Rays, *C. R. Acad. Sci. (Paris)* **180**, 65–8 (1925).

(*A27*) Auger, P., Compound Photoelectric Effect, *J. Phys. Radium* **6**, 205–8 (1925).

(*A28*) Axelrod, J. M., and I. Adler, X-Ray Spectrographic Determination of Cesium and Rubidium, *Anal. Chem.* **29**, 1280–1 (1957).

(*B1*) Backerud, L., An Evaluation of the Suitability of X-Ray Fluorescence Spectroscopy in the Analysis of Complex Alloy Systems, *X-Ray Spectrom.* **1**, 3–14 (1972).

(*B2*) Baird, A. K., A Pressed-Specimen Die for the Norelco Vacuum-Path X-Ray Spectrograph, *Norelco Rep.* **8**, 108–9 (1961).

(*B3*) Baldwin, J. M., *Q*-Switched Laser Sampling of Copper–Zinc Alloys, *Appl. Spectrosc.* **24**, 429–35 (1970).

(*B4*) Bambynek, W., B. Crasemann, R. W. Fink, H. U. Freund, H. Mark, C. D. Swift, R. E. Price, and P. V. Rao, X-Ray Fluorescence Yields, Auger, and Coster–Kronig Transition Probabilities, *Rev. Mod. Phys.* **44**, 716–813 (1972).

(*B5*) Banks, M., and E. T. Hall, X-Ray Fluorescent Analysis in Archaeology—The "Milliprobe," *Archaeometry* **6**, 31–6 (1963).

(*B6*) Barieau, R. E., X-Ray Absorption-Edge Spectrometry as an Analytical Tool, *Anal. Chem.* **29**, 348–52 (1957).

(*B7*) Barkla, C. G., Spectra of the Fluorescent Röntgen Radiation, *Phil. Mag.* [6] **22**, 396–412 (1911).

(*B8*) Barkla, C. G., and C. A. Sadler, Absorption of Roentgen Rays, *Phil. Mag.* [6] **17**, 739–60 (1909).

(*B9*) Barnhisel, R. I., W. R. Phillippe, and R. L. Blevins, Simple X-Ray Fluorescence Technique for Determination of Iron and Manganese in Soils and Concretions, *Soil Sci. Soc. Amer. Proc.* **33**, 811–13 (1969).

(*B10*) Barstad, G. E. B., and I. N. Refsdal, Sensitive Quantitative Recording X-Ray Spectrometer, *Rev. Sci. Instrum.* **29**, 343–8 (1958).

(*B11*) Batt, A., Use of a Calculation Method for X-Ray Emission Analysis, *Appl. Spectrosc.* **23**, 346–53 (1969).

(*B12*) Baumgartner, W. V., X-Ray Spectrometry Extends Film-Badge Dosimetry, *Nucleonics* **18**(8), 76, 78–9 (1960).

(*B13*) Baun, W. L., Use for X-Ray Spectrometry of the (200) Reflection of Ammonium Dihydrogen Phosphate, *Anal. Chem.* **41**, 830–1 (1969).

(*B14*) Baun, W. L., and E. W. White, Clinochlore—a Versatile New Analyzing Crystal for the X-Ray Region 5–27 Å, *Anal. Chem.* **41**, 831–2 (1969).

(*B15*) Beaman, D. R., and J. A. Isasi, Critical Examination of Computer Programs Used in Quantitative Electron Microprobe Analysis, *Anal. Chem.* **42**, 1540–68 (1970).

(*B16*) Beaman, D. R., and J. A. Isasi, Electron-Beam Microanalysis, *Mater. Res. Stand.* **11**(11), 8–35, 45, 48–68, 70–1, 73–5, 77–8; (12) 12–31, 51–6 (1971); *Amer. Soc. Test. Mater. Spec. Tech. Publ.* **STP-506**, 80 pp. (1972).

(*B17*) Bean, L., Method of Producing Sturdy Specimens of Pressed Powders for Use in X-Ray Spectrochemical Analysis, *Appl. Spectrosc.* **20**, 191–3 (1966).

(*B18*) Beard, D. W., and E. M. Proctor, A Method of Liquid Analysis Providing Increased Sensitivity for Light Elements, *Advan. X-Ray Anal.* **10**, 506–19 (1967).

(*B19*) Bearden, A. J., and F. N. Huffman, Precision Measurement of the Cleavage Plane Grating Spacing of Potassium Acid Phthalate, *Rev. Sci. Instrum.* **34**, 1233–4 (1963).

(*B20*) Beattie, H. J., and R. M. Brissey, Calibration Method for X-Ray Fluorescence Spectrometry, *Anal. Chem.* **26**, 980–3 (1954).

(*B21*) Beatty, R. T., Energy of Roentgen Rays, *Proc. Roy Soc.* (*London*) (*A*)**89**, 314–27 (1913).

(*B22*) Bedo, D. E., and D. H. Tomboulian, Photon Counting Spectrometer for Attenuation Measurements in the Soft X-Ray Region, *Rev. Sci. Instrum.* **32**, 184–8 (1961).

(*B23*) Beeghley, H. F., X-Ray Method for Determining Tin Coating Thickness, *J. Electrochem. Soc.* **97**, 152–7 (1950).

(*B24*) Beezhold, W., Ion-Induced Characteristic X-Ray Microanalysis; in "Semiconductor Silicon 1973" (Proc. Int. Symp. Silicon Mater. Sci. Technol., Second), H. R. Huff and R. R. Burgess, eds.; Electrochemical Society, Princeton, N. J., pp. 437–47 (1973).

(*B25*) Behr, F. A., and P. W. Zingaro, Notes on the Selection of Analyzing Crystals in X-Ray Spectrography, *Norelco Rep.* **1**, 26–8 1(953).

(*B26*) Belt, R. F., Synthetic Corundum as an X-Ray Monochromator, *X-Ray Spectrom.* **1**, 55–8 (1972).

(*B27*) Bernstein, F., Application of X-Ray Fluorescence Analysis to Process Control, *Advan. X-Ray Anal.* **5**, 486–99 (1962).

(*B28*) Bernstein, F., Particle-Size and Mineralogical Effects in Mining Applications, *Advan. X-Ray Anal.* **6**, 436–45 (1963).

(*B29*) Bernstein, F., Mineralogical Problems in X-Ray Emission Analysis of Sylvite Concentrates, *Advan. X-Ray Anal.* **7**, 555–65 (1964).

(*B30*) Bernstein, F., Relationship Between X-Ray Tube Target Materials and X-Ray Emission Intensities, *Develop. Appl. Spectrosc.* **4**, 45–55 (1965); *ISA* (*Instrum. Soc. Amer.*) *Trans.* **6**, 99–102 (1967).

(*B31*) Berry, P. F., Study of Particulate Heterogeneity Effects in Radioisotope X-Ray Spectrometry, *U. S. At. Energy Comm. Rep.* **ORO-3847-2**, 107 pp. (1971).

(*B32*) Bertin, E. P., Evaluation and Application of an Improved Slit Probe for the X-Ray Secondary-Emission Spectrometer, *Advan. X-Ray Anal.* **8**, 231–47 (1965).

(*B33*) Bertin, E. P., Comparison of Four Slit Apertures for Selected-Area Analysis with the X-Ray Secondary-Emission Spectrometer, *Advan. X-Ray Anal.* **10**, 462–73 (1967).

(*B34*) Bertin, E. P., Recent Advances in Quantitative X-Ray Spectrometric Analysis by Solution Techniques, *Advan. X-Ray Anal.* **11**, 1–22 (1968).

(*B35*) Bertin, E. P., X-Ray Probe with Slit Aperture in the Secondary Beam, *Anal. Chem.* **36**, 441–3 (1964).

(*B36*) Bertin, E. P., Intensity-Ratio Technique for X-Ray Spectrometric Analysis of Binary Samples, *Anal. Chem.* **36**, 826–32 (1964).

(*B37*) Bertin, E. P., Solution Techniques in X-Ray Spectrometric Analysis, *Norelco Rep.* **12**, 15–26 (1965).

(*B38*) Bertin, E. P., and R. J. Longobucco, Some Special Sample Mounting Devices for the X-Ray Fluorescence Spectrometer, *Advan. X-Ray Anal.* **5**, 447–56 (1962).

(*B39*) Bertin, E. P., and R. J. Longobucco, X-Ray Spectrometric Determination of Composition and Distribution of Sublimates in Receiving-Type Electron Tubes, *Advan. X-Ray Anal.* **7**, 566–83 (1964).

(*B40*) Bertin, E. P., and R. J. Longobucco, X-Ray Spectrometric Determination of Plate Metals on Plated Wires, *Anal. Chem.* **34**, 804–11 (1962).

(*B41*) Bertin, E. P., and R. J. Longobucco, X-Ray Spectrometric Analysis—Nickel, Copper, Silver, and Gold in Plating Baths, *Metal Finish.* **60**(3), 54–8 (1962).

(*B42*) Bertin, E. P., and R. J. Longobucco, X-Ray Methods for Determination of Plate Thickness, *Metal Finish.* **60**(8), 42–4 (1962).

(*B43*) Bertin, E. P., and R. J. Longobucco, Sample Preparation Methods for X-Ray Fluorescence Emission Spectrometry, *Norelco Rep.* **9**, 31–43 (1962).

(*B44*) Bertin, E. P., and R. J. Longobucco, Spectral-Line Interference in X-Ray Fluorescence Spectrometry, *Norelco Rep.* **9**, 64–72 (1962).

(*B45*) Bertin, E. P., R. J. Longobucco, and R. J. Carver, Simplified Routine Method for X-Ray Absorption-Edge Spectrometric Analysis, *Anal. Chem.* **36**, 641–55 (1964).

(*B46*) Bessen, I. I., Principles and Applications of Proportional Counters for X-Ray Spectrographs, *Advan. X-Ray Anal.* **1**, 455–68 (1958).

(*B47*) Betteridge, D., and A. D. Baker, Analytical Potential of Photoelectron Spectroscopy, *Anal. Chem.* **42**(1), 43A–44A, 46A, 48A, 50A, 52A, 54A, 56A (1970).

(*B48*) Bigelow, J. E., X-Ray Collimator Comprising a Plurality of Spaced Plastic Laminae with X-Ray Absorbent Material Coated Thereon, *U. S. Pat.* **3,143,652** (1964).

(*B49*) Birks, L. S., Apparatus for Vacuum X-Ray Fluorescence Analysis of Light Elements, *Rev. Sci. Instrum.* **22**, 891–1004 (1951).

(*B50*) Birks, L. S., Fluorescent X-Ray Excitation Efficiencies, *Spectrochim. Acta* **17**, 148–54 (1961).

(*B51*) Birks, L. S., Convex Curved-Crystal X-Ray Spectrograph, *Rev. Sci. Instrum.* **41**, 1129–32 (1970).

(*B52*) Birks, L. S., and A. P. Batt, Use of a Multichannel Analyzer for Electron-Probe Microanalysis, *Anal. Chem.* **35**, 778–82 (1963).

(*B53*) Birks, L. S., and E. J. Brooks, Hafnium–Zirconium and Tantalium–Columbium Systems—Quantitative Analysis by X-Ray Fluorescence, *Anal. Chem.* **22**, 1017–20 (1950).

(*B54*) Birks, L. S., and E. J. Brooks, Analysis of Uranium Solutions by X-Ray Fluorescence, *Anal. Chem.* **23**, 707–9 (1951).

(*B55*) Birks, L. S., and E. J. Brooks, Applications of Curved-Crystal X-Ray Spectrometers, *Anal. Chem.* **27**, 437–40 (1955).

(*B56*) Birks, L. S., and E. J. Brooks, Miniature Fluorescence X-Ray Spectrograph, *Anal. Chem.* **27**, 1147–9 (1955).

(*B57*) Birks, L. S., and D. M. Brown, Precision in X-Ray Spectrochemical Analysis—Fixed Time *vs.* Fixed Count, *Anal. Chem.* **34**, 240–1 (1962).

(*B58*) Birks, L. S., and J. V. Gilfrich, Development of X-Ray Fluorescence Spectroscopy for Elemental Analysis of Particulate Matter in the Atmosphere and in Source Emissions. II. Evaluation of Commercial Multiple-Crystal Spectrometer Instruments, *U. S. Nav. Res. Lab. Rep.* **7617**, 7 pp. (1973).

(*B59*) Birks, L. S., and D. L. Harris, Unusual Matrix Effects in Fluorescent X-Ray Spectrometry, *Anal. Chem.* **34**, 943–5 (1962).

(*B60*) Birks, L. S., and R. T. Seal, X-Ray Properties of Plastically Deformed LiF, *J. Appl. Phys.* **28**, 541–3 (1957).

(*B61*) Birks, L. S., and J. M. Siomkajlo, Long-Spacing Metal-Organic Crystals for X-Ray Spectrometry, *Spectrochim. Acta* **18**, 363–6 (1962).

(*B62*) Birks, L. S., E. J. Brooks, and H. Friedman, Fluorescent X-Ray Spectroscopy, *Anal. Chem.* **25**, 692–7 (1953).

(*B63*) Birks, L. S., E. J. Brooks, and G. W. Gourlay, Compact Curved-Crystal X-Ray Spectrometer, *Rev. Sci. Instrum.* **29**, 425–6 (1958).

(*B64*) Birks, L. S., E. J. Brooks, H. Friedman, and R. M. Roe, X-Ray Fluorescence Analysis of Ethyl Fluid in Aviation Gasoline, *Anal. Chem.* **22**, 1258–61 (1950).

(*B65*) Birks, L. S., R. E. Seebold, A. P. Batt, and J. S. Grosso, Excitation of Characteristic X-Rays by Protons, Electrons, and Primary X-Rays, *J. Appl. Phys.* **35**, 2578–81 (1964).

(*B66*) Blank, G. R., and H. A. Heller, X-Ray Spectrometric Determination of Copper, Tin, and Uranium in Bronze Heat-Treating Material, *Advan. X-Ray Anal.* **4**, 457–73 (1961).

(*B67*) Blank, G. R., and H. A. Heller, X-Ray Spectrometric Determination of Thorium in Refined Uranium Materials, *Norelco Rep.* **8**, 112–15 (1961); *Develop. Appl. Spectrosc.* **1**, 3–15 (1962).

(*B68*) Blank, G. R., and H. A. Heller, X-Ray Spectrometric Determination of Zirconium in Refined Uranium Materials, *Norelco Rep.* **9**, 23–7 (1962).

(*B69*) Blavier, P., H. Hans, P. Tyou, and I. Houbart, Elemental Analysis of Alnico-Type Alloys by X-Ray Fluorescence, *Cobalt* **1960**(7), 33–7; *Chem. Abstr.* **54**, 13952h (1960).

(*B70*) Blodgett, K. B., Films Built by Depositing Successive Unimolecular Layers on a Solid Surface, *J. Amer. Chem. Soc.* **57**, 1007–22 (1935).

(*B71*) Blount, C. W., D. E. Leyden, T. L. Thomas, and S. M. Guill, Application of Chelating Ion-Exchange Resins for Trace-Element Analysis of Geological Samples Using X-Ray Fluorescence, *Anal. Chem.* **45**, 1045–50 (1973).

(*B72*) Boehm, G., and K. Ulmer, Growth and Testing of Spectrometer Crystals for the Soft X-Ray Region (10–100 Å), *Z. Angew. Phys.* **29**, 287–91 (1970).

(*B73*) Bonar, L. G., and R. W. Strong, X-Ray Fluorescence Analysis in a Steel Foundry, *J. Metals* **18**, 1323–30 (1966).

(*B74*) Booker, G. R., and J. Norbury, Extraction Replica Method for Large Precipitates and Nonmetallic Inclusions in Steel, *Brit. J. Appl. Phys.* **8**, 109–13 (1957).

(*B75*) Bowman, H. R., E. K. Hyde, S. G. Thompson, and R. C. Jared, Application of High-Resolution Semiconductor Detectors in X-Ray Emission Spectrography, *Science*, **151**, 562–8 (1966).

(*B76*) Boyd, B. R., and H. T. Dryer, Analysis of Nonmetallics by X-Ray Fluorescence Techniques, *Develop. Appl. Spectrosc.* **2**, 335–49 (1963).

(*B77*) Brachfeld, B., Simple, Fast Technique for Sample Preparation of Composite Metal Powders for Analysis by X-Ray Fluorescence, *Appl. Spectrosc.* **27**, 289 (1973).

(*B78*) Bragg, W. H., and W. L. Bragg, Reflection of X-Rays by Crystals, *Proc. Phys. Soc.* (*London*) **A88**, 428–38 (1913).

(*B79*) Bragg, W. L., Diffraction of Short Electromagnetic Waves by a Crystal, *Proc. Cambridge Phil. Soc.* **17**, 43–57 (1912).

(*B80*) Brandt, W. and R. Laubert, Pauli Excitation of Atoms in Collision, *Phys. Rev. Lett.* **24**, 1037–40 (1970).

(*B81*) Brentano, J. C. M., and I. Ladany, Scintillation Counting of Low-Energy Photons and Its Use in X-Ray Diffraction, *Rev. Sci. Instrum.* **25**, 1028–9 (1954).

(*B82*) Brown, G., and R. Kanaris-Sotiriou, Thin Windows for Flow-Proportional Counters in X-Ray Fluorescence Spectrometers, *J. Sci. Instrum.* [2] **2**, 551–2 (1969).

(*B83*) Brown, G. A., Applications of an X-Ray Pulse Spectroscope; in "X-Ray and Electron Microscopy News," *Siemens Rev.* **34** (Special Issue), 69–73 (1967).

(*B84*) Brown, G. A., D. A. Nelson, and W. Ritzert, Two Approaches (Hardware and Software) to Automation of an X-Ray Spectroscopy System, *Siemens Rev.* **36** (Special Issue) 41–5 (1969).

(*B85*) Brown, O. E., H. L. Austin, and D. M. Schaefer, Preparation of Powdered Materials in Plastic Mounts for X-Ray Analysis, *Norelco Rep.* **7**, 140 (1960).

(*B86*) Bruch, J., Application of X-Ray Fluorescence Analysis to Various Ferrous Alloys, *Arch. Eisenhuettenw.* **33**(1), 5–15 (1962).

(*B87*) Buchanan, E. B., Jr., and T. D. Schroeder, Trace Lead Analysis Employing a Dual-Channel, Single-Crystal X-Ray Spectrometer, *Appl. Spectrosc.* **24**, 100–102 (1970).

(*B88*) Burbank, J., Identification and Characterization of Electrochemical Reaction Products by X-Ray Diffraction, *Nav. Res. Lab.* (*U. S.*) *Rep.* **6626**, 15 pp. (1967).

(*B89*) Burhop, E. H. S., Fluorescence Yield, *J. Phys. Radium* **16**, 625–9 (1955).

(*B90*) Burke, W. E., and D. L. Wood, Rare Earth Analysis by X-Ray-Excited Optical Fluorescence, *Advan. X-Ray Anal.* **11**, 204–13 (1968).

(*B91*) Burkhalter, P. G., Radioisotope X-Ray Analysis of Silver Ores Using Compton Scatter for Matrix Compensation, *Anal. Chem.* **43**, 10–17 (1971).

(*B92*) Burkhalter, P. G., and W. J. Campbell, Comparison of Detectors for Isotopic X-Ray Analyzers; in "Proceedings of the Second Symposium on Low-Energy X and Gamma Sources and Applications," P. S. Baker and M. Gerrard, eds.; *Oak Ridge Nat. Lab. Rep.* **ORNL-IIC-10**, vol. 1, pp. 393–423 (1967).

(*B93*) Burley, R. A., Diffraction Effects in X-Ray Fluorescence Analysis of Semiconductor-Grade Silicon and Other Monocrystalline Materials, *Appl. Spectrosc.* **21**, 416–19 (1967).

(*B94*) Burnham, H. D., J. Hower, and L. C. Jones, Generalized X-Ray Emission Spectrographic Calibration Applicable to Varying Compositions and Sample Forms, *Anal. Chem.* **29**, 1827–34 (1957).

(*C1*) Cairns, J. A., Characteristic X-Ray Production by Heavy-Ion Bombardment as a Technique for the Examination of Solid Surfaces, *Surface Sci.* **34**, 638–48 (1973).

(*C2*) Cairns, J. A., and D. F. Holloway, Selectivity of X-Ray Detection by a Variable-Geometry Proportional Counter, *Nature* **223**, 1262–3 (1969).

(*C3*) Cairns, J. A, and D. F. Holloway, Proportional Counter Window Filter for Preferential Detection of Boron *K* X-Rays, *Nucl. Instrum. Methods* **96**, 469–70 (1971).

(*C4*) Cairns, J. A., and R. S. Nelson, Selective X-Ray Generation by Heavy Ions. 1. Use of Energetic Heavy Ions to Generate Characteristic X-Rays from Elements in a Selective Manner, *Radiat. Eff.* **7**, 163–6 (1971).

(*C5*) Caley, J. L., Use of X-Ray Emission Spectrography for Petroleum Product Quality and Process Control, *Advan. X-Ray Anal.* **6**, 396–402 (1963).

(*C6*) Callendar, M. V., and D. F. White, Aspects of the Emission of X-Rays from Television Receivers, *J. Brit. Inst. Radio Eng.* **21**, 389–400 (1961).

(*C7*) Cameron, J. F., and J. R. Rhodes, Measurement of Tinplate Thickness Using Fluorescent X-Rays Excited by a Radioactive Source, *Brit. J. Appl. Phys.* **11**, 49–52 (1960).

(*C8*) Cameron, J. F., and J. R. Rhodes, Coating Thickness Measurement and Chemical Analysis by Radioisotope X-Ray Spectrometry; in "Encyclopedia of X-Rays and Gamma Rays," G. L. Clark, ed.; Reinhold Publishing Corp., N. Y., pp. 150–6 (1963).

(*C9*) Campbell, J. T., F. W. J. Garton, and J. D. Wilson, An X-Ray Milliprobe Analyzer with Photographic Recording of the Spatial Distribution of Elements, *Talanta* **15**, 1205–16 (1968).

(*C10*) Campbell, W. J., Fluorescent X-Ray Spectrographic Analysis of Trace Elements, Including Thin Films, *Amer. Soc. Test. Mater. Spec. Tech. Publ.* **349**, 48–69 (1964).

(*C11*) Campbell, W. J., Apparatus for Continuous Fluorescent X-Ray Spectrographic Analysis of Solutions, *Appl. Spectrosc.* **14**, 26–7 (1960).

(*C12*) Campbell, W. J., Fluorescent X-Ray Spectrographic Analysis—Studies of Low-Energy *K*, *L*, and *M* Spectral Lines, *U. S. Bur. Mines Rep. Invest.* **5538**, 20 pp. (1959).

(*C13*) Campbell, W. J., Energy-Dispersion X-Ray Analysis Using Radioactive Sources; in "X-Ray and Electron Methods of Analysis," H. Van Olphen and W. Parrish, eds.; Plenum Press, N. Y., pp. 36–54 (1968).

(*C14*) Campbell, W. J., and H. F. Carl, Quantitative Analysis of Niobium and Tantalum in Ores by Fluorescent X-Ray Spectroscopy, *Anal. Chem.* **26**, 800–5 (1954).

(*C15*) Campbell, W. J., and H. F. Carl, Combined Radiometric and Fluorescent X-Ray Spectrographic Method of Analyzing for Uranium and Thorium, *Anal. Chem.* **27**, 1884–6 (1955).

(*C16*) Campbell, W. J., and H. F. Carl, Fluorescent X-Ray Spectrographic Determination of Tantalium in Commercial Niobium Oxides, *Anal. Chem.* **28**, 960–2 (1956).

(*C17*) Campbell, W. J., and H. F. Carl, Quantitative Analysis by Fluorescent X-Ray Spectrography—Determination of Germanium in Coal and Coal Ash, *Anal. Chem.* **29**, 1009–17 (1957).

(*C18*) Campbell, W. J., and M. Leon, Fluorescent X-Ray Spectrograph for Dynamic Selective Oxidation Rate Studies—Design and Principles, *U. S. Bur. Mines Rep. Invest.* **5739**, 21 pp. (1961).

(*C19*) Campbell, W. J., and J. W. Thatcher, Determination of Calcium in Wolframite Concentrates by Fluorescent X-Ray Spectrography, *U. S. Bur. Mines Rep. Invest.* **5416**, 18 pp. (1958); *Advan. X-Ray Anal.* **2**, 313–32 (1959).

(*C20*) Campbell, W. J., and J. W. Thatcher, Fluorescent X-Ray Spectrography—Determination of Trace Elements, *U. S. Bur. Mines Rep. Invest.* **5966**, 29 pp. (1962); *Develop. Appl. Spectrosc.* **1**, 31–62 (1962).

(*C21*) Campbell, W. J., M. Leon, and J. W. Thatcher, Flat-Crystal X-Ray Optics, *Advan. X-Ray Anal.* **1**, 193–206 (1958).

(*C22*) Campbell, W. J., M. Leon, and J. W. Thatcher, Solution Techniques in Fluorescent X-Ray Spectrography, *U. S. Bur. Mines Rep. Invest.* **5497**, 24 pp. (1959).

(*C23*) Campbell, W. J., E. F. Spano, and T. E. Green, Micro and Trace Analysis by a Combination of Ion-Exchange Resin-Loaded Papers and X-Ray Spectrograhy, *Anal. Chem.* **38**, 987–96 (1966).

(*C24*) Canon, M., Polycrystalline X-Ray Analyzing Crystal, *Advan. X-Ray Anal.* **8**. 285–300 (1965).

(*C25*) Carl, H. F., and W. J. Campbell, Fluorescent X-Ray Spectrographic Analysis of Minerals, *Amer. Soc. Test. Mater. Spec. Tech. Publ.* **157**, 63–8 (1954).

(*C26*) Carman, C. J., X-Ray Fluorescent Determination of Major Constituents in Multielement Matrices by Use of Coherent to Incoherent Scattering Ratios, *Develop. Appl. Spectrosc.* **5**, 45–58 (1966).

(*C27*) Carr-Brion, K. G., X-Ray Fluorescence Determination of Zinc in Samples of Unknown Composition, *Analyst* (*London*) **89**, 346–9 (1964).

(*C28*) Carson, J. S., Timesaving with a Parallel Analyzing Channel Attachment, *Norelco Rep.* **10**, 84, 120 (1963).

(*C29*) Carter, G. F., X-Ray Fluorescence Analysis of Roman Coins, *Anal. Chem.* **36**, 1264–6 (1964).

(*C30*) Carter, G. F., Analysis of Copper and Brass Coins of the Early Roman Empire, *Science* **151**, 196–7 (1966).

(*C31*) Caruso, A. J., and H. H. Kim, Methods of Stretching Polypropylene for Use as Soft X-Ray Detector Windows, *Rev. Sci. Instrum.* **39**, 1059–60 (1968).

(*C32*) Caticha-Ellis, S., A. Ramos, and L. Saravia, Use of Primary Filters in X-Ray Spectrography—A New Method for Trace Analysis, *Advan. X-Ray Anal.* **11**, 105–13 (1968).

(*C33*) Cauchois, Y., Spectrograph Giving High Intensity from Transmission of Non-Collimated X-Rays through Curved Mica Sheets, *C. R. Acad. Sci.* (*Paris*) **194**, 362–5 (1932).

(*C34*) Cauchois, Y., Universal X-Ray Spectrograph for Use up to 20 Å, *J. Phys. Radium* **6**, 89–96 (1945).

(*C35*) Cavanagh, M. B., Application of X-Ray Fluorescence to Trace Analysis, *U. S. Nav. Res. Lab. Rep.* **4528**, 4 pp. (1955).

(*C36*) Champion, K. P., and R. N. Whittem, Utilization of Increased Sensitivity of X-Ray Fluorescence Spectrometry Due to Polarization of the Background Radiation, *Nature* **199**, 1082 (1963).

(*C37*) Chan, F. L., Detection, Confirmation, and Determination of Trace Amounts of Selenium by X-Ray Methods, *Advan. X-Ray Anal.* **7**, 542–54 (1964).

(*C38*) Chan, F. L., Study of X-Ray Fluorescence Method with Vacuum and Air-Path Spectrographs for the Determination of Film Thickness of II–VI Compounds, *Advan. X-Ray Anal.* **11**, 191–203 (1968).

(*C39*) Chan, F. L., Apparatus for Analysis of Liquid Samples by the X-Ray Fluorescence Method with a Vacuum Spectrograph, *Develop. Appl. Spectrosc.* **5**, 59–75 (1966).

(*C40*) Chan, F. L., and D. A. Brooks, Study of Homogeneity of Solid Solutions of Cadmium Sulfide and Cadmium Selenide by X-Ray Fluorescence, *Advan. X-Ray Anal.* **8**, 420–30 (1965).

(*C41*) Chappell, B. W., W. Compston, P. A. Arriens, and M. J. Vernon, Rubidium and Strontium Determination by X-Ray Fluorescence Spectrometry and Isotope Dilution below the Part per Million Level, *Geochim. Cosmochim. Acta* **33**, 1002–6 (1969).

(*C42*) Chesnin, L., and A. H. Beavers, Preparation of Plant Samples for X-Ray Emission Spectrography, *Agron. J.* **54**, 487–9 (1962).

(*C43*) Chessin, H., and B. E. Artz, Specimen Standards for X-Ray Spectrometric Analysis of Atmospheric Aerosols, *Advan. X-Ray Anal.* **17**, 225–35 (1974).

(*C44*) Chessin, H., and E. H. McLaren, X-Ray Spectrometric Determination of Atmospheric Aerosols, *Advan. X-Ray Anal.* **16**, 165–76 (1973).

(*C45*) Chodos, A. A., and C. G. Engel, Fluorescent X-Ray Spectrographic Analysis of Amphibolite Rocks, *Amer. Mineral.* **46**, 120–33 (1961).

(*C46*) Chodos, A. A., and W. Nichiporuk, Application of the Mutual Standards Concept to X-Ray Fluorescence Spectroscopy—Analysis of Meteoritic Sulfide Nodules for Eight Elements, *Advan. X-Ray Anal.* **2**, 247–55 (1959).

(*C47*) Claisse, F., Accurate X-Ray Fluorescence Analysis without Internal Standard, *Que. Prov. Dep. Mines. Prelim. Rep.* **327**, 24 pp. (1956); *Norelco Rep.* **4**, 3–7, 17, 19, 95–6 (1957).

(*C48*) Claisse, F., Overcoming the Particle-Size Effect in the Internal-Standard Method of X-Ray Fluorescence Analysis, *Spectrochim. Acta* **25B**, 209–18 (1970).

(*C49*) Claisse, F., and C. Samson, Heterogeneity Effects in X-Ray Analysis, *Advan. X-Ray Anal.* **5**, 335–54 (1962).

(*C50*) Clark, G. L., and W. Duane, Theory of the Tertiary Radiation Produced by Impacts of Photoelectrons, *Proc. Nat. Acad. Sci.* **10**, 41–7, 191–6 (1924).

(*C51*) Clark, G. L., and C. H. Liu, Quantitative Determination of Porosity by X-Ray Absorption, *Anal. Chem.* **29**, 1539–41 (1957).

(*C52*) Cockroft, A. L., and S. C. Curran, Elimination of the End Effects in Counters, *Rev. Sci. Instrum.* **22**, 37–42 (1951).

(*C53*) Colby, J. W., Quantitative Microprobe Analysis of Thin Insulating Films, *Advan. X-Ray Anal.* **11**, 287–305 (1968).

(*C54*) Cole, M., I. A. Bucklow, and C. W. B. Grigson, Technique for the Rapid, Accurate, and Strain-Free Machining of Metallic Single Crystals, *Brit. J. Appl. Phys.* **12**, 296–7 (1961).

(*C55*) Collin, R. L., X-Ray Fluorescence Analysis Using Ion Exchange Resin for Sample Support—Determination of Strontium in 0.1 *M* Calcium Acetate Solutions, *Anal. Chem.* **33**, 605–7 (1961).

(*C56*) Compton, A. H., Quantum Theory of the Scattering of X-Rays by Light Elements, *Phys. Rev.* **21**, 483–502 (1923).

(*C57*) Cook, G. B., C. E. Mellish, and J. A. Payne, Measurement of Thin Metal Layers —Fluorescent X-Ray Production by Radioisotope Sources, *Anal. Chem.* **32**, 590–3 (1960).

(*C58*) Cooke, C. J., and P. Duncumb, Comparison of a Nondispersive Method of X-Ray Microanalysis with Conventional Crystal Spectrometry; in "X-Ray Optics and Microanalysis," R. Castaing, P. Deschamps, and J. Philibert, eds.; Hermann et Cie. Paris, pp. 467–76 (1966).

(*C59*) Coolidge, W. D., A Powerful Roentgen-Ray Tube with a Pure Electron Discharge, *Phys. Rev.* [2] **2**, 409–30 (1913).

(*C60*) Cope, J. H., Quantitative X-Ray Spectrographic Analysis for Uranium with the Use of an Internal Standard, *Norelco Rep.* **3**, 41–3 (1956).

(*C61*) Cosslett, V. E., P. Duncumb, J. V. P. Long, and W. C. Nixon, Microanalysis by X-Ray Absorption, Fluorescence, Emission and Diffraction Using Ultra-Fine X-Ray Sources, *Advan. X-Ray Anal.* **1**, 329–37 (1958).

(*C62*) Coster, D., and G. von Hevesy, Missing Element of Atomic Number 72, *Nature* **111**, 79 (1923).

(*C63*) Coster, D., and G. von Hevesy, The New Element Hafnium, *Nature* **111**, 182 (1923).

(*C64*) Cowgill, U. M., X-Ray Emission Spectroscopy in the Chemical Analysis of Lake Sediments—Determining 41 Elements, *Develop. Appl. Spectrosc.* **5**, 3–23 (1966).

(*C65*) Criss, J. W., and L. S. Birks, Calculation Methods for Fluorescent X-Ray Spectrometry—Empirical Coefficients *vs* Fundamental Parameters, *Anal. Chem.* **40**, 1080–6 (1968).

(*C66*) Criss, J. W., and L. S. Birks, *Int. Spectrosc. Colloq.* 15th, Madrid (1969).

(*C67*) Croke, J. F., W. J. Pfozer, and M. J. Solazzi, New Automatic X-Ray Spectrometer in Industrial Control, *Norelco Rep.* **11**, 129–31 (1964).

(*C68*) Cullen, T. J., Potassium Pyrosulfate Fusion Technique—Determination of Copper in Mattes and Slags by X-Ray Spectroscopy, *Anal. Chem.* **32**, 516–17 (1960).

(*C69*) Cullen, T. J., Briqueted Copper Alloy Drillings as a Sample for X-Ray Spectroscopy, *Anal. Chem.* **33**, 1342–4 (1961).

(*C70*) Cullen, T. J., Coherent Scattered Radiation Internal Standardization in X-Ray Spectrometric Analysis of Solutions, *Anal. Chem.* **34**, 812–14 (1962).

(*C71*) Cullen, T. J., Addition of Sodium Fluoride to Potassium Pyrosulfate Fusions for X-Ray Spectrometric Analysis of Siliceous Samples, *Anal. Chem.* **34**, 862 (1962).

(*C72*) Cullen, T. J., Internal Differential X-Ray Absorption-Edge Spectrometry, *Anal. Chem.* **37**, 711–13 (1965).

(*C73*) Cullen, T. J., X-Ray Spectrometric Analysis of Solutions—Comparison of Techniques, *Pittsburgh Conf. Anal. Chem. Appl. Spectrosc.* (1963), unpublished; reviewed in detail in Reference (*B34*).

(*C74*) Cullen, T. J., X-Ray Methods for the Analysis of Solutions, *Develop. Appl. Spectrosc.* **8**, 76–83 (1970).

(*C75*) Curran, S. C., and E. R. Rae, Some Properties of Counters with Beaded Wires, *J. Sci. Instrum.* **24**, 233–8 (1947).

(*D1*) Dams, R., K. A. Rahn, and J. W. Winchester, Evaluation of Filter Materials and Impaction Surfaces for Nondestructive Neutron Activation Analysis of Aerosols, *Environ. Sci. Technol.* **6**, 441–8 (1972).

(*D2*) Das Gupta, K., H. Welch, P. F. Gott, J. F. Priest, S. Cheng, and E. Chu, Some New Methods of Precision X-Ray Spectrometry, *Advan. X-Ray Anal.* **16**, 251–9 (1973).

(*D3*) Daugherty, K. E., and R. J. Robinson, X-Ray Fluorescence Spectrometric Analysis of the Iron(II), Cobalt(II), Nickel(II), and Copper(II) Chelates of 8-Quinolinol, *Anal. Chem.* **36**, 1869-70 (1964).

(*D4*) Daugherty, K. E., R. J. Robinson, and J. I. Mueller, X-Ray Fluorescence Spectrometric Analysis of the Copper(II) and Mercury(II) Complexes of 6-Chloro-2-Methoxy-9-Thioacridine, *Anal. Chem.* **36**, 1098–1100 (1964).

(*D5*) Davies, R. D., and H. K. Herglotz, Total-Reflection X-Ray Spectrograph for Fluorescent Analysis of Light Elements, *Advan. X-Ray Anal.* **12**, 496–505 (1969).

(*D6*) Davis, C. M., and G. R. Clark, X-Ray Spectrographic Analysis of Nickel-Containing Alloys with Varied Sample Forms, *Advan. X-Ray Anal.* **1**, 351–66 (1958).

(*D7*) Davis, C. M., and M. M. Yanak, Performance of an Unattended Automated X-Ray Spectrograph, *Advan. X-Ray Anal.* **7**, 644-52 (1964).

(*D8*) Davis, C. M., K. E. Burke, and M. M. Yanak, X-Ray Spectrographic Analysis of Traces in Metals by Preconcentration Techniques, *Advan. X-Ray Anal.* **11**, 56–62 (1968).

(*D9*) Davis, E. N., and B. C. Hoeck, X-Ray Spectrographic Method for the Determination of Vanadium and Nickel in Residue Fuels and Charging Stocks, *Anal. Chem.* **27**, 1880–4 (1955).

(*D10*) Davis, E. N., and R. A. Van Nordstrand, Determination of Barium, Calcium, and Zinc in Lubricating Oils, *Anal. Chem.* **26**, 973–7 (1954).

(*D11*) De Jongh, W. K., Heterogeneity Effects in X-Ray Fluorescence Analysis, N. V. Philips Gloeilampenfabrieken, Eindhoven, Neth., *Sci. Anal. Equip. Bull.* 7000.38. 0266.11, 6 pp. (1970).

(*D12*) DeKalb, E. L., V. A. Fassel, T. Taniguchi, and T. R. Saranathan, Analytical Applications of X-Ray Excited Optical Fluorescence Spectra; the Internal Standard Principle, *Anal. Chem.* **40**, 2082–4 (1968).

(*D13*) Den Boggende, A. J. F., A. C. Brinkman, and W. de Graaff, Comments on the Ageing Effects of Gas-Filled Proportional Counters, *J. Sci. Instrum.* [2] **2**, 701–5 (1969).

(*D14*) Deslattes, R. D., J. L. Torgesen, B. Paretzkin, and A. T. Horton, Preliminary Studies on the Characterization of Solution-Grown ADP Crystals, *Advan. X-Ray Anal.* **8**, 315–24 (1965).

(*D15*) Despujols, J., H. Roulet, and G. Senemaud, X-Ray Fluorescence Analysis with a Focused Primary Beam; in "X-Ray Optics and X-Ray Microanalysis," H. H. Pattee, V. E. Cosslett, and A. Engstrom, eds.; Academic Press, N. Y., pp. 445–9 (1963).

(*D16*) Deutsch, Y., Direct X-Ray Spectrometric Determination of Bromine in Water, *Anal. Chem.* **46**, 437–40 (1974).

(*D17*) Dijksterhuis, P. R., Automatic X-Ray Analysis of Cement Materials, Including Sample Preparation by Fusion or Pressing, N. V. Philips Gloeilampenfabrieken, Eindhoven, Neth., *Sci. Anal. Equip. Bull.* 7000.38.0164.11, 4 pp. (1968).

(*D18*) Dinklage, J. B., X-Ray Diffraction by Multilayered Thin-Film Structures and Their Diffusion, *J. Appl. Phys.* **38**, 3781–5 (1967).

(*D19*) Dinklage, J., and R. Frerichs, X-Ray Diffraction and Diffusion in Metal-Film-Layered Structures, *J. Appl. Phys.* **34**, 2633–5 (1963).

(*D20*) Dodd, C. G., and D. J. Kaup, Optimum Conditions for Chemical Analysis by X-Ray Absorption-Edge Spectrometry, *Anal. Chem.* **36**, 2325–9 (1964).

(*D21*) Dolby, R. M., Methods for Analyzing Unresolved Proportional Counter Curves of X-Ray Line Spectra, *Proc. Phys. Soc.* (*London*) **73**, 81–96 (1959).

(*D22*) Dothie, H. J., "Filterscan"—Novel Technique in X-Ray Fluorescence Analysis, *Nature* **196**, 984–5 (1962).

(*D23*) Dothie, H. J., and B. Gale, Nondispersive X-Ray Fluorescence Absorption-Edge Spectroscopy, *Spectrochim. Acta* **20**, 1735–55 (1964).

(*D24*) Doughman, W. R., R. C. Hirt, and R. T. Langdon, Water-Cooled Liquid Sample Cell for the X-Ray Emission Spectrograph, *Norelco Rep.* **4**, 36, 40 (1957).

(*D25*) Dowling, P. H., C. F. Hendee, T. R. Kohler, and W. Parrish, Counters for X-Ray Analysis, *Philips Tech. Rev.* **18**, 262–75 (1956–57); *Norelco Rep.* **4**, 23–33 (1957).

(*D26*) Dozier, C. M., J. V. Gilfrich, and L. S. Birks, Quantitative Calibration of X-Ray Film Response in the 5-keV to 1.3-MeV Region, *Appl. Opt.* **6**, 2136–9 (1967).

(*D27*) Drummond, C. H., Preparation of Glass Blanks by Fusion with Lithium Tetraborate for X-Ray Spectrographic Analysis, *Appl. Spectrosc.* **20**, 252 (1966).

(*D28*) Duane, W., and F. L. Hunt, X-Ray Wave-Lengths, *Phys. Rev.* **6**, 166–71 (1915).

(*D29*) Duggan, J. L., W. L. Beck, L. Albrecht, L. Munz, and J. D. Spaulding, Studies of X-Rays Induced by Charged Particles, *Advan. X-Ray Anal.* **15**, 407–23 (1972).

(*D30*) Du Mond, J. W. M., A High Resolving-Power Curved-Crystal Focusing Spectrometer for Short-Wavelenghth X-Rays and Gamma Rays, *Rev. Sci. Instrum.* **18**, 626–38 (1947).

(*D31*) DuMond, J. W. M., and H. A. Kirkpatrick, Multiple-Crystal X-Ray Spectrograph, *Rev. Sci. Instrum.* **1**, 88–105 (1930).

(*D32*) Duncumb, P., Absorption and Emission Methods of Microanalysis, *Rev. Universelle Mones* **17**, 240–46 (1961).

(*D33*) Duncumb, P., and P. K. Shields, Present State of Quantitative X-Ray Microanalysis. I. Physical Basis, *Brit. J. Appl. Phys.* **14**, 617–25 (1963).

(*D34*) Dunn, H. W., X-Ray Absorption-Edge Analysis, *Anal. Chem.* **34**, 116–21 (1962).

(*D35*) Dunne, J. A., A New Focusing Vacuum X-Ray Macroprobe Analyzer, *Advan. X-Ray Anal.* **8**, 223–30 (1965).

(*D36*) Dunne, J. A., Components for X-Ray Fluorescence Spectroscopy in the 5–65 Å Wavelength Region, *Norelco Rep.* **11**, 109–12 (1964).

(*D37*) Dunne, J. A., and W. R. Muller, Demountable X-Ray Tube for Light-Element Fluorescence Analysis, *Develop. Appl. Spectrosc.* **4**, 33–44 (1965).

(*D38*) Dwiggins, C. W. Jr., Quantitative Determination of Low Atomic Number Elements Using Intensity Ratio of Coherent to Incoherent Scattering of X-Rays—Determination of Hydrogen and Carbon, *Anal. Chem.* **33**, 67–70 (1961).

(*D39*) Dwiggins, C. W., Jr., Automated Determination of Elements in Organic Samples Using X-Ray Emission Spectrometry, *Anal. Chem.* **36**, 1577–81 (1964).

(*D40*) Dwiggins, C. W., Jr., and H. N. Dunning, Quantitative Determination of Traces of Vanadium, Iron, and Nickel in Oils by X-Ray Spectrography, *Anal. Chem.* **32**, 1137–41 (1960).

(*D41*) Dwiggins, C. W., Jr., J. R. Lindley, and B. H. Eccleston, Cooled Sample Holder for the X-Ray Spectrograph, *Anal. Chem.* **31**, 1928 (1959).

(*D42*) Dyer, G. R., D. A. Gedke, and T. R. Harris, Fluorescence Analysis Using an Si(Li) X-Ray Energy Analysis System with Low-Power X-Ray Tubes and Radioisotopes, *Advan. X-Ray Anal.* **15**, 228–39 (1972).

(*D43*) Dyroff, G. V., and P. Skiba, Determination of Trace Amounts of Iron, Nickel, and Vanadium on Catalysts by Fluorescent X-Ray Spectrography, *Anal. Chem.* **26**, 1774–8 (1954).

(*E1*) Ebel, H., Thickness Determination of Planar Thin Films by X-Ray Fluorescence Analysis, *Z. Metallk.* **56**, 802–4 (1965).

(*E2*) Ebel, H., Quantitative X-Ray Fluorescence Analysis with Variable Take-Off Angle, *Advan. X-Ray Anal.* **13**, 68–79 (1970).

(*E3*) Ebel, H., and M. F. Ebel, Analogy between X-Ray Scatter and Fluorescence Analysis, *X-Ray Spectrom.* **1**, 15–21 (1972).

(*E4*) Ebel, H., and J. Klugel, Determination of Thickness of Thin Planar Multilayers by X-Ray Fluorescence, *Z. Naturforsch.* **21**, 2108–9 (1966).

(*E5*) Ebel, H., H. Dirschmid, A. Wagendristel, and G. Stehno, Determination of Particle-Size Distributions by Scattering of X-Rays of Various Wavelengths, *Acta Phys. Austr.* **31**, 363–72 (1970).

(*E6*) Ebert, F., and A. Wagner, Explanation of "Accidental Interferences" as Systematic Accompanying Phenomena in X-Ray Fluorescent Spectroscopic Measurements and the Possibility of Avoiding the Interferences, *Z. Metallk.* **48**, 646–9 (1957).

(*E7*) Eddy, C. E., and T. H. Laby, Analysis by X-Ray Spectroscopy, *Proc. Roy. Soc. (London)* **124A**, 249–68 (1929).

(*E8*) Eddy, C. E., and T. H. Laby, Quantitative Analysis by X-Ray Spectroscopy, *Proc. Roy. Soc. (London)* **127A**, 20–42 (1930).

(*E9*) Ehlert, R. C., Diffraction of X-Rays by Multilayer Stearate Soap Films, *Advan. X-Ray Anal.* **8**, 325–31 (1965).

(*E10*) Ehlert, R. C., and R. A. Mattson, Characteristic X-Rays from Boron and Beryllium, *Advan. X-Ray Anal.* **9**, 456-70 (1966).

(*E11*) Ehlert, R. C., and R. A. Mattson, Multilayer Soap Film Structures, *Advan. X-Ray Anal.* **10**, 389–97 (1967).

(*E12*) Ehrhardt, C. H., E. M. Banas, and J. L. Janik, Application of Spherically Curved Crystals for X-Ray Fluorescence, *Appl. Spectrosc.* **22**, 730–5 (1968).

(*E13*) Eichoff, H. J., and S. K. Kiefer, Universal Method for the Semiquantitative Analysis of Powdered Samples by X-Ray Fluorescence, *Method. Phys. Anal.* **1966**(1), 43–7.

(*E14*) Eick, J. D., H. J. Caul, D. L. Smith, and S. D. Rasberry, Analysis of Gold and Platinum Group Alloys by X-Ray Emission and Correction for Interelement Effects, *Appl. Spectrosc.* **21**, 324–8 (1967).

(*E15*) Elion, H. A., and R. E. Ogilvie, Reflecting Variable Bent-Crystal Spectrometer, *Rev. Sci. Instrum.* **33**, 753–5 (1962).

(*E16*) Ellis, J. H., R. E., Phillips, and R. A. Barnhisel, Diffusion of Iron in Montmorillonite as Determined by X-Ray Emission, *Soil Sci. Soc. Amer. Proc.* **34**, 591–5 (1970).

(*E17*) Engström, A., "X-Ray Microanalysis in Biology and Medicine," Elsevier Publishing Co., N. Y., 92 pp. (1962).

(*E18*) Esmail, E. I., C. J. Nicholls, and D. S. Urch, Detection of Light Elements by X-Ray Emission Spectroscopy with Use of Low-Energy Satellite Peaks, *Analyst* (*London*) **98**, 725–31 (1973).

(*F1*) Fabbi, B. P., Refined Fusion X-Ray Fluorescence Technique, and Determination of Major and Minor Elements in Silicate Standards, *Amer. Mineral.* **57**, 237–45 (1972).

(*F2*) Fabbi, B. P., Die for Pelletizing Samples for X-Ray Fluorescence Analysis, *X-Ray Spectrom*, **1**, 39–41 (1972).

(*F3*) Fabbi, B. P., X-Ray Fluorescence Determination of Sodium in Silicate Standards Using Direct Dilution and Dilution-Fusion Preparation Techniques, *X-Ray Spectrom.* **2**, 15–17 (1973).

(*F4*) Fadley, C. S., S. B. M. Hagstrom, J. M. Hollander, M. P. Klein, and D. A. Shirley, Chemical Bonding Information from Photoelectron Spectroscopy, *Science* **157**, 1571–3 (1967).

(*F5*) Faessler, A., and M. Goehring, X-Ray Spectrum and State of Combination from the *K*-Fluorescence Radiation of Sulfides, *Naturwissenschaften* **39**, 169–77 (1952).

(*F6*) Fagel, J. E., Jr., E. W. Balis, and L. B. Bronk, Organometallic Precipitates and Briqueting for X-Ray Spectrography, *Anal. Chem.* **29**, 1287–9 (1957).

(*F7*) Fagel, J. E., Jr., H. A. Liebhafsky, and P. D. Zemany, Determination of Tungsten or Molybdenum in X-Ray Emission Spectrography, *Anal. Chem.* **30**, 1918–20 (1958).

(*F8*) Fagot, C., and R. Tertian, X-Ray Fluorescence Determination of Impurities Present in Polyolefins, *Method. Phys. Anal.* **1966**(1), 30–7.

(*F9*) Fahlbusch, W. A., Preparation of Metallic Buttons and Analysis by X-Ray Spectrography, *Appl. Spectrosc.* **17**, 72–3 (1963).

(*F10*) Fano, U., Ionization Yield of Radiations. II. Fluctuations of the Number of Ions, *Phys. Rev.* **72**, 26–9 (1947).

(*F11*) Fatemi, M., and L. S. Birks, Obtaining Consistent Solutions of Empirical Equations in X-Ray Fluorescence, *Anal. Chem.* **45**, 1443–7 (1973).

(*F12*) Felten, E. J., I. Fankuchen, and J. Steigman, A Possible Solution to the Matrix Problem in X-Ray Fluorescence Spectrometry, *Anal. Chem.* **31**, 1771–6 (1959).

(*F13*) Fink, R. W., R. C. Jopson, H. Mark, and C. D. Swift, Atomic Fluorescence Yields, *Rev. Mod. Phys.* **38**, 513–40 (1966).

(*F14*) Finnegan, J. J., Thin-Film X-Ray Spectroscopy, *Advan. X-Ray Anal.* **5**, 500–11 (1962).

(*F15*) Fischer, D. W., and W. L. Baun, Experimental Dispersing Devices and Detection Systems for Soft X-Rays, *Advan. X-Ray Anal.* **7**, 489–96 (1964).

(*F16*) Fitzgerald, R., and P. Gantzel, X-Ray Energy Spectrometry in the 0.1- to 10-Å Range, *Amer. Soc. Test. Mater. Spec. Tech. Publ.* **STP-485**, 3–35 (1971).

(*F17*) Flikemma, D. S., and R. V. Schablaske, Direct Assay for Molybdenum and Ruthenium in Uranium Alloys by X-Ray Emission Spectrometry, *Advan. X-Ray Anal.* **1**, 387–98 (1958).

(*F18*) Florestan, J., X-Ray Fluorescence Determination of Ni and Cr in Stainless Steel, *Method. Phys. Anal.* **3**(2), 65–118 (1967).

(*F19*) Fonda, G. R., and G. B. Collins, The Cathode-Ray Tube in X-Ray Spectroscopy and Quantitative Analysis, *J. Amer. Chem. Soc.* **53**, 113–25 (1931).

(*F20*) Fox, J. G. M., X-Ray Spectrometric Determination of Light Elements Using Direct Electron Excitation, *J. Inst. Metals* **91**, 239–41 (1963).

(*F21*) Fraenkel, B. S., Twisted-Crystal X-Ray Spectrometer, *Rev. Sci. Instrum.* **40**, 732 (1969).

(*F22*) Francis, J. E., P. R. Bell, and J. C. Gundlach, Single-Channel Analyzer, *Rev. Sci. Instrum.* **22**, 133–7 (1951).

(*F23*) Franks, A., Optical Techniques Applied to the Study of X-Ray Wavelengths; in "X-Ray Optics and X-Ray Microanalysis," H. H. Pattee, V. E. Cosslett, and A. Engstrom, eds.; Academic Press, N. Y., pp. 199–213 (1963).

(*F24*) Franks, A., and R. F. Braybrook, Analysis of the Lighter Elements by Total Reflection of Their Characteristic X-Ray Wavelengths, *Brit. J. Appl. Phys.* **10**, 190–1 (1959).

(*F25*) Frerichs, R., Cadmium Sulfide X-Ray Detector, *J. Appl. Phys.* **21**, 312–7 (1950).

(*F26*) Friedman, H., Geiger Counter Tubes, *Proc. I.R.E.* (*Inst. Radio Eng.*) **37**, 791–808 (1949).

(*F27*) Friedman, H., and L. S. Birks, Geiger-Counter Spectrometer for X-Ray Fluorescence Analysis, *Rev. Sci. Instrum.* **19**, 323–30 (1948).

(*F28*) Friedman, H., L. S. Birks, and E. J. Brooks, Basic Theory and Fundamentals of Fluorescent X-Ray Spectrographic Analysis, *Amer. Soc. Test. Mater. Spec. Tech. Publ.* **157**, 3–26 (1954).

(*F29*) Friedrich, W., P. Knipping, and M. Von Laue, Interference Phenomena with Roentgen Rays, *Ber. Bayer. Akad. Wiss. Munchen* **1912**, 303–22; *Ann. Phys.* **41**, 971–88 (1913).

(*G1*) Gamage, C. F., The N900 Nondispersive Laboratory Analyzer, *X-Ray Spectrom.* **1**, 99–106 (1972).

(*G2*) Gamble, F. R., J. H. Osiecki, and F. J. DiSalvo, Some Superconducting Intercalation Complexes of TaS_2 and Substituted Pyridines, *J. Chem. Phys.* **55**, 3525–30 (1971).

(*G3*) Gamble, F. R., J. H. Osiecki, M. Cais, R. Pisharody, F. J. DiSalvo, and T. H. Geballe, Intercalation Complexes of Lewis Bases and Layered Sulfides—a Large Class of New Superconductors, *Science* **174**, 493–7 (1971).

(*G4*) Gaylor, D. W., Precision of Fixed-Time *vs.* Fixed-Count Measurements, *Anal. Chem.* **34**, 1670–1 (1962).

(*G5*) Gedke, D. A., The Si(Li) X-Ray Energy Analysis System: Operating Principles and Performance, *X-Ray Spectrom.* **1**, 129–41 (1972).

(*G6*) Geiger, H., and W. Muller, Mode of Action and Construction of a Counter Tube, *Phys. Z.* **29**, 839–41 (1928).

(*G7*) Gerold, V., X-Ray Thickness Gaging of Galvanized Layers with a Counter-Tube Diffractometer, *Z. Angew. Phys.* **4**, 247–54 (1952).

(*G8*) Gianelos, J., Rapid Direct X-Ray Fluorescence Method for Simultaneously Determining Brass Composition and Plating Weight for Brass-Plated Steel Tire-Cord Wires, *Advan. X-Ray Anal.* **17**, 325–36 (1974).

(*G9*) Giauque, R. D., F. S. Goulding, J. M. Jaklevic, and R. H. Pehl, Trace-Element Determination with Semiconductor-Detector X-Ray Spectrometers, *Anal. Chem.* **45**, 671–81 (1973).

(*G10*) Giessen, B. C., and G. E. Gordon, X-Ray Diffraction—New Technique Based on X-Ray Spectrography, *Science* **159**, 973–5 (1968).

(*G11*) Giessen, B. C., and G. E. Gordon, X-Ray Diffraction—New High-Speed Technique Based on X-Ray Spectrography, *Norelco Rep.* **17**(2), 17–21 (1970).

(*G12*) Gilberg, E., Device for Bending Crystal Slabs for Use in Focusing X-Ray Spectrometers, *Rev. Sci. Instrum.* **42**, 1189–90 (1971).

(*G13*) Gilfrich, J. V., and L. S. Birks, Spectral Distribution of X-Ray Tubes for Quantitative X-Ray Fluorescence Analysis, *Anal. Chem.* **40**, 1077–80 (1968).

(*G14*) Gilfrich, J. V., P. G. Burkhalter, and L. S. Birks, X-Ray Spectrometry for Particulate Air Pollution—Quantitative Comparison of Techniques, *Anal. Chem.* **45**, 2002–9 (1973).

(*G15*) Gilfrich, J. V., P. G. Burkhalter, R. R. Whitlock, E. S. Warden, and L. S. Birks, Spectral Distribution of a Thin-Window Rhodium–Target X-Ray Spectrographic Tube, *Anal. Chem.* **43**, 934–6 (1971).

(*G16*) Gilmore, J. T., Use of a Primary-Beam Filter in X-Ray Fluorescence Spectrometric Determination of Trace Arsenic, *Anal. Chem.* **40**, 2230–2 (1968).

(*G17*) Glade, G. H., Production Control of Gold and Rhodium Plating Thickness on Very Small Samples by X-Ray Spectroscopy, *Advan. X-Ray Anal.* **11**, 185–90 (1968).

(*G18*) Glade, G. H., and H. R. Post, Aging of Solder Standards, *Appl. Spectrosc.* **22**, 123 (1968).

(*G19*) Glocker, R., and W. Frohnmeyer, Roentgen-Spectroscopic Determination of the Quantity of an Element in Mixtures, *Ann. Phys.* **76**, 369–95 (1925).

(*G20*) Glocker, R., and W. Frohnmeyer, Use of X-Rays in Quantitative Chemical Analysis, *Fortschr. Geb. Röntgenstr.* **31**, 90–2 (1923).

(*G21*) Glocker, R., and H. Schreiber, Quantitative Roentgen Spectrum Analysis by Means of Cold Excitation of the Spectrum, *Ann. Phys.* **85**, 1089–1102 (1928).

(*G22*) Gohshi, Y., H. Yoshiro, and K. Hori, Wide-Range, Single-Axis, Vacuum, Two-Crystal Spectrometer for Fluorescent X-Ray Analysis, *Spectrochim. Acta* **27B**, 135–42 (1972).

(*G23*) Goodwin, P. S., and C. L. Winchester, Continuous Measurement of Plating Thickness, *Plating* **46**, 41–4 (1959).

(*G24*) Gould, R. W., and S. R. Bates, Some Applications of a Computer Program for Quantitative Spectrochemical Analysis, *X-Ray Spectrom.* **1**, 29–35 (1972).

(*G25*) Gould, R. W., S. R. Bates, and C. J. Sparks, Application of the Graphite Monochromator to Light-Element X-Ray Spectroscopy, *Appl. Spectrosc.* **22**, 549–51 (1968).

(*G26*) Goulding, F. S., and J. M. Jaklevic, Trace-Element Analysis by X-Ray Fluorescence, *U. S. At. Energy Comm. Rep.* **UCRL-20625**, 54 pp. (1971).

(*G27*) Goulding, F. S., J. M. Jaklevic, B. V. Jarrett, and D. A. Landis, Detector Background and Sensitivity of Semiconductor X-Ray Fluorescence Spectrometers, *Advan. X-Ray Anal.* **15**, 470–82 (1972).

(*G28*) Graeber, E. J., Direct Identification of X-Ray Spectra, *Appl. Spectrosc.* **16**, 24 (1962).

(*G29*) Graham, M. J., and C. S. Bray, Use of Evaporated Metal-Film Standards in Thin-Layer X-Ray Fluorescence Analysis of Mixed Oxides, *J. Sci. Instrum.* [2] **2**, 706–8 (1969).

(*G30*) Green, T. E., S. L. Law, and W. J. Campbell, Use of Selective Ion-Exchange Paper in X-Ray Spectrography and Neutron Activation—Application to the Determination of Gold, *Anal. Chem.* **42**, 1749–53 (1970).

(*G31*) Griffin, L. H., Iron-55 X-Ray Absorption Analysis of Organically Bound Chlorine Using Conventional Proportional Counting Facilities, *Anal. Chem.* **34**, 606–9 (1962).

(*G32*) Grodski, J. J., Photoelectric Detectors for 10–400-Å Photons, *Norelco Rep.* **14**, 107–11 (1967).

(*G33*) Grubb, W. T., and P. D. Zemany, X-Ray Emission Spectrography with Ion-Exchange Membrane, *Nature* **176**, 221 (1955).

(*G34*) Gunn, E. L., X-Ray Fluorescence of Suspended Particles in a Liquid Hydrocarbon, *Advan. X-Ray Anal.* **11**, 164–76 (1968).

(*G35*) Gunn, E. L., X-Ray Spectrometric Analysis of Liquids and Solutions, *Amer. Soc. Test. Mater. Spec. Tech. Publ.* **349**, 70–85 (1964).

(*G36*) Gunn, E. L., Fluorescent X-Ray Spectral Analysis of Powdered Solids by Matrix Dilution, *Anal. Chem.* **29**, 184–9 (1957).

(*G37*) Gunn, E. L., X-Ray Fluorescence Intensity of Elements Evaporated from Solution onto Thin Films, *Anal. Chem.* **33**, 921-7 (1961).

(*G38*) Gunn, E. L., Problems of Direct Determination of Trace Nickel in Oil by X-Ray Emission Spectrography, *Anal. Chem.* **36**, 2086–90 (1964).

(*G39*) Gunn, E. L., Determination of Lead in Gasoline by X-Ray Fluorescence Using an Internal Intensity Reference, *Appl. Spectrosc.* **19**, 99–102 (1965).

(*G40*) Gunn, E. L., Practical Methods of Solving Absorption and Enhancement Problems in X-Ray Emission Spectrography, *Develop. Appl. Spectrosc.* **3**, 69–96 (1964).

(*H1*) Hadding, A., Mineral Analysis by X-Ray Spectroscopic Methods, *Z. Anorg. Chem.* **122**, 195–200 (1922).

(*H2*) Haftka, F. J., X-Ray Fluorescence Analysis of Powders, *Rev. Universelle Mines* **15**, 549–54 (1959).

(*H3*) Hagg, G., and N. Karlsson, Aluminum Monochromator with Double Curvature for High-Intensity X-Ray Powder Photographs, *Acta Cryst.* **5**, 728–30 (1952).

(*H4*) Hakkila, E. A., and G. R. Waterbury, X-Ray Fluorescence Spectrometric Determination of Chromium, Iron, and Nickel in Ternary Alloys, *Anal. Chem.* **37**, 1773–5 (1965).

(*H5*) Hakkila, E. A., and G. R. Waterbury, Applications of X-Ray Absorption-Edge Analysis, *Develop. Appl. Spectrosc.* **2**, 297–307 (1963).

(*H6*) Hakkila, E. A., R. G. Hurley, and G. R. Waterbury, X-Ray Fluorescence Spectrometric Determination of Zirconium and Molybdenum in the Presence of Uranium, *Anal. Chem.* **36**, 2094–7 (1964).

(*H7*) Hakkila, E. A., R. G. Hurley, and G. R. Waterbury, A Three-Wavelength X-Ray Absorption-Edge Method for Determination of Plutonium in Nitrate Media, *Anal. Chem.* **38**, 425–8 (1966).

(*H8*) Hakkila, E. A., R. G. Hurley, and G. R. Waterbury, Solvent Extraction—X-Ray Spectrometric Measurement of Microgram Quantities of Tantalum, *Anal. Chem.* **40**, 818–20 (1940).

(*H9*) Hale, C. C. and W. H. King, Jr., Direct Nickel Determination in Petroleum Oils by X-Ray at the 0.1-ppm Level, *Anal. Chem.* **33**, 74–7 (1961).

(*H10*) Hall, E. T., X-Ray Fluorescent Analysis Applied to Archaeology, *Archaeometry* **3**, 29–35 (1960).

(*H11*) Hall, E. T., M. S. Banks, and J. M. Stern, Uses of X-Ray Fluorescent Analysis in Archaeology, *Archaeometry* **7**, 84–9 (1964).

(*H12*) Hall, T. Nondispersive X-Ray Fluorescence Unit for Analysis of Biological Tissue Sections, *Advan. X-Ray Anal.* **1**, 297–313 (1958).

(*H13*) Hall, T., X-Ray Fluorescence Analysis in Biology, *Science* **134**, 449–55 (1961).

(*H14*) Hanson, H. P., W. E. Flynt, and J. E. Dowdey, Improved Measuring Apparatus for X-Ray Absorption Spectrometry, *Rev. Sci. Instrum.* **29**, 1107–11 (1958).

(*H15*) Hanson, V. F., Quantitative Elemental Analysis of Art Objects by Energy-Dispersive X-Ray Fluorescence Spectroscopy, *Appl. Spectrosc.* **27**, 309–33 (1973).

(*H16*) Harms, J., Automatic Dead-Time Correction for Multichannel Pulse-Height Analyzers at Variable Counting Rates, *Nucl. Instrum. Methods* (*Neth.*) **53**, 192–6 (1967).

(*H17*) Harris, L. A., Secondary Electron Spectroscopy, *Ind. Res.* **10**(2), 52–6 (1968).

(*H18*) Harris, L. A., Analysis of Materials by Electron-Excited Auger Electrons, *J. Appl. Phys.* **39**, 1419–27 (1968).

(*H19*) Harris, T. J., and E. Mathieson, Pulse-Shape Discrimination in Proportional Counters—Experimental Results with an Optimized Electronic System, *Nucl. Instrum. Methods* (*Neth.*) **96**, 397–403 (1971).

(*H20*) Harvey, C. E., "A Method of Semiquantitative Spectrographic Analysis," Applied Research Labs., Glendale, Calif., 285 pp. (1947).

(*H21*) Harvey, P. K., D. M. Taylor, R. D. Hendry, and F. Bancroft, An Accurate Fusion Method for the Analysis of Rocks and Chemically Related Materials by X-Ray Fluorescence Spectrometry, *X-Ray Spectrom.* **2**, 33–44 (1973).

(*H22*) Hasler, M. F., and J. W. Kemp, Use of Multichannel Recording in X-Ray Fluorescent Analysis, *Amer. Soc. Test. Mater. Spec. Tech. Publ.* **157**, 34–42 (1954).

(*H23*) Hayakawa, K., H. Okano, S. Kawase, and S. Yamamoto, Auger-Electron Emission Micrography and Microanalysis of Solid Surfaces, *Advan. X-Ray Anal.* **17**, 498–508 (1974).

(*H24*) Hayashi, S., K, Sugahara, and M. Shimojo, Rapid Determination of Micro Amounts of Copper, Titanium, and Chromium in Polyethylene by X-Ray Fluorescence Spectrometry, *Jap. Anal.* **18**, 862–8 (1969).

(*H25*) Hayashi, S., K. Sugahara, and J. K. Teranishi, Rapid Determination of Micro Amounts of Aluminum and Chlorine in Polyethylene by X-Ray Fluorescence Spectrometry, *Jap. Anal.* **12**, 1502–8 (1968).

(*H26*) Heath, R. L., Application of High-Resolution Solid-State Detectors for X-Ray Spectrometry—a Review, *Advan. X-Ray Anal.* **15**, 1–35 (1972).

(*H27*) Heinrich, K. F. J., Pulse-Height Selection in X-Ray Fluorescence, *Advan. X-Ray Anal.* **4**, 370–81 (1961).

(*H28*) Heinrich, K. F. J., X-Ray Probe with Collimation of the Secondary Beam, *Advan. X-Ray Anal.* **5**, 516–26 (1962).

(*H29*) Heinrich, K. F. J., Instrumental Developments for Electron Microprobe Readout, *Advan. X-Ray Anal.* **7**, 382–94 (1964).

(*H30*) Heller, H. A., Analytical Control in Electroplating Technology, *Plating* **56**, 277–85 (1969).

(*H31*) Hendee, C. H., and S. Fine, Moseley's Law Applied to Proportional Counter Resolution of Adjacent Elements, *Phys. Rev.* **95**, 281–3 (1954).

(*H32*) Hendee, C. F., S. Fine, and W. B. Brown, Gas-Flow Proportional Counter for Soft X-Ray Detection, *Rev. Sci. Instrum.* **27**, 531–5 (1956).

(*H33*) Henke, B. L., Microanalysis with Ultrasoft X-Radiations, *Advan. X-Ray Anal.* **5**, 285–305 (1962).

(*H34*) Henke, B. L., X-Ray Fluorescence Analysis for Sodium, Fluorine, Oxygen, Nitrogen, Carbon, and Boron, *Advan. X-Ray Anal.* **7**, 460–88 (1964).

(*H35*) Henke, B. L., Some Notes on Ultrasoft X-Ray Fluorescence Analysis—10 to 100 Å Region, *Advan. X-Ray Anal.* **8**, 269–84 (1965).

(*H36*) Henke, B. L., Production, Detection, and Application of Ultrasoft X-Rays; in "X-Ray Optics and X-Ray Microanalysis," H. H. Pattee, V. E. Cosslett, and A. Engstrom, eds.; Academic Press, N. Y., pp. 157–72 (1963).

(*H37*) Henke, B. L., Spectroscopy in the 15–150 Å Ultrasoft X-Ray Region; in "X-Ray Optics and Microanalysis," R. Castaing, P. Deschamps, and J. Philibert, eds.; Hermann, Paris, pp. 440–53 (1966).

(*H38*) Henke, B. L., and R. E. Lent, Some Recent Work in Low-Energy X-Ray and Electron Analysis, *Advan. X-Ray Anal.* **12**, 480–95 (1969).

(*H39*) Hercules, D. M., Electron Spectroscopy, *Anal. Chem.* **42**(1), 20A, 21A–28A, 30A, 32A, 34A–35A, 38A–40A (1970).

(*H40*) Herglotz, H., Roentgen Tube with Transparent Anode for Secondary Excitation of Roentgen Spectra, *Osterr. Ing.-Arch.* **6**, 135–40 (1952).

(*H41*) Herglotz, H. K., Miniature X-Ray Equipment for Diffraction and Fluorescence Analysis, *Advan. X-Ray Anal.* **16**, 260–72 (1973).

(*H42*) Hermes, O. D., and P. C. Ragland, Quantitative Chemical Analysis of Minerals in Thin Section with the X-Ray Macroprobe, *Amer. Mineral.* **52**, 493–508 (1967).

(*H43*) Herzog, L. F., Age Determination by X-Ray Fluorescence Rubidium–Strontium Measurement in Lepidolite, *Science* **132**, 293–5 (1960).

(*H44*) Hinckley, D. N., and T. F. Bates, X-Ray Fluorescence Method for Quantitative Determination of Small Amounts of Montmorillonite in Kaolin Clays, *Amer. Mineral.* **45**, 239–41 (1960).

(*H45*) Hirokawa, K., Determination of Impurities in Some Nonferrous Metals by Fluorescent X-Ray Spectroscopy, *Sci. Rep. Res. Inst. Tohuku Univ., Ser. A* **13**, 263–70 (1961).

(*H46*) Hirokawa, K., Determination of Alloyed Elements in Special Steels by X-Ray Fluorescent Spectroscopy. I. Nickel and Chromium in Nickel-Chromium Steel and Tungsten in Tungsten Steel, *Sci. Rep. Res. Inst. Tohuku Univ., Ser. A* **14**, 278–87 (1962).

(*H47*) Hirokawa, K., and H. Goto, Determination of Small Amounts of Copper and Iron with X-Ray Fluorescent Spectroscopy—Combination with Organometallic Precipitation, *Bull. Chem. Soc. Jap.* **35**, 961–4 (1962).

(*H48*) Hirokawa, K., and H. Goto, Analysis of Alloys by Fluorescent X-Ray Spectroscopy; Nondestructive Addition Method, *Z. Anal. Chem.* **185**, 124–35 (1962).

(*H49*) Hirokawa, K., and H. Goto, Determination of Uranium in Thorium by X-Ray Fluorescent Spectroscopy—Organometallic Precipitation Method, *Bull. Chem. Soc. Jap.* **35**, 964–6 (1962).

(*H50*) Hirokawa, K., T. Shimanuki, and H. Goto, Determination of Manganese in Ferromanganese and Chromium in Ferrochromium by X-Ray Fluorescent Spectroscopy, *Sci. Rep. Res. Inst. Tokuku Univ., Ser. A* **15**, 124–32 (1963).

(*H51*) Hirokawa, K., T. Shimanuki, and H. Goto, Simultaneous Determination of Alloy Film Thickness and Its Composition with X-Ray Fluorescent Spectroscopy—Determination of the Composition and Thickness of Nickel–Iron Alloy Films on Copper, *Z. Anal. Chem.* **190**, 309–15 (1962).

(*H52*) Hirokawa, K., M. Suzuki, and H. Goto, Effect of Backing Metals in the X-Ray Fluorescence Spectral Analysis of Metal Films and Its Application—Absorption-Edge Effect, *Z. Anal. Chem.* **199**, 89–94 (1964).

(*H53*) Hirt, R. C., W. R. Doughman, and J. B. Gisclard, Application of X-Ray Emission Spectrography to Air-Borne Dusts in Industrial Hygiene Studies, *Anal. Chem.* **28**, 1649–51 (1956).

(*H54*) Hisano, K., and K. Oyama, Determination of Tantalum by X-Ray Fluorescence Analysis in Treated Tin Slag, *Jap. Anal.* **17**, 1373–7 (1968).

(*H55*) Hockings, W. A., Particle- and Grain-Size Measurement by X-Ray Fluorescence, *Powder Technol.* **3**(1), 29–40 (1969).

(*H56*) Hoffmeister, W., X-Ray Fluorescence Determination of Argon in Silicon Dioxide Films Made by Cathodic Sputtering, *Z. Anal. Chem.* **245**, 244–6 (1969).

(*H57*) Hofstadter, R., and J. A. McIntyre, Gamma-Ray Spectroscopy with Crystals of NaI(Tl), *Nucleonics* **7**(3), 32–7 (1950).

(*H58*) Hollander, J. M., and W. L. Jolly, X-Ray Photoelectron Spectroscopy, *Accounts Chem. Res.* **3**, 193–200 (1970).

(*H59*) Holliday, J. E., Interpretation of Changes in Shape of *K* Emission Bands of Light Elements with Chemical Combination, *Develop. Appl. Spectrosc.* **5**, 77–105 (1966).

(*H60*) Holliday, J. E., Soft X-Ray Spectroscopy in the 13 to 44 Å Region, *J. Appl. Phys.* **33**, 3259–65 (1962).

(*H61*) Holliday, J. E., Soft X-Ray Spectrometer Using a Flow Proportional Counter, *Rev. Sci. Instrum.* **31**, 891–5 (1960).

(*H62*) Hooper, P. R., Rapid Analysis of Rocks by X-Ray Fluorescence, *Anal. Chem.* **36**, 1271–6 (1964).

(*H63*) Horowitz, P., and J. A. Howell, Scanning X-Ray Microscope Using Synchrotron Radiation, *Science* **178**, 608–11 (1972).

(*H64*) Hoskins, J. H., Universal Vacuum X-Ray Spectrograph, *Norelco Rep.* **7**, 111–13, 124 (1960).

(*H65*) Houk, W. W., and L. Silverman, Determination of Oxygen to Uranium Ratio in Uranium Dioxides by Fluorescent X-Ray Analysis, *Pittsburgh Conf. Anal. Chem. Appl. Spectrosc., 10th* (1959); Abstr.: *Spectrochim. Acta* **15**, 308 (1959).

(*H66*) Houpt, P. M., An Element-Specific X-Ray Fluorescence Scanner for Thin-Layer Chromatograms, *X-Ray Spectrom.* **1**, 37–8 (1972).

(*H67*) Hower, J., Matrix Corrections in the X-Ray Spectrographic Trace-Element Analysis of Rocks and Minerals, *Amer. Mineral.* **44**, 19–32 (1959).

(*H68*) Huberman, H., G. Warner, and F. Widman, Trace Metals Analysis by X-Ray Emission Spectroscopy, *Norelco Rep.* **20**(3), 10–15 (1973).

(*H69*) Hudgens, C. R., and G. Pish, Gas Analysis by X-Ray Fluorescence Emission Spectrometry, *Anal. Chem.* **37**, 414 (1965).

(*H70*) Hudgens, C. R., and G. Pish, X-Ray Fluorescence Emission Analysis of Slurries, *Develop. Appl. Spectrosc.* **5**, 25–9 (1966).

(*H71*) Hughes, D. J., and J. E. Davey, X-Ray Fluorescence Determination of Light Elements in Volatile Liquids *in Vacuo* with a Newly Designed Sample Cell, *Analyst* (*London*) **95**, 992–4 (1970).

(*H72*) Hughes, H. K., and J. W. Wilczewski, *K*-Capture Spectroscopy—Iron-55 X-Ray Absorption Determination of Sulfur in Hydrocarbons, *Anal. Chem.* **26**, 1889–93 (1954).

(*H73*) Hunter, C. B., and J. R. Rhodes, Particle-Size Effects in X-Ray Emission Analysis—Formulae for Continuous Size Distributions, *X-Ray Spectrom.* **1**, 107–11 (1972).

(*H74*) Hurst, C. V., and G. L. Glen, Simultaneous Determination of I_0 and I for X-Ray Absorption Spectroscopy Using a Monitored-Beam Technique, *Advan. X-Ray Anal.* **14**, 352–9 (1971).

(*I1*) Ingersoll, R. M., and D. H. Derouin, Application of Color Photography to Electron Microbeam Probe Samples, *Rev. Sci. Instrum.* **40**, 637–9 (1969).

(*I2*) International Union of Pure and Applied Chemistry (IUPAC), Commission V4; Nomenclature, Symbols, Units, and Their Usage in Spectrochemical Analysis. Part. IV. X-Ray Spectrochemical Analysis; in preparation.

(*J1*) Jackwerth, E., and H. G. Kloppenburg, Quantitative Evaluation of Paper Chromatograms of Inorganic Ions by Means of X-Ray Fluorescence Analysis, *Naturwissenschaften* **47**, 444 (1960).

(*J2*) Jackwerth, E., and H. G. Kloppenburg, Quantitative Evaluation of Paper Chromatograms in Trace Analysis by X-Ray Fluorescence Spectroscopy, *Z. Anal. Chem.* **179**, 186–96 (1961).

(*J3*) Jackwerth, E., and H. G. Kloppenburg, X-Ray Fluorescence Spectroscopic Determination of Traces of Arsenic by Using a Modified Gutzeit Method, *Z. Anal. Chem.* **186**, 428–36 (1962).

(*J4*) Jacob, L., R. Noble, and H. Yee, Photomultiplier Soft X-Ray Spectrometer, *J. Sci. Instrum.* **37**, 460–3 (1960).

(*J5*) Jacquet, P. A., Electrolytic and Chemical Polishing, *Met. Rev.* **1**, 157–211 (1956).

(*J6*) Jaklevic, J. M., F. S. Goulding, and D. A. Landis, High-Rate X-Ray Fluorescence Analysis by Pulsed Excitation, *IEEE Trans. Nucl. Sci.* **NS-19**(3), 392–5 (1972).

(*J7*) Jaklevic, J. M., R. D. Giauque, D. F. Malone, and W. L. Searles, Small X-Ray Tubes for Energy-Dispersive Analysis Using Semiconductor Spectrometers, *Advan. X-Ray Anal.* **15**, 266–73 (1972).

(*J8*) Jecht, U., and I. Petersohn, Counters and Analyzer Crystals for the X-Ray Fluorescence Analysis of Elements of Low Atomic Number, *Siemens-Z.* **41**, 59–63 (1967).

(*J9*) Jenkins, R., Fractionation Effects in the Gas Supply of Flow Proportional Counters, *J. Sci. Instrum.* **41**, 696–7 (1964).

(*J10*) Jenkins, R., Modification of a Philips PW-1540 X-Ray Spectrometer for Non-dispersive Analysis, *J. Sci. Instrum.* **42**, 480–2 (1965).

(*J11*) Jenkins, R., Use of the X-Ray Spectrometer/Computer Combination for Analysis of Wide Concentration Range Alloys, *Norelco Rep.* **19**, 14–20 (1972).

(*J12*) Jenkins, R., Recent Developments in Analyzing Crystals for X-Ray Spectrometry, *X-Ray Spectrom.* **1**, 23–8 (1972).

(*J13*) Jenkins, R., Editorial [on Notation for X-Ray Spectral Lines], *X-Ray Spectrom.* **1**, 81 (1972).

(*J14*) Jenkins, R., and A. Campbell-Whitelaw, Determination of Interelement Correction Factors for Matrix Correction Procedures in X-Ray Fluorescence Spectrometry, *Can. Spectrosc.* **15**(2), 32–8, 52 (1970).

(*J15*) Jenkins, R., and D. J. Haas, Hazards in the Use of X-Ray Analytical Instrumentation, *X-Ray Spectrom.* **2**, 135–41 (1973).

(*J16*) Jenkins, R., and P. W. Hurley, Spark Erosion; in "Proceedings of the 5th Conference on X-Ray Analytical Methods," Swansea, Wales, U. K., J. Buwalda, ed.; Scientific and Analytical Equipment Dept., N. V. Philips Gloeilampenfabrieken, Eindhoven, Neth., pp. 88–9 (1966).

(*J17*) Jenkins, R., J. F. Croke, and R. G. Westberg, Present Status of Sorbitol Hexaacetate as an X-Ray Monochromator, *X-Ray Spectrom.* **1**, 59–68 (1972).

(*J18*) Johann, H. H., Intense X-Ray Spectra Obtained with Curved Crystals, *Z. Phys.* **69**, 185–206 (1931).

(*J19*) Johansson, T., New Focusing X-Ray Spectrometer, *Z. Phys.* **82**, 507–28 (1933).

(*J20*) Johnson, C. M., and P. R. Stout, Fluorescent X-Ray Spectrometry—Interferences from Compton Scattering from Matrices of Low Atomic Number, *Anal. Chem.* **30**, 1921–3 (1958).

(*J21*) Johnson, J. L., and B. E. Nagel, Microanalysis on Controlled Spot Test Paper by X-Ray Fluorescence, *Mikrochim. Acta* **3**, 525–31 (1963).

(*J22*) Johnson, M. P., P. R. Beeley, and J. Nutting, Application of X-Ray Fluorescence Probe Analysis to the Study of Segregation in Steels, *Advan. X-Ray Anal.* **8**, 259–68 (1965).

(*J23*) Johnson, M. P., P. R. Beeley, and J. Nutting, X-Ray Fluorescence Technique for the Study of Segregation in Steel, *J. Iron Steel Inst.* **201**, 32–5 (1967); in "X-Ray and Electron Microscopy News," *Siemens Rev.* **34** (Special Issue) 46–50, (1967).

(*J24*) Jones, M. P., J. Gavrilovic, and C. H. J. Beaven, Color Synthesis as an Aid to Electron-Probe X-Ray Microanalysis, *Trans. Inst. Mining. Met.* **75B**, 274–7 (1966).

(*J25*) Jones, R. A., Determination of Manganese in Gasoline by X-Ray Emission Spectrography, *Anal. Chem.* **31**, 1341–4 (1959).

(*J26*) Jones, R. A., Determination of Sulfur in Gasoline by X-Ray Emission Spectrography, *Anal. Chem.* **33**, 71–4 (1961).

(*J27*) Jönsson, A., Study of the Intensities of Soft X-Rays and Their Dependence on Potential, *Z. Phys.* **43**, 845–63 (1927).

(*K1*) Kaelble, E. F., and G. J. McEwan, X-Ray Spectrographic Measurement of Wax Film Thickness, *11th Ann. Symp. Spectrosc.*, Chicago (1960), unpublished; reviewed in detail in Reference (*25j*).

(*K2*) Kalman, Z. H., and L. Heller, Theoretical Study of X-Ray Fluorescent Determination of Traces of Heavy Elements in a Light Matrix, *Anal. Chem.* **34**, 946–51 (1962).

(*K3*) Kanaris-Sotiriou, R., and G. Brown, Diminution of Sulfur Contamination of Powder Specimens in X-Ray Fluorescence Analysis, *Analyst (London)* **94**, 780–1 (1969).

(*K4*) Kang, C. C., E. W. Keel, and E. Solomon, Determination of Traces of Vanadium, Iron, and Nickel in Petroleum Oils by X-Ray Emission Spectrometry, *Anal. Chem.* **32**, 221–5 (1960).

(*K5*) Karttunen, J. O., Separation and Fluorescent X-Ray Spectrometric Determination of Zirconium, Molybdenum, Ruthenium, Rhodium, and Palladium in Solution in Uranium-Base Fissium Alloys, *Anal. Chem.* **35**, 1044–9 (1963).

(*K6*) Karttunen, J. O., H. B. Evans, D. J. Henderson, P. J. Markovich, and R. L. Niemann, Portable Fluorescent X-Ray Instrument Utilizing Radioisotope Sources, *Anal. Chem.* **36**, 1277–82 (1964).

(*K7*) Keesaer, W. C., Establishment of *Q* Rating Factor for Decorative Chromium Plate by X-Ray Fluorescence, *Advan. X-Ray Anal.* **3**, 77–94 (1959).

(*K8*) Kehl, W. L., and R. G. Russell, Fluorescent X-Ray Spectrographic Determination of Uranium in Waters and Brines, *Anal. Chem.* **28**, 1350–1 (1956).

(*K9*) Kemp, J. W., M. F. Hasler, J. L. Jones, and L. Zeitz, Multichannel Instruments for Fluorescent X-Ray Spectroscopy, *Spectrochim. Acta* **7**, 141–8 (1955).

(*K10*) Kepros, J. G., E. M. Eyring, and F. W. Cagle, Jr., Experimental Evidence of an X-Ray Laser, *Proc. Nat. Acad. Sci. (U. S.)* **69**, 1744–5 (1972); *Phys. Today* **25**(10), 18–19 (1972).

(*K11*) Kilday, B. A., and R. E. Michaelis, Determination of Lead in Leaded Steels by X-Ray Spectroscopy, *Appl. Spectrosc.* **16**, 136–8 (1962).

(*K12*) Kiley, W. R., Function and Application of Counters and the Pulse-Height Analyzer, *Norelco Rep.* **7**, 143–9 (1960).

(*K13*) Kiley, W. R., and R. W. Deichert, Design Considerations for On-Stream X-Ray Analysis, *Advan. X-Ray Anal.* **6**, 429–35 (1963).

(*K14*) Kiley, W. R., and J. A. Dunne, X-Ray Detectors, *Amer. Soc. Test. Mater. Spec. Tech. Publ.* **STP-349**, 24–40 (1964).

(*K15*) Kimoto, S., M. Sato, H. Kamada, and T. Ui, The Primary X-Ray Analyzer, *Advan. X-Ray Anal.* **9**, 508–14 (1966).

(*K16*) Kirkpatrick, P., Theory and Use of Ross Filters, *Rev. Sci. Instrum.* **10**, 186–91 (1939); **15**, 223–9 (1944).

(*K17*) Klockenkämper, R., Influence of Spectral and Energy Resolution on Detection Power in X-Ray Spectral Analysis. I. Dependence of Detection Limit on Spectrometer Parameters, *Spectrochim. Acta* **26B**, 547–66 (1971).

(*K18*) Klockenkämper, R., Influence of Spectral and Energy Resolution on Detection Power in X-Ray Spectral Analysis. II. Determination of Optimal Spectrometer Parameters, *Spectrochim. Acta* **26B**, 567–76 (1971).

(*K19*) Klockenkämper, R., K. Laqua, and H. Massmann, Improvement of Detection Power in X-Ray Spectral Analysis with a Double-Crystal Spectrometer, *Spectrochim. Acta* **26B**, 577–93 (1971).

(*K20*) Knapp, K. T., R. H. Lindahl, and A. J. Mabis, An X-Ray Absorption Method for Elemental Analysis, *Advan. X-Ray Anal.* **7**, 318–24 (1964).

(*K21*) Koch, G. P., N. Christ, and J. L. Weber, Jr., Spectrochemical Standards for Aluminum Alloys, *Anal. Chem.* **36**, 1957–61 (1964).

(*K22*) Kodama, H., J. E. Brydon, and B. C. Stone, X-Ray Spectrochemical Analysis of Silicates Using Synthetic Standards with a Correction for Interelemental Effects by a Computer Method, *Geochim. Cosmochim. Acta* **31**, 649–59 (1967).

(*K23*) Koenig, J. H., J. A. Schoeffel, and G. J. Carron, X-Ray Spectrographic Camera, *Rev. Sci. Instrum.* **39**, 1635–8 (1968).

(*K24*) Koh, P. K., and B. Caugherty, Metallurgical Applications of X-Ray Fluorescent Analysis, *J. Appl. Phys.* **23**, 427–33 (1952).

(*K25*) Kohler, T. R., and W. Parrish, X-Ray Diffractometry of Radioactive Samples, *Rev. Sci. Instrum.* **26**, 374–9 (1955).

(*K26*) Kohler, T. R., and W. Parrish, Conversion of Quantum Counting Rate to Roentgens, *Rev. Sci. Instrum.* **27**, 705–6 (1956).

(*K27*) Kokotailo, G. T., and G. F. Damon, Determination of Bromine in Liquids by X-Ray Fluorescent Spectroscopy, *Anal. Chem.* **25**, 1185–7 (1963).

(*K28*) Kowalski, B. R., and T. L. Isenhour, X-Ray Emission Analysis of High Atomic Number Elements—a New Excitation Source Using Monoenergetic 477-keV γ-Rays from the ^{10}B (n,α) ^{7m}Li $\rightarrow$ ^{7}Li Reaction, *Appl. Spectrosc.* **23**, 1–4 (1969).

(*K29*) Kramers, H. A., Theory of X-Ray Absorption and of the Continuous X-Ray Spectrum, *Phil. Mag.* **46**, 836–71 (1923).

(*K30*) Krause, M. O., Photoelectron Spectrometry—a New Approach to X-Ray Analysis, *Advan. X-Ray Anal.* **16**, 74–89 (1973).

(*K31*) Kulenkampff, H., Continuous X-Ray Spectrum, *Ann. Phys.* **69**, 548–96 (1923).

(*K32*) Kullbom, S. D., W. K. Pollard, and H. F. Smith, Combined Infrared and X-Ray Spectrometric Method for Determining Sulfonate and Sulfate Concentration of Detergent Range Alkylbenzene Sulfonate Solutions, *Anal. Chem.* **37**, 1031–4 (1965).

(*K33*) Kuroha, T., and S. Shibuya, Determination of Trace Metals by Fluorescent X-Ray Spectrometry with Solvent-Extraction, Thin-Film Method, *Jap. Anal.* **17**, 801–5 (1968).

(*K34*) Kustner, H., Intense Monochromatic X-Rays Obtained from Ordinary Tubes without Spectroscopic Apparatus, *Z. Phys.* **70**, 324–47 (1931).

(*K35*) Kustner, H., Monochromatic X-Rays, *Z. Phys.* **77**, 52–9 (1932).

(*L1*) Lachance, G. R., and R. J. Traill, A Practical Solution to the Matrix Problem in X-Ray Analysis. I. Method, *Can. Spectrosc.* **11**(2), 43–8 (1966).

(*L2*) Ladell, J., and W. Parrish, Determination of Contamination of X-Ray Tubes, *Philips Res. Rep.* **14**, 401–20 (1959).

(*L3*) Ladell, J., and N. Spielberg, Laue Spectrometer for Multichannel X-Ray Spectrochemical Analysis, *Rev. Sci. Instrum.* **31**, 23–9 (1960).

(*L4*) Lamb, F. W., L. M. Niebylski, and E. W. Kieffer, Determination of Tetraethyllead in Gasoline by X-Ray Fluorescence, *Anal. Chem.* **27**, 129–32 (1955).

(*L5*) Lambert, M. C., Some Practical Aspects of X-Ray Spectrography, *Norelco Rep.* **6**, 37–51 (1959).

(*L6*) Lang, A. R., Design and Performance of X-Ray Proportional Counters, *J. Sci. Instrum.* **33**, 96–102 (1956).

(*L7*) Langheinrich, A. P., and J. W. Forster, Application of Radioisotope Nondispersive X-Ray Spectrometry to Analysis of Molybdenum, *Advan. X-Ray Anal.* **11**, 275–86 (1968).

(*L8*) Larson, J. A., and W. A. Ziemke, Increasing the Versatility of the General Electric Miniature X-Ray Probe, *Rev. Sci. Instrum.* **41**, 1576–7 (1970).

(*L9*) Larson, J. A., W. R. Pierson, and M. A. Short, Corrected Equation for the Standard-Addition Technique in X-Ray Fluorescence Spectrometric Analysis, *Anal. Chem.* **45**, 616 (1973).

(*L10*) Larson, J. O., R. A. Winkler, and J. C. Guffy, Glass-Fusion Method for X-Ray Fluorescence Analysis, *Advan. X-Ray Anal.* **10**, 489–93 (1967).

(*L11*) Lavrent'ev, Y. G., X-Ray Spectral Determination of Tin by Internal-Standard Methods, *Zavod. Lab.* **35**, 1118 (1969).

(*L12*) Law, S. L., and W. J. Campbell, Resin-Loaded Papers—a Versatile Medium for Sampling and Standardization, *Advan. X-Ray Anal.* **17**, 279–92 (1974).

(*L13*) Lazar, V. A., and K. C. Beeson, Determination of Copper and Molybdenum in Plants by X-Ray Spectrography, *J. Assoc. Offic. Agr. Chemists* **41**, 416–9 (1958).

(*L14*) Lee, F. S., and W. J. Campbell, Variation of X-Ray Spectral Line Position with Ambient Temperature Change—A Source of Error in X-Ray Spectrography, *Advan. X-Ray Anal.* **8**, 431–42 (1965).

(*L15*) Legrand, C., Measurement of the Thickness of Thin Layers by X-Ray Fluorescence, *J. Chim. Phys.* **53**, 587–92 (1956).

(*L16*) Legrand, C., and J. J. Trillat, Investigation of Surface Films by X-Ray Fluorescence, *Rev. Met.* **53**, 645–8 (1956).

(*L17*) Lehovillier, R., C. Samson, and F. Claisse, Selection of Diluents for the Analysis of Powders by X-Ray Fluorescence, *Can. J. Spectrosc.* **17**(5), 141–5 (1972).

(*L18*) Leroux, J., Method for Finding Mass-Absorption Coefficients by Empirical Equations and Graphs, *Advan. X-Ray Anal.* **5**, 153–60 (1962).

(*L19*) Leroux, J., Absorption Coefficients (Mass); Empirical Determination Between 1 and 50 keV; in "Encyclopedia of X-Rays and Gamma Rays," G. L. Clark, ed.; Reinhold Publishing Corp., N. Y., pp. 9–15 (1963).

(*L20*) Leroux, J., and M. Mahmud, X-Ray Quantitative Analysis by an Emission–Transmission Method, *Anal. Chem.* **38**, 76–82 (1966).

(*L21*) Leroux, J., P. A. Maffett, and J. L. Monkman, Microdetermination of Heavy Elements Such as Mercury and Iodine in Solution by X-Ray Absorption, *Anal. Chem.* **29**, 1089–92 (1957).

(*L22*) Le Sage, R. V., and V. Grubis, Use of the Thallium Phthalate Crystal for X-Ray Fluorescence Analysis in the Very Long Wavelength Region, *X-Ray Spectrom.* **2**, 189–205 (1973).

(*L23*) Lewis, C. L., W. L. Ott, and N. M. Sine, "The Analysis of Nickel," Pergamon Press. N. Y., ch. 4, pp. 85–102 (1966).

(*L24*) Lewis, R., S. Moll, R. E. Ogilvie, and S. Shur, A Method for Identification of Stratospheric Particles, *Advan. Metals Res. Corp., Res. Rep.* **AMR-1005**, 65 pp. (1960).

(*L25*) Leyden, D. E., Chelating Ion-Exchange Resins and X-Ray Fluorescence, *Advan. X-Ray Anal.* **17**, 293–301 (1974).

(*L26*) Liebhafsky, H. A., Analytical Methods Based on X-Ray Absorption, *Anal. Chem.* **25**, 689–92 (1953).

(*L27*) Liebhafsky, H. A., and P. D. Zemany, Film Thickness by X-Ray Emission Spectrography, *Anal. Chem.* **28**, 455–9 (1956).

(*L28*) Liebson, S. H., and H. Friedman, Self-Quenching Halogen-Filled Counters, *Rev. Sci. Instrum.* **19**, 303–6 (1948).

(*L29*) Linares, R. C., J. B. Schroeder, and L. A. Hurlbut, Applications of X-Ray-Excited Optical Fluorescence to Analytical Chemistry, *Spectrochim. Acta* **21**, 1915–20 (1965).

(*L30*) Lincoln, A. J., and E. N. Davis, Determination of Platinum in Alumina-Base Reforming Catalyst by X-Ray Spectroscopy, *Anal. Chem.* **31**, 1317–20 (1959).

(*L31*) Lloyd, L. C., Determination of Argon in RF-Sputtered Silicon Dioxide by X-Ray Emission, *Advan. X-Ray Anal.* **12**, 601–11 (1969).

(*L32*) Loch, G., Optimization of X-Ray Spectrometric Analysis by Application of Specific Detector Gases, *X-Ray Spectrom.* **2**, 125–8 (1973).

(*L33*) Lokay, J., and R. Shinn, Glass Analysis by X-Ray Fluorescence, *Glass Ind.* **44**, 431–3 (1963).

(*L34*) Long, A., Detectors for Low-Energy X-Radiation, *J. Brit. Inst. Radio Eng.* **19**, 273–87 (1959).

(*L35*) Long, J. V. P., Applications of and Some Sources of Error in X-Ray Microchemical Analysis: in "X-Ray Microscopy and Microradiography," V. E. Cosslett, A. Engström, and H. H. Pattee, eds; Academic Press, N. Y., pp. 628–35 (1957).

(*L36*) Long, J. V. P., and V. E. Cosslett, Some Methods of Microchemical Analysis; in "X-Ray Microscopy and Microradiography," V. E. Cosslett, A. Engström, and H. H. Pattee, eds.; Academic Press, N. Y., pp. 435–42 (1957).

(*L37*) Long, J. V. P., and H. O. E. Rockert, X-Ray Fluorescence Microanalysis and the Determination of Potassium in Nerve Cells; in "X-Ray Optics and X-Ray Microanalysis," H. H. Pattee, V. E. Cosslett, and A. Engström, eds.; Academic Press, N. Y., pp. 513–21 (1963).

(*L38*) Longobucco, R. J., X-Ray Spectrometric Determination of Major and Minor Constituents in Ceramic Materials, *Anal. Chem.* **34**, 1263–7 (1962).

(*L39*) Loomis, T. C., X-Ray Spectroscopy as an Analytical Tool, *Ann. N. Y. Acad. Sci.* **137**, 284–96 (1966).

(*L40*) Loomis, T. C., and K. H. Storks, X-Ray Microanalysis in Chemistry, *Bell. Labs. Rec.* **45**, 2–7 (1967).

(*L41*) Lopp, V. R., and C. G. Claypool, Direct Determination of Vanadium and Nickel in Crude Oils by X-Ray Fluorescence, *Advan. X-Ray Anal.* **3**, 131–7 (1960).

(*L42*) Lublin, P., Novel Approach to Discrimination in X-Ray Spectrographic Analysis, *Advan. X-Ray Anal.* **2**, 229–38 (1958).

(*L43*) Lublin, P., Determination of Cladding Thickness of Nuclear Fuel Elements by X-Rays, *Norelco Rep.* **6**, 57–9 (1959).

(*L44*) Lucas-Tooth, H. J., Description of the Solartron Vacuum Automatic X-Ray Spectrometer and Program Selection by Punched Card, *Rev. Universelle Mines* **17**, 180–3 (1961).

(*L45*) Lucas-Tooth, H. J., Excitation; in "Proceedings of the 5th Conference on X-Ray Analytical Methods," Swansea, Wales, U. K., J. Buwalda, ed.; Scientific and Analytical Equipment Dept., N. V. Philips Gloeilampenfabrieken, Eindhoven, Neth., pp. 1–5 (1966).

(*L46*) Lucas-Tooth, H. J., and B. J. Price, A Mathematical Method for Investigation of Interelement Effects in X-Ray Fluorescent Analysis, *Metallurgia* **64**, 149–52 (1961).

(*L47*) Lucas-Tooth, H. J., and E. C. Pyne, Accurate Determination of Major Constituents by X-Ray Fluorescent Analysis in the Presence of Large Interelement Effects, *Advan. X-Ray Anal.* **7**, 523–41 (1964).

(*L48*) Luke, C. L., Trace Analysis of Metals by Borax Disk X-Ray Spectrometry, *Anal. Chem.* **35**, 1551–2 (1963).

(*L49*) Luke, C. L., Ultratrace Analysis of Metals with a Curved-Crystal X-Ray Milliprobe, *Anal. Chem.* **36**, 318–22 (1964).

(*L50*) Luke, C. L., Determination of Trace Elements in Inorganic Materials by X-Ray Fluorescence Spectroscopy, *Anal. Chim. Acta* **41**, 237–50 (1968).

(*L51*) Luke, C. L., Determination of Traces of Lithium, Beryllium, or Phosphorus by X-Ray Analysis, *Anal. Chim. Acta* **45**, 365–76 (1969).

(*L52*) Luke, C. L., Determination of Traces of Boron or Sodium by X-Ray Analysis, *Anal. Chim. Acta* **45**, 377–81 (1969).

(*L53*) Lumb, P. G., Standards for the Analysis of Tin–Lead Alloys by X-Ray Spectrometry, *Metallurgia* **80**, 127–8 (1969).

(*L54*) Lund, P. K., and J. C. Mathies, X-Ray Spectroscopy in Biology and Medicine. I. Total Iron (Hemoglobin) Content in Whole Human Blood, *Norelco Rep.* **7**, 127–30 (1960).

(*L55*) Lundquist, L. R., Porosity Test for Thin Electrodeposits on Uranium by X-Ray Fluorescence, *Plating* **52**, 1316 (1965).

(*L56*) Lüschow, H. M., and H. U. Steil, X-Ray Analysis of Tin–Lead–Antimony Alloys, *Z. Anal. Chem.* **245**, 304–11 (1969).

(*L57*) Lynch, J. T., X-Ray Techniques as a Tool to Analyze Intermetallic Coatings, *Plating* **51**, 1173–7 (1964).

(*L58*) Lytle, F. W., J. I. Botsford, and H. A. Heller, X-Ray Emission Spectrographic Analysis of Bastnaesite Rare Earths, *Advan. X-Ray Anal.* **1**, 367–86 (1958).

(*L59*) Lytle, F. W., W. B. Dye, and H. J. Seim, Determination of Trace Elements in Plant Materials by Fluorescent X-Ray Analysis, *Advan. X-Ray Anal.* **5**, 433–46 (1962).

(*M1*) Mabis, A. J., and K. T. Knapp, *Pittsburgh Conf. Anal. Chem. Appl. Spectrosc.* (1964), unpublished.

(*M2*) MacDonald, G. L., Counting Statistics in X-Ray Spectrometry; in "Proceedings of the 4th Conference on X-Ray Analytical Methods," Sheffield, Eng.; Scientific and Analytical Equipment Dept., N. V. Philips Gloeilampenfabrieken, Eindhoven, Neth., pp. 1–10 (1964).

(*M3*) MacDonald, N. C., and J. R. Waldrop, Auger-Electron Spectroscopy in the Scanning-Electron Microscope—Auger-Electron Images, *Appl. Phys. Lett.* **19**, 315–18 (1971).

(*M4*) Mack, M., and N. Spielberg, Statistical Factors in X-Ray Intensity Measurements, *Spectrochim. Acta* **12**, 169–78 (1958).

(*M5*) MacKenzie, A. R., T. Hall, and W. F. Whitmore, Zinc Content of Expressed Human Prostatic Fluid, *Nature* **193**, 72–3 (1962).

(*M6*) MacNevin, W. M., and E. A. Hakkila, Fluorescent X-Ray Spectroscopic Estimations of Palladium, Platinum, Rhodium, and Iridium, *Anal. Chem.* **29**, 1019–22 (1957).

(*M7*) Maneval, D. R., and H. R. Lovell, Determination of Lanthanum, Cerium, Praseodymium, and Neodymium as Major Components by X-Ray Emission Spectroscopy, *Anal. Chem.* **32**, 1289–92 (1960).

(*M8*) Manners, V. J., J. V. Craig, and F. H. Scott, Effect of Specimen Surface Preparation on the Accuracy of X-Ray Fluorescence Analysis of Leaded Copper Alloys, *J. Inst. Metals Bull. Met. Rev.* **95**(6), 173–6 (1967).

(*M9*) Mantel, M., and S. Amiel, Application of High-Resolution X-Ray Spectrometry to Activation Analysis, *Anal. Chem.* **44**, 548–53 (1972).

(*M10*) Mantel, M., and S. Amiel, Simultaneous Determination of Uranium and Thorium by Instrumental Neutron Activation and High-Resolution X-Ray Spectrometry, *Anal. Chem.* **45**, 2393–9 (1973).

(*M11*) Margoshes, M., and S. D. Rasberry, Fitting of Analytical Functions with Digital Computers in Spectrochemical Analysis, *Anal. Chem.* **41**, 1163–72 (1969).

(*M12*) Marshall, D. J., and T. A. Hall, Method for Microanalysis of Thin Films; in "Optique des Rayons X et Microanalyse," R. Castaing, P. Deschamps, and J. Philibert, eds.; Hermann et Cie, Paris, pp. 374–81 (1966).

(*M13*) Marti, W., Determination of the Interelement Effect in the X-Ray Fluorescence Analysis of Steels, *Spectrochim. Acta* **18**, 1499–1504 (1962).

(*M14*) Martin, G. W., and A. S. Klein, Complete Instrumental System for Energy-Dispersive Diffractometry and Fluorescence Analysis, *Advan. X-Ray Anal.* **15**, 254–65 (1972).

(*M15*) Mathies, J. C., and P. K. Lund, X-Ray Spectroscopy in Biology and Medicine. II. Calcium Content of Human Blood Serum, *Norelco Rep.* **7**, 130–3 (1960).

(*M16*) Mathies, J. C., and P. K. Lund, X-Ray Spectroscopy in Biology and Medicine. III. Bromide (Total Bromine) in Human Blood, Serum, Urine, and Tissues, *Norelco Rep.* **7**, 134–5, 139 (1960).

(*M17*) Mathies, J. C., P. K. Lund, and W. Eide, X-Ray Spectroscopy in Biology and Medicine. IV. A Simple, Indirect, Sensitive Procedure for Determination of Nitrogen (Ammonia) at the Microgram and Submicrogram Level, *Anal. Biochem.* **3**, 408–14 (1962); *Norelco Rep.* **9**, 93–5 (1962).

(*M18*) Matocha, C. K., New Briqueting Technique, *Appl. Spectrosc.* **20**, 252–3 (1966).

(*M19*) McCall, J. M., Jr., D. E. Leyden, and C. W. Blount, Rapid Determination of Heavy Elements in Organometallic Compounds Using X-Ray Fluorescence, *Anal. Chem.* **43**, 1324–5 (1971).

(*M20*) McCrary, J. H., and L. D. Looney, Miniature Field-Emission X-Ray Tube, *Rev. Sci. Instrum.* **41**, 1095–6 (1970).

(*M21*) McCrary, J. H., and T. Van Vorous, Use of Field-Emission Tubes in X-Ray Analysis, *Advan. X-Ray Anal.* **15**, 285–94 (1972).

(*M22*) McCue, J. C., L. L. Bird, C. A. Ziegler, and J. J. O'Connor, X-Ray Rayleigh Scattering Method for Determination of Uranium in Solution, *Anal. Chem.* **33**, 41–3 (1961).

(*M23*) McCue, R. A., W. M. Mueller, and P. J. Dunton, Feasibility of the X-Ray Spectrograph as a Continuous Analytical Instrument for Process Control, *Advan. X-Ray Anal.* **1**, 399–407 (1958).

(*M24*) McGinness, J. D., R. W. Scott, and J. S. Mortensen, X-Ray Emission Analysis of Paints by Thin-Film Method, *Anal. Chem.* **41**, 1858–61 (1969).

(*M25*) McGrath, J. G., Jr., New Technique for the Design of an Extreme Ultraviolet Collimator, *Rev. Sci. Instrum.* **39**, 1036–8 (1968).

(*M26*) Mecke, P., Quantitative X-Ray Fluorescence Analysis of Small Amounts of Mixtures by a Filter Difference Method. *Z. Anal. Chem.* **193**, 241–7 (1963).

(*M27*) Mege, R. V., Standard Inert Dilution Method in X-Ray Fluorescence Analysis, *Anal. Chem.* **41**, 42–5 (1969).

(*M28*) Mellish, C. E., X-Ray Fluorescence Spectroscopy—Applications of a Simple Apparatus Using Radioactive Sources, *Research* (*London*) **12**, 212–17 (1959).

(*M29*) Menis, O., E. K. Halteman, and E. E. Garcia, X-Ray Emission Analysis of Plutonium and Uranium Compound Mixtures, *Anal. Chem.* **35**, 1049–52 (1963).

(*M30*) Merrit, J., and E. J. Agazzi, Casting Formvar Films on Water, *Rev. Sci. Instrum.* **38**, 127 (1967).

(*M31*) Metz, W. D., X-Rays from Heavy Ions—New Molecular Phenomena, *Science* **177**, 156–7 (1972).

(*M32*) Michaelis, R. E., "Report on Available Standard Samples, Reference Samples, and High-Purity Materials for Spectrochemical Analysis," *Amer. Soc. Test. Mater. Data Ser.* **DS-2** 156 pp. (1964).

(*M33*) Michaelis, R. E., and B. A. Kilday, Surface Preparation of Solid Metallic Samples for X-Ray Spectrochemical Analysis, *Advan. X-Ray Anal.* **5**, 405–11 (1962).

(*M34*) Miller, D. C., Norelco Portable Spectrometer (Portospec), *Advan. X-Ray Anal.* **3**, 57–67 (1960).

(*M35*) Miller, D. C., Norelco Pinhole Attachment, *Advan. X-Ray Anal.* **4**, 513–20 (1961).

(*M36*) Miller, D. C., Some Considerations in the Use of Pulse-Height Analysis with X-Rays, *Norelco Rep.* **4**, 37–40 (1957).

(*M37*) Miller, D. C., and P. W. Zingaro, Universal Vacuum Spectrograph and Comparative Data on the Intensities Observed in Air, Helium, and Vacuum Paths, *Advan. X-Ray Anal.* **3**, 49–56 (1960).

(*M38*) Mitchell, B. J., X-Ray Spectrographic Determination of Tantalum, Niobium, Iron, and Titanium Oxide Mixtures Using Simple Arithmetic Corrections for Interelement Effects, *Anal. Chem.* **30**, 1894–1900 (1958).

(*M39*) Mitchell, B. J., Complex Systems—Interelement Effects; in "Encyclopedia of Spectroscopy," G. L. Clark, ed.; Reinhold Publishing Corp, N. Y., pp. 736–45 (1960).

(*M40*) Mitchell, B. J., X-Ray Spectrographic Determination of Zirconium, Tungsten, Vanadium, Iron, Titanium, Tantalum, and Niobium Oxides—Application of the Correction Factor Method, *Anal. Chem.* **32**, 1652–6 (1960).

(*M41*) Mitchell, B. J., Prediction of X-Ray Fluorescent Intensities and Interelement Effects, *Anal. Chem.* **33**, 917–21 (1961).

(*M42*) Mitchell, B. J., and F. N. Hopper, Digital Computer Calculation and Correction of Matrix Effects in X-Ray Spectroscopy, *Appl. Spectrosc.* **20**, 172–80 (1966).

(*M43*) Mitchell, B. J., and J. E. Kellam, Unusual Matrix Effects in X-Ray Spectroscopy —A Study of the Range and Reversal of Absorption-Enhancement, *Appl. Spectrosc.* **22**, 742–8 (1968).

(*M44*) Mitchell, B. J., and H. J. O'Hear, General X-Ray Spectrographic Solution Method for Iron-, Chromium-, and/or Manganese-Bearing Materials, *Anal. Chem.* **34**, 1620–5 (1962).

(*M45*) Mitchell, J. W., C. L. Luke, and W. R. Northover, Techniques for Monitoring the Quality of Ultrapure Reagents—Neutron Activation and X-Ray Fluorescence, *Anal. Chem.* **45**, 1503–6 (1973).

(*M46*) Mitzner, B. M., Cell for X-Ray Examination of Compounds Requiring a Special Atmosphere, *Appl. Spectrosc.* **11**, 45–6 (1957).

(*M47*) Mizuike, A., Separation and Preconcentration Techniques; in "Modern Analytical Techniques for Metals and Alloys," Part 1, R. F. Bunshah, ed., Techniques of Metals Research Series, vol. 3, part 1; Interscience Publishers, N. Y., pp. 25–67 (1970).

(*M48*) Moak, W. D., Cooling Liquid Samples for X-Ray Fluorescence Analysis, *Anal. Chem.* **29**, 1906 (1957).

(*M49*) Momoki, K., Nomographic Method of X-Ray Fluorescence Routine Analysis for Daily Variations of Analytical Line Intensities with a Standard X-Ray Spectrograph, *Bunseki Kagaku* **10**, 40–7 (1961); *Chem. Abstr.* **58**, 6165*c* (1963).

(*M50*) Momoki, K., Nonlinear Calibration Curves in Nomographic Fluorescent X-Ray Spectrometry—Comparison of Various Calibration Curves, *Bunseki Kagaku* **10**, 330–6 (1961); *Chem. Abstr.* **55**, 25583*f* (1961).

(*M51*) Moon, R. J., Amplifying and Intensifying the Fluoroscopic Image by Means of a Scanning X-Ray Tube, *Science* **112**, 389–95 (1950).

(*M52*) Moore, J. E., G. P. Happ, and D. W. Stewart, Automatic Direct-Reading X-Ray Spectrometry—Application to Determinations of Silver, *Anal. Chem.* **33**, 61–4 (1961).

(*M53*) Moore, T. M., and D. J. McDonald, eds., Radiation Safety Recommendations for X-Ray Diffraction and Spectrographic Equipment, U. S. Dept. Health, Education, and Welfare (HEW), *DHEW Publ.* **MORP 68-14** (1968).

(*M54*) Moore, T. M., W. E. Gundaker, and J. W. Thomas, eds., Radiation Safety in X-Ray Diffraction and Spectroscopy, U. S. Dept. Health, Education, and Welfare (*HEW*), *DHEW Publ.* **(FDA)72-8009**, **BRH/DEP 72-3**, 239 pp. (1971).

(*M55*) Morrison, G. H., and H. Freiser, "Solvent Extraction in Analytical Chemistry," John Wiley and Sons, Inc., N. Y., 269 pp. (1966).

(*M56*) Mortimore, D. M., P. A. Romans, and J. L. Tews, X-Ray Spectroscopic Determination of Columbium and Tantalum in Rare Earth Ores, *Norelco Rep.* **1**, 107–10 (1954).

(*M57*) Moseley, H. G. J., High-Frequency Spectra of the Elements, *Phil. Mag.* **26**, 1024–34 (1913); **27**, 703–13 (1914).

(*M58*) Mueller, J. I., V. G. Scotti, and J. J. Little, Fluorescent X-Ray Analysis of Highly Radioactive Samples, *Advan. X-Ray Anal.* **2**, 157–66 (1959).

(*M59*) Muller, E. A. W., Attenuation of X-Rays in the Wavelength Range 0.01–10 mμ. I. General Principles, *Arch. Tech. Mess.* **9114-17**, 37–8 (1956); II. Mass Absorption Coefficients μ/ϱ and Position of Absorption Edges; *ibid.* **9114-18**, 107–10 (1956).

(*M60*) Mulligan, B. W., H. J. Caul, S. D. Rasberry, and B. F. Scribner, X-Ray Spectrometric Analysis of Noble-Metal Dental Alloys, *J. Res. Nat. Bur. Stand.* **68A**, 5–8 (1964).

(*M61*) Mulvey T., and A. J. Campbell, Proportional Counters in X-Ray Spectrochemical Analysis, *Brit. J. Appl. Phys.* **9**, 406–10 (1958).

(*M62*) Munch, R. H., Process Stream Analysis by X-Ray Spectroscopy, *Develop. Appl. Spectrosc.* **3**, 45–57 (1964).

(*M63*) Myers, R. H., D. Womeldorph, and B. J. Alley, Method for Adjusting for Particle-Size and Matrix Effects in the X-Ray Fluorescence Analysis Procedure, *Anal. Chem.* **39**, 1031–3 (1967).

(*N1*) Narbutt, K. I., X-Ray Fluorescence Microanalyzer, *Izv. Akad. Nauk SSSR, Ser. Fiz.* **26**, 423–8 (1962); *Chem. Abstr.* **57**, 7877*e* (1962).

(*N2*) Natelson, S., Microanalysis in the Modern Analytical Laboratory of Clinical Chemistry, *Anal. Chem.* **31**(3), 17A–30A (1959).

(*N3*) Natelson, S., Present Status of X-Ray Spectrometry in the Clinical Laboratory, *Trans. N. Y. Acad. Sci.* [2] **26**, 3–26 (1963).

(*N4*) Natelson, S., and B. Sheid, [X-Ray Spectroscopy in the Clinical Laboratory. V.] X-Ray Spectrometric Determination of Strontium in Human Serum and Bone, *Anal. Chem.* **33**, 396–401 (1961).

(*N5*) Natelson, S., and B. Sheid, X-Ray Spectroscopy in the Clinical Laboratory. II. Chlorine and Sulfur—Automatic Analysis of Ultramicro Samples, *Clin. Chem.* **6**, 299–313 (1960).

(*N6*) Natelson, S., and B. Sheid, X-Ray Spectroscopy in the Clinical Laboratory. III. Sulfur Distribution in the Electrophoretic Protein Fractions of Human Serum, *Clin. Chem.* **6**, 314–26 (1960).

(*N7*) Natelson, S., and B. Sheid, X-Ray Spectroscopy in the Clinical Laboratory. IV. Phosphorus—Total Blood Iron as a Measure of Hemoglobin Content, *Clin. Chem.* **7**, 115–29 (1961).

(*N8*) Natelson, S., and B. Sheid, X-Ray Spectroscopy in the Clinical Laboratory. VI. Acid-Labile Protein-Bound Iodine, *Clin. Chem.* **8**, 17–34 (1962).

(*N9*) Natelson, S., M. R. Richelson, B. Sheid, and S. L. Bender, X-Ray Spectroscopy in the Clinical Laboratory. I. Calcium and Potassium, *Clin. Chem.* **5**, 519–31 (1959).

(*N10*) National Bureau of Standards (U. S.), "Standard Materials Issued by the National Bureau of Standards," *Nat. Bur. Stand.* (*U. S.*) *Circ.* **552**, ed. 3, 27 pp. (1959).

(*N11*) National Bureau of Standards (U. S.), Standard Reference Materials—Sources of Information, *Nat. Bur. Stand.* (*U. S.*) *Misc. Publ.* **260-4**, 18 pp. (1965).

(*N12*) National Bureau of Standards (U. S.), Catalog of Standard Reference Materials, *Nat. Bur. Stand.* (*U. S.*) *Spec. Publ.* **260**, 77 pp. (1970).

(*N13*) Neeb, R., Analytical Chemistry of the Platinum Metals. I. X-Ray Fluorimetric Determination of Microgram Amounts of Osmium, *Z. Anal. Chem.* **179**, 21–9 (1961).

(*N14*) Needham, P. B., Jr., and B. D. Sartwell, X-Ray Production Efficiencies for *K*-, *L*-, *M*-, and *N*-Shell Excitation by Ion Impact, *Advan. X-Ray Anal.* **14**, 184–213 (1971).

(*N15*) Nelson, J. T., and R. T. Ellickson, Scintillation Counter for Detection of X-Rays, *J. Opt. Soc. Amer.* **45**, 984–6 (1955).

(*N16*) Nicholson, J. B., and D. B. Wittry, Comparison of the Performance of Gratings and Crystals in the 20–115-Å Region, *Advan. X-Ray Anal.* **7**, 497–511 (1964).

(*N17*) Nickey, D. A., and J. O. Rice, Modified Micro Sample Support for X-Ray Emission Spectrography, *Appl. Spectrosc.* **25**, 383 (1971).

(*N18*) Nikol'skii, A. P., New Principle for Constructing X-Ray and γ Spectrometers, *Prib. Sist. Upr.* **1968**(4), 42–3; *Chem. Abstr.* **69**, 40566 (1969).

(*N19*) Noakes, G. E., An Absolute Method of X-Ray Fluorescence Analysis Applied to Stainless Steels, *Amer. Soc. Test. Mater. Spec. Tech. Publ.* **STP-157**, 57–62 (1954).

(*N20*) Norrish, K., and J. T. Hutton, X-Ray Spectrographic Method for the Analysis of a Wide Range of Geological Samples, *Geochim. Cosmochim. Acta* **33**, 431–53 (1969).

(*O1*) Olson, E. C., and J. W. Shell, Simultaneous Determination of Traces of Selenium and Mercury in Organic Compounds by X-Ray Fluorescence, *Anal. Chim. Acta* **23**, 219–24 (1960).

(*O2*) Ong, P. S., Projection Microscopy and Microanalysis, *Advan. X-Ray Anal.* **5**, 324–34 (1962).

(*O3*) Ong, P. S., E. L. Cheng, and G. Sroka, Use of Multiple Standards for Absorption Correction and Quantitation in FRIEDA, *Advan. X-Ray Anal.* **17**, 269–78 (1974).

(*O4*) Oswin, E. H., Some Aspects of Sample Preparation and Presentation; in "Proceedings of the 5th Conference on X-Ray Analytical Methods," Swansea, Wales, U. K., J. Buwalda, ed.; Scientific and Analytical Equipment Dept., N. V. Philips Gloeilampenfabrieken, Eindhoven, Neth., pp. 67–71 (1966).

(*P1*) Paff, R. J., RCA Laboratories, Princeton, N. J.; unpublished work.

(*P2*) Panson, A. J., and M. Kuriyama, Monitored-Beam X-Ray Absorption Spectrometer, *Rev. Sci. Instrum.* **36**, 1488–9 (1965).

(*P3*) Pantony, D. A., and P. W. Hurley, Statistical and Practical Considerations of Limits of Detection in X-Ray Spectrometry, *Analyst* (*London*) **97**, 497–518 (1972).

(*P4*) Papariello, G. J., H. Letterman, and W. J. Mader, X-Ray Fluorescent Determination of Organic Substances through Inorganic Association, *Anal. Chem.* **34**, 1251–3 (1962).

(*P5*) Park, R. L., and J. E. Houston, Soft X-Ray Appearance-Potential Spectroscopy, *J. Vac. Sci. Technol.* **11**, 1–18 (1974).

(*P6*) Park, R. L., J. E. Houston, and D. G. Schreiner, Soft X-Ray Appearance-Potential Spectrometer for the Analysis of Solid Surfaces, *Rev. Sci. Instrum.* **41**, 1810–12 (1970).

(*P7*) Parker, D. L., An Electronic Partial-Vacuum Regulator, *Rev. Sci. Instrum.* **43**, 1103–6 (1972).

(*P8*) Parker, D. L., Spherically Bent Crystal X-Ray Spectrometer with Variable Curvature, *Advan. X-Ray Anal.* **17**, 521–30 (1974).

(*P9*) Parrish, W., Escape Peaks in X-Ray Diffractometry, *Advan. X-Ray Anal.* **8**, 118–33 (1965).

(*P10*) Parrish, W., X-Ray Intensity Measurements with Counter Tubes, *Philips Tech. Rev.* **17**, 206–21 (1956).

(*P11*) Parrish, W., X-Ray Spectrochemical Analysis, *Philips Tech. Rev.* **17**, 269–86 (1956); *Norelco Rep.* **3**, 24–36 (1956).

(*P12*) Parrish, W., and T. R. Kohler, Use of Counter Tubes in X-Ray Analysis, *Rev. Sci. Instrum.* **27**, 795–808 (1956).

(*P13*) Passell, T. O., J. B. Swedlund, and S. L. Taimuty, Development of X-Ray Fluorescence Analysis Techniques for Small Samples—A Prototype Atmospheric Electron Probe, *Stanford Res. Inst. Final Rep.* **AFCRL-62-933**, 60 pp. (1962).

(*P14*) Pattee, H. H., The Scanning X-Ray Microscope, *J. Opt. Soc. Amer.* **43**, 61–2 (1953).

(*P15*) Payne, K. W., Application of the Lithium Fluoride (220) Crystal in the Analysis of the Lanthanides; in "Proceedings of the 5th Conference on X-Ray Analytical Methods," Swansea, Wales, U. K., J. Buwalda, ed.; Scientific and Analytical Equipment Dept,. N. V. Philips Gloeilampenfabrieken, Eindhoven, Neth., pp. 51–3 (1966).

(*P16*) Pfeiffer, H. G., and P. D. Zemany, Trace Analysis by X-Ray Emission Spectrography, *Nature* **174**, 397 (1954).

(*P17*) Phifer, L. H., Use of X-Ray Fluorescence in the Analysis of Viscose, *Tappi* **52**, 671–3 (1969).

(*P18*) Philips Electronic Instruments, X-Ray Spectrograph, *Rev. Sci. Instrum.* **41**, 605–6 (1970).

(*P19*) Piccolo, B., D. Mitcham, V. W. Tripp, and R. T. O'Connor, Elemental Analysis of Chemically Modified Cotton Textiles, *Appl. Spectrosc.* **20**, 326–9 (1966).

(*P20*) Pike, E. R., Introduction to Soft X-Ray Spectroscopy, *Amer. J. Phys.* **28**, 235–42 (1960).

(*P21*) Pish, G., and A. A. Huffmann, Quantitative Determination of Thorium and Uranium in Solutions by Fluorescent X-Ray Spectrometry, *Anal. Chem.* **27**, 1875–8 (1955).

(*P22*) Plesch, R., Statistics and Probability in Radiation Measuring Technique; in "X-Ray and Electron Microscopy News," *Siemens Rev.* **34** (Special Issue), 73–9 (1967).

(*P23*) Pluchery, M., Absolute Method for X-Ray Fluorescence Analysis Applicable to the Films Obtained by Vacuum Evaporation of Iron–Nickel Alloys, *Spectrochim. Acta* **19**, 533–40 (1963).

(*P24*) Pollai, G., and H. Ebel, Tertiary Excitation in X-Ray Fluorescence Analysis, *Spectrochim. Acta* **26B**, 761–6 (1971).

(*P25*) Pollai, G., M. Mantler, and H. Ebel, Contribution of Scattering in the Measurement of X-Ray Fluorescence Intensities, *Spectrochim. Acta* **26B**, 733–46 (1971).

(*P26*) Pollai, G., M. Mantler, and H. Ebel, Secondary Excitation in X-Ray Fluorescence Analysis of Plane Thin Films, *Spectrochim. Acta* **26B**, 747–59 (1971).

(*P27*) Porter, D. E., and R. Woldseth, X-Ray Energy Spectrometry, *Anal. Chem.* **45**(7), 604A–606A, 608A, 610A, 612A, 614A (1973).

(*P28*) Preis, H., and A. Esenwein, X-Ray Fluorescence Spectrographic Method for Determination of Lead in Gasoline, *Schweiz. Arch. Angew. Wiss. Tech.* **26**, 317–20 (1960).

(*P29*) Priestly, E. F., New Method for Increasing X-Ray Reflection Power of Lithium Fluoride Crystals, *Brit. J. Appl. Phys.* **10**, 141–2 (1959).

(*P30*) Priestly, E. F., X-Ray Spectrographs for Rapid X-Ray Fluorescence Analysis with Photographic Recording, *J. Sci. Instrum.* **35**, 409–13 (1958).

(*P31*) Proctor, E. M., Picker X-Ray Corp., Cleveland, Ohio, unpublished work.

(*Q1*) Quebbeman, N. P., and C. W. Jones, Rotating Sample Holder for X-Ray Emission Spectrometry, *Appl. Spectrosc.* **23**, 67–9 (1969).

(*R1*) Raag, V., E. P. Bertin, and R. J. Longobucco, Distribution of Cathode Sublimation Deposits in a Receiving Tube as Determined by X-Ray Spectrometric Scanning, *Advan. Electron. Tube Tech.* **2**, 249–59 (1963).

(*R2*) Ranzetta, G. V. T., and V. D. Scott, Point-Anode Proportional Counters for the Detection of Soft X-Rays in Microanalysis, *J. Sci. Instrum.* **44**, 983–7 (1967).

(*R3*) Rasberry, S. D., Application of Computers in Electron-Probe and X-Ray Fluorescence Analysis, *Advan. X-Ray Anal.* **15**, 56–69 (1972).

(*R4*) Rasberry, S. D., X-Ray Fluorescence Spectrometry; in "Activities of the NBS Spectrochemical Section, July 1970 to June 1971," K. F. J. Heinrich and S. D. Rasberry, eds.; *Nat. Bur. Stand.* (*U. S.*) *Tech. Note* **582**, pp. 22–7 (1972).

(*R5*) Rasberry, S. D., and K. F. J. Heinrich, Computations for Quantitative X-Ray Fluorescence Analysis in the Presence of Interelement Effects, *Colloq. Spectrosc. Int.* 16th, Heidelberg (1971), **1**, 337–42.

(*R6*) Rasberry, S. D., and K. F. J. Heinrich, Calibration for Interelement Effects in X-Ray Fluorescence Analysis, *Anal. Chem.* **46**, 81–9 (1974)

(*R7*) Rasberry, S. D., H. J. Caul, and A. Yezer, X-Ray Fluorescence Analysis of Silver Dental Alloys with Correction for Line Interference, *Spectrochim. Acta* **23B**, 345–51 (1968).

(*R8*) Reynolds, R. C., Matrix Corrections in Trace Element Analysis by X-Ray Fluorescence—Estimation of Mass Absorption Coefficient by Compton Scattering, *Amer. Mineral.* **48**, 1133–43 (1963).

(*R9*) Rhodes, J. R., Radioisotope X-Ray Spectrometry, a Review, *Analyst* (*London*) **91**, 683–99 (1966).

(*R10*) Rhodes, J. R., and T. Furuta, Applications of a Portable Radioisotope X-Ray Fluorescence Spectrometer to Analysis of Minerals and Alloys, *Advan. X-Ray Anal.* **11**, 249–74 (1968).

(*R11*) Rhodes, J. R., Design and Application of X-Ray Emission Analyzers Using Radioisotope X-Ray or Gamma-Ray Sources, *Amer. Soc. Test. Mater. Spec. Tech. Publ.* **STP-485**, 243–85 (1971).

(*R12*) Rhodes, J. R., Energy-Dispersive X-Ray Spectrometry for Multielement Pollution Analysis, *Amer. Lab.* **5**(7), 57–8, 60, 62, 64, 66–8, 70–3 (1973).

(*R13*) Rhodes, J. R., and C. B. Hunter, Particle-Size Effects in X-Ray Emission Analysis—Simplified Formulae for Certain Practical Cases, *X-Ray Spectrom.* **1**, 113–17 (1972).

(*R14*) Rhodin, T. N., Chemical Analysis of Thin Films by X-Ray Emission Spectrography, *Anal. Chem.* **27**, 1857–61 (1955).

(*R15*) Riggs, F. B., Simple Aid to Pulse-Height Selection with Scanning X-Ray Spectrometers, *Rev. Sci. Instrum.* **34**, 312 (1963).

(*R16*) Rinaldi, F. F., and P. E. Aguzzi, Simple Technique for Casting Glass Disks for X-Ray Fluorescence Analysis, *Spectrochim. Acta* **23B**, 15–18 (1967).

(*R17*) Rizzo, F. E., Diffusion of Electroplated Couples, Ph. D. Thesis, Univ. Cincinnati, Cincinnati, Ohio, 168 pp (1964).

(*R18*) Robinson, J. M. M., and E. P. Gertiser, X-Ray Spectrographic Analysis of Raw Materials for the Cement Industry, *Mater. Res. Stand.* **4**, 228–31 (1964).

(*R19*) Roentgen, W. C., On a New Kind of Rays, *Sitzungsber. Phys.-Med. Ges. Würzburg* (137) (1895); *Ann. Phys. Chem.* **64**, 1–11 (1898).

(*R20*) Roentgen, W. C., On a New Kind of Rays—Second Communication, *Sitzungsber. Phys.-Med. Ges. Würzburg* (11) (1896); *Ann. Phys. Chem.* **64**, 12–17 (1898).

(*R21*) Roentgen, W. C., Further Observations on the Properties of the X-Rays, *Math. Naturwiss. Mitt. Sitzungsber. Preuss. Akad. Wiss., Phys.-Math. Kl.* **1897**, 392; *Ann. Phys. Chem.* **64**, 18–37 (1898).

(*R22*) Rogers, T. H., High-Intensity Radiation from Beryllium-Window X-Ray Tubes, *Radiology* **48**, 594–603 (1947).

(*R23*) Rose, H. J., Jr. and F. Cuttitta, X-Ray Fluorescence Spectroscopy in the Analysis of Ores, Minerals, and Waters, *Advan. X-Ray Anal.* **11**, 23–39 (1968).

(*R24*) Rose, H. J., Jr., I. Adler, and F. J. Flanagan, X-Ray Fluorescence Analysis of Light Elements in Rocks and Minerals, *Appl. Spectrosc.* **17**, 81–5 (1963).

(*R25*) Rose, H. J., Jr., F. Cuttitta, and R. R. Larson, Use of X-Ray Fluorescence in Determination of Selected Major Constituents in Silicates, *U. S. Geol. Surv. Prof. Pap.* **525-B**, 155–9 (1965).

(*R26*) Rose, H. J., Jr., R. P. Christian, J. R. Lindsay, and R. R. Larson, Microanalysis with the X-Ray Milliprobe, *U. S. Geol. Surv. Prof. Pap.* **650B**, 128–35 (1969).

(*R27*) Ross, P. A., New Method of Spectroscopy for Faint X-Radiations, *J. Opt. Soc. Amer.* **16**, 433–7 (1928).

(*R28*) Rothe, G., and A. Koster-Pflugmacher, The Gelatin Method, a New Sample Preparation Technique for X-Ray Fluorescence Spectroscopy, *Z. Anal. Chem.* **201**, 241–5 (1964).

(*R29*) Roulet, H., and J. Despujols, New Method of Analysis by X-Ray Fluorescence Applicable to Small Samples, *C. R. Acad. Sci.* (*Paris*) **253**, 641–3 (1961).

(*R30*) Rowe, W. A., and K. P. Yates, X-Ray Fluorescence Method for Trace Metals in Refinery Fluid Catalytic Cracking Feedstocks, *Anal. Chem.* **35**, 368–70 (1963).

(*R31*) Roy, O., R. Beique, and G. D'Ombrian, Method of Iodine Determination by Characteristic X-Ray Absorption, *I.R.E. Trans. Bio-Med. Electron.* **9**, 50–53 (1962).

(*R32*) Ruderman, I. W., and B. Michelman, Present Status of X-Ray Analyzer Crystals; presented at 5th Int. Conf. X-Ray Optics and Microanal., Tübingen, Ger. (1968); not published in the Conference Proceedings (*28*).

(*R33*) Ruderman, I. W., K. J. Ness, and J. C. Lindsay, Analyzer Crystals for X-Ray Spectroscopy in the Region 25–100 Å, *Appl. Phys. Lett.* **7**, 17–19 (1965).

(*R34*) Rudolph, J. S., and R. J. Nadalin, Determination of Microgram Quantities of Chloride in High-Purity Titanium by X-Ray Spectrochemical Analysis, *Anal. Chem.* **36**, 1815-17 (1964).

(*R35*) Rudolph, J. S., O. H. Kriege, and R. J. Nadalin, Applications of Chemical Precipitation Methods for Improving Sensitivity in X-Ray Fluorescent Analysis, *Develop. Appl. Spectrosc.* **4**, 57–64 (1965).

(*R36*) Russ, J. C., Background Subtraction for Energy-Dispersive X-Ray Spectra, *Proc. Nat. Conf. Electron-Probe Anal. Soc. Amer.*, 7th, Paper 76, 3 pp. (1972).

(*R37*) Ryder, R. J., and E. C. Taylor, X-Ray Analysis of Glass, *Norelco Rep.* **15**, 96–9 (1968).

(*R38*) Ryland, A. L., A. General Approach to the X-Ray Spectroscopic Analysis of Samples of Low Atomic Number, *Nat. Meet. Am. Chem. Soc.*, 147th, Phila. (1964); unpublished work.

(*S1*) Sahores, J. J., E. P. Larribau, and J. Mihura, New Results Obtained with a Regular XRF Spectrometer and a Portable Field Spectrometer Equipped with a Cold-Cathode Electron Tube, *Advan. X-Ray Anal.* **16**, 27–36 (1973).

(*S2*) Salmon, M. E., An X-Ray Fluorescence Method for Micro Samples, *Siemens Rev.* **38** (Special Issue), 12–18 (1971).

(*S3*) Salmon, M. L., Recent Developments in Fluorescent X-Ray Spectrographic-Absorptiometric Analysis of Mineral Systems, *Advan. X-Ray Anal.* **2**, 303–12 (1959).

(*S4*) Salmon, M. L., Effects of Operating Variables in the Application of Multi-Element Calibration Systems for Fluorescent X-Ray Spectrographic Analyses of Mineral Samples, *Advan. X-Ray Anal.* **4**, 433–56 (1961).

(*S5*) Salmon, M. L., Simple Multielement-Calibration System for Analysis of Minor and Major Elements in Minerals by Fluorescent X-Ray Spectrography, *Advan. X-Ray Anal.* **5**, 389-404 (1962).

(*S6*) Salmon, M. L., Practical Applications of Filters in X-Ray Spectrography, *Advan. X-Ray Anal.* **6**, 301–12 (1963).

(*S7*) Salmon, M. L., Improved Trace Analysis with the Use of Synchronized Electronic Discrimination in an X-Ray Scanning Procedure, *Advan. X-Ray Anal.* **7**, 604–14 (1964).

(*S8*) Salvage, L. R., Obtaining Color Images of Occlusions with a Microprobe, *Metallography* **2**, 101–5 (1969).

(*S9*) Samuels, L. E., and J. V. Craig, Abrasion Damage to Graphite Flakes in Cast Irons, *J. Iron Steel Inst.* **203**, 75–7 (1965).

(*S10*) Sarian, S., and H. W. Weart, X-Ray Fluorescence Analysis of Polyphase Metals—Use of Borax Glass Matrix, *Anal. Chem.* **35**, 115 (1963).

(*S11*) Sayce, L. A., and A. Franks, N. P. L. Gratings for X-Ray Spectrometry, *Proc. Roy. Soc. (London)* **282A**, 353–7 (1964).

(*S12*) Schaefer, E. A., P. F. Elliot, and J. O. Hibbits, X-Ray Fluorescence Measurements of Surface Uranium on Oxidized Fuel Elements, *Anal. Chim. Acta* **44**, 21–8 (1969).

(*S13*) Schindler, M. J., A. Month, and J. Antalec, An X-Ray Fluorescence Technique for Analysis of Iron–Nickel Films over an Extremely Wide Range of Composition and Thickness, *Pittsburgh Conf. Anal. Chem. Appl. Spectrosc.* (1961), unpublished; reviewed in detail in Reference (*B34*).

(*S14*) Schneider, H., Analysis of Small Quantities of Substances by Use of X-Ray Fluorescence Spectrometry, *Z. Anal. Chem.* **249**, 225–8 (1970).

(*S15*) Schnell E., Roentgen Fluorescence Analysis—Use of Secondary Radiation in the Detection of Halogen and Sulfur in Gas Samples, *Monatsch. Chem.* **96**, 1302–7 (1965).

(*S16*) Schreiber, H., Quantitative Chemical Analysis by Means of X-Ray Emission Spectra, *Z. Phys.* **58**, 619–50 (1929).

(*S17*) Schreiber, T. P., A. C. Ottolini, and J. L. Johnson, X-Ray Emission Analysis of Thin Films Produced by Lubricating Oil Additives, *Appl. Spectrosc.* **17**, 17–19 (1963).

(*S18*) Schumacher, B. W., and J. J. Grodiszewski, Research on a Versatile X-Ray Detector, *U. S. Air Force Mater. Lab. Rep.* **AFML-TR-64-414**, 51 pp. (1964).

(*S19*) Seaman, W., H. C. Lawrence, and H. C. Craig, Solid-Sample Techniques in *K*-Capture Spectroscopy—Determination of Chlorine and Bromine, *Anal. Chem.* **29**, 1631–2 (1957).

(*S20*) Seibel, G., Nondispersive X-Ray Spectrometry—Applications to Elementary Chemical Analysis, *Int. J. Appl. Radiat. Isotop.* **15**, 25–41 (1964).

(*S21*) Sellers, W. W., Jr. and K. G. Carroll, Gaging of Thin Nickel Coatings by X-Ray Fluorescence, *Tech. Proc. Amer. Electroplat. Soc.* **43**, 97–100 (1956).

(*S22*) Sewell, P. B., D. F. Mitchell, and M. Cohen, High-Energy Electron Diffraction and X-Ray Emission Analysis of Surfaces and Their Reaction Products, *Develop. Appl. Spectrosc.* **7A**, 61–79 (1969).

(*S23*) Shelby, W. D., and P. Cukor, X-Ray Fluorescence Analysis of Nickel–Iron Thin Films with Standards Prepared by Pyrolysis of Metal-Organic Compounds, *Anal. Chim. Acta* **49**, 275–8 (1970).

(*S24*) Shenberg, C., and S. Amiel, Critical Evaluation of Correction Methods for Interelement Effects in X-Ray Fluorescence Spectrometric Analysis, *Anal. Chem.* **46**, 1512–16 (1974).

(*S25*) Shenberg, C., A. B. Haim, and S. Amiel, Accurate Determination of Copper in Mixtures and Ores by Radioisotope-Excited X-Ray Fluorescence Spectrometric Analysis Using Peak Ratios, *Anal. Chem.* **45**, 1804–8 (1973).

(*S26*) Sherman, J., Theoretical Derivation of the Composition of Mixable Specimens from Fluorescent X-Ray Intensities, *Advan. X-Ray Anal.* **1**, 231–50 (1958).

(*S27*) Sherman, J., Correlation between Fluorescent X-Ray Intensity and Chemical Composition, *Amer. Soc. Test. Mater. Spec. Tech. Publ.* **STP-157**, 27–33 (1954).

(*S28*) Sherman, J., Theoretical Derivation of Fluorescent X-Ray Intensities from Mixtures, *Spectrochim. Acta* **7**, 283–306 (1955).

(*S29*) Shibuya, Y., E. Nishiyama, and K. Yanagase, Determination of Sulfur in Heavy Oils by X-Ray Fluorescence Analysis, *Jap. Anal.* **16**, 123–8 (1967).

(*S30*) Shiraiwa, T., and N. Fujino, Micro Fluorescent X-Ray Analyzer, *Advan. X-Ray Anal.* **11**, 95–104 (1968).

(*S31*) Shirley, D. A., Applications of Germanium Gamma-Ray Detectors, *Nucleonics* **23**(3), 62–6 (1965).

(*S32*) Shono, T., and K. Shinra, Determination of Metals in Airborne Dust by Fluorescent X-Ray Spectrometry with Filter-Paper Technique, *Jap. Anal.* **18**, 1032–4 (1969).

(*S33*) Short, M. A., Detection and Correction of Nonlinearity in X-Ray Counting Systems; in "Proceedings of the 5th Conference on X-Ray Analytical Methods," Swansea, Wales, U. K., J. Buwalda, ed.; Scientific and Analytical Equipment Dept., N. V. Philips Gloeilampenfabrieken, Eindhoven, Neth., pp. 60–62 (1966).

(*S34*) Shott, J. E., Jr., T. J. Garland, and R. O. Clark, Determination of Traces of Nickel and Vanadium in Petroleum Distillates—An X-Ray Emission Spectrographic Method Based on a New Rapid Ashing Procedure, *Anal. Chem.* **33**, 506–10 (1961).

(*S35*) Siemes, H., X-Ray Fluorescence Spectrometric Determination of Light Elements in Low Concentrations in Water Solutions, *Z. Anal. Chem.* **199**, 321–34 (1964).

(*S36*) Silver, M. D., and E. T.-K. Chow, Thickness Measurement of Thin Permalloy Films—Comparison of X-Ray Emission Spectroscopy, Interferometry, and Stylus Methods, *J. Vac. Sci. Technol.* **2**, 203–7 (1965).

(*S37*) Slack, C. M., Leonard-Ray Tube with Glass Window, *J. Opt. Soc. Amer. and Rev. Sci. Instrum.* **18**, 123–6 (1929).

(*S38*) Slaughter, M., and D. Carpenter, Modular Automatic X-Ray Analysis System, *Advan. X-Ray Anal.* **15**, 135–47 (1972).

(*S39*) Sloan, R. D., X-Ray Spectrographic Analysis of Thin Films by the Milliprobe Technique, *Advan. X-Ray Anal.* **5**, 512–15 (1962).

(*S40*) Smallbone, A. H., and R. Lathe, On-Stream X-Ray Analyzer and Digital Computer Simplify Ore Analysis, *Mining Eng.* (*N. Y.*) **21**(8), 66–8 (1969).

(*S41*) Smith, G. S., A Useful Technique in X-Ray Fluorescence Spectrography, *Chem. Ind.* (*London*) **1963**(22), 907–9.

(*S42*) Solazzi, M. J., The "Spectro-Rule"—a New Aid to X-Ray Spectroscopists, *Amer. Lab.* **1**(2), 89 (1969).

(*S43*) Soller, W., A New Precision X-Ray Spectrometer, *Phys. Rev.* **24**, 158–67.

(*S44*) Sommer, S. E., Selected-Area X-Ray Luminescence Spectroscopy with the X-Ray Milliprobe, *Appl. Spectrosc.* **26**, 557–8 (1972).

(*S45*) Spielberg, N., Tube Target and Inherent Filtration as Factors in the Fluorescence Excitation of X-Rays, *J. Appl. Phys.* **33**, 2033–5 (1962).

(*S46*) Spielberg, N., Intensities of Radiation from X-Ray Tubes and the Excitation of Fluorescence X-Rays, *Philips Res. Rep.* **14**, 215–36 (1959).

(*S47*) Spielberg, N., Orientation of Topaz Crystals for X-Ray Spectrochemical Analysis, *Rev. Sci. Instrum.* **36**, 1377–8 (1965).

(*S48*) Spielberg, N., and G. Abowitz, Calibration Techniques for X-Ray Fluorescence Analysis of Thin Nickel–Chromium Films, *Anal. Chem.* **38**, 200–3 (1966).

(*S49*) Spielberg, N., and M. Brandenstein, Instrumental Factors and Figure of Merit in the Detection of Low Concentrations by X-Ray Spectrochemical Analysis, *Appl. Spectrosc.* **17**, 6–9 (1963).

(*S50*) Spielberg, N., and J. Ladell, Crystallographic Aspects of Extra Reflections in X-Ray Spectrochemical Analysis, *J. Appl. Phys.* **31**, 1659–64 (1960).

(*S51*) Spielberg, N., and J. Ladell, Scanning Single-Crystal Multichannel X-Ray Spectrometer, *Rev. Sci. Instrum.* **34**, 1208–12 (1963).

(*S52*) Spielberg, N., W. Parrish, and K. Lowitzsch, Geometry of the Non-Focusing X-Ray Fluorescence Spectrograph, *Spectrochim. Acta* **13**, 564–83 (1959).

(*S53*) Splettstosser, H. R., and H. E. Seeman, Application of Fluorescence X-Rays to Metallurgical Microradiography, *J. Appl. Phys.* **23**, 1217–22 (1952).

(*S54*) Staats, G., and H. Bruck, Preparation of Oxide Samples for X-Ray Fluorescence Analysis, *Z. Anal. Chem.* **250**, 289–94 (1970).

(*S55*) Steele, W. K., Correction of X-Ray Fluorescence Analyses for Scattered Background Peaks of Target Elements, *Chem. Geol.* **11**, 149–56 (1973).

(*S56*) Steinhardt, R. G., Jr. and E. J. Serfass, X-Ray Photoelectron Spectrometer for Chemical Analysis, *Anal. Chem.* **23**, 1585–90 (1951).

(*S57*) Steinhardt, R. G., Jr. and E. J. Serfass, Surface Analysis with the X-Ray Photoelectron Spectrometer, *Anal. Chem.* **25**, 697–700 (1953).

(*S58*) Steinhardt, R. G., Jr., F. A. D. Granados, and G. L. Post, X-Ray Photoelectron Spectrometer with Electrostatic Deflection, *Anal. Chem.* **27**, 1046–50 (1955).

(*S59*) Stephenson, D. A., Improved Flux-Fusion Technique for X-Ray Emission Analysis, *Anal. Chem.* **41**, 966–7 (1969).

(*S60*) Stephenson, D. A., Multivariable Analysis of Quantitative X-Ray Emission Data, *Anal. Chem.* **43**, 310–18 (1971).

(*S61*) Sterk, A. A., X-Ray Generation by Proton Bombardment, *Advan. X-Ray Anal.* **8**, 189–97 (1965).

(*S62*) Stever, K. R., J. L. Johnson, and H. H. Heady, X-Ray Fluorescence Analysis of Tungsten–Molybdenum Metals and Electrolytes, *Advan. X-Ray Anal.* **4**, 474–87 (1961).

(*S63*) Stewart, J. H., Jr., Determination of Thorium by Monochromatic X-Ray Absorption, *Anal. Chem.* **32**, 1090–2 (1960).

(*S64*) Stoner, G. A., Rapid Automatic Analysis of Magnesium Alloys, *Anal. Chem.* **34**, 123–6 (1962).

(*S65*) Strasheim, A., and M. P. Brandt, Quantitative X-Ray Fluorescence Method of Analysis for Geological Samples Using a Correction Technique for Matrix Effects, *Spectrochim. Acta* **23B**, 183–96 (1967).

(*S66*) Strasheim, A., and F. T. Wybenga, Determination of Certain Noble Metals in Solution by Means of X-Ray Fluorescence Spectroscopy. I. Determination of Platinum, Gold, and Iridium, *Appl. Spectrosc.* **18**, 16–20 (1964).

(*S67*) Sun, S. C., Fluorescent X-Ray Spectrometric Estimation of Aluminum, Silicon, and Iron in Flotation Products of Clays and Bauxites, *Anal. Chem.* **31**, 1322–4 (1959).

(*S68*) Suoninen, E. J., Simultaneous Use of an Analyzer Crystal for Reflection and Transmission in X-Ray Fluorescence Spectrometry, *Suom. Kemistilehti* **34B**, 45–7 (1961); *Chem. Abstr.* **55**, 24223*e* (1961).

(*S69*) Sweatman, I. R., K. Norrish, and R. A. Durie, Assessment of X-Ray Fluorescence Spectrometry for Determination of Inorganic Constituents in Brown Coals, *Div. Coal Res. C.S.I.R.O.* (Commonwealth Scientific and Industrial Research Organization) *Misc. Rep.* **177**, 30 pp. (1963).

(*T1*) Tabikh, A. A., Use of Sample Rotators for Fluorescence Analysis of Light Elements in Cement Raw-Mix, *Anal. Chem.* **34**, 303–4 (1962).

(*T2*) Tanemura, T., Nondispersive X-Ray Spectroanalysis with Filter and Proportional Counter, *Rev. Sci. Instrum.* **32**, 364–6 (1961).

(*T3*) Taylor, A., B. J. Kagle, and E. W. Beiter, Geometrical Representation and X-Ray Fluorescence Analysis of Ternary Alloys, *Anal. Chem.* **33**, 1699–1706 (1961).

(*T4*) Taylor, D. L., and G. Andermann, Evaluation of an Isolated-Atom Model for Internal Standardization in X-Ray Fluorescence Analysis, *Anal. Chem.* **43**, 712–16 (1971).

(*T5*) Taylor, J., and W. Parrish, Absorption and Counting Efficiency Data for X-Ray Detectors, *Rev. Sci. Instrum.* **26**, 367–73 (1955); **27**, 108 (1956).

(*T6*) Tertian, R., Absolute and General Method of Elemental Quantitative Chemical Analysis, *C. R. Acad. Sci., Paris, Ser. A, B* **1968**, 266B(10), 617–19.

(*T7*) Tertian, R., Control of the Matrix Effect in X-Ray Fluorescence and Principle of a Quasi-Absolute Method of Quantitative Analysis in Solid and Liquid Solution, *Spectrochim. Acta* **23B**, 305–21 (1968).

(*T8*) Tertian, R., Quantitative X-Ray Fluorescence Analysis Using Solid-Solution Specimens—a Theoretical Study of the Influence of the Quality of Primary Radiation, *Spectrochim. Acta* **26B**, 71–94 (1971).

(*T9*) Thatcher, J. W., and W. J. Campbell, Fluorescent X-Ray Spectrographic Probe—Design and Applications, *U. S. Bur. Mines Rep. Invest.* **5500**, 23 pp. (1959).

(*T10*) Togel, K., X-Ray Fluorescence Analysis of Small Quantities and Areas, *Siemens-Z.* **36**, 497–501 (1962).

(*T11*) Togel, K., Preparation Technique for X-Ray Spectrometry; in "Zerstörungsfreie Materialprüfung" ("Nondestructive Material Testing"), E. A. W. Muller, ed.; R. Oldenbourg, Munich, Ger., sec. U152 (1961).

(*T12*) Tomaino, M., and A. De Pietro, [Philips] Inverted-Sample Three-Position Spectrograph, *Norelco Rep.* **3**, 57, 90, 94 (1956).

(*T13*) Tomkins, M. L., G. A. Borun, and W. A. Fahlbusch. Quantitative Determination of Tantalum, Tungsten, Niobium, and Zirconium in High-Temperature Alloys by X-Ray Fluorescent Solution Method, *Anal. Chem.* **34**, 1260–3 (1962).

(*T14*) Tomura, T., H. Okano, K. Hara, and T. Watanabe, Multistep Intensity Indication in Scanning Microanalysis, *Advan. X-Ray Anal.* **11**, 316–25 (1968).

(*T15*) Tousimis, A. J., and J. A. Nicolino, Universal X-Ray Detector—a Flow Proportional Counter for Analysis of all Elements from Beryllium to Uranium, *Proc. Nat. Conf. Electron-Probe Anal. Soc. Amer.*, 6th, Paper 60 (1971).

(*T16*) Toussaint, C. J., and G. Vos, Quantitative Determination of Carbon in Solid Hydrocarbons Using the Intensity Ratio of Incoherent to Coherent Scattering of X-Rays, *Appl. Spectrosc.* **18**, 171–4 (1964).

(*T17*) Traill, R. J., and G. R. Lachance, A New Approach to X-Ray Spectrochemical Analysis, *Geol. Surv. Can. Pap.* **64-57**, 22 pp. (1965).

(*T18*) Traill, R. J., and G. R. Lachance, A Practical Solution to the Matrix Problem in X-Ray Analysis. II. Application to a Multicomponent Alloy System, *Can. Spectrosc.* **11**(3), 63–71 (1966).

(*T19*) Trillat, J.-J., Metallurgical Aspects of Microradiography, *Met. Rev.* **1**, 3–30 (1956).

(*U1*) Umbarger, C. J., R. C. Bearse, D. A. Close, and J. J. Malanify, Sensitivity and Detectability Limits for Elemental Analysis by Proton-Induced X-Ray Fluorescence with a 3-MV Van de Graaff, *Advan. X-Ray Anal.* **16**, 102–10 (1973).

(*U2*) Underwood, J. H., An Improved Collimator for Extreme Ultraviolet and X-Rays, *Rev. Sci. Instrum.* **40**, 894–6 (1969).

(*V1*) Van Niekerk, J. N., and J. F. De Wet, Trace Analysis by X-Ray Fluorescence Using Ion Exchange Resins, *Nature*, **186**, 380–1 (1960).

(*V2*) Van Niekerk, J. N., J. F. De Wet, and F. T. Wybenga, Trace Analysis by Combination of Ion-Exchange and X-Ray Fluorescence—Determination of Uranium in Sulfate Effluents, *Anal. Chem.* **33**, 213–15 (1961).

(*V3*) Van Rennes, A. B., Pulse-Amplitude Analysis in Nuclear Research. I. Voltage Discriminators, *Nucleonics* **10**(7), 20–7 (1952).

(*V4*) Van Rennes, A. B., Pulse-Amplitude Analysis in Nuclear Research. II. Single-Channel Differential Analyzers, *Nucleonics* **10**(8), 22–8 (1952).

(*V5*) Van Rennes, A. B., Pulse-Amplitude Analysis in Nuclear Research. III, IV. Multichannel Analyzers, *Nucleonics* **10**(9), 32–8; (10) 50–6 (1952).

(*V6*) Vassos, B. H., R. F. Hirsch, and H. Letterman, X-Ray Microdetermination of Chromium, Cobalt, Copper, Mercury, Nickel, and Zinc in Water Using Electrochemical Preconcentration, *Anal. Chem.* **45**, 792–4 (1973).

(*V7*) Vassos, B. H., R. F. Hirsch, and H. Letterman, Seton Hall Univ., South Orange, N. J.; unpublished work.

(*V8*) Vassos, B. H., R. F. Hirsch, and D. G. Pachuta, Filters for X-Ray Spectrometry Prepared by Thin-Layer Electrodeposition, *Anal. Chem.* **43**, 1503–6 (1971).

(*V9*) Vassos, B. H., F. J. Berlandi, T. E. Neal, and H. B. Mark, Jr., Electrochemical Preparation of Thin Metal Films as Standards on Pyrolytic Graphite, *Anal. Chem.* **37**, 1653–6 (1965).

(*V10*) Verderber, R. R., X-Ray Fluorescence Analysis of the Composition of Ni–Fe Thin Films, *Norelco Rep.* **10**, 30–34 (1963).

(*V11*) Verhoeven, J., and E. Rieger, X-Ray Appearance-Potential Measurements with a Proportional Counter, *Ned. Tijdschr. Vacuumtech.* **10**, 80–3 (1972).

(*V12*) Victoreen, J. A., Calculation of X-Ray Mass-Absorption Coefficients, *J. Appl. Phys.* **20**, 1141–7 (1949).

(*V13*) Volborth, A., Total Instrumental Analysis of Rocks, *Nev. Bur. Mines Rep.* **6A**, 1–72 (1963).

(*V14*) Von Hamos, L., X-Ray Image Method of Chemical Analysis, *Amer. Mineral.* **23**, 215–26 (1938).

(*V15*) Von Hamos, L., Determination of Very Small Quantities of Substances by the X-Ray Microanalyzer, *Ark. Mat. Astron. Fys.* **31**, 1–11 (1945).

(*V16*) Von Hamos, L., X-Ray Micro-Analyzer, *J. Sci. Instrum.* **15**, 87–94 (1938).

(*V17*) Voparil, R., X-Ray Nondispersive Spectral Analysis Using Modulated Radiation, *J. Sci. Instrum.* [2] **3**, 798–800 (1970).

(*V18*) Vos, G., Correction Method for the X-Ray Fluorescent Determination of Iron, Manganese, Calcium, and Potassium in Light Matrices, *Spectrochim. Acta* **27B**, 89–92 (1972).

(*V19*) Vose, G. P., Determination of Organic–Inorganic Ratio in Osseous Tissue by X-Ray Absorption, *Anal. Chem.* **30**, 1819–21 (1958).

(*W1*) Waehner, K. A., Analyzing Refractory Metal Alloys—Fansteel 82 Metal, *Fansteel Met.* (Dec. 1961).

(*W2*) Wagner, J. C., and F. R. Bryan, X-Ray Spectrographic Analysis of Automotive Combustion Deposits without the Use of Calibration Curves, *Advan. X-Ray Anal.* **9**, 528–32 (1966).

(*W3*) Wang, M. S., Rapid Simple Fusion with Lithium Tetraborate for Emission Spectroscopy, *Appl. Spectrosc.* **16**, 141–2 (1962).

(*W4*) Wasilewska, M., and A. Robert, Determination of Silicon in Soluble Sodium Silicates by Nondispersive X-Ray Fluorescence Excited by a Radioactive Source, *Radiochem. Radioanal. Lett.* **1**, 263–72 (1969).

(*W5*) Weber, K., and J. Marchal, Device for Coupling a Pulse-Height Discriminator to a Scanning X-Ray Spectrometer, *J. Sci. Instrum.* **41**, 15–22 (1964).

(*W6*) Welday, E. E., A. K. Baird, D. B. McIntyre, and K. W. Madlem, Silicate Sample Preparation for Light-Element Analysis by X-Ray Spectrography, *Amer. Mineral.* **49**, 889–903 (1964).

(*W7*) Wenger, P. E., I. Kapetanidis, and W. Van Janstein, X-Ray Fluorescence as a Supplement to Paper Electrophoresis, *Pharm. Acta Helv.* **37**, 472–89 (1962).

(*W8*) West, H. I., Jr., W. E. Meyerhof, and R. Hofstadter, Detection of X-Rays by Means of NaI(Tl) Scintillation Counters, *Phys. Rev.* **81**, 141–2 (1951).

(*W9*) Weyl, R., Nondestructive Measurement of Composition and Layer Thickness of Thin Films by X-Ray Fluorescence, *Z. Angew. Phys.* **13**, 283–8 (1961).

(*W10*) White, E. W., H. A. McKinstry, and T. F. Bates, Crystal Chemical Studies by X-Ray Fluorescence, *Advan. X-Ray Anal.* **2**, 239–45 (1959).

(*W11*) White, J. E., X-Ray Diffraction by Elastically Deformed Crystals, *J. Appl. Phys.* **21**, 855–9 (1950).

(*W12*) Wilkinson, D. H., A Stable 99-Channel Pulse-Amplitude Analyzer for Slow Counting, *Proc. Cambridge Phil. Soc.* **46** (Part 3), 508–18 (1950).

(*W13*) Willgallis, A., and G. Schneider, Preparation of Silicate Samples for X-Ray Analysis, *Z. Anal. Chem.* **246**, 115–18 (1969).

(*W14*) Williams, R. W., Quantitative Elemental Analysis of Small Liquid Samples (Less Than 1 ml) by X-Ray Fluorescence, *Proc. Soc. Anal. Chem.* **7**(5), 80–83 (1970).

(*W15*) Winslow, E. H., H. M. Smith, H. E. Tanis, and H. A. Liebhafsky, Chemical Analysis Based on X-Ray Absorption Measurements with a Multiplier Phototube, *Anal. Chem.* **19**, 866–7 (1947).

(*W16*) Witmer, A. W., and N. H. W. Addink, Quantitative X-Ray Fluorescence Analysis without Standard Samples—Thin-Layer Method, *Sci. Ind.* **12**(2), 1 (1965).

(*W17*) Wittig, W. J., Uses of an X-Ray Fluorescence Semi-Microprobe Attachment in Metallurgy, *Advan. X-Ray Anal.* **8**, 248–58 (1965).

(*W18*) Wittig, W. J., Development and Use of a Semimicro X-Ray Fluorescence Attachment in Metallurgy, *Develop. Appl. Spectrosc.* **3**, 36–44 (1964).

(*W19*) Wolber, W. G., B. D. Klettke, and P. W. Graves, The Preamplified Spiraltron Electron Multiplier, *Rev. Sci. Instrum.* **41**, 724–8 (1970).

(*W20*) Wolfe, R. A., and R. G. Fowler, Application of Spectrographic Methods to Chemical Concentration of Trace Elements in Iron and Steel Analysis, *J. Opt. Soc. Amer.* **35**, 86–91 (1945).

(*W21*) Wybenga, F. T., and A. Strasheim, Determination of Certain Noble Metals in Solution by Means of X-Ray Fluorescence Spectroscopy. II. Determination of Palladium, Rhodium, and Ruthenium, *Appl. Spectrosc.* **20**, 247–50 (1966).

(*W22*) Wyckoff, R. W. G., and F. D. Davidson, Windowless X-Ray Tube Spectrometer for Light-Element Analysis, *Rev. Sci. Instrum.* **35**, 381–3 (1964).

(*W23*) Wytzes, S. A., Theoretical Considerations on the Collimators of an X-Ray Spectrograph, *Philips Res. Rep.* **16**, 201–24 (1961).

(*W24*) Wytzes, S. A., Automatic X-Ray Spectrometer, *Philips Tech. Rev.* **27**, 300–10 (1966).

(*W25*) Wytzes, S. A., Device for X-Ray Spectrochemical Analysis by Means of Fluorescent Radiation, *U. S. Pat.* **3,060,314** (1962).

(*Y1*) Yaguchi, K., and H. Shimomura, Fluorescent X-Ray Analysis of Antimony and Lead in Crude Copper, *Jap. Anal.* **18**, 1188–95 (1969).

(*Y2*) Yakowitz, H., and K. F. J. Heinrich, Color Representation of Electron Microprobe Area-Scan Images by a Color-Separation Process, *J. Res. Nat. Bur. Stand.* (*U. S.*) **73A**, 113–23 (1969).

(*Y3*) Yao, T. C., and F. W. Porsche, Determination of Sulfur and Chlorine in Petroleum Liquids by X-Ray Fluorescence, *Anal. Chem.* **31**, 2010–12 (1959).

(*Y4*) Yoneda, Y., and T. Horiuchi, Optical Flats for Use in X-Ray Spectrochemical Microanalysis, *Rev. Sci. Instrum.* **42**, 1069–70 (1971).

(*Y5*) Young, W. J., Applications of the Electron Microbeam Probe and Micro X-Rays in Nondestructive Analysis; in "Recent Advances in Conservation," G. Thomson, ed.; Butterworths, London, pp. 33–8 (1961).

(*Z1*) Zanin, S. J., and G. E. Hooser, Analysis of Solders by X-Ray Spectrometry, *Appl. Spectrosc.* **22**, 105–8 (1968).

(*Z2*) Zeitz, L., Potentialities of an X-Ray Microprobe, *Rev. Sci. Instrum.* **32**, 1423–4 (1961).

(*Z3*) Zeitz, L., X-Ray Emission Analysis in Biological Specimens; in "X-Ray and Electron-Probe Analysis in Biomedical Research," Progress in Analytical Chemistry, vol. 3; K. M. Earle and A. J. Tousimis, eds.; Plenum Press, N. Y., pp 35–73 (1969).

(*Z4*) Zeitz, L., and A. V. Baez, Microchemical Analysis by Emission Spectrographic and Absorption Methods; in "X-Ray Microscopy and Microradiography," V. E. Cosslett, A. Engström, and H. H. Pattee, eds.; Academic Press, N. Y., pp. 417–34 (1957).

(*Z5*) Zemany, P. D., Effects Due to Chemical State of the Samples in X-Ray Emission and Absorption, *Anal. Chem.* **32**, 595–7 (1960).

(*Z6*) Zemany, P. D., Use of Fe^{55} for Measuring Titanium Coating Thickness, *Rev. Sci. Instrum.* **30**, 292–3 (1959).

(*Z7*) Zemany, P. D., Line Interference Corrections for X-Ray Spectrographic Determination of Vanadium, Chromium, and Manganese in Low-Alloy Steels, *Spectrochim. Acta* **16**, 736–41 (1960).

(*Z8*) Zemany, P. D., and H. A. Liebhafsky, Plating Thickness by the Attenuation of Characteristic X-Rays, *J. Electrochem. Soc.* **103**, 157–9 (1956).

(*Z9*) Zemany, P. D., H. G. Pfeiffer, and H. A. Liebhafsky, Precision in X-Ray Emission Spectrography—Background Present, *Anal. Chem.* **31**, 1776–8 (1959).

(*Z10*) Zemany, P. D., W. W. Welbon, and G. L. Gaines, Determination of Microgram Quantities of Potassium by X-Ray Emission Spectrography of Ion-Exchange Membranes, *Anal. Chem.* **30**, 299–300 (1958).

(*Z11*) Zemany, P. D., E. H. Winslow, G. S. Poellmitz, and H. A. Liebhafsky, X-Ray Absorption Measurements—Comparative Method for Chemical Analysis Based on Measurements with a Multiplier Phototube, *Anal. Chem.* **21**, 493–7 (1949).

(*Z12*) Ziebold, T. O., and R. E., Ogilvie, Empirical Method for Electron Microanalysis, *Anal. Chem.* **36**, 322–7 (1964).

(*Z13*) Ziegler, C. A., L. L. Bird, and D. J. Chleck, X-Ray Rayleigh Scattering Method for Analysis of Heavy Elements in Low-*Z* Media, *Anal. Chem.* **31**, 1794–8 (1959).

(*Z14*) Zingaro, P. W., Helium-Chamber X-Ray Spectrograph, *Norelco Rep.* **1**, 45–6 (1953-54).

(*Z15*) Zimmerman, R. H., Industrial Applications of X-Ray Methods for Measuring Plating Thickness, *Advan. X-Ray Anal.* **4**, 335–50 (1961).

(*Z16*) Zimmerman, R. H., X-Rays Check Plating Thickness, *Iron Age* **186** (15), 84–7 (1960).

(*Z17*) Zimmerman, R. H., Measuring Plating Thickness—Industrial Applications of X-Ray Methods, *Metal Finish.* **59**(5), 67–73 (1961).

(*Z18*) Zimmerman, R. H., Industrial Applications of X-Ray Methods for Measuring Plating Thickness, *Qual. Assurance* **2**(3), 27–32 (1963).

Index